Precalculus
BUILDING CONCEPTS AND CONNECTIONS
Second Edition

Revathi Narasimhan

2ND PRINTING—JULY 2016
- Misc. corrections made

PAPERBACK PRINTING—JULY 2017

*xyz*textbooks

Precalculus

Building Concepts and Connections
Second Edition

Revathi Narasimhan

Publisher: XYZ Textbooks

Copy Editor: Margaret Parks & Associates

Project Manager: Jennifer Thomas

Developmental Editor: Katherine Heistand Shields

Composition: Jennifer Thomas

Sales: Amy Jacobs, Richard Jones, Bruce Spears

Cover Design: Rachel Hintz

ISBN-13: 978-1-63098-132-7 / ISBN-10: 1-63098-132-X

For product information and technology assistance, contact us at
XYZ Textbooks, 1-877-745-3499

For permission to use material from this text or product,
e-mail: **info@mathtv.com**

XYZ Textbooks
1339 Marsh Street
San Luis Obispo, CA 93401
USA

Printed in the United States of America

For your course and learning solutions, visit **www.xyztextbooks.com**

i

Brief Contents

Contents

6 Trigonometric Identities and Equations   503

7 Additional Topics in Trigonometry   553

8 Systems of Equations and Inequalities   641

Preface to the Instructor

Vision

This Second Edition of *Precalculus: Building Concepts and Connections* remains true to Revathi Narasimhan's original vision—to help students with varying levels of preparation, abilities, and motivation build a conceptual understanding of precalculus, increase awareness of how mathematical concepts connect and relate, and promote strong problem solving and mathematical reasoning skills.

Throughout the book, the author's reader-friendly, clear presentation combines mathematical rigor with an engaging writing style that maximizes student success and helps build a strong foundation for success in future mathematics courses.

Structured for Student Success

Early Algebra Review and Continued Reinforcement of Algebraic Skills

Based on feedback from instructors across the country who identified weak algebraic skills as one of the most significant challenges they face in teaching the course, the author moved algebra review from its previous position in an Appendix to the first chapter of the book. Using an easy-to-understand writing style supported by boxed definitions and detailed examples with clear explanations, this first chapter sets the stage for student success in the course. In addition, detailed examples in every chapter give clear explanations and often include *Just In Time* references to reinforce algebraic skills as students move through the book.

Functions as a Unifying Theme

Functions are introduced in Chapter 2, followed by related equations and ways of solving these equations in the context of their associated functions. By presenting equations in conjunction with their "functional" counterparts, students come away with a coherent picture of mathematics that will stay with them long after they complete the course.

Pedagogical Reinforcement

In every chapter, *Just in Time* references and exercises "recall" previous topics and skills. These consistent prompts of where important previous topics can be reviewed ensure that students have the skills and knowledge they need to learn the new material presented.

Online Videos for Every Example and Every Learning Style

Instructional videos available online at MathTV.com help students master the content and skills covered in every example in the book. Students first select from a number of online tutors and then can see and hear the examples solved in both English and Spanish. Each tutor approaches the problem in a slightly different way, ensuring that students with different learning styles and preferences can get the help they need when they need it.

Instruction And Pedagogy

Designed throughout to help students make sense of mathematics and build good study habits, this Second Edition includes innovative features, such as *Just in Time* exercises, *Notes*, and *Observations* that provide connections between various concepts. Technology and applications are incorporated throughout, and engaging exercise sets with a variety of problems round out the text. Strong pedagogy throughout engages students, helps them make stronger connections between concepts, and encourages frequent exploration and review.

Engage

Chapter Openers: To help students see the relevance of what they'll be learning, each chapter opens with compelling photos and/or illustrations that illustrate an interesting application of the chapter's content. In addition, a *Chapter Outline* provides an overview of the chapter's topics.

Applications: Real-world applications from business, economics, and the social and natural sciences demonstrate how mathematics is used in real-life situations. Both contemporary and classical applications are included in exercise sets and examples to provide a motivating real-world context for each topic. The frequent use of interesting applications helps students develop a stronger understanding of how mathematics is used to analyze problems in various disciplines and builds the skills they need to draw comparisons between discrete sets of data and make more informed decisions.

A Wealth of Clear Easy-to-Follow Examples: Well-marked examples illustrate topics in a straightforward manner, helping students develop problem-solving and mathematical thinking skills. When solutions involve complex, multiple steps, the solution is presented in a tabular format to better clarify the problem-solving process.

Video Tutorials: *QR Codes* in the margins beside each example allow students to access video tutorials instantly via smart phone or other internet-connected mobile device. For every video tutorial, students can *see* a tutor of their choice work the example on a whiteboard and *hear* detailed, easy-to-understand explanations, along with reminders of important concepts and potential pitfalls. Students can view the tutorials as many times as they want and can choose to switch back and forth between different tutors to enhance their understanding. Over time, students may develop a preferred tutor whose explanations are easiest for them to follow and understand.

Connect

Just in Time references: These references, found in the margins where appropriate, point to specific sections in the book where the referenced topics were first introduced, enabling students to turn back and review the material as needed to ensure they are prepared for the content that follows.

Just in Time exercises: Included as the first set of exercises for many sections, these exercises, help students recall and apply what they have previously learned to the current section's new concepts. If students have difficulty with these exercises, they can go back to the referenced section for study and review. Many *Just in Time* exercises refer students back to algebra concepts that must be mastered for success in precalculus.

Notes: Placed in the margin where appropriate, *Notes* speak to students in a conversational, one-on-one tone. They may be cautionary or informative, offer practical tips on avoiding common errors, or provide further useful information.

Explore

Using Technology boxes: Identified by a graphing calculator icon, these boxes and appear in the book margins to support the optional use of graphical calculator technology. The screen shots and instructions included have been carefully prepared to illustrate and support some of the subtle details of calculator use that can be overlooked. The limitations of graphing calculator technology are acknowledged and tips on ways to work through these limitations are included. Each box refers students to the appropriate section or sections in the Keystroke Appendix that can be downloaded from the textbook website.

Keystroke Appendix: This downloadable guide provides specific keystrokes for the TI-83/84 series of graphing calculators. The coverage of the guide parallels the order of topics in the book and includes detailed instruction on keystrokes, commands, and menus.

Review And Reinforce

Section Objectives: To help students read with a purpose, each section begins with a bulleted list of objectives that offer an at-a-glance overview of the knowledge and skills they need to master in that section.

Check It Out: Following every example, *Check It Out* exercises give students an opportunity to try a problem similar to the one given in the example. Answers are provided at the back of the book to allow students to check their work and assess their level of understanding. Answers to the *Check It Out* exercises can be found in the XYZ eBook.

Observations: Directly following the graphs of functions, *Observations* appear as short, bulleted lists that highlight key features of the graphs. They may also illustrate patterns that can help students organize their thinking. As students encounter the various observations throughout the text, they get into the habit of analyzing key features of functions, building their skills in interpreting and analyzing what they see.

Two-Column Chapter Summary: A detailed summary, organized by section, appears at the end of every chapter. The summary covers the section's main concepts and refers students to related exercises that will help them practice and review, ensuring more efficient, focused study and increased master of skills and concepts.

Exercises: Comprehensive exercise sets after every section help students master key concepts and skills. Many begin with *Just in Time* exercises to help students recall and apply what they have previously learned to the current section's new concepts. All section exercise sets include *Skills* exercises and *Applications* that ask students to apply section skills to real-world problems. Many sections also include *Concepts* exercises that draw on both the ideas presented in the section and the student's general math background.

End-of-Chapter Review Exercises and Chapter Tests: To further reinforce student understanding and skill development, *Review Exercises* that include both skills and applications appear at the end of each chapter. A comprehensive *Chapter Test* concludes every chapter, allowing students to assess their understanding of key chapter concepts and skills.

Additional Resources

Precalculus: Building Concepts and Connections, Second Edition is not only a printed textbook, it is a complete interactive solution for your classroom. Every new copy comes with a *free* one-year All-Access Pass that allows you and your students unlimited access to tutorial videos that cover every example in the book. Other online resources include an eBook, videos on study skills, interviews with student instructors, and downloadable matched-problem worksheets.

XYZ Homework provides a powerful, self-sufficient, online instructional system with real-time functionality for faculty and students. The unified learning environment combines online assessment with MathTV.com video lessons and *Precalculus: Building Concepts and Connections, Second Edition* to reinforce the concepts taught in the classroom. Ready-to-use assignments and randomized questions, correlated section by section to the printed book, provide unlimited practice and instant feedback with all the benefits of automatic grading. Other features include electronic access to the online book, an easy-to-use email system and discussion forum that allow for proper mathematical notation, and a course calendar that allows you or your students to see all upcoming assignments and tests.

About The Author

Revathi (Reva) Narasimhan is an Associate Professor of Mathematics at Kean University in Union, New Jersey. She received her Ph.D. in Applied Mathematics at the University of Maryland, College Park. She grew up in Mesa, Arizona and received her Bachelor's and Master's degrees at Arizona State University.

As a professional mathematician and a dedicated teacher, she is committed to helping her students understand "big picture" concepts in math and in helping them succeed in their math courses. In addition to this textbook, she has written scholarly articles for academic journals and technology supplements for other textbooks. She also presents regularly at regional and national mathematics conferences.

She and her husband, Prem Sreenivasan, a research microbiologist, have two sons, Vivek and Vijay. Reva likes to garden and sew and is an avid reader. She has been known to wake up before the crack of dawn to chase away the deer snacking in her garden.

Acknowledgements from Charles P. McKeague

XYZ Textbooks Crew

Production Team: Jennifer Thomas, Katherine Heistand Shields

Our fantastic production team shepherded this book from handwritten manuscript to the final form you see today. Their attention to detail and ideas for making this book user-friendly for students is greatly appreciated.

Editing and Proofreading Team: Judy Barclay, Katherine Heistand Shields, Andrew Pardo, Breylor Grout, Nathan Wicka, Donna Martin, Lauren Dvorak, Stephanie Freeman, Octabio Garcia, Ryan Gillett, Gordon Kirby, Tim McCaughey, Lauren Reeves

Their eye for detail and ability to ferret out even the most trivial error never ceased to amaze us. This book is better off both mathematically and grammatically due to their invaluable assistance.

Office and Account Management: Rachael Hillman, Katherine Hofstra

Our office staff is reliable, pleasant, and efficient. Plus they are lots of fun to work with.

Sales Department: Amy Jacobs, Rich Jones, Bruce Spears

Our award-winning, responsive sales staff is always conscientious and hard-working.

Directors of Web Development: Stephen Aiena, Lauren Barker

Our web development team is the brains behind XYZ Textbooks' family of websites, including XYZTextbooks.com, MathTV.com, and XYZHomework.com.

MathTV Peer Instructors: Breylor, Nathan, Shelby, Andrew, Gordon, Lauren R., Molly, Ryan, Stephanie, Penelope, Tim

These students and instructors have a genuine love for math. The videos they've filmed have helped countless students improve their math skills.

Focus Group Participants

Many thanks to the following instructors who participated in one of our online focus groups. Your suggestions were invaluable to us!

Leslie Banta, *Mendocino College*
Leah Wooden, *Delgado Community College*
Debra Lackey, *Odessa College*
Patrick Ward, *Illinois Central College*
Linda Tansil, *Southeast Missouri State University*
Shawna M Bynum, *Napa Valley College*
Dan Kleinfelter, *College of the Desert*
Carolyn Chapel, *Western Technical College*
Meghan McIntyre, *Wake Technical Community College*
Annik Martin, *Idaho State University*
Karen Pace, *Tarrant County College*
Damien Adams, *Cabrillo College*
Manuel Carames, *Miami Dade College North Campus*
Vochita Mihai, *Medaille College*
Daniel Rothe, *Alpena Community College*
Dr. Arcola W. Sullivan,
    Copiah-Lincoln Community College
Laurie Ann Schmidt, Ph.D.,
    Iowa Lakes Community College
Benjamin Falero, *Central Carolina Community College*
David French, *Tidewater Community College*
Bryan Ingham, *Finger Lakes Community College*
Hernando Tellez, *Cascadia College*
Ray DeWitt, *Eastern Wyoming College*
Lisa Hodge, *Wake Technical Community College*
Dr. Dan Canada, *Eastern Washington University*
Angela Everett, *Chattanooga State Community College*
Sylvia Brown, *Mountain Empire Community College*
Elsie M. Campbell, *Angelo State University*
Joe LaMontagne, *Davenport University*
James Eby, *Blinn College - Bryan Campus*
Jennifer Johnson,
    Metropolitan Community College - Longview

Jeganathan Sriskandarajah, *Madison College*
Cheryl Kane, *University of Nebraska-Lincoln*
Frederick Fritz, *Central Carolina Community College*
Jeff Kroemer, *Lone Star College Tomball*
Mary Ann Teel, *University of North Texas*
Christina Cornejo, *Erie Community College - North*
Michael Headley,
    Metropolitan Community College-Penn Valley
Joe D Heine, *Tarrant County College*
Harriet Kiser, *Georgia Highlands College*
Sylvia Gutowska,
    Community College of Baltimore County
Sonya McCook, *Alamance Community College*
Valsala Mohanakumar,
    Hillsborough Community College, Dale Mabry Campus
Kira Heater, *Laramie County Community College*
Jay Villanueva, *Florida Memorial University*
Nicole Francis, *Linn Benton Community College*
Sheeny Behmard, *Chemekea Community College*
Lee Widmer, *Miami University Hamilton*
Patricia Elliott-Traficante, *Holyoke Community College*
Wendy Pogoda, *Hillsborough Community College*
Dina Yagodich, *Frederick Community College*
Matt Blackmon, *Harrisburg Area Community College*
Dianne Twigger, *Evangel University*
Timothy J Bayer, *Virginia Western Community College*
Kelly Proffitt, *Patrick Henry Community College*
Mary Merchant, *Cedar Valley College*
Light Bryant, *Arizona Western College*
Joanne Manville, *Bunker Hill Community Collete*
Celeste Hernandez, *Richland College*
Mike Tieleman, *Anoka Technical College*

Algebra Review

I f you are planning to paint a room, you have to know how to apply problem solving skills to calculate how much paint you need to cover a given area. Problem solving skills are essential to understanding mathematics. This chapter will review algebra skills that are necessary for your study of precalculus, and show how they are used in various applications.

The perimeter of a rectangular fence is 15 feet. Write the width of the fence in terms of the length.

$$2l + 2w = 15$$
$$2w = 15 - 2l$$
$$w = \frac{1}{2}(15 - 2l)$$

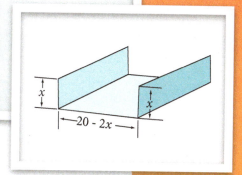

Outline

Objectives

- Understand basic properties of real numbers.
- Understand interval notation.
- Represent an interval of real numbers on the number line.
- Find the absolute value of a real number.
- Simplify expressions using order of operations.

The number system that you are familiar with is formally called the **real number system**. The real numbers evolved over time as the need arose to numerically represent various types of values. When people first started using numbers, they counted 1, 2, 3, and so on. It is customary to enclose a list of numbers in braces, and to formally call them a **set** of numbers. The counting numbers are called the set of **natural numbers**:

$$\{1, 2, 3, 4, 5, 6, \ldots\}$$

Soon a number was needed to represent nothing, and so zero was added to the set of natural numbers to get the set of **whole numbers**:

$$\{0, 1, 2, 3, 4, 5, \ldots\}$$

As commerce intensified, the concept of being in debt introduced negative numbers. The set consisting of the negatives of the natural numbers, the positive natural numbers, and zero is called the set of **integers**:

$$\{\ldots, -6, -5, -4, -3, -2, -1, 0, 1, 2, 3, 4, 5, 6, \ldots\}$$

The next step in the evolution of numbers is the set of **rational numbers**. A loaf of bread could be cut into pieces and a unit of time or money could be subdivided. Rational numbers are represented by the division of two integers $\frac{r}{s}$, with $s \neq 0$. They can be described by terminating or nonterminating, repeating decimal representations. The following are examples of rational numbers:

$$\frac{1}{3} = 0.3333\ldots, \quad -0.25 = -\frac{1}{4}, \quad 4\frac{2}{5} = \frac{22}{5}$$

When the ancient Greeks used triangles and circles in designing buildings, they discovered that some measurements were not represented by rational numbers. These new numbers were therefore called **irrational numbers**. The irrational number can be expressed as nonterminating and nonrepeating decimals. New symbols were introduced to represent the exact values of these numbers. Examples of irrational numbers include the following:

$$\sqrt{2}, \quad \pi, \quad \frac{\sqrt{5}}{2}$$

Properties of Real Numbers

The **associative property** states that numbers can be grouped in any order when adding or multiplying, and the result will be the same.

> **Associative Properties**
>
> Associative property of addition $\qquad a + (b + c) = (a + b) + c$
> Associative property of multiplication $\quad a(bc) = (ab)c$
> Here, a, b, and c are any real numbers.

The **commutative property** states that numbers can be added or multiplied in any order.

> **Commutative Properties**
>
> Commutative property of addition $\qquad a + b = b + a$
> Commutative property of multiplication $\quad ab = ba$
> Here, a and b are any real numbers.

Note A calculator display of 3.141592654 is just an approximation to π. Calculators and computers can only approximate irrational numbers since they can store only a finite number of digits.

The **distributive property** of multiplication over addition changes sums to products or products to sums.

> ### Distributive Property
>
> $ab + ac = a(b + c)$, where a, b, and c are any real numbers.

We also have the following definitions for additive and multiplicative identities.

> ### Additive and Multiplicative Identities
>
> There exists a unique real number 0 such that, for any real number a,
>
> $$a + 0 = 0 + a = a.$$
>
> The number 0 is called the **additive identity**.
> There exists a unique real number 1 such that, for any real number a,
>
> $$a \cdot 1 = 1 \cdot a = a.$$
>
> The number 1 is called the **multiplicative identity**.

Finally, we have the following definitions for additive and multiplicative inverses.

> ### Additive and Multiplicative Inverses
>
> The **additive inverse** of a real number a is $-a$, since $a + (-a) = 0$.
> The **multiplicative inverse** of a real number, $a \neq 0$, is $\frac{1}{a}$, since $a \cdot \frac{1}{a} = 1$.

VIDEO EXAMPLES

SECTION 1.1

Example 1   **Properties of Real Numbers**

What property does each of the following equations illustrate?

a. $4 + 6 = 6 + 4$ **b.** $3(5 - 8) = 3(5) - 3(8)$

c. $2 \cdot (3 \cdot 5) = (2 \cdot 3) \cdot 5$ **d.** $4 \cdot 1 = 4$

Solution

a. The equation $4 + 6 = 6 + 4$ illustrates the commutative property of addition.

b. The equation $3(5 - 8) = 3(5) - 3(8)$ illustrates the distributive property.

c. The equation $2 \cdot (3 \cdot 5) = (2 \cdot 3) \cdot 5$ illustrates the associative property of multiplication.

d. The equation $4 \cdot 1 = 4$ illustrates the multiplicative identity.

Check It Out 1   What property does the equality $7 \cdot 8 + 7 \cdot 5 = 7(8 + 5)$ illustrate?

Ordering of Real Numbers

The real numbers can be represented on a **number line**. Each real number corresponds to exactly one point on the number line. The number 0 corresponds to the **origin** of the number line. The positive numbers are to the right of the origin and the negative numbers are to the left of the origin. Figure 1 shows a number line.

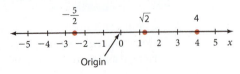

Figure 1

If we consider any two real numbers a and b, we can put them in a relative order. For example, $a < b$, is read as "a is less than b". It can also be read as "b is greater than a", that is $b > a$. They both mean that a is to the left of b on the number line.

The statement $a \leq b$ is read as "a is less than or equal to b". It means that *either $a < b$ or $a = b$.* Only one of these conditions needs to be satisfied for the entire statement to be true.

To express the set of real numbers that lie between two real numbers a and b, and include a and b, we write an **inequality** of the form $a \leq x \leq b$. This inequality can also be expressed in interval notation as $[a, b]$, called a **closed interval**. If the endpoints of the interval a and b are not included, we write the inequality $a < x < b$ as an **open interval** (a, b). The symbol ∞, **positive infinity**, is used to show that an interval extends forever in the positive direction. The symbol $-\infty$, **negative infinity**, is used to show that an interval extends forever in the negative direction.

Table 1 lists different types of inequalities, their corresponding interval notations, and their graphs on the number line. On a number line, when an endpoint of an interval is included, it is indicated by a filled-in, or closed circle. If an endpoint is not included, it is indicated by an open circle.

Note The symbol for infinity, ∞, is not a number. Therefore, you cannot use the bracket symbol next to it in interval notation. This is why any interval extending infinitely is denoted by the infinity symbol followed by a parenthesis.

Inequality	Interval Notation	Graph
$a \leq x \leq b$	$[a, b]$	
$a < x < b$	(a, b)	
$a \leq x < b$	$[a, b)$	
$a < x \leq b$	$(a, b]$	
$x \geq a$	$[a, \infty)$	
$x > a$	(a, ∞)	
$x \leq a$	$(-\infty, a]$	
$x < a$	$(-\infty, a)$	
All real numbers	$(-\infty, \infty)$	

Table 1

Example 2   Graphing an Interval

Graph each of the following intervals on the number line and give a verbal description.

a. $(-2, 3]$

b. $(-\infty, 4]$

Solution

a. The interval $(-2, 3]$ is graphed in Figure 2. Because -2 is not included in the set, it is represented by an open circle. Because 3 is included in the set, it is represented by a closed circle. The interval $(-2, 3]$ consists of all real numbers greater than -2 and less than or equal to 3.

Figure 2

b. The interval $(-\infty, 4]$ is graphed in Figure 3. Because 4 is included in the set, it is represented by a closed circle. The interval $(-\infty, 4]$ consists of all real numbers less than or equal to 4.

Figure 3

Check It Out 2 Graph the interval $[3, 5]$ on the number line and give a verbal description of the interval.

| **Example 3** | **Writing a Set in Interval Notation** |

Write the interval graphed in Figure 4 in interval notation.

Figure 4

Solution The interval consists of all numbers greater than or equal to -3 and less than 2. The number 2 is not included since it is represented by an open circle. Thus in interval notation, the set of points is written as $[-3, 2)$.

Check It Out 3 Write the interval graphed in Figure 5 in interval notation.

Figure 5

Absolute Value

The distance from the origin to a real number c is defined as the **absolute value** of c, denoted by $|c|$. The absolute value of a number is always nonnegative, and is also known as its **magnitude**. The algebraic definition of absolute value is given next.

Definition of Absolute Value

$$|x| = \begin{cases} x & \text{if } x \geq 0 \\ -x & \text{if } x < 0 \end{cases}$$

| **Example 4** | **Evaluating Absolute Value Expressions** |

Evaluate the following.

a. $|-3|$ **b.** $-4 - |4.5|$ **c.** $|-4 + 9| + 3$

Solution

a. Since $-3 < 0, |-3| = -(-3) = 3$.

b. Since $4.5 > 0, |4.5| = 4.5$. Thus, $-4 - |4.5| = -4 - 4.5 = -8.5$.

c. Perform the addition first and then evaluate the absolute value.

$$|-4 + 9| + 3 = |5| + 3 \qquad \text{Evaluate } -4 + 9 = 5 \text{ first}$$

$$= 5 + 3 = 8 \qquad \text{Since } 5 > 0, |5| = 5$$

Check It Out 4 Evaluate: $|4 - 6| - 3$.

Next we list some basic properties of absolute values that can be derived from the definition of absolute value.

Properties of Absolute Value

For any real numbers a and b, we have,

1. $|a| \geq 0$

2. $|-a| = |a|$

3. $|ab| = |a||b|$

4. $\left|\dfrac{a}{b}\right| = \dfrac{|a|}{|b|}, b \neq 0$

Using absolute value, we can determine the distance between two points on the number line. For instance, the distance between 7 and 12 is 5. We can write this using absolute value notation as follows:

$$|12 - 7| = 5 \quad \text{or} \quad |7 - 12| = 5$$

The distance is the same regardless of the order of subtraction.

Distance Between Two Points on the Real Number Line

Let a and b be two points on the number line. Then the distance between a and b is given by

$$|b - a| \quad \text{or} \quad |a - b|$$

Example 5   **Distance on the Real Number Line**

Find the distance between -6 and 4 on the real number line.

Solution Letting $a = -6$ and $b = 4$, we have

$$|a - b| = |-6 - 4| = |-10| = 10$$

Thus the distance between -6 and 4 is 10. We will obtain the same result if we compute $|b - a|$ instead.

$$|b - a| = |4 - (-6)| = |10| = 10$$

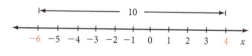

Figure 6

Check It Out 5   Find the distance between -4 and 5 on the real number line.

Order of Operations

When evaluating an arithmetic expression, the result must be the same regardless of who performs the operations. To ensure this, certain conventions for combining numbers must be followed. These conventions are outlined next.

Rules for Order of Operations

When evaluating a mathematical expression, perform the operations in the following order, beginning with the innermost parentheses and working outward.

- Simplify all numbers with exponents, working from left to right.
- Perform all multiplications and divisions, working from left to right.
- Perform all additions and subtractions, working from left to right.

When an expression is written as a quotient of two other expressions, the numerator and denominator are evaluated separately, and then the division is performed.

Using Technology

Calculators use order of operations to evaluate expressions. In Figure 1, note the use of parentheses to evaluate $\frac{3^2+1}{4-7}$. Using the **Frac** menu option, you can represent a repeating or terminating decimal as a fraction.

Keystroke Appendix:
Sections 2, 3, 4

```
3²+2(−3+7)
                    17
(3²+1)/(4−7)
           −3.333333333
Ans▶Frac
                 −10/3
■
```

Figure 1

Example 6   **Simplifying Using Order of Operations**

Simplify each expression.

a. $3^2 + 2(-3 + 7)$

b. $\dfrac{3^2 + 1}{4 - 7}$

c. $6 + 3(5 - (-3^2 + 2))$

Solution

a. Following the order of operations,

$$
\begin{aligned}
3^2 + 2(-3 + 7) &= 3^2 + 2(4) && \text{Evaluate within parentheses} \\
&= 9 + 2(4) && \text{Simplify number with exponent: } 3^2 = 9 \\
&= 9 + 8 && \text{Multiply } 2(4) \\
&= 17 && \text{Add}
\end{aligned}
$$

b. Because the expression $\frac{3^2+1}{4-7}$ is a quotient of two other expressions, evaluate the numerator and denominator separately, and then divide.

$$
\begin{aligned}
3^2 + 1 &= 10 && \text{Evaluate numerator} \\
4 - 7 &= -3 && \text{Evaluate denominator} \\
\frac{3^2 + 1}{4 - 7} &= -\frac{10}{3} && \text{Divide}
\end{aligned}
$$

c.
$$
\begin{aligned}
6 + 3(5 - (-3^2 + 2)) &= 6 + 3(5 - (-7)) && \text{Evaluate within innermost parentheses} \\
& && \text{Note that } -3^2 + 2 = -9 + 2 = -7 \\
&= 6 + 3(12) = 6 + 36 && \text{Evaluate within parentheses; then multiply} \\
&= 42 && \text{Add}
\end{aligned}
$$

Check It Out 6   Simplify the expression $5^2 - 3(15 - 12) + 4 \div 2$.

Understanding the order of operations is extremely important when entering and evaluating expressions in calculators.

Example 7   **Calculator Use and Order of Operations**

Evaluate the following expressions on your calculator. Check the calculator output with a hand calculation.

a. $4 + 10 \div 5 + 2$ **b.** $\dfrac{4 + 10}{5 + 2}$

Solution

a. Enter $4 + 10 \div 5 + 2$ in a calculator as

$$\boxed{4} \; \boxed{+} \; \boxed{10} \; \boxed{\div} \; \boxed{5} \; \boxed{+} \; \boxed{2}$$

The result is 8. This result checks with a hand calculation of

$$4 + 10 \div 5 + 2 = 4 + 2 + 2 = 8$$

b. To calculate $\frac{4+10}{5+2}$, we need to proceed with caution. The additions in the numerator and denominator must be performed first, before the division. This is entered in the calculator as

$$\boxed{(} \; \boxed{4} \; \boxed{+} \; \boxed{10} \; \boxed{)} \; \boxed{\div} \; \boxed{(} \; \boxed{5} \; \boxed{+} \; \boxed{2} \; \boxed{)}$$

The result is 2, which can be quickly checked by hand. The parentheses are a *must* in entering this expression. Omitting them will result in a wrong answer because the calculator will perform the operations in a different order.

 Check It Out 7   Evaluate the expression $\frac{3-10}{4+3} - 5$ using a calculator. Check your result by hand.

Note   See the XYZ eBook for answers to the *Check It Out* exercises.

Exercises 1.1

Skills

In Exercises 1–8, consider the following numbers.

$$\sqrt{2}, 0.5, \frac{4}{5}, -1, 0, 40, \pi, 10, -1.67$$

1. Which are integers?

2. Which are natural numbers?

3. Which are rational numbers?

4. Which are irrational numbers?

5. Which are whole numbers?

6. Which are rational numbers that are not integers?

7. Which are integers that are not positive?

8. Which are real numbers?

In Exercises 9–14, name the property illustrated by each equality.

9. $3 \cdot (8 \cdot 9) = (3 \cdot 8) \cdot 9$

10. $(5 + x) + z = 5 + (x + z)$

11. $-4(x - 2) = -4x + 8$

12. $3x + y = y + 3x$

13. $9(a) = a(9)$

14. $b(x + 2) = bx + 2b$

In Exercises 15–24, graph each interval on the number line.

15. $[-2, 4]$

16. $[-3, -1]$

17. $[-5, 0)$

18. $(-2, 4)$

19. $(-\infty, 3)$

20. $[2, \infty)$

21. $\left[\frac{1}{2}, 4\right]$

22. $\left[-\frac{3}{2}, 2\right]$

23. $(-2.5, 3)$

24. $[3.5, \infty)$

In Exercises 25–32, write the graphed interval in interval notation.

25.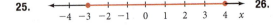

26.

27.

28.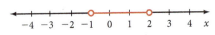

29.

30.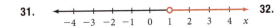

31.

32.

In Exercises 33–38, fill in the given table.

	Inequality	Interval Notation	Graph
33.	$4 \le x \le 10$		
34.		$(-\infty, 6)$	
35.	$-3 \le x < 0$		
36.		$(-5, 10]$	
37.			
38.			

In Exercises 39–54, evaluate each expression without using a calculator.

39. $|-3.2|$ **40.** $|-45.5|$ **41.** $|253|$ **42.** $|-37|$

43. $\left|\dfrac{5}{4}\right|$ **44.** $\left|-\dfrac{1}{2}\right|$ **45.** $\left|-\dfrac{4}{3}\right|$ **46.** $\left|\dfrac{7}{2}\right|$

47. $|-2|+4$ **48.** $-|5|$ **49.** $-|-4.5|$ **50.** $-|3.2|$

51. $|5-2|+4$ **52.** $4+|6-7|$ **53.** $5-|12-4|$ **54.** $-3-|6-10|$

In Exercises 55–66, find the distance between the numbers on the real number line.

55. $-2,4$ **56.** $-5,7$ **57.** $0,9$ **58.** $-12,0$

59. $-7.5,-12$ **60.** $6,-5.5$ **61.** $-4.3,7.9$ **62.** $6.7,13.4$

63. $-\dfrac{1}{2},\dfrac{5}{2}$ **64.** $-\dfrac{4}{3},\dfrac{7}{3}$ **65.** $\dfrac{1}{3},\dfrac{4}{5}$ **66.** $\dfrac{1}{4},-\dfrac{2}{3}$

In Exercises 67–76, evaluate each expression without using a calculator.

67. $9-3(2)-8$ **68.** $6+4(3)-14$ **69.** $(3-5)^2$ **70.** $8-(7-9)^2$

71. $3^2+3(5)-10$ **72.** $10+2^3\div4$ **73.** $(-3)^5+3$ **74.** $-6(1+2^2)$

75. $\dfrac{10-4^2}{2+3^2}$ **76.** $\dfrac{4(2)-6^2}{2^3-1}$

In Exercises 77–84, evaluate each expression using a calculator. Check your solution by hand.

77. -3^2 **78.** $(-3)^2$ **79.** $1+\dfrac{2}{3}-4^2$ **80.** $\left(-\dfrac{3}{5}\right)\left(\dfrac{25}{6}\right)$

81. $\dfrac{2}{3+5}$ **82.** $\dfrac{5+7}{3}$ **83.** $\dfrac{4^2+3}{5}$ **84.** $\dfrac{-6^2+3(7)}{5-2^3}$

Applications

85. Shopping The available shoe sizes at a store are whole numbers from 5 to 9, inclusive. List all the possible shoe sizes in the store.

86. Salary The annual starting salary for an administrative assistant at a university can range from \$30,000 to \$40,000, inclusive. Write the salary range in interval notation.

87. Weather On a particular winter day in Chicago, the temperature ranged from a low of 25°F to a high of 36°F. Write this range of temperatures in interval notation.

88. Travel On the Garden State Parkway in New Jersey, the distance between Exit A and Exit B, in miles, is given by $|A-B|$. How many miles are traveled between Exit 88 and Exit 127 on the Parkway?

89. Geography The highest point in the United States is Mount McKinley, Alaska, with an elevation of 20,320 feet. The lowest point in the U.S. is Death Valley, California, with an elevation of -282 feet (282 feet below sea level). Find the absolute value of the difference in elevation between the lowest and highest points in the U.S.
(Source: U.S. Geological Survey)

Concepts

90. Find two numbers a and b such that $|a+b|\neq|a|+|b|$. (Answers may vary.)

91. If a number is nonnegative, must it be positive? Explain.

92. Is the absolute value of a number always positive? Explain.

93. Find two points on the number line that are a distance of 4 units from -3.

94. Find a rational number less than π. (Answers may vary.)

SPOTLIGHT ON SUCCESS   *Student Instructor Lauren*

There are a lot of word problems in algebra and many of them involve topics that I don't know much about. I am better off solving these problems if I know something about the subject. So, I try to find something I can relate to. For instance, an example may involve the amount of fuel used by a pilot in a jet airplane engine. In my mind, I'd change the subject to something more familiar, like the mileage I'd be getting in my car and the amount spent on fuel, driving from my hometown to my college. Changing these problems to more familiar topics makes math much more interesting and gives me a better chance of getting the problem right. It also helps me to understand how greatly math affects and influences me in my everyday life. We really do use math more than we would like to admit—budgeting our income, purchasing gasoline, planning a day of shopping with friends—almost everything we do is related to math. So the best advice I can give with word problems is to learn how to associate the problem with something familiar to you.

You should know that I have always enjoyed math. I like working out problems and love the challenges of solving equations like individual puzzles. Although there are more interesting subjects to me, and I don't plan on pursuing a career in math or teaching, I do think it's an important subject that will help you in any profession.

Exponents, Roots, and Radicals

1.2

Objectives

- Simplify expressions with integer exponents.
- Write numbers using scientific notation.
- Find the nth root of a number.
- Simplify radical expressions.
- Add and subtract radical expressions.
- Simplify expressions with rational exponents.

In this section, we will discuss integers and rational numbers as exponents and the general rules for evaluating and simplifying expressions with exponents. We will also discuss scientific notation, a method used for writing very large and very small numbers.

Integer Exponents

If we want to multiply the same number by itself many times, writing out the multiplication becomes tedious, and so a new notation is needed. For example, $6 \cdot 6 \cdot 6$ can be compactly written as 6^3. The number 3 is called the **exponent** and the number 6 is called the **base**. The exponent tells you how many 6's are multiplied together. In general, we have the following definition.

> ### Definition of Positive Integer Exponents
>
> For any positive integer n,
> $$a^n = \underbrace{a \cdot a \cdot a \cdots a}_{n \text{ factors}}$$
> The number a is the **base** and the number n is the **exponent**.

Negative integer exponents are defined in terms of positive integer exponents as follows.

> ### Definition of Negative Integer Exponents
>
> Let a be any nonzero real number and m be a positive integer. Then
> $$a^{-m} = \frac{1}{a^m}$$

Using Technology

To calculate with exponents, use the x^y key in most scientific calculators. In a graphing calculator, use the ^ (hat) key.

Keystroke Appendix:
Section 4

Example 1   **Writing an Expression with Positive Exponents**

Write each expression using positive exponents.

a. 4^{-2} **b.** 1.45^{-3} **c.** $\dfrac{x^{-4}}{y^5}$, where $x, y \neq 0$

Solution

a. $4^{-2} = \dfrac{1}{4^2}$, using the rule for negative exponents.

b. $1.45^{-3} = \dfrac{1}{1.45^3}$, using the rule for negative exponents.

c. $\dfrac{x^{-4}}{y^5} = x^{-4} \cdot \dfrac{1}{y^5} = \dfrac{1}{x^4} \cdot \dfrac{1}{y^5} = \dfrac{1}{x^4 y^5}$

Check It Out 1   Write $x^{-2} y^3$ using positive exponents.

The following properties of exponents are used to simplify expressions containing exponents.

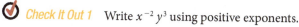

Properties of Integer Exponents

Let a and b be real numbers and let m and n be integers. Then the following properties hold.

Property	Illustration
1. $a^m \cdot a^n = a^{m+n}, a \neq 0$	$3^4 \cdot 3^{-2} = 3^{4+(-2)} = 3^2$
2. $(a^m)^n = a^{mn}, a \neq 0$	$(5^4)^3 = 5^{4 \cdot 3} = 5^{12}$
3. $(ab)^m = a^m b^m, a, b \neq 0$	$(4x)^3 = 4^3 x^3 = 64x^3$
4. $\left(\dfrac{a}{b}\right)^m = \dfrac{a^m}{b^m}, a, b \neq 0$	$\left(\dfrac{3}{7}\right)^2 = \dfrac{3^2}{7^2} = \dfrac{9}{49}$
5. $\dfrac{a^m}{a^n} = a^{m-n}, a \neq 0$	$\dfrac{5^4}{5^8} = 5^{4-8} = 5^{-4}$ or $\dfrac{1}{5^4}$
6. $a^1 = a$	$7^1 = 7$
7. $a^0 = 1, a \neq 0$	$10^0 = 1$

Example 2 shows how the properties of exponents can be combined to simplify expressions.

Example 2   **Simplifying Expressions Containing Exponents**

Simplify each expression and write it using positive exponents. Assume that variables represent nonzero real numbers.

a. $\dfrac{16x^{10}y^4}{4x^{10}y^8}$
  b. $\dfrac{(3x^{-2})^3}{x^{-5}}$
  c. $\left(\dfrac{54s^2t^{-3}}{6zt}\right)^{-2}$

Solution

a. $\dfrac{16x^{10}y^4}{4x^{10}y^8} = \dfrac{16}{4}x^{10-10}y^{4-8} = 4x^0y^{-4} = \dfrac{4}{y^4}$

The fact $x^0 = 1$ was used in the final step. It is usually a good idea to wait until the last step to write the expression using only positive exponents.

b. $\dfrac{(3x^{-2})^3}{x^{-5}} = \dfrac{3^3 x^{-2 \cdot 3}}{x^{-5}} = \dfrac{27x^{-6}}{x^{-5}} = 27x^{-6-(-5)} = 27x^{-1} = \dfrac{27}{x}$

c. $\left(\dfrac{54s^2t^{-3}}{6zt}\right)^{-2} = \left(\dfrac{9s^2t^{-3-1}}{z}\right)^{-2}$ Simplify within parentheses

$\qquad = \left(\dfrac{9s^2t^{-4}}{z}\right)^{-2}$

$\qquad = \dfrac{9^{-2}s^{-4}t^8}{z^{-2}}$ Use Property (4)

$\qquad = \dfrac{z^2t^8}{81s^4}$ Note that $9^{-2} = \dfrac{1}{81}$

Check It Out 2 Simplify the expression $\dfrac{(4y^{-3})^4}{(xy)^{-2}}$ and write it using positive exponents.

Scientific Notation

It is often more convenient to write very large or very small numbers using a system called **scientific notation**, which we now discuss.

> ### Scientific Notation
>
> A nonzero number x is written in scientific notation in the form
>
> $$a \times 10^b$$
>
> where $1 \le a < 10$ if $x > 0$, and $-10 < a \le -1$ if $x < 0$, and b is an integer.
>
> The number b is sometimes called the **order of magnitude** of x.

 Using Technology

Scientific notation is entered in most calculators using the EE key.

Keystroke Appendix:
Section 4

Example 3   **Expressing a Number in Scientific Notation**

Express each of the following numbers in scientific notation.

a. 328.5   **b.** 4.69   **c.** 0.00712

Solution

a. Because 328.5 is greater than 0, $1 \le a < 10$. In this case, $a = 3.285$. In going from 328.5 to 3.285, we shifted the decimal point two places to the left. In order to start *from* 3.285 and "get back" to 328.5, we have to shift the decimal point two places to the right. Thus $b = 2$. Thus, in scientific notation, we have

$$328.5 = 3.285 \times 10^2$$

b. Since $1 \le 4.69 < 10$, there is no need to shift the decimal point, and so $a = 4.69$ and $b = 0$. In scientific notation, 4.69 is written as 4.69×10^0.

c. Because 0.00712 is greater than 0, $1 \le a < 10$. Thus, $a = 7.12$. To get from 0.00712 to 7.12, we had to shift the decimal point three places to the right. To get back to 0.00712 from 7.12, we have to shift the decimal point three places to the left. Thus, $b = -3$. In scientific notation,

$$0.00712 = 7.12 \times 10^{-3}$$

Check It Out 3   Express 0.0315 in scientific notation.

Sometimes, we may need to convert a number in scientific notation to decimal form.

> ### Converting From Scientific Notation to Decimal Form
>
> To convert a number in the form $a \times 10^b$ into decimal form, proceed as follows.
>
> - If $b > 0$, the decimal point has to be shifted b places *to the right*.
> - If $b = 0$, the decimal point is not shifted at all.
> - If $b < 0$, the decimal is shifted $|b|$ places *to the left*.

Example 4   **Converting From Scientific Notation to Decimal Form**

Write the following numbers in decimal form.

a. 2.1×10^5 **b.** 3.47×10^{-3}

Solution

a. To express 2.1×10^5 in decimal form, move the decimal point in 2.1 **five** places to the **right**. You will need to append four zeros.

$$2.1 \times 10^5 = 210,000$$

b. To express 3.47×10^{-3} in decimal form, move the decimal point in 3.47 **three** places to the **left**. So we need to attach two zeros to the right of the decimal point.

$$3.47 \times 10^{-3} = 0.00347$$

Check It Out 4   Write 7.05×10^{-4} in decimal form.   ■

To multiply and divide numbers using scientific notation, we apply the rules of exponents as illustrated in Example 5.

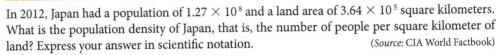

Example 5   **Population Density**

In 2012, Japan had a population of 1.27×10^8 and a land area of 3.64×10^5 square kilometers. What is the population density of Japan, that is, the number of people per square kilometer of land? Express your answer in scientific notation.   (*Source*: CIA World Factbook)

Solution   To compute the number of people per square kilometer, divide the total population by the total land area.

$$\frac{\text{Total population}}{\text{Total land area}} = \frac{1.27 \times 10^8}{3.64 \times 10^5}$$

$$= \frac{1.27}{3.64} \cdot \frac{10^8}{10^5}$$

$$\approx 0.349 \times 10^3 \quad \text{Use properties of exponents}$$

$$= 3.49 \times 10^2 \quad \text{Write in scientific notation}$$

Thus the population density is approximately 349 people per square kilometer.

Check It Out 5   Find the population density of Italy, with a population of 6.13×10^7 and land area of 2.94×10^5 square kilometers (2012 estimates).   ■

Roots and Radicals

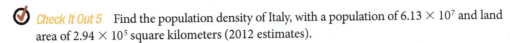

The square root of 25 is defined to be the positive number 5, since $5^2 = 25$. The square root of 25 is indicated by $\sqrt{25}$. The fourth root of 25 is denoted by $\sqrt[4]{25}$. Note that we can take even roots of only nonnegative numbers.

Similarly, the cube root of -8 is indicated by $\sqrt[3]{-8} = -2$, since $(-2)^3 = -8$. We can take odd roots of all real numbers, negative and nonnegative. This leads to the following definition.

The nth Root of a

Let n be an even positive integer and a be a nonnegative real number. Then

$$\sqrt[n]{a}$$ denotes the nonnegative number whose nth power is a.

Let n be an odd positive integer greater than 1 and a be any real number. Then

$$\sqrt[n]{a}$$ denotes the number whose nth power is a.

The number $\sqrt[n]{a}$ is called the **nth root of a**. The $\sqrt[n]{}$ symbol is called a **radical**. If the n is omitted in the radical symbol, it is assumed to be a square root.

Example 6   **Evaluating nth Root of a Number**

Determine $\sqrt[3]{64}$, $\sqrt[4]{16}$, and $\sqrt[5]{-32}$.

Solution   First, $\sqrt[3]{64} = 4$, since $4^3 = 64$. Next, $\sqrt[4]{16} = 2$, since $2^4 = 16$. Finally, $\sqrt[5]{-32} = -2$, since $(-2)^5 = -32$.

Check It Out 6   Determine $\sqrt[4]{81}$.

We now introduce the following rules for radicals.

Rules for Operations with Radicals

Suppose a and b are real numbers such that their nth roots are defined.

Product Rule: $\sqrt[n]{a} \cdot \sqrt[n]{b} = \sqrt[n]{ab}$

Quotient Rule: $\dfrac{\sqrt[n]{a}}{\sqrt[n]{b}} = \sqrt[n]{\dfrac{a}{b}}$, $b \neq 0$

It is conventional to leave radicals only in the numerator of a fraction. This is known as **rationalizing the denominator**, a technique that is illustrated in the following three examples.

Example 7   **Rationalizing the Denominator**

Rationalize the denominator of each expression.

a. $\sqrt{\dfrac{18}{5}}$

b. $\dfrac{\sqrt[3]{10}}{\sqrt[3]{3}}$

Solution

a. $\sqrt{\dfrac{18}{5}} = \dfrac{\sqrt{18}}{\sqrt{5}} \cdot \dfrac{\sqrt{5}}{\sqrt{5}} = \dfrac{\sqrt{90}}{5} = \dfrac{3\sqrt{10}}{5}$ $\sqrt{90} = \sqrt{9} \cdot \sqrt{10} = 3\sqrt{10}$

b. $\dfrac{\sqrt[3]{10}}{\sqrt[3]{3}} = \dfrac{\sqrt[3]{10}}{\sqrt[3]{3}} \cdot \dfrac{\sqrt[3]{9}}{\sqrt[3]{9}}$ Multiply numerator and denominator by $\sqrt[3]{9}$ to obtain a perfect cube in the denominator

$\phantom{\dfrac{\sqrt[3]{10}}{\sqrt[3]{3}}} = \dfrac{\sqrt[3]{90}}{\sqrt[3]{27}} = \dfrac{\sqrt[3]{90}}{3}$ $\sqrt[3]{27} = 3$

Note that $\sqrt[3]{90}$ cannot be further simplified since there are no perfect cubes which are factors of 90.

Check It Out 7   Rationalize the denominator of $\dfrac{5}{\sqrt{2}}$, and then simplify, if possible.

The product and quotient rules for radicals can be used to simplify expressions involving radicals. It is conventional to write radical expressions such that for the nth root, there are no powers under the radical greater than or equal to n. We have the following rule for finding the nth root of a^n, $a \neq 0$.

Note   Unless otherwise stated, variables are assumed to be positive to avoid the issue of taking absolute values.

Finding nth root of a^n

If n is odd, then $\sqrt[n]{a^n} = a$, $a \neq 0$.
If n is even, then $\sqrt[n]{a^n} = |a|$, $a \neq 0$.
For instance, $\sqrt[3]{(-4)^3} = -4$ whereas $\sqrt{(-4)^2} = |-4| = 4$.

Example 8   Simplifying Radicals

Simplify the following.

a. $\sqrt{75}$

b. $\sqrt[3]{\dfrac{125}{108}}$

c. $\sqrt{x^5 y^7}$, $\quad x, y > 0$

Solution

a. Because $75 = 25 \cdot 3$, and 25 is a perfect square, we use the product rule for radicals to obtain

$$\sqrt{75} = \sqrt{25 \cdot 3} = \sqrt{25}\sqrt{3} = 5\sqrt{3}$$

b. Apply the quotient rule for radicals. Since $125 = 5^3$ and $108 = 27 \cdot 4$, we can write

$$\sqrt[3]{\frac{125}{108}} = \frac{\sqrt[3]{125}}{\sqrt[3]{108}} = \frac{5}{\sqrt[3]{27 \cdot 4}} = \frac{5}{3\sqrt[3]{4}}$$

Next, we clear the radical $\sqrt[3]{4}$ in the denominator by multiplying the numerator and denominator by $\sqrt[3]{2}$.

$$\frac{5}{3\sqrt[3]{4}} = \frac{5}{3\sqrt[3]{4}} \cdot \frac{\sqrt[3]{2}}{\sqrt[3]{2}} = \frac{5\sqrt[3]{2}}{3\sqrt[3]{8}} = \frac{5\sqrt[3]{2}}{6}$$

c. Because $x^5 = x \cdot x^4 = x(x^2)^2$ and $y^7 = y \cdot y^6 = y(y^3)^2$, we have

$$\sqrt{x^5 y^7} = \sqrt{x(x^2)^2 y(y^3)^2} = x^2 y^3 \sqrt{xy}$$

Because we were given that $x, y > 0$, the square root of the quantity under the radical is always defined.

✓ *Check It Out 8*   Simplify $\sqrt{45}$ and $\sqrt[3]{x^5 y^6}$. ■

If the denominator of an expression contains two terms of the form $\sqrt{a} + \sqrt{b}$, eliminate the radical(s) by multiplying the numerator and denominator by $\sqrt{a} - \sqrt{b}$, $a, b > 0$. We see that

$$(\sqrt{a} + \sqrt{b})(\sqrt{a} - \sqrt{b}) = (\sqrt{a} + \sqrt{b})(\sqrt{a}) - (\sqrt{a} + \sqrt{b})(\sqrt{b})$$

$$= \sqrt{a} \cdot \sqrt{a} + \sqrt{b} \cdot \sqrt{a} - \sqrt{a} \cdot \sqrt{b} - \sqrt{b} \cdot \sqrt{b}$$

$$= \sqrt{a} \cdot \sqrt{a} - \sqrt{b} \cdot \sqrt{b} = a - b$$

Therefore, the radical(s) in the denominator have been eliminated. If the denominator is of the form $\sqrt{a} - \sqrt{b}$, then multiply both numerator and denominator by $\sqrt{a} + \sqrt{b}$. The same approach can be used for denominators of the form $a \pm \sqrt{b}$ and $\sqrt{a} \pm b$.

Example 9   Rationalizing a Denominator Containing Two Terms

Rationalize the denominator:

$$\frac{5}{4 - \sqrt{2}}$$

Solution To remove the radical in the denominator, multiply the numerator and denominator by $4 + \sqrt{2}$.

$$\frac{5}{4 - \sqrt{2}} = \frac{5}{4 - \sqrt{2}} \cdot \frac{4 + \sqrt{2}}{4 + \sqrt{2}}$$

$$= \frac{5(4 + \sqrt{2})}{4^2 - (\sqrt{2})^2} = \frac{20 + 5\sqrt{2}}{14}$$

Check It Out 9 Rationalize the denominator of $\dfrac{4}{1+\sqrt{3}}$.

Adding and Subtracting Radical Expressions

Two or more radicals of the form $\sqrt[n]{a}$ can be combined provided that they all have the same expression under the radical and all have the same value of n. For instance, $3\sqrt{2} - 2\sqrt{2} = \sqrt{2}$ and $\sqrt[3]{x} + 2\sqrt[3]{x} = 3\sqrt[3]{x}$.

Example 10   **Combining Radical Expressions**

Simplify the following radical expressions. Assume $x \geq 0$.

a. $\sqrt{48} + \sqrt{27} - \sqrt{12}$

b. $(3 + \sqrt{x})(4 - 2\sqrt{x})$

Solution

a. $\sqrt{48} + \sqrt{27} - \sqrt{12} = \sqrt{16 \cdot 3} + \sqrt{9 \cdot 3} - \sqrt{4 \cdot 3}$

$$= 4\sqrt{3} + 3\sqrt{3} - 2\sqrt{3} = 5\sqrt{3}$$

b. $(3 + \sqrt{x})(4 - 2\sqrt{x}) = 12 - 6\sqrt{x} + 4\sqrt{x} - 2\sqrt{x}\,\sqrt{x}$

$$= 12 - 2\sqrt{x} - 2x$$

 Check It Out 10   Simplify $\sqrt{16x} - \sqrt{9x} + \sqrt{x^3}$. Assume $x \geq 0$.

Rational Exponents

We have already discussed integer exponents in this section. We can also define exponents that are rational.

> **Definition of $a^{1/n}$**
>
> If a is a real number and n is a positive integer greater than 1, then
> $$a^{1/n} = \sqrt[n]{a}, \quad \text{where } a \geq 0 \text{ when } n \text{ is even.}$$
> The quantity $a^{1/n}$ is called the **nth root of a**.

Example 11   **Expressions Involving nth Roots**

Evaluate the following.

a. $27^{1/3}$

b. $-36^{1/2}$

Solution

a. Applying the definition of $a^{1/n}$ with $n = 3$,

$$27^{1/3} = \sqrt[3]{27} = 3.$$

b. Applying the definition of $a^{1/n}$ with $n = 2$, $-36^{1/2} = -(\sqrt{36}) = -6$. Note that only the number 36 is raised to the 1/2 power, and the result is multiplied by -1.

 Check It Out 11   Evaluate $(-8)^{1/3}$.

Next we give the definition of $a^{m/n}$, where m/n is in lowest terms and $n \geq 2$. The definition is given in a way that the laws of exponents given earlier also hold for rational exponents.

Definition of $a^{m/n}$

Let a be a positive real number and m and n be integers such that m/n is in lowest terms and $n \geq 2$. We then have

$$a^{m/n} = \sqrt[n]{a^m} = \left(\sqrt[n]{a}\right)^m$$

For example $64^{2/3} = \left(\sqrt[3]{64}\right)^2 = 4^2 = 16$. Also, $64^{2/3} = \sqrt[3]{64^2} = \sqrt[3]{4096} = 16$.

The rules for integer exponents given earlier also hold for rational exponents, with some restrictions. These rules are summarized below.

Properties of Rational Exponents

Assume that a and b are real numbers and m and n are rational numbers. Whenever m and n indicate even roots, we assume that a and b are *nonnegative* real numbers. Then the following properties hold.

1. $a^m \cdot a^n = a^{m+n}$ **2.** $(a^m)^n = a^{mn}$

3. $(ab)^m = a^m b^m$ **4.** $a^{-m} = \dfrac{1}{a^m}, a \neq 0$

5. $\left(\dfrac{a}{b}\right)^m = \dfrac{a^m}{b^m}, b \neq 0$ **6.** $\dfrac{a^m}{a^n} = a^{m-n}, a \neq 0$

| **Example 12** | **Expressions Containing Rational Exponents** |

Evaluate the following expressions without using a calculator.

a. $(4)^{3/2}$ **b.** $\left(\dfrac{1}{3}\right)^{-2}$ **c.** $(23.1)^0$

Solution

a. Using the rules for rational exponents,

$$(4)^{3/2} = (4^{1/2})^3 = 2^3 = 8$$

b. Using the rules for exponents,

$$\left(\dfrac{1}{3}\right)^{-2} = \left(\left(\dfrac{1}{3}\right)^{-1}\right)^2 = (3)^2 = 9 \quad \text{Recall that } \left(\dfrac{1}{3}\right)^{-1} = 3$$

c. Note that $(23.1)^0 = 1$, since any nonzero number raised to the zero power is defined to be equal to 1.

Check It Out 12   Evaluate the following expressions without using a calculator.

a. $(8)^{2/3}$ **b.** $\left(\dfrac{1}{4}\right)^{-3}$ **c.** $(16)^{-1/2}$

Example 13   **Simplifying Expressions Containing Rational Exponents**

Simplify and write with positive exponents.

a. $(25)^{3/2}$ **b.** $(16s^{4/3}\,t^{-3})^{3/2}$,   where $s, t > 0$ **c.** $\dfrac{(x^{-7/3}\,y^{5/2})^3}{y^{-1/2}\,x^2}$,   where $x, y > 0$

Solution

a. $(25)^{3/2} = (25^{1/2})^3 = \left(\sqrt{25}\right)^3 = 5^3 = 125$

b. First use the third Property of rational exponents to rewrite the power of a product.

$$(16s^{4/3}\,t^{-3})^{3/2} = 16^{3/2}(s^{4/3})^{3/2}(t^{-3})^{3/2} \qquad \text{Use Property (3)}$$

$$= 16^{3/2}s^{(4/3)\,(3/2)}t^{-3(3/2)} \qquad \text{Use Property (2)}$$

$$= 16^{3/2}s^2t^{-9/2} \qquad \tfrac{4}{3}\cdot\tfrac{3}{2} = 2 \text{ and } (-3)\left(\tfrac{3}{2}\right) = -\tfrac{9}{2}$$

$$= \frac{64s^2}{t^{9/2}} \qquad 16^{3/2} = (16^{1/2})^3 = 64$$

c. Use the third Property to simplify the numerator. This gives

$$\frac{(x^{-7/3}\,y^{5/2})^3}{y^{-1/2}\,x^2} = \frac{x^{(-7/3)\cdot 3}\,y^{(5/2)\cdot 3}}{y^{-1/2}\,x^2} \qquad \text{Use Property (3)}$$

$$= \frac{x^{-7}\,y^{15/2}}{y^{-1/2}\,x^2} \qquad -\tfrac{7}{3}\cdot 3 = -7 \text{ and } \tfrac{5}{2}\cdot 3 = \tfrac{15}{2}$$

$$= x^{-7-2}\,y^{15/2-(-1/2)} = x^{-9}y^8 \qquad \text{Use Property (6)}$$

$$= \frac{y^8}{x^9} \qquad \text{Use Property (4) to write with positive exponents}$$

Check It Out 13   Simplify the following.

a. $(16)^{5/4}$

b. $(8s^{5/3}\,t^{-1/6})^3$

Exercises 1.2

Skills

In Exercises 1–8, simplify each expression without using a calculator.

1. -3^2

2. $(-3)^2$

3. 4^{-3}

4. 6^{-2}

5. -2^0

6. $(-3)^0$

7. $\left(\frac{3}{4}\right)^{-2}$

8. $\left(\frac{5}{2}\right)^{-2}$

In Exercises 9–24, simplify each expression and write using positive exponents. Assume that all variables represent nonzero numbers.

9. $-(4x^2y^4)^2$

10. $(-4xy^3)^2$

11. $(2x^5)^{-2}$

12. $(4y^2)^{-3}$

13. $-2(a^2b^5)^2$

14. $-3(a^4b^2)^2$

15. $(-4xy^2)^{-2}$

16. $(-3x^2y^2)^{-3}$

17. $\dfrac{7x^6y^2}{21x^3}$

18. $\dfrac{6x^3y^4}{24x^2y}$

19. $\dfrac{(2y^{-3})^2}{y^{-5}}$

20. $\dfrac{(-4x^{-1})^2}{x^{-3}}$

21. $\left(\dfrac{2x^{-2}y^3}{xy^4}\right)^2$

22. $\left(\dfrac{3x^3y}{xy^2}\right)^3$

23. $\left(\dfrac{-4s^5t^{-2}}{12s^3t}\right)^{-2}; s, t \neq 0$

24. $\left(\dfrac{27s^3z^5t^{-2}}{3z^2t}\right)^{-3}; z, t \neq 0$

In Exercises 25–32, write each number in scientific notation.

25. 0.0051

26. 23.37

27. 5600

28. 497

29. 0.0000567

30. 0.0000032

31. 1,760,000

32. 5,341,200

In Exercises 33–40, write the following numbers in decimal form.

33. 3.71×10^2

34. 4.26×10^4

35. 2.8×10^{-2}

36. 6.25×10^{-3}

37. 5.96×10^5

38. 2.5×10^3

39. 4.367×10^7

40. 3.105×10^{-2}

In Exercises 41–44, compute and write the answer in scientific notation.

41. $(2.1 \times 10^3)(4.3 \times 10^4)$

42. $(3.7 \times 10^{-1})(5.1 \times 10^3)$

43. $\dfrac{9.4 \times 10^2}{4.7 \times 10^3}$

44. $\dfrac{1.3 \times 10^{-4}}{3.9 \times 10^{-2}}$

In Exercises 45–52, evaluate without using a calculator.

45. $\sqrt{49}$

46. $\sqrt[3]{64}$

47. $\sqrt[3]{\dfrac{1}{8}}$

48. $-\sqrt{\dfrac{9}{4}}$

49. $(49)^{3/2}$

50. $(27)^{2/3}$

51. $\left(\dfrac{16}{625}\right)^{1/4}$

52. $\left(\dfrac{-8}{125}\right)^{1/3}$

In Exercises 53–86, simplify the radical expressions. Assume that all variables represent positive real numbers.

53. $\sqrt{32}$

54. $\sqrt{75}$

55. $\sqrt[3]{250}$

56. $\sqrt[3]{80}$

57. $\sqrt{32} \cdot \sqrt{8}$

58. $\sqrt{27} \cdot \sqrt{12}$

59. $\sqrt[3]{\dfrac{-16}{125}}$

60. $\sqrt[3]{\dfrac{32}{125}}$

61. $\sqrt{\dfrac{3}{5}}$

62. $\sqrt{\dfrac{2}{7}}$

63. $\sqrt[3]{\dfrac{7}{9}}$

64. $\sqrt[3]{\dfrac{5}{2}}$

65. $\sqrt{\dfrac{50}{147}}$

66. $\sqrt{\dfrac{32}{125}}$

67. $\sqrt[3]{\dfrac{-48}{81}}$

68. $\sqrt[3]{\dfrac{-375}{32}}$

69. $\sqrt{x^3 y^4}, \quad x, y > 0$

70. $\sqrt{s^6 y^3}, \quad s, y > 0$

71. $\sqrt[3]{x^8 y^4}$

72. $\sqrt[3]{s^{10} t^7}$

73. $\sqrt{3x^2 y} \cdot \sqrt{15xy^3}$

74. $\sqrt{5yz^3} \cdot \sqrt{8y^2 z^2}$

75. $\sqrt[3]{6x^3 y^2} \cdot \sqrt[3]{4x^2 y}$

76. $\sqrt[3]{9xy^4} \cdot \sqrt[3]{6x^2 y^2}$

77. $\sqrt{98} + 3\sqrt{32}$

78. $2\sqrt{200} - \sqrt{72}$

79. $\sqrt{216} - 4\sqrt{24} + \sqrt{3}$

80. $-\sqrt{125} + \sqrt{20} - \sqrt{50}$

81. $(1 + \sqrt{5})(1 - \sqrt{5})$

82. $(3 - \sqrt{2})(3 + \sqrt{2})$

83. $\dfrac{4}{1 - \sqrt{5}}$

84. $\dfrac{3}{2 + \sqrt{6}}$

85. $\dfrac{1}{\sqrt{3} - \sqrt{2}}$

86. $\dfrac{2}{\sqrt{5} + \sqrt{2}}$

In Exercises 87–98, simplify and write with positive exponents. Assume that all variables represent positive real numbers.

87. $3^{2/3} \cdot 3^{-4/3}$

88. $2^{1/2} \cdot 2^{-1/3}$

89. $5^{-1/2} \cdot 5^{3/2}$

90. $3^{2/3} \cdot 3^{1/3}$

91. $\dfrac{7^{-1/4}}{7^{1/2}}$

92. $\dfrac{5^{1/2}}{5^{1/3}}$

93. $\dfrac{4^{-1/3}}{4^{1/4}}$

94. $\dfrac{2^{3/2}}{2^{1/4}}$

95. $(x^2 y^3)^{-1/2}$

96. $(s^4 y^5)^{-1/3}$

97. $\dfrac{x^{1/3} \cdot x^{1/2}}{x^2}$

98. $\dfrac{y^{2/3} \cdot y^{3/2}}{y^3}$

Applications

99. Astronomy The moon orbits the earth at 36,800 kilometers per hour. Express the given quantity in scientific notation.

100. Biology The length of a large amoeba is 0.005 millimeters. Express the given quantity in scientific notation.

101. Land Area The total land area of the United States is approximately 3,540,000 square miles. Express the given quantity using scientific notation.

102. Population The United States population estimate for 2012 is 3.14×10^8 people. If the total land area of the United States is approximately 3,540,000 square miles, how many people are there per square mile of land in the U.S.? Express your answer in decimal form.

103. Geometry The length of a diagonal of a square with a side of length s is $s\sqrt{2}$. Find the sum of the lengths of the two diagonals of a square whose side is five inches long.

104. Physics If an object is dropped from a height of h meters, it will take $\sqrt{\dfrac{h}{4.9}}$ seconds to hit the ground. How long will it take a ball dropped from a height of 30 meters to hit the ground? Round to the nearest tenth.

105. Ecology The number of tree species in a forested area in Malaysia is given approximately by the expression $386a^{1/4}$, where a is the area of the forested region in square kilometers. Determine the number of tree species in an area of 20 square kilometers. Round to the nearest whole species.

(*Source*: Plotkin et al., Proceedings of the National Academy of Sciences, Sept. 2000).

Concepts

106. For what value(s) of x is the expression $\dfrac{3x^{-2}}{5}$ defined?

107. For what value(s) of x and y is the expression $\dfrac{3x^2 y^{-3}}{y}$ defined?

108. For what values of b is the expression $\sqrt[2]{-b}$ a real number?

109. Show with a numerical example that $\sqrt{x^2 + y^2} \neq x + y$. (Answers may vary.)

110. Without using a calculator, explain why $\sqrt{10}$ must be greater than 3.

Polynomials and Factoring

Objectives

- Define a polynomial.
- Write a polynomial in descending order.
- Add and subtract polynomials.
- Multiply polynomials.
- Factor by grouping.
- Factor trinomials.
- Factor difference of squares and perfect square trinomials.
- Factor sums and differences of cubes.

Algebraic Expressions

In algebra, letters such as x and y are known as **variables**. You can use a variable to represent an unknown quantity. When you combine variables and numbers using multiplication, division, addition, and subtraction, as well as powers and roots, you get an **algebraic expression**.

In this section, we discuss a specific type of algebraic expression known as a **polynomial**. Some examples of polynomials are

$$2x + 7, \quad 3y^2 + 8y - \frac{3}{2}, \quad 10x^7 + x\sqrt{5}, \quad \text{and} \quad 5$$

Polynomials consist of sums of individual expressions, where each expression is the product of a real number and a variable raised to a nonnegative integer power.

Polynomial Expression in One Variable

A **polynomial in one variable** is an algebraic expression of the form

$$a_n x^n + a_{n-1} x^{n-1} + a_{n-2} x^{n-2} + \ldots + a_1 x + a_0,$$

where n is a nonnegative integer, a_n, $a_{n-1}, \ldots, a_0$ are real numbers, and $a_n \neq 0$.

All polynomials discussed in this section are in one variable only. They will form the basis for our work with quadratic and polynomial functions in Chapters 2 and 3.

Note The variable x in the definition of a polynomial can be consistently replaced by any other variable.

Polynomial Terminology

The following is a list of important definitions related to polynomials.

- The **degree** of the polynomial is n, the highest power to which a variable is raised.
- The parts of a polynomial separated by plus signs are called **terms**.
- The numbers a_n, $a_{n-1}, \ldots, a_0$ are called **coefficients** and the **leading coefficient** is a_n.
- The **constant term** is a_0.
- Polynomials are usually written in **descending order**, with the exponents decreasing from left to right.

VIDEO EXAMPLES

SECTION 1.3

Example 1   **Identifying Features of a Polynomial**

Write the following polynomial in descending order and find its degree, terms, coefficients, and constant term.

$$-4x^2 + 3x^5 - 2x - 7$$

Solution Writing the polynomial in descending order, with the exponents decreasing from left to right, we have

$$3x^5 - 4x^2 - 2x - 7$$

The degree of the polynomial is 5 because that is the highest power to which a variable is raised. The terms of the polynomial are

$$3x^5, \quad -4x^2, \quad -2x, \quad \text{and} \quad -7$$

The coefficients of the polynomial are $a_5 = 3$, $a_2 = -4$, $a_1 = -2$, and $a_0 = -7$. Note that $a_4 = a_3 = 0$. The constant term is $a_0 = -7$.

Check It Out 1   Write the following polynomial in descending order and find its degree, terms, coefficients, and constant term.

$$-3x^2 + 7 + 4x^5$$

Special Names for Polynomials

Polynomials with one, two, or three terms have specific names.

- A polynomial with one term, such as $2y^3$, is called a **monomial**.

- A polynomial with two terms, such as $-3z^4 + \frac{1}{5}z^2$, is called a **binomial**.

- A polynomial with three terms, such as $4t^6 - \frac{3}{4}t^2 - \sqrt{3}$, is called a **trinomial**.

Addition and Subtraction of Polynomials

Terms of an expression that have the same variable raised to the same power are called **like terms**. To add or subtract polynomials, we *combine* or *collect like terms* by adding their respective coefficients using the distributive property. This is illustrated in Example 2.

   **Adding and Subtracting Polynomials**

Add or subtract each of the following.

a. $(3x^3 + 2x^2 - 5x + 7) + (x^3 - x^2 + 5x - 2)$ **b.** $\left(s^4 + \frac{3}{4}s^2\right) - (s^4 - s^2)$

Solution

a. Rearranging terms so that the terms with the same powers are grouped together gives

$$(3x^3 + 2x^2 - 5x + 7) + (x^3 - x^2 + 5x - 2)$$
$$= (3x^3 + x^3) + (2x^2 - x^2) + (-5x + 5x) + (7 - 2)$$

Using the distributive property,

$$= (3 + 1)x^3 + (2 - 1)x^2 + (-5 + 5)x + 7 - 2$$
$$= 4x^3 + x^2 + 5 \qquad\qquad\qquad \text{Simplify}$$

b. We must distribute the minus sign throughout the second polynomial.

$$\left(s^4 + \frac{3}{4}s^2\right) - (s^4 - s^2) = s^4 + \frac{3}{4}s^2 - s^4 + s^2$$

$$= (s^4 - s^4) + \left(\frac{3}{4}s^2 + s^2\right) \qquad \text{Collect like terms}$$

$$= (1 - 1)s^4 + \left(\frac{3}{4} + 1\right)s^2 \qquad \text{Use distributive property}$$

$$= \frac{7}{4}s^2$$

 Check It Out 2   Add the following:

$$(5x^4 - 3x^2 + x - 4) + (-4x^4 + x^3 + 6x^2 + 4x)$$

Multiplication of Polynomials

We will first explain the multiplication of monomials, since these are the simplest of the polynomials. To multiply two monomials, we multiply the coefficients of the monomials and then multiply the variable expressions.

Example 3   **Multiplication of Monomials**

Multiply: $(-2x^2)(4x^7)$.

Solution

$$(-2x^2)(4x^7) = (-2)(4)(x^2 x^7) \qquad \text{Commutative and associate properties of multiplication}$$
$$= -8x^9 \qquad\qquad\quad \text{Multiply coefficients and add exponents of same base}$$

Check It Out 3   Multiply: $(4x^4)(-3x^5)$.

To multiply binomials, apply the distributive property twice and then apply the rules for multiplying monomials.

Example 4   **Multiplying Binomials**

Multiply: $(3x + 2)(-2x - 3)$.

Solution

$$(3x + 2)(-2x - 3) = \mathbf{3x}(-2x - 3) + \mathbf{2}(-2x - 3) \qquad \text{Apply distributive property}$$

Note that the terms $3x$ and 2 are *each* multiplied by $(-2x - 3)$.

$$= -6x^2 - 9x - 4x - 6 \qquad \text{Remove parentheses}$$
$$= -6x^2 - 13x - 6 \qquad\quad \text{Combine like terms}$$

Check It Out 4   Multiply: $(x + 4)(2x - 3)$.

Examining our work, we see that to multiply binomials, each term in the first polynomial is multiplied by each term in the second polynomial. To make sure you have multiplied all combinations of the terms, use the memory aid **FOIL**: multiply the First terms, then the **O**uter terms, then the **I**nner terms, and then the **L**ast terms. Collect like terms and simplify, if possible.

Example 5   **Using FOIL to Multiply Binomials**

Multiply: $(-7x + 4)(5x - 1)$.

Solution   Since we are multiplying two binomials, apply FOIL to get

$$(-7x + 4)(5x - 1) = \underbrace{(-7x)(5x)}_{\text{First}} + \overbrace{(-7x)(-1)}^{\text{Outer}} + \underbrace{(4)(5x)}_{\text{Inner}} + \overbrace{(4)(-1)}^{\text{Last}}$$

$$= -35x^2 + 7x + 20x - 4$$
$$= -35x^2 + 27x - 4$$

Check It Out 5   Use FOIL to multiply $(2x - 5)(3x + 1)$.

To multiply general polynomials, use the distributive property repeatedly, as illustrated in Example 6.

Example 6   **Multiplying General Polynomials**

Multiply: $(4y^3 - 3y + 1)(y - 2)$.

Solution   We have

$$(4y^3 - 3y + 1)(y - 2) = (4y^3 - 3y + 1)(y) - (4y^3 - 3y + 1)(2) \qquad \text{Use the distributive property}$$

Making sure to distribute the second negative sign throughout, we have

$$= 4y^4 - 3y^2 + y - 8y^3 + 6y - 2 \qquad \text{Use distributive property again}$$

$$= 4y^4 - 8y^3 - 3y^2 + 7y - 2 \qquad \text{Combine like terms}$$

Check It Out 6   Multiply $(3y^2 - 6y + 5)(y + 3)$.

The products of binomials given in Table 1 occur often enough that they are worth committing to memory. They will be used later in this section to help factor polynomial expressions.

Special products of binomials	Illustration
Square of a sum $(A + B)^2 = A^2 + 2AB + B^2$	$(3y + 4)^2 = (3y)^2 + 2(3y)(4) + 4^2$ $\qquad\qquad = 9y^2 + 24y + 16$
Square of a difference $(A - B)^2 = A^2 - 2AB + B^2$	$(4x^2 - 5)^2 = (4x^2)^2 - 2(4x^2)(5) + (5)^2$ $\qquad\qquad = 16x^4 - 40x^2 + 25$
Product of a sum and a difference $(A + B)(A - B) = A^2 - B^2$	$(7y + 3)(7y - 3) = (7y)^2 - (3)^2$ $\qquad\qquad = 49y^2 - 9$

Table 1

Factoring and Common Factors

The process of factoring a polynomial reverses the process of multiplication. That is, we find *factors* that can be multiplied together to produce the original polynomial expression. Factoring skills are of great importance in understanding the quadratic, polynomial and rational functions discussed in Chapter 3.

Common Factors

When factoring, the first step is to look for the *greatest common factor* in all the terms of the polynomial and then factor it out using the distributive property. The **greatest common factor** is a monomial, whose constant part is an integer with the largest absolute value common to all terms. Its variable part is the variable with the largest exponent common to all terms.

Example 7   **Factoring the Greatest Common Factor**

Factor the greatest common factor from each of the following.

a. $3x^4 + 9x^3 + 18x^2$ **b.** $-8y^2 - 6y + 4$

Solution

a. Because $3x^2$ is common to all the terms, we use the distributive property to write,

$$3x^4 + 9x^3 + 18x^2 = \mathbf{3x^2}(x^2 + 3x + 6)$$

Inside the parentheses, there are no further factors common to all the terms. Therefore, $3x^2$ is the greatest common factor of all the terms in the polynomial.

To check the factoring, multiply $3x^2(x^2 + 3x + 6)$ to see that it gives the original polynomial expression.

b. Because 2 is common to all the terms, we have

$$-8y^2 - 6y + 4 = \mathbf{2}(-4y^2 - 3y + 2), \text{ or } -\mathbf{2}(4y^2 + 3y - 2)$$

There are no variable terms to factor out. You can check that the factoring is correct by multiplying. Also note there can be more than one way to factor.

 Check It Out 7   Factor the greatest common factor from $5y + 10y^2 - 25y^3$.

Factoring by Grouping

Suppose we have an expression of the form $pA + qA$, where p, q and A can be any expression. Using the distributive property, we can write

$$pA + qA = (p + q)A, \text{ or } A(p + q)$$

This is the key to a technique called **factoring by grouping**.

Example 8   **Factoring by Grouping**

Factor $x^3 - x^2 + 2x - 2$ by grouping.

Solution Group the terms as follows.

$$\begin{aligned} x^3 - x^2 + 2x - 2 &= (x^3 - x^2) + (2x - 2) \\ &= x^2(\mathbf{x - 1}) + 2(\mathbf{x - 1}) \quad \text{\small Common factor in both groups is } (x - 1) \end{aligned}$$

Using the distributive property to factor out the term $(x - 1)$, we have

$$= (\mathbf{x - 1})(x^2 + 2)$$

 Check It Out 8   Factor $x^2 + 3x + 4x + 12$ by grouping.

When using factoring by grouping, it is essential to group together terms with the *same* common factor. Not all polynomials can be factored by grouping.

Factoring Trinomials of the Form $ax^2 + bx + c$

Note Factoring does not change an expression; It simply puts it in a different form. The factored form of a polynomial is quite useful when solving equations and when graphing quadratic and other polynomial functions.

One of the most common factoring problems involves trinomials of the form $ax^2 + bx + c$. In such problems, we assume that a, b and c have no common factors other than 1 or -1. If they do, simply factor out the greatest common factor first.

To factor, we simply reverse the FOIL method for multiplying polynomials. We try to find integers P, Q, R and S such that

$$(Px + Q)(Rx + S) = \underbrace{PR}x^2 + \overbrace{(PS + QR)}^{\text{Outer + Inner}}x + \underbrace{QS}_{\text{Last}} = ax^2 + bx + c.$$
$$\underset{\text{First}}{}$$

We see that $PR = a$, $PS + QR = b$, and $QS = c$. That is, we find factors of a and factors of c and choose only those factor combinations such that the inner and outer terms add to bx. This method is illustrated in Example 9.

> **Example 9**   **Factoring Trinomials**

Factor each of the following.

a. $x^2 - 2x - 8$

b. $8x^3 - 10x^2 - 12x$

Solution

a. First, note that there is no common factor to factor out. The factors of 1 are ± 1. The factors of -8 are ± 1, ± 2, ± 4, ± 8. Since $a = 1$, we must have a factorization of the form

$$x^2 - 2x - 8 = (x + \square)(x + \square)$$

where the numbers in the boxes are yet to be determined. The factors of $c = -8$ must be chosen so that the coefficient of x in the product is -2. Since 2 and -4 satisfy this condition, we have

$$x^2 - 2x - 8 = (x + \mathbf{2})(x + \mathbf{-4}) = (x + 2)(x - 4)$$

You can multiply the factors to check your answer.

b. Factor out the greatest common factor $2x$ to get

$$8x^3 - 10x^2 - 12x = \mathbf{2x}(4x^2 - 5x - 6)$$

The expression in parentheses is a trinomial of the form $ax^2 + bx + c$. We want to factor it as follows:

$$4x^2 - 5x - 6 = (\square x + \triangle)(\square x + \triangle)$$

The factors of $a = 4$ are placed in the boxes and the factors of $c = -6$ are placed in the triangles.

$$a = 4 \qquad \text{Factors: } \pm 1, \pm 2, \pm 4$$
$$c = -6 \qquad \text{Factors: } \pm 1, \pm 2, \pm 3, \pm 6$$

Find a pair of factors each for a and c such that the middle term of the trinomial is $-5x$. Note also that the two factors of -6 must be opposite in sign. We try different possibilities until we get the correct result.

Binomial factors		ax^2	bx	c
$(2x + 3)(2x - 2)$	$\longrightarrow$	$4x^2$	$2x$	-6
$(4x + 1)(x - 6)$	$\longrightarrow$	$4x^2$	$-23x$	-6
$(4x - 3)(x + 2)$	$\longrightarrow$	$4x^2$	$5x$	-6
$(4x + 3)(x - 2)$	$\longrightarrow$	$4x^2$	$-5x$	-6

Table 2

The last factorization is the correct one. Thus,

$$8x^3 - 10x^2 - 12x = 2x(4x^2 - 5x - 6) = 2x(4x + 3)(x - 2)$$

You should multiply out the factors to check.

✅ *Check It Out 9*   Factor: $2x^2 + 8x - 10$.

Special Factorization Patterns

One of the most efficient ways to factor is to remember special factorization patterns that occur frequently. We have categorized them into two groups: quadratic factoring patterns and cubic factoring patterns. The quadratic factoring patterns follow directly from the special products of binomials mentioned earlier. Tables 3 and 4 list the quadratic and cubic factoring patterns respectively.

Quadratic factoring pattern	Illustration
Difference of squares $A^2 - B^2 = (A + B)(A - B)$	$9x^2 - 5 = (3x)^2 - (\sqrt{5})^2$ $= (3x + \sqrt{5})(3x - \sqrt{5})$ where $A = 3x$ and $B = \sqrt{5}$
Perfect square trinomial $A^2 + 2AB + B^2 = (A + B)^2$	$25t^2 + 30t + 9 = (5t)^2 + 2(5t)(3) + 3^2$ $= (5t + 3)^2$ where $A = 5t$ and $B = 3$
Perfect square trinomial $A^2 - 2AB + B^2 = (A - B)^2$	$16s^2 - 8s + 1 = (4s)^2 - 2(4s)(1) + 1^2$ $= (4s - 1)^2$ where $A = 4s$ and $B = 1$

Table 3

Cubic factoring pattern	Illustration
Difference of cubes $A^3 - B^3 = (A - B)(A^2 + AB + B^2)$	$y^3 - 27 = y^3 - 3^3$ $= (y - 3)(y^2 + 3y + 3^2)$ $= (y - 3)(y^2 + 3y + 9)$ where $A = y$ and $B = 3$
Sum of cubes $A^3 + B^3 = (A + B)(A^2 - AB + B^2)$	$8x^3 + 125 = (2x)^3 + 5^3$ $= (2x + 5)((2x)^2 - 2x(5) + 5^2)$ $= (2x + 5)(4x^2 - 10x + 25)$ where $A = 2x$ and $B = 5$

Table 4

Example 10 **Special Factorization Patterns**

Factor using one of the special factorization patterns.

a. $27x^3 - 64$ **b.** $8x^2 + 32x + 32$

Solution

Note Not all polynomial expressions can be factored with techniques covered in this section. A more detailed study of the factorization of polynomials is given in Chapter 3.

a. Because $27x^3 - 64 = (3x)^3 - 4^3$, we can use the formula for the difference of cubes.

$27x^3 - 64 = (3x)^3 - 4^3$ Use $A = 3x$ and $B = 4$
$= (3x - 4)((3x)^2 + (3x)(4) + 4^2)$
$= (3x - 4)(9x^2 + 12x + 16)$

b. $8x^2 + 32x + 32 = 8(x^2 + 4x + 4)$ Factor out 8
$= 8(x + 2)^2$ Perfect square trinomial with $A = x$ and $B = 2$

Check It Out 10 Factor $4y^2 - 100$ using a special factorization pattern.

Exercises 1.3

Skills

In Exercises 1–6, collect like terms and arrange the polynomial in descending order. Give the degree of the polynomial.

1. $3y + 16 + 2y + 10$
2. $5z + 3 - 6z + 2$
3. $2t^2 - 2t + 5 + t^2$

4. $v^2 + v + v^2 - 1$
5. $-7s^2 - 6s + 3 + 3s^2 + 4$
6. $3s^3 + 4s - 2 + 5s^2 - 6s + 18$

In Exercises 7–56, perform the given operations. Express your answer as a single polynomial in descending order.

7. $(z + 6) + (5z + 8)$
8. $(2x - 3) - (x + 6)$
9. $(-9x^2 - 32x + 14) + (-2x^2 + 15x - 6)$

10. $(-9x^3 + 6x^2 - 20x + 3) + (x^2 - 5x - 6) + (x^5 + 7x^3 - 3x^2 + 5)$

11. $(z^5 - 4z^4 + 7) - (-z^3 + 15z^2 - 8z)$
12. $(x^5 - 3x^4 + 7x) - (4x^5 - 3x^3 + 8x + 1)$

13. $(3v^4 - 6v + 5) + (25v^5 - 16v^3 + 3v^2)$
14. $(-2x^3 + 4x^2 - 2x) - (7x^5 + 4x^3 - 1)$

15. $(-21t^2 + 21t + 21) + (9t^3 + 2)$
16. $(x^2 + 3) - (3x^2 + x - 10) + (2x^3 - 3x)$

17. $s(2s + 1)$
18. $-v(3v - 4)$
19. $3z(-6z^2 - 5)$

20. $5u(4u^2 - 10)$
21. $-t(-7t^2 + 3t + 9)$
22. $-5z(9z^2 - 2z - 4)$

23. $7z^2(z^2 + 9z - 8)$
24. $-6v^2(5v^2 - 3v + 7)$
25. $(y + 6)(y + 5)$

26. $(x - 4)(x + 7)$
27. $(-v - 12)(v - 3)$
28. $(x + 8)(x - 3)$

29. $(5 + 4v)(-7v - 6)$
30. $(-12 + 6z)(3z - 7)$
31. $(u^2 - 9)(u + 3)$

32. $(s^2 + 5)(s - 1)$
33. $(x + 4)^2$
34. $(t - 5)^2$

35. $(s + 6)^2$
36. $(-v + 3)^2$
37. $(5t + 4)^2$

38. $(7x + 1)^2$
39. $(v + 9)(v - 9)$
40. $(z - 7)(z + 7)$

41. $(9s + 7)(7 - 9s)$
42. $(-6 + 5t)(-5t - 6)$
43. $(v^2 + 3)(v^2 - 3)$

44. $(7 - z^2)(7 + z^2)$
45. $(5y^2 - 4)(5y^2 + 4)$
46. $(2x^2 + 3)(2x^2 - 3)$

47. $(4z^2 + 5)(4z^2 - 5)$
48. $(6 - 7u^2)(6 + 7u^2)$
49. $(x + 2z)(x - 2z)$

50. $(u - 3v)(u + 3v)$
51. $(-t^2 - 5t + 1)(t + 6)$
52. $(v^2 + 3v - 7)(-v - 2)$

53. $(x - 2)(x^2 + 3x - 7)$
54. $(u + 3)(6u^2 - 4u + 5)$
55. $(5u^3 - 6u^2 - 7u + 9)(-4u + 7)$

56. $(-8v^3 + 7v^2 + 5v - 4)(3v - 9)$

In Exercises 57–62, factor the greatest common factor from each expression.

57. $2x^3 + 6x^2 - 8x$
58. $4x^4 - 8x^3 + 12$
59. $-2t^6 - 4t^5 + 10t^2$

60. $12x^5 - 6x^3 - 18x^2$
61. $-5x^7 + 10x^5 - 15x^3$
62. $-14z^5 + 7z^3 + 28$

In Exercises 63–68, factor each expression by grouping.

63. $3(x + 1) + x(x + 1)$
64. $x(x - 2) + 4(x - 2)$
65. $s^3 - 5s^2 - 9s + 45$

66. $-27v^3 - 36v^2 + 3v + 4$
67. $12u^3 + 4u^2 - 3u - 1$
68. $75t^3 + 25t^2 - 12t - 4$

In Exercises 69–78, factor each trinomial.

69. $x^2 + 4x + 3$
70. $x^2 + 2x - 35$
71. $x^2 - 6x - 16$

72. $x^2 - 10x + 24$
73. $3s^2 + 15s + 12$
74. $4y^2 - 20y - 24$

75. $-6t^2 + 24t + 72$
76. $-5z^2 - 20z + 60$
77. $9u^2 - 27u + 18$

78. $2x^2 - 4x - 6$

In Exercises 79–86, factor each polynomial using one of the special factorization patterns.

79. $x^2 - 16$
80. $t^2 - 25$
81. $4x^2 + 4x + 1$

82. $9x^2 - 6x + 1$
83. $y^3 + 64$
84. $8x^3 - 27$

85. $8y^3 + 1$
86. $64x^3 - 8$

32

In Exercises 87–132, factor each expression completely, using any of the methods from this section.

87. $z^2 + 13z + 42$ **88.** $z^2 + z - 30$ **89.** $x^2 + 12x + 36$

90. $x^2 + 8x + 16$ **91.** $-y^2 + 4y - 4$ **92.** $-y^2 - 6y - 9$

93. $z^2 - 16z + 64$ **94.** $z^2 - 8z + 16$ **95.** $-2y^2 + 7y - 3$

96. $-3y^2 - 2y + 8$ **97.** $9y^2 + 12y + 4$ **98.** $4x^2 + 12x + 9$

99. $6z^2 - 3z - 18$ **100.** $8v^2 + 20v - 12$ **101.** $15t^2 - 70t - 25$

102. $14y^2 - 7y - 21$ **103.** $-10u^2 - 45u - 20$ **104.** $-6x^2 + 27x - 30$

105. $v^2 - 4$ **106.** $y^2 - 9$ **107.** $-s^2 + 49$

108. $-u^2 + 36$ **109.** $-25t^2 + 4$ **110.** $-49v^2 + 16$

111. $t^3 - 16t^2$ **112.** $x^3 - 9x^2$ **113.** $12u^3 + 4u^2 - 40u$

114. $6x^3 - 15x^2 + 9x$ **115.** $-10t^3 + 5t^2 + 15t$ **116.** $-8y^3 - 44y^2 - 20y$

117. $-15z^3 - 5z^2 + 20z$ **118.** $6x^3 + 14x^2 - 12x$ **119.** $2y^3 + 3y^2 - 8y - 12$

120. $4x^4 + 20x^3 + 24x^2$ **121.** $-18z^3 + 27z^2 + 32z - 48$

122. $-10s^4 - 25s^3 + 15s^2$ **123.** $3y^4 + 18y^3 + 24y^2$ **124.** $21v^4 - 28v^3 + 7v^2$

125. $7x^5 - 63x^3$ **126.** $5y^5 - 20y^3$ **127.** $-6s^5 - 30s^3$

128. $15u^5 + 18u^3$ **129.** $8x^3 + 64$ **130.** $27x^3 + 1$

131. $-8y^3 + 1$ **132.** $-64z^3 + 27$

Applications

133. Home Improvement The amount of paint needed to cover the walls of a bedroom is $132x$, where x is the thickness of the coat of paint. The amount of same paint that is needed to cover the walls of the den is $108x$. How much more paint is needed for the bedroom than for the den? Express your answer as a monomial in terms of x.

134. Geometry Two circles have a common center. Let r denote the radius of the smaller circle. What is the area of the region between the two circles if the area of the larger circle is $9\pi r^2$ and the area of the smaller circle is $4\pi r^2$? Express your answer as a monomial in r.

135. Geometry The *perimeter* of a square is the sum of the lengths of all four sides.
  a. If one side is of length s, find the perimeter of the square in terms of s.
  b. If each side of the square in part (a) is doubled, find the perimeter of the new square.

136. Shopping At the Jolly Ox, a gallon of milk sells for $3.30, and apples go for $0.49 per pound. Suppose Tania bought x gallons of milk and y pounds of apples.
  a. How much did she spend altogether (in dollars)? Express your answer as a binomial in x and y.
  b. If Tania gave the cashier a $20 bill, how much would she receive in change? Express in terms of x and y, and assume her purchases do not exceed $20.

137. Investment Suppose an investment of $1000 is worth $1000(1 + r)^2$ after two years, where r is the interest rate. Assume that no additional deposits or withdrawals are made.
  a. Write $1000(1 + r)^2$ as a polynomial in descending order.
  b. If the interest rate is 5%, use a calculator to determine how much the $1000 investment is worth after two years. (In the formula $1000(1 + r)^2$, r is assumed to be in decimal form.)

138. Investment Suppose an investment of $500 is worth $500(1 + r)^3$ after three years, where r is the interest rate. Assume that no additional deposits or withdrawals are made.
  a. Write $500(1 + r)^3$ as a polynomial in descending order.
  b. If the interest rate is 4%, use a calculator to determine how much the $500 investment is worth after three years. (In the formula $500(1 + r)^3$, r is assumed to be in decimal form.)

Concepts

---

139. If two polynomials of degree 3 are added, is their sum necessarily a polynomial of degree 3? Explain.

140. A student writes the following on an exam: $(x + 2)^2 = x^2 + 4$. Explain the student's error and give the correct answer for $(x + 2)^2$.

141. What is the constant term in the product of $5x^2 - 3x + 2$ and $6x^2 - 9x$?

142. If a polynomial of degree 2 is multiplied by a polynomial of degree 3, what is the degree of their product?

143. For what value(s) of a is $-8x^3 + 5x^2 + ax$ a binomial?

144. Is $(x^2 - 4)(x + 5)$ completely factored? Explain.

145. Give an example of a polynomial of degree 2 that can be expressed as the square of a binomial, and then express it accordingly.

146. Can $y^2 + a^2$ be factored as $(y + a)^2$? Explain.

A quotient of two polynomial expressions is called a **rational expression**. A rational expression is defined whenever the denominator is not equal to zero.

Example 1   **Values for Which a Rational Expression is Defined**

For what values of x is the following rational expression defined?

$$\frac{x+1}{(x-3)(x-5)}$$

Solution   The rational expression is defined only when the denominator is *not* zero. This happens whenever

$$x - 3 \neq 0 \Rightarrow x \neq 3 \qquad \text{or} \qquad x - 5 \neq 0 \Rightarrow x \neq 5$$

Thus, the rational expression is defined whenever x is *not* equal to 3 or 5. We can also say that 3 and 5 are *excluded values* of x.

✓ *Check It Out 1*   For what values of x is the following rational expression $\dfrac{x}{x^2 - 1}$ defined? ■

Simplifying Rational Expressions

Recall that if you have a fraction such as $\frac{4}{12}$, you simplify it by first factoring the numerator and the denominator and then dividing out the common factors:

$$\frac{4}{12} = \frac{2 \cdot 2}{2 \cdot 2 \cdot 3} = \frac{1}{3}$$

When simplifying rational expressions containing variables, you completely factor polynomials instead of numbers. Familiarity with the many factoring techniques is the most important tool in manipulating rational expressions.

Example 2   **Simplifying a Rational Expression**

Simplify: $\dfrac{x^2 - 2x + 1}{3x^2 - 4x + 1}$.

Solution

$$\frac{x^2 - 2x + 1}{3x^2 - 4x + 1} = \frac{(x-1)(x-1)}{(3x-1)(x-1)} \qquad \text{Factor completely}$$

$$= \frac{x-1}{3x-1} \qquad \text{Divide out } x-1, \text{ a common factor}$$

✓ *Check It Out 2*   Simplify:  $\dfrac{x^2 - 4}{x^2 + 5x + 6}$. ■

Next, we discuss the arithmetic of rational expressions, which is very similar to the arithmetic of rational numbers.

Multiplication and Division of Rational Expressions

Multiplication of rational expressions is straightforward. You multiply the numerators, multiply the denominators, and then simplify your answer.

> **Example 3**   **Multiplication of Rational Expressions**

Multiply the following and express your answer in lowest terms. For what values of the variable is the expression meaningful?

a. $\dfrac{3a}{8} \cdot \dfrac{24}{6a^3}$

b. $\dfrac{x^2 + x - 6}{x^2 - 4} \cdot \dfrac{(x + 2)^2}{x^2 + 9}$

Solution

a. $\dfrac{3a}{8} \cdot \dfrac{24}{6a^3} = \dfrac{(3a)(24)}{(8)(6a^3)}$ Write as a single fraction

$\qquad\qquad = \dfrac{3 \cdot a \cdot 6 \cdot 4}{4 \cdot 2 \cdot 6 \cdot a^3}$ Factor

$\qquad\qquad = \dfrac{3}{2a^2}$ Divide out common factors

The expression is meaningful for $a \neq 0$.

b. $\dfrac{x^2 + x - 6}{x^2 - 4} \cdot \dfrac{(x + 2)^2}{x^2 + 9} = \dfrac{(x^2 + x - 6)(x + 2)^2}{(x^2 - 4)(x^2 + 9)}$

$\qquad\qquad = \dfrac{(x + 3)(x - 2)\,(x + 2)^2}{(x + 2)(x - 2)\,(x^2 + 9)}$ Factor

$\qquad\qquad = \dfrac{(x + 3)(x + 2)}{x^2 + 9}$ Divide out common factors, assuming $x \neq 2, -2$

The expression is meaningful for $x \neq 2, -2$, because they appeared in the denominator of the original expression. Observe that $x^2 + 9$ cannot be factored any further using real numbers, and is never zero.

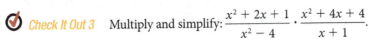 **Check It Out 3**   Multiply and simplify: $\dfrac{x^2 + 2x + 1}{x^2 - 4} \cdot \dfrac{x^2 + 4x + 4}{x + 1}$.

When dividing two rational expressions, the expression following the division symbol is called the **divisor**. To divide rational expressions, multiply the first expression by the reciprocal of the divisor.

> **Example 4**   **Dividing Rational Expressions**

Divide and simplify: $\dfrac{3x^2 - 5x - 2}{x^2 - 4x + 4} \div \dfrac{9x^2 - 1}{x + 5}$.

Solution Taking the reciprocal of the divisor and multiplying, we have

$$\frac{3x^2 - 5x - 2}{x^2 - 4x + 4} \div \frac{9x^2 - 1}{x + 5} = \frac{3x^2 - 5x - 2}{x^2 - 4x + 4} \cdot \frac{x + 5}{9x^2 - 1}$$

Factor, divide out common factors, and multiply to get

$$= \frac{(3x + 1)\,(x - 2)}{(x - 2)^2} \cdot \frac{x + 5}{(3x + 1)\,(3x - 1)}$$

$$= \frac{x + 5}{(x - 2)\,(3x - 1)}$$

Check It Out 4   Divide and simplify: $\dfrac{7x + 14}{x^2 - 4} \div \dfrac{7x}{x^2 + x - 6}$.

Addition and Subtraction of Rational Expressions

To add and subtract rational expressions, follow the same procedure used for adding and subtracting rational numbers. Before we can add rational expressions, we must write them in terms of the same denominator, known as the **least common denominator**. For instance, to compute $\frac{1}{4} + \frac{1}{6}$, we find the least common multiple of 4 and 6, which is 12. The number $12 = 2 \cdot 2 \cdot 3$ is the smallest number whose factors includes the factors of 4, which are 2 and 2, *and* the factors 6, which are 2 and 3.

> **Definition of the Least Common Denominator**
>
> The **least common denominator (LCD)** of a set of rational expressions is the simplest expression that includes all the factors of each of the denominators.

When adding or subtracting rational expressions, you factor *polynomials* instead of numbers to find the least common denominator.

Example 5 **Adding and Subtracting Rational Expressions**

Add or subtract the following and express your answer in lowest terms.

a. $\dfrac{x+4}{3x+6} + \dfrac{2x+1}{x^2+7x+10}$

b. $\dfrac{3x}{4-2x} - \dfrac{x+5}{x^2-4}$

Solution

a. Factor the denominators and find the least common denominator.

$$\frac{x+4}{3x+6} + \frac{2x+1}{x^2+7x+10} = \frac{x+4}{3(x+2)} + \frac{2x+1}{(x+5)(x+2)}$$

The LCD is: $\mathbf{3(x+2)(x+5)}$. Write both expressions as equivalent rational expressions using the LCD.

$$= \frac{x+4}{3(x+2)} \cdot \frac{\mathbf{x+5}}{\mathbf{x+5}} + \frac{2x+1}{(x+5)(x+2)} \cdot \frac{\mathbf{3}}{\mathbf{3}}$$

Simplify the numerators and add the two fractions.

$$= \frac{x^2+9x+20}{3(x+2)(x+5)} + \frac{6x+3}{3(x+2)(x+5)} = \frac{x^2+15x+23}{3(x+2)(x+5)}$$

The expression cannot be simplified any further.

b. Factor the denominators and find the least common denominator.

$$\frac{3x}{4-2x} - \frac{x+5}{x^2-4} = \frac{3x}{2(2-x)} - \frac{x+5}{(x+2)(x-2)}$$

$$= \frac{3x}{-2(x-2)} - \frac{x+5}{(x+2)(x-2)}$$

The LCD is: $-2(x - 2)(x + 2)$. Note that $2(2 - x) = -2(x - 2)$. Write both expressions as equivalent rational expressions using the LCD.

$$= \frac{3x}{-2(x - 2)} \cdot \frac{x + 2}{x + 2} - \frac{x + 5}{(x + 2)(x - 2)} \cdot \frac{-2}{-2}$$

$$= \frac{3x^2 + 6x}{-2(x - 2)(x + 2)} - \frac{-2x - 10}{-2(x - 2)(x + 2)} \qquad \text{Simplify the numerators}$$

$$= \frac{(3x^2 + 6x) - (-2x - 10)}{-2(x - 2)(x + 2)} = \frac{3x^2 + 8x + 10}{-2(x - 2)(x + 2)} \qquad \text{Subtract, taking care to distribute the minus sign}$$

The expression cannot be simplified any further.

 Check It Out 5   Subtract and express your answer in lowest terms: $\dfrac{-2}{x + 2} - \dfrac{3}{x^2 - 4}$. ■

Complex Fractions

A **complex fraction** is one where the numerator and/or denominator of the fraction contains a rational expression. Complex fractions are also commonly referred to as **complex rational expressions**.

> **Example 6**   **Simplifying a Complex Fraction**

Simplify $\dfrac{\dfrac{x^2 - 4}{2x + 1}}{\dfrac{x^2 + x - 6}{x - 1}}$.

Solution   Because we have a quotient of two rational expressions, we can write

$$\frac{\dfrac{x^2 - 4}{2x + 1}}{\dfrac{x^2 + x - 6}{x - 1}} = \frac{x^2 - 4}{2x + 1} \div \frac{x^2 + x - 6}{x - 1}$$

$$= \frac{x^2 - 4}{2x + 1} \cdot \frac{x - 1}{x^2 + x - 6}$$

$$= \frac{(x + 2)(x - 2)}{2x + 1} \cdot \frac{x - 1}{(x + 3)(x - 2)} \qquad \text{Factor.}$$

$$= \frac{(x + 2)(x - 1)}{(2x + 1)(x + 3)} \qquad \text{Divide out } x - 2, \text{ the common factor.}$$

There are no more common factors, so the expression is simplified.

 *Check It Out 6*   Simplify: $\dfrac{\dfrac{x + y}{y}}{\dfrac{x^2 - y^2}{x}}$. ■

Another way to simplify a complex fraction is to multiply the numerator and denominator by the least common denominator of all the denominators.

Example 7   **Simplifying a Complex Fraction**

Simplify: $\dfrac{\dfrac{1}{x} + \dfrac{1}{xy}}{\dfrac{3}{y^2} + \dfrac{1}{y}}$.

Solution First find the LCD of the four rational expressions. The denominators are x, xy, y^2, and y.

Thus, the LCD is xy^2. We then can write

$$\frac{\dfrac{1}{x} + \dfrac{1}{xy}}{\dfrac{3}{y^2} + \dfrac{1}{y}} = \frac{\dfrac{1}{x} + \dfrac{1}{xy}}{\dfrac{3}{y^2} + \dfrac{1}{y}} \cdot \frac{xy^2}{xy^2}$$

$$= \frac{\left(\dfrac{1}{x} + \dfrac{1}{xy}\right)xy^2}{\left(\dfrac{3}{y^2} + \dfrac{1}{y}\right)xy^2}$$

$$= \frac{\dfrac{1}{x}(xy^2) + \dfrac{1}{xy}(xy^2)}{\dfrac{3}{y^2}(xy^2) + \dfrac{1}{y}(xy^2)} \qquad \text{Distribute } xy^2$$

$$= \frac{y^2 + y}{3x + xy} \qquad \text{Simplify each term}$$

$$= \frac{y(y + 1)}{x(3 + y)} \qquad \text{Factor to see if any factors can be removed}$$

There are no common factors, so the expression is simplified.

Check It Out 7   Simplify: $\dfrac{\dfrac{2}{x} + \dfrac{1}{xy}}{\dfrac{1}{y} - \dfrac{2}{x}}$.

Exercises 1.4

 Skills

In Exercises 1–10, simplify each rational expression and indicate, where applicable, the values of the variable for which the expression is defined.

1. $\dfrac{57}{24}$

2. $\dfrac{56}{49}$

3. $\dfrac{x^2 - 4}{6(x + 2)}$

4. $\dfrac{3(x - 3)}{x^2 - 9}$

5. $\dfrac{x^2 - 2x - 3}{x^2 - 9}$

6. $\dfrac{z^2 - 1}{z^2 + 2z + 1}$

7. $\dfrac{x^4 - x^2}{x + 1}$

8. $\dfrac{y^3 - y}{y - 1}$

9. $\dfrac{x^3 - 1}{x^2 - 1}$

10. $\dfrac{y^3 + 8}{y^2 - 4}$

In Exercises 11–26, multiply or divide. Express your answer in lowest terms, if possible.

11. $\dfrac{3x}{6y^2} \cdot \dfrac{2xy}{x^3}$

12. $\dfrac{x^2y}{2y^2} \cdot \dfrac{4y^4}{x^3}$

13. $\dfrac{x + 2}{x^2 - 9} \cdot \dfrac{x + 3}{x^2 + 4x + 4}$

14. $\dfrac{x - 3}{x^2 - 2x + 1} \cdot \dfrac{x - 1}{2x - 6}$

15. $\dfrac{3x + 9}{x^2 + x - 6} \cdot \dfrac{2x - 4}{x + 6}$

16. $\dfrac{x - 3}{4x + 16} \cdot \dfrac{3x + 12}{x^2 - 5x + 6}$

17. $\dfrac{6x - 12}{3x^3 - 12x} \cdot \dfrac{x^2 - 4x + 4}{x^2 + 3x - 10}$

18. $\dfrac{4x^4 - 36x^2}{8x - 8} \cdot \dfrac{x^2 - 2x + 1}{x^2 + 2x - 15}$

19. $\dfrac{x^3 + 1}{x^2 - 1} \cdot \dfrac{2x^2 - x - 1}{x + 2}$

20. $\dfrac{a^3 - 1}{a^2 - 2a + 1} \cdot \dfrac{(a - 1)^2}{a^2 + a + 1}$

21. $\dfrac{5x - 20}{x^2 - 4x - 5} \div \dfrac{x^2 - 8x + 16}{x - 5}$

22. $\dfrac{3x - 6}{2x + 2} \div \dfrac{3x^2 - 5x - 2}{x^2 - 5x + 6}$

23. $\dfrac{6x^3 - 24x}{3x^2 - 3} \div \dfrac{2x^2 + 4x}{x^2 - 2x + 1}$

24. $\dfrac{5x^4 - 45x^2}{7x - 14} \div \dfrac{3x^2 + 9x}{x^2 + 3x - 10}$

25. $\dfrac{x^3 - 8}{2x^2 - 3x - 2} \div \dfrac{x^2 - 4}{2x + 1}$

26. $\dfrac{a^3 + 27}{a^2 - 1} \div \dfrac{a^2 + 6a + 9}{a^2 + 2a + 1}$

In Exercises 27–42, add or subtract. Express your answer in lowest terms, if possible.

27. $\dfrac{2}{x} + \dfrac{3}{x^2}$

28. $\dfrac{-3}{y^2} + \dfrac{4}{y}$

29. $\dfrac{1}{x + 1} + \dfrac{4}{x - 1}$

30. $\dfrac{5}{x - 3} + \dfrac{6}{x + 2}$

31. $\dfrac{4x}{x^2 - 9} + \dfrac{2x^2}{3x + 9}$

32. $\dfrac{2z}{5z - 10} + \dfrac{z + 1}{z^2 - 4z + 4}$

33. $\dfrac{x}{x + 1} - \dfrac{x - 4}{x - 1}$

34. $\dfrac{x + 1}{x - 3} - \dfrac{6}{x + 2}$

35. $\dfrac{-3x}{x^2 - 16} - \dfrac{3x^2}{3x + 12}$

36. $\dfrac{z}{3z - 15} - \dfrac{z - 1}{z^2 - 10z + 25}$ **37.** $\dfrac{3}{x - 1} + \dfrac{4}{1 - x}$ **38.** $\dfrac{6}{2x - 1} + \dfrac{4}{1 - 2x}$

39. $\dfrac{4}{x + 2} - \dfrac{2}{x - 2} + \dfrac{1}{x^2 - 4}$ **40.** $\dfrac{-1}{x - 1} + \dfrac{2}{x + 1} - \dfrac{3}{x^2 - 1}$

41. $\dfrac{7}{3 - x} - \dfrac{1}{x + 2} + \dfrac{4}{x^2 - x - 6}$ **42.** $\dfrac{3}{y - 4} + \dfrac{2}{y^2 - 5y + 4} + \dfrac{2}{1 - y}$

In Exercises 43–57, simplify each complex fraction.

43. $\dfrac{\dfrac{x + 1}{x}}{\dfrac{x^2 - 1}{x^2}}$ **44.** $\dfrac{\dfrac{a^2 - 1}{a}}{\dfrac{a - 1}{a^3}}$ **45.** $\dfrac{\dfrac{1}{x} + \dfrac{1}{y}}{\dfrac{1}{y^2} - \dfrac{2}{x}}$

46. $\dfrac{\dfrac{1}{y} - \dfrac{1}{x^2}}{\dfrac{1}{x} + \dfrac{2}{y}}$ **47.** $\dfrac{1}{\dfrac{1}{r} + \dfrac{1}{s} + \dfrac{1}{t}}$ **48.** $\dfrac{2}{\dfrac{1}{x^2} + \dfrac{1}{xy} + \dfrac{1}{y^2}}$

49. $\dfrac{1 + x^{-1}}{x^{-2} - 1}$ **50.** $\dfrac{a^{-1} + b^{-1}}{a + b}$ **51.** $\dfrac{\dfrac{1}{x - 1} - \dfrac{1}{x - 3}}{\dfrac{2}{x - 1} + \dfrac{3}{x + 1}}$

52. $\dfrac{\dfrac{2}{x - 2} + \dfrac{1}{x - 1}}{\dfrac{3}{x + 3} - \dfrac{1}{x - 2}}$ **53.** $\dfrac{\dfrac{1}{x + h} - \dfrac{1}{x}}{h}$ **54.** $\dfrac{\dfrac{1}{x} - \dfrac{1}{a}}{x - a}$

55. $\dfrac{\dfrac{2}{x^2 - 4} + \dfrac{1}{x - 2}}{\dfrac{4}{x + 2}}$ **56.** $\dfrac{\dfrac{3}{x^2 - 9} - \dfrac{1}{x + 3}}{\dfrac{2}{x - 3}}$ **57.** $\dfrac{\dfrac{a}{a^2 - b^2} + \dfrac{b}{a + b}}{\dfrac{1}{a - b}}$

Applications

58. Average Cost The average cost per book for printing x booklets is $\dfrac{300 + 0.5x}{x}$. Evaluate this expression for $x = 100$ and interpret the result.

59. Driving Speed If it takes t hours to drive a distance of 400 miles, then the average driving speed is given by $\dfrac{400}{t}$. Evaluate this expression for $t = 8$ and interpret the result.

60. Work Rate One pump can fill a pool in 4 hours and another can fill it in 3 hours. Working together, it takes the pumps $t = \dfrac{1}{\frac{1}{4} + \frac{1}{3}}$ hours to fill the pool. Find t.

61. Physics In an electrical circuit, if three resistors are connected in parallel, then their total resistance is given by

$$R = \dfrac{1}{\dfrac{1}{R_1} + \dfrac{1}{R_2} + \dfrac{1}{R_3}}$$

Simplify the expression for R.

Concepts

62. Find two numbers x and y such that $\frac{1}{x} + \frac{1}{y} \neq \frac{2}{x+y}$. (Answers may vary.)

63. The expression $\frac{x^2 - 1}{x + 1}$ simplifies to $x - 1$. What value(s) of x must be excluded when performing the simplification?

64. Is $\frac{x^2}{x} = x$ for all values of x? Explain.

65. In an answer to an exam question, $\frac{x^2 + 4}{x + 2}$ is simplified as $x + 2$. Is this correct? Explain.

Linear and Quadratic Equations 1.5

Objectives

- Solve simple linear equations.
- Solve equations involving fractions.
- Solving fractions involving decimals.
- Solve equations for one variable in terms of another.
- Solve a quadratic equation by factoring.
- Solve a quadratic equation by completing the square.
- Solve a quadratic equation by using the quadratic formula.

In previous sections, we manipulated various algebraic expressions to simplify them. In this section, we will review some basic equation-solving skills that you learned in your previous algebra courses. When you set two algebraic expressions equal to each other, you form an **equation**. If you can find a value of a variable which makes the equation true, you have **solved the equation**. The following strategies can help you solve an equation.

Equation–Solving Strategies

When solving an equation, you may need to use one or more of the following steps.

- Simplify an expression by removing parentheses. Then combine **like terms**—that is, combine real numbers or expressions with the same variable names.

- Add or subtract the *same* real number or expression to, or from, *both sides* of the equation.

$$a = b \text{ is equivalent to } a + c = b + c$$

- Multiply or divide *both sides* of the equation by the *same nonzero* real number.

$$a = b \text{ is equivalent to } ac = bc, c \neq 0$$

We first examine solutions of a **linear equation**, that is, an equation that can be written in the form $ax + b = 0$, where a, b are real numbers and $a \neq 0$. This is the most elementary type of equation.

VIDEO EXAMPLES

SECTION 1.5

Example 1 **Solving an Equation**

Solve the following equation for x.

$$3(x + 2) - 2 = 4x$$

Solution Proceed as follows.

$3(x + 2) - 2 = 4x$	Given equation
$3x + 6 - 2 = 4x$	Remove parentheses
$3x + 6 - 2 - \mathbf{4x} = 4x - \mathbf{4x}$	Subtract $4x$ from both sides
$-x + 4 = 0$	Combine like terms
$-x = -4$	Isolate term containing x
$\mathbf{x = 4}$	Multiply both sides of the equation by -1

Thus $\mathbf{x = 4}$ is the solution to the given equation. Check the solution by substituting $x = 4$ in the original equation:

$$3(4 + 2) - 2 = 4(4) \quad \Rightarrow \quad 16 = 16$$

✓ *Check It Out 1*   Solve the equation $2(x + 4) = 3x + 2$ for x.   ■

When an equation involves fractions, it is easier to solve if the denominators are cleared first, as illustrated in the next example.

> **Example 2**   **Solving an Equation Involving Fractions**

Solve the equation.

$$\frac{x+5}{2} + \frac{2x-1}{5} = 5$$

Solution Clear the denominators by multiplying *both* sides of the equation by the least common denominator, which is **10**.

$$10\left(\frac{x+5}{2} + \frac{2x-1}{5}\right) = 5 \cdot 10 \qquad \text{Multiply both sides by LCD}$$

$$5(x+5) + 2(2x-1) = 50 \qquad \text{Simplify each term}$$

$$5x + 25 + 4x - 2 = 50 \qquad \text{Remove parentheses}$$

$$9x + 23 = 50 \qquad \text{Combine like terms}$$

$$9x = 27 \qquad \text{Subtract 23 from both sides}$$

$$\boldsymbol{x = 3} \qquad \text{Divide both sides by 9 to solve for } x$$

✓ *Check It Out 2* Solve the equation.

$$\frac{x+3}{4} + \frac{x+5}{2} = 7$$ ■

We can also clear decimals in an equation to make it easier to work with. This procedure is illustrated in Example 3.

> **Example 3**   **Solving an Equation Involving Decimals**

Solve the equation.

$$0.3(x+2) - 0.02x = 0.5$$

Solution There are two decimal coefficients, 0.3 and 0.02. Multiply both sides of the equation by the smallest power of 10 that will eliminate the decimals. In this case, multiply both sides by 100.

$$0.3(x+2) - 0.02x = 0.5 \qquad \text{Given equation}$$

$$\boldsymbol{100}(0.3(x+2) - 0.02x) = 0.5(\boldsymbol{100}) \qquad \text{Multiply both sides by 100}$$

$$30(x+2) - 2x = 50 \qquad 100(0.3) = 30, 100(0.02) = 2, \text{ and } 100(0.5) = 50$$

$$30x + 60 - 2x = 50 \qquad \text{Remove parentheses}$$

$$28x + 60 = 50 \qquad \text{Combine like terms}$$

$$28x = -10 \qquad \text{Subtract 60 from both sides}$$

$$x = -\frac{10}{28} = -\frac{5}{14} \qquad \text{Divide each side by 28 and reduce the fraction}$$

✓ *Check It Out 3* Solve the equation $0.6(2x+1) - 0.03(x-1) = 0.15$ ■

In Example 4, we solve for one variable in terms of another. In this case, the solution is not just a number.

Example 4 **Solving for one Variable in Terms of Another**

The perimeter of a rectangular fence is 15 feet. Write the width of the fence in terms of the length.

Solution The perimeter formula for a rectangle is $P = 2l + 2w$. Thus we have

$$2l + 2w = 15 \qquad P = 15$$

$$2w = 15 - 2l \qquad \text{Isolate } w \text{ term}$$

$$w = \frac{1}{2}(15 - 2l) \qquad \text{Divide by 2 to solve for } w$$

 Check It Out 4 Rework Example 4 if the perimeter is 20 feet.

Quadratic Equations

We now turn our attention to solving equations which include variables raised to the second power, known as quadratic equations.

> **Definition of a Quadratic Equation**
>
> A **quadratic equation** is an equation which can be written in the **standard form**
>
> $$ax^2 + bx + c = 0$$
>
> where a, b, and c are real numbers with $a \neq 0$.

Solving a Quadratic Equation by Factoring

One of the simplest ways to solve a quadratic equation is by factoring. We need the following rule to justify our procedure for solving equations by factoring.

> **Zero Product Rule**
>
> If a product of real numbers is zero, then at least one of the factors is zero. That is,
>
> If $cd = 0$, then $c = 0$ **or** $d = 0$.

Just in Time
Review factoring
in Section 1.3.

Example 5 **Solving an Equation by Factoring**

Solve $2x^2 - 7x + 3 = 0$ by factoring.

Solution Factoring the left-hand side gives

$$(2x - 1)(x - 3) = 0$$

According to the Zero Product Rule, if the product of two factors equals zero then at least one of the factors is equal to zero. Thus, we set each factor equal to zero and solve for x.

$$2x - 1 = 0 \quad \Rightarrow \quad x = \frac{1}{2}$$

$$\text{or}$$

$$x - 3 = 0 \quad \Rightarrow \quad x = 3$$

The solutions to the equation are $x = \frac{1}{2}$ and $x = 3$. You should check that these values satisfy the original equation.

 Check It Out 5 Solve $5x^2 - 3x - 2 = 0$ by factoring.

Solving a Quadratic Equation by Completing the Square

In Example 5, we found the solutions of $2x^2 - 7x + 3 = 0$ by factoring. Unfortunately, not all equations of the form $ax^2 + bx + c = 0$ can be solved by factoring. In this section, we will discuss a general method that can be used to solve all quadratic equations. We first need the following rule regarding square roots.

Principle of Square Roots

If $x^2 = c$, where $c \geq 0$, then $x = \sqrt{c}$ or $x = -\sqrt{c}$.

Example 6   **Using the Principle of Square Roots**

Solve $-3x^2 + 9 = 0$.

Solution We use the principle of square roots to solve this equation.

$$-3x^2 + 9 = 0 \qquad \text{Given equation}$$
$$-3x^2 = -9 \qquad \text{Subtract 9 from both sides}$$
$$x^2 = 3 \qquad \text{Divide by } -3 \text{ on both sides}$$
$$x = \sqrt{3} \text{ or } x = -\sqrt{3} \qquad \text{Apply the principle of square roots}$$

 Check It Out 6   Solve $4x^2 - 20 = 0$.

We now introduce the method of *completing the square* as a tool to solve *any* type of quadratic equation.

Example 7   **Solving by Completing the Square**

Solve the equation $3x^2 - 6x - 1 = 0$ by using the technique of completing the square.

Solution

$$3x^2 - 6x - 1 = 0 \qquad \text{Given equation}$$
$$3x^2 - 6x = 1 \qquad \text{Move the constant to right side}$$
$$x^2 - 2x = \frac{1}{3} \qquad \text{Divide by 3 to get coefficient of } x^2 \text{ to equal 1}$$
$$x^2 - 2x + 1 = \frac{1}{3} + 1 \qquad \text{Complete the square by taking half of } -2 \text{ and}$$
$$\text{squaring it: } \left(\tfrac{1}{2}(-2)\right)^2 = 1. \text{ Add 1 to both sides.}$$
$$(x - 1)^2 = \frac{4}{3} \qquad \text{Write the left hand side as a perfect square}$$
$$x - 1 = \pm \sqrt{\frac{4}{3}} \qquad \text{Use the principle of square roots}$$
$$x = 1 \pm \sqrt{\frac{4}{3}} \qquad \text{Solve for } x$$
$$= 1 \pm 2\frac{\sqrt{3}}{3} \qquad \text{Simplify the radical}$$

Thus, $x = 1 + 2\frac{\sqrt{3}}{3} \approx 2.155$ and $x = 1 - 2\frac{\sqrt{3}}{3} \approx -0.155$ are the two solutions of the equation.

 Check It Out 7   Solve the equation $2x^2 - 4x - 1 = 0$ by using the technique of completing the square.

Note The principle of square roots can be used to solve a quadratic equation works only when the quadratic equation can be rewritten in the form $x^2 = c$. One side of the equation must be a constant and the other side must be a perfect square.

Just in Time
Review perfect square trinomials in Section 1.3.

Solving a Quadratic Equation Using the Quadratic Formula

We now derive a general formula for solving quadratic equations. In the following derivation, we assume $a > 0$. If $a < 0$, we can multiply the equation by -1 and obtain a positive coefficient for x^2.

$$ax^2 + bx + c = 0 \qquad \text{Quadratic equation}$$

$$a\left(x^2 + \frac{b}{a}x\right) + c = 0 \qquad \text{Factor } a \text{ out of the first two terms on left side}$$

$$a\left(x^2 + \frac{b}{a}x + \left(\frac{b}{2a}\right)^2\right) - a\left(\frac{b}{2a}\right)^2 + c = 0 \qquad \text{Complete the square on } x^2 + \frac{b}{a}x$$

Since $x^2 + \frac{b}{a}x + \left(\frac{b}{2a}\right)^2 = \left(x + \frac{b}{2a}\right)^2$ and $a\left(\frac{b}{2a}\right)^2 = \frac{b^2}{4a}$, we have

$$a\left(x + \frac{b}{2a}\right)^2 + \left(c - \frac{b^2}{4a}\right) = 0$$

$$a\left(x + \frac{b}{2a}\right)^2 = -\left(c - \frac{b^2}{4a}\right)$$

$$a\left(x + \frac{b}{2a}\right)^2 = \frac{b^2 - 4ac}{4a} \qquad \text{Simplify the right-hand side}$$

$$\left(x + \frac{b}{2a}\right)^2 = \frac{b^2 - 4ac}{4a^2} \qquad \text{Divide by } a$$

$$x + \frac{b}{2a} = \pm\sqrt{\frac{b^2 - 4ac}{4a^2}} \qquad \text{Take the square roots of both sides}$$

$$x = -\frac{b}{2a} \pm \sqrt{\frac{b^2 - 4ac}{4a^2}} \qquad \text{Subtract } \frac{b}{2a}$$

$$x = -\frac{b}{2a} \pm \frac{\sqrt{b^2 - 4ac}}{2a} \qquad \begin{array}{l}\text{Simplify under the radical:} \\ \sqrt{4a^2} = 2a \text{ since } a > 0\end{array}$$

The Quadratic Formula

The solutions of $ax^2 + bx + c = 0$, with $a \neq 0$, is given by the **quadratic formula**

$$x = \frac{-b \pm \sqrt{b^2 - 4ac}}{2a}$$

In summary, to solve a **quadratic equation**, perform the following steps.

- Write the equation in the form $ax^2 + bx + c = 0$.

- If possible, factor the left-hand side of the equation to find the solution(s).

- If it is not possible to factor, use the quadratic formula to solve the equation. Alternatively, you can complete the square.

Note Since the quadratic formula gives all the solutions of a quadratic equation, we see that a quadratic equation can have at most two solutions.

Example 8   **Solving an Equation by Using the Quadratic Formula**

Solve the equation $-4x^2 + 3x + \frac{1}{2} = 0$ by using the quadratic formula.

Solution Since the expression $-4x^2 + 3x + \frac{1}{2}$ cannot be readily factored, we use the quadratic formula.

$$x = \frac{-(3) \pm \sqrt{(3)^2 - 4(-4)\left(\frac{1}{2}\right)}}{2(-4)}$$ Substitute $a = -4$, $b = 3$, and $c = \frac{1}{2}$ in the formula

$$= \frac{-3 \pm \sqrt{9 + 8}}{-8}$$ $-(4)(-4)\left(\frac{1}{2}\right) = 8$

$$= \frac{-3 \pm \sqrt{17}}{-8}$$ Simplify under the radical

$$= \frac{3}{8} \pm \frac{\sqrt{17}}{8}$$

The solutions of the equation are $x = \frac{3}{8} + \frac{\sqrt{17}}{8}$ and $x = \frac{3}{8} - \frac{\sqrt{17}}{8}$.

 Check It Out 8   Solve $2x^2 - 4x - 1 = 0$ using the quadratic formula.   ■

In the quadratic formula, the quantity under the radical, $b^2 - 4ac$, can be positive, negative, or zero. The characteristics of the solutions in each of these three cases will be different, as summarized below.

Types of Solutions to the Quadratic Equation and the Discriminant

The quadratic formula contains the quantity $b^2 - 4ac$ under the radical; this quantity is known as the **discriminant**. Solutions to quadratic equations that are real numbers are called **real solutions**. The number of real solutions to a particular quadratic equation depends on the sign of the discriminant:

- If $b^2 - 4ac > 0$, there will be *two* distinct, real solutions.

- If $b^2 - 4ac = 0$, there will be *one* real solution.

- If $b^2 - 4ac < 0$, there will be *no* real solutions.

Solutions that are not real numbers are called *nonreal* solutions; these are part of the complex number system discussed in Chapter 3.

Example 9   **The Quadratic Formula and Nonreal Solutions**

Use the quadratic formula to find the real solutions of the equation $-2t^2 + 3t = 5$. Find the value of the discriminant.

Solution Writing the given equation as $-2t^2 + 3t - 5 = 0$ and applying the quadratic formula gives

$$t = \frac{-(3) \pm \sqrt{(3)^2 - 4(-2)(-5)}}{2(-2)}$$ Substitute $a = -2$, $b = 3$ and $c = -5$

$$= \frac{-3 \pm \sqrt{9 - 40}}{-4}$$

$$= \frac{-3 \pm \sqrt{-31}}{-4}$$ Simplify, using care with the signs

The discriminant, $b^2 - 4ac = -31$, is negative. Thus, **there are no real numbers as solutions** to this equation. Solutions of quadratic equations in which the discriminant is negative can be found only by using complex numbers, discussed in Chapter 3.

 Check It Out 9   Use the quadratic formula to solve $x^2 - x + 3 = 0$. Find the value of the discriminant.

Applications

Quadratic equations arise frequently in applications, some of which we now discuss.

Example 10   **Quadratic Model for the Height of a Baseball in Flight**

The height of a ball after being thrown vertically upward from a point 80 feet above the ground, with a velocity of 40 feet per second, is given by $h = -16t^2 + 40t + 80$, where t is the time in seconds since the ball was thrown and h is in feet.

a. When will the ball be 50 feet above the ground?

b. When will the ball reach the ground?

c. For what values of t does this problem make sense (from a physical stand point)?

Solution

a. Setting the $h = 50$ and solving for t, we have

$$-16t^2 + 40t + 80 = 50$$
$$-16t^2 + 40t + 30 = 0 \qquad \text{Write in standard form}$$
$$t = \frac{-(40) \pm \sqrt{(40)^2 - 4(-16)(30)}}{2(-16)} \approx 3.10 \text{ or } -0.604 \qquad \text{Use the quadratic formula}$$

Since a negative number makes no sense for a value of time, the ball will be 50 feet above the ground in $t \approx 3.10$ seconds.

b. When the ball reaches the ground, the height h will be zero. Thus, setting $h = 0$ and solving for t gives

$$-16t^2 + 40t + 80 = 0$$

$$t = \frac{-(40) \pm \sqrt{(40^2) - 4(-16)(80)}}{2(-16)} \approx 3.81 \text{ or } -1.31$$

Since a negative number makes no sense for a value of time, the ball will reach the ground in $t \approx 3.81$ seconds.

c. The problem makes sense for values of t in the interval $[0, 3.81]$. The time t must be greater than or equal to 0. Once the ball hits the ground, the motion of the ball is no longer governed by the given expression.

 Check It Out 10   In Example 10, when will the ball be 40 feet above the ground?

Exercises 1.5

In Exercises 1–30, solve the equation.

1. $3x + 5 = 8$
2. $4x + 1 = 17$
3. $-2x - 5 = 3x + 10$
4. $4x - 2 = 2x + 8$
5. $-3(x - 1) = 12$
6. $5(x + 2) = 20$
7. $-2(x + 4) - 3 = 7$
8. $5(x - 2) + 4 = 19$
9. $-3(x - 4) = -(x + 1) - 6$
10. $6(2x + 1) = 3(x - 3) + 7$
11. $-2(5 + x) - (x - 2) = 10(x + 1)$
12. $3(4 + x) + 2(x + 2) = 2(2x - 1)$
13. $\dfrac{1}{2} + \dfrac{x}{3} = \dfrac{7}{6}$
14. $-\dfrac{1}{3} + \dfrac{x}{5} = \dfrac{2}{3}$
15. $\dfrac{x + 3}{4} + \dfrac{x}{3} = 6$
16. $\dfrac{x - 1}{5} + \dfrac{x}{2} = 4$
17. $\dfrac{2x - 3}{3} - \dfrac{x}{2} = -\dfrac{2}{3}$
18. $\dfrac{3x + 1}{2} - \dfrac{2x}{3} = \dfrac{3}{2}$
19. $\dfrac{3x + 4}{2} + x = 4$
20. $\dfrac{7x - 1}{3} - x = 1$
21. $0.4(x - 1) + 1 = 0.5x$
22. $-0.3(2x + 1) - 3 = 0.2x$
23. $1.2(x + 5) = 3.1x$
24. $2.6(x - 1) = 4.5x$
25. $0.01(x - 3) - 0.02 = 0.05$
26. $-0.03(x + 4) + 0.05 = 0.03$
27. $0.5(2x - 1) - 0.02x = 0.3$
28. $0.4(x - 2) - 0.05x = 0.7$
29. $\pi x + 3 = 4\pi x$
30. $\sqrt{2}\,(x + 1) - 1 = 3\sqrt{2}$

In Exercises 31–38, solve each equation for y in terms of x.

31. $x + y = 5$
32. $-x + y = 3$
33. $-4x + 2y = 6$
34. $6x + 3y = 12$
35. $5x + 4y = 10$
36. $3x + 2y = 12$
37. $4x + y - 5 = 0$
38. $-5x + y + 4 = 0$

In Exercises 39–50, solve the quadratic equation by factoring.

39. $x^2 - 25 = 0$
40. $x^2 - 16 = 0$
41. $x^2 - 7x + 12 = 0$
42. $x^2 - 4x - 21 = 0$
43. $-3x^2 + 12 = 0$
44. $-5x^2 + 45 = 0$
45. $6x^2 - x - 2 = 0$
46. $5x^2 - 7x - 6 = 0$
47. $4x^2 + 1 = 4x$
48. $9x^2 + 1 = -6x$
49. $2t^2 = t + 3$
50. $-10t + 8 = -3t^2$

In Exercises 51–58, solve the quadratic equation by completing the square.

51. $x^2 + 4x = -3$
52. $x^2 - 6x = 7$
53. $x^2 - 2x = 4$
54. $x^2 + 8x = 6$
55. $x^2 + x = 2$
56. $x^2 - x = 3$
57. $2x^2 + 8x - 1 = 0$
58. $3x^2 - 6x + 2 = 0$

In Exercises 59–72, solve the quadratic equations using the quadratic formula. Find only real solutions.

59. $x^2 + 2x - 1 = 0$

60. $x^2 + x - 5 = 0$

61. $-2x^2 + 2x + 1 = 0$

62. $2t^2 + 4t - 5 = 0$

63. $3 - x - x^2 = 0$

64. $-2 + t^2 + t = 0$

65. $2x^2 + x + 2 = 0$

66. $-3x^2 + 2x - 1 = 0$

67. $-l^2 + 40l = 100$

68. $-x^2 + 50x = 300$

69. $\frac{1}{2}t^2 - 4t - 3 = 0$

70. $-\frac{1}{3}x^2 - 3x + 9 = 0$

71. $-0.75x^2 + 2 = 2x$

72. $0.25x^2 - 0.5x = 1$

In Exercises 73–82, solve the quadratic equations using any method.

73. $x^2 - 4 = 0$

74. $x^2 - 9 = 0$

75. $-x^2 + 2x = 1$

76. $x^2 - 4x = -4$

77. $-2x^2 - 1 = 3x$

78. $-3x^2 - 2 = 7x$

79. $x^2 - 2x = 9$

80. $-x^2 - 3x = 1$

81. $(x - 1)(x + 2) = 1$

82. $(x + 1)(x - 2) = 2$

For each of the equations in Exercises 83–86, find the discriminant, $b^2 - 4ac$, and use it to determine the number of real solutions to the equation. You need not solve the equation.

83. $x^2 - 2x - 1 = 0$

84. $-x^2 + x + 3 = 0$

85. $x^2 + 2x = -3$

86. $-x^2 + 4x = 4$

Applications

87. Commerce The profit, in dollars, from selling x units of Blu-ray players is given by $40x - 200$. Set up and solve an equation to find out how many Blu-ray players must be sold to obtain a profit of $800.

88. Construction A contractor builds a square fence with 50 feet of fencing material. Find the length of a side of the square.

89. Art A rectangular frame for a painting has a perimeter of 96 inches. If the length of the frame is 30 inches, find the width of the frame.

90. Physics: Ball Height The height of a ball after being dropped from a point 100 feet above the ground is given by $h = -16t^2 + 100$, where t is the time in seconds since the ball was dropped and h is in feet.

   a. When will the ball be 60 feet above the ground?

   b. When will the ball reach the ground?

   c. For what values of t does this problem make sense (from a physical standpoint)?

91. Performing Arts Revenue from Broadway shows in New York can be modeled by $p = 0.0489t^2 - 0.7815t + 10.31$, where t is the number of years since 1981 and p is in millions of dollars. The model is based on data for the years 1981–2000. When will the revenue be $12 million?

                                            (Source: The League of American Theaters and Producers, Inc.)

92. Leisure The average amount of money spent on books and magazines per household in the United States can be modeled by $r = -0.2837t^2 + 5.547t + 136.7$. Here, r is in dollars and t is the number of years since 1985. The model is based on data for the years 1985–2000. According to this model, in what year(s) was the average expenditure per household for books and magazines equal to $160?

                                                  (Source: U.S. Bureau of Labor Statistics)

93. Manufacturing A rain gutter is to be fabricated with an open top and a rectangular cross-section by bending up a flat piece of metal that is 18 feet long and 20 inches wide. The top of the gutter is open. How much metal has to be bent upward to obtain a cross-sectional area of 30 square inches?

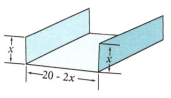

Concepts

94. Can the equation $x + 2 = x$ be solved for x? Explain.

95. The width of a fence is given by $w = \frac{1}{2}(15 - 2l)$. Evaluate w when $l = 2.5$ feet. If you try to evaluate w for $l = 10$ feet, do you get a realistic value for w? Explain.

96. Find the mistake in the following "solution" of the equation $\frac{x+1}{4} + 1 = 4$.

$$\frac{x+1}{4} + 1 = 4 \qquad \text{Multiplying by 4}$$

$$(x + 1) + 1 = 4$$

$$x = 2$$

97. For what values of a does the equation $ax + x = 5$ have a solution?

98. Find a quadratic equation whose solutions are 3 and -3. (Answers may vary.)

99. For what value(s) of k does $x^2 - k = 0$ have two distinct real solutions?

Linear Inequalities

Objectives

- Solve a linear inequality.
- Solve a compound inequality.
- Use inequalities to model and solve real world problems.

When you use mathematics in a real-world situation, you will often need to compare two different quantities. For example, you may need to compare two types of telephone calling plans to see which one is more economical. We solve this problem at the end of this section, but first you need to know how to solve linear inequalities.

Linear Inequalities

In this section, we explain how to solve a **linear inequality**, which is an inequality where a variable appears only to the first power. For example, $3x - 2 < 7$ is a linear inequality, but $2x^2 - 1 < 7$ is not. First, we review some properties of inequalities.

Properties of Inequalities

Let a, b, and c be any real numbers.

Addition principle: If $a < b$, then $a + c < b + c$.

Multiplication principle for $c > 0$: If $a < b$, then $ac < bc$ if $c > 0$.

Multiplication principle for $c < 0$: If $a < b$, then $ac > bc$ if $c < 0$. Note that the *direction* of the inequality is *reversed* when both sides are multiplied by a negative number.

Similar statements hold true for $a \leq b$, $a > b$, and $a \geq b$.

VIDEO EXAMPLES

SECTION 1.6

Example 1 shows how to use these properties to solve an inequality.

Example 1   **Solving a Linear Inequality**

Solve $x - 4 > -2x + 2$

Solution

$$
\begin{aligned}
x - 4 &> -2x + 2 &&\text{Given inequality}\\
-6 &> -3x &&\text{Collect like terms}\\
-\left(\frac{1}{3}\right)(-6) &< -\left(\frac{1}{3}\right)(-3x) &&\text{Multiply by } -\tfrac{1}{3}; \text{ switch direction of inequality}\\
2 &< x &&\text{Solve for } x\\
x &> 2 &&\text{Rewrite solution}
\end{aligned}
$$

The set of values of x such that $x > 2$ is called the solution set of the inequality. In interval notation, this set is written as $(2, \infty)$.

 Check It Out 1   Solve the inequality $-3x + 5 > -x + 1$.

Note Unlike solving an equation, solving an inequality gives an infinite number of solutions. You cannot really check your solution the same way you do for an equation; but you can get an idea of whether your solution is correct by substituting some values from your solution set into the inequality.

Compound Inequalities

If two inequalities are joined by the word *and*, then the conditions for both inequalities must be satisfied. Such inequalities are called **compound inequalities**. For example, $-2 < x + 4$ *and* $x + 4 \leq 9$ is a compound inequality that can be abbreviated as $-2 < x + 4 \leq 9$.

Example 2 illustrates additional techniques for solving inequalities, including compound inequalities.

Example 2   **Solving Additional Types of Inequalities**

Solve the following inequalities.

a. $2x + \dfrac{5}{2} > 3x - 6$ **b.** $-4 \le 3x - 2 < 7$

Solution

a. Solving this inequality involves clearing the fraction. Otherwise, all steps are similar to those used in the previous example.

$$2x + \frac{5}{2} > 3x - 6 \qquad \text{Given inequality}$$

$$2\left(2x + \frac{5}{2}\right) > 2(3x - 6) \qquad \text{Clear fraction: multiply each side by 2}$$

$$4x + 5 > 6x - 12 \qquad \text{Simplify each side}$$

$$-2x > -17 \qquad \text{Collect like terms}$$

$$x < \frac{17}{2} \qquad \text{Divide by } -2\text{: reverse inequality}$$

Thus, the solution set is the set of all real numbers that are less than $\frac{17}{2}$. In interval notation, this is $\left(-\infty, \frac{17}{2}\right)$.

b. We solve a compound inequality by working with all parts at once.

$$-4 \le 3x - 2 < 7$$

$$-2 \le 3x < 9 \qquad \text{Add 2 to each part}$$

$$-\frac{2}{3} \le x < 3 \qquad \text{Multiply by } \frac{1}{3}$$

Thus the solution set is the set of all real numbers greater than or equal to $-\frac{2}{3}$ and less than 3. In interval notation, this is $\left[-\frac{2}{3}, 3\right)$.

 Check It Out 2   Solve the inequality $-\dfrac{2}{3}x + 4 \le 3x + 5$.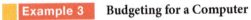

To summarize, we see that by using the properties of inequalities, we can algebraically solve any linear inequality in a manner similar to that used to solve a linear equation.

Applications

Example 3 illustrates how an inequality can be used in making budget decisions.

Example 3   **Budgeting for a Computer**

Alicia has a total of \$1000 to spend on a new computer system. If the sales tax is 8%, what is the retail price range of computers that she should consider?

Solution   Let p denote the price of the computer system. The sales tax is then 8% of p, or $0.08p$. The problem can be written and solved as an inequality, as follows:

$$\text{Price} + \text{sales tax} \le 1000 \qquad \text{From problem statement}$$

$$p + 0.08p \le 1000 \qquad \text{Substitute for price and sales tax}$$

$$1.08p \le 1000 \qquad \text{Collect like terms}$$

$$p \le 925.93 \qquad \text{Solve for } p$$

Thus, Alicia can purchase any computer system that has a retail price of less than or equal to \$925.93, without having the combination of price and sales tax exceed her budget of \$1000.

 Check It Out 3   Rework Example 3 if Alicia has a total of \$1200 to spend on a new computer system.

Example 4 analyzes the comparison of phone plans mentioned at the beginning of this section.

Example 4   Comparing Rate Plans

The Verizon phone company in New Jersey has the following two plans for local toll calls:

- Plan A charges $4.00 per month plus 8 cents per minute for every local toll call.
- Plan B charges a flat rate of $20 per month for local toll calls regardless of the number of minutes of use.

a. Express the monthly cost for Plan A in terms of the number of minutes used.

b. Express the monthly cost for Plan B in terms of the number of minutes used.

c. How many minutes would you have to use per month for Plan B to be cheaper than Plan A?

Solution

a. Let t be the number of minutes used, and A be the monthly cost of plan A. From the wording of the problem, we have

$$A = 4 + 0.08t \quad \text{Total monthly cost for Plan A}$$

b. Let t be the number of minutes used, and B be the monthly cost of plan B. Since the monthly cost is the same regardless of the number of minutes, we have

$$B = 20 \quad \text{Total monthly cost for Plan B}$$

c. To find out when Plan B will be cheaper than Plan A, we set $B < A$ and solve for t:

$$20 < 4 + 0.08t \quad \text{Set } B < A$$
$$16 < 0.08t \quad \text{Subtract 4 from both sides}$$
$$200 < t \quad \text{Divide by 0.08}$$

You would have to use more than 200 minutes of local toll calls per month for the cost of Plan B to be cheaper than Plan A. A table of values of the costs of both plans is given below. Note that for $t = 200$ minutes, both plans cost the same.

t	$A = 4 + 0.08t$	$B = 20$
0	4	20
50	8	20
100	12	20
150	16	20
200	20	20
250	24	**20**
300	28	**20**

Check It Out 4   In Example 4, suppose Plan B cost $28 per month and Plan A remained the same. How many minutes would you have to use per month for Plan B to be cheaper than Plan A?

An important application of linear equations and inequalities occurs in business models, when dealing with the production or operating costs of a product and the revenue earned from selling the product. We would like to determine the "break-even point", defined as the point where the production cost equals revenue. Example 5 explores this topic.

Example 5   **Cost and Revenue**

To operate a gourmet coffee booth in a shopping mall, it costs $500 (the fixed cost) plus $6 for each pound of coffee bought at wholesale price. The coffee is then sold to customers for $10 per pound.

a. Find an expression for the operating cost of selling q pounds of coffee.

b. Find an expression for the revenue earned by selling q pounds of coffee.

c. Find the break-even point.

d. How many pounds of coffee must be sold for the revenue to be greater than the total cost?

Solution

a. Let C represent the cost of selling q pounds of coffee. From the wording of the problem, we have

$$C = 500 + 6q$$

b. Since the coffee is sold for $10 per pound, the revenue R is

$$R = 10q$$

c. To find the break-even point, we set the expressions for the cost and revenue equal to each other to get

$$500 + 6q = 10q \qquad \text{Set cost equal to revenue}$$
$$500 = 4q \qquad \text{Collect like terms}$$
$$125 = q \qquad \text{Solve for } q$$

Thus, the store owner must sell 125 pounds of coffee for the operating cost to equal revenue. In that case, the production cost is $1250, and so is the revenue.

d. To calculate the amount of coffee sold so that the revenue is greater than the cost, we solve the inequality $R > C$:

$$10q > 500 + 6q \qquad \text{Substitute expressions for cost and revenue}$$
$$4q > 500 \qquad \text{Collect like terms}$$
$$q > 125 \qquad \text{Solve for } q$$

More than 125 pounds must be sold for the revenue to be greater than the cost.

✓ *Check It Out 5*   Rework Example 5, now assuming the coffee is sold for $11 per pound. The cost remains unchanged. Comment on the differences between the new result and the one in Example 5. ◼

Exercises 1.6

Skills

In Exercises 1–4, check whether the indicated value of the variable satisfies the given inequality.

1. Value: $x = 1$; Inequality: $x + 1 < 2$

2. Value: $x = \dfrac{1}{2}$; Inequality: $3x + 1 > -1$

3. Value: $s = 3.2$; Inequality: $2s - 1 \le 10$

4. Value: $t = \sqrt{2}$; Inequality: $5 > -t - 1$

In Exercises 5–30, solve the inequalities. Express your answer in interval notation.

5. $-3x + 2 \le 5x + 10$

6. $2t + 1 > 3t + 4$

7. $8s - 9 \ge 2s + 15$

8. $-x + 6 \le 2x + 9$

9. $2x < 3x - 10$

10. $x \ge 3x - 6$

11. $2x + 3 \ge 0$

12. $x - 4 < 0$

13. $4x - 5 > 3$

14. $3x + 1 \le 7$

15. $2 - 2x \ge x - 1$

16. $x + 4 < x - 1$

17. $-4(x + 2) \ge x + 5$

18. $-3(x - 3) < 7x + 1$

19. $\dfrac{x}{3} \le \dfrac{2x}{3} - 1$

20. $-\dfrac{x}{2} > \dfrac{3x}{2} + 3$

21. $\dfrac{1}{3}(x + 1) < x + 3$

22. $-2x - 1 \ge \dfrac{(x + 5)}{2}$

23. $\dfrac{1}{3}x + 2 \le \dfrac{3}{2}x - 1$

24. $\dfrac{2}{3}x + 3 \le 5$

25. $-2 \le 2x + 1 \le 3$

26. $-4 \le 3x - 2 \le 2$

27. $0 < -x + 5 < 4$

28. $-1 < -2x + 1 < 5$

29. $0 < \dfrac{x + 3}{2} < 3$

30. $1 \le \dfrac{2x - 1}{3} \le 4$

Applications

Cost and Revenue For each set of cost and revenue expressions in Exercises 31–34, (a) find the break-even point and (b) calculate the values of q for which revenue exceeds cost.

31. $C = 2q + 10$; $R = 4q$

32. $C = 3q + 21$; $R = 6q$

33. $C = 10q + 200$; $R = 15q$

34. $C = 8q + 150$; $R = 10q$

35. **Manufacturing** To manufacture boxes, it costs \$750 (the fixed cost) plus \$2 for each box produced. The boxes are then sold for \$4 each.

    a. Find an expression for the production cost of q boxes.

    b. Find an expression for the revenue earned by selling q boxes.

    c. For what values of q will the revenue be greater than the production cost?

36. Film Industry Films with plenty of special effects are very expensive to produce. For example, *Terminator 3* cost $55 million to make and another $30 million to market. Suppose an average movie ticket costs $8, and only half of this amount goes to the studio that made the film. How many tickets must be sold for the movie studio to break even for this film?   (Source: Stanford Graduate School of Business)

37. Health and Fitness A jogger on a pre-set treadmill burns 3.2 Calories per minute. How long must she jog to burn at least 200 Calories?

38. Compensation A salesperson earns $100 a week in salary plus 20% percent commission on total sales. How much must the salesperson generate in sales in one week to earn a total of at least $400 for the week?

39. Exam Scores In a math class, a student has scores of 94, 86, 84, and 97 on the first four exams. Assuming that 100 is the maximum number of points on each test, what must the range of scores be on the fifth exam so that the average of the five tests is greater than or equal to 90?

40. Sales Tax The total cost of a certain type of laptop computer ranges from $1200 to $2000. The total cost includes sales tax of 6%. Set up and solve an inequality to find the range of prices for the laptop before tax.

41. Communications A telephone company offers two different long distance calling plans. Plan A charges a fee of $4.95 per month plus $0.07 for each minute used. Plan B costs $0.10 per minute of use, but has no monthly fee.

  a. Find the total monthly cost of using Plan A in terms of the number of minutes used.

  b. Find the total monthly cost of using Plan B in terms of the number of minutes used.

  c. Calculate the number of minutes of long-distance calling at which the two plans will cost the same. What will be the monthly charge at that level of usage?

42. Cost Comparison Rental car company A charges a flat rate of $45 per day to rent a car, with unlimited mileage. Company B charges $25 per day plus $0.25 per mile.

  a. Find an expression for the cost of a car rental for one day from Company A in terms of the number of miles driven.

  b. Find an expression for the cost of a car rental for one day from Company B in terms of the number of miles driven.

  c. Determine how many miles must be driven so that Company A charges the same amount as Company B. What is the daily charge at this number of miles?

43. Meteorology At a dew point of 70°F, the relative humidity, in percentage points, can be approximated by the expression

$$RH = -2.58x + 280$$

where x represents the actual temperature, in degrees Fahrenheit. We assume that $x \geq 70$, the dew point temperature. What is the range of temperatures for which the relative humidity is greater than or equal to 50%?

Concepts

44. Does the inequality $x > x + 2$ have a solution? Explain.

45. Solve the inequality $2x + 4 \geq 2x$. What do you observe?

46. Explain why the following is incorrect: $3 \leq x \leq -4$.

Objectives

- Express absolute value of a number in terms of distance on the number line.
- Solve equations involving absolute value.
- Solve inequalities involving absolute value.
- Solve an applied problem involving absolute value.

The absolute value of a real number can be defined as follows:

$$|x| = \begin{cases} x \text{ if } x \geq 0 \\ -x \text{ if } x < 0 \end{cases}$$

Note that $|x|$ has two different expressions: x if $x \geq 0$ and $-x$ if $x < 0$. You use *only one* of the two expressions depending on the value of x. To solve equations and inequalities involving absolute value, it is useful to think of the absolute value of a number in terms of distance from the origin on the number line. Figure 1 illustrates an example of this.

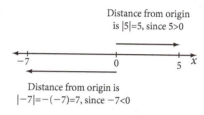

Distance from origin
is $|5|=5$, since $5>0$

Distance from origin is
$|-7|=-(-7)=7$, since $-7<0$

Figure 1

In this section, we discuss general methods to find solutions to equations and inequalities involving absolute value.

Equations Involving Absolute Value

From the definition of absolute value, we have the following statement.

> **Absolute Value Equations**
>
> Let $a > 0$. Then the expression $|X| = a$ is equivalent to $X = a$ or $X = -a$.

In the above statement, X can be *any* expression, not just a single variable. The set of all numbers which satisfy the equation $|X| = a$ is called its **solution set**. We can use this statement to solve equations involving absolute value, as shown in the next example.

Example 1   **Equations Involving Absolute Value**

Solve the following equations:

a. $|2x - 3| = 7$ **b.** $|x| = -3$ **c.** $-|3x + 1| - 3 = -8$

Solution

a. We have the following two equations that, taken together, correspond to the single equation $|2x - 3| = 7$:

$$2x - 3 = 7 \quad \text{or} \quad 2x - 3 = -7$$

The word *or* means that a number x is a solution of the equation $|2x - 3| = 7$ if and only if x is a solution of *at least one* of the two equations $2x - 3 = 7$ or $2x - 3 = -7$. Each of these two equations must be solved separately.

$2x - 3 = 7$	or $2x - 3 = -7$	Write down both equations
$2x = 10$	$2x = -4$	Add 3 to both sides
$x = 5$	$x = -2$	Divide by 2

Thus, the solution set is $\{-2, 5\}$.

b. Since the absolute value of any number must be greater than or equal to zero, the equation $|x| = -3$ has **no solution**.

c. In order to solve the equation $-|3x + 1| - 3 = -8$, we must first isolate the absolute value term.

$$-|3x + 1| - 3 = -8 \quad \text{Given equation}$$
$$-|3x + 1| = -5 \quad \text{Add 3 to both sides}$$
$$|3x + 1| = 5 \quad \text{Isolate absolute value term}$$

We next apply the definition of absolute value to get the following two equations that, taken together, correspond to the single equation $|3x + 1| = 5$.

$$3x + 1 = 5 \quad \text{or} \quad 3x + 1 = -5 \quad \text{Write down both equations}$$
$$3x = 4 \qquad\qquad 3x = -6 \quad \text{Subtract 1 from both sides}$$
$$x = \frac{4}{3} \qquad\qquad x = -2 \quad \text{Divide by 3}$$

Thus the solution set is $\left\{ -2, \frac{4}{3} \right\}$.

 Check It Out 1   Solve the equation $|-5x + 2| = 12$.

Inequalities Involving Absolute Value

Just in Time
Review linear inequalities in Section 1.6

Solving inequalities involving absolute value is straightforward if you keep in mind the definition of the absolute value. Thinking of the absolute value of a number as its distance from the origin (on the number line) leads to the following statements about inequalities.

Absolute Value Inequalities

Let $a > 0$. Then the inequality $|X| < a$ is equivalent to $-a < X < a$

$$-a < X < a$$

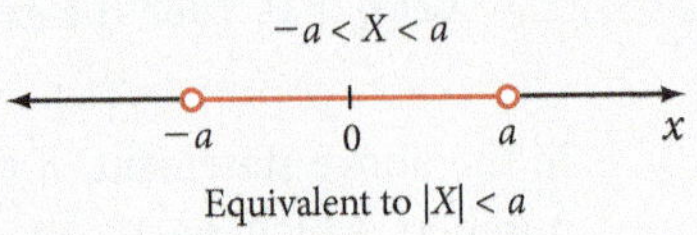

Figure 2

Similarly, $|X| \leq a$ is equivalent to $-a \leq X \leq a$.

Let $a > 0$. Then the inequality $|X| > a$ is equivalent to $X < -a \quad \textbf{or} \quad X > a$

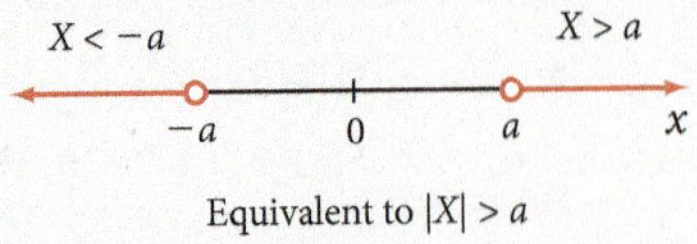

Figure 3

Similarly, $|X| \geq a$ is equivalent to $X \leq -a \quad \textbf{or} \quad X \geq a$.

Observations:
- In the above statements, X can be *any* expression, not just a single variable.
- The pair of equivalent inequalities for $|X| > a$ *must* be written as two *separate* inequalities, and similarly for $|X| \geq a$.

We next show how to solve inequalities involving absolute value.

> **Example 2** **Inequalities Involving Absolute Value**

Solve the following inequalities and indicate the solution set on a number line.

a. $|2x - 3| > 7$ 　　　　**b.** $\left|-\dfrac{2}{3}x + 4\right| \le 5$ 　　　　**c.** $-4 + |3 - x| > 5$

Solution

a. To solve the inequality $|2x - 3| > 7$, we proceed as follows. Since this is a "greater than" absolute value inequality, we must rewrite it as two separate inequalities without an absolute value:

$$2x - 3 < -7 \quad \text{or} \quad 2x - 3 > 7 \qquad \text{Rewrite as two separate inequalities}$$
$$2x < -4 \qquad\qquad 2x > 10 \qquad\quad \text{Add 3 to both sides}$$
$$x < -2 \qquad\qquad\quad x > 5 \qquad\qquad \text{Divide by 2}$$

Thus the solution of $|2x - 3| > 7$ is the set of all x such that $x < -2$ or $x > 5$. In interval notation, the solution set is $(-\infty, -2) \cup (5, \infty)$, which is graphed on the number line in Figure 4.

Figure 4

b. To solve $\left|-\dfrac{2}{3}x + 4\right| \le 5$ algebraically, we first write the inequality as an equivalent expression without an absolute value:

$$-5 \le -\frac{2}{3}x + 4 \le 5$$

We now solve the inequality for x.

$$-5 \ \le \ -\frac{2}{3}x + 4 \le 5 \qquad \text{Write equivalent expression}$$

$$-15 \ \le \ -2x + 12 \le 15 \qquad \text{Multiply by 3 to clear fraction}$$

$$-27 \ \le \ -2x \le 3 \qquad\qquad \text{Subtract 12 from each part}$$

$$\frac{27}{2} \ \ge \ x \ge -\frac{3}{2} \qquad\qquad \text{Divide by } -2; \text{ inequalities are reversed}$$

$$-\frac{3}{2} \ \le \ x \le \frac{27}{2} \qquad\qquad \text{Rewrite inequality}$$

The solution of $\left|-\dfrac{2}{3}x + 4\right| \le 5$ is the set of all x such that $-\dfrac{3}{2} \le x \le \dfrac{27}{2}$. In the last step, we turned the inequality around and rewrote it so that $-\dfrac{3}{2}$, the smaller of the two numbers $-\dfrac{3}{2}$ and $\dfrac{27}{2}$, comes first. So the solution in interval notation is $\left[-\dfrac{3}{2}, \dfrac{27}{2}\right]$. The solution set is graphed on the number line in Figure 5.

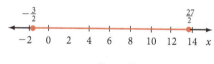

Figure 5

c. To solve the inequality $-4 + |3 - x| > 5$, first isolate the term containing the absolute value.

$$-4 + |3 - x| > 5 \qquad \text{Given inequality}$$
$$|3 - x| > 9 \qquad\qquad \text{Add 4 to each side}$$

Since this is a "greater than" inequality, we must rewrite it as two separate inequalities without the absolute value:

$$3 - x < -9 \quad \text{or} \quad 3 - x > 9 \qquad \text{Rewrite as two separate inequalities}$$
$$-x < -12 \qquad\qquad -x > 6 \qquad \text{Subtract 3 from both sides}$$
$$x > 12 \qquad\qquad x < -6 \qquad \text{Multiply by } -1; \text{ inequalities are reversed}$$

Therefore, the solution is $(-\infty, -6) \cup (12, \infty)$. The solution set is graphed on the number line in Figure 6.

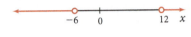

Figure 6

Check It Out 2   Solve the inequality $|-2x - 1| \leq 6$. Express your answer in interval notation and graph the solution set on the number line.

Understanding absolute value as distance plays an important role in applications and in more advanced math courses such as calculus. The quantity $|a - b|$, which is the same as $|b - a|$, represents the distance between a and b. The following examples explore the relationship between distance and absolute value.

Example 3   **Distance and Absolute Value**

Graph the following on a number line, and write each set using an absolute value inequality.

a. The set of all x whose distance from 4 is less than 5.
b. The set of all x whose distance from 4 is greater than 5.

Solution

a. The set of all x whose distance from 4 is less than 5 is indicated on the number line in Figure 7.

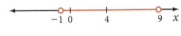

Figure 7

If $x > 4$, then $x - 4$ gives the distance from 4. If $x < 4$, then $4 - x = -(x - 4)$ gives the distance from 4. By the definition of absolute value, $|x - 4|$ gives the distance of x from 4. Thus, the set of all such x that are within 5 units of 4 is given by the inequality

$$|x - 4| < 5$$

b. To write an inequality which represents the set of all x whose distance from 4 is greater than 5, we simply set the distance expression, $|x - 4|$, to be greater than 5. The corresponding points on the number line are shown in Figure 8.

$$|x - 4| > 5$$

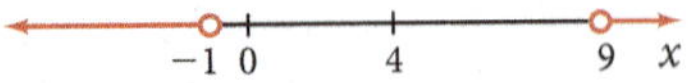

Figure 8

Check It Out 3   Graph the following on a number line and write the set using an absolute value inequality: the set of all x whose distance from 6 is greater than or equal to 3.

In applications, absolute value can be used to indicate intervals of uncertainty, as shown in Example 4.

Example 4   **Distance and Absolute Value**

A thermometer measures temperature with an uncertainty of 0.25°F. If a person's body temperature is measured at 98.3°F, use absolute value notation to write an inequality for the range of possible body temperatures.

Solution Since the uncertainty is 0.25°F, the person's temperature, x, can be anywhere in the following interval:

$$98.3 - 0.25 \leq x \leq 98.3 + 0.25$$

Thus x represents the set of all values that are a distance of at most 0.25 from 98.3. Written using absolute value, we have

$$|x - 98.3| \leq 0.25$$

 Check It Out 4   A scale measures weight with an uncertainty of 0.5 pounds. The scale reads 140 pounds while a certain person is standing on it. Use absolute value notation to write an inequality for the range of possible values of that person's weight.

Exercises 1.7

In Exercises 1–16, solve the equations.

1. $|x + 4| = 6$

2. $|x - 2| = 7$

3. $|2x - 4| = 8$

4. $|3x + 6| = 9$

5. $\left|2s - \dfrac{3}{2}\right| = 10$

6. $\left|3s + \dfrac{4}{3}\right| = 9$

7. $4\,|t + 5| = 16$

8. $-|3t + 2| = -5$

9. $2\,|2x + 1| = 10$

10. $4\,|x - 5| = 12$

11. $|x - 1| + 5 = 9$

12. $|x - 2| + 3 = 8$

13. $1 + |-2x + 5| = 3$

14. $-2 + |4x - 3| = 7$

15. $|x^2 - 8| = 1$

16. $|x^2 - 1| = 3$

In Exercises 17–20, determine whether the given value of x satisfies the inequality.

17. $|x - 2| > 4$; $x = 3$

18. $|x + 2| < 4$; $x = -2$

19. $|3x - 2| \le 4$; $x = \dfrac{3}{2}$

20. $|3x - 2| \ge 2$; $x = 4.1$

In Exercises 21–28, graph the solution set of each inequality on the real number line.

21. $x > -3$

22. $t < 4$

23. $-1 \le s \le 2$

24. $-4 \le x \le -1$

25. $|x| \le \dfrac{4}{3}$

26. $|x| \le 3$

27. $|x| > 7$

28. $|x| > 5$

In Exercises 29–50, solve the inequality. Express your answer in interval notation and graph the solution set on the number line.

29. $|2x| > 8$

30. $|3x| > 9$

31. $|x + 3| \le 4$

32. $|x - 4| \le 11$

33. $|x - 10| > 6$

34. $|x - 4| < 7$

35. $|2s - 7| > 3$

36. $|3s + 2| \ge 6$

37. $|2 - 3x| \le 10$

38. $|-1 + 7x| \le 13$

39. $\left|\dfrac{1}{2}x + 6\right| \le 5$

40. $\left|\dfrac{2}{3}x - 2\right| < 9$

41. $\left|\dfrac{x + 7}{6}\right| < 5$

42. $\left|\dfrac{x + 5}{8}\right| > 3$

43. $|x - 4| - 2 \ge 6$

44. $|x + 3| - 1 \le 4$

45. $|3x + 7| - 2 < 8$

46. $|4x + 2| + 4 \ge 9$

47. $|t - 6| \ge 0$

48. $|t - 6| > 0$

49. $|x - 4| < 0.001$

50. $|x - 3| < 0.01$

In Exercises 51–56, use absolute value notation to write an appropriate equation or inequality for each set of numbers.

51. All numbers whose distance from -7 is equal to 3.

52. All numbers whose distance from 8 is equal to $\dfrac{5}{4}$.

53. All numbers whose distance from 8 is less than 5.

54. All numbers whose distance from -4 is less than 7.

55. All numbers whose distance from -6.5 is greater than 8.

56. All numbers whose distance from 5 is greater than 12.3.

Applications

57. **Weather** The average temperatures in Frostbite Falls, over the course of a year are given by $|T + 10| < 20$, in degrees Fahrenheit. Solve this inequality and interpret it.

58. **Weather** Over the course of a year, the average daily temperature in Honolulu, Hawaii, varies from 65°F to 80°F. Express this range of temperatures using an inequality with an absolute value.

59. **Geography** You are located at the center of Omaha, Nebraska. Write an inequality, using an absolute value, which gives all the points within 30 miles north or south of the center of Omaha. Indicate what point you would use as the origin.

60. **Geography** You are located at the center of Hartford, Connecticut. Write an inequality, using an absolute value, which gives all the points more than 65 miles east or west of the center of Hartford. Indicate what point you would use as the origin.

61. **Temperature Measurement** A room thermostat is set at 68°F and measures the temperature of the room with an uncertainty of ± 1.5°F. Assuming that the temperature is uniform throughout the room, use absolute value notation to write an inequality for the range of possible temperatures in the room.

62. **Length Measurement** A ruler measures an object with an uncertainty of $\frac{1}{16}$ inch. If a pencil is measured to be 8 inches, use absolute value notation to write an inequality for the range of possible lengths of the pencil.

Concepts

63. Explain why $|-3(x + 2)|$ is *not* the same as $-3|x + 2|$.

64. Explain why the expression "$x > 3$ *or* $x < -2$" *cannot* be written as $3 < x < -2$.

65. Show that $|x - k| = |k - x|$, where k is any real number.

66. Can you think of an absolute value equation with no solution.

67. Explain why $|x| < 0$ has no solution.

Objectives

- Solve polynomial equations by reduction to quadratic form.

- Solve equations containing rational expressions.

- Solve equations containing radical expressions.

- Solve applied problems.

VIDEO EXAMPLES

SECTION 1.8

In this section, we will solve equations containing polynomial, rational, and radical expressions by algebraically manipulating them to resemble quadratic equations. We note that the techniques covered in this section apply only to a select class of equations. Chapter 3 on polynomial and rational functions discusses general graphical solutions of such equations in more detail.

Solving Equations by Reducing Them to Quadratic form

Recall that to solve a quadratic equation, you could either factor a quadratic polynomial or use the quadratic formula. Some types of polynomial equations can be solved by using a substitution technique to reduce them to quadratic form. Then either the quadratic formula or factoring is used to solve the equation. We illustrate this technique in the following examples.

Example 1   **Solving Using Substitution**

Solve the equation $3x^4 + 5x^2 - 2 = 0$ for real values of x.

Solution   Since the powers of x in this equation are all even, we can use the substitution

$$u = x^2$$

This gives

$$3x^4 + 5x^2 - 2 = 0$$
$$3(x^2)^2 + 5(x^2) - 2 = 0$$
$$3u^2 + 5u - 2 = 0$$

The equation involving u is quadratic and is easily factored.

$$3u^2 + 5u - 2 = (3u - 1)(u + 2) = 0$$

This implies that

$$u = -2 \text{ or } u = \frac{1}{3}$$

We now go back and find the values of x that correspond to these solutions for u.

$$u = x^2 = -2 \Rightarrow \text{no real solutions for } x$$

$$u = x^2 = \frac{1}{3} \Rightarrow x = \pm\sqrt{\frac{1}{3}} = \pm\frac{\sqrt{3}}{3}$$

Therefore, the original equation has two real solutions, $x = \dfrac{\sqrt{3}}{3}$ and $x = -\dfrac{\sqrt{3}}{3}$.

✅ *Check It Out 1*   Solve the equation $2x^4 + x^2 - 3 = 0$.

> **Example 2**   **Solving Using Substitution**

Solve the equation $t^6 - t^3 = 2$ for real values of t.

Solution Here, too, we will try to find a substitution that will reduce the equation to a quadratic. We note that only cubic powers of t appear in the equation, so we can use the substitution

$$u = t^3$$

Thus, we have

$$(t^3)^2 - t^3 = 2$$
$$u^2 - u = 2$$
$$u^2 - u - 2 = 0$$
$$(u - 2)(u + 1) = 0$$
$$u = -1 \text{ or } u = 2$$

Converting back to t, we have

$$u = t^3 = -1 \Rightarrow t = -1$$
$$u = t^3 = 2 \Rightarrow t = \sqrt[3]{2}$$

Thus the real valued solutions are $t = -1$ and $t = \sqrt[3]{2}$. Both of the equations $t^3 = -1$ and $t^3 = 2$ have complex-valued solutions as well, but we will not discuss those here.

 Check It Out 2   Solve the equation $3t^6 + 5t^3 = 2$ for real values of t. ■

Just in Time
Review rational
expressions in
Section 1.4.

Solving Equations Containing Rational Expressions

Many equations involving rational expressions can be reduced to linear or quadratic equations. To do so, we multiply both sides of the equation by the least common denominator (LCD) of each term in the equation. The process is illustrated in the following example.

> **Example 3**   **Solving an Equation with Rational Expressions**

Solve $\dfrac{1}{x} = \dfrac{2}{x - 2} + 3$.

Solution The LCD of the three terms in this equation is $x(x - 2)$. Note that the term 3 just has a denominator of 1. We proceed as follows.

$$\frac{1}{x} = \frac{2}{x - 2} + 3 \qquad \text{Original equation}$$

$$x(x - 2)\frac{1}{x} = x(x - 2)\frac{2}{x - 2} + x(x - 2)3 \qquad \begin{array}{l}\text{Multiply both sides of the equation} \\ \text{by the LCD}\end{array}$$

$$(x - 2)(1) = 2x + 3x(x - 2) \qquad \text{Divide out common factors}$$

$$x - 2 = 2x + 3x^2 - 6x \qquad \text{Expand products}$$

$$-3x^2 + 5x - 2 = 0 \qquad \text{Standard form of a quadratic equation}$$

$$(-3x + 2)(x - 1) = 0 \qquad \text{Factor}$$

Apply the Zero Product Rule to get

$$-3x + 2 = 0 \Rightarrow x = \frac{2}{3} \text{ or } x - 1 = 0 \Rightarrow x = 1$$

Thus the possible solutions are $x = \frac{2}{3}$ and $x = 1$. We must check both possibilities in the *original* equation.

Check $x = \frac{2}{3}$:

$$\frac{1}{x} = \frac{2}{x-2} + 3 \qquad \text{Original equation}$$

$$\frac{1}{\frac{2}{3}} = \frac{2}{\frac{2}{3} - 2} + 3 \qquad \text{Let } x = \frac{2}{3}$$

$$\frac{3}{2} = \frac{2}{-\frac{4}{3}} + 3$$

$$\frac{3}{2} = -\frac{3}{2} + 3$$

$$\frac{3}{2} = \frac{3}{2} \qquad\qquad x = \frac{2}{3} \text{ checks}$$

Check $x = 1$:

$$\frac{1}{x} = \frac{2}{x-2} + 3 \qquad \text{Original equation}$$

$$\frac{1}{1} = \frac{2}{1-2} + 3 \qquad \text{Let } x = 1$$

$$1 = 1 \qquad\qquad x = 1 \text{ checks}$$

Thus, $x = \frac{2}{3}$ and $x = 1$ are solutions.

✓ *Check It Out 3*   Solve $-\dfrac{2}{x+3} = \dfrac{1}{x} - \dfrac{3}{2}$.   ■

Solving Equations with Radical Expressions

The next example shows how to solve an equation involving variables under a radical symbol. The main idea is to isolate the term with the radical and then square both sides to eliminate the radical.

> **Example 4**   **Solving an Equation Containing One Radical**

Solve $\sqrt{3x + 1} + 2 = x - 1$.

Solution

$$\sqrt{3x + 1} + 2 = x - 1 \qquad \text{Original equation}$$

$$\sqrt{3x + 1} = x - 3 \qquad \text{Isolate the radical term by subtracting 2}$$

$$3x + 1 = x^2 - 6x + 9 \qquad \text{Square both sides}$$

$$0 = x^2 - 9x + 8 \qquad \text{Quadratic equation in standard form}$$

$$0 = (x - 8)(x - 1) \qquad \text{Factor}$$

Note When solving equations with radicals, we often obtain extraneous solutions. This occurs as a result of modifying the original equation in the course of the solution process. Therefore, it is very important to check all possible solutions by substituting each of them into the original equation.

The only possible solutions are $x = 1$ and $x = 8$. To determine whether they actually are solutions, we substitute them for x—one at a time—in the *original* equation:

    Check $x = 8$: $\sqrt{3x + 1} + 2 = \sqrt{3(8) + 1} + 2 = 7$ and $x - 1 = 8 - 1 = 7$

Therefore $x = 8$ is a solution. Now check $x = 1$:

    Check $x = 1$: $\sqrt{3x + 1} + 2 = \sqrt{3(1) + 1} + 2 = 4$ but $x - 1 = 1 - 1 = 0$

Since $4 \neq 0$, $x = 1$ is not a solution of the original equation. Therefore, the only solution is $x = 8$.

✓ *Check It Out 4*   Solve $\sqrt{4x + 5} - 1 = x + 1$.   ■

In the case of Example 4, the extraneous solution crept in when we squared both sides of the equation $\sqrt{3x + 1} = x - 3$, which gave $3x + 1 = x^2 - 6x + 9$. Note that $x = 1$ is a solution of the latter but not of the former.

If a radical equation contains two terms with variables under the radicals, isolate one of the radicals and raise both sides to an appropriate power. If a radical term with a variable still remains, repeat the process. The next example illustrates the technique.

Example 5   **Solving an Equation Containing Two Radicals**

Solve $\sqrt{3x+1} - \sqrt{x+4} = 1$.

Solution

$\sqrt{3x+1} - \sqrt{x+4} = 1$	Original equation
$\sqrt{3x+1} = 1 + \sqrt{x+4}$	Isolate a radical
$3x+1 = 1 + 2\sqrt{x+4} + (x+4)$	Square both sides
$3x+1 = x+5 + 2\sqrt{x+4}$	Combine like terms
$2x-4 = 2\sqrt{x+4}$	Isolate the radical
$4x^2 - 16x + 16 = 4x + 16$	Square both sides
$4x^2 - 20x = 0$	Quadratic equation in standard form
$4x(x-5) = 0$	Factor the left-hand side

Apply the Zero Product Rule to get

$$4x = 0 \Rightarrow x = 0 \text{ or } x - 5 = 0 \Rightarrow x = 5$$

The only possible solutions are $x = 0$ and $x = 5$. To determine whether they actually are solutions, we substitute them for x—one at a time—in the *original* equation:

$$\textbf{Check } x = 0: \sqrt{3x+1} - \sqrt{x+4} = \sqrt{3(0)+1} - \sqrt{0+4} = -1 \neq 1$$

Therefore $x = 0$ is **not** a solution. Now check $x = 5$:

$$\textbf{Check } x = 5: \sqrt{3(5)+1} - \sqrt{(5)+4} = 4 - 3 = 1$$

Thus $x = 5$ is a solution of the original equation. Therefore, the only solution is $x = 5$.

Check It Out 5   Solve $\sqrt{2x-1} - \sqrt{x+3} = -1$.

Applications

Rational and radical equations occur in a variety of applications. In the following examples, we examine two such applications.

 Example 6   **Average Cost**

A theater club arranged for a chartered bus trip to a play at a cost of $350. To lower costs, 10 nonmembers were invited to join the trip. The bus fare per person then decreased by $4. How many theater club members were going on the trip?

Solution First, identify the variable and the relationships among the many quantities mentioned in the problem.

Variable: the number of club members going on the trip, denoted by x

Total cost: $350

Number of people on trip: $x + 10$

Original cost per club member: $\frac{350}{x}$

New cost per person: $\frac{350}{x} - 4$

Equation: (New cost per person) (number of people on trip) = Total cost

Thus, have the equation

$$\left(\frac{350}{x} - 4\right)(x + 10) = 350 \qquad \text{(Cost per person) (number of people)} = \text{total cost}$$

$$\left(\frac{350 - 4x}{x}\right)(x + 10) = 350 \qquad \text{Write the first factor as a single fraction}$$

$$(350 - 4x)(x + 10) = 350x \qquad \text{Multiply by } x$$

$$350x - 4x^2 - 40x + 3500 = 350x \qquad \text{Expand the left side}$$

$$-4x^2 - 40x + 3500 = 0 \qquad \text{Make the right side of the equation zero}$$

$$-4(x^2 + 10x - 875) = 0 \qquad \text{Factor out } -4$$

$$x^2 + 10x - 875 = 0 \qquad \text{Divide by } -4 \text{ on both sides}$$

$$(x + 35)(x - 25) = 0 \qquad \text{Factor}$$

Setting each factor to zero,

$$x + 35 = 0 \Rightarrow x = -35$$

$$x - 25 = 0 \Rightarrow x = 25$$

Only the positive value of x makes sense, and so there are 25 members of the theater club going on the trip. You should check this solution in the original equation.

✔ *Check It Out 6*   Check the solution to Example 6. ■

The next problem will illustrate an application of a radical equation.

Example 7   **Distance and Rate**

Jennifer is standing on one side of a river that is 3 kilometers wide. Her bus is located on the opposite side of the river. Jennifer plans to cross the river by rowboat and then jog the rest of the way to reach the bus, which is 10 kilometers along the river from a point B directly across the river from her current location (point A). If she can row 5 kilometers per hour and jog 7 kilometers per hour, at which point on the other side of the river should she dock her boat so that it will take her a total of exactly 2 hours to reach her bus? Assume that Jennifer's path on each leg of the trip is a straight line, and that there is no river current or wind speed.

Solution   First, we draw a figure illustrating the problem. From Figure 1, the distance rowed is $\sqrt{x^2 + 9}$ and the distance jogged is $10 - x$.

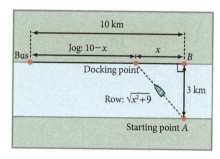

Figure 1

Recall that Distance = Speed × Time, so Time = $\dfrac{\text{Distance}}{\text{Speed}}$.

$$\text{Time to row} = \frac{\text{Distance rowed}}{\text{Rowing speed}} = \frac{\sqrt{x^2 + 9}}{5}$$

$$\text{Time to jog} = \frac{\text{Distance jogged}}{\text{Jogging speed}} = \frac{10 - x}{7}$$

Since the total time must equal two hours, we have

$$\frac{\sqrt{x^2 + 9}}{5} + \frac{10 - x}{7} = 2 \qquad \text{Time rowed + time jogged = 2 hours}$$

$$7\left(\sqrt{x^2 + 9}\right) + 5(10 - x) = 70 \qquad \text{Multiply by 35 to clear fractions}$$

$$7\left(\sqrt{x^2 + 9}\right) + 50 - 5x = 70 \qquad \text{Distribute the 5}$$

$$7\left(\sqrt{x^2 + 9}\right) = 20 + 5x \qquad \text{Isolate the radical expression}$$

$$49(x^2 + 9) = 400 + 200x + 25x^2 \qquad \text{Square both sides}$$

$$49x^2 + 441 = 400 + 200x + 25x^2 \qquad \text{Distribute the 49}$$

$$24x^2 - 200x + 41 = 0 \qquad \text{Write the quadratic equation in standard form}$$

Using the quadratic formula to solve the equation, we have

$$x = \frac{-(-200) \pm \sqrt{(-200)^2 - 4(24)(41)}}{2(24)} \approx 0.2103 \text{ or } 8.123$$

In order to reach the bus in 2 hours, Jennifer should dock the boat either 0.2103 km along the river from Point B or 8.123 km along the river from point B.

You can check that both solutions satisfy the original equation, within rounding error.

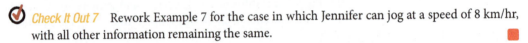 **Check It Out 7**   Rework Example 7 for the case in which Jennifer can jog at a speed of 8 km/hr, with all other information remaining the same.

In practice, problems such as the one in Example 7 would ask for the *minimum time* it takes to reach a destination. You can solve such problems by hand only by using calculus.

Exercises 1.8

Just in Time Exercises

These exercises correspond to the Just in Time references in this section. Complete them to review topics relevant to the remaining exercises.

In Exercises 1–4, multiply and simplify.

1. $(x - 2)\left(\dfrac{3}{x - 2}\right)$ **2.** $x^2\left(\dfrac{5}{x}\right)$ **3.** $x(x + 2)\left(\dfrac{2}{x} + \dfrac{x}{x + 2}\right)$ **4.** $(x^2 - 9)\left(\dfrac{5}{x + 3} + \dfrac{x}{x - 3}\right)$

Skills

In Exercises 5–16, find all real solutions of the polynomial equations. Check your solutions.

5. $x^4 - 49 = 0$ **6.** $x^4 - 25 = 0$ **7.** $x^4 - 10x^2 = -21$ **8.** $x^4 - 5x^2 = 24$

9. $6s^4 - s^2 - 2 = 0$ **10.** $4s^4 + 11s^2 - 3 = 0$ **11.** $4x^4 - 7x^2 = 2$ **12.** $-x^4 + 2x^2 - 1 = 0$

13. $x^6 - 4x^3 = 5$ **14.** $x^6 = x^3 + 6$ **15.** $3t^6 + 14t^3 = -8$ **16.** $2x^6 - 7x^3 = 7$

In Exercises 17–32, find all real solutions of the rational equations. Check your solutions.

17. $\dfrac{2}{3} + \dfrac{3}{5} = \dfrac{2}{x}$

18. $\dfrac{1}{4} - \dfrac{3}{2} = \dfrac{3}{x}$

19. $-\dfrac{2}{3x} + \dfrac{1}{x} = \dfrac{1}{4}$

20. $\dfrac{1}{2x} + \dfrac{4}{5} = \dfrac{3}{x}$

21. $\dfrac{1}{x^2} - \dfrac{3}{x} = 10$

22. $\dfrac{1}{x^2} - \dfrac{7}{x} = 18$

23. $\dfrac{2x}{x - 1} - \dfrac{3}{x} = 2$

24. $-\dfrac{3x}{x + 2} + \dfrac{1}{x} = 2$

25. $\dfrac{3}{x + 1} + \dfrac{2}{x - 3} = 4$

26. $\dfrac{1}{2x - 3} - \dfrac{x}{x - 1} = 2$

27. $\dfrac{1}{x^2 - x - 6} + \dfrac{3}{x + 2} = \dfrac{-4}{x - 3}$

28. $\dfrac{1}{x^2 + 4x - 5} + \dfrac{6}{x + 5} = \dfrac{1}{x - 1}$

29. $\dfrac{x}{2x^2 + x - 3} + \dfrac{1}{x - 1} = \dfrac{3}{2x + 3}$

30. $\dfrac{x}{3x^2 + 5x - 2} - \dfrac{5}{x + 2} = -\dfrac{1}{3x - 1}$

31. $\dfrac{x - 3}{2x - 4} + \dfrac{1}{x^2 - 4} = \dfrac{1}{x + 2}$

32. $\dfrac{x - 1}{3x + 3} - \dfrac{9}{x^2 - 1} = \dfrac{2}{x + 1}$

In Exercises 33–46, solve the radical equation to find all real solutions. Check your solutions.

33. $\sqrt{x + 3} = 5$

34. $\sqrt{x + 2} = 6$

35. $\sqrt{x^2 + 1} = \sqrt{17}$

36. $\sqrt{x^2 + 3} = \sqrt{28}$

37. $\sqrt{x^2 + 6x} - 1 = 3$

38. $\sqrt{x^2 - 5x} + 4 = 10$

39. $\sqrt{x + 1} + 2 = x$

40. $\sqrt{2x - 1} + 2 = x$

41. $\sqrt{x - 4} + \sqrt{x} = 2$

42. $\sqrt{x + 3} + \sqrt{x - 5} = 4$

43. $\sqrt{2x + 3} - \sqrt{x - 2} = 2$

44. $\sqrt{x - 1} - \sqrt{2x - 1} = 3$

45. $\sqrt[3]{x + 3} = 5$

46. $\sqrt[3]{5x - 3} = 5$

In Exercises 47–50, find all real solutions of the given equations. Check your solutions.

47. $x - 4\sqrt{x} = -3$ (*Hint:* use $u = \sqrt{x}$)   **48.** $x - 6\sqrt{x} = -5$ (*Hint:* use $u = \sqrt{x}$)

49. $3x^{2/3} + 2x^{1/3} - 1 = 0$ (*Hint:* use $u = x^{1/3}$)   **50.** $2x^{2/3} - 5x^{1/3} - 3 = 0$ (*Hint:* use $u = x^{1/3}$)

Applications

51. **Average Cost** Four students plan to rent a minivan for a weekend trip and equally share the rental cost of the van. By adding two more people, each person can save $10 on his or her share of the cost. How much is the total rental cost of the van?

52. **Work Rate** Two painters are available to paint a room. Working alone, the first painter can paint the room in 5 hours. The second painter can paint the room in 4 hours working by herself. If they work together, they can paint the room in t hours. To find t, we note that in 1 hour, the first painter paints $\frac{1}{5}$ of the room and the second one paints $\frac{1}{4}$ of the room. If they work together, they paint $\frac{1}{t}$ portion of the room. The equation is thus

$$\frac{1}{5} + \frac{1}{4} = \frac{1}{t}$$

Find t, the time it takes both painters to paint the room working together.

53. **Work Rate** Two water pumps work together to fill a storage tank. If the first pump can fill the tank in 6 hours and the two pumps working together can fill the tank in 4 hours, how long would it take to fill the storage tank using just the second pump? (Hint: To set up an equation, refer to preceding problem.)

54. **Engineering** In electrical circuit theory, the formula

$$\frac{1}{R} = \frac{1}{R_1} + \frac{1}{R_2}$$

is used to find the total resistance R of a circuit when two resistors with resistance R_1 and R_2 are connected in parallel. In such a parallel circuit, if the total resistance R is $\frac{8}{3}$ ohms and R_2 is twice R_1, find the resistances R_1 and R_2.

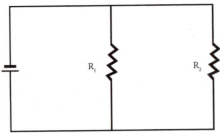

Use the following figure for Problems 55–56.

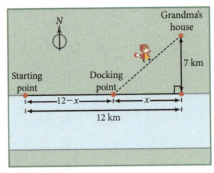

55. **Distance and Rate** To get to her grandmother's house, Little Red Riding Hood first rows a boat along a river at a speed of 10 km/hr and then walks through the woods at 4 km/hr. If she starts from a point that is 7 km south and 12 km west of Grandma's house, at what point along the river should she dock her boat so that she can reach the house in a total of exactly 3 hours? Assume that all paths are straight paths.

56. **Distance and Rate** The Big Bad Wolf would like to beat Little Red Riding Hood to her grandmother's house. The Big Bad Wolf can row a boat at a speed of 12 km/hr and can walk through the woods at 5 km/hr. If the wolf starts from the same point as Little Red Riding Hood, at what point along the river should the wolf dock its boat so that it can reach the house in a total of exactly 2.5 hours? Assume that all paths are straight paths.

Concepts

57. Explain what is wrong with the following steps for solving a radical equation.

$$\sqrt{x + 1} - 2 = 0$$

$$(x + 1) + 4 = 0$$

$$x = -5$$

58. Without any calculations, explain why $\sqrt{x + 1} = -2$ does *not* have a real number as a solution.

59. Without any calculations, explain why $\frac{1}{x^2 + 1} = 0$ does *not* have a real number as a solution.

60. Write an equation that has $2, -2, 3, -3$ as solutions. (There are many correct answers.)

Chapter 1 Summary

Section 1.1 The Real Number System

Properties of real numbers
[Review exercises 1–6]

Any real number is either a rational or an irrational number. All real numbers satisfy the following properties.

- The **associative properties** of addition and multiplication:

$$a + (b + c) = (a + b) + c \quad \text{and} \quad a(bc) = (ab)c$$

- The **commutative properties** of addition and multiplication:

$$a + b = b + a \quad \text{and} \quad ab = ba$$

- The **distributive property** of multiplication:

$$ab + ac = a(b + c)$$

where a, b, and c are any real numbers.

Ordering of real numbers
[Review exercises 7–10]

The real numbers are ordered increasingly from left to right on the number line.
- An **inequality** of the form $a \leq x \leq b$ can be expressed in interval notation as $[a, b]$. This interval is called a **closed interval**.
- If the endpoints of the interval, a and b, are not included, we write the inequality $a < x < b$ as an **open interval** (a, b).

Absolute value of a number
[Review exercises 11 and 12]

$$|x| = \begin{cases} x & \text{if } x \geq 0 \\ -x & \text{if } x < 0 \end{cases}$$

The distance between two points a and b on the number line is given by $|b - a|$ or $|a - b|$.

Rules for order of operations
[Review exercises 13–16]

When evaluating a numerical expression, the **order of operations** is to remove parentheses first, then exponentiate, multiply, divide, add and subtract, in that order, proceeding from left to right.

Section 1.2 Exponents, Roots, and Radicals

Positive and negative integer exponents
[Review exercises 17–22]

Positive integer exponents   For any positive integer n,

$$a^n = \underbrace{a \cdot a \cdot a \cdots a}_{n \text{ factors}}$$

The number a is the **base** and the number n is the **exponent**.

Negative integer exponents

Let a be any nonzero real number and m a positive integer. Then,

$$a^{-m} = \frac{1}{a^m}$$

Properties of integer exponents [Review exercises 17–22]

Let a and b be real numbers and let m and n be integers. Then the following properties hold.

1. $a^m \cdot a^n = a^{m+n}$

2. $(a^m)^n = a^{mn}$

3. $(ab)^m = a^m b^m$

4. $\left(\dfrac{a}{b}\right)^m = \dfrac{a^m}{b^m}, b \neq 0$

5. $\dfrac{a^m}{a^n} = a^{m-n}, a \neq 0$

6. $a^1 = a$

7. $a^0 = 1, a \neq 0$

Scientific Notation [Review exercises 23–29]

A nonzero number x is written in **scientific notation** as

$$a \times 10^b$$

where $1 \leq a < 10$, if $x > 0$ and $-10 < a \leq -1$, if $x < 0$ and b is an integer.

The nth root of a number [Review exercises 30–37]

For n an integer, $\sqrt[n]{a}$ is called the **nth root of a**. It denotes the number whose nth power is a. If n is even, then $a \geq 0$. If n is odd, then a can be any real number.

Rules for radicals [Review exercises 30–37]

Suppose a and b are real numbers such that their nth roots are defined.

Product Rule:　$\sqrt[n]{a} \cdot \sqrt[n]{b} = \sqrt[n]{ab}$

Quotient Rule:　$\dfrac{\sqrt[n]{a}}{\sqrt[n]{b}} = \sqrt[n]{\dfrac{a}{b}}, b \neq 0$

Rational exponents [Review exercises 38–46]

If a is a real number and n is a positive integer greater than 1, then

$$a^{1/n} = \sqrt[n]{a}, \quad \text{where } a \geq 0 \text{ when } n \text{ is even}$$

Let a be a positive real number and m and n integers. Then

$$a^{m/n} = \sqrt[n]{a^m} = \left(\sqrt[n]{a}\right)^m$$

Section 1.3 Polynomials and Factoring

Definition of polynomials [Review exercises 47–56]

A **polynomial in one variable** is an algebraic expression of the form

$$a_n x^n + a_{n-1} x^{n-1} + a_{n-2} x^{n-2} + \cdots + a_1 x + a_0$$

where n is a nonnegative integer and $a_n, a_{n-1}, \ldots, a_0$ are real numbers and $a_n \neq 0$.

Addition of polynomials [Review exercises 47–50]

To add or subtract polynomials, *combine* or *collect like terms* by adding or subtracting their respective coefficients.

Products of polynomials [Review exercises 51–56]

To multiply polynomials, apply the distributive property and then apply the rules for multiplying monomials. The **degree** of the polynomial is n, the highest power to which the variable is raised.

Special products of polynomials   [Review exercises 57–60]

Special products of polynomials are listed below.

$$(A + B)^2 = A^2 + 2AB + B^2$$
$$(A - B)^2 = A^2 - 2AB + B^2$$
$$(A + B)(A - B) = A^2 - B^2$$

General factoring techniques   [Review exercises 61–76]

A trinomial can be factored either by reversing the FOIL method of multiplication or by working backwards and trying various factors,

$$x^2 - 2x - 8 = (x - 4)(x + 2)$$

Quadratic and cubic factoring patterns   [Review exercises 61–76]

These are some common factoring patterns for some specific forms of polynomials.

Quadratic factoring patterns:
$$A^2 - B^2 = (A + B)(A - B)$$
$$A^2 + 2AB + B^2 = (A + B)^2$$
$$A^2 - 2AB + B^2 = (A - B)^2$$

Cubic factoring patterns:
$$A^3 - B^3 = (A - B)(A^2 + AB + B^2)$$
$$A^3 + B^3 = (A + B)(A^2 - AB + B^2)$$

Section 1.4 Rational Expressions

Arithmetic of rational expressions   [Review exercises 77–88]

A quotient of two polynomial expressions is called a **rational expression**. A rational expression is defined whenever the denominator is not equal to zero.

To find the **product** of two rational expressions, multiply the numerators, multiply the denominators, and then simplify the answer.

To **add** and **subtract** rational expressions, first write them in terms of the same denominator, known as the **least common denominator**. Then add and simplify.

Section 1.5 Linear and Quadratic Equations

Solving linear equations   [Review exercises 89–93]

To solve an equation, simplify it using basic operations until you arrive at the form $x = c$ for some number c.

If an equation contains decimals or fractions, multiply the equation by the LCD to clear the decimal or fraction. This makes the equation easier to work with.

If an equation contains two variables, solve for one variable in terms of another.

Solving quadratic equations   [Review exercises 94–111]

A **quadratic equation** is an equation that can be written in the **standard form**

$$ax^2 + bx + c = 0$$

where a, b, and c are real numbers with $a \neq 0$.

A quadratic equation can be solved by factoring, completing the square, or by using the quadratic formula.

The solution of $ax^2 + bx + c = 0$, with $a \neq 0$, is given by the **quadratic formula**

$$x = \frac{-b \pm \sqrt{b^2 - 4ac}}{2a}$$

Section 1.6 **Linear Inequalities**

Properties of inequalities   [Review exercises 112–117]

Let a, b, and c be any real numbers. The following properties are useful when solving inequalities.

Addition principle: If $a < b$, then $a + c < b + c$.
Multiplication principle for $c > 0$: If $a < b$, then $ac < bc$ if $c > 0$.
Multiplication principle for $c < 0$: If $a < b$, then $ac > bc$ if $c < 0$. Note that the *direction* of the inequality is *reversed* when both sides are multiplied by a negative number.

Similar statements hold true for $a \leq b$, $a > b$, and $a \geq b$.

Section 1.7 **Equations and Inequalities Involving Absolute Value**

Equations Involving Absolute Value   [Review exercises 118–123]

Let $a > 0$. Then the expression $|X| = a$ is equivalent to $X = a$ or $X = -a$.

Inequalities involving absolute value   [Review exercises 124–129]

Let $a > 0$. Then the following hold:
$|X| < a$ is equivalent to $-a < X < a$. Similarly, $|X| \leq a$ is equivalent to $-a \leq X \leq a$.
$|X| > a$ is equivalent to $X < -a$ **or** $X > a$.
Similarly, $|X| \geq a$ is equivalent to $X \leq -a$ **or** $X \geq a$.

Section 1.8 **Other Types of Equations**

Equations reducible to quadratic form   [Review exercises 130–133]

Use a substitution such as $u = x^2$ or $u = x^3$ to reduce the given equation to quadratic form. Then solve by using factoring or the quadratic formula. You should *always* check your solutions.

Equations with rational expressions   [Review exercises 134–137]

When an equation involves rational expressions, multiply both sides of the equation by the least common denominator (LCD) of all terms in the equation. This results in a linear or quadratic equation.

Equations containing radical expressions   [Review exercises 138–141]

When an equation contains a square root symbol, isolate the radical term and square both sides. Solve the resulting quadratic equation and check your solution(s).

Chapter 1 Review Exercises

Section 1.1

In Exercises 1–4, consider the following numbers.

$$\sqrt{3}, 1.2, 3, -1.006, \frac{3}{2}, -5, 8$$

1. Which are integers?

2. Which are irrational numbers?

3. Which are integers that are not negative?

4. Which are rational numbers that are not integers?

In Exercises 5 and 6, name the property illustrated by each equality.

5. $4 + (5 + 7) = (4 + 5) + 7$ 6. $2(x + 5) = 2x + 10$

In Exercises 7–10, graph each interval on the real number line.

7. $[-4, 1)$ 8. $\left(-3, \frac{3}{2}\right)$ 9. $(-1, \infty)$ 10. $(-\infty, -3]$

In Exercises 11 and 12, find the distance between the numbers on the real number line.

11. $-6, 4$ 12. $4.7, -3.5$

In Exercises 13–16, evaluate each expression without using a calculator.

13. $-|-3.7|$ 14. $2^3 - 5(4) + 1$ 15. $\dfrac{6 + 3^2}{-2^2 + 1}$ 16. $-7 + 4^2 \div 8$

Section 1.2

In Exercises 17–22, simplify the expression and write it using positive exponents. Assume all variables represent nonzero numbers.

17. $6x^{-2}y^4$

18. $-(7x^3y^2)^2$

19. $\dfrac{4x^3y^{-2}}{x^{-1}y}$

20. $\dfrac{xy^4}{3^{-1}x^3y^{-2}}$

21. $\left(\dfrac{16x^4y^{-2}}{4x^{-2}y}\right)^2$

22. $\left(\dfrac{5x^{-2}y^3}{15x^3y}\right)^{-1}$

In Exercises 23 and 24, express the number in scientific notation.

23. $-4{,}670{,}000$ 24. 0.000317

In Exercises 25 and 26, express the number in decimal form.

25. 3.001×10^4 26. 5.617×10^{-3}

In Exercises 27 and 28, compute and write the answer in scientific notation.

27. $3.2 \times 10^5 \times 2.0 \times 10^{-3}$ 28. $\dfrac{4.8 \times 10^{-2}}{1.6 \times 10^{-1}}$

29. **Chemistry**   If 1 liter of a chemical solution contains 5×10^{-3} grams of arsenic, how many grams of arsenic are in 3.2 liters of the same solution?

In Exercises 30–35, simplify the expression. Assume that all variables represent positive real numbers.

30. $\sqrt{5x} \cdot \sqrt{10x^2}$ **31.** $\sqrt{3x^2} \cdot \sqrt{15x}$ **32.** $\sqrt{\dfrac{50}{36}}$

33. $\sqrt[3]{\dfrac{-96}{125}}$ **34.** $\sqrt{25x} - \sqrt{36x} + \sqrt{16}$ **35.** $\sqrt[3]{24} - \sqrt[3]{81} + \sqrt[3]{-64}$

In Exercises 36 and 37, rationalize the denominator.

36. $\dfrac{5}{3 - \sqrt{2}}$ **37.** $-\dfrac{2}{1 + \sqrt{3}}$

In Exercises 38–41, evaluate the expression.

38. $-16^{1/2}$ **39.** $(-125)^{1/3}$ **40.** $64^{3/2}$ **41.** $(-27)^{2/3}$

In Exercises 42–45, simplify and write your answer using positive exponents.

42. $3x^{1/3} \cdot 12x^{1/4}$ **43.** $5x^{1/2} y^{1/2} \cdot 4x^{2/3} y$ **44.** $\dfrac{12x^{2/3}}{4x^{1/2}}$ **45.** $\dfrac{16x^{1/3} y^{1/2}}{8x^{2/3} y^{3/2}}$

46. **Physics** If an object is dropped from a height of h feet, it will take $\sqrt{\dfrac{h}{16}}$ seconds to hit the ground. How long will it take a ball dropped from a height of 50 feet to hit the ground?

Section 1.3

In Exercises 47–56, perform the indicated operations and write your answer as a polynomial in descending order.

47. $(13y^2 + 19y - 9) + (6y^3 + 5y - 3)$

48. $(-11z^2 - 4z - 8) - (6z^3 + 25z + 10)$

49. $(3t^4 - 8) - (-9t^5 + 2)$ **50.** $(17u^5 + 8u) + (-16u^4 - 21u + 6)$

51. $(4u + 1)(3u - 10)$ **52.** $(2y + 7)(2 - y)$

53. $(8z - 9)(-3z + 8)$ **54.** $(-3y + 5)(9 + y)$

55. $(3z + 5)(2z^2 - z + 8)$ **56.** $(4t + 1)(7t^2 - 6t - 5)$

In Exercises 57–60, find the special product.

57. $(3x + 2)(3x - 2)$ **58.** $(2x + 5)^2$

59. $(5 - x)^2$ **60.** $(x + \sqrt{3})(x - \sqrt{3})$

In Exercises 61–76, factor each expression.

61. $8z^3 + 4z^2$ **62.** $125u^3 - 5u^2$ **63.** $y^2 + 11y + 28$

64. $-y^2 + 2y + 15$ **65.** $3x^2 - 7x - 20$ **66.** $2x^2 + 3x - 9$

67. $5x^2 - 8x - 4$ **68.** $-3x^2 - 10x + 8$ **69.** $9u^2 - 49$

70. $4y^2 - 25$ **71.** $z^3 + 8z$ **72.** $4z^2 - 16$

73. $2x^2 + 4x + 2$ **74.** $3x^3 - 18x^2 + 27x$ **75.** $4x^3 + 32$

76. $5y^3 - 40$

Section 1.4

In Exercises 77 and 78, simplify each rational expression and indicate the values of the variable for which the expression is defined.

77. $\dfrac{x^2 - 9}{x - 3}$

78. $\dfrac{x^2 + 2x - 15}{x^2 - 25}$

In Exercises 79–82, multiply or divide. Express your answer in lowest terms.

79. $\dfrac{x^2 + 2x + 1}{x^2 - 1} \cdot \dfrac{x^2 - x - 2}{x + 1}$

80. $\dfrac{y^2 - y - 12}{y^2 - 9} \cdot \dfrac{y + 3}{y^2 - 4y}$

81. $\dfrac{3x + 6}{x^2 - 4} \div \dfrac{3x}{x^2 + 4x + 4}$

82. $\dfrac{4x + 12}{x^2 - 9} \div \dfrac{x^2 + 1}{x + 3}$

In Exercises 83–86, add or subtract. Express your answer in lowest terms.

83. $\dfrac{1}{x + 1} + \dfrac{4}{x - 3}$

84. $\dfrac{3}{x - 4} - \dfrac{2}{x^2 - x - 12}$

85. $\dfrac{2x}{x - 3} + \dfrac{1}{x + 3} - 2x^2 - 9$

86. $\dfrac{2x + 1}{x^2 + 3x + 2} - \dfrac{3x - 1}{2x^2 + 3x - 2}$

In Exercises 87 and 88, simplify the complex fraction.

87. $\dfrac{\dfrac{a^2 - b^2}{ab}}{\dfrac{a - b}{b}}$

88. $\dfrac{\dfrac{3}{x - 2} - \dfrac{1}{x + 1}}{\dfrac{2}{x - 1} + \dfrac{3}{x + 1}}$

Section 1.5

In Exercises 89–91, solve the equation.

89. $3(x + 4) - 2(2x + 1) = 13$

90. $\dfrac{3x - 1}{5} + 1 = \dfrac{1}{2}$

91. $0.02(x + 4) - 0.1(x - 2) = 0.2$

In Exercises 92 and 93, solve the equation for y in terms of x.

92. $3x + y = 5$

93. $2(x - 1) = y - 7$

In Exercises 94–99, solve the quadratic equation by factoring.

94. $x^2 - 9 = 0$

95. $2x^2 - 8 = 0$

96. $x^2 - 9x + 20 = 0$

97. $x^2 + x - 12 = 0$

98. $6x^2 - x - 12 = 0$

99. $-6x^2 - 5x + 4 = 0$

In Exercises 100–103, solve the quadratic equation by completing the square.

100. $x^2 - 4x - 2 = 0$

101. $-x^2 - 2x = -5$

102. $x^2 + 3x - 7 = 0$

103. $-2x^2 + 8x = 1$

In Exercises 104–109, solve the quadratic equation using the quadratic formula. Find only real solutions.

104. $-3x^2 - x + 3 = 0$

105. $-x^2 + 2x + 2 = 0$

106. $-3t^2 - 2t + 4 = 0$

107. $\dfrac{4}{3}x^2 + x = 2$

108. $-s^2 - \sqrt{2}s = -\dfrac{1}{2}$

109. $-(x + 1)(x - 4) = -6$

110. **Economics** The volume of toys, games, and sporting goods sold by a toy store chain can be modeled by the quadratic expression

$$s = 0.1525t^2 + 0.3055t + 18.66$$

where t is the number of years since 2008 and s is the dollar amount of the imports, in millions of dollars. Use this model to find the dollar amount of toys, games, and sporting goods sold for the year 2012.

111. **Business Expenditures** The percentage of total operating expenses incurred by airlines for airline food can be modeled by the expression

$$f = -0.0055t^2 + 0.116t + 2.90$$

where t is the number of years since 1980. The model is based on data for selected years from 1980 to 2010. (*Source*: Statistical Abstract of the United States)

a. In what year after 1980 was the expenditure for airline food 2% of the total operating expenses? Do not convert the percentage to decimal form.

b. Is this model reliable as a long-term indicator of airline expenditures for airline food as a percentage of total operating expenses? Justify your answer.

Section 1.6

In Exercises 112–117, solve the inequality. Express your answer in interval notation.

112. $-7x + 3 > 5x - 2$ **113.** $5x - 2 \le 3x + 7$ **114.** $\dfrac{1}{3}x - 6 \ge 4x - 1$

115. $x - 4 \le \dfrac{2}{5}x$ **116.** $4 \le \dfrac{2x + 2}{3} \le 7$ **117.** $-1 \le \dfrac{x - 4}{3} \le 4$

Section 1.7

In Exercises 118–123, solve the equation.

118. $|x - 5| = 6$ **119.** $|x + 6| = 7$ **120.** $\left|2s - \dfrac{1}{2}\right| = 8$

121. $\left|4s + \dfrac{3}{2}\right| = 10$ **122.** $1 + 3|2x - 5| = 4$ **123.** $-3 + 2|x - 4| = 7$

In Exercises 124–129, solve the inequality. Write your answer in interval notation and graph the solution set on the number line.

124. $|3x + 10| > 5$ **125.** $|x + 6| < 7$ **126.** $\left|\dfrac{1}{2}x + 6\right| \le 5$

127. $\left|\dfrac{2x + 1}{5}\right| \le 1$ **128.** $3\left|\dfrac{3}{2}x + 1\right| < 9$ **129.** $|-2x - 7| + 4 \le 8$

Section 1.8

In Exercises 130–133, find all real solutions of the polynomial equation.

130. $x^4 - 11x^2 + 24 = 0$

131. $x^6 - 4x^3 = 21$

132. $-3x^4 + 11x^2 = -4$

133. $6x^2 - 7x = 3$

In Exercises 134–137, find all real solutions of the rational equation.

134. $\dfrac{3x}{x+1} + \dfrac{1}{x} = \dfrac{5}{2}$

135. $\dfrac{5}{x+2} + x = -8$

136. $\dfrac{1}{x^2 + 2x - 3} + \dfrac{4}{x+3} = 1$

137. $\dfrac{2}{x^2 + 3x - 4} + \dfrac{1}{x-1} = -\dfrac{6}{x+4}$

In Exercises 138–141, find all real solutions of the radical equation.

138. $\sqrt{2x+1} - x = -1$

139. $\sqrt{3x+4} + 2 = x$

140. $\sqrt{3x-5} - \sqrt{x-3} = 2$

141. $\sqrt{2x+3} + \sqrt{x+6} = 6$

Chapter 1 Test

1. Which of the numbers in the set $\left\{ \sqrt{2}, -1, 1.55, \pi, 41 \right\}$ are rational?

2. Name the property illustrated by the equality $3(x + 4) = 3x + 12$.

3. Graph the interval $[-4, 2)$ on the number line.

4. Find the distance between 4.6 and -5.7 on the number line.

5. Evaluate without using a calculator: $\dfrac{2^3 - 6 \cdot 4 - 2}{-3^2 - 5}$

6. Evaluate the expression $-3x^2 + 6x - 1$ for $x = -2$.

7. Express $8{,}903{,}000$ in scientific notation.

In Exercises 8–12, simplify each expression. Write your answer using positive exponents.

8. $-(6x^2y^5)^2$

9. $\left(\dfrac{36x^5y^{-1}}{9x^{-4}y^3} \right)^3$

10. $\sqrt{6x}\ \sqrt{8x^2}, x \geq 0$

11. $-5x^{1/3} \cdot 6x^{1/5}$

12. $\dfrac{45x^{2/3}y^{-1/2}}{5x^{1/3}y^{5/2}}, x, y > 0$

In Exercises 13 and 14, simplify without using a calculator.

13. $\sqrt[3]{54} - \sqrt[3]{16}$

14. $(-125)^{2/3}$

15. Rationalize the denominator: $\dfrac{7}{1 + \sqrt{5}}$

In Exercises 16–21, factor each expression completely.

16. $25 - 49y^2$

17. $4x^2 + 20x + 25$

18. $6x^2 - 7x - 5$

19. $4x^3 - 9x$

20. $3x^2 + 8x - 35$

21. $2x^3 + 16$

In Exercises 22–26, perform the operation, and simplify.

22. $\dfrac{2x + 4}{x^2 - 9} \cdot \dfrac{2x^2 - 5x - 3}{x^2 - 4}$

23. $\dfrac{5x + 1}{x^2 + 4x + 4} \div \dfrac{5x^2 - 9x - 2}{x^2 + x - 2}$

24. $\dfrac{5}{x^2 - 4} - \dfrac{7}{x + 2}$

25. $\dfrac{3x - 1}{2x^2 - x - 1} - \dfrac{1}{x^2 + 2x - 3}$

26. $\dfrac{\dfrac{5}{x - 2} + \dfrac{3}{x}}{\dfrac{1}{x} - \dfrac{4}{x - 2}}$

27. Solve for x: $\dfrac{2x + 1}{2} - \dfrac{3x - 2}{5} = 2$

28. Solve the inequality $-2 \leq \dfrac{5x - 1}{2} < 4$, and express your answer in interval notation.

In Exercises 29 and 30, solve the equation.

29. $\left| 6x + \dfrac{4}{3} \right| = 5$

30. $|4x - 7| - 3 = 6$

31. Solve $x^2 - 5x - 6 = 0$ by factoring.

32. Solve $2x^2 - 4x - 3 = 0$ by completing the square.

In Exercises 33 and 34, solve the equations using the quadratic formula.

33. $3x^2 + x - 1 = 0$ **34.** $-2x^2 + 2x + 3 = 0$

In Exercises 35 and 36, find all real solutions using any method.

35. $3x^2 - x - 4 = 0$ **36.** $2x^2 + 2x - 5 = 0$

In Exercises 37 and 38, solve the inequality. Write your answer in interval notation.

37. $2|2x - 5| \geq 6$ **38.** $|5x + 2| + 6 < 13$

In Exercises 39–41, find all real solutions of the equation.

39. $6x^4 - 5x^2 - 4 = 0$ **40.** $\dfrac{1}{2x + 1} + \dfrac{3}{x - 2} = \dfrac{5}{2x^2 - 3x - 2}$

41. $\sqrt{2x - 1} + \sqrt{x + 4} = 6$

42. If 1 liter of a chemical solution contains 5×10^{-6} grams of sodium, how many grams are in 5.7 liters of the same solution? Express your answer in scientific notation.

43. Julia is comparing two rate plans for cell phones. Plan A charges $0.18 per minute with no monthly fee. Plan B charges $8 per month plus $0.10 per minute. What is the minimum number of minutes per month that Julia must use her cell phone for the cost of Plan B to be less than or equal to that of Plan A?

44. The height of a ball after being dropped from a point 256 feet above the ground is given by $h = -16t^2 + 256$, where t is the time in seconds since the ball was dropped and h is in feet. When will the ball reach the ground?

Functions and Graphs

Recent blockbuster films have been extremely profitable, according to the Motion Picture Association of America. Such data can be studied mathematically using the language of functions. See Exercise 75 in Section 2.3.

This chapter will define what functions are, show you how to work with them, and illustrate how they are used in various applications.

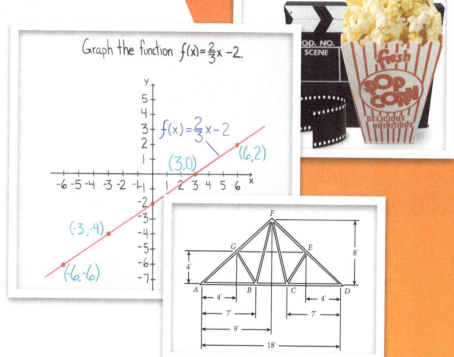

Outline

Objectives

- Find the slope of a line.

- Find the equation of a line in both slope-intercept and point-slope forms.

- Find the equation of line in general form.

- Find equations of parallel and perpendicular lines.

- Solve applied problems using equations of lines.

In Section 1.1, you saw that real numbers can be plotted as points on a number line. To get from one point to another on the real number line, you need to move either to the left or to the right. However, if you wanted to describe how to get from one point to another where you have to first move left, and then move down, you will need two numbers—one to describe the horizontal distance and direction, and the other to describe the vertical distance and direction. The **coordinate system** allows us to describe points using horizontal and vertical directions.

Coordinate System

On a plane, we draw horizontal and vertical lines, known as **axes**. The point of intersection of these lines is denoted by the **origin**, $(0, 0)$. The horizontal axis is called the **x-axis** and the vertical axis is called the **y-axis**, although other variable names can be used.

The axes divide the plane into four regions called **quadrants**. The convention for labeling quadrants is to begin with Roman numeral I for the quadrant in the upper right, and continue numbering counter-clockwise. See Figure 1.

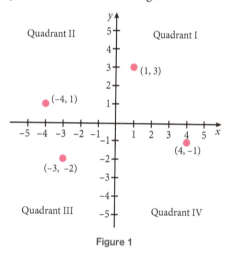

Figure 1

To locate a point in the plane, we use an ordered pair of numbers (x, y). The x value gives the horizontal location of the point and the y value gives the vertical location. The first number x is called the **x-coordinate**, **first coordinate**, or **abscissa**. The second number y is called the **y-coordinate**, **second coordinate**, or **ordinate**. If an ordered pair is given by another pair of labels, such as (u, v), it is assumed that the first coordinate represents the horizontal location and the second coordinate represents the vertical location. The xy-coordinate system is also called the **Cartesian coordinate system** or the **rectangular coordinate system.**

Lines and Their Slopes

The xy-coordinate system is useful for describing geometric objects, such as lines and circles. In this section, we discuss how a line can be described using a relationship between the x- and y-coordinates of the points on the line.

One way to describe a line is to measure its steepness. That is, measure how quickly the line rises (or falls) as we move from left to right.

Definition of Slope

The **slope** of a nonvertical line containing the points (x_1, y_1) and (x_2, y_2) is given by

$$m = \frac{\text{Rise}}{\text{Run}} = \frac{y_2 - y_1}{x_2 - x_1}$$

where $x_1 \neq x_2$. See Figure 2.

The slope of a vertical line is undefined.

Note Remember the following when calculating the slope of a line:

- It does **not** matter in which order a point is called.
- As long as two points lie on the same line, the slope will be the same regardless of which two points are used.

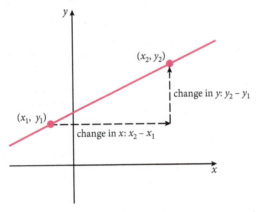

Figure 2

Example 1   **Finding the Slope of a Line**

Find the slope of the line passing through the points $(-1, 2)$ and $(-3, 4)$. Plot the points and indicate the slope on your plot.

Solution The points and the line passing through them are shown in Figure 3. Letting $(x_1, y_1) = (-1, 2)$ and $(x_2, y_2) = (-3, 4)$, we have

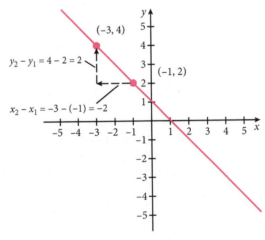

Figure 3

$$m = \frac{y_2 - y_1}{x_2 - x_1} \qquad \text{Formula for slope}$$

$$= \frac{4 - 2}{-3 - (-1)} \qquad \text{Substitute values}$$

$$= \frac{2}{-2} = -1 \qquad \text{Simplify}$$

Check It Out 1 Find the slope of the line passing through the points $(-5, 3)$ and $(9, 4)$.

Graphically, the sign of m shows you how the line slants: if $m > 0$, the line slopes upward as the value of x increases. If $m < 0$, the line slopes downward as the value of x increases. This is illustrated in Figure 4.

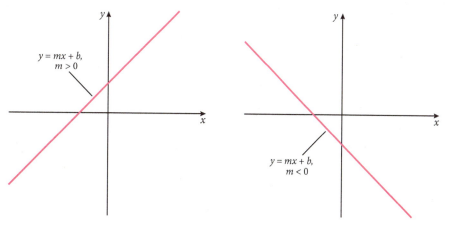

Figure 4

Equation of a Line: Point-Slope Form

To derive an equation of a line, first sketch the graph of a line containing a point (x_1, y_1). On this graph, we indicate (x, y) to be any point on the line, as shown in Figure 5.

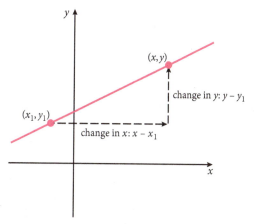

Figure 5

We can derive the equation of the line in Figure 5 by following these steps:

- Write down the formula for the slope of the line passing through the points (x, y) and (x_1, y_1).

$$m = \frac{y - y_1}{x - x_1}$$

- Multiply both sides by $x - x_1$.

$$m(x - x_1) = y - y_1$$

Equation of a Line in Point-Slope Form

The equation of a line with slope m and containing the point (x_1, y_1) is given by

$$y - y_1 = m(x - x_1)$$

or, equivalently,

$$y = m(x - x_1) + y_1$$

> **Example 2**   **Equation of a Line in Point-Slope Form**

Find the equation of the line passing through $(2, 6)$ and $(-1, -2)$ in point-slope form.

Solution The points and the line passing through them are plotted in Figure 6.

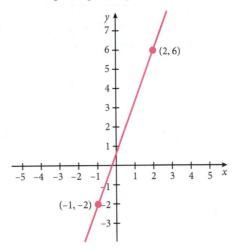

Figure 6

First, we must find the slope. Letting $(x_1, y_1) = (-1, -2)$ and $(x_2, y_2) = (2, 6)$, we have

$$m = \frac{y_2 - y_1}{x_2 - x_1} \qquad \text{Formula for slope}$$

$$= \frac{6 - (-2)}{2 - (-1)} \qquad \text{Substitute values}$$

$$= \frac{8}{3}$$

The point-slope formula gives

$$y - y_1 = m(x - x_1)$$

$$y - (-2) = \frac{8}{3}(x - (-1)) \qquad \text{Substitute } m = \frac{8}{3}, x_1 = -1, y_1 = -2$$

$$y + 2 = \frac{8}{3}(x + 1)$$

Note that we could have used the point $(2, 6)$ instead of $(-1, -2)$. The equation, when written in slope-intercept form, would still be the same.

✅ *Check It Out 2*   Find the equation of the line passing through $\left(-\frac{3}{2}, 1\right)$ and $\left(\frac{1}{2}, 2\right)$ in point-slope form.

Equation of a Line: Slope-Intercept Form

When we use $(0, b)$ as a point in the point-slope form of an equation of a line, we get another form of an equation of a line known as the **slope-intercept** form of an equation of a line. The point where the graph of the line crosses the y axis, $(0, b)$, is called the **y-intercept**. Notice that the x coordinate of the y-intercept is 0.

> ### Slope-Intercept Form of the Equation of a Line
>
> The **slope-intercept** form of the equation of a line with slope m and y-intercept $(0, b)$ is given by
>
> $$y = mx + b$$

Other terminology related to lines appears below.

- A point (x_1, y_1) is said to **lie on the line** $y = mx + b$ if $y_1 = mx_1 + b$. That is, (x_1, y_1) satisfies the equation $y = mx + b$.

- The point where the graph of a non-horizontal line crosses the x-axis is called the **x-intercept**. Since the y-coordinate of the x-intercept is 0, the x-intercept is found by setting y to 0 and solving for x.

> **Example 3**   **Equation of a Line in Slope-Intercept Form**

Write the equation of a line with a slope of -4 and y-intercept of $\left(0, \sqrt{2}\right)$ in slope-intercept form. Does $(0, 1)$ lie on this line? Find the x-intercept of this line.

Solution Since the slope, m, is -4 and b is $\sqrt{2}$, the equation of the line in slope-intercept form is

$$y = -4x + \sqrt{2}$$

Note that b is irrational. Slopes and intercepts of a line can be any real number.

Substituting $(x, y) = (0, 1)$, we see that $1 \neq -4(0) + \sqrt{2}$. Thus, $(0, 1)$ does not lie on the line $y = -4x + \sqrt{2}$.

To find the x-intercept of this line, we set $y = -4x + \sqrt{2} = 0$. Solving for x, we have $x = \frac{\sqrt{2}}{4}$. Thus, $\left(\frac{\sqrt{2}}{4}, 0\right)$ is the x-intercept.

✓ *Check It Out 3*   Write the equation of a line with a slope of 2 and a y-intercept of $\left(0, -\frac{1}{3}\right)$ in slope-intercept form. Does $\left(\frac{1}{6}, 0\right)$ lie on this line? ■

> **Example 4**   **Finding an Equation Given a Graph**

Find the equation, in slope-intercept form, of the line whose graph is given in Figure 7.

Just in Time
Review equation solving in Section 1.5.

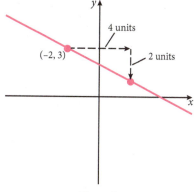

Figure 7

Solution From the point $(-2, 3)$, move 4 units to the right and then 2 units down to a second point on the line. Therefore,

$$\text{Slope} = m = \frac{\text{Change in } y}{\text{Change in } x} = \frac{-2}{+4} = -\frac{1}{2}$$

Note that the change in y is *negative*, since we move down 2 units. We can now easily use the point-slope form for the equation of the line to get

$$y - y_1 = m(x - x_1)$$

$$y - 3 = -\frac{1}{2}(x - (-2)) \qquad \text{Substitute } m = -\frac{1}{2}, x_1 = -2, y_1 = 3$$

$$y = -\frac{1}{2}(x + 2) + 3 \qquad \text{Add 3 to both sides}$$

$$y = -\frac{1}{2}x + 2 \qquad \text{Simplify}$$

Note that we started with the point-slope form for the equation of the line and rearranged that equation to obtain the slope-intercept form. This is a common technique that you will use when working with equations of lines.

Check It Out 4 Find the equation of the line whose graph is given in Figure 8.

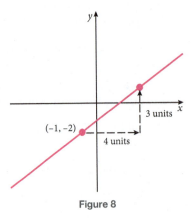

Figure 8

Horizontal and Vertical Lines

Two special cases of equations of lines are worth noting: equations of horizontal lines and equations of vertical lines.

Horizontal Line Two points lying on a horizontal line will have the same y-coordinate. Let (x_1, b) and (x_2, b) denote two points on this line. See Figure 9.

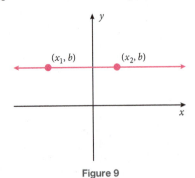

Figure 9

The slope of this line will be

$$m = \frac{y_2 - y_1}{x_2 - x_1} = \frac{b - b}{x_2 - x_1} = 0$$

Substituting 0 for m in the point-slope formula, we get

$$y - b = 0(x - x_1)$$
$$y = b \qquad \text{Solve for } y$$

Vertical Line To find the equation of a vertical line passing through a point (a, y_1), we can first sketch its graph as shown in Figure 10.

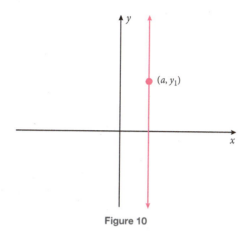

Figure 10

A second point, (a, y_2), on this vertical line will have the same x-coordinate. The slope is then $\frac{y_2 - y_1}{a - a}$, which is undefined because the denominator is zero. Therefore, we cannot use the formula for an equation of a line involving slope. The equation of a vertical line is simply given by the constant value of the x-coordinate and is written as $x = a$.

Equation of Horizontal Line

The **equation of the horizontal line** passing through (a, b) is

$$y = b$$

Equation of Vertical Line

The **equation of the vertical line** passing through (a, b) is

$$x = a$$

Example 5 **Equation of a Horizontal Line and a Vertical Line**

Find the equations of the lines described below.

a. A horizontal line passing through the point $(2, 3)$.

b. A vertical line passing through the point $(1, -2)$.

Solution

a. The y-coordinate of the point $(2, 3)$ is 3. Since any two points on a horizontal line have the same y-coordinate, the equation of the horizontal line is $y = 3$. See Figure 11.

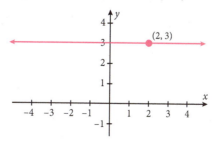

Figure 11

b. The x-coordinate of the point $(1, -2)$ is 1. Since any two points on a vertical line have the same x-coordinate, the equation of the vertical line is $x = 1$. See Figure 12.

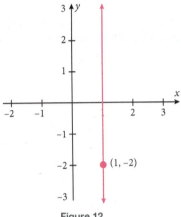

Figure 12

Check It Out 5   Find the equation of the horizontal line passing through the point $(-3, -1)$. Find the equation of the vertical line passing through the point $(4, 2)$.

General Form for the Equation of a Line

In this section, we discuss another form for the equation of a line, known as the **general form** for the equation of a line.

> **General Form of the Equation of a Line**
>
> The **general form of the equation of a line** is
>
> $$Ax + By + C = 0$$
>
> where A, B, are not both zero. Here, A, B, C are constants that are real numbers.

Note that if $A = 0$ in the general form, we get the equation of a horizontal line, $y = -\frac{C}{B}$. If $B = 0$, then we get the equation of a vertical line, $x = -\frac{C}{A}$.

Example 6   **General Form of the Equation of a Line**

Write the equation $3x + 2y - 4 = 0$ in slope-intercept form and graph the line.

Solution Solve for y to write the equation in slope-intercept form.

$$3x + 2y - 4 = 0 \qquad \text{Given equation}$$
$$2y = -3x + 4 \quad \text{Isolate the } y \text{ term}$$
$$y = -\frac{3}{2}x + 2 \quad \text{Divide by 2}$$

Thus, $y = -\frac{3}{2}x + 2$ is the equation of the line in slope-intercept form. This is a line with slope $-\frac{3}{2}$ and y-intercept $(0, 2)$.

To graph the line, locate the y-intercept $(0, 2)$ in the coordinate plane. Since the slope is $-\frac{3}{2} = \frac{\text{Rise}}{\text{Run}}$, move 3 units down and 2 units to the right from $(0, 2)$ to obtain the second point, $(2, -1)$. See Figure 13. Draw the line through the two points.

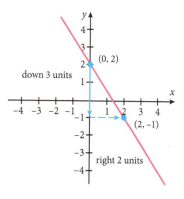

Figure 13

Check It Out 6 Write the equation $-3x + 5y = 15$ in slope-intercept form and graph the line.

Parallel and Perpendicular Lines

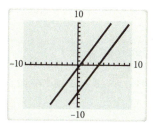
If two nonvertical lines are parallel, they have the same slope. Two vertical lines are always parallel. Also, if you know that two lines have the same slope, then they are parallel. This leads to the following statement about parallel lines.

Slopes of Parallel Lines

Two nonvertical lines are parallel if and only if they have the same slope. Vertical lines are always parallel.

If two lines are perpendicular, it can be shown that the following result holds.

Slopes of Perpendicular Lines

Two lines with slopes m_1 and m_2 are perpendicular if and only if
$$m_1 m_2 = -1 \quad \text{or, equivalently,} \quad m_2 = -\frac{1}{m_1}$$

Thus, perpendicular lines have slopes that are **negative reciprocals** of each other. Vertical and horizontal lines are always perpendicular to each other. The following examples illustrate these properties.

Example 7 **Finding the Equation of a Parallel Line**

Find an equation of the line parallel to $2x - y = 1$ and passing through $(2, -3)$. Write the equation in slope-intercept form.

Solution First, find the slope of the original line by writing it in slope-intercept form:

$$2x - y = 1$$
$$-y = 1 - 2x \quad \text{Subtract } 2x$$
$$y = 2x - 1 \quad \text{Multiply by } -1 \text{ and rearrange terms}$$

Thus, the slope of the original line is $m = 2$.

Since the new line is parallel to the given line, it has the same slope. Using the point-slope form of the equation of a line,

$$y - y_1 = m(x - x_1)$$

$$y - (-3) = 2(x - 2) \qquad \text{Use } m = 2, (x_1, y_1) = (2, -3)$$

$$y + 3 = 2x - 4 \qquad \text{Remove parentheses}$$

$$y = 2x - 7 \qquad \text{Subtract 3 from both sides}$$

Thus, the equation of the parallel line, in slope-intercept form, is $y = 2x - 7$.

Check It Out 7 Find an equation of the line parallel to $-4x + y = 8$ and passing through $(-1, 2)$. Write the equation in slope-intercept form.

Using Technology

Graph the lines
$Y_1 = -\frac{2}{3}x + \frac{1}{3}$ and
$Y_2 = \frac{3}{2}x - 7$ using a
SQUARE window to
see that the lines are
perpendicular
(Figure 15).

Keystroke Appendix:
Section 7

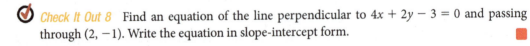

Figure 15

Example 8   Finding the Equation of a Perpendicular Line

Find an equation of the line perpendicular to $2x + 3y - 1 = 0$ and passing through $(4, -1)$. Write the equation in slope-intercept form.

Solution First, find the slope of the original line by writing the equation in slope-intercept form.

$$2x + 3y - 1 = 0$$

$$3y = -2x + 1 \qquad \text{Isolate the } y \text{ term}$$

$$y = -\frac{2}{3}x + \frac{1}{3} \qquad \text{Divide by 3}$$

The slope of the original line is $-\frac{2}{3}$. The line perpendicular to it has slope

$$m = -\frac{1}{-\frac{2}{3}} = \frac{3}{2} \qquad \text{Negative reciprocal}$$

Using the point-slope form for the equation of a line:

$$y - y_1 = m(x - x_1)$$

$$y - (-1) = \frac{3}{2}(x - 4) \qquad \text{Use } m = \frac{3}{2}, (x_1, y_1) = (4, -1)$$

$$y + 1 = \frac{3}{2}x - 6 \qquad \text{Remove parentheses}$$

$$y = \frac{3}{2}x - 7 \qquad \text{Subtract 1 from each side}$$

Thus, the equation of the perpendicular line is $y = \frac{3}{2}x - 7$ in slope-intercept form.

Check It Out 8 Find an equation of the line perpendicular to $4x + 2y - 3 = 0$ and passing through $(2, -1)$. Write the equation in slope-intercept form.

Applications of Equations of Lines

The following two examples will use linear equations to solve applied problems.

Example 9   Application to Depreciation

A 2012 Honda Civic's value over time can be approximated by the equation $v = -2200t + 22000$, where t denotes the number of years after its purchase. Answer the following questions:

a. What will be the value of the car 6 years after purchase?

b. What are the slope and v-intercept, and what do they represent?

c. For what value of t will the value of the car be zero?

Solution

a. To find the value of the car 6 years after its purchase, substitute $t = 6$ in the equation.

$$v = -2200(6) + 22000 \qquad \text{Substitute } t = 6$$
$$v = -13200 + 22000 = 8800$$

The car will have a value of $8800 six years after its purchase.

b. The slope is -2200, which indicates that the car loses its value by $2200 every year. The v-intercept is $(0, 22000)$. It signifies the value of the car at $t = 0$, which is the original purchase price of $22,000.

c. To find the value of t for which the value of the car will be zero, set $v = 0$ in the equation and solve for t.

$$0 = -2200t + 22000 \qquad \text{Set } v = 0$$
$$2200t = 22000 \qquad \text{Add } 2200t \text{ to each side}$$
$$t = 10 \qquad \text{Divide by 2200}$$

So, in 10 years, the car will have zero value.

 Check It Out 9   Repeat the above example if $v = -1500t + 30000$.

Example 10   **Application to Ecology**

In 2010, there were 1,400 Canada Geese in a wildlife refuge. This population has been increasing by 70 geese each year.

a. Write an equation for the population of Canada Geese, p, in terms of t, the number of years since 2010.

b. What are the slope and p-intercept and what do they represent?

c. When will the goose population reach 1,960?

Solution

a. From the problem statement, after t years, the population would have increased from the initial population by $70t$. Since the initial population was 1,400, we have

$$p = 70t + 1400$$

b. The slope is 70, and represents the increase in population each year. The p-intercept is $(0, 1400)$ and represents the population at $t = 0$, corresponding to the year 2010.

c. To find out when the goose population will reach 1,960, substitute 1960 for p in the equation and solve.

$$p = 70t + 1400 \qquad \text{Given equation}$$
$$1960 = 70t + 1400 \qquad \text{Substitute } p = 1960$$
$$560 = 70t \qquad \text{Subtract 1400 from each side}$$
$$8 = t \qquad \text{Divide by 70}$$

The goose population will reach 1,960 after 8 years, in 2018.

 Check It Out 10   Use the equation in Example 10 to predict the goose population in the year 2022.

Exercises 2.1

In Exercises 1–4, solve for y.

1. $y - 4 = 2(y + 1)$　　**2.** $y + 6 = 3(y - 2)$　　**3.** $\frac{1}{2}y + 3 = 4$　　**4.** $\frac{1}{4}y - 1 = -2$

In Exercises 5 and 6, solve for y in terms of x.

5. $4y + 2x - 8 = 0$　　**6.** $3y - 6x + 6 = 0$

 Skills

7. Which of the following represent equations of lines? Explain your answers.

　a. $y = 1 + 3x$　　**b.** $y = \frac{1}{x} - 1$　　**c.** $y = -5x$　　**d.** $h = \sqrt{s + 1}$

8. Which of the following represent equations of lines? Explain your answers.

　a. $h = \frac{1}{3}s + 1$　　**b.** $y = \frac{2}{x^2} + 1$　　**c.** $y = 3$　　**d.** $y = -3\sqrt{t}$

In Exercises 9–22, find the slope of the line passing through each given pair of points, if the slope is defined.

9. $(1, -3)$ and $(0, 4)$　　　**10.** $(-1, 2)$ and $(0, -2)$　　　**11.** $(1, 3)$ and $(2, 3)$

12. $(4, -1)$ and $(4, 2)$　　　**13.** $(0, 1)$ and $(-2, 0)$　　　**14.** $(3, 0)$ and $(0, -4)$

15. $(-5, -2)$ and $(-5, 1)$　　**16.** $(4, 1)$ and $(2, 4)$　　　**17.** $(0, -2)$ and $(0, 1/2)$

18. $\left(-\frac{1}{2}, 3\right)$ and $(-4, 3)$　　**19.** $\left(\frac{2}{3}, -1\right)$ and $\left(-\frac{1}{3}, -2\right)$　　**20.** $\left(1, \frac{4}{3}\right)$ and $\left(\frac{1}{2}, \frac{2}{3}\right)$

21. $\left(\pi, \frac{\pi}{2}\right)$ and $\left(\frac{\pi}{4}, \frac{\pi}{3}\right)$　　**22.** $\left(\frac{\pi}{2}, \pi\right)$ and $\left(\frac{\pi}{3}, \frac{\pi}{4}\right)$

In Exercises 23 and 24, use the graph of the line to find the slope of the line.

23.

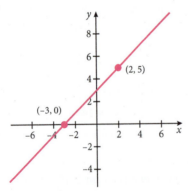

24.

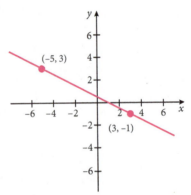

25. Plot the points given in Exercise 9, and indicate the direction of the changes in x and y on your plot.

26. Plot the points given in Exercise 10, and indicate the direction of the changes in x and y on your plot.

In Exercises 27–30, check whether each point lies on the line having the equation $y = -2x + 5$.

27. $(2, 1)$ **28.** $(0, 3)$ **29.** $(-1, 0)$ **30.** $(1, 3)$

In Exercises 31–40, find the equation of the line in point-slope form, passing through the given pair of points.

31. $(2, -1)$ and $(1, 4)$ **32.** $(-1, 3)$ and $(0, 1)$ **33.** $(-1, -4)$ and $(2, -3)$

34. $(-3, 2)$ and $(5, 0)$ **35.** $(4, 5)$ and $(7, 5)$ **36.** $(10, 8)$ and $(5, 8)$

37. $(1.5, 3.6)$ and $(2.5, 4.5)$ **38.** $(4.2, 1.2)$ and $(6.2, 5.7)$ **39.** $\left(\frac{1}{2}, -1\right)$ and $\left(3, \frac{1}{2}\right)$

40. $\left(\frac{1}{3}, 2\right)$ and $\left(4, \frac{1}{4}\right)$

In Exercises 41–46, rewrite the equation in the form $y = mx + b$ *or* $x = c$. *Identify the slope, if defined, and the y-intercept, if it exists.*

41. $-3x + 6y - 2 = 0$ **42.** $4x + 2y - 10 = 0$ **43.** $3(x - 2) + 4y - 7 = 0$

44. $-2(x + 2) + 4y - 6 = 0$ **45.** $2(x - 3) = 0$ **46.** $3(y + 1) = -1$

In Exercises 47–56, write each equation of a line in the form $y = mx + b$. *Find the slope and the x- and y-intercepts, if any. Graph the lines.*

47. $2x + y = 6$ **48.** $-3x + y = 4$ **49.** $4x - 3y = -2$

50. $3x - 4y = 1$ **51.** $y - 3 = -2(x - 6)$ **52.** $y + 5 = -1(x + 1)$

53. $-5x + 3y - 9 = 0$ **54.** $-2x + 4y - 8 = 0$ **55.** $2y - 4 = 0$

56. $3(y + 1) = 0$

In Exercises 57–62, find the equation in the form $y = mx + b$ *or* $x = c$, *of each of the lines pictured.*

57.
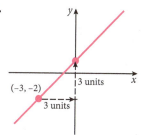

58.
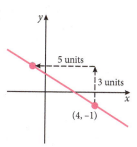

59.
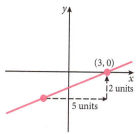

60.
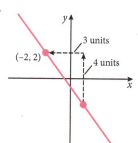

61.
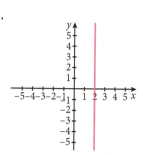

62.
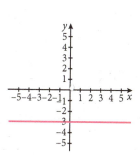

In Exercises 63–68, use a graphing utility to graph each line. Choose an appropriate window to display the graph clearly.

63. $y = 5.7x + 13.8$ **64.** $y = -3.2x - 12.6$ **65.** $3x - y = 15$

66. $0.1x + 0.2y = 2$ **67.** $y - 0.5 = 2.8(x + 1.7)$ **68.** $y + 0.9 = -1.4(x - 1.2)$

In Exercises 69–102, find the equation of the line, in the form $y = mx + b$, or $x = c$, having the given properties.

69. Slope: -1; y-intercept: $(0, 2)$

70. Slope: 3; y-intercept: $(0, 5)$

71. x-intercept: $(-3, 0)$; y-intercept: $(0, 1)$

72. x-intercept: $(1, 0)$; y-intercept: $(0, 4)$

73. Slope: $-\frac{1}{2}$; passes through $(1, 2)$

74. Slope: $\frac{2}{3}$; passes through $(2, -3)$

75. Slope: -3; passes through $(3, -1)$

76. Slope: -2; passes through $(1, -1)$

77. Slope: $\frac{3}{4}$; y-intercept: $\left(0, -\frac{3}{2}\right)$

78. Slope: $-\frac{1}{3}$; y-intercept: $(0, 3)$

79. x-intercept: $\left(\frac{1}{2}, 0\right)$; y-intercept: $(0, 3)$

80. x-intercept: $(-3, 0)$; y-intercept: $\left(0, \frac{3}{2}\right)$

81. Vertical line through $(4, 6)$

82. Horizontal line through $(-1, -4)$

83. Horizontal line through $(2, -1)$

84. Vertical line through $(0, 3)$

85. Horizontal line through $(4, 0.5)$

86. Vertical line through $(-2, 0)$

87. Parallel to the line $y = -3x$ and passing through $(0, -1)$

88. Parallel to the line $y = 2x + 5$ and passing through $(0, 3)$

89. Parallel to the line $y = -\frac{1}{2}x + 2$ and passing through $(4, -1)$

90. Parallel to the line $y = -\frac{1}{3}x - 1$ and passing through $(3, 2)$

91. Parallel to the line $2x + 3y = 6$ and passing through $(-3, 1)$

92. Parallel to the line $-3x + 4y = 8$ and passing through $(8, -2)$

93. Perpendicular to the line $y = \frac{2}{3}x$ and passing through $(0, -1)$

94. Perpendicular to the line $y = -\frac{1}{4}x$ and passing through $(0, -2)$

95. Perpendicular to the line $y = 2x - 1$ and passing through $(-2, 1)$

96. Perpendicular to the line $y = -3x + 2$ and passing through $(1, 4)$

97. Perpendicular to the line $x + 2y = 1$ and passing through $(2, 1)$

98. Perpendicular to the line $2x - y = 1$ and passing through $(-1, 0)$

99. Perpendicular to the line $y = 4$ and passing through $(0, 2)$

100. Perpendicular to the line $x = 3$ and passing through $(2, 1)$

101. Parallel to the line $x = 1$ and passing through $(-2, -5)$

102. Parallel to the line $y = 3$ and passing through $(1, -2)$

Applications

103. Salary A computer salesperson earns $650 per week plus $50 for each computer sold.

   a. Express the salesperson's earnings for one week in terms of the number of computers sold.

   b. Find the values of m and b and interpret them.

104. Salary An appliance salesperson earns $800 per week plus $75 for each appliance sold.

   a. Express the salesperson's earnings for one week in terms of the number of appliances sold.

   b. Find the values of m and b and interpret them.

105. Film Industry The yearly amount of global box office revenue, R, in billions of U.S. dollars, has been rising according to the following equation

$$R = 1.82t + 27.46$$

where t is the number of years since 2008. *(Source: Motion Picture Association of America, 2013)*

   a. According to this equation, what will be the box office revenue for the year 2015?

   b. What does the R-intercept represent for this problem?

106. Travel Amy is driving her car to a conference. The expression $d = 65t + 100$ represents her total distance from home in miles, and t is the number of hours since 8:00 AM.

   a. How many miles from home will Amy be at 12:00 noon?

   b. What does the d-intercept represent in this problem?

107. Sales The number of handbags sold per year by Tres Chic Boutique since 2009 is given by the equation $h = 160t + 500$. Here, t is the number of years since 2009, and h is in millions of handbags.

   a. How many handbags were sold in 2012?

   b. What is the h-intercept and what does it represent?

   c. According to this equation, in what year will 1,200 million handbags be sold?

108. Sales The number of computers sold per year by T.J.'s Computers since 2008 is given by the equation $c = 25t + 350$. Here, t is the number of years since 2008.

   a. How many computers were sold in 2012?

   b. What is the c-intercept of this equation, and what does it represent?

   c. According to this equation, in what year will 600 computers be sold?

109. Leisure The admission price to Wonderland Amusement Park has been increasing by $1.50 each year. If the price of admission was $25.50 in 2009, find an equation that gives the price of admission in terms of t, where t is the number of years since 2009.

110. Pricing The total cost of a car is the sum of the sales price of the car, a sales tax of 6% of the sales price, and $500 for title and tags. Express the total cost of the car in terms of the sales price of the car.

111. Temperature Scales Temperatures are commonly measured using the Celsius scale or the Fahrenheit scale. A temperature of 0° Celsius is equivalent to 32° Fahrenheit, and a temperature of 100°C is equivalent to 212°F. If a temperature, C, is given in degrees Celsius, find an equation of the form $F = mC + b$ that will yield the equivalent Fahrenheit temperature.

112. Manufacturing To manufacture watches, it costs $5 for each watch, over and above the fixed cost of $5,000.

   a. Express the cost of manufacturing watches in terms of the number of watches made.

   b. Using the answer you found in part (a), find the total cost of manufacturing 1250 watches.

Concepts

113. What happens when you graph $y = x + 100$ in the standard viewing window of your graphing utility? How can you change the window so that you can see a clearer graph?

114. What happens when you graph $y = 100x$ in the standard viewing window of your graphing utility? How can you change the window so that you can see a clearer graph?

115. Sketch by hand the graph of the line with slope $\frac{-4}{5}$ and y-intercept $(0, -1)$. Find the equation of this line.

116. Sketch by hand the graph of the line with slope $\frac{-3}{2}$ and y-intercept $(0, -2)$. Find the equation of this line.

117. Sketch by hand the graph of the line with slope $\frac{5}{3}$ that passes through the point $(-2, 6)$. Find the equation of this line.

118. Sketch by hand the graph of the line with slope $\frac{2}{5}$ that passes through the point $(1, -3)$. Find the equation of this line.

2.2

Coordinate Geometry, Circles and Other Equations

Objectives

- Use the distance formula.
- Use the midpoint formula.
- Write the standard form of the equation of a circle and sketch the circle.
- Find the center and radius of a circle from an equation in general form and sketch the circle.
- Sketch the graph of an equation.

The coordinate system allows us to tie together concepts from geometry and algebra. In Section 2.1, we saw how to describe a line using the xy-coordinate system. In this section, we will describe circles and some other curves using the coordinate system.

Distance Between Two Points

In this section, you will learn how to find the distance between two points and use it to find the equation of a circle. The formula for the distance between two points is based on the Pythagorean Theorem from geometry. In a right triangle, the square of the length of the side opposite the right angle equals the sum of the squares of the lengths of the other two sides.

We place the right triangle on an xy-coordinate system with the other two sides parallel to the x- and y-axes, and label the sides and vertices as in Figure 1.

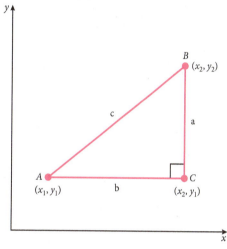

Figure 1

We then use the Pythagorean Theorem to find c, the length of the side opposite the right angle, which is the distance between the points A and B.

$$\text{Length of side } AC = b = x_2 - x_1$$

$$\text{Length of side } BC = a = y_2 - y_1$$

$$\text{Length of side } AB = c = \sqrt{b^2 + a^2} = \sqrt{(x_2 - x_1)^2 + (y_2 - y_1)^2}$$

Distance Formula

The distance d between the points (x_1, y_1) and (x_2, y_2) is given by

$$d = \sqrt{(x_2 - x_1)^2 + (y_2 - y_1)^2}$$

Midpoint of a Line Segment

The midpoint of a line segment is the point that is equidistant from the endpoints of the segment.

Midpoint of a Line Segment

The coordinates of the **midpoint** of the line segment joining the points (x_1, y_1) and (x_2, y_2) are

$$\left(\frac{x_1 + x_2}{2}, \frac{y_1 + y_2}{2} \right)$$

Notice that the x-coordinate of the midpoint is the average of the x-coordinates of the endpoints and the y-coordinate of the midpoint is the average of the y-coordinates of the endpoints.

VIDEO EXAMPLES

SECTION 2.2

Example 1   **Calculating Distance and Midpoint**

a. Find the distance between the points $(3, -5)$ and $(6, 1)$.

b. Find the midpoint of the line segment joining the points $(3, -5)$ and $(6, 1)$.

Solution

a. Using the distance formula with $(x_1, y_1) = (3, -5)$ and $(x_2, y_2) = (6, 1)$,

$$d = \sqrt{(x_2 - x_1)^2 + (y_2 - y_1)^2} \qquad \text{Distance formula}$$

$$d = \sqrt{(6 - 3)^2 + (1 - (-5))^2} \qquad \text{Substitute values}$$

$$= \sqrt{3^2 + 6^2} = \sqrt{45} = 3\sqrt{5}$$

b. Using the midpoint formula with $(x_1, y_1) = (3, -5)$ and $(x_2, y_2) = (6, 1)$, the coordinates of the midpoint are

$$\left(\frac{x_1 + x_2}{2}, \frac{y_1 + y_2}{2} \right) = \left(\frac{3 + 6}{2}, \frac{-5 + 1}{2} \right) = \left(\frac{9}{2}, -2 \right)$$

 Check It Out 1   (a) Find the distance between the points $(1, 2)$ and $(-4, 7)$. (b) Find the midpoint of the line segment joining the points $(1, 2)$ and $(-4, 7)$. ■

The distance formula is useful in describing the equations of some basic figures. Next we apply the distance formula to find the standard form of the equation of a circle.

Equation of a Circle

Recall from geometry that a circle is the set of all points in a plane whose distance from a fixed point is a constant. The fixed point is called the **center** of the circle and the distance from the center to any point on the circle is called the **radius** of the circle. A **diameter** of a circle is a line segment through the center of the circle with its endpoints on the circle. The length of a diameter is twice the radius of the circle. See Figure 2.

Next we find the equation of a circle with its center at the origin.

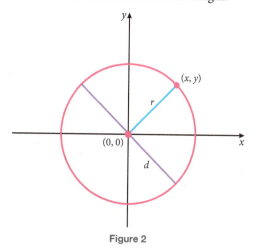

Figure 2

The distance from the center to any point (x, y) on the circle is the radius of the circle, r. This gives us the following.

Distance from center $(0, 0)$ to (x, y) = radius of circle

$$\sqrt{(x - 0)^2 + (y - 0)^2} = r \qquad \text{Apply distance formula}$$

$$x^2 + y^2 = r^2 \qquad \text{Square both sides of equation}$$

Similarly, we can find the equation of a circle with the center at the point (h, k) and radius r, shown in Figure 3. To do this, we use the distance formula to express r in terms of x, y, h, and k.

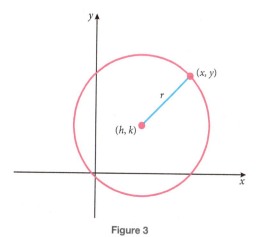

Figure 3

Distance from center (h, k) to (x, y) = radius of circle

$$\sqrt{(x - h)^2 + (y - k)^2} = r \qquad \text{Apply distance formula}$$

$$(x - h)^2 + (y - k)^2 = r^2 \qquad \text{Square both sides of equation}$$

Equation of a Circle in Standard Form

The circle with center at $(0, 0)$ and radius r is the set of all points (x, y) satisfying the equation

$$x^2 + y^2 = r^2$$

The circle with center at (h, k) and radius r is the set of all points (x, y) satisfying the equation

$$(x - h)^2 + (y - k)^2 = r^2$$

Example 2   **Finding the Standard Form of the Equation of a Circle**

Write the standard form of the equation of the circle with center at $(3, -2)$ and radius 5. Sketch the circle.

Solution The center $(3, -2)$ corresponds to the point (h, k) in the equation of the circle. We have

$$(x - h)^2 + (y - k)^2 = r^2 \qquad \text{Equation of circle}$$
$$(x - 3)^2 + (y - (-2))^2 = 5^2 \qquad \text{Use } h = 3, k = -2, r = 5$$
$$(x - 3)^2 + (y + 2)^2 = 25 \qquad \text{Standard form of equation}$$

To sketch the circle, first plot the center $(3, -2)$. Since the radius is 5, we can plot four points on the circle that are 5 units to the left, to the right, up, and down from the center. Using these points as a guide, we can sketch the circle as shown in Figure 4.

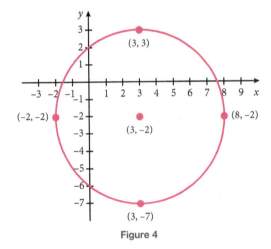

Figure 4

Check It Out 2 Write the standard form of the equation of the circle with center at $(4, -1)$ and radius 3. Sketch this circle.

Example 3   **Finding the Equation of a Circle Given a Point on the Circle**

Write the standard form of the equation of the circle with center at $(-1, 2)$ and containing the point $(1, 5)$. Sketch the circle.

Solution Since the radius is the distance from $(-1, 2)$ to $(1, 5)$,

$$r = \sqrt{(1 - (-1))^2 + (5 - 2)^2} = \sqrt{4 + 9} = \sqrt{13}$$

The equation is then

$$(x - h)^2 + (y - k)^2 = r^2 \qquad \text{Equation of circle}$$
$$(x - (-1))^2 + (y - 2)^2 = (\sqrt{13})^2 \qquad \text{Use } h = -1, k = 2, r = \sqrt{13}$$
$$(x + 1)^2 + (y - 2)^2 = 13 \qquad \text{Standard form of equation}$$

The circle is sketched in Figure 5.

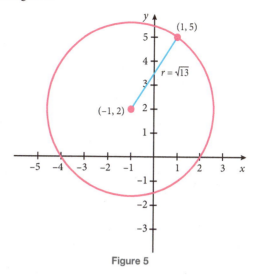

Figure 5

 **Check It Out 3** Write the standard form of the equation of the circle with center at $(3, -1)$ and containing the point $(0, 2)$.

The General Form of the Equation of a Circle

The equation of a circle can also be written in another form, known as the general form of the equation of the circle. To do so, we start with the standard form of the equation and expand the terms.

$$(x - h)^2 + (y - k)^2 = r^2 \quad \text{Equation of circle}$$

$$x^2 - 2hx + h^2 + y^2 - 2ky + k^2 = r^2 \quad \text{Expand terms on the left side}$$

$$x^2 + y^2 - 2hx - 2ky + h^2 + k^2 - r^2 = 0 \quad \text{Rearrange terms in decreasing powers of } x \text{ and } y$$

Letting $D = -2h$, $E = -2k$, and $F = h^2 + k^2 - r^2$, we get the general form of the equation of a circle.

General Form of the Equation of a Circle

The **general form of the equation of a circle** with center (h, k) and radius r is given by
$$x^2 + y^2 + Dx + Ey + F = 0$$
where $D = -2h$, $E = -2k$, and $F = h^2 + k^2 - r^2$.

If you are given an equation of a circle in general form, you can use the technique of **completing the square** to rewrite the equation in standard form. You can then quickly identify the center and radius of the circle. To complete the square on an expression of the form $x^2 + bx$, add an appropriate number c so that $x^2 + bx + c$ is a perfect square trinomial.

For example, to complete the square on $x^2 + 8x$, you add 16. This gives

$$x^2 + 8x + 16 = (x + 4)^2$$

In general, if the expression is of the form $x^2 + bx$, you add $c = \left(\frac{b}{2}\right)^2$ to make $x^2 + bx + c$ a perfect square trinomial. Table 1 gives more examples.

Just in Time
Review trinomials that are perfect squares in Section 1.3.

Begin With	Then Add	To Get
$x^2 + 10x$	$\left(\dfrac{10}{2}\right)^2 = 5^2 = 25$	$x^2 + 10x + 25 = (x + 5)^2$
$y^2 - 8y$	$\left(\dfrac{-8}{2}\right)^2 = (-4)^2 = 16$	$y^2 - 8y + 16 = (y - 4)^2$
$x^2 + 3x$	$\left(\dfrac{3}{2}\right)^2 = \dfrac{9}{4}$	$x^2 + 3x + \dfrac{9}{4} = \left(x + \dfrac{3}{2}\right)^2$
$x^2 + bx$	$\left(\dfrac{b}{2}\right)^2$	$x^2 + bx + \left(\dfrac{b}{2}\right)^2 = \left(x + \dfrac{b}{2}\right)^2$

Table 1

Example 4   **Completing the Square to Write the Equation of a Circle**

Write the equation of the circle, $x^2 + y^2 + 8x - 2y - 8 = 0$, in standard form. Find the coordinates of the center of the circle and find its radius. Sketch the circle.

Solution To put the equation in standard form, we complete the square on both x and y.

Step 1 Group together the terms containing x, and then the terms containing y. Move the constant to the right side of the equation. This gives

$$x^2 + y^2 + 8x - 2y - 8 = 0 \quad \text{Original equation}$$

$$(x^2 + 8x) + (y^2 - 2y) = 8. \quad \text{Group } x, y \text{ terms}$$

Step 2 Complete the square for each expression in parentheses by using $\left(\dfrac{8}{2}\right)^2 = 16$ and $\left(\dfrac{-2}{2}\right)^2 = 1$. Remember that any number added to the left side of the equation must also be added to the right side.

$$(x^2 + 8x + \mathbf{16}) + (y^2 - 2y + \mathbf{1}) = 8 + \mathbf{16} + \mathbf{1}$$

$$(x^2 + 8x + 16) + (y^2 - 2y + 1) = 25$$

Step 3 Factoring to get $x^2 + 8x + 16 = (x + 4)^2$ and $y^2 - 2y + 1 = (y - 1)^2$, we have

$$(x + 4)^2 + (y - 1)^2 = 25$$

The equation is now in standard form. The coordinates of the center are $(-\mathbf{4}, \mathbf{1})$ and the radius is **5**. Using the center and the radius, we can sketch the circle shown in Figure 6.

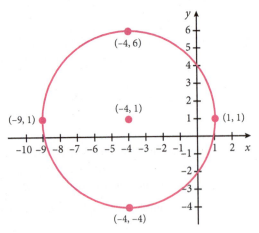

Figure 6

Ⓥ **Check It Out 4**   Write the equation of the circle, $x^2 + y^2 + 2x - 6y - 6 = 0$, in standard form. Find the coordinates of the center of the circle and find its radius. Sketch the circle.

Graphs of Other Equations

In your prior algebra courses, you learned to graph by plotting points on the coordinate plane. To do so, you solved for the variable y in a given equation, and substituted suitable values for the x variable. We briefly review those techniques here.

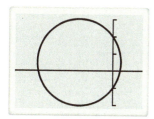

Using Technology

The APPS button in a graphing calculator will bring up a CONICS menu item for graphing circles. Choose the option for Circle, and enter the equation to get a graph of the circle. See Figure 7.

Keystroke Appendix:
Section 17

Figure 7

Example 5   **Sketching the Graph of an Equation**

Sketch the graph of the equation $y - x^2 = -3$.

Solution   First, solve for y: $y = x^2 - 3$. Then make a table of x and y values. See Table 2.

x	$y = x^2 - 3$
-2	$(-2)^2 - 3 = 1$
-1	$(-1)^2 - 3 = -2$
0	$(0)^2 - 3 = -3$
1	$(1)^2 - 3 = -2$
2	$(2)^2 - 3 = 1$

Table 2

Plot the points and connect them with a smooth curve. See Figure 8. A curve with this shape is called a **parabola**.

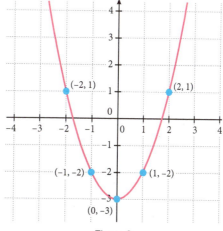

Figure 8

Check It Out 5   Sketch the graph of the equation $y - 2x^2 = 0$.

Example 6   **Sketching the Graph of an Equation**

Sketch the graph of the equation $y^2 + x = 0$.

Solution   First, solve for y.

$$y^2 + x = 0 \qquad \text{Given equation}$$

$$y^2 = -x \qquad \text{Subtract } x \text{ from both sides}$$

$$|y| = \sqrt{-x} \qquad \begin{array}{l}\text{Take square root of both sides}\\ \text{Recall } \sqrt{y^2} = |y|\end{array}$$

$$y = \pm\sqrt{-x} \qquad \text{Definition of absolute value}$$

Note that $\sqrt{-x}$ is a real number only when $-x \geq 0$, which means $x \leq 0$. Also, there are two values of y corresponding to each allowable nonzero value for x. We make two separate tables of x and y values.

x	$y = \sqrt{-x}$
-9	$\sqrt{-(-9)} = 3$
-4	$\sqrt{-(-4)} = 2$
-1	$\sqrt{-(-1)} = 1$
0	$\sqrt{-(0)} = 0$

Table 3

x	$y = -\sqrt{-x}$
-9	$-\sqrt{-(-9)} = -3$
-4	$-\sqrt{-(-4)} = -2$
-1	$-\sqrt{-(-1)} = -1$
0	$-\sqrt{-(0)} = 0$

Table 4

Plot the points and connect them with a smooth curve. See Figure 9. This curve is a parabola that opens to the left.

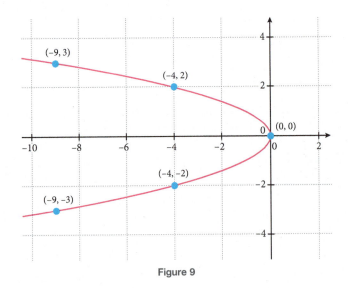

Figure 9

Check It Out 6   Sketch the graph of the equation $y^2 - x = 0$.

Sections 2.4, 2.5, and 2.7 will cover graphs in greater detail, and most of the subsequent chapters will deal with specific graphing techniques and analysis of graphs.

Exercises 2.2

Just in Time Exercises

1. True or False: $(x + 3)^2 = x^2 + 9$
2. Expand: $(x + 1)^2$
3. Factor: $x^2 + 8x + 16$
4. Factor: $x^2 - x + \dfrac{1}{4}$

Skills

In Exercises 5–16, find the distance between each pair of points and the midpoint of the line segment joining them.

5. $(6, 4), (-8, 11)$
6. $(-5, 8), (-10, 14)$
7. $(-4, 20), (-10, 14)$
8. $(4, 3), (-5, 13)$

9. $(1, -1), (5, 5)$
10. $(-5, -2), (6, 10)$
11. $(6, -3), (6, 11)$
12. $(4, 7), (-10, 7)$

13. $\left(-\dfrac{1}{2}, 1\right), \left(-\dfrac{1}{4}, 0\right)$
14. $\left(-\dfrac{3}{4}, 0\right), (-5, 3)$
15. $(a_1, a_2), (b_1, b_2)$
16. $(a_1, 0), (0, b_2)$

In Exercises 17–26, write the standard form of the equation of the circle having the given radius and center. Sketch the circle.

17. $r = 5$; center: $(0, 0)$
18. $r = 3$; center: $(0, 0)$
19. $r = 3$; center: $(-1, 0)$

20. $r = 4$; center: $(0, -2)$
21. $r = 5$; center: $(3, -1)$
22. $r = 3$; center: $(-2, 4)$

23. $r = \dfrac{3}{2}$; center: $(1, 0)$
24. $r = \dfrac{5}{3}$; center: $(0, -2)$
25. $r = \sqrt{3}$; center: $(1, 1)$

26. $r = \sqrt{5}$; center: $(-2, -1)$

In Exercises 27–34, write the standard form of the equation of the circle having the given center and containing the given point.

27. Center: $(0, 0)$; point: $(1, 3)$
28. Center: $(0, 0)$; point: $(-2, 1)$

29. Center: $(2, 0)$; point: $(2, 5)$
30. Center: $(0, 3)$; point: $(-2, 3)$

31. Center: $(1, -2)$; point: $(5, 1)$
32. Center: $(-3, 2)$; point: $(3, -2)$

33. Center: $\left(\dfrac{1}{2}, 0\right)$; point: $(1, 3)$
34. Center: $\left(0, \dfrac{1}{3}\right)$; point: $(4, 2)$

In Exercises 35–40, what number must be added to complete the square?

35. $x^2 + 12x$
36. $x^2 - 10x$
37. $y^2 - 5y$
38. $y^2 + 7y$
39. $x^2 - 3x$
40. $y^2 + 5y$

In Exercises 41–54, find the center and radius of the circle having the given equation.

41. $x^2 + y^2 = 36$
42. $x^2 + y^2 = 49$
43. $(x - 1)^2 + (y + 2)^2 = 36$

44. $(x + 3)^2 + (y - 5)^2 = 121$
45. $(x - 8)^2 + y^2 = \dfrac{1}{4}$
46. $x^2 + (y - 12)^2 = \dfrac{1}{9}$

47. $x^2 + y^2 - 6x + 4y - 3 = 0$
48. $x^2 + y^2 + 8x + 2y - 8 = 0$

49. $x^2 + y^2 - 2x + 2y - 7 = 0$
50. $x^2 + y^2 + 8x - 2y + 8 = 0$

51. $x^2 + y^2 - 6x - 4y - 5 = 0$
52. $x^2 + y^2 + 4x - 2y - 7 = 0$
53. $x^2 + y^2 - x = 2$

54. $x^2 + y^2 + 3y = 4$

In Exercises 55–58, find the equation, in standard form, of each of the circles pictured.

55.

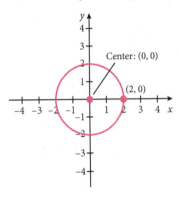

56.

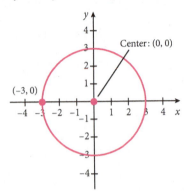

57.

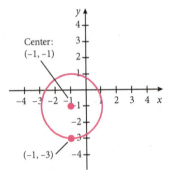

58.

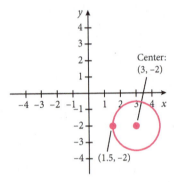

In Exercises 59–62, find the center and radius of the circle having the given equation. Use a graphing utility to graph the circle.

59. $x^2 + y^2 = 6.25$

60. $x^2 + y^2 = 12.25$

61. $(x - 3.5)^2 + y^2 = 10$

62. $(y - 2)^2 + (x + 4.2)^2 = 30$

In Exercises 63–70, sketch the graph of the equation.

63. $y = 2x^2 + 1$

64. $y = -x^2 - 2$

65. $y - |x| = 3$

66. $y + 2|x| = 1$

67. $y^2 = 2x$

68. $y^2 = x + 1$

69. $|y| - x = 0$

70. $|y| - 2x = 0$

Applications

71. Construction A circular walkway is to be built around a monument, with the monument as the center. The distance from the monument to any point on the inner boundary of the walkway is 30 feet.

Concentric circles

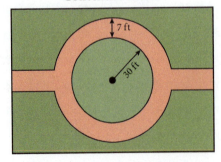

a. What is the equation of the inner boundary of the walkway? Use a coordinate system with the monument at $(0, 0)$.

b. If the walkway is 7 feet wide, what is the equation of the outer boundary of the walkway?

72. Distance Measurement Two people are standing at the same road intersection. One walks directly east at 3 miles per hour. The other walks directly north at 4 miles per hour. How far apart will they be from each other after half an hour?

73. Engineering The Howe Truss was developed in about 1840 by the Massachusetts bridge builder William Howe. It is illustrated below and constitutes a section of a bridge. The points labeled *A*, *B*, *C*, *D*, and *E* are equally spaced. The points *F*, *G*, and *H* are also equally spaced.

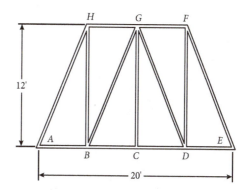

a. Let the point labeled *A* be the origin for a coordinate system for this problem. What are the coordinates of the points *A* through *H*?

b. Trusses such as the one illustrated here were used to build wooden bridges in the nineteenth century. Use the distance formula to find the total length of all the lumber required to build this truss.

74. Engineering The following drawing illustrates a type of roof truss found in many homes.

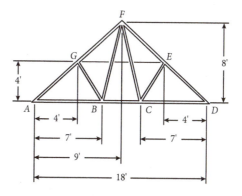

a. Let the point labeled *A* be the origin for a coordinate system for this problem. What are the coordinates of the points *A* through *G*?

b. Use the distance formula to find the total length of all the lumber required to build this truss.

Concepts

75. Write the equation of the circle whose diameter has endpoints at $(-3, 4)$ and $(1, 2)$ and sketch the circle.

76. Find the equation of the circle with center at the origin and a circumference of 7π units.

77. Find the equation of the circle with center at $(-5, -1)$ and an area of 64π.

78. Find the point(s) on the *x*-axis that is(are) at a distance of 7 units from the point $(3, 4)$.

79. Find the point(s) on the *y*-axis that is(are) at a distance of 8 units from the point $(1, -2)$.

SPOTLIGHT ON SUCCESS   *Instructor Anna N. Tivy*

Dear Student,

I was born in Russia, in a town named Astrakhan. I was born into a family of teachers. My mother Lyudmila was a middle school principal for more than ten years. She also was a physics and astronomy teacher. My father Nikolay was a PE teacher. Even my darling grandmother Claudia was a teacher of mathematics. For my entire childhood, I was challenged to be a good example for other children in local schools, and be a top student in every discipline. I was tutored in Mathematics, Russian composition, and literacy during every summer break! Even when my family took vacation trips to the Black Sea, half of my suitcase was full of textbooks and assignments. As a result, I excelled in mathematics and all other subjects. You may ask me, did you love math? It is hard to say. If you are good at something, I mean, really good … does it mean you love it or you love the feeling of being good? Well, it was the latter for me. However, I now realize that I was also in love with mathematics.

You may know that Perestroika and Glasnost were very big historical events in Russia from 1985 through 1990. It was the crash of the Soviet Union, and the birth of the Russian Federation. There was chaos in the streets, and in education. I was about to become a high school graduate with all A's toward my diploma. In spite of the chaos, I made a decision to become an elementary school teacher. I graduated from the State Pedagogical University in Astrakhan with honors, and became an elementary school teacher. But my country was still unsettled, so after a few years, I made a personal decision to relocate to the United States.

When I arrived in the United States, I was not able to speak and understand the English language. I was not familiar with the culture and traditions. My hard earned diploma from Russia did not allow me to start to work using my specialization. But I did not give up! I began to attend English and Math classes at Ventura Community College to study language and refresh my memory in Calculus. Soon, I realized that I was ready to apply for a Master program in Applied Mathematics at California State University of Northridge. This was a time when I fully understood that I love mathematics and I love to teach. I have continued to teach mathematics since my graduation, and my love for teaching has not diminished. My one regret is that my grandmother, Claudia, did not see my success. She was the one who truly believed in me and helped me to become what I am right now.

My advice to you, dear Student, is to follow your dreams, listen to your heart. Mathematics is not an easy subject, but you can learn it with a lot of practice and dedication. Invite Mathematics, Queen of Sciences, to be your friend and not your enemy. Together you can achieve your dreams and success will follow.

Sincerely,

Anna N. Tivy
Full-time lecturer of Mathematics at California State University of Channel Islands, Adjunct Faculty of Mathematics at Ventura Community College, CA

Functions

Objectives

- Define a function.
- Evaluate a function at a certain input value.
- Interpret tabular and graphical representations of a function.
- Define the domain and range of a function.

A function describes a relationship between two quantities of interest. For example, the distance driven depends on, or is a function of, the amount of time driven. A salesperson's earnings depends on, or is a function of, the amount of sales he or she generates. In this section, we will define precisely what a function is, and examine some of its applications.

Definition of a Function

Definition of a Function

A **function** establishes a correspondence between a set A and a set B such that for each element in A, there is *exactly one* corresponding element in B. See Figure 1.

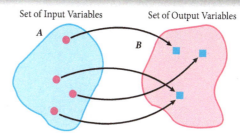

Set of Input Variables   Set of Output Variables

Figure 1

We can think of A as a set of all input values, and B as the set containing all output values. If x is an element of A, there is exactly one corresponding output value, $f(x)$, which is an element of B. The letter f is just the name of the function. See Figure 2. The quantity $f(x)$ is pronounced as "f of x" or "f at x," and the parentheses in $f(x)$ do **not** represent multiplication.

Input x Function f Output $f(x)$

Figure 2

A variable, such as x, can refer to any one of the possible input values and is usually referred to as the **independent variable**. The corresponding output, $f(x)$, is usually referred to as the **dependent variable**.

Functions need not always be defined by formulas. They can be given as a table, pictured as a graph, or described simply in words. As long as the definition for a function is satisfied, it does not matter how the function itself is described.

VIDEO EXAMPLES

SECTION 2.3

Input	Output
-4	0.50
-3	0.50
0	0.50
2.5	0.50
$\dfrac{13}{2}$	0.50

Table 1

Example 1   **Checking the Definition of a Function**

Which of the following correspondences satisfy the definition of a function?

a. The input value is one of the letters from the set {J, F, M, A, S, O, N, D} and the output value is a name of a month beginning with that letter.

b. The correspondence defined by Table 1.

c. The input value is the radius of a circle and the output value is the area of the circle.

Solution

a. If the letter J was input, the output could be January, June or July. Thus, **this correspondence is not a function**.

b. Table 1 represents a function since each input value has only one corresponding output value. It does *not* matter that the *output* values are *repeated*.

c. The area A, of a circle is related to its radius, r, by the formula $A = \pi r^2$. For each value of r, only one value of A is output by the formula. Therefore, **this correspondence is a function**.

✓ *Check It Out 1*   Which of the following represent functions?

a. The input variable is the number of days in a month and the output variable is the name of the corresponding month(s).

b.

Input	Output
−6	1
−4	4.3
−3	−1
1	3.1
6	−2

Table 2

c. The input variable is the diameter of a circle and the output variable is the circumference of the circle.

The following examples will illustrate the usefulness of the function notation.

Example 2   **Evaluating a Function**

Let $f(x) = x^2 - 1$. Evaluate the following:

a. $f(-2)$ b. $f\left(\dfrac{1}{2}\right)$ c. $f(x + 1)$ d. $f(\sqrt{5})$ e. $f(x^3)$

Solution

a. $f(-2) = (-2)^2 - 1 = 4 - 1 = 3$

b. $f\left(\dfrac{1}{2}\right) = \left(\dfrac{1}{2}\right)^2 - 1 = \dfrac{1}{4} - 1 = -\dfrac{3}{4}$

c. $f(x + 1) = (x + 1)^2 - 1 = x^2 + 2x + 1 - 1 = x^2 + 2x$

d. $f(\sqrt{5}) = (\sqrt{5})^2 - 1 = 5 - 1 = 4$

e. $f(x^3) = (x^3)^2 - 1 = x^6 - 1$

✓ *Check It Out 2*   Let $f(t) = -t^2 + 2$. Evaluate the following:

a. $f(-1)$ b. $f(a + 1)$ c. $f(x^2)$

Example 3 provides additional examples of evaluating functions.

Example 3   **Evaluation of Functions**

Evaluate $g(-3)$ and $g(a^2)$ for the following functions.

a. $g(x) = \dfrac{\sqrt{1 - x}}{2}$ b. $g(x) = \dfrac{x + 4}{x - 2}$

 Using Technology

To evaluate function values for a particular numerical value, first store the value of x and then evaluate the function at that value. Figure 3 shows how to evaluate $x^2 - 1$ at $x = -2$.

Keystroke Appendix:

Section 4

−2→X	
	−2
X²−1	
	3

Figure 3

Solution

a. $g(-3) = \dfrac{\sqrt{1-(-3)}}{2} = \dfrac{\sqrt{4}}{2} = 1$

$g(a^2) = \dfrac{\sqrt{1-(a^2)}}{2}$. This expression cannot be simplified further.

b. $g(-3) = \dfrac{(-3)+4}{(-3)-2} = -\dfrac{1}{5}$

$g(a^2) = \dfrac{(a^2)+4}{(a^2)-2}$. This expression cannot be simplified further.

 Check It Out 3   Let $f(x) = \dfrac{x-3}{x^2+1}$. Evaluate $f(-3)$.

Observations:

- A function can be assigned any name. Functions do not always have to be called f.
- The variable x in $f(x)$ is a placeholder. It can be replaced by any quantity, as long as the same replacement occurs in the entire expression for the function.

Piecewise-Defined Functions

Functions which are defined using different expressions, corresponding to different conditions satisfied by the independent variable, are called **piecewise-defined functions**. For instance, the absolute value function is defined piecewise as follows:

$$|x| = \begin{cases} x & \text{if } x \geq 0 \\ -x & \text{if } x < 0 \end{cases}$$

So if $x = -3, |-3| = -(-3) = 3$, since $-3 < 0$. But if $x = 3$, then $|3| = 3$, since $3 > 0$.

Example 4 shows how to evaluate another piecewise-defined function.

Example 4   **Evaluating a Piecewise-Defined Function**

Define $H(x)$ as follows:

$$H(x) = \begin{cases} 1 & \text{if } x < 0 \\ -x+1 & \text{if } 0 < x \leq 2 \\ -3 & \text{if } x > 2 \end{cases}$$

Evaluate the following, if defined:

a. $H(2)$ **b.** $H(-6)$ **c.** $H(0)$

Solution   Note that the function H is not given by just one formula. The expression for $H(x)$ will depend on whether $x < 0, 0 < x \leq 2$, or $x > 2$.

a. To evaluate $H(2)$, first note that $x = 2$ and thus, you must use the expression for $H(x)$ corresponding to $0 < x \leq 2$. Therefore, $H(2) = -(2) + 1 = -1$.

b. To evaluate $H(-6)$, note that $x = -6$. Since x is less than zero, we use the value for $H(x)$ corresponding to $x < 0$. Therefore, $H(-6) = 1$.

c. Now we are asked to evaluate $H(0)$. We see that $H(x)$ is defined only for $x < 0$, or $0 < x \leq 2$, or $x > 2$, and that $x = 0$ satisfies none of these three conditions. Therefore, $H(0)$ **is not defined**.

 Check It Out 4   For $H(x)$ defined in Example 4, find

a. $H(4)$ **b.** $H(-3)$

Domain and Range of a Function

It is important to know when a function is defined and when it is not. For example, the function $f(x) = \frac{1}{x}$ is not defined for $x = 0$. The set of all input values for which a function is defined has a special name, called the **domain**. Throughout this text, mathematical functions which output only real numbers are considered, and so we restrict our definition of domain to that effect.

Definition of Domain

The **domain** of a function is the set of all input values for which the function will produce a real number.

Similarly, we can consider a set of all output values of a function, called the **range**.

Definition of Range

The **range** of a function is the set of all output values which correspond to at least one element of the domain of the function.

Example 5   **Finding the Domain of the Function**

Find the domain of each of the following functions. Write your answer using interval notation.

a. $g(t) = t^2 + 4$

b. $f(x) = \sqrt{4 - x}$

c. $h(s) = \dfrac{1}{s - 1}$

d. $h(t) = \dfrac{1}{\sqrt{4 - t}}$

Solution

Just in Time
Review square roots in Section 1.2 and inequalities in Section 1.6.

a. For the function $g(t) = t^2 + 4$, any value can be substituted for t and a real number will be output by the function. **Thus, the domain for g is all real numbers**, or $(-\infty, \infty)$ in interval notation,

b. For $f(x) = \sqrt{4 - x}$, we recall that the square root of a number is a real number only when the number under the square root sign is greater than or equal to zero. Therefore, we need to have

$4 - x \geq 0$ Expression under square root sign must be greater than or equal to zero

$4 \geq x$ Solve for x

$x \leq 4$ Rewrite inequality

Thus, the domain is **the set of all real numbers less than or equal to 4**, or $(-\infty, 4]$, in interval notation.

Just in Time
Review rational expressions in Section 1.4.

c. For $h(s) = \frac{1}{s-1}$, we see that this expression is defined only when the denominator, $s - 1$ is *not equal to zero*. So, $s \neq 1$. Thus, the domain is $(-\infty, 1) \cup (1, \infty)$.

d. For $h(t) = \frac{1}{\sqrt{4-t}}$, we see that this expression is not defined when the denominator is equal to zero, which happens when $t = 4$. However, since the square root is not defined for negative numbers, we must have

$4 - t > 0$ Expression under radical is greater than 0

$4 > t$ Solve for t

$t < 4$ Rewrite inequality

Thus, the domain, in interval notation, is $(-\infty, 4)$.

○ *Check It Out 5*   Find the domain of the following functions. Write your answer using interval notation.

a. $H(t) = -t^2 - 1$

b. $g(x) = \sqrt{x - 4}$

c. $h(x) = \dfrac{1}{2x + 1}$

d. $f(t) = \dfrac{2}{\sqrt{t - 4}}$

■

Note Finding the range of a function involves techniques discussed in later chapters, and so we will not discuss them here. However, the next section will show how you can determine the range graphically.

More Examples of Functions

Function notation is extremely useful in applications where a problem stated verbally must be restated in mathematical terms, as seen in Example 6.

Example 6   **Modeling Weekly Pay**

Eduardo is a part-time salesperson at Digitex Audio, a sound equipment store. Each week, he is paid a salary of $200 plus a commission of 10% of the amount of sales he generates that week, in dollars.

a. What are the input and output variables for this problem?

b. Express Eduardo's pay for one week as a function of the sales he generates that week.

c. What was his pay for a week in which he generated $4000 worth of sales?

Solution

a. Let the variables be defined as follows:

Input variable: x (amount of sales generated in one week, in dollars)
Output variable: $P(x)$ (pay for that week, in dollars)

b. Eduardo's pay for a given week consists of a **fixed portion, $200**, plus a commission based on the amount of sales generated that week. Since he receives 10% of the sales generated, **the commission portion of his pay is given by 0.10x**. Hence his total pay for the week is given by

$$\text{Pay} = \text{Fixed portion} + \text{Commission portion}$$
$$P(x) = 200 + 0.10x$$

c. If Eduardo generated $4000 worth of weekly sales, his pay would be $P(4000)$. Substituting $x = 4000$ into the expression for $P(x)$,

$$P(4000) = 200 + 0.10(\mathbf{4000}) = 600$$

Thus, he will be paid $600 for the week.

○ *Check It Out 6*   Rework Example 6, but now assume that Eduardo's weekly salary is increased to $500 and his commission is 15% of total sales generated.

■

In everyday life, information such as postal rates or income tax is often given in tables. The following example lists postal rates as a function of weight.

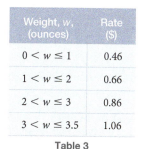

Weight, w, (ounces)	Rate ($)
$0 < w \le 1$	0.46
$1 < w \le 2$	0.66
$2 < w \le 3$	0.86
$3 < w \le 3.5$	1.06

Table 3

Example 7   **Postal Rate Table**

Table 3 gives the U.S. Postal Service's rate for a first-class letter in 2013 as a function of the weight of the letter.

(*Source*: United States Postal Service)

Letters heavier than 3.5 ounces are classified separately, and are not listed here.

a. Identify the independent variable and the dependent variable.

b. Explain why this table represents a function.

c. What is the rate for a first-class letter weighing 3.2 ounces?

Solution

a. Reading the problem again, the rate for the first-class letter is a *function of the weight* of the letter. Thus, the independent variable is the weight of the first-class letter. The dependent variable is the rate charged.

b. This table represents a function, because for each input weight, only one rate will be output.

c. Since 3.2 is between 3 and 3.5, it will cost $1.06 to mail this letter.

Check It Out 7   Use the table in Example 7 to determine the rate for a first-class letter weighing 2.3 ounces. ■

Functions can also be depicted graphically. In newspapers and magazines, you will see many graphical representations of relationships between quantities. The input variable is on the horizontal axis of the graph and the output variable is on the vertical axis of the graph. The following example shows a graphical representation of a function. We will discuss graphs of functions in more detail in the next section.

Example 8   **Graphical Depiction of a Function**

Figure 4 depicts the average high temperature in Fargo, North Dakota, in degrees Fahrenheit, as a function of the month of the year.   (*Source*: weather.yahoo.com)

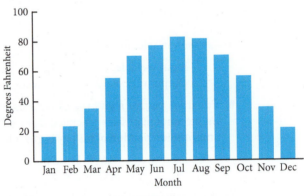

Figure 4

a. Let $T(m)$ be the function represented by the graph, where m is the month of the year. What is $T(\text{May})$?

b. What are the valid input values for this function?

Solution

a. Estimating from the graph, $T(\text{May}) \approx 70°\text{F}$. Thus, the average high temperature for May in Fargo is approximately 70°F.

b. Here, the valid input values are the names of months of the year from January through December. Note that input values and output values do not necessarily have to be numbers for functions that are not defined by a mathematical expression.

Check It Out 8   In Example 8, estimate $T(\text{August})$. ■

Exercises 2.3

Just in Time Exercises

1. True or False: $\sqrt{-4} = -2$

2. $\sqrt{x}$ is a real number when   a) $x < 0$ b) $x \geq 0$ c) $x \leq 0$

3. Solve: $2x + 1 > 0$

4. For what values of x is the denominator of $\frac{1}{2x+1}$ equal to 0?

Skills

In Exercises 5–16, evaluate $f(3)$, $f(-1)$, and $f(0)$.

5. $f(x) = 5x + 3$

6. $f(x) = -4x - 1$

7. $f(x) = -\frac{7}{2}x + 2$

8. $f(x) = \frac{2}{3}x - 1$

9. $f(x) = x^2 + 2$

10. $f(x) = -x^2 - 4$

11. $f(x) = -2(x+1)^2 - 4$

12. $f(x) = 3(x-3)^2 + 5$

13. $f(t) = \sqrt{3t + 4}$

14. $f(t) = \sqrt{2t + 5}$

15. $f(t) = \frac{t^2 - 1}{t + 3}$

16. $f(t) = \frac{t^2 + 1}{t - 2}$

In Exercises 17–24, evaluate $f(a)$, $f(a + 1)$, and $f\left(\frac{1}{2}\right)$.

17. $f(x) = 4x + 3$

18. $f(x) = -2x - 1$

19. $f(x) = -x^2 + 4$

20. $f(x) = -2x^2 + 1$

21. $f(x) = \sqrt{3x - 1}$

22. $f(x) = \sqrt{x + 1}$

23. $f(x) = \frac{1}{x + 1}$

24. $f(x) = \frac{1}{2x + 1}$

In Exercises 25–34, evaluate $g(-x)$, $g(2x)$, and $g(a + h)$.

25. $g(x) = \sqrt{6}$

26. $g(x) = \sqrt{5}$

27. $g(x) = 2x - 3$

28. $g(x) = -\frac{1}{2}x + 1$

29. $g(x) = 3x^2$

30. $g(x) = -x^2$

31. $g(x) = \frac{1}{x}$

32. $g(x) = -\frac{3}{x}$

33. $g(x) = -x^2 - 3x + 5$

34. $g(x) = x^2 + 6x - 1$

In Exercises 35–42, evaluate $f(-2)$, $f(0)$, and $f(1)$, if possible, for each function. If a function value is undefined, state so.

35. $f(x) = \begin{cases} 1 & \text{if } x < -2 \\ \frac{1}{2} & \text{if } x > -2 \end{cases}$

36. $f(x) = \begin{cases} \frac{2}{3} & \text{if } x < 1 \\ -2 & \text{if } x > 1 \end{cases}$

37. $f(x) = \begin{cases} x & \text{if } x < 0 \\ 1 & \text{if } x \geq 0 \end{cases}$

38. $f(x) = \begin{cases} -2 & \text{if } x < 1 \\ x^2 & \text{if } x \geq 1 \end{cases}$

39. $f(x) = \begin{cases} -1 & \text{if } x \leq -2 \\ 2 & \text{if } -2 < x < 1 \\ 4 & \text{if } x \geq 1 \end{cases}$

40. $f(x) = \begin{cases} 0 & \text{if } x < 0 \\ 2 & \text{if } 0 \leq x < 1 \\ 4 & \text{if } x \geq 1 \end{cases}$

41. $f(x) = \begin{cases} \sqrt{x} & 0 \leq x < 1 \\ -x + 4 & x \geq 1 \end{cases}$

42. $f(x) = \begin{cases} 2 & x < -2 \\ |x| & x \geq 2 \end{cases}$

43. Let $g(t)$ be defined by the following table:

t	$g(t)$
-2	$\frac{4}{3}$
-1	4.5
0	2
2	-1
5	-1

a. Evaluate $g(5)$.
b. Evaluate $g(0)$.
c. Is $g(3)$ defined? Explain.

44. Let $h(t)$ be defined by the following table:

t	$h(t)$
-2	4
$-\frac{1}{2}$	4.5
0	2
3	$\frac{4}{3}$
4	-1

a. Evaluate $h(-2)$.
b. Evaluate $h(4)$.
c. Is $h(5)$ defined? Explain.

Paying Attention to Instructions: *Exercises 45 and 46 are intended to give you practice reading, and paying attention to, the instructions that accompany the problems you are working.*

45. a. Solve for x: $2x + 3 = 9$.

    b. Evaluate $f(9)$, where $f(x) = 2x + 3$.

    c. Find x such that $f(x) = 11$, where $f(x) = 2x + 3$.

46. a. Solve for x: $x + 5 = 13$.

    b. Evaluate $f(10)$, where $f(x) = x + 5$.

    c. Find x such that $f(x) = 15$, where $f(x) = x + 5$.

In Exercises 47–54, determine whether a function is being described.

47. The length of a side of a square is the input variable and the perimeter of the square is the output variable.

48. A person's height is the input variable and his/her weight is the output variable.

49. The price of a store product is input and the name of the product is output.

50. The perimeter of a rectangle is input and its length is output.

51. The input variable is the denomination of a U.S. paper bill (1-dollar bill, 5-dollar bill, etc.) and the output variable is the length of the bill.

52. The input variable is the bar code on a product at a store and the output variable is the name of the product.

53. The following input–output table:

Input	Output
-2	$\frac{1}{3}$
-1	-2
0	2
2	-1
5	-1

54. The following input–output table:

Input	Output
-2	$-\frac{1}{3}$
-1	0
-1	2
2	5
5	4

In Exercises 55–68, find the domain of each function.

55. $f(x) = x^2 - 4$

56. $g(x) = -x^3 - 2$

57. $f(s) = \dfrac{1}{s + 1}$

58. $h(y) = \dfrac{1}{y + 2}$

59. $f(w) = \dfrac{5}{w - 3}$

60. $H(t) = \dfrac{3}{1 - t}$

61. $h(x) = \dfrac{1}{x^2 - 4}$

62. $f(x) = \dfrac{2}{x^2 - 9}$

63. $g(x) = \sqrt{2 - x}$

64. $F(w) = \sqrt{-4 - w}$

65. $f(x) = \dfrac{1}{x^2 + 1}$

66. $h(s) = \dfrac{3}{s^2 + 3}$

67. $f(x) = \dfrac{2}{\sqrt{x + 7}}$

68. $g(x) = \dfrac{3}{\sqrt{8 - x}}$

Applications

69. Geometry The volume of a sphere is given by

$$V(r) = \frac{4}{3}\pi r^3$$

where r is the radius. Find the volume, in cubic inches, when the radius is 3 inches.

70. Geometry The hypotenuse of a right triangle, having sides of lengths 4 and x, is given by

$$H(x) = \sqrt{16 + x^2}$$

Find and interpret the quantity $H(2)$.

71. Sales A commissioned salesperson's earnings can be determined by the function

$$S(x) = 1000 + 20x$$

where x is the number of items sold. Find and interpret $S(30)$.

72. **Engineering** The distance between the n supports of a horizontal beam 10 feet long is given by

$$d(n) = \frac{10}{\sqrt{n^2 - 1}}$$

if bending is to be kept to a minimum.

   a. Find $d(2)$, $d(3)$, and $d(5)$.

   b. Can n be a negative integer? Explain.

73. **Motion and Distance** A car travels 45 miles per hour.

   a. Express the distance traveled by the car (in miles) as a function of the time (in hours).

   b. How far does the car travel in two hours? Write this using function notation.

   c. Find the domain of this function.

74. **Business** The following graph gives the total revenues from consumer electronics shipments, in billions of dollars, in the United States for various years.   (Source: Consumer Electronics Association, 2013)

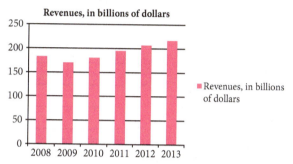

   a. If the dollar value of shipments, S, is a function of the year, estimate $S(2010)$ and interpret it.

   b. What is the general trend you notice in revenues from consumer electronics shipments?

75. **Film Industry** The following tables gives the amount of money grossed (in millions of dollars) by certain top movie hits.

   (*Source*: movies.yahoo.com)

Movie	Earnings (in millions)
The Hobbit: An Unexpected Journey	$303
The Dark Knight Rises	$448
Marvel's The Avengers	$623
The Amazing Spider-Man	$262
The Twilight Saga: Breaking Dawn Part 2	$292
The Hunger Games	$408

   a. If the dollar amount grossed, D, is a function of the movie title, find D(The Hunger Games) and interpret it.

   b. What is the domain of this function?

76. **Manufacturing** It costs a watchmaker $20 for each watch manufactured, over and above the watchmaker's fixed cost of $5000.

   a. Express the total manufacturing cost, in dollars, as a function of the number of watches produced.

   b. How much does it cost for 35 watches to be manufactured? Write this information in function notation.

   c. Find the domain of this function which makes sense in the real world.

77. **Geometry** If the length of a rectangle is three times its width, express the area of the rectangle as a function of its width.

78. Geometry If the height of a triangle is twice the length of the base, find the area of the triangle in terms of the length of the base.

79. Physics A ball is dropped from a height of 100 feet. The height of the ball at t seconds after it is dropped is given by the function $h(t) = -16t^2 + 100$.

    a. Find $h(0)$ and interpret it.

    b. Find the height of the ball after 2 seconds.

80. Marine Biology The amount of coral, in kilograms, harvested in North American waters is shown on the following graph.   (*Source*: United Nations, FAOSTAT data)

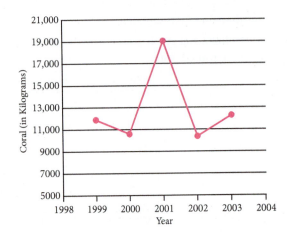

    a. From the graph, in what year(s) was the amount of coral harvested approximately 12,000 kilograms?

    b. From the data in the graph, in what year was the maximum amount of coral harvested?

81. Consumer Behavior The following table lists the per capita consumption of high-fructose corn syrup, a sweetener found in many foods and beverages, for selected years between 1970 and 2009.

    (*Source*: Statistical Abstract of the United States, 2013)

Year	Per capita consumption (in pounds)
1970	0.5
1980	19.0
1990	49.6
2000	62.6
2005	59.2
2008	53.0
2009	50.1

    a. If the input value is the year, and the output value is the per capita consumption of high fructose corn syrup, explain why this table represents a function. Denote this function by S.

    b. Find $S(2005)$ and interpret it.

    c. High fructose corn syrup was developed in the early 1970's and gained popularity as a cheaper alternative to sugar in processed foods. How is this reflected in the given table?

    d. What trend do you notice in the use of high fructose corn syrup from 2000 to 2009?

82. Automobile Mileage An automobile called the Ford Fusion Hybrid (2013 model), has a combined city and highway mileage rating of 47 miles per gallon. Write a formula for the distance it can travel (in miles) as a function of the amount of gasoline used (in gallons).

    (*Source*: www.fueleconomy.gov)

83. **Rate Plan** A long distance telephone company advertises that it charges $1.00 for the first twenty minutes and 7 cents a minute for every minute beyond the first 20 minutes. Let $C(t)$ denote the total cost of a telephone call lasting t minutes. Assume that the minutes are nonnegative integers.

    a. Many people will assume that it will cost only $0.50 to talk for 10 minutes. Why is this incorrect?
    b. Write an expression for the function $C(t)$.
    c. How much will it cost to talk for 5 minutes? 20 minutes? 30 minutes?

84. **Online Shopping** On the online auction site eBay, the minimum amount that one may bid on an item is based on the current bid, as shown in the table below.   (*Source*: www.ebay.com)

Current Bid	Minimum Bid Increment
$1.00 − $4.99	$0.25
$5.00 − $24.99	$0.50
$25.00 − $99.99	$1.00
$100.00 − $249.99	$2.50

For example, if the current bid on an item is $7.50, then the next bid must be *at least* $0.50 higher.

    a. Explain why the *minimum* bid increment, I, is a function of the current bid, b.
    b. Find $I(2.50)$ and interpret it.
    c. Find $I(175)$ and interpret it.
    d. Can you find $I(400)$ using this table? Why or why not?

Concepts

85. Let $f(x) = ax^2 + 5$. Find a if $f(1) = 2$.

86. Let $f(x) = -2x + c$. Find c if $f(3) = 1$.

Find the domain of the following functions.

87. $f(t) = k$, k is a fixed real number.

88. $g(s) = 1 + \sqrt{s^2 + 1}$

Using Technology

To graph $f(x) = \frac{2}{3}x - 2$, enter expression for $f(x)$ in the Y = editor. Use a table of values (Figure 1) to choose an appropriate window for the graph (see Figure 2).

Keystroke Appendix:
Sections 6 and 7

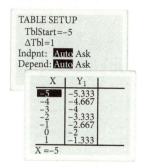

Figure 1

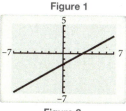

Figure 2

A graph is one of the most common ways to represent a function. Many newspaper articles and magazines summarize pertinent data in a graph. In this section, you will learn how to sketch the graphs of various functions which were discussed in Section 2.3. As you learn more about various types of functions in later chapters, you will build on the ideas presented in this section.

Graphs of Functions Defined by a Single Expression

One of the main features of a function is its graph. Recall that if we are given a value for x, then $f(x)$ is the value that is output by the function f. In the xy-plane, the input value, x, is the x-coordinate and the output value, $f(x)$, is the y-coordinate. The **graph** of f is the set of all points $(x, f(x))$ such that x is in the domain of f.

The following example shows the connection between a set of (x, y) values and the definition of a function.

Example 1   **Satisfying the Function Definition**

Does the following set of points define a function?

$$S = \{(-1, -1), (0, 0), (1, 2), (2, 4)\}$$

Solution   When a point is written in the form (x, y), x is the input variable, or the independent variable; y is the output variable, or the dependent variable. Table 1 shows the correspondence between the x and y values.

x	y
-1	-1
0	0
1	2
2	4

Table 1

The definition of a function states that for each value of x, there must be exactly one value of y. This definition is satisfied, and thus, the set of points S defines a function.

Check It Out 1   Does the following set of points define a function?

$$S = \{(-3, -1), (1, 0), (1, 2), (3, 4)\}$$

To graph a function by hand, we first make a table of x and $f(x)$ values, choosing various values for x. The set of coordinates given by $(x, f(x))$ are then plotted in the xy-plane. This process is illustrated in Example 2.

Example 2   **Graph of a Function**

Graph the function $f(x) = \dfrac{2}{3}x - 2$.

Solution Make a table of values of x and $f(x)$ (Table 2) and then plot the points on the xy-plane. We have chosen multiples of 3 for the x values to make the arithmetic easier. In principle, we can choose any value of x as long as the function is defined for that value of x. Recall that the y-coordinate corresponding to x is given by $f(x)$. Since the points lie along a line, we draw the line passing through the points to get the graph shown in Figure 3.

x	$f(x) = \dfrac{2}{3}x - 2$
-6	-6
-3	-4
0	-2
3	0
6	2

Table 2

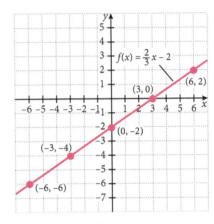

Figure 3

This is just the graph of a line, which you have seen in Section 2.1. The difference here is that we use the function notation, which states the dependence on x.

 Check It Out 2   Graph the function $g(x) = -3x + 1$.

Example 3   **Finding Domain and Range from a Graph**

Graph the function $g(t) = -2t^2$. Use the graph to find the domain and range of g.

Solution Make a table of values of t and $g(t)$ (Table 3) and then plot the points on the ty-plane as shown in Figure 4. Recall that the y-coordinate corresponding to t is given by $g(t)$.

This curve is a parabola that opens downward (Figure 4). We will explore functions like this in more detail in Section 3.1.

t	$g(t)$
-2	-8
-1	-2
0	0
1	-2
2	-8

Table 3

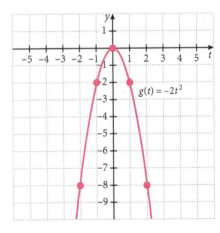

Figure 4

**Using Technology**

To manually generate a table of values, use the **Ask** option in the TABLE feature (Figure 5). The graph of $Y_1 = \sqrt{4-x}$ is shown in Figure 6.

Keystroke Appendix:
Section 6

Figure 5

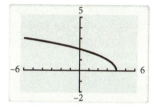

Figure 6

To find the domain of this function, we see that $g(t) = -2t^2$ is defined for all values of t. Graphically, there is a y-coordinate on the curve corresponding to *any* t that you choose. Thus, the **domain** is the set of *all real numbers*, $(-\infty, \infty)$.

Since the curve lies only in the bottom half of the ty-plane and touches the t axis at the origin, the y-coordinates take on values which are *less than or equal to zero*. Hence the **range** is $(-\infty, 0]$.

By looking at the expression for $g(t)$, which is $-2t^2$, we can come to the same conclusion about the range: t^2 will always be greater than or equal to zero, and when it is multiplied by -2, the end result will always be less than or equal to zero.

✓ **Check It Out 3** Graph the function $f(x) = x^2 - 2$ and use the graph to find its domain and range. ■

Example 4 **Graphing More Functions**

Graph each of the following functions. Use the graph to find the domain and range of f.

a. $f(x) = \sqrt{4-x}$ 　　　　　　　**b.** $f(x) = |x|$

Solution

a. Make a table of values of x and $f(x)$ (Table 4), and then plot the points on the xy-plane. Connect the dots to get a smooth curve. See Figure 7.

x	$f(x) = \sqrt{4-x}$
-5	3
-1	$\sqrt{5} \approx 2.236$
0	2
2	$\sqrt{2} \approx 1.414$
3	1
4	0

Table 4

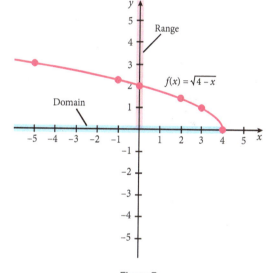

Figure 7

Just in Time
Review absolute value in Section 1.1.

From Figure 7, we see the x values for which the function is defined are $x \le 4$. For these values of x, $f(x) = \sqrt{4-x}$ will be a real number. Thus, the **domain** is $(-\infty, 4]$.

The y-coordinates of the points on the graph of $f(x) = \sqrt{4-x}$ take on all values greater than or equal to zero. Thus, the **range** is $[0, \infty)$.

b. The absolute value function is given by $f(x) = |x|$. Make a table of values of x and $f(x)$ (Table 5) and plot the points. The points form a "V" shape — the graph comes in sharply at the origin, as shown in Figure 8. This can be confirmed by choosing additional points near the origin. From the graph, we see that the **domain** is the set of all real numbers, $(-\infty, \infty)$. The **range** is the set of all real numbers greater than or equal to zero, $[0, \infty)$.

x	$f(x) = |x|$
-4	4
-2	2
-1	1
0	0
1	1
2	2
4	4

Table 5

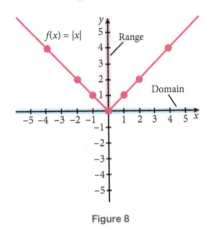

Figure 8

Check It Out 4 Graph the function $g(x) = \sqrt{x - 4}$ and use the graph to find its domain and range.

When a function is to be graphed with some x values excluded, then the corresponding points are denoted by an open circle on the graph. For example, Figure 9 shows the graph of $f(x) = x^2 + 1$, $x > 0$. To indicate that $(0, 1)$ is *not* part of the graph, it is indicated by an open circle.

 Using Technology

In a graphing utility, you must give each piece of the piecewise function a different name in the $Y=$ editor, along with the conditions that x must satisfy. See Figure 10. The calculator should be in Dot mode so that the pieces are not joined. See Figure 11.

Keystroke Appendix:
Section 7

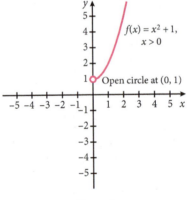

Figure 9

Graphing Piecewise-Defined Functions

We can graph piecewise-defined functions by essentially following the same procedures given thus far. However, you have to be careful about "jumps" in the function that may occur at points at which the function expression changes. Example 5 illustrates this situation.

Example 5   **Graphing a Piecewise-Defined Function**

Graph the function $H(x)$, defined as follows:

$$H(x) = \begin{cases} x^2 & \text{if } x < 0 \\ x + 1 & \text{if } 0 \le x < 3 \\ -1 & \text{if } x \ge 3 \end{cases}$$

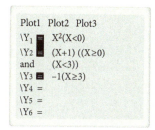

Figure 10

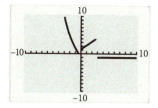

Solution This function is given by three different expressions, depending on the value of x. The graph will thus be constructed in three steps, as follows.

Step 1 If $x < 0$, then $H(x) = x^2$. So we first graph $H(x) = x^2$ on the interval $(-\infty, 0)$. The value $x = 0$ is not included since $H(x) = x^2$ holds only for $x < 0$. Since $x = 0$ is not part of the graph, we indicate it by an open circle. See Part I of the graph in Figure 12.

Step 2 If $0 \le x < 3$, then $H(x) = x + 1$. Thus, we graph the line $H(x) = x + 1$ on $[0, 3)$, indicated by Part II of the graph in Figure 12. Note that $x = 3$ is not part of the graph.

Step 3 If $x \ge 3$, then $H(x) = -1$. Thus, we graph the horizontal line $H(x) = -1$ on the interval $[3, \infty)$, indicated by Part III of the graph in Figure 12.

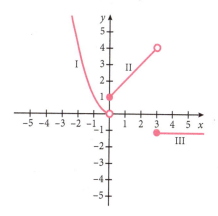

Figure 12

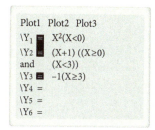

Figure 11

Check It Out 5   Graph the function $H(x)$, defined as follows:

$$H(x) = \begin{cases} -5, & \text{if } x \le 0 \\ 4, & \text{if } x > 0 \end{cases}$$

The Vertical Line Test for Functions

We can examine the graph of a set of ordered pairs to determine whether that graph describes a function.

Recall that the definition of a function states that for each value in the domain of the correspondence, there can be only one value in the range. Graphically, this means that any vertical line can intersect the graph of a function at most once.

Figure 13 is the graph of a function, because any vertical line crosses the graph at no more than one point. Some sample vertical lines are drawn for reference.

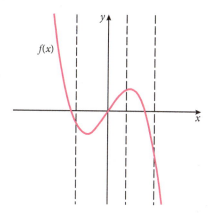

Figure 13

Figure 14 does *not* represent the graph of a function because there are vertical lines that cross the graph at more than one point.

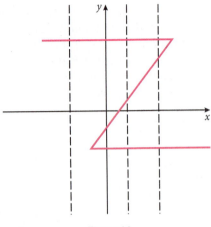

Figure 14

| **Example 6** | **Vertical Line Test** |

Use the vertical line test to determine which of the graphs in Figure 15 are graphs of functions.

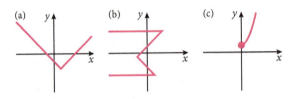

Figure 15

Solution Graphs (a) and (c) both represent functions, since any vertical line intersects each graph at most once. Graph (b) does not represent a function because a vertical line intersects the graph at more than one point.

Check It Out 6   Does the graph in Figure 16 represent a function? Explain.

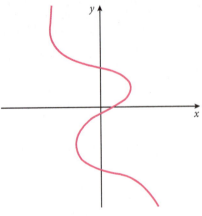

Figure 16

Intercepts and Zeros of Functions

When graphing a function, it is important to know where the graph crosses the *x*- and *y*-axes. An ***x*-intercept** is a point where the graph of a function crosses the *x*-axis. In terms of function terminology, the first coordinate of an *x*-intercept is a value of *x* such that $f(x) = 0$. The second coordinate of an *x*-intercept is 0. Values of *x* satisfying $f(x) = 0$ are called **zeros** of the function *f*. The ***y*-intercept** is a point where the graph of a function crosses the *y*-axis. Thus, the first coordinate of the *y*-intercept is 0 and its second coordinate is simply $f(0)$. Graphs of functions need not always have *x*- and *y*-intercepts.

Examining the graph of a function helps us to understand the various features of the function that may not be evident from its algebraic expression alone.

> **Example 7**   **Function Values and Intercepts from a Graph**
>
> Consider the graph of the function *f* shown in Figure 17.

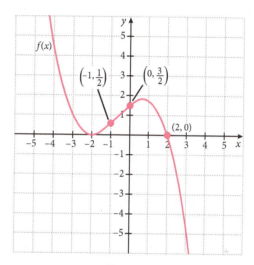

Figure 17

a. Find $f(-1), f(0)$ and $f(2)$.

b. Find the domain of *f*.

c. What are the *x*- and *y*-intercepts of the graph of *f*?

Solution

a. Since $\left(-1, \frac{1}{2}\right)$ lies on the graph of *f*, the *y*-coordinate $\frac{1}{2}$ corresponds to $f(-1)$. Thus, $f(-1) = \frac{1}{2}$. Similarly, $f(0) = \frac{3}{2}$ and $f(2) = 0$.

b. From the graph, the domain of *f* seems to be all real numbers. Unless otherwise stated, graphs are assumed to extend beyond the actual region shown.

c. The graph of *f* intersects the *x*-axis at $x = 2$ and $x = -2$. Thus, **the *x*-intercepts are $(-2, 0)$ and $(2, 0)$**. The graph of *f* intersects the *y*-axis at $y = \frac{3}{2}$ and so **the *y*-intercept is $\left(0, \frac{3}{2}\right)$**.

✓ *Check It Out 7*   In Example 7, estimate $f(1)$ and $f(-3)$.

Equations that Describe Functions

In Sections 2.1 and 2.2, you sketched graphs of lines and circles, respectively. A non-vertical line is the graph of a function since it satisfies the vertical line test, but the graph of a circle does not satisfy the vertical line test and so it does not represent a function. However, they can both be written as *equations* involving x and y. Thus, not all equations result in the variable y being a function of the variable x.

Example 8   **Equations Describing Functions**

Which of the following equations describe y as a function of x?

a. $2x - y = 1$ **b.** $|y| = x$

Solution

a. Solve the equation $2x - y = 1$ for y to get

$$y = 2x - 1$$

For each value of the variable x, there is only one value for the variable y. So y represents a function of x. The corresponding graph satisfies the vertical line test, as seen in Figure 18.

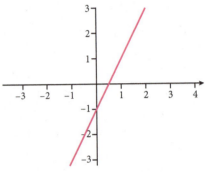

Figure 18

b. Using the definition of absolute value, solve $|y| = x$ for y to get

$$y = x \quad \text{or} \quad y = -x, x \geq 0$$

Note that $x \geq 0$ since $|y|$ cannot be negative. For each positive value of the variable x, there are two values for the variable y. So y cannot be written as a function of x. The corresponding graph does not satisfy the vertical line test, as seen in Figure 19.

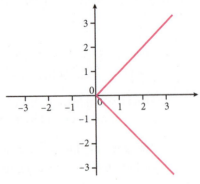

Figure 19

✔ *Check It Out 8*   Determine if the equation $y - |x| = 1$ describes y as a function of x. ■

Exercises 2.4

Just in Time Exercises

In Exercises 1–6, evaluate.

1. $-|2|$

2. $-|-4|$

3. $|-3| - 1$

4. $|(2)(-5)|$

5. $\sqrt{-(-9)}$

6. $-\sqrt{4}$

Skills

In Exercises 7–12, determine whether each of the sets of points defines a function.

7. $S = \{(-2, 1), (1, 5), (1, 2), (6, 1)\}$

8. $S = \{(-4, -1), (1, -1), (2, 0), (3, -1)\}$

9. $S = \left\{\left(-\dfrac{3}{2}, 1\right), (0, 4), (1.4, -2), (0, 1.3)\right\}$

10. $S = \left\{\left(\dfrac{2}{3}, 3\right), (6.7, 1.2), (3.1, 1.4), (4.2, 3.5)\right\}$

11. $S = \{(-5, 2.3), (-4, 3.1), (3, 2.5), (-5, 1.3)\}$

12. $S = \{(-3, -3), (-2, 2), (0, 0), (1, 1)\}$

In Exercises 13–16, fill in the table with function values for the given function. Then sketch its graph.

13.

x	-4	-2	0	2	4
$f(x) = -\dfrac{1}{2}x - 4$					

14.

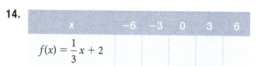

x	-6	-3	0	3	6
$f(x) = \dfrac{1}{3}x + 2$					

15.

x	0	2	$\dfrac{9}{2}$	8	18
$f(x) = \sqrt{2x}$					

16.

x	-16	-9	-4	-1	0
$f(x) = \sqrt{-x} + 3$					

In Exercises 17–46, graph the function without using a graphing utility, and determine the domain and range. Write your answer in interval notation.

17. $f(x) = 2x - 1$

18. $g(x) = -3x + 4$

19. $f(x) = 4x + 5$

20. $g(x) = -5x - 2$

21. $f(x) = -\dfrac{1}{3}x - 4$

22. $f(x) = \dfrac{3}{2}x + 3$

23. $f(x) = -2x + 1.5$

24. $f(x) = 3x - 4.5$

25. $f(x) = 4$

26. $h(x) = 7$

27. $g(x) = 4x^2$

28. $h(x) = -x^2 + 1$

29. $h(s) = -3s^2 + 4$

30. $g(s) = s^2 - 2$

31. $h(x) = \sqrt{x + 4}$

32. $g(t) = \sqrt{t - 3}$

33. $f(x) = \sqrt{3x}$

34. $f(x) = -\sqrt{3x}$

35. $f(x) = \sqrt{x} - 2$

36. $f(x) = \sqrt{x} + 1$

37. $f(x) = 2|x|$

38. $f(x) = -3|x|$

39. $f(x) = -|x|$

40. $f(x) = |x| + 4$

41. $f(x) = |-2x|$

42. $f(x) = |-4x|$

43. $f(x) = \sqrt{-2x}$

44. $f(x) = \sqrt{-3x}$

45. $h(s) = -s^3 + 1$

46. $f(x) = x^3 - 3$

In Exercises 47–54, graph the function by hand.

47. $f(x) = \begin{cases} 0 & \text{if } x \le 1 \\ 2 & \text{if } x > 1 \end{cases}$

48. $h(x) = \begin{cases} -1 & \text{if } x < 0 \\ 4 & \text{if } x \ge 0 \end{cases}$

49. $f(x) = \begin{cases} x + 2 & x \le 2 \\ 4 & x > 2 \end{cases}$

50. $f(x) = \begin{cases} 3 & x \le -1 \\ -x + 2 & x > -1 \end{cases}$

51. $g(x) = \begin{cases} x + 1 & x \le 0 \\ x & x > 0 \end{cases}$

52. $g(x) = \begin{cases} x + 1 & x < 0 \\ -x + 1 & x \ge 0 \end{cases}$

53. $f(x) = \begin{cases} x^2 & -1 \le x \le 2 \\ -2 & 2 < x \le 3 \\ x + 1 & x > 3 \end{cases}$

54. $f(x) = \begin{cases} 3 & -2 \le x \le 1 \\ -x^2 & 1 < x \le 2 \\ 5 & x > 2 \end{cases}$

In Exercises 55–58, determine whether the graph depicts the graph of a function. Explain your answer.

55.

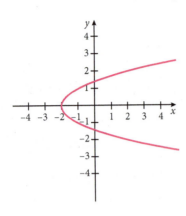

56.

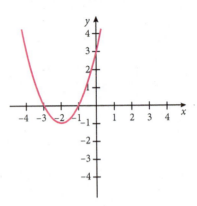

57.

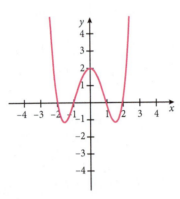

58.

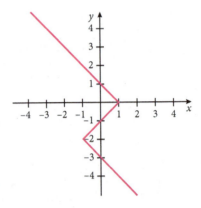

In Exercises 59–64, find the domain and range for each function whose graph is given. Write your answer in interval notation.

59.

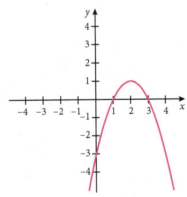

60.

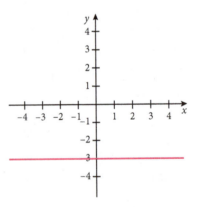

61.

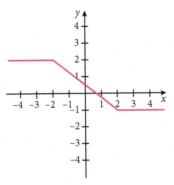

62.

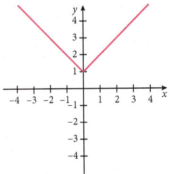

63.

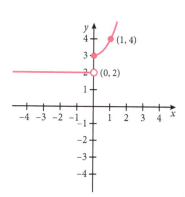

64.

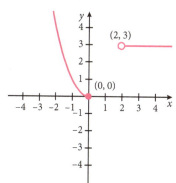

In Exercises 65–70, use a graphing utility to graph each function. Be sure to adjust your window size to see a complete graph.

65. $f(x) = 2.5|x| + 10$

66. $f(x) = -|1.4x| - 15.2$

67. $f(x) = 0.4\sqrt{0.4x - 4.5}$

68. $f(x) = 1.6\sqrt{2.6 - 0.3x}$

69. $f(x) = -3.6x^2 - 9$

70. $f(x) = 2.4x^2 + 8.5$

In Exercises 71 and 72, graph the piecewise-defined function using a graphing utility. The display should be in DOT mode.

71. $f(x) = \begin{cases} 0.5x^2 & \text{if } x \le 0 \\ -x^2 & \text{if } x > 0 \end{cases}$

72. $f(x) = \begin{cases} x^2 & \text{if } -2 \le x < 0 \\ -x + 1 & \text{if } 0 \le x < 2.5 \\ x - 3.5 & \text{if } x \ge 2.5 \end{cases}$

In Exercises 73–78, for each function f given by its graph, approximate the following.

a. $f(-1), f(0),$ and $f(2)$ **b.** The domain of f **c.** x- and y-intercepts of the graph of f

73.

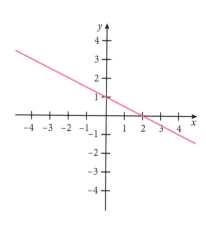

74.

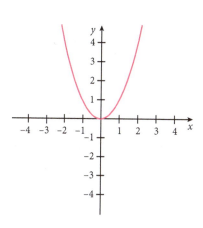

75.

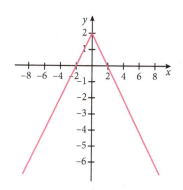

76.

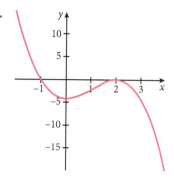

77.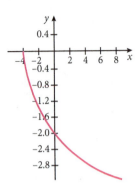

78.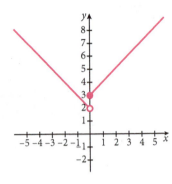

In Exercises 79–84, determine if the equation defines y as a function of x.

79. $3x + y - 1 = 0$

80. $-2x - y + 5 = 0$

81. $y^2 = x$

82. $2y = x^2$

83. $y + 2x^2 = 1$

84. $x^2 + y^2 = 1$

Applications

85. NASA Budget The following graph gives the budget for the National Aeronautics and Space Administration (NASA) for the years 2008–2013.   (*Source*: NASA)

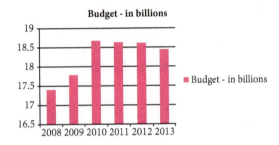

a. Estimate the amount allotted for 2010.

b. By approximately how much did the budget increase from 2009 to 2010?

86. Geometry The area of a circle of radius r is given by the function $A(r) = \pi r^2$. Sketch a graph of the function A using values of $r > 0$. Why are only positive values of r used?

87. Geometry The volume of a sphere of radius r is given by the function $V(r) = \frac{4}{3}\pi r^3$. Sketch the graph of the function V using values of $r > 0$. Why are only positive values of r used?

88. Rate and Distance The time it takes for a person to row a boat 50 miles is given by $T(s) = \frac{50}{s}$, where s is the speed at which the boat is rowed. For what values of s does this function make sense? Sketch the graph of the function T using these values of s.

89. Rate and Distance A car travels 55 miles per hour. Find and graph the distance traveled by the car (in miles) as a function of the time (in hours). For what values of the input variable will your function values be greater than or equal to zero?

90. Construction Costs It costs $85 per square foot of area to build a house. Find and graph the total cost of building a house as a function of area (in square feet).

 91. Ecology A coastal region with an area of 250 square miles in 2009 has been losing 2.5 square miles of land per year due to erosion. Thus, the area A of the region t years after 2009 is $A(t) = 250 - 2.5t$.

a. Sketch the graph of A for $0 \le t \le 100$.

b. What does the t-intercept represent for this problem?

92. **Environmental Science** The sports utility vehicle (SUV) Toyota Highlander emits 5.2 tons of greenhouse gases per year (2013 model), while the SUV Volvo XC90 releases 7.8 tons of greenhouse gases per year (2013 model). (*Source:* www.fueleconomy.gov)

   a. Express the amount of greenhouse gases released by the Highlander as a function of time and graph the function. What are the units of the input and output variables?

   b. On the same set of coordinate axes as in part (a), graph the amount of greenhouse gases released by the Volvo XC90 as a function of time.

   c. Compare the two graphs. What do you observe?

Concepts

In Exercise 93–96, graph the pair of functions on the same set of coordinate axes and explain the differences between the two graphs.

93. $f(x) = 2$ and $g(x) = 2x$

94. $h(x) = -2x$ and $g(x) = 2x$

95. $f(x) = -3x^2$ and $g(x) = 3x^2$

96. $f(x) = 3x^2$ and $g(x) = 3x^2 + 1$

97. Explain why the graph of $x = |y|$ is not a function.

98. Is the graph of $f(x) = \sqrt{x + 4}$ the same as the graph of $g(x) = \sqrt{x} + 4$? Explain by sketching their graphs.

In Exercises 99–102, graph the pair of functions on the same set of coordinate axes and find their respective ranges.

99. $f(x) = |x|, g(x) = |x| - 3$

100. $f(x) = x^2 + 4, g(x) = x^2 - 4$

101. $f(x) = \sqrt{2x}, g(x) = \sqrt{x}$

102. $f(x) = 3x + 4, g(x) = 3x + 7$

Analyzing the Graph of a Function

2.5

In this section, we study further properties of functions that will be useful in later chapters. These properties will provide additional tools for understanding functions and their graphs.

Even and Odd Functions

You may have observed that the graph of $f(x) = x^2$ is a mirror image of itself when reflected across the y-axis. This is referred to as **symmetry with respect to the y-axis**. See Figure 1.

Another type of symmetry that occurs is defined as **symmetry with respect to the origin**. The graph of $f(x) = x^3$ displays this type of symmetry. See Figure 2.

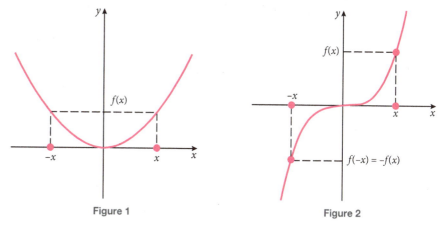

Figure 1   Figure 2

Observations

- In Figure 1, $f(x) = x^2$. The points $(x, f(x))$ and $(-x, f(-x))$ are symmetric with respect to the y-axis, and $f(x) = f(-x)$. For example $f(-3) = (-3)^2 = 9$ and $f(3) = 3^2 = 9$.

- In Figure 2, $f(x) = x^3$. The points $(x, f(x))$ and $(-x, f(-x))$ are symmetric with respect to the origin, and $f(-x) = -f(x)$.

The following definitions extend our observations.

Definition of an Even Function

The **domain** of a function is the set of all input values for which the function will produce a real number.

A function is **symmetric with respect to the y-axis** if

$$f(x) = f(-x), \text{ for each } x \text{ in the domain of } f$$

Functions with this property are called **even functions**.

Definition of an Odd Function

A function is **symmetric with respect to the origin** if

$$f(-x) = -f(x), \text{ for each } x \text{ in the domain of } f$$

Functions with this property are called **odd functions**.

Observations:

- A function cannot be both odd and even at the same time unless it is the function $f(x) = 0$ for all x.

- There are various other symmetries besides the ones we have discussed here. For example, a function can be symmetric with respect to a vertical line other than the y-axis or with respect to a point other than the origin. However, these types of symmetries are beyond the scope of the current discussion.

Objectives

- Determine if a function is even, odd, or neither.

- Given a graph, determine intervals where a function is increasing, decreasing, or constant.

- Determine the average rate of change of a function over an interval.

- Find the difference quotient of a function.

Using Technology

You can graph a function to see if it is odd, even, or neither and then check your conjecture algebraically. The graph of $h(x) = -x^3 + 3x$ seems as though it is symmetric with respect to the origin (Figure 3). This is checked algebraically in Example 1(c).

Keystroke Appendix:
Section 7

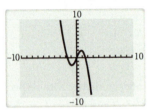

Figure 3

Example 1   **Determining Odd or Even Functions**

Using the definitions of odd and even functions, classify the following functions as odd, even, or neither.

a. $f(x) = |x| + 2$ **b.** $g(x) = (x - 4)^2$ **c.** $h(x) = -x^3 + 3x$

Solution

a. First check to see if f is an even function:

$$f(-x) = |-x| + 2 = |-1||x| + 2 = |x| + 2 = f(x)$$

Since $f(x) = f(-x)$, **f is an even function.** The graph of f is symmetric with respect to the y-axis, as shown in Figure 4. This graph verifies what we found by the use of algebra alone.

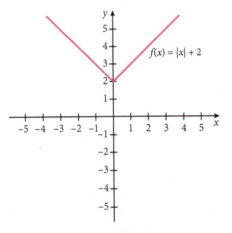

Figure 4

b. For $g(x) = (x - 4)^2$, it would help to first expand the expression for the function, which gives

$$g(x) = (x - 4)^2 = x^2 - 8x + 16$$

We can then see that

$$g(-x) = (-x)^2 - 8(-x) + 16 = x^2 + 8x + 16$$

Since $g(x) \neq g(-x)$, g is not even. Using the expression for $g(-x)$, which we have already found, we see that

$$g(-x) = x^2 + 8x + 16 \quad \text{and} \quad -g(x) = -(x^2 - 8x + 16) = -x^2 + 8x - 16$$

Since $g(-x) \neq -g(x)$, g is not odd. The fact that this function is neither even nor odd can also be seen from its graph. Figure 5 shows no symmetry with respect to the y-axis or with respect to the origin.

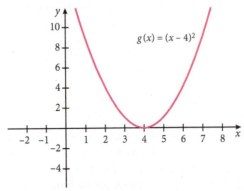

Figure 5

c. Since $h(x)$ has an odd-powered term, we will first check to see if it is an odd function:

$$h(-x) = -(-x)^3 + 3(-x) = -(-x^3) - 3x = -(-x^3 + 3x) = -h(x)$$

Since $h(-x) = -h(x)$, **h is an odd function**. Thus, it is symmetric with respect to the origin as verified by the graph in Figure 6.

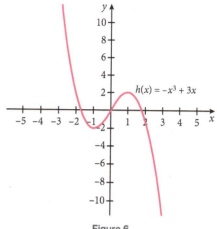

Figure 6

  Check It Out 1 Decide whether the following functions are even, odd, or neither.

a. $h(x) = 2|x|$

b. $f(x) = (x + 1)^2$

Increasing and Decreasing Functions

An important idea in studying functions is figuring out how the function value, $f(x)$, changes as x changes. You should already have some intuitive ideas about this quality of a function. The following definition about increasing and decreasing functions makes these ideas precise.

> ### Increasing, Decreasing, and Constant Functions
>
> - A function f is **increasing** on an open interval I if, for any a, b in the interval with $a < b$, we have $f(a) < f(b)$. See Figure 7.
>
> - A function f is **decreasing** on an open interval I if, for any a, b in the interval with $a < b$, we have $f(a) > f(b)$. See Figure 8.
>
> - A function f is **constant** on an open interval I if, for any a, b in the interval, we have $f(a) = f(b)$. See Figure 9.

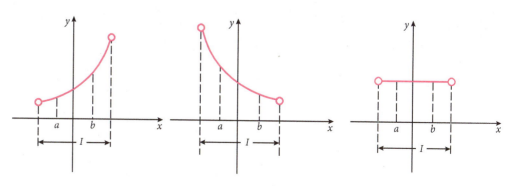

Figure 7   Figure 8   Figure 9

| **Example 2**   **Increasing and Decreasing Functions**

For the function f given in Figure 10, find the interval(s) on which the following is true:

a. f is increasing.   **b.** f is decreasing.   **c.** f is constant.

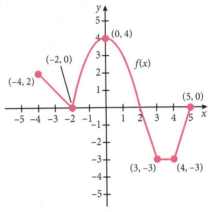

Figure 10

Solution

a. From the graph, the function is **increasing on the intervals $(-2, 0)$ and $(4, 5)$.**

b. From the graph, the function is **decreasing on the intervals $(-4, -2)$ and $(0, 3)$.**

c. From the graph, the function is **constant on the interval $(3, 4)$.**

✓ *Check It Out 2*   Find the interval(s) where f is decreasing for f given in Figure 11.

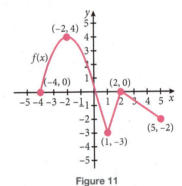

Figure 11

In Chapters 3 and 4, we will discuss increasing and decreasing functions in more detail. In those chapters, we will also examine points where the graph of a function "turns" from increasing to decreasing or vice versa.

Average Rate of Change

While determining whether a function is increasing or decreasing is of some value, it is of greater interest to figure out how quickly a function changes. One quantity that tells us how quickly a function changes is called the **average rate of change**.

| **Average Rate of Change**

The average rate of change of a function f on an interval $[x_1, x_2]$ is given by

$$\text{Average rate of change} = \frac{f(x_2) - f(x_1)}{x_2 - x_1}$$

Example 3   **Determining the Average Rate of Change**

Find the average rate of change of $f(x) = 2x^2 + 1$ on the following intervals:

a. $[-3, -2]$ **b.** $[0, 2]$

Solution

a. Using $x_1 = -3$ and $x_2 = -2$ in the definition of the average rate of change, we have

$$\text{Average rate of change} = \frac{f(x_2) - f(x_1)}{x_2 - x_1}$$

$$= \frac{f(-2) - f(-3)}{-2 - (-3)} = \frac{9 - 19}{1} = -10$$

b. Using $x_1 = 0$ and $x_2 = 2$ in the definition of the average rate of change, we have

$$\text{Average rate of change} = \frac{f(x_2) - f(x_1)}{x_2 - x_1} = \frac{f(2) - f(0)}{2 - 0}$$

$$= \frac{9 - 1}{2} = \frac{8}{2} = 4$$

Note that the average rate of change in part (a) is negative, but in part (b), it is positive. In this example, the average rate of change varies, depending on the interval chosen.

 Check It Out 3   Find the average rate of change of $f(x) = 2x^2 + 1$ on the interval $[3, 4]$. ◾

Difference Quotient

An interval $[a, b]$ can be rewritten in the form $[a, a + h]$, where $h = b - a$. Thus, the average rate of change of f on the interval $[a, a + h]$ is

$$\frac{f(a + h) - f(a)}{(a + h) - a} = \frac{f(a + h) - f(a)}{h}$$

This quantity is known as a **difference quotient** and it plays an important role in calculus.

Example 4   **Rate of Change and Difference Quotient**

A ball is dropped from a height of 100 meters. At time t, in seconds, the height of the ball from the ground is given by $f(t) = -4.9t^2 + 100$. Calculate the following.

a. The average rate of change of f on the interval $[1, 3]$

b. The difference quotient $\dfrac{f(a + h) - f(a)}{h}$

Solution

a. We first calculate $f(1)$ and $f(3)$:

$$f(1) = -4.9(1)^2 + 100 = 95.1$$

$$f(3) = -4.9(3)^2 + 100 = 55.9$$

Using $t_1 = 1$ and $t_2 = 3$ in the definition of the average rate of change, we have

$$\text{Average rate of change} = \frac{f(t_2) - f(t_1)}{t_2 - t_1}$$

$$= \frac{f(3) - f(1)}{3 - 1} = \frac{55.9 - 95.1}{2} = -19.6$$

The average rate of change of the height on the interval $[1, 3]$ is $-19.6 \frac{\text{m}}{\text{sec}}$.

b. We first calculate $f(a + h)$ and $f(a)$:

$$f(a + h) = -4.9(a + h)^2 + 100 = -4.9(a^2 + 2ah + h^2) + 100$$

$$= -4.9a^2 - 9.8ah - 4.9h^2 + 100$$

$$f(a) = -4.9a^2 + 100$$

$$\text{Difference quotient} = \frac{f(a + h) - f(a)}{h}$$

$$= \frac{(-4.9a^2 - 9.8ah - 4.9h^2 + 100) - (-4.9a^2 + 100)}{h}$$

$$= \frac{-4.9a^2 - 9.8ah - 4.9h^2 + 100 + 4.9a^2 - 100}{h} \qquad \text{Distribute negative sign}$$

$$= \frac{-9.8ah - 4.9h^2}{h} = -9.8a - 4.9h \qquad \text{Simplify numerator and divide out } h$$

The difference quotient is given in terms of a and h, since we did not specify their numerical values. You will explore the significance of this quantity in a calculus course.

✓ *Check It Out 4*   In Example 4, find the rate of change of f on the interval $[0, 2]$.   ■

We conclude this section with facts about lines and linear functions. Since the equation $y = mx + b$ defines y as a function of x, we can write $f(x) = mx + b$. This is called a **linear function**. Its graph is just the line with slope m and y-intercept $(0, b)$. Unlike the functions in Examples 3 and 4, the average rate of change of a linear function is the same, on *any* interval, as shown below.

$$\text{Rate of change} = \frac{f(x_2) - f(x_1)}{x_2 - x_1} \qquad \text{Definition of rate of change}$$

$$= \frac{mx_2 + b - (mx_1 + b)}{x_2 - x_1} \qquad \text{Substitute } x_2, x_1 \text{ into } f(x)$$

$$= \frac{mx_2 + b - mx_1 - b}{x_2 - x_1} \qquad \text{Remove parentheses}$$

$$= \frac{mx_2 - mx_1}{x_2 - x_1} \qquad \text{Simplify}$$

$$= \frac{m(x_2 - x_1)}{x_2 - x_1} \qquad \text{Factor out } m$$

$$= m \qquad \text{Cancel the term } (x_2 - x_1), \text{ since } x_2 \neq x_1$$

Thus, we see that for a *linear* function, the rate of change is exactly the slope m, regardless of the values of x_1 and x_2. This connection between the slope and the average rate of change is very important in calculus.

Skills

In Exercises 1–6, classify each function given by its graph as odd, even, or neither.

1.

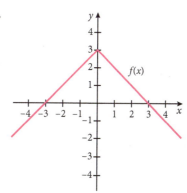

2.

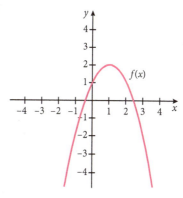

3.

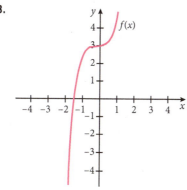

4.

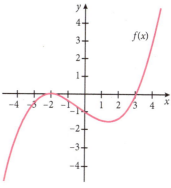

5.

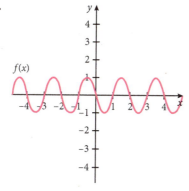

6.

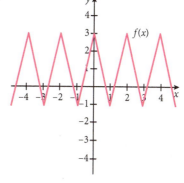

Questions 7–13 apply to the function f whose graph is as follows.

7. Find the domain of *f*.

8. Find the range of *f*.

9. Find the *y*-intercept.

10. Find the interval(s) on which *f* is increasing.

11. Find the interval(s) on which *f* is decreasing.

12. Find the interval(s) on which *f* is constant.

13. Is this function odd, even, or neither?

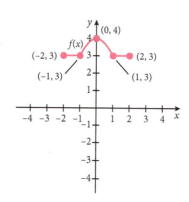

Questions 14–21 apply to the function f whose graph is as follows.

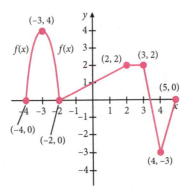

14. Find the interval(s) on which f is increasing.

15. Find the interval(s) on which f is decreasing.

16. Find the interval(s) on which f is constant.

17. Find the y-intercept.

18. Find the average rate of change of f on the interval $[-3, -2]$.

19. Find the average rate of change of f on the interval $[-2, 2]$.

20. Find the average rate of change of f on the interval $[2, 3]$.

21. Find the average rate of change of f on the interval $[4, 5]$.

Questions 22–26 apply to the function f whose graph is as follows.

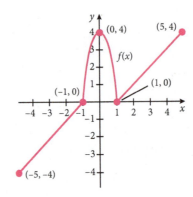

22. Find the interval(s) on which f is increasing.

23. Find the interval(s) on which f is decreasing.

24. Find the average rate of change of f on the interval $[-1, 0]$.

25. Find the average rate of change of f on the interval $[0, 1]$.

26. Is this function odd, even, or neither?

In Exercises 27–38, decide if each function is odd, even, or neither, using the definitions.

27. $f(x) = x + 3$

28. $f(x) = -2x$

29. $f(x) = |3x| - 2$

30. $f(x) = -4x^2 + x$

31. $f(x) = -3x^2 + 1$

32. $f(x) = -x^3 + 1$

33. $f(x) = -|x| + 1$

34. $f(x) = 2x$

35. $f(x) = |x - 1|$

36. $f(x) = x^5 - 2x$

37. $f(x) = (x^2 + 1)(x - 1)$

38. $f(x) = (x^2 - 3)(x^2 - 4)$

In Exercises 39–54, find the average rate of change of each function on the given interval.

39. $f(x) = -2x^2 + 5$, interval: $[-2, -1]$

40. $f(x) = 3x^2 - 1$, interval: $[2, 3]$

41. $f(x) = x^3 + 1$, interval: $[0, 2]$

42. $f(x) = -2x^3$, interval: $[-2, 0]$

43. $f(x) = 2x^2 + 3x - 1$, interval: $[-2, -1]$

44. $f(x) = 3x^3 + x^2 + 4$, interval: $[-2, 0]$

45. $f(x) = -x^4 + 6x^2 - 1$, interval: $[1, 2]$

46. $f(x) = -4x^3 - 3x^2 - 1$, interval: $[0, 2]$

47. $f(x) = 2|x| + 4$, interval: $[3, 5]$

48. $f(x) = |x| - 5$, interval: $[-4, -2]$

49. $f(x) = \sqrt{-x}$, interval: $[-4, -3]$

50. $f(x) = \sqrt{x + 3}$, interval: $[2, 4]$

51. $f(x) = 2x - 4$, interval: $[a, a + h]$

52. $f(x) = -3x + 1$, interval: $[a, a + h]$

53. $f(x) = x^2 - 1$, interval: $[a, a + h]$

54. $f(x) = -2x^2$, interval: $[a, a + h]$

In Exercises 55–60, use a graphing utility to decide if the function is odd, even, or neither.

55. $f(x) = x^2 - 4x + 1$

56. $f(x) = -2x^2 + 2x + 3$

57. $f(x) = 2x^3 - x$

58. $f(x) = (x + 1)(x - 2)(x + 3)$

59. $f(x) = x^4 - 5x^2 + 4$

60. $f(x) = -x^4 + 4x^2$

Applications

61. Demand Function The demand for a product, in thousands of units, is given by $d(x) = \frac{100}{x}$, where x is the price of the product. Is this an increasing or a decreasing function? Explain.

62. Revenue The revenue for a company is given by $R(x) = 30x$, where x is the number of units sold, in thousands. Is this an increasing or a decreasing function? Explain.

63. Depreciation The value of a computer t years after purchase is given by $v(t) = 2000 - 300t$, where $v(t)$ is in dollars. Find the average rate of change of the value of the computer on the interval $[0, 3]$ and interpret it, explaining why the average rate of change of v on any interval will be the same.

64. Stamp Collecting The value of a commemorative stamp t years after purchase appreciates according to the function $v(t) = 0.37 + 0.05t$, where $v(t)$ is in dollars. Find the average rate of change of the value of the stamp in the interval $[0, 4]$ and interpret it.

65. Commerce The following table lists the annual sales of DVDs at a small store for selected years.

Year	Number of units sold
2007	10,000
2009	30,000
2010	33,000

Find the average rate of change in sales from 2007 to 2009. Also, find the average rate of change in sales from 2009 to 2010. Is the average rate of change the same for the two intervals?

Concepts

66. Fill in the following table for $f(x) = 3x^2 - 2$.

Interval	Average rate of change
[1, 2]	
[1, 1.1]	
[1, 1.05]	
[1, 1.01]	
[1, 1.001]	

What do you notice about the average rate of change as the right endpoint of the interval gets closer to the left endpoint of the interval?

67. Fill in the following table for $f(x) = -x^2 + 1$.

Interval	Average rate of change
[1, 2]	
[1.9, 2]	
[1.95, 2]	
[1.99, 2]	
[1.999, 2]	

What do you notice about the average rate of change as the left endpoint of the interval gets closer to the right endpoint of the interval?

68. Suppose f is constant on an interval $[a, b]$. Show that the average rate of change of f on $[a, b]$ is zero.

69. If the average rate of change of a function in an interval is zero, does that mean that the function is constant on that interval?

70. Let f be decreasing on an interval (a, b). Show that the average rate of change of f on $[c, d]$ is negative, where $a < c < d < b$.

The Algebra of Functions

2.6

Objectives

- Find the sum or difference of two functions and the corresponding domain.

- Find the product or quotient of two functions and the corresponding domain.

- Find the composition of functions.

- Find the domain of a composite function.

- Write a function as the composition of two functions.

- Calculate the difference quotient.

Multinational firms, such as the courier service DHL, must deal with conversions between different currencies and/or different systems of weights and measures. Computations involving these conversions use "operations" on functions—operations such as addition and composition. These topics will be studied in detail in this section. They have a wide variety of uses in both practical and theoretical situations.

Arithmetic Operations on Functions

We can add, subtract, multiply, or divide two functions. These operations are defined as follows.

> **Arithmetic Operations on Functions**
>
> Given functions f and g, for each x in the domain of *both* f and g, the sum, difference, product, and quotient of f and g are defined as follows.
>
> $$(f + g)(x) = f(x) + g(x)$$
> $$(f - g)(x) = f(x) - g(x)$$
> $$(fg)(x) = f(x) \cdot g(x)$$
> $$\left(\frac{f}{g}\right)(x) = \frac{f(x)}{g(x)}, \quad \text{where } g(x) \neq 0$$

In Example 1, we will compute the composition of functions that are given by algebraic expressions.

VIDEO EXAMPLES

SECTION 2.6

Example 1   **Arithmetic Operations on Two Functions**

Let f and g be two functions defined as follows:

$$f(x) = \frac{2}{x - 4} \quad \text{and} \quad g(x) = 3x - 1$$

Find the following and determine the domain of each.

a. $(f + g)(x)$ **b.** $(f - g)(x)$ **c.** $(fg)(x)$ **d.** $\left(\dfrac{f}{g}\right)(x)$

Solution

a. From the definition of the sum of two functions,

$$(f + g)(x) = f(x) + g(x) = \frac{2}{x - 4} + 3x - 1$$

We can simplify as follows.

$$(f + g)(x) = \frac{2}{x - 4} + (3x - 1)$$

$$= \frac{2 + (3x - 1)(x - 4)}{x - 4} \quad \text{LCD is } x - 4$$

$$= \frac{2 + (3x^2 - 13x + 4)}{x - 4}$$

$$= \frac{3x^2 - 13x + 6}{x - 4} \quad \text{Combine like terms}$$

Just in Time
Review rational
expressions in
Section 1.4.

The domain of f is $(-\infty, 4) \cup (4, \infty)$ and the domain of g consists of all real numbers. Thus, **the domain of $f + g$ consists of $(-\infty, 4) \cup (4, \infty)$**, since these values are in the domains of both f and g.

b. $(f - g)(x) = f(x) - g(x) = \dfrac{2}{x - 4} - (3x - 1)$

$\qquad = \dfrac{2 - (3x - 1)(x - 4)}{x - 4}$ LCD is $x - 4$

$\qquad = \dfrac{2 - (3x^2 - 13x + 4)}{x - 4}$

$\qquad = \dfrac{-3x^2 + 13x - 2}{x - 4}$ Collect like terms; be careful with the negative sign

The **domain of $f - g$ consists of $(-\infty, 4) \cup (4, \infty)$**, since these values are in the domains of both f and g.

c. $(fg)(x) = f(x) \cdot g(x) = \left(\dfrac{2}{x - 4}\right)(3x - 1) = \dfrac{6x - 2}{x - 4}$

The **domain of fg consists of $(-\infty, 4) \cup (4, \infty)$**, since these values are in the domains of both f and g.

d. $\left(\dfrac{f}{g}\right)(x) = \dfrac{f(x)}{g(x)} = \dfrac{\dfrac{2}{x - 4}}{3x - 1}$

$\qquad = \dfrac{2}{x - 4} \cdot \dfrac{1}{3x - 1} = \dfrac{2}{(x - 4)(3x - 1)}$

The set of x values common to both functions is $(-\infty, 4) \cup (4, \infty)$. In addition, $g(x) = 0$ for $x = \frac{1}{3}$. Thus, **the domain of $\frac{f}{g}$ is the set of all real numbers such that $x \neq 4$ and $x \neq \frac{1}{3}$**. In interval notation, we have

$$\left(-\infty, \frac{1}{3}\right) \cup \left(\frac{1}{3}, 4\right) \cup (4, \infty)$$

 Check It Out 1   Let f and g be two functions defined as follows:

$$f(x) = \frac{1}{x + 2} \quad \text{and} \quad g(x) = 2x + 1$$

a. Find $(f + g)(x)$.   **b.** Find $\left(\dfrac{f}{g}\right)(x)$.

Once we find the arithmetic combination of two functions, we can evaluate the new function at any point in its domain, as illustrated in the following example.

Example 2   **Evaluating a Sum, Product, and a Quotient of Two Functions**

Let f and g be two functions defined as follows:

$$f(x) = 2x^2 - x \quad \text{and} \quad g(x) = \sqrt{x + 1}$$

Evaluate the following.

a. $(f + g)(-1)$ **b.** $\left(\dfrac{f}{g}\right)(3)$ **c.** $(fg)(3)$

Solution

a. Compute the sum of the two functions as follows.

$$(f + g)(x) = f(x) + g(x) = 2x^2 - x + \sqrt{x + 1}$$

Next, evaluate $(f + g)(x)$ at $x = -1$.

$$(f + g)(-1) = 2(-1)^2 - (-1) + \sqrt{-1 + 1}$$

$$= 2 - (-1) + 0 = 3$$

b. Compute the quotient of the two functions as follows:

$$\left(\frac{f}{g}\right)(x) = \frac{f(x)}{g(x)} = \frac{2x^2 - x}{\sqrt{x + 1}}$$

Next, evaluate $\left(\dfrac{f}{g}\right)(x)$ at $x = 3$.

$$\left(\frac{f}{g}\right)(3) = \frac{2(3)^2 - (3)}{\sqrt{(3) + 1}}$$

$$= \frac{2(9) - 3}{2}$$

$$= \frac{15}{2}$$

c. Compute the product of the two functions as follows:

$$(fg)(x) = f(x) \cdot g(x) = (2x^2 - x)(\sqrt{x + 1})$$

Thus, $(fg)(3) = (2(3)^2 - 3)(\sqrt{3 + 1}) = 30$.

 Check It Out 2   Let f and g be two functions defined as follows:

$$f(x) = 3x - 4 \quad \text{and} \quad g(x) = x^2 + x$$

Evaluate the following:

a. $(f + g)(2)$ **b.** $\left(\dfrac{f}{g}\right)(2)$

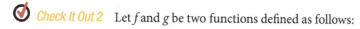

The next example shows an application of arithmetic involving functions.

Example 3   **Calculating Profit**

The GlobalEx Corporation has revenues modeled by the function $R(t) = 40 + 2t$, where t is the number of years since 2010 and $R(t)$ is in millions of dollars. Its operating costs are modeled by the function $C(t) = 35 + 1.6t$, where t is the number of years since 2010 and $C(t)$ is in millions of dollars. Find the profit function $P(t)$ for GlobalEx Corporation.

Solution Since profit is equal to revenue minus cost, we can write

$$P(t) = R(t) - C(t)$$

Substituting the expressions for $R(t)$ and $C(t)$ gives

$$P(t) = (40 + 2t) - (35 + 1.6t) = 40 + 2t - 35 - 1.6t = 5 + 0.4t$$

Thus, the profit function is $P(t) = 5 + 0.4t$, where t is the number of years since 2010.

 Check It Out 3   Find the profit function for GlobalEx Corporation in Example 3 if the revenue and cost functions are given by $R(t) = 42 + 2.2t$ and $C(t) = 34 + 1.5t$.

Composition of Functions

When converting between currencies or weights and measures, a function expressed in terms of a given unit must be restated in terms of a new unit. For example, if profit is given in terms of dollars, another function must convert the profit function to a different currency. Successive evaluation of this series of two functions, is known as a **composition of functions** and is of both practical and theoretical importance.

Definition of a Composite Function

The **composition of functions** f and g is a function that is denoted by $f \circ g$ and defined as

$$(f \circ g)(x) = f(g(x))$$

The domain of $f \circ g$ is the set of all x in the domain of g such that $g(x)$ is in the domain of f.

The function $f \circ g$ is called a **composite function**.

Example 4   **Finding and Evaluating Composite Functions**

Let $f(s) = s^2 + 1$ and $g(s) = -2s$.

a. Find an expression for $(f \circ g)(s)$ and give the domain of $f \circ g$.

b. Find an expression for $(g \circ f)(s)$ and give the domain of $g \circ f$.

c. Evaluate $(f \circ g)(-1)$.

d. Evaluate $(g \circ f)(-1)$.

Solution

a. The composite function $(f \circ g)$ is defined as $(f \circ g)(s) = f(g(s))$. Computing the quantity $f(g(s))$ is often the most confusing part. To make things easier, think of $f(s)$ as $f(\square)$, where the box can contain anything. Then proceed as follows:

$$f(\square) = (\square)^2 + 1 \qquad \text{Definition of } f$$

$$f(\boxed{g(s)}) = (\boxed{g(s)})^2 + 1 \qquad \text{Place } g(s) \text{ in the box}$$

$$= (\boxed{-2s})^2 + 1 \qquad \text{Substitute expression for } g(s)$$

$$= 4s^2 + 1 \qquad \text{Simplify: } (-2s)^2 = 4s^2$$

Thus, $(f \circ g)(s) = 4s^2 + 1$. Since the domain of g is all real numbers and the domain of f is also all real numbers, the domain of $f \circ g$ is all real numbers, $(-\infty, \infty)$.

b. The composite function $g \circ f$ is defined as $(g \circ f)(s) = g(f(s))$. Now, we compute $g(f(s))$. Thinking of $g(s)$ as $g(\square)$, we have

$$g(\square) = -2(\square) \qquad \text{Definition of } g$$

$$g(\boxed{f(s)}) = -2(\boxed{f(s)}) \qquad \text{Place } f(s) \text{ in the box}$$

$$= -2(\boxed{s^2 + 1}) \qquad \text{Substitute expression for } f(s)$$

$$= -2s^2 - 2 \qquad \text{Simplify}$$

Thus, $(g \circ f)(s) = -2s^2 - 2$. Note that $(f \circ g)(s)$ is **not** equal to $(g \circ f)(s)$. Since the domain of f is all real numbers and the domain of g is also all real numbers, the domain of $g \circ f$ is all real numbers, $(-\infty, \infty)$.

c. Since $(f \circ g)(s) = 4s^2 + 1$, the value of $(f \circ g)(-1)$ is

$$(f \circ g)(\mathbf{-1}) = 4(\mathbf{-1})^2 + 1 = 4(1) + 1 = 4 + 1 = 5$$

Alternatively, using the expressions for the individual functions,

$$(f \circ g)(-1) = f(g(-1))$$

$$= f(2) \qquad g(-1) = -2(-1) = 2$$

$$= 5 \qquad f(2) = 2^2 + 1 = 4 + 1 = 5$$

d. Since $(g \circ f)(s) = -2s^2 - 2$, the value of $(g \circ f)(-1)$ is

$$(g \circ f)(\mathbf{-1}) = -2(\mathbf{-1})^2 - 2 = -2(1) - 2 = -2 - 2 = -4$$

Alternatively, using the expressions for the individual functions,

$$(g \circ f)(-1) = g(f(-1))$$

$$= g(2) \qquad f(-1) = (-1)^2 + 1 = 1 + 1 = 2$$

$$= -4 \qquad g(2) = -2(2) = -4$$

✅ *Check It Out 4*   Let $f(s) = s^2 - 2$ and $g(s) = 3s$. Evaluate $(f \circ g)(-1)$ and $(g \circ f)(-1)$.   ◼

Using Technology

For Example 5(a), enter
$Y_1(x) = \frac{1}{x}$ and
$Y_2(x) = x^2 - 1$. Then
$Y_3(x) = Y_1(Y_2(x))$ defines the composite function
$Y_1 \circ Y_2$. Note that the table of values gives "ERROR" as the value of $Y_3(x)$ for $x = -1$
and $x = 1$. These numbers are not in the domain of
$Y_3 = Y_1 \circ Y_2$. See Figure 1.

Keystroke Appendix:

Sections 4 and 6

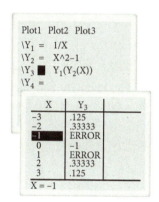

Figure 1

Note In some instances, there can be more than one way to write a function as a composition of two functions.

The domain of definition of a composite function can differ from the domain of either or both of the two functions from which it is composed, as illustrated in Example 5.

Example 5   **Domain of Composite Functions**

Let $f(x) = \frac{1}{x}$ and $g(x) = x^2 - 1$.

a. Find $f \circ g$ and its domain.   **b.** Find $g \circ f$ and its domain.

Solution We first note that the domain of f is $(-\infty, 0) \cup (0, \infty)$ and the domain of g is all real numbers.

a. To find $f \circ g$, proceed as follows:

$$(f \circ g)(x) = f(g(x)) \quad \text{Definition of } f \circ g$$
$$= f(x^2 - 1) \quad \text{Substitute expression for } g(x)$$
$$= \frac{1}{x^2 - 1} \quad \text{Use definition of } f$$

The domain of $f \circ g$ is the set of all x in the domain of g such that $g(x)$ is in the domain of f. The domain of f is $(-\infty, 0) \cup (0, \infty)$. Therefore, every value output by g must be a number other than 0. Thus, we find the numbers for which $x^2 - 1 = 0$, and then *exclude* them. By factoring,

$$x^2 - 1 = (x + 1)(x - 1)$$

Hence

$$x^2 - 1 = 0 \Rightarrow x + 1 = 0 \text{ or } x - 1 = 0 \Rightarrow x = -1 \text{ or } x = 1$$

The domain of $f \circ g$ is $(-\infty, -1) \cup (-1, 1) \cup (1, \infty)$.

b. To find $g \circ f$, proceed as follows.

$$(g \circ f)(x) = g(f(x)) \quad\quad\quad \text{Definition of } g \circ f$$
$$= g\!\left(\frac{1}{x}\right) \quad\quad\quad\quad \text{Substitute expression for } f(x)$$
$$= \left(\frac{1}{x}\right)^2 - 1 = \frac{1}{x^2} - 1 \quad \text{Use definition of } g$$

The domain of $g \circ f$ is the set of all x in the domain of f such that $f(x)$ is in the domain of g. Since the domain of g is all real numbers, any value output by f is acceptable. Thus, by definition, the domain of $g \circ f$ is the set of all x in the domain of f, which is $(-\infty, 0) \cup (0, \infty)$.

 Check It Out 5 Let $f(x) = \sqrt{x}$ and $g(x) = x + 1$. Find $f \circ g$ and its domain.   ■

The next example illustrates how a given function can be written as a composition of two other functions.

Example 6   **Writing a Function as a Composition**

If $h(x) = \sqrt{3x^2 - 1}$, find two functions f and g such that $h(x) = (f \circ g)(x) = f(g(x))$.

Solution The function h takes the square root of the quantity $3x^2 - 1$. Since $3x^2 - 1$ must be calculated before its square root can be taken, let $g(x) = 3x^2 - 1$. Then let $f(x) = \sqrt{x}$. Thus, two functions that can be used for the composition are

$$f(x) = \sqrt{x} \text{ and } g(x) = 3x^2 - 1$$

We check this by noting that

$$h(x) = (f \circ g)(x) = f(g(x))$$
$$= f(3x^2 - 1) = \sqrt{3x^2 - 1}$$

 Check It Out 6 If $h(x) = (x^3 + 9)^5$, find two functions f and g such that $h(x) = (f \circ g)(x) = f(g(x))$.

The next example illustrates an application of composite function.

Example 7   **Evaluating Two Functions Successively with Tables**

The cost of fuel for running a fleet of vehicles owned by GlobalEx Corporation is given in Table 1 in terms of the number of gallons used. However, the European branch of GlobalEx Corporation records its fuel consumption in units of liters. So Table 2 lists the equivalent quantity of fuel in gallons.

a. Find the cost of 55 gallons of fuel.   **b.** Find the cost of 113.55 liters of fuel.

Solution

a. From Table 1, it is clear that 55 gallons of fuel costs $82.50.

b. For 113.55 liters of fuel, we must find the equivalent quantity of fuel in gallons before looking up the price.

$$\text{Quantity (liters)} \rightarrow \text{Quantity (gallons)} \rightarrow \text{Cost}$$

From Table 2, we see that 113.5 liters is equal to 30 gallons. We then refer to the first table to find that 30 gallons of fuel costs the company $45.

To answer the second question, we had to use *two* different tables to look up the value of the cost function. The lookup process was a composition of the cost function with the unit conversion function.

 Check It Out 7   In Example 7, find the cost of 264.95 liters of fuel. ■

Quantity (gallons)	Cost ($)
30	45
45	67.5
55	82.5
70	105

Table 1

Quantity (liters)	Quantity (gallons)
113.55	30
170.325	45
208.175	55
264.95	70

Table 2

Difference Quotient

Combining functions is a technique that is often used in calculus. For example, it is used in calculating the **difference quotient** of a function f, which is an expression of the form $\frac{f(x + h) - f(x)}{h}$, $h \neq 0$. This is illustrated in the following example.

Example 8   **Computing a Difference Quotient**

Compute $\dfrac{f(x + h) - f(x)}{h}$, $h \neq 0$, for $f(x) = 2x^2 + 1$.

Solution Compute each component, step by step.

$$f(x + h) = 2(x + h)^2 + 1$$
$$= 2(x^2 + 2xh + h^2) + 1 \qquad \text{Expand } (x + h)^2$$
$$= 2x^2 + 4xh + 2h^2 + 1$$

Next,

$$f(x + h) - f(x) = 2x^2 + 4xh + 2h^2 + 1 - (2x^2 + 1)$$
$$= 4xh + 2h^2$$

Finally,

$$\frac{f(x + h) - f(x)}{h} = \frac{4xh + 2h^2}{h} = 4x + 2h$$

 Check It Out 8   Compute $\dfrac{f(x + h) - f(x)}{h}$, $h \neq 0$, for $f(x) = -x^2 + 4$. ■

Exercises 2.6

 Just in Time Exercises

1. A quotient of two polynomial expressions is called a _____ and is defined whenever the denominator is not equal to _____.

2. True or False: The variable x in $f(x)$ is a placeholder and can be replaced by any quantity as long as the same replacement occurs in the expression for the function.

3. What is the domain of the function $f(x) = x^2 - 3x$?

4. What is the domain of the function $f(x) = \sqrt{x - 1}$?

5. What is the domain of the function $f(x) = \sqrt{x^2 - 9}$?

6. What is the domain of the function $f(x) = \dfrac{x + 2}{x - 1}$?

Skills

In Exercises 7–16, for the functions f and g, find each of the following and identify its domain.

a. $(f + g)(x)$ b. $(f - g)(x)$ c. $(fg)(x)$ d. $\left(\dfrac{f}{g}\right)(x)$

7. $f(x) = 3x - 5; g(x) = -x + 3$ 8. $f(x) = 2x + 1; g(x) = -5x - 1$

9. $f(x) = x - 3; g(x) = x^2 + 1$ 10. $f(x) = x^3; g(x) = 3x^2 + 4$

11. $f(x) = \dfrac{1}{x}; g(x) = \dfrac{1}{2x - 1}$ 12. $f(x) = \dfrac{2}{x + 1}; g(x) = \dfrac{-1}{x^2}$

13. $f(x) = \sqrt{x}; g(x) = -x + 1$ 14. $f(x) = 2x - 1; g(x) = \sqrt{x}$

15. $f(x) = |x|; g(x) = \dfrac{1}{2x + 5}$ 16. $f(x) = \dfrac{2}{x - 4}; g(x) = -|x|$

In Exercises 17–32, let $f(x) = -x^2 + x$, $g(x) = \dfrac{2}{x + 1}$ and $h(x) = -2x + 1$. Evaluate each of the following.

17. $(f + g)(1)$ 18. $(f + g)(0)$ 19. $(g + h)(0)$ 20. $(g + h)(1)$ 21. $(f - g)(2)$

22. $(f - g)(-3)$ 23. $(g - h)(-2)$ 24. $(g - h)(3)$ 25. $(fg)(3)$ 26. $(fg)(-3)$

27. $(gh)(-3)$ 28. $(gh)(0)$ 29. $\left(\dfrac{f}{g}\right)(-2)$ 30. $\left(\dfrac{f}{g}\right)(3)$ 31. $\left(\dfrac{f}{h}\right)(1)$

32. $\left(\dfrac{h}{f}\right)(2)$

In Exercises 33–40, use f and g given by the following tables of values.

x	−1	0	3	6
$f(x)$	−2	3	4	2

x	−2	1	2	4
$g(x)$	0	6	−2	3

33. Evaluate $f(-1)$.   34. Evaluate $g(4)$.   35. Evaluate $(f \circ g)(-2)$.   36. Evaluate $(f \circ g)(4)$.

37. Evaluate $(g \circ f)(-1)$.   38. Evaluate $(g \circ f)(6)$.   39. Is $(g \circ f)(0)$ defined? Why or why not?

40. Is $(f \circ g)(2)$ defined? Why or why not?

In Exercises 41–52, let $f(x) = x^2 + x$, $g(x) = \sqrt{x}$ and $h(x) = -3x$. Evaluate each of the following.

41. $(f \circ h)(5)$ 42. $(f \circ h)(1)$ 43. $(h \circ g)(4)$ 44. $(h \circ g)(0)$

45. $(f \circ g)(4)$ 46. $(f \circ g)(9)$ 47. $(g \circ f)(2)$ 48. $(g \circ f)(-3)$

49. $(h \circ f)(2)$ 50. $(h \circ f)(-3)$ 51. $(h \circ f)\left(\dfrac{1}{2}\right)$ 52. $(h \circ f)\left(\dfrac{3}{2}\right)$

In Exercises 53-72, find expressions for $(f \circ g)(x)$ and $(g \circ f)(x)$. Give the domains of $f \circ g$ and $g \circ f$.

53. $f(x) = -x^2 + 1, g(x) = x + 1$

54. $f(x) = 2x + 5, g(x) = 3x^2$

55. $f(x) = 4x - 1, g(x) = \dfrac{x + 1}{4}$

56. $f(x) = 2x + 3, g(x) = \dfrac{x - 3}{2}$

57. $f(x) = 3x^2 + 4x, g(x) = x + 2$

58. $f(x) = -2x + 1, g(x) = 2x^2 - 5x$

59. $f(x) = \dfrac{1}{x}, g(x) = 2x + 5$

60. $f(x) = 3x + 1, g(x) = \dfrac{2}{x}$

61. $f(x) = \dfrac{3}{2x + 1}, g(x) = 2x^2$

62. $f(x) = 3x^2 + 1, g(x) = \dfrac{2}{x + 5}$

63. $f(x) = \sqrt{x + 1}, g(x) = -3x - 4$

64. $f(x) = 5x + 1, g(x) = \sqrt{x - 3}$

65. $f(x) = |x|, g(x) = \dfrac{2x}{x - 1}$

66. $f(x) = |x|, g(x) = \dfrac{x}{x - 3}$

67. $f(x) = x^2 - 2x + 1, g(x) = x + 1$

68. $f(x) = x - 2, g(x) = 2x^2 - x + 3$

69. $f(x) = \dfrac{x^2 + 1}{x^2 - 1}, g(x) = |x|$

70. $f(x) = |x|, g(x) = \dfrac{x^2 + 3}{x^2 - 4}$

71. $f(x) = \dfrac{1}{x^2 + 1}, g(x) = \dfrac{2x + 1}{3x - 1}$

72. $f(x) = \dfrac{-x + 1}{2x + 3}, g(x) = \dfrac{1}{x^2 + 1}$

In Exercises 73–82, find two functions f and g such that $h(x) = (f \circ g)(x) = f(g(x))$. Answers may vary.

73. $h(x) = (3x - 1)^2$

74. $h(x) = (-2x + 5)^2$

75. $h(x) = \sqrt[3]{4x^2 - 1}$

76. $h(x) = \sqrt[5]{-x^3 + 8}$

77. $h(x) = \dfrac{1}{2x + 5}$

78. $h(x) = \dfrac{3}{x^2 + 1}$

79. $h(x) = \sqrt{x^2 + 1} + 5$

80. $h(x) = 4(2x + 9)^5 - (2x + 9)^8$

81. $h(x) = \sqrt[3]{5x + 7} - 2$

82. $h(x) = (3x - 7)^{10} + 5(3x - 7)^2$

In Exercises 83–86, let $f(t) = -t^2$ and $g(x) = x^2 - 1$.

83. Evaluate $(f \circ f)(-1)$.

84. Evaluate $(g \circ g)\left(\dfrac{2}{3}\right)$.

85. Find an expression for $(f \circ f)(t)$, and give the domain of $f \circ f$.

86. Find an expression for $(g \circ g)(x)$, and give the domain of $g \circ g$.

In Exercises 87–90, let $f(t) = 3t + 1$ and $g(x) = x^2 + 4$.

87. Evaluate $(f \circ f)(2)$.

88. Evaluate $(g \circ g)\left(\dfrac{1}{2}\right)$.

89. Find an expression for $(f \circ f)(t)$, and give the domain of $f \circ f$.

90. Find an expression for $(g \circ g)(x)$, and give the domain of $g \circ g$.

In Exercises 91–96, find the difference quotient $\dfrac{f(x + h) - f(x)}{h}$, $h \neq 0$, for the given function f.

91. $f(x) = 3x - 1$

92. $f(x) = -2x + 3$

93. $f(x) = -x^2 + x$

94. $f(x) = 3x^2 + 2x$

95. $f(x) = \dfrac{1}{x - 3}, x \neq 3$

96. $f(x) = \dfrac{1}{x + 1}, x \neq -1$

Applications

97. Sports The Washington Redskins' revenue can be modeled by the function $R(t) = 327 + 10t$, where t is the number of years since 2008 and $R(t)$ is in millions of dollars. Their operating costs are modeled by the function $C(t) = 246 + 20t$, where t is the number of years since 2008 and $C(t)$ is in millions of dollars. Find the profit function $P(t)$.

(*Source*: Forbes.com)

98. Commerce The following two tables give revenues, in dollars, from two stores for various years. Compute the table for $(f + g)(x)$ and explain what it represents.

Year x	Revenue: Store 1 $f(x)$
2008	200,000
2009	210,000
2010	195,000
2011	230,000

Year x	Revenue: Store 2 $g(x)$
2008	300,000
2009	320,000
2010	295,000
2011	330,000

99. Commerce The number of copies of a popular mystery writer's newest release sold at a local bookstore during each month after its release is given by $n(x) = -5x + 100$. The price of the book during each month after its release is given by $p(x) = -1.5x + 30$. Find $(np)(3)$. Interpret your results.

100. Real Estate A realtor sells new homes in a housing development for \$400,000 each. Her commission is 6% of total sales generated.

a. What is the total amount of sales, $S(x)$, if x is the number of homes sold?

b. What is the commission, $C(x)$, if x is the number of homes sold?

c. Interpret the amount $S(x) - C(x)$.

101. Currency Exchange The exchange rate from U.S. dollars to Euros on a particular day is given by the function $f(x) = 0.75x$, where x is in U.S. dollars. If GlobalEx Corporation has revenue given by the function $R(t) = 40 + 2t$, where t is the number of years since 2009 and $R(t)$ is in millions of dollars, find $(f \circ R)(t)$ and explain what it represents.

(*Source*: www.xe.com)

102. Unit Conversion The conversion of temperature units from degrees Fahrenheit to degrees Celsius is given by the equation $C(x) = \frac{5}{9}(x - 32)$, where x is given in degrees Fahrenheit. Let $T(x) = 70 + 4x$ denote the temperature, in degrees Fahrenheit, in Phoenix, Arizona on a typical July day, where x is the number of hours after 6 A.M. Assume the temperature models holds until 4 P.M. of the same day. Find $(C \circ T)(x)$ and explain what it represents.

103. Geometry The surface area of a sphere is given by $A(r) = 4\pi r^2$, where r is in inches and $A(r)$ is in square inches. The function $C(x) = 6.4516x$ takes x square inches as input and outputs the equivalent result in square centimeters. Find $(C \circ A)(r)$ and explain what it represents.

104. Geometry The perimeter of a square is $P(s) = 4s$, where s is the length of a side in inches. The function $C(x) = 2.54x$ takes x inches as input and outputs the equivalent result in centimeters. Find $(C \circ P)(s)$ and explain what it represents.

Concepts

105. Is it true that $(fg)(x)$ is the same as $(f \circ g)(x)$ for any functions f and g? Explain.

106. Give an example to show that $(f \circ g)(x) \neq (g \circ f)(x)$.

107. Let $f(x) = ax + b$ and $g(x) = cx + d$, where a, b, c, and d are constants. Show that $(f + g)(x)$ and $(f - g)(x)$ also represent linear functions.

108. Find $\frac{f(x + h) - f(x)}{h}$, $h \neq 0$ for $f(x) = ax + b$, where a and b are constants.

Transformations of the Graph of a Function

2.7

Objectives

- Graph vertical and horizontal shifts of the graph of a function.

- Graph a vertical compression or stretch of the graph of a function.

- Graph reflections across the x-axis of the graph of a function.

- Graph a horizontal compression or a stretch of the graph of a function.

- Graph reflections across the y-axis of a graph of a function.

- Graph combinations of transformations of the graph of a function.

- Identify an appropriate transformation of the graph of a function from a given expression for the function.

In this section, you will see how to create graphs of new functions from the graph of an existing function by simple geometric **transformations**. The general properties of these transformations are very useful for sketching the graphs of various functions.

Throughout this section, we will investigate transformations of the graphs of the functions x^2, $|x|$, and $\sqrt{x}$ (see Figure 1), and state the general rules of transformations. Since these rules apply to the graph of any function, they will be applied to transformations of the graphs of other functions in later chapters.

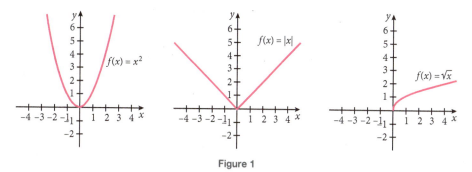

Figure 1

Vertical and Horizontal Shifts of the Graph of a Function

The simplest transformations we can consider are the ones where you take the graph of a function and simply shift it vertically or horizontally. Such shifts are also called *translations*.

For example, if $f(x) = x^2$, and $g(x) = x^2 - 3$, the corresponding y-value of $g(x)$, for each x, will always be 3 units less than the value for $f(x)$. So the graph of g will be the graph of f shifted down by 3 units. We have the following general statement about **vertical shifts** of the graph of a function.

Vertical Shifts of the Graph of $f(x)$

Let f be a function and c a positive constant.

- The graph of $g(x) = f(x) + c$ is the graph of $f(x)$ shifted c units **upward**.

- The graph of $g(x) = f(x) - c$ is the graph of $f(x)$ shifted c units **downward**.

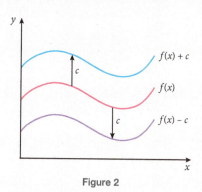

Figure 2

Example 1   **Graphing** $f(x) = |x|$ **and** $g(x) = |x| - 2$

Make a table of values of the functions $f(x) = |x|$ and $g(x) = |x| - 2$, for $x = -3, -2, -1, 0,$ 1, 2, and 3. Use your table to sketch the graphs of the two functions. What are the domain and range of f and g?

Solution   We make a table of function values as shown in Table 1, and sketch the graphs as shown in Figure 3.

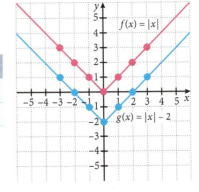

x	−3	−2	−1	0	1	2	3
$f(x) = |x|$	3	2	1	0	1	2	3
$g(x) = |x| - 2$	1	0	−1	−2	−1	0	1

Table 1

Figure 3

Examining the table of function values, we see that the values of $g(x) = |x| - 2$ are two units less than the corresponding values of $f(x) = |x|$. The graph of $g(x) = |x| - 2$ has the same shape as the graph of $f(x) = |x|$, but is shifted *down by 2 units*.

The domain of both f and g is the set of all real numbers. From the graphs, we see that the range of f is $[0, \infty)$, while the range of g is $[-2, \infty)$.

Check It Out 1   Make a table of values of the functions $f(x) = |x|$ and $g(x) = |x| + 3$, for $x = -3, -2, -1, 0, 1, 2,$ and 3. Use your table to sketch the graphs of the two functions. What are the domain and range of f and g?

If $f(x) = x^2$ and $g(x) = (x - 3)^2$, we see that $f(0)$ has the same value as $g(3)$, $f(1)$ has the same value as $g(4)$, and so on. Thus, the graph of g is the same as the graph of f shifted to the right by 3 units. We have the following general statement about **horizontal shifts** of the graph of a function.

**Using Technology**

Vertical shifts can be easily seen with a graphing calculator. Figure 4 shows the graphs of $f(x) = |x|$ and $g(x) = |x| - 2$ on the same set of axes using a decimal window.

Keystroke Appendix:
Section 7

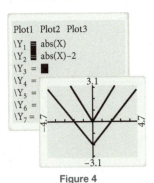

Figure 4

Horizontal Shifts of the Graph of $f(x)$

Let f be a function and c a positive constant.

- The graph of $g(x) = f(x - c)$ is the graph of $f(x)$ shifted c units **to the right**.

- The graph of $g(x) = f(x + c)$ is the graph of $f(x)$ shifted c units **to the left**.

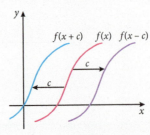

Figure 5

Example 2   Graphing $f(x) = |x|$ and $g(x) = |x - 2|$

Make a table of values of the functions $f(x) = |x|$ and $g(x) = |x - 2|$, for $x = -3, -2, -1, 0, 1, 2, 3$. Use your table to sketch the graphs of the two functions. What are the domain and range of f and g?

Solution We make a table of function values as shown in Table 2 and sketch the graphs as shown in Figure 6.

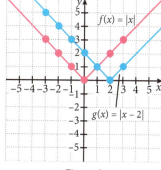

x	-3	-2	-1	0	1	2	3
$f(x) = |x|$	3	2	1	0	1	2	3
$g(x) = |x - 2|$	5	4	3	2	1	0	1

Table 2

Figure 6

Examining the table of function values, we see that the values of $g(x) = |x - 2|$ are the values of $f(x) = |x|$ *shifted to the right by 2 units*. This is illustrated by the numbers in red in the rows labeled $f(x) = |x|$ and $g(x) = |x - 2|$. The graph of $g(x) = |x - 2|$ is the same as the graph of $f(x) = |x|$, but is *shifted to the right by 2 units*.

The domain of both f and g is the set of all real numbers. From the graphs, we see that the range of both f and g is $[0, \infty)$.

✅ *Check It Out 2* Make a table of values of the functions $f(x) = |x|$ and $g(x) = |x - 1|$, for $x = -3, -2, -1, 0, 1, 2, 3$. Use your table to sketch the graphs of the two functions. What are the domain and range of f and g?

Vertical and horizontal shifts can also be combined to create the graph of a new function.

Example 3   Combining Vertical and Horizontal Shifts of a Function

Use vertical and/or horizontal shifts, along with a table of values, to graph the following functions.

a. $g(x) = \sqrt{x + 2}$

b. $g(x) = |x + 3| - 2$

Solution

a. The graph of $g(x) = \sqrt{x + 2}$ is a horizontal shift of the graph of $f(x) = \sqrt{x}$ by 2 units to the *left*, since $f(x + 2) = f(x - (-2)) = \sqrt{x + 2}$. Make a table of function values as shown in Table 3 and use it to sketch the graphs of both functions, as shown in Figure 8.

x	-3	-2	-1	0	1	2	3
$f(x) = \sqrt{x}$	error	error	error	0	1	$\sqrt{2} \approx 1.414$	$\sqrt{3} \approx 1.732$
$g(x) = f(x + 2)$ $= \sqrt{x + 2}$	error	0	1	$\sqrt{2} \approx 1.414$	$\sqrt{3} \approx 1.732$	2	$\sqrt{5} \approx 2.236$

Table 3

Using Technology

Horizontal shifts can be seen easily with a graphing calculator. Figure 7 shows $f(x) = |x|$ and $g(x) = |x - 2|$ on the same set of axes using a decimal window.

Keystroke Appendix:
Section 7

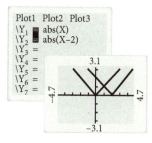

Figure 7

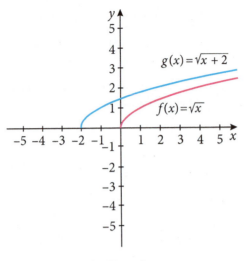

Figure 8

b. The graph of the function $g(x) = |x + 3| - 2$ consists of both a horizontal shift of the graph of $f(x)$ by 3 units *to the left* and a vertical shift by 2 units *down*. We make a table of function values as shown in Table 4.

x	−4	−3	−2	−1	0	1	2
$f(x) = |x|$	4	3	2	1	0	1	2
$y = |x + 3|$	1	0	1	2	3	4	5
$g(x) = |x + 3| - 2$	−1	−2	−1	0	1	2	3

Table 4

Using Technology

Combinations of horizontal and vertical shifts can be seen easily with a graphing calculator. Figure 9 shows $Y_1 = |x|$, $Y_2 = |x + 3|$, and $Y_3 = |x + 3| - 2$ on the same set of axes, using a decimal window.

Keystroke Appendix:

Section 7

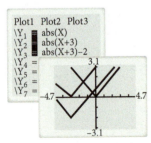

Figure 9

The values of $y = |x + 3|$ occur in the same order as the values as $|x|$, but are shifted to the *left by 3 units*, as illustrated by the numbers in red.

To find the values of $g(x) = |x + 3| - 2$, we next shift the values of $y = |x + 3|$ *down by 2 units*. We can graph the function $g(x) = |x + 3| - 2$ in two stages—first the horizontal shift and then the vertical shift. See Figure 10.

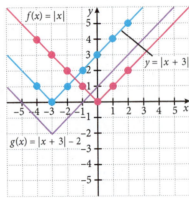

Figure 10

The order of the horizontal and vertical translations does not matter.

✅ *Check It Out 3*   Graph the function $f(x) = \sqrt{x - 2} + 1$ using transformations.   ▪

Vertical Scalings and Reflections Across the Horizontal Axis

In this subsection, we will examine what happens when an expression for a function is multiplied by a nonzero constant. For instance, if $f(x) = x^2$, $g(x) = 2x^2$ and $h(x) = \frac{1}{2}x^2$, we see that $g(x)$ doubles the function values from $f(x)$. The function $h(x) = \frac{1}{2}x^2$ halves the function values of $f(x)$.

We next have a general statement about **vertical scalings** of the graph of a function.

Vertical Scalings of the Graph of $f(x)$

Let f be a function and c be a positive constant.

- If $c > 1$, the graph of $g(x) = cf(x)$ is the graph of $f(x)$ **stretched vertically** away from the x-axis, with the y-coordinates of $f(x)$ multiplied by c.

- If $0 < c < 1$, the graph of $g(x) = cf(x)$ is the graph of $f(x)$ **compressed vertically** toward the x-axis, with the y-coordinates of $g(x) = f(x)$ multiplied by c.

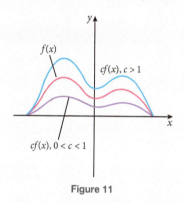

Figure 11

We can also consider the graph of $g(x) = -f(x)$. In this case, the y-coordinate of $(x, g(x))$ will be the negative of the the y-coordinate of $(x, f(x))$. Graphically, this results in a reflection of the graph of $f(x)$ across the x-axis.

Reflection of the Graph of $f(x)$ Across the x-Axis

Let f be a function.

- The graph of $g(x) = -f(x)$ is the graph of $f(x)$ **reflected across the x-axis**.

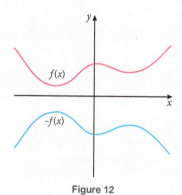

Figure 12

Many graphing problems involve some combination of vertical and horizontal translations as well as vertical scalings and reflections. These actions on the graph of a function are collectively known as **transformations** of the graph of the function and are explored in the following examples.

Example 4   **Sketching Graphs Using Transformations**

Identify the basic function $f(x)$ which is transformed to obtain $g(x)$. Then use transformations to sketch the graphs of both $f(x)$ and $g(x)$.

a. $g(x) = 3\sqrt{x}$ **b.** $g(x) = -2|x|$ **c.** $g(x) = -2|x + 1| + 3$

Solution

a. The graph of the function $g(x) = 3\sqrt{x}$ is a vertical stretch of the graph of $f(x) = \sqrt{x}$, since the function values of $f(x)$ are multiplied by a factor of 3. Both functions have domain $[0, \infty)$. Table 5 gives function values for the two functions, and their corresponding graphs are shown in Figure 13.

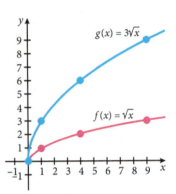

x	$f(x) = \sqrt{x}$	$g(x) = 3\sqrt{x}$
0	0	0
1	1	3
2	1.414	4.243
4	2.000	6.000
9	3	9

Table 5

Figure 13

b. The graph of the function $g(x) = -2|x|$ consists of a vertical stretch and then a reflection across the axis of the graph of $f(x) = |x|$, since the values of $f(x)$ are not only doubled but also negated. Table 6 gives function values for the two functions, and their corresponding graphs are shown in Figure 14.

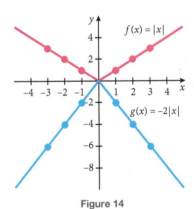

x	$f(x) = |x|$	$g(x) = -2|x|$
-3	3	-6
-2	2	-4
-1	1	-2
0	0	0
1	1	-2
2	2	-4
3	3	-6

Table 6

Figure 14

c. We can view the graph of $g(x)$ as resulting from transformations of the graph of $f(x) = |x|$ in the following manner:

$f(x) = |x|$ $\rightarrow$ $y_1 = -2|x|$ $\rightarrow$ $y_2 = -2|x + 1|$ $\rightarrow$ $g(x) = -2|x + 1| + 3$

Original   Vertical scaling   Horizontal shift   Vertical shift
function   by 2 and   to the left by   upward by
                 reflection   1 unit   3 units
                 across the
                 x-axis

The transformation from $f(x) = |x|$ to $y_1 = -2|x|$ was already discussed and graphed in part (b). We now take the graph of $y_1 = -2|x|$ and shift it to the left by 1 unit and then upward by 3 units, as shown in Figure 15.

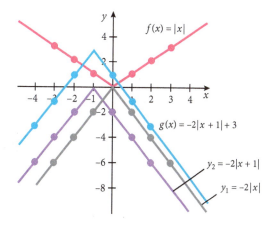

Figure 15

Using Technology

When entering transformations of a function into the Y = editor, you can define each transformation in terms of the function name Y_1. This is convenient if you want to observe the effects of the same transformation on a different function. You need only change the expression for Y_1. See Figure 16.

Keystroke Appendix:
Sections 4 and 7

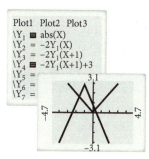

Figure 16

✔ *Check It Out 4*   Use transformations to sketch the graphs of the following functions:

a. $g(x) = -3x^2$

b. $g(x) = -|x - 1| + 2$

Example 5   **Using Transformations to Sketch a Graph**

Suppose the graph of a function $g(x)$ is produced from the graph of $f(x) = x^2$ by vertically compressing the graph of f by a factor of $\frac{1}{3}$, then shifting it to the left by 1 unit, and finally shifting it downward by 2 units. Give an expression for $g(x)$ and sketch the graphs of both f and g.

Solution   The series of transformations can be summarized as follows:

$$f(x) = x^2 \quad \rightarrow \quad y_1 = \frac{1}{3}x^2 \quad \rightarrow \quad y_2 = \frac{1}{3}(x + 1)^2 \quad \rightarrow \quad g(x) = \frac{1}{3}(x + 1)^2 - 2$$

Original function → Vertical scaling by $\frac{1}{3}$ → Shift left by 1 → Shift down by 2

The function g is then given by

$$g(x) = \frac{1}{3}(x + 1)^2 - 2$$

The graphs of f and g are given in Figure 17, which also shows the graph of $\frac{1}{3}x^2$. The horizontal and vertical shifts are indicated by arrows.

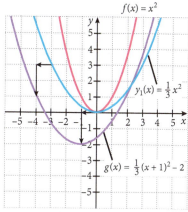

Figure 17

Check It Out 5   Let the graph of $g(x)$ be produced from the graph of $f(x) = \sqrt{x}$ by vertically stretching the graph of f by a factor of 3, then shifting it to the left by 2 units, and finally shifting it upward by 1 unit. Give an expression for $g(x)$, and sketch the graphs of both f and g.   ■

Horizontal Scalings and Reflections Across the Vertical Axis

The final set of transformations involve stretching and compressing the graph of a function along the horizontal axis. In function notation, we examine the relationship between the graph of $f(x)$ and the graph of $f(cx)$, $c > 0$.

If $f(x) = \sqrt{x}$ and $g(x) = \sqrt{2x}$, we see that $f(4) = g(2)$. So $g(x)$ obtains the same y values for smaller values of x, resulting in a shrinking effect of the graph of $f(x)$. We have the following general statement about **horizontal scalings** of the graph of a function.

Horizontal Scalings of the Graph of $f(x)$

Let f be a function and c be a positive constant.

- If $c > 1$, the graph of $g(x) = f(cx)$ is the graph of $f(x)$ **compressed horizontally** toward the y-axis, scaled by a factor of $\frac{1}{c}$.

- If $0 < c < 1$, the graph of $g(x) = f(cx)$ is the graph of $f(x)$ **stretched horizontally** away from y-axis, scaled by a factor of $\frac{1}{c}$.

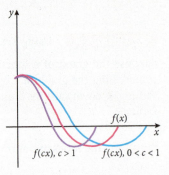

Figure 18

We can also consider the graph of $g(x) = f(-x)$, which is a reflection of the graph of $f(x)$ across the y-axis.

Reflection of the Graph of $f(x)$ Across the y-Axis

Let f be a function.

- The graph of $g(x) = f(-x)$ is the graph of $f(x)$ **reflected across the y-axis**.

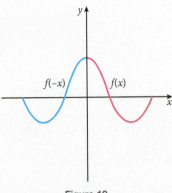

Figure 19

| **Example 6** | Using Transformations to Sketch a Graph |

The graph of $f(x)$ is shown in Figure 20. Use it to sketch the graphs of

a. $f(3x)$

b. $f(-x) + 1$

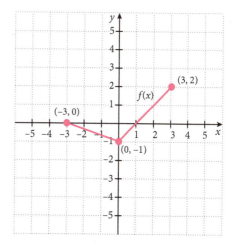

Figure 20

Solution

a. The graph of $f(3x)$ is a *horizontal compression* of the graph of $f(x)$. The x-coordinates of $f(x)$ are scaled by a factor of $\frac{1}{3}$. Table 7 summarizes how each of the key points in the graph of $f(x)$ is transformed to the corresponding point on the graph of $f(3x)$. The points are then used to sketch the graph of $f(3x)$. See Figure 21.

Point on graph of $f(x)$		Point on graph of $f(3x)$
$(-3, 0)$	$\rightarrow$	$\left(\frac{1}{3}(-3), 0\right) = (-1, 0)$
$(0, -1)$	$\rightarrow$	$\left(\frac{1}{3}(0), -1\right) = (0, -1)$
$(3, 2)$	$\rightarrow$	$\left(\frac{1}{3}(3), 0\right) = (1, 2)$

Table 7

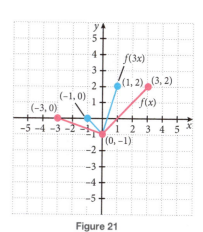

Figure 21

b. The graph of $f(-x) + 1$ is a *reflection across the y-axis* of the graph of $f(x)$ followed by a *vertical shift upward* of 1 unit. To obtain the graph of $f(-x)$, the x-coordinates of $(x, f(x))$ are negated. For the vertical shift, 1 is then added to the y-coordinate.

Table 8 summarizes how each of the key points in the graph of $f(x)$ is transformed to the corresponding point on the graph of $f(-x)$, and finally to the points on the graph of $f(-x) + 1$. The points are then used to sketch the graph of $f(-x) + 1$. See Figure 22.

Point on graph of $f(x)$		Point on graph of $f(-x)$		Point on graph $f(-x) + 1$
$(-3, 0)$	$\rightarrow$	$(-(-3), 0) = (3, 0)$	$\rightarrow$	$(3, 0 + 1) = (3, 1)$
$(0, -1)$	$\rightarrow$	$(-(0), -1) = (0, -1)$	$\rightarrow$	$(0, -1 + 1) = (0, 0)$
$(3, 2)$	$\rightarrow$	$(-(3), 0) = (-3, 2)$	$\rightarrow$	$(-3, 2 + 1) = (-3, 3)$

Table 8

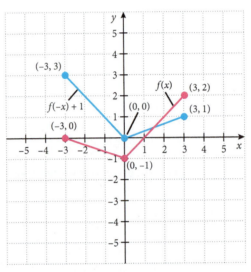

Figure 22

Note It is important to see that these horizontal scalings work in an opposite manner as compared to vertical scalings. For example, if $x = 1$, $f(3x) = f(3)$. The value of $f(3)$ is reached "earlier" at $x = 1$ by the function $f(3x)$. This accounts for the shrinking effect.

Check It Out 6 For f given in the above example, graph $f\left(\frac{1}{2}x\right)$.

Exercises 2.7

Skills

In Exercises 1–30, identify the underlying basic function, and use transformations of the basic function to sketch the graph of the given function.

1. $g(t) = t^2 + 1$

2. $g(t) = t^2 - 3$

3. $f(x) = \sqrt{x} - 2$

4. $g(x) = \sqrt{x} + 1$

5. $h(x) = |x - 2|$

6. $h(x) = |x + 4|$

7. $f(s) = (s + 5)^2$

8. $g(s) = (s - 3)^2$

9. $f(x) = \sqrt{x - 4}$

10. $f(x) = \sqrt{x + 3}$

11. $h(x) = |x - 2| + 1$

12. $g(x) = \sqrt{x + 1} - 2$

13. $s(x) = (x + 3)^2 - 1$

14. $g(x) = (x - 2)^2 + 5$

15. $h(t) = 3t^2$

16. $g(x) = 2\sqrt{x}$

17. $s(x) = -4|x|$

18. $h(x) = -2x^2$

19. $h(s) = -|s| - 3$

20. $f(x) = -\sqrt{x + 4}$

21. $h(x) = -\dfrac{1}{2}|x + 1| - 3$

22. $h(x) = -2|x - 4| + 1$

23. $g(x) = -3(x + 2)^2 - 4$

24. $h(x) = -\dfrac{1}{3}(x - 2)^2 - \dfrac{3}{2}$

25. $f(x) = |2x|$

26. $f(x) = \left|\dfrac{x}{2}\right|$

27. $f(x) = (2x)^2$

28. $f(x) = \left(\dfrac{1}{2}x\right)^2$

29. $g(x) = \sqrt{3x}$

30. $f(x) = \sqrt{2x}$

In Exercises 31–34, explain how each graph is a transformation of the graph of $f(x) = |x|$ and find a suitable expression for the function represented by the graph.

31.

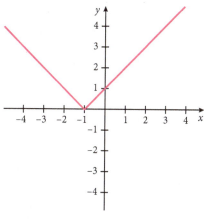

32.

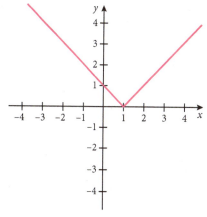

33.

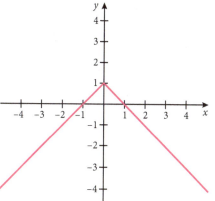

34.

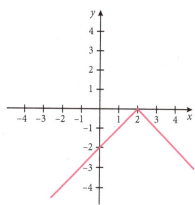

In Exercises 35–38, explain how each graph is a transformation of the graph of $f(x) = x^2$ and find a suitable expression for the function represented by the graph.

35.

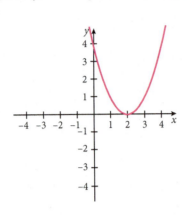

36.

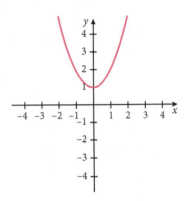

37.

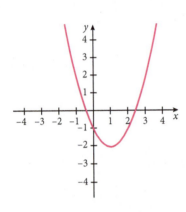

38.

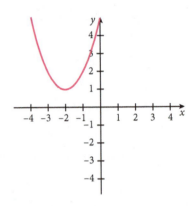

In Exercises 39–44, use the verbal description to find an algebraic expression for the function.

39. The graph of the function $g(t)$ is formed by translating the graph of $f(t) = |t|$ to the left 4 units and down 3 units.

40. The graph of the function $f(t)$ is formed by translating the graph of $h(t) = t^2$ to the right 2 units and upward 6 units.

41. The graph of the function $g(t)$ is formed by vertically scaling the graph of $f(t) = t^2$ by a factor of -3 and moving it to the right by 1 unit.

42. The graph of the function $g(t)$ is formed by vertically scaling the graph of $f(t) = |t|$ by a factor of -2 and moving it to the left by 5 units.

43. The graph of the function $h(x)$ is formed by scaling the graph of $f(x) = x^2$ horizontally by a factor of $\frac{1}{2}$ and moving it down 4 units.

44. The graph of the function $g(x)$ is formed by scaling the graph of $f(x) = x^2$ vertically by a factor of $\frac{1}{2}$ and moving it up 4 units.

In Exercises 45–52, use the given function f to sketch a graph of the indicated transformation of f. First copy the graph of f onto a sheet of graph paper.

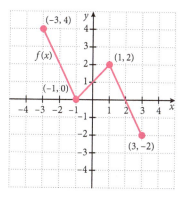

45. $2f(x)$

46. $\dfrac{1}{2}f(x)$

47. $f(x) + 2$

48. $-f(x) - 3$

49. $f(2x)$

50. $f\left(\dfrac{1}{2}x\right)$

51. $f(x - 1) + 2$

52. $-f(x + 2) - 1$

In Exercises 53–60, use the given function f to sketch a graph of the indicated transformation of f. First copy the graph of f onto a sheet of graph paper.

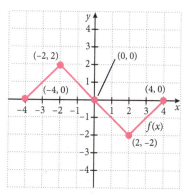

53. $f(x + 2)$

54. $f(x - 1)$

55. $2f(x) + 1$

56. $-f(x) - 1$

57. $f(2x)$

58. $f\left(\dfrac{1}{3}x\right)$

59. $f(x + 1) - 3$

60. $-2f(x + 3)$

In Exercises 61 and 62, use the table giving values for f(x) and g(x) = f(x) + k to find the appropriate value of k.

61.

x	f(x)	g(x) = f(x) + k
−2	−4	−6
−1	−6	−8
0	−9	−11
1	−6	−8
2	−4	−6

62.

x	f(x)	g(x) = f(x) + k
10	3	4.5
9	2	3.5
8	1	2.5
7	2	3.5
6	3	4.5

In Exercises 63 and 64, use the table giving values for $f(x)$ and $g(x) = f(x - k)$ to find the appropriate value of k.

63.

x	-2	-1	0	1	2
$f(x)$	4	6	8	10	12
$g(x) = f(x - k)$	2	4	6	8	10

64.

x	-2	-1	0	1	2
$f(x)$	-4	-5	-6	-5	-4
$g(x) = f(x - k)$	-5	-6	-5	-4	-3

In Exercises 65 and 66, fill in the missing values for the function f in the table.

65.

x	$f(x)$	$g(x) = f(x) - 3$
-2		33
-1		22
0		13
1		6
2		1

66.

x	$f(x)$	$g(x) = f(x) + 2$
-2		20
-1		18.5
0		16
1		17.2
2		13

 In Exercises 67–74, use a graphing utility to solve the problem.

67. Graph $f(x) = |x + 3.5|$ and $g(x) = |x| + 3.5$. Describe each graph in terms of transformations of the graph of $h(x) = |x|$.

68. Graph $f(x) = (x - 4.5)^2$ and $g(x) = x^2 + 4.5$. Describe each graph in terms of transformations of the graph of $h(x) = x^2$.

69. If $f(x) = \sqrt{x}$, graph $f(x)$ and $f(x - 4.5)$ in the same window. What is the relationship between the two graphs?

70. If $f(x) = |x|$, graph $f(x)$ and $f(0.3x)$ in the same window. What is the relationship between the two graphs?

71. If $f(x) = |x|$, graph $-2f(x)$ and $f(-2x)$ in the same window. Do they produce the same graphs? Explain.

72. If $f(x) = \sqrt{x}$, graph $3f(x)$ and $f(3x)$ in the same window. Do they produce the same graphs? Explain.

73. Graph $f(x) = x^3$ and $g(x) = (x - 7)^3$. How can the graph of g be described in terms of the graph of f?

74. Graph the functions $f(x) = |x - 4|$ and $g(x) = f(-x) = |(-x) - 4|$. What relationship do you observe between the graphs of the two functions? Do the same with $f(x) = (x - 2)^2$ and $g(x) = f(-x) = ((-x) - 2)^2$. What type of reflection of the graph of $f(x)$ gives the graph of $g(x) = f(-x)$?

Applications

75. Coffee Sales Let $P(x)$ represent the price of x pounds of coffee. Assuming the entire amount is taxed at 6%, find an expression, in terms of $P(x)$, of just the sales tax for x pounds of coffee.

76. Salary Let $S(x)$ represent the weekly salary of a sales person, where x is the weekly dollar amount of sales generated by the salesperson. If she pays 15% of her salary in federal taxes, express her after-tax salary in terms of $S(x)$. Assume there are no other deductions to her salary.

77. Printing The production cost, in dollars, for x number of color brochures is $C(x) = 500 + 3x$. The fixed cost is $500, since that is the amount of money need to start production even if no brochures are printed.

   a. If the fixed cost is decreased by $50, find the new cost function.

   b. Graph both cost functions and interpret the effect of the decreased fixed cost.

78. Geometry The area of a square is given by $A(s) = s^2$, where s is the length of a side in inches. Compute the expression for $A(2s)$ and explain what it represents.

79. Physics The height of a ball thrown upward with an initial velocity of 30 feet/second from an initial height of h feet is given by

$$s(t) = -16t^2 + 30t + h$$

where t is the time, in seconds.

   a. If $h = 0$, how high is the ball at time $t = 1$?

   b. If $h = 20$, how high is the ball at time $t = 1$?

   c. In terms of shifts, what is the effect of h in the function $s(t)$?

80. Unit Conversion Let $T(x)$ be the temperature, in degrees Celsius, of the point on a long rod located x cm from the end of the rod, which corresponds to $x = 0$. Temperature for the absolute temperature scales can be measured in kelvin by adding 273 to the temperature in degrees Celsius. Let $t(x)$ be the temperature function in kelvin, and write an expression for the temperature function $t(x)$ in kelvin in terms of the function $T(x)$.

Concepts

81. The point $(2, 4)$ on the graph of $f(x) = x^2$ has been shifted horizontally to the point $(-3, 4)$. Identify the shift and write a new function $g(x)$ in terms of $f(x)$.

82. The point $(-2, 2)$ on the graph of $f(x) = |x|$ has been shifted horizontally *and* vertically to the point $(3, 4)$. Identify the shifts and write a new function $g(x)$ in terms of $f(x)$.

83. When using transformations with both vertical scaling and vertical shifts, the order in which you perform the transformations matters. Let $f(x) = |x|$.

   a. Find the function $g(x)$ whose graph is obtained by first vertically stretching $f(x)$ by a factor of 2 and then shifting upward by 3 units. A table of values and/or a sketch of the graph will be helpful.

   b. Find the function $g(x)$ whose graph is obtained by first shifting $f(x)$ upward by 3 units and then multiplying the result by a factor of 2. A table of values and/or a sketch of the graph will be helpful.

   c. Compare your answers to parts (a) and (b). Explain why they are different.

84. Let $f(x) = 2x + 5$ and $g(x) = f(x + 2) - 4$. Graph both functions on the same set of coordinate axes. Describe the effect of the transformation. What do you observe?

Linear Functions and Models; Variation

2.8

Objectives

- Identify the input variable and the output variable for a given application.

- Extract suitable data from the statement of a word problem.

- Find a linear function that models a real-world application.

- Explain the significance of the slope and *x*- and *y*-intercepts for an application.

- Examine situations involving direct and inverse variation.

VIDEO EXAMPLES

SECTION 2.8

Many problems that arise in business, marketing, and the social and physical sciences are *modeled* using linear functions. This means that we *assume* that the quantities we are observing have a linear relationship. In this section, we explore means to analyze such problems. That is, you will learn to take the mathematical ideas in the previous section and *apply* them to a problem. In particular, you will learn to do the following:

- Identify input and output variables for a given situation.

- Find a function which describes the relationship between the input and output variables.

- Use the function you found to further analyze the problem.

In addition to linear models, we will briefly examine models where the input and output variables have a relationship that is not linear.

Linear Models

A **linear function** is a function of the form $f(x) = mx + b$, where m and b are constants. It can also be written in the point-slope form, $f(x) = m(x - x_1) + y_1$, where m is the slope, and (x_1, y_1) is a point on the line.

A **linear model** is an application whose mathematical analysis results in a linear function. The following example illustrates how such a model may arise.

> **Example 1** **Modeling Yearly Bonus**

At the end of each year, Jocelyn's employer gives her an annual bonus of $1,000 plus $200 for each year she has been employed by the company. Answer the following questions.

a. Find a linear function that relates Jocelyn's bonus to the number of years of her employment with the company.

b. Use the function you found in part (a) to calculate the annual bonus Jocelyn will receive after she has worked for the company for 8 years.

c. Use the function you found in part (a) to calculate how long Jocelyn would have to work at the company for her annual bonus to amount to $3,200.

d. Interpret the slope and *y*-intercept for this problem, both verbally and graphically.

Solution

a. We first identify the variables for this problem.

- t is the input, or independent, variable, denoting the number of years worked.

- B is the output, or dependent, variable, denoting Jocelyn's bonus.

  Next we proceed to write the equation.

$$\text{Bonus} = 1000 + 200 \cdot \text{\# of years employed} \quad \text{From problem statement}$$
$$B = 1000 + 200t \quad \text{Substitute the variable letters}$$

Since B depends on t, we can express the bonus using function notation as follows:

$$B(t) = 1000 + 200t$$

Note that $B(t)$ could also be written as $B(t) = 200t + 1000$. These two forms are equivalent.

b. Jocelyn's bonus after 8 years of employment, is given by $B(8)$.

$$B(8) = 1000 + 200(8) = 1000 + 1600 = 2600$$

Thus, she will receive a bonus of **$2,600** after eight years of employment.

c. To find out when Jocelyn's bonus would amount to \$3,200, we have to find the value of t such that $B(t) = 3200$.

$$3200 = 1000 + 200t \quad \text{Substitute \$3,200 for the bonus}$$
$$2200 = 200t \qquad\quad \text{Isolate the } t \text{ variable (subtract 1000 from both sides)}$$
$$11 = t \qquad\qquad\; \text{Solve for } t \text{ (divide both sides by 200)}$$

We see that Jocelyn will need to work for the company for **11 years** to receive a bonus of \$3,200.

d. Since the function describing Jocelyn's bonus is $B(t) = 1000 + 200t$, the slope of the line $B = 1000 + 200t$ is 200. Recall that the slope is the ratio of the change in B to the change in t.

$$\text{Slope} = 200 = \frac{200}{1} = \frac{\text{Change in bonus}}{\text{Change in years employed}}$$

The slope of 200 signifies that Jocelyn's bonus will increase by \$200 each year that she works for the company.

The y-intercept for this problem is **(0, 1000)**. It means that Jocelyn will receive a bonus of \$1,000 at the start of her employment with the company, which corresponds to $t = 0$.

The graphical interpretations of the slope and y-intercept are indicated in Figure 1.

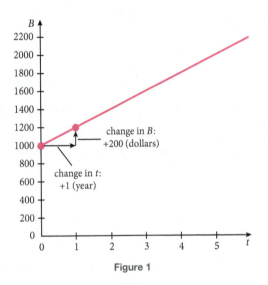

Figure 1

✅ *Check It Out 1*   Rework Example 1 above for the case where Jocelyn's bonus is \$1,400 plus \$300 for every year she has been with the company. ■

In many models using linear functions, you will often be given two data points and asked to find the equation of the line passing through them. We review the procedure in the following example.

x	f(x)
1.3	4.5
2.6	1.9

Table 1

Example 2   **Finding a Linear Function Given Two Points**

Find the linear function f whose input and output values are given in Table 1. Evaluate $f(5)$.

Solution From the information in the table, we see that the following two points must lie on a line.

$$(x_1, y_1) = (1.3, 4.5) \quad (x_2, y_2) = (2.6, 1.9)$$

Recall that a y-coordinate is the function value corresponding to a given x-coordinate.

Next, calculate the slope.

$$m = \frac{y_2 - y_1}{x_2 - x_1} = \frac{1.9 - 4.5}{2.6 - 1.3} = \frac{-2.6}{1.3} = -2$$

Using the point-slope form for a linear function with $m = -2$ and $(x_1, y_1) = (1.3, 4.5)$,

$$f(x) = m(x - x_1) + y_1 = -2(x - 1.3) + 4.5 = -2x + 2.6 + 4.5$$
$$f(x) = -2x + 7.1$$

Evaluating $f(5)$, we have

$$f(5) = -2(5) + 7.1 = -10 + 7.1 = -2.9$$

✅ *Check It Out 2* Find the linear function f whose input and output values are given in Table 2. Evaluate $f(-1)$.

x	f(x)
1	4.5
2.5	7.5

Table 2

Our next example approaches a real-world application in a completely different manner. It is important to keep in mind that these types of problems never present themselves in a uniform fashion in the real world. One of the challenges of solving them is to take the various facts that are given and put them into the context of the mathematics that you have learned.

Example 3   **Depreciation Model**

Table 3 gives the value of a commercial printer at two different times after its purchase. Answer the following questions.

a. Identify the input and output variables.

b. Express the value of the printer as a linear function of the number of years after its purchase.

c. Using the function found in part (c), find the original purchase price.

d. Assuming that the value of the printer is a linear function of the number of years after its purchase, when will the printer's value reach \$0?

Time after purchase (years)	Value (dollars)
2	10,000
3	9,000

Table 3

Solution

a. The input variable, t, is the number of years after purchase of the printer. The output variable, v, is the value of the printer after t years.

b. We first compute the slope with the two data points $(2, 10000)$, $(3, 9000)$.

$$\text{Slope} = m = \frac{9000 - 10000}{3 - 2} = \frac{-1000}{1} = -1000$$

Then, using the point-slope form for a linear function, with $(2, 10000)$ as the point, we have

$$v(t) = -1000(t - 2) + 10000 \quad \text{Substitute into point-slope form of a linear function}$$
$$v(t) = -1000t + 2000 + 10000 \quad \text{Remove parentheses}$$
$$v(t) = -1000t + 12000 \quad \text{Simplify}$$

c. The original purchase price would correspond to the value of the printer 0 years after purchase. Substituting $t = 0$ into the equation above, we get

$$v(0) = -1000(0) + 12000 = 12000$$

Thus, the printer originally cost \$12000. Note that this is also the v-intercept of the linear function $v(t) = -1000t + 12000$.

d. To find out when the printer's value will reach \$0, we must set $v(t)$ equal to 0 and solve for t.

$$0 = -1000t + 12000 \quad \text{Set } v(t) \text{ equal to zero}$$
$$-12000 = -1000t \quad \text{Isolate the } t \text{ term}$$
$$t = 12 \quad \text{Solve for } t$$

Thus, it will take 12 years for the printer to reach a value of \$0. To illustrate the decrease in value, we can make a table of values of t and $v(t)$. See Table 4.

Note that for each 2-year increase in its age, the value of the printer goes down by the same amount, \$2000.

t	v(t)
0	12,000
2	10,000
4	8,000
6	6,000
8	4,000
10	2,000
12	0

Table 4

✓ *Check It Out 3* Rework Example 3 if the value of the printer is given in Table 5.

Time after purchase (years)	Value (dollars)
2	10,000
5	4,000

Table 5

Linear Models Using Curve-Fitting

Using mathematics to model real-world situations involves a variety of techniques. We have already used one such technique: taking a problem statement given in English and translating it into a mathematical expression. In Example 1, the linear function that arose was an exact representation of the problem at hand. However, many real world problems can only be *approximately* represented by a linear function.

In this section, we will analyze trends in a set of actual data, and we will gain some experience in generating a model that approximates a set of data. One common technique used in modeling a real-world situation involves finding a suitable mathematical function whose graph closely resembles the plot of a set of data points. This technique is known as **fitting data points to a curve**, or just **curve-fitting**. It is also referred to as **regression**.

Guidelines for Curve-Fitting

- Examine the given set of data and decide which variable would be most appropriate as the **input variable** and which as the **output variable**.

- Set up a coordinate system, and plot the given points, either by hand or by using a graphing utility.

- Observe the trend in the data points—does it look like the graph of a function you are familiar with?

- If you find that your data represents a linear function, $y = mx + b$, then you must find the values of m and b that best approximate the data. The mathematics behind finding the "best" values for m and b is beyond the scope of this text. However, you can use your graphing utility to find these values.

Using Technology

Curve fitting features can be accessed under the statistics options in your graphing utility.

Keystroke Appendix: Section 12

Body weight (g)	Heart weight (g)
281.58	1.0353
285.03	1.0534
290.03	1.0726
295.16	1.1034
300.63	1.1842
313.46	1.2673

Table 6

Example 4 **Modeling the Relation of Body Weight to Organ Weight**

Table 6 gives the body weights of laboratory rats and the weights of their hearts, in grams. All data points are given to five significant digits. (*Source:* NASA life sciences data archive)

a. Let x denote the body weight. Make a scatter plot of the heart weight h vs. x.

b. State, in words, any general observations you can make about the data.

c. Find an expression for the *linear* function that best fits the given data points.

d. Compare the actual heart weights for the given body weights with the heart weights predicted by your function. What do you observe?

e. If a rat weighs 308 grams, use your model to predict its heart weight.

Solution

a. The scatter plot is given in Figure 2.

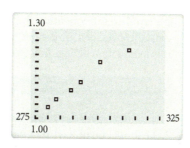

Figure 2

b. From the scatter plot, it appears that the weight of the heart increases with the weight of the rat.

c. Using the curve-fitting features of a graphing utility, we obtain the following equation for the "best-fit" line through these points:

$$h(x) = 0.0075854x - 1.1131$$

We have retained five significant digits in the output since the given data values were given to that accuracy. See Figure 3 and Figure 4 for graphing calculator output.

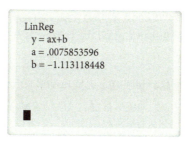

Figure 3

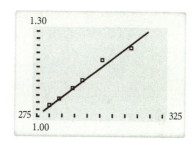

Figure 4

d. Table 7 gives the actual heart weights and the weights of the heart that were predicted by the model. We see that the values are close, though not exact. When generating a model from a set of data, you should always compare the given values to those predicted by the model to see if the model is valid.

Body weight (g)	Actual heart weight (g)	Predicted heart weight (g)
281.58	1.0353	1.0228
285.03	1.0534	1.0490
290.03	1.0726	1.0869
295.16	1.1034	1.1258
300.63	1.1842	1.1673
313.46	1.2673	1.2646

Table 7

Note Linear models should be used only as an approximation for predicting values close to those for which actual data is available. Generally, linear models will be less accurate as we move further away from the given data points.

e. Substituting the weight of 308 grams into $h(x)$, we have

$$h(308) = 0.0075854(308) - 1.1131 = 1.2232$$

Thus, the rat's heart weight is 1.2232 grams, to five significant digits.

Check It Out 4 Use the model in Example 4 to predict the heart weight of a rat weighing 275 grams.

Direct Variation

Linear models of the form $f(x) = kx$, $k > 0$, are often referred to as models involving **direct variation**. That is, the ratio of the output value to the input value is a positive constant. Models of this type occur frequently in business and scientific applications. For example, if a car travels at a constant speed of 45 miles per hour, then the distance traveled in x hours is given by $f(x) = 45x$. This a model involving direct variation.

Direct Variation

If a model gives rise to a linear function of the form $f(x) = kx$ or $y = kx$, $k > 0$, we say that the model involves **direct variation**. Other ways of stating this are:

- y varies directly as x.

- y is directly proportional to x.

The number k is called the **variation constant** or the **constant of proportionality**.

Example 5   Temperature-Volume Relation

The volume of a fixed mass of gas is directly proportional to its temperature. If the volume of a gas is 40 cc (cubic centimeters) at 25°C (degrees Celsius), find the following.

a. The variation constant k in the equation

$$V = kT$$

b. The volume of the same gas at 50°C.

c. The temperature of the same gas if the volume is 30 cc.

Solution

a. To find k, substitute the values for volume and temperature to get

$$40 = k(25) \quad \text{Use } V = 40 \text{ and } T = 25$$

$$\frac{40}{25} = k \quad \text{Divide by 25}$$

$$\frac{8}{5} = k \quad \text{Simplify}$$

Thus, $k = \frac{8}{5}$ and the equation is

$$V = \frac{8}{5}T$$

b. Substituting $T = 50$ in the equation $V = \frac{8}{5}T$,

$$V = \frac{8}{5}T \quad \Rightarrow \quad V = \frac{8}{5}(50) = 80$$

Thus, at 50°C, the volume is 80 cc.

c. Now substitute $V = 30$ into the $V = \frac{8}{5}T$.

$$30 = \frac{8}{5}T \quad \Rightarrow \quad T = \frac{5}{8}(30) = 18.75$$

Thus, the volume is 30 cc at at 18.75°C.

 Check It Out 5 Find the constant k in the equation $V = kT$ if $V = 35$ cc and $T = 45$°C.

Inverse Variation

Our focus so far in this section has been on linear models. Situations that are not modeled by linear functions are called **nonlinear models**. We briefly discuss one such nonlinear model which occurs frequently in science and business.

Suppose a bus travels a distance of 200 miles. The time t it takes to reach its destination depends on its speed, r, and the distance traveled, and can be determined by the equation $rt = 200$, or $t = \frac{200}{r}$. The greater the speed, the less time it will take to travel 200 miles. This is an example of **inverse variation**.

Inverse Variation

If a model gives rise to a function of the form $f(x) = \frac{k}{x}$ or $xy = k$, $k > 0$, we say that the model involves **inverse variation**. Other ways of stating this are

- y varies inversely as x, and

- y is inversely proportional to x.

The number k is called the **variation constant** or the **constant of proportionality**.

A complete discussion of functions of the form $f(x) = \frac{k}{x}$ will be presented in Section 3.7. We next examine a model involving inverse variation.

Example 6　Price-Demand Relation

The price of a product is inversely proportional to its demand. That is, $P = \frac{k}{q}$, where P is the price per unit and q is the number of products demanded. If 3,000 units are demanded at \$10 per unit, how many units are demanded at \$6 per unit?

Solution　Since k is a constant, we have

$$k = Pq = (10)(3000) = 30{,}000$$

If $P = 6$, we then have

$$6 = \frac{30{,}000}{q} \qquad \text{Substitute } P = 6, k = 30{,}000$$

$$6q = 30{,}000 \qquad \text{Multiply by } q$$

$$q = \frac{30{,}000}{6} = 5000 \quad \text{Solve for } q$$

Thus, 5,000 units can be demanded at \$6 per unit.

Check It Out 6　If $P = \frac{k}{q}$, where P is the price per unit and q is the number of products demanded, find k if $q = 2500$ and $P = 5$.

Exercises 2.8

 Skills

In Exercises 1–6, for each table of values, find the linear function f having the given input and output values.

1.

x	f(x)
3	5
4	9

2.

x	f(x)
2	10
6	7

3.

x	f(x)
3.1	−2.5
5.6	3.5

4.

x	f(x)
1.7	15
3.2	10

5.

x	f(x)
30	600
50	900

6.

x	f(x)
60	1000
80	1500

In Exercises 7–22, find the variation constant and the corresponding equation for each situation.

7. Let y vary directly as x, and $y = 40$ when $x = 10$.

8. Let y vary directly as x, and $y = 100$ when $x = 25$.

9. Let y vary inversely as x, and $y = 5$ when $x = 2$.

10. Let y vary inversely as x, and $y = 6$ when $x = 8$.

11. Let y vary directly as x, and $y = 35$ when $x = 10$.

12. Let y vary directly as x, and $y = 80$ when $x = 15$.

13. Let y vary inversely as x, and $y = 2.5$ when $x = 6$.

14. Let y vary inversely as x, and $y = 4.2$ when $x = 10$.

15. Let y vary directly as x, and $y = \frac{1}{3}$ when $x = 2$.

16. Let y vary directly as x, and $y = \frac{1}{5}$ when $x = 3$.

17. Let y vary inversely as x, and $y = \frac{3}{4}$ when $x = 8$.

18. Let y vary inversely as x, and $y = \frac{5}{2}$ when $x = 6$.

19. The variable y is directly proportional to x, and $y = 35$ when $x = 7$.

20. The variable y is directly proportional to x, and $y = 48$ when $x = 8$.

21. The variable y is inversely proportional to x, and $y = 14$ when $x = 7$.

22. The variable y is inversely proportional to x, and $y = 4$ when $x = 12$.

Applications

23. Utility Bill A monthly long-distance bill is $4.50 plus $0.07 for each minute of use. Express the amount of the long-distance bill as a linear function of the number of minutes of use.

24. Printing The total cost of printing small booklets is $500 (the fixed cost) plus $2 for each booklet printed. Express the total cost of producing booklets as a function of the number of booklets produced.

25. Temperature Scales If 32° Fahrenheit corresponds to 0° Celsius and 212° Fahrenheit corresponds to 100° Celsius, find a linear function that converts a Fahrenheit temperature to a Celsius temperature.

26. Demand Function At $5 each, 300 hats will be sold. But at $3 each, 800 hats will be sold. Express the number of hats sold as a linear function of the price per hat.

27. Physics The volume of a gas varies directly with the temperature. Find k if the volume is 50 cc at 40°C. What is the volume if the temperature is 60°C?

28. **Tax Rules** According to Internal Revenue Service rules, the depreciation amount of an item is directly proportional to its purchase price. If the depreciation amount of a $2,500 piece of equipment is $500, what is the depreciation amount of a piece of equipment purchased at $4,000?

   (*Source*: Internal Revenue Service)

29. **Construction** The rise of a roof (y) is directly proportional to its run (x). If the rise is 5 feet when the run in 8 feet, find the rise when the run is 20 feet.

30. **Electronics Demand** It is known that 10,000 units of a computer chip is demanded at $50 per chip. How many units are demanded at $60 per chip, if price varies inversely as the number of chips?

31. **Work and Rate** The time required to do a job, t, varies inversely as the number of people, p, who work on the job. Assume all work on the job at the same rate. If it takes 10 people to paint the inside of an office building in 5 days, how long will it take 15 people to finish the same job?

32. **Travel** The time t required to drive a fixed distance varies inversely as the speed of the car. If it takes two hours to drive from New York to Philadelphia at a speed of 55 miles per hour, how long will it take to drive the same route at a speed of 50 miles per hour?

33. **Rental Cost** A 10-foot U-Haul truck for in-town use rents for $19.95 per day plus $0.99 per mile. You are planning to rent the truck for just one day.

   (*Source*: www.uhaul.com)

   a. Write the total cost of rental as a linear function of the number of miles driven.
   b. Give the slope and y-intercept of the graph of this function and explain their significance.
   c. How much will it cost to rent the truck if you drive a total of 56 miles?

34. **Depreciation** The following table gives the value of a personal computer purchased in 2008 at two different times after its purchase.

Time after purchase (years)	Value (dollars)
2	1000
4	500

   a. Express the value of the computer as a linear function of the number of years after its purchase.
   b. Using the function found in (a), find the original purchase price.
   c. Assuming that the value of the computer is a linear function of the number of years after its purchase, when will the computer's value reach $0?
   d. Sketch a graph of this function, indicating the x- and y-intercepts.

35. **Website Traffic** The number of visitors to a popular web site grew from 40 million in September 2011 to 58 million in March 2012.

   a. Let t be the number of *months* since September 2011. Plot the given data points, with N, the number of visitors, on the vertical axis and t on the horizontal axis. You may find it easier to represent N in millions; that is, represent 40 million visitors as 40 and set the scale for the vertical axis accordingly.
   b. From your plot, find the slope of the line between the two points you plotted.
   c. What does the slope represent?
   d. Find the expression that gives the number of visitors as a linear function of t.
   e. How many visitors are expected in July 2012?

36. **Communications** A certain piece of communications equipment cost $123 to manufacture in 2010. Since then, its manufacturing cost has been decreasing by $4.50 year.

   a. If the input variable, t, is the number of years since 2010, find a linear function that gives the manufacturing cost as a function of t.
   b. If the trend continues, what is the cost of manufacturing the equipment in 2013?
   c. When will the manufacturing cost of the equipment reach $78?

37. **Beverage Sales** In 2001, Americans consumed an average of 18.2 gallons of bottled water per person. In 2011, that number climbed to 29.2 gallons per person.   (*Source:* Beverage Marketing Association)

    a. Find a linear function that describes the gallons of bottled water consumed per person as a function of t. Let t denote the number of years since 2001.
    b. What is the slope of the corresponding line, and what does it signify?
    c. What is the y-intercept of the corresponding line, and what does it signify?

38. **Automobile Costs** A 2013 Subaru Outback wagon costs $23,500 and gets 22 miles per gallon, according to the website, fueleconomy.gov. Assume that gasoline costs $4 per gallon.

    a. What is the cost of gasoline per mile for the Outback wagon?
    b. Assume that the total cost of owning the car consists of the price of the car and the cost of gasoline. (In reality, the total cost is much more than this.) For the Subaru Outback, find a linear function describing the total cost, with the input variable being the number of miles driven.
    c. What is the slope of the graph of the function in part (b), and what does it signify?
    d. What is the y-intercept of the graph of the function in part (b), and what does it signify?

39. **Consumer Behavior** Linear models can be used to predict buying habits of consumers. Suppose that a survey found that in 2008, 20% of a surveyed group bought designer frames for their eyeglasses. In 2011, the percentage climbed to 29%.

    a. Assuming that the percentage of people buying frames is a linear function of time, find an equation for the percentage of people buying designer frames. Let t correspond to the number of years since 2008.
    b. Use your equation to predict the percentage of people buying designer frames in 2014.
    c. Use your equation to predict when the percentage of people who buy designer frames will reach 50%.
    d. Do you think you can use this model to predict the percentage of people buying designer frames in the year 2038? Why or why not?
    e. From your answer to the previous question, what do you think are some limitations to this model?

40. **Economy** In the year 2000, the average hourly earnings of production workers nationwide rose steadily from $13.50 per hour in January to $14.03 per hour in December.   (*Source:* Bureau of Labor Statistics)

    a. Create a linear model that expresses average hourly earnings as a function of time, t. How would you define t?
    b. Using your function, how much was the average worker earning in March? in October?
    c. How fast is the average hourly wage increasing per month?

41. **Traffic Flow** In 2007, the average weekday volume of traffic on a particular stretch of the Princess Parkway was 175,000 vehicles. By 2011, the volume had increased to 200,000 per weekday.

    a. By how much did the traffic increase per year? Mathematically, what does this quantity represent?
    b. Create a linear model for the volume of traffic as a function of time and use it to determine the average weekday traffic flow for 2013.

42. **Sports Revenue** The revenues for the hockey teams Dallas Stars and New York Rangers are given in the following table for the years 2008 and 2010.   (*Source:* Forbes Magazine)

Year	Stars' Revenue ($ million)	Rangers' Revenue ($ million)
2008	105	137
2010	95	154

    a. Express the revenue, R, for the Stars as a linear function of time, t. Let t correspond to the number of years since 2008.
    b. Express the revenue, R, for the Rangers as a linear function of time, t. Let t correspond to the number of years since 2008.
    c. Project the Stars' revenue for the year 2012.
    d. Project the Rangers' revenue for the year 2012.

43. Population Mobility The following table lists historical mobility rates (the percentage of people who had a change of residence) for selected years between 1960 and 2000. (*Source:* U.S. Census Bureau)

Year	Percentage
1960	20.6
1970	18.7
1980	18.6
1990	16.3
2000	16.1

a. What general trend do you notice in these figures?

b. Fit a linear function to this set of points, using the number of years since 1960 as the independent variable.

 44. Aging The following table lists the population of those in the U.S. who are 65 years of age or older, in millions. (*Source: Statistical Abstract of the United States*)

Year	Population, 65 or older (in millions)
1995	31.7
2000	32.6
2005	35.2
2010	38.6

a. What general trend do you notice in these figures?

b. Fit a linear function to this set of points, using the number of years since 1995 as the independent variable.

c. Use your function to predict the number of people over 65 in the year 2013.

Concepts

45. Can $y = 2x + 5$ represent an equation for direct variation?

46. Does the following table of values represent a linear model? Explain.

x	y
−1	4
0	7
1	16
2	30

Chapter 2 Summary

Section 2.1 The Coordinate System; Lines and Their Graphs

Definition of slope [Review Exercises 1–4]

The **slope** of a line containing the points (x_1, y_1) and (x_2, y_2) is given by

$$m = \frac{y_2 - y_1}{x_2 - x_1}$$

where $x_1 \neq x_2$.

Equations of lines [Review Exercises 5–18]

The **point-slope form** of the equation of a line with slope m and passing through (x_1, y_1) is $y - y_1 = m(x - x_1)$.

The **slope-intercept form** of the equation of a line with slope m and y-intercept $(0, b)$ is $y = mx + b$.

The **general form** of the equation of a line is $Ax + By + C = 0$, where A, B, C are real numbers, and A, B are not both zero.

The equation of a **horizontal line** passing through (a, b) is given by $y = b$. The equation of a **vertical line** through (a, b) is given by $x = a$.

Parallel and perpendicular lines [Review Exercises 11–14]

Nonvertical **parallel lines** have the *same slope*. All vertical lines are parallel to each other.

Perpendicular lines have slopes that are *negative reciprocals* of each other. Vertical and horizontal lines are always perpendicular to each other.

Section 2.2 Coordinate Geometry, Circles, and Other Equations

Distance between two points [Review Exercises 19–22, 34]

The **distance** d between the points (x_1, y_1) and (x_2, y_2) is given by

$$d = \sqrt{(x_2 - x_1)^2 + (y_2 - y_1)^2}$$

Midpoint of a line segment [Review Exercises 19-22]

The coordinates of the **midpoint** of the line segment joining the points (x_1, y_1) and (x_2, y_2) are

$$\left(\frac{x_1 + x_2}{2}, \frac{y_1 + y_2}{2} \right)$$

Equation of a circle in standard form [Review Exercises 23–30]

The circle with center at (h, k) and radius r is the set of all points (x, y) satisfying the equation
$$(x - h)^2 + (y - k)^2 = r^2$$

The general form of the equation of a circle [Review Exercises 31, 32]

The **general form of the equation of a circle** is given by
$$x^2 + y^2 + Dx + Ey + F = 0$$

By completing the square, we can rewrite the general equation in standard form.

Graphs of equations [Review Exercises 33, 34]

To graph an equation, solve the given equation for y in terms of x. Then generate a table of x and y values to help you sketch the graph of the equation.

Section 2.3 **Functions**

Definition of a function   [Review Exercises 35–38]

A **function** establishes a correspondence between a set A and a set B such that for each element in A, there is *exactly one* corresponding element in B.

Piecewise-defined functions   [Review Exercises 39–42]

A **piecewise-defined function** is a function that is defined using different expressions, corresponding to different conditions satisfied by the independent variable.

Domain and range of a function   [Review Exercises 43–50]

The **domain** of a function is the set of all allowable input values for which the function is defined. For a mathematical function in this textbook, the function is defined only when the output is a real number.

The **range** of a function is the set of all output values that correspond to at least one element of the domain of the function.

Section 2.4 **Graphs of Functions**

Graph of a function   [Review Exercises 53–60]

The **graph** of f is the set of all points $(x, f(x))$ such that x is in the domain of f.

The vertical line test for functions   [Review Exercise 61]

Any vertical line can intersect the graph of a function at most once.

Intercepts and zeros of functions   [Review Exercise 62]

The x-**intercept** is a point at which the graph of a function crosses the x-axis. The first coordinate of an x-intercept is a value of x such that $f(x) = 0$. Values of x satisfying $f(x) = 0$ are called **zeros** of the function f. The second coordinate of the x-intercept is 0.

The y-**intercept** is a point at which the graph of a function crosses the y-axis. The coordinates of the y-intercept are $(0, f(0))$.

Equations and functions   [Review Exercises 63, 64]

To see if an equation defines y as a function of x, first solve for y. Then check if there is only one y value for each allowable value of x.

Section 2.5 **Analyzing the Graph of a Function**

Odd and even functions   [Review Exercises 65–72]

A function is **symmetric with respect to the y-axis** if

$$f(x) = f(-x), \text{ for each } x \text{ in the domain of } f$$

Functions with this property are called **even functions**.

A function is **symmetric with respect to the origin** if

$$f(-x) = -f(x), \text{ for each } x \text{ in the domain of } f$$

Functions with this property are called **odd functions**.

Increasing, decreasing, and constant functions   [Review Exercises 73–78]

A function f is **increasing** on an open interval I, if, for any a, b in the interval with $a < b$, we have $f(a) < f(b)$.

A function f is **decreasing** on an open interval I, if, for any a, b in the interval with $a < b$, we have $f(a) > f(b)$.

A function f is **constant** on an open interval I, if, for any a, b in the interval, we have $f(a) = f(b)$.

Average rate of change   [Review Exercises 79–82]

The **average rate of change** of a function f on an interval $[x_1, x_2]$ is given by

$$\text{Average rate of change} = \frac{f(x_2) - f(x_1)}{x_2 - x_1}$$

Section 2.6 The Algebra of Functions

Arithmetic operations on functions   [Review Exercises 83–88, 93–96]

Given functions f and g, then for each x in the domain of *both* f and g, the **sum**, **difference**, **product**, and **quotient** of f and g are defined as follows.

$$(f + g)(x) = f(x) + g(x)$$
$$(f - g)(x) = f(x) - g(x)$$
$$(fg)(x) = f(x) \cdot g(x)$$
$$\left(\frac{f}{g}\right)(x) = \frac{f(x)}{g(x)}, \text{ where } g(x) \neq 0$$

Definition of a composite function   [Review Exercises 89–92, 97–108]

Let f and g be functions. The **composite function** $f \circ g$ is a function defined as

$$(f \circ g)(x) = f(g(x))$$

The domain of $f \circ g$ is the set of all x in the domain of g such that $g(x)$ is in the domain of f.

Difference quotient   [Review Exercises 109, 110]

The **difference quotient** of a function f is an expression of the form $\dfrac{f(x + h) - f(x)}{h}$, $h \neq 0$.

Section 2.7 Transformations of the Graph of a Function

Vertical and horizontal shifts of the graph of a function   [Review Exercises 111–114, 123, 124, 128, 131]

Let f be a function and c be a positive constant.

Vertical shifts of the graph of $f(x)$
The graph of $g(x) = f(x) + c$ is the graph of $f(x)$ shifted c units **upward**.
The graph of $g(x) = f(x) - c$ is the graph of $f(x)$ shifted c units **downward**.

Horizontal shifts of the graph of $f(x)$
The graph of $g(x) = f(x - c)$ is the graph of $f(x)$ shifted c units **to the right**.
The graph of $g(x) = f(x + c)$ is the graph of $f(x)$ shifted c units **to the left**.

Vertical scalings and reflections across the horizontal axis   [Review Exercises 115–119, 125, 127, 130, 132]

Let f be a function and c be a positive constant.

Vertical scalings of the graph of $f(x)$

If $c > 1$, the graph of $g(x) = cf(x)$ is the graph of $f(x)$ **stretched vertically** away from the x-axis, with the y-coordinates of $f(x)$ multiplied by c.

If $0 < c < 1$, the graph of $g(x) = cf(x)$ is the graph of $f(x)$ **compressed vertically** toward the x-axis, with the y-coordinates of $f(x)$ multiplied by c.

Reflection of the graph of $f(x)$ across the x-axis

The graph of $g(x) = -f(x)$ is the graph of $f(x)$ **reflected across the x-axis**.

Horizontal scalings and reflections across the vertical axis   [Review Exercises 120–122, 126, 129]

Let f be a function and c be a positive constant.

Horizontal scalings of the graph of $f(x)$

If $c > 1$, the graph of $g(x) = f(cx)$ is the graph of $f(x)$ **compressed horizontally** toward the y-axis, scaled by a factor of $\frac{1}{c}$.

If $0 < c < 1$, the graph of $g(x) = f(cx)$ is the graph of $f(x)$ **stretched horizontally** away from y-axis, scaled by a factor of $\frac{1}{c}$.

Reflection of the graph of $f(x)$ across the y-axis

The graph of $g(x) = f(-x)$ is the graph of $f(x)$ **reflected across the y-axis**.

Section 2.8 Linear Functions and Models; Variation

Guidelines for finding a linear model   [Review Exercises 133, 134, 141, 142]

Begin by reading the problem a couple of times to get an idea of what is going on.

Identify the *input* and *output* variables.

Sometimes, you will be able to write down the linear function for the problem by just reading the problem and "translating" the words into mathematical symbols.

At other times, you will have to look for two data points within the problem to find the slope of your line. Only after you perform this step can you find the linear function. Interpret the slope and y-intercept both verbally and graphically.

Direct and inverse variation   [Review Exercises 135–140]

Direct variation: a model giving rise to a linear function of the form $f(x) = kx$ or $y = kx$, $k > 0$.

Inverse variation: a model giving rise to a function of the form $f(x) = \frac{k}{x}$ or $xy = k$, $k > 0$. In both models, $k > 0$ is called a **variation constant** or **constant of proportionality**.

Chapter 2 Review Exercises

Section 2.1

In Exercises 1–4, for each pair of points, find the slope of the line passing through the points (if the slope is defined). If the slope is undefined, state so.

1. $(-2, 0), (0, 5)$

2. $(1, -6), (-4, 5)$

3. $\left(\frac{2}{3}, 1\right), \left(-\frac{1}{2}, 3\right)$

4. $(4.1, 5.5), (2.1, -3.5)$

In Exercises 5–14, find an equation of the line with the given properties, express the equation in the form $y = mx + b$ or $x = c$, and then graph the line.

5. Passing through the point $(4, -1)$ and with slope -2

6. Vertical line through the point $(5, 0)$

7. x-intercept: $(-2, 0)$; y-intercept: $(0, 3)$

8. x-intercept: $(1, 0)$; y-intercept: $(0, -2)$

9. Passing through the points $(-8, -3)$ and $(12, -7)$

10. Passing through the points $(-3, -5)$ and $(0, 5)$

11. Perpendicular to the line $x - y = 1$ and passing through the point $(-1, 2)$

12. Perpendicular to the line $-3x + y = 4$ and passing through the point $(2, 0)$

13. Parallel to the line $x + y = 3$ and passing through the point $(3, -1)$

14. Parallel to the line $-2x + y = -1$ and passing through the point $(0, 3)$

In Exercises 15 and 16, express the given equations in the form $y = mx + b$ or $x = c$. Graph the line.

15. $3x + 6 = 0$

16. $3x + 4y + 8 = 0$

17. Salary A commissioned salesperson's earnings can be determined by

$$S = 800 + 0.1x$$

where x is the total amount of sales generated by the salesperson per week. How much sales should be generated in order to earn a salary of $1500?

18. Sales The number of gift-boxed pens sold per year since 2008 by The Pen and Quill Shop is given by $h = 400 + 80t$. Here, t is the number of years since 2008.

  a. How many gift-boxed pens were sold in 2011?

  b. What is the value of h when $t = 0$, and what does it represent?

  c. When will 1120 gift-boxed pens be sold?

Section 2.2

In Exercises 19–22, find the distance between the points and the midpoint of the line segment joining them.

19. $(0, -2), (-3, 3)$ **20.** $(-4, 6), (8, -10)$

21. $(6, 9), (11, 13)$ **22.** $(-7, -6), (13, 9)$

In Exercises 23–26, write the standard form of the equation of the circle having the given radius and center. Sketch the circle.

23. $r = 6$; center: $(-1, 2)$ **24.** $r = 3$; center: $(4, 1)$

25. $r = \dfrac{1}{2}$; center: $(0, -1)$ **26.** $r = \sqrt{3}$; center: $(-2, 1)$

In Exercises 27–30, find the center and the radius of circle having the given equation. Sketch the circle.

27. $(x - 3)^2 + (y - 2)^2 = \dfrac{1}{4}$ **28.** $x^2 + (y - 12)^2 = \dfrac{1}{9}$

29. $x^2 + y^2 - 2x + 2y - 7 = 0$ **30.** $x^2 + y^2 - 2x + 6y + 1 = 0$

In Exercises 31 and 32, sketch the graph of the given equation.

31. $-4x + y^2 = 0$ **32.** $|x| - y = 3$

33. Gardening A circular flower border is to be planted around a statue, with the statue as the center. The distance from the statue to any point on the inner boundary of the flower border is 20 feet. What is the equation of the outer boundary of the border, if the flower border is two feet wide? Use a coordinate system with the statue at $(0, 0)$.

34. Distance Two boats start from the same point. One travels directly west at 30 miles per hour. The other travels directly north at 44 miles per hour. How far apart will they be from each other after an hour and a half?

Section 2.3

In Exercises 35–38, evaluate the following for each function.

a. $f(4)$ **b.** $f(-2)$ **c.** $f(a)$ **d.** $f(a + 1)$

35. $f(x) = 2x^2 - 1$ **36.** $f(x) = \dfrac{1}{x^2 + 1}$ **37.** $f(x) = \sqrt{x^2 - 4}$ **38.** $f(x) = \dfrac{x + 1}{x - 1}$

In Exercises 39–42, let $f(x) = \begin{cases} -2x + 1, & \text{if } x < 0 \\ x, & \text{if } x \geq 0 \end{cases}$ *. Evaluate the following.*

39. $f(0)$ **40.** $f(-1)$ **41.** $f(3)$ **42.** $f\left(\dfrac{1}{2}\right)$

In Exercises 43–50, find the domain of the function. Write your answer in interval notation.

43. $h(x) = x^3 + 2$ **44.** $h(x) = \sqrt{6 - x}$ **45.** $f(x) = \dfrac{7}{x - 2}$

46. $g(x) = \dfrac{1}{x^2 - 4}$ **47.** $f(x) = \dfrac{-3}{x^2 + 2}$ **48.** $f(x) = -\dfrac{1}{(x + 5)^2}$

49. $f(x) = |x| + 1$ **50.** $g(x) = \dfrac{x}{(x + 2)(x - 3)}$

51. Geometry If the surface area of a sphere is $S(r) = 4\pi r^2$, find and interpret $S(3)$. Why would only positive values of r make sense for this function?

52. Pediatrics The length of an infant 21 inches long at birth can be modeled by the linear function $h(t) = 21 + 1.5t$, where t is the age of the infant, in months, and $h(t)$ is in inches.   (*Source*: Growth Charts, Centers for Disease Control)

    a. What is the slope of the line and what does it represent?

    b. What is the length of the infant at 6 months?

Section 2.4

In Exercises 53–58, graph the function and determine its domain and range.

53. $h(x) = 3x + \dfrac{1}{2}$ **54.** $f(x) = \sqrt{5 - x}$ **55.** $g(x) = -2x^2 + 3$

56. $g(x) = x^2 - 4$ **57.** $f(x) = -2|x|$ **58.** $f(x) = |x| - 2$

In Exercises 59 and 60, sketch a graph of the following piecewise-defined functions.

59. $g(x) = \begin{cases} x, & \text{if } x \le -2 \\ 2.5, & \text{if } x > -2 \end{cases}$ **60.** $f(s) = \begin{cases} 4, & \text{if } x < 0 \\ s^2, & \text{if } 0 \le x < 2 \\ -3, & \text{if } 2 \le x \le 3 \end{cases}$

In Exercise 61, determine which of the following are graphs of a function. Explain your answer.

61. a.   **b.**   **c.**   **d.**

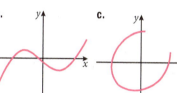

In Exercise 62, evaluate $f(-2)$, $f(1)$ and find the x- and y-intercepts for f given by the graph.

62.

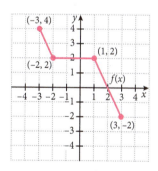

In Exercises 63 and 64, determine if the equation describes y as a function of x.

63. $2x + 3y = 6$ **64.** $y^2 + 9x = 0$

Section 2.5

In Exercises 65–70, decide if the function is odd, even, or neither, using the definitions.

65. $f(x) = -x$ **66.** $f(x) = 3x + 4$ **67.** $f(x) = -x^2 + 5$

68. $f(x) = x^5 + x^2$ **69.** $f(x) = -3x^5 - x$ **70.** $f(x) = 2x^4 + x^2 + 1$

In Exercises 71 and 72, use a graphing utility to decide if each of the following functions is odd, even, or neither. Confirm your result using the definitions.

71. $f(x) = 2x^3 + 3x - 5$ **72.** $f(x) = -x^4 + x^2 + 3$

Exercises 73–78 apply to the function f given by the following graph.

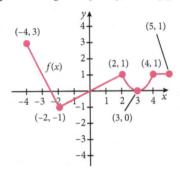

73. Find the interval(s) where f is increasing.

74. Find the interval(s) where f is decreasing.

75. Find the interval(s) where f is constant.

76. Find the average rate of change of f in the interval $[-1, 1]$.

77. Find the average rate of change of f in the interval $[3, 4]$.

78. Find the y-intercept.

In Exercises 79–82, for each function, find the average rate of change on the given interval.

79. $f(x) = -2x + 4$, interval: $[2, 3]$

80. $f(x) = -2x^3 + 3x$, interval: $[-3, -1]$

81. $f(x) = 2x - 3$, interval: $[a, a + h]$, $h > 0$

82. $f(x) = x^2 - 2x$, interval: $[a, a + h]$, $h > 0$

Section 2.6

In Exercises 83–88, for the given functions f and g, find each combination of functions and identify its domain.

a. $(f + g)(x)$ **b.** $(f - g)(x)$ **c.** $(fg)(x)$ **d.** $\left(\dfrac{f}{g}\right)(x)$

83. $f(x) = 4x^2 + 1; g(x) = x + 1$ **84.** $f(x) = 3x - 1; g(x) = x^2 - 4$

85. $f(x) = \left(\dfrac{1}{2x}\right); g(x) = \dfrac{1}{x^2 + 1}$ **86.** $f(x) = \sqrt{x}; g(x) = \dfrac{1}{\sqrt{x}}$

87. $f(x) = \dfrac{2}{x - 4}; g(x) = 3x^2$ **88.** $f(x) = \dfrac{2}{x + 3}; g(x) = \dfrac{x + 1}{x - 4}$

In Exercises 89–92, use f and g given by the following tables of values.

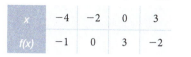

x	−4	−2	0	3
f(x)	−1	0	3	−2

x	−1	0	3
g(x)	−2	0	5

89. Evaluate $(f \circ g)(-1)$.

90. Evaluate $(f \circ g)(0)$.

91. Evaluate $(g \circ f)(0)$.

92. Evaluate $(g \circ f)(-4)$.

In Exercises 93–100, let $f(x) = 3x - 1$, $g(x) = -2\sqrt{x}$, and $h(x) = 4x$. Evaluate each of the following.

93. $(f + g)(4)$

94. $(g - h)(9)$

95. $\left(\dfrac{f}{h}\right)(2)$

96. $(fh)(3)$

97. $(f \circ h)(-1)$

98. $(h \circ f)(2)$

99. $(f \circ g)(9)$

100. $(g \circ f)(3)$

In Exercises 101–108, find expressions for $(f \circ g)(x)$ and $(g \circ f)(x)$, and give the domain of $f \circ g$ and the domain of $g \circ f$.

101. $f(x) = -x^2 + 4$, $g(x) = x - 2$

102. $f(x) = 2x + 5$, $g(x) = \dfrac{x - 5}{2}$

103. $f(x) = -x^2 + 3x$, $g(x) = x - 3$

104. $f(x) = -\dfrac{2}{x}$, $g(x) = x + 5$

105. $f(x) = \dfrac{1}{x - 2}$, $g(x) = x^2 + x$

106. $f(x) = \sqrt{x + 2}$, $g(x) = 2x + 1$

107. $f(x) = \dfrac{x}{x + 3}$, $g(x) = |x|$

108. $f(x) = x^2 - 4x + 4$, $g(x) = \dfrac{1}{x}$

In Exercises 109 and 110, find the difference quotient $\dfrac{f(x + h) - f(x)}{h}$, $h \neq 0$, for each function.

109. $f(x) = 4x - 3$

110. $f(x) = -3x^2$

Section 2.7

In Exercises 111–122, use transformations to sketch the graph of each function.

111. $g(x) = |x| - 6$

112. $f(s) = (s - 5)^2$

113. $h(x) = |x - 1| + 2$

114. $g(x) = (x + 4)^2 - 3$

115. $f(x) = 2\sqrt{x}$

116. $h(s) = -|s|$

117. $f(s) = -(s + 4)^2$

118. $p(x) = -\sqrt{x} + 1$

119. $f(x) = -3(x + 2)^2 + 1$

120. $h(x) = \sqrt{3x}$

121. $h(x) = |2x| - 3$

122. $h(x) = \left|\dfrac{1}{3}x\right|$

In Exercises 123–126, use the verbal description to find an algebraic expression for each function.

123. The graph of the function $g(x)$ is formed by translating the graph of $f(x) = |x|$ three units to the right and one units up.

124. The graph of the function $f(x)$ is formed by translating the graph of $h(x) = \sqrt{x}$ two units to the right and three units up.

125. The graph of the function $g(x)$ is formed by vertically scaling the graph of $f(x) = x^2$ by a factor of 2 and moving to the left by 1 unit.

126. The graph of the function $g(x)$ is formed by horizontally compressing the graph of $f(x) = |x|$ by a factor of $\frac{1}{2}$ and moving to up by 2 units.

In Exercises 127–130, use the given graph of f to graph each expression.

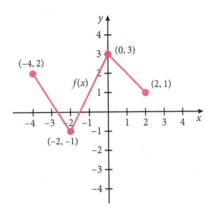

127. $2f(x) - 3$ **128.** $f(x - 1)$ **129.** $f(2x)$ **130.** $-f(x) + 1$

 In Exercises 131 and 132, use a graphing utility for each problem.

131. Graph the functions $y_1(x) = (x - 1.5)^2$ and $y_2(x) = x^2 - 1.5$ in the same window and comment on the difference between the two in terms of transformations of $f(x) = x^2$.

132. Graph the function $y_1(x) = 4|x + 1.5| - 2.5$ and describe the graph in terms of transformations of $f(x) = |x|$.

Section 2.8

In Exercises 133 and 134, for each table of values, find the linear function f with the given input and output values.

133.

x	f(x)
0	4
3	6

134.

x	f(x)
−1	3
4	−2

In Exercises 135–138, find the variation constant and the corresponding equation.

135. Let y vary directly as x, and $y = 25$ when $x = 10$.

136. Let y vary directly as x, and $y = 40$ when $x = 8$.

137. Let y vary inversely as x, and $y = 9$ when $x = 6$.

138. Let y vary inversely as x, and $y = 12$ when $x = 8$.

139. Business If the revenue of a wallet manufacturer varies directly with the quantity of wallets sold, find the revenue function if the revenue from selling 5000 wallets is $30,000. What would be the revenue if 800 wallets were sold?

140. Economics The demand for a product is inversely proportional to its price. If 400 units are demanded at a price of $3 per unit, how many units are demanded at a price of $2 per unit?

141. Depreciation The following table gives the value of a computer printer purchased in 2008 at two different times after its purchase.

Time after purchase (years)	Value (dollars)
2	300
4	200

a. Express the value of the printer as a linear function of the number of years after its purchase.

b. Using the function found in (a), find the original purchase price.

c. Assuming that the value of the printer is a linear function of the number of years after its purchase, when will the printer's value reach $0?

 142. Music The following table shows the number of music compact discs (CDs), in millions, sold in the U.S. between the years 2009 and 2012.

(*Source*: Recording Industry Association of America)

Year	Units sold (in millions)
2009	293
2010	226
2011	240
2012	211

a. What general trend do you notice in these figures?

b. Fit a linear function to this set of points, using the number of years since 2009 as the independent variable.

c. Use your function to predict the number of units of CDs that will be sold in the U.S. in 2015.

Chapter 2 Test

1. Find the slope of the line, if defined, passing through the pair if points. If undefined, state so.

   a. $(2, -5)$ and $(4, 2)$

   b. $(-2, 4)$ and $(-2, 6)$

2. Find the slope-intercept form of the equation of the line, passing through the point $(-1, 3)$ and with slope -4.

3. Find the general form of the equation of the line, passing through the points $(5, -2)$ and $(3, 0)$.

4. Find the equation of the line perpendicular to the line $2y - x = 3$ and passing through $(1, 4)$. Write the equation in slope-intercept form.

5. Find the equation of the line parallel to the line $-4x - y = 6$ and passing through $(-3, 0)$. Write the equation in slope-intercept form.

6. Find the equation of the horizontal line through $(4, -5)$.

7. Find the equation of the vertical line through $(7, -1)$.

8. Given the points $(1, 2)$ and $(-4, 3)$, find the distance between them and the midpoint of the line segment joining them.

9. Write the standard for the equation of the circle with center $(2, 5)$ and radius 6.

10. Find the center and the radius of the circle with equation $x^2 + 2x + y^2 - 4y = 4$. Sketch the circle.

11. Graph the equation $|y| = 4x$. Determine if y is a function of x.

12. Let $f(x) = -x^2 + 2x$ and $g(x) = \sqrt{x + 6}$. Evaluate each of the following.

    a. $f(-2)$
    b. $f(a - 1)$
    c. $g(3)$
    d. $g(10)$

13. Find the domain in interval form of $f(x) = -3x$.

14. Find the domain in interval form of $f(x) = \frac{1}{x - 5}$.

15. Sketch the graph of $f(x) = -2x - 3$ and find its domain.

16. Sketch the graph of $f(x) = \sqrt{x + 3}$ and find its domain.

17. Sketch the graph of $f(x) = -x^2 + 4$ and find its domain and range.

18. Determine whether the set of points $S = \{(1, -1), (0, 1), (1, 2), (2, 3)\}$ defines a function.

19. Determine whether the following is the graph of a function.

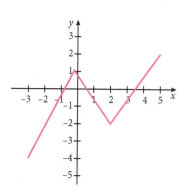

20. Let f be defined as follows.

$$f(x) = \begin{cases} -x + 1, & \text{if } x \le 1 \\ x - 1, & \text{if } x > 1 \end{cases}$$

   a. Sketch a graph of f.

   b. Find a single expression for f in terms of absolute value.

In Exercises 21–23, decide if the function is odd, even, or neither, using the appropriate definitions.

21. $f(x) = 2x + 1$ **22.** $f(x) = -3x^2 - 5$ **23.** $f(x) = 2x^5 + x^3$

In Exercises 24 and 25, use the following graph.

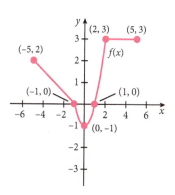

24. Find the interval(s) on which

   a. f is increasing.   **b.** f is decreasing.   **c.** f is constant.

25. Find the average rate of change of f on the interval $[-2, 1]$.

26. Find the difference quotient $\dfrac{f(x + h) - f(x)}{h}$ for $f(x) = 4x^2 - 5$.

In Exercises 27–34, let $f(x) = x^2 + 2x$ and $g(x) = 2x - 1$.

27. Find $(f + g)(x)$ and identify its domain.

28. Find $(g - f)(x)$ and identify its domain.

29. Evaluate $\left(\dfrac{f}{g}\right)(3)$.

30. Evaluate $(fg)(-2)$.

31. Evaluate $(f \circ g)(0)$.

32. Evaluate $(g \circ f)(-1)$.

33. Find $(fg)(x)$ and identify its domain.

34. Find $\left(\dfrac{f}{g}\right)(x)$ and identify its domain.

35. Let $f(x) = \dfrac{1}{2x}$ and $g(x) = x^2 - 1$. Find $(f \circ g)(x)$ and identify its domain.

In Exercises 36–39, use transformations to sketch the graph of each function.

36. $f(x) = \left|\dfrac{1}{2}x\right|$ **37.** $f(x) = |3x| - 1$

38. $f(x) = -2\sqrt{x} + 3$ **39.** $f(x) = -(x + 2)^2 - 2$

In Exercises 40 and 41, use the verbal description of the function to find its corresponding algebraic expression.

40. The graph of the function $g(x)$ is formed by translating the graph of $f(x) = |x|$ two units to the left and one units up.

41. The graph of the function $g(x)$ is formed by horizontally compressing the graph of $f(x) = x^2$ by a factor of $\dfrac{1}{2}$ and moving down by 1 unit.

42. If y varies directly as x, and $y = 36$ when $x = 8$, find the variation constant and the corresponding equation.

43. If y varies inversely as x, and $y = 10$ when $x = 7$, find the variation constant and the corresponding equation.

44. Housing A house purchased for $300,000 in 2006 increases in value by $15,000 each year.

    a. Express the value of the house as linear function of t, the number of years after its purchase.

    b. According to your function, when will the price of the house reach $420,000?

Chapter 3

Polynomial and Rational Functions

Boxes can be manufactured in many shapes and sizes. A polynomial function can be used in constructing a box to meet a volume specification. See Example 10 in Section 3.2 for such an application. In this section, we study quadratic, polynomial, and rational functions. These functions have many mathematical properties that play an important role in advanced mathematics and applications.

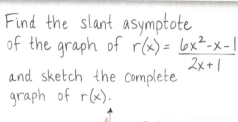

Find the slant asymptote of the graph of $r(x) = \dfrac{6x^2 - x - 1}{2x + 1}$ and sketch the complete graph of $r(x)$.

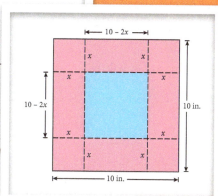

Outline

Polynomial and
Rational Functions

Quadratic Functions and Their Graphs

Objectives

- Define a quadratic function.
- Use transformations to graph a quadratic function in standard form.
- Write a quadratic function in vertex form by completing the square.
- Find the vertex and axis of symmetry of the graph of a quadratic function.
- Identify the maximum or minimum value of a quadratic function.
- Graph a quadratic function in standard form.
- Find a quadratic function given its zeros.
- Solve applied problems using maximum and minimum function values.

In Chapter 2, you examined linear functions, which are of the form $f(x) = mx + b$. The graph of a linear function is simply a line. In this section, we examine **quadratic functions**, in which the independent variable x is raised to the second power.

Definition of a Quadratic Function

A function f is a **quadratic function** if it can be expressed in the form

$$f(x) = ax^2 + bx + c$$

where a, b, and c are real numbers and $a \neq 0$. The **domain** of a quadratic function is the set of all real numbers.

Throughout our discussion, a is the coefficient of the x^2 term; b is the coefficient of the x term; and c is the constant term. To better understand quadratic functions, it is helpful to look at their graphs.

We study the graphs of general quadratic functions by first examining the graph of the function $f(x) = ax^2$, $a \neq 0$. We will later see that the graph of *any* quadratic function can be produced by a suitable combination of transformations of this graph.

Consider the quadratic functions $f(x) = x^2$, $g(x) = -x^2$, and $h(x) = 2x^2$. We make a table of values (Table 1) and graph the three functions on the same set of coordinate axes (Figure 1).

x	$f(x) = x^2$	$g(x) = -x^2$	$h(x) = 2x^2$
-2	4	-4	8
-1	1	-1	2
0	0	0	0
1	1	-1	2
2	4	-4	8

Table 1

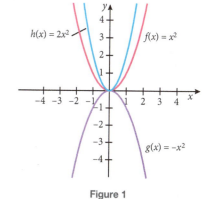

Figure 1

Observations:

- The graph of $f(x) = x^2$ opens upward, while the graph of $g(x) = -x^2$ opens downward.
- The graph of $h(x) = 2x^2$ is vertically stretched by a factor of 2 compared to the graph of $f(x) = x^2$, but the graphs of both f and h open upward.

The next section will generalize these observations.

Graphing Quadratic Functions Using the Form $f(x) = a(x - h)^2 + k$

In many instances, it is easier to analyze a quadratic function if it is written in the form $f(x) = a(x - h)^2 + k$. This is known as the **standard form** of the quadratic function.

> ### Quadratic Function in Standard Form
>
> A quadratic function $f(x) = ax^2 + bx + c$ can be rewritten in the form
>
> $$f(x) = a(x - h)^2 + k$$
>
> known as the **standard form**. The graph of $f(x) = ax^2 + bx + c$ is called a **parabola**. The **vertex** of the parabola is given by (h, k).
>
> - If $a > 0$, the parabola opens upward, and the vertex is the lowest point on the graph at which f has its **minimum** value. See Figure 2.
>
> - If $a < 0$, the parabola opens downward, and the vertex is the highest point on the graph at which f has its **maximum** value. See Figure 3.
>
>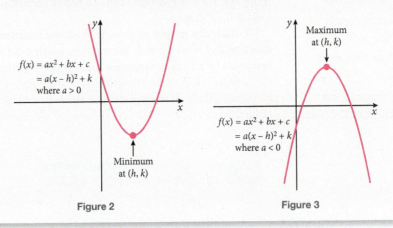
>
> **Figure 2**   **Figure 3**

When a quadratic function is expressed in the form $a(x - h)^2 + k$, we can use transformations to help sketch its graph, as seen in the following example.

VIDEO EXAMPLES

SECTION 3.1

> **Example 1**   **Graphing a Quadratic Function Using Transformations**
>
> Let $f(x) = -2(x - 1)^2 + 3$.
>
> **a.** Use transformations to graph f.   **b.** What is the vertex of the associated parabola?

Solution

a. Note that the graph of $f(x) = -2(x - 1)^2 + 3$ can be written as a series of transformations of the graph of $y = x^2$ as follows:

$$x^2 \rightarrow -2x^2$$
Vertical stretch by factor of 2 and reflection across x-axis

$$\rightarrow -2(x - 1)^2$$
Horizontal shift 1 unit to the right

$$\rightarrow -2(x - 1)^2 + 3$$
Vertical shift 3 units up

Just in Time
Review transformations in Section 2.7.

The transformation of the graph of $y = x^2$ to the graph of $y = -2x^2$ is sketched in Figure 4.

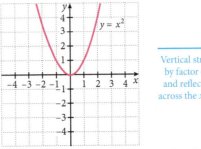

Figure 4

The shifts of $-2x^2$ to the right by 1 unit and up by 3 units are shown in the series of graphs in Figure 5.

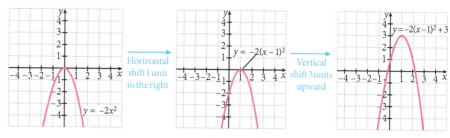

Figure 5

b. The vertex of the parabola is $(h, k) = (1, 3)$. Because $a = -2 < 0$, the parabola opens downward.

Just in Time
Review factoring of trinomials that are perfect squares in Section 1.3.

Check It Out 1 Rework Example 1 for $g(x) = 2(x - 2)^2 - 1$.

We can write *any* quadratic function $f(x) = ax^2 + bx + c$ in the form $f(x) = a(x - h)^2 + k$, where a, h, and k are real numbers, by using a technique known as **completing the square**. For example, to complete the square on $x^2 + 8x$, you add 16. This gives

$$x^2 + 8x + 16 = (x + 4)^2.$$

In general, if the expression is of the form $x^2 + bx$, you add $c = \left(\frac{b}{2}\right)^2$ to make $x^2 + bx + c$ a perfect square. If the expression is of the form $ax^2 + bx$, $a \neq 1$, you factor out a to get $a\left(x^2 + \frac{b}{a}x\right)$ and then complete the square on the expression *inside* the parentheses.

> **Example 2** **Writing a Quadratic Function in Standard Form**

Let $f(x) = 3x^2 + 12x + 8$.

a. Use the technique of completing the square to write $f(x) = 3x^2 + 12x + 8$ in the form $f(x) = a(x - h)^2 + k$.

b. What is the vertex of the associated parabola? Does f reach a maximum or a minimum value at the vertex?

Solution

Just in Time
See more examples on completing the square in Section 2.2.

a. To complete the square on $3x^2 + 12x + 8$, we examine the first two terms: $3x^2 + 12x$. Factor out the 3 to get $3(x^2 + 4x)$. Taking half of 4, the coefficient of x, gives 2. Squaring it gives $(2)^2 = 4$. Thus, the number 4 will complete the square on $x^2 + 4x$.

Putting all this together, we have

$$3x^2 + 12x + 8 = 3(x^2 + 4x) + 8 \qquad \text{Factor 3 out of the } x^2 \text{ and } x \text{ terms}$$

$$= 3(x^2 + 4x + \mathbf{4} - \mathbf{4}) + 8 \qquad \begin{array}{l}\text{Add within parentheses to complete the square} \\ \text{Subtract within parentheses to maintain the} \\ \text{value of the original expression}\end{array}$$

$$= 3(x^2 + 4x + 4) - 3(4) + 8 \qquad \text{Regroup terms}$$

$$= 3(x + 2)^2 - 12 + 8 \qquad \text{Rewrite } x^2 + 4x + 4 \text{ as } (x + 2)^2$$

$$= 3(x + 2)^2 - 4 \qquad \text{Combine } -12 \text{ and } 8$$

$$= 3(x - (-2))^2 - 4 \qquad \text{Write as } (x - h)^2$$

This now is in the form $a(x - h)^2 + k$, with $a = 3$, $h = -2$, and $k = -4$.

b. The vertex of the associated parabola is $(h, k) = (-2, -4)$. Because $a = 3 > 0$, the parabola opens upward, and so f reaches a minimum value of -4 at the vertex. See Figure 6.

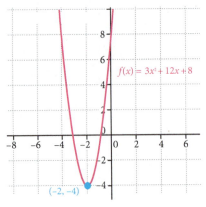

$f(x) = 3x^2 + 12x + 8$

$(-2, -4)$

Figure 6

✓ *Check It Out 2*   Use the technique of completing the square to write $f(x) = 3x^2 - 12x + 7$ in the form $f(x) = a(x - h)^2 + k$. Find the vertex of the associated parabola. Does f reach a maximum or a minimum value at the vertex? ◼

Vertex, Axis of Symmetry, and *x*-Intercepts

It can be tedious to write a quadratic function in the form $f(x) = a(x - h)^2 + k$ in order to find the vertex. By completing the square for a general quadratic function $f(x) = ax^2 + bx + c$, it is possible to write it in the form $f(x) = a(x - h)^2 + k$ and obtain a general formula for the values of h and k.

Formula for Vertex of a Quadratic Function

The **vertex** of the parabola associated with the quadratic function $f(x) = ax^2 + bx + c$ is given by (h, k), where $h = -\dfrac{b}{2a}$ and $k = c - \dfrac{b^2}{4a}$.

Because (h, k) is a point on the graph of f, we must have $f(h) = k$. Thus the vertex is given by

$$(h, k) = \left(-\frac{b}{2a}, c - \frac{b^2}{4a}\right) \text{ or } (h, k) = \left(-\frac{b}{2a}, f\left(-\frac{b}{2a}\right)\right)$$

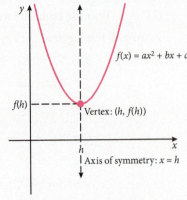

$f(x) = ax^2 + bx + c$

$f(h)$

Vertex: $(h, f(h))$

h

Axis of symmetry: $x = h$

Figure 7

Note that a parabola is symmetric about the vertical line passing through the vertex. That is, the part of the parabola that lies to the left of that line is the mirror image of the part that lies to the right of it. This line is known as the **axis of symmetry**.

Axis of Symmetry

The equation of the **axis of symmetry** of the parabola associated with the quadratic function $f(x) = ax^2 + bx + c$ is given by

$$x = \frac{-b}{2a}$$

Example 3   **Finding the Vertex and Axis of Symmetry of the Graph of a Quadratic Function**

Find the vertex and axis of symmetry of the graph of the quadratic function $f(x) = 2x^2 - 8x + 1$.

Solution The vertex is given by $(h, k) = \left(-\dfrac{b}{2a}, f\left(-\dfrac{b}{2a}\right)\right)$.

$$h = \frac{-b}{2a} = \frac{-(-8)}{2(2)} = 2$$

$$k = f(2) = 2(2)^2 - 8(2) + 1$$

$$= 8 - 16 + 1 = -7$$

The vertex is therefore $(2, -7)$.

The axis of symmetry is given by the equation of the vertical line $x = -\dfrac{b}{2a}$, which is $x = 2$. Because the vertical line passes through the vertex, the line has a constant x-value that is the same as the x-coordinate of the vertex. See Figure 8.

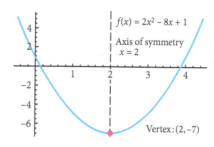

Figure 8

Check It Out 3 Find the vertex and axis of symmetry of the graph of the quadratic function $f(x) = -5x^2 - 10x + 1$.

Certain properties of a quadratic function are related to the solution of an associated quadratic equation. A **real zero** of a function is defined as follows.

Definition of Real Zeros

The real numbers x at which $f(x) = 0$ are called **real zeros** of the function f.

Recall that the points where the graph of a function crosses the x-axis are called **x-intercepts**. The x-coordinate of an x-intercept is a value of x such that $f(x) = 0$. Thus, a **real zero** of a function f is the first coordinate (the x-coordinate) of an x-intercept of the graph of f.

Just in Time
Review the
quadratic formula
and discriminants
in Section 1.5.

If $f(x) = ax^2 + bx + c$, then solving $f(x) = 0$ is the same as solving the quadratic equation $ax^2 + bx + c = 0$. The solution to this equation is given by the quadratic formula, $x = \frac{-b \pm \sqrt{b^2 - 4ac}}{2a}$. The connection between the sign of the discriminant, $b^2 - 4ac$, and the number of x-intercepts is illustrated in Figure 9.

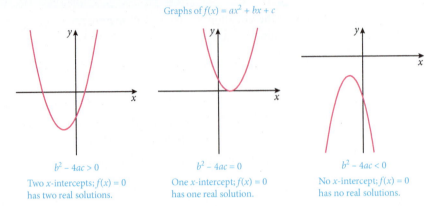

Graphs of $f(x) = ax^2 + bx + c$

$b^2 - 4ac > 0$
Two x-intercepts; $f(x) = 0$ has two real solutions.

$b^2 - 4ac = 0$
One x-intercept; $f(x) = 0$ has one real solution.

$b^2 - 4ac < 0$
No x-intercept; $f(x) = 0$ has no real solutions.

Figure 9

Just in Time
Review factoring
in Section 1.3.

Example 4   **Relating x-Intercepts to Real Zeros**

Find the x-intercepts of the parabola associated with the function $f(x) = 3x^2 - 5x - 2$. Also, find the real zeros of f.

Solution   We set $f(x) = 0$ and solve the resulting equation.

$$3x^2 - 5x - 2 = 0 \quad \text{Set the function expression equal to 0}$$

$$(3x + 1)(x - 2) = 0 \quad \text{Factor the left-hand side}$$

The product of two factors will equal zero if at least one of the factors is equal to zero. Thus, we set each factor equal to zero and solve for x.

$$3x + 1 = 0 \Rightarrow x = -\frac{1}{3}$$

or

$$x - 2 = 0 \Rightarrow x = 2$$

Hence, $3x^2 - 5x - 2 = 0$ if $x = -\frac{1}{3}$ or $x = 2$. The x-intercepts are therefore $\left(-\frac{1}{3}, 0\right)$ and $(2, 0)$. The graph of the function crosses the x-axis at just these points and no others, as seen in Figure 10. The zeros of f are the values of x such that $f(x) = 0$. We see that these values are $x = -\frac{1}{3}$ and $x = 2$.

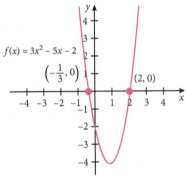

Figure 10

Check It Out 4   Rework Example 4 for the function $f(x) = 2x^2 - 3x - 2$.

In Example 4, each of the zeros of the function f is also the first coordinate of an x-intercept of the graph of f. However, as we shall see in a later section, there are quadratic functions whose graphs have no x-intercepts, even though the functions themselves have zeros. Such zeros are not "real zeros." In order to treat quadratic functions of this type, we will introduce a set of numbers known as *complex numbers* in Section 3.5.

The next example will show you how to put together everything you learned in this section to create the graph of a quadratic function.

Using Technology

The maximum value of f can be found using the MAXIMUM feature of your graphing utility.

Keystroke Appendix:
Section 10

> **Example 5**   **Graphing a Quadratic Function Using the Vertex and Intercepts**

Let $f(x) = -3x^2 - 6x + 2$.

a. Find the maximum value of f.

b. Find the axis of symmetry of the parabola.

c. Find the x-intercepts, if they exist.

d. Find two additional points on the graph, and then sketch the graph of f by hand.

e. Use the graph to find intervals where f is increasing or decreasing and find the range.

Solution

a. The vertex is given by $(h, k) = \left(\dfrac{-b}{2a}, f\left(\dfrac{-b}{2a} \right) \right)$.

$$h = \frac{-b}{2a} = \frac{-(-6)}{2(-3)} = -1$$

$$k = f(-1) = -3(-1)^2 - 6(-1) + 2$$

$$= -3 + 6 + 2 = 5$$

The vertex is therefore $(-1, 5)$. Because $a = -3 < 0$, the parabola opens downward. The maximum value of f is the y-coordinate of the vertex, which is 5.

b. The axis of symmetry is given by $x = -1$.

c. The x-intercepts can be found by solving the equation $f(x) = 0$, resulting in the equation

$$-3x^2 - 6x + 2 = 0$$

The discriminant, $b^2 - 4ac = (-6)^2 - 4(-3)(2) = 60$, is greater than 0, and so there are two x-intercepts. The equation cannot be readily factored. Using the quadratic formula,

$$x = \frac{-b \pm \sqrt{b^2 - 4ac}}{2a} = \frac{-(-6) \pm \sqrt{(-6)^2 - 4(-3)(2)}}{2(-3)} \qquad \text{Substitute } a = -3, b = -6, \text{ and } c = 2$$

$$= \frac{6 \pm \sqrt{36 + 24}}{-6} = -1 \pm \frac{\sqrt{60}}{6}$$

$$= -1 \pm \frac{\sqrt{15}}{3}$$

So $x = -1 + \dfrac{\sqrt{15}}{3} \approx 0.29$, and $x = -1 - \dfrac{\sqrt{15}}{3} \approx -2.29$. The x-intercepts are approximately $(0.29, 0)$ and $(-2.29, 0)$.

d. The y-intercept can be easily calculated by substituting $x = 0$ into the expression for $f(x)$, giving $(0, 2)$ as the y-intercept. The point $(0, 2)$ is 1 unit to the right of the axis of symmetry, so its mirror image will be 1 unit to the left of that line, with coordinates $(-2, 2)$. Putting all this together, we get the graph in Figure 11.

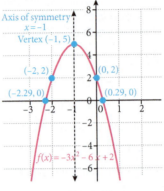

Figure 11

> **Note** When finding points on a parabola, if the vertex also happens to be the y-intercept, then you must find a different point on the parabola and then use symmetry to find one more point.

Note that the axis of symmetry is **not** part of the graph of the function. It is sketched merely to indicate the symmetry of the graph.

e. From the graph, we see that the function is increasing on $(-\infty, -1)$ and decreasing on $(-1, \infty)$. The domain is $(-\infty, \infty)$ and the range is $(-\infty, 5]$.

 Check It Out 5 Rework Example 5 for the function $g(t) = -3t^2 + 6t - 2$. ■

The next example illustrates an important connection between the real zeros and factors of a quadratic function and the x-intercepts of its graph. This connection will resurface later in the study of polynomial functions.

Example 6   **Real Zeros and Factors of a Quadratic Function**

Find a possible expression for a quadratic function $f(x)$ with zeros at $x = -3$ and $x = 1$.

Solution Because the zeros are obtained by factoring the expression for $f(x)$ and setting it to zero, we can work backward by noting that

$$x = -3 \Rightarrow x + 3 = 0 \quad \text{and} \quad x = 1 \Rightarrow x - 1 = 0$$

Thus the factors for $f(x)$ are $x + 3$ and $x - 1$.

$$f(x) = (x + 3)(x - 1) = x^2 + 2x - 3$$

This is not the only possible expression for f. Any function of the form $a(x + 3)(x - 1)$, $a \neq 0$, would have the same zeros as f. Two possible functions are pictured in Figure 12.

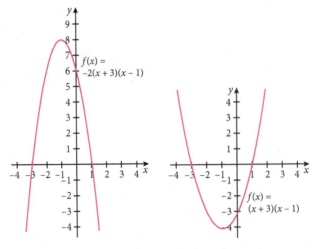

Figure 12

✅ *Check It Out 6*   Find a possible expression for a quadratic function $f(x)$ with zeros at $x = 2$ and $x = -4$. ◼

Applications

This section illustrates a number of applications involving quadratic functions.

Example 7   **Maximizing Area**

Traffic authorities have 100 feet of rope to cordon off a rectangular region to form a ticket arena specifically for concert goers who are waiting to purchase tickets.

a. Express the area of this region as a function of the length of just one of the four sides of the region.

b. Find the dimensions of the enclosed region that will give the maximum area, and determine the maximum area.

Solution

a. Because this is a geometric problem, first draw a diagram of the cordoned-off region with x as the length of the region and w as its width. See Figure 13.

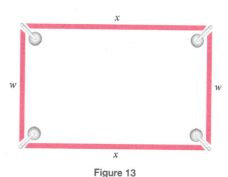

Figure 13

Using Technology

You can use a graphing utility to generate a table of values for the area function and see how the values for the area vary as the length is changed. See Figure 14.

Keystroke Appendix:
Section 6

X	Y₁
5	225
10	400
25	525
20	600
25	625
30	600
35	525
X = 25	

Figure 14

From geometry, we know that Area $= A = wx$.

To express the area as a function of just *one* of the two variables x, w, find a relationship between the length and the width. Because 100 feet of rope is available, the perimeter of the enclosed region must equal 100

$$\text{Perimeter} = 2x + 2w = 100$$

Solve for w in terms of x

$$2w = 100 - 2x \qquad \text{Isolate the } w \text{ term}$$

$$w = \frac{1}{2}(100 - 2x) \qquad \text{Solve for } w$$

$$w = 50 - x \qquad \text{Simplify}$$

Now, we can write the area as

$$A(x) = xw = x(50 - x) = 50x - x^2 = -x^2 + 50x$$

b. Since $a = -1 < 0$, the graph of the area function is a parabola opening downward. Therefore, the second coordinate of the vertex will yield a maximum value. See Figure 15.

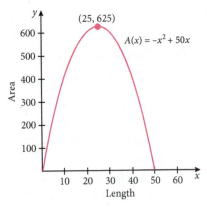

Figure 15

The first coordinate of the vertex is

$$\frac{-b}{2a} = \frac{-50}{2(-1)} = 25$$

The maximum area occurs when the **length** is 25 ft. Since $2x + 2w = 100$, we find that $w = 25$, and so the **width** is 25 ft. as well. Substituting into the area function, we have

$$A(\mathbf{25}) = -(\mathbf{25})^2 + 50(\mathbf{25}) = 625 \text{ sq. ft.}$$

Thus, the **maximum area** is 625 sq.ft., corresponding to roping off a 25-foot-square region. See Figure 16.

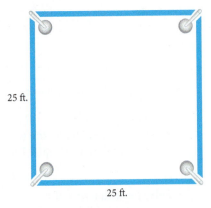

25 ft.

25 ft.

Figure 16

✓ *Check It Out 7*   If the area function of a rectangular region cordoned off with 120 feet of rope is given by $A(x) = -x^2 + 60x$, find the dimensions of the region that will give the maximum area, and determine the maximum area.

Example 8 **Modeling Car Mileage**

The following table lists the speed at which a car is driven, in miles per hour, and the corresponding gas mileage obtained, in miles per gallon. (*Source*: Environmental Protection Agency)

Speed (x) (miles per hour)	Gas Mileage (m) (miles per gallon)
5	12
10	17
25	27
45	30
65	25
75	22

a. Make a scatter plot of *m*, the gas mileage, vs. *x*, the speed at which a car is driven. What type of trend do you observe—linear or quadratic? Find an expression for the function that best fits the given data points.

b. Use your function to find the gas mileage obtained when a car is driven at 35 miles per hour.

c. Find the speed at which the car's gas mileage is at its maximum, using a graphing utility.

Solution

a. Enter the data into your graphing utility and display the scatter plot, as shown in Figure 17.

Note that the gas mileage first increases with speed and then decreases. This suggests that a quadratic model would be appropriate. Using the *regression* feature of your graphing utility, the expression for the function best fits this data is

$$m(x) = -0.0108x^2 + 0.981x + 8.05$$

The graph of the function along with the data points in the same window (Figure 18) shows that the function approximates the data reasonably well. Store this function in your calculator as Y_1 for subsequent analysis.

Using Technology

These curve fitting features are available on most graphing utilities, under the *regression* option.

Keystroke Appendix:
Sections 9, 12

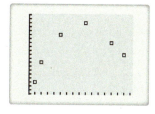

Figure 17

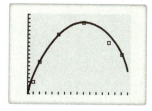

Figure 18

b. Evaluating the function at $x = 35$ gives a gas mileage of approximately 29.2 miles per gallon. See Figure 19.

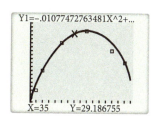

Figure 19

c. Using the MAXIMUM feature of your graphing utility, the maximum point is approximately (45.5, 30.4), as seen in Figure 20. Thus, at a speed of 45.5 miles per hour, the gas mileage is at a maximum of 30.4 miles per gallon.

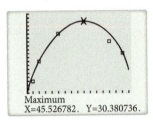

Maximum
X=45.526782. Y=30.380736.

Figure 20

✅ *Check It Out 8*   Use the TABLE feature in ASK mode to evaluate the gas mileage function in Example 8 at the values of the speeds given in the table. Compare with the actual gas mileage listed for those speeds and comment on the validity of the model. ▪

Note   You can fit a linear or quadratic function to *any* set of points. It is up to you to decide whether this is an appropriate choice or not. Choosing a reasonable function to model a set of data requires that you be familiar with the properties of different types of functions. As you work with the data sets for the problems in this and other sections, make sure you understand why you are choosing one particular type of function over another.

Exercises 3.1

Just in Time Exercises

1. The graph of $g(x) = f(x) + 2$ is the graph of f shifted _____ by 2 units.

2. The graph of $g(x) = f(x + 3)$ is the graph of f shifted to the _____ by 3 units.

3. Factor: $x^2 - 13x + 40$.

4. Solve using the quadratic formula: $x^2 - 2x - 1 = 0$.

5. Find c so that $x^2 - 3x + c$ is a perfect square trinomial.

6. Write the expression in the form $(x - h)^2$: $x^2 - 8x + 16$.

Skills

In Exercises 7–10, graph each pair of functions on the same set of coordinate axes, and find the domain and range of each function.

7. $f(x) = -2x^2, g(x) = -x^2$

8. $f(x) = 3x^2, g(x) = x^2$

9. $f(x) = x^2 + 1, g(x) = x^2 - 1$

10. $f(x) = -x^2 + 2, g(x) = -x^2 - 2$

Paying Attention to Instructions: *Exercises 11 and 12 are intended to give you practice reading and paying attention to the instructions that accompany the problems you are solving.*

11. Refer to graph (d) in this exercise set.

   a. Find the domain and range of f.   b. For what value of x does f attain a maximum?

   c. What is the maximum value attained by f?   d. For what value(s) of x is $f(x) = 0$?

12. Refer to graph (g) in this exercise set.

   a. Find the domain and range of f.   b. For what value of x does f attain a minimum?

   c. What is the minimum value attained by f?   d. For what value(s) of x is $f(x) = 0$?

In Exercises 13–20, each of the graphs represents a quadratic function. Match the graph with its corresponding expression or description.

a.

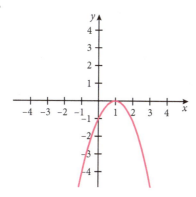

b.

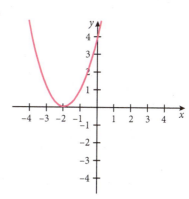

c.

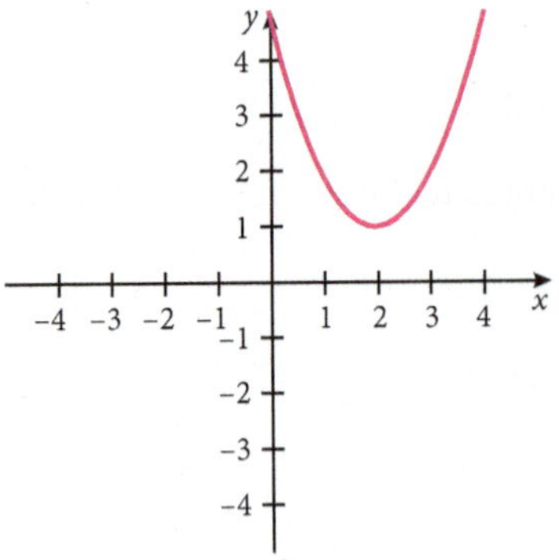

d.

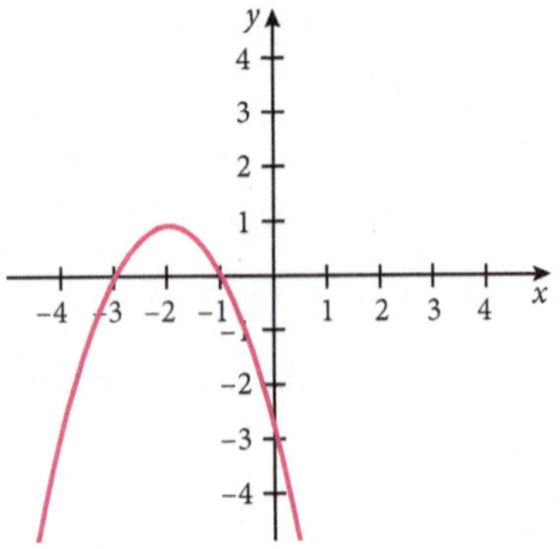

e.

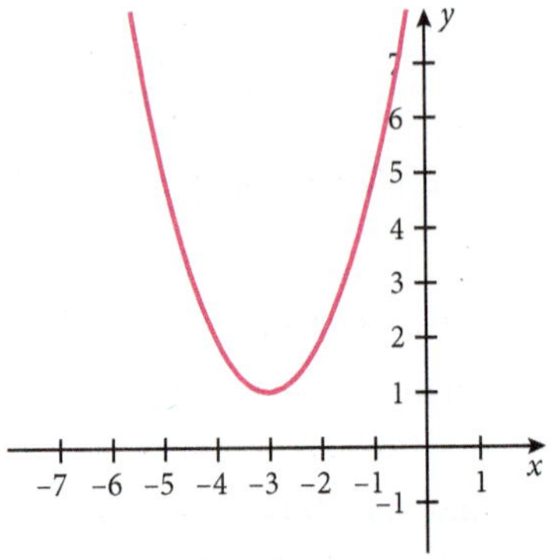

f.

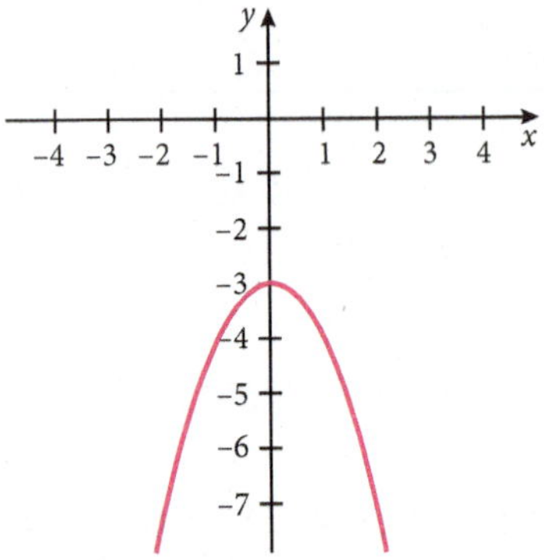

g.

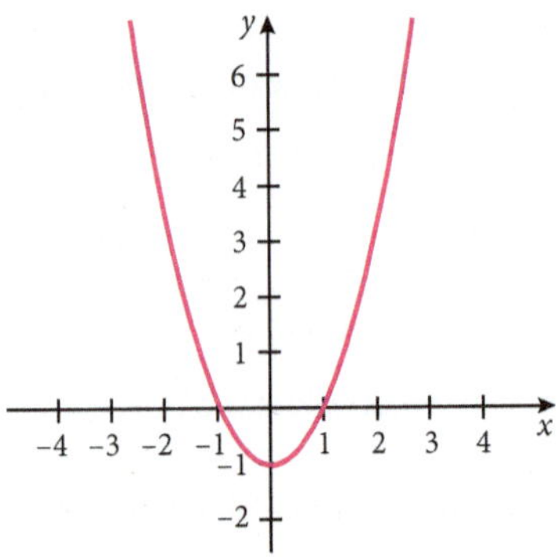

h.

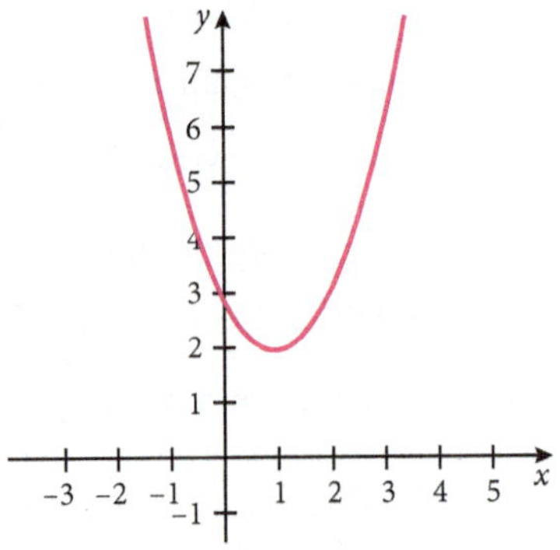

13. $f(x) = (x + 2)^2$

14. $f(x) = -(x + 2)^2 + 1$

15. The graph of $f(x) = x^2$ shifted 1 unit to the right and reflected across the x-axis

16. The graph of $f(x) = x^2$ shifted 2 units to the right and 1 unit up

17. $f(x) = (x + 3)^2 + 1$

18. $f(x) = (x - 1)^2 + 2$

19. The graph of $f(x) = x^2$ reflected across the x-axis and shifted 3 units down

20. The graph of $f(x) = x^2$ shifted 1 unit down

In Exercises 21–26, use transformations to graph each of the quadratic functions and find the vertex of the associated parabola.

21. $f(x) = (x + 2)^2 - 1$

22. $h(x) = (x - 3)^2 + 2$

23. $f(x) = -(x + 1)^2 - 1$

24. $g(s) = -(s - 2)^2 + 2$

25. $h(x) = -3(x + 4)^2 - 2$

26. $g(t) = 2(t - 3)^2 + 3$

In Exercises 27–34, write each quadratic function in the form $a(x - h)^2 + k$ by completing the square. Also find the vertex of the associated parabola and sketch the graph.

27. $g(x) = x^2 + 2x + 5$

28. $h(x) = x^2 - 4x + 6$

29. $w(x) = -x^2 + 6x + 4$

30. $f(x) = -x^2 - 4x + 5$

31. $h(x) = x^2 + x - 3$

32. $g(x) = -x^2 + x - 7$

33. $f(x) = 3x^2 + 6x - 4$

34. $f(x) = -2x^2 + 8x + 3$

In Exercises 35–40, for each function f, find the discriminant and use it to determine the number of x-intercepts of the graph of f. Also, determine the number of real solutions to the equation $f(x) = 0$.

35. $f(x) = x^2 - 2x - 1$

36. $f(x) = -x^2 + x + 3$

37. $f(x) = 2x^2 + x + 1$

38. $f(x) = 3x^2 - 4x + 4$

39. $f(x) = x^2 + 2x + 1$

40. $f(x) = -x^2 + 4x - 4$

In Exercises 41–56, find the vertex, axis of symmetry, y-intercept, and any x-intercepts of the associated parabola. Find at least two additional points on the parabola and sketch.

41. $f(x) = -2x^2 + 4x - 1$

42. $f(x) = x^2 - 6x + 1$

43. $g(x) = -x^2 + 4x - 3$

44. $f(x) = 3x^2 - 12x + 4$

45. $h(x) = x^2 - 3x + 5$

46. $h(x) = -x^2 + x - 2$

47. $f(t) = -16t^2 + 100$

48. $f(x) = 10x^2 - 65$

49. $G(x) = -6x + x^2 + 5$

50. $h(t) = -5t + 3 - t^2$

51. $f(s) = -2s^2 + 3s + 1$

52. $g(t) = 3t^2 - 6t - \dfrac{3}{4}$

53. $g(t) = t^2 + t + 1$

54. $f(t) = -t^2 - 1$

55. $f(t) = \dfrac{1}{3} - 3t + t^2$

56. $h(t) = 1 - \dfrac{1}{2}t - t^2$

In Exercises 57–64, find the vertex and axis of symmetry of the associated parabola. Sketch the parabola. Find the intervals where the function is increasing or decreasing and find the range.

57. $f(x) = -x^2 + 10x - 8$

58. $h(x) = x^2 + 6x - 7$

59. $f(x) = 2x^2 + 4x - 3$

60. $g(x) = -4x^2 - 8x + 5$

61. $f(x) = -0.2x^2 + 0.4x - 2.2$

62. $f(x) = 0.3x^2 + 0.6x + 1.3$

63. $h(x) = \dfrac{1}{4}x^2 + \dfrac{1}{2}x - 2$

64. $g(x) = -\dfrac{1}{6}x^2 + \dfrac{1}{3}x + 1$

In Exercises 65–68, find a quadratic function with the given properties. There can be more than one answer.

65. The graph of f opens downward, and has x-intercepts $(3, 0)$ and $(-1, 0)$.

66. The graph of f opens upward, and has x-intercepts $(-2, 0)$ and $(4, 0)$.

67. The function f has zeros at $x = -1$ and $x = 4$. The vertex is the lowest point on its graph.

68. The function f has zeros at $x = 2$ and $x = -2$. The vertex is the highest point on its graph.

In Exercises 69–73, graph each quadratic function by finding a suitable viewing window with the help of the TABLE feature. Also, find the vertex of the associated parabola using the graphing utility.

69. $y_1(x) = 0.4x^2 + 20$ **70.** $g(s) = -s^2 - 15$ **71.** $h(x) = (\sqrt{2})x^2 + x + 1$

72. $h(x) = x^2 + 5x - 20$ **73.** $s(t) = -16t^2 + 40t + 120$

74. Graph the function $f(t) = t^2 - 4$ in a decimal window. Using your graph, determine the values of t for which $f(t) \geq 0$.

75. Suppose the vertex of the parabola associated with a certain quadratic function is $(2, 1)$ and another point on this parabola is $(3, -1)$.
   a. Find the equation of the axis of symmetry of the parabola.
   b. Use symmetry to find a third point on the parabola.
   c. Sketch the parabola.

76. Examine the following table of values of a quadratic function f.

x	$f(x)$
-2	3
-1	0
0	-1
1	0
2	3

   a. What is the equation of the axis of symmetry of the associated parabola? Justify your answer.
   b. Find the minimum or maximum value of the function and the value of x at which it occurs.
   c. Sketch a graph of the function from the values given in the table and find an expression for the function.

77. Let $g(s) = -2s^2 + bs$. Find the value of b so that the vertex of the parabola associated with this function is $(1, 2)$.

78. Let $f(x) = x^2 + c$.
   a. Use a graphing utility to sketch the graph of f by choosing different values of c: some positive, some negative, and 0.
   b. For what value(s) of c will the graph of f have two x-intercepts? Justify your answer.
   c. For what value(s) of c will f have two real zeros? How is your answer related to part (b)?
   d. Repeat parts (b) and (c) for the case of *one* x-intercept and *one* real zero, respectively.
   e. Repeat parts (b) and (c) for the case of *no* x-intercepts and *no* real zeros, respectively.

79. Use the *intersection* feature of your graphing calculator to explore the real solution(s), if any, of $x^2 = x + k$ for $k = 0$, $k = -\frac{1}{4}$, and $k = -3$. Also use the *zero* feature to explore the solution(s). Relate your observations to the quadratic formula.

Applications

80. Landscaping A rectangular garden plot is to be enclosed, with a fence on three of its sides and a brick wall on the fourth side. If 100 feet of fencing material is available, what dimensions will yield the maximum area?

81. Physics: Ball Height The height of a ball that is thrown directly upward from a point 200 feet above the ground with an initial velocity of 40 feet per second is given by $h(t) = -16t^2 + 40t + 200$, where t is the amount of time elapsed since the ball was thrown. Here, t is in seconds and $h(t)$ is in feet.
   a. Sketch a graph of h.
   b. When will the ball reach its maximum height, and what is the maximum height?

82. Manufacturing A rain gutter with a rectangular cross-section is to be fabricated by bending up a flat piece of metal that is 18 feet long and 20 inches wide. The top of the gutter is open.

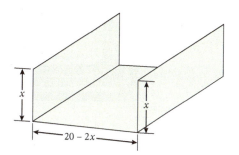

   a. Write an expression for the cross-sectional area in terms of x, the length of metal which is bent upwards.
   b. How much metal has to be bent upwards to maximize the cross- sectional area? What is the maximum cross-sectional area?

Use the information below to work Exercises 83 and 84.

The quadratic function $p(x) = -0.387(x - 45)^2 + 2.73(x - 45) - 3.89$ gives the percentage (in decimal form) of puffin eggs that hatch during a breeding season in terms of x, the mean sea surface temperature of the surrounding area, in degrees Fahrenheit.

(*Source*: Gjerdum et al., "Tufted puffin reproduction reveal ocean climate variability,"
Proceedings of the National Academy of Sciences, August 2003.)

83. Ecology What is the percentage of puffin eggs that will hatch at 49°F? at 47°F?

84. Ecology For what mean temperature is the percentage of hatched puffin eggs a maximum? Find the percentage of hatched eggs at this temperature.

85. Performing Arts Attendance at Broadway shows in New York can be modeled by the quadratic function $p(t) = 0.0489t^2 - 0.7815t + 10.31$, where t is the number of years since 1981 and $p(t)$ is in millions. The model is based on data for the years 1981–2000.

(*Source:* The League of American Theaters and Producers, Inc.)

   a. Use this model to estimate the attendance in the year 1995. Compare it to the actual value of 9 million.
   b. Use this model to estimate the attendance for the year 2006.
   c. What is the vertex of the parabola associated with the function p and what does it signify in relation to this problem?
   d. Would this model be suitable for predicting the attendance at Broadway shows for the year 2025? Why or why not?

86. **Maximizing Revenue** A chartered bus company has the following price structure. A single bus ticket costs $30. For each *additional* ticket sold to a group of travelers, the price per ticket is reduced by $0.50. The reduced price applies to all the tickets sold to the group.
    a. Calculate the total cost for one, two, and five tickets.
    b. Using your calculations in part (a) as a guide, find a quadratic function that gives the total cost of the tickets, in terms of the number of tickets sold.
    c. How many tickets must be sold to maximize the revenue for the bus company?

87. **Construction Analysis** A farmer has 400 feet of fencing material available to make two identical, adjacent rectangular corrals for the farm animals, as pictured. If the farmer wants to maximize the total enclosed area, what will be the dimensions of each corral?

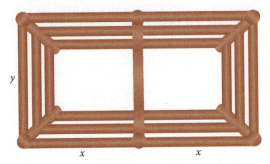

88. **Architecture** There is 24 feet of material available to trim the outer frame of a window shaped as in the diagram.

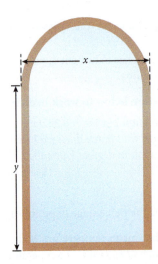

    a. Write an expression for the length of the semicircular arc (just the curved portion) in terms of x.
    b. Set up an equation relating the amount of available trim material to the lengths of the parts of the window that are to be covered with the trim. Solve this equation for y.
    c. Write an expression for the area of the window in terms of x.
    d. What are the dimensions of the window that will maximize the area of the window? What is the maximum area of the window?

89. **Engineering** When designing buildings, engineers must pay careful attention to how different factors affect the load a structure can bear. The following table gives the load in terms of the weight of concrete that can be borne when threaded rod anchors of various diameters are used to form joints.

(*Source*: Simpson Anchor Systems)

Diameter (in.)	Load (lbs.)
0.3750	2105
0.5000	3750
0.6250	5875
0.7500	8460
0.8750	11500

   a. Examine the table and explain why the relationship between the diameter and the load is *not* linear.
   b. The function

$$f(x) = 14926x^2 + 148x - 51$$

   gives the load (in pounds of concrete) that can be borne when rod anchors of diameter x (in inches) are employed. Use this function to determine the load for an anchor with a diameter of 0.8 inch.

90. **Design** The parabolic arc of a fountain has a maximum height of 5 ft. The horizontal range of the arc is 3 ft. The diagram below shows the arc placed in the xy-coordinate system. Find an expression for the quadratic function associated with the parabola shown in the diagram. Graph your function as a check.

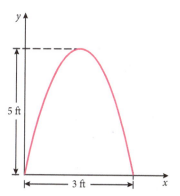

91. **Physics** A ball is thrown directly upward from ground level at time $t = 0$ (t is in seconds). At $t = 3$, the ball reaches its maximum distance from the ground, which is 144 feet. Assume that the distance of the ball from the ground (in feet) at time t is given by a quadratic function d as a function of t. Find an expression for $d(t)$ in the form $a(t - h)^2 + k$ by performing the following steps.
   a. From the given information, find the values of h and k and substitute them into the expression $a(t - h)^2 + k$.
   b. Now find a. To do this, use the fact that at time $t = 0$ the ball is at ground level. That is, $d(0) = 0$. This will give you an equation with just a as a variable. Solve for a.
   c. Now, substitute the value you found for a into the expression you found in part (a).
   d. Check your answer. Is $(3, 144)$ the vertex of the associated parabola? Does it pass through $(0, 0)$?

92. Sociology The percentage of employed mothers of preschoolers who used a nursery school or preschool as the primary source of child care is listed below for selected years between 1990 and 2000.

(*Source:* U.S. Census Bureau)

Year	Percentage Using Nursery School/ Preschool for Child Care
1990	6.9
1991	7.3
1995	5.9
1997	4.2
2000	3.8

a. Make a scatter plot of the data, using the number of years since 1990 as the independent variable. Explain why a quadratic function would be a better choice than a linear function for fitting this set of data.

b. Find the quadratic function that best fits the given data points.

c. Use the function to estimate the percentage of mothers using a preschool or nursery for child care in the year 2001.

d. Compare the actual percentages for the given years with the values predicted by your function. Comment on the reliability of your function as a model.

93. Accounting The table below gives the number of electronically filed tax returns for the calendar years listed.

(*Source:* Internal Revenue Service)

Year	1995	1998	2000	2003	2005
Number of returns (in millions)	11.8	24.6	35.4	52.9	68.2

a. Make a scatter plot of the data, using the number of years since 1995 as the independent variable.

b. Find the quadratic function that best fits the given data points.

c. Compare the actual numbers of electronically filed tax returns for the given years with the numbers predicted by your function.

d. Use your function to estimate the number of tax returns filed electronically in the year 2004.

e. The IRS had a total of 61.2 million tax returns filed electronically in 2004. How does that compare with your answer in part (d)?

Concepts

94. Why must we have $a \neq 0$ in the definition of a quadratic function?

95. What are at least two features of a quadratic function that differ from those of a linear function?

96. Which of the following points lie on the graph of the function $f(s) = -s^2 + 6$? Justify your answer.

   a. $(3, -1)$ **b.** $(0, 6)$ **c.** $(2, 1)$

97. Suppose the vertex and an x-intercept of the parabola associated with a certain quadratic function are given by $(-1, 2)$ and $(4, 0)$, respectively.

   a. Find the other x-intercept.

   b. Find the equation of the parabola.

   c. Check your answer by graphing.

98. The range of a quadratic function $g(x) = ax^2 + bx + c$ is given by $(-\infty, 2]$. Is a positive or negative? Justify your answer.

99. A parabola associated with a certain quadratic function f has the point $(2, 8)$ as its vertex and passes through the point $(4, 0)$. Find an expression for $f(x)$ in the form $a(x - h)^2 + k$.

   a. From the given information, find the values of h and k.

   b. Substitute the values you found for h and k into the expression $a(x - h)^2 + k$.

   c. Now find a. To do this, use the fact that the parabola passes through the point $(4, 0)$. That is, $f(4) = 0$. You should get an equation with just a as a variable. Solve for a.

   d. Substitute the value you found for a into the expression you found in (b).

   e. Graph the function using a graphing utility and check your answer. Is $(2, 8)$ the vertex of the parabola? Does it pass through $(4, 0)$?

100. Is it possible for a quadratic function to have the set of all real numbers as its range? Explain. (Hint: Examine the graph of a general quadratic function.)

Polynomial Functions and Their Graphs

3.2

Objectives

- Define a polynomial function.
- Determine end behavior.
- Find x-intercepts and zeros by factoring.
- Sketch a complete graph of a polynomial function.
- Relate zeros, x-intercepts, and factors of a polynomial.
- Solve applied problems using polynomials.

Just in Time
Review polynomials in Section 1.3.

VIDEO EXAMPLES

SECTION 3.2

In Chapter 2 and Section 3.1 we discussed linear and quadratic functions in detail. Recall that linear functions are of the form $f(x) = mx + b$, and quadratic functions are of the form $f(x) = ax^2 + bx + c$.

We can extend this notion to define functions involving x raised to nonnegative integer powers. The resulting functions are known as **polynomial functions**. Linear and quadratic functions are special types of polynomial functions.

> **Definition of a Polynomial Function**
>
> A function f is said to be a **polynomial function** if it can be written in the form
>
> $$f(x) = a_n x^n + a_{n-1} x^{n-1} + \cdots + a_1 x + a_0$$
>
> where $a_n \neq 0$, n is a nonnegative integer, and $a_0, a_1, \ldots, a_n$ are real-valued constants. The domain of f is the set of all real numbers.

Some of the constants that appear in the definition of a polynomial function have specific names associated with them.

- The nonnegative integer n is called the **degree** of the polynomial. Polynomials are usually written in **descending order**, with the exponents decreasing from left to right.
- The expressions $a_0, a_1 x, \ldots, a_n x^n$ are called **terms**, and the constants $a_0, a_1, \ldots, a_n$ are called **coefficients**.
- The term $a_n x^n$ is called the **leading term**, and the coefficient a_n is called the **leading coefficient**.
- A function of the form $f(x) = a_0$ is called a **constant polynomial** or a **constant function**.

Example 1   **Identifying Polynomial Functions**

Which of the following are polynomial functions? For the ones that are, find the degree and the coefficients, and identify the leading coefficient.

  a. $g(x) = 3 + 5x$ **b.** $h(s) = 2s(s^2 - 1)$ **c.** $f(x) = \sqrt{x^2 + 1}$

Solution

a. The function $g(x) = 3 + 5x = 5x + 3$ is a polynomial function of degree 1, with **coefficients $a_0 = 3$ and $a_1 = 5$.** The leading coefficient is $a_1 = 5$. This is a linear function.

b. Simplifying gives $h(s) = 2s(s^2 - 1) = 2s^3 - 2s$. This is a polynomial of degree 3. **The coefficients are $a_3 = 2$, $a_2 = 0$, $a_1 = -2$, and $a_0 = 0$.** The leading coefficient is $a_3 = 2$.

c. The function $f(x) = \sqrt{x^2 + 1} = (x^2 + 1)^{1/2}$ is not a polynomial function because the expression $x^2 + 1$ is raised to a fractional exponent, and $(x^2 + 1)^{1/2}$ cannot be written as a sum of terms in which x is raised to nonnegative integer powers.

✓ *Check It Out 1*   Rework Example 1 for the following functions.

  a. $f(x) = 6$ **b.** $g(x) = (x + 1)(x - 1)$ **c.** $h(t) = \sqrt{t} + 3$

The rest of this section will be devoted to graphs of polynomial functions. These graphs have no breaks or holes, and no sharp corners. Figure 1 illustrates graphs of several functions, and indicates which are the graphs of polynomial functions.

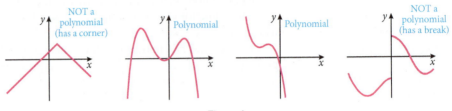

Figure 1

Polynomials of the Form x^n and Their Transformations

x	$f(x) = x^3$	$g(x) = x^4$
-100	-10^6	10^8
-10	-10^3	10^4
-2	-8	16
-1	-1	1
0	0	0
1	1	1
2	8	16
10	10^3	10^4
100	10^6	10^8

Table 1

The graphs of polynomial functions can be quite varied. The simplest polynomial functions are those with just one term, x^n. Consider the functions $f(x) = x^3$ and $g(x) = x^4$. Table 1 gives a few of their function values.

It is useful to examine the trend in the function value as the value of x increases to positive infinity ($x \to \infty$) or when the value of x decreases to negative infinity ($x \to -\infty$). This is known as **end behavior** and is summarized in Table 2.

Function	Behavior of Function as $x \to -\infty$	Behavior of Function as $x \to \infty$
x^3	$x^3 \to -\infty$	$x^3 \to \infty$
x^4	$x^4 \to \infty$	$x^4 \to \infty$

Table 2

The graphs and their end behaviors are shown in Figure 2.

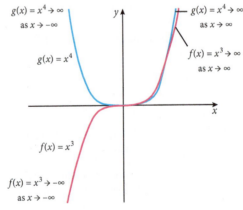

Figure 2

We now summarize properties of the general function $f(x) = x^n$, n a positive integer.

Properties of $f(x) = x^n$	
For $f(x) = x^n$, n an odd, positive integer, the following properties hold:	For $f(x) = x^n$, n an even, positive integer, the following properties hold:
■ **Domain:** $(-\infty, \infty)$	■ **Domain:** $(-\infty, \infty)$
■ **Range:** $(-\infty, \infty)$	■ **Range:** $[0, \infty)$
■ **End behavior:** As $x \to \infty$, $x^n \to \infty$ 　　　　　　As $x \to -\infty$, $x^n \to -\infty$	■ **End behavior:** As $x \to \infty$, $x^n \to \infty$ 　　　　　　As $x \to -\infty$, $x^n \to \infty$

Just in Time
Review transformations in Section 2.7.

In Example 2, we graph some transformations of the basic graph of $f(x) = x^n$.

Example 2 **Transformations of $f(x) = x^n$**

Graph the following functions using transformations.

a. $g(x) = (x - 1)^4$ **b.** $g(x) = -x^3 + 1$

Solution

a. The graph of $g(x) = (x - 1)^4$ is obtained by horizontally shifting the graph of $f(x) = x^4$ to the right by 1 unit, as shown in Figure 3.

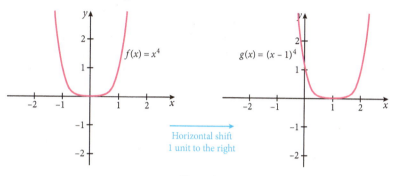

Figure 3

b. The graph of $g(x) = -x^3 + 1$ can be written as a series of transformations of the graph of $f(x) = x^3$. See Figure 4.

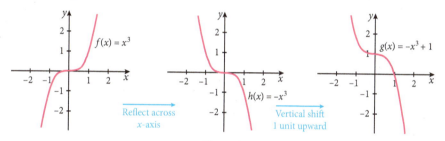

Figure 4

 Check It Out 2   Graph $f(x) = x^4 + 2$ using transformations.

The Leading Term Test for End Behavior

So far we have discussed the end behavior of polynomials with just one term. What about polynomial functions with more than one term? Specifically, we will examine the end behavior of two functions, $f(x) = -2x^3$ and $g(x) = -2x^3 + 8x$. Their graphs are shown in Figure 5.

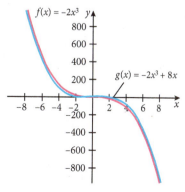

Figure 5

The graphs of the two functions are nearly indistinguishable. For very large values of $|x|$, the $8x$ term makes a very small contribution to the value of $g(x)$, compared to the $-2x^3$ term. So the values of $f(x)$ and $g(x)$ are almost the same. For small values of $|x|$, the values of $f(x)$ and $g(x)$ are indistinguishable on the graph because of the choice of vertical scale.

We now generalize our observations about the end behavior of polynomial functions.

Leading Term Test for End Behavior

Given a polynomial function of the form

$$f(x) = a_n x^n + a_{n-1} x^{n-1} + \cdots + a_1 x + a_0, \; a_n \neq 0, \; n \geq 1$$

the **end behavior** of f is determined by the **leading term** of the polynomial, $a_n x^n$.

The shape of the graph of $a_n x^n$ will resemble the shape of the graph of $y = x^n$ if $a_n > 0$, and it will resemble the shape of the graph of $y = -x^n$ if $a_n < 0$. Refer back to Example 2 for the graphs of $f(x) = x^3$ and $g(x) = -x^3 + 1$.

Figure 6 summarizes our discussion of the end behavior of non-constant polynomials.

Note Only the *end behavior* of the function can be sketched by using the leading term test. The shape of the graph for small and moderate values of $|x|$ *cannot* be determined from this test.

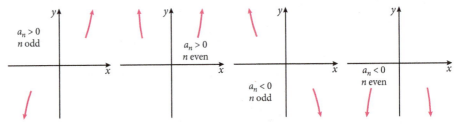

The shape of the graph in the middle region cannot be determined using the leading term test.

Figure 6

Example 3   **Determining End Behavior**

Determine the end behavior of the following functions by examining the leading term.

a. $f(x) = x^3 + 3x^2 + x$ **b.** $g(t) = -2t^4 + 8t^2$

Solution

a. For $f(x) = x^3 + 3x^2 + x$, the leading term is x^3. Thus, for large values of $|x|$, we expect $f(x)$ to behave like $y = x^3$: $f(x) \to \infty$ as $x \to \infty$ and $f(x) \to -\infty$ as $x \to -\infty$.

b. For $g(t) = -2t^4 + 8t^2$, the leading term is $-2t^4$. Thus for large values of $|t|$, we expect $g(t)$ to behave like $y = -2t^4$: $g(t) \to -\infty$ as $t \to \infty$ and $g(t) \to -\infty$ as $t \to -\infty$.

Check It Out 3   Determine the end behavior of the following functions by examining the leading term.

a. $h(x) = -3x^3 + x$ **b.** $s(x) = 2x^2 + 1$

Using Technology

Use caution with window settings when examining end behavior with a graphing utility. Figure 7 shows the graphs of a values of $Y_1(x) = -2x^4 + 8x^2$ and $Y_2(x) = -2x^4$ in a standard window. The graphs look very different. However, a window setting of $[-100, 100]$ $(10) \times [0, 20000](1000)$, shows two nearly identical graphs. The choice of scale suppresses the differences between the two graphs.

Keystroke Appendix:
Sections 7, 8

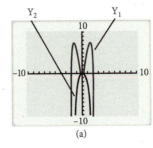

(a)

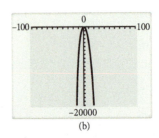

(b)

Figure 7

Finding Zeros and *x*-Intercepts by Factoring

In addition to their end behavior, an important feature of the graphs of polynomial functions is the location of their *x*-intercepts. Recall from Section 3.1 the following connection between the real zeros of a function and the *x*-intercepts of its graph: the real number values of *x* satisfying $f(x) = 0$ are called *real zeros* of the function *f*. Each of these values of *x* is the first coordinate of an *x*-intercept of the graph of the function.

Just in Time
Review factoring in Section 1.3 and *x*-intercepts and zeros in Section 3.1.

Example 4   **Finding Real Zeros and *x*-Intercepts**

Find the real zeros of $f(x) = 2x^3 - 18x$ and the corresponding *x*-intercepts of the graph of *f*.

Solution   To find the real zeros of *f*, solve the equation $f(x) = 0$:

$$2x^3 - 18x = 0 \qquad \text{Set expression for } f \text{ equal to } 0$$

$$2x(x^2 - 9) = 0 \qquad \text{Factor out } 2x$$

$$2x(x + 3)(x - 3) = 0 \qquad \text{Factor } x^2 - 9 = (x + 3)(x - 3)$$

$$2x = 0 \Rightarrow x = 0$$

$$x + 3 = 0 \Rightarrow x = -3 \quad \text{Set each factor equal to zero and solve for } x$$
$$x - 3 = 0 \Rightarrow x = 3$$

The real zeros of *f* are $x = 0$, $x = -3$, and $x = 3$. The *x*-intercepts of the graph of *f* are $(0, 0)$, $(-3, 0)$, and $(3, 0)$.

 Check It Out 4   Find the real zeros of $f(x) = -3x^3 + 12x$ and the *x*-intercepts of the graph of *f*.

Hand Sketching the Graph of a Polynomial Function

If a polynomial function can easily be written in factored form, then we can find the *x*-intercepts, and sketch the function by hand using the following procedure.

Hand Sketching the Graph of a Polynomial Function

1. Determine the end behavior of the function.
2. Find the *y*-intercept and plot it.
3. Find and plot the *x*-intercepts of the graph of the function. These points will divide the *x*-axis into smaller intervals.
4. Find the sign and value of $f(x)$ for a test value *x* in each of these intervals. Plot these points.
5. Use the plotted points and the end behavior to sketch a smooth graph of the function. Plot any additional points, if needed.

> **Example 5**   **Sketching a Polynomial Function**

Determine the end behavior of $f(x) = x^3 - x$. Find the x- and y-intercepts of its graph. Use this information to sketch the graph of f by hand.

Solution

Step 1 Determine the end behavior. For $|x|$ large, $f(x)$ behaves like $y = x^3$; $f(x) \to \infty$ as $x \to \infty$ and $f(x) \to -\infty$ as $x \to -\infty$. See Figure 8.

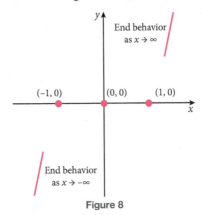

Figure 8

Step 2 Find the y-intercept. Since $f(0) = 0$, the y-intercept is $(0, 0)$.

Step 3 Find the x-intercepts. The zeros of this function are found by setting $f(x)$ to 0 and solving for x:

$$x^3 - x = 0 \qquad \text{Set expression to zero}$$

$$x(x^2 - 1) = x(x + 1)(x - 1) = 0 \qquad \text{Factor left side completely}$$

$$x = 0$$

$$x + 1 = 0 \Rightarrow x = -1 \qquad \text{Use the Zero Product Rule}$$

$$x - 1 = 0 \Rightarrow x = 1$$

Thus the zeros are $x = 0$, $x = -1$, and $x = 1$, and the x-intercepts are $(0, 0)$, $(-1, 0)$, and $(1, 0)$. See Figure 8.

Step 4 Determine signs of function values. Next, we graph the function between the x-intercepts, and observe that the three x-intercepts break up the x-axis into four intervals:

$$(-\infty, -1), (-1, 0), (0, 1), (1, \infty)$$

Table 3 lists the value of $f(x)$ for at least one value of x, called a *test value*, in each of these intervals. We can choose just one test value, since the sign of the function value is unchanged within each interval.

Interval	Test Value x	Function Value $f(x) = x^3 - x$	Sign of $f(x) = x^3 - x$
$(-\infty, -1)$	-2	-6	$-$
$(-1, 0)$	-0.5	0.375	$+$
$(0, 1)$	0.5	-0.375	$-$
$(1, \infty)$	2	6	$+$

Table 3

The signs given in Table 3 are summarized on the number line in Figure 9.

Figure 9

Plotting the x-intercepts and the test values, we have the partial sketch shown in Figure 10 (a).

Step 5 Sketch the entire graph. By plotting the points given in the table and using the sign of the function value in each of the intervals, we can sketch the graph of the function, as shown in Figure 10 (b).

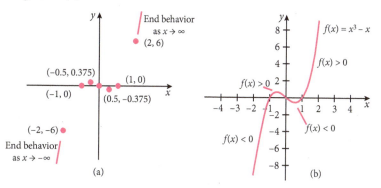

Figure 10

 Check It Out 5   Rework Example 5 for the function $f(x) = -3x^3 + 27x$.

Multiplicities of Real Zeros

We will now examine the finer properties of the graphs of polynomial functions. These include:

- Examining the behavior of the polynomial function at its x-intercepts

- Observing any types of symmetry in the graph of the polynomial function

- Locating the local maxima and minima of the graph of a polynomial function by using a graphing utility

A complete analysis of the graph of a polynomial function involves calculus, which is beyond the scope of this book.

The number of times a linear factor $x - a$ occurs in the completely factored form of a polynomial expression is known as the **multiplicity** of the real zero a associated with that factor. For example, $f(x) = (x + 1)(x - 3)^2 = (x + 1)(x - 3)(x - 3)$ has two real zeros: $x = -1$ and $x = 3$. The zero at $x = -1$ has multiplicity 1 and the zero at $x = 3$ has multiplicity 2, since their corresponding factors are raised to the powers 1 and 2, respectively. A formal definition of the multiplicity of a zero of a polynomial is given in Section 3.5.

The multiplicity of a real zero of a polynomial and the graph of the polynomial at the corresponding x-intercept have a close connection, as discussed next.

Multiplicities of Real Zeros and Behavior at the *x*-Intercept

- If the multiplicity of a real zero of a polynomial function is **odd**, the graph of the function **crosses** the *x*-axis at the corresponding *x*-intercept.

- If the multiplicity of a real zero of a polynomial function is **even**, the graph of the function **touches**, but does not cross, the *x*-axis at the corresponding *x*-intercept. See Figure 11.

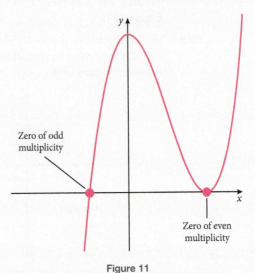

Zero of odd multiplicity

Zero of even multiplicity

Figure 11

| **Example 6** | **Multiplicity of Real Zeros** |

Determine the multiplicities of the real zeros of the following functions. Does the graph cross the *x*-axis or just touch it at the *x*-intercepts?

a. $f(x) = (x - 6)^3(x + 2)^2$ **b.** $h(t) = t^3 + 2t^2 + t$

Solution

a. Recall that the zeros of f are found by setting $f(x)$ to 0 and solving for x.

$$(x - 6)^3(x + 2)^2 = 0$$

$$(x - 6)^3 = 0 \quad \text{or} \quad (x + 2)^2 = 0$$

Thus, $x = 6$ or $x = -2$. The zero at $x = 6$ has multiplicity 3 since the corresponding factor, $x-6$, is raised to the third power. Because the multiplicity is odd, the graph will cross the *x*-axis at $(6, 0)$.

The zero at $x = -2$ has multiplicity 2. Because this zero has even multiplicity, the graph of f only touches the *x*-axis at $(-2, 0)$. This is verified by the graph of f in Figure 12.

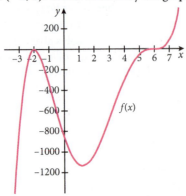

$f(x)$

Figure 12

b. To find the zeros of h, we first need to factor the expression. This gives

$$t^3 + 2t^2 + t = t(t^2 + 2t + 1) = t(t + 1)^2 = 0$$

$$t = 0 \text{ or } (t + 1)^2 = 0$$

Thus, $t = 0$ or $t = -1$. We see that the zero at $t = 0$ has multiplicity 1, and so the graph will cross the t-axis at $(0, 0)$. The zero at $t = -1$ has multiplicity 2, and will simply touch the t axis at $(-1, 0)$. This is verified by the graph of h in Figure 13.

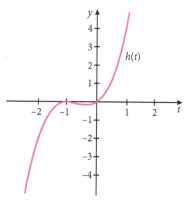

Figure 13

 Check It Out 6   Rework Example 6 for $g(x) = x^2(x - 5)^2$.

Symmetry of Polynomial Functions

Just in Time
Review symmetry in Section 2.5.

Recall that a function is *even* if $f(x) = f(-x)$ for all x in the domain of f. The graph of an even function is symmetric with respect to the y-axis. A function is *odd* if $f(x) = -f(-x)$ for all x in the domain of f. The graph of an odd function is symmetric with respect to the origin. If a polynomial function happens to be odd or even, we can use that fact to help sketch its graph.

Example 7   **Checking for Symmetry**

Check whether $f(x) = x^4 + x^2 + x$ is odd, even, or neither.

Solution   Because $f(x)$ has some even-powered terms, we will check to see if it is an even function:

$$f(-\pmb{x}) = (-\pmb{x})^4 + (-\pmb{x})^2 + (-\pmb{x}) = x^4 + x^2 - x$$

So $f(x) \neq f(-x)$.
Now check if it is odd

$$-f(-x) = -(f(-x)) = -(x^4 + x^2 - x) = -x^4 - x^2 + x$$

So $f(x) \neq -f(-x)$. Thus, f is neither even nor odd. Its graph is not symmetric with respect to the y-axis or the origin.

 Check It Out 7   Decide whether the following functions are even, odd or neither.

a. $h(t) = -t^4 + t$ **b.** $g(s) = s^3 + 8s$

Finding Local Extrema and Sketching a Complete Graph

The graphs of most of the polynomials we have examined have peaks and valleys, known as **local maxima** and **local minima**, respectively. They are also known as **local extrema** or **turning points**. The term "local" is used since the values are not necessarily the maximum and minimum values of the function over its entire domain.

Finding the precise locations of local extrema requires the use of calculus. If you sketch a graph by hand, you can get a rough idea of the locations of the local extrema by plotting additional points in the intervals where such local extrema may exist. If you are using a graphing utility, you can accurately find the local extrema.

In addition to the techniques presented earlier, we can now use multiplicity of zeros and symmetry to sketch graphs of polynomials. A sketch of the graph of a polynomial is **complete** if it shows all the x-intercepts and the y-intercept and illustrates the correct end behavior of the function.

Example 8   **Sketching a Complete Graph**

Sketch a complete graph of $f(x) = -2x^4 + 8x^2$.

Solution

Step 1 This function's end behavior is similar to that of $y = -2x^4$: as $x \to \pm\infty$, $f(x) \to -\infty$.

Step 2 The y-intercept is $(0, 0)$.

Step 3 Find the x-intercepts of the graph of the function.

$$-2x^4 + 8x^2 = 0 \qquad \text{Set expression equal to zero}$$

$$-2x^2(x^2 - 4) = -2x^2(x + 2)(x - 2) = 0 \qquad \text{Factor left side completely}$$

$$x^2 = 0 \Rightarrow x = 0 \qquad x = 0 \text{ is a zero of multiplicity 2}$$

$$x + 2 = 0 \Rightarrow x = -2$$

$$x - 2 = 0 \Rightarrow x = 2$$

Thus the x-intercepts of the graph of this function are $(0, 0)$, $(-2, 0)$, and $(2, 0)$. Because $x = 0$ is a zero of multiplicity 2, the graph *touches*, but *does not cross* the x-axis at $(0, 0)$. The zeros at $x = -2$ and $x = 2$ are of multiplicity 1, so the graph *crosses* the x-axis at $(-2, 0)$ and $(2, 0)$.

Step 4 Check for symmetry, if any. Since $f(x)$ has even-powered terms, we will check to see if it is an even function:

$$f(-x) = -2(-x)^4 + 8(-x)^2 = -2x^4 + 8x^2 = f(x)$$

Because $f(x) = f(-x)$, f is an even function. The graph is symmetric with respect to the y-axis.

Step 5 Make a table of test values for x in the intervals $(0, 2)$ and $(2, \infty)$ to determine the sign of f (See Table 4). The positive values of x are sufficient since the graph is symmetric with respect to the y-axis. We simply reflect the graph across the y axis to complete the graph.

Interval	Test Value x	Function Value $f(x) = -2x^4 + 8x^2$	Sign of $f(x)$
$(0, 2)$	1	6	$+$
$(2, \infty)$	3	-90	$-$

Table 4

Putting the information from all these steps together, we obtain the graph in Figure 14.

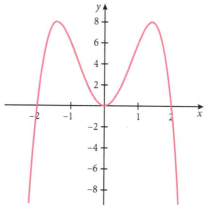

Figure 14

Using Technology

Examining Figure 14, we see that there must be one local maximum between $x = -2$ and $x = 0$, and another local maximum between $x = 0$ and $x = 2$. The value $x = 0$ gives rise to a local minimum. Using the MAXIMUM and MINIMUM features of your graphing utility, you can determine the local extrema. One of the maximum values is shown in Figure 15. The local extrema are summarized in Table 5.

✅ **Check It Out 8**   Rework Example 8 for $f(x) = x^5 - x^3$.

x	$f(x)$	Type of Extremum
-1.4142	8	Local maximum
0	0	Local minimum
1.4142	8	Local maximum

Table 5

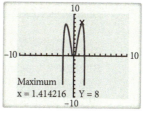

Figure 15

Relationship Between x-Intercepts and Factors

Thus far, we have used the factored form of a polynomial to find the x-intercepts of its graph. Next, we try to find a possible expression for a polynomial function *given* the x-intercepts of the graph. To do so, the following important fact is needed. It will be discussed in detail in Section 3.6.

Number of Real Zeros

The **number of real zeros** of a polynomial function f of degree n is less than or equal to n, counting multiplicities. The graph of f can cross the x-axis at most n times.

We also have the following relationship between a real zero of a polynomial, its factor, and the corresponding x-intercept of its graph.

Zeros, x-Intercepts and factors

If c is a real zero of a polynomial function $f(x)$, that is, $f(c) = 0$, and c is a zero of multiplicity k, then

a. $(c, 0)$ is an x-intercept of the graph of f, and

b. $(x - c)^k$ is a factor of $f(x)$, but the factor $(x - c)^{k+1}$ is not.

Example 9 **Finding a Polynomial from Its Zeros**

Find a polynomial $f(x)$ of degree 4 that has zeros at -1 and 3, each with multiplicity 1, and a zero at -2 of multiplicity 2.

Solution Use the fact that if c is a zero of f, then $x - c$ is a factor of f. Because -1 and 3 are zeros with multiplicity 1, the corresponding factors of f are $x - (-1)$ and $x - 3$. Because -2 is a zero with multiplicity 2, the corresponding factor is $(x - (-2))^2$. A possible fourth-degree polynomial satisfying all of the given conditions is

$$f(x) = (x + 1)(x - 3)(x + 2)^2$$

This is not the only possible expression. Any nonzero multiple of $f(x)$ will satisfy the same conditions. Thus any polynomial of the form

$$f(x) = a(x + 1)(x - 3)(x + 2)^2, \quad a \neq 0$$

will suffice.

 Check It Out 9 Find a polynomial $f(x)$ of degree 3 that has zeros at 1, 2, and -1, each with multiplicity 1.

Application of Polynomials

Example 10 **Graphing a Volume Function**

Gift Horse, Inc., manufactures various types of decorative gift boxes. The bottom portion of one such box is made by cutting a small square of length x inches from each corner of a 10-inch by 10-inch piece of cardboard and folding up the sides of the box. See Figure 16.

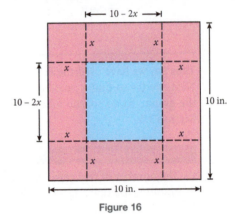

Figure 16

a. Find an expression for the volume of the resulting box.

b. Graph the volume function. Use your graph to determine the values of x for which the expression makes realistic sense.

Solution

a. The volume of a rectangular box is given by the formula

$$V = \text{length} \times \text{width} \times \text{height}$$

From Figure 16, the expressions for the length, width, and height of the box are given by

$$l = 10 - 2x$$

$$w = 10 - 2x$$

$$h = x$$

Substituting into the expression for the volume, we get,

$$V(x) = \text{length} \times \text{width} \times \text{height} = (10 - 2x)(10 - 2x)x = x(10 - 2x)^2$$

We leave it in factored form, because it is easier to analyze.

b. The x-intercepts of the graph of this function are found by setting $V(x)$ equal to 0:

$$x(10 - 2x)^2 = 0 \quad \text{Set } V(x) \text{ equal to zero}$$

$$x = 0 \quad \text{Use the Zero Product Rule}$$

$$(10 - 2x)^2 = 0 \Rightarrow x = 5$$

Thus the x-intercepts are $(0, 0)$ and $(5, 0)$. Because $x = 0$ is a zero of multiplicity 1, the graph will cross the axis at $x = 0$. Since $x = 5$ is a zero of multiplicity 2, the graph will touch the x-axis at $(5, 0)$ and not cross it. The x-intercepts break up the x-axis into three intervals. We next find the value of the function at one point in each of these intervals. See Table 6.

Interval	Test Value (x)	Function Value $V(x)$	Sign of $V(x)$
$(-\infty, 0)$	-1	-144	$-$
$(0, 5)$	2	72	$+$
$(5, \infty)$	6	24	$+$

Table 6

Using the end behavior along with the x-intercepts, information about the multiplicity of the zeros, and the data given in the table, we obtain the graph in Figure 17.

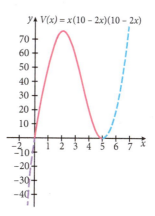

Figure 17

We see that x cannot be negative since it is a length. From the graph, we see that $V(x)$ is positive for values of x such that $0 < x < 5$ or $x > 5$. Because the piece of cardboard is only 10 inches by 10 inches, we cannot cut out a square more than 5 inches on a side from each corner. Thus, the only allowable values for x are $x \in (0, 5)$. We exclude $x = 0$ and $x = 5$, because a solid object with zero volume is meaningless.

Mathematically, the function $V(x)$ is defined for all values of x, whether or not it makes realistic sense.

✓ *Check It Out 10*   Graph the volume function of the open box obtained by cutting a square of length x from each corner of a 8 inches by 8 inches piece of cardboard and then folding up the sides.

Exercises 3.2

1. The graph of $f(x + 2)$ is the same as the graph of f shifted to the _____ by _____ units.
2. What is the zero of $f(x) = 5x - 15$?
3. Find the x- and y-intercepts of the graph of $f(x) = -2x + 4$.
4. Factor the following: **a.** $x^3 + 2x^2 + x$ **b.** $-3x^3 + 12x$
5. If $f(x) = f(-x)$ for all x in the domain of f, what type of symmetry does the graph of f have?
6. True or False: If $f(x) = -f(-x)$ for all x in the domain of f, then the function is even.

 Skills

Paying Attention to Instructions: Exercises 7 and 8 are intended to give you practice reading and paying attention to the instructions that accompany the problems you are working.

7. Let $f(x) = -2x^3 + 18x$
   a. Find the factors of f.
   b. Find the real zeros of f.
   c. Find the x-intercepts of the graph of f.

8. Let $f(x) = x^3 - 2x^2 + x$
   a. Find the factors of f.
   b. Find the real zeros of f.
   c. Find the x-intercepts of the graph of f.

In Exercises 9–12, determine whether the graph represents a polynomial. Explain your reasoning.

9.

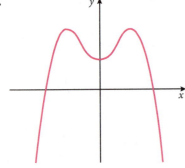

10.

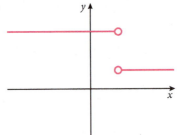

11.

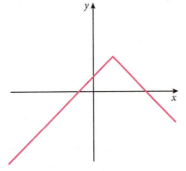

12.

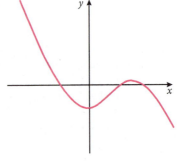

In Exercises 13–20 determine whether the function is a polynomial function. If so, find the degree. If not, state the reason.

13. $f(x) = -x^3 + 3x^2 + 1$ 14. $f(s) = 4s^5 - 5s^3 + 6s - 1$ 15. $f(t) = \sqrt{t}$ 16. $g(t) = \dfrac{1}{t}$

17. $f(x) = 5$ 18. $g(x) = -2$ 19. $f(x) = -(x + 1)^3$ 20. $g(x) = (x - 1)^2$

In Exercises 21–30, determine the end behavior of the function.

21. $f(x) = 7x$

22. $g(x) = -2x$

23. $f(x) = -2x^3 + 4x - 1$

24. $g(x) = 3x^4 + 2x^2 - 1$

25. $H(x) = -5x^4 + 3x^2 + x - 1$

26. $h(x) = 5x^6 - 3x^3$

27. $g(x) = -10x^3 + 3x^2 + 5x - 2$

28. $h(x) = 3x^3 - 4x^2 + 5$

29. $f(s) = \dfrac{7}{2}s^5 - 14s^3 + 10s$

30. $g(s) = -\dfrac{3}{4}s^4 + 8s^2 - 3s - 16$

In Exercises 31–42, sketch the following polynomial functions using transformations.

31. $f(x) = x^3 - 2$

32. $f(x) = x^4 - 1$

33. $f(x) = \dfrac{1}{2}x^3$

34. $g(x) = -\dfrac{1}{2}x^4$

35. $g(x) = (x - 2)^3$

36. $h(x) = (x + 1)^4$

37. $h(x) = -2x^5 - 1$

38. $f(x) = 3x^4 + 2$

39. $f(x) = -(x + 1)^3 - 2$

40. $f(x) = (x - 2)^4 + 1$

41. $h(x) = -\dfrac{1}{2}(x + 1)^3 - 2$

42. $h(x) = \dfrac{1}{2}(x - 2)^4 - 1$

In Exercises 43–46, find a function of the form $y = cx^k$ that has the same end behavior. Confirm your results with a graphing utility.

43. $g(x) = -5x^3 - 4x^2 + 4$

44. $h(x) = 6x^3 - 4x^2 + 7x$

45. $f(x) = 1.5x^5 - 10x^2 + 14x$

46. $g(x) = -3.6x^4 + 4x^2 + x - 20$

In Exercises 47–62, for each polynomial function:

   a. *Find a function of the form $y = cx^k$ which has the same end behavior.*
   b. *Find the x- and y-intercept(s) of the graph.*
   c. *Find the interval(s) where the value of the function is positive.*
   d. *Find the interval(s) where the value of the function is negative.*
   e. *Use the information in parts (a)–(d) to sketch a graph of f or g.*

47. $f(x) = -2x^3 + 8x$

48. $f(x) = 3x^3 - 27x$

49. $g(x) = (x - 3)(x + 4)(x - 1)$

50. $f(x) = (x + 1)(x - 2)(x + 3)$

51. $f(x) = -\dfrac{1}{2}(x^2 - 4)(x^2 - 1)$

52. $f(x) = (x^2 - 4)(x + 1)(x - 3)$

53. $f(x) = x^3 - 2x^2 - 3x$

54. $g(x) = x^3 + x^2 - 3x$

55. $f(x) = -x(2x + 1)(x - 3)$

56. $g(x) = 2x(x - 2)(2x - 1)$

57. $f(x) = -(x^2 - 1)(x - 2)(x + 3)$

58. $f(x) = x(x^2 - 4)(x + 1)$

59. $g(x) = 2x^2(x + 3)$

60. $f(x) = -3x^2(x - 1)$

61. $f(x) = (2x + 1)(x - 3)(x^2 + 1)$

62. $g(x) = -(x - 2)(3x - 1)(x^2 + 1)$

In Exercises 63–66, for each polynomial function graphed below, find
   a. *The x-intercepts of the graph of the function, if any*
   b. *The y-intercept of the graph of the function*
   c. *Whether the leading term is odd- or even-powered*
   d. *The sign of the leading coefficient*

63.

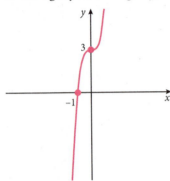

64.

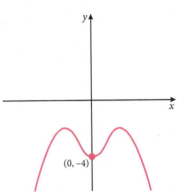

65.

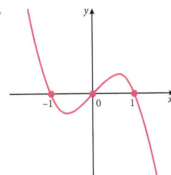

66.

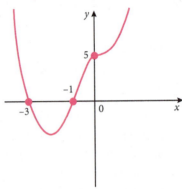

In Exercises 67–74, determine the real zeros of the function and their multiplicities. Comment on the behavior of the graph at the x-intercepts: Does it cross or just touch the x-axis? You may check your results with a graphing utility.

67. $f(x) = (x - 2)^2(x + 5)^5$ **68.** $g(s) = (s + 6)^4(s - 3)^3$ **69.** $h(t) = t^2(t - 1)(t + 2)$

70. $g(x) = x^3(x + 2)(x - 3)$ **71.** $f(x) = x^2 + 2x + 1$ **72.** $h(s) = s^2 - 2s + 1$

73. $g(s) = 2s^3 + 4s^2 + 2s$ **74.** $h(x) = 2x^3 - 4x^2 + 2x$

In Exercises 75–82, classify the function as odd, even, or neither. Determine the symmetry, if any, of its graph.

75. $g(x) = x^4 + 2x^2 - 1$ **76.** $h(x) = 2x^4 - x^2 + 2$ **77.** $f(x) = -3x^3 + 1$ **78.** $g(x) = x^3 - 2$

79. $f(x) = -x^3 + 2x$ **80.** $g(x) = x^3 - 3x$ **81.** $h(x) = -2x^4 + 3x^2 - 1$

82. $g(x) = 3x^4 - 2x^2 + 1$

In Exercises 83–86, use the graph of the polynomial function to find the real zeros of the function and to determine whether they are of even or odd multiplicity.

83.

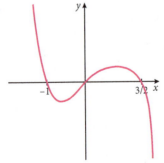

84.

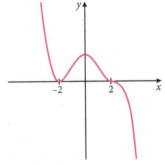

85.

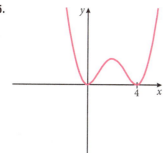

86.

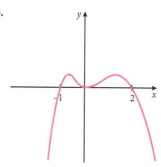

In Exercises 87–98, for each polynomial function, find

    a. *The end behavior*

    b. *The y-intercept of its graph*

    c. *The x-intercept(s) of the graph of the function and the multiplicity of each of the real zeros*

    d. *Symmetry of the graph of the function, if any*

    e. *The intervals where the function is positive or negative.*

Use this information to sketch a graph of the function. Factor first if the expression is not in factored form.

87. $f(x) = x^2(x - 1)$ **88.** $h(x) = x(x - 2)^2$ **89.** $f(x) = (x - 2)^2(x + 2)$

90. $g(x) = (x + 1)(x - 2)^2$ **91.** $g(x) = (x + 1)^2 (x - 2)(x + 3)$

92. $f(x) = (x - 1)(x + 2)^2(x + 1)$ **93.** $g(x) = -2(x + 1)^2(x - 3)^2$

94. $f(x) = -3(x - 2)^2(x + 1)^2$ **95.** $f(x) = x^3 + 4x^2 + 4x$ **96.** $f(x) = -x^3 - 2x^2 - x$

97. $h(x) = -2x^4 + 4x^3 + 2x^2$ **98.** $f(x) = 3x^4 - 6x^3 + 3x^2$

In Exercises 99–106, find an expression for a polynomial function f(x) with the given properties. There can be more than one correct answer.

99. Degree 3; zeros at $-2, 5, 6$, each with multiplicity 1

100. Degree 3; zeros at $-6, 0, 3$, each with multiplicity 1

101. Degree 4; zeros at 2 and 4, each with multiplicity 2

102. Degree 4; zeros at 2 and -3, each with multiplicity 1; zero at 5 with multiplicity 2

103. Degree 3; zero at 2, with multiplicity 1; zero at -3 with multiplicity 2

104. Degree 3; zero at 5, with multiplicity 3

105. Degree 5; zeros at -2 and -1, each with multiplicity 1; zero at 5 with multiplicity 3

106. Degree 5; zeros at -3 and 1, each with multiplicity 2; zero at 4 with multiplicity 1

In Exercises 107–110, graph each polynomial function using a graphing utility. Then

    a. *approximate the x-intercept(s) of the graph of the function.*

    b. *find the intervals where the function is positive or negative.*

    c. *approximate the values of x at which a local maximum or minimum occurs.*

    d. *discuss symmetry of the graph, if any.*

107. $f(x) = -x^3 + 3x + 1$ **108.** $f(x) = x^3 + x^2 + \dfrac{1}{2}$ **109.** $f(x) = x^4 + 2x^3 - 1$

110. $f(x) = -x^4 + 3x - 1$

111. Consider the function $f(x) = 0.001x^3 + 2x^2$.

    a. Graph the function in a standard window of a graphing utility. Explain why this does not give a complete graph of the function.

    b. Using the x-intercepts of the graph of this function and the end behavior of the function, sketch an approximate graph of this function by hand.

    c. Find a graphing window that shows a correct graph for this function.

Applications

112. Manufacturing An open box is to be made by cutting four squares of equal size out of a 10-inch by 15-inch rectangular piece of cardboard (one at each corner) and then folding up the sides.

    a. Let x be the length of the side of the square cut from each corner. Find an expression for the volume of the box in terms of x. Leave the expression in factored form.

    b. What is a realistic range of values for x? Explain.

113. Construction A cylindrical container is to be constructed so that the *sum* of its height and its diameter is 10 feet.

    a. Write an equation relating the height of the cylinder, h, to its radius, r. Solve the equation for h in terms of r.

    b. The volume of a cylinder is given by $V = \pi r^2 h$. Use your answer from part (a) to express the volume in terms of r alone. Leave your expression in factored form so it will be easier to analyze.

    c. What are the values of r for which this problem makes realistic sense? Explain.

114. Manufacturing A rectangular container with a square base is constructed so that the *sum* of the height and the perimeter of the base is 20 feet.

    a. Write an equation relating the height, h, to the length of a side of the base, s. Solve the equation for h in terms of s.

    b. Use your answer from part (a) to express the volume of the container in terms of s alone. Leave it in factored form so it will be easier to analyze.

    c. What are the values of s for which this problem makes sense? Explain.

115. Geometry A rectangular solid has height h and a square base. One side of the square base is 3 inches greater than the height.

    a. Find an expression for the volume of the solid in terms of h.

    b. Sketch a graph of the volume function.

    c. For what values of h does the volume function make sense?

116. Manufacturing An open box is to be made by cutting four squares of equal size out of a 12-inch by 12-inch square piece of cardboard, one at each corner, and then folding up the sides of the box.

    a. Let x be the length of the side of the square cut from each corner. Find an expression for the volume of the box in terms of x.

    b. Sketch a graph of the volume function.

    c. Find the value of x that gives the maximum volume for the box.

117. Economics Gross Domestic Product (GDP) is the market value of all final goods and services produced within a country during a given time period. The following fifth degree polynomial approximates the per capita GDP (the average GDP per person) for the United States for the years from 1933 to 1950.

$$g(x) = 0.294x^5 - 12.2x^4 + 169x^3 - 912x^2 + 2025x + 4508$$

where $g(x)$ is in 1996 dollars and x is the number of years since 1933. Note that when dollar amounts are measured over time, they are converted to the dollar value for a specific base year. In this case, the base year is 1996.   (*Source:* Economic History Services)

    a. Use this model to calculate the per capita GDP (in 1996 dollars) for the years 1934, 1942 and 1949. What do you observe?

    b. Explain why this model may not be suitable as a predictor of the per capita GDP for the year 2002.

    c. Use your graphing utility to find the year(s), during the period 1933–1950, when the GDP reached a local maximum.

Concepts

118. Explain why the following graph is **not** a complete graph of the function $p(x) = 0.01x^3 + x^2$.

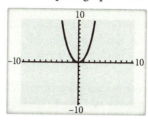

119. Sketch the graph of a cubic polynomial function with *exactly* 2 real zeros.

120. Find a polynomial function whose zeros are $x = 0, 1, -1$. Is your answer the only correct answer? Why or why not? You may confirm your answer with a graphing utility.

121. Find a polynomial function whose graph crosses the x-axis at $(2, 0)$ and $(1, 0)$. Is your answer the only correct answer? Why or why not? You may confirm your answer with a graphing utility.

 In Exercises 122–125, use the given information to

    a. *Sketch a possible graph of the polynomial function.*

    b. *Indicate on your graph roughly where the local maxima and minima, if any, may occur.*

    c. *Find a possible expression for the polynomial.*

    d. *Use a graphing utility to check your answers to parts (a)–(c).*

122. A polynomial $p(x)$ has real zeros at $x = -1$ and $x = 3$, and the graph crosses the x-axis at both these zeros. As $x \to \pm\infty$, $p(x) \to \infty$.

123. The only points at which the graph of the polynomial $f(s)$ crosses the s-axis are $(-1, 0)$ and $(2, 0)$, and the only point at which it just touches the s-axis is $(0, 0)$. The function is positive on the intervals $(-\infty, -1)$ and $(2, \infty)$.

124. The real zeros of a polynomial $h(x)$ are $x = 3$ and $x = 0.5$, each with multiplicity 1, and $x = \sqrt{2}$, which has multiplicity 2. As $|x|$ gets larger, the value of $h(x) \to \infty$.

125. A polynomial $q(x)$ has exactly one real zero and no local maxima or minima.

SPOTLIGHT ON SUCCESS   *Student Instructor Shelby*

The price of success is hard work, dedication to the job at hand, and the determination that whether we win or lose, we have applied the best of ourselves to the task at hand.

—*Vince Lombardi*

My earliest memory of math is from elementary school, having to continuously take the division and times tables test because I couldn't finish it in the allotted time. I have never been naturally gifted at math, but I enjoy a challenge. I wouldn't allow my setbacks to stop me from succeeding in my classes. As a high school freshman I struggled in my geometry class, and as a senior, I thrived in my AP Calculus class. The difference in those short four years was the effort I put in, as well as having great teachers, who provided the necessary knowledge and support along the way.

However, to continue to thrive in math, I had to realize that my success wasn't solely in the hands of the teacher. I saw students fail classes taught by some of the best teachers on campus. It all stemmed from the personal goals each student held. There were multiple times when I told myself, "You can do this. You can learn the concepts." I'm not sure I would have moved past prealgebra without those personal words of encouragement.

Having confidence in yourself is necessary to be successful in all aspects of life. It's easy to give up on something that doesn't come naturally, but the reward of achieving something you thought to be impossible is worth the hardship.

Objectives

- Perform long division of polynomials.

- Perform synthetic division of polynomials.

- Apply the Remainder and Factor Theorems.

In previous courses, you may have learned how to factor polynomials using various techniques. Many of these techniques apply only to special kinds of polynomial expressions. For example, the previous two sections of this chapter dealt only with polynomials that could easily be factored to find the zeros and x-intercepts.

In order to be able to find zeros and x-intercepts of polynomials that cannot readily be factored, we first introduce long division of polynomials. We will then use the long division algorithm to make a general statement about the factors of a polynomial.

Long Division of Polynomials

Long division of polynomials is similar to long division of numbers. When dividing polynomials, we obtain a quotient and a remainder. Just as with numbers, if a remainder is 0, then the divisor is a factor of the dividend.

VIDEO EXAMPLES

SECTION 3.3

Example 1 **Determining Factors by Division**

Divide to determine whether $x - 1$ is a factor of $x^2 - 3x + 2$.

Solution Here, $x^2 - 3x + 2$ is the *dividend*, and $x - 1$ is the *divisor*.

Step 1 Set up the division as follows.

$$\text{Divisor} \rightarrow \quad x - 1 \overline{)x^2 - 3x + 2} \quad \leftarrow \text{Dividend}$$

Step 2 Divide the leading term of the dividend (x^2) by the leading term of the divisor (x). The result (x) is the first term of the quotient, as illustrated below.

$$\begin{array}{r} x \quad\quad\quad \leftarrow \text{First term of quotient} \\ x - 1 \overline{)x^2 - 3x + 2} \end{array}$$

Step 3 Take the first term of the quotient, x, and multiply it by the divisor, which gives $x^2 - x$. Put this result in the second row.

$$\begin{array}{r} x \quad\quad\quad \\ x - 1 \overline{)x^2 - 3x + 2} \\ \underline{x^2 - x} \quad\quad \leftarrow \text{Multiply } x \text{ by divisor} \end{array}$$

Step 4 Subtract the second row from the first row and treat the resulting expression, $-2x + 2$, as though it were a new dividend, just as in long division of numbers.

$$\begin{array}{r} x \quad\quad\quad \\ x - 1 \overline{)x^2 - 3x + 2} \\ \underline{x^2 - x} \quad\quad \\ -2x + 2 \quad \leftarrow (x^2 - 3x + 2) - (x^2 - x) \text{ (Watch your signs!)} \end{array}$$

Step 5 Continue as in Steps 1 − 4, but in Step 2 divide the leading term of the expression, $-2x + 2$, in the bottom row by the leading term of the divisor. This result, (-2), is the second term of the quotient.

$$\begin{array}{r} x - 2 \quad\quad \\ x - 1 \overline{)x^2 - 3x + 2} \\ \underline{x^2 - x} \quad\quad \\ \text{Leading term is } -2x \rightarrow \quad -2x + 2 \\ \text{Multiply } -2 \text{ by divisor} \rightarrow \quad \underline{-2x + 2} \\ (-2x + 2) - (-2x + 2) = 0 \rightarrow \quad 0 \end{array}$$

Thus, dividing $x^2 - 3x + 2$ by $x - 1$ gives $x - 2$ as the *quotient* and 0 as the *remainder*. This tells us that $(x - 1)$ is a factor of $(x^2 - 3x + 2)$.

Check your answer: $(x - 2)(x - 1) = x^2 - 3x + 2$.

✓ *Check It Out 1* Find the quotient and remainder when the polynomial $x^2 + x - 6$ is divided by $x - 2$.

We can make the following statement about the relationship between the dividend, the divisor, the quotient, and the remainder:

$$(\text{Divisor} \times \text{Quotient}) + \text{Remainder} = \text{Dividend}$$

This is a very important result and is stated formally as follows.

The Division Algorithm

Let $p(x)$ be a polynomial divided by a nonzero polynomial $d(x)$. Then, there is a quotient polynomial $q(x)$ and a remainder polynomial $r(x)$ such that

$$p(x) = d(x)q(x) + r(x) \text{ or, equivalently, } \frac{p(x)}{d(x)} = q(x) + \frac{r(x)}{d(x)}$$

where either $r(x) = 0$ or the degree of $r(x)$ is less than the degree of $d(x)$.

The following result illustrates the relationship between factors and remainders.

Factors and Remainders

Let a polynomial $p(x)$ be divided by a nonzero polynomial $d(x)$, with a **quotient polynomial** $q(x)$ and a **remainder polynomial** $r(x)$. If $r(x) = 0$, then $d(x)$ and $q(x)$ are both *factors* of $p(x)$.

Example 2 shows how polynomial division can be used to factor a polynomial that cannot be factored by straightforward methods that you may already know.

Example 2   Long Division of Polynomials

Find the quotient and remainder when $2x^4 + 7x^3 + 4x^2 - 7x - 6$ is divided by $2x + 3$.

Solution We follow the same steps as before, but condense them in this example.

Step 1
- Divide the leading term of the dividend $(2x^4)$ by the leading term of the divisor $(2x)$. The result, x^3, is the first term of the quotient.

- Multiply the first term of the quotient, x^3, by the divisor, and put the result, $2x^4 + 3x^3$, in the second row.

- Subtract the second row from the first row, just as in the division of numbers.

$$
\begin{array}{r}
x^3 \\
2x+3\,\overline{)2x^4 + 7x^3 + 4x^2 - 7x - 6} \\
\underline{2x^4 + 3x^3 } \quad \leftarrow \text{Multiply } x^3 \text{ by the divisor} \\
4x^3 + 4x^2 - 7x - 6 \quad \leftarrow \text{Subtract}
\end{array}
$$

Step 2 Divide the leading term of the expression in the bottom row, $4x^3$, by the leading term of the divisor. Multiply the result, $2x^2$, by the divisor and subtract.

$$
\begin{array}{r}
x^3 \ + 2x^2 \\
2x+3\,\overline{)2x^4 + 7x^3 + 4x^2 - 7x - 6} \\
\underline{2x^4 + 3x^3 } \\
4x^3 + 4x^2 - 7x - 6 \\
\underline{4x^3 + 6x^2 } \quad \leftarrow \text{Multiply } 2x^2 \text{ by divisor} \\
-2x^2 - 7x - 6 \quad \leftarrow \text{Subtract}
\end{array}
$$

Step 3 Divide the leading term of the expression in the bottom row, $-2x^2$, by the leading term of the divisor. Multiply the result, $-x$, by the divisor and subtract.

$$
\begin{array}{r}
x^3 + 2x^2 - x \phantom{{}- 6} \\
2x + 3\overline{)2x^4 + 7x^3 + 4x^2 - 7x - 6} \\
\underline{2x^4 + 3x^3} \phantom{{}+ 4x^2 - 7x - 6} \\
4x^3 + 4x^2 - 7x - 6 \\
\underline{4x^3 + 6x^2} \phantom{{}- 7x - 6} \\
-2x^2 - 7x - 6 \\
\underline{-2x^2 - 3x} \phantom{{}- 6} \quad \leftarrow \text{Multiply } -x \text{ by divisor} \\
-4x - 6 \quad \leftarrow \text{Subtract. (Be careful with signs!)}
\end{array}
$$

Step 4 Divide the leading term of the expression in the bottom row, $-4x$, by the leading term of the divisor. Multiply the result, -2, by the divisor, and subtract.

$$
\begin{array}{r}
x^3 + 2x^2 - x - 2 \\
2x + 3\overline{)2x^4 + 7x^3 + 4x^2 - 7x - 6} \\
\underline{2x^4 + 3x^3} \phantom{{}+ 4x^2 - 7x - 6} \\
4x^3 + 4x^2 - 7x - 6 \\
\underline{4x^3 + 6x^2} \phantom{{}- 7x - 6} \\
-2x^2 - 7x - 6 \\
\underline{-2x^2 - 3x} \phantom{{}- 6} \\
-4x - 6 \\
\underline{-4x - 6} \quad \leftarrow \text{Multiply } -2 \text{ by divisor} \\
0 \quad \leftarrow \text{Subtract. (Be careful with signs!)}
\end{array}
$$

Thus, dividing $2x^4 + 7x^3 + 4x^2 - 7x - 6$ by $2x + 3$ gives a quotient, $q(x)$, of $x^3 + 2x^2 - x - 2$ with a remainder, $r(x)$, of 0. Equivalently,

$$
\text{Check:} \quad \frac{2x^4 + 7x^3 + 4x^2 - 7x - 6}{2x + 3} = x^3 + 2x^2 - x - 2
$$

You can check that $(2x + 3)(x^3 + 2x^2 - x - 2) = 2x^4 + 7x^3 + 4x^2 - 7x - 6$.

 Check It Out 2   Find the quotient and remainder when $6x^3 - x^2 - 3x + 1$ is divided by $2x - 1$.

Thus far, none of the long division problems illustrated have produced a remainder. Example 3 illustrates long division with a remainder.

> **Example 3**   **Long Division with Remainder**

Find the quotient and remainder when $p(x) = 6x^3 + x - 1$ is divided by $d(x) = x + 2$. Write your answer in the form $\frac{p(x)}{d(x)} = q(x) + \frac{r(x)}{d(x)}$.

Solution   The steps for long division should now be fairly clear to you.

Step 1 Note that the expression $6x^3 + x - 1$ does not have an x^2 term. Thus, when we set up the division, we write the x^2 term as $0x^2$. We then divide $6x^3$ by the leading term of the divisor.

$$
\begin{array}{r}
6x^2 \phantom{{}+ x - 1} \\
x + 2\overline{)6x^3 + 0x^2 + x - 1} \\
\underline{6x^3 + 12x^2} \phantom{{}+ x - 1} \quad \leftarrow \text{Multiply } 6x^2 \text{ by divisor} \\
-12x^2 + x - 1 \quad \leftarrow \text{Subtract}
\end{array}
$$

Step 2 Next, we divide $-12x^2$ by the leading term of the divisor.

$$
\begin{array}{r}
6x^2 - 12x \phantom{{}- 1} \\
x + 2\overline{)6x^3 + 0x^2 + x - 1} \\
\underline{6x^3 + 12x^2} \phantom{{}+ x - 1} \\
-12x^2 + x - 1 \\
\underline{-12x^2 - 24x} \phantom{{}- 1} \quad \leftarrow \text{Multiply } -12x \text{ by divisor} \\
25x - 1 \quad \leftarrow \text{Subtract}
\end{array}
$$

Step 3 Finally, we divide $25x$ by the leading term of the divisor.

$$
\begin{array}{r}
6x^2 - 12x + 25 \\
x + 2\overline{)6x^3 + 0x^2 + x - 1} \\
\underline{6x^3 + 12x^2} \\
-12x^2 + x - 1 \\
\underline{-12x^2 - 24x} \\
25x - 1 \\
\underline{25x + 50} \quad \leftarrow \text{Multiply 25 by divisor} \\
-51 \quad \leftarrow \text{Subtract}
\end{array}
$$

Thus, the quotient is $q(x) = 6x^2 - 12x + 25$, and the remainder, $r(x)$, is -51, or equivalently,

$$
\frac{6x^3 + x - 1}{x + 2} = 6x^2 - 12x + 25 - \frac{51}{x + 2}
$$

In this example, long division yields a nonzero remainder. Thus $x + 2$ is *not* a factor of $6x^3 + x - 1$.

✓ *Check It Out 3*   Find the quotient and remainder when $3x^3 + x^2 - 1$ is divided by $x + 1$. Write your answer in the form $\frac{p(x)}{d(x)} = q(x) + \frac{r(x)}{d(x)}$.   ■

Synthetic Division

Synthetic division is a compact way of dividing polynomials when the divisor is of the form $x - c$. Instead of writing out all the terms of the polynomial, we work only with the coefficients. We illustrate this shorthand form of polynomial division with the problem from Example 3.

Example 4   **Synthetic Division**

Use synthetic division to divide $6x^3 + x - 1$ by $x + 2$.

Solution Because the divisor is of the form $x - c$, we can use synthetic division. Note that $c = -2$.

Step 1 Write down the coefficients of the dividend in a row, from left to right, and then place the value of c, which is -2, in that same row, to the left of the leading coefficient of the dividend.

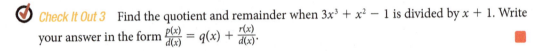

Value of $c \rightarrow$ -2⌋ 6   0   1 -1 $\leftarrow$ Coefficients of the dividend

Step 2 Bring down the leading coefficient of the dividend, 6, and then multiply it by c, which is -2.

$$
\begin{array}{r}
-2\,\rfloor\; 6 \quad 0 \quad 1 \quad -1 \\
\downarrow \; -12 \qquad\qquad \leftarrow \text{b) Then multiply 6 by } -2 \\
\text{a) First bring down 6} \rightarrow \quad 6 \;\nearrow
\end{array}
$$

Step 3 Place the result, -12, below the coefficient of the next term of the dividend, 0, and add.

$$
\begin{array}{r}
-2\,\rfloor\; 6 \quad 0 \quad 1 \quad -1 \\
\downarrow \; -12 \\
\hline
6 \quad -12 \qquad \leftarrow \text{Add 0 and } -12
\end{array}
$$

Step 4 Apply Steps 2(b) and 3 to the result, which is -12.

$$
\begin{array}{r}
-2\,\rfloor\; 6 \quad 0 \qquad 1 \quad -1 \\
\downarrow \; -12 \quad \boxed{24} \qquad \leftarrow \text{Multiply } -12 \text{ by } -2 \\
\hline
6 \quad \boxed{-12} \nearrow \; 25 \qquad \leftarrow \text{Add 1 and 24}
\end{array}
$$

Step 5 Apply Steps 2(b) and 3 to the result, which is 25.

$$
\begin{array}{r}
-2\,\rfloor\; 6 \quad 0 \quad 1 \qquad -1 \\
\downarrow \; -12 \;\; 24 \quad \boxed{-50} \qquad \leftarrow \text{Multiply 25 by } -2 \\
\hline
6 \quad -12 \;\boxed{25} \nearrow \; -51 \qquad \leftarrow \text{Add } -1 \text{ and } -50
\end{array}
$$

Step 6 The last row, except for the last number, consists of the coefficients of the quotient polynomial. The degree of the quotient is one degree less than the degree of the dividend. So, the 6 in the last row represents $6x^2$; the -12 represents $-12x$; and the 25 represents the constant term. Thus we have

$$q(x) = 6x^2 - 12x + 25$$

The remainder, $r(x)$, is -51, the last number in the bottom row.

 Check It Out 4 Use synthetic division to divide $-2x^3 + 3x^2 - 1$ by $x - 1$. ◼

The Remainder and Factor Theorems

We now examine an important connection between a polynomial $p(x)$ and the remainder obtained when $p(x)$ is divided by $x - c$. In Example 3, division of $p(x) = 6x^3 + x - 1$ by $x + 2$ yielded a remainder of -51. Also, $p(-2) = -51$. This is a consequence of the Remainder Theorem, formally stated as follows.

> **The Remainder Theorem**
>
> When a polynomial $p(x)$ is divided by $x - c$, the remainder is equal to the value of $p(c)$.

Because the remainder is equal to $p(c)$, synthetic division provides a quick way to evaluate $p(c)$. This is illustrated in the next example.

Example 5 **Applying the Remainder Theorem**

Let $p(x) = -2x^4 + 6x^3 + 3x - 1$. Use synthetic division to evaluate $p(2)$.

Solution From the Remainder Theorem, $p(2)$ is the remainder obtained when $p(x)$ is divided by $x - 2$. Following the steps outlined in Example 4, we have the following.

$$
\begin{array}{r|rrrrr}
2 & -2 & 6 & 0 & 3 & -1 \\
  & \downarrow & -4 & 4 & 8 & 22 \\
\hline
  & -2 & 2 & 4 & 11 & 21
\end{array}
$$

Because the remainder is 21, we know from the Remainder Theorem that $p(2) = 21$.

 Check It Out 5 Let $p(x) = 3x^4 - x^2 + 3x - 1$. Use synthetic division to evaluate $p(-2)$. ◼

The **Factor Theorem** is a direct result of the Remainder Theorem.

> **The Factor Theorem**
>
> The term $x - c$ is a ***factor*** of a polynomial $p(x)$ if and only if $p(c) = 0$.

The Factor Theorem makes an important connection between zeros and factors. It states that if we have a ***linear factor*** of a polynomial, that is a factor of the form $x - c$, then $p(c) = 0$. That is, *c is a zero of the polynomial* $p(x)$. It also works the other way around: if c is a zero of the polynomial $p(x)$, then $x - c$ is a factor of $p(x)$.

Example 6 **Applying the Factor Theorem**

Determine whether $x + 3$ is a factor of $2x^3 + 3x - 2$.

Solution Let $p(x) = 2x^3 + 3x - 2$. Because $x + 3$ is in the form $x - c$, we can apply the Factor Theorem with $c = -3$.

Evaluating, $p(-3) = -65$. By the Factor Theorem, $x + 3$ is *not* a factor of $2x^3 + 3x - 2$ since $p(-3) \neq 0$.

 Check It Out 6 Determine whether $x - 1$ is a factor of $2x^3 - 4x + 2$. ◼

Exercises 3.3

Skills

In Exercises 1–14, find the quotient and remainder when the first polynomial is divided by the second. You may use synthetic division wherever applicable.

1. $2x^2 + 13x + 15; x + 5$ 2. $2x^2 - 7x + 3; x - 3$ 3. $2x^3 - x^2 - 8x + 4; 2x - 1$

4. $3x^3 + 2x^2 - 3x - 2; 3x + 2$ 5. $x^3 - 3x^2 + 2x - 4; x + 2$ 6. $x^3 + 2x^2 - x - 3; x - 3$

7. $-3x^4 + x^2 - 2; 3x - 1$ 8. $2x^4 - x^3 + x^2 - x; 2x + 1$ 9. $x^6 + 1; x + 1$

10. $-x^3 + x; x - 5$ 11. $x^3 + 2x^2 - 5; x^2 - 2$ 12. $-x^3 - 3x^2 + 6; x^2 + 1$

13. $x^5 - x^4 + 2x^3 + x^2 - x + 1; x^3 + x - 1$ 14. $-2x^5 + x^4 - x^3 + 2x^2 - 1; x^3 + x^2 + 1$

In Exercises 15–20, write each polynomial in the form $p(x) = d(x)q(x) + r(x)$. You may use synthetic division wherever applicable.

15. $p(x) = x^2 + x + 1; d(x) = x + 1$ 16. $p(x) = x^2 + x + 1; d(x) = x - 1$

17. $p(x) = 3x^3 + 2x - 8; d(x) = x - 4$ 18. $p(x) = 4x^3 - x + 4; d(x) = x - 2$

19. $p(x) = x^6 - 3x^5 + x^4 - 2x^2 - 5x + 6; d(x) = x^2 + 2$

20. $p(x) = -x^6 + 4x^5 - x^3 + x^2 + x - 8; d(x) = x^2 + 4$

In Exercises 21–28, use synthetic division to find the function values.

21. $f(x) = x^3 - 7x + 5$; find $f(3)$ and $f(5)$.   22. $f(x) = -2x^3 + 4x^2 - 7$; find $f(4)$ and $f(-3)$.

23. $f(x) = -2x^4 - 10x^3 - 3x + 10$; find $f(-1)$ and $f(2)$.

24. $f(x) = -x^4 + 3x^3 - 2x - 4$; find $f(-2)$ and $f(3)$.   25. $f(x) = x^5 - 2x^3 + 12$; find $f(3)$ and $f(-2)$.

26. $f(x) = -2x^5 + x^4 + x^2 - 2$; find $f(-3)$ and $f(4)$.   27. $f(x) = x^4 - 2x^2 + 1$; find $f\left(\dfrac{1}{2}\right)$.

28. $f(x) = -x^4 + 3x^2 - 2x$; find $f\left(\dfrac{3}{2}\right)$.

In Exercises 29–38, determine whether $q(x)$ is a factor of $p(x)$ and justify your answer.

29. $p(x) = x^3 - 7x + 6; q(x) = x - 3$ 30. $p(x) = x^3 - 5x^2 + 8x - 4; q(x) = x + 2$

31. $p(x) = x^3 - 7x + 6; q(x) = x + 3$ 32. $p(x) = x^3 - 5x^2 + 8x - 4; q(x) = x - 2$

33. $p(x) = x^5 - 3x^3 + 2x - 8; q(x) = x - 4$ 34. $p(x) = -2x^4 - 7x^3 + 5; q(x) = x + 2$

35. $p(x) = x^4 - 50; q(x) = x - 5$ 36. $p(x) = 2x^5 - 1; q(x) = x - 2$

37. $p(x) = 3x^3 - 48x - 4x^2 + 64; q(x) = x + 4$ 38. $p(x) = x^3 + 9x + x^2 + 9; q(x) = x + 1$

Concepts

39. Given the following graph of a polynomial function $p(x)$, find a linear factor of $p(x)$.

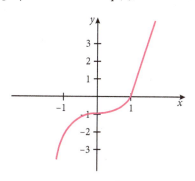

40. Consider the following graph of a polynomial function $p(x)$.

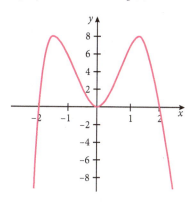

    a. Evaluate $p(2)$.

    b. Is $x - 2$ a factor of $p(x)$? Explain.

    c. Find the remainder when $p(x)$ is divided by $x - 2$.

41. Find the remainder when $x^7 + 7$ is divided by $x - 1$.

42. Find the remainder when $x^8 - 3$ is divided by $x + 1$.

43. Let $x - \frac{1}{2}$ be a factor of a polynomial function $p(x)$. Find $p\left(\frac{1}{2}\right)$.

44. For what value(s) of k do you get a remainder of 15 when you divide $kx^3 + 2x^2 - 10x + 3$ by $x + 2$?

45. For what value(s) of k do you get a remainder of -2 when you divide $x^3 - x^2 + kx + 3$ by $x + 1$?

46. Why is the Factor Theorem a direct result of the Remainder Theorem?

Real Zeros of Polynomials; Solutions of Equations

Objectives

- Find the rational zeros of a polynomial.

- Solve polynomial equations by finding zeros.

- Know and apply Descartes' Rule of Signs.

VIDEO EXAMPLES

SECTION 3.4

This section will cover some additional properties of polynomial functions. These properties, together with the ones we have encountered earlier in this chapter, will in turn be used to solve polynomial equations.

To find the zeros of polynomials of degree 2, you can use the quadratic formula. However, there are no easy to use formulas for finding the zeros of a polynomial of degree three or four. For polynomials of degree five or greater, formulas for finding zeros do not even exist! We can find zeros of polynomials algebraically if they can readily be factored or are of a special type. We can also find rational zeros of polynomials with integer coefficients. We will study these types of polynomials and corresponding polynomial equations in this section.

A graphing calculator or computer can be used to find the zeros of a general polynomial function. If you use technology, you will be able to find the zeros of a wider range of polynomials and solve a wider range of polynomial equations.

Example 1   **Using a Known Zero to Factor a Polynomial**

Show that $x = 4$ is a zero of $p(x) = 3x^3 - 4x^2 - 48x + 64$. Use this fact to completely factor $p(x)$.

Solution   Evaluate $p(4)$ to get

$$p(4) = 3(4)^3 - 4(4)^2 - 48(4) + 64 = 0$$

We see that $x = 4$ is a zero of $p(x)$. By the Factor Theorem, $x - 4$ is a factor of $p(x)$. Using synthetic division, we have $p(x) = (x - 4)(3x^2 + 8x - 16)$. The second factor, $3x^2 + 8x - 16$, is a quadratic expression that can be factored further:

$$3x^2 + 8x - 16 = (3x - 4)(x + 4)$$

Therefore,

$$p(x) = 3x^3 - 4x^2 - 48x + 64 = (x - 4)(3x - 4)(x + 4)$$

Check It Out 1   Show that $x = -2$ is a zero of $p(x) = 2x^3 + x^2 - 5x + 2$. Use this fact to completely factor $p(x)$.

The following summarizes the key connections among the factors and real zeros of a polynomial and the x-intercepts of its graph.

Zeros, Factors and x-Intercepts of a Polynomial

Let $p(x)$ be a polynomial function. Let c be a real number. Then the following are equivalent statements. That is, if one of the following statements is true, then the other two statements are also true. Similarly, if one of the following statements is false, then the other two statements are also false.

- $p(c) = 0$.

- $x - c$ is a factor of $p(x)$.

- $(c, 0)$ is an x-intercept of the graph of $p(x)$.

Example 2   **Relating Zeros, Factors and x-Intercepts**

Fill in Table 1, where p, h, and g are some polynomial functions.

	Function	Zero	x-Intercept	Factor
(a)	$p(x)$		$(5, 0)$	
(b)	$h(x)$			$x - 3$
(c)	$g(x)$	-1		

Table 1

Solution

a. Because $(5, 0)$ is an x-intercept of the graph of $p(x)$, the corresponding zero is 5 and the corresponding factor is $x - 5$.

b. Because $x - 3$ is a factor of $h(x)$, corresponding zero is 3, and the corresponding x-intercept is $(3, 0)$.

c. Because -1 is a zero of $g(x)$, the corresponding x-intercept is $(-1, 0)$, and the corresponding factor is $x - (-1)$, or equivalently, $x + 1$.

These results are shown in Table 2.

	Function	Zero	x-Intercept	Factor
(a)	$p(x)$	5	$(5, 0)$	$x - 5$
(b)	$h(x)$	3	$(3, 0)$	$x - 3$
(c)	$g(x)$	-1	$(-1, 0)$	$x + 1$

Table 2

 *Check It Out 2*   Fill in Table 3, where p, h, and g are some polynomial functions.

Function	Zero	x-Intercept	Factor
$p(x)$			$x + 6$
$h(x)$	4		
$g(x)$		$(2, 0)$	

Table 3

The Rational Zero Theorem

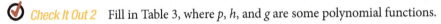

Although there is a relationship between the factors of a polynomial and its zeros, we still do not know how to *find* the zeros of a given polynomial. The following fact can be helpful in this regard.

Just in Time
Review the real number system in Section 1.1.

Number of Real Zeros of a Polynomial
A nonconstant polynomial function $p(x)$ of degree n has at most n real zeros where each zero of multiplicity k is counted k times.

In general, finding *all* the zeros of any given polynomial by hand is not possible. However, we can use a theorem called the Rational Zero Theorem to find out if a polynomial with *integer coefficients* has any *rational zeros* — that is, rational numbers that are zeros of the polynomial.

Consider $p(x) = 10x^2 - 29x - 21 = (5x + 3)(2x - 7)$. The zeros of $p(x)$ are $\frac{7}{2}$ and $-\frac{3}{5}$. Note that the numerator of each of the zeros is a factor of the constant term of the polynomial, -21. Also, the denominator of each of the zeros is a factor of the leading coefficient of the polynomial, 10. This observation can be generalized to the rational zeros of *any* polynomial of degree n with integer coefficients. This is summarized in the **Rational Zero Theorem**.

> ### The Rational Zero Theorem
>
> If $f(x) = a_n x^n + a_{n-1} x^{n-1} + \cdots + a_1 x + a_0$ is a polynomial with integer coefficients, and $\frac{p}{q}$ is a rational zero of f with p and q having no common factor other than 1, then p is a factor of a_0 and q is a factor of a_n.

The Rational Zero Theorem can be used in conjunction with long division to help find real zeros of a polynomial, as discussed in the next example.

Example 3   **Applying the Rational Zero Theorem**

Find all the real zeros of $p(x) = 3x^3 - 6x^2 - x + 2$.

Solution

Step 1 First, list all the possible rational zeros.

We consider all possible combinations of factors of 2 for the numerator of a rational zero and all possible factors of 3 for the denominator. The factors of 2 are ± 1 and ± 2, and the factors of 3 are ± 1 and ± 3. So, the possible rational zeros are

$$\pm 1, \ \pm 2, \ \pm \frac{1}{3}, \ \pm \frac{2}{3}$$

Step 2 The value of $p(x)$ at each of the possible rational zeros is summarized in Table 4.

x	-2	-1	$-\frac{2}{3}$	$-\frac{1}{3}$	$\frac{1}{3}$	$\frac{2}{3}$	1	2
$p(x)$	-44	-6	-0.88889	1.55556	1.11111	-0.44444	-2	0

Table 4

Only $x = 2$ is an actual zero. Thus, $x - 2$ is a factor of $p(x)$.

Step 3 Divide $p(x)$ by $x - 2$ using synthetic division.

$$
\begin{array}{r|rrrr}
2 & 3 & -6 & -1 & 2 \\
  &   & 6 & 0 & -2 \\
\hline
  & 3 & 0 & -1 & 0
\end{array}
$$

Thus, $p(x) = 3x^3 - 6x^2 - x + 2 = (x - 2)(3x^2 - 1)$.

Step 4 To find the other zeros, solve the quadratic equation

$$3x^2 - 1 = 0$$

which gives $x = \pm \frac{\sqrt{3}}{3}$. Note that the Rational Zero Theorem does *not* give these two zeros, since they are *irrational*.

Thus, the three zeros of $p(x)$ are $2, \frac{\sqrt{3}}{3}, -\frac{\sqrt{3}}{3}$. These are the only zeros, because a cubic polynomial function can have at most three zeros. The graph of $p(x)$ (Figure 1) indicating the locations of the zeros can be sketched using the techniques presented in Section 3.2.

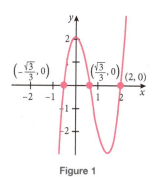

Figure 1

✓ *Check It Out 3*   Find all the real zeros of $p(x) = 4x^3 + 4x^2 - x - 1$.

Example 4   **Using a Graphing Utility to Locate Zeros**

Use a graphing utility to find all the real zeros of $p(x) = 3x^3 - 6x^2 - x + 2$.

Solution We can quickly locate all the *rational* zeros of the polynomial by using a graphing utility in conjunction with the Rational Zero Theorem. First, list all the possible rational zeros.

$$\pm 1, \pm 2, \pm\frac{1}{3}, \pm\frac{2}{3}$$

The zeros range from -2 to 2. When we graph the function $p(x) = 3x^3 - 6x^2 - x + 2$ in a decimal window, we immediately see that 2 is a probable zero (Figure 2). The possibilities $x = \pm 1$ can be excluded right away. (See Figure 2.)

Using the graphing calculator to evaluate the function, we obtain $p(2) = 0$. Evaluating the function at $\pm\frac{1}{3}$ shows that they are *not* zeros. It may happen that the polynomial has an irrational zero very close to the suspected rational zero.

Using the ZERO feature of your graphing utility, you will find that the other zeros are approximately ± 0.5774. See Figure 3. Thus, we have the following zeros: $x = 2$, $x \approx -0.5774$, and $x \approx 0.5774$. The numbers ± 0.5774 are approximations to the exact values $\pm\frac{\sqrt{3}}{3}$ found algebraically in Example 3. Because this is a cubic polynomial, we cannot have more than three zeros. So we have found all the zeros of $p(x) = 3x^3 - 6x^2 - x + 2$.

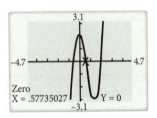

Figure 2

Figure 3

 Check It Out 4   Use a graphing utility to find all the real zeros of $x^3 - x^2 - 7x - 2$.

Solving Polynomial Equations by Finding Zeros

Because any equation can be rewritten so that the right-hand side is zero, solving an equation is identical to finding the zeros of a suitable function.

Example 5   **Solving a Polynomial Equation**

Solve the equation $3x^4 - 8x^3 - 9x^2 + 22x = -8$.

Solution **Algebraic Approach:** We first write the equation in the form $p(x) = 0$.

$$3x^4 - 8x^3 - 9x^2 + 22x + 8 = 0$$

The task now is to find all zeros of $p(x) = 3x^4 - 8x^3 - 9x^2 + 22x + 8$. First, use the Rational Zero Theorem to list all the possible rational zeros.

$$\frac{\text{Factors of 8}}{\text{Factors of 3}} = \frac{\pm 1, \pm 2, \pm 4, \pm 8}{\pm 1, \pm 3}$$

$$= \pm 1, \pm 2, \pm 4, \pm 8, \pm\frac{1}{3}, \pm\frac{2}{3}, \pm\frac{4}{3}, \pm\frac{8}{3}$$

Next, we evaluate $p(x)$ at each of these possibilities, by either direct evaluation or synthetic division. We try the integer possibilities first, from the smallest in magnitude to the largest. We find that $p(2) = 0$, and so $x - 2$ is a factor of $p(x)$. Using synthetic division, we factor out the term $(x - 2)$.

$$
\begin{array}{r|rrrrr}
2 & 3 & -8 & -9 & 22 & 8 \\
  & \downarrow & 6 & -4 & -26 & -8 \\
\hline
  & 3 & -2 & -13 & -4 & 0 \\
\end{array}
$$

Thus,

$$3x^4 - 8x^3 - 9x^2 + 22x + 8 = (x - 2)(3x^3 - 2x^2 - 13x - 4)$$

Next, we try to factor $q(x) = 3x^3 - 2x^2 - 13x - 4$. Checking the possible rational zeros listed earlier, we find that none of the other integers on the list is a zero of $q(x)$. We try the fractions, and see that $q\left(-\frac{1}{3}\right) = 0$. Then we use synthetic division with $q(x)$ as the dividend in order to factor out the term $\left(x + \frac{1}{3}\right)$.

$$
\begin{array}{r|rrrr}
-\dfrac{1}{3} & 3 & -2 & -13 & -4 \\
 & \downarrow & -1 & 1 & 4 \\
\hline
 & 3 & -3 & -12 & 0
\end{array}
$$

Thus,

$$3x^3 - 2x^2 - 13x - 4 = \left(x + \frac{1}{3}\right)(3x^2 - 3x - 12) = 3\left(x + \frac{1}{3}\right)(x^2 - x - 4).$$

Note that we factored the quadratic expression $3x^2 - 3x - 12$ as $3(x^2 - x - 4)$. To find the remaining zeros, solve the equation $x^2 - x - 4 = 0$. Using the quadratic formula, we find that

$$x^2 - x - 4 = 0 \Rightarrow x = \frac{1}{2} \pm \frac{1}{2}\sqrt{17}.$$

Because these two zeros are *irrational*, they did not appear in the list of possible rational zeros. Thus, the solutions to the equation $3x^4 - 8x^3 - 9x^2 + 22x = -8$ are

$$x = 2, \ x = -\frac{1}{3}, \ x = \frac{1}{2} + \frac{\sqrt{17}}{2}, \ x = \frac{1}{2} - \frac{\sqrt{17}}{2}$$

Graphical Approach: To solve the equation $3x^4 - 8x^3 - 9x^2 + 22x = -8$, graph the function $p(x) = 3x^4 - 8x^3 - 9x^2 + 22x + 8$ and find its zero(s). Using a window size of $[-4.7, 4.7] \times [-15, 25]$, YScl $= 5$, we obtain the graph in Figure 4.

It looks as though there is a zero at $x = 2$. Using the graphing calculator, we can verify that, indeed, $p(2) = 0$.

Using the ZERO feature, we find that there is another zero, close to 2, at $x \approx 2.5616$, as shown in Figure 5. The two negative zeros, which can be found by using the graphical solver twice, are $x \approx -1.5616$ and $x \approx -0.33333$. Thus, the four zeros are

$$x = 2, \ x \approx -0.33333, \ x \approx 2.5616, \ x \approx -1.5616$$

One of the zeros is highlighted in Figure 5.

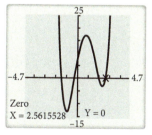

Figure 4

Zero
X = 2.5615528   Y = 0

Figure 5

✓ *Check It Out 5*   Solve the equation $2x^4 + 3x^3 - 6x^2 = 5x - 6$. ■

Descartes' Rule of Signs

An nth degree polynomial can have *at most* n real zeros. But many nth degree polynomials have fewer real zeros. For example, $p(x) = x(x^2 + 1)$ has only one real zero, and $p(x) = x^4 + 16$ has no real zeros. To get a better idea of the number of real zeros of a polynomial, a rule using the signs of the coefficients was developed by the French mathematician Rene Descartes around 1637. For a polynomial written in descending order, the number of **variations of sign** is the number of times that successive coefficients are of different signs. It plays a key role in Descartes' rule.

For instance, the polynomial $p(x) = -3x^4 + 6x^3 + x^2 - x + 1$ has *three* variations in signs, illustrated as follows.

$$p(x) = -3x^4 + 6x^3 + x^2 - x + 1$$

We now state Descartes' Rule of Signs, without proof.

> ### Descartes' Rule of Signs
>
> Let $p(x)$ be a polynomial function with real coefficients and a nonzero constant term. Let k be the number of variations of sign of $p(x)$, and let m be the number of variations of sign of $p(-x)$.
>
> 1. The number of **positive zeros** of p is either equal to k or less than k by an even integer.
> 2. The number of **negative zeros** of p is either equal to m or less than m by an even integer.

Note In Descartes' Rule, the number of positive and negative zeros includes multiplicity. For example, if a zero has multiplicity 2, it counts as two zeros.

Example 6   **Applying Descartes' Rule of Signs**

Use Descartes' Rule of Signs to determine the number of positive and negative zeros of $p(x) = -3x^4 + 4x^2 - 3x + 2$.

Solution First, determine the variations of sign of $p(x)$.

$$p(x) = -3x^4 + 4x^2 - 3x + 2$$
$$\underbrace{\qquad}_{1} \underbrace{\qquad}_{2} \underbrace{\qquad}_{3}$$

Because $p(x)$ has three variations of sign, the number of *positive zeros* of p is equal to either 3 or less than 3 by an even integer. Therefore, the number of positive zeros is **3** or **1**, since a negative number of zeros does not make sense.

Next, determine the variations of sign of $p(-x)$.

$$p(-x) = -3(-x)^4 + 4(-x)^2 - 3(-x) + 2 = -3x^4 + 4x^2 + 3x + 2$$
$$\underbrace{\qquad}_{1}$$

Because $p(-x)$ has one variation of sign, **the number of negative zeros of p is equal to 1.**

 Check It Out 6 Use Descartes' Rule of Signs to determine the number of positive and negative zeros of $p(x) = 4x^4 - 3x^3 + 2x - 1$.

Exercises 3.4

Just in Time Exercises

1. A rational number is a number that can be expressed as a quotient of two _____.
2. True or False: $\sqrt{3}$ is a rational number.
3. True or False: $0.1111\ldots$ is an irrational number.
4. True or False: 0.25 is a rational number.

Skills

In Exercises 5–10, for each polynomial, determine which of the numbers listed next to it are zeros of the polynomial.

5. $p(x) = (x - 10)^8$, $x = 6, -10, 10$

6. $p(x) = (x + 6)^{10}$, $x = 6, -6, 0$

7. $g(s) = s^2 + 4$, $s = -2, 2$

8. $f(x) = x^2 + 9$, $x = -3, 3$

9. $f(x) = x^3 + 2x^2 - 3x - 6$; $x = \sqrt{3}, -\sqrt{2}$

10. $f(x) = x^3 + 2x^2 - 2x - 4$; $x = \sqrt{2}, -\sqrt{3}$

In Exercises 11–18, show that the given value of x is a zero of the polynomial. Use the zero to completely factor the polynomial.

11. $p(x) = x^3 - 5x^2 + 8x - 4$; $x = 2$

12. $p(x) = x^3 - 7x + 6$; $x = 2$

13. $p(x) = -x^4 - x^3 + 18x^2 + 16x - 32$; $x = 1$

14. $p(x) = 2x^3 - 11x^2 + 17x - 6$; $x = \dfrac{1}{2}$

15. $p(x) = 3x^3 - 2x^2 + 3x - 2$; $x = \dfrac{2}{3}$

16. $p(x) = 2x^3 - x^2 + 6x - 3$; $x = \dfrac{1}{2}$

17. $p(x) = 3x^3 + x^2 + 24x + 8$; $x = -\dfrac{1}{3}$

18. $p(x) = 2x^5 + x^4 - 2x - 1$; $x = -\dfrac{1}{2}$

In Exercises 19–22, fill in the following table, where f, p, h, and g are some polynomial functions.

	Function	Zero	x-Intercept	Factor
19.	$f(x)$		$(-2, 0)$	
20.	$p(x)$			$x + 5$
21.	$h(x)$	-4		
22.	$g(x)$	6		

In Exercises 23–34, find all the real zeros of the polynomial.

23. $P(x) = x^3 + 2x^2 - 5x - 6$

24. $P(x) = 2x^3 + 3x^2 - 8x + 3$

25. $P(x) = x^4 - 13x^2 - 12x$

26. $Q(s) = s^4 - s^3 + s^2 - 3s - 6$

27. $P(s) = 4s^4 - 25s^2 + 36$

28. $P(t) = 6t^3 - 4t^2 + 3t - 2$

29. $F(x) = -4x^4 - 11x^3 - x^2 - 11x + 3$

30. $g(x) = -2x^3 - x^2 + 16x + 15$

31. $f(x) = x^4 + 2x^3 - 5x^2 - 4x + 6$

32. $g(x) = x^4 + x^3 + 3x^2 + 5x - 10$

33. $h(x) = x^4 + 3x^3 - 8x^2 - 22x - 24$

34. $f(x) = x^5 - 7x^4 + 10x^3 + 14x^2 - 24x$

In Exercises 35–42, find all real solutions of the polynomial equation.

35. $x^3 + 2x^2 + 2x = -1$

36. $3x^3 - 7x^2 = -5x + 1$

37. $x^3 - 6x^2 + 5x = -12$

38. $4x^3 - 16x^2 + 19x = -6$

39. $2x^3 - 3x^2 = 11x - 6$

40. $2x^3 - x^2 - 18x = -9$

41. $x^4 + x^3 - x = 1$

42. $6x^4 + 11x^3 - 3x^2 = 2x$

In Exercises 43–52, use Descartes' Rule of Signs to determine the number of positive and negative zeros of p. You need not find the zeros.

43. $p(x) = 4x^4 - 5x^3 + 6x - 3$

44. $p(x) = x^4 + 6x^3 - 7x^2 + 2x - 1$

45. $p(x) = -2x^3 + x^2 - x + 1$

46. $p(x) = -3x^3 + 2x^2 - x - 1$

47. $p(x) = 2x^4 - x^3 - x^2 + 2x + 5$

48. $p(x) = 3x^4 - 2x^3 + 3x^2 - 4x + 1$

49. $p(x) = x^5 + 3x^4 - 4x^2 + 10$

50. $p(x) = 2x^5 - 6x^3 + 7x^2 - 8$

51. $p(x) = x^6 + 4x^3 - 3x + 7$

52. $p(x) = 5x^6 - 7x^5 + 4x^3 - 6$

 In Exercises 53–59, graph the function using a graphing utility, and find its zeros.

53. $f(x) = x^3 - 3x^2 - 3x - 4$

54. $g(x) = 2x^5 + x^4 - 2x - 1$

55. $h(x) = 4x^3 - 12x^2 + 5x + 6$

56. $p(x) = -x^4 - x^3 + 18x^2 + 16x - 32$

57. $p(x) = -2x^4 + 13x^3 - 23x^2 + 3x + 9$

58. $f(x) = x^3 + x^2 + x - 3.1x^2 - 2.5x - 4$

59. $p(x) = x^3 + (3 + \sqrt{2})x^2 + 4x + 6.7$

Applications

60. Geometry A rectangle has length $x^2 - x + 6$ units and width $x + 1$ units. Find x such that the area of the rectangle is 24 square units.

61. Geometry The length of a rectangular box is 10 inches more than its height, and its width is 5 inches more than its height. Find the dimensions of the box if the volume is 168 cubic inches.

 62. Manufacturing An open rectangular box is constructed by cutting a square of length x from each corner of a 12-inch by 15-inch rectangular piece of cardboard and then folding up the sides.

   a. What is the length of the square that must be cut from each corner if the volume of the box is 112 cubic inches?

   b. What is the length of the square that must be cut from each corner if the volume of the box is 150 cubic inches?

 63. Manufacturing The height of a right circular cylinder is 5 inches more than its radius. Find the dimensions of the cylinder if its volume is 1000 cubic inches. (The volume of a cylinder is given by $V = \pi r^2 h$, where r is the radius and h is the height.)

Concepts

64. The following is the graph of a cubic polynomial function. Find an expression for the polynomial function with leading coefficient 1 that corresponds to this graph. You may check your answer by using a graphing utility.

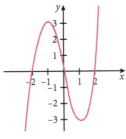

65. Find at least two different cubic polynomials whose only real zero is -1. Graph your answers to check them.

 66. Let $p(x) = x^5 + x^3 - 2x$.

   a. Show that p is symmetric with respect to the origin.

   b. Find a zero of p by inspection of the polynomial expression.

   c. Use a graphing utility to find the other zeros.

   d. How do you know that you have found *all* the zeros of p?

Complex Numbers

Objectives

- Define a complex number.
- Perform arithmetic with complex numbers.
- Find the complex zeros of a quadratic function.
- Find the complex solutions of a quadratic equation.

In order to complete our analysis of zeros of polynomial functions, we must introduce a set of numbers known as the *complex numbers*. Because finding the zeros of a polynomial function is equivalent to solving a related polynomial equation, the introduction of complex numbers will also complete our discussion of polynomial equations.

If we tried to solve the equation

$$x^2 + 4 = 0$$

we would get

$$x^2 = -4$$

which has no real numbers as its solution. The problem is that we cannot take the square root of a negative number and get a real number as an answer.

We need a special number that will produce -1 when it is squared. One such special number is denoted by the letter i, which is an example of what we call an **imaginary number**.

Definition of the Imaginary Number *i*

The **imaginary number** i is defined as the number such that

$$i = \sqrt{-1} \quad \text{and} \quad i^2 = -1$$

Numbers of the form bi, where b is a real number, are called **pure imaginary numbers**.

VIDEO EXAMPLES

SECTION 3.5

Example 1   Pure Imaginary Numbers

Write the following as pure imaginary numbers.

a. $\sqrt{-36}$　　　　　**b.** $\sqrt{-8}$　　　　　**c.** $\sqrt{-\dfrac{1}{4}}$

Solution

a. $\sqrt{-36} = i\sqrt{36} = 6i$, using the fact that $i = \sqrt{-1}$.

b. $\sqrt{-8} = i\sqrt{8} = 2i\sqrt{2}$.

c. $\sqrt{-\dfrac{1}{4}} = i\sqrt{\dfrac{1}{4}} = \dfrac{1}{2}i$.

Note that when i is to be multiplied by a radical, we place the i in front of the radical so that it is clear that it is not under the radical.

 Check It Out 1　Write $\sqrt{-25}$, $\sqrt{-108}$, and $\sqrt{-\dfrac{4}{9}}$ as pure imaginary numbers. ■

Real numbers and pure imaginary numbers are a subset of a set of numbers called the **complex numbers**, defined as follows.

Definition of a Complex Number

A **complex number** is a number of the form $a + bi$, where a and b are real numbers.

Observations:

- If $b = 0$ in the above definition, then $a + bi = a$, which is a real number.
- If $a = 0$ in the above definition, then $a + bi = bi$, which is a pure imaginary number.

Example 2   **Writing Numbers in the Form $a + bi$**

Write the following numbers in the form $a + bi$, and identify a and b.

a. $\sqrt{2}$

b. $\dfrac{1}{3}i$

c. $1 + \sqrt{3}$

Solution

a. $\sqrt{2} = \sqrt{2} + 0i$. Note that $a = \sqrt{2}$, $b = 0$.

b. $\dfrac{1}{3}i = 0 + i\dfrac{1}{3}$. Note that $a = 0$, $b = \dfrac{1}{3}$.

c. $1 + \sqrt{3} = \left(1 + \sqrt{3}\right) + 0i$. Note that $a = 1 + \sqrt{3}$, $b = 0$.

✓ *Check It Out 2*   Write the following numbers in the form $a + bi$: $\sqrt[3]{5}, -1 + \sqrt{2}, \dfrac{i}{4}$. ■

> **Parts of a Complex Number**
>
> For a complex number $a + bi$, a is called the **real part** and b is called the **imaginary part**.

Example 3   **Real and Imaginary Parts of a Complex Number**

What are the real and imaginary parts of the following complex numbers?

a. $-2 + 3i$

b. $-i + 4$

c. $-\sqrt{3}$

Solution

a. Note that $-2 + 3i$ is in the form $a + bi$. Thus the real part is $a = -2$, and the imaginary part is $b = 3$.

b. First rewrite $-i + 4$ as $4 - i$. This gives 4 as the real part and -1 as the imaginary part.

c. Note that $-\sqrt{3} = -\sqrt{3} + 0i$. This gives $-\sqrt{3}$ as the real part and 0 as the imaginary part.

✓ *Check It Out 3*   What are the real and imaginary parts of the following complex numbers? $\sqrt[3]{5}, -1 + \sqrt{2}, \dfrac{i}{4}$. ■

Addition and Subtraction of Complex Numbers

📱 **Using Technology**

Arithmetic with complex numbers can be performed in a graphing utility in the $a + bi$ mode. Some models will use the ordered pair (a, b) to represent a complex number. See Figure 1.

Keystroke Appendix:

Section 11

To add two complex numbers, we simply add their real parts to get the real part of their sum, and we add their imaginary parts to get the imaginary part of their sum. To subtract two complex numbers, we subtract their real and imaginary parts instead of adding them.

Example 4   **Adding and Subtracting Complex Numbers**

Perform the following operations.

a. $(1 + 2i) + (3 - 5i)$

b. $\left(\sqrt{2} + i\right) + \left(-\sqrt{2} - i\right)$

c. $i + (-1)$

d. $(1 + i) - (2 - i)$

Solution

a. Grouping the real terms together and the imaginary terms together, we have

$$(1 + 2i) + (3 - 5i) = (1 + 3) + (2i - 5i) = 4 - 3i$$

b. Once again, we group the real terms together and the imaginary terms together to get

$$\left(\sqrt{2} + i\right) + \left(-\sqrt{2} - i\right) = \left(\sqrt{2} - \sqrt{2}\right) + (i - i) = 0 + 0i = 0$$

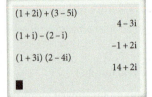

(1 + 2i) + (3 − 5i)
                    4 − 3i
(1 + i) − (2 − i)
                    −1 + 2i
(1 + 3i) (2 − 4i)
                    14 + 2i

Figure 1

c. Note that $i + (-1)$ cannot be simplified any further, and so the answer is $-1 + i$, rewritten in the form $a + bi$.

d. To calculate $(1 + i) - (2 - i)$, we first distribute the negative sign over the real and imaginary parts of the second complex number and then add

$$(1 + i) - (2 - i) = 1 + i - 2 + i = -1 + 2i$$

 Check It Out 4   Perform the following operations:

a. $3 + 2i - 4 + i$ **b.** $-2 + i$ **c.** $\sqrt{3} + i - \sqrt{3}$ ■

Multiplication of Complex Numbers

 Just in Time
Review multiplication of binomials in Section 1.3.

To multiply two complex numbers, we apply the rules of multiplication of binomials. This is illustrated in the following example.

Example 5   **Multiplying Complex Numbers**

Multiply

a. $(1 + 3i)(2 - 4i)$ **b.** $\sqrt{-4}\,\sqrt{-9}$

Note As illustrated in Example 5 (b), you must be careful when multiplying if there are negative numbers under the radical. In this example, we cannot simply multiply the numbers under the radical first. The rule $\sqrt{x}\,\sqrt{y} = \sqrt{xy}$ is valid when x or y is positive. It does not hold when *both* x and y are negative. To avoid this potential source of error, always write square roots of negative numbers in terms of i *before* simplifying.

Solution

a. $(1 + 3i)(2 - 4i) = 2 - 4i + 6i - 12i^2$ Multiply (using FOIL)

$\qquad\qquad\qquad\quad = 2 + 2i - 12i^2$ Add the real and the imaginary parts

$\qquad\qquad\qquad\quad = 2 + 2i - 12(-1)$ Note that $i^2 = -1$

$\qquad\qquad\qquad\quad = 14 + 2i$ Simplify

b. $\sqrt{-4}\,\sqrt{-9} = (2i)\,(3i)$ Write as imaginary numbers

$\qquad\qquad\quad = 6i^2 = -6$ Use $i^2 = -1$

Note that we wrote $\sqrt{-4}$ and $\sqrt{-9}$ using imaginary numbers *before* simplifying. The reason for this is explained in the note on the left.

 Check It Out 5   Multiply $(-3 + 4i)(5 - 2i)$.   ■

Division of Complex Numbers

Before we can define division of complex numbers, we must define the **complex conjugate** of a complex number.

Definition of Complex Conjugate

The **complex conjugate** of a complex number $a + bi$ is given by $a - bi$. The complex conjugate of a complex number has the *same real part* as the original number, but the *negative of the imaginary part*.

The next example will illustrate this definition.

Example 6   **Conjugate of a Complex Number**

Find the complex conjugates of the following numbers.

a. $1 + 2i$ **b.** $-3i$ **c.** 2

Solution

a. The complex conjugate of $1 + 2i$ is $1 - 2i$. We simply take the negative of the imaginary part and keep the real part of the original number.

b. The complex conjugate of $-3i$ is $3i$. The real part here is zero, and so we just negate the imaginary part. Every pure imaginary number is equal to the negative of its complex conjugate.

c. The complex conjugate of 2 is 2. The real part is 2, and it remains the same. The imaginary part is zero, and will remain zero when negated. Every real number is equal to its complex conjugate.

 *Check It Out 6*   Find the complex conjugate of $-3 - 7i$.

Complex conjugates are often abbreviated as simply *conjugates*. The next example will illustrate why conjugates are useful.

Example 7   **Multiplying a Number by Its Conjugate**

Multiply $-3 + 2i$ by its conjugate. What type of number results from this operation?

Solution The conjugate of $-3 + 2i$ is $-3 - 2i$. Following the rules of multiplication, we have

$$(-3 + 2i)(-3 - 2i) = 9 + 6i - 6i - 4i^2 \quad \text{Use FOIL to multiply}$$

$$= 9 + 0i - 4i^2 \quad \text{Add the real and the imaginary parts}$$

$$= 9 - 4(-1) \quad \text{Note that } i^2 = -1 \text{ and } 0i = 0$$

$$= 13 \quad \text{Simplify}$$

Thus we see that the product of $-3 + 2i$ and its conjugate is a positive real number.

 *Check It Out 7*   Multiply $-3 - 7i$ by its conjugate. What type of number results from this operation?

It can be shown that the product of *any* nonzero complex number and its conjugate is a *positive* real number. This fact is extremely useful in the division of complex numbers, as illustrated in the next example.

Example 8   **Dividing Complex Numbers**

Find $\dfrac{2}{-3 + 2i}$.

Solution The idea is to multiply both the numerator and the denominator of the expression $\frac{2}{-3 + 2i}$ by the complex conjugate of the denominator. This will give a real number in the denominator. Thus, we have

$$\frac{2}{-3 + 2i} = \frac{2}{-3 + 2i} \cdot \frac{-3 - 2i}{-3 - 2i} \quad \begin{array}{l}\text{Multiply the numerator and denominator by}\\ \text{the conjugate of } -3 + 2i\end{array}$$

$$= \frac{2(-3 - 2i)}{13} \quad \begin{array}{l}(-3 + 2i)(-3 - 2i) = 13 \text{ from the previous}\\ \text{example}\end{array}$$

$$= \frac{2}{13}(-3 - 2i) = \frac{-6}{13} - \frac{4}{13}i \quad \text{Simplify}$$

 *Check It Out 8*   Find $\dfrac{4}{-3 - 7i}$.

Zeros of Quadratic Functions and Solutions of Quadratic Equations

By expanding from the real number system to the complex number system, we can find the nonreal zeros of a quadratic function by using the quadratic formula. We can also find complex-valued solutions to quadratic equations.

Just in Time
Review the quadratic formula Section 1.5.

Example 9   **Imaginary Solutions of a Quadratic Equation**

Use the definition of i to solve the equation $x^2 = -4$.

Solution Since $i^2 = -1$, we can write

$$x^2 = (-1)(4) = i^2 2^2 \qquad \text{Use the definition of } i$$

$$x^2 = (2i)^2 \qquad\qquad \text{Use the properties of exponents}$$

$$x = \pm 2i \qquad\qquad \text{Solve for } x$$

Checking the solutions, we see that

$$x^2 = (2i)^2 = 4i^2 = -4 \quad \text{and} \quad x^2 = (-2i)^2 = 4i^2 = -4$$

We can therefore conclude that

$$x^2 = -4 \text{ for } x = \pm 2i$$

 Check It Out 9   Use the definition of i to solve the equation $x^2 = -9$.

Example 10   **Complex Zeros of Quadratic Functions**

Compute the zeros of the quadratic function $f(x) = 3x^2 + x + 1$. Use the zeros to find the x-intercepts, if any, of the graph of the function. Verify your results by graphing the function.

Solution We solve $f(x) = 3x^2 + x + 1 = 0$ for x. Noting that the expression cannot be factored easily, we use the quadratic formula to solve for x. Thus

$$x = \frac{-b \pm \sqrt{b^2 - 4ac}}{2a} \qquad \text{The quadratic formula}$$

$$= \frac{-(1) \pm \sqrt{(1)^2 - 4(3)(1)}}{2(3)} \qquad a = 3, b = 1, \text{ and } c = 1$$

$$= \frac{-1 \pm \sqrt{-11}}{6} = -\frac{1}{6} \pm \frac{\sqrt{11}}{6} i \quad \text{Simplify}$$

We see that this function has two nonreal zeros: $x = -\frac{1}{6} + \frac{\sqrt{11}}{6} i$ and $x = -\frac{1}{6} - \frac{\sqrt{11}}{6} i$. Since the zeros are not real numbers, the graph of the function has no x-intercepts. These results are confirmed by examining the graph of $f(x) = 3x^2 + x + 1$ in Figure 2.

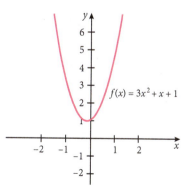

Figure 2

 Check It Out 10   Rework Example 10 using the quadratic function $g(s) = -3s^2 + 2s - 1$.

Example 11   **Complex Valued Solutions of a Quadratic Equation**

Find all solutions of the quadratic equation $2t^2 - 2t = -\frac{3}{2}$. Relate the solutions of this equation to the zeros of an appropriate quadratic function.

Solution   First write the equation in standard form

$$2t^2 - 2t + \frac{3}{2} = 0$$

Apply the quadratic formula to solve for t:

$$t = \frac{-b \pm \sqrt{b^2 - 4ac}}{2a} \qquad \text{The quadratic formula}$$

$$= \frac{-(-2) \pm \sqrt{(-2)^2 - 4(2)(3/2)}}{2(2)} \qquad a = 2, b = -2, \text{ and } c = \frac{3}{2}$$

$$= \frac{2 \pm \sqrt{4 - 12}}{4} = \frac{2 \pm \sqrt{-8}}{4} = \frac{2 \pm i\sqrt{8}}{4} \qquad \text{Use } \sqrt{-8} = i\sqrt{8}$$

$$= \frac{2 \pm 2i\sqrt{2}}{4} = \frac{1}{2} \pm i\frac{\sqrt{2}}{2} \qquad \text{Simplify}$$

We see that this quadratic equation has two solutions, $t = \frac{1}{2} + i\frac{\sqrt{2}}{2}$ and $t = \frac{1}{2} - i\frac{\sqrt{2}}{2}$, both of which are nonreal. An associated quadratic function is $f(t) = 2t^2 - 2t + \frac{3}{2}$, which has two nonreal zeros: $t = \frac{1}{2} + i\frac{\sqrt{2}}{2}$ and $t = \frac{1}{2} - i\frac{\sqrt{2}}{2}$. **The graph of f will have no x-intercepts, since f has no real zeros.** This is confirmed by the graph of $f(t) = 2t^2 - 2t + \frac{3}{2}$ in Figure 4.

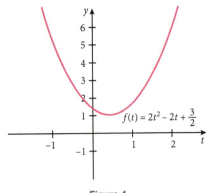

Figure 4

✓ *Check It Out 11*   Find all solutions of the quadratic equation $-2t^2 + 3t - 2 = 0$.   ■

Complex numbers play a central role in the complete factorization of polynomials and will be discussed in the next section. They also have applications in engineering, physics, and advanced mathematics. However, they require a certain amount of technical background that is beyond the scope of this book, and so we will not discuss them here.

Exercises 3.5

Just in Time Exercises

1. Multiply: $(2 + 3x)(-3 + x)$.

2. Multiply: $(x + 2)(x - 2)$.

3. Multiply: $(x + \sqrt{2})(x - \sqrt{3})$.

4. Multiply: $(x + 2y)(x - 3y)$.

In Exercises 5–8, use the quadratic formula to solve the equation.

5. $x^2 - 3x = 0$

6. $x^2 + 4x - 2 = 0$

7. $2x^2 - x = 1$

8. $x^2 = x + 1$

Skills

In Exercises 9–14, write the number as a pure imaginary number.

9. $\sqrt{-16}$

10. $\sqrt{-64}$

11. $\sqrt{-12}$

12. $\sqrt{-24}$

13. $\sqrt{-\dfrac{4}{25}}$

14. $\sqrt{\dfrac{-9}{4}}$

In Exercises 15–22, find the real and imaginary parts of the complex number.

15. 2

16. -3

17. $-\pi i$

18. $i\sqrt{3}$

19. $1 + \sqrt{5}$

20. $\sqrt{7} - 1$

21. $1 + \sqrt{-5}$

22. $\sqrt{-7} - 1$

In Exercises 23–30, find the complex conjugate of each number.

23. -2

24. -5

25. $i - 1$

26. $-2i + 4$

27. $3 + \sqrt{2}$

28. $9 - \sqrt{3}$

29. i^2

30. i^3

In Exercises 31–42, find $x + y$, $x - y$, xy, and $\dfrac{x}{y}$.

31. $x = 3i; y = 2 - i$

32. $x = -2i; y = 5 + i$

33. $x = -3 + 5i; y = 2 - 3i$

34. $x = 2 - 9i; y = -4 + 6i$

35. $x = 4 - 5i; y = 3 + 2i$

36. $x = 2 - 7i; y = 11 + 2i$

37. $x = \dfrac{1}{2} - 3i; y = \dfrac{1}{5} + \dfrac{4}{3}i$

38. $x = \dfrac{1}{2} - 2i; y = \dfrac{1}{3} - \dfrac{2}{5}i$

39. $x = -\dfrac{1}{3} + i\sqrt{5}; y = -\dfrac{1}{2} - 2i\sqrt{5}$

40. $x = \dfrac{1}{2} + i\sqrt{2}; y = -\dfrac{1}{5} + i\sqrt{2}$

41. $x = -3 + i; y = i + \dfrac{1}{2}$

42. $x = -2 - i; y = i + 2$

In Exercises 43–48, use the definition of i to solve the equation.

43. $x^2 = -16$

44. $x^2 = -25$

45. $-x^2 = 8$

46. $-x^2 = 12$

47. $3x^2 = -30$

48. $5x^2 = -60$

Paying Attention to Instructions: *Exercises 49–52 are intended to give you practice reading and paying attention to the instructions that accompany the problems you are working.*

49. Let $f(x) = 2x^2 + 4$.
    a. Find the real zeros of f, if any.
    b. Find all zeros of f.

50. Let $f(x) = 3x^2 + 9$.
    a. Find the real zeros of f, if any.
    b. Find all zeros of f.

51. Consider the equation $x^2 + x + 1 = 0$.
    a. Find the real solutions of this equation, if any.
    b. Find all solutions of this equation.

52. Consider the equation $2x^2 + x + 1 = 0$.
    a. Find the real solutions of this equation, if any.
    b. Find all solutions of this equation.

In Exercises 53–60, compute the zeros of each quadratic function.

53. $f(x) = 2x^2 + 9$

54. $h(x) = -3x^2 - 10$

55. $f(x) = -x^2 - x - 1$

56. $g(x) = x^2 - x + 1$

57. $h(t) = 3t^2 - 2t - 9$

58. $f(x) = -2x^2 - 2x + 11$

59. $h(x) = 3x^2 + 8x - 16$

60. $f(t) = 2t^2 + 11t + 9$

In Exercises 61–76, find all solutions of the quadratic equation. Relate the solutions of the equation to the zeros of an appropriate quadratic function.

61. $x^2 + 2x + 3 = 0$

62. $-x^2 + x - 5 = 0$

63. $-3x^2 + 2x - 4 = 0$

64. $-2x^2 + 3x - 1 = 0$

65. $5x^2 - 2x + 3 = 0$

66. $-7x^2 + 2x - 1 = 0$

67. $5x^2 = -2x - 3$

68. $7x^2 = -x - 1$

69. $-3x^2 + 8x = 16$

70. $2t^2 + 8t = -9$

71. $-4t^2 + t - \dfrac{1}{2} = 0$

72. $-6t^2 + 2t - \dfrac{1}{3} = 0$

73. $\dfrac{2}{3}x^2 + x = -1$

74. $-\dfrac{3}{4}x^2 - x = 2$

75. $(x + 1)^2 = -25$

76. $(x - 2)^2 = -16$

In Exercises 77–80, use the following definition. A complex number $a + bi$ is often denoted by the letter z. Its conjugate, $a - bi$, is then denoted by $\bar{z}$.

77. Show that $z + \bar{z} = 2a$ and that $z - \bar{z} = 2bi$.

78. Show that $z\bar{z} = a^2 + b^2$.

79. Show that the real part of z is equal to $\dfrac{z + \bar{z}}{2}$.

80. Show that the imaginary part of z is equal to $\dfrac{z - \bar{z}}{2i}$.

In Exercises 81–84, refer to $f(x) = ax^2 + 2x + 1$, where a is a real number.

81. Find the discriminant $b^2 - 4ac$.

82. For what value(s) of a will f have two real zeros?

83. For what value(s) of a will f have one real zero?

84. For what value(s) of a will f have no real zeros?

In Exercises 85–88, solve the quadratic equations by entering the quadratic formula in the home screen of your graphing utility. See the Using Technology sidebar next to Example 11.

85. $-0.25x^2 + 1.14x - 2.5 = 0$

86. $0.62t^2 - 1.29t + 1.5 = 0$

87. $3t^2 + \sqrt{19} = 2t$

88. $2x^2 + \sqrt{11} = x$

Concepts

89. Consider a parabola that opens upward and has vertex (0, 4).

   a. Why does the quadratic function associated with such a parabola have no real zeros?

   b. Show that $f(x) = 2x^2 + 4$ is a possible quadratic function associated with such a parabola. Is this the only possible quadratic function associated with such a parabola? Explain.

   c. Find the zeros of the function f given in part (b).

90. For the functions y_1 and y_2 graphed below, explain why the equation $y_1(x) = y_2(x)$ has no real-valued solutions. Assuming that y_1 is a quadratic function with real coefficients and y_2 is a linear function, explain why the equation $y_1(x) = y_2(x)$ has at least one complex-valued solution.

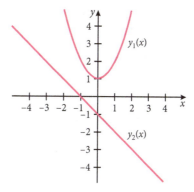

91. In this problem, you will explore the relationship between factoring a quadratic expression over the complex numbers and finding the zeros of the associated quadratic function. This topic will be explained in greater detail in Section 3.6.

   a. Multiply $(x + i)(x - i)$.

   b. What are the zeros of $f(x) = x^2 + 1$?

   c. What is the relationship between your answers to parts (a) and (b)?

   d. Using your answers to parts (a)–(c) as a guide, how would you factor $x^2 + 9$?

   e. Using your answers to parts (a)–(d) as a guide, how would you factor $x^2 + c^2$, where c is a positive real number?

92. We know that $i^2 = -1$, but is there a complex number z such that $z^2 = i$? We answer that question in this exercise.

   a. Calculate $\left(\dfrac{\sqrt{2}}{2}(1 + i) \right)\left(\dfrac{\sqrt{2}}{2}(1 + i) \right)$.

   b. Use your answer in part (a) to find a complex number z such that $z^2 = i$.

93. Examine the following table of values of a quadratic function.

x	f(x)
−2	9
−1	3
0	1
1	3
2	9

a. What is the equation of the axis of symmetry of the associated parabola? Explain how you got your answer.

b. Find the minimum or maximum value of the function and the value of x at which it occurs.

c. Sketch a graph of the function from the values given in the table.

d. Does this function have real or nonreal zeros? Explain.

94. Is it possible for a quadratic function with real coefficients to have one real zero and one nonreal zero? Explain. (*Hint*: Examine the quadratic formula.)

The Fundamental Theorem of Algebra; Complex Zeros

3.6

Objectives

- Understand the statement and consequences of the Fundamental Theorem of Algebra.

- Be able to factor polynomials with real coefficients over the complex numbers.

- Understand the connection among real zeros, x-intercepts, and factors of a polynomial.

In previous sections of this chapter, we examined ways to find zeros of a polynomial function. We also saw that there was a close connection between the zeros of a polynomial and its factors. In this section, we will make these observations precise by presenting some known facts about the zeros of a polynomial.

The Fundamental Theorem of Algebra

Recall that the solutions of the equation $P(x) = 0$ are known as **zeros** of the polynomial function P. Another name for a solution of a polynomial equation is *root*. A famous theorem about the existence of a solution to a polynomial equation was proved by a mathematician named Karl Friedrich Gauss in 1799. It is stated as follows.

> **The Fundamental Theorem of Algebra**
>
> Every nonconstant polynomial function with real or complex coefficients has at least one complex zero.

Many proofs of this theorem are known, but they are beyond the scope of this text.

In order to find the exact number of zeros of a polynomial, the following precise definition is needed for the multiplicity of a zero of a polynomial function. Recall that multiplicity was briefly discussed in Section 3.2.

> **Definition of Multiplicity of a Zero**
>
> A zero c of a polynomial P of degree $n > 0$ has **multiplicity** k if $P(x) = (x - c)^k Q(x)$, where $Q(x)$ is a polynomial of degree $n - k$ and c is not a zero of $Q(x)$.

Note The Fundamental Theorem of Algebra states only that a solution *exists*. It does not tell you how to find the solution.

The following example will help you unravel the notation used in the definition of multiplicity.

VIDEO EXAMPLES

SECTION 3.6

Example 1   **Determining Multiplicity of a Zero**

Let $h(x) = x^3 + 2x^2 + x$.

a. What is the value of the multiplicity, k, of the zero at $x = -1$?

b. Write $h(x)$ in the form $h(x) = (x + 1)^k Q(x)$. What is $Q(x)$?

Solution

a. Factoring $h(x)$, we obtain

$$x^3 + 2x^2 + x = x(x^2 + 2x + 1) = x(x + 1)^2$$

Thus, the zero at $x = -1$ is of multiplicity **2**, that is $k = 2$.

b. We have

$$h(x) = (x + 1)^2 Q(x) = (x + 1)^2 x$$

where $Q(x) = x$. Note that the degree of $Q(x)$ is $n - k = 3 - 2 = 1$. Hence we see that the various aspects of the definition of multiplicity are verified.

✓ *Check It Out 1*   For the function $h(x) = x^4 - 2x^3 + x^2$, what is the value of the multiplicity at $x = 0$?

Factorization and Zeros of Polynomials with Real Coefficients

The Factorization Theorem gives information on factoring a polynomial with real coefficients.

> **The Factorization Theorem**
>
> Any polynomial P with **real** coefficients can be factored uniquely into linear factors and/or **irreducible quadratic factors**, where an irreducible quadratic factor is one that cannot be factored any further using real numbers. The two zeros of each irreducible quadratic factor are **complex conjugates** of each other.

Example 2   **Factorization of a Polynomial**

Using the fact that $x = -2$ is a zero of f, factor $f(x) = x^3 + 2x^2 + 7x + 14$ into linear and irreducible quadratic factors.

Solution Because $x = -2$ is a zero of f, we know that $x + 2$ is a factor of $f(x)$. Dividing $f(x) = x^3 + 2x^2 + 7x + 14$ by $x + 2$, we have

$$f(x) = (x + 2)(x^2 + 7)$$

Because $x^2 + 7$ cannot be factored any further using real numbers, the factorization above is complete as far as real numbers are concerned. The factor $x^2 + 7$ is an example of an irreducible quadratic factor.

Check It Out 2   Using the fact that $t = 6$ is a zero of h, factor $h(t) = t^3 - 6t^2 + 5t - 30$ into linear and irreducible quadratic factors.   ▇

If we allow factorization over the complex numbers, then we can use the Fundamental Theorem of Algebra to write a polynomial

$$p(x) = a_n x^n + a_{n-1} x^{n-1} + a_{n-2} x^{n-2} + \cdots + a_1 x + a_0$$

in terms of factors of the form $x - c$, where c is a complex zero of $p(x)$. To do so, let c_1 be a complex zero of the polynomial $p(x)$. The existence of c_1 is guaranteed by the Fundamental Theorem of Algebra. Since $p(c_1) = 0$, $x - c_1$ is a factor of $p(x)$ by the Factorization Theorem, and

$$p(x) = (x - c_1)q_1(x)$$

where $q_1(x)$ is a polynomial of degree less than n.

Assuming that the degree of $q_1(x)$ is greater than or equal to one, $q_1(x)$ has a complex zero c_2. Then,

$$q_1(x) = (x - c_2)q_2(x)$$

Thus,

$$p(x) = (x - c_1)q_1(x)$$
$$= (x - c_1)(x - c_2)q_2(x) \quad \text{Substitute } q_1(x) = (x - c_2)q_2(x)$$

This process can be continued until we get a complete factored form:

$$p(x) = a_n(x - c_1)(x - c_2) \cdots (x - c_n)$$

In general, the c_i's may not be distinct.
We have thus established the following result.

> **The Linear Factorization Theorem**
>
> Let $p(x) = a_n x^n + a_{n-1} x^{n-1} + a_{n-2} x^{n-2} + \cdots + a_1 x + a_0$, where $n \geq 1$ and $a_n \neq 0$. Then
>
> $$p(x) = a_n(x - c_1)(x - c_2) \cdots (x - c_n)$$
>
> The numbers $c_1, c_2, \ldots, c_n$ are complex, possibly real, and not necessarily distinct.
>
> Thus, every polynomial $p(x)$ of degree $n \geq 1$ has exactly n zeros, if multiplicities and complex zeros are counted.

Example 3 illustrates factoring over the complex numbers and finding complex zeros.

> **Example 3**   **Factorization Over the Complex Numbers**

Factor $f(x) = 2x^5 + 12x^3 + 18x$ over the complex numbers.

Solution Using standard factoring procedures,

$$f(x) = 2x(x^4 + 6x^2 + 9) \quad \text{Factor out } 2x$$

$$= 2x(x^2 + 3)^2 \qquad \text{Factor the perfect square trinomial}$$

But $x^2 + 3 = (x + i\sqrt{3})(x - i\sqrt{3})$. Thus, the factorization over the complex numbers is given by

$$f(x) = 2x\big((x + i\sqrt{3})(x - i\sqrt{3})\big)^2$$

$$= 2x(x + i\sqrt{3})^2(x - i\sqrt{3})^2$$

Note that $x = 0$ is a zero of multiplicity 1. The zeros $x = i\sqrt{3}$ and $x = -i\sqrt{3}$ each have multiplicity 2 and are complex conjugates of each other. Thus, there are five zeros for this fifth degree polynomial, counting multiplicities and complex zeros.

✓ *Check It Out 3*   Factor $h(t) = t^3 - 6t^2 + 5t - 30$ over the complex numbers. Use the fact that $t = 6$ is a zero of h. ■

> *Note* The statements discussed thus far regarding the zeros and factors of polynomials do *not* tell us *how to find* the factors or zeros.

Finding a Polynomial Given Its Zeros

So far, we have been given a polynomial and have been asked to factor it and find its zeros. If the zeros of a polynomial are given, we can reverse the process and find a factored form of the polynomial using the Linear Factorization Theorem.

> **Example 4**   **Finding an Expression for a Polynomial**

Find a polynomial $p(x)$ of degree 4 with $p(0) = -9$ and zeros $x = -3$, $x = 1$ and $x = 3$, with $x = 3$ a zero of multiplicity 2. For this polynomial, is it possible for the zeros other than 3 to have a multiplicity greater than 1?

Solution By the Factorization Theorem, $p(x)$ is of the form

$$p(x) = a(x - (-3))(x - 1)(x - 3)^2 = a(x + 3)(x - 1)(x - 3)^2$$

where a is the leading coefficient, which is still to be determined. Since we are given that $p(0) = -9$, we write down this equation first.

$$p(0) = -9$$

$$a(0 + 3)(0 - 1)(0 - 3)^2 = -9 \quad \text{Substitute 0 for } x \text{ in expression for } p$$

$$a(3)(-1)(-3)^2 = -9 \quad \text{Simplify}$$

$$-27a = -9$$

$$a = \frac{1}{3} \quad \text{Solve for } a$$

Thus, the desired polynomial is $p(x) = \frac{1}{3}(x + 3)(x - 1)(x - 3)^2$. It is not possible for the other zeros to have multiplicities greater than 1 since the number of zeros already adds up to four, counting the multiplicity of the zero at $x = 3$, and p is a polynomial of degree 4.

✓ *Check It Out 4*   Rework Example 4 for a polynomial of degree 5 with $p(0) = 32$ and zeros -2, 4, and 1, where -2 is a zero of multiplicity 2 and 1 is a zero of multiplicity 2. ■

We have already discussed how to find any possible *rational* zeros of a polynomial. If rational zeros exist, we can use synthetic division to help factor the polynomial. We can use a graphing utility when a polynomial of degree greater than two has only irrational or complex zeros.

Exercises 3.6

In Exercises 1–4, list the zeros of each polynomial and state the multiplicity of each zero.

1. $g(x) = (x - 1)^3 (x - 4)^5$

2. $f(t) = t^5(t - 3)^2$

3. $f(s) = (s - \pi)^{10} (s + \pi)^3$

4. $h(x) = (x - \sqrt{2})^{13} (x + \sqrt{2})^7$

In Exercises 5–16, find all the zeros, real and nonreal, of the polynomial. Then express $p(x)$ as a product of linear factors.

5. $p(x) = 2x^2 - 5x + 3$

6. $p(x) = 2x^2 - x - 6$

7. $p(x) = x^2 - \pi^2$

8. $p(x) = x^2 - 2$

9. $p(x) = x^2 + 9$

10. $p(x) = x^2 + 4$

11. $p(x) = x^2 - 3$

12. $p(x) = x^2 - 7$

13. $p(x) = x^3 + 3x$

14. $p(x) = x^3 + 5x$

15. $p(x) = x^4 - 9$ (*Hint*: Factor first as difference of squares.)

16. $p(x) = x^4 - 16$ (*Hint*: Factor first as difference of squares.)

In Exercises 17–22, one zero of each polynomial is given. Use it to express the polynomial as a product of linear and irreducible quadratic factors.

17. $x^3 - 2x^2 + x - 2$; zero: $x = 2$

18. $x^3 - x^2 + 4x - 4$; zero: $x = 1$

19. $2x^3 - 9x^2 - 11x + 30$; zero: $x = 5$

20. $2x^3 - 9x^2 + 7x + 6$; zero: $x = 2$

21. $x^4 - 5x^3 + 7x^2 - 5x + 6$; zero: $x = 3$

22. $x^4 + 2x^3 - 2x^2 + 2x - 3$; zero: $x = -3$

In Exercises 23–30, one zero of each polynomial is given. Use it to express the polynomial as a product of linear factors over the complex numbers. You may have already factored some of these polynomials into linear and irreducible quadratic factors in the previous group of exercises.

23. $x^3 - 2x^2 + x - 2$; zero: $x = 2$

24. $x^3 - x^2 + 4x - 4$; zero: $x = 1$

25. $x^4 + 4x^3 - x^2 + 16x - 20$; zero: $x = -5$

26. $x^4 - 6x^3 + 9x^2 - 24x + 20$; zero: $x = 5$

27. $x^4 - 5x^3 + 7x^2 - 5x + 6$; zero: $x = 3$

28. $x^4 + 2x^3 - 2x^2 + 2x - 3$; zero: $x = -3$

29. $2x^4 - 5x^3 + 8x^2 - 15x + 6$; zero: $x = 2$ (*Hint*: Use Rational Zero Theorem after division.)

30. $2x^4 - x^3 + 5x^2 - 3x - 3$; zero: $x = 1$ (*Hint*: Use Rational Zero Theorem after division.)

In Exercises 31–36, find an expression for a polynomial $p(x)$ with real coefficients satisfying the given conditions. There may be more than one possible answer.

31. Degree 2; $x = 2$ and $x = -1$ are zeros.

32. Degree 2; $x = \dfrac{1}{2}$ and $x = \dfrac{3}{4}$ are zeros.

33. Degree 3; $x = 1$ is a zero of multiplicity 2 and the origin is the y-intercept.

34. Degree 3; $x = -2$ is a zero of multiplicity 2 and the origin is an x-intercept.

35. Degree 4; $x = 1$ and $x = \dfrac{1}{3}$ are both zeros of multiplicity 2.

36. Degree 4; $x = -1$ and $x = -3$ are zeros of multiplicity 1 and $x = \dfrac{1}{3}$ is a zero of multiplicity 2.

Concepts

37. One of the zeros of a certain quadratic polynomial with real coefficients is $1 + i$. What is its other zero?

38. The graph of a certain cubic polynomial function, f, has one x-intercept at $(1, 0)$ that crosses the x-axis, and another x-intercept at $(-3, 0)$ that touches the x-axis but does not cross it. What are the zeros of f and their multiplicities?

39. Explain why there cannot be two different points at which the graph of a cubic polynomial touches the x-axis without crossing it.

40. Why can't the numbers i, $2i$, 1, 2 be the set of zeros of some fourth-degree polynomial with real coefficients?

41. The graph of a polynomial function is given below. What is the lowest possible degree of this polynomial? Explain. Find a possible expression for the function.

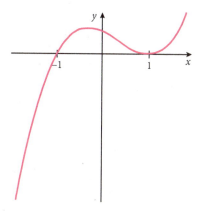

42. The graph of a polynomial function is given below. What is the lowest possible degree of this polynomial? Explain. Find a possible expression for the function.

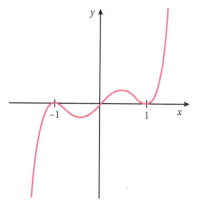

Rational Functions

Objectives

- Define a rational function.
- Examine the end behavior of a rational function.
- Find vertical asymptotes and intercepts.
- Find horizontal asymptotes.
- Sketch a complete graph of a rational function.

Just in Time
Review rational expressions in Section 1.4.

VIDEO EXAMPLES

SECTION 3.7

x	$f(x) = \dfrac{1}{x-1}$
-100	-0.009901
-10	-0.090909
0	-1
0.5	-2
0.9	-10
1	Undefined
1.1	10
1.5	2
2	1
10	0.111111
100	0.010101

Table 1

Thus far, we have studied linear, quadratic, and other polynomial functions. In this section, we extend our study of functions to include rational functions. This type of function is defined by a rational expression.

> ### Definition of a Rational Function
>
> A **rational function** $r(x)$ is defined as a quotient of two polynomials $p(x)$ and $h(x)$,
> $$r(x) = \frac{p(x)}{h(x)}$$
> where $h(x)$ is not the constant zero function.

The **domain** of a rational function consists of all real numbers for which the denominator is not equal to zero. We will be especially interested in the behavior of the rational function very close to the value(s) of x at which the denominator is zero.

Before examining rational functions in general, we look at some specific examples.

Example 1 **Analyzing a Simple Rational Function**

Let $f(x) = \dfrac{1}{x-1}$.

a. What is the domain of f?

b. Make a table of values of x and $f(x)$. Include values of x that are near 1 as well as larger values of x.

c. Graph f by hand.

d. Comment on the behavior of the graph.

Solution

a. The domain of f is the set of all values of x such that the denominator, $x - 1$, is not equal to zero. This is true for all $x \neq 1$. In interval notation, the domain is $(-\infty, 1) \cup (1, \infty)$.

b. Table 1 is a table of values of x and $f(x)$. Note that it contains some values of x that are close to 1 as well as some larger values of x.

c. Graphing the data in Table 1 gives us Figure 1. Note that there is no value for $f(x)$ at $x = 1$.

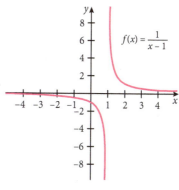

Figure 1

d. Examining the graph of $f(x) = \dfrac{1}{x-1}$, we see that as the absolute value of x gets large, the value of the function approaches zero, even though it never actually reaches zero. Also, as the value of x approaches 1, the absolute value of the *function* gets large.

✓ *Check It Out 1* What is the domain for $f(x) = \dfrac{1}{x}$? Sketch a graph of f.

When considering rational functions, we often use a pair of facts about the relationship between numbers that are large in absolute value and their reciprocals. These facts are informally stated as follows.

LARGE-Small Principle

$$\frac{1}{\text{LARGE}} = \text{small}; \quad \frac{1}{\text{small}} = \text{LARGE}$$

We will refer to this pair of facts as the **LARGE-small principle**.

Vertical Asymptotes

In Example 1, the absolute value of $f(x) = \frac{1}{x-1}$ near $x = 1$ is very large; we also noted that $f(x)$ is *not defined* at $x = 1$. We say that the line with equation $x = 1$ is a **vertical asymptote** of the graph of f. The line $x = 1$ is indicated by a dashed line. It is *not* part of the graph of the function $f(x) = \frac{1}{x-1}$. See Figure 2.

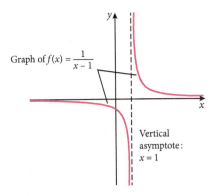

Graph of $f(x) = \frac{1}{x-1}$

Vertical asymptote: $x = 1$

Figure 2

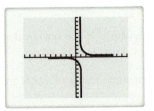

Vertical Asymptote

To find the **vertical asymptote(s)** of a rational function $r(x) = \frac{p(x)}{q(x)}$, first make sure that $p(x)$ and $q(x)$ have no common factors. Then the vertical asymptotes occur at values of x where $q(x) = 0$. At these values of x, the function values will approach positive or negative infinity.

Note A case in which $p(x)$ and $q(x)$ have common factors is given in Example 7.

For now, we will use algebraic methods to find the vertical asymptotes of the graphs of several rational functions. Later, we will use them to sketch graphs.

Example 2   **Finding Vertical Asymptotes**

Find all vertical asymptotes of the following functions.

a. $f(x) = \dfrac{2x}{x+1}$ **b.** $f(x) = \dfrac{x+2}{x^2-1}$ **c.** $f(x) = \dfrac{x^2-3}{2x+1}$

Solution

a. For the function $f(x) = \frac{2x}{x+1}$, we see that the numerator and denominator have no common factors, so we can set the denominator equal to zero and find the vertical asymptote(s):

$$x + 1 = 0 \Rightarrow x = -1$$

Thus **the line $x = -1$** is the only vertical asymptote of the function.

b. The numerator and denominator of $f(x) = \frac{x+2}{x^2-1}$ have no common factors, so we can set the denominator equal to zero and find the vertical asymptote(s). In this case, the denominator can be factored, so we will apply the Zero Product Rule:

$$x^2 - 1 = 0 \Rightarrow (x+1)(x-1) = 0 \Rightarrow x = 1, -1$$

This function has *two* vertical asymptotes: **the line $x = 1$ and the line $x = -1$.**

c. Once again, the numerator and denominator have no common factors. The **vertical asymptote is the line $x = -\frac{1}{2}$,** because $x = -\frac{1}{2}$ is the solution of the equation $2x + 1 = 0$.

 *Check It Out 2*   Find all vertical asymptotes of $f(x) = \frac{3x}{x^2 - 9}$.

End Behavior of Rational Functions and Horizontal Asymptotes

Just as we did with polynomial functions, we can also examine the end behavior of rational functions. We will use this information later to help sketch complete graphs of rational functions.

We can examine what happens to the values of a rational function $r(x)$ as $|x|$ gets large. This is the same as determining the end behavior of the rational function. For example, as $x \to \infty$, $f(x) = \frac{1}{x-1} \to 0$, because the denominator becomes large in magnitude, but the numerator stays constant at 1. Similarly, as $x \to -\infty$, $f(x) = \frac{1}{x-1} \to 0$. These are instances of the LARGE-small principle. When such a behavior occurs, we say that $y = 0$ is a **horizontal asymptote** of the function $f(x) = \frac{1}{x-1}$.

The following gives the necessary conditions for a rational function to have a horizontal asymptote.

Horizontal Asymptotes of Rational Functions

Let $r(x)$ be a rational function given by

$$r(x) = \frac{p(x)}{q(x)} = \frac{a_n x^n + a_{n-1} x^{n-1} + \cdots + a_1 x + a_0}{b_m x^m + b_{m-1} x^{m-1} + \cdots + b_1 x + b_0}$$

Here, $p(x)$ is a polynomial of degree n and $q(x)$ is a polynomial of degree m. Assume that $p(x)$ and $q(x)$ have no common factors

- If $n < m$, $r(x)$ approaches zero for large values of $|x|$. The line $y = 0$ is the **horizontal asymptote** of the graph of $r(x)$.

- If $n = m$, $r(x)$ approaches a nonzero constant, $\frac{a_n}{b_m}$, for large values of $|x|$. The line $y = \frac{a_n}{b_m}$ is the **horizontal asymptote** of the graph of $r(x)$.

- If $n > m$, $r(x)$ has no horizontal asymptote.

> **Example 3**   **Finding Horizontal Asymptotes**

Find the horizontal asymptote, if it exists, for each of the following rational functions. Use a table and a graph to discuss the end behavior of each function.

a. $f(x) = \dfrac{x + 2}{x^2 - 1}$ **b.** $f(x) = \dfrac{2x}{x + 1}$ **c.** $f(x) = \dfrac{x^2 - 3}{2x + 1}$

Solution

a. The degree of the numerator is 1, and the degree of the denominator is 2. Because $1 < 2$, the line $y = 0$ is a horizontal asymptote of the graph of f, as shown in Figure 4.

We can generate a table of values of $f(x) = \dfrac{x + 2}{x^2 - 1}$ for $|x|$ large, as shown in Table 2.

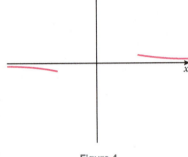

x	$f(x)$
-1000	-0.00099800
-100	-0.0098009
-50	-0.019207
50	0.020808
100	0.010201
1000	0.0010020

Table 2

Figure 4

Note that as $x \to \infty$, $f(x)$ gets close to 0 and will always be slightly above 0. As $x \to -\infty$, $f(x)$ gets close to 0 again, but this time it will be slightly below 0.

Recalling the end behavior of polynomials for large values of $|x|$, the numerator $x + 2$ is about the same as x, and the denominator $x^2 - 1$ is approximately the same as x^2. We then have

$$f(x) = \frac{x + 2}{x^2 - 1} \approx \frac{x}{x^2} = \frac{1}{x} \to 0, \text{ for large values of } |x|$$

by the LARGE-small principle. This is what we observed in Table 2 and the corresponding graph.

b. The degree of the numerator is 1, and the degree of the denominator is 1. Since $1 = 1$, the line $y = \dfrac{a_1}{b_1} = 2$ is a horizontal asymptote of the graph of f as shown in Table 3. We can generate a table of values of $f(x) = \dfrac{2x}{x + 1}$ for large values of $|x|$, as shown in Figure 5. We include both positive and negative values of x.

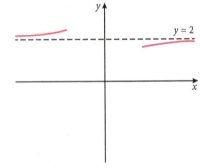

x	$f(x)$
-1000	2.0020
-100	2.0202
-50	2.0408
50	1.9607
100	1.9801
1000	1.9980

Table 3

Figure 5

From Table 3 and Figure 5, we observe that as $x \to \infty$, $f(x)$ will be slightly below 2. As $x \to -\infty$, $f(x)$ gets close to 2 again, but this time it will be slightly above 2.

To justify the end behavior in for large values of $|x|$, the numerator $2x$ is just $2x$, and the denominator $x + 1$ is approximately the same as x. We then have $f(x) = \dfrac{2x}{x + 1} \approx \dfrac{2x}{x} = 2$ for large values of $|x|$. This is what we observed in Table 3 and the corresponding graph.

x	$f(x)$
-1000	-500.25
-500	-250.25
-100	-50.236
-50	-25.222
50	24.722
100	49.736
500	249.75
1000	499.75

Table 4

c. The degree of the numerator is 2, and the degree of the denominator is 1. Because $2 > 1$, the graph of $f(x)$ has no horizontal asymptote. Table 4 gives values of $f(x) = \frac{x^2 - 3}{2x + 1}$ for $|x|$ large.

Unlike in parts (a) and (b), the values in the table do not seem to tend to any *one* particular number for $|x|$ large. However, you will notice that the values of $f(x)$ are very close to the values of $\frac{1}{2}x$ as $|x|$ gets large. We will discuss the behavior of functions such as this later in this section.

Check It Out 3 Find the horizontal asymptote of the rational function $f(x) = \frac{3x}{x^2 - 9}$.

Graphs of Rational Functions

The features of rational functions that we have discussed so far, combined with some additional information, can be used to sketch the graphs of rational functions. The procedure for doing so is summarized below, followed by examples.

Sketching the Graph of a Rational Function

Step 1 Find the vertical asymptote(s), if any, and indicate them on the graph.

Step 2 Find the horizontal asymptote(s), if any, and indicate them on the graph.

Step 3 Find the x- and y-intercepts and plot these points on the graph. For rational functions, the x-intercepts occur at those points in the domain of f at which

$$f(x) = \frac{p(x)}{q(x)} = 0$$

This means that $p(x) = 0$ at the x-intercepts.
To calculate the y-intercept, evaluate $f(0)$, if $f(0)$ is defined.

Step 4 Use the information in Steps 1–3 to sketch a partial graph. That is, find function values for points near the vertical asymptote(s) and sketch the behavior near the vertical asymptote(s). Also sketch the end behavior.

Step 5 Determine whether the function has any symmetries.

Step 6 Plot some other points to help you complete the graph.

Example 4   **Graphing Rational Functions**

Sketch the graph of $f(x) = \frac{2x}{x + 1}$.

Solution

Steps 1 and 2 The vertical and horizontal asymptotes of this function were computed in Examples 2 and 3 as follows.

Vertical Asymptote	$x = -1$
Horizontal Asymptote	$y = 2$

Table 5

Step 3 To find the y-intercept, evaluate $f(0) = \frac{2(0)}{(0) + 1} = 0$. The y-intercept is $(0, 0)$. To find the x-intercept, we find the points where the numerator, $2x$, is equal to zero. This happens at $x = 0$. Thus the x-intercept is also $(0, 0)$.

x-Intercept	$(0, 0)$
y-Intercept	$(0, 0)$

Table 6

Step 4 We now find values of $f(x)$ near the vertical asymptote $x = -1$. Note that we have chosen some values of x that are slightly to the right of $x = -1$ and some that are slightly to the left of $x = -1$. See Table 7.

x	-1.5	-1.1	-1.01	-1.001	-1	-0.999	-0.99	-0.9	-0.5
$f(x)$	6	22	202	2002	Undefined	-1998	-198	-18	-2

Table 7

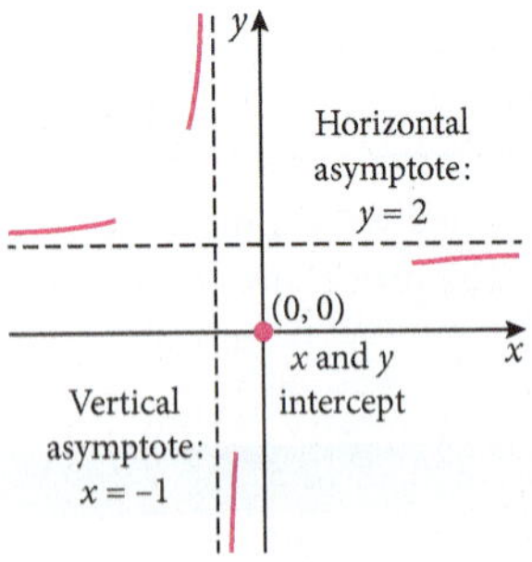

From Table 7, we see that the value of $f(x)$ increases to ∞ as x approaches -1 from the left, and that it decreases to $-\infty$ as x approaches -1 from the right.

The end behavior of f is as follows: $f(x) \to 2$ as $x \to \pm\infty$. Next we sketch the information collected so far. See Figure 6.

Step 5 Check for symmetry. Because

$$f(-x) = \frac{2(-x)}{(-x) + 1} = -\frac{2x}{-x + 1}$$

$f(-x) \neq f(x)$ and $f(-x) \neq -f(x)$. Thus, the graph of this function has no symmetries.

Step 6 There is an x-intercept at $(0, 0)$ and a vertical asymptote at $x = -1$, so we choose values in the intervals $(-\infty, -1)$, $(-1, 0)$, $(0, \infty)$ to fill out the graph. See Table 8.

x	-5	$-\frac{1}{2}$	1	2	5
$f(x)$	2.5	-2	1	1.3333	1.6667

Table 8

Plotting these points and connecting them with a smooth curve gives the graph shown in Figure 7. The horizontal and vertical asymptotes are *not* part of the graph of f. They are shown on the plot to indicate the behavior of the graph of f.

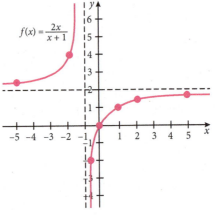

$$f(x) = \frac{2x}{x + 1}$$

Figure 7

Check It Out 4   Sketch the graph of $f(x) = \frac{x - 1}{3x + 1}$.

(Figure 6 — side margin)

Horizontal asymptote: $y = 2$

$(0, 0)$
x and y intercept

Vertical asymptote: $x = -1$

Figure 6

Example 5 **Graphing Rational Functions**

Sketch the graph of $f(x) = \dfrac{x+2}{x^2-1}$.

Solution

Steps 1 and 2 The vertical and horizontal asymptotes of this function were computed in Examples 3 and 4 as follows.

Vertical Asymptotes	$x = 1, x = -1$
Horizontal Asymptote	$y = 0$

Table 9

Step 3 The y-intercept is at $(0, -2)$ because

$$f(0) = \frac{0+2}{0^2-1} = -2$$

x-Intercept	$(-2, 0)$
y-Intercept	$(0, -2)$

Table 10

The x-intercept is at $(-2, 0)$ since $x + 2 = 0$ at $x = -2$. See Table 10.

Step 4 Find some values of $f(x)$ near the vertical asymptotes $x = -1$ and $x = 1$. See Table 11.

x	$f(x)$	x	$f(x)$
-1.1	4.2857	0.9	-15.2632
-1.01	49.2537	0.99	-150.2513
-1.001	499.2504	0.999	-1500.2501
-1	undefined	1	undefined
-0.999	-500.7504	1.001	1499.7501
-0.99	-50.7538	1.01	149.7512
-0.9	-5.7895	1.1	14.7619

Table 11

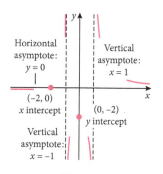

Figure 8

Observations:

- The value of $f(x)$ increases to ∞ as x approaches -1 from the left.
- The value of $f(x)$ decreases to $-\infty$ as x approaches -1 from the right.
- The value of $f(x)$ decreases to $-\infty$ as x approaches 1 from the left.
- The value of $f(x)$ increases to ∞ as x approaches 1 from the right.

Sketch the portions of the graph with the information collected thus far. See Figure 8.

Step 5 Check for symmetry. Because

$$f(-x) = \frac{(-x)+2}{(-x)^2-1} = \frac{-x+2}{x^2-1}$$

$f(-x) \neq f(x)$ and $f(-x) \neq -f(x)$. So the graph of this function has no symmetries.

Step 6 Now choose some additional values to fill out the graph. Since there is an x-intercept at $(-2, 0)$ and vertical asymptotes at $x = \pm 1$, choose at least one value in each of the intervals $(-\infty, -2), (-2, -1), (-1, 1), (1, \infty)$. See Table 12.

x	-3	-1.5	-0.5	0.5	2
$f(x)$	-0.1250	0.4000	-2.0000	-3.3333	1.3333

Table 12

Plotting these points and connecting them with a smooth curve gives the graph in Figure 9.

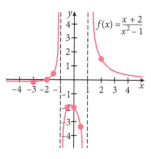

Figure 9

✓ *Check It Out 5* Sketch the graph of $f(x) = \dfrac{3x}{x^2-9}$.

Rational Functions with Slant Asymptotes

If the degree of the numerator of a rational function is 1 greater than the degree of the denominator, the graph of the function will have what is known as a **slant asymptote**. Using long division, we can write

$$r(x) = \frac{p(x)}{q(x)} = ax + b + \frac{s(x)}{q(x)}$$

where ax and b are the first two terms of the quotient and $s(x)$ is the remainder. Because the degree of $s(x)$ is less than the degree of $q(x)$, the value of the function $\frac{s(x)}{q(x)}$ approaches 0 as $|x|$ goes to infinity. Thus $r(x)$ will resemble the line $y = ax + b$ as $x \to \infty$ or $x \to -\infty$. The end behavior analysis performed earlier, without using long division, gives only the ax expression for the line. Next, we show how to find the equation of the asymptotic line $y = ax + b$.

| **Example 6** | **Rational Functions with Slant Asymptotes** |

Find the slant asymptote of the graph of $r(x) = \frac{6x^2 - x - 1}{2x + 1}$ and sketch the complete graph of $r(x)$.

Solution Performing the long division, we can write

$$r(x) = \frac{6x^2 - x - 1}{2x + 1} = 3x - 2 + \frac{1}{2x + 1}$$

For large values of $|x|$, $\frac{1}{2x + 1} \to 0$, and so the graph of $r(x)$ resembles the graph of the line $y = 3x - 2$. The equation of the slant asymptote is thus $y = 3x - 2$. Properties of the graph are summarized in Table 13.

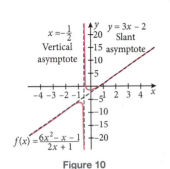

$x = -\frac{1}{2}$
Vertical
asymptote

$y = 3x - 2$
Slant
asymptote

$f(x) = \frac{6x^2 - x - 1}{2x + 1}$

Figure 10

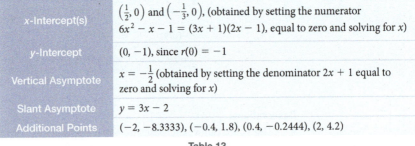

x-Intercept(s)	$\left(\frac{1}{2}, 0\right)$ and $\left(-\frac{1}{3}, 0\right)$, (obtained by setting the numerator $6x^2 - x - 1 = (3x + 1)(2x - 1)$, equal to zero and solving for x)
y-Intercept	$(0, -1)$, since $r(0) = -1$
Vertical Asymptote	$x = -\frac{1}{2}$ (obtained by setting the denominator $2x + 1$ equal to zero and solving for x)
Slant Asymptote	$y = 3x - 2$
Additional Points	$(-2, -8.3333), (-0.4, 1.8), (0.4, -0.2444), (2, 4.2)$

Table 13

The function $r(x) = \frac{6x^2 - x - 1}{2x + 1}$ is graphed in Figure 10.

✓ *Check It Out 6* Find the slant asymptote of the graph of $r(x) = \frac{x^2 - 3x - 4}{x - 5}$ and sketch the complete graph of $r(x)$. ■

Rational Function with Common Factors

We now examine the graph of a rational function in which the numerator and the denominator have a common factor.

| **Example 7** | **Rational Function with Common Factors** |

Sketch the complete graph of $r(x) = \frac{x - 1}{x^2 - 3x + 2}$.

Solution Factor the numerator and denominator of $r(x)$ to obtain

$$r(x) = \frac{x - 1}{x^2 - 3x + 2} = \frac{x - 1}{(x - 2)(x - 1)}$$

This function is undefined at $x = 1$ and $x = 2$, because those values of x give rise to a zero denominator. If $x \neq 1$, then the factor $x - 1$ can be divided out to obtain

$$r(x) = \frac{1}{x - 2}, \quad x \neq 1, 2$$

x-Intercept	None
y-Intercept	$\left(0, -\frac{1}{2}\right)$
Vertical Asymptote	$x = 2$
Horizontal Asymptote	$y = 0$
Undefined at	$x = 1$ $x = 2$

Table 14

The features of the graph of $r(x)$ are summarized in Table 14.

Even though $r(x)$ is not defined at $x = 1$, it does *not* have a vertical asymptote there. To see why, examine Table 15, which gives the values of $r(x)$ at some values of x close to 1.

x	0.9	0.99	0.999	1.001	1.01	1.1
r(x)	-0.9091	-0.9901	-0.999	-1.001	-1.01	-1.111

Table 15

The values of $r(x)$ near $x = 1$ do not tend to infinity and suggest that $r(x)$ gets close to -1 as x gets close to 1. Thus the graph of $r(x) = \frac{1}{x - 2}$, $x \neq 1, 2$ will have a *hole* at $(1, -1)$, as shown in Figure 11.

You can obtain the value -1 by substituting 1 for x into $r(x) = \frac{1}{x - 2}$. This substitution can be justified by using theorems of calculus, but is beyond the scope of this discussion.

Check It Out 7 Sketch the complete graph of $r(x) = \frac{x - 3}{x^2 - 5x + 6}$. ▪

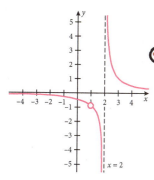

Figure 11

An Application of a Rational Function

Example 8 **Average Cost**

Suppose it costs \$45 a day to rent a car with unlimited mileage.

a. What is the expression for the average cost per mile per day?

b. Make a table and graph of the average cost function.

c. What happens to the average cost per mile per day as the number of miles driven per day increases?

Solution

a. Let x be the number of miles driven per day. The average cost per mile per day will depend on the number of miles driven per day, as follows:

$$A(x) = \text{Average cost per mile per day} = \frac{\text{Total cost per day}}{\text{Miles driven per day}} = \frac{45}{x}$$

b. Table 16 gives a table of values and Figure 12 gives the corresponding graph.

c. We see that as the number of miles driven per day increases, the average cost per mile per day goes down. Note that this function is defined only when $x > 0$, since you cannot drive a negative number of miles and the average cost of driving zero miles is not defined.

Check It Out 8 What is the average cost per mile per day if the daily cost of renting a car with unlimited mileage is \$50? ▪

Miles Driven	Average Cost
0	Undefined
$\frac{1}{2}$	90
1	45
5	9
10	4.5
100	0.45
1000	0.045

Table 16

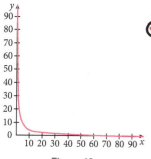

Figure 12

Exercises 3.7

1. True or False: The expression $\frac{1}{1 + x^2}$ is not defined for $x = -1$.

2. Evaluate $f(2)$ if $f(x) = \frac{2x}{x^2 - 1}$.

3. True or False: $\frac{x}{x^2 - 1} = \frac{1}{x} - x$.

4. True or False: $\frac{x - 1}{x^2 - 1} = \frac{1}{x + 1}$, for $x \neq 1$

Skills

In Exercises 5–18, for each function, find the domain and the vertical and horizontal asymptotes (if any).

5. $h(x) = \dfrac{-2}{x + 6}$

6. $F(x) = \dfrac{4}{x - 3}$

7. $g(x) = \dfrac{3}{x^2 - 4}$

8. $f(x) = \dfrac{2}{x^2 - 9}$

9. $f(x) = \dfrac{-x^2 + 9}{-2x^2 + 8}$

10. $h(x) = \dfrac{-3x^2 + 12}{x^2 - 9}$

11. $h(x) = \dfrac{1}{(x - 2)^2}$

12. $G(x) = \dfrac{-2}{(x + 4)^2}$

13. $h(x) = \dfrac{3x^2}{x + 1}$

14. $f(x) = \dfrac{-2x^2}{x - 1}$

15. $f(x) = \dfrac{2x + 7}{2x^2 + 5x - 3}$

16. $f(x) = \dfrac{3x + 5}{x^2 - x - 2}$

17. $f(x) = \dfrac{x + 1}{x^2 + 1}$

18. $h(x) = \dfrac{x + 2}{4 + x^2}$

In Exercises 19-22, find the domain, the vertical and horizontal asymptotes (if any), and the x- and y-intercepts (if any), of their graphs.

19.

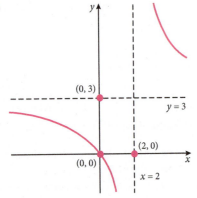

20.

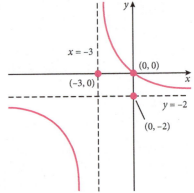

21.

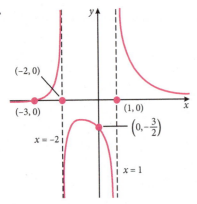

22.

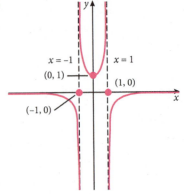

In Exercises 23–26, for each function, fill in the given tables.

23. Let $f(x) = \dfrac{2}{x+1}$.

    a. Fill in the following table for values of x near -1. What do you observe about the value of $f(x)$ as x approaches -1 from the right? From the left?

x	−1.5	−1.1	−1.01	−0.99	−0.9	−0.5
f(x)						

    b. Complete the following table. What happens to the value of $f(x)$ as $x \to \infty$?

x	10	50	100	1000
f(x)				

    c. Complete the following table. What happens to the value of $f(x)$ as $x \to -\infty$?

x	−1000	−100	−50	−10
f(x)				

24. Let $f(x) = \dfrac{1}{3-x}$.

    a. Fill in the following table for values of x near 3. What do you observe about the value of $f(x)$ as x approaches 3 from the right? From the left?

x	2.5	2.9	2.99	3.01	3.1	3.5
f(x)						

    b. Complete the following table. What happens to the value of $f(x)$ as $x \to \infty$?

x	10	50	100	1000
f(x)				

    c. Complete the following table. What happens to the value of $f(x)$ as $x \to -\infty$?

x	−1000	−100	−50	−10
f(x)				

25. Let $f(x) = \dfrac{2x^2 - 1}{x^2}$.

    a. Fill in the following table for values of x near 0. What do you observe about the value of $f(x)$ as x approaches 0 from the right? From the left?

x	−0.5	−0.1	−0.01	0.01	0.1	0.5
f(x)						

    b. Complete the following table. What happens to the value of $f(x)$ as $x \to \infty$?

x	10	50	100	1000
f(x)				

    c. Complete the following table. What happens to the value of $f(x)$ as $x \to -\infty$?

x	−1000	−100	−50	−10
f(x)				

26. Let $f(x) = \dfrac{-3x^2 + 4}{x^2}$.

    a. Fill in the following table for values of x near 0. What do you observe about the value of $f(x)$ as x approaches 0 from the right? From the left?

x	−0.5	−0.1	−0.01	0.01	0.1	0.5
f(x)						

    b. Complete the following table. What happens to the value of $f(x)$ as $x \to \infty$?

x	10	50	100	1000
f(x)				

    c. Complete the following table. What happens to the value of $f(x)$ as $x \to -\infty$?

x	−1000	−100	−50	−10
f(x)				

In Exercises 27–46, sketch the graph of the rational function. Indicate any vertical and horizontal asymptote(s), and all intercepts.

27. $f(x) = \dfrac{1}{x-2}$ **28.** $f(x) = \dfrac{1}{x+3}$ **29.** $f(x) = \dfrac{-12}{x+6}$ **30.** $f(x) = \dfrac{-10}{x+2}$

31. $f(x) = \dfrac{8}{4-x}$ **32.** $f(x) = \dfrac{12}{3-x}$ **33.** $f(x) = \dfrac{3}{(x+1)^2}$ **34.** $h(x) = \dfrac{-9}{(x-3)^2}$

35. $g(x) = \dfrac{3-x}{x+4}$ **36.** $g(x) = \dfrac{x+5}{x-2}$ **37.** $h(x) = \dfrac{-2x}{(x-1)(x+4)}$

38. $f(x) = \dfrac{x}{(x-3)(x-1)}$ **39.** $f(x) = \dfrac{3x^2}{x^2-x-2}$ **40.** $f(x) = \dfrac{-4x^2}{x^2-x-6}$

41. $f(x) = \dfrac{x-1}{2x^2-5x-3}$ **42.** $f(x) = \dfrac{x-2}{2x^2+x-3}$ **43.** $f(x) = \dfrac{x^2+x-6}{x^2-1}$

44. $f(x) = \dfrac{x^2+3x+2}{x^2-9}$ **45.** $h(x) = \dfrac{1}{x^2+1}$ **46.** $h(x) = \dfrac{2}{x^2+4}$

In Exercises 47–58, sketch a graph of the rational function. Indicate all intercepts and asymptotes.

47. $g(x) = \dfrac{x^2}{x+4}$ **48.** $g(x) = \dfrac{x^2}{x-2}$ **49.** $h(x) = \dfrac{-x^2}{x-3}$ **50.** $h(x) = \dfrac{-x^2}{x+1}$

51. $h(x) = \dfrac{4-x^2}{x}$ **52.** $h(x) = \dfrac{x^2-9}{x}$ **53.** $h(x) = \dfrac{x^2+x+1}{x-1}$ **54.** $h(x) = \dfrac{x^2+2x+1}{x+3}$

55. $h(x) = \dfrac{3x^2+5x-2}{x+1}$ **56.** $h(x) = \dfrac{2x^2+11x+5}{x-3}$ **57.** $h(x) = \dfrac{x^3+1}{x^2+3x}$ **58.** $h(x) = \dfrac{x^3-1}{x^2-2x}$

In Exercises 59–64, sketch a graph of the rational function involving common factors and find all asymptotes and intercepts. Indicate them on the graph.

59. $f(x) = \dfrac{3x+9}{x^2-9}$ **60.** $f(x) = \dfrac{2x-4}{x^2-4}$ **61.** $f(x) = \dfrac{x^2+x-2}{x^2+2x-3}$

62. $f(x) = \dfrac{2x^2-5x+2}{x^2-5x+6}$ **63.** $f(x) = \dfrac{x^2+3x-10}{x-2}$ **64.** $f(x) = \dfrac{x^2+2x+1}{x+1}$

Applications

65. Drug Concentration The concentration, $C(t)$, of a drug in a patient's bloodstream t hours after administration is given by

$$C(t) = \frac{10t}{1+t^2}$$

where $C(t)$ is in milligrams per liter.
a. What is the drug concentration in the bloodstream 8 hours after administration?
b. Find the horizontal asymptote of $C(t)$ and explain its significance.

66. Environmental Costs The annual cost in the United States, in millions of dollars, of removing arsenic from drinking water in the U.S. can be modeled by the function

$$C(x) = \frac{1900}{x}$$

where x is the concentration of arsenic remaining in the water, in micrograms per liter. A microgram is 10^{-6} grams.   (*Source*: Environmental Protection Agency)
a. Evaluate $C(10)$ and explain its significance.
b. Evaluate $C(5)$ and explain its significance.
c. What happens to the cost function as x gets closer to zero?

67. Rental Costs A truck rental company charges a daily rate of $15 plus $0.25 per mile driven. What is the average cost per mile of driving x miles per day? Use this expression to find the average cost per mile of driving 50 miles per day.

68. Printing To print booklets, it costs $300 and an additional $0.50 per booklet. What is the average cost per booklet of printing x books? Use this expression to find the average cost per booklet when printing 1000 booklets.

69. Phone Plans A wireless phone company has a pricing scheme that includes 250 minutes' worth of phone usage in the basic monthly fee of $30. Each minute over and above the first 250 minutes of usage is charged at $0.60 per minute.

   a. Let x be the number of minutes of phone usage per month. What is the expression for the average cost per minute if the value of x is between 0 and 250?

   b. What is the expression for the average cost per minute if the value of x is not in the interval (0, 250)?

   c. If phone usage in a certain month is 600 minutes, what is the average cost per minute?

70. Health The body-mass index (BMI) is a measure of body fat based on height and weight that applies to both adult males and adult females. It is calculated using the following formula:

$$\text{BMI} = \frac{703w}{h^2}$$

where w is the person's weight, in pounds, and h is the person's height, in inches. A BMI in the range 18.5–24.9 is considered normal. (*Source*: National Institutes of Health)

   a. Calculate the BMI for a person who is 5 feet 5 inches tall and weighs 140 pounds. Is this person's BMI within the normal range?

   b. Calculate the weight of a person who is 6 feet tall and has a BMI of 24.

   c. Calculate the height of a person who weighs 170 pounds and has a BMI of 24.3.

71. Metallurgy How much pure gold should be added to a 2-ounce alloy that is presently 25% gold to make it 60% percent gold?

72. Manufacturing A packaging company wants to design an open box with a square base and a volume of exactly 30 cubic feet.

   a. Let x denote the length of a side of the base of the box, and let y denote the height of the box. Express the total surface area of the box in terms of x and y.

   b. Write an equation relating x and y to the total volume of 30 cubic feet.

   c. Solve the equation from part (b) for y in terms of x.

   d. Now, write an expression for the surface area in terms of just x. Call this function $S(x)$.

   e. Fill in the following table with the value of the surface area for the given values of x.

x	1	2	3	4	5	6
$S(x)$						

   f. What do you observe about the total surface area as x increases? From your table, approximately what value of x would give the minimum surface area?

   g. Use a graphing utility to find the value of x that would give the minimum surface area.

73. Manufacturing A gift box company wishes to make a small open box by cutting four equal squares from a 3 inch by 5 inch card, one from each corner.

   a. Let x denote the length of the square cut from each corner. Write an expression for the volume of the box in terms of x. Call this function $V(x)$. What is the realistic domain of this function?

   b. Write an expression for the surface area of the box in terms of x. Call this function $S(x)$.

   c. Write an expression for the ratio of the volume of the box to the surface area, in terms of x. Call this function $r(x)$.

   d. Fill in the following table with the value of $r(x)$ for the given values of x.

x	0.2	0.4	0.6	0.8	1.0	1.2	1.4
$r(x)$							

   e. What do you observe about the ratio of the volume to the surface area as x increases? From your table, approximately what value of x would give the maximum ratio of volume to surface area?

   f. Use a graphing utility to find the value of x that would give the maximum ratio of volume to surface area.

Concepts

74. Sketch a possible graph of a rational function $r(x)$ of the following description: the graph of r has a horizontal asymptote $y = -2$ and a vertical asymptote $x = 1$, with y-intercept at $(0, 0)$.

75. Sketch a possible graph of a rational function $r(x)$ of the following description: the graph of r has a horizontal asymptote $y = -2$ and a vertical asymptote $x = 1$, with y-intercept at $(0, 0)$ and x-intercept at $(2, 0)$.

76. Give a possible expression for a rational function $r(x)$ of the following description: the graph of r has a horizontal asymptote $y = 2$ and a vertical asymptote $x = 1$, with y-intercept at $(0, 0)$. It may be helpful to sketch its graph first. You may check your answer with a graphing utility.

77. Give a possible expression for a rational function $r(x)$ of the following description: the graph of r has a horizontal asymptote $y = 0$ and a vertical asymptote $x = 0$, with no x- or y-intercepts. It may be helpful to sketch its graph first. You may check your answer with a graphing utility.

78. Give a possible expression for a rational function $r(x)$ of the following description: the graph of r is symmetric with respect to the y-axis; it has a horizontal asymptote $y = 0$ and a vertical asymptote $x = 0$, with no x- or y- intercepts. It may be helpful to sketch its graph first. You may check your answer with a graphing utility

79. Explain why the following output from a graphing utility is not a complete graph of the function $f(x) = \frac{1}{(x - 10)(x + 3)}$.

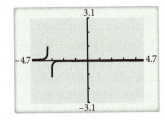

Objectives

- Solve a quadratic inequality.
- Solve a polynomial inequality.
- Solve a rational inequality.

VIDEO EXAMPLES

SECTION 3.8

In this section, we will show how to solve inequalities involving polynomial and rational expressions. Example 1 will show how to solve a quadratic inequality.

Example 1   **Solving a Quadratic Inequality**

Solve the inequality $2x^2 - 3x - 2 > 0$.

Solution

Step 1   Factor the expression $2x^2 - 3x - 2$ to get

$$(2x + 1)(x - 2) > 0$$

Step 2   Determine the values where $(2x + 1)(x - 2)$ equals zero.

$$2x + 1 = 0 \Rightarrow x = -\frac{1}{2}$$

$$x - 2 = 0 \Rightarrow x = 2$$

Step 3   Since the expression $(2x + 1)(x - 2)$ can change sign only at $x = -\frac{1}{2}$ or at $x = 2$, we form the following intervals:

$$\left(-\infty, -\frac{1}{2}\right), \left(-\frac{1}{2}, 2\right), (2, \infty).$$

Step 4   Make a table with these intervals in the first column. Choose a test value in each interval and determine the sign of each factor of the quadratic expression in that interval. See Table 1.

Interval	Test value	Sign of $2x + 1$	Sign of $x - 2$	Sign of $(2x + 1)(x - 2)$
$\left(-\infty, -\frac{1}{2}\right)$	-1	$-$	$-$	$+$
$\left(-\frac{1}{2}, 2\right)$	0	$+$	$-$	$-$
$(2, \infty)$	3	$+$	$+$	$+$

Table 1

Step 5   Since the inequality is satisfied on either of the intervals, $\left(-\infty, -\frac{1}{2}\right)$ or $(2, \infty)$, the solution set is $\left(-\infty, -\frac{1}{2}\right) \cup (2, \infty)$. The endpoints are **not** included since the inequality is strictly "greater than". Graphically, this is the set of all x values for which the graph of $f(x) = 2x^2 - 3x - 2$ lies above the x-axis. See Figure 1.

Note   Unlike solving an equation, solving an inequality gives an infinite number of solutions. You can get an idea of whether your solution is correct by substituting some values from your solution set into the inequality.

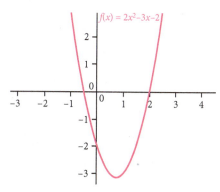

Figure 1

✓ *Check It Out 1*   Solve the inequality $-2x^2 + 5x < 3$.

Polynomial Inequalities

A **polynomial inequality** can be written in the form $a_n x^n + a_{n-1} x^{n-1} + \cdots + a_0 \,\square\, 0$, where the symbol inside the box can be $>$, $\geq$, $<$, or $\leq$. By factoring, we can also solve certain polynomial inequalities, as shown in Example 2.

Example 2   **Solving a polynomial inequality**

Solve the inequality.

$$x^3 - 2x^2 - 3x \leq 0$$

Solution

Step 1 One side of the inequality is already zero. Therefore, factor the nonzero side.

$$x(x^2 - 2x - 3) \leq 0 \quad \text{Factor out } x$$

$$x(x - 3)(x + 1) \leq 0 \quad \text{Factor inside parentheses}$$

Step 2 Determine the values where $x(x - 3)(x + 1)$ equals zero. These are $x = 0$, $x = 3$, $x = -1$. Since the expression $x(x - 3)(x + 1)$ can change sign only at these three values, we form the following intervals:

$$(-\infty, -1), (-1, 0), (0, 3), (3, \infty)$$

Step 3 Make a table with these intervals in the first column. Choose a test value in each interval and determine the sign of each factor of the polynomial expression in that interval. See Table 2.

Interval	Test value	Sign of x	Sign of $x - 3$	sign of $x + 1$	Sign of $x(x - 3)(x + 1)$
$(-\infty, -1)$	-2	$-$	$-$	$-$	$-$
$(-1, 0)$	$-\dfrac{1}{2}$	$-$	$-$	$+$	$+$
$(0, 3)$	1	$+$	$-$	$+$	$-$
$(3, \infty)$	4	$+$	$+$	$+$	$+$

Table 2

Step 4 From Table 2, we observe that $x(x - 3)(x + 1) \leq 0$ for all x in the intervals

$$(-\infty, -1] \cup [0, 3]$$

Thus the solution of the inequality is $(-\infty, -1] \cup [0, 3]$. The endpoints of the interval are included because we want values of x that make the expression $x(x - 3)(x + 1)$ *less than or equal to* zero.

Confirm the results by graphing the function $f(x) = x^3 - 2x^2 - 3x$ and observing where the graph lies above the x-axis and where it intersects the x-axis. See Figure 2.

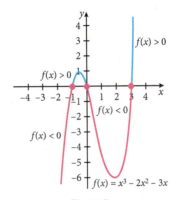

Figure 2

✅ *Check It Out 2*   Solve the inequality $x(x^2 - 4) \geq 0$.

Example 3　An Application of a Polynomial Inequality

A box with a square base and height that is 3 inches less than the length of one side of the base is to be built. What lengths of the base will produce a volume greater than or equal to 16 cubic inches?

Solution Let x be the length of the square base. The volume is given by

$$V(x) = x \cdot x \cdot (x - 3) = x^2(x - 3)$$

Because we want the volume to be greater than or equal to 16 cubic inches, we solve the inequality as shown below.

$$x^2(x - 3) \geq 16$$

$$x^3 - 3x^2 \geq 16 \quad \text{Remove parentheses}$$

$$x^3 - 3x^2 - 16 \geq 0 \quad \text{Make right-hand side equal to 0.}$$

Next we factor the nonzero side of the inequality. The expression does not seem to be factorable by any of the elementary techniques, so we will use the Rational Zero Theorem to factor, if possible.

The possible rational zeros are

$$x = \pm 16, \pm 8, \pm 4, \pm 2, \pm 1$$

We see that $x = 4$ is a zero. Using synthetic division, we have

$$p(x) = x^3 - 3x^2 - 16 = (x - 4)(x^2 + x + 4)$$

Now, $x^2 + x + 4$ cannot be factored further over the real numbers. Thus we have only the intervals $(-\infty, 4)$ and $(4, \infty)$ to check. For x in $(-\infty, 4)$, a test value of 0 yields $p(0) < 0$. For x in $(4, \infty)$, a test value of 5 yields $p(5) > 0$. We find that $p(x) = (x - 4)(x^2 + x + 4) \geq 0$ for all x in $[\mathbf{4}, \infty)$.

Thus a box with a square base whose side **is greater than or equal to 4 inches** will produce a volume **greater than or equal to 16 cubic inches**. The corresponding height will be 3 inches less than the length of the side.

 Check It Out 3　Rework Example 2 if the volume of the box is 50 cubic inches.

 **Using Technology**

A graphing utility can be used to confirm the results of Example 4. The graph of $f(x) = \frac{x^2 - 9}{x + 5}$ lies above the x axis for x in $(-5, -3)$ or $(3, \infty)$. See Figure 3.

Keystroke Appendix:
Sections 7 and 8

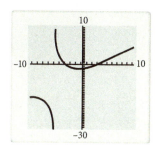

Figure 3

Rational Inequalities

In some applications and in more advanced mathematics courses, it is important to know how to solve an inequality involving a rational function, referred to as a **rational inequality**. Let $p(x)$ and $q(x)$ be polynomial functions with $q(x)$ not identically equal to zero. A rational inequality can be written in the form $\frac{p(x)}{q(x)} \,\square\, 0$, where the symbol inside the box can be $>$, $\geq$, $<$, or $\leq$. Example 4 shows you how to solve such an inequality.

Example 4　Solving a Rational Inequality

Solve the inequality $\frac{x^2 - 9}{x + 5} > 0$.

Solution

Step 1 The right-hand side is already 0. We factor the numerator on the left side.

$$\frac{(x + 3)(x - 3)}{x + 5} > 0$$

Step 2 Determine the values where the numerator, $(x + 3)(x - 3)$, equals zero. These are $x = 3$, and $x = -3$. Also, the denominator is equal to zero at $x = -5$. Because the expression $\frac{(x + 3)(x - 3)}{x + 5}$ can change sign only at these three values, we form the following intervals

$$(-\infty, -5), (-5, -3), (-3, 3), (3, \infty)$$

Step 3 Make a table with these intervals in the first column. Choose a test value in each interval and determine the sign of each factor of the inequality in that interval. See Table 3.

Interval	Test Value	Sign of $x + 3$	Sign of $x - 3$	Sign of $x + 5$	Sign of $\dfrac{(x+3)(x-3)}{(x+5)}$
$(-\infty, -5)$	-6	$-$	$-$	$-$	$-$
$(-5 -3)$	-4	$-$	$-$	$+$	$+$
$(-3, 3)$	0	$+$	$-$	$+$	$-$
$(3, \infty)$	4	$+$	$+$	$+$	$+$

Table 3

Step 4 The inequality $\dfrac{(x+3)(x-3)}{(x+5)} > 0$ is satisfied for all x in the intervals

$$(-5, -3) \cup (3, \infty).$$

The numbers 3 and -3 are not included because the inequality symbol is *greater than*, not *greater than or equal to*.

Check It Out 4   Solve the inequality $\dfrac{(x-1)}{(x+2)} < 0$. ■

Example 5   **Solving a Rational Inequality**

Solve the inequality.

$$\frac{2x+5}{x-2} \geq x$$

Solution

Using Technology

Using the INTERSECTION feature of a graphing utility, find the intersection points of $Y_1 = \frac{2x+5}{x-2}$ and $Y_2 = x$. The values of x for which $Y_1 \geq Y_2$ will be the solution set. One of the intersection points is given in Figure 4.

Keystroke Appendix:

Sections 7, 8, 9

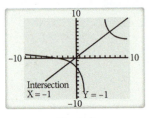

Figure 4

Step 1 Make the right-hand side zero by subtracting x from both sides. Then simplify.

$$\frac{2x+5}{x-2} - x \geq 0$$

$$\frac{2x+5-x(x-2)}{x-2} \geq 0 \quad \text{Combine terms using } x-2 \text{ as the LCD}$$

$$\frac{2x+5-x^2+2x}{x-2} \geq 0 \quad \text{Simplify}$$

$$\frac{-x^2+4x+5}{x-2} \geq 0 \quad \text{Collect like terms}$$

$$\frac{x^2-4x-5}{x-2} \leq 0 \quad \text{Multiply by } -1. \text{ Note reversal of inequality.}$$

Step 2 Factor the numerator on the left side.

$$\frac{(x+1)(x-5)}{x-2} \leq 0$$

Step 3 Determine the values at which the numerator $(x+1)(x-5)$ equals zero. These are $x = -1$, $x = 5$. Also, the denominator is equal to zero at $x = 2$. Because the expression $\frac{(x+1)(x-5)}{(x-2)}$ can change sign only at these three values, we form the following intervals:

$$(-\infty, -1), (-1, 2), (2, 5), (5, \infty)$$

Step 4 Make a table of these intervals. Choose a test value in each interval and determine the sign of each factor of the inequality in that interval. See Table 4.

Interval	Test Value	Sign of $x + 1$	Sign of $x - 2$	Sign of $x - 5$	Sign of $\dfrac{(x + 1)(x - 5)}{(x - 2)}$
$(-\infty, -1)$	-2	$-$	$-$	$-$	$-$
$(-1, 2)$	0	$+$	$-$	$-$	$+$
$(2, 5)$	3	$+$	$+$	$-$	$-$
$(5, \infty)$	6	$+$	$+$	$+$	$+$

Table 4

Step 5 The inequality $\dfrac{(x + 1)(x - 5)}{(x - 2)} \leq 0$ is satisfied for all x in the intervals.

$$(-\infty, -1] \cup (2, 5]$$

We have included the endpoints -1 and 5 because the inequality states *less than or equal to*. The endpoint 2 is *not* included because division by zero is not defined.

✅ *Check It Out 5* Solve the inequality.

$$\frac{2x - 1}{x + 3} \leq 0$$

◼

To summarize, follow these general steps when solving quadratic, polynomial, or rational inequalities.

1. First rewrite the inequality with a 0 on the right side.

2. Factor the nonzero side and find the zero of each factor in the inequality.

3. The resulting zeros divide the x-axis into multiple intervals. Make a table and test the sign of the inequality in each of these intervals by choosing appropriate test values.

4. Choose the intervals that satisfy the inequality.

Exercises 3.8

Skills

In Exercises 1–8 solve the quadratic inequality.

1. $x^2 - 1 \leq 0$

2. $x^2 - 9 < 0$

3. $2x^2 + 3x - 5 \geq 0$

4. $6x^2 - 5x - 6 < 0$

5. $2x^2 < x + 1$

6. $6x^2 \geq 13x - 5$

7. $5x^2 - 8x \geq 4$

8. $-3x^2 \leq -7x - 6$

In Exercises 9–26, solve the polynomial inequality.

9. $2x(x + 5)(x - 3) \geq 0$

10. $(x + 1)^2(x - 2) \leq 0$

11. $x^3 - 16x < 0$

12. $x^3 - 9x > 0$

13. $x^3 - 4x > 0$

14. $x^3 - 16x \leq 0$

15. $(x - 2)(x^2 - 4) < 0$

16. $(x - 5)(x^2 - 9) > 0$

17. $(x + 2)(x^2 - 5x + 4) \geq 0$

18. $(x + 3)(x^2 - 3x + 2) \geq 0$

19. $x^4 - x^2 > 3$

20. $x^4 - 3x^2 < 10$

21. $x^3 - 4x \leq -x^2 + 4$

22. $x^3 - 7x \leq -6$

23. $x^3 \leq 4x$

24. $x^3 \geq x$

25. $x^3 < 2x^2 + 3x$

26. $x^3 < 4x^2 - 4x$

In Exercises 27–46, solve the rational inequality.

27. $\dfrac{x + 2}{x - 1} \leq 0$

28. $\dfrac{x - 4}{2x + 1} > 0$

29. $\dfrac{x^2 - 4}{x - 3} \leq 0$

30. $\dfrac{4x^2 - 9}{x + 2} < 0$

31. $\dfrac{x(x + 1)}{1 + x^2} \geq 0$

32. $\dfrac{x^2 - 2x - 3}{x^2 + 2x + 1} \leq 0$

33. $\dfrac{4 - x}{x - 1} > x$

34. $\dfrac{-8}{x + 3} < -2x$

35. $\dfrac{1}{x} \leq \dfrac{1}{2x - 1}$

36. $\dfrac{2}{x + 1} > \dfrac{1}{x - 2}$

37. $\dfrac{3}{x - 1} \leq 2$

38. $\dfrac{-1}{2x + 1} \geq 1$

39. $\dfrac{x - 1}{x + 2} \geq 0$

40. $\dfrac{3x + 6}{x - 3} < 0$

41. $\dfrac{1}{2x + 1} \leq 0$

42. $\dfrac{-1}{3x - 1} > 0$

43. $\dfrac{x + 1}{x^2 - 9} < 0$

44. $\dfrac{x^2 - 4}{x + 5} \geq 0$

45. $\dfrac{x + 1}{x - 3} \leq \dfrac{x - 2}{x + 4}$

46. $\dfrac{x - 2}{x + 2} > \dfrac{x + 5}{x - 1}$

Applications

47. Geometry A rectangular solid has a square base and a height that is 2 inches less than the length of one of sides of the base. What lengths of the base will produce a volume greater than or equal to 32 cubic inches?

48. Manufacturing A rectangular box with a rectangular base is to be built. The length of one side of the rectangular base is 3 inches more than the height, while the length of the other side of the rectangular base is 1 inch more than the height of the box. For what values of the height will the volume be greater than equal to 40 cubic inches?

49. Drug Concentration The concentration, $C(t)$, of a drug in a patient's bloodstream t hours after administration is given by

$$C(t) = \frac{4t}{3 + t^2}$$

where $C(t)$ is in milligram per liter. During what time interval will the concentration be greater than 1 milligrams per liter?

50. Printing Costs To print booklets, it costs $400 and an additional $0.50 per booklet. What is the minimum number of booklets that must be printed so that the average cost per booklet is less than $0.55?

Concepts

51. To solve the inequality $x(x + 1)(x - 1) < 2$, a student starts by setting up the following inequalities:

$$x < 2, \quad x + 1 < 2, \quad x - 1 < 2$$

Why is this the *wrong* way to start the problem? What is the correct way to start this problem?

52. To solve the inequality $\frac{x}{x + 1} \geq 2$ a student first "simplifies" the problem by multiplying both sides by $x + 1$ to get

$$x \geq 2(x + 1)$$

Why is this an incorrect way to start the problem?

53. Find a polynomial $p(x)$ such that $p(x) > 0$ has the solution set $(0, 1) \cup (3, \infty)$. There may be more than one correct answer.

54. Find polynomials $p(x)$ and $q(x)$, with $q(x)$ nonconstant, such that $\frac{p(x)}{q(x)} \geq 0$ has the solution set $[3, \infty)$. There may be more than one correct answer.

Chapter 3 Summary

Section 3.1 Quadratic Functions and Their Graphs

Definition of a quadratic function [Review Exercises 1–14]

A function f is a **quadratic function** if it can be expressed in the form $f(x) = ax^2 + bx + c$, where a, b, and c are real numbers and $a \neq 0$.

Graph of a quadratic function [Review Exercises 1–14]

The graph of the function $f(x) = ax^2 + bx + c$ is called a **parabola**. If $a > 0$, the parabola opens upward. If $a < 0$, the parabola opens downward.

Standard form of a quadratic function [Review Exercises 1–8]

The **standard form** of a quadratic function is given by $f(x) = a(x - h)^2 + k$. The graph of any quadratic function can be represented as a series of transformations of the basic function $y = x^2$.

Vertex of a parabola [Review Exercises 9–14, 19, 20]

The **vertex** (h, k) of the parabola is given by $h = \frac{-b}{2a}$ and $k = f(h) = f\left(\frac{-b}{2a}\right) = c - \frac{b^2}{4a}$. The axis of **symmetry** of the parabola is given by $x = \frac{-b}{2a}$.

Zeros of a quadratic function and x-intercepts of its graph [Review Exercises 15–18]

The real number values of x at which $f(x) = 0$ are called **real zeros** of the function f. The x-coordinate of an x-intercept is a value of x such that $f(x) = 0$.

Section 3.2 Polynomial Functions and Their Graphs

Polynomial functions and terminology [Review Exercises 21–24]

A function f is a **polynomial function** if it can be expressed in the form $f(x) = a_n x^n + a_{n-1} x^{n-1} + \cdots + a_1 x + a_0$, where $a_n \neq 0$, n is a nonnegative integer, and $a_0, a_1, \cdots, a_n$ are real numbers.

In the definition of a polynomial function,

- The nonnegative integer n is called the **degree** of the polynomial.

- The constants $a_0, a_1, \cdots, a_n$ are called **coefficients**.

- The term $a_n x^n$ is called the **leading term**, and the coefficient a_n is called the **leading coefficient**.

The leading term test for end behavior [Review Exercises 25–30]

For sufficiently large values of $|x|$, the leading term of a polynomial function $f(x)$ will be much larger in magnitude than any of the subsequent terms of $f(x)$.

Hand sketching a polynomial function [Review Exercises 31–36]

For a polynomial written in factored form, sketch the function using the following procedure.

Step 1 Determine the end behavior of the function.

Step 2 Find and plot the y-intercept.

Step 3 Find and plot the x-intercepts of the graph of the function; these points divide the x-axis into smaller intervals.

Step 4 Find the sign and value of $f(x)$ for a test value x in each of these intervals. Plot these points.

Step 5 Use the plotted points and the end behavior to sketch a smooth graph of the function. Plot additional points, if needed.

Multiplicities of zeros and behavior at the *x*-intercept [Review Exercises 37–42]

The number of times a linear factor $x - a$ occurs in the completely factored form of a polynomial expression is known as the **multiplicity** of the real zero a associated with that factor.

- If the multiplicity of a real zero of a polynomial function is **odd**, the graph of the function **crosses** the *x*-axis at the corresponding *x*-intercept.

- If the multiplicity of a real zero of a polynomial function is **even**, the graph of the function **touches**, but does not cross, the *x*-axis at the corresponding *x*-intercept.

The number of real zeros of a polynomial $f(x)$ of degree n is less than or equal to n, counting multiplicity.

Finding local extrema and sketching a complete graph [Review Exercises 43–52]

The peaks and valleys of the graphs of most polynomial functions are known as **local maxima** and **local minima**, respectively. Together, they are known as **local extrema** and can be located by using a graphing utility.

Connection between zeros and *x*-intercepts [Review Exercises 37–42]

The real number values of x satisfying $f(x) = 0$ are called **real zeros** of the function f. Each real zero x is the first coordinate of an *x*-intercept of the graph of the function.

Section 3.3 Division of Polynomials; the Remainder and Factor Theorems

The division algorithm and synthetic division [Review Exercises 53–56]

Synthetic division is a compact way of dividing polynomials when the divisor is of the form $x - c$.

Let $p(x)$ be a polynomial divided by a nonzero polynomial $d(x)$. Then, there exists a quotient polynomial $q(x)$ and a remainder polynomial $r(x)$ such that

$$p(x) = d(x)q(x) + r(x)$$

The remainder $r(x)$ is either equal to 0 or its degree is less than the degree of $d(x)$.

The remainder and factor theorems [Review Exercises 57–60]

The Remainder Theorem: When a polynomial $p(x)$ is divided by $x - c$, the remainder is equal to the value of $p(c)$.

The Factor Theorem: The term $x - c$ is a **factor** of a polynomial $p(x)$ if and only if $p(c) = 0$.

Section 3.4 Real Zeros of Polynomials; Solutions of Equations

Zeros, factors, and *x*-intercepts of a polynomial [Review Exercises 61–64]

Let $p(x)$ be a polynomial function and let c be a real number. The following are equivalent statements.

- $p(c) = 0$.

- $x - c$ is a factor of $p(x)$.

- $(c, 0)$ is an *x*-intercept of the graph of $p(x)$.

The rational zero theorem and polynomial equations [Review Exercises 65–72, 75]

If $f(x) = a_n x^n + a_{n-1} x^{n-1} + \cdots + a_1 x + a_0$ is a polynomial with integer coefficients and $\frac{p}{q}$ is a rational zero of f with p and q having no common factor other than 1, then p is a factor of a_0 and q is a factor of a_n.

Because any equation can be rewritten so that the right-hand side is zero, solving an equation is identical to finding the zeros of a suitable function.

Descartes' Rule of Signs [Review Exercises 73, 74]

The number of *positive* real zeros of a polynomial $p(x)$, counting multiplicity, is either equal to the number of variations of sign in $p(x)$ or less than that number by an even integer.

The number of *negative* real zeros of a polynomial $p(x)$, counting multiplicity, is either equal to the number of variations of sign in $p(-x)$ or less than that number by an even integer.

Section 3.5 Complex Numbers

Definition of a complex number [Review Exercises 76–79]

The number i is defined as $\sqrt{-1}$. **A complex number** is a number of the form $a + bi$, where a and b are real numbers. If $a = 0$, then the number is a **pure imaginary number**.

Addition and subtraction of complex numbers [Review Exercises 84–89]

To add two complex numbers, add their corresponding real and imaginary parts. To subtract two complex numbers, subtract their corresponding real and imaginary parts.

Multiplication of complex numbers [Review Exercises 84–89]

To multiply two complex numbers, apply the rules of multiplication of binomials.

Conjugates and division of complex numbers [Review Exercises 80–89]

The **complex conjugate** of a complex number $a + bi$ is given by $a - bi$. The quotient of two complex numbers, written as a fraction, can be found by multiplying the numerator and denominator by the complex conjugate of the denominator.

Zeros of quadratic functions and solutions of quadratic equations [Review Exercises 90–93]

By using complex numbers, you can find the nonreal zeros of a quadratic function and the nonreal solutions of a quadratic equation by using the quadratic formula.

Section 3.6 The Fundamental Theorem of Algebra; Complex Zeros

The fundamental theorem of algebra [Review Exercises 94–97]

Every nonconstant polynomial function with real or complex coefficients has at least one complex zero. This theorem does not tell you what the zero is—only that a complex zero *exists*.

Definition of multiplicity of zeros [Review Exercises 94–97]

A zero c of a polynomial $P(x)$ of degree $n > 0$ has **multiplicity** k if $P(x) = (x - c)^k Q(x)$ where $Q(x)$ is a polynomial of degree $n - k$ and c is not a zero of $Q(x)$.

Polynomials with real coefficients [Review Exercises 94–97]

Any polynomial $P(x)$ with *real* coefficients can be factored uniquely into linear factors and/or **irreducible quadratic factors**.

Factorization over the complex numbers [Review Exercises 94–97]

Every polynomial $p(x)$ of degree $n > 0$ has exactly n zeros if multiplicities and complex zeros are counted and can be factored uniquely into linear factors over the complex numbers.

Section 3.7 **Rational Functions**

Definition of a rational function [Review Exercises 98–108]

A **rational function** $r(x)$ is defined as a quotient of two polynomials $p(x)$ and $h(x)$, $r(x) = \frac{p(x)}{h(x)}$, where $h(x)$ is not the constant zero function. The **domain** of a rational function consists of all real numbers for which $h(x)$ is not equal to zero.

Vertical asymptotes [Review Exercises 98, 99]

The vertical asymptotes of $r(x) = \frac{p(x)}{h(x)}$ occur at the values of x at which $h(x) = 0$ assuming $p(x)$ and $h(x)$ have no common factors.

Horizontal asymptote and end behavior [Review Exercises 100–105, 108]

Let $r(x) = \frac{p(x)}{h(x)}$, where $p(x)$ and $h(x)$ are polynomials of degrees n and m, respectively.

- If $n < m$, $r(x)$ approaches zero for large values of $|x|$. The line $y = 0$ is the **horizontal asymptote** of the graph of $r(x)$.

- If $n = m$, $r(x)$ approaches a nonzero constant for large values of $|x|$. The line $y = \frac{a_n}{b_n}$ is the horizontal asymptote of the graph of $r(x)$.

- If $n > m$, then $r(x)$ does not have a horizontal asymptote.

Slant asymptotes [Review Exercises 106, 107]

If the degree of the numerator of a rational function is 1 greater than the degree of the denominator, the graph of the rational function has a **slant asymptote**.

Section 3.8 **Quadratic, Polynomial, and Rational Inequalities**

Quadratic and polynomial inequalities [Review Exercises 109–116,121]

A **polynomial inequality** can be written in the form $a_n x^n + a_{n-1} x^{n-1} + \cdots + a_0 \,\square\, 0$, where the symbol inside the box can be $>, \geq, <,$ or $\leq$. When $n = 2$, we obtain a quadratic inequality.

Rational inequalities [Review Exercises 117-120]

Let $p(x)$ and $q(x)$ be polynomial functions with $q(x)$ not identically equal to zero. A **rational inequality** can be written in the form $\frac{p(x)}{q(x)} \,\square\, 0$, where the symbol inside the box can be $>, \geq, <,$ or $\leq$.

Chapter 3 Review Exercises

Section 3.1

In Exercises 1–4, use transformations to graph the quadratic function and find the vertex of the associated parabola.

1. $f(x) = (x + 3)^2 - 1$

2. $g(x) = (x - 1)^2 + 4$

3. $f(x) = -2x^2 - 1$

4. $g(x) = 3(x + 1)^2 + 3$

In Exercises 5–8, write each quadratic function in the form $a(x - h)^2 + k$ by completing the square. Also find the vertex of the associated parabola.

5. $f(x) = x^2 - 4x + 3$

6. $g(x) = 3 - 6x + x^2$

7. $f(x) = 4x^2 + 8x - 1$

8. $g(x) = -3x^2 + 12x + 5$

In Exercises 9–14, for each quadratic function, find the vertex, x- and y-intercepts, and axis of symmetry of the associated parabola. Sketch the parabola. Find the intervals where the function is increasing and decreasing and find the range.

9. $f(x) = x^2 - 2x + 1$

10. $g(x) = -2x^2 - 3x$

11. $f(x) = -x^2 - 3x + 1$

12. $f(x) = 1 - 4x + 3x^2$

13. $g(x) = \frac{1}{2}x^2 - 2x + 5$

14. $f(x) = -\frac{2}{3}x^2 + x - 1$

In Exercises 15–18, for each function of the form $f(x) = ax^2 + bx + c$, find the discriminant, $b^2 - 4ac$, and use it to determine the number of x-intercepts of the graph of f. Also, determine the number of real solutions to the equation $f(x) = 0$.

15. $f(x) = x^2 - 6x + 4$

16. $f(x) = -2x^2 - 7x$

17. $f(x) = x^2 - 6x + 9$

18. $f(x) = -x^2 - x - 2$

19. Geometry A rectangular play yard is to be enclosed, with a fence on three of its sides and a brick wall on the fourth side. If 120 feet of fencing material is available, what dimensions will yield the maximum area?

20. Airline Industry The percentage of total operating expenses incurred by airlines for airline food can be modeled by the function

$$f(t) = -0.0055t^2 + 0.116t + 2.90$$

where *t* is the number of years since 1980. The model is based on data for selected years from 1980 to 2000. (*Source*: Statistical Abstract of the United States)

a. What is the *y*-intercept of the graph of this function and what does it signify in relation to this problem?

b. In what year between 1980 and 2000 was the expenditure for airline food at a maximum?

c. Is this model reliable as a long-term indicator of airline expenditures for airline food as a percentage of total operating expenses? Justify your answer.

Section 3.2

In Exercises 21–24, determine whether the function is a polynomial function. If so, find its degree, its coefficients, and the leading coefficient.

21. $f(x) = -x^3 - 6x^2 + 5$

22. $f(s) = s^5 + 6s - 1$

23. $f(t) = \sqrt{t + 1}$

24. $g(t) = \frac{1}{t^2}$

In Exercises 25–30, determine the end behavior of the following functions.

25. $f(x) = -3x^3 + 5x + 9$ **26.** $g(t) = 5t^4 - 6t^2 + 1$

27. $H(s) = -6s^4 - 3s$ **28.** $g(x) = -x^3 + 2x - 1$

29. $h(s) = 10s^5 - 2s^2$ **30.** $f(x) = 7x^2 - 4$

In Exercises 31–36, for each polynomial function, find

     a. *the end behavior*

     b. *the x- and y-intercepts of the graph of the function*

     c. *the interval(s) on which the value of the function is positive*

     d. *the interval(s) on which the value of the function is negative*

Use this information to sketch a graph of the function. Factor first if the expression is not in factored form.

31. $f(x) = -(x - 1)(x + 2)(x + 4)$ **32.** $g(x) = (x - 3)(x - 4)(x - 1)$

33. $f(t) = t(3t - 1)(t + 4)$ **34.** $g(t) = 2t(t + 4)\left(t + \frac{3}{2}\right)$

35. $f(x) = 2x^3 + x^2 - x$ **36.** $g(x) = -x^3 - 6x^2 + 7x$

In Exercises 37–42, determine the multiplicities of the real zeros of the functions. At the x-intercepts, does the graph cross or just touch the x-axis?

37. $f(x) = (x + 2)^3(x + 7)^2$ **38.** $g(s) = (s + 8)^5(s - 1)^2$

39. $h(t) = -t^2(t + 1)(t - 2)$ **40.** $g(x) = -x^3(x^2 - 16)$

41. $f(x) = x^3 + 2x^2 + x$ **42.** $h(s) = s^7 - 16s^3$

In Exercises 43–50, for each polynomial function, find

     a. *the x- and y-intercepts of the graph of the function*

     b. *the multiplicities of each of the real zeros*

     c. *the end behavior*

     d. *the intervals where the function is positive and the intervals where it is negative*

Use this information to sketch a graph of the function.

43. $f(x) = x^2(2x + 1)$ **44.** $h(t) = -t(t + 4)^2$

45. $f(x) = \left(x - \frac{1}{2}\right)^2(x - 4)$ **46.** $f(x) = (x - 7)^2(x + 2)(x - 3)$

47. $f(t) = (t + 2)(t - 1)(t^2 + 1)$ **48.** $g(s) = \left(s - \frac{1}{2}\right)(s + 3)(s^2 + 4)$

49. $g(x) = x^4 - 3x^3 - 18x^2$ **50.** $h(t) = -2t^5 + 4t^4 + 2t^3$

51. Manufacturing An open box is to be made by cutting four squares of equal size out of a 8-inch by 11-inch rectangular piece of cardboard, one at each corner, and then folding up the sides of the box.

     a. Let x be the length of the side of the square cut from each corner. Find an expression for the volume of the box in terms of x. Leave the expression in factored form.

     b. What is a realistic set of values for x? Explain.

      c. Use a graphing utility to find an approximate value of x that will yield the maximum volume.

52. Design A pencil holder in the shape of a right circular cylinder is to be designed with the specification that the sum of its radius and height must equal 8 inches. The volume of a right circular cylinder is given by $V = \pi r^2 h$, where r is the radius and h is the height.

     a. Let r denote the radius. Find an expression for the volume of the cylinder in terms of r. Leave the expression in factored form.

     b. What is a realistic set of values for r? Explain.

     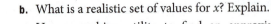 **c.** Use a graphing utility to find an approximate value of r that will yield the maximum volume.

Section 3.3

In Exercises 53–56, write each polynomial as $p(x) = d(x)q(x) + r(x)$, where $p(x)$ is the first polynomial and $d(x)$ is the second polynomial. You may use synthetic division wherever applicable.

53. $-4x^2 + x - 7; x - 4$ **54.** $5x^3 + 2x + 4; x + 2$

55. $x^5 - x^4 + x^2 - 3x + 1; x^2 + 3$ **56.** $-4x^3 - x^2 + 2x + 1; 2x + 1$

In Exercises 57–60, find the remainder when the first polynomial, $p(x)$, is divided by the second polynomial, $d(x)$. Determine whether $d(x)$ is a factor of $p(x)$ and justify your answer.

57. $-x^3 + 7x + 6; x + 1$ **58.** $2x^3 + x^2 + 8x; x + 2$

59. $x^3 - 7x + 6; x + 3$ **60.** $x^{10} - 1; x - 1$

Section 3.4

In Exercises 61–64, show that the given value of x is a zero of the polynomial. Use the zero to completely factor the polynomial.

61. $p(x) = x^3 - 6x^2 + 3x + 10; x = 2$ **62.** $p(x) = -x^3 - 7x - 8; x = -1$

63. $p(x) = -x^4 + x^3 + 4x^2 + 5x + 3; x = 3$ **64.** $p(x) = x^4 - x^2 + 6x; x = -2$

In Exercises 65–68, find all the real zeros of the polynomial.

65. $P(x) = 2x^3 - 3x^2 - 2x + 3$ **66.** $P(t) = -t^3 - 5t^2 + 4t + 20$

67. $h(x) = x^3 - 3x^2 + x - 3$ **68.** $f(x) = x^3 + 3x^2 - 9x + 5$

In Exercises 69–72, solve the equation for real values of x.

69. $2x^3 + 9x^2 - 6x = 5$ **70.** $x^3 = 21x - 20$

71. $x^3 - 7x^2 = -14x + 8$ **72.** $x^4 - 9x^2 - 2x^3 + 2x = -8$

In Exercises 73 and 74, use Descartes' Rule of Signs to determine the number of positive and negative zeros of p. You need not find the zeros.

73. $p(x) = -x^4 + 2x^3 - 7x - 4$ **74.** $p(x) = x^5 + 3x^4 - 8x^2 - x - 3$

75. Geometry The length of a rectangular solid is 3 inches more than its height and its width is 4 inches more than its height. Write and solve a polynomial equation to determine the height of the solid such that the volume is 60 cubic inches.

Section 3.5

In Exercises 76–79, find the real and imaginary parts of the complex number.

76. $\sqrt{3}$ **77.** $-\frac{3}{2}i$ **78.** $7 - 2i$ **79.** $-1 - \sqrt{-5}$

In Exercises 80–83, find the complex conjugate of each number.

80. $\frac{1}{2}$ **81.** $3 - i$ **82.** $i + 4$ **83.** $1 - \sqrt{2} + 3i$

In Exercises 84–89, find $x + y$, $x - y$, xy, and $\frac{x}{y}$.

84. $x = 1 + 4i, y = 2 - 3i$ **85.** $x = 3 + 2i, y = -4 + 3i$

86. $x = 1.5 - 3i, y = 2i - 1$ **87.** $x = -\sqrt{2} + i, y = -3$

88. $x = \frac{1}{2}i - 1, y = i + \frac{3}{2}$ **89.** $x = -i, y = -\sqrt{-3} + 1$

In Exercises 90–93, find all solutions of the quadratic equation. Relate the solutions of each equation to the zeros of an appropriate quadratic function.

90. $-x^2 + x - 3 = 0$

91. $-2x^2 = -x + 1$

92. $-\dfrac{4}{5}t - 1 = t^2$

93. $2t^2 - \sqrt{13} = t$

Section 3.6

In Exercises 94–97, find all the zeros, real and nonreal, of the polynomial. Then express $p(x)$ as a product of linear factors.

94. $p(x) = x^2 + 25$

95. $p(x) = x^3 + x - 4x^2 - 4$

96. $p(x) = x^3 + 2x^2 + 4x + 8$

97. $p(x) = x^4 - 8x^2 - 9$

Section 3.7

98. Let $f(x) = \dfrac{1}{(x+1)^2}$.

 a. Fill in the following table for values of x near -1. What do you observe about the value of $f(x)$ as x approaches -1 from the right? From the left?

x	-1.5	-1.1	-1.01	-0.99	-0.9	-0.5
$f(x)$						

 b. Complete the following table. What happens to the value of $f(x)$ as $x \to \infty$?

x	10	50	100	1000
$f(x)$				

 c. Complete the following table. What happens to the value of $f(x)$ as $x \to -\infty$?

x	-1000	-100	-50	-10
$f(x)$				

99. Let $f(x) = \dfrac{1}{(3-x)^2}$.

 a. Fill in the following table for values of x near 3. What do you observe about the value of $f(x)$ as x approaches 3 from the right? From the left?

x	2.5	2.9	2.99	3.01	3.1	3.5
$f(x)$						

 b. Complete the following table. What happens to the value of $f(x)$ as $x \to \infty$?

x	10	50	100	1000
$f(x)$				

 c. Complete the following table. What happens to the value of $f(x)$ as $x \to -\infty$?

x	-1000	-100	-50	-10
$f(x)$				

In Exercises 100–107, for each rational function, find all asymptotes and intercepts, and sketch a graph.

100. $f(x) = \dfrac{2}{x - 1}$

101. $f(x) = \dfrac{3x}{x + 5}$

102. $h(x) = \dfrac{1}{x^2 - 4}$

103. $g(x) = \dfrac{2x^2}{x^2 - 1}$

104. $g(x) = \dfrac{x - 2}{x^2 - 2x - 3}$

105. $h(x) = \dfrac{x^2 - 2}{x^2 - 4}$

106. $r(x) = \dfrac{x^2 - 1}{x + 2}$

107. $p(x) = \dfrac{x^2 - 4}{x - 1}$

108. Average Cost A truck rental company charges a daily rate of \$20 plus \$0.25 per mile driven.

   a. What is the average cost per mile of driving x miles per day?

   b. Use this expression to find the average cost per mile of driving 100 miles per day.

   c. Find the horizontal asymptote of this function and explain its significance.

Section 3.8

In Exercises 109–120, solve the inequality.

109. $4x^2 + 21x + 5 \le 0$

110. $x^2 - 5x - 6 > 0$

111. $-6x^2 - 5x > -4$

112. $2x^2 - 11x \le -12$

113. $-x(x + 1)(x^2 - 9) > 0$

114. $x^2(x + 2)(x - 3) \le 0$

115. $x^3 + 4x^2 \le -x + 6$

116. $9x^3 - x > -9x^2 + 1$

117. $\dfrac{x^2 - 1}{x + 1} \le 0$

118. $\dfrac{x^2 + 2x - 3}{x - 3} \le 0$

119. $\dfrac{4x - 2}{3x - 1} > 2$

120. $-\dfrac{2}{x + 3} \le x$

121. Geometry A rectangular solid has a square base and a height that is 1 inch less than the length of one side of the base. Set up and solve a polynomial inequality to determine the lengths of the base that will produce a volume greater than or equal to 48 cubic inches.

Chapter 3 Test

1. Write $f(x) = 2x^2 - 4x + 1$ in the form $f(x) = a(x - h)^2 + k$. Find the vertex of the associated parabola and determine if f attains a maximum or minimum at that point.

In Exercises 2–4, find the vertex and axis of symmetry of the parabola represented by $f(x)$. Sketch the graph of f and find its range.

2. $f(x) = -(x - 1)^2 + 2$ 3. $f(x) = x^2 + 4x + 2$ 4. $f(x) = -2x^2 + 8x - 4$

5. Sketch a graph of $f(x) = 3x^2 + 6x$. Find the vertex, axis of symmetry, x- and y-intercepts, and intervals on which the function is increasing and decreasing.

6. Determine the degree, coefficients, and the leading coefficient of the polynomial $p(x) = 3x^5 + 4x^2 - x + 7$.

7. Determine the end behavior of $p(x) = -8x^4 + 3x - 1$.

8. Determine the real zeros of $p(x) = -2x^2(x^2 - 9)$ and their multiplicities. Also, find the x-intercepts and determine whether the graph of p crosses or touches the x-axis.

In Exercises 9–12, find

a. *the x- and y-intercepts of the graph of each polynomial*

b. *the multiplicities of each of the real zeros*

c. *the end behavior*

d. *the intervals on which the function is positive and the intervals on which it is negative*

Use this information to sketch a graph of the function.

9. $f(x) = -2x(x - 2)(x + 1)$ 10. $f(x) = (x + 1)(x - 2)^2$

11. $f(x) = -3x^3 - 6x^2 - 3x$ 12. $f(x) = 2x^4 + 5x^3 + 2x^2$

In Exercises 13 and 14, write each polynomial as $p(x) = d(x)q(x) + r(x)$, where $p(x)$ is the first polynomial and $d(x)$ is the second polynomial.

13. $3x^4 - 6x^2 + x - 1; x^2 + 1$ 14. $-2x^5 + x^4 - 4x^2 + 3; x - 1$

15. Find the remainder when $p(x) = x^4 + x - 2x^3 - 2$ is divided by $x - 2$. Is $x - 2$ a factor of $p(x)$? Explain.

16. Use the fact that $x = 3$ is a zero of $p(x) = x^4 - x - 3x^3 + 3$ to completely factor $p(x)$.

In Exercises 17 and 18, find all the real zeros of the given polynomial.

17. $p(x) = x^3 - 3x - 2x^2 + 6$ 18. $p(x) = 2x^4 + 9x^3 + 14x^2 + 9x + 2$

In Exercises 19 and 20, find all the real solutions of the given equation.

19. $2x^3 + 5x^2 = 2 - x$

20. $x^4 - 4x^3 + 2x^2 + 4x = 3$

21. Use Descartes' Rule of Signs to determine the possible number of positive and negative zeros of $p(x) = -x^5 + 4x^4 - 3x^2 + x + 8$. You need not find the zeros.

22. Find the real and imaginary parts of the complex number $4 - \sqrt{-2}$.

In Exercises 23–25, perform the indicated operations and write in the form $a + bi$.

23. $5 + 4i - (6 + 2i)$ 24. $(3 - 4i)(-2 + i)$ 25. $\dfrac{2 + i}{3 - 2i}$

In Exercises 26 and 27, find all the solutions, real or complex, of each equation.

26. $x^2 + 2x + 3 = 0$

27. $-2x^2 + x - 1 = 0$

In Exercises 28 and 29, find all the zeros, real and nonreal, of each polynomial. Then express $p(x)$ as a product of linear factors.

28. $p(x) = x^5 - 16x$

29. $p(x) = x^4 - x^3 - 2x^2 - 4x - 24$

In Exercises 30–32, find all asymptotes and intercepts and sketch a graph of each rational function.

30. $f(x) = \dfrac{-3}{(x + 3)}$

31. $f(x) = \dfrac{-2x}{x - 2}$

32. $f(x) = \dfrac{2}{2x^2 - 3x - 2}$

In Exercises 33–36, solve each inequality.

33. $3x^2 - 4x - 15 < 0$

34. $(x^2 - 4)(x + 3) \le 0$

35. $\dfrac{x^2 - 4x - 5}{x + 2} > 0$

36. $\dfrac{4}{3x + 1} \ge 2$

37. The radius of a right circular cone is 2 inches more than its height. Write and solve a polynomial equation to determine the height of the cone such that the volume is 48π cubic inches. The volume of a cone is given by $V = \frac{1}{3}\pi r^2 h$, where r is the radius and h is the height.

38. A couple rents a moving van at a daily rate of $50 plus $0.25 per mile driven.

   a. What is the average cost per mile of driving x miles per day?

   b. If the couple drive 250 miles per day, find the average cost per mile.

Exponential and Logarithmic Functions

The price of computers and other electronics has been decreasing steadily over the past twenty years. This phenomenon can be modeled by an *exponential function,* a type of function in which the independent variable appears in the *exponent.* A simple illustration of this model is given in Example 5 in Section 4.4. This chapter will explore exponential functions, and their inverse, logarithmic functions. These functions are invaluable in the study of more advanced mathematics and have numerous applications in a variety of fields, including engineering, the life sciences, business, physics and computer science.

Chapter 4

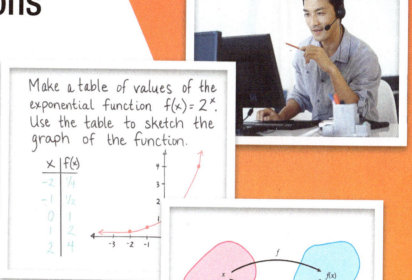

Outline

Inverse Functions

Objectives

- Define the inverse of a function.
- Verify that two functions are inverses of each other.
- Define a one-to-one function.
- Define the conditions for the existence of an inverse function.
- Find the inverse of some functions.

Inverse Functions

In Section 2.6, we discussed the composition of functions, which involves using the output of one function as the input to another. Using this idea of composition of functions, we can sometimes find a function which will *undo* the action of another function, f. Such a function is called the **inverse** of f.

For example, take a number x and multiply it by 6, giving $6x$. To get back to the original number x, you multiply $\frac{1}{6}$ by $6x$. This is an instance of undoing the action of a function. We next give the formal definition of the inverse of a function.

Inverse of a Function

Let f be a function. A function g is said to be the **inverse function** of f if the domain of g is equal to the range of f and, for every x in the domain of f and every y in the domain of g,

$$g(y) = x \quad \text{if and only if} \quad f(x) = y.$$

The notation for the inverse function of f is f^{-1}. Equivalently,

$$f^{-1}(y) = x \quad \text{if and only if} \quad f(x) = y.$$

The notation f^{-1} does NOT mean $\frac{1}{f}$.

Note The definition does not tell you how to find the inverse of a function f, or if such an inverse function even exists. It only states that if f has an inverse function, then the inverse function must have the given properties.

The idea of an inverse can be illustrated graphically as follows. Consider evaluating $f^{-1}(4)$ using the graph of a function f in Figure 1. From the definition of an inverse function, $f^{-1}(y) = x$ if and only if $f(x) = y$. Thus, to evaluate $f^{-1}(4)$, we have to determine the value of x that produces $f(x) = 4$. From the graph of f, we see that $f(2) = 4$, so $f^{-1}(4) = 2$.

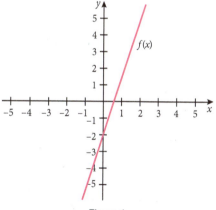

Figure 1

We next examine the action of a function on x, followed by the action of its inverse. These two actions, in succession, will simply result in getting back to x. Recall that actions of functions in sequence are accomplished by using compositions of functions. The following definition generalizes this notion.

Just in Time
Review the composition of functions in Section 2.6

Composition of a Function and Its Inverse

If f is a function with an inverse function f^{-1}, then

- for every x in the domain of f, $f^{-1}(f(x))$ is defined and $f^{-1}(f(x)) = x$;
- for every x in the domain of f^{-1}, $f(f^{-1}(x))$ is defined and $f(f^{-1}(x)) = x$.

See Figure 2. If g is any function with the same properties with respect to f, as the ones stated here for f^{-1}, then f and g are inverse functions of one another.

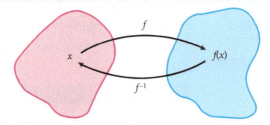

Figure 2 Relationship between f and f^{-1}

Example 1   **Verifying Inverse Functions**

Verify that the following functions are inverses of each other.

$$f(x) = 3x - 4; \quad g(x) = \frac{x}{3} + \frac{4}{3}$$

Solution To check that f and g are inverses of each other, we verify that $(f \circ g)(x) = x$ and $(g \circ f)(x) = x$. We begin with $(f \circ g)(x)$.

$$(f \circ g)(x) = f(g(x)) = 3\left(\frac{x}{3} + \frac{4}{3}\right) - 4 = x + 4 - 4 = x$$

Next, we calculate $(g \circ f)(x)$.

$$(g \circ f)(x) = g(f(x)) = \frac{3x - 4}{3} + \frac{4}{3} = x - \frac{4}{3} + \frac{4}{3} = x$$

Thus, by definition, f and g are inverses of each other.

 Check It Out 1   Verify that $f(x) = 2x - 9$ and $g(x) = \frac{x}{2} + \frac{9}{2}$ are inverses of each other.   ■

Example 2 shows how to determine the inverse of a function.

Example 2   **Finding the Inverse of a Function**

Find the inverse of the function $f(x) = -2x + 3$ and check that it is the inverse.

Solution Before we illustrate an algebraic method for finding the inverse of f, let us examine what f does: it takes a number x, multiplies it by -2 and adds 3 to the result. The inverse function will *undo* this sequence, in order to get back to x. It will start with the value output by f, then *subtract* 3 and *divide the result* by -2.

We next show how this can be done using a set of algebraic steps.

Steps	Example
1. Start with the expression for the given function f.	$f(x) = -2x + 3$
2. Replace $f(x)$ with y.	$y = -2x + 3$
3. Interchange the variables x and y so that the input variable for the inverse function f^{-1} is x and its output variable is y.	$x = -2y + 3$
4. Solve for y. This gives the same result that was explained before in words.	$x = -2y + 3$ $x - 3 = -2y$ $\dfrac{x-3}{-2} = y$
5. The inverse function f^{-1} is now given by y, so replace y with $f^{-1}(x)$ and simplify the expression for $f^{-1}(x)$.	$f^{-1}(x) = \dfrac{x-3}{-2} = -\dfrac{1}{2}x + \dfrac{3}{2}$

To check that the function $f^{-1}(x) = -\frac{1}{2}x + \frac{3}{2}$ is the inverse of $f(x) = -2x + 3$, we find expressions for $(f \circ f^{-1})(x)$ and $(f^{-1} \circ f)(x)$.

$$(f \circ f^{-1})(x) = f(\boldsymbol{f^{-1}(x)}) = -2\left(-\frac{1}{2}\boldsymbol{x} + \frac{3}{2}\right) + 3 = x - 3 + 3 = x$$

$$(f^{-1} \circ f)(x) = f^{-1}(\boldsymbol{f(x)}) = -\frac{1}{2}(\boldsymbol{-2x + 3}) + \frac{3}{2} = x - \frac{3}{2} + \frac{3}{2} = x$$

Since $(f \circ f^{-1})(x) = x = (f^{-1} \circ f)(x)$, the functions f and f^{-1} are inverses of each other.

 Check It Out 2   Find the inverse of the function $f(x) = 4x - 5$ and check that it is the inverse.

x	$f(x) = x^2$
-2	4
-1	1
0	0
1	1
2	4

Table 1

One-to-One Functions

Thus far in this section, we have dealt only with functions that have an inverse, and so you may have the impression that every function has an inverse or that an inverse function can be easily found. Neither of these statements is true. We will now study conditions under which a function can have an inverse.

Table 1 lists the values of the function $f(x) = x^2$ for selected values of x.

The inverse function of f would take as input a value of y in the range of f, and output the value of x such that $f(x) = y$. But in this case, the inverse of f would output *both* 2 and -2, for the same input value of 4. Since *no* function is permitted to do that, we see that the function $f(x) = x^2$ does *not* have an inverse function.

It seems reasonable to assert that the only functions that have inverses are those for which different input values always produce different output values, since each value output by f would correspond to only *one* value input by f.

We now make the above discussion mathematically precise.

Definition of a One-to-One Function

For a function f to have an inverse function, f must be **one-to-one**. That is, if $f(a) = f(b)$, then $a = b$.

Graphically, any horizontal line can cross the graph of a one-to-one function at most once. The reasoning is as follows: if there exists a horizontal line that crosses the graph of f more than once, a single output value of f corresponds to two different input values of f. Thus, f is no longer one-to-one.

The above comment gives rise to the *horizontal line test*.

The Horizontal Line Test

A function f is one-to-one if every horizontal line intersects the graph of f at most once.

In Figure 3, the function f does *not* have an inverse because there are horizontal lines that intersect the graph of f more than once. The function h *has* an inverse because *every* horizontal line intersects the graph of h exactly once.

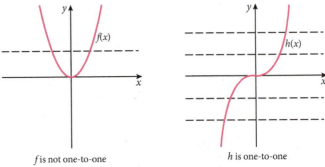

f is not one-to-one h is one-to-one

Figure 3 Horizontal line test

Example 3   **Checking Whether a Function is One-to-One**

Which of the following functions are one-to-one and therefore have an inverse?

a. $f(t) = -3t + 1$

b. The function f given graphically in Figure 4.

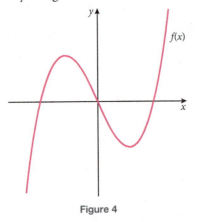

Figure 4

t	$f(t)$
-4	17
-3	19
-1	0
1	21
5	0

Table 2

c. The function f given by the values in Table 2.

Solution

a. To show that f is one-to-one, we show that if $f(a) = f(b)$, then $a = b$.

$$f(a) = f(b) \qquad \text{Assumption}$$
$$-3a + 1 = -3b + 1 \qquad \text{Evaluate } f(a) \text{ and } f(b)$$
$$-3a = -3b \qquad \text{Subtract 1 from each side}$$
$$a = b \qquad \text{Divide each side by } -3$$

We have shown that if $f(a) = f(b)$, then $a = b$. Thus f is one-to-one and does have an inverse.

b. In Figure 5, we have drawn a horizontal line which intersects the graph of f more than once, and it is easy to see that there are other such horizontal lines. Thus, according to the horizontal line test, f is not one-to-one and does not have an inverse.

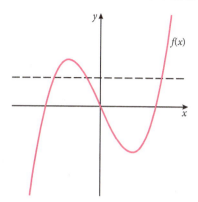

Figure 5 Horizontal line test

c. Note that $f(-1) = 0 = f(5)$. Since two different inputs yield the same output, f is not one-to-one and does not have an inverse.

 Check It Out 3 Show that the function $f(x) = 4x - 6$ is one-to-one.

Graph of a Function and Its Inverse

We have seen how to obtain some simple inverse functions algebraically. Visually, the graph of a function f and its inverse are *mirror images* of each other, with the line $y = x$ being the mirror. This is stated formally as follows.

> **Graph of a Function and Its Inverse**
>
> The graphs of a function f and its inverse function f^{-1} are symmetric with respect to the line $y = x$.

This fact is illustrated in Example 4.

Example 4 **Graphing a Function and Its Inverse**

Graph the function $f(x) = -2x + 3$ and its inverse, $f^{-1}(x) = -\frac{1}{2}x + \frac{3}{2}$, on the same set of axes, using the same scale for both axes. What do you observe?

Solution The graphs of f and its inverse f^{-1} are shown in Figure 7. The inverse of f was computed in Example 2.

Recall that the inverse of a function is found by interchanging the input and output values. For instance, the points $(3, -3)$ and $(-1, 5)$ on the graph of $f(x) = -2x + 3$ are reflected to the points $(-3, 3)$ and $(5, -1)$, respectively, on the graph of $f^{-1}(x) = -\frac{1}{2}x + \frac{3}{2}$. It is this interchange of inputs and outputs that causes the graphs of f and f^{-1} to be symmetric with respect to the line $y = x$.

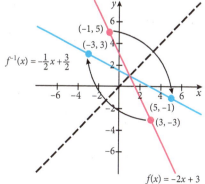

Figure 7

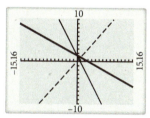

✓ *Check It Out 4*   Graph the function $f(x) = 3x - 2$ and its inverse, $f^{-1}(x) = \frac{1}{3}x + \frac{2}{3}$, on the same set of axes, using the same scale for both axes. What do you observe?

Example 5   **Finding an Inverse Function and Its Graph**

Let $f(x) = x^3 + 1$.

a. Show that f is one-to-one using the horizontal line test.

b. Find the inverse of f.

c. Graph f and its inverse on the same set of axes.

Solution

a. The graph of $f(x) = x^3 + 1$, along with a few horizontal lines, is shown in Figure 8. It passes the horizontal line test, and so f is one-to-one.

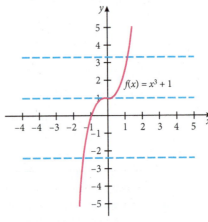

$f(x) = x^3 + 1$

Figure 8

Just in Time
Review cube roots in Section 1.2.

b. To find the inverse, we proceed as follows.

Step 1 Start with the definition of the given function, letting $y = f(x)$.

$$y = x^3 + 1$$

Step 2 Interchange the variables x and y.

$$x = y^3 + 1$$

Step 3 Solve for y.

$$x - 1 = y^3$$

$$\sqrt[3]{x - 1} = y$$

Step 4 The expression for the inverse $f^{-1}(x)$ is now given by y.

$$f^{-1}(x) = \sqrt[3]{x - 1}$$

Thus the inverse of $f(x) = x^3 + 1$ is $f^{-1}(x) = \sqrt[3]{x - 1}$.

x	f(x) = x³ + 1
−1.5	−2.375
−1	0
0	1
1.5	4.375

Table 3

c. Points on the graph of f^{-1} can be found by interchanging the x and y coordinates of points on the graph of f. Table 3 lists several points on the graph of f.

The graph of f and its inverse are shown in Figure 9.

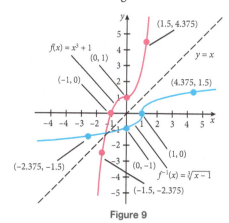

Figure 9

✓ *Check It Out 5* Rework Example 5 for the function $f(x) = -x^3 + 2$.

Restriction of Domain to Find an Inverse Function

We saw earlier that the function $f(x) = x^2$ does not have an inverse. We can, however, define a new function g from f, by restricting the domain to only non-negative numbers x, so that g will have an inverse. This is shown in the next example.

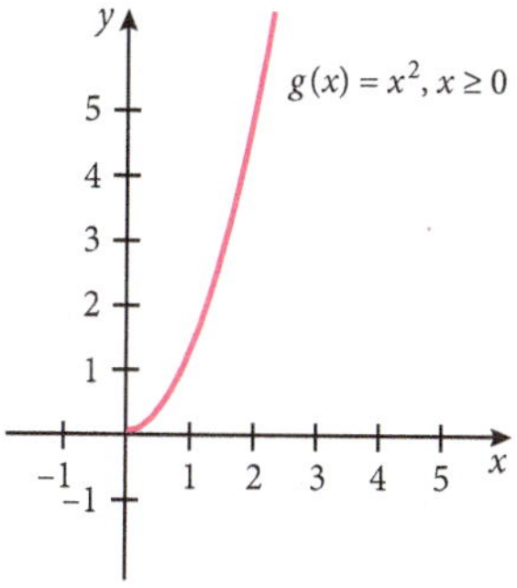

Example 6 **Restriction of Domain to Find an Inverse Function**

Show that the function $g(x) = x^2$, $x \geq 0$ has an inverse and find it.

Solution
Recall that the function $f(x) = x^2$, whose domain consists of the set of all real numbers has *no* inverse, because f is not one-to-one. However, here we are examining the function g that has the same function expression as f, but is defined *only* for $x \geq 0$. From the graph of g in Figure 10, we see that g is one-to-one by the horizontal line test and thus has an inverse.
  To find the inverse, we proceed as follows.

$$y = x^2, \quad x \geq 0 \qquad \text{Definition of the given function}$$
$$x = y^2, \quad x \geq 0 \qquad \text{Interchange the variables } x \text{ and } y$$
$$\sqrt{x} = y, \quad x \geq 0 \qquad \text{Take the square root of both sides}$$

Because $x \geq 0$, $\sqrt{x}$ is a real number. The inverse function g^{-1} is now given by y.

$$g^{-1}(x) = \sqrt{x} = x^{1/2}, \quad x \geq 0$$

✓ *Check It Out 6* Show that the function $f(x) = x^4$, $x \geq 0$ has an inverse and find the inverse.

Figure 10

g(x) = x², x ≥ 0

An Application of Inverse Functions

The next example discusses how inverse functions are used in converting units.

Example 7   **Unit Conversion**

Table 4 lists certain quantities of fuel in gallons and the corresponding quantities in liters.

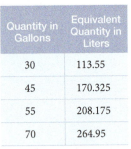

Quantity in Gallons	Equivalent Quantity in Liters
30	113.55
45	170.325
55	208.175
70	264.95

Table 4

a. There are 3.785 liters in 1 gallon. Find an expression for a function $L(x)$ which will take the number of gallons of fuel as its input and give the number of liters of fuel as its output.

b. Rewrite Table 4 so that the number of liters is the input and the number of gallons is the output.

c. Find an expression for a function $G(x)$ that will take the number of liters of fuel as its input and give the number of gallons of fuel as its output.

d. Show that $L(x)$ and $G(x)$ are inverses of each other.

Solution

a. To convert from gallons to liters, we *multiply* the number of gallons by 3.785. The function $L(x)$ is thus given by

$$L(x) = 3.785x$$

where x is the number of *gallons* of fuel.

b. Because we already have the amount of fuel in gallons and the corresponding amount in liters, we simply interchange the columns of the table, so that the *input* is the quantity of fuel in *liters* and the *output* is the equivalent quantity of that fuel in *gallons*. See Table 5.

Liters x	Gallons $G(x)$
113.55	30
170.325	45
208.175	55
264.95	70

Table 5

c. To convert from liters to gallons, we *divide* the number of liters by 3.785. The function $G(x)$ is thus given by

$$G(x) = \frac{x}{3.785}$$

where x is the number of *liters* of fuel.

d. To show that the functions L and G are inverses of each other, we must show that $(L \circ G)(x) = x$ and $(G \circ L)(x) = x$.

$$(L \circ G)(x) = L(G(x)) = 3.785 \left(\frac{x}{3.785} \right) = x$$

$$(G \circ L)(x) = G(L(x)) = \left(\frac{3.785x}{3.785} \right) = x$$

Thus, the functions L and G are inverses of each other. In applied problems, inverse functions are denoted by another letter. In these contexts, the f^{-1} notation is not usually used.

✓ *Check It Out 7* Find a function, $C(x)$, which takes the number of inches as input and produces the number of centimeters as output. Also find $N(x)$, the inverse of $C(x)$. ■

Exercises 4.1

Just in Time Exercises

1. True or False: $\sqrt[3]{x}$ is defined for real numbers.

2. True or False: $\sqrt[2]{x-2}$ is defined for real numbers.

3. Find $f \circ g$ if $f(x) = x^2$ and $g(x) = x - 1$.

4. Find $f \circ g$ if $f(x) = \dfrac{1}{x}$ and $g(x) = \dfrac{1}{x^2}$.

5. Simplify: $2\left(\dfrac{1}{2}x + 1\right) - 2$.

6. Simplify: $\sqrt[3]{27x^3}$.

Skills

Paying Attention to Instructions: Exercises 7 and 8 are intended to give you practice reading and paying attention to the instructions that accompany the problems you are working.

7. Let $f(x) = x^3$ and $f^{-1}(x) = \sqrt[3]{x}$.
   Evaluate the following:
   a. $f(3)$
   b. $f^{-1}(3)$
   c. $(f(3))(f^{-1}(3))$
   d. $(f \circ f^{-1})(3)$

8. Let $f(x) = x + 1$ and $f^{-1}(x) = x - 1$.
   Evaluate the following:
   a. $f(5)$
   b. $f^{-1}(5)$
   c. $(f^{-1}(5))(f(5))$
   d. $(f^{-1} \circ f)(5)$

In Exercises 9–18, verify that the given functions are inverses of each other.

9. $f(x) = x - 2;\quad g(x) = x + 2$

10. $f(x) = x + 7;\quad g(x) = x - 7$

11. $f(x) = 6x;\quad g(x) = \dfrac{1}{6}x$

12. $f(x) = -8x;\quad g(x) = -\dfrac{1}{8}x$

13. $f(x) = -3x + 8;\quad g(x) = -\dfrac{1}{3}x + \dfrac{8}{3}$

14. $f(x) = \dfrac{1}{2}x + 1;\quad g(x) = 2x - 2$

15. $f(x) = x^3 + 2;\quad g(x) = \sqrt[3]{x - 2}$

16. $f(x) = x^3 - 4;\quad g(x) = \sqrt[3]{x + 4}$

17. $f(x) = \dfrac{1}{x};\quad g(x) = \dfrac{1}{x}$

18. $f(x) = x;\quad g(x) = x$

In Exercises 19–22, state whether each function given by a table is one-to-one. Explain your reasoning.

19.

x	$f(x)$
-3	6
-2	-8
0	0
1	8
3	-6

20.

x	$f(x)$
-3	4
-1	7
0	4
1	5
3	12

21.

x	$f(x)$
-2	-6
-1	5
0	9
1	4
2	9

22.

x	$f(x)$
-2	-9
-1	-8
0	-7
1	-6
2	-5

In Exercises 23–28, state whether each function given graphically is one-to-one.

23.

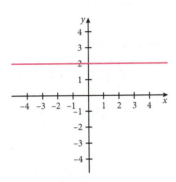

24.

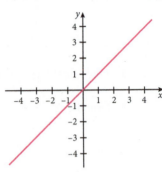

25.

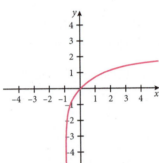

26.

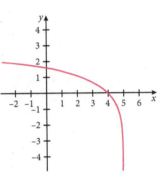

27.

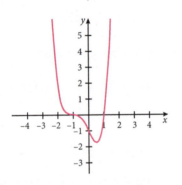

28.

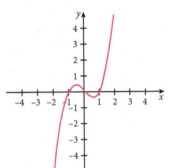

In Exercises 29–34, state whether each of the functions is one-to-one.

29. $f(x) = -3x + 2$

30. $f(x) = \dfrac{4}{3}x + 1$

31. $f(x) = 2x^2 - 3$

32. $f(x) = -3x^2 + 1$

33. $f(x) = -2x^3 + 4$

34. $f(x) = -\dfrac{1}{3}x^3 - 5$

In Exercises 35–54, find the inverse of the given function and graph both on the same set of axes.

35. $f(x) = -\dfrac{2}{3}x$

36. $g(x) = \dfrac{4}{3}x$

37. $f(x) = -4x + \dfrac{1}{5}$

38. $f(s) = 2s - \dfrac{9}{5}$

39. $f(x) = x^3 - 6$

40. $f(x) = -x^3 + 4$

41. $g(x) = -x^2 + 8, x \geq 0$

42. $g(x) = x^2 - 6, x \geq 0$

43. $f(x) = -2x^3 + 7$

44. $g(x) = 3x^3 - 5$

45. $f(x) = -4x^5 + 9$

46. $g(x) = 2x^5 - 6$

47. $f(x) = \dfrac{1}{3}x$

48. $g(x) = \dfrac{-1}{2x}$

49. $g(x) = (x - 1)^2, x \geq 1$

50. $g(x) = (x + 2)^2, x \geq -2$

51. $f(x) = \sqrt{x + 3}, x \geq -3$

52. $f(x) = \sqrt{x - 4}, x \geq 4$

53. $f(x) = \dfrac{2x}{x - 1}$

54. $f(x) = \dfrac{x + 3}{x}$

In Exercises 55–62, find the inverse of the given function. Then graph the given function and its inverse on the same set of axes.

55. $f(x) = 2x$ **56.** $f(x) = -3x$ **57.** $f(x) = -3x + 3$ **58.** $f(x) = 4x + 4$

59. $f(x) = 8x^3$ **60.** $f(x) = -x^3 + 1$ **61.** $f(x) = x^2 + 2, x \geq 0$ **62.** $f(x) = x^2 - 1, x \geq 0$

In Exercises 63–66, a graph of a one-to-one function f is given. Draw the graph of the inverse function f^{-1}. Copy the given graph onto a piece of graph paper and use the line $y = x$ to help you sketch the inverse. State the domain and range of f and f^{-1}.

63.

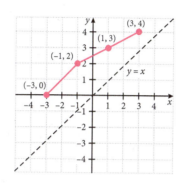

64.

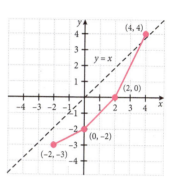

65.

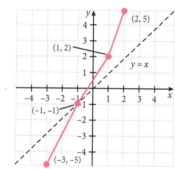

66.

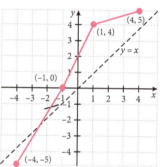

Evaluate the quantities indicated in Exercises 67–70, by referring to the function f given by the following table.

67. $f^{-1}(1)$

68. $f^{-1}(2)$

69. $f^{-1}(f^{-1}(-2))$

70. $f^{-1}(f^{-1}(1))$

x	f(x)
-2	1
-1	2
0	0
1	-1
2	-2

Evaluate the quantities indicated in Exercises 71–74, by referring to the function f given by the following graph.

71. $f^{-1}(1)$

72. $f^{-1}(3)$

73. $f^{-1}(f^{-1}(-3))$

74. $f^{-1}(f^{-1}(3))$

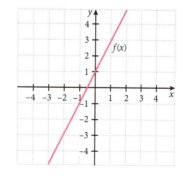

Applications

75. Converting Liquid Measures Find a function which converts x gallons into quarts. Find its inverse and explain what it does.

76. Shopping When you buy products at a store, the Universal Product Code (UPC) is scanned and the price is output by a computer. The price is a function of the UPC. Why? Does this function have an inverse? Why or why not?

77. Economics In economics, the demand function gives the price p as a function of the quantity q. One example of a demand function is $p = 100 - 0.1q$. However, mathematicians tend to think of the price as the input variable and the quantity as the output variable. How can you take this example of a demand function and express q as a function of p?

78. Physics After t seconds, the height of an object dropped from an initial height of 100 feet is given by $h = -16t^2 + 100, t \geq 0$.

  a. Why does h have an inverse?

  b. Write t as a function of h and explain what it represents.

79. Fashion A woman's dress size in the United States can be converted to a woman's dress size in France by using the function $f = s + 30$, where s takes on all even values from 2 to 24, inclusive.

                                        (*Source:* www.onlineconversion.com)

  a. What is the range of f?

  b. Find the inverse of f and interpret it.

80. Temperature When measuring temperature, $100°$ Celsius(C) is equivalent to $212°$ Fahrenheit (F). Also $0°$C is equivalent to $32°$F.

  a. Find a linear function which converts Celsius temperatures to Fahrenheit temperatures.

  b. Find the inverse of the function you found in part (a). What does this inverse function accomplish?

Concepts

81. The following is the graph of a function f.

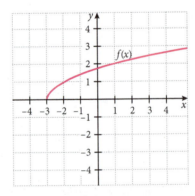

  a. Find x such that $f(x) = 1$. You will have to approximate the value of x from the graph.

  b. Let g be the inverse of f. Approximate $g(2)$ from the graph.

  c. Sketch the graph of g.

82. Two students have an argument. One says that the inverse of the function f given by the expression $f(x) = 6$ is the function g given by the expression $g(x) = \frac{1}{6}$; the other claims that f has no inverse. Who is correct and why?

83. If the graph of a function f is symmetric with respect to the y-axis, can f be one-to-one? Explain.

84. Give an example of an odd function that is not one-to-one.

85. Give an example of a function which is its own inverse.

Exponential Functions

Objectives

- Define an exponential function.
- Sketch the graph of an exponential function.
- Identify the main properties of an exponential function.
- Define the natural exponential function.
- Find an exponential function suitable to a given application.

In this section, we will study a class of functions known as **exponential functions**. They play an important role in mathematics and in various applied problems. For example, bacteria such as *E. coli* reproduce by splitting into two identical pieces. If there are no other constraints to their growth, their population over time can be modeled by a function known as an **exponential function**.

If we start with one bacterium, and the bacterium splits into two bacteria every hour, the population is *doubled* every hour. Let $P(t)$ denote the population of bacteria after t hours. Table 1 gives some values for the number of bacteria after t hours.

t (hours)	0	1	2	3	4	5	6	7	8
$P(t)$ (number of bacteria)	1	2	4	8	16	32	64	128	256

Table 1

Note the following about $P(t)$:

$$P(1) = 2(1) = 2^1; \quad P(2) = 2\,(P(1)) = 2(2) = 4 = 2^2;$$

$$P(3) = 2(P(2)) = 2(2^2) = 2^3; \quad P(4) = 2(P(3)) = 2(2^3) = 2^4; \ldots$$

Following this pattern, we find that $P(t) = 2^t$. Here, the *independent variable, t, is in the exponent,* unlike the functions examined in the previous chapters, where the independent variable was raised to a *fixed* power. The function $P(t)$ is an example of an exponential function.

Next, we give the formal definition of an exponential function, and examine its properties.

Just in Time
Review properties of exponents in Section 1.2.

Definition of an Exponential Function

We now briefly recall some properties of exponents. You already know how to calculate quantities such as 2^3 or $1.5^{1/2}$ or $3^{2/3}$. In each of these expressions, the exponent is either an integer or a rational number. If $a > 0$, any real number can be used as an exponent in an expression of the form a^x.

For example, expressions such as $3^{\sqrt{2}}$ or 2^π represent a real number. We will take these general properties of exponents for granted, since their verification is beyond the scope of this discussion. All the properties of integer and rational exponents apply to real-number exponents as well.

We now present a definition of an exponential function.

> **Definition of an Exponential Function**
>
> An **exponential function** is a function of the form
>
> $$f(x) = Ca^x$$
>
> where a and C are constants such that $a > 0$, $a \neq 1$ and $C \neq 0$. The domain of the exponential function is the set of all real numbers, $(-\infty, \infty)$. The range will vary, according to the values of C and a.
>
> The number a is known as the **base** of the exponential function.

VIDEO EXAMPLES

SECTION 4.2

Example 1   **Evaluating an Exponential Expression**

Evaluate each expression to four decimal places using a calculator

a. $0.5^{1/3}$ **b.** $2.8^{2.5}$ **c.** $4^{\sqrt{2}}$

Solution The expressions are evaluated as follows, with keystrokes given for a graphing calculator.

Expression	Calculator Keystrokes	Result
a. $0.5^{1/3}$	$0.5\ \boxed{\wedge}\ \boxed{(}\ \boxed{1}\ \boxed{\div}\ \boxed{3}\ \boxed{)}\ \boxed{\text{ENTER}}$	0.7937
b. $2.8^{2.5}$	$2.8\ \boxed{\wedge}\ 2.5\ \boxed{\text{ENTER}}$	13.1188
c. $4^{\sqrt{2}}$	$4\ \boxed{\wedge}\ \boxed{\sqrt{\ }}\ 2\ \boxed{\text{ENTER}}$	7.1030

✓ *Check It Out 1*   Evaluate 3^{π} to four decimal places using a calculator. ■

In the following two examples, we graph some exponential functions and make observations.

Example 2   **Graphing an Exponential Function**

Make a table of values of the exponential function $f(x) = 2^x$. Use the table to sketch the graph of the function. What happens to the value of the function as $x \to \pm\infty$?

Solution We first make a table of values of $f(x)$. See Table 2.
We then plot the points and connect them with a smooth curve. See Figure 2.

🖩 **Using Technology**

When graphing an exponential function, you will need to adjust the window size so that you can see how rapidly the y-value increases. In Figure 1, the graph of $f(x) = 2^x$ uses a window size of $[-5, 5]$ by $[0, 35](5)$.

Keystroke Appendix:
Section 7

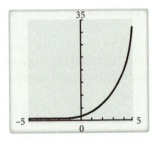

Figure 1

x	$f(x) = 2^x$
-10	$2^{-10} = \frac{1}{2^{10}} \approx 0.000977$
-5	$2^{-5} = \frac{1}{2^5} = 0.03125$
-2	$2^{-2} = 0.25$
-1	$2^{-1} = 0.5$
0	$2^0 = 1$
1	$2^1 = 2$
2	$2^2 = 4$
5	$2^5 = 32$
10	$2^{10} = 1024$

Table 2

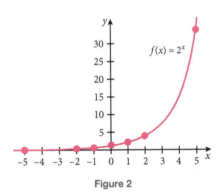

Figure 2

Observations:

- As $x \to \infty$, the value of $f(x)$ gets very large. This is an example of exponential *growth*.

- As $x \to -\infty$, the value of $f(x)$ gets extremely small but never reaches zero.

For example, $2-1000 = \frac{1}{2^{1000}}$, which is quite small but still positive.
   Note that when you use a calculator, you may sometimes get 0 instead of an extremely small value. This is because of the limited precision of the calculator. It does not mean that the actual value is zero.

- The graph of $f(x) = 2^x$ has a **horizontal asymptote** at $y = 0$. That means the graph of f gets very close to the line $y = 0$, but never touches it.

✓ *Check It Out 2*   Rework Example 2 for the function $g(x) = 3^x$. ■

We now make the following general observation about the graphs of exponential functions.

Properties of Exponential Functions

Given an **exponential function** $f(x) = Ca^x$ with $C > 0$, the function will exhibit one of the following two types of behavior, depending on the value of the base a:

If $a > 1$ and $C > 0$:

$f(x) = Ca^x \to \infty$ as $x \to \infty$

The *domain* is the set of *all real numbers*, $(-\infty, \infty)$.

The *range* is the set of *all positive real numbers*, $(0, \infty)$.

The *x*-axis is a horizontal asymptote; the graph of *f* approaches the *x*-axis as $x \to -\infty$, but does not touch or cross it.

The function is *increasing* on $(-\infty, \infty)$ and illustrates *exponential growth*. See Figure 3.

If $0 < a < 1$ and $C > 0$:

$f(x) = Ca^x \to 0$ as $x \to \infty$

The *domain* is the set of *all real numbers*, $(-\infty, \infty)$.

The *range* is the set of *all positive numbers*, $(0, \infty)$.

The *x*-axis is a horizontal asymptote; the graph of *f* approaches the *x*-axis as $x \to \infty$, but does not touch or cross it.

The function is *decreasing* on $(-\infty, \infty)$. and illustrates *exponential decay*. See Figure 4.

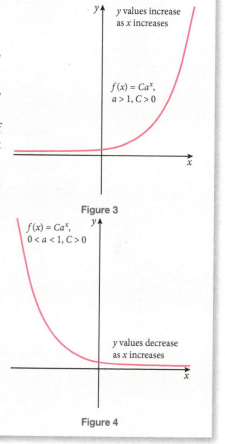

Figure 3

Figure 4

Example 3   **Graphing an Exponential Function**

Make a table of values of the exponential function $f(x) = 5\left(\frac{1}{3}\right)^x = 5(3^{-x})$. Use the table to sketch the graph of the function. What happens to the value of the function as $x \to \pm\infty$? Determine the range of the function from the graph.

Solution Use a calculator to generate a table of values for $f(x)$. See Table 3.

We then plot the points and connect them with a smooth curve. See Figure 5.

x	$f(x) = 5\left(\dfrac{1}{3}\right)^x$
-10	$5\left(\frac{1}{3}\right)^{-10} = 295245$
-5	$5\left(\frac{1}{3}\right)^{-5} = 1215$
-2	$5\left(\frac{1}{3}\right)^{-2} = 45$
-1	$5\left(\frac{1}{3}\right)^{-1} = 15$
0	$5\left(\frac{1}{3}\right)^{0} = 5$
1	$5\left(\frac{1}{3}\right)^{1} \approx 1.6667$
2	$5\left(\frac{1}{3}\right)^{2} \approx 0.5556$
5	$5\left(\frac{1}{3}\right)^{5} \approx 0.0208$
10	$5\left(\frac{1}{3}\right)^{10} \approx 0.0000847$

Table 3

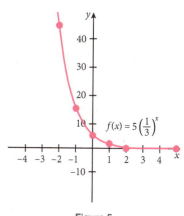

Figure 5

Observations:

- This function is an example of exponential *decay*, because the value of the function decreases as x increases.

- Note that as $x \to -\infty$, the value of $f(x)$ gets very large. For example, if $x = -1000$, then $5\left(\frac{1}{3}\right)^{-1000} = 5(3^{1000})$, which is quite large.

- As $x \to \infty$, the value of $f(x)$ gets extremely small but never reaches zero. For example, $5\left(\frac{1}{3}\right)^{1000} = \frac{5}{3^{1000}}$, which is quite small but still positive. This is an example of *exponential decay*, because the function values decrease as x increases.

- As $x \to \infty, f(x) \to 0$. Thus the *horizontal asymptote* is $y = 0$, because the graph of f approaches the line $y = 0$, but never touches it.

Range of Function

Because the graph of f is always above the line $y = 0$, but comes closer to it as the value of x increases, the range of f is the set of all positive numbers, or $(0, \infty)$.

 Check It Out 3   Rework Example 3 for the function $g(x) = 6^{-x}$.

Example 4   **Graphing an Exponential Function**

Make a table of values of the function $h(x) = -(2)^{2x}$ and sketch the graph of the function. Find the y-intercept, domain, and range. Describe the behavior of the function as x approaches $\pm\infty$.

Solution   Note that

$$h(x) = -2^{2x} = -(2^2)^x = -4^x$$

We can make a table of values of $h(x)$, as shown in Table 4. We then plot the points and connect them with a smooth curve. See Figure 6.

Be careful when calculating the values of the function. For example,

$$h(-2) = -[(4)^{-2}] = -\left(\frac{1}{4^2}\right) = -0.0625$$

The negative sign in front of the 4 is applied only after the exponentiation is performed.

x	$h(x) = -2^{2x} = -4^x$
-10	-9.536×10^{-7}
-5	-0.000977
-2	-0.0625
0	-1
1	-4
2	-16
5	-1024
10	-1048576

Table 4

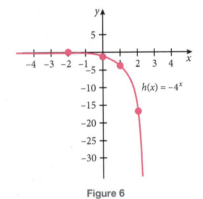

Figure 6

Observations:

- The y-intercept is $(0, -1)$.

- The domain of h is the set of all real numbers, or $(-\infty, \infty)$.

- From the sketch of the graph, we see that the *range of h is the set of all negative real numbers*, or $(-\infty, 0)$ in interval notation.

- As $x \to \infty, h(x) \to -\infty$.

- As $x \to -\infty, h(x) \to 0$. Thus, the *horizontal asymptote* is the line $y = 0$.

 Check It Out 4   Rework Example 4 for the function $h(s) = -(3)^s$.

The Number *e* and the Natural Exponential Function

There are some special numbers which occur frequently in the study of mathematics. For instance, you may be familiar with the irrational number π from geometry. Recall that an irrational number is one which cannot be written in the form of a terminating decimal or a repeating decimal.

Another irrational number that occurs frequently is the number *e*, which is defined as the number that the quantity $\left(1 + \frac{1}{x}\right)^x$ approaches as *x* approaches infinity. The non-terminating, non-repeating decimal representation of the number *e* is

$$e = 2.7182818284\ldots$$

The fact that the quantity $\left(1 + \frac{1}{x}\right)^x$ levels off as *x* increases can be seen by examining the graph of $A(x) = \left(1 + \frac{1}{x}\right)^x$ in Figure 7.

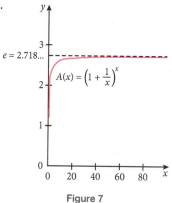

Figure 7

Calculations involving *e* are generally handled with a calculator.

> **Example 5**   **Evaluating an Exponential Expression Using *e***

Evaluate each expression to four decimal places using a calculator.

a. $e^{1/3}$ **b.** $e^{2.5}$ **c.** e^{-2}

Solution   Use the $\boxed{e^x}$ button on your calculator to evaluate the expressions. The right parentheses is already provided when you press the $\boxed{e^x}$ button. Other calculators may have different keystrokes.

Expressions	Calculator Keystrokes	Result
a. $e^{1/3}$	$\boxed{e^x}\ \boxed{1}\ \boxed{\div}\ \boxed{3}\ \boxed{)}\ \boxed{\text{ENTER}}$	1.3956
b. $e^{2.5}$	$\boxed{e^x}\ \boxed{2.5}\ \boxed{)}\ \boxed{\text{ENTER}}$	12.1825
c. e^{-2}	$\boxed{e^x}\ \boxed{(-)}\ \boxed{2}\ \boxed{)}\ \boxed{\text{ENTER}}$	0.1353

✓ *Check It Out 5*   Evaluate $e^{-1.8}$ to four decimal places using a calculator.   ■

We can define an exponential function with *e* as the base. The graph of the exponential function $f(x) = e^x$ has the same general shape as that of $f(x) = 2^x$ or $f(x) = 3^x$. See Figure 8.

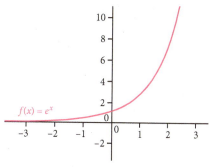

$f(x) = e^x$

Figure 8

| Example 6 | **Exponential Function with Base *e*** |

Make a table of values for the function $g(x) = e^x + 3$ and sketch the graph of the function. Find the *y*-intercept, domain, and range. Describe the behavior of the function as *x* approaches $\pm\infty$.

Solution Make a table of values for $g(x)$ by choosing various values of *x*, as shown in Table 5. Then plot the points and connect them with a smooth curve. See Figure 9.

x	$g(x) = e^x + 3$
−10	3.000
−5	3.007
−2	3.135
−1	3.368
0	4.000
1	5.718
2	10.389
5	151.413
10	22029.466

Table 5

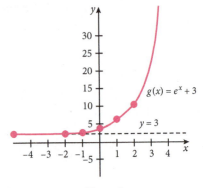

Figure 9

Observations:

- The *y*-intercept is (0, 4).

- The *domain* of g is the set of *all real numbers*, or $(-\infty, \infty)$. Because the graph of g is always above the line $y = 3$ but approaches it as the value of *x* decreases, the range of g is the set of *all real numbers strictly greater than 3*, or $(3, \infty)$.

- As $x \to \infty$, $g(x) \to \infty$.

- As $x \to -\infty$, $g(x) \to 3$. Thus, the *horizontal asymptote* is $y = 3$.

- The graph of $g(x) = e^x + 3$ is the graph of e^x shifted upward by 3 units.

 Check It Out 6   Rework Example 6 for the function $h(x) = e^{-x}$. ◼

Applications of Exponential Functions

Exponential functions are extremely useful in a variety of fields, including finance, biology, and the physical sciences. This section will illustrate the usefulness of exponential functions in analyzing some real-world problems. We will study additional applications in later sections on exponential equations and exponential and logarithmic models.

The calculation of compounded interest is an important application of exponential functions. To *compound* interest means to calculate interest on the sum of original amount and the interest earned up to a point. To calculate the total amount of money in the account after *t* years, it is useful to make a table because the process above will be repeated many times. We write the expressions in exponential form so that we can observe a pattern.

Year	Amount at Start of Year	Interest Earned During Year	Amount at End of Year
1	100	$(0.05)(100)$	$100(1 + 0.05) = 100(1.05)$
2	$100(1.05)$	$(0.05)(100(1.05))$	$100(1.05)(1 + 0.05) = \mathbf{100(1.05)^2}$
3	$100(1.05)^2$	$(0.05)(100(1.05)^2)$	$100(1.05)^2(1 + 0.05) = \mathbf{100(1.05)^3}$
4	$100(1.05)^3$	$(0.05)(100(1.05)^3)$	$100(1.05)^3(1 + 0.05) = \mathbf{100(1.05)^4}$
...	...	...	...

Table 6

The dots indicate that the table can be continued indefinitely. From the table of values, we can conclude from this pattern that the *amount of money in the account after t years* will be $100(1.05)^t$ dollars. This is another instance of **exponential growth**. Note that the amount each year was *multiplied* by a *constant* of 1.05 to obtain the amount for the start of the following year.

If we replace 0.05 with a variable r for the annual rate, and replace 100 with a variable P, for principal, we have the formula $A(t) = P(1 + r)^t$, for the amount after t years.

In the preceding discussion, interest was compounded annually. It can also be compounded in many other ways—quarterly, monthly, daily, and so on. The following is a general formula that applies to interest compounded n times a year.

Compounded Interest

Suppose an amount P is invested in an account that pays interest at rate r, and the interest is compounded n times a year. Then, after t years, the amount in the account will be

$$A(t) = P\left(1 + \frac{r}{n}\right)^{nt}$$

When interest is compounded as soon as it is earned, it is said to be compounded *continuously*, rather than being compounded at intervals such as once a month or once a year. A different formula, using e as the base, is used to calculate the total amount of money in the account in the account when interest is continuously compounded.

Continuously Compounded Interest

Suppose an amount P is invested in an account that pays interest at rate r, and the interest is compounded *continuously*. Then, after t years, the amount in the account will be

$$A(t) = Pe^{rt}$$

Example 7   Computing the Value of a Savings Account

Suppose $2500 is invested in a savings account. Find the following quantities.

a. Amount in the account after 4 years if the interest rate is 5.5% compounded monthly.

b. Amount in the account after 4 years if the interest rate is 5.5% compounded continuously.

Solution

a. Here, $P = 2500$, $r = 0.055$, $t = 4$ and $n = 12$. Substituting these values, we obtain

$$A(t) = P\left(1 + \frac{r}{n}\right)^{nt}; A(4) = \mathbf{2500}\left(1 + \frac{\mathbf{0.055}}{\mathbf{12}}\right)^{(12)(4)} \approx 3113.63$$

There will be $3113.63 in the account after four years, if the interest is compounded monthly.

b. Here, $P = 2500$, $r = 0.055$ and $t = 4$. Since the interest is compounded continuously, we have

$$A(t) = Pe^{rt}; \; A(4) = \mathbf{2500}e^{0.055(4)} \approx 3115.19$$

There will be \$3115.19 in the account after four years, if the interest is compounded continuously. Note that this amount is just slightly more than that obtained in part (a).

 Check It Out 7   Suppose \$3000 is invested in a savings account. Find the following quantities

a. Amount in the account after 3 years if the interest rate is 6.5% compounded monthly.

b. Amount in the account after 3 years if the interest rate is 6.5% compounded continuously.

Using Technology

Use a table of values to find a suitable window to graph $Y_1(x) = 20000(0.92)^x$. One possible window size is [0, 30](5) by [0, 21000] (1000). See Figure 10.

Keystroke Appendix:

Sections 6 and 7

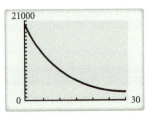

Figure 10

| **Example 8** | **Depreciation of an Automobile** |

A Honda Civic (2-Door coupe) depreciates at a rate of about 8% per year. This means that each year it will *lose* 8% of the value it had the previous year. If the Honda Civic was purchased at \$20,000, make a table of its value over the first 5 years after purchase. Find a function that gives its value t years after purchase, and sketch its graph.

(*Source:* Kelley Blue Book)

Solution Note that if the car loses 8% of its value, then it retains 92% percent of its value from the previous year. Using this fact, we can generate Table 7.

Years Since Purchase	Expression for Value	Value
0	20000	\$20,000
1	$\mathbf{0.92} \cdot 20000$	\$18,400
2	$0.92(0.92 \cdot 20000) = \mathbf{0.92^2}(20000)$	\$16,928
3	$0.92(0.92^2 \cdot 20000) = \mathbf{0.92^3}(20000)$	\$15,574
4	$0.92(0.92^3 \cdot 20000) = \mathbf{0.92^4}(20000)$	\$14,328
5	$0.92(0.92^4 \cdot 20000) = \mathbf{0.92^5}(20000)$	\$13,182

Table 7

We would like to find a function of the form $v(t) = Ca^t$, where $v(t)$ is the value of the Honda Civic t years after purchase. Note that $v(0) = C = \$20,000$. From the table, we see that $a = 0.92$, since the car's value in a certain year is multiplied by 0.92 to obtain its value in the following year. Thus

$$v(t) = Ca^t = 20000(0.92)^t$$

Because $0 < 0.92 < 1$, we can expect this function to decrease over time. This is an example of **exponential decay**, which is confirmed by sketching the graph of $v(t)$. See Figure 11.

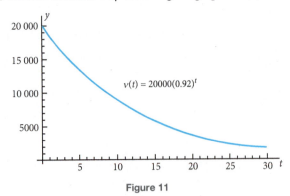

Figure 11

 Check It Out 8   If a Ford Focus was purchased for \$22,000 and depreciates at a rate of 10% per year, find an exponential function which gives its value t years after purchase, and sketch the graph of the function.

Observations:

Note the following differences between a linear model of depreciation (see Section 2.8) and an exponential model of depreciation:

- In a linear model of depreciation, a fixed dollar amount is *subtracted* each year from the previous year's value. In an exponential model, a positive constant less than 1 is *multiplied* each year by the previous year's value.

- With an exponential model of depreciation, the value *never* reaches zero; with a linear model, the value eventually *does* reach zero. In this sense, an exponential model of depreciation is more realistic than a linear model.

We next give an example where a graphing utility is used to determine a solution of an equation.

Example 9   **Finding Doubling Time for an Investment**

Use a graphing utility to find out how long it will take an investment of $2,500 to double if the interest rate is 5.5% compounded monthly.

Solution Because $r = 0.055$ and $n = 12$, the expression for the amount in the account after t years is given by

$$A(t) = P\left(1 + \frac{r}{n}\right)^{nt} = 2500\left(1 + \frac{0.055}{12}\right)^{12t}$$

We must find the value of t so that the total amount in the account will be equal to $5,000, which is twice the initial investment of $2,500. We must therefore solve the equation

$$5000 = 2500\left(1 + \frac{0.055}{12}\right)^{12t}$$

This equation cannot be solved by any of the algebraic means studied so far. However, you can solve it by using the INTERSECT feature of your graphing utility with $y_1 = 5000$ and $y_2 = 2500\left(1 + \frac{0.055}{12}\right)^{12x}$. You will need to choose your horizontal and vertical scales appropriately by using a table of values. As shown in Figure 12, a window size of [0, 15] by [2500, 5500](250) works well.

> **Note** The powerful features of your graphing utility allow you to solve equations for which you may not yet know an algebraic technique. However, you need good analytical skills to get the approximate solution. These skills include setting up a good viewing window so that you can see the point(s) of intersection of a pair of graphs. Knowledge of the behavior of the functions involved is crucial to finding a suitable window.

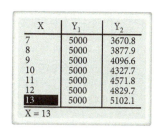
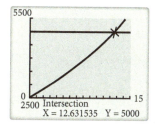

Figure 12

The solution is $t \approx 12.63$. Thus, it will take approximately 12.63 years for the initial investment of $2,500 to double, at a rate of 5.5% compounded monthly.

Check It Out 9   Use a graphing utility to find out how long it will take an investment of $3,500 to double if the interest rate is 6% compounded monthly.

Exercises 4.2

In Exercises 1–6, evaluate the expression.

1. 4^3

2. $27^{1/3}$

3. 3^{-2}

4. -3^2

5. $4(2^3)$

6. $3(-2)^3$

 Skills

In Exercises 7–16, evaluate each expression to four decimal places, using a calculator.

7. $2.1^{1/3}$

8. $3.2^{1/2}$

9. $4^{1.6}$

10. $6^{2.5}$

11. $3^{\sqrt{2}}$

12. $2^{\sqrt{3}}$

13. e^3

14. e^6

15. $e^{-2.5}$

16. $e^{-3.2}$

Paying Attention to Instructions: Exercises 17 and 18 are intended to give you practice reading, and paying attention to, the instructions that accompany the problems you are working.

17. Let $f(x) = 2^x$ and $g(x) = x^2$.

   a. Evaluate $f(3)$ and $g(3)$.

   b. Evaluate $f(-2)$ and $g(-2)$.

   c. Which function, f or g, has a horizontal asymptote?

   d. Which function, f or g, has the range $[0, \infty)$?

18. Let $f(x) = 3^{-x}$ and $g(x) = x^{-3}$.

   a. Evaluate $f(2)$ and $g(2)$.

   b. Evaluate $f(-2)$ and $g(-2)$.

   c. Which function, f or g, has the domain $(-\infty, \infty)$?

   d. Which function, f or g, can have negative function values?

In Exercises 19–36, sketch the graph of each function.

19. $f(x) = 4^x$

20. $f(x) = 5^x$

21. $g(x) = \left(\dfrac{1}{4}\right)^x$

22. $g(x) = \left(\dfrac{1}{5}\right)^x$

23. $f(x) = 2(3)^{-x}$

24. $f(x) = 4(2)^{-x}$

25. $f(x) = 2e^x$

26. $g(x) = 5e^x$

27. $f(x) = 2 + 3e^x$

28. $f(x) = 5 + 2e^x$

29. $g(x) = 10(2)^x$

30. $h(x) = -5(3)^x$

31. $f(x) = -2\left(\dfrac{1}{3}\right)^x$

32. $h(x) = 4\left(\dfrac{2}{3}\right)^x$

33. $f(x) = 3^{2x}$

34. $g(x) = 2^{3x}$

35. $f(x) = 2^{-x} - 1$

36. $f(x) = 3^{-x} + 1$

In Exercises 37–46, sketch the graph of each function and find the following.

a. the y-intercept

b. the domain and range

c. the horizontal asymptote

d. behavior of the function as x approaches $\pm\infty$

37. $f(x) = -5^x$

38. $f(x) = 2^{-x}$

39. $f(x) = 3 - 2^x$

40. $g(x) = 6 - 5^x$

41. $f(x) = 7e^x$

42. $g(x) = -4e^{2x}$

43. $g(x) = 3e^{-x} - 4$

44. $h(x) = 10e^{-x} + 2$

45. $f(x) = -4(3)^x + 1$

46. $f(x) = -2(3)^x + 1$

In Exercises 47–50, state whether each graph represents an exponential function of the form Ca^x, $a \neq 0$, $a \neq 1$, $C \neq 0$. Explain your reasoning.

47.

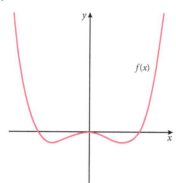

48.

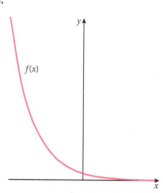

49.

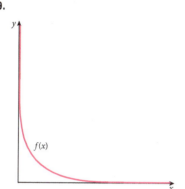

50.

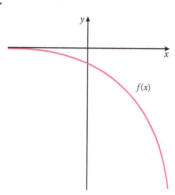

In Exercises 51–56, use a graphing utility to solve each equation for x.

51. $5 = 3^x$ **52.** $7 = 4^x$ **53.** $10 = 2^{-x}$

54. $20 = 100(5)^{-x}$ **55.** $100 = 50e^{0.06x}$ **56.** $25 = 50e^{-0.05x}$

57. Consider the function $f(x) = xe^{-x}$.

a. Use a graphing utility to graph this function, with x ranging from -5 to 5. You may need to scroll through the table of values to set an appropriate scale for the vertical axis.

b. What are the domain and range of f?

c. What are the x- and y-intercepts, if any, of the graph of this function?

d. Describe the behavior of the function as x approaches $\pm\infty$.

58. Consider the function $f(x) = e^{-x^2}$.

a. Use a graphing utility to graph this function, with x ranging from -5 to 5. You may need to scroll through the table of values to set an appropriate scale for the vertical axis.

b. What are the domain and range of f?

c. Does f have any symmetries?

d. What are the x- and y-intercepts, if any, of the graph of this function?

e. Describe the behavior of the function as x approaches $\pm\infty$.

Applications

Compound Interest For an initial deposit of $1500, find the total amount in a bank account after 5 years for the interest rates and compounding frequencies given.

59. 6% compounded annually

60. 3% compounded semiannually

61. 6% compounded monthly

62. 3% compounded quarterly

Banking For an initial deposit of $1500, with interest compounded continuously, find the total amount in a bank account after t years for the interest rates and values of t given.

63. 6% interest; $t = 3$

64. 7% interest; $t = 4$

65. 3.25% interest; $t = 5.5$

66. 4.75% interest; $t = 6.5$

In Exercises 67–70, fill in the tables according to the given rule and find an expression for the function represented by each rule.

67. Salary The annual salary of an employee of a certain company starts at $10,000 and is increased by 5% at the end of every year.

Years at Work	Annual Salary
0	
1	
2	
3	
4	

68. Population Growth A population of cockroaches starts out at 100 and doubles every month.

Month	Population
0	
1	
2	
3	
4	

69. Depreciation An automobile purchased at $20,000 depreciates at 10% per year.

Years Since Purchase	Value
0	
1	
2	
3	
4	

70. **Ecology** A rain forest with a current area of 10,000 square kilometers loses 5% of its area every year.

Years in the Future	Area of Rain Forest
0	
1	
2	
3	
4	

71. **Depreciation** The depreciation rate of a Mercury Sable is about 30% per year. If the Sable was purchased at $18,000, make a table of its value over the first 5 years after purchase. Find a function which gives its value t years after purchase, and sketch the graph of the function.

(*Source:* Kelley Blue Book)

72. **Depreciation** The depreciation rate of a Toyota Camry is about 8% per year. If the Camry was purchased at $25,000, make a table of its value over the first 4 years after purchase. Find a function which gives its value t years after purchase, and sketch the graph of the function.

(*Source:* Kelley Blue Book)

73. **Savings Bonds** U.S. Savings Bonds, Series I, pay interest at a rate of 3% compounded quarterly. How much would a bond purchased for $1000 be worth in 10 years? These bonds stop paying interest after 30 years. Why do you think this is so? (*Hint:* Think about how much this bond would be worth in 80 years.)

74. **Salary** The hourly wage for construction workers was $17.48 in 2000 and has risen at a rate of 2.7% annually.

(*Source:* Bureau of Labor Statistics)

  a. Find an expression for the hourly wage as a function of time t. Measure t in years since 2000.

  b. Using your answer to part (a), make a table of predicted values of the hourly wage for the years 2000–2007.

  c. The actual hourly wage for 2003 was $18.95. How does this compare with the predicted value?

75. **Pharmacology** When a drug is administered orally, the amount of the drug present in the bloodstream of the patient can be modeled by a function of the form

$$C(t) = ate^{-bt}$$

where $C(t)$ is the concentration of the drug (in milligrams per liter), t is the number of hours since the drug was administered, and a and b are positive constants. For a 300 mg dose of the asthma drug aminophylline, this function is

$$C(t) = 4.5te^{-0.275t}$$

(*Source:* Merck Manual of Diagnosis and Therapy)

  a. How much of this drug is present in the bloodstream at $t = 0$?

  b. How much of this drug is present in the bloodstream after 1 hour?

  c. Sketch the graph of this function, either by hand or with a graphing utility, with t ranging from 0 to 20.

  d. What happens to the value of the function as $t \to \infty$? Does this make sense in the context of this problem? Why?

  e. Use a graphing utility to find the time at which the concentration of this drug reaches its maximum.

  f. Use a graphing utility to determine when the concentration of this drug reaches 3 mg/L for the second time. (This will occur after the concentration peaks.)

76. Design The height, in feet, of the Gateway Arch in Saint Louis can be written as a function of the horizontal distance, x, in feet, from the midpoint of the base of the arch using a combination of exponential functions. The height $h(x)$ is given by

$$h(x) = -34.38\,(e^{-0.01x} + e^{0.01x}) + 693.76$$

(*Source:* National Park Service)

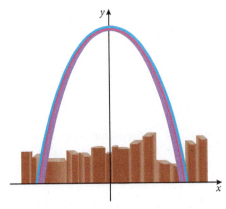

a. What is the maximum value of this function?

b. Evaluate $h(100)$.

c. Graph the function $h(x)$, using a graphing utility. Choose a suitable window size so that you can see the entire arch. For what value(s) of x is $h(x)$ equal to 300 feet?

Concepts

77. In the definition of the exponential function, why is $a = 1$ excluded?

78. Consider the function $f(x) = 2 + e^{-x}$.

a. What number does $f(x)$ approach as $x \to \infty$?

b. How could you use the graph of this function to confirm the answer to part (a)?

79. The graph of the function $f(x) = Ca^x$ passes through the points $(0, 12)$ and $(2, 3)$.

a. Use $f(0)$ to find C.

b. Is this function increasing or decreasing? Explain.

c. Now that you know C, use $f(2)$ to find a. Does your value of a confirm your answer to part (b)?

80. Consider the two functions $f(x) = 2x$ and $g(x) = 2^x$.

a. Make a table of values of $f(x)$ and $g(x)$, with x ranging from -1 to 4 in steps of 0.5.

b. Find the interval(s) where $2x < 2^x$.

c. Find the interval(s) where $2x > 2^x$.

d. Using your table from part (a) as an aid, state what happens to the value of $f(x)$ if x is increased by 1 unit.

e. Using your table from part (a) as an aid, state what happens to the value of $g(x)$ if x is increased by 1 unit.

f. Using your answers from parts (c) and (d) as an aid, explain why the value of $g(x)$ is increasing much faster than the value of $f(x)$.

81. Explain why the function $f(x) = 2^x$ has no vertical asymptotes. (Review Section 3.7.)

Logarithmic Functions 4.3

Objectives

- Define a logarithm of a positive number.

- Convert between logarithmic and exponential statements.

- Define common logarithm and natural logarithm.

- Use the change-of-base formula.

- Define a logarithmic function.

- Identify the logarithm and exponential functions as inverses.

- Sketch the graph of a logarithmic function.

- Use a logarithmic function in an application.

When you are given the output of an exponential function and asked to find the exponent, or the corresponding input, you are taking the inverse of the exponential function. This inverse function is called the **logarithmic function**.

For instance, we can ask, "For what value of x is $2^x = 32$?" The answer is $x = 5$. The exponent, 5, is called a **logarithm**, and the corresponding inverse functions is called the logarithmic function.

Definition of Logarithm

The following is the formal definition of a logarithm.

> **Definition of Logarithm**
>
> Let $a > 0$, $a \neq 1$. If $x > 0$, then the **logarithm of x with respect to base a** is denoted by $y = \log_a x$ and defined by
>
> $$y = \log_a x \quad \text{if and only if} \quad x = a^y$$
>
> The number a is known as the **base**. Thus the functions $f(x) = a^x$ and $g(x) = \log_a x$ are inverses of each other. That is,
>
> $$a^{\log_a x} = x \quad \text{and} \quad \log_a a^x = x$$

This formal definition of a logarithm does *not* tell us how to calculate the value of $\log_a x$; it simply gives a definition for such a number.

Observations

- The number denoted by $\log_a x$ is defined to be the unique exponent y that satisfies the equation $a^y = x$.

- Substituting for y, the definition of the logarithm gives

$$a^y = a^{\log_a x} = x$$

Thus, a *logarithm is an exponent*. To understand the definition of the logarithm, it is helpful to go back and forth between a logarithmic statement and its corresponding exponential statement. For instance, $\log_2 8 = 3$ is equivalent to $2^3 = 8$.

Example 1   **Equivalent Exponential Statements**

Complete the following table by filling in the exponential statements that are equivalent to the given logarithmic statements.

	Logarithmic Statement	Exponential Statement
a.	$\log_3 9 = 2$	
b.	$\log_5 \sqrt{5} = \dfrac{1}{2}$	
c.	$\log_2 \dfrac{1}{4} = -2$	
d.	$\log_a b = k, a > 0, a \neq 1$	

Solution To find the exponential statement, we use the fact that the logarithmic equation $y = \log_a x$ is equivalent to the exponential equation $a^y = x$.

	Logarithmic Statement	Question to Ask Yourself	Exponential Statement
a.	$\log_3 9 = 2$	To what power must 3 be raised to produce 9? The answer is **2**.	$3^2 = 9$
b.	$\log_5 \sqrt{5} = \dfrac{1}{2}$	Note that square root of a number is the same as raising a number to the $\frac{1}{2}$ power. To what power must 5 be raised to produce $\sqrt{5} = 5^{1/2}$? The answer is $\frac{1}{2}$.	$5^{1/2} = \sqrt{5}$
c.	$\log_2 \dfrac{1}{4} = -2$	To what power must 2 be raised to produce $\frac{1}{4}$? The answer is $-$**2**.	$2^{-2} = \dfrac{1}{4}$
d.	$\log_a b = k, a > 0, a \neq 1$	To what power must a be raised to produce b? The answer is **k**.	$a^k = b$

✓ *Check It Out 1*   Rework Example 1 for the following logarithmic statements.

	Logarithmic Statement	Exponential Statement
a.	$\log_3 27 = 3$	
b.	$\log_4 \dfrac{1}{4} = -1$	

Example 2   **Equivalent Logarithmic Statements**

Complete the following table by filling in the logarithmic statements that are equivalent to the given exponential statements.

	Exponential Statement	Logarithmic Statement
a.	$4^0 = 1$	
b.	$10^{-1} = 0.1$	
c.	$6^{1/3} = \sqrt[3]{6}$	
d.	$a^k = v, a > 0, a \neq 1$	

Solution To find the logarithmic statements, we use the fact that the logarithmic equation $y = \log_a x$ is equivalent to the exponential equation $a^y = x$.

	Exponential Statement	Question to Ask Yourself	Logarithmic Statement
a.	$4^0 = 1$	What is the logarithm of 1 with respect to base 4? The answer is **0**.	$\log_4 1 = 0$
b.	$10^{-1} = 0.1$	What is the logarithm of 0.1 with respect to base 10? The answer is $-$**1**.	$\log_{10} 0.1 = -1$
c.	$6^{1/3} = \sqrt[3]{6}$	What is the logarithm of $\sqrt[3]{6}$ with respect to base 6? The answer is $\frac{1}{3}$.	$\log_6 \sqrt[3]{6} = \dfrac{1}{3}$
d.	$a^k = v, a > 0, a \neq 1$	What is the logarithm of v with respect to base a? The answer is **k**.	$\log_a v = k$

✓ *Check It Out 2* Rework Example 2 for the following exponential statements.

	Exponential Statement	Logarithmic Statement
a.	$4^3 = 64$	
b.	$10^{1/2} = \sqrt{10}$	

You must understand the definition of a logarithm and its relationship to an exponential expression in order to evaluate logarithms without a calculator. The next example will show how to evaluate logarithms using the definition.

Example 3 **Evaluating Logarithms Without Using a Calculator**

Evaluate the following without using a calculator. If there is no solution, so state.

a. $\log_5 125$ **b.** $\log_{10} \frac{1}{100}$ **c.** $\log_a a^4, a > 0, a \neq 1$

d. $3^{\log_3 5}$ **e.** $\log_{10}(-1)$

Solution

a. Let $y = \log_5 125$. The equivalent exponential equation is $5^y = 125$. To find y note that $125 = 5^3$, so

$$5^y = 125 \Rightarrow y = 3$$

Thus, $\log_5 125 = 3$.

b. Let $y = \log_{10} \frac{1}{100}$; equivalently, $10^y = \frac{1}{100}$. Because $\frac{1}{100} = 10^{-2}$, we find that $y = -2$. Thus $\log_{10} \frac{1}{100} = -2$.

c. Let $y = \log_a a^4, a > 0, a \neq 1$; equivalently, $a^y = a^4$. Thus $y = 4$ and $\log_a a^4 = 4$.

d. To evaluate $3^{\log_3 5}$, we note from the definition of the logarithm that $a^{\log_a x} = x$. Using $a = 3$ and $x = 5$, we see that $3^{\log_3 5} = 5$. This is an illustration of the fact that the exponential and logarithmic functions are inverses of each other.

e. Let $y = \log_{10}(-1)$; equivalently, $10^y = -1$. However, 10 raised to any real number is always positive. Thus the equation $10^y = -1$ has no solution. Thus $\log_{10}(-1)$ does not exist. It is not possible to take the logarithm of a negative number.

✓ *Check It Out 3* Evaluate the following without a calculator: $\log_6 36, \log_b b^{1/3}$ $(b > 0), b \neq 1$, and $10^{\log_{10} 9}$.

Example 4 **Solving an Equation Involving a Logarithm**

Use the definition of the logarithm to find the value of x.

a. $\log_4 16 = x$ **b.** $\log_3 x = -2$

Solution

a. Using the definition of the logarithm, the equation $\log_4 16 = x$ can be written as

$$4^x = 16$$

Because $16 = 4^2$, we see that $x = 2$.

b. Using the definition of the logarithm, the equation $\log_3 x = -2$ can be written as

$$3^{-2} = x$$

Because $3^{-2} = \frac{1}{9}$, we obtain $x = \frac{1}{9}$.

✓ *Check It Out 4* Solve the equation $\log_5 x = 3$.

Common Logarithms and Natural Logarithms

Certain bases for logarithms occur so often that they have special names. The logarithm with respect to base 10 is known as the **common logarithm**, and is abbreviated as **log**, without the subscript 10.

Definition of a Common Logarithm

$$y = \log x \quad \text{if and only if } x = 10^y$$

The LOG key on your calculator evaluates the common logarithm of a number. For example, using a calculator, $\log 4 \approx 0.6021$, rounded to four decimal places.

Just as the exponential function 10^x is the inverse of the logarithmic function with respect to base 10, the exponential function e^x is the inverse of the logarithmic function with respect to base e. This base occurs so often in mathematics that the logarithm with respect to base e is called the **natural logarithm**, abbreviated as **ln**.

Definition of a Natural Logarithm

$$y = \ln x \quad \text{if and only if } x = e^y$$

The natural logarithm of a number can be found by pressing the **LN** key on your calculator. For example, using a calculator, $\ln 3 \approx 1.0986$, rounded to four decimal places.

Example 5   Evaluating Common and Natural Logarithms

Without using a calculator, evaluate the following expressions.

a. $\log 10{,}000$ **b.** $\ln e^{1/2}$ **c.** $e^{\ln a},\, a > 0$

Solution

a. To find $\log 10{,}000$, we find the power to which 10 must be raised to get 10,000. Because $10{,}000 = 10^4$, we get $\log 10{,}000 = 4$.

b. Once again, we ask the question "to what power must e be raised to get $e^{1/2}$?" The answer is $\frac{1}{2}$. Thus, $\ln e^{1/2} = \frac{1}{2}$.

c. By the definition of the natural logarithm,

$$e^{\ln a} = a, \quad a > 0$$

 Check It Out 5 Without using a calculator, find $\log 10^{2/3}$ and $\ln e^{4/3}$. ■

The inverse relationship between exponents and logarithms can be clearly seen with the help of a graphing calculator, as illustrated in the following example.

Example 6   **Evaluating a Logarithm Graphically**

Use the definition of the natural logarithm and the graph of the exponential function $f(x) = e^x$ to find an approximate value for ln 10. Use a window size of $[0, 3]$ by $[0, 12]$. Compare this solution with the answer obtained with the LN key on your calculator.

Solution   To evaluate ln 10 means that we must find an exponent, x, such that

$$e^x = 10$$

We are given the output value of 10 and asked to find the input value of x. This is exactly the process of finding an inverse. To solve for x, use the INTERSECT feature of your graphing calculator, with $Y_1(x) = e^x$ and $Y_2(x) = 10$. The solution is $x \approx 2.3026$. To reiterate, $e^{2.3026} \approx 10$, implying that ln $10 \approx 2.3026$. The LN key gives the same answer. See Figure 1.

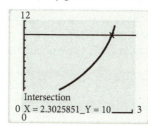
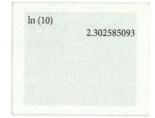

Figure 1

Check It Out 6   Use the definition of the natural logarithm and the graph of the exponential function $f(x) = e^x$ to find an approximate value for ln 8 and compare it with the answer obtained with the **ln** key on your calculator.

Change-of-Base Formula

Because your scientific or graphing calculator has special keys for only common logarithms and natural logarithms, you must use a change-of-base formula to calculate logarithms with respect to other bases.

Change-of-Base Formula

To write a logarithm with base a in terms of a logarithm with base 10 or base e, we use the formulas

$$\log_a x = \frac{\log_{10} x}{\log_{10} a}$$

$$\log_a x = \frac{\ln x}{\ln a}$$

where $x > 0$, $a > 0$, and $a \neq 1$.

It makes no difference whether you choose the change-of-base formula with the common logarithm or the natural logarithm. The two formulas will give the same result for the value of $\log_a x$.

Example 7   **Using the Change-of-Base Formula**

Use the change-of-base formula with the indicated logarithm to calculate the following:

a. $\log_6 15$, using common logarithm   **b.** $\log_7 0.3$, using natural logarithm

Solution

a. Using the change-of-base formula with the common logarithm, with $x = 15$ and $a = 6$,

$$\log_6 15 = \frac{\log_{10} 15}{\log_{10} 6} \approx \frac{1.176}{0.7782} \approx 1.511$$

b. Using the change-of-base formula with the natural logarithm, with $x = 0.3$ and $a = 7$,

$$\log_7 0.3 = \frac{\ln 0.3}{\ln 7} \approx \frac{-1.2040}{1.9459} \approx -0.6187$$

 Check It Out 7   Compute $\log_6 15$ using the change-of-base formula with the natural logarithm and show that you get the same result as in Example 7(a). ▪

Graphs of Logarithmic Functions

Consider the function $f(x) = \log x$. We will graph this function after making a table of function values. To fill in Table 1, ask yourself the question "10 raised to *what* power equals x?" First, let $x = 0$. Since 10 raised to any power does not equal zero, log 0 is undefined. Next, suppose $x = 0.001$. Since $0.001 = 10^{-3}$, $\log 0.001 = -3$. Fill in the rest of the table in a similar manner. By plotting the points in Table 1, we obtain the graph in Figure 2.

x	$f(x) = \log x$
0	Undefined
10^{-10}	-10
0.001	-3
$\dfrac{1}{10}$	-1
1	0
10	1
100	2
1.00×10^7	7

Table 1

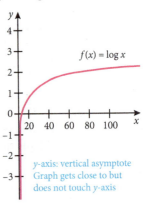

Figure 2

Observations:

- Although the x-values in the table range in magnitude from 10^{-10} to 10^7, the y-values range only from -10 to 7. Thus, the logarithmic function can take inputs of widely varying magnitude and yield output values that are much closer together in magnitude. This is reflected in the horizontal and vertical scales of the graph.

- As x approaches 0, the graph of the function approaches the line $x = 0$, or the y-axis, but does not touch it. The function is *not* defined at $x = 0$.

- Since we cannot solve the equation $10^y = x$ for y when x is negative or zero, $y = \log_{10} x$ is defined only for *positive real numbers*.

- From the graph, we see that $f(x) \to -\infty$ as $x \to 0$. Thus, the y-axis is a vertical asymptote of the graph of f.

- Is there a horizontal asymptote? It can be shown that $\log x$ increases without any upper limit, although the value of $\log x$ grows fairly slowly. Therefore, the graph has no horizontal asymptote.

- If the two columns of the table were interchanged, we would have a table of values of the function $g(x) = 10^x$, because the functions $g(x) = 10^x$ and $f(x) = \log x$ are inverses of each other.

We next summarize the properties of the logarithmic function with respect to any base $a > 0, a \neq 1$.

Properties of Logarithmic Functions

$f(x) = \log_a x, a > 1$

Domain: all *positive real numbers*; $(0, \infty)$

Range: *all real numbers*; $(-\infty, \infty)$

Vertical asymptote: $x = 0$ (the y-axis)

Increasing on $(0, \infty)$

Inverse function of $g(x) = a^x$

See Figure 3.

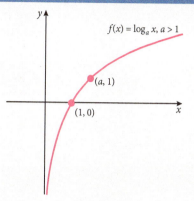

Figure 3

$f(x) = \log_a x, 0 < a < 1$

Domain: all *positive real numbers*; $(0, \infty)$

Range: *all real numbers*; $(-\infty, \infty)$

Vertical asymptote: $x = 0$ (the y-axis)

Decreasing on $(0, \infty)$

Inverse function of $g(x) = a^x$

See Figure 4.

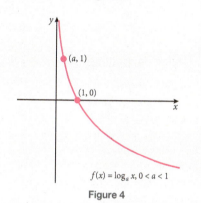

Figure 4

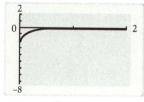

Using Technology

If you graph $y = \log x$ on a calculator, you will find that the graph will stop at a certain point near the vertical asymptote, $x = 0$. See Figure 6. This is because the calculator can plot only a finite number of points and cannot go beyond a certain limit. However, from the foregoing discussion, you know that the value of $y = \log x$ approaches $-\infty$ as x gets close to zero.

Keystroke Appendix:

Section 7

Figure 6

Logarithmic functions with bases between 0 and 1 are rarely used in practice.

Because $f(x) = \log_a x$ and $g(x) = a^x$ are inverses of each other, the graph of $f(x) = \log_a x$ can be obtained by reflecting the graph of $g(x) = a^x$ across the line $y = x$. See Figure 5.

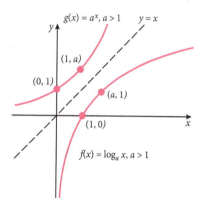

Figure 5

Example 8 **Graphing Logarithmic Functions**

Find the domain of each of the following logarithmic functions. Sketch a graph of the function and find its range. Indicate the vertical asymptote.

a. $g(t) = \ln(-t)$

b. $f(x) = 3\log_2(x - 1)$

Solution

a. The function $g(t) = \ln(-t)$ is defined only when $-t > 0$, which is equivalent to $t < 0$. Thus, its domain is the set of all negative real numbers, or $(-\infty, 0)$. We generate a table (Table 2) by choosing a suitable set of t values from the interval $(-\infty, 0)$, and then evaluate the corresponding function values. We then sketch the graph shown in Figure 7.

t	$g(t) = \ln(-t)$
0	Undefined
-0.025	-3.689
-0.5	-0.693
-1	0.000
$-e$	1.000
-5	1.609

Table 2

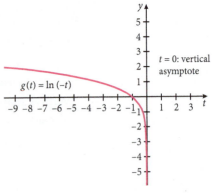

Figure 7

The range of g is the set of all real numbers, or $(-\infty, \infty)$ in interval notation. We see from the graph that as t approaches 0 from the left, the value of $\ln(-t)$ approaches $-\infty$. Therefore, the graph has a vertical asymptote at $t = 0$.

b. The function $f(x) = 3\log_2(x - 1)$ is defined only when $x - 1 > 0$, which is equivalent to $x > 1$. Thus its domain is the set of all real numbers greater than 1, or $(1, \infty)$. We can use transformations to graph this function. Note that the graph of $f(x) = 3\log_2(x - 1)$ is the same as the graph of $y = \log_2 x$ shifted to the right by 1 unit and then vertically stretched by a factor of 3. See Figure 8.

 Using Technology

In order to graph $f(x) = 3\log_2(x - 1)$ on a calculator, you must first use the change-of-base formula to rewrite the logarithm with respect to base 10. In the function editor, you must enter $Y1 = 3\,\boxed{\text{LOG}}\,\boxed{(}(X - 1\boxed{)}$ $\boxed{\div}\,\boxed{\text{LOG}}\,\boxed{(}2\boxed{)}$. See Figure 9.

Keystroke Appendix:

Sections 4, 7

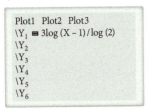

```
Plot1  Plot2  Plot3
\Y₁ ■ 3log (X – 1)/log (2)
\Y₂
\Y₃
\Y₄
\Y₅
\Y₆
```

Figure 9

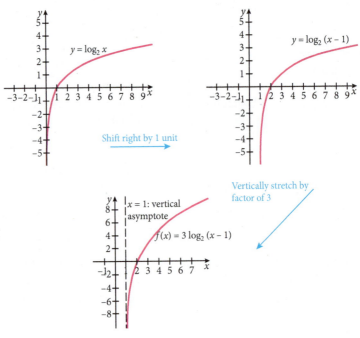

Figure 8

The range of f is the set of all real numbers, or $(-\infty, \infty)$. As x approaches 1 from the right, the value of $f(x) = 3\log_2(x - 1)$ approaches $-\infty$. Therefore, $x = 1$ is a vertical asymptote.

 Check It Out 8   Rework Example 8 for the function $g(x) = -\log x$.   ■

Applications of Logarithmic Functions

The fact that logarithmic functions grow very slowly is an attractive feature for modeling certain applications. We examine some of these in the following examples. Additional applications of logarithmic functions will be studied in the last two sections of this chapter.

Example 9   **Earthquakes and the Richter Scale**

Since the intensities of earthquakes vary widely, they are measured on a logarithmic scale known as the Richter scale, using the formula

$$R(I) = \log\left(\frac{I}{I_0}\right)$$

where I represents the actual intensity of the earthquake, and I_0 is a baseline intensity used for comparison. The Richter scale gives the *magnitude* of the earthquake.

Because of the logarithmic nature of this function, an *increase* of a *single unit* in the value of $R(I)$ represents a *tenfold increase* in the intensity of the earthquake. A recording of 7, for example, corresponds to an intensity that is 10 times as large as the intensity of an earthquake with a recording of 6.   (*Source*: U.S. Geological Survey)

a. If the intensity of an earthquake is 100 times the baseline intensity I_0, what is its magnitude on the Richter scale?

b. A 2003 earthquake in San Simeon, CA registered 6.5 on the Richter scale. Express its intensity in terms of I_0.

c. A 2004 earthquake in central Japan registered 5.4 on the Richter scale. Express its intensity in terms of I_0. What is the ratio of the intensity of the 2003 San Simeon quake to the intensity of this quake?

Solution

a. If the intensity of an earthquake is 100 times the baseline intensity I_0, then $I = 100I_0$. Substituting this expression for I in the formula for $R(I)$, we have

$$R(I) = R(\mathbf{100\,I_0}) = \log\left(\frac{\mathbf{100\,I_0}}{I_0}\right) = \log 100 = 2$$

Thus, the earthquake has a magnitude of 2 on the Richter Scale.

b. Substituting 6.5 for $R(I)$ in the formula $R(I) = \log\left(\frac{I}{I_0}\right)$ gives

$6.5 = \log\left(\frac{I}{I_0}\right)$. Rewriting this equation in exponential form, we get

$$10^{6.5} = \frac{I}{I_0}$$

from which it follows that

$$I = 10^{6.5} I_0 \approx 3{,}162{,}278 I_0$$

Therefore, the San Simeon earthquake had an intensity of nearly 3.2 million times that of the baseline intensity I_0.

c. Substituting 5.4 for $R(I)$ in the formula $R(I) = \log\left(\frac{I}{I_0}\right)$ gives

$5.4 = \log\left(\frac{I}{I_0}\right)$. Rewriting this in exponential form, we find that

$$10^{5.4} = \frac{I}{I_0}$$

so $I = 10^{5.4} I_0 \approx 251{,}189 I_0$. Therefore, the Japan earthquake had an intensity of about 250,000 times that of the baseline intensity I_0. Comparing the intensity of this earthquake with that of the San Simeon earthquake, we find that the ratio is

$$\frac{\text{Intensity of San Simeon quake}}{\text{Intensity of Japan quake}} = \frac{3{,}162{,}278 I_0}{251{,}189 I_0} \approx 12.6$$

The San Simeon quake was therefore 12.6 times as intense as the Japan quake.

✓ *Check It Out 9*   Find the intensity in terms of I_0 of a quake which measures 7.2 on the Richter scale.

Just in Time
Review scientific notation in Section 1.2.

Example 10 Distance of Planets

Table 3 lists the distance from the sun of various planets in the solar system, as well as the nearest star, Alpha Centauri. Find the common logarithm of each distance.

Planet or Star	Distance from Sun (miles)
Earth	9.350×10^7
Jupiter	4.862×10^8
Neptune	2.798×10^9
Alpha Centauri	2.543×10^{13}

Table 3

Solution We first compute the common logarithms of the distances given in the table. For example,

$$\log(9.350 \times 10^7) \approx 7.971$$

Table 4 summarizes the results.

Planet or Star	Distance from Sun (miles)	Logarithm of Distance
Earth	9.350×10^7	7.971
Jupiter	4.862×10^8	8.687
Neptune	2.798×10^9	9.447
Alpha Centauri	2.543×10^{13}	13.41

Table 4

The distances from the sun of these planets or star vary widely—the longest distance given in the table exceeds the shortest distance by several powers of 10. But the common logarithms of the distances vary only from 7.971 to 13.41.

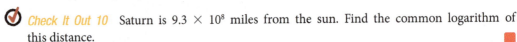 *Check It Out 10*   Saturn is 9.3×10^8 miles from the sun. Find the common logarithm of this distance.

Exercises 4.3

 ## Just in Time Exercises

In Exercises 1–4, write each number using a rational exponent.

1. $\sqrt{13}$ 2. $\sqrt[5]{4}$ 3. $\sqrt[3]{e}$ 4. $\sqrt{e^3}$

In Exercises 5–8, f and g are inverses of each other.

5. True or False: If $f(2) = 3$, then $f^{-1}(3) = 2$. Assume f is one-to-one.

6. True or False: If $g^{-1}(8) = 2$, then $g(2) = 8$. Assume g is one-to-one.

7. True or False: If $f(x) = 4$, then its inverse exists, and is $f^{-1}(x) = \frac{1}{4}$.

8. True or False: If f and g are inverses, then $g(x) = \frac{1}{f(x)}$.

In Exercises 9 and 10, write each number using scientific notation.

9. 0.036   10. 102,000,000

 ## Skills

In Exercises 11 and 12, complete the table by filling in the exponential statements that are equivalent to the given logarithmic statements.

11.

Logarithmic Statement	Exponential Statement
$\log_3 1 = 0$	
$\log 10 = 1$	
$\log_5 \frac{1}{5} = -1$	
$\log_a x = b, a > 0, a \neq 1$	

12.

Logarithmic Statement	Exponential Statement
$\log_2 4 = 2$	
$\log 100 = 2$	
$\log_7 \frac{1}{49} = -2$	
$\log a = b, a > 0$	

In Exercises 13 and 14, complete the table by filling in the logarithmic statements that are equivalent to the given exponential statements.

13.

Exponential Statement	Logarithmic Statement
$3^4 = 81$	
$5^{1/3} = \sqrt[3]{5}$	
$6^{-1} = \frac{1}{6}$	
$a^v = u, a > 0, a \neq 1$	

14.

Exponential Statement	Logarithmic Statement
$3^5 = 243$	
$7^{1/2} = \sqrt{7}$	
$6^{-2} = \frac{1}{36}$	
$10^a = b$	

In Exercises 15–34, evaluate each expression without using a calculator. Round answers to four decimal places.

15. $\log 10000$
16. $\log 0.001$
17. $\log \sqrt[3]{10}$
18. $\log \sqrt{10}$

19. $\ln e^2$
20. $\ln \sqrt{e}$
21. $\ln e^{1/3}$
22. $\ln \dfrac{1}{e}$

23. $\log 10^{x+y}$
24. $\ln e^{x-z}$
25. $\log 10^k$
26. $\ln e^w$

27. $\log_2 \sqrt{2}$
28. $\log_7 49$
29. $\log_3 \dfrac{1}{81}$
30. $\log_7 \dfrac{1}{49}$

31. $\log_{1/2} 4$
32. $\log_{1/3} 9$
33. $\log_4 4^{x^2+1}$
34. $\log_6 6^{6x}$

In Exercises 35–42, evaluate the expression to four decimal places using a calculator. Round answers to four decimal places.

35. $2\log 4$
36. $-3\log 6$
37. $\ln \sqrt{2}$
38. $\ln \pi$

39. $\log 1400$
40. $\log 2500$
41. $2\ln \dfrac{1}{5}$
42. $-\ln \dfrac{2}{3}$

In Exercises 43–50, use the change-of-base formula to evaluate the following using a calculator. Round answers to four decimal places.

43. $\log_3 1.25$
44. $\log_3 2.75$
45. $\log_5 0.5$
46. $\log_5 0.65$

47. $\log_2 12$
48. $\log_2 20$
49. $\log_7 150$
50. $\log_7 230$

In Exercises 51–56, use the definition of a logarithm to solve for x.

51. $\log_2 x = 3$
52. $\log_5 \sqrt{5} = x$
53. $\log_3 x = \dfrac{1}{3}$

54. $\log_6 x = -2$
55. $\log_x 216 = 3$
56. $\log_x 9 = \dfrac{1}{2}$

In Exercises 57–72, find the domain of each function. Use your answer to help graph the function, and label all asymptotes and intercepts.

57. $f(x) = 2\log x$
58. $f(x) = 4\ln x$
59. $f(x) = 4\log_3 x$
60. $f(x) = 3\log_5 x$

61. $g(x) = (\log x) - 3$
62. $h(x) = (\ln x) + 2$
63. $f(x) = \log_4(x + 1)$
64. $f(x) = \log_5 (x - 2)$

65. $f(x) = \ln(x + 4)$
66. $f(x) = \log(x - 3)$
67. $g(x) = 2\log_3 (x - 1)$
68. $f(x) = -\log_2(x + 3)$

69. $f(t) = \log_{1/3} t$
70. $g(s) = \log_{1/2} s$
71. $f(x) = \log|x|$
72. $g(x) = \ln(x^2)$

73. Use the following graph of $f(x) = 10^x$ to estimate $\log 7$. Explain how you obtained your answer.

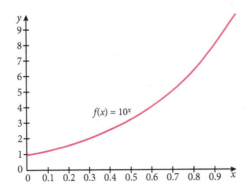

74. Use the following graph of $f(x) = e^x$ to estimate ln 10. Explain how you obtained your answer.

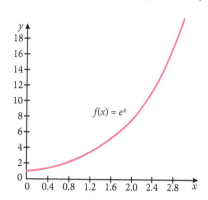

In Exercises 75–78, match the description with the correct graph. Each description applies to exactly one of the four graphs pictured.

a.

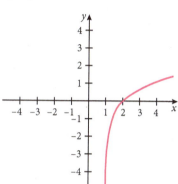

b.

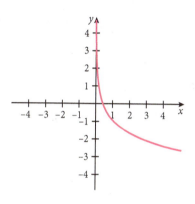

c.

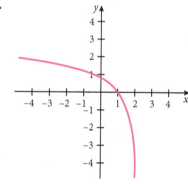

d.

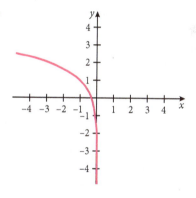

75. Graph of a logarithmic function with vertical asymptote at $x = 2$ and domain $x < 2$.

76. Graph of a logarithmic function with vertical asymptote at $x = 1$ and domain $x > 1$.

77. Graph of $f(x) = \ln(x)$ reflected across the x-axis and shifted down 1 unit.

78. Graph of $f(x) = \ln(x)$ reflected across the y-axis and shifted up 1 unit.

In Exercises 79–82, solve each equation graphically, and then express each solution as an appropriate logarithm to four decimal places. If a solution does not exist, explain why.

79. $10^t = 7$ **80.** $e^t = 6$ **81.** $4(10^x) = 20$ **82.** $e^t = -3$

Applications

Exercises 83–87 refer to the following: The magnitude of an earthquake is measured on the Richter Scale, using the formula

$$R(I) = \log\left(\frac{I}{I_0}\right)$$

where I represents the actual intensity of the earthquake, and I_0 is a baseline intensity used for comparison.

83. **Richter Scale** If the intensity of an earthquake is 10000 times the baseline intensity I_0, what is its magnitude on the Richter scale?

84. **Richter Scale** If the intensity of an earthquake is a million times the baseline intensity I_0, what is its magnitude on the Richter scale?

85. **Great Earthquakes** The great San Francisco earthquake of 1906, the most powerful earthquake in Northern California's recorded history, is estimated to have registered 7.8 on the Richter scale. Express its intensity in terms of I_0. *(Source: U.S. Geological Survey)*

86. **Great Earthquakes** In 1984, another significant earthquake in San Francisco registered 6.1 on the Richter scale. Express its intensity in terms of I_0.

87. **Earthquake Intensity** What is the ratio of the intensity of a quake that measures 7.1 on the Richter scale to the intensity of one that measures 4.2?

Exercises 88 and 89 refer to the following: The pH of a chemical solution is given by

$$pH = -\log[H^+]$$

where $[H^+]$ is the concentration of hydrogen ions in the solution, in units of moles per liter. (One mole is 6.02×10^{23} molecules.)

88. **Chemistry** Find the pH of a solution for which $[H^+] = 0.001$ moles per liter.

89. **Chemistry** Find the pH of a solution for which $[H^+] = 10^{-4}$ moles per liter.

90. **Astronomy** The brightness of a star is designated on a numerical scale called *magnitude*, which is defined by the formula

$$M(I) = -\log_{2.5}\frac{I}{I_0}$$

where I is the energy intensity of the star and I_0 is the baseline intensity used for comparison. A decrease of one unit in magnitude represents an increase in energy intensity by a factor of 2.5. *(Source: National Aeronautics and Space Agency)*

   a. If the star Spica has magnitude 1, find its intensity in terms of I_0.

   b. The star Sirius, the brightest star other than the sun, has magnitude -1.46. Find its intensity in terms of I_0. What is the ratio of the intensity of Sirius to that of Spica?

91. **Computer Science** Computer programs perform many kinds of sorting. It is preferable to use the least amount of computer time to do the sorting, where the measure of computer time is the number of operations that the computer needs to perform. We discuss two methods of sorting: the **bubble sort** and the **heap sort**. It is known that the bubble sort algorithm requires approximately n^2 operations to sort a list of n items, while the heap sort algorithm requires approximately $n\log_{10} n$ operations to sort n items.

   a. To sort 100 objects, how many operations are required by the bubble sort? by the heap sort?

   b. Make a table of the number of operations required for the bubble sort to sort a list of n items, with n ranging from 5 to 20 in steps of 5. If the number of items sorted is doubled from 10 to 20, what is the corresponding increase in the number of operations?

   c. Rework part (b) for the heap sort.

   d. Which algorithm, the bubble sort or the heap sort, is more efficient? Why?

   e. In the same window, graph the functions that give the number of operations for the bubble sort and the heap sort. Let n range from 1 to 20. Which function is growing faster and why? Note that you will have to choose the vertical scale carefully so that the $n \log n$ function does not get "squashed."

92. Ecology The pH scale measures the level of acidity of a solution on a logarithmic scale. A pH of 7.0 is considered neutral. If the pH is less than 7.0, then the solution is acidic. The lower the pH, the more acidic a solution.

Since the pH scale is logarithmic, a single unit *decrease* in pH represents a *tenfold increase* in the acidity level.

a. The average pH of rainfall in the northeastern part of the United States is 4.5. Normal rainfall has a pH of 5.5. Compared to normal rainfall, how many times more acidic is the rainfall in the northeastern U.S. on average? Explain. (*Source:* U.S. Environmental Protection Agency)

b. Because of increases in the acidity of the rain, many lakes in the northeastern U.S. have become more acidic. The degree to which acidity can be tolerated by fish in these lakes depends on the species. The Yellow Perch can easily tolerate a pH of 4.0, while the Common Shiner cannot easily tolerate pH levels below 6.0. Which species is more likely to survive in a more acidic environment and why? What is the ratio of the acidity levels that are easily tolerated by the Yellow Perch and the Common Shiner? Explain. (*Source:* U.S. Environmental Protection Agency)

Concepts

93. Explain why log 400 is between 2 and 3, without using a calculator.

94. Explain why ln 4 is between 1 and 2, without using a calculator.

In Exercises 95–98, explain how you would use the following table of values for the function $f(x) = 10^x$ *to find the given quantity.*

x	$f(x) = 10^x$
0.4771	3
0.5	$\sqrt{10}$
3	1000
−0.3010	0.5
$\sqrt{10}$	1452

95. log 1000   **96.** log 3   **97.** log 0.5   **98.** log $\sqrt{10}$

99. The graph of $f(x) = a\log x$ passes through the point $(10, 3)$. Find a and thus the complete expression for f. You can check your answer by graphing f.

100. The graph of $f(x) = A\ln x + B$ passes through the points $(1, 2)$ and $(e, 4)$.

a. Find A and B using the given points.

b. Check your answer by graphing f.

101. Use the change-of-base formula to show that $\log_{1/2} x = -\log_2 x$. Sketch the graphs of the two functions to check that this is true.

102. Find the domains of $f(x) = 2\ln x$ and $g(x) = \ln x^2$. Graph these functions in separate windows. Where are the graphs identical? Explain in terms of the domain you found for each function.

SPOTLIGHT ON SUCCESS *Student Instructor Breylor*

There are three ingredients in the good life: learning, earning and yearning.

—*Christopher Morley*

It can be hard to improve yourself in life, no matter what you are doing. To succeed and prosper, it is helpful to think about Christopher Morley's quote above. I love to learn new things and value what I earn from learning, but sometimes life can get busy and I think, "I know enough. I can slow down." The real key to improvement is perseverance and yearning for more. Training in martial arts is a passion of mine, and it continues to enthuse me to this day. However, obstacles often pop up that can distract me from my training. In the moment, I find it easy to think, "I'll just skip today." Then I ask myself, "Where would that thinking take me?" I strive for improvement, I yearn for it, and skipping a day of learning will not help me reach my goals and earn the success I seek. This thinking relates to all aspects of life, math included. Improvement only happens if someone has a desire to get better, or a yearning for the knowledge to come.

4.4

Properties of Logarithms

Objectives

- Define the various properties of logarithms.
- Combine logarithmic expressions.
- Use properties of logarithms in an application.

In the previous section, you were introduced to logarithms and discussed logarithmic functions. We continue our study of logarithms by examining some of their special properties.

Product Property of Logarithms

If you compute log 3.6 and log 36 using a calculator, you will note that the value of log 36 exceeds the value of log 3.6 by only one unit, even though 36 is ten times as large as 3.6. This curious fact is actually the result of a more general property of logarithms, which we now present.

Product Property of Logarithms

Let $x, y > 0$ and $a > 0$, $a \neq 1$. Then,

$$\log_a(xy) = \log_a x + \log_a y.$$

We derive the product property of common logarithms as follows:

$$xy = a^{\log_a(xy)} \qquad \text{Logarithmic and exponential functions are inverses}$$

Again using the inverse relationship of logarithmic and exponential functions, we have $x = a^{\log_a x}$ and $y = a^{\log_a y}$, so we can also write xy as $a^{\log_a x} \times a^{\log_a y}$. Carrying out the multiplication, we obtain

$$a^{\log_a x} \times a^{\log_a y} = a^{\log_a x + \log_a y} \qquad \text{Add exponents, since the bases are the same}$$

Now, we equate the two expressions for xy.

$$a^{\log_a x + \log_a y} = a^{\log_a(xy)} \qquad \text{Equate expressions for } xy$$

$$\log_a x + \log_a y = \log_a(xy) \qquad \text{Equate exponents, since the bases are the same}$$

VIDEO EXAMPLES

SECTION 4.4

Example 1   **Using the Product Property to Calculate Logarithms**

Given that log 2.5 ≈ 0.3979 and log 3 ≈ 0.4771, first calculate the following logarithms *without* the use of a calculator. Then check your answers with a calculator.

a. log 25   **b.** log 75

Solution

a. Because we can write 25 as 2.5 × 10, and we are given an approximate value for log 2.5, we have

$$\log 25 = \log(\mathbf{2.5 \times 10}) \qquad \text{Write 25 as a product}$$
$$= \log \mathbf{2.5} + \log \mathbf{10} \qquad \text{Use product property}$$
$$\approx 0.3979 + 1 = 1.3979 \qquad \text{Substitute and simplify}$$

b. First, write

$$75 = 7.5 \times 10 = (3 \times 2.5) \times 10$$

Using the approximate values of log 2.5 and log 3, we have

$$\log 75 = \log((\mathbf{3 \times 2.5}) \times \mathbf{10}) \qquad \text{Write 75 as a product}$$

$$= \log(\mathbf{3 \times 2.5}) + \log \mathbf{10} = \log 3 + \log 2.5 + \log 10 \qquad \text{Use product property twice}$$

$$\approx 0.4771 + 0.3979 + 1 = 1.8750 \qquad \text{Substitute data and simplify}$$

You can check the results with a calculator.

Check It Out 1   Using the approximate value of log 2.5 from Example 1, calculate $\log 2500$ without using your calculator. Compare with the answer from your calculator.   ■

Before the widespread use of calculators, the product property of logarithms was used to calculate products of large numbers. Precalculus textbooks would contain tables of logarithms in the appendix to aid in the calculation. Today, properties of logarithms are discussed mainly to help students develop an understanding of the nature of logarithms and their applications.

Power Property of Logarithms

We next present the power property of logarithms.

> **Power Property of Logarithms**
>
> Let $x > 0$, $a > 0$, $a \neq 1$, and let k be any real number. Then,
> $$\log_a x^k = k \log_a x$$

It is important to note that the power property holds only when $x > 0$. We can illustrate this using the case where $a = e$ and $k = 2$, so that the power property gives $\ln x^2 = 2 \ln x$.

Consider the functions $f(x) = \ln x^2$ and $g(x) = 2 \ln x$, which are graphed in Figure 1.

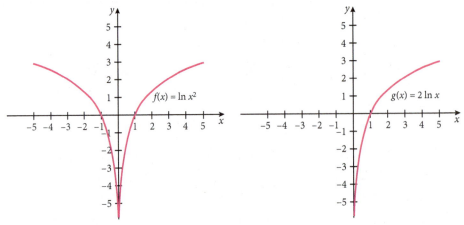

Figure 1

The domain of f is $(-\infty, 0) \cup (0, \infty)$, whereas the domain of g is only $(0, \infty)$, since $2\ln x$ is undefined if x is negative. From the graphs, we observe that these functions are equal, but only on their common domain, which is $(0, \infty)$. Thus, the power property $\ln x^2 = 2 \ln x$ holds only when $x > 0$.

Just in Time
Review rational exponents in Section 1.2.

Example 2   **Simplifying Logarithmic Expressions**

Simplify the following expressions, if possible, by eliminating exponents and radicals. Assume $x, y > 0$.

a. $\log(xy^{-3})$ **b.** $\ln(3x^{1/2} \sqrt[3]{y})$ **c.** $(\ln x)^{1/3}$

Solution

a. We use the product property, followed by the power property.

$$\log(xy^{-3}) = \log x + \log y^{-3} \qquad \text{Product property}$$
$$= \log x - 3 \log y \qquad \text{Power property}$$

b. Again we begin by applying the product property:

$$\ln(3x^{1/2}\sqrt[3]{y}) = \ln 3 + \ln x^{1/2} + \ln \sqrt[3]{y}$$

Using the fact that $\sqrt[3]{y} = y^{1/3}$ and applying the power property, we find that

$$\ln(3x^{1/2}\sqrt[3]{y}) = \ln 3 + \frac{1}{2}\ln x + \frac{1}{3}\ln y$$

c. We are asked to simplify $(\ln x)^{1/3}$. However, the power property applies *only* to logarithms of the form $\ln x^a$, and the given expression is in the form $(\ln x)^a$. Therefore, we cannot simplify $(\ln x)^{1/3}$. Applying the power property to expressions such as this is a common mistake.

 Check It Out 2 Simplify $\log(4x^{-1/3}\sqrt{y})$ by eliminating exponents and radicals. Assume $x, y > 0$.

Quotient Property of Logarithms

We derive another property of logarithms, which is known as the quotient property. Let $x, y > 0$. Then

$$\log_a \frac{x}{y} = \log_a(xy^{-1}) \qquad \text{Write } \frac{x}{y} \text{ as } xy^{-1}$$

$$= \log_a x + \log_a y^{-1} \quad \text{Product property of logarithms}$$

$$= \log_a x - \log_a y \quad \text{Power property of logarithms}$$

Thus, we see that $\log_a \frac{x}{y} = \log_a x - \log_a y$. This is known as the **quotient property of logarithms**.

Quotient Property of Logarithms

Let $x, y > 0$ and $a > 0$, $a \neq 1$. Then

$$\log_a \frac{x}{y} = \log_a x - \log_a y$$

Example 3   **Writing an Expression as a Sum or Difference of Logarithms**

Write the following as a sum and/or difference of logarithmic expressions. Eliminate exponents and radicals wherever possible.

a. $\log\left(\dfrac{x^3 y^2}{100}\right), x, y > 0$ **b.** $\log_a\left(\dfrac{\sqrt{x^2 + 2}}{(x+1)^3}\right), x > -1$

Solution

a. Use the quotient property, the product property, and the power property, in that order:

$$\log\left(\frac{x^3 y^2}{100}\right) = \log(x^3 y^2) - \log 100 \qquad \text{Quotient property}$$

$$= \log x^3 + \log y^2 - \log 100 \qquad \text{Product property}$$

$$= 3\log x + 2\log y - 2 \qquad \text{Power property; also } \log 100 = 2$$

b. Use the quotient property, followed by the power property:

$$\log_a\left(\frac{\sqrt{x^2 + 2}}{(x+1)^3}\right) = \log_a\sqrt{x^2+2} - \log_a(x+1)^3 \qquad \text{Quotient property}$$

$$= \frac{1}{2}\log_a(x^2 + 2) - 3\log_a(x+1) \qquad \text{Power property; also } \sqrt{x^2 + 2} = (x^2 + 2)^{1/2}$$

Because the logarithm of a sum *cannot* be simplified further, we cannot eliminate the exponent in the expression $(x^2 + 2)$.

 Check it Out 3   Write $\log\left(\dfrac{\sqrt{x-1}}{x^2 + 4}\right)$ as a sum and/or difference of logarithmic expressions.

Eliminate exponents and radicals wherever possible. Assume $x > 1$.

Combining Logarithmic Expressions

The properties of logarithms can also be used to combine sums and differences of logarithms into a single expression. This will be useful in the next section, where we solve exponential and logarithmic equations.

> **Example 4** **Writing an Expression as a Single Logarithm**

Write each of the following as the logarithm of a single quantity.

a. $\log_a 3 + \log_a 6, a > 0$

b. $\dfrac{1}{3} \ln 64 + \dfrac{1}{2} \ln x, x > 0$

c. $3 \log 5 - 1$

d. $\log_a x + \dfrac{1}{2} \log_a (x^2 + 1) - \log_a 3, a > 0, x > 0$

Solution

a. Using the product property, write the sum of logarithms as the logarithm of a product.

$$\log_a 3 + \log_a 6 = \log_a (3 \times 6) = \log_a 18$$

b. Use the power property first, and then the product property.

$$\frac{1}{3} \ln 64 + \frac{1}{2} \ln x = \ln 64^{1/3} + \ln x^{1/2} \qquad \text{Power property}$$
$$= \ln 4 + \ln x^{1/2} \qquad \text{Write } 64^{1/3} \text{ as } 4$$
$$= \ln(4x^{1/2}) \qquad \text{Product property}$$

c. Write 1 as log 10 (so that every term is expressed as a logarithm), and then apply the power property followed by the quotient property.

$$3 \log 5 - 1 = 3 \log 5 - \log 10 \qquad \text{Write 1 as log 10}$$
$$= \log 5^3 - \log 10 \qquad \text{Power property}$$
$$= \log \left(\frac{5^3}{10} \right) \qquad \text{Quotient property}$$
$$= \log \frac{125}{10} = \log 12.5 \qquad \text{Simplify}$$

d. Use the power property, the product property, and the quotient property, in that order.

$$\log_a x + \frac{1}{2} \log_a (x^2 + 1) - \log_a 3 = \log_a x + \log_a (x^2 + 1)^{1/2} - \log_a 3 \quad \text{Power property}$$
$$= \log_a (x(x^2 + 1)^{1/2}) - \log_a 3 \quad \text{Product property}$$
$$= \log_a \left(\frac{x(x^2+1)^{1/2}}{3} \right) \quad \text{Quotient property}$$

✔ *Check It Out 4* Write $3 \log_a x - \log_a (x^2 + 1)$ as the logarithm of a single quantity. Assume $x > 0$.

Applications of Logarithms

Logarithms occur in a variety of applications. Example 5 explores an application of logarithms that occurs frequently in chemistry and biology.

| **Example 5** | **Measuring the pH of a Solution** |

The pH of a solution is a measure of the concentration of hydrogen ions in the solution. This concentration, which is denoted by $[H^+]$, is given in units of moles per liter, where one mole is 6.02×10^{23} molecules.

Because the concentration of hydrogen ions can vary by several powers of 10 from one solution to another, the pH scale was introduced to express the concentration in more accessible terms. The pH of a solution is defined as

$$pH = -\log[H^+]$$

a. Find the pH of solution A, whose hydrogen ion concentration is 10^{-4} moles/liter.

b. Find the pH of solution B, whose hydrogen ion concentration is 4.1×10^{-8} moles/liter.

c. If a solution has a pH of 9.2, what is its concentration of hydrogen ions?

Solution

a. Using the definition of pH, we have

$$pH = -\log 10^{-4} = -(-4 \log 10) = 4 \log 10 = 4(1) = 4$$

b. Again, using the definition of pH, we have

$$pH = -\log(4.1 \times 10^{-8}) = -(\log 4.1 + \log 10^{-8})$$
$$= -(\log 4.1 - 8 \log 10) \approx -(0.613 - 8) \approx 7.387$$

Note that solution B has a higher pH, but a smaller concentration of hydrogen ions, than solution A. As the concentration of hydrogen ions *decreases* in a solution, the solution is said to become more *basic*. Likewise, if the concentration of hydrogen ions *increases* in a solution, the solution is said to become more *acidic*.

c. Here we are given the pH of a solution and must find $[H^+]$. We have

$$9.2 = -\log[H^+] \quad \text{Set pH to 9.2 in definition of pH}$$
$$-9.2 = \log[H^+] \quad \text{Isolate log expression}$$
$$10^{-9.2} = [H^+] \quad \text{Use definition of logarithm}$$

Thus, the concentration of hydrogen is $10^{-9.2} \approx 6.310 \times 10^{-10}$ moles/liter. Note how we used the definition of the logarithm to solve the logarithmic equation $-9.2 = \log[H^+]$ in a single step.

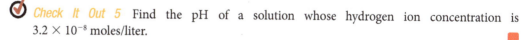

Check It Out 5 Find the pH of a solution whose hydrogen ion concentration is 3.2×10^{-8} moles/liter.

Exercises 4.4

Just in Time Exercises

In Exercises 1–4, rewrite using rational exponents. Assume all variables are non-negative.

1. $\sqrt[3]{x}$ **2.** $\sqrt[4]{z}$ **3.** $\sqrt[4]{x^3}$ **4.** $\sqrt[5]{z^3}$

Skills

In Exercises 5–12, use $\log 2 \approx 0.3010$, $\log 5 \approx 0.6990$ and $\log 7 \approx 0.8451$, to evaluate each logarithm without the use of a calculator. Then check your answer with a calculator.

5. $\log 35$ **6.** $\log 14$ **7.** $\log \dfrac{2}{5}$ **8.** $\log \dfrac{5}{7}$

9. $\log \sqrt{2}$ **10.** $\log \sqrt{5}$ **11.** $\log 125$ **12.** $\log 8$

In Exercises 13–18, use the properties of logarithms to simplify each expression by eliminating all exponents and radicals. Assume that $x, y > 0$.

13. $\log(xy^3)$ **14.** $\log(x^3 y^2)$ **15.** $\log\left(\sqrt[3]{x}\,\sqrt[4]{y}\right)$ **16.** $\log\left(\sqrt[5]{x^2}\,\sqrt{y^5}\right)$

17. $\log \dfrac{\sqrt[4]{x}}{y^{-1}}$ **18.** $\log \dfrac{\sqrt[3]{x}}{y^2}$

In Exercises 19–28, write each logarithm as a sum and/or difference of logarithmic expressions. Eliminate exponents and radicals, and evaluate logarithms wherever possible. Assume that $a, x, y, z > 0$, and $a \neq 1$.

19. $\log \dfrac{x^2 y^5}{10}$ **20.** $\log \dfrac{x^5 y^4}{1000}$ **21.** $\ln \dfrac{\sqrt[3]{x^2}}{e^2}$ **22.** $\ln \dfrac{\sqrt[4]{y^3}}{e^5}$

23. $\log_a \dfrac{\sqrt{x^2 + y}}{a^3}$ **24.** $\log_a \dfrac{\sqrt{x^3 y + 1}}{a^4}$ **25.** $\log_a \sqrt{\dfrac{x^6}{y^3 z^5}}$ **26.** $\log_a \sqrt{\dfrac{z^5}{xy^4}}$

27. $\log \sqrt[3]{\dfrac{xy^3}{z^5}}$ **28.** $\log \sqrt[3]{\dfrac{x^3 z^5}{10 y^2}}$

In Exercises 29–44, write each expression as a logarithm of a single number or expression, and then simplify if possible. Assume each of the given variable expressions is defined for appropriate values of the variable(s) contained in it. Do not use a calculator.

29. $\log 6.3 - \log 3$ **30.** $\log 4.1 + \log 3$ **31.** $\log 3 + \log x + \log \sqrt{y}$

32. $\ln y - \ln 2 + \ln \sqrt{x}$ **33.** $\ln 4 - 1$ **34.** $\log 8 + 1$

35. $3 \log x + 2$ **36.** $2 \ln y + 3$ **37.** $\dfrac{1}{3} \log_4 8x^9 - \log_4 x^2$

38. $\dfrac{1}{4} \log_3 81 y^8 + \log_3 y^3$ **39.** $\ln(x^2 - 9) - \ln(x + 3)$ **40.** $\ln(x^2 - 1) - \ln(x - 1)$

41. $\dfrac{1}{2}\left[\log(x^2 - 1) - \log(x + 1)\right] + \log x$ **42.** $\dfrac{1}{3}\left[\log(x^2 - 9) - \log(x - 3)\right] - \log x$

43. $\dfrac{3}{2} \log 16x^4 - \dfrac{1}{2} \log y^8$ **44.** $\dfrac{4}{3} \log 8x^6 - \dfrac{1}{3} \log 27 y^9$

In Exercises 45–50, let $b = \log k$. Write the following expressions in terms of b. Assume $k > 0$.

45. $\log 10k$ **46.** $\log 100k$ **47.** $\log k^3$ **48.** $\log k^4$

49. $\log \dfrac{1}{k}$ **50.** $\log \dfrac{1}{k^3}$

In Exercises 51–60, simplify each expression. Assume that each variable expression is defined for appropriate values of x. Do not use a calculator.

51. $\log 10^{\sqrt{2}}$

52. $\ln e^{\sqrt{3}}$

53. $\ln e^{x+2}$

54. $\log 10^{2x}$

55. $10^{\log(3x+1)}$

56. $e^{\ln(5x-1)}$

57. $e^{\ln(x^2+1)}$

58. $10^{\log(2x^2+3)}$

59. $\log_a \sqrt[5]{a^2}, a > 0, a \neq 1$

60. $\log_b \sqrt[3]{b}, b > 0, b \neq 1$

In Exercises 61–64, use a graphing utility with a decimal window.

61. Graph $f(x) = \log 10x$ and $g(x) = \log x$ on the same set of axes. Explain the relationship between the two graphs in terms of the properties of logarithms.

62. Graph $f(x) = \log 0.1x$ and $g(x) = \log x$ on the same set of axes. Explain the relationship between the two graphs in terms of the properties of logarithms.

63. Graph $f(x) = \ln e^2 x$ and $g(x) = \ln x$ on the same set of axes. Explain the relationship between the two graphs in terms of the properties of logarithms.

64. Graph $f(x) = \log x - \log(x - 1)$ and $g(x) = \log \frac{x}{x-1}$ on the same set of axes

   a. What are the domains of the two functions?

   b. For what values of x do these two functions agree?

   c. To what extent does this pair of functions exhibit the quotient property of logarithms?

Applications

Chemistry *Refer to the definition of pH in Example 5 to solve Exercises 65–69.*

65. Suppose solution A has a pH of 5 and solution B has a pH of 9. What is the ratio of the concentration of hydrogen ions in Solution A to the concentration of hydrogen ions in Solution B?

66. Find the pH of a solution with $[H^+] = 4 \times 10^{-5}$.

67. Find the pH of a solution with $[H^+] = 6 \times 10^{-8}$.

68. Find the hydrogen ion concentration of a solution with a pH of 7.2.

69. Find the hydrogen ion concentration of a solution with a pH of 3.4.

Noise levels *Use the following information for Exercises 70–72. The decibel (dB) is a unit that is used to express the relative loudness of two sounds. One application of this is the relative value of the output power of an amplifier with respect to the input power. Since power levels can vary greatly in magnitude, the relative value D of power level P_1 with respect to power level P_2 is given (in units of dB) in terms of the logarithm of their ratio as follows.*

$$D = 10 \log \frac{P_1}{P_2}.$$

The values of P_1 and P_2 are expressed in the same units, such as watts (W).

70. If $P_1 = 20$ W and $P_2 = 0.3$ W, find the relative value of P_1 with respect to P_2, in units of dB.

71. If an amplifier output power is 10 W and the input power is 0.5 W, what is the relative value of the output with respect to the input, in units of dB?

72. Use the properties of logarithms to show that the relative value of one power level with respect to another, expressed in units of dB, is actually a *difference* of two quantities.

Concepts

73. Consider the function $f(x) = 2^x$.
    a. Sketch the graph of f.
    b. What are the domain and range of f?
    c. Graph the inverse function.
    d. What are the domain and range of the inverse?

74. Consider the function $f(x) = x^3$.
    a. Sketch the graph of f.
    b. What are the domain and range of f?
    c. Graph the inverse function.
    d. What are the domain and range of the inverse?

75. Graph $f(x) = e^{\ln x}$ and $g(x) = x$ on the same set of axes.
    a. What are the domains of the two functions?
    b. For what values of x are these two functions equal?

76. Graph $f(x) = \ln e^x$ and $g(x) = x$ on the same set of axes.
    a. What are the domains of the two functions?
    b. For what values of x are these two functions equal?

77. Let $a > 1$. Can $(-3, 1)$ lie on the graph of $\log_a x$? Why or why not?

Exponential and Logarithmic Equations

4.5

Objective

- Solve exponential equations.
- Solve applied problems using exponential equations.
- Solve logarithmic equations.
- Solve applied problems using logarithmic equations.

Just in Time
Review one-to-one functions in Section 4.1.

VIDEO EXAMPLES

SECTION 4.5

Exponential Equations

Equations with variables in the exponents occur quite frequently and are called **exponential equations**. In this section, we will illustrate some algebraic techniques for solving these types of equations by using logarithms.

Since the exponential and logarithmic functions are inverses of each other, they are one-to-one functions. We will use the following one-to-one property to solve exponential and logarithmic equations.

One-to-one Property

For any $a > 0$, $a \neq 1$,

$$a^x = a^y \quad \text{implies} \quad x = y$$

Example 1 **Solving an Exponential Equation**

Solve the equation $2^t = 128$.

Solution Because 128 can be written as a power of 2, we have

$2^t = 128$	Original equation
$2^t = 2^7$	Write 128 as power of 2
$t = 7$	Equate exponents using the one-to-one property

The solution of the equation is $t = 7$.

 Check It Out 1 Solve the equation $2^t = 512$.

In some cases, the two sides of an exponential equation cannot be written easily in the form of exponential expressions with the same base. In such cases, we take the logarithm of both sides with respect to a suitable base, and then use the one-to-one property to solve the equation.

Example 2 **Solving an Exponential Equation**

Solve the equation $10^{2x-1} = 3^x$.

Solution We begin by taking the logarithm of both sides. Since 10 is a base of one of the expressions, we will take the logarithm with base 10 on each side.

$10^{2x-1} = 3^x$	Original equation
$\log 10^{2x-1} = \log 3^x$	Take common logarithm of both sides
$(2x - 1)\log 10 = x \log 3$	Power property of logarithms
$2x - 1 = x \log 3$	$\log 10 = 1$
$2x - x \log 3 = 1$	Collect like terms
$x(2 - \log 3) = 1$	Factor out x
$x = \dfrac{1}{2 - \log 3} \approx 0.6567$	Divide both sides by $(2 - \log 3)$

You can use your calculator to verify that this is indeed the solution of the equation.

 Check It Out 2 Solve the equation in Example 2 by taking $\log_3$ of each side in the second step and making other modifications as appropriate. If you solve it correctly, you will obtain the same answer that was found in Example 2.

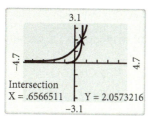

Using Technology

To solve the equation in Example 2 with a calculator, graph $Y_1(x) = 10^{2x-1}$ and $Y_2(x) = 3x$ and use the INTERSECT feature. See Figure 1.

Keystroke Appendix:
Section 9

Intersection
X = .6566511   Y = 2.0573216

Figure 1

The next example will show how to manipulate an equation before taking the logarithm of both sides of the equation.

> **Example 3**   **Solving an Exponential Equation**

Solve the equation $3e^{2t} + 6 = 24$.

Solution   To solve this equation, we first isolate the exponential term.

$$3e^{2t} + 6 = 24 \qquad \text{Original equation}$$

$$3e^{2t} = 18 \qquad \text{Subtract 6 from both sides to isolate the exponential expression}$$

$$e^{2t} = 6 \qquad \text{Divide both sides by 3}$$

Because e is the base appearing in the exponential expression on the left-hand side of this equation, we will use natural logarithms.

$$\ln e^{2t} = \ln 6 \qquad \text{Take natural logarithm of both sides}$$

$$2t \ln e = \ln 6 \qquad \text{Power property of logarithms}$$

$$2t = \ln 6 \qquad \ln e = 1$$

$$t = \frac{\ln 6}{2} \approx 0.8959 \qquad \text{Divide both sides by 2}$$

 Check It Out 3   Solve the equation $4e^{3t} - 10 = 26$.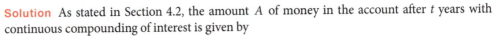

Applications of Exponential Equations

Exponential equations occur frequently in applications. We'll explore some of these applications in the examples that follow.

> **Example 4**   **Continuous Compound Interest**

Suppose a bank pays interest at a rate of 5%, compounded continuously, on an initial deposit of $1000. How long does it take for an investment of $1000 to grow to a total of $1200, assuming that no withdrawals or additional deposits are made?

Solution   As stated in Section 4.2, the amount A of money in the account after t years with continuous compounding of interest is given by

$$A = Pe^{rt}$$

where P is the initial deposit and r is the interest rate. Thus, we start by substituting the given data

$$1200 = 1000e^{0.05t} \qquad A = 1200, P = 1000, r = 0.05$$

$$1.2 = e^{0.05t} \qquad \text{Divide both sides by 1000 to isolate exponential expression}$$

$$\ln 1.2 = 0.05t \ln e \qquad \text{Take natural log of both sides; use power property}$$

$$\ln 1.2 = 0.05t \qquad \ln e = 1$$

$$t = \frac{\ln 1.2}{0.05} \approx 3.646 \qquad \text{Solve for } t$$

Thus, it will take about 3.65 years for the amount of money in the account to reach $1200.

 Check It Out 4   In Example 4, how long will it take for the amount of money in the account to reach $1400?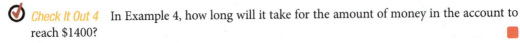

Using Technology

To solve the problem in Example 4, first create a table of values of the amount over time. The amount of $1200 will be reached somewhere between 3 and 4 years. Graph the functions $y_1(t) = 1000e^{0.05t}$ and $y_2(t) = 1200$ and find the intersection. Use the table to choose a suitable window size, such as [0, 10] by [1000, 1500](25). See Figure 2.

Keystroke Appendix:
Section 9

X	Y_1	Y_2
0	1000	1200
1	1051.3	1200
2	1105.2	1200
3	1161.8	1200
4	1221.4	1200
5	1284	1200
6	1349.9	1200

X = 0

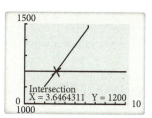

Figure 2

It is fairly easy to make algebraic errors when solving exponential equations. For an application problem involving compound interest, for example, you can use common sense: you know that the number of years cannot be negative or be very large. The total interest earned is $200 in Example 4. This is 20% of $1000, and at even 5% simple interest (not compounded), this amount can be reached in 4 years. Compounding continuously will lessen the time somewhat, and so 3.65 years is a reasonable solution. Using a table of values is another way to see if your answer is reasonable.

The following example examines a model where the exponential function decreases over time.

Example 5   **Cost of Computer Disk Storage**

Computer storage and memory are calculated using a *byte* as a unit. A kilobyte (KB) is 1000 bytes and a megabyte (MB) is 1,000,000 bytes. The costs of computer storage and memory have decreased exponentially since the 1990s. For example, the cost of computer storage over time can be modeled by the exponential function

$$C(t) = 10.7(0.48)^t$$

where t is the number of years since 1997 and $C(t)$ is the cost at time t, given in cents per megabyte.

(*Source:* www.microsoft.com)

a. How much did a megabyte of computer storage cost in 2002?

b. When did the cost of computer storage decrease to 1 cent per megabyte?

Solution Note that the cost function is of the form $y = Ka^t$, with $a = 0.48 < 1$, making it a decreasing function.

a. Because 2002 is five years after 1997, substitute 5 for t in the cost function.

$$C(5) = 10.7(0.48)^5 \approx 0.273$$

Thus, in 2002, one megabyte of storage cost approximately 0.273 of a cent.

b. To compute when the cost of one megabyte of computer storage reached 1 cent, set the cost function equal to 1 and solve for t.

$$1 = 10.7(0.48)^t \qquad \text{Set cost function equal to 1}$$

$$\frac{1}{10.7} = (0.48)^t \qquad \text{Isolate exponential expression}$$

$$\log \frac{1}{10.7} = t \log 0.48 \qquad \text{Take common logarithm of both sides}$$

$$t = \frac{\log \frac{1}{10.7}}{\log 0.48} \approx 3.23 \qquad \text{Solve for } t$$

Therefore, computer storage costs only 1 cent per megabyte approximately 3.23 years after 1997, or some time in the beginning of the year 2000.

 Check It Out 5   Using the model in Example 5, how much will a megabyte of storage cost in 2006?

The next example uses logarithms to find the parameters for a function that models bacterial population growth.

Using Technology

You can generate a table of values for the cost function in Example 5. See Figure 3. Observe that the cost is roughly halved each year, since 0.48 is close to 0.5.

Keystroke Appendix:

Section 6

X	Y$_1$
0	10.7
1	5.136
2	2.4653
3	1.1833
4	.568
5	.27264
6	.13087

X = 0

Figure 3

Example 6   Bacterial Growth

Suppose a colony of bacteria doubles its initial population of 10,000 in 10 hours. Assume the function that models this growth is given by $P(t) = P_0 e^{kt}$, where t is given in hours, and P_0 is the initial population.

a. Find the population at time $t = 0$.

b. Find the value of k.

c. What is the population at time $t = 20$?

Solution

a. The population at $t = 0$ is simply the initial population of 10,000. Thus, the function modeling this growth is $P(t) = 10,000e^{kt}$. We still have to find k, which is done in the next step.

b. To find k, we need to write an equation with k as the only variable. Using the fact that the population doubles in 10 hours, we have

$$10,000e^{k(10)} = \mathbf{20,000} \qquad \text{Using } t = 10 \text{ and } P(10) = 20{,}000$$

$$e^{10k} = 2 \qquad \text{Divide both sides by } 10{,}000$$

$$10k \ln e = \ln 2 \qquad \text{Take natural logarithm of both sides}$$

$$10k = \ln 2 \qquad \ln e = 1$$

$$k = \frac{\ln 2}{10} \approx 0.0693 \qquad \text{Solve for } k$$

c. Using the expression for $P(t)$ from part (a) and the value of k from part (b), we have

$$P(t) = 10000e^{0.0693t}$$

Evaluating the function at $t = 20$ gives

$$P(\mathbf{20}) = 10000e^{0.0693(20)} \approx 40{,}000$$

Note that this is twice 20,000, the population at $t = 10$. So, the population doubles every 10 hours.

✔ *Check It Out 6*   When will the population of the bacteria in Example 6 reach 50,000?

Equations Involving Logarithms

When an equation involves logarithms, we can use the inverse relationship between exponents and logarithms to solve it. Recall the following properties:

$$e^{\ln a} = a \quad \text{and} \quad 10^{\log a} = a, \quad a > 0$$

To solve an equation involving logarithms, we exponentiate both sides of the equation by choosing a suitable base, and then raise that base to the expression on each side of the equation. We then apply the inverse relationship between exponents and logarithms to complete the problem. Example 7 illustrates this method.

Example 7   Solving a Logarithmic Equation

Solve the equation $4 + \log_3 x = 6$.

Solution

$$4 + \log_3 x = 6 \qquad \text{Original equation}$$

$$\log_3 x = 2 \qquad \text{Isolate logarithmic expression}$$

$$3^{\log_3 x} = 3^2 \qquad \text{Exponentiate both sides (base 3)}$$

$$x = 9 \qquad \text{Inverse property: } 3^{\log_3 x} = x$$

The solution is $x = 9$. You can check this by substituting 9 for x in the original equation.

✔ *Check It Out 7*   Solve the equation $-2 + \log_2 x = 3$.

Keystroke Appendix:
Section 9

Using Technology

To solve the equation in Example 8 with a calculator, graph $Y_1(x) = \log 2x + \log(x + 4)$, $Y_2(x) = 1$ and use the INTERSECT feature. See Figure 4. There are no extraneous solutions, because we directly solve the original equation, rather than the quadratic equation obtained through algebra in Example 8.

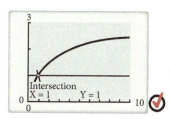

Figure 4

Example 8   **Solving an Equation Containing Two Logarithmic Expressions**

Solve the equation $\log 2x + \log(x + 4) = 1$.

Solution

$$\log 2x + \log(x + 4) = 1 \qquad \text{Original equation}$$
$$\log 2x(x + 4) = 1 \qquad \text{Combine using the product property}$$
$$10^{\log 2x(x + 4)} = 10^1 \qquad \text{Exponentiate both sides (base 10)}$$
$$2x(x + 4) = 10 \qquad \text{Inverse property: } 10^{\log a} = a$$
$$2x^2 + 8x - 10 = 0 \qquad \text{Write as a quadratic equation in standard form}$$
$$2(x^2 + 4x - 5) = 2(x + 5)(x - 1) = 0 \quad \text{Factor}$$

Setting each factor equal to 0, we find that the only possible solutions are $x = -5$ and $x = 1$. We next check each of these possible solutions by substituting them into the original equation.

Check $x = -5$:   Since $\log(2(-5)) + \log(-5 + 4) = \log(-10) + \log(-1)$, and logarithms of negative numbers are not defined, $x = -5$ is *not* a solution.

Check $x = 1$:

$$\log(2(1)) + \log(1 + 4) = \log 2 + \log 5$$
$$= \log(2 \times 5)$$
$$= \log 10 = 1$$

Thus $x = 1$ satisfies the equation and is the only solution.

Check It Out 8   Solve the equation $\log x + \log(x - 3) = 1$ and check your solution(s).

Example 9   **Solving an Equation Involving a Natural Logarithm**

Solve the equation $\ln x = 2 + \ln(x - 1)$.

Solution

$$\ln x = 2 + \ln(x - 1) \qquad \text{Original equation}$$
$$\ln x - \ln(x - 1) = 2 \qquad \text{Gather logarithmic expressions on one side}$$
$$\ln\left(\frac{x}{x - 1}\right) = 2 \qquad \text{Quotient property of logarithms}$$
$$e^{\ln(x/(x - 1))} = e^2 \qquad \text{Exponentiate both sides (base } e\text{)}$$
$$\frac{x}{x - 1} = e^2 \qquad \text{Inverse property}$$

We now solve for x.

$$x = e^2(x - 1) \qquad \text{Clear fraction: multiply both sides by } x - 1$$
$$x - e^2 x = -e^2 \qquad \text{Gather } x \text{ terms on one side}$$
$$x(1 - e^2) = -e^2 \qquad \text{Factor out } x$$
$$x = -\frac{e^2}{1 - e^2} \approx 1.1565 \quad \text{Solve for } x$$

You can check this answer by substituting it into the *original* equation.

Check It Out 9   Solve the equation $\ln x = 1 + \ln(x - 2)$ and check your solution.

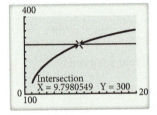
Application of Logarithmic Equations

Logarithmic functions can be used to model phenomena where the growth is rapid at first and then slows down. For instance, the total revenue from ticket sales for a movie grows rapidly at first, and then continues to grow, but at a slower rate. This is illustrated in Example 10.

Example 10   **Box Office Revenues**

The cumulative box office revenue from the movie *Finding Nemo* can be modeled by the logarithmic function

$$R(x) = 78.05 \ln(x + 1) + 114.3$$

where x is the number of weeks since the movie opened and $R(x)$ is given in millions of dollars. How many weeks after the opening of the movie was the cumulative revenue equal to $300 million?   (*Source:* movies.yahoo.com)

Solution   We set $R(x)$ to 300 and solve the resulting logarithmic equation.

$300 = 78.05 \ln(x + 1) + 114.3$	Original equation
$185.7 = 78.05 \ln(x + 1)$	Subtract 114.3 from both sides
$2.379 \approx \ln(x + 1)$	Divide both sides by 78.05 and round the result
$e^{2.379} \approx e^{\ln(x + 1)}$	Exponentiate both sides
$e^{2.379} \approx x + 1$	Inverse property
$x \approx e^{2.379} - 1 \approx 9.794$	Solve for x

Thus, by around 9.794 weeks after the opening of the movie, $300 million in total revenue had been generated.

✓ *Check It Out 10*   In Example 10, when did the cumulative revenue reach $200 million?   ■

Exercises 4.5

Just in Time Exercises

1. True or False: $f(x) = 2 \ln x$ is a one-to-one function.

2. True or False: The equation $x^3 + 2 = 4$ has two solutions.

3. True or False: Let f be a function. If the equation $f(x) = 4$ has two distinct solutions for x, then f is not one-to-one.

4. True or False: Since $f(x) = 10^x$ is a one-to-one function, the equation $10^x = 5$ has only one solution.

Skills

In Exercises 5–30, solve the exponential equation. Round to three decimal places when needed.

5. $5^x = 125$ 6. $7^{2x} = 49$ 7. $10^x = 10,000$ 8. $10^x = 0.0001$

9. $4^x = \dfrac{1}{16}$ 10. $6^x = \dfrac{1}{216}$ 11. $4e^x = 36$ 12. $5e^x = 60$

13. $3(1.3^x) = 5$ 14. $6(0.9^x) = 7$ 15. $10^x = 2^{-x+4}$ 16. $3^{-x} = 10^{-4x+1}$

17. $3^{-2x-1} = 2^x$ 18. $5^{x+5} = 3^{-2x+1}$ 19. $1000e^{0.04x} = 2000$ 20. $250e^{0.05x} = 400$

21. $5e^x + 7 = 32$ 22. $4e^x + 6 = 22$ 23. $2(0.8^x) - 3 = 8$ 24. $4(1.2^x) - 4 = 9$

25. $e^{x^2+1} - 2 = 3$ 26. $10^{2x^2+1} - 8 = 4$

 27. $1.7e^{0.5x} = 3.26$

 28. $4e^x = -x + 3$

 29. $xe^{-x} + e^x = 2$

 30. $e^x + e^{-x} = -x + 4$

In Exercises 31–56, solve the logarithmic equation and eliminate any extraneous solutions. If there are no solutions, so state.

31. $\log x = 0$ 32. $\ln x = 1$

33. $\ln(x - 1) = 2$ 34. $\ln(x + 1) = 3$

35. $\log(x + 3) = 1$ 36. $\log(x - 1) = 2$

37. $\log_3(x + 4) = 2$ 38. $\log_5(x + 3) = 1$

39. $\log(x + 1) + \log(x - 1) = 0$ 40. $\log(x + 3) + \log(x - 3) = 0$

41. $\log x + \log(x + 3) = 1$ 42. $\log x + \log(2x - 1) = 1$

43. $\log_2 x = 2 - \log_2(x - 3)$ 44. $\log_5 x = 1 - \log_5(x - 4)$

45. $\ln(2x) = 1 + \ln(x + 3)$ 46. $\log_3 x = 2 + \log_3(x - 2)$

47. $\log(3x + 1) - \log(x^2 + 1) = 0$ 48. $\log(x + 5) - \log(4x^2 + 5) = 0$

49. $\log(2x + 5) + \log(x + 1) = 1$ 50. $\log(3x + 1) + \log(x + 1) = 1$

51. $\log_2(x + 5) = \log_2(x) + \log_2(x - 3)$ 52. $\ln 2x - \ln(x^2 + 1) = \ln 1$

 53. $2 \ln x + \ln(x - 1) = 3.1$

 54. $-\ln x - \ln(x + 2) = 2.5$

55. $\log|x - 2| + \log|x| = 1.2$

56. $\ln x = (x - 2)^2$

Applications

Banking *In Exercises 57–62, determine how long it takes to double the given investment if r is the interest rate and the interest is compounded continuously. Assume that no withdrawals or further deposits are made.*

57. Initial Amount: $1,500; $r = 6\%$

58. Initial Amount: $3,000; $r = 4\%$

59. Initial Amount: $4,000; $r = 5.75\%$

60. Initial Amount: $6,000; $r = 6.25\%$

61. Initial Amount: $2,700; $r = 7.5\%$

62. Initial Amount: $3,800; $r = 5.8\%$

Banking *An initial deposit is made in a bank account. In Exercises 63–68, find the interest rate r if the interest is compounded continuously and no withdrawals or further deposits are made.*

63. Initial Amount: $1,500; Amount in 5 years: $2,000

64. Initial Amount: $3,000; Amount in 3 years: $3,600

65. Initial Amount: $4,000; Amount in 8 years: $6,000

66. Initial Amount: $6,000; Amount in 10 years: $12,000

67. Initial Amount: $8,500; Amount in 5 years: $10,000

68. Initial Amount: $12,000; Amount in 20 years: $25,000

69. Bacterial Growth Suppose a colony of bacteria doubles in 12 hours from an initial population of 1 million. Find the growth constant k if the population is modeled by the function $P_0 e^{kt}$. When will the population reach 4 million? 8 million?

70. Bacterial Growth Suppose a colony of bacteria doubles in 20 hours from an initial population of 1 million. Find the growth constant k if the population is modeled by the function $P_0 e^{kt}$. When will the population reach 4 million? 8 million?

71. Computer Science In 1965, Gordon Moore, then director of Intel research, conjectured that the number of transistors that fit on a computer chip doubles every few years. This came to be known as *Moore's Law*. Analysis of data from Intel Corporation yields the following model of the number of transistors per chip over time:

$$s(t) = 2297.1 e^{0.3316t}$$

where $s(t)$ is the number of transistors per chip, and t is the number of years since 1971.

(Source: Intel Corporation)

a. What is the number of transistors per chip in 1971 according to this model?

b. How long does it take to double the number of transistors?

72. Depreciation The value of a 2003 Toyota Corolla is given by the function

$$v(t) = 14000(0.93)^t$$

where t is the number of years since its purchase and $v(t)$ is its value in dollars. *(Source: Kelley Blue Book)*

a. What was its initial purchase price?

b. What percentage of its value does the Toyota Corolla lose each year?

c. How long will it take for the value of the Toyota Corolla to reach $12,000?

73. Film Industry The cumulative box office revenue from the movie *Terminator 3* can be modeled by the logarithmic function

$$R(x) = 26.203 \ln x + 90.798$$

where x is the number of weeks since the movie opened and $R(x)$ is given in millions of dollars. How many weeks after the opening of the movie did the cumulative revenue reach $140 million?

(Source: movies.yahoo.com)

74. Physics Plutonium is a radioactive element which has a *half-life* of 24,360 years. The half-life of a radioactive substance is the time it takes for *half* of the substance to decay. Find an exponential function of the form Ae^{kt} that gives the amount of plutonium left after t years if the initial amount of plutonium is 10 pounds. How long will it take for the plutonium to decay to 2 pounds?

Chemistry *Exercises 75 and 76 refer to the following. The pH of a solution is defined as*

$$pH = -\log[H^+]$$

The concentration of hydrogen ions, $[H^+]$, is given in moles per liter, where one mole is 6.02×10^{23} molecules.

75. What is the concentration of hydrogen ions in a solution that has a pH of 6.2?

76. What is the concentration of hydrogen ions in a solution that has a pH of 1.5?

77. Acoustics The decibel (dB) is a a unit that is used to express the relative loudness of two sounds. One application of this is the relative value of the output power of an amplifier with respect to the input power. Because power levels can vary greatly in magnitude, the *relative value D* of power level P_1 with respect to power level P_2 is given (in units of dB) in terms of the logarithm of their ratio as follows:

$$D = 10 \log \frac{P_1}{P_2}$$

where the values of P_1 and P_2 are expressed in the same units, such as watts (W). If $P_2 = 75$ W, find the value of P_1 at which $D = 0.7$.

78. Depreciation A new car that costs $25,000 depreciates to 80% of its value in 3 years.

   a. Assume that the depreciation is linear. What is the linear function that models the value of this car t years after purchase?

   b. Assume that the value of the car is given by an exponential function Ae^{kt}, where A is the initial price of the car. Find the value of the constant k.

   c. For the linear model used in part (a), find the value of the car 5 years after purchase, then do the same for the exponential model used in part (b).

   d. Graph both models using a graphing utility. Which model do you think is more realistic and why?

Investment *In Exercises 79–82, use the following table, which illustrates the growth over time of an amount of money deposited in a bank account.*

t (years)	Amount ($)
0	5000
1	5309.18
2	5637.48
4	6356.25
6	7166.65
10	9110.59
12	10,272.17

79. What is the amount of the initial deposit?

80. Approximately how long does it take to earn a total of $600 in interest?

81. Approximately how long does it take for the amount of money in the account to double?

82. Assume that the amount of money in the account at time t (in years) is given by $P_0 e^{rt}$, where P_0 is the initial deposit and r is the interest rate. Find the value of r.

Concepts

83. Do $\ln x^2 = 1$ and $2 \ln x = 1$ have the same solution? Explain.

84. Explain why the equation $2e^x = -1$ has no solution.

85. What is wrong with the following step?

$$\log x + \log(x + 1) = 0 \quad \Rightarrow \quad (x(x + 1)) = 0$$

86. What is wrong with the following step?

$$2^{x+5} = 3^{4x} \quad \Rightarrow \quad x + 5 = 4x$$

In Exercises 87–90, solve using any method, and eliminate extraneous solutions.

87. $\ln(\log x) = 1$

88. $e^{\log x} = e$

89. $\log_5 |x - 2| = 2$

90. $\ln|2x - 3| = 1$

Exponential, Logistic, and Logarithmic Models

4.6

Objectives

- Construct an exponential decay model.

- Use curve-fitting for an exponential model.

- Use curve-fitting for an logarithmic model.

- Define a logistic model.

- Use curve-fitting for a logistic model.

Just in Time
Review properties of exponential functions in Section 4.2.

In the previous sections, you examined applications of exponential and logarithmic functions where you were given the expression for the function. In this section, we will study how an appropriate model can be selected when we are given some *data* about a certain problem. This is how real-world problems often arise.

In simple cases, it is possible to come up with an appropriate model through paper-and-pencil work. For more complicated data sets, we will need to use technology to find a suitable model. We will examine both types of problems in this section. In addition, we will investigate other functions, closely related to the exponential function, that are useful in many applications.

Exponential Growth and Decay

Recall the following facts from Section 4.2 about exponential functions.

Properties of Exponential Functions

- An exponential function of the form $f(x) = Ca^x$, where $C > 0$ and $a > 1$ models **exponential growth**. See Figure 1.

- An exponential function of the form $f(x) = Ca^x$, where $C > 0$ and $0 < a < 1$ models **exponential decay**. See Figure 2.

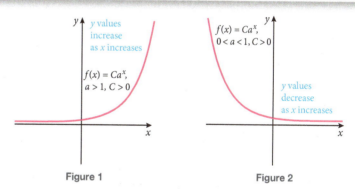

Figure 1   Figure 2

When modeling exponential growth and decay without the use of technology, it is more convenient to state the functions modeling exponential growth and decay using the base e. If $a > 1$, we can write $a = e^k$, where k is some constant such that $k > 0$. If $0 < a < 1$, we can write $a = e^k$, where k is some constant such that $k < 0$. We then have the following.

- An exponential function of the form $f(x) = Ce^{kx}$, where $C > 0$ and $k > 0$ models **exponential growth**.

- An exponential function of the form $f(x) = Ce^{kx}$, where $C > 0$ and $k < 0$ models **exponential decay**.

The exponential function is well suited to many models in the social, life, and physical sciences. The following example discusses the decay of the radioactive metal strontium-90. This metal has a variety of commercial and research uses. For example, it is used in fireworks displays to produce the red flame color. (*Source*: Argonne National Laboratories)

VIDEO EXAMPLES

SECTION 4.6

Example 1   **Modeling Radioactive Decay**

It takes 29 years for an initial amount A_0 of strontium-90 to break down into half the initial amount, $\frac{A_0}{2}$. That is, the *half-life* of strontium-90 is 29 years.

a. Given an initial amount of A_0 grams of strontium-90, at $t = 0$, find an exponential decay model, $A(t) = A_0 e^{kt}$ that gives the amount of strontium-90 at time t, $t \geq 0$.

b. Calculate the time required for strontium-90 to decay to $\frac{1}{10} A_0$.

Solution

a. The exponential model is given by $A(t) = A_0 e^{kt}$. After 29 years, the amount of strontium-90 is $\frac{1}{2} A_0$. Putting this information together, we have

$$A(t) = A_0 e^{kt} \qquad \text{Given model}$$

$$A(29) = \frac{1}{2} A_0 = A_0 e^{k(29)} \qquad \text{At } t = 29, A(29) = \frac{1}{2} A_0$$

$$\frac{1}{2} = e^{k(29)} \qquad \text{Divide both sides of equation by } A_0$$

$$\ln\left(\frac{1}{2}\right) = 29k \qquad \text{Take natural logarithm of both sides}$$

$$\frac{\ln \frac{1}{2}}{29} = k \qquad \text{Solve for } k$$

$$k \approx -0.0239 \qquad \text{Approximate } k \text{ to four decimal places}$$

Because this is a decay model, we know that k is negative. Thus, the amount of strontium-90 at time t is given by

$$A(t) = A_0 e^{-0.0239t}$$

b. We must calculate t such that $A(t) = \frac{1}{10} A_0$. We proceed as follows.

$$A(t) = A_0 e^{-0.0239t} \qquad \text{Given model}$$

$$\frac{1}{10} A_0 = A_0 e^{-0.0239t} \qquad \text{Substitute } A(t) = \frac{1}{10} A_0$$

$$\frac{1}{10} = e^{-0.0239t} \qquad \text{Divide both sides of the equation by } A_0$$

$$\ln\left(\frac{1}{10}\right) = -0.0239t \qquad \text{Take natural logarithm of both sides}$$

$$\frac{\ln \frac{1}{10}}{-0.02390} = t \qquad \text{Solve for } t$$

$$t \approx 96.34 \qquad \text{Approximate } t \text{ to 2 decimal places}$$

Thus, in approximately 96.34 years, one-tenth of the original amount of strontium-90 will remain.

✓ *Check It Out 1*   Radium-228 is a radioactive metal with a half-life of 6 years. Find an exponential decay model, $A(t) = A_0 e^{kt}$ that gives the amount of radium-228 at time t, $t \geq 0$.

The next example discusses population growth assuming an exponential model.

Example 2   **Modeling Population Growth**

The population of the United States is expected to grow from 282 million in 2000 to 335 million in 2020.   *(Source: U.S. Census Bureau)*

a. Find a function of the form $P(t) = Ce^{kt}$ that which models the population growth. Here, t is the number of years after 2000, and $P(t)$ is in millions.

b. Use your model to predict the population of the United States in 2016.

Solution

a. If t is the number of years after 2000, we have the following two data points:

$$(0, 282) \text{ and } (20, 335)$$

First, we find C:

$P(t) = Ce^{kt}$	Given equation
$P(0) = Ce^{k(0)} = 282$	Population at $t = 0$ is 282 million
$C = 282$	Because $Ce^{k(0)} = C(1) = C$

Thus, the model is $P(t) = 282e^{kt}$. Next, we find k.

$P(t) = 282e^{kt}$	Given equation
$P(20) = 282e^{k(20)} = 335$	Population at $t = 20$ is 335 million
$282e^{20k} = 335$	
$e^{20k} = \dfrac{335}{282}$	Isolate the exponential term.
$20k = \ln\dfrac{335}{282}$	Take natural logarithm of both sides
$k = \dfrac{\ln\frac{335}{282}}{20} \approx 0.00861$	Solve for k, and approximate to five decimal places

Thus the function is

$$P(t) = 282e^{0.00861t}$$

b. We substitute $t = 16$, because 2016 is 16 years after 2000.

$$P(16) = 282e^{0.00861(16)} \approx 324$$

Thus, in 2016, the population of the United States is predicted to be 324 million, according to this model.

 Check It Out 2   Use the function in Example 2 to estimate the population of the United States in 2015.

 **Using Technology**

Curve-fitting features are available under the Statistics option on most graphing calculators.

Keystroke Appendix:

Section 12

Years Since 1975	National Debt (in billions)
0	576.6
5	930.2
10	1946
15	3233
20	4974
25	5674
30	7933

Table 1

Models Using Curve Fitting

This section will discuss real-world problems that can be analyzed using only the curve-fitting, or regression, capabilities of your graphing utility. We will discuss examples of exponential, logistic, and logarithmic models.

 Example 3   **Growth of the National Debt**

Table 1 shows the United States national debt (in billions of dollars) for selected years from 1975–2005.   *(Source: U.S. Department of the Treasury)*

a. Make a scatter plot of the data and find an exponential function of the form $f(x) = Ca^x$ that best fits this data.

b. From the model in part (a) what was the estimated national debt in the year 2010? Compare your answer with the actual value of $13,000 billion dollars.

Solution

a. Figure 3 shows a scatter plot of the data, along with the exponential curve.

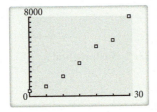 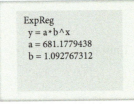 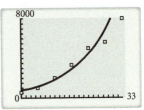

ExpReg
y = a*b^x
a = 681.1779438
b = 1.092767312

Figure 3

The exponential function that best fits this data is given by

$$d(x) = 681.2(1.093)^x$$

b. To find the estimated debt in 2010, we calculate $d(35)$:

$$d(x) = 681.2(1.093)^{35} \approx 15{,}310$$

Thus, the estimated debt in 2010 was approximately $15,310 *billion* dollars, or $15.31 *trillion* dollars. The actual debt in 2010 was approximately $13,000 billion dollars, and so our model overestimated the amount.

 Check It Out 3   Use the model in Example 3 to estimate the national debt in the year 2012.

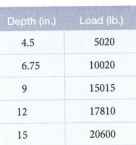

Depth (in.)	Load (lb.)
4.5	5020
6.75	10020
9	15015
12	17810
15	20600

Table 2

Example 4   **Modeling Loads in a Structure**

When designing buildings, engineers must pay careful attention to how different factors affect the load a structure can carry. Table 2 gives the load, in pounds of concrete, when a 1-inch-diameter anchor is used as a joint. The table summarizes the relation between the load and how deep the anchor is drilled into the concrete.   *(Source: Simpson Anchor Systems)*

a. From examining the table, what is the general relationship between the depth of the anchor and the load?

b. Make a scatterplot of the data and find a natural logarithmic function that best fits the data.

c. If an anchor were drilled 10 inches deep, what is the resulting load that can be carried?

d. What is the minimum depth an anchor should be drilled in order to sustain a load of 9000 pounds?

Solution

a. From examining the table, we see that deeper the anchor is drilled, the more load that can be sustained. However, the sustainable load increases rapidly at first and then increases slowly. Thus a logarithmic model seems appropriate.

b. Figure 4 shows a scatterplot of the data, along with the logarithmic curve.

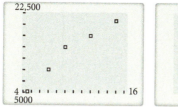 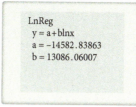 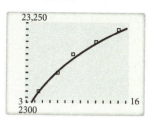

Figure 4

The logarithmic function that best fits this data is given by

$$L(x) = 13,086 \ln(x) - 14,583$$

where x is the depth the anchor is drilled.

c. To find the sustainable load when an anchor is drilled 10 inches deep, we evaluate $L(10)$.

$$L(10) = 13,086 \ln(10) - 14,583 \approx 15,548$$

The resulting load is approximately 15,548 pounds. As a check, we see that this value is reasonable by comparing it with the data in the table.

d. To find the minimum depth, we set the expression for the load to 9,000 and solve for x:

$$13,086 \ln(x) - 14,583 = 9,000 \qquad \text{Set load expression equal to 9,000}$$

$$13,086 \ln x = 23,583 \qquad \text{Add 14,583 to both sides}$$

$$\ln x = \frac{23,583}{13,086} \approx 1.8021 \qquad \text{Divide by 13,086}$$

$$x = e^{1.8021} \approx 6.0623 \qquad \text{Exponentiate both sides}$$

Thus, the anchor must be drilled at least to a depth of 6.06 inches to sustain a load of 9,000 pounds. Drilling to a greater depth simply means the anchor will sustain more than 9,000 pounds.

We could have also found the solution with a graphing utility by storing the logarithmic function as y_1 and finding its intersection with $y_2 = 9,000$.

 Check It Out 4   Find the solution to part (d) in the above example using a graphing utility.

The Logistic Model

We examined population growth models by using an exponential function in the previous section. However, it seems unrealistic that any population would simply tend to infinity over a long period of time. Other factors, such as the ability of the environment to support the population, would eventually come into play and level off the population. Thus, we need a more refined model of population growth that takes such issues into account.

One function that models this behavior discussed is known as a **logistic function**. It is defined as

$$f(x) = \frac{c}{1 + ae^{-bx}}$$

where a, b and c are constants determined from a given set of data. Finding these constants involves using the *logistic regression* feature of your graphing utility. The graph of the logistic function is shown in Figure 5.

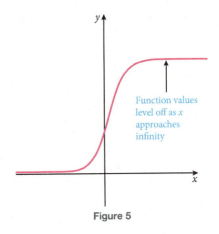

Figure 5

Let us examine some properties of this function by letting $f(x) = \frac{3}{1 + 2e^{-0.5x}}$.
Table 3 lists some values for $f(x)$.

x	-10	-5	-1	0	1	2	5	10	25
$f(x) = \dfrac{3}{1 + 2e^{-0.5x}}$	0.0101	0.1183	0.6981	1.0000	1.3556	1.7284	2.5769	2.9601	2.9999

Table 3

Observations:

- As x increases, the function values approach 3. The term $2e^{-0.5x}$ becomes very small in magnitude as x increases to ∞, making the denominator of $f(x)$, $1 + 2e^{-0.5x}$, very close to 1. Because $f(x) = \frac{3}{1 + 2e^{-0.5x}}$, the values of $f(x)$ will approach 3 as x increases to ∞.

- As x decreases to $-\infty$, the function values approach 0. The term $2e^{-0.5x}$ increases as x decreases to $-\infty$, making the denominator of $f(x)$, $1 + 2e^{-0.5x}$, very large in magnitude. Since $f(x) = \frac{3}{1 + 2e^{-0.5x}}$, the values of $f(x)$ will approach 0 as x decreases to $-\infty$.

In Example 5, a logistic function is used to analyze a set of data.

 Example 5   **Logistic Population Growth**

Table 4 gives the population of South America for selected years from 1970 to 2000.

(*Source:* U.S. Census Bureau)

Year	Population (millions)
1970	191
1980	242
1990	296
2000	347

Table 4

a. Use a graphing utility to make a scatter plot of the data and find the logistic function of the form $f(x) = \frac{c}{1 + ae^{-bx}}$ that best fits the data from 1970 to 2000. Let x be the number of years from 1970.

b. From this model, what is the projected population in 2020? How does it compare with the projection of 421 million given by the U.S. Census Bureau?

Solution

a. Figure 6 shows a scatter plot of the data along with the logistic curve.

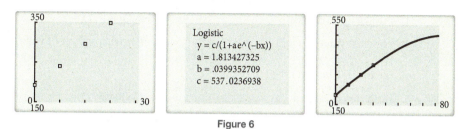

Logistic
y = c/(1+ae^(−bx))
a = 1.813427325
b = .0399352709
c = 537.0236938

Figure 6

The logistic function that best fits this data is given by

$$p(x) = \frac{537}{1 + 1.813e^{-0.04x}}$$

b. To find the projected population in 2020, we calculate $p(50)$:

$$p(50) = \frac{537}{1 + 1.813e^{-0.04(50)}} \approx 431$$

This model predicts that there will be approximately 431 million people in South America in 2020. The statisticians who study these types of data use a more sophisticated analysis, of which curve-fitting is only a part. Their projection of 421 million is relatively close to that predicted by our model.

✔ *Check It Out 5* From this model, what is the projected population in 2040? How does it compare with the projection of 468 million from the U.S. Census Bureau?

Exercises 4.6

Skills

In Exercises 1–4, use $f(t) = 10e^{-t}$.

1. Evaluate $f(0)$.

2. Evaluate $f(2)$.

3. For what value of t will $f(t) = 5$?

4. For what value of t will $f(t) = 2$?

In Exercises 5–8, use $f(t) = 4e^{t}$.

5. Evaluate $f(1)$.

6. Evaluate $f(3)$.

7. For what value of t will $f(t) = 8$?

8. For what value of t will $f(t) = 10$?

In Exercises 9–12, use $f(x) = \dfrac{0.3}{1 + 10e^{-2x}}$.

9. *Evaluate $f(0)$.* **10.** Evaluate $f(1)$. **11.** Evaluate $f(10)$. **12.** Evaluate $f(12)$.

In Exercises 13–16, use $f(x) = 3 \ln x - 4$.

13. Evaluate $f(e)$.

14. Evaluate $f(1)$.

15. For what value of x will $f(x) = 2$?

16. For what value of x will $f(x) = 3$?

In Exercises 17–20, match the description to one of the graphs (a)–(d).

a.

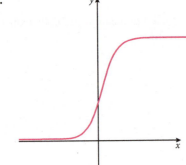

b.

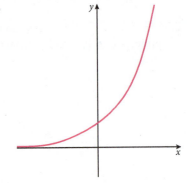

c.

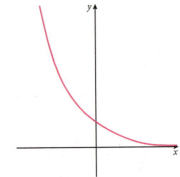

d.

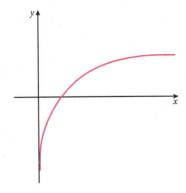

17. Exponential decay model

18. Logarithmic growth model

19. Logistic growth model

20. Exponential growth model

Applications

21. **Chemistry** It takes 5700 years for an initial amount A_0 of carbon-14 to break down into half the amount, $\frac{A_0}{2}$.

    a. Given an initial amount of A_0 grams of carbon-14 at $t = 0$, find an exponential decay model, $A(t) = A_0 e^{kt}$ that gives the amount of carbon-14 at time t, $t \geq 0$.

    b. Calculate the time required for carbon-14 to decay to $\frac{1}{3}A_0$.

22. **Chemistry** The half-life of plutonium-238 is 88 years.

    a. Given an initial amount of A_0 grams of plutonium-238, at $t = 0$, find an exponential decay model, $A(t) = A_0 e^{kt}$ that gives the amount of plutonium-238 at time t, $t \geq 0$.

    b. Calculate the time required for plutonium-238 to decay to $\frac{1}{3}A_0$.

23. **Population Growth** The population of the United States is expected to grow from 282 million in 2000 to 335 million in 2020.
    (*Source:* U.S. Census Bureau)

    a. Find a function of the form $P(t) = Ce^{kt}$ that models the population growth of the United States. Here, t is the number of years since 2000, and $P(t)$ is in millions.

    b. Assuming the trend in part (a) continues, in what year will the population of United States be 350 million?

24. **Population Growth** The population of Florida grew from 17.8 million in 2005 to 19.3 million in 2012.
    (*Source:* U.S. Census Bureau)

    a. Find a function of the form $P(t) = Ce^{kt}$ that models the population growth. Here, t is the number of years since 2005, and $P(t)$ is in millions.

    b. Use your model to predict the population of Florida in 2015.

25. **Housing Prices** The median price of a home in the United States rose from \$175,000 in 2000 to \$280,000 in 2007.
    (*Source:* National Association of Home Builders)

    a. Find an exponential function $P(t) = Ce^{kt}$ that models the growth of housing prices, where t is the number of years since 2000.

    b. The median price of a home in the U.S. dropped to \$180,000 in 2012. Why is the model in part (a) inadequate for predicting the price in 2012?

26. **Economics** Due to inflation, a dollar in the year 1994 is worth \$1.28 in 2005 dollars. Find an exponential function $v(t) = Ce^{kt}$ that models the value of a 1994 dollar t years after 1994.
    (*Source:* Inflationdata.com)

27. **Wildlife Conservation** The population of white-tailed deer in a wildlife refuge t months after their introduction into the refuge can be modeled by the logistic function

    $$N(t) = \frac{300}{1 + 14e^{-0.05t}}$$

    a. How many deer were initially introduced into the refuge?

    b. How many deer will be in the wildlife refuge 10 months after introduction?

28. **Health Sciences** The spread of the flu in an elementary school can be modeled by a logistic function. The number of children infected with the flu virus t days after the first infection is given by

    $$N(t) = \frac{150}{1 + 4e^{-0.5t}}$$

    a. How many children were initially infected with the flu?

    b. How many children were infected with the flu virus after 5 days? After 10 days?

29. Horticulture Pesticides decay at different rates depending on the pH level of the water that is used to mix the pesticide solution. The pH scale measures the acidity of a solution. The lower the pH value, the more acidic the solution.

When mixed with water that has a pH of 6.0, the pesticide chemical known as malathion has a *half-life* of 8 days; that is, *half* the initial amount will remain after 8 days. However, if it is mixed with water that has a pH of 7.0, the half-life decreases to 3 days.   (*Source:* Cooperative Extension Program, University of Missouri)

a. Assume that the initial amount of malathion is 5mg. Find an exponential function of the form Ae^{kt} that gives the amount of malathion that remains after t days if it is mixed with water that has a pH of 6.0.

b. Assume that the initial amount of malathion is 5mg. Find an exponential function of the form Ae^{kt} that gives the amount of malathion that remains after t days if it is mixed with water that has a pH of 7.0.

c. How long will it take for the amount of malathion in each of the solutions to to decay to 3 mg?

d. If the malathion is to be stored for a few days before use, which of the two solutions would be more effective and why?

e. Graph the two exponential functions in the same window and describe how the graphs illustrate the differing decay rates.

30. Tourism The following table shows the tourism revenue in China for selected years from 2000–2012, in billions of dollars.   (*Source:* World Tourism Organization)

Year	Revenue (billions of dollars)
2000	16.2
2003	17.3
2005	29.3
2010	45.8
2011	48.5
2012	50.0

a. Make a scatter plot of the data and find an exponential function of the form $f(x) = Ca^x$ that best fits this data. Let x be the number of years since 2000.

b. From this model, what is the projected revenue from tourism in the year 2014?

c. Do you think it is realistic for this model to be accurate over the long term? Explain.

31. Commerce The following table gives the sales, in billions of current dollars, at restaurants in the United States.   (*Source:* National Restaurant Association Fact Sheet)

Year	1970	1985	1995	2005	2013
Sales	42.8	173.7	295.7	475.8	660.5

a. Make a scatter plot of the data and find an exponential function of the form $f(x) = Ca^x$ that best fits this data. Let x be the number of years since 1970.

b. Why must a be greater than 1 in your model?

c. From this model, what is the projected sales for restaurants in the year 2015?

d. Do you think it is realistic for this model to be accurate over the long term? Explain.

32. Car Racing The following table lists the qualifying speed of the Indianapolis 500 car race winners for selected years from 1941–2013.

(*Source:* www.indy500.com)

Year	1941	1951	1961	1971	1981	1991	2005	2013
Qualifying Speed (miles per hour)	121	135	145	174	200	224	228	227

   a. Explain why a logistic function would fit this data.

   b. Make a scatter plot of the data and find a logistic function of the form $f(x) = \frac{c}{1 + ae^{-bx}}$ that best fits this data. Let x be the number of years since 1941.

   c. What does c signify in your model?

   d. From this model, what is the projected qualifying speed for the winner in 2016?

33. Health Sciences The spread of disease can be modeled by the logistic function. For example, in early 2003, there was an outbreak of an illness called SARS (Severe Acute Respiratory Syndrome) in many parts of the world. The following table gives the *total number of cases* in Canada for weeks following March 20, 2003.

(*Source:* World Health Organization)

Weeks Since March 20, 2003	0	1	2	3	4	5	6	7	8
Total Cases	9	62	132	149	140	216	245	252	250

Note: The total number dropped from 149 to 140 between weeks 3 and 4 because some of the infected cases thought to be SARS were reclassified as other diseases.

   a. Explain why a logistic function would suit this data.

   b. Make a scatter plot of the data and find a logistic function of the form $f(x) = \frac{c}{1 + ae^{-bx}}$ that best fits this data.

   c. What does c signify in your model?

   d. The World Health Organization declared in July 2003 that SARS no longer posed a threat in Canada. From analyzing this data, explain why that would be so.

34. **Environmental Science** The cost of removing chemicals from drinking water depends on how much or how little of the chemical can be safely left behind in the water. The following table lists the annual removal costs for arsenic in terms of the concentration of arsenic in the drinking water.
(*Source:* Environmental Protection Agency)

Arsenic Concentration (micrograms per liter)	Annual cost (millions of dollars)
3	645
5	379
10	166
20	65

a. Interpret the data in the table. What is the relation between the amount of arsenic left behind in the removal process and the annual cost? (One microgram is equal to 10^{-6} grams.)

b. Make a scatter plot of the data and find an exponential function of the form $C(x) = ab^x$ that best fits this data. Here, x is the arsenic concentration.

c. Why must a be less than 1 in your model?

d. From this model, what is the annual cost for having an arsenic concentration of 12 micrograms per liter?

e. It would be best to have the smallest amount of arsenic in the drinking water, but the cost may not be feasible. Use your model to calculate the annual cost of processing so that the concentration of arsenic is only 2 micrograms per liter of water. Interpret your result.

35. **Prenatal Care** The following data gives the percentage of women who smoked during pregnancy for selected years from 1994 to 2002.
(*Source:* National Center for Health Statistics)

Year	Percent Smoking During Pregnancy
1994	14.6
1996	13.6
1998	12.9
2000	12.2
2001	12.0
2002	11.4

a. From examining the table, what is the general relationship between time and the percentage of women who smoke during pregnancy?

b. Let t be the number of years since 1993. Make a scatterplot of the data and find a natural logarithmic function of the form $a \ln t + b$ that best fits the data. Why must a be negative?

c. Project the percentage of women who smoke during pregnancy for the year 2007.

Concepts

36. Refer to Example 1. Without solving an equation, how would you figure out when strontium-90 would reach one-fourth the amount A_0?

37. The value c in the logistic function $f(x) = \frac{c}{1 + ae^{-bx}}$ is sometimes called the "carrying capacity." Can you give a reason why this term might be used?

38. Explain why the function $f(t) = e^{\frac{1}{2}t}$ cannot model exponential decay.

39. For the logistic function, $f(x) = \frac{c}{1 + ae^{-bx}}$, show that $f(x) > 0$ for all x if a and c are positive.

Chapter 4 Summary

Section 4.1 Inverse Functions

Definition of an inverse function [Review Exercises 1–4]

Let f be a function. A function g is said to be the **inverse function** of f if the domain of g is equal to the range of f and, for every x in the domain of f and every y in the domain of g,

$$g(y) = x \quad \text{if and only if} \quad f(x) = y$$

The notation for the inverse function of f is f^{-1}.

Composition of a function and its inverse [Review Exercises 1–4]

If f is a function with an inverse function f^{-1}, then

- for every x in the domain of f, $f^{-1}(f(x))$ is defined and $f^{-1}(f(x)) = x$
- for every x in the domain of f^{-1}, $f(f^{-1}(x))$ is defined and $f(f^{-1}(x)) = x$

One-to-one function [Review Exercises 5–16]

A function f is **one-to-one** if $f(a) = f(b)$ implies $a = b$. For a function to have an inverse, it must be one-to-one.

Graph of a function and its inverse [Review Exercises 13–16]

The graphs of a function f and its inverse function f^{-1} are symmetric with respect to the line $y = x$.

Restriction of domain to find inverse [Review Exercises 11, 12, 16]

Many functions that are not one-to-one can be restricted to an interval in which they *are* one-to-one. Their inverses are then defined on this restricted interval.

Section 4.2 Exponential Functions

Definition of an exponential function [Review Exercises 17–24]

An **exponential function** is a function of the form

$$f(x) = Ca^x$$

where a and C are constants such that $a > 0$, $a \neq 1$ and $C \neq 0$. The domain of the exponential function is the set of all real numbers.

Properties of exponential functions [Review Exercises 17–24, 29, 30]

- If $a > 1$ and $C > 0$, $f(x) = Ca^x \to \infty$ as $x \to \infty$. The function is **increasing** on $(-\infty, \infty)$ and illustrates **exponential growth.**
- If $0 < a < 1$ and $C > 0$, $f(x) = Ca^x \to 0$ as $x \to \infty$. The function is **decreasing** on $(-\infty, \infty)$ and represents **exponential decay.**

Application: Periodic compounded interest [Review Exercises 25, 26]

Suppose an amount P is invested in an account that pays interest at rate r, and the interest is compounded n times a year. Then, after t years, the amount in the account will be

$$A(t) = P\left(1 + \frac{r}{n}\right)^{nt}$$

Application: Continuous compounded interest   [Review Exercises 27, 28]

Suppose an amount P is invested in an account that pays interest at rate r, and the interest is compounded *continuously*. Then, after t years, the amount in the account will be

$$A(t) = Pe^{rt}$$

The number e is defined as the number that the quantity $\left(1 + \frac{1}{n}\right)^n$ approaches as n approaches infinity. The non-terminating, non-repeating decimal representation of the number e is

$$e = 2.7182818284\ldots$$

Section 4.3 Logarithmic Functions

Definition of logarithm   [Review Exercises 31, 32]

Let $a > 0$, $a \neq 1$. If $x > 0$, then the **logarithm of x with respect to base a** is denoted by $y = \log_a x$ and defined by

$$y = \log_a x \quad \text{if and only if} \quad x = a^y$$

Common logarithms and natural logarithms   [Review Exercises 33−46, 55, 56]

If the base of the logarithm is 10, the logarithm is a common logarithm:

$$y = \log x \quad \text{if and only if} \quad x = 10^y$$

If the base of the logarithm is e, the logarithm is a natural logarithm:

$$y = \ln x \quad \text{if and only if} \quad x = e^y$$

Change of base formula   [Review Exercises 47−50]

To write a logarithm with base a in terms of a logarithm with base 10 or base e, use

$$\log_a x = \frac{\log_{10} x}{\log_{10} a}$$

$$\log_a x = \frac{\ln x}{\ln a}$$

where $x > 0$, $a > 0$, and $a \neq 1$.

Graphs of logarithmic functions   [Review Exercises 51−54]

$f(x) = \log_a x, a > 1$

Domain: all *positive* real numbers; $(0, \infty)$
Range: all real numbers; $(-\infty, \infty)$
Vertical asymptote: $x = 0$ (the y-axis)
Increasing on $(0, \infty)$
Inverse function of $y = a^x$

$f(x) = \log_a x, 0 < a < 1$

Domain: all *positive* real numbers; $(0, \infty)$
Range: all real numbers; $(-\infty, \infty)$
Vertical asymptote: $x = 0$ (the y-axis)
Decreasing on $(0, \infty)$
Inverse function of $y = a^x$

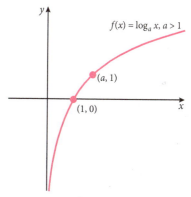

Figure 3

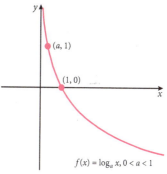

Figure 4

Section 4.4 **Properties of Logarithms**

Product property of logarithms   [Review Exercises 57−62]

Let $x, y > 0$ and $a > 0, a \neq 1$. Then

$$\log_a(xy) = \log_a x + \log_a y$$

Power property of logarithms   [Review Exercises 61−66]

Let $x > 0, a > 0, a \neq 1$, and let k be any real number. Then

$$\log_a x^k = k \log_a x$$

Quotient property of logarithms   [Review Exercises 59, 63−66, 67−72]

Let $x, y > 0$, and $a > 0, a \neq 1$. Then

$$\log_a \frac{x}{y} = \log_a x - \log_a y$$

Section 4.5 Exponential and Logarithmic Equations

One-to-one property　　　　　　　　　[Review Exercises 73−75]

For any $a > 0$, $a \neq 1$,

$$a^x = a^y \quad \text{implies} \quad x = y$$

Solving exponential equations　　　　　　[Review Exercises 73−82]

To solve an exponential equation, take the logarithm of both sides of the equation and use the power property of logarithms and solve the resulting equation.

Solving logarithmic equations　　　　　　[Review Exercises 83−92]

To solve an logarithmic equation, isolate the logarithmic term, exponentiate both sides of the equation, and use the inverse relationship between logarithms and exponents and solve the resulting equation.

Section 4.6 Exponential, Logistic, and Logarithmic Models

Exponential Models　　　　　　[Review Exercises 93, 94, 99, 100]

- An exponential function of the form $f(x) = Ce^{kx}$, where $C > 0$ and $k > 0$ models **exponential growth.**

- An exponential function of the form $f(x) = Ce^{kx}$, where $C > 0$ and $k < 0$ models **exponential decay.**

Logistic Models　　　　　　[Review Exercises 97, 98, 101]

A **logistic function,** $f(x) = \dfrac{c}{1 + ae^{-bx}}$, is used to model a growth phenomenon which must eventually level off.

Logarithmic Models　　　　　　[Review Exercises 95, 96]

A **logarithmic function** can be used to model a phenomenon that grows rapidly at first and then grows more slowly.

Chapter 4 Review Exercises

Section 4.1

In Exercises 1–4, verify that the functions are inverses of each other.

1. $f(x) = 2x + 7$; $g(x) = \dfrac{x - 7}{2}$

2. $f(x) = -x + 3$; $g(x) = -x + 3$

3. $f(x) = 8x^3$; $g(x) = \dfrac{\sqrt[3]{x}}{2}$

4. $f(x) = -x^2 + 1$, $\quad x \geq 0$; $\quad g(x) = \sqrt{1 - x}$

In Exercises 5–12, find the inverse of each one-to-one function.

5. $f(x) = -\dfrac{4}{5}x$

6. $g(x) = \dfrac{2}{3}x$

7. $f(x) = -3x + 6$

8. $f(x) = -2x - \dfrac{5}{3}$

9. $f(x) = x^3 + 8$

10. $f(x) = -2x^3 + 4$

11. $g(x) = -x^2 + 8, x \geq 0$

12. $g(x) = 3x^2 - 5, x \geq 0$

In Exercises 13–16, find the inverse of each one-to-one function. Graph the function and its inverse on the same set of axes, making sure the scales on both axes are the same.

13. $f(x) = -x - 7$

14. $f(x) = 2x + 1$

15. $f(x) = -x^3 + 1$

16. $f(x) = x^2 - 3, x \geq 0$

Section 4.2

In Exercises 17–24, sketch the graph of each function. Label the y-intercept and a few other points (by giving their coordinates). Determine the domain and range and describe the behavior of the function as $x \to \pm\infty$.

17. $f(x) = -4^x$
18. $f(x) = -3^x$
19. $g(x) = \left(\dfrac{2}{3}\right)^x$
20. $g(x) = \left(\dfrac{3}{5}\right)^x$

21. $f(x) = 4e^x$
22. $g(x) = -3e^x + 2$
23. $g(x) = 2e^{-x} + 1$
24. $h(x) = 5e^{-x} - 3$

Banking In Exercises 25 and 26, for an initial deposit of $1500, find the total amount in a bank account after 6 years for the interest rates and compounding frequencies given.

25. 5% compounded quarterly

26. 8% compounded semiannually

Banking In Exercises 27 and 28, for an initial deposit of $1500, find the total amount in a bank account after t years for the interest rates and values of t given. Assume continuous compounding of interest.

27. 8% interest; $t = 4$

28. 3.5% interest; $t = 5$

29. Depreciation The depreciation rate of a Ford Focus is about 25% per year. If the Focus was purchased at $17,000, make a table of its values over the first 5 years after purchase. Find a function which gives its value t years after purchase, and sketch a graph of the function. (*Source:* www.edmunds.com)

30. Tuition savings To save for her newborn daughter's education, Jennifer invests $4,000 at an interest rate of 4% compounded monthly. What is the value of this investment in 18 years, assuming there are no additional deposits or withdrawals to this investment?

Section 4.3

31. Complete the table by filling in the exponential statements that are equivalent to the given logarithmic statements.

Logarithmic Statement	Exponential Statement
$\log_3 9 = 2$	
$\log 0.1 = -1$	
$\log_5 \dfrac{1}{25} = -2$	

32. Complete the table by filling in the logarithmic statements that are equivalent to the given exponential statements.

Exponential Statement	Logarithmic Statement
$3^5 = 243$	
$4^{1/5} = \sqrt[5]{4}$	
$8^{-1} = \dfrac{1}{8}$	

In Exercises 33–42, evaluate each expression without using a calculator.

33. $\log_5 625$ **34.** $\log_6 \dfrac{1}{36}$ **35.** $\log_9 81$ **36.** $\log_7 \dfrac{1}{7}$

37. $\log \sqrt{10}$ **38.** $\ln e^{1/2}$ **39.** $\ln \sqrt[3]{e}$ **40.** $\ln e^{-1}$

41. $\log 10^{x+2}$ **42.** $\ln e^{5x}$

In Exercises 43–46, evaluate each expression to four decimal places using a calculator.

43. $4 \log 2$ **44.** $-6 \log 7.3$ **45.** $\ln \sqrt{8}$ **46.** $\ln\left(\dfrac{\pi}{2}\right)$

In Exercises 47–50, use the change-base formula to evaluate each expression using a calculator. Round your answer to four decimal places.

47. $\log_3 4.3$ **48.** $\log_4 6.52$ **49.** $\log_6 0.75$ **50.** $\log_5 0.85$

In Exercises 51–54, find the domain of each function. Use your answer to help you graph the function, and label all asymptotes and intercepts.

51. $f(x) = \log x - 6$ **52.** $f(x) = \ln(x - 4)$

53. $f(x) = 3\log_4 x$ **54.** $f(x) = \log_5 x + 4$

Geology *The magnitude of an earthquake is measured on the Richter Scale, using the formula*

$$R(I) = \log\left(\dfrac{I}{I_0}\right)$$

where I *represents the actual intensity of the earthquake and* I_0 *is a baseline intensity used for comparison.*

55. If the intensity of an earthquake is 10 times the baseline intensity I_0, what is its magnitude on the Richter scale?

56. In October 8, 2005, a devastating earthquake affected areas of northern Pakistan. It registered a magnitude of 7.6 on the Richter scale. Express its intensity in terms of I_0.

(*Source:* U.S. Geological Survey)

Section 4.4

In Exercises 57–60, use the following. Given that $\log 3 \approx 0.4771$, $\log 5 \approx 0.6990$ *and* $\log 7 \approx 0.8451$, *first evaluate the following logarithms without using a calculator. Then check your answer using a calculator.*

57. $\log 21$ **58.** $\log 15$ **59.** $\log\left(\dfrac{5}{3}\right)$ **60.** $\log \sqrt{3}$

In Exercises 61–66, write each expression as a sum and/or difference of logarithmic expressions. Eliminate exponents and radicals and evaluate logarithms, wherever possible. Assume that a, x, y, z > 0 and a ≠ 1.

61. $\log\left(\sqrt[4]{x}\,\sqrt[3]{y}\right)$ **62.** $\ln\left(\sqrt[3]{x^5}\,\sqrt[3]{y^3}\right)$ **63.** $\log_a \sqrt{\dfrac{x^6}{y^3 z^5}}$

64. $\log_a \sqrt{\dfrac{z^5}{xy^4}}$ **65.** $\ln \sqrt[3]{\dfrac{xy^3}{z^5}}$ **66.** $\log \sqrt[3]{\dfrac{x^3 z^5}{10y^2}}$

In Exercises 67–72, write each expression as a logarithm of a single quantity, and then simplify if possible. Assume that each variable expression is defined for appropriate values of the variable(s).

67. $\ln(x^2 - 3x) - \ln(x - 3)$ **68.** $\log_a(x^2 - 4) - \log_a(x + 2), a > 0, a \neq 1$

69. $\dfrac{1}{4}[\log(x^2 - 1) - \log(x + 1)] + 3\log x$ **70.** $\dfrac{2}{3}\log 9x^3 - \dfrac{1}{4}\log 16y^8$

71. $2\log_3 x^2 - \dfrac{1}{3}\log_3 \sqrt{x}$ **72.** $2\ln(x^2 + 1) + \dfrac{1}{2}\ln x^4 - 3\ln x$

Section 4.5

In Exercises 73–82, solve each exponential equation.

73. $5^x = 625$ **74.** $6^{2x} = 1296$ **75.** $7^x = \dfrac{1}{49}$ **76.** $4e^x + 6 = 38$

77. $25e^{0.04x} = 100$ **78.** $3(1.5^x) - 2 = 9$ **79.** $4^{2x+3} = 16$ **80.** $5^{3x+2} = \dfrac{1}{5}$

81. $e^{2x+1} = 4$ **82.** $2^{x-1} = 10$

In Exercises 83–92, solve each logarithmic equation and eliminate any extraneous solutions.

83. $\ln(2x - 1) = 0$ **84.** $\log(x + 3) - \log(2x - 4) = 0$

85. $\ln x^{1/2} = 2$ **86.** $2\log_3 x = -4$

87. $\log x + \log(2x - 1) = 1$ **88.** $\log(x + 6) = \log x^2$

89. $\log(3x + 1) - \log(x^2 + 1) = 0$ **90.** $\log_3 x + \log_3(x + 8) = 2$

91. $\log_4 x + \log_4(x + 3) = 1$ **92.** $\log(3x + 10) = 2\log x$

Section 4.6

In Exercises 93–98, evaluate $f(0)$ and $f(3)$ for each function. Round your answer to four decimal places.

93. $f(x) = 4e^{-2.5x}$ **94.** $f(x) = 30e^{1.2x}$ **95.** $f(x) = 20\ln(x + 2) + 1$

96. $f(x) = 10\ln(2x + 1) - 2$ **97.** $f(x) = \dfrac{0.2}{1 + 100e^{-4x}}$ **98.** $f(x) = \dfrac{0.5}{1 + 200e^{-5x}}$

99. Global Economy One measure of a country's economy is its collective purchasing power. In 2000, China had a purchasing power of 2.5 trillion dollars. If the purchasing power was forecasted to grow at a rate of 7% a year, find a function of the form $P(t) = Ca^t$, $a > 1$ that models China's purchasing at time t. Here, t is the number of years since 2000.   (*Source:* Proceedings of the National Academy of Sciences)

100. Music In the first few years of introduction of a popular product, the number of units sold per year increases exponentially. Consider the following table which shows the sales of portable music players for the years 2000–2005.

(*Source:* Consumer Electronics Association)

Year	Number of units sold (in millions)
2000	0.510
2001	0.724
2002	1.737
2003	3.031
2004	6.972
2005	10.052

a. Make a scatter plot of the data and find the exponential function of the form $f(x) = Ca^x$ that best fits this data. Let x denote the number of years since 2000.

b. Use the function from part (a) to find the number of portable music players sold in 2007.

c. Use the function from part (a) to determine the year when the number of portable music players sold equals 21 million units.

101. Ecology The number of trout in a pond t months after their introduction into the pond can be modeled by the logistic function

$$N(t) = \frac{450}{1 + 9e^{-0.3t}}$$

a. How many trout were initially introduced into the pond?

b. How many trout will be in the pond 15 months after introduction?

c. Graph this function for $0 \le t \le 30$. What do you observe as t increases?

d. How many months after introduction will the number of trout in the pond be equal to 400?

Chapter 4 Test

1. Verify that the functions $f(x) = 3x - 1$ and $g(x) = \frac{x+1}{3}$ are inverses of each other.

2. Find the inverse of the one-to-one function $f(x) = 4x^3 - 1$.

3. Find $f^{-1}(x)$ given $f(x) = x^2 - 2$, $x \geq 0$. Graph f and f^{-1} on the same set of axes.

In Exercises 4–6, sketch a graph of the function and describe its behavior as $x \to \pm\infty$.

4. $f(x) = -3^x + 1$
5. $f(x) = 2^{-x} - 3$
6. $f(x) = e^{-2x}$

7. Write in exponential form: $\log_6 \dfrac{1}{216} = -3$.

8. Write in logarithmic form: $2^5 = 32$.

In Exercises 9–10, evaluate the expression without using a calculator.

9. $\log_8 \dfrac{1}{64}$

10. $\ln e^{3.2}$

11. Use a calculator to evaluate $\log_7 4.91$ to four decimal places.

12. Sketch the graph of $f(x) = \ln(x + 2)$. Find all asymptotes and intercepts.

In Exercises 13–14, write the expression as a sum or difference of logarithmic expressions. Eliminate exponents and radicals when possible.

13. $\log \sqrt[3]{x^2 y^4}$

14. $\ln(e^2 x^2 y)$

In Exercises 15–16, write each expression as a logarithm of a single quantity, and simplify when possible.

15. $\ln(x^2 - 4) - \ln(x - 2) + \ln x$
16. $4 \log_2 x^{1/3} + 2 \log_2 x^{1/3}$

In Exercises 17–22, solve.

17. $6^{2x} = 36^{3x-1}$
18. $4^x = 7.1$
19. $4e^{x+2} - 6 = 10$

20. $200e^{0.2t} = 800$
21. $\ln(4x + 1) = 0$
22. $\log x + \log(x + 3) = 1$

23. For an initial deposit of $3000, find the total amount in a bank account after 6 years if the interest rate is 5% compounded quarterly.

24. Find the value in three years of an initial investment of $4000 at an interest rate of 7%, compounded continuously.

25. The depreciation rate of a laptop computer is about 40% per year. If a new laptop computer was purchased for $900, find a function that gives its value t years after purchase.

26. The magnitude of a earthquake is measured on the Richter scale using the formula $R(I) = \log\left(\frac{I}{I_0}\right)$, where I represents the actual intensity of the earthquake and I_0 is a baseline intensity used for comparison. If an earthquake registers 6.2 on the Richter scale, express its intensity on terms of I_0.

27. The number of college students infected with a cold virus in a dormitory can be modeled by the logistic function $N(t) = \frac{120}{1 + 3e^{-0.4t}}$, where t is the number of days after the first infection.

   a. How many students were initially infected?

   b. Approximately how many students will be infected after 10 days?

28. The population of a small town grew from 28,000 in 2010 to 32,000 in 2012. Find a function of the form $P(t) = Ce^{kt}$ that models this growth, where t is the number of years since 2010.

Trigonometric Functions

Trigonometry is one of the most practical branches of mathematics. For instance, Ferris wheels and carousels travel in circular paths. Their speed and distance traversed can be calculated using trigonometry. In addition, trigonometry is used in advanced mathematics and can be applied in many professional fields including electronics, medical imaging, surveying, and acoustics.

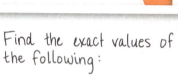

Find the exact values of the following:

a) $\cos 225° = -\dfrac{\sqrt{2}}{2}$

b) $\sin \dfrac{5\pi}{6} = \dfrac{1}{2}$

c) $\tan(-45) = -1$

$(x,y) = (\cos t, \sin t)$

Outline

Objectives

- Know terminology related to angles.
- Know radian and degree measure of angles.
- Convert between radian measure and degree measure of angles.
- Find the length of an arc of a circle.
- Compute angular and linear speed.

Trigonometric functions are used to describe phenomena such as sound waves, pendulum motion, and planetary orbits. To help study these functions, we cover some basic material on angles and discuss their relation to circular motion.

Angles and the *xy*-Coordinate System

An **angle** is formed by rotating a ray, or half-line, about its endpoint. The **initial side** of the angle is the starting position of the ray, and the **terminal side** of the angle is the final position of the ray after rotation. The **vertex** of the angle is the point about which the ray is rotated. Angles are usually denoted by lowercase Greek letters such as α (alpha), β (beta), or θ (theta).

Describing angles using the *xy*-coordinate system will simplify our discussion and will provide a standard way of referring to angles. An angle is said to be in **standard position** if its initial side is on the positive *x*-axis and its vertex is at the origin. The angles in Figure 1 and Figure 2 are both in standard position.

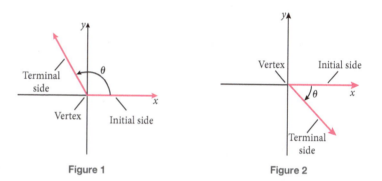

Figure 1   Figure 2

The **measure of an angle** is determined by the amount of rotation of the ray as it goes from the initial side to the terminal side. A complete revolution sweeps out 360°, and so an angle of 1° is equivalent to a rotation of $\frac{1}{360}$ of a complete revolution. A **positive angle** in standard position is generated by a counterclockwise rotation and a **negative angle** is generated by a clockwise rotation. See Figure 3.

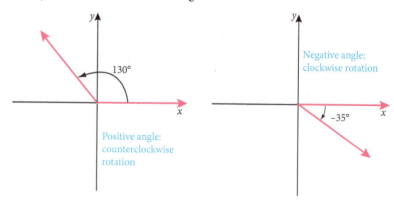

Figure 3

Some types of angles are given special names. An angle whose measure is greater than 0° and less than 90° is called an **acute** angle. An angle whose measure is greater than 90° and less than 180° is called an **obtuse** angle. A 90° angle is called a **right angle**, and a 180° angle is called a **straight angle**. See Figure 4. Figure 5 shows the quadrants where the terminal sides of the angles between 0° and 360° lie.

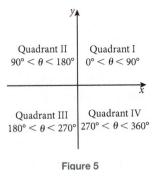

Figure 5

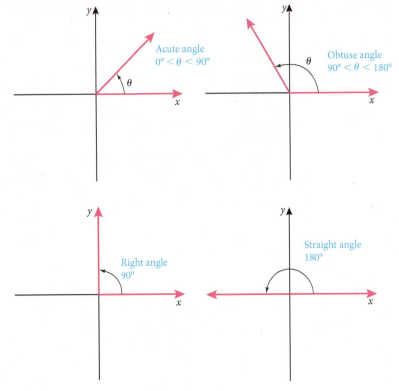

Figure 4

VIDEO EXAMPLES

SECTION 5.1

Example 1   Sketching an Angle

Sketch each of the following angles in standard position.

a. 120°   **b.** −45°   **c.** 450°

Solution

a. Note that 120° = 90° + 30°. Starting from the initial side, an angle of 120° consists of a counterclockwise rotation of 90° followed by a rotation of 30°. See Figure 6.

b. Starting from the initial side, an angle of −45° consists of a clockwise rotation terminating midway in the fourth quadrant, since 45° is $\frac{1}{2}$ of 90°. See Figure 7.

c. Note that 450° = 360° + 90°. Starting from the initial side, make one complete counterclockwise revolution followed by a 90° rotation. See Figure 8.

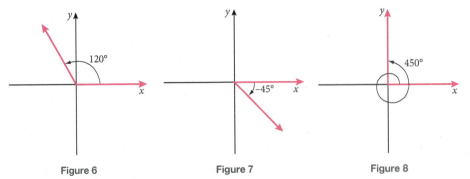

Figure 6   **Figure 7**   **Figure 8**

 Check It Out 1   Sketch an angle of −135° in standard position.

Two angles that have the same initial *and* terminal sides are said to be **coterminal,** as shown in Figure 9.

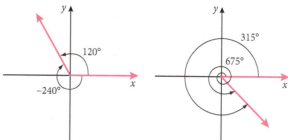

Figure 9

> **Example 2**   **Coterminal Angles**

For each angle, find two angles that are coterminal with it.

a. 135°   **b.** −60°

Solution To find a coterminal angle, we add an integer multiple of 360° to the original angle. This produces another angle that has the same initial and terminal sides, but is generated by a different amount of rotation.

a. Adding 360° to 135° and adding 720° to 135° give, respectively,

$$135° + \mathbf{360°} = 495° \text{ and } 135° + \mathbf{720°} = 855°$$

Thus, angles of measure 495° and 855° are coterminal with an angle of measure 135°. See Figure 10.

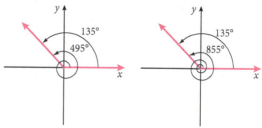

Figure 10

b. Adding 360° to −60° and adding −360° to −60° give, respectively,

$$-60° + \mathbf{360°} = 300° \text{ and } -60° + (\mathbf{-360°}) = -420°$$

Thus, angles of measure 300° and −420° are coterminal with an angle of measure −60°. See Figure 11.

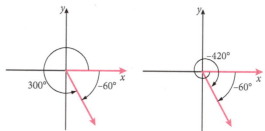

Figure 11

Check It Out 2   Find two angles that are coterminal with an angle of 150°.

Definition of Complementary and Supplementary Angles

- Two acute angles are called **complementary** (complements of each other) if the sum of their measures is 90°.
- Two positive angles are called **supplementary** (supplements of each other) if the sum of their measures is 180°.

Example 3   **Complementary and Supplementary Angles**

a. Find the complement of a 63° angle.   **b.** Find the supplement of a 105° angle.

Solution

a. To find the complement of an angle of 63°, subtract 63° from 90°.

$$90° - 63° = 27°$$

Thus, angles of measure 27° and 63° are complementary.

b. To find the supplement of an angle of 105°, subtract 105° from 180°.

$$180° - 105° = 75°$$

Thus, angles of measure 75° and 105° are supplementary.

✔ *Check It Out 3*   Find the complement and the supplement of an angle of 72°.

Radian Measure of Angles and the Unit Circle

In order to study trigonometry using functions, we measure angles with a unit called a **radian.** To define radian measure, we will use the circle that has radius 1 and is centered at the origin, called the **unit circle.** See Figure 12.

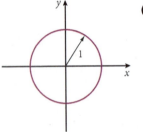

Figure 12

Definition of a Radian

An angle of 1 radian is an angle in standard position that is generated in the counterclockwise direction and cuts off an arc of length 1 on the unit circle. See Figure 13.

We can apply the definition of a radian to find the measure of other angles. For example, one complete counterclockwise revolution of the terminal side of an angle cuts off the entire circumference of the unit circle, given by

$$C = 2\pi r = 2\pi (1) = 2\pi$$

Thus, the radian measure of 360° corresponds to 2π radians. An angle produced by half a revolution cuts off half of the circumference, which is $\frac{2\pi}{2} = \pi$. Similarly, a quarter revolution produces an angle of $\frac{2\pi}{4} = \frac{\pi}{2}$, and so on. See Figure 14.

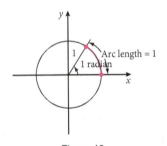

Figure 13

Note There is no special symbol used for the measure of an angle in radians. However, the word radian is sometimes used after an angle to emphasize that it is in radians.

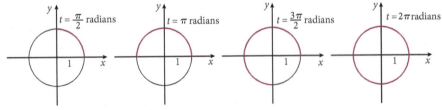

Figure 14

Converting Between Radian and Degree Measure

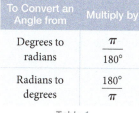

To Convert an Angle from	Multiply by
Degrees to radians	$\dfrac{\pi}{180°}$
Radians to degrees	$\dfrac{180°}{\pi}$

Table 1

To convert between radian and degree measure, use the fact that, in any circle, the length of arc cut off by an angle of 360° is the circumference of that circle. Because the circumference of the unit circle is $2\pi(1)$, the definition of radian measure yields $360° = 2\pi$ radians. Dividing by 2π gives

$$1 \text{ radian} = \frac{360°}{2\pi} = \frac{180°}{\pi} \approx 57.296°$$

Thus, we have the conversion factors shown in Table 1.

Figure 15 shows the quadrants where the terminal sides of the angles between 0 and 2π lie.

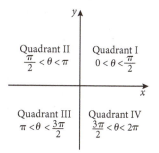

Figure 15

Example 4 **Converting Angles from Degrees to Radians**

Convert each angle from degrees to radians.

a. 225° **b.** 60° **c.** −135°

Solution To convert an angle from degrees to radians, multiply it by $\frac{\pi}{180°}$.

a. $225° = 225° \left(\dfrac{\pi}{180°} \right) = \dfrac{5\pi}{4}$ radians

b. $60° = 60° \left(\dfrac{\pi}{180°} \right) = \dfrac{\pi}{3}$ radians

c. $-135° = -135° \left(\dfrac{\pi}{180°} \right) = -\dfrac{3\pi}{4}$ radians

✅ *Check It Out 4*   Convert 210° from degrees to radians.

Example 5 **Converting Angles from Radians to Degrees**

Convert each angle from radians to degrees.

a. $\dfrac{\pi}{6}$ **b.** $-\dfrac{3\pi}{4}$ **c.** 3

Solution To convert an angle from radians to degrees, multiply it by $\frac{180°}{\pi}$.

a. $\dfrac{\pi}{6} = \dfrac{\pi}{6} \left(\dfrac{180°}{\pi} \right) = 30°$

b. $-\dfrac{3\pi}{4} = -\dfrac{3\pi}{4} \left(\dfrac{180°}{\pi} \right) = -135°$

c. $3 = 3 \left(\dfrac{180°}{\pi} \right) = \dfrac{540°}{\pi} \approx 171.89°$

✅ *Check It Out 5*   Convert $\frac{\pi}{3}$ from radians to degrees.

Arc Length

Arc length: $s = r\theta$

Figure 16

The length of an arc of a circle is directly proportional to the angle θ that cuts off that arc. If θ is given in radians, the **arc length,** shown in Figure 16, is

$$s = r\theta$$

In this formula, s is expressed in the same units of length as r.

Example 6 **Arc Length**

Find the length of an arc of a circle of radius 3 inches that is cut off by an angle of $\frac{\pi}{6}$ radians.

Solution Because $s = r\theta$, we have

$$s = (3) \left(\frac{\pi}{6} \right) = \frac{\pi}{2} \text{ inches} \approx 1.571 \text{ inches}$$

Check It Out 6   Find the length of an arc of a circle of radius 5 meters that is cut off by an angle of $\frac{3\pi}{4}$ radians.   ■

Any object that revolves about a fixed axis traces out a path that is circular in shape, so the formula for arc length can be used to compute the distance traversed by that object.

> **Example 7**   **Distance Traversed on a Circular Path**
>
> Find the distance traversed by the tip of the second hand on a clock as it moves from the 0-second mark to the 20-second mark. The length of the second hand is 5 inches.
>
> **Solution**   First compute θ, the angle that is swept out in going from the 0-second mark to the 20-second mark. Because the second hand on a clock makes a complete revolution in 60 seconds, it makes one-third of a revolution in 20 seconds. Converting to the corresponding angle θ in radians, we get
>
> $$\theta = \frac{1}{3}\,\text{revolution} \cdot \frac{2\pi\,\text{radians}}{\text{revolution}} = \frac{2\pi}{3}\,\text{radians}$$
>
> Thus, the tip of the second hand sweeps out an angle of $\frac{2\pi}{3}$ radians, and so
>
> $$s = r\theta = (5)\left(\frac{2\pi}{3}\right) = \frac{10\pi}{3}\,\text{inches}$$

 Check It Out 7   Rework Example 7 for the case where the tip of the second hand moves from the 20-second mark to the 50-second mark.   ■

Angular Speed and Linear Speed

An object that revolves about a fixed axis and at a fixed distance from it is said to be in **circular motion**, because it traces out a circular path as it revolves. The rate at which the angle is swept out per unit time is called its **angular speed**, usually expressed as the *angle through which it rotates* per unit time. If the angular speed is constant throughout the motion, the object is said to be in **uniform circular motion**. The distance traversed by the object per unit time is called its **linear speed**.

> **Definition of Linear and Angular Speed**
>
> If a point on a circle of radius r travels in uniform circular motion through an angle θ in time t, its **linear speed** v is
>
> $$v = \frac{s}{t}$$
>
> where s is the arc length given by $s = r\theta$.
> The corresponding **angular speed** ω (the Greek letter "omega") is
>
> $$\omega = \frac{\theta}{t}$$
>
> where θ is in radians.

For an object in circular motion, we can write the linear speed v in terms of the angular speed ω by using the arc length formula. That is,

$$v = \frac{s}{t} = \frac{r\theta}{t} = rw, \text{ since } s = r\theta$$

> **Linear Speed in Terms of Angular Speed**
>
> The linear speed v of a point that is a distance r from the axis of rotation is given by
>
> $$v = r\omega,$$
>
> where ω is the angular speed in radians per unit of time. The linear speed, v, is expressed in units of distance per unit time.

Figure 17

| Example 8 | **Converting Angular Speed to Linear Speed** |

A vintage vinyl record makes 45 revolutions per minute. What is the linear speed of a point 3 inches from the center? See Figure 17.

Solution First, we convert the angular speed from revolutions per minute to ω radians per minute:

$$\omega = \frac{45 \text{ revolutions}}{\text{minute}} \cdot \frac{2\pi \text{ radians}}{\text{revolution}}$$

$$= 90\pi \frac{\text{radians}}{\text{minute}}$$

Therefore, the angular speed is 90π radians per minute, so

$$v = r\omega$$

$$= (3 \text{ inches}) \cdot \frac{90\pi}{\text{minute}} = 270\pi \frac{\text{inches}}{\text{minute}}$$

✅ *Check It Out 8*   Rework Example 8 for a point 2 inches from the center of the record. ■

| Example 9 | **Converting Linear Speed to Angular Speed** |

In a bicycle race, one of the bicyclists travels at a speed of 40 kilometers per hour. If the diameter of the bicycle wheel is 0.66 meters, find the angular speed of the wheel in

a. radians per hour   **b.** revolutions per hour

Solution

a. To find the angular speed, use the formula $v = r\omega$ and solve for ω.

$$v = r\omega \quad \Rightarrow \quad \omega = \frac{v}{r}$$

Because v is the linear speed at a point on the outside of the wheel, we have $v = 40$ kilometers per hour, the speed of the bicycle. Because the diameter of the wheel is in meters, convert the linear speed from kilometers per hour to meters per hour.

$$v = \frac{40 \text{ kilometers}}{\text{hour}} \cdot \frac{1000 \text{ meters}}{1 \text{ kilometer}} = \frac{40{,}000 \text{ meters}}{\text{hour}}$$

Also, the radius r is half the diameter d, and so

$$r = \frac{d}{2} = \frac{0.66 \text{ meters}}{2} = 0.33 \text{ meters}$$

Therefore,

$$\omega = \frac{v}{r} = \frac{40{,}000 \text{ meters}}{\text{hour}} \div 0.33 \text{ meters} = \frac{40{,}000 \text{ meters}}{\text{hour}} \cdot \frac{1}{0.33 \text{ meters}}$$

$$= \frac{40{,}000 \text{ radians}}{0.33 \text{ hours}} \approx 121{,}212 \text{ radians per hour}$$

b. To find the angular speed in revolutions per hour, use the fact that there is 1 revolution for every 2π radians.

$$\omega \approx \frac{121{,}212 \text{ radians}}{\text{hour}} \cdot \frac{1 \text{ revolution}}{2\pi \text{ radians}}$$

$$\approx 19{,}291 \text{ revolutions per hour}$$

✅ *Check It Out 9*   Rework Example 9 for the case where the speed of the bicycle racer is 35 kilometers per hour. ■

Exercises 5.1

 Just in Time Exercises

In Exercises 1–4, identify the quadrant where each point lies.

1. $(-2, -1)$ **2.** $(1, 5)$ **3.** $(3, -4)$ **4.** $(-2, 6)$

 Skills

Paying Attention to Instructions: *Exercises 5–8 are intended to give you practice in reading and paying attention to the instructions that accompany the problems you are working.*

5. True or False: An angle of measure 3 is in degree measure.

6. True or False: An angle of measure 3° is in degree measure.

7. In what quadrant does the terminal side of a 135° angle lie?

8. In what quadrant does the terminal side of a −135° angle lie?

In Exercises 9–20, sketch the angle in standard position.

9. $135°$ **10.** $210°$ **11.** $-135°$ **12.** $-225°$ **13.** $270°$ **14.** $450°$

15. $\dfrac{7\pi}{4}$ **16.** $\dfrac{5\pi}{3}$ **17.** $\dfrac{9\pi}{4}$ **18.** $\dfrac{8\pi}{3}$ **19.** $-\dfrac{9\pi}{4}$ **20.** $-\dfrac{4\pi}{3}$

For each angle in Exercises 21–32, find two angles that are coterminal with it.

21. $140°$ **22.** $160°$ **23.** $-55°$ **24.** $-100°$ **25.** $60°$ **26.** $75°$

27. $210°$ **28.** $240°$ **29.** $\dfrac{\pi}{3}$ **30.** $\dfrac{3\pi}{4}$ **31.** $-\dfrac{\pi}{6}$ **32.** $-\dfrac{5\pi}{4}$

For each angle in Exercises 33–38, find a positive angle and a negative angle that are coterminal with it.

33. $140°$ **34.** $270°$ **35.** $-65°$ **36.** $-110°$ **37.** $\dfrac{\pi}{6}$ **38.** $\dfrac{5\pi}{3}$

For each angle in Exercises 39–44, find its complement.

39. $57°$ **40.** $35°$ **41.** $48°$ **42.** $33°$ **43.** $15°$ **44.** $23°$

For each angle in Exercises 45–50, find its supplement.

45. $105°$ **46.** $67°$ **47.** $89°$ **48.** $112°$ **49.** $130°$ **50.** $49°$

Convert each angle in Exercises 51–62 from degrees to radians.

51. $240°$ **52.** $210°$ **53.** $-150°$ **54.** $-225°$ **55.** $270°$ **56.** $300°$

57. $-270°$ **58.** $-180°$ **59.** $720°$ **60.** $540°$ **61.** $390°$ **62.** $-405°$

Convert each angle in Exercises 63–68 from radians to degrees.

63. 3π **64.** $-\dfrac{5\pi}{4}$ **65.** $\dfrac{\pi}{180}$ **66.** $\dfrac{\pi}{45}$ **67.** $-\dfrac{2\pi}{5}$ **68.** $-\dfrac{\pi}{5}$

Applications

69. **Time** What is the angle swept out by the second hand of a clock in a 10-second interval? Express your answer in both degrees and radians.

70. **Time** What is the angle swept out by the second hand of a clock in a 15-second interval? Express your answer in both degrees and radians.

71. **Robotics** A robotic arm pinned at one end makes a complete revolution in 2 minutes. What is the angle swept out by the robotic arm in 1.5 minutes? Express your answer in both degrees and radians.

72. **Robotics** A robotic arm pinned at one end makes a complete revolution in half a minute. What is the angle swept out by the robotic arm in 20 seconds? Express your answer in both degrees and radians.

73. **Geography** Earth makes one full rotation about its axis every 24 hours. How many degrees does the Earth rotate in 1 hour?

74. **Clocks** Find the distance traversed by the tip of the minute hand on a clock between 5:12 P.M. and 6:27 P.M. on any given day if the length of the minute hand is 4 inches. Express your answer in inches.

75. **History** In the 1800s, women often carried pleated fans. One of the fans on display at the Smithsonian is 7 inches long and, when fully open, sweeps out an angle of 80°. How long is the trim, to the nearest tenth of an inch, on the curved edge of the fan?

76. **Landscaping** Sam has a fountain in his backyard. He has decided to make a bed of flowers around the fountain in the shape of a disc with diameter 6 feet. As a border around the bed, he is going to place shrubs around $\frac{2}{5}$ of it and a stone border around the remaining part. How many feet of stone border does he need?

The latitude of a point P on the surface of Earth is defined as the angle formed by a ray from P to C, the center of the Earth, and a ray from C to the equator. Latitude is measured from the equator, with values going north or south.

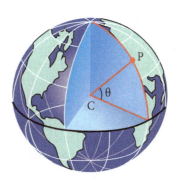

77. **Flight Path** Earth's radius is approximately 3900 miles. Find the distance traveled by a low-flying bird in getting from a location at 40° north latitude to a location at 20° north latitude if the bird flies due south. Express your answer in miles.

78. **Distance between cities** Chicago, Illinois (42° north latitude) is due north of Birmingham, Alabama (33° north latitude). If Earth's radius is approximately 3900 miles, find the approximate distance between the two cities, to two decimal places.

79. **Pet Exercise** A hamster "runs in place" on a circular treadmill, causing the treadmill to rotate at constant angular speed. What is the linear speed of the treadmill in feet per second if it makes one revolution every 3 seconds and has a diameter of 8 inches?

80. **Transportation** A car with tires 24 inches in diameter travels at a speed of 60 mph. What is the angular speed of the tires? Express your answer in both degrees per second and radians per second.

81. **Recreation** A bicycle with a tire radius of 18-inches travels at a speed of 20 mph. What is the angular speed of the tires? Express your answer in both degrees per second and radians per second.

82. **Music Box** A circular music box rotates at a constant rate while the music is playing. What is the linear speed of a fly that is perched on the music box at a point 2 inches from its center if it takes the music box 6 seconds to make one revolution? Express your answer in inches per second.

83. **Entertainment** In 2013, the singer Josh Groban toured "in the round." This means that the stage was round and rotated during the performance. If the stage made one complete rotation every 10 minutes, and Mr. Groban stood at a distance of 20 feet from the center of the stage during one of the songs, what was his linear speed in feet per minute?

84. **Pulleys** A weight is moved upward through the use of a pulley, 10 inches in radius. If the pulley is rotated counterclockwise through an angle of 45°, approximate the height, in inches, that the weight will rise. Round your answer to two decimal places.

85. **Pulleys** Refer to the figure in the previous exercise. Through what angle, in degrees, should the pulley be rotated so that the weight is raised 6 inches from its position shown in the figure? Round your answer to two decimal places.

86. **Sewing** A seamstress secures one end of a piece of thread to a spool. Then she uses an attachment on her sewing machine to cause the spool to spin around, which in turn causes the thread to wind around the spool. If the spool has a diameter of 1.6 cm and spins at a rate of 3 revolutions per second, what length of thread (in centimeters) is wound around the spool in 1 second?

87. **Leisure** In one of the rides at an amusement park, you sit in a circular "car" and cause it to rotate by turning a wheel in the center. The faster you turn the wheel, the faster the car rotates. How far from the center of the car are you sitting if your car makes one revolution every 3 seconds and your linear speed is 5 feet per second? Express your answer in feet.

88. **History** The first commercially successful steamboat was Robert Fulton's *Clermont*. It had side paddles that were 15 feet in diameter. On its maiden voyage from New York City, NY, to Albany, NY, in 1807, the boat traveled at an average linear speed of 5 mph. How many revolutions per minute were the paddles turning?

89. **Games** A game played by many children involves placing a cuff around one ankle that has a ball attached to it by a 2-foot-long string. The ball is spun around the child's leg while he or she jumps over the rope with the other foot. Suppose the ball is making one revolution per second. Calculate the linear speed of the ball in feet per second.

90. **Entertainment** The first Ferris wheel was 250 feet in diameter. It was invented by John Ferris in 1893. Assuming it made one revolution every 30 seconds, what was the angular speed of a passenger (assume the passenger is on the edge of the wheel) in degrees per second? What was the passenger's linear speed in feet per minute?

91. Earth Rotation Earth rotates about an axis through its poles, making one revolution per day.

    a. What is the exact angular speed of Earth about its axis? Express your answer in both degrees per hour and radians per hour.

    b. The radius of Earth is approximately 3900 miles. What distance is traversed by a point on Earth's surface at the equator during any 8-hour interval as a result of Earth's rotation about its axis? Express your answer in miles.

    c. What is the linear speed (in miles per hour) of the point in part (b)?

92. Area of a Sector Consider an angle θ in standard position whose vertex coincides with the center of a circle of radius r. The portion of the circle bounded by the initial and terminal side of the angle θ is called a *sector* of the circle.

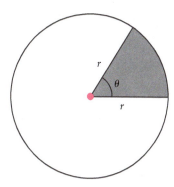

    a. If A is the area of the circle, then $A_s = A\frac{\theta}{2\pi}$ represents the area of the sector, because $\frac{\theta}{2\pi}$ gives the fraction of the area covered by the sector. Show that the area of a sector, A_s, is $A_s = \frac{r^2\theta}{2}$. Here, *theta* is in radians.

    b. Find A_s if $\theta = \frac{\pi}{3}$ and $r = 12$ inches.

Concepts

93. Which is the larger angle, $1°$ or 1 radian? Explain.

94. What is the area of the portion of the unit circle swept out by an angle of $\dfrac{\pi}{6}$ radians?

In Exercises 95–100 find the radian measure of an angle in standard position that is generated by the specified rotation.

95. Quarter of a full revolution clockwise

96. Half of a full revolution counterclockwise

97. One-third of a full revolution counterclockwise

98. Two-thirds of a full revolution clockwise

99. Two full revolutions clockwise

100. Three full revolutions counterclockwise

Trigonometric Functions Using the Unit Circle

Objectives

- Define trigonometric functions using the unit circle.

- Know basic relationships among the trigonometric functions.

- Find exact values of trigonometric functions of certain special angles.

- Use reference angles to find values of trigonometric functions.

- Apply the Pythagorean Identities.

- Apply the Negative Angle Identities.

Historically, trigonometry was a study of relationships between the sides and angles of right triangles. It arose out of the need for precise measurements in navigation and surveying. However, the right triangle can only accommodate measurement of acute angles, and calculus and physics require trigonometric relationships to be defined on a larger domain. With that in mind, we will use the unit circle to define trigonometric functions for any real number.

Let s be a real number. Starting at the point $(1, 0)$, we can measure an arc of length s on the unit circle, proceeding counterclockwise if $s > 0$ and clockwise if $s < 0$. The endpoint is a point (x, y) on the unit circle. Note the terminal side of the angle t intersects the unit circle at the point (x, y). See Figure 1 and Figure 2. If $|s| > 2\pi$, we simply wind around the circle enough times so that a distance of $|s|$ is covered.

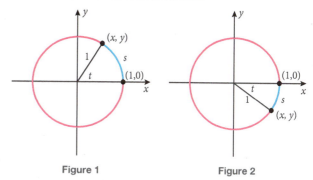

Figure 1   Figure 2

Using the arc length formula, $s = r\theta$, we have

$$s = r\theta = (1)(t) = t$$

We see that the arc length is equal to the measure of the angle t, in radians. So for each real number t, there is a corresponding point (x, y) on the unit circle. This correspondence will be used to define the trigonometric functions on the unit circle *for any real number*.

Just in Time
Review the definitions of the unit circle and radian measure in Section 5.1.

Definitions of Sine and Cosine in Terms of the Unit Circle

Let (x, y) be the point where the terminal side of an angle of t radians in standard position intersects the unit circle. See Figure 3.
The value of the **cosine** function at t, abbreviated as cos t, is defined as cos $t = x$.
The value of the **sine** function at t, abbreviated as sin t, is defined as sin $t = y$.

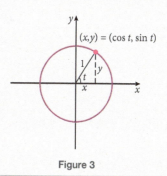

Figure 3

In the following example, we will use these definitions to find the values of the sines and cosines of some special angles.

Example 1 Sine and Cosine of 0 and $\frac{3\pi}{2}$

Find the sine and cosine of the following angle t.

a. $t = 0$ b. $t = \frac{3\pi}{2}$

Solution

a. The terminal side of the angle $t = 0$ in standard position intersects the unit circle at $(x, y) = (1, 0)$. Thus, $\cos 0 = x = 1$ and $\sin 0 = y = 0$. See Figure 4.

b. The terminal side of the angle $t = \frac{3\pi}{2}$ in standard position intersects the unit circle at $(x, y) = (0, -1)$. Thus, $\cos \frac{3\pi}{2} = x = 0$ and $\sin \frac{3\pi}{2} = y = -1$. See Figure 5.

✓ *Check It Out 1* Find the sine and cosine of $t = \frac{\pi}{2}$.

Note You may find it a bit hard to "think in radians," since you are more familiar with degree measure. But radian measure is standard for working with trigonometric functions. Table 1 lists some special angles in both radians and degrees.

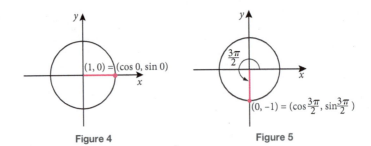

Figure 4   Figure 5

Radian Measure	0	$\frac{\pi}{6}$	$\frac{\pi}{4}$	$\frac{\pi}{3}$	$\frac{\pi}{2}$	π	$\frac{3\pi}{2}$	2π
Degree Measure	0°	30°	45°	60°	90°	180°	270°	360°

Table 1

Table 2 and Figure 6 give the sines and cosines of all the angles whose terminal side lies on the x- or y-axis. These angles are known as **quadrantal angles**.

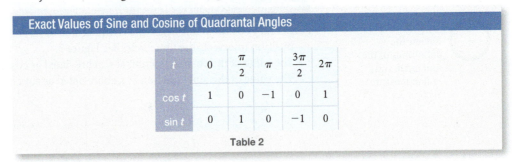

Exact Values of Sine and Cosine of Quadrantal Angles

t	0	$\frac{\pi}{2}$	π	$\frac{3\pi}{2}$	2π
$\cos t$	1	0	−1	0	1
$\sin t$	0	1	0	−1	0

Table 2

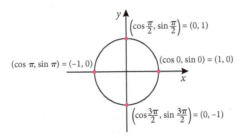

Figure 6

Before evaluating sines and cosines of more special angles, we define some additional trigonometric functions.

Definitions of Other Trigonometric Functions

Four additional trigonometric functions, defined in terms of the sine and cosine functions, are given as follows.

Definitions of Other Trigonometric Functions

$$\text{Tangent: } \tan t = \frac{\sin t}{\cos t} \qquad \text{Cotangent: } \cot t = \frac{\cos t}{\sin t}$$

$$\text{Secant: } \sec t = \frac{1}{\cos t} \qquad \text{Cosecant: } \csc t = \frac{1}{\sin t}$$

Because these functions involve quotients and reciprocals, they are undefined whenever the denominator is zero. Calculating these quantities is straightforward once you know the sine and cosine of an angle.

Example 2 **Calculating Values of Trigonometric Functions**

Calculate the exact value of the following, if defined

a. $\tan 0$

b. $\csc \dfrac{3\pi}{2}$

c. $\cot \pi$

Solution Refer to Table 2 and Figure 6 for the values of sine and cosine of these angles.

a. Using the definition of tangent, $\tan 0 = \frac{\sin 0}{\cos 0} = \frac{0}{1} = 0$.

b. Using the definition of cosecant, $\csc \frac{3\pi}{2} = \frac{1}{\sin \frac{3\pi}{2}} = \frac{1}{-1} = -1$.

c. Using the definition of cotangent $\cot \pi = \frac{\cos \pi}{\sin \pi} = \frac{-1}{0}$. Because there is a zero in the denominator, $\cot \pi$ is undefined.

Check It Out 2 Calculate the exact value of $\sec \pi$, if defined.

Note that $\cot t$ is the reciprocal of $\tan t$. Similar relationships exist between $\sin t$ and $\csc t$, and between $\cos t$ and $\sec t$. Thus, we have the following **reciprocal identities**.

Reciprocal Identities

$$\sin t = \frac{1}{\csc t} \qquad \cos t = \frac{1}{\sec t} \qquad \tan t = \frac{1}{\cot t}$$

$$\csc t = \frac{1}{\sin t} \qquad \sec t = \frac{1}{\cos t} \qquad \cot t = \frac{1}{\tan t}$$

These identities hold whenever the denominator is nonzero.

Just in Time
Refer to Section 2.1 for the definition of quadrants.

Sines and Cosines of Special Angles in the First Quadrant

To find the sine and cosine of $\frac{\pi}{4}$, proceed as follows. Observe that the line $y = x$ makes an angle of $\frac{\pi}{4}$ and intersects the unit circle. See Figure 7. To find the point of intersection, substitute $y = x$ into $x^2 + y^2 = 1$, the equation of the unit circle. This gives

$$x^2 + x^2 = 1 \Rightarrow 2x^2 = 1 \Rightarrow x^2 = \frac{1}{2}$$

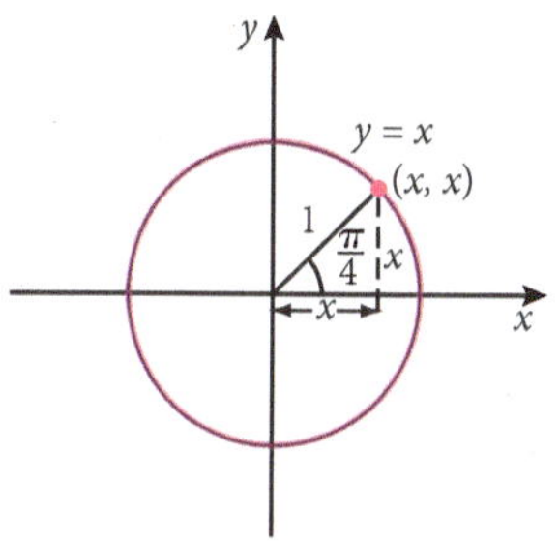

Figure 7

Solving for x by taking the square root of both sides, we have

$$x = \sqrt{\frac{1}{2}} = \frac{\sqrt{2}}{2}$$

Because $\frac{\pi}{4}$ lies in the first quadrant, only the positive square root is taken. Thus, we have

$$x = \cos\frac{\pi}{4} = \frac{\sqrt{2}}{2}$$

$$y = \sin\frac{\pi}{4} = \frac{\sqrt{2}}{2}$$

Exact values of the sine and cosine of $\frac{\pi}{6}$ can also be found, as shown in the following example.

Example 3   **Exact Values of $\cos\frac{\pi}{6}$ and $\sin\frac{\pi}{6}$**

Find the exact values $\cos\frac{\pi}{6}$ and $\sin\frac{\pi}{6}$.

Solution On the coordinate plane, draw the unit circle and then draw the angle $\frac{\pi}{6}$ in standard position. See Figure 8.

From geometry, $x = (\sqrt{3})y$ for this special case. Substituting $x = (\sqrt{3})y$ into the equation for the unit circle, we have,

$$x^2 + y^2 = 1 \Rightarrow ((\sqrt{3})y)^2 + y^2 = 1 \Rightarrow 3y^2 + y^2 = 1$$

This gives

$$4y^2 = 1 \Rightarrow y^2 = \frac{1}{4}$$

Because $\frac{\pi}{6}$ is in the first quadrant, take only the positive square root to get

$$y = \sin\frac{\pi}{6} = \frac{1}{2}$$

Because $x = (\sqrt{3})y$

$$x = \cos\frac{\pi}{6} = \sqrt{3}\left(\frac{1}{2}\right) = \frac{\sqrt{3}}{2}$$

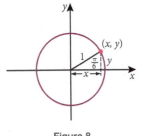

Figure 8

✓ *Check It Out 3*   Find the exact values of $\cos\frac{\pi}{3}$ and $\sin\frac{\pi}{3}$. ■

Table 3 and Figure 9 summarize the exact values of the sine and cosine of $\frac{\pi}{6}$, $\frac{\pi}{4}$, and $\frac{\pi}{3}$.

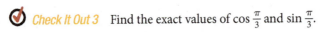

Exact Values of Sine and Cosine in the First Quadrant

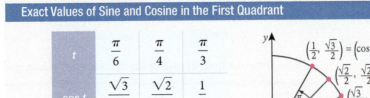

t	$\dfrac{\pi}{6}$	$\dfrac{\pi}{4}$	$\dfrac{\pi}{3}$
$\cos t$	$\dfrac{\sqrt{3}}{2}$	$\dfrac{\sqrt{2}}{2}$	$\dfrac{1}{2}$
$\sin t$	$\dfrac{1}{2}$	$\dfrac{\sqrt{2}}{2}$	$\dfrac{\sqrt{3}}{2}$

$\left(\frac{1}{2}, \frac{\sqrt{3}}{2}\right) = \left(\cos\frac{\pi}{3}, \sin\frac{\pi}{3}\right)$

$\left(\frac{\sqrt{2}}{2}, \frac{\sqrt{2}}{2}\right) = \left(\cos\frac{\pi}{4}, \sin\frac{\pi}{4}\right)$

$\left(\frac{\sqrt{3}}{2}, \frac{1}{2}\right) = \left(\cos\frac{\pi}{6}, \sin\frac{\pi}{6}\right)$

Table 3   **Figure 9**

Sines and Cosines of Special Angles in Other Quadrants

To calculate sines and cosines of angles in other quadrants, we apply the concept of a **reference angle,** which is defined as follows.

Definition of Reference Angle

For an angle t in standard position whose terminal side lies in one of the four quadrants, the **reference angle** for t is the *acute* angle that the terminal side of t makes with the x-axis. See Figure 10.

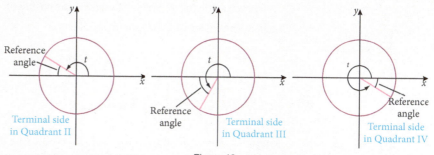

Figure 10

| | **Example 4** | **Reference Angles** |

Find the reference angle for each of the following angles.

a. $t = \dfrac{5\pi}{4}$ **b.** $t = -\dfrac{4\pi}{3}$

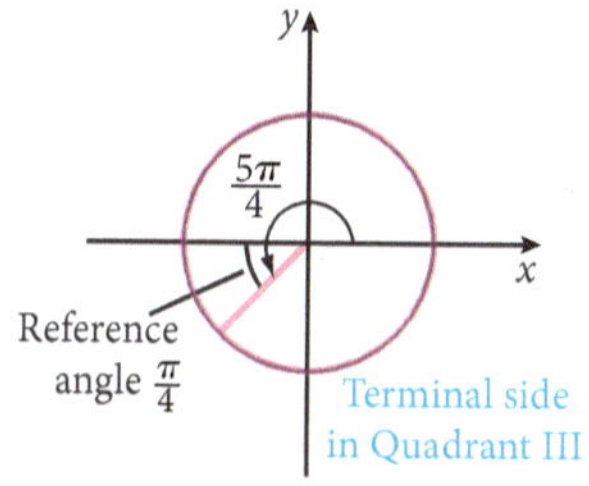

Solution

a. Note that $t = \frac{5\pi}{4} = \pi + \frac{\pi}{4}$. Because $\frac{5\pi}{4}$ is a positive angle, we measure it in the *counter-clockwise* direction from the positive x-axis. Its terminal side makes an acute angle of $\frac{\pi}{4}$ with the horizontal axis. Thus, the reference angle is $\frac{\pi}{4}$. See Figure 11.

b. Note that $t = -\frac{4\pi}{3} = -\left(\pi + \frac{\pi}{3}\right)$. Because $-\frac{4\pi}{3}$ is a negative angle, we measure it in the *clockwise* direction from the positive x-axis. Its terminal side makes an acute angle of $\frac{\pi}{3}$ with the horizontal axis. Thus, the reference angle is $\frac{\pi}{3}$. See Figure 12.

✔ *Check It Out 4*   Find the reference angle for $t = \frac{5\pi}{6}$. ■

| | **Example 5** | **Sines and Cosines of Special Angles in Other Quadrants** |

Find

a. $\sin \dfrac{5\pi}{4}$

b. $\cos \dfrac{2\pi}{3}$

c. $\cos\left(-\dfrac{\pi}{6}\right)$

Solution Before studying these examples, refer to Table 3 of sines and cosines of special angles in the first quadrant.

a. The terminal side of $\frac{5\pi}{4}$ lies in the third quadrant, so $\sin \frac{5\pi}{4}$ will be negative. The reference angle for $\frac{5\pi}{4}$ is $\frac{\pi}{4}$. Therefore, the sine of $\frac{5\pi}{4}$ will have the same magnitude as the sine of $\frac{\pi}{4}$, but will *differ in sign*. See Figure 13. Hence

$$\sin \frac{5\pi}{4} = -\frac{\sqrt{2}}{2}$$

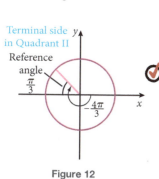

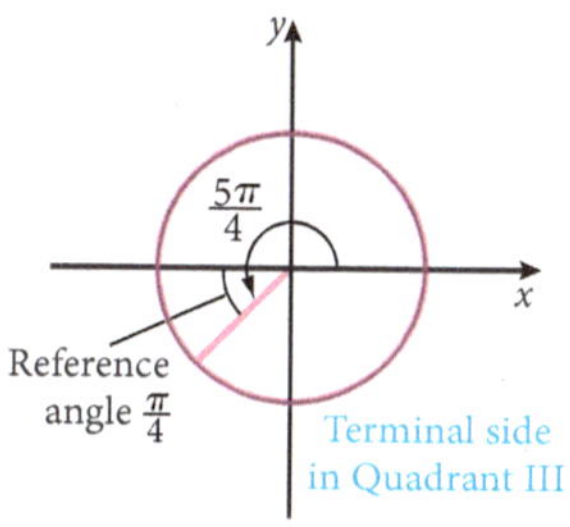

Figure 11

Figure 12

Figure 13

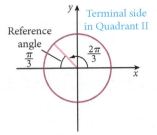

Figure 14

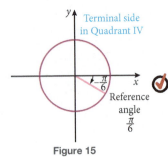

Figure 15

b. The terminal side of $\frac{2\pi}{3}$ lies in the second quadrant, so $\cos \frac{2\pi}{3}$ will be negative. The reference angle for $\frac{2\pi}{3}$ is $\frac{\pi}{3}$. Therefore, the cosine of $\frac{2\pi}{3}$ will have the same magnitude as the cosine of $\frac{\pi}{3}$, but will *differ in sign*. See Figure 14. Hence

$$\cos \frac{2\pi}{3} = -\frac{1}{2}$$

c. The terminal side of $-\frac{\pi}{6}$ lies in the fourth quadrant, so $\cos\left(-\frac{\pi}{6}\right)$ will be positive. The reference angle for $-\frac{\pi}{6}$ is $\frac{\pi}{6}$. Therefore, the cosine of $-\frac{\pi}{6}$ will have the *same magnitude and sign* as the cosine of $\frac{\pi}{6}$. See Figure 15. Hence

$$\cos\left(-\frac{\pi}{6}\right) = \frac{\sqrt{3}}{2}$$

✓ *Check It Out 5* Find $\sin\left(-\frac{4\pi}{3}\right)$ and $\cos\left(-\frac{4\pi}{3}\right)$. ■

Figure 16 shows the signs of the sine and cosine functions in each of the four quadrants. The signs for $\sin t$ and $\cos t$ agree with the signs of the x- and y-coordinates, respectively, of a point on the unit circle in that quadrant.

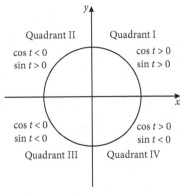

Figure 16

The exact values of sine and cosine of all special angles in the interval $(0, 2\pi)$ are shown in Figure 17 and Figure 18.

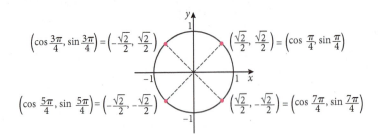

Figure 17

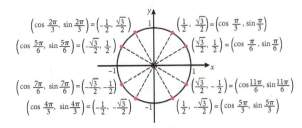

Figure 18

Other Trigonometric Functions

Now we find the values of the other trigonometric functions for each of the special angles.

Example 6   **Exact Values of the Other Trigonometric Functions**

Find the values of the trigonometric functions of $\frac{\pi}{3}$ to complete Table 4.

$\cos \frac{\pi}{3}$	$\sin \frac{\pi}{3}$	$\tan \frac{\pi}{3}$	$\cot \frac{\pi}{3}$	$\sec \frac{\pi}{3}$	$\csc \frac{\pi}{3}$
$\frac{1}{2}$	$\frac{\sqrt{3}}{2}$				

Table 4

Solution   Using the definitions of the other trigonometric functions, we have the following.

$$\tan \frac{\pi}{3} = \frac{\left(\sin \frac{\pi}{3} \right)}{\left(\cos \frac{\pi}{3} \right)} = \frac{\left(\frac{\sqrt{3}}{2} \right)}{\left(\frac{1}{2} \right)} = \sqrt{3}$$

$$\cot \frac{\pi}{3} = \frac{\left(\cos \frac{\pi}{3} \right)}{\left(\sin \frac{\pi}{3} \right)} = \frac{\left(\frac{1}{2} \right)}{\left(\frac{\sqrt{3}}{2} \right)} = \frac{1}{\sqrt{3}} = \frac{\sqrt{3}}{3}$$

$$\sec \frac{\pi}{3} = \frac{1}{\left(\cos \frac{\pi}{3} \right)} = \frac{1}{\left(\frac{1}{2} \right)} = 2$$

$$\csc \frac{\pi}{3} = \frac{1}{\left(\sin \frac{\pi}{3} \right)} = \frac{1}{\left(\frac{\sqrt{3}}{2} \right)} = \frac{2}{\sqrt{3}} = \frac{2\sqrt{3}}{3}$$

The table is then completed as follows.

$\cos \frac{\pi}{3}$	$\sin \frac{\pi}{3}$	$\tan \frac{\pi}{3}$	$\cot \frac{\pi}{3}$	$\sec \frac{\pi}{3}$	$\csc \frac{\pi}{3}$
$\frac{1}{2}$	$\frac{\sqrt{3}}{2}$	$\sqrt{3}$	$\frac{\sqrt{3}}{3}$	2	$\frac{2\sqrt{3}}{3}$

Table 5

Using Technology

When using a calculator to find the sine and cosine of an angle, make sure it is set in the correct mode — RADIAN or DEGREE. Figure 19 shows a calculator in radian mode.

Keystroke Appendix:
Section 15

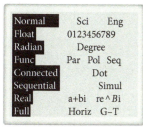

Figure 19

Check It Out 6   Find the values of the trigonometric functions of $\frac{\pi}{6}$ to complete the table.

$\cos \frac{\pi}{6}$	$\sin \frac{\pi}{6}$	$\tan \frac{\pi}{6}$	$\cot \frac{\pi}{6}$	$\sec \frac{\pi}{6}$	$\csc \frac{\pi}{6}$
$\frac{\sqrt{3}}{2}$	$\frac{1}{2}$				

Table 6

Values of Trigonometric Functions Using a Calculator

You can find exact values of the trigonometric functions for only a few special angles. For all others, a calculator is needed to evaluate trigonometric function values, as shown in the next example.

> **Example 7**   **Using a Calculator to Approximate Trigonometric Functions**

Use a calculator to approximate each of the following to three decimal places.

a. $\sin 4$ **b.** $\cot \dfrac{\pi}{8}$ **c.** $\cos(-2)$

Solution

a. Because the 4 is not followed by a degree symbol, the angle is assumed to be in radians. With your calculator in RADIAN mode, and using the $\boxed{SIN}$ key, you will find that $\sin 4 \approx -0.757$.

b. The angle here is also in radians. There is no key for the cotangent function in your calculator, and so you must input $1/\boxed{TAN}\,(\pi/8)$. This gives $\cot \dfrac{\pi}{8} \approx 2.414$.

c. The angle here is in radians. Using the COS key, you will find that $\cos(-2) = -0.416$.

 Check It Out 7   Use a calculator to approximate each of the following to three decimal places.

a. $\sin(-1.3)$

b. $\tan\left(\dfrac{5\pi}{2}\right)$

Pythagorean Identities

Using the equation for the unit circle, we can establish relationships between $\sin t$ and $\cos t$. Let $(x,y) = (\cos t, \sin t)$ be a point on the unit circle. Because the equation of a unit circle is $x^2 + y^2 = 1$, it follows that

$$x^2 + y^2 = (\cos t)^2 + (\sin t)^2 = 1$$

which is known as a **Pythagorean identity**. It is conventional to write $(\sin t)^2$ as $\sin^2 t$, and $(\cos t)^2$ as $\cos^2 t$. Thus, this Pythagorean identity is usually written as

$$\cos^2 t + \sin^2 t = 1$$

We can use this identity to derive two additional Pythagorean identities involving the other trigonometric functions.

$$1 + \tan^2 t = \sec^2 t$$
$$1 + \cot^2 t = \csc^2 t$$

To derive the first of these two identities, start with the original Pythagorean identity and use algebraic manipulation.

$\sin^2 t + \cos^2 t = 1$	Original Pythagorean identity
$\dfrac{\sin^2 t}{\cos^2 t} + \dfrac{\cos^2 t}{\cos^2 t} = \dfrac{1}{\cos^2 t}$	Divide each term by $\cos^2 t$
$\left(\dfrac{\sin t}{\cos t}\right)^2 + 1 = \left(\dfrac{1}{\cos t}\right)^2$	Rewrite each term
$\tan^2 t + 1 = \sec^2 t$	Substitute $\dfrac{\sin t}{\cos t} = \tan t$ and $\dfrac{1}{\cos t} = \sec t$

The identity $1 + \cot^2 t = \csc^2 t$ can be derived in a similar manner.

> **The Pythagorean Identities**
>
> $$\cos^2 t + \sin^2 t = 1$$
> $$1 + \tan^2 t = \sec^2 t$$
> $$1 + \cot^2 t = \csc^2 t$$

Finding Values of Trigonometric Functions Using Pythagorean Identities

We can find the values of other trigonometric functions when given the value of just one of the functions, as shown in the following example.

> **Example 8**   **Using Pythagorean Identities to Find Function Values**

Given that $\tan t = -\frac{2}{3}$ and that the terminal side of t lies in Quadrant II, find $\cos t$.

Solution We apply one of the Pythagorean identities as follows.

$$1 + \tan^2 t = \sec^2 t \qquad \text{Pythagorean identity}$$

$$1 + \left(-\frac{2}{3}\right)^2 = \sec^2 t \qquad \text{Substitute } \tan t = -\frac{2}{3}$$

$$1 + \frac{4}{9} = \sec^2 t \qquad \text{Simplify}$$

$$\frac{13}{9} = \sec^2 t$$

$$\sec t = \pm \frac{\sqrt{13}}{3} \qquad \text{Take square root of both sides}$$

Because the terminal side of t lies in Quadrant II, $\sec t = \frac{1}{\cos t}$ is negative, because $\cos t$ is negative. Thus,

$$\sec t = -\frac{\sqrt{13}}{3}$$

Because $\cos t = \frac{1}{\sec t}$,

$$\cos t = \frac{1}{-\dfrac{\sqrt{13}}{3}} = -\frac{3}{\sqrt{13}} = -\frac{3\sqrt{13}}{13}$$

✓ **Check It Out 8** Given that $\cot t = \frac{1}{3}$ and that the terminal side of t lies in Quadrant III, find $\sin t$.

■

Negative Angle Identities

We can establish relationships between $\sin t$, $\cos t$ and $\sin(-t)$ and $\cos(-t)$ by examining corresponding points on the unit circle. From Figure 20, the coordinates of points R and S are

$$R: (\cos t, \sin t), \quad S: (\cos(-t), \sin(-t))$$

In Figure 20, note that the x-coordinates of R and S are the same. Because $x = \cos t$, we have

$$\cos t = \cos(-t)$$

The y-coordinate of S is the negative of the y-coordinate of R. Because $y = \sin t$, we have

$$\sin t = -\sin(-t)$$

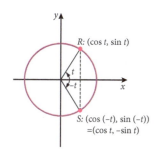

Figure 20

From these facts, and the definitions of the other trigonometric functions, we obtain the following relationships, which are sometimes referred to as **negative-angle identities.**

Negative-Angle Identities	
$\sin(-t) = -\sin t$	$\csc(-t) = -\csc t$
$\cos(-t) = \cos t$	$\sec(-t) = \sec t$
$\tan(-t) = -\tan t$	$\cot(-t) = -\cot t$

Recall that a function is even if $f(t) = f(-t)$, and it is odd if $f(-t) = -f(t)$. Thus, the cosine and secant functions are *even*, while the sine, tangent, cotangent, and cosecant functions are *odd*. These properties of the trigonometric functions will be quite useful in sketching graphs in Sections 5.4 and 5.5.

Example 9   **Applying the Negative-Angle Identities**

Use the negative-angle identities to find the exact values of the following.

a. $\cos\left(-\dfrac{3\pi}{4}\right)$ **b.** $\tan\left(-\dfrac{\pi}{3}\right)$

Solution

a. Because $\cos t = \cos(-t)$,

$$\cos\left(-\frac{3\pi}{4}\right) = \cos\left(\frac{3\pi}{4}\right) = -\frac{\sqrt{2}}{2}$$

b. Because $\tan(-t) = -\tan t$,

$$\tan\left(-\frac{\pi}{3}\right) = -\tan\left(\frac{\pi}{3}\right) = -\sqrt{3}$$

✓ *Check It Out 9*   Find the exact value of $\sin\left(-\frac{5\pi}{4}\right)$.

Repetitive Behavior of Sine, Cosine, and Tangent

If $t \in [0, 2\pi)$, then any integer multiple of 2π added to t produces a coterminal angle. Therefore, the terminal sides of t and $t + 2n\pi$ intersect at the same point on the unit circles and so must have the same values of sine and cosine.

Repetitive Behavior of Sine and Cosine

Let t be a real number and let n be an integer. Then

$$\sin(t + 2n\pi) = \sin t \text{ and } \cos(t + 2n\pi) = \cos t$$

The tangent function also repeats, but does so by an integer multiple of π.

Repetitive Behavior of Tangent

Let t be a real number and let n be an integer. Then

$$\tan(t + n\pi) = \tan t$$

Example 10   **Finding Sine and Cosine of Angles of the Form $t + 2n\pi$**

Determine the exact value of each of the following.

a. $\cos\left(\dfrac{9\pi}{4}\right)$ **b.** $\tan\left(-\dfrac{13\pi}{3}\right)$

Solution

a. The terminal side of $\frac{9\pi}{4}$ lies in Quadrant I, as shown in Figure 21. Because

$$\frac{9\pi}{4} = \frac{\pi}{4} + \frac{8\pi}{4} = \frac{\pi}{4} + 2\pi$$

$$\cos\left(\frac{9\pi}{4}\right) = \cos\left(\frac{\pi}{4} + 2\pi\right) = \cos\left(\frac{\pi}{4}\right) = \frac{\sqrt{2}}{2}$$

by the repetitive behavior of cosine.

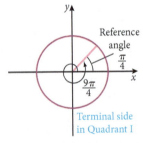

Reference angle

Terminal side in Quadrant I

Figure 21

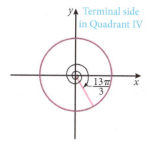

y Terminal side in Quadrant IV

$-\dfrac{13\pi}{3}$ *x*

Figure 22

b. The terminal side of $-\dfrac{13\pi}{3}$ lies in Quadrant IV, as shown in Figure 22. Because

$$-\frac{13\pi}{3} = -\frac{\pi}{3} - \frac{12\pi}{3} = -\frac{\pi}{3} - 4\pi$$

$$\tan\left(-\frac{13\pi}{3}\right) = \tan\left(-\frac{\pi}{3} - 4\pi\right) = \tan\left(-\frac{\pi}{3}\right) = -\sqrt{3}$$

by the repetitive behavior of sine.

✓ *Check It Out 10*   Determine the exact value of $\sin\left(\frac{11\pi}{4}\right)$.

Summary of Exact Values of the Sine and Cosine Function

For quick reference, Table 7, Table 8, and Table 9 summarize the exact values of the sine and cosine functions for *t* in $[0, 2\pi)$.

- Terminal side of *t* along one of the coordinate axes

t	0	$\dfrac{\pi}{2}$	π	$\dfrac{3\pi}{2}$	2π
cos *t*	1	0	−1	0	1
sin *t*	0	1	0	−1	0

Table 7

- Measure of *t* is an integer multiple of $\frac{\pi}{4}$.

t	$\dfrac{\pi}{4}$	$\dfrac{3\pi}{4}$	$\dfrac{5\pi}{4}$	$\dfrac{7\pi}{4}$
cos *t*	$\dfrac{\sqrt{2}}{2}$	$-\dfrac{\sqrt{2}}{2}$	$-\dfrac{\sqrt{2}}{2}$	$\dfrac{\sqrt{2}}{2}$
sin *t*	$\dfrac{\sqrt{2}}{2}$	$\dfrac{\sqrt{2}}{2}$	$-\dfrac{\sqrt{2}}{2}$	$-\dfrac{\sqrt{2}}{2}$

Table 8

- Measure of *t* is an integer multiple of $\frac{\pi}{6}$.

t	$\dfrac{\pi}{6}$	$\dfrac{\pi}{3}$	$\dfrac{2\pi}{3}$	$\dfrac{5\pi}{6}$	$\dfrac{7\pi}{6}$	$\dfrac{4\pi}{3}$	$\dfrac{5\pi}{3}$	$\dfrac{7\pi}{6}$
cos *t*	$\dfrac{\sqrt{3}}{2}$	$\dfrac{1}{2}$	$-\dfrac{1}{2}$	$-\dfrac{\sqrt{3}}{2}$	$-\dfrac{\sqrt{3}}{2}$	$-\dfrac{1}{2}$	$\dfrac{1}{2}$	$\dfrac{\sqrt{3}}{2}$
sin *t*	$\dfrac{1}{2}$	$\dfrac{\sqrt{3}}{2}$	$\dfrac{\sqrt{3}}{2}$	$\dfrac{1}{2}$	$-\dfrac{1}{2}$	$-\dfrac{\sqrt{3}}{2}$	$-\dfrac{\sqrt{3}}{2}$	$-\dfrac{1}{2}$

Table 9

Exercises 5.2

Just in Time Exercises

In Exercises 1–4, determine whether the given points lie on the unit circle. That is, determine whether they satisfy the equation $x^2 + y^2 = 1$.

1. $(0, -1)$

2. $(0.5, 0.5)$

3. $(-0.3, 0.7)$

4. $\left(\dfrac{1}{2}, \dfrac{\sqrt{3}}{2}\right)$

In Exercises 5–8, determine the quadrant where the terminal side of the given angle lies.

5. $\dfrac{4\pi}{3}$

6. $-\dfrac{5\pi}{4}$

7. $-\dfrac{11\pi}{3}$

8. $\dfrac{13\pi}{6}$

Skills

Paying Attention to Instructions: *Exercises 9 and 10 are intended to give you practice reading and paying attention to the instructions that accompany the problems you are working.*

9. Let $t = \frac{5\pi}{3}$.

   a. Find the reference angle for t.

   b. Find an angle that is coterminal to t.

10. Let $t = -\frac{\pi}{3}$.

   a. Find the reference angle for t.

   b. Find an angle that is coterminal to t.

In Exercises 11–22, find the reference angle for each given angle.

11. $\dfrac{7\pi}{6}$ **12.** $\dfrac{11\pi}{4}$ **13.** $-\dfrac{7\pi}{6}$ **14.** $-\dfrac{5\pi}{4}$ **15.** $\dfrac{3\pi}{4}$ **16.** $\dfrac{4\pi}{3}$

17. $-\dfrac{5\pi}{6}$ **18.** $-\dfrac{7\pi}{4}$ **19.** $\dfrac{7\pi}{8}$ **20.** $\dfrac{11\pi}{8}$ **21.** $-\dfrac{\pi}{5}$ **22.** $-\dfrac{8\pi}{5}$

In Exercises 23–38, find the exact values of all the trigonometric functions for the following values of t. If a certain value is undefined, state so. Do not use a calculator.

23. $t = 5\pi$

24. $t = 6\pi$

25. $t = -3\pi$

26. $t = -\dfrac{3\pi}{2}$

27. $t = -\dfrac{\pi}{3}$

28. $t = -\dfrac{\pi}{6}$

29. $t = -\dfrac{\pi}{4}$

30. $t = -\dfrac{3\pi}{4}$

31. $t = \dfrac{11\pi}{4}$

32. $t = \dfrac{9\pi}{4}$

33. $t = \dfrac{13\pi}{6}$

34. $t = -\dfrac{15\pi}{4}$

35. $t = \dfrac{13\pi}{3}$

36. $t = \dfrac{15\pi}{4}$

37. $t = -\dfrac{10\pi}{3}$

38. $t = -\dfrac{14\pi}{3}$

In Exercises 39–60, use a calculator in radian mode to evaluate the given trigonometric function to four decimal places.

39. $\cos 3$

40. $\sin 1$

41. $\sin(-3)$

42. $\cos(-1.5)$

43. $\sin\dfrac{\pi}{12}$

44. $\cot\dfrac{\pi}{5}$

45. $\tan 2$

46. $\tan(-1)$

47. $\tan\dfrac{3\pi}{5}$

48. $\cot\dfrac{5\pi}{12}$

49. $\tan(-0.5)$

50. $\tan(-1.5)$

51. $\sec\dfrac{7\pi}{5}$

52. $\sec\dfrac{4\pi}{5}$

53. $\sec(3.2)$

54. $\cot(2.7)$

55. $\csc(-2.5)$

56. $\sec(-1.4)$

57. $\sec(1.5)$

58. $\csc(2.8)$

59. $\cot(-3.9)$

60. $\cot(-6.6)$

In Exercises 61–72, fill in the following table with the missing information. Approximate all non-exact answers to four decimal places.

	Quadrant	$\sin t$	$\cos t$	$\tan t$
61.	I	$\frac{1}{2}$		
62.	IV		$\frac{1}{2}$	
63.	III			1
64.	II			-1
65.	II		$-\frac{1}{2}$	
66.	II		$-\frac{\sqrt{3}}{2}$	
67.	IV	-0.6		
68.	III	-0.8		
69.	II		$-\frac{5}{13}$	
70.	IV		$\frac{12}{13}$	
71.	IV			-2
72.	II			$-\frac{2}{3}$

In Exercises 73–78, use the negative-angle identities to compute the exact value of each of the following trigonometric functions.

73. $\sin\left(-\dfrac{2\pi}{3}\right)$ **74.** $\cos\left(-\dfrac{10\pi}{3}\right)$ **75.** $\sec\left(-\dfrac{4\pi}{3}\right)$

76. $\csc\left(-\dfrac{5\pi}{4}\right)$ **77.** $\tan\left(-\dfrac{7\pi}{3}\right)$ **78.** $\cot\left(-\dfrac{11\pi}{6}\right)$

In Exercises 79–90, find the exact value of each expression without using a calculator.

79. $\sin\dfrac{\pi}{2} + \cos\pi$ **80.** $\sin\dfrac{3\pi}{2} + \cos\dfrac{\pi}{2}$ **81.** $3\sin\dfrac{\pi}{4} + 2\cos\dfrac{3\pi}{4}$ **82.** $2\sin\dfrac{\pi}{6} - \cos\dfrac{\pi}{3}$

83. $\sin\dfrac{\pi}{4}\cos\dfrac{\pi}{4}$ **84.** $\sin\dfrac{\pi}{3}\cos\dfrac{\pi}{6}$ **85.** $\tan\dfrac{\pi}{4}\sec\dfrac{\pi}{4}$ **86.** $\cot\dfrac{\pi}{3}\csc\dfrac{\pi}{6}$

87. $\csc\dfrac{\pi}{2} - 4\cot\dfrac{\pi}{2}$ **88.** $3\tan\pi + 5\sec\pi$ **89.** $\tan\dfrac{\pi}{3} - \cos\dfrac{\pi}{6}$ **90.** $\sin\dfrac{\pi}{6} + \cot\dfrac{\pi}{6}$

In Exercises 91–96, find the sine and cosine of the angle t in $(0, 2\pi)$, in standard position, whose terminal side intersects the unit circle at the given point.

91. $(-1, 0)$ **92.** $(0, 1)$ **93.** $\left(-\dfrac{4}{5}, \dfrac{3}{5}\right)$

94. $(0.8, -0.6)$ **95.** $\left(-\dfrac{12}{13}, -\dfrac{5}{13}\right)$ **96.** $\left(-\dfrac{1}{2}, \dfrac{\sqrt{3}}{2}\right)$

In Exercises 97–100, find an angle s such that $s \neq t$, $0 \leq s < 2\pi$, and $\cos s = \cos t$.

97. $t = \dfrac{\pi}{4}$ **98.** $t = \dfrac{\pi}{2}$ **99.** $t = \dfrac{4\pi}{3}$ **100.** $t = \dfrac{11\pi}{6}$

In Exercises 101–104, find an angle s such that s ≠ t, 0 ≤ s < 2π, and sin s = sin t.

101. $t = \pi$ **102.** $t = \dfrac{5\pi}{4}$ **103.** $t = \dfrac{2\pi}{3}$ **104.** $t = \dfrac{3\pi}{2}$

In Exercises 105–108, find exact values of cos 3t and cos $\left(\frac{t}{3}\right)$ for the following values of t.

105. $t = 0$ **106.** $t = \dfrac{\pi}{2}$ **107.** $t = -\dfrac{\pi}{2}$ **108.** $t = -\pi$

Concepts

109. Does the equation $\sin(t + \pi) = \sin t + \sin \pi$ hold for all t? Explain.

110. Does the equation $\cos\left(\frac{t}{2}\right) = \frac{\cos t}{2}$ hold for all t? Explain.

111. Find all values of t in $(0, 2\pi)$ such that $\sin t = \cos t$.

112. Find all values of t in $(0, 2\pi)$ such that $\cos t = -\frac{1}{2}$.

113. Suppose t is in $\left(0, \frac{\pi}{2}\right)$. Express $\cos(t + \pi)$ in terms of $\cos t$. (*Hint:* It is helpful to sketch a figure.)

114. Suppose t is in $\left(0, \frac{\pi}{2}\right)$. Express $\sin\left(t + \frac{\pi}{2}\right)$ in terms of $\sin t$. (*Hint:* It is helpful to sketch a figure.)

115. Derive the Pythagorean identity $1 + \cot^2 t = \csc^2 t$.

Right Triggle Trigonometry $\quad$ **5.3**

5.3

Objectives

- Define trigonometric functions using a right triangle.
- Find unknown sides and angles of a right triangle.
- Determine exact values of trigonometric functions of special angles.
- Solve applied problems using right triangles.

In many instances, it is convenient to view trigonometric functions in terms of ratios of the sides of right triangles. For example, by using right triangle trigonometry, you can determine the height of an object using lengths and angles that are easier to measure. You will see how to do this later in this section.

Trigonometric Functions Using a Right Triangle

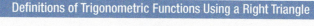

In this section, we will study trigonometry by examining right triangles. Figure 1 shows a right triangle and its sides relative to angle θ. The *ratio* of the lengths of any pair of sides in a triangle depends only on the measure of the acute angle θ. Each of these ratios is given a special name, and they are collectively called **trigonometric functions.**

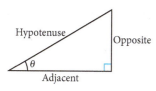

Figure 1

Note The right triangle definition of trigonometric functions is consistent with the unit circle definition when the values of the trigonometric functions are restricted to the first quadrant. That is,

$$0 < t < \frac{\pi}{2}.$$

Definitions of Trigonometric Functions Using a Right Triangle			
Sine:	$\sin \theta = \dfrac{\text{opposite}}{\text{hypotenuse}}$	Cosecant:	$\csc \theta = \dfrac{\text{hypotenuse}}{\text{opposite}}$
Cosine:	$\cos \theta = \dfrac{\text{adjacent}}{\text{hypotenuse}}$	Secant:	$\sec \theta = \dfrac{\text{hypotenuse}}{\text{adjacent}}$
Tangent:	$\tan \theta = \dfrac{\text{opposite}}{\text{adjacent}}$	Cotangent:	$\cot \theta = \dfrac{\text{adjacent}}{\text{opposite}}$

We next state the Pythagorean Theorem, which is used to solve many problems in trigonometry.

The Pythagorean Theorem
Let a and b be the legs of a right triangle, and c be the length of its hypotenuse. See Figure 2. Then $$a^2 + b^2 = c^2$$

VIDEO EXAMPLES

SECTION 5.3

Example 1 $\quad$ Evaluating Trigonometric Functions

Use the triangle in Figure 3 to find the values of the six trigonometric functions of θ.

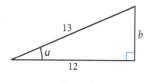

Figure 3

Solution Figure 3 gives the lengths of the hypotenuse and the side adjacent to θ. Use the Pythagorean Theorem to find the length of the other side:

$$12^2 + b^2 = 13^2$$
$$144 + b^2 = 169$$
$$b^2 = 25$$
$$b = 5$$

Hypotenuse
c

b
Leg

90°

a
Leg

Figure 2

The lengths of all three sides of the triangle are

side adjacent to θ (adj) = 12

side opposite θ (opp) = 5

hypotenuse (hyp) = 13

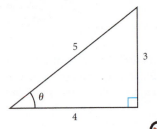

Figure 4

The six trigonometric functions of θ are

$$\sin \theta = \frac{\text{opp}}{\text{hyp}} = \frac{5}{13} \qquad \csc \theta = \frac{\text{hyp}}{\text{opp}} = \frac{13}{5} \qquad \tan \theta = \frac{\text{opp}}{\text{adj}} = \frac{5}{12}$$

$$\cos \theta = \frac{\text{adj}}{\text{hyp}} = \frac{12}{13} \qquad \sec \theta = \frac{\text{hyp}}{\text{adj}} = \frac{13}{12} \qquad \cot \theta = \frac{\text{adj}}{\text{opp}} = \frac{12}{5}$$

 Check It Out 1 Use the triangle in Figure 4 to find the values of the six trigonometric functions of θ. ■

Example 2   **Evaluating Trigonometric Functions**

Let θ be an acute angle such that $\cos \theta = \frac{2}{5}$. Find the values of

a. $\sin \theta$ **b.** $\tan \theta$

Solution

a. In order to solve this problem, we need to relate the given information to an appropriate right triangle that has θ as one of its acute angles. Because $\cos \theta = \frac{2}{5}$, we have

$$\cos \theta = \frac{\text{adj}}{\text{hyp}} = \frac{2}{5}$$

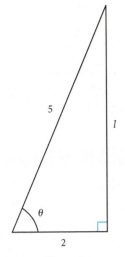

Figure 5

We will thus let adj $= 2$ and hyp $= 5$. Because $\sin \theta = \frac{\text{opp}}{\text{hyp}}$, we need to determine the length, l, of the side opposite angle θ. See Figure 5. We apply the Pythagorean Theorem to find l

$$2^2 + l^2 = 5^2$$
$$4 + l^2 = 25$$
$$l^2 = 25 - 4$$
$$l^2 = 21$$
$$l = \sqrt{21} \approx 4.583$$

Using this value of l, we find that

$$\sin \theta = \frac{\text{opp}}{\text{hyp}} = \frac{l}{5} \approx \frac{4.583}{5} \approx 0.917$$

b. Because we now know the lengths of the opposite and adjacent sides of the triangle, we can find $\tan \theta$.

$$\tan \theta = \frac{\text{opp}}{\text{adj}} = \frac{l}{2} \approx \frac{4.583}{2} \approx 2.292$$

 Check It Out 2 Let θ be an acute angle such that $\sin \theta = 0.7$. Find $\cos \theta$. ■

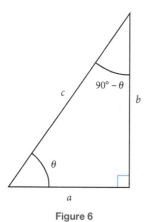

Figure 6

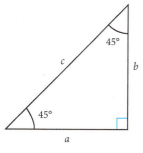

Figure 7

Cofunction Identities

The sine and the cosine are a pair of functions known as **cofunctions**; likewise for tangent and cotangent, and for secant and cosecant. For example, if θ is an acute angle, then value of $\sin\theta$ is equal to the value of $\cos(90° - \theta)$. To see this, we first sketch a right triangle and apply the definitions of sine and cosine.

From Figure 6,

$$\sin\theta = \frac{b}{c} \text{ and } \cos(90° - \theta) = \frac{b}{c} \text{ so } \cos(90° - \theta) = \sin\theta$$

In fact, we can show that all of the following relationships hold.

Cofunction Identities	
$\sin(90° - \theta) = \cos\theta$	$\cos(90° - \theta) = \sin\theta$
$\tan(90° - \theta) = \cot\theta$	$\cot(90° - \theta) = \tan\theta$
$\sec(90° - \theta) = \csc\theta$	$\csc(90° - \theta) = \sec\theta$

Example 3   Using a Cofunction Identity

If $\sin 20° \approx 0.3420$, find $\cos(70°)$ without using a calculator.

Solution Since $70° = 90° - 20°$, applying the cofunction identity for $\cos(90° - \theta)$ gives

$$\cos(70°) = \sin(20°) \approx 0.3420$$

 Check It Out 3 If $\cos 40° \approx 0.766$, find $\sin(50°)$ without using a calculator.

Sines and Cosines of Special Angles

Using geometry, we can determine the sine and cosine of angles with measures of 30°, 45°, and 60°. We begin with the 45° angle.

Sine and cosine of 45°

Draw a right triangle with one of the acute angles measuring 45°. Then the other acute angle also has measure 45°. See Figure 7. Recall from geometry that sides opposite congruent angles are equal, so $a = b$.

By the Pythagorean Theorem,

$$a^2 + b^2 = c^2 \qquad \text{Pythagorean Theorem}$$
$$a^2 + a^2 = c^2 \qquad \text{Use the fact that } a = b$$
$$2a^2 = c^2$$
$$\frac{a^2}{c^2} = \frac{1}{2}$$
$$\frac{a}{c} = \frac{\sqrt{2}}{2} \qquad \text{Solve for } \frac{a}{c}$$
$$\sin 45° = \frac{a}{c} = \frac{\sqrt{2}}{2}$$

Therefore,

$$\sin 45° = \cos 45° = \frac{\sqrt{2}}{2}$$

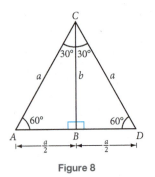

Figure 8

Sine and Cosine of 30° and 60°

To find the sines and cosines of 30° and 60° angles, first position two $30-60-90$ right triangles, ABC and BCD, side by side, as shown in Figure 8.

This gives us a larger triangle, ACD, all of whose angles are of measure 60°. Hence all of its sides are of equal length. Call this length a. By symmetry, segments AB and BD are each of length $\frac{a}{2}$. Let b denote the length of segment BC. By the Pythagorean Theorem applied to triangle ABC,

$$\left(\frac{a}{2}\right)^2 + b^2 = a^2 \qquad \text{Pythagorean Theorem}$$

$$b^2 = a^2 - \left(\frac{a}{2}\right)^2 \qquad \text{Subtract } \left(\frac{a}{2}\right)^2$$

$$b^2 = a^2 - \frac{a^2}{4} \qquad \text{Simplify}$$

$$b^2 = a^2\left(\frac{3}{4}\right) \qquad \text{Combine like terms}$$

$$b = a\,\frac{\sqrt{3}}{\sqrt{4}} = a\,\frac{\sqrt{3}}{2} \qquad \text{Take square root}$$

Now use this result and triangle ABC to determine the values of sin 60° and cos 60°.

$$\sin 60° = \frac{b}{a} = \frac{\left(\dfrac{a\sqrt{3}}{2}\right)}{a} = \frac{\sqrt{3}}{2}$$

$$\cos 60° = \frac{\left(\dfrac{a}{2}\right)}{a} = \frac{1}{2}$$

Now find the values of sin 30° and cos 30° by using the cofunction identities.

$$\sin 30° = \cos 60° = \frac{1}{2}$$

$$\cos 30° = \sin 60° = \frac{\sqrt{3}}{2}$$

You can use the trigonometric functions to help you find missing sides and angles of a triangle. This is commonly referred to as **solving a triangle.**

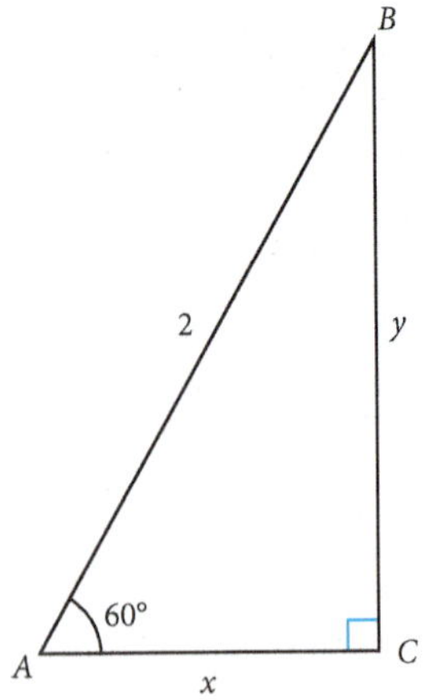

| **Example 4** | **Solving the Triangle** |

Find the unknown angle measure and the unknown lengths of sides of the triangle in Figure 9.

Solution The measure of A is 60°. Because this is a right triangle, the other acute angle, B, must have measure 30°.
Proceed as follows:

$$\frac{y}{2} = \frac{\text{opposite}}{\text{hypotenuse}} = \sin 60°$$

$$\frac{y}{2} = \frac{\sqrt{3}}{2} \qquad \sin 60° = \frac{\sqrt{3}}{2}$$

$$y = (2)\left(\frac{\sqrt{3}}{2}\right) = \sqrt{3} \qquad \text{Solve for } y$$

Figure 9

Now find x using the Pythagorean Theorem.

$$x^2 + (\sqrt{3})^2 = 2^2$$

$$x^2 + 3 = 4$$

$$x^2 = 1 \qquad \text{Subtract 3 from each side}$$

$$x = 1 \qquad \text{Take square root}$$

Therefore,

$$x = 1, \quad y = \sqrt{3}, \quad \text{and} \quad B = 30°$$

✓ *Check It Out 4*   Find the unknown angle measure and the unknown lengths of sides of the triangle in Figure 10.

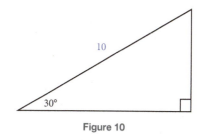

Figure 10

**Using Technology**

When using a calculator to find the sine and cosine of an angle, make sure it is set in the correct mode — DEGREE or RADIAN. Figure 11 shows a graphing calculator in degree mode.

Keystroke Appendix:

Section 15

Figure 11

Determining Sine and Cosine Using a Calculator

You can find exact values of the sine and cosine functions for only a few special angles. For other angles, a calculator is needed to evaluate their sines and cosines.

Example 5   **Using a Calculator to Evaluate Sines and Cosines**

Use a calculator to evaluate each of the following to four decimal places.

a. $\cos 42°$ **b.** $\sin 67.5°$

Solution

a. With your calculator in DEGREE mode, and using the ⎡COS⎤ key, you will find that cos $42° \approx 0.7431$.

b. With your calculator in DEGREE mode, and using the ⎡SIN⎤ key, you will find that sin $67.5° \approx 0.9239$.

✓ *Check It Out 5*   Use a calculator to evaluate sin 58° to four decimal places.

Applications of Right Triangle Trigonometry

The next three examples will show how trigonometric functions can be used in applications involving right triangles.

One type of angle that occurs in many applications is the **angle of elevation**. It is the angle from a given point to a point that is at a higher elevation.

Another type of angle is the **angle of depression**. It is the angle from a given point to a point that is at a lower elevation. See Figure 12.

Figure 12

Figure 13

| | Example 6 | **Application to Surveying** |

A surveyor stands 100 feet from the base of a building. From that point, the angle of elevation to the top of the building is 78°. See Figure 13. Find the height of the building. Round your answer to two decimal places.

Solution From the figure, we see that the lengths of the sides opposite and adjacent to the 78° angle are involved. The side opposite that angle is the height h of the building, and the side adjacent to that angle is 100 feet. Thus, using the definition of the tangent function, we have

$$\tan 78° = \frac{\text{opp}}{\text{adj}} \qquad \text{Definition of tangent function}$$

$$\tan 78° = \frac{h}{100} \qquad \text{Substitute opp} = h \text{ and adj} = 100$$

$$100 \tan 78° = h \qquad \text{Solve for } h$$

$$100(4.7046) \approx h \qquad \text{Use calculator to find } \tan 78°$$

$$470.46 \approx h \qquad \text{Simplify}$$

Thus, the height of the building is about 470.46 feet, rounded to two decimal places.

Check It Out 6   A surveyor stands 120 feet from the base of a building. From that point, the angle of elevation to the top of the building is 60°. Find the height of the building.

| | Example 7 | **Traffic Monitoring** |

A traffic helicopter hovering 500 feet above ground observes a police car on the ground at an angle of depression of 25°. See Figure 14. What is the distance of the helicopter from the police car, to the nearest foot?

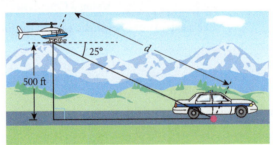

Figure 14

Solution The distance of the helicopter from the police car is given by d in Figure 14. Since d is the hypotenuse of the right triangle pictured, we have

$$\sin 25° = \frac{500}{d} \qquad \text{Definition of sine}$$

$$d \sin 25° = 500 \qquad \text{Multiply both sides of equations by } d$$

$$d = \frac{500}{\sin 25°} \qquad \text{Solve for } d$$

$$d = \frac{500}{0.4226} \approx 1183 \quad \sin 25° \approx 0.4226$$

The distance of the helicopter from the police car is approximately 1183 feet, rounded to the nearest foot.

Check It Out 7   In Example 7, find the horizontal distance of the helicopter from the police car.

Example 8	**Application to Navigation**

An observer measures an angle of elevation of 41° to the top of a lighthouse from a certain point on the ground. She then walks 15 feet closer to the lighthouse and measures an angle of elevation of 44°. See Figure 15. How tall is the lighthouse? Round your answer to two decimal places.

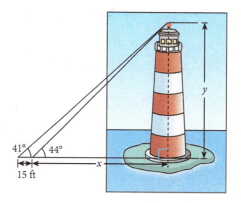

Figure 15

Solution   Referring to the figure, we have the following relationships:

$$\tan 44° = \frac{y}{x}$$

$$\tan 41° = \frac{y}{x+15}$$

Note that we have two unknown quantities, x and y. Rewrite the equations as

$$y = x \tan 44°$$

$$y = (x + 15) \tan 41°$$

Setting the two expressions for y equal to each other and solving for x.

$$x \tan 44° = (x + 15) \tan 41°$$

$$x \tan 44° = x \tan 41° + 15 \tan 41° \qquad \text{Expand right side}$$

$$x(\tan 44° - \tan 41°) = 15 \tan 41° \qquad \text{Subtract } x \tan 41° \text{ and factor left side}$$

$$x = \frac{15 \tan 41°}{\tan 44° - \tan 41°} \qquad \text{Solve for } x$$

$$\approx \frac{15(0.9657)}{0.9657 - 0.8693} \qquad \text{Use calculator for tangent values}$$

$$x \approx 135.26 \qquad \text{Use calculator to evaluate } x$$

Now calculate y.

$$y = x \tan 44°$$

$$\approx (135.26)(0.9657) \approx 130.62$$

The height of the lighthouse is about 130.62 feet, rounded to two decimal places.

Check It Out 8   An observer measures an angle of elevation of 30° from a certain point on the ground to the top of a tower. She then walks 10 feet closer to the tower and measures an angle of elevation of 35°. How tall is the tower?

Exercises 5.3

In Exercises 1–6, find the missing dimension of a right triangle with sides a and b and hypotenuse c.

1. $a = 3, b = 4, c = $ _____
2. $a = 6, b = 8, c = $ _____
3. $a = 2, b = 3, c = $ _____

4. $a = 1, b = 4, c = $ _____
5. $a = 3, c = 6, b = $ _____
6. $b = 4, c = 10, a = $ _____

 Skills

In Exercises 7–14, find the exact values of the six trigonometric functions of θ.

7.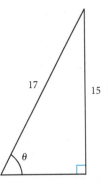

8.

9.

10.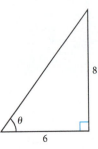

11.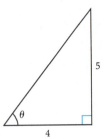

12.

13.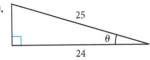

14.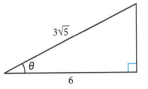

In Exercises 15–26, use the given value of a trigonometric function of θ to find the values of the other five trigonometric functions. Assume θ is an acute angle.

15. $\cos \theta = \dfrac{3}{5}$
16. $\sin \theta = \dfrac{12}{13}$
17. $\tan \theta = \dfrac{3}{2}$
18. $\tan \theta = \dfrac{1}{2}$

19. $\sin \theta = \dfrac{3}{5}$
20. $\cos \theta = \dfrac{4}{5}$
21. $\cot \theta = 1.5$
22. $\cos \theta = 0.7$

23. $\csc \theta = 2$
24. $\sec \theta = 3$
25. $\cot \theta = 4$
26. $\tan \theta = 5$

In Exercises 27–34, find the exact value of each of the following trigonometric functions.

27. $\cos 30°$
28. $\sin 45°$
29. $\sec 30°$
30. $\csc 45°$

31. $\tan 60°$
32. $\cot 45°$
33. $\csc 60°$
34. $\tan 30°$

In Exercises 35–44, use a calculator, in DEGREE mode to evaluate the following. Round to four decimal places.

35. sin 38°   **36.** cos 57°   **37.** tan 65°   **38.** tan 17°   **39.** cos 83°

40. sin 78°   **41.** sec 15°   **42.** csc 36°   **43.** cot 67°   **44.** cot 34°

In Exercises 45–52, find a cofunction that has the same value as the given quantity.

45. sin 35°   **46.** cos 42°   **47.** tan 40°   **48.** tan 20°

49. sec 47°   **50.** csc 31°   **51.** cot 67°   **52.** cot 72°

In Exercises 53–58, find the lengths of the sides and angle measure of each right triangle. Round to four decimal places.

53.
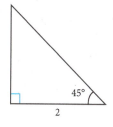

54.
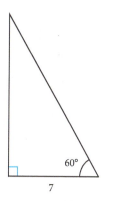

55.
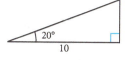

56.
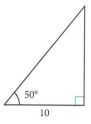

57.
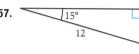

58.
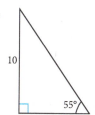

In Exercises 59–64, find the measure of the acute angle θ whose sine or cosine is given.

59. $\sin \theta = \dfrac{1}{2}$ **60.** $\cos \theta = \dfrac{\sqrt{3}}{2}$ **61.** $\cos \theta = \dfrac{\sqrt{2}}{2}$

62. $\sin \theta = \dfrac{\sqrt{3}}{2}$ **63.** $\sin \theta = \dfrac{\sqrt{2}}{2}$ **64.** $\cos \theta = \dfrac{1}{2}$

65. Let α be an acute angle with $\sin \alpha = a$. Find $\csc \alpha$ and $\cos (90° - \alpha)$ in terms of a.

66. Let β be an acute angle with $\cos \beta = b$. Find $\sec \beta$ and $\sin(90° - \beta)$ in terms of b.

Applications

Unless otherwise indicated, round answers to four decimal places.

67. **Construction** A 20-foot-long piece of wire is attached to the top of a pole at one end and nailed to the ground at the other end. If the wire makes an angle of 30° with the ground, find the height of the pole.

68. **Construction** A pole 10 feet high is supported by a taut wire staked to the ground at a 35° angle. How long is the wire?

69. **Ballooning** From the basket of a hot-air balloon 100 feet above the ground, the angle of depression of a point A on the ground is 10.5°. What is the distance from point A to the basket of the balloon?

70. **Navigation** The angle of depression from the top of a lighthouse to a small boat near the coast is 8°. If the lighthouse is 120 feet high, what is the distance from the top of the lighthouse to the boat?

71. **Surveying** The angle of elevation from a certain point on the ground to the top of a tower of is 38°. From a point that is 15 feet closer to the tower, the angle of elevation is 42°. Find the height of the tower.

72. **Surveying** From a certain point on a 10-foot-high platform, the angle of depression to the base of a building is 15°, and the angle of elevation to the top of the building is 55°. How high is the building?

73. **Ladders** A 16-foot-long ladder leans against a vertical wall. The base of the ladder makes an angle of 68° with the lawn on which the foot of the ladder rests. How high above the surface of the lawn is the top of the ladder?

74. **Design** A 30-foot-long driveway slopes downward, away from a house, at an angle of 8° with respect to the adjacent street. How far below the house is the lowest point of the driveway?

75. **Surveying** The angle of elevation of the top of a hill from a certain point on the surrounding level ground is 10°. If the hill is 9 feet high, what is the horizontal distance of the top of the hill from that point?

76. **Design** A ramp makes an angle of 15° with the ground. If the top of the ramp is 4 feet above the ground, how long is the ramp?

77. **Building Code** In Milwaukee, WI, the building code states that for a ramp to qualify as handicapped accessible, it can rise only 1 foot for every 8 feet of length. What is the degree of incline for the ramp?
(*Source:* www.mkedcd.org)

78. **Quilting** Sara wants to make a quilt square by using right triangles of varying colors. The triangles need to have an hypotenuse of length 10 centimeters. If the triangles are isosceles right triangles, what is the length of each side of the triangle?

79. **Surveying** When the Eiffel Tower was dedicated in 1889, the height, including the flagpole on top, was 324 meters. If the angle of elevation of the top of the tower from a certain point from the ground is 50°, how far is the point from the center of the base?
(*Source:* www.tour-eiffel.fr/teiffel/uk/)

80. **Distance** Standing on the south rim of the Grand Canyon near Grand Canyon Village, the vertical distance to the canyon floor is approximately 5000 feet. Standing at this point you can see Phantom Ranch by looking down at an angle of depression of 30°. Find the distance to the camp from the point at the base of the rim directly under where you stand.
(*Source:* www.nps.gov/grca/)

81. **Surveillance** A security camera is mounted on the wall at a height of 10 feet. At what angle of depression should the camera be set if the camera is to be pointed at a door 50 feet from the point on the floor directly under the camera?

82. **Statue Height** During a hike in Mexico, Sam discovered a large stone statue. To estimate the height of the object, he stood 20 feet from the statue and measured the angle of elevation to the top of the statue to be 70°. What is the height of the statue to the nearest tenth of a foot?

83. **Surfing** Jake has a Surftech Softop surfboard. When he stands it up in the sand, it casts a shadow that is 84 inches long. If the angle of elevation of the sun is 45°, how long is the board?

84. **Ancient Buildings** The Great Pyramid of Giza in Egypt is 481 feet high. The distance from the point directly under the highest point to the edge of the pyramid is 375.5 feet. What is the angle of elevation of the sides of the pyramid?

85. **Physics** A force F can be decomposed into its horizontal component F_x and its vertical component F_y. The ratio of F_y to F_x is given by $\tan \theta$, where θ is the angle whose initial side is the positive x-axis and whose terminal side is oriented in the direction of F. Find the value of that ratio if $\theta = 36°$.

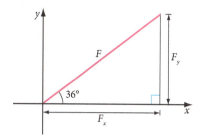

86. **Sewing** A piece of fabric that will serve as the background for a monogram on a sweater is to be cut in the form of a right triangle. What is the ratio of the lengths of the two shortest edges of the piece of fabric if one of the shortest edges is 3.5 inches long and the angle opposite that edge is 45°?

87. **Surveying** A surveyor stands 180 feet from the base of a building. The angle of elevation to the top of the building is 57°. Find the height of the building.

88. **Swimming** The angle of depression of a certain point on the surface of the water in a swimming pool from the end of the diving board is 25° degrees. That point is at a horizontal distance of 15 feet from the end of the diving board. A 6-foot-tall swimmer aims for the point in question and makes a straight-line dive, head first. If the top of his head makes contact with the water at that point, find the distance d traversed by his feet between the time they left the end of the diving board and the time his head hit the water.

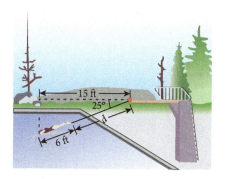

89. **Monuments** The Washington Monument in Washington, DC, is 555 feet high. If the angle of elevation of the top of the monument from a certain point from the ground is 60°, how far is that point from the center of the base? Assume that the base of the monument is on the ground.

90. **Leisure** A 45-foot-long hill makes an angle of 20° with the level ground at the bottom of it. After a big snowfall, people like to sled down the hill. How high above the surrounding level ground do they begin their descent?

91. Illumination A light pole is located next to a house. The light rays that graze the top of the roof make an angle of 27° with respect to the front lawn. If the top of the pole is 25 feet above the lawn, what is the distance traversed by those light rays between the time they are emitted and the time they reach the surface of the lawn?

92. Gardening A garden consists of four sections of equal dimensions. Each section is in the shape of a right triangle. One of the acute angles in each triangle measures 24°, and the side adjacent to that angle is 16 feet long.

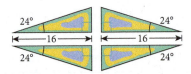

a. Find the length of the hypotenuse.

b. Find the length of the side opposite the 24° angle.

c. Compute the total area of the garden (all four sections combined).

93. Surveying The known dimensions of a parcel of land adjacent to a road are shown in the figure. Find the lengths of the missing sides L and H. Round your answers to two decimal places.

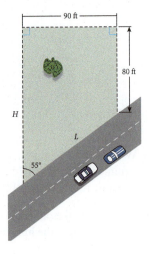

94. **Geometry** Compute the area of the shaded triangle. Round to two decimal places.

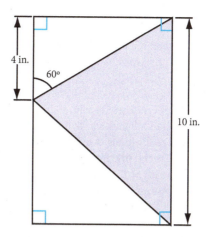

95. **Geometry** The perimeter of a right triangle is 30 inches and one of the angles has a measure of 35°.

a. Find the lengths of all the sides of the triangle. Round your answer to two decimal places.

b. Find the area of the triangle. Round your answer to two decimal places.

Concepts

96. Show that $\tan(90° - \theta) = \cot \theta$.

97. Show that $\csc(90° - \theta) = \sec \theta$.

98. Can a right triangle be used to define sin 90°?

99. Before the widespread use of calculators, values for sine and cosine of selected angles in the interval [0°, 45°] were given in tables. Why was there no need to list the sine and cosine of angles in the interval (45°, 90°)?

SPOTLIGHT ON SUCCESS   *Student Instructor Ryan*

You do not determine a man's greatness by his talent or wealth, as the world does, but rather by what it takes to discourage him.
— *Dr. Jerry Falwell*

From very early in school, I seemed to have a knack for math, but I also had a knack for laziness and procrastination. In rare times, I would really focus in class and nail all the material, but more often I would spend my time goofing off with friends and coasting on what came easily to me. My parents tried their hardest to motivate me, to do anything they could to help me succeed, but when it came down to it, it was on me to control the outcome.

At one point, my dad voiced his frustrations with having to pay for such a good education when I wasn't taking advantage of it. He told me that if I didn't get above a 3.8 GPA, that I would be attending a different school the following fall. It was the motivation I needed. Finally, I understood the importance of trying my best in school. That semester I achieved the goal my dad had given to me.

But it wasn't that easy. I encountered numerous things in high school that could have derailed my academic journey. I felt abandoned by people I considered family, I suffered multiple injuries requiring surgery, and I experienced the death of a classmate and teammate. There was so much going on that I could have just shut down and gone back to coasting by, but luckily, I had already learned my lesson. I managed to stay positive and work hard through all the challenges. Everything paid off senior year when I was accepted to California Polytechnic State University and managed to pass every AP test I took. This journey has taught me that one of the most important things we can do is to work hard no matter what the circumstances; if we refuse to be discouraged, then we can achieve greatness.

Trigonometric Functions of Any Angle Using Right Triangles

5.4

Objectives

- Define trigonometric functions for any angle.
- Find exact values of trigonometric functions of certain special angles.
- Use reference angles to find values of trigonometric functions.
- Relate trigonometric function through the Pythagorean Theorem.
- Use a calculator to find trigonometric function values.

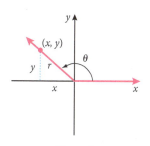

Because many applications involve angles that are not acute, we need to extend the definition of the trigonometric functions to these types of angles. To do so, we consider any angle θ, in standard position, and use a point (x, y) on its terminal side to construct a triangle. See Figure 1. We then use this triangle to define the six trigonometric functions of any angle θ.

Figure 1

Trigonometric Functions of Any Angle θ

Let (x, y) be a point and θ be the angle, in standard position, whose terminal side is the ray from the origin passing through (x, y). Let $r = \sqrt{x^2 + y^2}$. Then

$$\cos\theta = \frac{x}{r} \qquad \sin\theta = \frac{y}{r}$$

$$\tan\theta = \frac{y}{x} \qquad \cot\theta = \frac{x}{y}$$

$$\sec\theta = \frac{r}{x} \qquad \csc\theta = \frac{r}{y}$$

It is useful to define the reciprocal functions, secant, cosecant, and cotangent, in terms of cosine, sine, and tangent, as follows.

Reciprocal Functions

$$\sec\theta = \frac{1}{\cos\theta} \qquad \csc\theta = \frac{1}{\sin\theta} \qquad \cot\theta = \frac{1}{\tan\theta}$$

VIDEO EXAMPLES

SECTION 5.4

Example 1 Evaluating Trigonometric Functions

In Figure 2 and Figure 3, find the exact values of all the trigonometric functions of each value of θ.

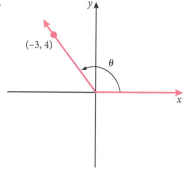

a.

Figure 2

b.

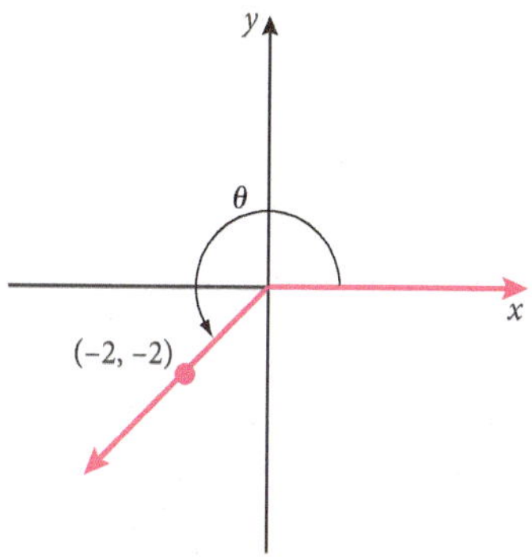

Figure 3

Solution

a. To evaluate the trigonometric functions of θ, first compute r.

$$r = \sqrt{x^2 + y^2} = \sqrt{(-3)^2 + 4^2} = \sqrt{25} = 5$$

Using the definitions of the trigonometric functions,

$$\cos\theta = \frac{x}{r} = \frac{-3}{5} = -\frac{3}{5} \qquad \sin\theta = \frac{y}{r} = \frac{4}{5}$$

$$\tan\theta = \frac{y}{x} = \frac{4}{-3} = -\frac{4}{3} \qquad \cot\theta = \frac{x}{y} = \frac{-3}{4} = -\frac{3}{4}$$

$$\sec\theta = \frac{r}{x} = \frac{5}{-3} = -\frac{5}{3} \qquad \csc\theta = \frac{r}{y} = \frac{5}{4}$$

b. First compute r.

$$r = \sqrt{x^2 + y^2}$$

$$= \sqrt{(-2)^2 + (-2)^2} = \sqrt{8} = 2\sqrt{2}$$

Using the definitions of the trigonometric functions,

$$\cos\theta = \frac{x}{r} = \frac{-2}{2\sqrt{2}} = -\frac{\sqrt{2}}{2} \qquad \sin\theta = \frac{y}{r} = \frac{-2}{2\sqrt{2}} = -\frac{\sqrt{2}}{2}$$

$$\tan\theta = \frac{y}{x} = \frac{-2}{-2} = 1 \qquad \cot\theta = \frac{x}{y} = \frac{-2}{-2} = 1$$

$$\sec\theta = \frac{r}{x} = \frac{2\sqrt{2}}{-2} = -\sqrt{2} \qquad \csc\theta = \frac{r}{y} = \frac{2\sqrt{2}}{-2} = -\sqrt{2}$$

✅ *Check It Out 1*   Let $(x, y) = (4, -3)$. Find the exact values of all the trigonometric functions of θ.

The quadrant in which the terminal side of the angle θ lies will tell us whether the values of the trigonometric functions are positive or negative. Because $r > 0$, the signs of the trigonometric functions will depend only on the signs or x and y. See Figure 4.

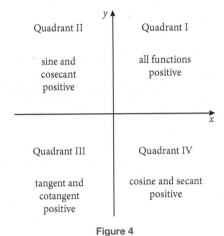

Figure 4

Example 2   **Determining the Quadrant in Which an Angle Lies**

Suppose $\sec\theta < 0$ and $\sin\theta > 0$. In what quadrant does the terminal side of θ lie?

Solution Since $\sec\theta < 0$ and $\sec\theta = \frac{1}{\cos\theta}$, $\cos\theta < 0$. Because $\sin\theta > 0$, and $\cos\theta < 0$, the terminal side of θ must lie in Quadrant II.

✅ *Check It Out 2*   Suppose $\cos\theta > 0$ and $\csc\theta < 0$. In what quadrant does the terminal side of θ lie?

Quadrantal Angles

When the terminal side of an angle lies on the horizontal or vertical axes, it is referred to as a **quadrantal angle.** The values of the trigonometric functions for these angles can still be calculated from the given definition, but some values may be undefined.

> **Example 3**   **Trigonometric functions of $\theta = 90°$ and $\theta = 180°$**
>
> Find the exact values of all six trigonometric function of

a. $\theta = 90°$ **b.** $\theta = 180°$

Solution

a. The terminal side of $\theta = 90°$ lies along the positive y-axis. Choose $(0, 2)$ as a point on this axis. Thus, $x = 0$, $y = 2$, and $r = 2$. Using the definitions of the trigonometric functions,

$$\cos \theta = \frac{x}{r} = \frac{0}{2} = 0 \qquad\qquad \sin \theta = \frac{y}{r} = \frac{2}{2} = 1$$

$$\tan \theta = \frac{y}{x} = \frac{2}{0} \text{ is undefined.} \qquad \cot \theta = \frac{x}{y} = \frac{0}{2} = 0$$

$$\sec \theta = \frac{r}{x} = \frac{2}{0} \text{ is undefined.} \qquad \csc \theta = \frac{r}{y} = \frac{2}{2} = 1$$

The values of trigonometric functions remain the same regardless of which point on the positive y-axis is chosen.

b. The terminal side of $\theta = 180°$ lies along the negative x-axis. Choose $(-2, 0)$ as a point on this axis. Thus $x = -2$, $y = 0$, and $r = 2$. Using the definitions of the trigonometric functions,

$$\cos \theta = \frac{x}{r} = \frac{-2}{2} = -1 \qquad \sin \theta = \frac{y}{r} = \frac{0}{2} = 0$$

$$\tan \theta = \frac{y}{x} = \frac{0}{-2} = 0 \qquad \cot \theta = \frac{x}{y} = \frac{-2}{0} \text{ is undefined.}$$

$$\sec \theta = \frac{r}{x} = \frac{2}{-2} = -1 \qquad \csc \theta = \frac{r}{y} = \frac{-2}{0} \text{ is undefined.}$$

The values of trigonometric functions remain the same regardless of which point on the negative x-axis is chosen.

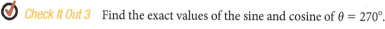 **Check It Out 3**   Find the exact values of the sine and cosine of $\theta = 270°$.

Table 1 gives the sines and cosines of all the angles whose terminal side lies on the x- or y-axis.

Exact Values of Sine and Cosine of Quadrantal Angles					
θ	0°	90°	180°	270°	360°
(in Radians)	(0)	$\left(\dfrac{\pi}{2}\right)$	(π)	$\left(\dfrac{3\pi}{2}\right)$	(2π)
$\cos \theta$	1	0	−1	0	1
$\sin \theta$	0	1	0	−1	0

Table 1

Trigonometric Functions of Special Angles in Other Quadrants

In Section 5.3, we computed the values of the trigonometric functions of acute angles. These angles have their terminal sides in the first quadrant. To calculate the trigonometric function values of angles whose terminal sides lie in the other quadrants, we apply the concept of a **reference angle,** defined as follows.

Reference Angle

For an angle θ in standard position whose terminal side lies in one of the four quadrants, the **reference angle** for θ is the *acute* angle that the terminal side of θ makes with the x-axis. See Figure 5.

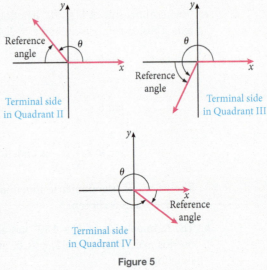

Figure 5

Example 4   **Reference Angles**

Find the reference angle for each of the following angles.

a. $\theta = 135°$ **b.** $\theta = -110°$

Solution

a. Because 135° is a positive angle, we measure it in the *counterclockwise* direction from the positive x-axis. Its terminal side makes an acute angle of 45° with the horizontal axis, and thus the reference angle is 45°. See Figure 6.

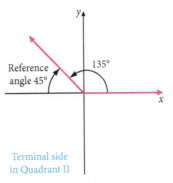

Figure 6

b. Because $-110°$ is a negative angle, we measure it in the *clockwise* direction from the positive x-axis. Its terminal side makes an acute angle of $70°$ with the horizontal axis, and thus the reference angle is $70°$. See Figure 7.

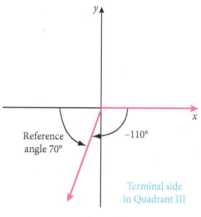

Figure 7

✅ *Check It Out 4*   Find the reference angle for $\theta = 330°$.

A reference angle can be used to find the values of the trigonometric functions in other quadrants. The values of the trigonometric functions of $30°$, $45°$ and $60°$ are given in Table 2.

θ	30°	45°	60°
$\cos\theta$	$\dfrac{\sqrt{3}}{2}$	$\dfrac{\sqrt{2}}{2}$	$\dfrac{1}{2}$
$\sin\theta$	$\dfrac{1}{2}$	$\dfrac{\sqrt{2}}{2}$	$\dfrac{\sqrt{3}}{2}$
$\tan\theta$	$\dfrac{\sqrt{3}}{3}$	1	$\sqrt{3}$

Table 2

These values will help you calculate trigonometric functions of angles whose terminal sides lie in other quadrants.

Example 5   **Trigonometric Function Values of Special Angles in other Quadrants**
Find the exact values of the following.

a. $\cos 225°$ 　　　　　　**b.** $\sin\dfrac{5\pi}{6}$ 　　　　　　**c.** $\tan(-45°)$

Solution

a. The terminal side of $225°$ lies in the third quadrant, so $\cos 225°$ will be negative. See Figure 8. The reference angle for $225°$ is $45°$. Therefore, the cosine of $225°$ will have the same magnitude as the cosine of $\cos 45° = \dfrac{\sqrt{2}}{2}$, but will *differ in sign*. Hence

$$\cos 225° = -\frac{\sqrt{2}}{2}$$

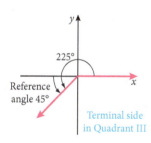

Figure 8

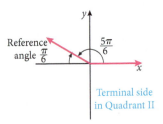

Reference angle $\frac{\pi}{6}$

$\frac{5\pi}{6}$

Terminal side in Quadrant II

Figure 9

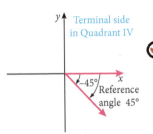

Terminal side in Quadrant IV

$-45°$ Reference angle $45°$

Figure 10

Just in Time
Review coterminal angles in Section 5.1.

b. The terminal side of $\frac{5\pi}{6}$, or 150°, lies in the second quadrant, so $\sin\frac{5\pi}{6}$ will be positive. See Figure 9. The reference angle for $\frac{5\pi}{6}$ is 30°, or $\frac{\pi}{6}$. Therefore, the sine of $\frac{5\pi}{6}$ will have the *same* magnitude and sign as $\sin\frac{\pi}{6} = \frac{1}{2}$. Hence

$$\sin\frac{5\pi}{6} = \frac{1}{2}$$

c. The terminal side of $-45°$ lies in the fourth quadrant, so $\tan(-45°)$ will be negative. The reference angle for $-45°$ is 45°. See Figure 10. Therefore, the tangent of $-45°$ will have the same magnitude as $\tan 45° = 1$, but *differ* in sign. Hence

$$\tan(-45°) = -1$$

Check It Out 5   Find $\sin\left(-\frac{4\pi}{3}\right)$ and $\cos\left(-\frac{4\pi}{3}\right)$. ■

Trigonometric Functions of Coterminal Angles

Coterminal angles share the same terminal side, and thus they have the same reference angles. We can use this fact to calculate the values of trigonometric functions of angles whose measure is greater than 360° or less than 0°.

Example 6   **Trigonometric Functions of Coterminal Angles**

Find the following:

a. $\sin 480°$

b. $\cos(-90°)$

Solution

a. Because the terminal sides of 480° and 120° are the same, the reference angle is 60°. See Figure 11. Thus,

$$\sin 480° = \sin 120° = \frac{\sqrt{3}}{2}$$

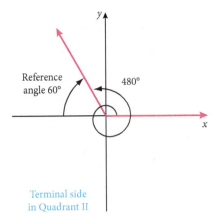

Reference angle 60°

480°

Terminal side in Quadrant II

Figure 11

b. Because the terminal sides of $-90°$ and $270°$ are the same,

$$\cos(-90°) = \cos 270° = 0$$

See Figure 12.

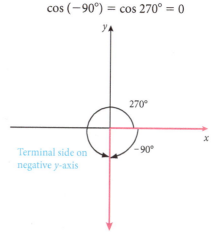

Figure 12

Check It Out 6 Find the exact value of $\cos 495°$

Relating the Trigonometric Functions

Given the value of a single trigonometric function of θ, we can determine the values of the other related trigonometric functions of θ by applying the Pythagorean Theorem. This is shown in the following example.

Just in Time
Review the Pythagorean Theorem in Section 5.3.

Example 7 **Evaluating Trigonometric Functions**

Suppose $\sin \theta = -\frac{1}{3}$ and the terminal side of θ lies in the fourth quadrant. Find $\cos \theta$ and $\tan \theta$.

Solution Since $\sin \theta = -\frac{1}{3} = \frac{y}{r}$, $y = -1$ and $r = 3$. See Figure 13. Note that r must always be positive.

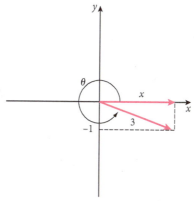

Figure 13

Now, find x using the Pythagorean Theorem:

$$x^2 + y^2 = r^2 \Rightarrow (x)^2 + (-1)^2 = 3^2$$

Solving for x^2 gives

$$x^2 = 8$$

So $x = \pm 2\sqrt{2}$. Because the terminal side of θ lies in the fourth quadrant, choose $x = 2\sqrt{2}$. Thus,

$$\cos \theta = \frac{x}{r} = \frac{2\sqrt{2}}{3}, \qquad \tan \theta = \frac{y}{x} = \frac{-1}{2\sqrt{2}} = -\frac{\sqrt{2}}{4}$$

Check It Out 7 Suppose $\sin \theta = -\frac{2}{3}$ and the terminal side of θ lies in the third quadrant. Find $\cos \theta$ and $\tan \theta$.

Evaluating Trigonometric Functions Using a Calculator

Only the trigonometric functions of special angles can be calculated exactly. For other angles, a calculator is needed to calculate the values of trigonometric functions.

> **Example 8**   **Using a Calculator to Evaluate Trigonometric Functions**

Use a calculator to find the following. Round your answers to four decimal places.

a. $\sin 242°$

b. $\cot 318°$

Solution

a. Setting the calculator in DEGREE mode, we have

$$\sin 242° \approx -0.8829$$

b. Calculators do not generally have a key for the cotangent function. So we use the fact that $\cot x = \frac{1}{\tan x}$, and use the key for the tangent function.

Thus,

$$\cot 318° = \frac{1}{\tan 318°} \approx -1.106$$

✓ *Check It Out 8*   Use a calculator to find $\cos 197°$ and $\sec 197°$. Round your answers to four decimal places.   ■

Exercises 5.4

Just in Time Exercises

In Exercises 1–4, find at least two angles, one positive and one negative, that are coterminal to the given angle. Answers may vary.

1. 50° **2.** 130° **3.** 250° **4.** 320°

In Exercises 5 and 6, use the Pythagorean Theorem to find the value of the missing leg, a or b, or the hypotenuse, c.

5. $a = 1, b = 4, c = $ _____

6. $a = 2, b = $ _____, $c = 5$

Skills

In Exercises 7–14, find the exact values of all the trigonometric functions for each value of θ.

7.
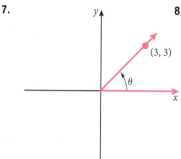

8.
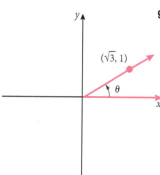

9.
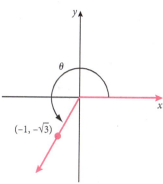

10.
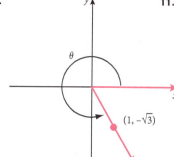

11.
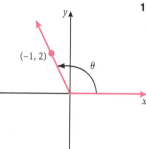

12.
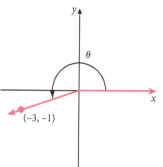

13. **14.**
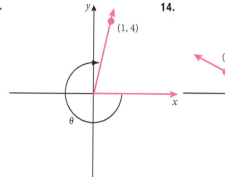

In Exercises 15–20, for each angle θ in standard position, determine the quadrant where the terminal side of θ lies.

15. $\cos \theta > 0$ and $\sin \theta < 0$ **16.** $\sin \theta > 0$ and $\cos \theta < 0$ **17.** $\sec \theta < 0$ and $\tan \theta > 0$

18. $\csc \theta < 0$ and $\tan \theta < 0$ **19.** $\cot \theta > 0$ and $\csc \theta > 0$ **20.** $\cot \theta < 0$ and $\sin \theta > 0$

In Exercises 21–32, find the reference angle for each of the given angles.

21. $120°$ **22.** $150°$ **23.** $240°$ **24.** $290°$ **25.** $-105°$ **26.** $-210°$

27. $400°$ **28.** $530°$ **29.** $\dfrac{7\pi}{4}$ **30.** $\dfrac{7\pi}{3}$ **31.** $-\dfrac{5\pi}{3}$ **32.** $-\dfrac{4\pi}{3}$

In Exercises 33–60, find the exact value of the given trigonometric function. If a certain value is undefined, state so.

33. $\sin 150°$ **34.** $\cos 240°$ **35.** $\sin 225°$ **36.** $\sin 135°$ **37.** $\cos 120°$

38. $\sin 330°$ **39.** $\sin \dfrac{2\pi}{3}$ **40.** $\cos \dfrac{7\pi}{6}$ **41.** $\tan 225°$ **42.** $\cot 315°$

43. $\tan 330°$ **44.** $\tan 270°$ **45.** $\sec 150°$ **46.** $\csc 210°$ **47.** $\cot \dfrac{3\pi}{2}$

48. $\cot \pi$ **49.** $\tan 420°$ **50.** $\tan 450°$ **51.** $\cos (-210°)$ **52.** $\sin (-60°)$

53. $\tan (-45°)$ **54.** $\tan (-150°)$ **55.** $\sec (-135°)$ **56.** $\csc (-210°)$ **57.** $\csc 180°$

58. $\sec 540°$ **59.** $\sec 2\pi$ **60.** $\csc 3\pi$

In Exercises 61–78, find the exact value of each expression. Simplify your answer.

61. $\cos 45° + \sin 30°$ **62.** $\sin 45° - \cos 30°$ **63.** $\cos 120° - \sin 210°$ **64.** $\sin 45° + \cos 135°$

65. $\tan 120° + \cot 135°$ **66.** $\cot 45° + \tan 120°$ **67.** $\cot 210° - \tan 330°$ **68.** $\tan 180° + \cot 90°$

69. $\sec 180° + \csc 270°$ **70.** $\csc 45° - \sec 135°$ **71.** $\sin \dfrac{\pi}{3} + \cos \dfrac{2\pi}{3}$ **72.** $\cos \pi + \sin \dfrac{3\pi}{2}$

73. $\tan \dfrac{5\pi}{4} - \cot \dfrac{\pi}{4}$ **74.** $\tan 2\pi + \cot \dfrac{3\pi}{4}$ **75.** $\cos 45° \cos 30° + \sin 45° \sin 30°$

76. $\sin 45° \cos 30° + \cos 45° \sin 30°$ **77.** $\sin 60° \cos 45° - \cos 45° \sin 60°$

78. $\cos 30° \cos 45° - \sin 30° \sin 45°$

In Exercises 79–88, find the exact value of the indicated trigonometric function, using the given information.

79. $\sin \theta$ if $\cos \theta = \frac{3}{5}$; terminal side of θ in Quadrant IV

80. $\cos \theta$ if $\sin \theta = -\frac{4}{5}$; terminal side of θ in Quadrant III

81. $\sec \theta$ if $\sin \theta = -\frac{5}{13}$; terminal side of θ in Quadrant III

82. $\csc \theta$ if $\cos \theta = -\frac{12}{13}$; terminal side of θ in Quadrant II

83. $\tan \theta$ if $\cos \theta = -\frac{1}{3}$; terminal side of θ in Quadrant III

84. $\cot \theta$ if $\sin \theta = \frac{2}{5}$; terminal side of θ in Quadrant II

85. $\cos \theta$ if $\sin \theta = -\frac{1}{5}$; terminal side of θ in Quadrant IV

86. $\tan \theta$ if $\cos \theta = -\frac{2}{3}$; terminal side of θ in Quadrant II

87. $\sec \theta$ if $\sin \theta = -0.8$; terminal side of θ in Quadrant IV

88. $\csc \theta$ if $\cos \theta = -0.6$; terminal side of θ in Quadrant II

In Exercises 89–98, evaluate the trigonometric function using a calculator. Round your answers to four decimal places.

89. $\sin 157°$ **90.** $\cos 234°$ **91.** $\cos 435.4°$ **92.** $\sin 385.7°$ **93.** $\tan 125°$

94. $\tan 192°$ **95.** $\cot 214°$ **96.** $\sec 143°$ **97.** $\csc 315.4°$ **98.** $\cot 167.9°$

Application

99. Distance In the given illustration, find the distance from the top of the pole to Point *A*, directly below it.

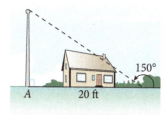

100. Navigation A ship starts at Point *O*, the origin, and travels 50 miles in a straight line path, that makes an angle of 120° with the horizontal. Find the distance *d* shown in the illustration.

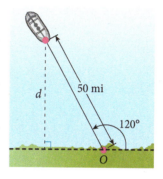

101. Maps Three locations on a map are shown. Find the distance from Lakecrest to Woodbrook.

Concepts

102. Why are sin 90° and sin 450° the same?

103. Explain why sec 270° is undefined.

104. Let $(3, -4)$ and $(6, -8)$ be two points on the terminal side of an angle θ. Compute $\sin \theta$ using each pair of coordinates and show that you get the same answer.

105. Let (x, y) and (kx, ky), $k > 0$ be two points on the terminal side of an angle θ. Compute $\cos \theta$ using each pair of coordinates and show that you get the same answer.

Graphs of the Sine and Cosine Functions

5.5

Objectives

- Sketch the graphs of the basic sine and cosine functions.
- Know the properties of the sine and cosine functions.
- Sketch transformations of the graphs of the sine and cosine functions.

In Section 5.2, we used the properties of the unit circle to define the sines and cosines of all *real numbers*. We now consider the graphs of these functions. Start by constructing a table (Table 1) of values of sin t for selected values of t from 0 to 2π.

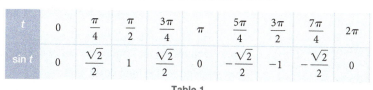

t	0	$\dfrac{\pi}{4}$	$\dfrac{\pi}{2}$	$\dfrac{3\pi}{4}$	π	$\dfrac{5\pi}{4}$	$\dfrac{3\pi}{2}$	$\dfrac{7\pi}{4}$	2π
sin t	0	$\dfrac{\sqrt{2}}{2}$	1	$\dfrac{\sqrt{2}}{2}$	0	$-\dfrac{\sqrt{2}}{2}$	-1	$-\dfrac{\sqrt{2}}{2}$	0

Table 1

Connecting the points with a smooth curve, we have the graph of sin t on $[0, 2\pi]$, as shown in Figure 1. Because the sine function is defined for all values of t, the complete graph of sin t extends infinitely in both directions of the t-axis. Moreover, the graph of the sine function on the interval $[0, 2\pi]$ repeats itself on successive intervals of length 2π in both directions, as shown in Figure 2.

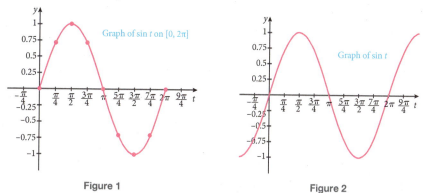

Figure 1

Figure 2

Similarly, for the cosine function, we get the following table of values (Table 2) and the corresponding graph of cos t on $[0, 2\pi]$, as shown in Figure 3. The complete graph of cos t also extends infinitely in both directions of the t-axis, and the graph of cos t on the interval $[0, 2\pi]$ repeats itself on successive intervals of length 2π in both directions, as shown in Figure 4.

t	0	$\dfrac{\pi}{4}$	$\dfrac{\pi}{2}$	$\dfrac{3\pi}{4}$	π	$\dfrac{5\pi}{4}$	$\dfrac{3\pi}{2}$	$\dfrac{7\pi}{4}$	2π
cos t	1	$\dfrac{\sqrt{2}}{2}$	0	$-\dfrac{\sqrt{2}}{2}$	-1	$-\dfrac{\sqrt{2}}{2}$	0	$\dfrac{\sqrt{2}}{2}$	1

Table 2

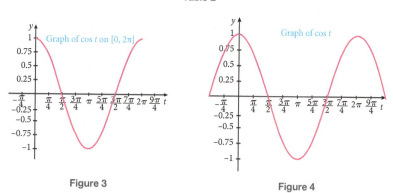

Figure 3

Figure 4

*Note* So far, we have used the variable t to represent the independent variable to avoid confusion with the coordinates (x, y) of a point on the unit circle. But the cosine function can also be referred to as $\cos x$ or $\cos \theta$, since any letter or symbol can be used as the independent variable. In what follows, we will use $\cos x$ and $\sin x$ to refer to the cosine and sine functions, because that is common usage in calculus.

A function f that repeats its values at regular intervals is called **periodic.** For any periodic function, the length of the shortest interval over which the function repeats itself is known as the **period,** p. In function notation, this means $f(x + p) = f(x)$.

Properties of the Sine and Cosine Functions

- The sine and cosine functions are both periodic with period 2π.

$$\sin(x + 2\pi) = \sin x \text{ and } \cos(x + 2\pi) = \cos x$$

  The function is said to go through one **cycle** between every pair of repetitions.
- Domain of $\sin x$ and $\cos x$: all real numbers, $(-\infty, \infty)$
- Range of $\sin x$ and $\cos x$: $[-1, 1]$
- The functions $\sin x$ and $\cos x$ oscillate between -1 and 1, and so have **amplitude** 1.

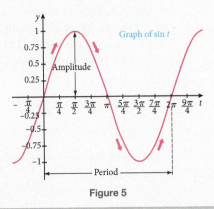

Figure 5

Observations:

- The zeros of $f(x) = \sin x$ are $x = n\pi$, n is an integer. The zeros of $f(x) = \cos x$ are $x = \frac{(2n + 1)\pi}{2}$, where n is an integer.
- From the negative-angle identities in Section 5.2, we have $\sin(-x) = -\sin x$. Thus, $f(x) = \sin x$ is an *odd function.*
- From the negative-angle identities in Section 5.2, we have $\cos(-x) = \cos x$. Thus, $f(x) = \cos x$ is an *even function.*

Just in Time
Review transformations in Section 2.7.

Shifts of the Graphs of the Sine and Cosine Functions

The facts about transformations in Chapter 2 play an important role in the graphs of the trigonometric functions. In addition, we will also need the values of x in $[0, 2\pi]$ where $\sin x$ and $\cos x$ equal 1, 0, or -1. See Figures 2 and 4. These values of x will be referred to as *key values* of x. See Table 3.

List of Key Values of x

x	0	$\frac{\pi}{2}$	π	$\frac{3\pi}{2}$	2π
$\cos x$	1	0	-1	0	1
$\sin x$	0	1	0	-1	0

Table 3

The next two examples illustrate how the basic graphs of the sine and cosine functions can be transformed by vertical or horizontal shifts.

Using Technology

Use a trigonometric graphing window to graph $f(x) = \cos x - 2$, as shown in Figure 6. Make sure the calculator is in RADIAN mode.

Keystroke Appendix:
Section 15

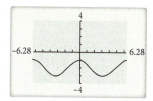

Figure 6

Example 1   **Vertical Shift of a Cosine Graph**

Sketch the graph of $f(x) = \cos x - 2$.

Solution   The graph of the function $f(x) = \cos x - 2$ is obtained by taking the graph of $\cos x$ and shifting it 2 units *downward*. To help graph the function, first create a table of values of $\cos x$ for key values of x from 0 to 2π. These are the values of x where $\cos x$ attains a maximum, minimum, or zero. Refer to Table 4. Then make a separate row of values of $\cos x - 2$.

x	0	$\dfrac{\pi}{2}$	π	$\dfrac{3\pi}{2}$	2π	
$\cos x$	1	0	−1	0	1	← Compute $\cos x$
$\cos x - 2$	−1	−2	−3	−2	−1	← and subtract 2

Table 4

Using the table of values for $f(x) = \cos x - 2$, sketch the graph of f on the interval $[0, 2\pi]$. Since the cosine function is periodic, the rest of the graph is obtained by repeating the cycle. See Figure 7.

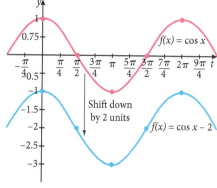

Figure 7

✓ *Check It Out 1*   Sketch the graph of $f(x) = \sin x + 1$.

Example 2   **Horizontal Shifts of the Sine and Cosine Graphs**

Use transformations to graph the following functions

a. $f(x) = \cos\left(x - \dfrac{\pi}{2}\right)$ **b.** $g(x) = \sin\left(x + \dfrac{\pi}{4}\right)$

Solution

a. To graph this function, first create a table of key values of $\cos x$ for x from 0 to 2π. See Table 5. Now, the graph of the function $f(x) = \cos\left(x - \dfrac{\pi}{2}\right)$ is obtained by taking the graph of $\cos x$ and shifting it to the *right* by $\dfrac{\pi}{2}$ units. Therefore, the values of $\cos x$ for x from 0 to 2π will coincide with the values of $\cos\left(x - \dfrac{\pi}{2}\right)$ for x from $\dfrac{\pi}{2}$ to $\dfrac{5\pi}{2}$. See Table 5 and Table 6.

x	0	$\dfrac{\pi}{2}$	π	$\dfrac{3\pi}{2}$	2π
$\cos x$	1	0	−1	0	1

Table 5

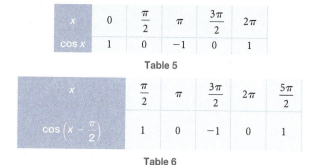

x	$\dfrac{\pi}{2}$	π	$\dfrac{3\pi}{2}$	2π	$\dfrac{5\pi}{2}$
$\cos\left(x - \dfrac{\pi}{2}\right)$	1	0	−1	0	1

Table 6

Using the values from Table 5, sketch the graph of $f(x) = \cos\left(x - \frac{\pi}{2}\right)$ on the interval $\left[\frac{\pi}{2}, \frac{5\pi}{2}\right]$. Then use the repeating behavior of the cosine function to extend the graph a little on each side. See Figure 8. Observe that the graph of $\cos\left(x - \frac{\pi}{2}\right)$ looks exactly like the graph of $\sin x$, a fact that will be proven in Chapter 6.

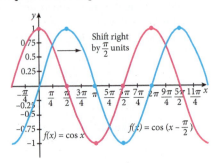

Figure 8

b. First, we make a table of key values of $\sin x$ for x from 0 to 2π. The graph of $g(x) = \sin\left(x + \frac{\pi}{4}\right)$ is obtained by taking the graph of $y = \sin x$ and shifting it *to the left* by $\frac{\pi}{4}$ units. Therefore, the values of $\sin x$ for x from 0 to 2π will coincide with the values of $\sin\left(x + \frac{\pi}{4}\right)$ for x from $-\frac{\pi}{4}$ to $\frac{7\pi}{4}$. See Table 7 and Table 8.

X	0	$\frac{\pi}{2}$	π	$\frac{3\pi}{2}$	2π
sin x	0	1	0	−1	0

Table 7

X	$-\frac{\pi}{4}$	$\frac{\pi}{4}$	$\frac{3\pi}{4}$	$\frac{5\pi}{4}$	$\frac{7\pi}{4}$
sin (x + π/4)	0	1	0	−1	0

Table 8

Using the table of values from Table 8, sketch the graph of the function g on the interval $\left[-\frac{\pi}{4}, \frac{7\pi}{4}\right]$, and then extend it a little on each side. The effect of the horizontal shift is illustrated in Figure 9.

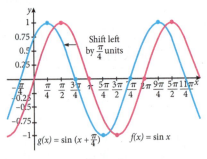

Figure 9

✓ *Check It Out 2*   Sketch the graph of $f(x) = \sin\left(x + \frac{\pi}{2}\right)$. ∎

Stretches and Compressions

When the functions $\cos x$ and $\sin x$ are shifted vertically or horizontally, the amplitude and the period of the functions do not change. However, when the graphs of sine and cosine functions are transformed by horizontal and vertical stretches and compressions, the period of the function and/or the amplitude of the function will change. This is illustrated in the following two examples.

Example 3 **Graph of *a* sin *x***

Use a transformation to sketch the graph of $g(x) = 3 \sin x$.

Solution The graph of $g(x) = 3 \sin x$ is obtained by *vertically stretching* the graph of sin *x* by a factor of 3. Thus, the amplitude is 3. This is seen from a table of values (Table 9) of sin *x* and 3 sin *x* at the key values of *x*, and from the resulting graph of the function $g(x) = 3 \sin x$, which is shown in Figure 10.

x	0	$\frac{\pi}{2}$	π	$\frac{3\pi}{2}$	2π
sin x	0	1	0	−1	0
3 sin x	0	3	0	−3	0

Table 9

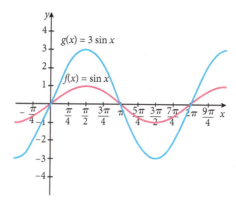

Figure 10

From Table 9 and Figure 10, the period of $g(x) = 3 \sin x$ is 2π, the same as that of sin *x*.

 Check It Out 3 Use a transformation to sketch the graph of $g(x) = 2 \cos x$.

The graphs of sine and cosine functions can be transformed through **horizontal stretches and compressions,** as seen in Example 4.

Example 4 **Graph of cos *bx***

Sketch the graph of $f(x) = \cos 2x$.

Solution Start by tabulating values of cos *x* at key values of *x* in the interval $[0, 2\pi]$. See Table 10.

x	0	$\frac{\pi}{2}$	π	$\frac{3\pi}{2}$	2π
cos x	1	0	−1	0	1

Table 10

Then tabulate values of cos 2*x*. Note that when we evaluate this function, we first multiply the value of *x* by 2 and then take the cosine. Thus, we need only generate values of cos 2*x* at key values of *x* in the interval $[0, \pi]$. The key *x* values are now in increments of $\frac{\pi}{4}$. See Table 11.

x	0	$\frac{\pi}{4}$	$\frac{\pi}{2}$	$\frac{3\pi}{4}$	π
cos 2x	1	0	−1	0	1

Table 11

The graphs of both cos x and cos $2x$ are given in Figure 11.

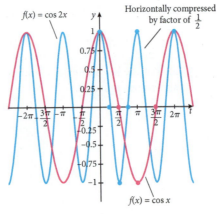

Figure 11

The *period* of $f(x) = \cos 2x$ is π. Thus, we see that the graph of cos $2x$ can be obtained from the graph of cos x by applying a *horizontal compression* of a factor of $\frac{1}{2}$.

✅ *Check It Out 4*   Sketch the graph of sin $2x$.   ■

Using Technology

Use a trigonometric graphing window to graph $f(x) = \cos 2x$, as shown in Figure 12. Make sure the calculator is in RADIAN mode.

Keystroke Appendix:
Section 15

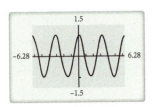

Figure 12

Functions of the Form $a \sin b(x - c) + d$ and $a \cos b(x - c) + d$

We now study general functions of the form

$$f(x) = a \sin b(x - c) + d \text{ and } f(x) = a \cos b(x - c) + d$$

where a, b, c, and d are constants such that $a \neq 0$ and $b > 0$. We can study these functions as transformations of the functions $a \sin bx$ and $a \cos bx$, respectively.

Functions of the Form $f(x) = a \sin b(x - c) + d$ and $f(x) = a \cos b(x - c) + d$

Assume that $a \neq 0$ and $b > 0$. We then have the following properties of f.

Amplitude: $|a|$ **Period:** $\frac{2\pi}{b}$

The graphs of these functions are found by applying the following transformations to the graphs of the functions sin x and cos x, respectively.

Horizontal stretch by a factor of $\frac{1}{b}$ if $b < 1$

Horizontal compression by a factor of $\frac{1}{b}$ if $b > 1$

Vertical stretch by a factor of $|a|$ if $|a| > 1$

Vertical compression by a factor of $|a|$ if $|a| < 1$

Reflection across the x-axis if $a < 0$

Horizontal shift of size $|c|$:   to the right if $c > 0$

(also known as phase shift)   to the left if $c < 0$

Vertical shift of size $|d|$:   upward if $d > 0$

          downward if $d < 0$

Example 5 Graph of $-3\cos\left(\frac{1}{2}x\right)$

Find the amplitude and period of $f(x) = -3\cos\frac{1}{2}x$, and sketch the graph of f.

Solution Here, $a = -3$ and $b = \frac{1}{2}$. The constants c and d are zero. The amplitude of this function is $|a| = |-3| = 3$. The period of this function is

$$P = \frac{2\pi}{b} = \frac{2\pi}{\frac{1}{2}} = 4\pi$$

So the graph of $-3\cos\frac{1}{2}x$ repeats itself every 4π units. Because this is a horizontal *stretching* of the graph the $\cos x$ function, the key values of x are in the interval $[0, 4\pi]$ in increments of π. See Table 12.

x	0	π	2π	3π	4π
$-3\cos\frac{1}{2}\pi$	-3	0	3	0	-3

Table 12

Using these points gives the graph in Figure 13. We can then extend the graph on either side because the function is periodic. See Figure 14.

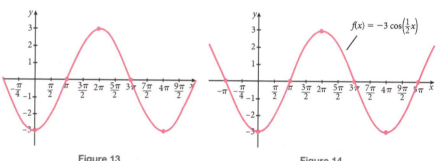

Figure 13

Figure 14

 Check It Out 5 Find the amplitude, and period of $f(x) = -\sin\left(\frac{1}{2}x\right)$, and sketch the graph of f.

Example 6 Amplitude, Period, and Horizontal Shift

Find the amplitude, period, and horizontal shift of the function $f(x) = 4\sin\left(\frac{x}{3} + \frac{\pi}{2}\right)$. Sketch the graph of f.

Solution First, write the function as

$$f(x) = 4\sin\left(\frac{x}{3} + \frac{x}{2}\right) = 4\sin\left(\frac{1}{3}\left(x + \frac{3\pi}{2}\right)\right) = 4\sin\left(\frac{1}{3}\left(x - \left(-\frac{3\pi}{2}\right)\right)\right)$$

Here, $a = 4$, $b = \frac{1}{3}$, and $c = -\frac{3\pi}{2}$. We have the following:

$$\text{Amplitude: } |a| = |4| = 4$$

$$\text{Period: } \frac{2\pi}{b} = \frac{2\pi}{\frac{1}{3}} = 6\pi$$

Horizontal shift (or phase shift): $\frac{3\pi}{2}$ units to the *left*, because $c = -\frac{3\pi}{2} < 0$.

Note that this function will repeat only every 6π units, since its graph has been horizontally stretched by a factor of 3.

We now use the following series of transformations to graph the function.

$$\sin x \longrightarrow \sin\left(\frac{1}{3}x\right) \longrightarrow 4\sin\left(\frac{1}{3}x\right) \longrightarrow 4\sin\left(\frac{1}{3}\left(x + \frac{3\pi}{2}\right)\right)$$

 **Using Technology**

Understanding transformations is critical to graphing, even when using a graphing utility. Figure 15 shows the graph of $f(x) = 4\sin\left(\frac{x}{3} + \frac{x}{2}\right)$ in the default trig window. Less than one complete period is visible. Using the information from Example 6, we obtain a window that shows two complete periods of the function (Figure 16).

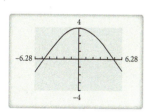

Figure 15

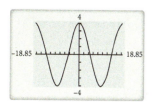

Figure 16

Step 1 The transformation $\sin x \longrightarrow \sin\left(\frac{1}{3}x\right)$:

This transformation *horizontally stretches* the graph of $\sin x$ by a factor of 3. Thus, the period is 6π, so we make a table of values of $\sin\left(\frac{1}{3}x\right)$ by using key values of x in the interval $[0, 6\pi]$. See Table 13. The graph is given in Figure 17.

x	0	$\frac{3\pi}{2}$	3π	$\frac{9\pi}{2}$	6π
$\sin\left(\frac{1}{3}x\right)$	0	1	0	-1	0

Table 13

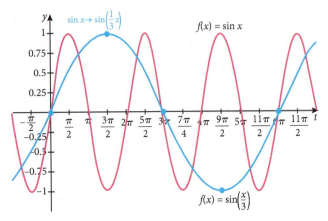

Figure 17

Step 2 The transformation $\sin\left(\frac{1}{3}x\right) \longrightarrow 4\sin\left(\frac{1}{3}x\right)$:

This transformation *vertically stretches* the graph of $\sin\left(\frac{1}{3}x\right)$ by a factor of 4, as shown in Figure 18.

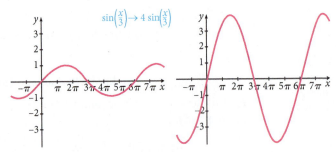

Figure 18

Step 3 The transformation $4\sin\left(\frac{1}{3}x\right) \longrightarrow 4\sin\left(\frac{1}{3}\left(x + \frac{3\pi}{2}\right)\right)$:

This *horizontally shifts* the graph of $4\sin\left(\frac{1}{3}x\right)$ *to the left* by $\frac{3\pi}{2}$ units, as shown in Figure 19.

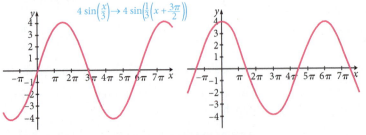

Figure 19

✓ *Check It Out 6*   Rework Example 6 for $f(x) = 3\sin(2x - \pi)$.

Graphing utilities can be quite useful in graphing sine and cosine functions. However, using them requires some care, as shown in Example 7.

Example 7   **Graphing a Sine Function with a Graphing Utility**

Use a graphing utility to graph at least two periods of $f(x) = \sin(20\pi x)$.

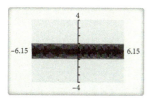

Figure 20

Solution Using a standard trigonometric window, we get the graph in Figure 20. It looks like a set of squiggly lines and doesn't give much information. We need to modify the window size to get a better graph. The period of the function $f(x) = \sin(20\pi x)$ is $\frac{2\pi}{20\pi} = 0.1$. Thus, the graph completes one period in a length of just 0.1.

One possible setting is to adjust the window size to XMin = -0.1, XMax = 0.1 and XScl = 0.01. Because the amplitude is 1, we can set YMin = -1.5, YMax = 1.5 and YScl = 1. See Figure 21. Now, we clearly see two periods of the graph. Understanding the role of period and amplitude is essential to getting reasonable graphs of sine and cosine functions with a graphing utility.

 Check It Out 7   Use a graphing utility to graph at least two periods of $f(x) = \cos(10\pi x)$.

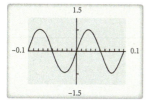

Figure 21

An Application of the Sine Function

The periodic properties of the sine and cosine functions make them attractive for modeling phenomena exhibiting repetitive behavior. Consider a block attached to the end of a spring that is suspended from the ceiling. Once it is pulled downward, it begins to bob up and down in a pattern known as *simple harmonic motion*, which we analyze next.

Example 8   **Simple Harmonic Motion**

The position d of a block attached to a spring is given by the formula $d(t) = 4\sin\frac{\pi}{3}t$, where t is in seconds and d is in centimeters.

a. Find the period of the motion and interpret it.

b. The frequency is given by $\frac{1}{p}$ where p is the period. Find the frequency of the motion and interpret it.

c. What is the maximum distance of the block from its equilibrium position (the position at which $d = 0$)?

Solution

a. The function is of the form $a\sin bt$, with $a = 4$ and $b = \frac{\pi}{3}$. The period is

$$\frac{2\pi}{b} = \frac{2\pi}{\frac{\pi}{3}} = 6$$

Therefore, in 6 seconds, the distance function completes one complete period and the block returns to its starting point.

b. The frequency is equal to $\frac{1}{6}$, the reciprocal of the period. In terms of this problem, the distance function completes $\frac{1}{6}$ of a period in 1 second. Frequency is usually given in the unit sec^{-1}.

c. Because the amplitude of the distance function is

$$|a| = |4| = 4$$

the maximum distance that the block travels from its equilibrium position is 4 centimeters.

 Check It Out 8   Rework Example 8 for the case where $d(t) = 6\sin\frac{\pi}{2}t$. ■

Exercises 5.5

Just in Time Exercises

In Exercises 1–4, fill in the blank with one of the following: upward, downward, to the left, to the right.

1. The graph of $f(x) + 3$ is obtained by shifting the graph of $f(x)$ _____ by 3 units.

2. The graph of $f(x) - 2$ is obtained by shifting the graph of $f(x)$ _____ by 2 units.

3. The graph of $f(x + 1)$ is obtained by shifting the graph of $f(x)$ _____ by 1 unit.

4. The graph of $f(x - 4)$ is obtained by shifting the graph of $f(x)$ _____ by 4 units.

In Exercises 5–8, fill in the blank with one of the following: horizontal, vertical.

5. The graph of $2f(x)$ is obtained by a _____ stretch of the graph of $f(x)$ by a factor of 2.

6. The graph of $f(3x)$ is obtained by a _____ compression of the graph of $f(x)$ by a factor of $\frac{1}{3}$.

7. The graph of $f(\frac{1}{2}x)$ is obtained by a _____ stretch of the graph of $f(x)$ by a factor of 2.

8. The graph of $\frac{1}{4}f(x)$ is obtained by a _____ compression of the graph of $f(x)$ by a factor of $\frac{1}{4}$.

Skills

In Exercises 9–12, use vertical translations to graph at least two periods of the following functions.

9. $f(x) = \cos x + 3$ 10. $f(x) = \sin x - 2$ 11. $g(x) = \cos x - \frac{1}{2}$ 12. $g(x) = \sin x + \frac{3}{2}$

In Exercises 13–16, use horizontal translations to graph at least two periods of the following functions.

13. $f(x) = \cos\left(x - \frac{\pi}{4}\right)$ 14. $f(x) = \sin(x + \pi)$ 15. $g(x) = \cos\left(\frac{3\pi}{4} + x\right)$ 16. $g(x) = \sin\left(\frac{5\pi}{4} + x\right)$

In Exercises 17–20, use vertical stretches and compressions to graph at least two periods of the following functions.

17. $f(x) = -2\sin x$ 18. $f(x) = -3\cos x$ 19. $g(x) = \frac{3}{2}\cos x$ 20. $g(x) = -\frac{1}{2}\sin x$

In Exercises 21–26, use horizontal stretches and compressions to graph at least two periods of the following functions.

21. $f(x) = \sin(3x)$ 22. $f(x) = \cos(4x)$ 23. $f(x) = \sin\left(\frac{1}{2}x\right)$

24. $f(x) = \cos\left(\frac{1}{2}x\right)$ 25. $g(x) = \cos(3x)$ 26. $g(x) = \sin(4x)$

In Exercises 27–36, graph at least two periods of the following functions.

27. $f(x) = -2\sin\left(x - \frac{\pi}{4}\right)$ 28. $f(x) = 3\cos\left(x + \frac{\pi}{2}\right)$ 29. $g(x) = \frac{1}{2}\sin(2x - \pi)$

30. $g(x) = \frac{3}{2}\cos(2x + \pi)$ 31. $h(x) = -3\sin(4x - \pi) + 2$ 32. $h(x) = 2\cos\left(2x + \frac{\pi}{2}\right) - 1$

33. $r(x) = -\cos(2\pi x) + 2$ 34. $r(x) = \sin(4\pi x) - 1$ 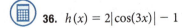 35. $h(x) = 2|\sin(2x)| + 1$

36. $h(x) = 2|\cos(3x)| - 1$

In Exercises 37–42, graph the given pair of functions on the same set of axes. Are the graphs of f and g identical or not?

37. $f(x) = \sin(3x); g(x) = 3\sin(x)$

38. $f(x) = \sin(x + \pi); g(x) = -\sin x$

39. $f(x) = \cos(x + \pi); g(x) = \cos(x) + \cos\pi$

40. $f(x) = \cos(x + \pi); g(x) = -\cos(x)$

41. $f(x) = \sin(3x); g(x) = \sin\left(\frac{1}{3}x\right)$

42. $f(x) = \cos\left(\frac{1}{2}x\right); g(x) = \cos(2x)$

In Exercises 43–46, for each graph, find its corresponding expression of the form $\sin(x - a) + b$, a, b, any real number.

43.

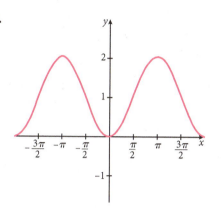

44.

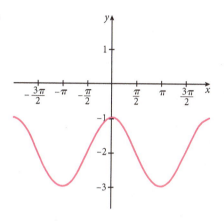

45.

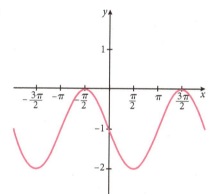

46.

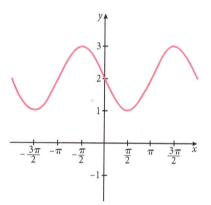

In Exercises 47–54, for each graph, find its corresponding expression of the form $a \sin bx$ or $a \cos bx$, $b > 0$.

47.

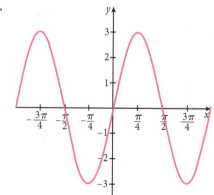

48.

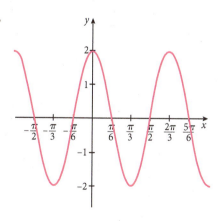

49.

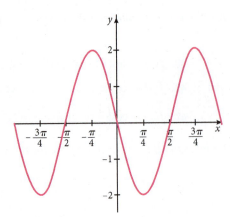

50.

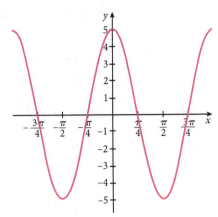

51.

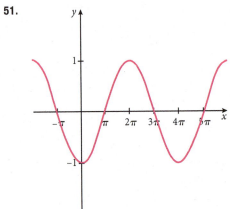

52.

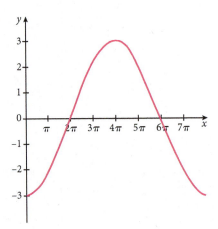

53.

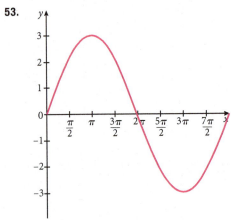

54.

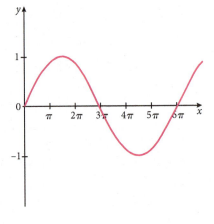

In Exercises 55–60, match the given set of properties with the appropriate function.

a. $h(x) = 2 \cos\left(\dfrac{3}{2}x + \pi\right) - 2$

b. $h(x) = 3 \cos\left(\dfrac{2}{3}x + \dfrac{\pi}{2}\right) + 3$

c. $h(x) = -2 \sin\left(2x - \dfrac{\pi}{2}\right) - 2$

d. $h(x) = -2 \cos\left(3x - \dfrac{\pi}{2}\right) - 3$

e. $h(x) = -3 \sin\left(2x + \dfrac{\pi}{3}\right) + 2$

f. $h(x) = 3 \sin\left(\dfrac{3}{2}x - \dfrac{\pi}{2}\right) + 3$

55. Amplitude: 2; vertical shift: 2 units downward; horizontal shift: $\frac{\pi}{4}$ units to the right; period: π

56. Amplitude: 3; vertical shift: 2 units upward; horizontal shift: $\frac{\pi}{6}$ units to the left; period: π

57. Amplitude: 2; vertical shift: 2 units downward; horizontal shift: $\frac{2\pi}{3}$ units to the left; period: $\frac{4\pi}{3}$

58. Amplitude: 3; vertical shift: 3 units upward; horizontal shift: $\frac{3\pi}{4}$ units to the left; period: 3π

59. Amplitude: 3; vertical shift: 3 units upward; horizontal shift: $\frac{\pi}{3}$ units to the right; period: $\frac{4\pi}{3}$

60. Amplitude: 2; vertical shift: 3 units downward; horizontal shift: $\frac{\pi}{6}$ units to the right; period: $\frac{2\pi}{3}$

Applications

61. Simple Harmonic Motion The position d of a block attached to a spring is given by the formula $d = 7 \cos \frac{\pi}{2} t$, where t is in seconds. Find d when $t = \frac{1}{2}$ and when $t = 4$.

62. Simple Harmonic Motion The position d of a block attached to a spring is given by the formula $d = 5 \sin \frac{\pi}{4} t$, where t is in seconds. What is the maximum distance of the block from its equilibrium position (the position at which $d = 0$)? Find the period of the motion.

63. Sales Toy sales at a department store t months after the month of December can be modeled by the function $s(t) = 210 + 150 \cos \frac{\pi}{6} t$, where s is in thousands of dollars. What is the value of $s(4)$, and what does it represent? Find the period of this function.

64. Radio Waves The *frequency* of a sine function is given by $\frac{1}{p}$, where p is the period. Find the frequency of an AM radio station whose wave form is given by $f(t) = A \sin(1.2 \times 10^6 \pi t)$, where A is some positive constant.

65. Electricity The voltage in an electrical circuit is given by the function $V(t) = \sin\left(3t - \frac{\pi}{2}\right)$. What is the smallest nonnegative value of t at which the voltage is equal to 0?

66. Circuit Theory The charge on an electrical capacitor is given by the function $q(t) = Q \cos\left(3t + \frac{\pi}{12}\right)$, where Q is a constant. What is the smallest positive value of t at which the charge is equal to $q(0)$?

67. Sound Wave The form of a sound wave is given by the function $f(x) = 25 \sin(4x + \pi)$. Find the amplitude, period, and frequency of the wave. The frequency of a sine function is given by $\frac{1}{p}$, where p is the period.

68. Light Wave The form of a light wave is given by the function $f(x) = 3 \cos\left(5x - \frac{\pi}{2}\right) + 4$. What are the maximum and minimum values of this function, and what is the smallest positive value of x at which the function attains its minimum value?

69. Flight Path A plane approaching an airport is told to maintain a holding pattern before being given clearance to land. The formula

$$d(t) = 80 \sin(0.75t) + 200$$

can be used to determine the distance in miles of the plane from the airport at time t. What is the maximum distance from the airport the plane travels while it is in the holding pattern?

70. Leisure The formula

$$h(t) = 125 \sin\left(2\pi t - \frac{\pi}{2}\right) + 125$$

represents the height, in feet, above the ground at time t, in minutes, of a person who is riding a Ferris wheel.

a. What is the period of the function? What does it mean?

b. What is the maximum height reached?

71. Biorhythm The function

$$P(t) = 50 \sin \frac{2\pi}{23} t + 50$$

is used in biorhythm theory to predict an individual's physical potential, as a percentage of the maximum, on a particular day, with $t = 0$ corresponding to birth.

a. What is the period of the function?

b. What is an individual's physical potential on his or her 3rd birthday (day 1095)?

Concepts

72. How can you show graphically that $\cos\left(\frac{\pi}{2} - x\right) = \sin x$?

73. Using appropriate graphs, show that $\cos(-x) = \cos x$.

74. Using appropriate graphs, show that $\sin(-x) = -\sin x$.

Objectives

- Sketch the graphs of the basic tangent and cotangent functions.
- Sketch the graphs of the basic secant and cosecant functions.
- Know the properties of the tangent, cotangent, secant, and cosecant functions.
- Sketch transformations of the graphs of tangent, cotangent, secant, and cosecant functions.

In this section we will construct the graphs of the tangent, cotangent, secant and cosecant functions. Because these functions are defined as quotients, we can expect that there will be breaks in the graphs at places where the denominator is zero.

Graph of the Tangent Function

The tangent function is defined as $\tan x = \frac{\sin x}{\cos x}$. Using this fact, we tabulate values of $\sin x$ and $\cos x$ for specific values of x in the interval $[0, 2\pi]$, and then compute the values of $\tan x$. Note that the tangent function is undefined whenever the value of $\cos x$ is 0. This is denoted by a dash (—) in Table 1.

x	0	$\frac{\pi}{4}$	$\frac{\pi}{2}$	$\frac{3\pi}{4}$	π	$\frac{5\pi}{4}$	$\frac{3\pi}{2}$	$\frac{7\pi}{4}$	2π
$\sin x$	0	$\frac{\sqrt{2}}{2}$	1	$\frac{\sqrt{2}}{2}$	0	$-\frac{\sqrt{2}}{2}$	-1	$-\frac{\sqrt{2}}{2}$	0
$\cos x$	1	$\frac{\sqrt{2}}{2}$	0	$-\frac{\sqrt{2}}{2}$	-1	$-\frac{\sqrt{2}}{2}$	0	$\frac{\sqrt{2}}{2}$	1
$\tan x$	0	1	—	-1	0	1	—	-1	0

Table 1

To see what happens to the tangent function between $x = \frac{\pi}{4}$ and $\frac{\pi}{2}$, we can use a calculator to compute additional values of $\tan x$. As x gets close to $\frac{\pi}{2}$ through values of x that are *less than* $\frac{\pi}{2}$, $\tan x$ approaches ∞. Similarly, as x gets close to $\frac{\pi}{2}$ through values of x that are *greater than* $\frac{\pi}{2}$, $\tan x$ approaches $-\infty$. Thus, the graph of the tangent function has a vertical asymptote at $x = \frac{\pi}{2}$, as shown in Figure 1. Similarly, a vertical asymptote exists at $x = \frac{3\pi}{2}$.

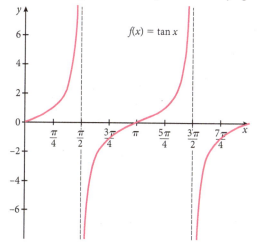

Figure 1

Moreover, we see that the graph of the tangent function on the interval $\left[0, \frac{\pi}{2}\right)$ looks exactly like that on the interval $\left[\pi, \frac{3\pi}{2}\right)$; likewise, the graph of $\tan x$ on the interval $\left(\frac{\pi}{2}, \pi\right]$ is identical to that on the interval $\left(\frac{3\pi}{2}, 2\pi\right]$. In fact, the tangent function is periodic and has period π. Because $\tan x$ is defined at every point at which the value of $\cos x$ is not equal to 0, the graph of $\tan x$ extends horizontally in both directions and has a vertical asymptote at every odd-integer multiple of $\frac{\pi}{2}$. See Figure 2.

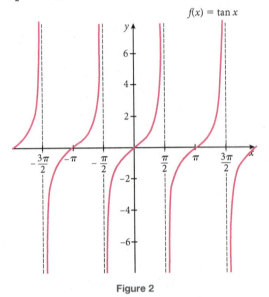

$f(x) = \tan x$

Figure 2

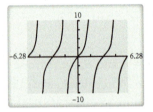

Properties of the Tangent Function

- Domain: All real numbers except $\frac{(2n+1)\pi}{2}$, where n is an integer
- Range: $(-\infty, \infty)$
- Period: π
- Vertical asymptotes: $x = \frac{(2n+1)\pi}{2}$, where n is an integer

Example 1   **Graph of $f(x) = \tan(x - c)$**

Sketch the graph of at least two full periods of the function, $f(x) = \tan\left(x - \frac{\pi}{4}\right)$.

Solution The graph of $\tan\left(x - \frac{\pi}{4}\right)$ is obtained by horizontally shifting the graph of $\tan x$ by $\frac{\pi}{4}$ units to the *right*. First, tabulate values of $\tan x$ for key values of x from $-\frac{\pi}{2}$ to $\frac{\pi}{2}$, and then tabulate values of $\tan\left(x - \frac{\pi}{4}\right)$ for key values of x from $-\frac{\pi}{4}$ to $\frac{3\pi}{4}$. The values of x in the Table 3 were obtained by adding $\frac{\pi}{4}$ to each of the values of x in Table 2.

x	$-\frac{\pi}{2}$	$-\frac{\pi}{3}$	$-\frac{\pi}{4}$	$-\frac{\pi}{6}$	0	$\frac{\pi}{6}$	$\frac{\pi}{4}$	$\frac{\pi}{3}$	$\frac{\pi}{2}$
$\tan x$	—	$-\sqrt{3}$	-1	$-\frac{\sqrt{3}}{3}$	0	$\frac{\sqrt{3}}{3}$	1	$\sqrt{3}$	—

Table 2

x	$-\frac{\pi}{4}$	$-\frac{\pi}{12}$	0	$\frac{\pi}{12}$	$\frac{\pi}{4}$	$\frac{5\pi}{12}$	$\frac{\pi}{2}$	$\frac{7\pi}{12}$	$\frac{3\pi}{4}$
$\tan\left(x - \frac{\pi}{4}\right)$	—	$-\sqrt{3}$	-1	$-\frac{\sqrt{3}}{3}$	0	$\frac{\sqrt{3}}{3}$	1	$\sqrt{3}$	—

Table 3

Graphing both functions on the same set of axes in Figure 4, we see the effect of the horizontal shift on $f(x) = \tan\left(x - \frac{\pi}{4}\right)$.

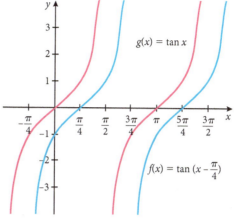

Figure 4

 Check It Out 1   Sketch the graph of at least two full periods of the function $f(x) = \tan\left(x + \frac{\pi}{2}\right)$.

Graph of the Cotangent Function

Because the cotangent function is defined as $\cot x = \frac{\cos x}{\sin x}$, its properties are very similar to those of $\tan x$. Create a table of values for the cotangent function (Table 4) using the definition. Note that the points at which the cotangent function is undefined are different from those at which the tangent function is undefined.

x	0	$\frac{\pi}{4}$	$\frac{\pi}{2}$	$\frac{3\pi}{4}$	π	$\frac{5\pi}{4}$	$\frac{3\pi}{2}$	$\frac{7\pi}{4}$	2π
cos x	1	$\frac{\sqrt{2}}{2}$	0	$-\frac{\sqrt{2}}{2}$	-1	$-\frac{\sqrt{2}}{2}$	0	$\frac{\sqrt{2}}{2}$	1
sin x	0	$\frac{\sqrt{2}}{2}$	1	$\frac{\sqrt{2}}{2}$	0	$-\frac{\sqrt{2}}{2}$	-1	$-\frac{\sqrt{2}}{2}$	0
cot x	—	1	0	-1	—	1	0	-1	—

Table 4

Properties of the Cotangent Function

- Domain: All real numbers except $n\pi$, where n is an integer
- Range: $(-\infty, \infty)$
- Period: π
- Vertical asymptotes: $x = n\pi$, where n is an integer

See Figure 5 for the graph of the cotangent function.

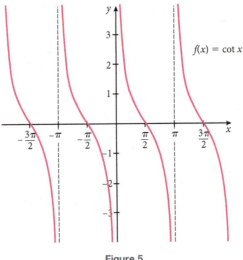

Figure 5

| Example 2 | **Graph of $f(x) = \cot(ax)$** |

Sketch the graph of at least two full periods of the function $f(x) = \cot 3x$.

Solution   The graph of $f(x) = \cot 3x$ is obtained by horizontally compressing the graph of $\cot x$ by a factor of $\frac{1}{3}$. Thus, the period of $\cot 3x$ is $\frac{\pi}{3}$.

First, make a table of values of $\cot x$ for key values of x from 0 to π. To obtain the table of values for $\cot 3x$, divide each of the key values of x in the first table by 3. See Table 5. The values of $\cot(3x)$ in the second table are the same as the values of $\cot x$, for corresponding values of x, in the first table. The graph of $\cot 3x$ in shown in Figure 6.

x	0	$\frac{\pi}{4}$	$\frac{\pi}{2}$	$\frac{3\pi}{4}$	π
$\cot x$	—	1	0	-1	—

x	0	$\frac{\pi}{12}$	$\frac{\pi}{6}$	$\frac{3\pi}{12}$	$\frac{\pi}{3}$
$\cot 3x$	—	1	0	-1	—

Table 5

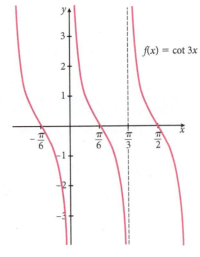

Figure 6

✓ *Check It Out 2*   Sketch the graph of at least two full periods of the function $f(x) = \cot\left(\frac{x}{2}\right)$. ■

Using Technology

To graph the $f(x) = \cot 3x$ in a graphing calculator, you must enter it as $Y_1 = 1/\tan(3x)$. See Figure 7 and Figure 8. Do *not* use the $\tan^{-1}$ function, because that refers to the inverse tangent function, which is discussed later in this chapter.

Keystroke Appendix:

Sections 8 and 15

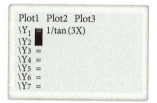

Figure 7

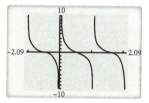

Figure 8

Graphs of the Secant and Cosecant Functions

The secant and cosecant functions are defined in terms of the cosine and sine functions, respectively, as follows.

$$\sec x = \frac{1}{\cos x} \qquad \csc x = \frac{1}{\sin x}$$

The secant function is undefined for values of x for which $\cos x = 0$, and the cosecant function is undefined for values of x such that $\sin x = 0$. We can expect the graphs of $\sec x$ and $\csc x$ to have vertical asymptotes whenever $\cos x = 0$ or $\sin x = 0$, respectively.

To sketch the graph of $\sec x$, first make a table of values (Table 6) of $\cos x$. Then take their reciprocals and use those points to sketch the graph, which is shown in Figure 9. The graph of the cosine function is also given for reference.

x	0	$\frac{\pi}{4}$	$\frac{\pi}{2}$	$\frac{3\pi}{4}$	π	$\frac{5\pi}{4}$	$\frac{3\pi}{2}$	$\frac{7\pi}{4}$	2π
$\cos x$	1	$\frac{\sqrt{2}}{2}$	0	$-\frac{\sqrt{2}}{2}$	-1	$-\frac{\sqrt{2}}{2}$	0	$\frac{\sqrt{2}}{2}$	1
$\sec x$	1	$\sqrt{2}$	—	$-\sqrt{2}$	-1	$-\sqrt{2}$	—	$\sqrt{2}$	1

Table 6

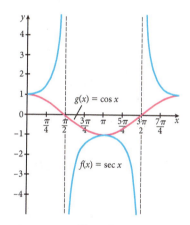

Figure 9

Because the cosine function is periodic and has period 2π, the secant function also has period 2π. See Figure 10 for a graph of $\sec x$ over two full periods.

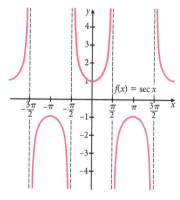

Figure 10

Using Technology

The cosecant function must be entered as $Y_1 = 1/\sin(x)$. See Figure 11. Do *not* use the $\sin^{-1}$ function, because that refers to the inverse sine function, which is discussed later in the chapter.

Keystroke Appendix:

Sections 8 and 15

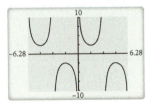

Figure 11

Properties of the Secant Function

- Domain: All real numbers except $\dfrac{(2n+1)\pi}{2}$, n an integer

- Range: $(-\infty, -1] \cup [1, \infty)$

- Period: 2π

- Vertical asymptotes: $\dfrac{(2n+1)\pi}{2}$, n an integer

The graph of the cosecant function, shown in Figure 12, can be obtained in a manner similar to the one used for the secant function. Because the sine function is periodic and has period 2π, the cosecant function also has period 2π.

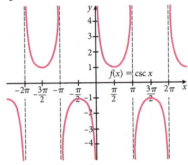

$f(x) = \csc x$

Figure 12

Properties of the Cosecant Function

- Domain: All real numbers except $n\pi$, where n is an integer.

- Range: $(-\infty, -1] \cup [1, \infty)$

- Period: 2π

- Vertical asymptotes: $x = n\pi$, where n is an integer.

Example 3   **Graph of $a\sec(x - c)$**

Sketch the graph of one full period of the function $f(x) = -2\sec\left(x - \dfrac{\pi}{2}\right)$.

Solution   Viewing the graph of f as a series of transformations of the graph of $\sec x$, we have the following:

$$\sec x \quad \longrightarrow \quad -2\sec x \quad \longrightarrow \quad -2\sec\left(x - \dfrac{\pi}{2}\right)$$

Vertically stretch by factor of 2   Horizontally shift to
and reflect across x-axis   the right by $\dfrac{\pi}{2}$

These steps are graphed in Figure 13.

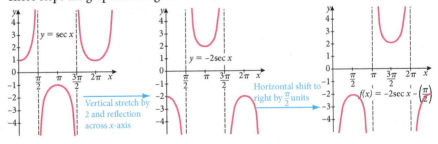

Figure 13

The period of the function is 2π, since the transformation does not involve any horizontal stretching or compression.

 Check It Out 3   Sketch the graph of one full period of the function $f(x) = 3\sec\left(x + \dfrac{\pi}{4}\right)$. ■

> **Example 4**   **Graph of $f(x) = \csc(ax) + c$**

Sketch the graph of two full periods of the function $f(x) = \csc(2x) - 1$.

Solution   Viewing the graph of f as a series of transformations of the graph of $\csc x$, we have the following:

$$\csc x \quad\longrightarrow\quad \csc(2x) \quad\longrightarrow\quad \csc(2x) - 1$$

Horizontally compression by factor of $\frac{1}{2}$ Vertically shift 1 unit downward

These steps are graphed in Figure 14.

The horizontal compression in the first step has halved the period of the original function $\csc x$, so the period of $f(x) = \csc(2x) - 1$ is π.

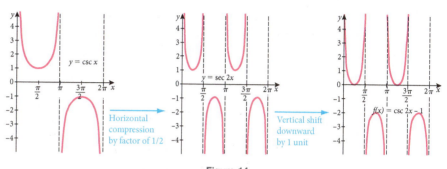

Figure 14

✓ **Check It Out 4**   Sketch the graph of two full periods of the function $f(x) = \csc\left(\dfrac{3x}{4}\right) + 5$.

Exercises 5.6

Skills

In Exercises 1–6, use vertical translations to graph at least two periods of the given function

1. $f(x) = \tan x - 3$

2. $f(x) = \tan x + 2$

3. $f(x) = \sec x + 1$

4. $f(x) = \csc x - 2$

5. $g(x) = \cot x + \dfrac{3}{2}$

6. $g(x) = \csc x - \dfrac{1}{2}$

In Exercises 7–12, use horizontal translations to graph at least two periods of the given functions.

7. $f(x) = \tan\left(x + \dfrac{\pi}{4}\right)$

8. $f(x) = \tan\left(x - \dfrac{\pi}{2}\right)$

9. $f(x) = \sec\left(x - \dfrac{\pi}{2}\right)$

10. $f(x) = \csc\left(x + \dfrac{\pi}{3}\right)$

11. $g(x) = \cot\left(\dfrac{3\pi}{4} + x\right)$

12. $g(x) = \sec\left(\dfrac{\pi}{2} + x\right)$

In Exercises 13–18, use vertical stretches to graph at least two periods of the given functions.

13. $f(x) = 4\tan x$

14. $f(x) = -3\tan x$

15. $f(x) = 2\csc x$

16. $f(x) = 3\sec x$

17. $g(x) = -2\cot x$

18. $g(x) = -3\csc x$

In Exercises 19–24, use horizontal stretches and compressions to graph at least two periods of the given functions.

19. $f(x) = \tan(2x)$

20. $f(x) = \tan(0.5x)$

21. $f(x) = \csc(2x)$

22. $f(x) = \sec(3x)$

23. $g(x) = \csc\left(\dfrac{\pi}{3}x\right)$

24. $g(x) = \cot\left(\dfrac{\pi}{2}x\right)$

In Exercises 25–34, graph at least two periods of the given functions.

25. $f(x) = \tan\left(x + \dfrac{\pi}{4}\right) + 1$

26. $f(x) = -\tan\left(x - \dfrac{\pi}{3}\right) - 1$

27. $f(x) = 2\sec\left(x + \dfrac{\pi}{4}\right)$

28. $f(x) = -3\csc\left(x - \dfrac{\pi}{3}\right)$

29. $g(x) = \dfrac{1}{2}\cot(2x)$

30. $g(x) = \dfrac{3}{2}\sec(3x)$

31. $h(x) = -3\csc(x - \pi) + 2$

32. $h(x) = 2\cot\left(x + \dfrac{\pi}{2}\right) - 1$

33. $r(x) = -\sec\left(\dfrac{2\pi}{3}x\right) + 2$

34. $r(x) = \cot\left(\dfrac{\pi}{3}x\right) - 1$

In Exercises 35–38, find an expression of the form a tan bx or a cot bx corresponding to each of the given graphs.

35.

36.

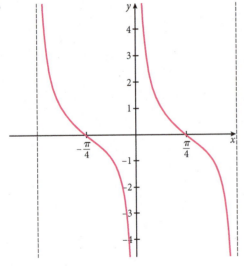

37.

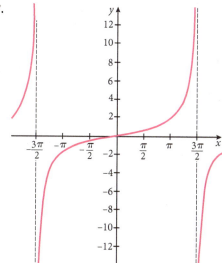

38.

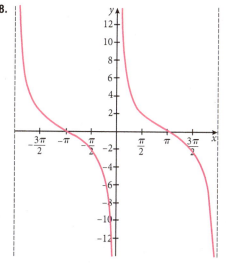

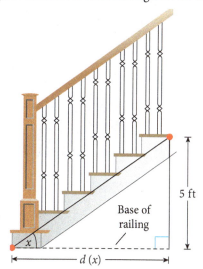 *In Exercises 39–44, graph the given pair of functions in the same window. Graph at least two periods of each function, and describe the similarities and differences between the graphs*

39. $f(x) = \tan(x); f(x) = -\tan(x)$

40. $f(x) = \tan(3x); f(x) = 3\tan(x)$

41. $f(x) = \sec(2x); f(x) = 2\sec(x)$

42. $f(x) = \frac{1}{2}\csc(x); f(x) = \csc\left(\frac{1}{2}x\right)$

43. $f(x) = \cot(3\pi x); f(x) = \cot\left(\frac{\pi}{3}x\right)$

44. $f(x) = \sec\left(\frac{\pi}{2}x\right); f(x) = \sec(2\pi x)$

Applications

45. Design The base of a railing for a set of steps leading up to the front door of a house makes an angle of x degrees with the horizontal. Let $d(x)$ be the horizontal distance between the two ends of the base of the railing. If the upper end of the railing is 5 feet higher than the lower end, find the positive number A such that $d(x) = A \cot x$, and use your function to find the length of the base of the railing if $x = 35°$.

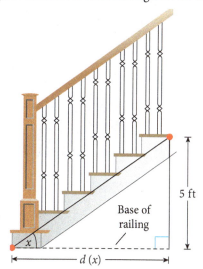

46. **Playground construction** A triangular-shaped playground with vertices at points P, Q, and R is to be enlarged to form a triangle with vertices P, R, and a new point, S (see the figure). Segment PQ is twice as long as segment RQ, and the length of segment PR is 45 feet. Let $d(x)$ be the length of segment PS, where x is the measure of angle QPS in radians. Find constants A and C such that A is positive, $0 < C < \frac{\pi}{2}$, and $d(x) = A \sec(x + C)$. Use your function to determine the length of segment PS if $x = \frac{2\pi}{15}$.

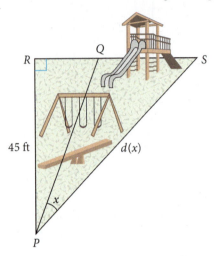

47. **Sailing** A triangular sail is to be constructed for a sailboat. The sail, with vertices at points P, Q, and R, is to be strengthened by two thin strips of heavy-duty material: one from point P to point S, the other from point Q to point S (see the figure). The length of segment PT is to be equal to that of segment ST, where point T is the midpoint of the base of the sail. The base is 16 feet long. Let $h(x)$ be the height of the sail, where x is the measure of angle RPS in radians. Find constants A and C such that A is positive, $0 < C < \frac{\pi}{2}$, and $h(x) = A \tan(x + C)$. Use your function to determine the height of the sail if $x = \frac{5\pi}{36}$.

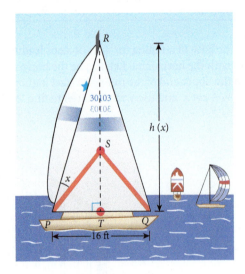

48. Fishing An angler is standing on the wharf and sees a fish in the water. Let $l(x)$ be the length of fishing line that would have to be let out in order to catch the fish, where x (in radians) is the angle of depression of the location of the fish with respect to the tip of the fishing rod. If the horizontal distance of the fish from the tip of the rod is 12 feet, find constants A and C such that $0 < C < \pi$ and $l(x) = A \csc(C - x)$. Use your function to determine the length of fishing line that would have to be let out if $x = \frac{\pi}{3}$. (*Hint:* First express $l(x)$ as a product of a constant and a trigonometric function of just x.)

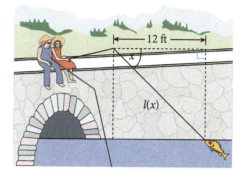

Concepts

49. What are the domain and range of the function $f(x) = \csc 2x + 3$?

50. What are the domain and range of the function $f(x) = \sec 3x - 1$?

51. What are the vertical asymptotes of the graph of $f(x) = \tan x + \cot x$?

Inverse Trigonometric Functions

5.7

Objectives

- Know the properties of the arcsine, arccosine and arctangent functions.
- Sketch the graphs of the inverse trigonometric functions.
- Calculate exact values of inverse trigonometric functions.
- Know how to evaluate the composite of a trigonometric function with the inverse of a trigonometric function.
- Use inverse trigonometric functions to solve applied problems.

In many situations, we are given the value of the sine of an angle, or number, and/or the value of its cosine, and we are asked to find the angle, or number. This involves finding the *inverse* of one of the trigonometric functions, which we now discuss.

Inverse of Sine, Cosine, and Tangent

Recall from Chapter 4 that inverses are only defined for functions that are one-to-one, a property that the sine, cosine, and tangent functions lack. Graphically, these functions fail the horizontal line test, as shown in Figure 1.

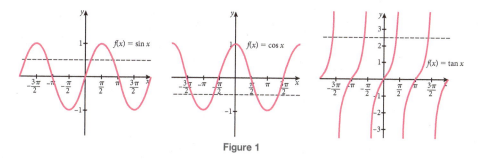

Figure 1

To define an inverse for any of these functions, we must first restrict the domain of the original function to an interval on which it is one-to-one. By convention, the sine function is restricted to the interval $\left[-\frac{\pi}{2}, \frac{\pi}{2}\right]$, because the sine function is one-to-one on this interval and also takes on all the values in the interval $[-1, 1]$, as shown in Figure 2. Similarly, the domain of the cosine function is restricted to the interval $[0, \pi]$, as illustrated in Figure 3.

Just in Time

Review inverse of functions in Section 4.1.

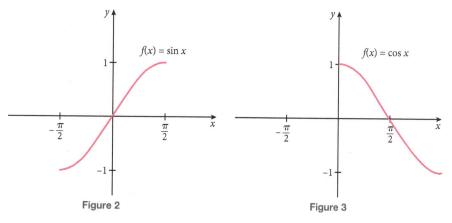

Figure 2

Figure 3

The inverse of the sine function is referred to as the **arcsine** function, and the inverse of the cosine function is referred to as the **arccosine** function. These inverse functions are defined as follows.

The Arcsine Function

The **arcsine** function, arcsin x, is defined by

$$\text{arcsin } x = y \text{ if and only if } x = \sin y$$

- The domain of arcsin x is $[-1, 1]$.
- The range of arcsin x is $\left[-\dfrac{\pi}{2}, \dfrac{\pi}{2}\right]$.

See Figure 4 for a graph of the arcsine function. The arcsine function is also denoted as $\sin^{-1} x$.

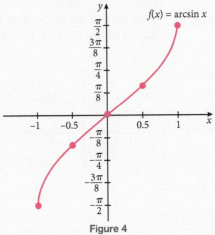

Figure 4

The Arccosine Function

The **arccosine** function, arccos x, is defined by

$$\text{arccos } x = y \text{ if and only if } x = \cos y$$

- The domain of arccos x is $[-1, 1]$.
- The range of arccos x is $[0, \pi]$.

See Figure 5 for a graph of the arccosine function. The arccosine function is also denoted as $\cos^{-1} x$.

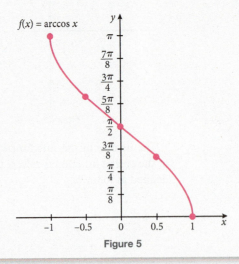

Figure 5

The range of the arcsine and the arccosine functions with reference to the unit circle are given in Figure 6.

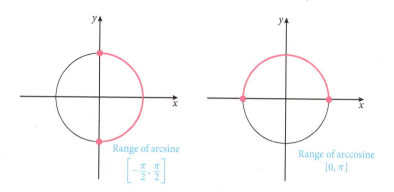

Range of arcsine
$\left[-\frac{\pi}{2}, \frac{\pi}{2}\right]$

Range of arccosine
$[0, \pi]$

Figure 6

VIDEO EXAMPLES

SECTION 5.7

| **Example 1** | **Evaluating Inverse Sine and Cosine Functions** |

Find the exact value of

a. $\arccos \dfrac{1}{2}$ **b.** $\arcsin\left(-\dfrac{\sqrt{3}}{2}\right)$ **c.** $\arccos\left(-\dfrac{\sqrt{2}}{2}\right)$

Solution

a. To determine $\arccos \dfrac{1}{2}$, we must find *the* angle or number in the interval $[0, \pi]$ whose cosine is $\dfrac{1}{2}$. That number is $\dfrac{\pi}{3}$, because $\cos \dfrac{\pi}{3} = \dfrac{1}{2}$. Thus,

$$\arccos \frac{1}{2} = \frac{\pi}{3}$$

b. To determine $\arcsin\left(-\dfrac{\sqrt{3}}{2}\right)$, we must find *the* angle or number in the interval $\left[-\dfrac{\pi}{2}, \dfrac{\pi}{2}\right]$ whose sine is $-\dfrac{\sqrt{3}}{2}$. Because $\sin\left(-\dfrac{\pi}{3}\right) = -\dfrac{\sqrt{3}}{2}$ and $-\dfrac{\pi}{3}$ is in $\left[-\dfrac{\pi}{2}, \dfrac{\pi}{2}\right]$,

$$\arcsin\left(-\frac{\sqrt{3}}{2}\right) = -\frac{\pi}{3}$$

c. To determine $\arccos\left(-\dfrac{\sqrt{2}}{2}\right)$, we must find *the* angle or number in the interval $[0, \pi]$ whose cosine is $-\dfrac{\sqrt{2}}{2}$. Because $\cos \dfrac{3\pi}{4} = -\dfrac{\sqrt{2}}{2}$, and $\dfrac{3\pi}{4}$ is in $[0, \pi]$,

$$\arccos\left(-\frac{\sqrt{2}}{2}\right) = \frac{3\pi}{4}$$

Using Technology

You can use the *SIN*$^{-1}$ or the *COS*$^{-1}$ keys on your calculator to approximate the inverse sine and cosine functions. See Figure 7.

Keystroke Appendix:
Section 15

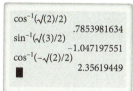

Figure 7

✅ *Check It Out 1* Find $\arcsin \dfrac{1}{2}$.

The inverse of the tangent function is called the **arctangent** function. It is defined by restricting the domain of the tangent function to the interval $\left(-\dfrac{\pi}{2}, \dfrac{\pi}{2}\right)$, which gives rise to a one-to-one function. The range of the tangent function on $\left(-\dfrac{\pi}{2}, \dfrac{\pi}{2}\right)$ is $(-\infty, \infty)$, the same as that of the unrestricted tangent function.

The Arctangent Function

The **arctangent** function, arctan x, is defined by

$$\arctan x = y \text{ if and only if } x = \tan y$$

- The domain of arctan x is $(-\infty, \infty)$.
- The range of arctan x is $\left(-\dfrac{\pi}{2}, \dfrac{\pi}{2}\right)$.

The graph of the arctangent function is given in Figure 8. The arctangent function so denoted by $\tan^{-1} x$.

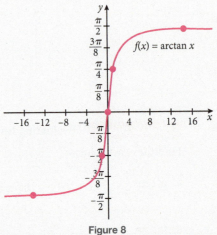

Figure 8

The range of arctangent function with reference to the unit circle is given in Figure 9.

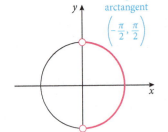

Figure 9

Example 2   Evaluating the Inverse Tangent Function

Find the exact value of

a. $\arctan \sqrt{3}$

b. $\arctan(-1)$

Solution

a. To evaluate arctan $\sqrt{3}$, we must find *the* angle or number in the interval $\left(-\frac{\pi}{2}, \frac{\pi}{2}\right)$ whose tangent is $\sqrt{3}$. That number is $\frac{\pi}{3}$. Thus,

$$\arctan \sqrt{3} = \frac{\pi}{3}$$

b. The number in $\left(-\frac{\pi}{2}, \frac{\pi}{2}\right)$ whose tangent is -1 is $-\frac{\pi}{4}$. Thus,

$$\arctan(-1) = -\frac{\pi}{4}$$

 Check It Out 2   Evaluate arctan 1.

Composition of Functions

From our work with functions and their inverses, we know that $f \circ f^{-1}(x) = x$ for all x in the domain of $f \circ f^{-1}$. Also, $f^{-1} \circ f(x) = x$ for all x in the domain of f. When sine, cosine, or tangent functions and their inverses are composed with each other, you must give careful consideration to the domain of definition of the inverse trigonometric functions, as summarized below.

Trigonometric Functions and Their Inverses

Sine and arcsin

- For x in $[-1, 1]$, $\sin(\arcsin x) = x$, or $\sin(\sin^{-1}(x)) = x$.
- For x in $\left[-\frac{\pi}{2}, \frac{\pi}{2}\right]$, $\arcsin(\sin x) = x$, or $\sin^{-1}(\sin(x)) = x$.

Cosine and arccosine

- For x in $[-1, 1]$, $\cos(\arccos x) = x$, or $\cos(\cos^{-1}(x)) = x$.
- For x in $[0, \pi]$, $\arccos(\cos x) = x$, $\cos^{-1}(\cos(x)) = x$.

Tangent and arctangent

- For x in $(-\infty, \infty)$, $\tan(\arctan x) = x$, $\tan(\tan^{-1}(x)) = x$.
- For x in $\left(-\frac{\pi}{2}, \frac{\pi}{2}\right)$, $\arctan(\tan x) = x$, $\tan^{-1}(\tan(x)) = x$.

The following example illustrates the rules for composing trigonometric functions.

Example 3   **Composition of Functions**

Find the following:

a. $\arcsin\left(\sin\frac{\pi}{6}\right)$ **b.** $\arccos\left(\cos\frac{5\pi}{4}\right)$ **c.** $\sin^{-1}\left(\cos\frac{2\pi}{3}\right)$

Solution In all these parts of the problem, the final answer will be the number whose sine or cosine is given.

a. Because $\sin\frac{\pi}{6} = \frac{1}{2}$, we have

$$\arcsin\left(\sin\frac{\pi}{6}\right) = \arcsin\frac{1}{2} = \frac{\pi}{6}$$

Another way to see this is from the definition. Because $\frac{\pi}{6}$ is in $\left[-\frac{\pi}{2}, \frac{\pi}{2}\right]$, $\arcsin\left(\sin\frac{\pi}{6}\right) = \frac{\pi}{6}$.

b. Because $\cos\frac{5\pi}{4} = -\frac{\sqrt{2}}{2}$, we have

$$\arccos\left(\cos\frac{5\pi}{4}\right) = \arccos\left(-\frac{\sqrt{2}}{2}\right) = \frac{3\pi}{4}$$

Note that $\arccos\left(\cos\frac{5\pi}{4}\right)$ is *not* equal to $\frac{5\pi}{4}$, because $\frac{5\pi}{4}$ is *not* in the interval $[0, \pi]$, the range of the arccosine function.

c. Because $\cos\frac{2\pi}{3} = -\frac{1}{2}$, we have

$$\sin^{-1}\left(\cos\frac{2\pi}{3}\right) = \sin^{-1}\left(-\frac{1}{2}\right) = -\frac{\pi}{6}$$

✓ *Check It Out 3* Find $\arccos\left(\cos\frac{4\pi}{3}\right)$.

Example 4   **Finding Values of Trigonometric Functions**

Find the following:

a. $\cos\left(\arcsin\left(\frac{3}{5}\right)\right)$

b. $\sin\left(\cos^{-1}\left(-\frac{12}{13}\right)\right)$

Solution

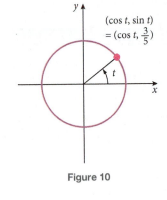

$(\cos t, \sin t)$
$= (\cos t, \frac{3}{5})$

Figure 10

a. Because $\frac{3}{5} > 0$, $\arcsin\frac{3}{5}$ is in $\left(0, \frac{\pi}{2}\right)$. We must now find the cosine of the angle whose sine is $\frac{3}{5}$. Let $t = \arcsin\frac{3}{5}$. Thus, $\sin t = \frac{3}{5}$. See Figure 10.

From one of the Pythagorean identities,

$$\sin^2 t + \cos^2 t = 1$$

$$\left(\frac{3}{5}\right)^2 + \cos^2 t = 1 \qquad\qquad \sin t = \frac{3}{5}$$

$$\cos^2 t = 1 - \frac{9}{25} \qquad\qquad \left(\frac{3}{5}\right)^2 = \frac{9}{25}$$

$$\cos^2 t = \frac{16}{25}$$

$$\cos t = \pm\frac{4}{5}$$

Because t is in $\left(0, \frac{\pi}{2}\right)$, the value of $\cos t$ is positive. Thus,

$$\cos\left(\arcsin\frac{3}{5}\right) = \cos t = \frac{4}{5}.$$

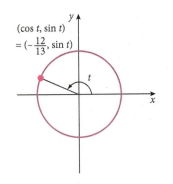

$(\cos t, \sin t)$
$= (-\frac{12}{13}, \sin t)$

Figure 11

b. Because $-\frac{12}{13} < 0$, $\cos^{-1}\left(-\frac{12}{13}\right)$ is in $\left(\frac{\pi}{2}, \pi\right)$. Let $t = \cos^{-1}\left(-\frac{12}{13}\right)$. Then, $\cos t = -\frac{12}{13}$. See Figure 11.

To find $\sin t$, use the Pythagorean identity:

$$\sin^2 t + \cos^2 t = 1$$

$$\sin^2 t + \left(-\frac{12}{13}\right)^2 = 1 \qquad\qquad \cos t = \left(-\frac{12}{13}\right)$$

$$\sin^2 t = 1 - \frac{144}{169} \qquad\qquad \left(-\frac{12}{13}\right)^2 = \frac{144}{169}$$

$$\sin^2 t = \frac{25}{169}$$

$$\sin t = \pm\frac{5}{13}.$$

Because t is in $\left(\frac{\pi}{2}, \pi\right)$, the value of $\sin t$ is positive. Thus,

$$\sin\left(\cos^{-1}\left(-\frac{12}{13}\right)\right) = \sin t = \frac{5}{13}$$

✓ *Check It Out 4*   Find $\sin\left(\arccos\left(-\frac{3}{5}\right)\right)$

Applications of Inverse Trigonometric Functions

Example 5   **Application: Slope of a Ramp**

A ramp is to be constructed with a slope of $\frac{1}{10}$. See Figure 12. Find the acute angle that the ramp makes with the horizontal line. Express the angle in degrees.

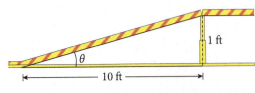

1 ft

θ

10 ft

Figure 12

Solution Let θ denote the acute angle that the ramp makes with the horizontal line. The slope of the ramp is given by $\frac{\text{Rise}}{\text{Run}}$. The slope can also be expressed as $\tan \theta$. Thus, $\tan \theta = \frac{1}{10}$. Using the $\tan^{-1}$ key on a calculator, in DEGREE mode, we have

$$\theta = \tan^{-1}\left(\frac{1}{10}\right) \approx 5.71°$$

 Check It Out 5 Rework Example 5 for the case where the slope of the ramp is $\frac{1}{2}$.

Example 6 **Application: Firing Angle of a Projectile**

When a projectile is fired with an initial velocity v_0 and at an acute angle θ, the horizontal range is given by

$$R = \frac{v_0^2 \sin 2\theta}{g}$$

where g is the gravitational constant, 9.8 meters per second squared. If a projectile is fired with an initial velocity of 80 meters per second, at what angle must it be fired in order to have a horizontal range of 500 meters? Express the angle in degrees.

Solution Substituting $v_0 = 80$, $R = 500$, and $g = 9.8$ into the equation for R gives

$$500 = \frac{(80)^2 \sin 2\theta}{9.8}$$

Isolating the quantity $\sin 2\theta$:

$$500(9.8) = (80)^2 \sin 2\theta$$

$$\frac{500\,(9.8)}{(80)^2} = \sin 2\theta$$

$$0.765625 = \sin 2\theta.$$

Using the $\sin^{-1}$ key on a calculator, in DEGREE mode, we have

$$2\theta = \sin^{-1}(0.765625) \approx 49.96°$$

$$\theta \approx 24.98°$$

Thus, the projectile must be fired at an angle of **24.98°** to have a horizontal range of 500 meters.

 Check It Out 6 Rework Example 6 for the case where the projectile is to have a horizontal range of 550 meters.

Other Inverse Trigonometric Functions

The inverse of cotangent, cosecant, and secant functions are defined as follows.

Function	Domain	Range
$\cot^{-1} x$	$(-\infty, \infty)$	$\left(-\frac{\pi}{2}, 0\right) \cup \left(0, \frac{\pi}{2}\right]$
$\csc^{-1} x$	$(-\infty, -1] \cup [1, \infty)$	$\left[-\frac{\pi}{2}, 0\right) \cup \left(0, \frac{\pi}{2}\right]$
$\sec^{-1} x$	$(-\infty, -1] \cup [1, \infty)$	$\left[0, \frac{\pi}{2}\right) \cup \left(\frac{\pi}{2}, \pi\right]$

These definitions are motivated by examining the graphs of the cotangent, cosecant, and secant functions, respectively. For instance, the graph of $\sec x$ is one-to-one on the interval $\left[0, \frac{\pi}{2}\right) \cup \left(\frac{\pi}{2}, \pi\right]$. Thus, the the definition for $\sec^{-1} x$ can be made with $|x| \geq 1$ as the domain and $\left[0, \frac{\pi}{2}\right) \cup \left(\frac{\pi}{2}, \pi\right]$ as the range.

Example 7   **Finding the Exact Value of an Inverse Secant Function**

Find the exact value of $\sec^{-1} 2$.

Solution Let $y = \sec^{-1} 2$. We look for $y \in \left[0, \frac{\pi}{2} \right) \cup \left(\frac{\pi}{2}, \pi \right]$ such that $\sec y = 2$.

This is equivalent to finding $y \in \left[0, \frac{\pi}{2} \right) \cup \left(\frac{\pi}{2}, \pi \right]$ such that $\cos y = \frac{1}{2}$.

From our work with exact values of cosine in Section 5.2, the only value of $y \in \left[0, \frac{\pi}{2} \right) \cup \left(\frac{\pi}{2}, \pi \right]$ with $\cos y = \frac{1}{2}$ is $y = \frac{\pi}{3}$. Thus, $\sec^{-1} 2 = \frac{\pi}{3}$.

 Check It Out 7   Find the exact value of $\csc^{-1} 2$. ■

Calculators do not have keys for the inverse secant, cosecant, and cotangent functions, so you must use their equivalent relationships to inverse cosine, sine, and tangent, respectively.

Evaluating Inverse Secant, Cosecant, and Cotangent Functions

$$\sec^{-1} x = \cos^{-1} \left(\frac{1}{x} \right), x \neq 0$$

$$\csc^{-1} x = \sin^{-1} \left(\frac{1}{x} \right), x \neq 0$$

$$\cot^{-1} x = \begin{cases} \frac{\pi}{2} - \tan^{-1}(x), & \text{for } x \geq 0 \\ -\frac{\pi}{2} - \tan^{-1}(x), & \text{for } x < 0 \end{cases}$$

The following example illustrates how to evaluate these inverse functions.

Example 8   **Calculator Approximation of Inverse, Cosecant, and Cotangent Functions**

Use a calculator to approximate the following in radians to four decimal places.

a. $\csc^{-1} \dfrac{3}{2}$ **b.** $\cot^{-1}(-3)$

Solution

a. Let $y = \csc^{-1} \frac{3}{2}$. This is equivalent to finding $y = \sin^{-1} \frac{2}{3}$. Thus,

$$\csc^{-1} \frac{3}{2} = \sin^{-1} \frac{2}{3} \approx 0.7297$$

The last step was performed using a calculator in RADIAN mode.

b. Let $y = \cot^{-1}(-3)$. We then have

$$\cot^{-1}(-3) = -\frac{\pi}{2} - \tan^{-1}(-3) \approx -0.3218$$

The last step was performed using a calculator in RADIAN mode.

 Check It Out 8   Approximate $\csc^{-1} 3$. ■

Exercises 5.7

For Exercises 1–4, use the definition of f(x) *as given by the following table.*

x	f(x)
−2	5
−1	3
1	−2
4	−1

1. Find $f^{-1}(-2)$ **2.** Find $f^{-1}(-1)$ **3.** Find $(f \circ f^{-1})(3)$ **4.** Find $(f^{-1} \circ f)(4)$

 Skills

In Exercises 5–16, find the exact values of the given trigonometric functions without using a calculator.

5. arcsin 1   **6.** arccos 1   **7.** arccos(−1)   **8.** arcsin 0

9. arctan 0   **10.** arctan 1   **11.** $\arccos \dfrac{1}{2}$ **12.** $\arcsin\left(-\dfrac{1}{2}\right)$

13. $\cos^{-1}\left(-\dfrac{\sqrt{3}}{2}\right)$ **14.** $\sin^{-1}\left(-\dfrac{\sqrt{2}}{2}\right)$ **15.** $\tan^{-1}(-\sqrt{3})$ **16.** $\tan^{-1}\left(-\dfrac{\sqrt{3}}{3}\right)$

In Exercises 17–28, use a calculator to evaluate each trigonometric function. Make sure that the calculator is in RADIAN mode.

17. $\arcsin\left(-\dfrac{1}{4}\right)$ **18.** $\arccos\left(-\dfrac{1}{5}\right)$ **19.** arccos 0.75   **20.** arcsin 0.15

21. arctan 5   **22.** arctan (−0.7)   **23.** $\cos^{-1} 0.125$ **24.** $\sin^{-1} 0.8$

25. $\cos^{-1}(-0.95)$ **26.** $\sin^{-1}(-0.05)$ **27.** $\tan^{-1}(-5)$ **28.** $\tan^{-1}\left(-\dfrac{1}{3}\right)$

In Exercises 29–38, evaluate each trigonometric function without using a calculator.

29. $\arccos\left(\cos\left(-\dfrac{\pi}{3}\right)\right)$ **30.** $\arcsin(\sin(5\pi))$ **31.** $\sin(\arcsin(0.3))$ **32.** $\cos(\arccos(0.8))$

33. $\tan(\tan^{-1}(4))$ **34.** $\tan(\tan^{-1}(-5))$ **35.** $\sin\left(\cos^{-1}\left(\dfrac{12}{13}\right)\right)$ **36.** $\cos\left(\sin^{-1}\left(-\dfrac{3}{5}\right)\right)$

37. $\cos\left(\sin^{-1}\left(-\dfrac{12}{13}\right)\right)$ **38.** $\sin\left(\cos^{-1}\left(-\dfrac{4}{5}\right)\right)$

In Exercises 39–44, sketch the graph of each function.

39. $f(x) = \arctan x - \dfrac{\pi}{2}$ **40.** $f(x) = \arcsin x + \pi$ **41.** $f(x) = 2 \arccos x$

42. $f(x) = -3 \arcsin x$ **43.** $g(x) = \sin^{-1} 2x - \pi$ **44.** $g(x) = \cos^{-1}\left(\dfrac{x}{2}\right) + \pi$

In Exercises 45–48, explain why each of the following expressions is undefined.

45. arccos 4   **46.** arcsin −3   **47.** sin(arcsin(5))   **48.** cos(arccos(−6))

In Exercises 49–56, find the exact values of the given expressions in radian measure.

49. $\csc^{-1}\sqrt{2}$

50. $\csc^{-1}\dfrac{2\sqrt{3}}{3}$

51. $\sec^{-1}\left(-\dfrac{2\sqrt{3}}{3}\right)$

52. $\sec^{-1}\left(-\sqrt{2}\right)$

53. $\csc^{-1}(-2)$

54. $\sec^{-1}(-2)$

55. $\cot^{-1}\sqrt{3}$

56. $\cot^{-1}(-1)$

In Exercises 57–64, evaluate the given expressions to four decimal places with a calculator.

57. $\sec^{-1}2.5$

58. $\csc^{-1}5$

59. $\csc^{-1}(-3.6)$

60. $\sec^{-1}(-4.2)$

61. $\cot^{-1}2.4$

62. $\cot^{-1}3.6$

63. $\cot^{-1}(-3.2)$

64. $\cot^{-1}(-1.8)$

Applications

65. Surveying A surveyor finds that $\tan\theta = \frac{19}{5}$, where θ is the angle that a straight line from the ground makes with the top of a platform. Find θ in degrees.

66. Design A ramp makes an acute angle θ with respect to the horizontal. Any two points along either edge of the ramp that are 10 feet apart differ in their horizontal positions by 9.5 feet. Give the value of θ in degrees.

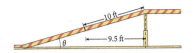

67. Construction A 15-foot pole is to be stabilized by two wires of equal length, one on each side of the pole. One end of each wire is to be attached to the top of the pole; the other end is to be staked to the ground at an acute angle θ with respect to the horizontal. Because of safety factors, the ratio of the length of either wire to the height of the pole is to be no more than $\frac{4}{3}$. What is the limiting value of θ in degrees? Is this limiting value a maximum value of θ or a minimum value of θ? Explain.

68. Robotics A robotic arm is hinged at one end. The acute angle that the robotic arm makes with the horizontal is given by $\theta = \arcsin\frac{a}{3}$, where a is the vertical distance of the end of the arm from the horizontal. Find θ in radians when $a = 1$ and when $a = 2$.

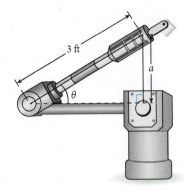

69. **Roofing** A smokestack cut from a right circular cylinder is to be attached to a roof of slope $\frac{4}{5}$. At what acute angle to the horizontal should the cylinder be cut so that it matches the angle of the roof? Express your answer in degrees.

70. **Construction** The pitch of a roof is its slope, which is given as $\frac{\text{Rise}}{\text{Run}}$. If the pitch of a roof is $\frac{2}{5}$, what acute angle does it make with the horizontal? Express your answer in radians.

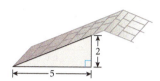

71. **Physics** The position of a block that is attached to one end of a spring oscillates according to the formula $d = 5 \sin 2t$ for t in the interval $\left[-\frac{\pi}{4}, \frac{\pi}{4} \right]$. Express t as a function of d, and state the domain of your function.

72. **Sound Waves** A sound wave has the form $y = 2 \cos \left(3x - \frac{\pi}{4} \right)$ for x in the interval $\left[\frac{\pi}{12}, \frac{5\pi}{12} \right]$. Express x as a function of y, and state the domain of your function.

73. **Physics** The horizontal range of a projectile that is fired with an initial velocity v_0 at an acute angle θ with respect to the horizontal is given by

$$R = \frac{(v_0)^2 \sin 2\theta}{g}$$

where g is the gravitational constant (9.8 m/sec²). If $v_0 = 30$ m/sec, find the angle at which the projectile must be fired if it is to have a horizontal range of 80 meters. Express your answer in degrees.

Concepts

In Exercises 74–77, use the given graph of sin x on the interval $\left[-\frac{\pi}{2}, \frac{\pi}{2}\right]$ to approximate the following expressions.

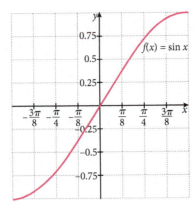

74. arcsin 0.6　　　　**75.** arcsin(−0.2)　　　　**76.** sin⁻¹ 0.4　　　　**77.** sin⁻¹ (−0.8)

In Exercises 78–81, use the given graph of tan x on the interval $\left(-\frac{\pi}{2}, \frac{\pi}{2}\right)$ to approximate the following expressions.

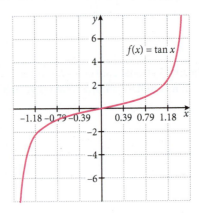

78. arctan 2　　　　**79.** arctan 4　　　　**80.** tan⁻¹ (−3)　　　　**81.** tan⁻¹ 6

Chapter 5 Summary

Section 5.1 Angles and Their Measures

Angles and their measures   [Review Exercises 1–16]

Degree measure: 1 revolution $= 360°$
Radian measure: 1 revolution $= 2\pi$ radians
Converting between radian and degree measure:

To Convert	Multiply by
Degrees to radians	$\dfrac{\pi}{180°}$
Radians to degrees	$\dfrac{180°}{\pi}$

Arc length   [Review Exercises 18]

The length of an arc of a circle of radius r subtended by an angle θ is $s = r\theta$, where θ is in radians.

Linear and angular speed   [Review Exercises 17–18]

If a point on a circle of radius r travels in uniform circular motion through an angle θ in time t, its **linear speed** v is $v = \dfrac{s}{t}$, where s is the arc length given by $s = r\theta$.

The corresponding **angular speed** is $\omega = \dfrac{\theta}{t}$. The linear speed v is related to angular speed ω by the formula $v = r\omega$, where r is the distance from the axis of rotation.

Section 5.2 Trigonometric Functions Using the Unit Circle

Definition of trigonometric functions using the unit circle   [Review Exercises 19–26]

Let (x, y) be the point where the terminal side of an angle of t radians in standard position intersects the unit circle.

The value of the **cosine** function at t, abbreviated as $\cos t$, is defined as $\cos t = x$.

The value of the **sine** function at t, abbreviated as $\sin t$, is defined as $\sin t = y$.

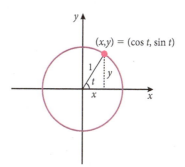

Definitions of the other trigonometric functions   [Review Exercises 23–26]

$$\text{Tangent: } \tan t = \frac{\sin t}{\cos t} \qquad \text{Cotangent: } \cot t = \frac{\cos t}{\sin t}$$

$$\text{Secant: } \sec t = \frac{1}{\cos t} \qquad \text{Cosecant: } \csc t = \frac{1}{\sin t}$$

Pythagorean identities   [Review Exercises 27–30]

For any real number t in the function's domain, the Pythagorean identities are

$$\cos^2 t + \sin^2 t = 1$$
$$1 + \tan^2 t = \sec^2 t$$
$$1 + \cot^2 t = \csc^2 t$$

Negative-angle identities   [Review Exercises 31 and 32]

For any real number t in the function's domain,

$$\sin(-t) = -\sin t \qquad \csc(-t) = -\csc t$$
$$\cos(-t) = \cos t \qquad \sec(-t) = \sec t$$
$$\tan(-t) = -\tan t \qquad \cot(-t) = -\cot t$$

Repetitive behavior of sine, cosine, and tangent   [Review Exercises 25 and 26]

Let t be a real number in the function's domain, and let n be an integer. Then

$$\sin(t + 2n\pi) = \sin t$$
$$\cos(t + 2n\pi) = \cos t$$
$$\tan(t + n\pi) = \tan t$$

Section 5.3 Right Triangle Trigonometry

Definitions of trigonometric functions using right triangles   [Review Exercises 34–44]

$$\text{Sine:} \quad \sin\theta = \frac{\text{opp}}{\text{hyp}} \qquad \text{Cosecant:} \quad \csc\theta = \frac{\text{hyp}}{\text{opp}}$$

$$\text{Cosine:} \quad \cos\theta = \frac{\text{adj}}{\text{hyp}} \qquad \text{Secant:} \quad \sec\theta = \frac{\text{hyp}}{\text{adj}}$$

$$\text{Tangent:} \quad \tan\theta = \frac{\text{opp}}{\text{adj}} \qquad \text{Cotangent:} \quad \cot\theta = \frac{\text{adj}}{\text{opp}}$$

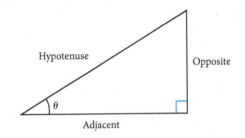

Cofunction identities   [Review Exercises 34–44]

The following cofunction identities hold for all acute angles.

$$\sin(90° - \theta) = \cos\theta \qquad \cos(90° - \theta) = \sin\theta$$
$$\tan(90° - \theta) = \cot\theta \qquad \cot(90° - \theta) = \tan\theta$$
$$\sec(90° - \theta) = \csc\theta \qquad \csc(90° - \theta) = \sec\theta$$

The arccosine function   [Review Exercises 66, 69, 74, 75, 77]

The **arccosine** function, arccos x, is defined by

$$\text{arccos } x = y \text{ if and only if } x = \cos y$$

- The domain of arccos x is $[-1, 1]$.
- The range of arccos x is $[0, \pi]$.

The arctangent function   [Review Exercises 71, 72, 79]

The **arctangent** function, arctan x, is defined by

$$\text{arctan } x = y \text{ if and only if } x = \tan y$$

- The domain of arctan x is $(-\infty, \infty)$.
- The range of arctan x is $\left(-\frac{\pi}{2}, \frac{\pi}{2}\right)$.

The other inverse trigonometric functions   [Review Exercises 67, 68]

The inverse of cotangent, cosecant, and secant function are defined as follows.

- The domain of $\cot^{-1} x$ is $(-\infty, \infty)$ and the range is $\left(-\frac{\pi}{2}, 0\right) \cup \left(0, \frac{\pi}{2}\right]$.
- The domain of $\csc^{-1} x$ is $(-\infty, -1] \cup [1, \infty)$ and the range is $\left[-\frac{\pi}{2}, 0\right) \cup \left(0, \frac{\pi}{2}\right]$.
- The domain of $\sec^{-1} x$ is $(-\infty, -1] \cup [1, \infty)$ and the range is $\left[0, \frac{\pi}{2}\right) \cup \left(\frac{\pi}{2}, \pi\right]$.

Chapter 5 Review Exercises

Section 5.1

In Exercises 1–6, sketch the given angles in standard position.

1. $225°$ **2.** $-60°$ **3.** $120°$ **4.** $\dfrac{3\pi}{4}$ **5.** $-\dfrac{5\pi}{4}$ **6.** $\dfrac{11\pi}{3}$

In Exercises 7 and 8, find the complement and supplement of each angle.

7. $35°$ **8.** $47°$

In Exercises 9–12, convert the given degree measure to radians.

9. $240°$ **10.** $210°$ **11.** $-150°$ **12.** $-225°$

In Exercises 13–16, convert the given radian measure to degrees.

13. $\dfrac{5\pi}{6}$ **14.** $-\dfrac{3\pi}{5}$ **15.** $\dfrac{\pi}{90}$ **16.** $\dfrac{7\pi}{5}$

17. Robotics A robotic arm pinned at one end makes a complete revolution in 1.5 minutes. What is the angle swept out in 1 minute by the robotic arm rotating counterclockwise? Express your answer in both degrees and radians. Assume that the robotic arm starts in the standard position.

18. Robotics A robotic arm with radius 2 feet and pinned at one end makes a complete revolution in 3 minutes.

    a. Find its angular speed in radians per minute, assuming it is constant.

    b. Find the distance traveled by the tip in 1 minute.

Section 5.2

In Exercises 19–22, find the reference angle corresponding the given angle.

19. $\dfrac{4\pi}{3}$ **20.** $-\dfrac{11\pi}{3}$ **21.** $\dfrac{5\pi}{3}$ **22.** $-\dfrac{7\pi}{6}$

In Exercises 23–26, find the exact values of sin t, cos t, and tan t for each value of t.

23. $t = -\dfrac{5\pi}{2}$ **24.** $t = \dfrac{11\pi}{6}$ **25.** $t = \dfrac{7\pi}{3}$ **26.** $t = -\dfrac{17\pi}{6}$

In Exercises 27–30, fill in the given table with the missing information.

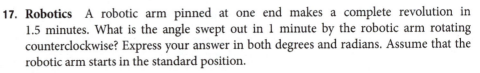
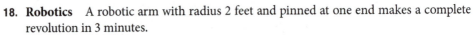

	Quadrant	sin t	cos t	tan t
27.	II		$-\dfrac{\sqrt{3}}{2}$	
28.	IV	$-\dfrac{\sqrt{2}}{2}$		
29.	IV			-1
30.	III		$-\dfrac{1}{2}$	

In Exercises 31 and 32, use the negative angle identities to evaluate the given functions.

31. $\tan\left(-\dfrac{\pi}{4}\right)$

32. $\cos\left(-\dfrac{5\pi}{4}\right)$

33. Physics A spring with a mass attached to its end moves a distance d from the origin according to the formula $d = 10 \sin \frac{\pi}{3} t$, where t is in seconds. What is the maximum distance from the origin that the spring travels? Find the period of the motion.

Section 5.3

In Exercises 34–37, you are given one of the values of a trigonometric function of θ. Find the values of the other five trigonometric functions. Assume θ is an acute angle.

34. $\cos \theta = \dfrac{1}{5}$ **35.** $\sin \theta = \dfrac{3}{10}$ **36.** $\tan \theta = 2$ **37.** $\cot \theta = 0.7$

In Exercises 38 and 39, find the lengths of the unknown sides in the right triangle.

38.
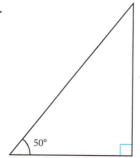

39.
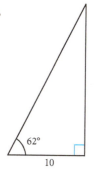

40. Surveying A surveyor stands 150 feet from the base of a tower. The angle of elevation to the top of the building is 63°. Find the height of the tower.

41. Surveying An observer measures an angle of elevation of 35° from a certain point on the ground to the top of a tower. She then walks 12 feet closer to the tower and measures an angle of elevation of 38° to the top of the tower. How tall is the tower?

42. Aviation A plane takes off at an angle of 16° with respect to the horizontal. How many miles has it traveled by the time it reaches an altitude of 5,000 feet?

43. Manufacturing A conveyor belt transports merchandise up an incline that makes an angle of 22° with respect to the horizontal. If the merchandise traverses a distance of 30 feet while on the incline, how much higher is the top of the incline than its base?

44. Earth Rotation Earth rotates about an axis through its poles, making one revolution per day. Earth's radius is approximately 3,900 miles. What is the length of the circular path traced out each day by a point on Earth's surface at 25° south latitude (i.e., the distance traversed by that point each day as a result of the Earth's rotation)? Express your answer in miles. (*Hint*: First find the radius of the path.)

Section 5.4

In Exercises 45–48, a point (x, y) on the terminal side of an angle θ is given. Find the exact value of the indicated trigonometric function.

45. $(-12, 5)$; $\cos \theta$ **46.** $(-4, -3)$; $\sin \theta$ **47.** $(3, -3)$; $\tan \theta$ **48.** $(1, -2)$; $\sec \theta$

In Exercises 49–52, find the reference angle and the exact value of the indicated trigonometric function for each of the given angles.

49. $\theta = 315°$; $\cos \theta$

50. $\theta = 150°$; $\tan \theta$

51. $\theta = -120°$; $\sec \theta$

52. $\theta = 420°$; $\sin \theta$

Section 5.5

In Exercises 53–58, graph at least two periods of the given function.

53. $f(x) = 3 \sin x - 1$

54. $f(x) = -2 \cos x + 3$

55. $f(x) = \cos 2x - 1$

56. $f(x) = \sin \left(\frac{x}{2} \right) + 3$

57. $g(x) = 2 \sin \left(x + \frac{\pi}{4} \right)$

58. $g(x) = \cos(2x - \pi) - 3$

Section 5.6

In Exercises 59–64, graph at least one period of the given function.

59. $f(x) = 2 \tan x + 1$

60. $f(x) = -\csc x$

61. $f(x) = -\cot 2x$

62. $f(x) = \sec \left(x - \frac{\pi}{2} \right)$

63. $f(x) = 2 \tan(x + \pi)$

64. $f(x) = -\cot \left(x - \frac{\pi}{4} \right) + 3$

Section 5.7

In Exercises 65–68, evaluate the given functions without using a calculator.

65. $\arcsin (-1)$

66. $\arccos 0$

67. $\cot^{-1} 0$

68. $\csc^{-1} \left(-\sqrt{2} \right)$

In Exercises 69–76, evaluate the given functions without using a calculator.

69. $\arccos \left(\cos \left(-\frac{\pi}{4} \right) \right)$

70. $\arcsin (\sin(3\pi))$

71. $\tan (\arctan (7))$

72. $\tan (\arctan (-3))$

73. $\sin (\arcsin (0.4))$

74. $\cos (\arccos (0.7))$

75. $\sin \left(\arccos \left(-\frac{12}{13} \right) \right)$

76. $\cos \left(\arcsin \left(\frac{4}{5} \right) \right)$

In Exercises 77 and 78, sketch a graph of each function.

77. $f(x) = -3 \arccos x$

78. $g(x) = \sin^{-1} x - \frac{\pi}{2}$

79. Roofing The roof of a house makes an angle θ with the horizontal such that $\tan \theta = \frac{1}{6}$. Find θ in degrees.

80. Physics If a projectile is fired with an initial velocity 70 m/sec and at an angle θ, the horizontal range is given by

$$R = \frac{v_0^2 \sin 2\theta}{g}$$

where g is the gravitational constant of 9.8 m/sec².

Find an angle that a projectile must be fired at 70 m/sec to cover a horizontal range of 400 meters. Express your answer in degrees.

Chapter 5 Test

In Exercises 1–4, sketch the angle in standard position.

1. $-135°$ **2.** $450°$ **3.** $\dfrac{9\pi}{4}$ **4.** $-\dfrac{8\pi}{3}$

In Exercises 5 and 6, convert the given degree measure to radians.

5. $75°$ **6.** $-315°$

In Exercises 7 and 8, convert the given radian measure to degrees.

7. $\dfrac{\pi}{18}$ **8.** $-\dfrac{\pi}{30}$

In Exercises 9 and 10, find the reference angle corresponding to each angle, and compute the exact values of their sine and cosine.

9. $-\dfrac{5\pi}{6}$ **10.** $\dfrac{8\pi}{3}$

11. Given that $\cos t = -\dfrac{\sqrt{2}}{2}$ and that the terminal side of t is in the third quadrant, find the exact value of $\sin t$ and $\tan t$.

In Exercises 12 and 13, given one of the values of a trigonometric function of θ, find the exact values of the other five trigonometric functions. Assume θ is an acute angle.

12. $\sin \theta = \dfrac{2}{5}$ **13.** $\tan \theta = 3$

14. Find the exact values of the lengths of the unknown sides in the right triangle.

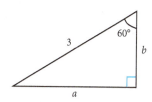

15. Given the point $(-1, 3)$ on the terminal side of θ, compute the exact value of $\cos \theta$ and $\csc \theta$.

16. Find the reference angle corresponding to $240°$ and compute the exact values of $\sin 240°$ and $\sec 240°$.

In Exercises 17–19, graph at least two periods of each function.

17. $f(x) = -\sin 2x + 1$ **18.** $f(x) = 3 \sin \left(x - \dfrac{\pi}{4}\right)$ **19.** $f(x) = 2\cos (2x + \pi)$

In Exercises 20–22, graph at least one period of each function.

20. $f(x) = \tan 2x + 1$ **21.** $f(x) = \sec \left(x + \dfrac{\pi}{2}\right)$ **22.** $f(x) = -3 \cot x$

In Exercises 23–25, evaluate without using a calculator.

23. $\arccos \left(\cos \left(\dfrac{3\pi}{2}\right)\right)$ **24.** $\tan (\arctan (8))$ **25.** $\sin \left(\arccos \left(\dfrac{4}{5}\right)\right)$

26. Sketch a graph of $f(x) = 2 \arcsin x + \pi$.

27. A Ferris wheel 200 feet in diameter makes one revolution every 2 minutes. What is the angular speed of a passenger in degrees per minute, assuming the passenger is on the edge of the wheel? What is the passenger's linear speed in feet per minute?

28. A wire of length 30 feet is attached to the roof of a building and to a stake 10 feet from the building. Find the angle the wire makes with the side of the building.

29. Find the dimensions of the square screen of a 20-inch television. *Note*: 20 inches refers to the length of the diagonal of the screen.

30. The function

$$P(t) = 50 \sin \frac{2\pi}{33} t + 50$$

is used in biorhythm theory to predict an individual's intellectual potential, as a percentage of the maximum, on a particular day, with $t = 0$ corresponding to birth.

a. What is the period of the function?

b. What percentage of an individual's intellectual potential is being met on his 3rd birthday (day 1095)?

Trigonometric Identities and Equations

The number of daylight hours varies during different times of the year and can be modeled by a trigonometric function. You can solve a related trigonometric equation to calculate the time of year when you have a specified number of daylight hours at a particular location. See Exercises 98 and 99 in Section 6.4.

In this chapter, we will study relations between the different trigonometric functions by establishing identities and using them to solve equations.

Use the sum identities to find the values of

a) $\sin\left(\dfrac{7\pi}{12}\right) = \dfrac{\sqrt{2} + \sqrt{6}}{4}$

b) $\cos\left(\dfrac{7\pi}{12}\right) = \sqrt{2}$

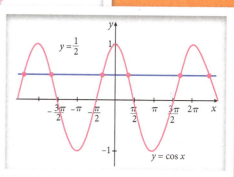

Outline

Verifying Identities

Objectives

- Know basic identities.
- Apply various strategies for verifying identities.

When working with trigonometric functions, it is useful to transform expressions containing these functions into an equivalent form. Consider the equation $\cos^2 x = 1 - \sin^2 x$, where $\cos^2 x$ is written in terms of $\sin^2 x$. This is an example of an **identity**, because it holds true for *all* values of x. The process of demonstrating that an equation holds true for all values of a variable for which the terms of the equation are defined is called **verifying an identity**. The process is also referred to as *establishing* or *proving* an identity.

Basic Identities

Table 1 summarizes some familiar trigonometric identities.

Identities as Ratios	$\tan x = \dfrac{\sin x}{\cos x}$	$\cot x = \dfrac{\cos x}{\sin x}$	
Reciprocal Identities	$\sec x = \dfrac{1}{\cos x}$	$\csc x = \dfrac{1}{\sin x}$	$\cot x = \dfrac{1}{\tan x}$
Pythagorean Identities	$\cos^2 x + \sin^2 x = 1$	$1 + \tan^2 x = \sec^2 x$	$1 + \cot^2 x = \csc^2 x$
Negative-Angle Identities	$\cos(-x) = \cos x$	$\sin(-x) = -\sin x$	$\tan(-x) = -\tan x$

Table 1

To show that an equation *is* an identity, we must show that the equations are true *for all values* of the variable for which the expressions are defined. To show that an equation is *not an identity*, we find values of x for which the terms of the equation are defined, but for which the equation will not hold true.

VIDEO EXAMPLES

SECTION 6.1

Using Technology

For Example 1(b), you can graph $Y_1 = \left(\dfrac{1}{\tan(x)}\right)(\tan x)$ and it seems you get the line $y = 1$ (Figure 1). This holds only for values of x where $\tan x$ is defined and not equal to zero—that is, for $x \neq \dfrac{n\pi}{2}$, n an integer. The calculator graph is not clear enough to show that Y_1 is defined only at these values of x. See Figure 1.

Keystroke Appendix:
Section 15

> **Example 1**   **Deciding Whether an Equation Is an Identity**

Determine which of the following are identities. For those that are not an identity, give a value of x for which every term in the equation is defined but the equation does not hold. For those that are identities, verify the identity.

a. $\sin x = 1$ **b.** $\cot x \tan x = 1$

Solution

a. If $x = 0$, the term $\sin x$ is defined. But the equation $\sin x = 1$ does not hold true, because $\sin 0 = 0 \neq 1$, and so the equation $\sin x = 1$ is *not* an identity. It is called a conditional equation, because it may hold for some, but not all, values of x.

b. To determine whether $\cot x \tan x = 1$ for all x, we can use the basic identities listed in Table 1 to rewrite the left-hand side of the equation:

$$\cot x \tan x = \frac{\cos x}{\sin x}\frac{\sin x}{\cos x}$$

Dividing out the factors $\cos x$ and $\sin x$, we have

$$\cot x \tan x = 1$$

for all x for which $\cot x \tan x$ is defined. Therefore, we have verified that $\cot x \tan x = 1$ is an identity.

Check It Out 1   Verify that $\sin x \csc x = 1$ for all x. ∎

Figure 1

The objective of this section is to determine whether certain trigonometric equations are identities. In general, there are two basic approaches to verifying identities, as shown below.

General Approaches for Verifying Identities

- **Approach 1: One Side of the Equation** Focus on the side of the equation with the more complicated expression and transform it until it is identical to the expression on the other side of the equation.

- **Approach 2: Both Sides of the Equation** Transform the expressions on both sides of the equation until the resulting expressions on both sides of the equation are identical.

As you work to verify identities, you will always use one of these two methods. In addition to these two general approaches, we will also use some specific strategies to verify identities.

Strategy 1   Write in Terms of Sine and/or Cosine

Write a given expression in terms of sine and/or cosine, and then simplify the resulting expression if necessary.

Example 2   **Verify an Identity by Reducing to Sine and Cosine Functions**

Verify the following identity.

$$(\csc x \tan x)(\sec x) = \sec^2 x$$

Solution Begin with the left-hand side of the equation, because that expression looks more complex:

$$(\csc x \tan x)(\sec x)$$

$$= \left(\frac{1}{\sin x} \cdot \frac{\sin x}{\cos x}\right)\left(\frac{1}{\cos x}\right) \qquad \text{Use } \csc x = \frac{1}{\sin x}, \tan x = \frac{\sin x}{\cos x}, \text{ and } \sec x = \frac{1}{\cos x}$$

$$= \left(\frac{1}{\cos x}\right)\left(\frac{1}{\cos x}\right) \qquad \text{Simplify inside parentheses}$$

$$= \frac{1}{\cos^2 x} = \left(\frac{1}{\cos x}\right)^2 = \sec^2 x \qquad \text{Use } \frac{1}{\cos x} = \sec x$$

Thus, we have verified that the left side expression, $(\csc x \tan x)(\sec x)$, can be transformed to the expression on the right, $\sec^2 x$.

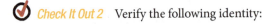 **Check It Out 2**   Verify the following identity:

$$(\sec x \cot x)(\sin^2 x) = \sin x$$

Example 3   **Verify an Identity by Reducing to Sine and Cosine Functions**

Verify the following identity.

$$(\sec x + \tan x)(1 - \sin x) = \cos x$$

Solution We begin with the left-hand side of the equation:

$$(\sec x + \tan x)(1 - \sin x)$$

$$= \left(\frac{1}{\cos x} + \frac{\sin x}{\cos x}\right)(1 - \sin x) \qquad \text{Use } \sec x = \frac{1}{\cos x} \text{ and } \tan x = \frac{\sin x}{\cos x}$$

$$= \left(\frac{1 + \sin x}{\cos x}\right)(1 - \sin x) \qquad \text{Add expression with like denominators}$$

$$= \frac{1 - \sin^2 x}{\cos x} \qquad \text{Multiply}$$

$$= \frac{\cos^2 x}{\cos x} = \cos x \qquad \text{Use } 1 - \sin^2 x = \cos^2 x$$

We have verified that the left-hand side expression, $(\sec x + \tan x)(1 - \sin x)$, can be transformed to the right side expression, $\cos x$.

 Check It Out 3   Verify the following identity.

$$(\csc x + \cot x)(1 - \cos x) = \sin x$$

Factoring Methods for Verifying Identities

Note You cannot verify an identity by substituting just a few numbers and noting that the equation holds for those numbers. The identity must be verified for all values of x in the domain of definition, and this can only be done algebraically.

Another useful strategy for verifying identities is to factor an expression in order to simplify it.

Strategy 2   Factor
Apply a factoring technique, if possible, to a given expression.

 Just in Time
Review factoring techniques in Section 1.3.

Example 4   **Factoring to Help Verify an Identity**

Verify the following identities.

a. $\cos^2 x \sin x - \cos^4 x \sin x = \cos^2 x \sin^3 x$ **b.** $\tan^3 x = \tan x \sec^2 x - \tan x$

Solution

a. Starting with the left-hand side of the equation,

$$\cos^2 x \sin x - \cos^4 x \sin x = \cos^2 x \sin x \,(1 - \cos^2 x) \qquad \text{Factor out } \cos^2 x \sin x$$

$$= \cos^2 x \sin x(\sin^2 x) \qquad \text{Use } 1 - \cos^2 x = \sin^2 x$$

$$= \cos^2 x \sin^3 x \qquad \text{Multiply}$$

Thus, we have verified the identity by transforming the expression on the left, $\cos^2 x \sin x - \cos^4 x \sin x$, to the expression on the right, $\cos^2 x \sin^3 x$.

b. Starting with the right-hand side of the equation,

$$\tan x \sec^2 x - \tan x = \tan x(\sec^2 x - 1) \qquad \text{Factor out } \tan x$$

$$= \tan x \tan^2 x \qquad \text{Use } \sec^2 x - 1 = \tan^2 x$$

$$= \tan^3 x \qquad \text{Multiply}$$

Thus, we have verified the identity by transforming the expression on the right, $\tan x \sec^2 x - \tan x$, to the expression on the left, $\tan^3 x$.

 Check It Out 4   Verify that $\csc^3 x = \csc x \cot^2 x + \csc x$.

The next strategy consists of transforming the numerator and denominator of a fraction.

> **Strategy 3   Obtaining a Difference of Squares**
>
> If an expression consists of a fraction whose numerator or denominator is of the form $a + b$ or $a - b$, then multiply the numerator and the denominator by $a - b$ or $a + b$, respectively. This will give a numerator or denominator of the form $a^2 - b^2$, to which one of the Pythagorean identities may be applied.

Just in Time
Review rational expressions in Section 1.4.

Example 5   **Obtaining a Difference of Squares**

Verify the identity.

$$\frac{\cos x}{1 + \sin x} = \frac{1 - \sin x}{\cos x}$$

Solution The expressions are already written in terms of sine and cosine. Because $1 - \sin^2 x = \cos^2 x$, multiply the numerator and denominator of the expression on the left-hand side by $1 - \sin x$.

$$\frac{\cos x}{1 + \sin x} = \frac{\cos x}{1 + \sin x} \cdot \frac{1 - \sin x}{1 - \sin x}$$

$$= \frac{\cos x \,(1 - \sin x)}{(1 + \sin x)(1 - \sin x)} \qquad \text{Write as a single fraction}$$

$$= \frac{\cos x \,(1 - \sin x)}{1 - \sin^2 x} \qquad \text{Multiply}$$

$$= \frac{\cos x \,(1 - \sin x)}{\cos^2 x} \qquad \text{Use } 1 - \sin^2 x = \cos^2 x$$

$$= \frac{1 - \sin x}{\cos x} \qquad \text{Divide out } \cos x \text{ term}$$

So, we have verified the identity by transforming the expression on the left, $\frac{\cos x}{1 + \sin x}$, to the expression on the right, $\frac{1 - \sin x}{\cos x}$.

 Check It Out 5   Verify the identity

$$\frac{\tan x}{\sec x + 1} = \frac{\sec x - 1}{\tan x}$$

The next example illustrates Approach 2—working with the expression on each side of the proposed identity separately, until the two sides have been transformed into the same expression.

Example 6 **Working with Both Sides of the Equation**

Verify the identity

$$\sec x + \tan x = \frac{\cos x}{1 - \sin x}$$

Solution We will work with both sides of the equation in this example. Write the left-hand side in terms of sine and cosine (Strategy 1).

$$\sec x + \tan x = \frac{1}{\cos x} + \frac{\sin x}{\cos x}$$

$$= \frac{1 + \sin x}{\cos x}$$

Now, transform the right-hand side by multiplying the numerator and denominator by $1 + \sin x$ (Strategy 3).

$$\frac{\cos x}{1 - \sin x} = \frac{\cos x}{1 - \sin x} \cdot \frac{(1 + \sin x)}{(1 + \sin x)}$$

$$= \frac{\cos x(1 + \sin x)}{1 - \sin^2 x} \qquad \text{Write as a single fraction}$$

$$= \frac{\cos x(1 + \sin x)}{\cos^2 x} \qquad \text{Use } 1 - \sin^2 x = \cos^2 x$$

$$= \frac{1 + \sin x}{\cos x} \qquad \text{Simplify}$$

Because both sides are equal to $\frac{1 + \sin x}{\cos x}$, the identity is verified.

 Check It Out 6 Verify the identity

$$\frac{1 - \cos t}{\cos t} = \frac{\tan^2 t}{1 + \sec t}$$

We have given some general and specific strategies to verify identities. Many problems may not neatly fit into just one category, and you may need to combine some of these strategies, as in Example 6.

Also, there are several correct ways to verify an identity, as long as each step is valid, and you proceed logically. It is important to remember that you *cannot* simply move terms from one side of the equation to the other just because you see an equals sign, because we do not know yet that the equation holds.

Just in Time Exercises

In Exercises 1–4, completely factor each expression.

1. $4x^2 - 9$ **2.** $x^2 + 4x + 4$ **3.** $9x^2 - 81$ **4.** $2x^2 - 16x + 32$

In Exercises 5–8, completely simplify each expression.

5. $\dfrac{2}{1+x} - \dfrac{1}{1-x}$ **6.** $\dfrac{3}{x-2} + \dfrac{4}{x+2}$ **7.** $\dfrac{5}{x-1} + \dfrac{3}{1-x}$ **8.** $\dfrac{7}{x-2} + \dfrac{4}{2-x}$

Skills

In Exercises 9–14, show that the given equation is not an identity.

9. $\cos x = 0.5$ **10.** $\tan x = 1$ **11.** $\sin(x + \pi) = \sin x + \sin \pi$

12. $\cos\left(x - \dfrac{\pi}{2}\right) = \cos x - \cos \dfrac{\pi}{2}$ **13.** $\sin 2x = 2 \sin x$ **14.** $\cos 2x = 2 \cos x$

In Exercises 15–20, write the given expression in terms of sin x and/or cos x only.

15. $\cot x \csc x$ **16.** $\csc^2 x$ **17.** $\dfrac{\tan^2 x}{\sin x}$ **18.** $\sec x \cot x$ **19.** $\sec^2 x - 1$ **20.** $1 - \tan x$

In Exercises 21–26, completely factor the given trigonometric expression.

21. $\sin x + \sin x \cos x$ **22.** $\tan^2 x - \sec x \tan x$ **23.** $1 - \sin^2 x$ **24.** $\sec^2 x - \tan^2 x$

25. $\sin^2 x - \cos^2 x$ **26.** $\sin^4 x - 1$

In Exercises 27–80, verify the given identity.

27. $\tan x \csc x = \sec x$

28. $\cot x \sec x = \csc x$

29. $\sin x + \cos x \cot x = \csc x$

30. $\cos x + \tan x \sin x = \sec x$

31. $\sec^2 x \, (1 - \sin^2 x) = 1$

32. $\csc^2 x \, (1 - \cos^2 x) = 1$

33. $\sin^2 (-x) + \cos^2 (-x) = 1$

34. $\sin^2 (-x) + \cos^2 x = 1$

35. $(\sec^2 x - 1) \cot^2 x = 1$

36. $(\csc^2 x - 1) \tan^2 x = 1$

37. $\dfrac{\cot x}{\csc x} = \cos x$

38. $\dfrac{\tan x}{\sec x} = \sin x$

39. $\sin^2 x \sec x = \sec x - \cos x$

40. $\cos^2 x \csc x = \csc x - \sin x$

41. $(\cos x + \sin x)^2 - 2 \sin x \cos x = 1$

42. $(\cos x - \sin x)^2 + 2 \sin x \cos x = 1$

43. $\sec x + \tan x = \dfrac{1 + \sin x}{\cos x}$

44. $\csc x + \cot x = \dfrac{\sin x}{1 - \cos x}$

45. $\cos^3 x = \cos x - \cos x \sin^2 x$

46. $\sin^3 x = \sin x - \sin x \cos^2 x$

47. $\sec x \cos^3 x = 1 - \sin^2 x$

48. $\sec^3 x = \dfrac{1 + \tan^2 x}{\cos x}$

49. $\dfrac{1}{1 + \cos x} + \dfrac{1}{1 - \cos x} = 2 \csc^2 x$

50. $\dfrac{1}{1 + \cos x} - \dfrac{1}{1 - \cos x} = -2 \csc^2 x \cos x$

51. $\dfrac{\sin x - \sin(-x)}{1 - \cos^2 x} = 2 \csc x$

52. $\dfrac{\cos x + \cos(-x)}{1 - \sin^2 x} = 2 \sec x$

53. $\dfrac{\sec^2 x}{1 + \sin x} = \dfrac{\sec^2 x - \sec x \tan x}{\cos^2 x}$

54. $\dfrac{\csc^2 x}{1 + \cos x} = \dfrac{\csc^2 x - \csc x \cot x}{\sin^2 x}$

55. $\cos^2 x - \sin^2 x = 1 - 2\sin^2 x$

56. $\sec^2 x + \tan^2 x = 1 + 2 \tan^2 x$

57. $\tan x + \cot x = \sec x \csc x$

58. $\tan x - \cot x = \sec x \csc x - 2 \cos x \csc x$

59. $\cos^4 x - \sin^4 x = \cos^2 x - \sin^2 x$

60. $\cot^2 x - \tan^2 x = \csc^2 x - \sec^2 x$

61. $\dfrac{\tan^2 x - 1}{1 + \tan^2 x} = 2\sin^2 x - 1$

62. $\dfrac{\cot^2 x - 1}{1 + \cot^2 x} = 2 \cos^2 x - 1$

63. $\csc^2 x + \sec^2 x = \csc^2 x \sec^2 x$

64. $1 - \tan^4 x = \sec^2 x(1 - \tan^2 x)$

65. $\sec^4 x - \tan^4 x = \sec^2 x + \tan^2 x$

66. $1 - \sin^4 x = \cos^2 x + \cos^2 x \sin^2 x$

67. $\dfrac{1}{1 + \cos x} = \csc^2 x - \cot x \csc x$

68. $\dfrac{1}{1 - \sin x} = \sec^2 x + \tan x \sec x$

69. $(\csc x - \cot x)^2 = \dfrac{1 - \cos x}{1 + \cos x}$

70. $(\sec x - \tan x)^2 = \dfrac{1 - \sin x}{1 + \sin x}$

71. $\cot x + \tan x = \sec^2 x \cot x$

72. $\sin x + \cos x + \dfrac{\sin x}{\cot x} = \sec x + \csc x - \dfrac{\cos x}{\tan x}$

73. $\dfrac{\sec^2 x - 1}{\tan x} = \dfrac{\sec x}{\csc x}$

74. $\dfrac{\csc^2 x - 1}{\cot x} = \dfrac{\csc x}{\sec x}$

75. $\dfrac{\sin x}{\csc x - \cot x} = 1 + \cos x$

76. $\dfrac{\cos x - \sin x}{\cos x + \sin x} = \dfrac{\cot x - 1}{\cot x + 1}$

77. $a \csc^2 x (1 + \cos x)(1 - \cos x) = a$

78. $b(\sec x - \tan x)(\sec x + \tan x) = b$

79. $\ln|\tan x| = -\ln|\cot x|$

80. $\ln|\sec x| = -\ln|\cos x|$

In Exercises 81–86, use a graphing utility to graph each side of the equation and decide if the equation is a possible identity. You need not verify the ones that are identities.

81. $\sin(x + \pi) = \sin x + \pi$

82. $\cos 2x = 2 \cos x$

83. $\cos 2x = 1 - 2 \sin^2 x$

84. $\sin 2x = 2 \sin x \cos x$

85. $\sin(x - \pi) = \sin x$

86. $\cos(x + \pi) = -\cos x$

Concepts

87. Explain why $\sin x = \sqrt{1 - \cos^2 x}$ is NOT an identity.

88. What values of $x \in [0, 2\pi)$ satisfy the equation $\sin x = \cos x$?

89. Does the identity $\tan x \cos x = \sin x$ hold for $x = \frac{\pi}{2}$? Why or why not?

90. Does the identity $\csc x \sin x = 1$ hold for all real values of x? Why or why not?

Math takes time. This fact holds true in the smallest of math problems as much as it does in the most math intensive careers. I see proof in each video I make. My videos get progressively better with each take, though I still make mistakes and find aspects I can improve on with each new video. In order to keep trying to improve in spite of any failures or lack of improvement, something else is needed. For me it is the sense of a specific goal in sight, to help me maintain the desire to put in continued time and effort.

When I decided on the number one university I wanted to attend, I wrote the name of that school in bold block letters on my door, written to remind myself daily of my ultimate goal. Stuck in the back of my head, this end result pushed me little by little to succeed and meet all of the requirements for the university I had in mind. And now I can say I'm at my dream school bringing with me that skill.

I recognize that others may have much more difficult circumstances than my own to endure, with the goal of improving or escaping those circumstances, and I deeply respect that. But that fact demonstrates to me how easy but effective it is, in comparison, to "stay with the problems longer" with a goal in mind of something much more easily realized, like a good grade on a test. I've learned to set goals, small or big, and to stick with them until they are realized.

Sum and Difference Identities

6.2

Objectives

- Know sum and difference identities for sine, cosine, and tangent.

- Use sum and difference identities to find exact values of trigonometric functions.

- Use sum and difference identities to derive other identities.

Note The sum and difference identities for sine and cosine are important enough to commit to memory. Note that $\sin(a + b)$ is written as a *sum* of two quantities, while $\cos(a + b)$ is written as a *difference* of two quantities. To prevent confusion between the two, a useful memory aid is "sine keeps the sign."

VIDEO EXAMPLES

SECTION 6.2

 **Using Technology**

We can confirm the result of Example 1 by seeing that the graphs of the functions $Y_1 = \cos(x + \pi)$ and $Y_2 = -\cos x$ are identical. See Figure 1.

Keystroke Appendix:

Section 15

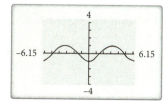

Figure 1

Some of the most important identities give the relationships between the sine, cosine, or tangent of a sum (or difference) of a pair of real numbers. These identities are in turn used to derive other identities, as seen in later sections.

Sum and Difference Identities

$$\sin (a + b) = \sin a \cos b + \cos a \sin b \qquad \tan (a + b) = \frac{\tan a + \tan b}{1 - \tan a \tan b}$$

$$\sin (a - b) = \sin a \cos b - \cos a \sin b \qquad \tan (a - b) = \frac{\tan a - \tan b}{1 + \tan a \tan b}$$

$$\cos (a + b) = \cos a \cos b - \sin a \sin b$$

$$\cos (a - b) = \cos a \cos b + \sin a \sin b$$

These identities are proved at the end of this section using techniques in analytic geometry. Our focus now is on using the identities in various contexts.

Example 1 **Rewriting $\cos(x + \pi)$**

Write $\cos(x + \pi)$ in terms of only $\cos x$.

Solution Using the addition identity, with $a = x$ and $b = \pi$, we have

$$\cos(x + \pi) = \cos x \cos \pi - \sin x \sin \pi$$

$$= (\cos x)(-1) - (\sin x)(0) \qquad \text{Substitute } \cos \pi = -1 \text{ and } \sin \pi = 0$$

$$= -\cos x \qquad \text{Simplify}$$

✓ *Check It Out 1* Write $\sin(x - \pi)$ in terms of $\sin x$. ■

The sum and difference identities can be used to compute the exact values of $\sin(a \pm b)$ or $\cos(a \pm b)$, where the exact values of $\sin a$, $\sin b$, $\cos a$, and $\cos b$ are known.

Example 2 **Finding Exact Values of Sine and Cosine (Radians)**

Use the sum identities to find the exact values of $\sin\left(\frac{7\pi}{12}\right)$ and $\cos\left(\frac{7\pi}{12}\right)$.

Solution Write $\frac{7\pi}{12}$ as a sum of two quantities whose exact values of sine and cosine are known.

$$\frac{7\pi}{12} = \frac{4\pi}{12} + \frac{3\pi}{12} = \frac{\pi}{3} + \frac{\pi}{4}$$

Apply the sum identities for sine and cosine, with $a = \frac{\pi}{3}$ and $b = \frac{\pi}{4}$.

$$\sin\left(\frac{7\pi}{12}\right) = \sin\left(\frac{\pi}{3} + \frac{\pi}{4}\right) \qquad \frac{7\pi}{12} = \frac{\pi}{3} + \frac{\pi}{4}$$

$$= \sin\frac{\pi}{3}\cos\frac{\pi}{4} + \cos\frac{\pi}{3}\sin\frac{\pi}{4} \qquad \text{Sum identity for sine}$$

$$= \left(\frac{\sqrt{3}}{2}\right)\left(\frac{\sqrt{2}}{2}\right) + \left(\frac{1}{2}\right)\left(\frac{\sqrt{2}}{2}\right) \qquad \text{Substitute values for sine and cosine}$$

$$= \frac{\sqrt{3} \cdot \sqrt{2}}{4} + \frac{\sqrt{2}}{4} \qquad \text{Multiply and add}$$

$$= \frac{\sqrt{2}\left(\sqrt{3}+1\right)}{4} \qquad \text{Factor out } \sqrt{2}$$

Just in Time
Review cosine and sine of special angles in Section 5.2.

$$\cos\left(\frac{7\pi}{12}\right) = \cos\left(\frac{\pi}{3} + \frac{\pi}{4}\right)$$

$$= \cos\frac{\pi}{3}\cos\frac{\pi}{4} - \sin\frac{\pi}{3}\sin\frac{\pi}{4} \qquad \text{Sum identity for cosine}$$

$$= \left(\frac{1}{2}\right)\left(\frac{\sqrt{2}}{2}\right) - \left(\frac{\sqrt{3}}{2}\right)\left(\frac{\sqrt{2}}{2}\right) \qquad \text{Substitute values for sine and cosine}$$

$$= \frac{\sqrt{2} - \sqrt{3}\cdot\sqrt{2}}{4} \qquad \text{Multiply and add}$$

$$= \frac{\sqrt{2}(1 - \sqrt{3})}{4} \qquad \text{Factor out } \sqrt{2}$$

Check It Out 2   Find the exact value of $\tan\frac{7\pi}{12}$.

Example 3   **Finding Exact Values of Cosine and Tangent (Degrees)**

Find the exact values of $\cos 15°$ and $\tan 15°$.

Solution   Write $15°$ as

$$15° = 45° - 30°$$

Apply the difference identity for cosine with $a = 45°$ and $b = 30°$.

$$\cos 15° = \cos(45° - 30°)$$

$$= \cos 45° \cos 30° + \sin 45° \sin 30° \qquad \text{Difference identity for cosine}$$

$$= \left(\frac{\sqrt{2}}{2}\right)\left(\frac{\sqrt{3}}{2}\right) + \left(\frac{\sqrt{2}}{2}\right)\left(\frac{1}{2}\right) \qquad \text{Substitute values for sine and cosine}$$

$$= \frac{\sqrt{6}}{4} + \frac{\sqrt{2}}{4} \qquad \text{Multiply and add}$$

$$= \frac{\sqrt{6} + \sqrt{2}}{4} \qquad \text{Combine terms}$$

Using the difference identity for tangent,

$$\tan 15° = \tan(45° - 30°)$$

$$= \frac{\tan 45° - \tan 30°}{1 + \tan 45° \tan 30°} \qquad \text{Difference identity for tangent}$$

$$= \frac{1 - \frac{\sqrt{3}}{3}}{1 + (1)\left(\frac{\sqrt{3}}{3}\right)} \qquad \text{Use } \tan 45° = 1 \text{ and } \tan 30° = \frac{\sqrt{3}}{3}$$

$$= \frac{\left(\frac{3 - \sqrt{3}}{3}\right)}{\left(\frac{3 + \sqrt{3}}{3}\right)} = \frac{3 - \sqrt{3}}{3 + \sqrt{3}}$$

The final answer can also be written as $2 - \sqrt{3}$ by rationalizing the denominator.

 Check It Out 3   Find the exact value of $\sin 15°$.

The next example shows how to find $\sin(a + b)$ and $\cos(a + b)$ when a and b are not given explicitly.

Example 4   **Finding Exact Values of Sine and Tangent**

Given (a, b) in Quadrant III with $\sin a = -\frac{1}{3}$ and $\cos b = -\frac{3}{5}$, find the exact value of each of the following:

a. $\sin(a + b)$ **b.** $\tan(a + b)$

Solution

a. From the sum identity,

$$\sin(a + b) = \sin a \cos b + \cos a \sin b.$$

Now determine $\cos a$ and $\sin b$. Using the Pythagorean identity, we find that

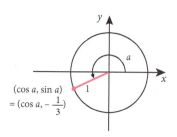

$(\cos a, \sin a)$
$= (\cos a, -\frac{1}{3})$

Figure 2

$$\cos^2 a + \sin^2 a = 1$$

$$\cos^2 a + \left(-\frac{1}{3}\right)^2 = 1 \qquad\qquad \text{Use } \sin a = -\frac{1}{3}$$

$$\cos^2 a + \frac{1}{9} = 1 \qquad\qquad \text{Simplify}$$

$$\cos^2 a = \frac{8}{9} \qquad\qquad \text{Subtract } \frac{1}{9} \text{ from both sides}$$

$$\cos a = \pm\sqrt{\frac{8}{9}} = \pm\frac{2\sqrt{2}}{3} \qquad \text{Take the square root}$$

Because a lies in Quadrant III, as seen in Figure 2, $\cos a < 0$, and so $\cos a = -\frac{2\sqrt{2}}{3}$. Similarly,

$$\cos^2 b + \sin^2 b = 1 \qquad\qquad \text{Pythagorean identity}$$

$$\left(-\frac{3}{5}\right)^2 + \sin^2 b = 1 \qquad\qquad \text{Use } \cos b = -\frac{3}{5}$$

$$\left(\frac{9}{25}\right)^2 + \sin^2 b = 1 \qquad\qquad \text{Simplify}$$

$$\sin^2 b = \frac{16}{25} \qquad\qquad \text{Subtract } \frac{9}{25}$$

$$\sin b = -\frac{4}{5} \qquad\qquad \text{Take the square root}$$

We chose the negative sign, because b lies in Quadrant III, as seen in Figure 3. We now have all the values necessary to calculate $\sin(a + b)$.

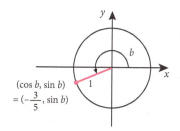

$(\cos b, \sin b)$
$= (-\frac{3}{5}, \sin b)$

Figure 3

$$\sin(a + b) = \sin a \cos b + \cos a \sin b \qquad \text{Sum identity for sine}$$

$$= \left(-\frac{1}{3}\right)\left(-\frac{3}{5}\right) + \left(-\frac{2\sqrt{2}}{3}\right)\left(-\frac{4}{5}\right) \qquad \text{Substitute values for sine and cosine}$$

$$= \frac{1}{5} + \frac{8\sqrt{2}}{15} = \frac{3}{15} + \frac{8\sqrt{2}}{15} \qquad \text{Simplify}$$

$$= \frac{3 + 8\sqrt{2}}{15}$$

b. $\tan(a + b) = \dfrac{\tan a + \tan b}{1 - \tan a \tan b}$

Now,

$$\tan a = \frac{\sin a}{\cos a} = \frac{-\frac{1}{3}}{-\frac{2\sqrt{2}}{3}} = \frac{1}{2\sqrt{2}} = \frac{\sqrt{2}}{4}$$

and

$$\tan b = \frac{\sin b}{\cos b} = \frac{-\frac{4}{5}}{-\frac{3}{5}} = \frac{4}{3}$$

Thus,

$$\tan(a + b) = \frac{\tan a + \tan b}{1 - \tan a \tan b} \qquad \text{Sum identity for tangent}$$

$$= \frac{\frac{\sqrt{2}}{4} + \frac{4}{3}}{1 - \left(\frac{\sqrt{2}}{4}\right)\left(\frac{4}{3}\right)} \qquad \text{Substitute values for tan } a \text{ and tan } b$$

$$= \frac{\frac{3\sqrt{2}+16}{12}}{1 - \left(\frac{\sqrt{2}}{3}\right)} \qquad \text{Simplify numerator}$$

$$= \frac{\frac{3\sqrt{2} + 16}{12}}{\frac{3 - \sqrt{2}}{3}} \qquad \text{Simplify denominator}$$

$$= \frac{3\sqrt{2} + 16}{12} \cdot \frac{3}{3 - \sqrt{2}} \qquad \text{Rewrite division as multiplication}$$

$$= \frac{3\sqrt{2} + 16}{4(3 - \sqrt{2})} \qquad \text{Simplify}$$

Rationalizing denominator:

$$\frac{(3\sqrt{2} + 6)}{4(3 - \sqrt{2})} \cdot \frac{(3 + \sqrt{2})}{(3 + \sqrt{2})} = \frac{24 + 15\sqrt{2}}{28}$$

 Check It Out 4   For a, b given in Example 4, find $\cos(a + b)$.

Cofunction Identities

The sum and difference identities can be used to establish the following cofunction identities.

Cofunction Identities
$\sin\left(\dfrac{\pi}{2} - a\right) = \cos a \qquad \cos\left(\dfrac{\pi}{2} - a\right) = \sin a$
$\tan\left(\dfrac{\pi}{2} - a\right) = \cot a \qquad \cot\left(\dfrac{\pi}{2} - a\right) = \tan a$
$\sec\left(\dfrac{\pi}{2} - a\right) = \csc a \qquad \csc\left(\dfrac{\pi}{2} - a\right) = \sec a$

We show that $\sin\left(\dfrac{\pi}{2} - a\right) = \cos a$. The other cofunction identities are proved similarly and are left as exercises.

Using the difference identity,

$$\sin\left(\frac{\pi}{2} - a\right) = \sin\frac{\pi}{2}\cos a - \cos\frac{\pi}{2}\cos a$$

$$= (1)(\cos a) - (0)(\cos a) \qquad \text{Substitute } \sin\frac{\pi}{2} = 1 \text{ and } \cos\frac{\pi}{2} = 0$$

$$= \cos a \qquad \text{Simplify}$$

Example 5   **Verifying an Identity**

Verify the identity

$$\frac{\sin\left(x - \frac{\pi}{2}\right)}{\sin x} = -\cot x$$

Solution   Start with the expression on the left-hand side. Writing $x - \frac{\pi}{2}$ as $-\left(\frac{\pi}{2} - x\right)$, we have

$$\sin\left(x - \frac{\pi}{2}\right) = \sin\left(-\left(\frac{\pi}{2} - x\right)\right)$$

$$= -\sin\left(\frac{\pi}{2} - x\right) \qquad \text{Use } \sin(-u) = -\sin u$$

$$= -\cos x \qquad\qquad \text{Cofunction identity}$$

Thus,

$$\frac{\sin\left(x - \frac{\pi}{2}\right)}{\sin x} = -\frac{\cos x}{\sin x} = -\cot x$$

✓ *Check It Out 5*   Verify the identity

$$\frac{\csc\left(\frac{\pi}{2} - x\right)}{\cos x} = \sec^2 x$$

Sums of Sines and Cosines

The sum and difference identities can be used to rewrite a sum of cosine and sine functions in terms of just the sine function as follows.

Note   We restrict θ to $[0, 2\pi)$ to obtain a unique solution.

> **Sums of Sines and Cosines**
>
> Let A and B be real numbers. Then
>
> $$A \sin x + B \cos x = C \sin(x + \theta)$$
>
> where $C = \sqrt{A^2 + B^2}$ and $\theta \in [0, 2\pi)$ satisfies
>
> $$\cos\theta = \frac{A}{C} \qquad \text{and} \qquad \sin\theta = \frac{B}{C}$$

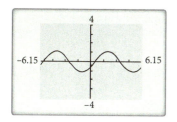 **Using Technology**

You can graph both functions in Example 6 and see that they are identical. See Figure 4. Change the line marker for the second graph so that a small circle traces nicely over the first graph.

Keystroke Appendix:
Section 15

Figure 4

Example 6   **Writing as a Sum of Sine and Cosine**

Write the function $f(x) = \sin x - \frac{1}{2}\cos x$ in the form $C\sin(x + \theta)$, where $\theta \in [0, 2\pi)$.

Solution   By the preceding identity for the sum of sines and cosines,

$$C = \sqrt{A^2 + B^2}$$

$$= \sqrt{1 + \left(-\frac{1}{2}\right)^2} = \sqrt{1 + \frac{1}{4}} = \frac{\sqrt{5}}{2}$$

The angle θ is determined so that

$$\cos\theta = \frac{A}{C} = \frac{1}{\left(\frac{\sqrt{5}}{2}\right)} = \frac{2}{\sqrt{5}}$$

$$\sin\theta = \frac{B}{C} = \frac{\left(-\frac{1}{2}\right)}{\left(\frac{\sqrt{5}}{2}\right)} = -\frac{1}{\sqrt{5}}$$

Because $\cos \theta > 0$ and $\sin \theta < 0$, θ is in Quadrant IV. Using a calculator, we find that $\sin^{-1}\left(-\frac{1}{\sqrt{5}}\right) \approx -0.4636$, in radians. Note that the inverse sine gives the angle in the fourth quadrant. Because $\theta \in [0, 2\pi)$, we have $\theta = 2\pi - 0.4636 \approx 5.820$ Thus $f(x) = \frac{\sqrt{5}}{2} \sin(x + 5.820)$.

 Check It Out 6   Write the function $f(x) = \sin x + \cos x$ in the form $C \sin (x + \theta)$, where $\theta \in [0, 2\pi)$.   ■

A Calculus Application

In calculus, the quantity $\frac{f(x + h) - f(x)}{h}$, for any real valued function f and any number $h \neq 0$, is known as a *difference quotient*. It measures the *average rate of change* on the interval $[x, x + h]$, which is one of the fundamental quantities studied in calculus. The sum identity can be used to simplify this quantity when f is a trigonometric function.

| Example 7 | **Calculus Application** |

Let $f(x) = \cos x$ and $h \neq 0$. Express $\frac{f(x + h) - f(x)}{h}$ in terms of $\sin x$, $\cos x$, $\sin h$, $\cos h$ and h.

Solution   Using the sum identity, we have

$$f(x + h) = \cos(x + h) = \cos x \cos h - \sin x \sin h$$

Therefore,

$$\frac{f(x + h) - f(x)}{h} = \frac{\cos x \cos h - \sin x \sin h - \cos x}{h}$$

$$= \frac{\cos x(\cos h - 1) - \sin x \sin h}{h}$$

$$= \frac{\cos x (\cos h - 1)}{h} - \frac{\sin x \sin h}{h}$$

 Check It Out 7   Rework Example 7 for the case where $f(x) = \sin x$.   ■

Proof of Sum and Difference Identities

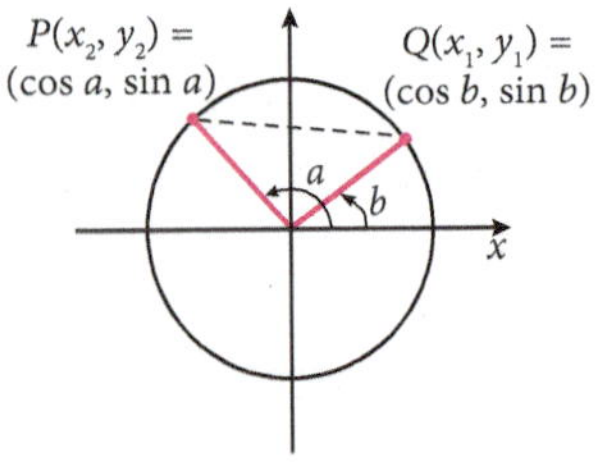

We first verify that $\cos(a - b) = \cos a \cos b + \sin a \sin b$.

Difference Identity for Cosine

From Figure 5 and the distance formula, we find that the distance PQ is

$$PQ = \sqrt{(x_2 - x_1)^2 + (y_2 - y_1)^2}$$

$$= \sqrt{(\cos a - \cos b)^2 + (\sin a - \sin b)^2} \quad \text{Use } (x_1, y_1) = (\cos b, \sin b), \text{ and } (x_2, y_2) = (\cos a, \sin a)$$

Now,

$$(\cos a - \cos b)^2 = \cos^2 a - 2 \cos a \cos b + \cos^2 b$$

and

$$(\sin a - \sin b)^2 = \sin^2 a - 2 \sin a \sin b + \sin^2 b$$

Substituting these results into the expression for PQ and then rearranging terms, we have

$$PQ = \sqrt{(\cos^2 a + \sin^2 a) - 2 \cos a \cos b - 2 \sin a \sin b + (\cos^2 b + \sin^2 b)}$$

Substituting $\cos^2 a + \sin^2 a = 1$ and $\cos^2 b + \sin^2 b = 1$ and then simplifying further, we obtain

$$PQ = \sqrt{2 - 2(\cos a \cos b + \sin a \sin b)}$$

We can rotate the circle so that point Q is at $(1, 0)$. The distance PQ will be the same, and P will be the point $(\cos(a - b), \sin(a - b))$. See Figure 6. Now we will be able to obtain an expression for $\cos(a - b)$. Letting $t = a - b$, we have

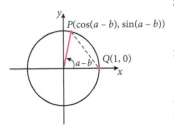

Figure 5

$P(x_2, y_2) = (\cos a, \sin a)$ $Q(x_1, y_1) = (\cos b, \sin b)$

Figure 6

$P(\cos(a - b), \sin(a - b))$ $Q(1, 0)$

$$PQ = \sqrt{(\cos t - 1)^2 + (\sin t - 0)^2}$$

$$= \sqrt{\cos^2 t - 2\cos t + 1 + \sin^2 t}$$

$$= \sqrt{2 - 2\cos t} \qquad \text{Use } \sin^2 t + \cos^2 t = 1$$

Equating the two expressions for PQ and solving for $\cos t$ gives

$$\sqrt{2 - 2\cos t} = \sqrt{2 - 2(\cos a \cos b + \sin a \sin b)}$$

$$2 - 2\cos t = 2 - 2(\cos a \cos b + \sin a \sin b) \quad \text{Square both sides of the equation}$$

$$-2\cos t = -2(\cos a \cos b + \sin a \sin b) \quad \text{Subtract 2 from both sides}$$

$$\cos(a - b) = \cos a \cos b + \sin a \sin b \quad \text{Substitute } t = a - b$$

Thus, we have verified the difference identity for cosine.

Sum Identity for Cosine

The sum identity for cosine follows by writing $a + b$ as $a - (-b)$ and using the difference identity for cosine. Thus,

$$\cos(a + b) = \cos(a - (-b)) = \cos a \cos(-b) + \sin a \sin(-b)$$

$$= \cos a \cos b - \sin a \sin b \qquad \text{Use } \cos(-b) = \cos b \text{ and } \sin(-b) = -\sin(b)$$

Sum Identity for Sine

We next prove the sum identity for sine. We start from the cofunction identity

$$\sin u = \cos\left(\frac{\pi}{2} - u\right)$$

which follows easily from the difference identity for cosine.

Letting $u = a + b$, we get

$$\sin(a + b) = \cos\left(\frac{\pi}{2} - (a + b)\right) = \cos\left(\frac{\pi}{2} - a - b\right) = \cos\left(\left(\frac{\pi}{2} - a\right) - b\right)$$

Now, we use the difference identity for cosine, which gives

$$\cos\left(\left(\frac{\pi}{2} - a\right) - b\right) = \cos\left(\frac{\pi}{2} - a\right)\cos b + \sin\left(\frac{\pi}{2} - a\right)\sin b$$

Now, by the cofunction identities,

$$\cos\left(\frac{\pi}{2} - a\right) = \sin a$$

$$\sin\left(\frac{\pi}{2} - a\right) = \cos\left(\frac{\pi}{2} - \left(\frac{\pi}{2} - a\right)\right) = \cos a$$

Substituting these results, we obtain the sum identity for sine.

$$\sin(a + b) = \sin a \cos b + \cos a \sin b$$

Difference Identity for Sine

Writing $a - b = a + (-b)$, we also get the difference identity for sine.

$$\sin(a - b) = \sin a \cos b - \cos a \sin b$$

The details are left as an exercise at the end of this section.

Exercises 6.2

Just in Time Exercises

In Exercises 1–6, find the exact value of each function.

1. $\sin \dfrac{\pi}{6}$ **2.** $\tan \dfrac{\pi}{3}$ **3.** $\cos\left(-\dfrac{5\pi}{4}\right)$ **4.** $\sin\left(-\dfrac{7\pi}{3}\right)$ **5.** $\cos 330°$ **6.** $\cos 240°$

Skills

Paying Attention to Instructions: *Exercises 7 and 8 are intended to give you practice reading and paying attention to the instructions that accompany the problems you are working.*

7. Find the exact value of each expression.

   a. $\cos \dfrac{\pi}{3} + \cos \dfrac{\pi}{6}$

   b. $\cos\left(\dfrac{\pi}{3} + \dfrac{\pi}{6}\right)$

8. Find the exact value of each expression.

   a. $\sin \dfrac{\pi}{3} - \sin \dfrac{\pi}{6}$

   b. $\sin\left(\dfrac{\pi}{3} - \dfrac{\pi}{6}\right)$

In Exercises 9–20, find the exact value of each expression.

9. $\sin \pi \sin \dfrac{\pi}{3}$ **10.** $\cos \dfrac{\pi}{2} \cos \dfrac{\pi}{6}$ **11.** $\sin\left(-\dfrac{\pi}{3}\right) \sin \dfrac{\pi}{3}$ **12.** $\cos\left(-\dfrac{\pi}{6}\right) \cos \dfrac{\pi}{4}$

13. $\sin(45°) \sin(30°)$ **14.** $\cos(45°) \cos(30°)$ **15.** $\sin(-45°) \cos(30°)$ **16.** $\cos(-30°) \sin(45°)$

17. $\cos \pi \cos \dfrac{\pi}{3} + \sin \pi \sin \dfrac{\pi}{3}$

18. $\cos \dfrac{\pi}{3} \sin \dfrac{\pi}{2} - \sin \dfrac{\pi}{3} \cos \dfrac{\pi}{2}$

19. $\sin(30°) \cos(45°) - \cos(30°) \sin(45°)$

20. $\sin(45°) \cos(30°) + \cos(90°) \sin(90°)$

In Exercises 21–38, find the exact value of the sine, cosine, and tangent of each angle.

21. $\dfrac{11\pi}{12} = \dfrac{2\pi}{3} + \dfrac{\pi}{4}$ **22.** $\dfrac{5\pi}{12} = \dfrac{\pi}{4} + \dfrac{\pi}{6}$ **23.** $-\dfrac{\pi}{12} = \dfrac{\pi}{4} - \dfrac{\pi}{3}$ **24.** $-\dfrac{5\pi}{12} = \dfrac{\pi}{4} - \dfrac{2\pi}{3}$

25. $\dfrac{3\pi}{4} = \pi - \dfrac{\pi}{4}$ **26.** $-\dfrac{4\pi}{3} = -\pi - \dfrac{\pi}{3}$ **27.** $-\dfrac{13\pi}{12}$ **28.** $\dfrac{7\pi}{6}$

29. $-\dfrac{7\pi}{12}$ **30.** $\dfrac{19\pi}{12}$ **31.** $240°$ **32.** $-135°$ **33.** $75°$ **34.** $-210°$

35. $105°$ **36.** $-165°$ **37.** $-195°$ **38.** $-75°$

In Exercises 39–42, verify the cofunction identity.

39. $\cos\left(\dfrac{\pi}{2} - a\right) = \sin a$ **40.** $\tan\left(\dfrac{\pi}{2} - a\right) = \cot a$ **41.** $\sec\left(\dfrac{\pi}{2} - a\right) = \csc a$

42. $\csc\left(\dfrac{\pi}{2} - a\right) = \sec a$

In Exercises 43–48, find the exact value of each function, given a, b in Quadrant IV with $\sin a = -\dfrac{3}{5}$ and $\cos b = \dfrac{5}{13}$.

43. $\cos(a + b)$ **44.** $\sin(a + b)$ **45.** $\tan(a + b)$ **46.** $\cot(a + b)$

47. $\cos\left(a + \dfrac{\pi}{2}\right)$ **48.** $\csc\left(b - \dfrac{\pi}{2}\right)$

In Exercises 49–54, given a, b in Quadrant II with $\sin a = \frac{1}{3}$ *and* $\cos b = -\frac{1}{4}$, *find the exact value of each expression.*

49. $\cos(a-b)$ **50.** $\sin(a-b)$ **51.** $\tan(a-b)$ **52.** $\cot(a-b)$

53. $\sin\left(a+\dfrac{\pi}{2}\right)$ **54.** $\sec\left(b-\dfrac{\pi}{2}\right)$

In Exercises 55–60, write each function in the form $C \sin(x+\theta)$, *where* $\theta \in [0, 2\pi)$.

55. $f(x) = \dfrac{\sqrt{2}}{2}\sin x + \dfrac{\sqrt{2}}{2}\cos x$ **56.** $f(x) = -\dfrac{\sqrt{3}}{2}\sin x + \dfrac{1}{2}\cos x$

57. $f(x) = -\sin x + \cos x$ **58.** $f(x) = -2\sin x - 2\cos x$

59. $f(x) = \sin x - \sqrt{3}\cos x$ **60.** $f(x) = \sqrt{2}\sin x + \sqrt{7}\cos x$

In Exercises 61–66, write in terms of a single trigonometric function of just x.

61. $\sin(x-\pi)$ **62.** $\cos\left(x+\dfrac{3\pi}{2}\right)$ **63.** $\sin(x+n\pi)$, *n is odd*

64. $\cos(x+n\pi)$, *n is even*   **65.** $\tan\left(x+\dfrac{3\pi}{2}\right)$ **66.** $\tan\left(x-\dfrac{\pi}{2}\right)$

In Exercises 67–80, verify the identity.

67. $\cos(x+x) = \cos^2 x - \sin^2 x$ **68.** $\sin(x+x) = 2\sin x \cos x$ **69.** $\tan(\pi - x) = -\tan x$

70. $\tan(\pi + x) = \tan x$ **71.** $\tan\left(x+\dfrac{\pi}{4}\right) = \dfrac{\tan x + 1}{1 - \tan x}$ **72.** $\tan\left(\dfrac{\pi}{4} - x\right) = \dfrac{1 - \tan x}{1 + \tan x}$

73. $\cos\left(x-\dfrac{\pi}{2}\right) = \cos\left(\dfrac{\pi}{2} - x\right)$ **74.** $\sin\left(x-\dfrac{\pi}{2}\right) = -\sin\left(\dfrac{\pi}{2} - x\right)$ **75.** $\sin\left(\dfrac{\pi}{3} - x\right) = -\sin\left(x-\dfrac{\pi}{3}\right)$

76. $\cos\left(\dfrac{\pi}{6} - x\right) = \cos\left(x-\dfrac{\pi}{6}\right)$ **77.** $\cos(x+y) + \cos(x-y) = 2\cos x \cos y$

78. $\sin(x+y) + \sin(x-y) = 2\sin x \cos y$ **79.** $\sin(x+y)\sin(x-y) = \sin^2 x \cos^2 y - \cos^2 x \sin^2 y$

80. $\cos(x-y)\cos(x+y) = \cos^2 x \cos^2 y - \sin^2 x \sin^2 y$

In Exercises 81–86, find the exact value of each expression.

81. $\sin\left(\cos^{-1}0 - \sin^{-1}\dfrac{1}{2}\right)$ **82.** $\cos\left(\sin^{-1}0 + \cos^{-1}\dfrac{1}{2}\right)$ **83.** $\cos\left(\sin^{-1}\dfrac{3}{5} + \dfrac{\pi}{2}\right)$

84. $\sin\left(\cos^{-1}\dfrac{4}{5} - \dfrac{\pi}{2}\right)$ **85.** $\tan\left(\dfrac{\pi}{4} + \cos^{-1}\dfrac{4}{5}\right)$ **86.** $\tan\left(\sin^{-1}\dfrac{3}{5} - \dfrac{\pi}{4}\right)$

Applications

87. Simple Harmonic Motion The displacement of a mass suspended on a spring, at time t, is given by $g(t) = -\dfrac{\sqrt{3}}{2} \sin t + \dfrac{1}{2} \cos t$. Find c in the interval $[0, 2\pi)$ such that $g(t)$ can be written in the form $g(t) = \sin(t + c)$.

88. Simple Harmonic Motion When a mass is suspended on a spring, its displacement at time t is given by $g(t) = \dfrac{1}{2} \sin t - \dfrac{\sqrt{3}}{2} \cos t$. Find c in the interval $[0, 2\pi)$ such that $g(t)$ can be written in the form $g(t) = \sin(t + c)$.

89. Radio Waves The wave form for a radio device is $f(x) = \sin\left(300\pi x + \dfrac{\pi}{4}\right)$. Find A, B, C, and D such that $f(x) = A \sin Bx + C \cos Dx$.

90. Radio Waves The wave form for a radio device is $f(x) = \sin\left(200\pi x + \dfrac{\pi}{3}\right)$. Find A, B, C, and D such that $f(x) = A \sin Bx + C \cos Dx$.

91. Calculus Let $f(x) = \cos 2x$ and $h \neq 0$. Express $\dfrac{f(x + h) - f(x)}{h}$ in terms of $\sin 2x$, $\cos 2x$, $\sin 2h$, $\cos 2h$, and h.

92. Calculus Let $f(x) = \sin 2x$ and $h \neq 0$. Express $\dfrac{f(x + h) - f(x)}{h}$ in terms of $\sin 2x$, $\cos 2x$, $\sin 2h$, $\cos 2h$, and h.

93. Circuit Theory When current in an electrical circuit is in the form of a sine or cosine wave, it is called alternating current. Two alternating current waves, with wave forms $y_1(x) = 10 \sin(50\pi x)$ and $y_2(x) = 10 \cos(50\pi x)$, respectively, interfere with each other to produce a third wave whose wave form is $y(x) = y_1(x) + y_2(x)$. Find the exact values of the positive number A and the number c in $[0, 2\pi)$ such that $y(x) = A \sin(50\pi x + c)$.

94. Circuit Theory In an electrical circuit, voltages are in the form of a sine or cosine wave. Two voltages $V_1(t) = 100 \sin(110\pi t)$ and $V_2(t) = 150 \cos(110\pi t)$ are applied to the same electrical circuit. Find the positive number A and the number c in $[0, 2\pi)$ such that

$$V(t) = V_1(t) + V_2(t) = A \sin(110\pi t + c)$$

Concepts

95. Using $a = b = x$, find a formula for $\sin 2x$.

96. Using $a = b = x$, find a formula for $\cos 2x$.

97. Simplify: $\sin(a + b) - \sin(a - b)$.

98. Simplify: $\cos(a + b) - \cos(a - b)$.

99. Using the identity for $\sin(a + b)$, verify that $\sin(a - b) = \sin a \cos b - \cos a \sin b$.

100. Using the identities for $\sin(a + b)$ and $\cos(a + b)$, verify that $\tan(a + b) = \dfrac{\tan a + \tan b}{1 - \tan a \tan b}$.

Multiple-Angle Identities; Sum and Product Identities

Objectives

- Know and use double-angle identities.
- Know and use power-reducing identities.
- Know and use half-angle identities.
- Know and use identities relating products and sums.

Note From the double angle identities, we see that $\sin 2x \neq 2 \sin x$. Also, $\cos 2x \neq 2 \cos x$ and $\tan 2x \neq 2 \tan x$.

VIDEO EXAMPLES

SECTION 6.3

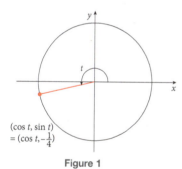

$(\cos t, \sin t)$
$= (\cos t, -\frac{1}{4})$

Figure 1

One of the most important applications of the sum identities from the previous section is the derivation of formulas for expressions such as $\sin 2x$ and $\cos 2x$. These are called **double-angle identities**. We begin by showing how the identity for $\sin 2x$ is derived. First, write $\sin 2x$ as $\sin(x + x)$. Applying the sum identity, we have

$$\sin 2x = \sin(x + x) = \sin x \cos x + \cos x \sin x = 2 \sin x \cos x$$

A similar approach yields identities for $\cos 2x$ and $\tan 2x$.

Double Angle Identities

$$\sin 2x = 2 \sin x \cos x \qquad \cos 2x = \cos^2 x - \sin^2 x$$
$$= 1 - 2 \sin^2 x$$
$$\tan 2x = \frac{2 \tan x}{1 - \tan^2 x} \qquad = 2 \cos^2 x - 1$$

Example 1 **Calculating sin 2t**

Find $\sin 2t$ if $\sin t = -\frac{1}{4}$ and $\pi < t < \frac{3\pi}{2}$.

Solution First, locate t on the unit circle as shown in Figure 1. Using the double angle identity, we have

$$\sin 2t = 2 \sin t \cos t$$

Next, determine $\cos t$ using the Pythagorean identity $\cos^2 t = 1 - \sin^2 t$.

$$\cos^2 t = 1 - \sin^2 t \qquad \textcolor{green}{\text{Pythagorean identity}}$$

$$= 1 - \left(-\frac{1}{4}\right)^2 \qquad \textcolor{green}{\text{Substitute } \sin t = -\frac{1}{4}}$$

$$= 1 - \frac{1}{16} = \frac{15}{16} \qquad \textcolor{green}{\text{Simplify}}$$

$$\cos t = \pm \frac{\sqrt{15}}{4} \qquad \textcolor{green}{\text{Solve for } \cos t}$$

Because $\pi < t < \frac{3\pi}{2}$, $\cos t$ is negative, and so

$$\cos t = -\frac{\sqrt{15}}{4}$$

Therefore,

$$\sin 2t = 2 \sin t \cos t \qquad \textcolor{green}{\text{Double-angle identity for } \sin 2t}$$

$$= 2\left(-\frac{1}{4}\right)\left(-\frac{\sqrt{15}}{4}\right) \qquad \textcolor{green}{\text{Substitute values}}$$

$$= \frac{\sqrt{15}}{8} \qquad \textcolor{green}{\text{Simplify}}$$

✔ *Check It Out 1* Rework Example 1 for $\cos 2t$.

Power-Reducing Identities

Earlier, we stated the double-angle identity $\cos 2x = 1 - 2\sin^2 x$. Solving for $\sin^2 x$ gives

$$\sin^2 x = \frac{1 - \cos 2x}{2}$$

Because this identity expresses a sine function raised to the *second* power in terms of a cosine function raised to the *first* power, it is called a **power-reducing identity**. Similarly, we can derive power-reducing identities for $\cos^2 x$ and $\tan^2 x$.

Note The power-reducing identities come up quite often in calculus, so you should commit them to memory.

Power-Reducing Identities
$\sin^2 x = \dfrac{1 - \cos 2x}{2}$ $\cos^2 x = \dfrac{1 + \cos 2x}{2}$ $\tan^2 x = \dfrac{1 - \cos 2x}{1 + \cos 2x}$

Example 2   **Power-Reducing Identity for $\cos^4 x$**

Express $f(x) = \cos^4 x$ in terms of constants and cosine functions raised to the first power.

Solution   Writing $\cos^4 x$ as $(\cos^2 x)(\cos^2 x)$ and applying the power-reducing identity, we get

$$f(x) = \cos^4 x = (\cos^2 x)(\cos^2 x)$$

$$= \left(\frac{1 + \cos 2x}{2}\right)\left(\frac{1 + \cos 2x}{2}\right) \quad \text{Apply power reducing identity to each factor}$$

$$= \frac{1 + 2\cos 2x + \cos^2 2x}{4} \quad \text{Multiply and simplify}$$

Because we want each cosine function in the result to be raised only to the first power, we apply the power-reducing identity to the function $\cos^2 2x$.

$$\cos^2 2x = \frac{1 + \cos(2(2x))}{2} = \frac{1 + \cos 4x}{2}$$

Substituting this result into the expression for $f(x)$ yields

$$f(x) = \frac{1 + 2\cos 2x + \frac{1}{2}(1 + \cos 4x)}{4}$$

$$= \frac{1}{4} + \frac{1}{2}\cos 2x + \frac{1}{8}(1 + \cos 4x) \quad \text{Separate terms}$$

$$= \frac{1}{4} + \frac{1}{2}\cos 2x + \frac{1}{8} + \frac{1}{8}\cos 4x \quad \text{Distribute}$$

$$= \frac{3}{8} + \frac{1}{2}\cos 2x + \frac{1}{8}\cos 4x \quad \text{Combine like terms}$$

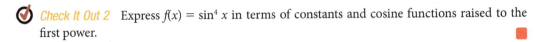 **Check It Out 2**   Express $f(x) = \sin^4 x$ in terms of constants and cosine functions raised to the first power.

Half-Angle Identities

The power-reducing identities can be used to derive another set of identities, known as the **half-angle identities**. For example,

$$\sin^2 u = \frac{1 - \cos 2u}{2} \quad \Rightarrow \quad \sin u = \pm \sqrt{\frac{1 - \cos 2u}{2}}$$

If we let $2u = x$. then $u = \frac{x}{2}$. Thus,

$$\sin\left(\frac{x}{2}\right) = \pm \sqrt{\frac{1 - \cos x}{2}}$$

Also,

$$\cos^2 u = \frac{1 + \cos 2u}{2} \quad \Rightarrow \quad \cos u = \pm \sqrt{\frac{1 + \cos 2u}{2}}$$

If we let $2u = x$, then $u = \frac{x}{2}$. Thus,

$$\cos\left(\frac{x}{2}\right) = \pm \sqrt{\frac{1 + \cos x}{2}}$$

We can derive a similar identity for $\tan\left(\frac{x}{2}\right)$.

$$\tan\left(\frac{x}{2}\right) = \frac{\sin\left(\frac{x}{2}\right)}{\cos\left(\frac{x}{2}\right)}$$

$$= \frac{\sin\left(\frac{x}{2}\right)\left(2\cos\left(\frac{x}{2}\right)\right)}{\cos\left(\frac{x}{2}\right)\left(2\cos\left(\frac{x}{2}\right)\right)} \qquad \text{Multiply numerator and denominator by } 2\cos\left(\frac{x}{2}\right)$$

$$= \frac{\sin 2\left(\left(\frac{x}{2}\right)\right)}{2\cos^2\left(\frac{x}{2}\right)} \qquad \text{Use double-angle identity for sine}$$

$$= \frac{\sin x}{2\left(\frac{1 + \cos x}{2}\right)} \qquad \text{Use power-reducing identity for cosine}$$

$$= \frac{\sin x}{1 + \cos x} \qquad \text{Simplify}$$

In the second step of the preceding derivation, if we instead multiply the numerator and denominator by $2\sin\left(\frac{x}{2}\right)$, we get the identity $\tan\left(\frac{x}{2}\right) = \frac{1 - \cos x}{\sin x}$.

Note In the half-angle identities for sine and cosine, only one of the signs, plus or minus, is chosen, depending on the quadrant in which the terminal side of $\frac{x}{2}$ lies.

Half-Angle Identities

$$\sin\left(\frac{x}{2}\right) = \pm \sqrt{\frac{1 - \cos x}{2}}$$

$$\cos\left(\frac{x}{2}\right) = \pm \sqrt{\frac{1 + \cos x}{2}}$$

$$\tan\left(\frac{x}{2}\right) = \frac{\sin x}{1 + \cos x} = \frac{1 - \cos x}{\sin x}$$

| **Example 3** | **Use Half-Angle Identities to Find Exact Values** |

Use the half-angle identity to find the exact value of

a. $\cos 75°$

b. $\sin \dfrac{17\pi}{12}$

Solution

a. Because $75° = \frac{1}{2}(150°)$, use the half-angle identity for cosine with $x = 150°$. We next choose the appropriate sign in front of the radical. The terminal side of the $75°$ angle lies in the first quadrant, where cosine is positive. Using the positive sign in front of the radical, we find that

$$\cos 75° = \cos\left(\frac{1}{2}(150°)\right)$$

$$= \sqrt{\frac{1 + \cos 150°}{2}}$$

$$= \sqrt{\frac{1 + \left(-\frac{\sqrt{3}}{2}\right)}{2}} \qquad \cos 150° = -\frac{\sqrt{3}}{2}$$

$$= \sqrt{\frac{\frac{2 - \sqrt{3}}{2}}{2}} \qquad \text{Combine terms in numerator}$$

$$= \sqrt{\frac{2 - \sqrt{3}}{4}} \qquad \text{Divide}$$

$$= \frac{\sqrt{2 - \sqrt{3}}}{2} \qquad \text{Simplify radical}$$

b. Because $\frac{17\pi}{12} = \frac{1}{2}\left(\frac{17\pi}{6}\right)$, use the half-angle identity for sine with $x = \frac{17\pi}{6}$. The terminal side of $\frac{17\pi}{12}$ lies in the third quadrant, so sine is negative. Thus,

$$\sin\frac{17\pi}{12} = \sin\left(\frac{1}{2}\left(\frac{17\pi}{6}\right)\right) = -\sqrt{\frac{1 - \cos\left(\frac{17\pi}{6}\right)}{2}}$$

Now,

$$\cos\left(\frac{17\pi}{6}\right) = \cos\left(\frac{12\pi}{6} + \frac{5\pi}{6}\right) = \cos\left(\frac{5\pi}{6}\right) = -\frac{\sqrt{3}}{2}$$

Therfore,

$$\sin\left(\frac{17\pi}{12}\right) = -\sqrt{\frac{1 - \left(-\frac{\sqrt{3}}{2}\right)}{2}} \qquad \text{Substitute } \cos\frac{17\pi}{6} = -\frac{\sqrt{3}}{2}$$

$$= -\sqrt{\frac{\frac{2 + \sqrt{3}}{2}}{2}} \qquad \text{Combine terms in numerator}$$

$$= -\sqrt{\frac{2 + \sqrt{3}}{4}} \qquad \text{Divide}$$

$$= -\frac{\sqrt{2 + \sqrt{3}}}{2} \qquad \text{Simplify radical}$$

✓ *Check It Out 3*   Find the exact value of $\sin\frac{5\pi}{12}$.

Identities Relating Products and Sums

In some applications of trigonometric functions, it is convenient to write a product of trigonometric functions as a sum. In other applications, it is desirable to write a sum of trigonometric functions as a product. Thus, we introduce the *product-to-sum identities* and the *sum-to-product* identities.

We first derive the product to sum identity for $\sin a \cos b$. To do this, we write the sum and difference identities for sine and then add them:

$$\sin (a + b) = \sin a \cos b + \cos a \sin b$$

$$\underline{\sin (a - b) = \sin a \cos b - \cos a \sin b}$$

$$\sin (a + b) + \sin (a - b) = 2 \sin a \cos b$$

Solving for $\sin a \cos b$, we have

$$\sin a \cos b = \frac{1}{2} (\sin (a + b) + \sin (a - b))$$

We can derive identities for $\cos a \sin b$, $\cos a \cos b$, and $\sin a \sin b$ similarly.

Product-to-Sum Identities
$$\sin a \cos b = \frac{1}{2} (\sin (a + b) + \sin (a - b))$$
$$\cos a \sin b = \frac{1}{2} (\sin (a + b) - \sin (a - b))$$
$$\cos a \cos b = \frac{1}{2} (\cos (a + b) + \cos (a - b))$$
$$\sin a \sin b = \frac{1}{2} (\cos (a - b) - \cos (a + b))$$

Using Technology

In Example 4, graph
$Y_1 = \sin 2x \cos 3x$ and
$Y_2 = \left(\frac{1}{2}\right) (\sin 5x - \sin x)$ to
see that they are identical.
See Figure 2. Change the
line marker for the second
graph so that a small circle
traces over the first graph.

Keystroke Appendix:
Section 15

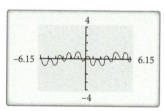

Figure 2

Example 4   **Writing as a Sum of Sine and Cosine**

Write $\sin 2x \cos 3x$ in terms of sine functions only.

Solution   Using the product-sum identity with $a = 2x$ and $b = 3x$,

$$\sin 2x \cos 3x = \frac{1}{2} (\sin (2x + 3x) + \sin(2x - 3x))$$

$$= \frac{1}{2} (\sin 5x + \sin(-x))$$

$$= \frac{1}{2} (\sin 5x - \sin x) \qquad \text{Use } \sin (-x) = -\sin x$$

Check It Out 4   Write $\sin 3x \sin 2x$ in terms of cosine functions only.

Example 5   **Verifying an Identity**

Verify the identity

$$\cos 4x \cos 2x = \frac{1}{2}(\cos 6x + \cos 2x)$$

Solution   Using the product-sum identity for $\cos a \cos b$, with $a = 4x$ and $b = 2x$, we find that

$$\cos 4x \cos 2x = \frac{1}{2}(\cos(4x + 2x) + \cos(4x - 2x))$$

$$= \frac{1}{2}(\cos 6x + \cos 2x)$$

Check It Out 5   Verify the identity

$$\sin 3x \cos x = \frac{1}{2}(\sin 4x + \sin 2x)$$

We can also derive identities which express a sum of trigonometric functions as a product. These are called **sum-to-product identities**. To obtain the sum-to-product identity for $\cos a + \cos b$, we start with the product-sum identity for cosine as follows:

$$\cos(x + y) + \cos(x - y) = 2\cos x \cos y$$

Next, make a careful choice of x and y in terms of a and b so that the left-hand side can be written as $\cos a + \cos b$:

Let
$$x = \frac{a + b}{2} \text{ and } y = \frac{a - b}{2}$$

Then,

$$x + y = \frac{a + b}{2} + \frac{a - b}{2} = \frac{(a + b) + (a - b)}{2} = \frac{a + b + a - b}{2} = a$$

$$x - y = \frac{a + b}{2} - \frac{a - b}{2} = \frac{(a + b) - (a - b)}{2} = \frac{a + b - a + b}{2} = b$$

Substituting these results, we get

$$\cos a + \cos b = 2\cos\left(\frac{a + b}{2}\right)\cos\left(\frac{a - b}{2}\right)$$

We can similarly derive other sum-to-product identities.

Sum-to-Product Identities
$$\sin a + \sin b = 2\sin\left(\frac{a + b}{2}\right)\cos\left(\frac{a - b}{2}\right)$$
$$\cos a + \cos b = 2\cos\left(\frac{a + b}{2}\right)\cos\left(\frac{a - b}{2}\right)$$
$$\sin a - \sin b = 2\cos\left(\frac{a + b}{2}\right)\sin\left(\frac{a - b}{2}\right)$$
$$\cos a - \cos b = -2\sin\left(\frac{a + b}{2}\right)\sin\left(\frac{a - b}{2}\right)$$

| **Example 6** | **Writing as a Product of Cosine and Sine** |

Write $\cos 2x - \cos 3x$ as a product of two trigonometric functions.

Solution Using the sum to product identity for cosine, with $a = 2x$ and $b = 3x$, we get

$$\cos 2x - \cos 3x = -2\sin\left(\frac{2x + 3x}{2}\right)\sin\left(\frac{2x - 3x}{2}\right)$$

$$= -2\sin\left(\frac{5x}{2}\right)\sin\left(\frac{-x}{2}\right)$$

$$= 2\sin\left(\frac{5x}{2}\right)\sin\left(\frac{x}{2}\right) \quad \text{Use } \sin\left(\frac{-x}{2}\right) = -\sin\left(\frac{x}{2}\right)$$

Check It Out 6 Write $\sin 3x + \sin 2x$ as a product of two trigonometric functions.

Exercises 6.3

 Skills

In Exercises 1–8, let x be the angle (in radians) that satisfies the conditions $0 < x < \frac{\pi}{2}$ and $\sin x = \frac{1}{3}$. Find the exact value of each function.

1. $\cos x$ **2.** $\tan x$ **3.** $\sin 2x$ **4.** $\cos 2x$ **5.** $\tan 2x$ **6.** $\csc 2x$ **7.** $\sec 2x$ **8.** $\cot 2x$

In Exercises 9–16, for the given angle x (in radians), find the exact values of sin 2x, cos 2x, and tan 2x using double-angle identities.

9. $\cos x = -\dfrac{3}{5}$ and $\pi < x < \dfrac{3\pi}{2}$

10. $\sin x = \dfrac{5}{13}$ and $\dfrac{\pi}{2} < x < \pi$

11. $\sin x = -\dfrac{1}{2}$ and $\pi < x < \dfrac{3\pi}{2}$

12. $\cos x = \dfrac{1}{4}$ and $\dfrac{3\pi}{2} < x < 2\pi$

13. $\sec x = 3$ and $0 < x < \dfrac{\pi}{2}$

14. $\csc x = -\dfrac{5}{3}$ and $\dfrac{3\pi}{2} < x < 2\pi$

15. $\sec x = -\dfrac{6}{5}$ and $\pi < x < \dfrac{3\pi}{2}$

16. $\csc x = \dfrac{9}{7}$ and $0 < x < \dfrac{\pi}{2}$

In Exercises 17–24, express the given function in terms of constants and sine and/or cosine functions to the first power.

17. $\cos^3 x$ **18.** $\sin^3 x$ **19.** $\sin x \cos^2 x$ **20.** $\cos x \sin^2 x$ **21.** $\sin^2 x \cos^3 x$

22. $\cos^2 x \sin^2 x$ **23.** $\dfrac{\tan^2 x}{\sec^2 x}$ **24.** $\sin x + \cos^2 x$

In Exercises 25–32, use the half-angle identities to find the exact value of the given function.

25. $\cos \dfrac{\pi}{12}$ **26.** $\sin \dfrac{\pi}{8}$ **27.** $\cos \dfrac{13\pi}{12}$ **28.** $\sin \dfrac{11\pi}{12}$ **29.** $\sin\left(-\dfrac{3\pi}{8}\right)$ **30.** $\cos\left(-\dfrac{\pi}{12}\right)$

31. $\tan \dfrac{3\pi}{12}$ **32.** $\tan \dfrac{5\pi}{8}$

In Exercises 33–38, given $\sin \theta = \frac{3}{5}$ and $0 < \theta < \frac{\pi}{2}$, find the exact value of each function.

33. $\sin \dfrac{\theta}{2}$ **34.** $\cos \dfrac{\theta}{2}$ **35.** $\tan \dfrac{\theta}{2}$ **36.** $\csc \dfrac{\theta}{2}$ **37.** $\cot \dfrac{\theta}{2}$ **38.** $\sec \dfrac{\theta}{2}$

In Exercises 39–44, given $\cos \theta = -\frac{3}{5}$ and $\pi < \theta < \frac{3\pi}{2}$, find the exact value of each function.

39. $\sin \dfrac{\theta}{2}$ **40.** $\cos \dfrac{\theta}{2}$ **41.** $\tan \dfrac{\theta}{2}$ **42.** $\csc \dfrac{\theta}{2}$ **43.** $\cot \dfrac{\theta}{2}$ **44.** $\sec \dfrac{\theta}{2}$

In Exercises 45–50, write each expression as a sum of two trigonometric functions.

45. $\sin 4x \cos 3x$ **46.** $\cos 5x \sin 2x$ **47.** $\sin 2x \sin 6x$ **48.** $\cos 6x \cos x$ **49.** $\sin x \sin 3x$

50. $\cos 3x \cos 5x$

In Exercises 51–56, write each expression as a product of two trigonometric functions.

51. $\sin 2x - \sin 3x$ **52.** $\cos 4x + \cos 3x$ **53.** $\cos 3x - \cos 5x$ **54.** $\sin 2x + \sin x$

55. $\cos 5x - \cos 2x$ **56.** $\cos x + \cos 3x$

In Exercises 57–76, verify the given identity.

57. $\sec^2 x = \dfrac{2}{1 + \cos 2x}$

58. $\csc^2 x = \dfrac{2}{1 - \cos 2x}$

59. $2 \cos^2 2x = 1 + \cos 4x$

60. $2 \sin^2 2x = 1 - \cos 4x$

61. $\sec^2 2x = \dfrac{2}{1 + \cos 4x}$

62. $\csc^2 2x = \dfrac{2}{1 - \cos 4x}$

63. $\cos 6x = 1 - 2 \sin^2 3x$

64. $\cos 8x = 2 \cos^2 4x - 1$

65. $\sin 6x = 2 \sin 3x \cos 3x$

66. $\sin 8x = 2 \sin 4x \cos 4x$

67. $\sin (2x + \pi) = -2 \sin x \cos x$

68. $\cos \left(2x - \dfrac{\pi}{2} \right) = 2 \sin x \cos x$

69. $\cos 4x = 1 - 8 \sin^2 x + 8 \sin^4 x$

70. $\sin 4x = 4 \sin x \cos x \,(1 - 2 \sin^2 x)$

71. $\sin 3x = \sin x \,(4 \cos^2 x - 1)$

72. $\cos 3x = \cos x \,(1 - 4 \sin^2 x)$

73. $4 \cos^3 x - 2 \cos x = \cos 3x + \cos x$

74. $4 \sin x \cos^2 x = \sin 3x + \sin x$

75. $\tan \left(-\dfrac{x}{2} \right) = \cot x - \csc x$

76. $\cot \left(\dfrac{x}{2} \right) = \cot x + \csc x$

In Exercises 77–84, find the exact value of each expression.

77. $\cos \left(2 \cos^{-1} \dfrac{3}{5} \right)$

78. $\sin \left(2 \cos^{-1} \dfrac{3}{5} \right)$

79. $\sin \left(2 \sin^{-1} \dfrac{1}{2} \right)$

80. $\cos \left(2 \sin^{-1} \dfrac{1}{2} \right)$

81. $\sin^2 \left(\dfrac{1}{2} \cos^{-1} \dfrac{4}{5} \right)$

82. $\cos^2 \left(\dfrac{1}{2} \cos^{-1} \dfrac{1}{2} \right)$

83. $\tan^2 \left(\dfrac{1}{2} \sin^{-1} \dfrac{\sqrt{3}}{2} \right)$

84. $\tan^2 \left(\dfrac{1}{2} \cos^{-1} \dfrac{\sqrt{3}}{2} \right)$

Applications

85. Physics The horizontal range of a projectile fired with an initial velocity of 40 meters per second at an angle θ is given by $R = \dfrac{40^2 \sin 2\theta}{9.8}$. Find R to four decimal places if it is known that $\sin \theta = 0.3$ and θ is in the first quadrant.

86. Traveling Waves The expression $\sin (x + ct) + \sin (x - ct)$ represents a traveling wave that is moving at speed c.

   a. Write the expression in terms of a product of functions.

   b. The function $f(x, t) = \sin (x - t)$ is a function of two variables, x and t, where x stands for position and t represents time. For a fixed value of t, $\sin (x - t)$ is a function of x alone. For each of three fixed values of t, $t = 0$, $t = 1$, and $t = 2$, graph this function. What happens to the graph as t increases?

87. Geometry Consider the triangle ABC as shown. Sides AB and AC are both of length s, α is the angle between them, b is the base of the triangle, and h is the height.

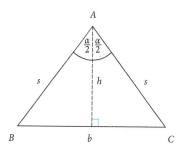

   a. Express b in terms of s and $\sin \left(\dfrac{\alpha}{2} \right)$.

   b. Express h in terms of s and $\cos \left(\dfrac{\alpha}{2} \right)$.

   c. Use your answers to parts (a) and (b) to express the area of the triangle in terms of s and a product of two trigonometric functions of the angle $\left(\dfrac{\alpha}{2} \right)$.

   d. Use your answer to part (c) to express the area of the triangle in terms of s and $\sin \alpha$.

   e. Use your answer to part (d) to find the exact value of the area if $s = 6$ and $\tan(\alpha) = \dfrac{4}{3}$.

88. **Construction** Michelle and Rick Bonneau are building a patio along the entire back wall of their home. The patio is designed in such a way that its outer border is an arc of a circle with center at point A and radius r, as shown in the accompanying figure. Points C and D are the ends of the back wall of the house, α is the angle between radii AC and AD, point B is the midpoint of CD (which is of length l), and BE (which is of length w) is perpendicular to CD.

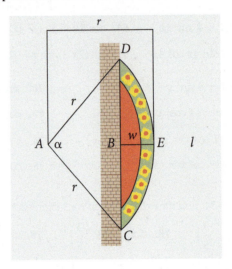

   a. Express $\sin\left(\frac{\alpha}{2}\right)$ in terms of r and l. (*Hint*: Use an appropriate right triangle.)
   b. Express r in terms of $\sin\left(\frac{\alpha}{2}\right)$ and l.
   c. Use your answer to part (b) to find the value of r if $l = 50$, $\cos\alpha = -0.4$, and α is in the second quadrant.
   d. Express $\cos\left(\frac{\alpha}{2}\right)$ in terms of r and w. (*Hint*: Use an appropriate right triangle.)
   e. Express r in terms of $\cos\left(\frac{\alpha}{2}\right)$ and w.
   f. Use your answer to part (e) to find the value of r if $w = 15$, $\tan\alpha = 12$, and α is in the first quadrant.
   g. What is the value of α (in degrees) if $l = 40$ and the distance from point A to point B is 20?

Concepts

89. Determine the constant A such that $\sin(\pi + x) = A\sin(\pi - x)$ is an identity.

90. Determine the constant A such that $\cos(\pi + x) = A\cos(\pi - x)$ is an identity.

91. Derive the following product-to-sum identity.

$$\cos a \cos b = \frac{1}{2}(\cos(a + b) + \cos(a - b))$$

92. Derive the following sum-to-product identity.

$$\sin a - \sin b = 2\cos\left(\frac{a + b}{2}\right)\sin\left(\frac{a - b}{2}\right)$$

Trigonometric Equations

6.4

Objectives

- Solve basic trigonometric equations exactly.

- Find approximate solutions of trigonometric equations.

- Use identities to solve trigonometric equations.

- Use substitution to solve trigonometric equations.

- Solve applied problems involving trigonometric equations.

- Solve a trigonometric equation using a graphing utility.

VIDEO EXAMPLES

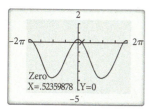

SECTION 6.4

 Using Technology

You can use the ZERO feature to find the approximate zeros of $Y_1 = 2 \cos t - \sqrt{3}$. Figure 2 shows the zero corresponding to the exact solution $x = \frac{\pi}{6}$. You can also approximate another solution corresponding to $x = \frac{11\pi}{6}$. However, a graphing utility cannot produce an infinite number of solutions.

Keystroke Appendix:
Sections 9, 15

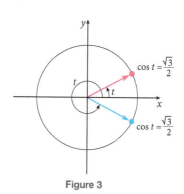

Figure 2

In this section, we study methods to solve equations that contain one or more trigonometric functions. Solutions to trigonometric equations have properties that are not shared by many of the algebraic equations you have solved so far. Unless you are looking for a solution in a specific interval, a trigonometric equation usually has an infinite number of solutions, because the trigonometric functions are periodic. Figure 1 shows the graphs of $y = \cos x$ and $y = \frac{1}{2}$. The x values of the intersection of the two graphs are the solutions of the equation $\cos x = \frac{1}{2}$.

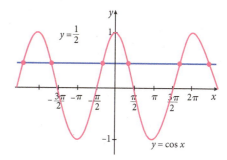

Figure 1

To solve trigonometric equations, you can apply familiar algebraic techniques, such as factoring and substitution, together with trigonometric identities.

Example 1 **Solving a Basic Equation Involving Cosine**

Find all solutions of the equation $2 \cos t - \sqrt{3} = 0$.

Solution Solve for $\cos t$ as follows.

$$2 \cos t - \sqrt{3} = 0 \qquad \text{Original equation}$$

$$2 \cos t = \sqrt{3} \qquad \text{Add } \sqrt{3} \text{ to each side}$$

$$\cos t = \frac{\sqrt{3}}{2} \qquad \text{Divide each side by 2}$$

We next find all values of t such that $\cos t = \frac{\sqrt{3}}{2}$. Because $0 < \cos t < 1$, we see that t must lie in the first quadrant or the fourth quadrant. Moreover, the equation $\cos t = \frac{\sqrt{3}}{2}$ will have exactly one solution in the interval $\left(0, \frac{\pi}{2}\right)$, and exactly one solution in the interval $\left(\frac{3\pi}{2}, 2\pi\right)$. See Figure 3.

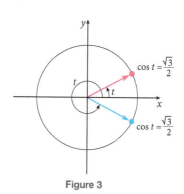

Figure 3

In the first quadrant, $t = \frac{\pi}{6}$ is a solution, because $\cos\frac{\pi}{6} = \frac{\sqrt{3}}{2}$. In the fourth quadrant, $t = \frac{11\pi}{6}$ is a solution, because

$$\cos\frac{11\pi}{6} = \cos\left(-\frac{\pi}{6}\right) = \cos\frac{\pi}{6} = \frac{\sqrt{3}}{2}$$

Because the cosine function is periodic with period 2π, there are infinitely many solutions. Thus, the solutions of the equation $2\cos t - \sqrt{3} = 0$ can be summarized as

$$t = \frac{\pi}{6} + 2n\pi \quad \text{and} \quad t = \frac{11\pi}{6} + 2n\pi$$

where n is an integer. You should check that these are solutions of the equation.

 Check It Out 1   Find all solutions of the equation $2\sin t - 1 = 0$.

Example 2   **Solving a Basic Equation Involving Tangent**

Find all solutions of the equation $\tan t + 2 = -\tan t$.

Solution First, isolate the $\tan t$ term as follows.

$\tan t + 2 = -\tan t$	Original equation
$\tan t + \tan t + 2 = 0$	Add $\tan t$ to each side
$2\tan t + 2 = 0$	Combine like terms
$2\tan t = -2$	Subtract 2 from each side
$\tan t = -1$	Divide each side by 2

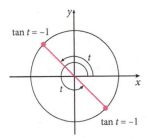

Figure 4

Because $\tan t < 0$, we see that t must lie in the second quadrant or the fourth quadrant. See Figure 4. Thus, $t = \frac{3\pi}{4}$ and $t = \frac{7\pi}{4}$ are solutions. The tangent function has period π, so the general solution can be written as

$$t = \frac{3\pi}{4} + n\pi, \quad n \text{ an integer}$$

Note that $t = \frac{7\pi}{4}$ is included in the general solution because $\frac{7\pi}{4} = \frac{3\pi}{4} + \pi$. You should check that these are solutions of the equation.

 Check It Out 2   Find all solutions of the equation $2\csc x - 4 = 0$.

Example 3   **Solving an Equation with Multiple Angles**

Find all solutions of the equation $2\sin 3x - 1 = 0$.

Solution First isolate the $\sin 3x$ term, then solve for $\sin 3x$.

$2\sin 3x = 1$	Given equation
$\sin 3x = \frac{1}{2}$	Divide by 2
$\sin u = \frac{1}{2}$	Substitute $u = 3x$

One value of u that satisfies $\sin u = \frac{1}{2}$ is $u = \frac{\pi}{6}$. Also, $\sin\frac{5\pi}{6} = \frac{1}{2}$. Because the sine function is periodic with period 2π, we have

$$u = 3x = \frac{\pi}{6} + 2n\pi, \quad u = 3x = \frac{5\pi}{6} + 2n\pi, n \text{ is an integer}$$

Solving for x, we have

$$x = \frac{\pi}{18} + \frac{2n\pi}{3}, \quad x = \frac{5\pi}{18} + \frac{2n\pi}{3}, n \text{ is an integer}$$

 Check It Out 3   Find all solutions of the equation $\cos 2x + 1 = 0$.

Using Technology
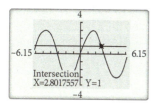

Graph $Y_1(x) = 3 \sin x$ and $Y_2 = 1$, in radian mode, and use the INTERSECT feature to solve the equation in Example 4. One solution is shown in Figure 5. Because a graphing calculator finds only a finite number of solutions, use periodicity to find all the solutions.

Keystroke Appendix:
Sections 9, 15

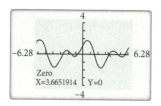

Figure 5

Note Example 4 shows that you must be careful when using the $\boxed{SIN^{-1}}$ key on your calculator. The calculator will produce only the answer within the range of the inverse function. Other solutions, if they exist, must be found by locating angles in other quadrants.

Using Technology

For Example 5, the graph of $Y_1 = \sin 2x + \cos x$ shows that there are four zeros in $[0, 2\pi)$. They can be found using the ZERO feature. In Figure 7, the zero at $\frac{7\pi}{6}$ is shown.

Keystroke Appendix:
Sections 9 and 15

Figure 7

The exact values of trigonometric functions are known for only a handful of angles in any finite interval. In many cases, you will need to use a scientific calculator to find a solution, as shown in Example 4.

Example 4   **Solving an Equation with Approximate Solutions**

Find all solutions of the equation $3 \sin t = 1$.

Solution   Isolating the $\sin t$ term by dividing both sides by 3 gives

$$\sin t = \frac{1}{3}$$

The value of t can be found only by using the $\boxed{SIN^{-1}}$ key on a scientific calculator:

$$t \approx 0.3398 \qquad \text{In radians}$$

The sine function is periodic with period 2π, so for every integer n, $t \approx 0.3398 + 2n\pi$ satisfies $\sin t = \frac{1}{3}$.

There are still other solutions. Because $\sin(\pi - t) = \sin t$, one of these is found by evaluating the expression $\pi - t$ for $t \approx 0.3398$, giving $\pi - 0.3398 \approx 2.802$. See Figure 6.

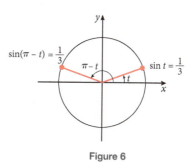

Figure 6

$$\pi - t \approx \pi - 0.3398 \approx 2.802$$

Thus, for every integer n, $t \approx 2.802 + 2n\pi$ is a solution of the equation $\sin t = \frac{1}{3}$. The solutions of the equation $3 \sin t = 1$ can be summarized as

$$t \approx 0.3398 + 2n\pi, \quad t \approx 2.802 + 2n\pi, \; n \text{ an integer}$$

You should check that these are solutions of the equation.

✓ *Check It Out 4*   Find all solutions of the equation $3 \cos t = 1$. ■

Using Identities and Factoring to Solve Equations

We next show the use of trigonometric identities to solve equations.

Example 5   **Using a Double Angle Identity to Solve an Equation**

Find the solutions of the equation

$$\sin 2x + \cos x = 0$$

that lie in the interval $[0, 2\pi)$.

Solution

$\sin 2x + \cos x = 0$	Original equation
$2 \sin x \cos x + \cos x = 0$	Use the identity $\sin 2x = 2 \sin x \cos x$
$\cos x (2 \sin x + 1) = 0$	Factor out $\cos x$

Using the zero product rule, we get

$$\cos x = 0 \Rightarrow x = \frac{\pi}{2}, \frac{3\pi}{2}$$

$$2\sin x + 1 = 0 \Rightarrow \sin x = -\frac{1}{2} \Rightarrow x = \frac{7\pi}{6}, \frac{11\pi}{6}$$

Thus, the solutions of the equation $\sin 2x + \cos x = 0$ that lie in the interval $[0, 2\pi)$ are $x = \frac{\pi}{2}, \frac{3\pi}{2}, \frac{7\pi}{6}, \frac{11\pi}{6}$. You should check that these are solutions of the equation.

 Check It Out 5   Find the solutions of the equation

$$\sin 2x - \sin x = 0$$

that lie in the interval $[0, 2\pi)$.

Example 6   **An Area Problem**

A metal plate is in the shape of a right triangle with a hypotenuse of 10 inches, as shown in Figure 8. Find θ if the area of the triangle is 20 square inches.

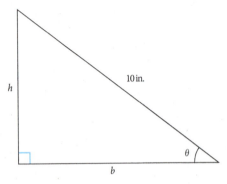

Figure 8

Solution   The area formula for a triangle is $A = \frac{1}{2}bh$. From Figure 8, we have

$$\cos \theta = \frac{b}{10} \Rightarrow b = 10 \cos \theta$$

$$\sin \theta = \frac{h}{10} \Rightarrow h = 10 \sin \theta$$

The area of the triangle is then given by

$$A = \frac{1}{2}bh = \frac{1}{2}(10 \cos \theta)(10 \sin \theta) = 50 \cos \theta \sin \theta$$

Because the area is 20 square inches, we have the following equation:

$$50 \cos \theta \sin \theta = 20$$

$50 \sin \theta \cos \theta = 20$	Original equation
$25(2 \sin \theta \cos \theta) = 20$	Rewrite to apply double-angle identity
$25 \sin 2\theta = 20$	Substitute $\sin 2\theta = 2 \sin \theta \cos \theta$
$\sin 2\theta = \frac{4}{5}$	Divide both sides of equation by 25
$2\theta \approx 53.13°$	Use $\boxed{SIN^{-1}}$ key on calculator in degree mode
$\theta \approx 26.565°$	Solve for θ

 Check It Out 6   Solve the equation

$$144 \sin \theta \cos \theta = 36$$

where $0 < \theta < 90°$.

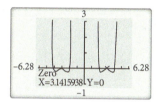

Figure 9

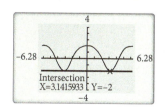

Figure 10

If the given equation is in the form of a quadratic equation, use an appropriate substitution to solve it, as shown in the following example.

Example 7   **Using Factoring to Solve an Equation**

Find all solutions of the equation

$$\sec^2 x + 3 \sec x + 2 = 0$$

Solution   Note that this equation is in the form $u^2 + 3u + 2 = 0$, where $u = \sec x$. Thus, we first solve the equation $u^2 + 3u + 2 = 0$.

$$u^2 + 3u + 2 = 0$$

$$(u + 2)(u + 1) = 0 \qquad \text{Factor}$$

$$u + 2 = 0 \quad \Rightarrow \quad u = -2 \qquad \text{Set each factor equal to 0}$$

$$u + 1 = 0 \quad \Rightarrow \quad u = -1$$

If $u = -2$, then $\sec x = -2$. Because $\sec x = \frac{1}{\cos x}$, we have $\frac{1}{\cos x} = -2 \Rightarrow \cos x = -\frac{1}{2}$.

The solutions of the equation $\cos x = -\frac{1}{2}$ can be summarized as $x = \frac{2\pi}{3} + 2n\pi, x = \frac{4\pi}{3} + 2n\pi$, n is an integer.

If $u = -1$, then $\sec x = -1$. Thus,

$$\frac{1}{\cos x} = -1 \Rightarrow \cos x = -1$$

The general solution of the equation $\cos x = -1$ is $x = \pi + 2n\pi$, n an integer. Thus, the solutions of the equation $\sec^2 x + 3 \sec x + 2 = 0$ can be summarized as

$$x = \frac{2\pi}{3} + 2n\pi, \quad x = \frac{4\pi}{3} + 2n\pi, \quad x = \pi + 2n\pi$$

n an integer. You should check that these are solutions of the equation.

 Check It Out 7   Find all solutions of the equation $\cos^2 x - \cos x - 2 = 0$.

The next example illustrates the use of a Pythagorean identity to write an equation in terms of just one function.

Example 8   **Using a Pythagorean Identity to Solve an Equation**

Find the solutions of the equation $2 \cos t + \sin^2 t = -2$ that lie in the interval $[0, 2\pi)$.

Solution   Use a Pythagorean identity to write the equation in terms of $\cos t$.

$$2 \cos t + \sin^2 t = -2 \qquad \text{Original equation}$$

$$2 \cos t + (1 - \cos^2 t) = -2 \qquad \text{Use } \sin^2 t = 1 - \cos^2 t$$

$$2 \cos t + (1 - \cos^2 t) + 2 = 0 \qquad \text{Add 2 to both sides}$$

$$-\cos^2 t + 2 \cos t + 3 = 0 \qquad \text{Simplify}$$

$$\cos^2 t - 2 \cos t - 3 = 0 \qquad \text{Multiply by } -1$$

Let $u = \cos t$. We then have the equation $u^2 - 2u - 3 = 0$. Solving this equation,

$$u^2 - 2u - 3 = 0$$

$$(u - 3)(u + 1) = 0 \qquad \text{Factor}$$

Setting each factor to 0,

$$u - 3 = 0 \quad \Rightarrow \quad u = 3$$

$$u + 1 = 0 \quad \Rightarrow \quad u = -1$$

In terms of cos t, we have

$$\cos t = 3 \quad \Rightarrow \quad \text{No solution, because } \cos t \text{ cannot be greater than } 1$$

$$\cos t = -1 \quad \Rightarrow \quad t = \pi$$

Thus, $t = \pi$ is the only solution of the equation $2\cos t + \sin^2 t = -2$ that lies in the interval $[0, 2\pi)$. You should check that this is a solution of the equation.

Check It Out 8   Find the solutions of the equation $\sin t - \cos^2 t = 1$ that lie in the interval $[0, 2\pi)$.

The next example illustrates how a half-angle identity can be used to solve an equation.

Example 9   **Using a Half-Angle Identity to Solve an Equation**

Find all solutions in the interval $[0, 2\pi)$ for the equation $\tan \frac{x}{2} = \sin x$.

Solution   Substitute $\tan\left(\frac{x}{2}\right) = \frac{\sin x}{1 + \cos x}$ in the equation to get

$$\frac{\sin x}{1 + \cos x} = \sin x \qquad\qquad \tan \frac{x}{2} = \frac{\sin x}{1 + \cos x}$$

$$\frac{\sin x}{1 + \cos x} - \sin x = 0 \qquad\qquad \text{Subtract } \sin x \text{ from both sides}$$

$$\sin x \left(1 - \frac{1}{1 + \cos x}\right) = 0 \qquad\qquad \text{Factor } \sin x$$

$$\sin x = 0 \text{ or } 1 - \frac{1}{1 + \cos x} = 0 \qquad\qquad \text{Use the Zero Product Rule}$$

For the equation $\sin x = 0$, the only solutions in $[0, 2\pi)$ are $x = 0, \pi$. Solve the second equation as follows.

$$1 - \frac{1}{1 + \cos x} = 0 \qquad\qquad \text{Given equation}$$

$$\frac{(1 + \cos x) - 1}{1 + \cos x} = 0 \qquad\qquad \text{Combine terms with common denominator}$$

$$\frac{\cos x}{1 + \cos x} = 0 \qquad\qquad \text{Simplify numerator}$$

$$\cos x = 0 \qquad\qquad \text{Set numerator to zero to solve}$$

$$x = \frac{\pi}{2}, \frac{3\pi}{2} \qquad\qquad \text{Find solutions in } [0, 2\pi)$$

So we have $x = 0, \frac{\pi}{2}, \pi, \frac{3\pi}{2}$ as four possible solutions to the equation $\tan \frac{x}{2} = \sin x$. However, $x = \pi$ cannot be a solution because $\tan \frac{\pi}{2}$ is undefined. The other three x values satisfy the equation. Thus, the solutions of the equation in $[0, 2\pi)$ are $x = 0, \frac{\pi}{2}, \frac{3\pi}{2}$.

Check It Out 9   Find all solutions in the interval $[0, 2\pi)$ for the equation $\cot \frac{x}{2} = \csc x$.

Using a Graphing Utility to Solve Equations

Thus far, the equations discussed could be solved by algebraic methods. However, many equations involving trigonometric functions can only be solved using a graphing utility, as shown in Example 10.

Example 10   **Solving a Trigonometric Equation with a Graphing Utility**

Solve $\cos x = x^2 + 2x$ with a graphing utility.

Solution   Graphing $Y_1 = \cos x$ and $Y_2 = x^2 + 2x$ in the same window and using the INTERSECT feature to find their intersections gives two solutions: $x \approx -1.8507$ and $x \approx 0.3877$. See Figure 11.

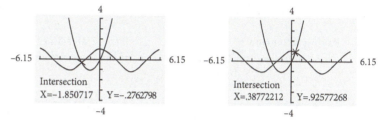

Figure 11

The graphs intersect in only two places, because the graph of the cosine function simply oscillates between 1 and -1, while the graph of $Y_2 = x^2 + 2x$ extends upward to ∞. Thus there are only two solutions to the equation $\cos x = x^2 + 2x$.

 Check It Out 10   Solve $\cos x = x$ with a graphing utility.

Exercises 6.4

Just in Time Exercises

In Exercises 1–4, complete each Pythagorean identity.

1. $\underline{} + \cos^2 x = 1$ **2.** $\sin^2 x = 1 - \underline{}$ **3.** $\sec^2 x - \underline{} = 1$ **4.** $1 + \underline{} = \csc^2 x$

In Exercises 5 and 6, solve each equation by factoring.

5. $x^2 - 3x - 4 = 0$ **6.** $2x^2 - x - 1 = 0$

Skills

Paying Attention to Instructions: *Exercises 7 and 8 are intended to give you practice reading and paying attention to the instructions that accompany the problems you are working.*

7. **a.** Verify the identity $\sin x \cot x = \cos x$.
 b. Solve the equation $\sin x \cot x = 1$ for $x \in [0, 2\pi)$.

8. **a.** Verify the identity $\sin x \csc x = 1$.
 b. Solve the equation $\sin x = 1$ for $x \in [0, 2\pi)$.

In Exercises 9–22, find the exact solutions of the given equation, in radians.

9. $2 \sin x = -1$ **10.** $-2 \cos x = \sqrt{3}$ **11.** $\tan x - 1 = 0$ **12.** $\sec x = -\sqrt{2}$

13. $\csc x = 2$ **14.** $\cot x - \sqrt{3} = 0$ **15.** $\sin^2 x = 1$ **16.** $\cos^2 x = \dfrac{1}{2}$

17. $\sin 2x = \dfrac{1}{2}$ **18.** $\cos 2x = -\dfrac{\sqrt{3}}{2}$ **19.** $\cos 4x + 1 = 0$ **20.** $\sin 4x + 2 = 3$

21. $\tan 2x = -1$ **22.** $\tan 3x = 1$

In Exercises 23–30, use a calculator to find the solutions of the given equation, in radians.

23. $4 \cos x = -3$ **24.** $3 \sin x = 2$ **25.** $2 \tan x = \tan x + 3$ **26.** $\sec x = \dfrac{3}{2}$

27. $12 + 3 \csc x = 7$ **28.** $\cot x - 2 = 2 \cot x$ **29.** $4 \csc x - 5 = 1$ **30.** $6 \cot x + 5 = 0$

In Exercises 31–66, find the exact solutions of each equation, in radians, that lie in the interval $[0, 2\pi)$.

31. $\sin^2 x - \sin x = 0$ **32.** $\cos^2 x + \cos x = 0$ **33.** $\sin^3 x = \sin x$ **34.** $\cos^3 x + \cos x = 0$

35. $\sin^2 x = \cos x + 1$ **36.** $\cos^2 x = \dfrac{1}{2}$ **37.** $\cos x - \sec x = 0$ **38.** $\sec x + \csc x = 0$

39. $\sin\left(x - \dfrac{\pi}{4}\right) = 0$ **40.** $\cos\left(x + \dfrac{\pi}{4}\right) = 1$ **41.** $\sin\left(2x + \dfrac{\pi}{2}\right) + 1 = 0$ **42.** $\cos\left(3x - \dfrac{\pi}{2}\right) - 1 = 0$

43. $\sin^2 x + 2 \cos x = 1$ **44.** $\sin x + \cos^2 x = 1$ **45.** $\cos x = \sec x$ **46.** $\sin x - \csc x = 0$

47. $\sin 2x = 2 \cos x$ **48.** $2 \sin x = \sin 2x$ **49.** $\cos 2x + 3 \cos x - 1 = 0$

50. $\cos 2x - 5 \cos x - 2 = 0$ **51.** $\sec^2 x = 2 \tan^2 x$ **52.** $\csc^2 x = 2 \cot^2 x$

53. $\tan^2 x + \sec x = -1$ **54.** $\sec^2 x - \tan x = 1$ **55.** $\cos 2x = -\sin^2 x$

56. $\cos^2 x - \sin^2 x = 0$

57. $\tan\left(\dfrac{x}{2}\right) = \csc x$

58. $\tan\left(\dfrac{x}{2}\right) = \dfrac{1}{2}\csc x$

59. $2\cos^2 x - 3\cos x + 1 = 0$ **60.** $4\sin^2 x + 4\sin x + 1 = 0$ **61.** $\tan^2 x - 2\tan x + 1 = 0$

62. $\sec^2 x + 4\sec x + 4 = 0$ **63.** $\sec^2 x - \sec x = 2$ **64.** $\csc^2 x + \csc x = 2$

65. $\cot^2 x + \cot x - \sqrt{3}\cot x - \sqrt{3} = 0$ **66.** $\sin^2 x + \sin x - \dfrac{\sqrt{2}}{2}\sin x - \dfrac{\sqrt{2}}{2} = 0$

In Exercises 67–82, use a calculator to find the solutions of the given equations, in radians, that lie in the interval $[0, 2\pi)$. Round your answers to four decimal places.

67. $4\tan^2 x = 1$ **68.** $3\sec^2 x = 7$ **69.** $3\sin^2 x - 5\sin x - 2 = 0$ **70.** $5\cos^2 x + 4\cos x - 1 = 0$

71. $3\cos^2 x + \cos x - 2 = 0$ **72.** $4\sin^2 x - 7\sin x + 3 = 0$ **73.** $2\sin 2x + \cos x = 0$

74. $3\sin 2x - 2\cos x = 0$ **75.** $\sin 2x = \sin^2 x$ **76.** $\cos 2x = \cos^2 x$

77. $\csc^2 x - \csc x = 20$ **78.** $\sec^2 x - \sec x = 6$ **79.** $4\tan^2 x - 4\tan x = 3$

80. $5\cot^2 x - 3\cot x = 2$ **81.** $\sin^4 x - \cos^4 x = -\dfrac{1}{3}$ **82.** $10\tan^4 x - 7\sec^4 x + 23 = 0$

 In Exercises 83–90, use a graphing utility to find the solutions of the following equations, in radians, that lie in the interval $[0, 2\pi)$. Round your answers to four decimal places.

83. $\cos 2x = -x^2$ **84.** $\sin 2x = 1 - x$ **85.** $\tan x = x + 2$ **86.** $\sec x = -x + 1$

87. $\sin 2x = \cos 2x$ **88.** $-\sin x + x = \cos x$ **89.** $\cos^2 x = \sin x$ **90.** $\sin^2 x = \cos x$

Applications

91. Physics The horizontal range of a projectile fired with an initial velocity of 70 meters per second at an angle θ is given by

$$R = \dfrac{70^2 \sin\theta \cos\theta}{4.9}$$

where R is in meters. At what acute angle must the projectile be fired so that the range is 300 meters?

92. Air Travel A plane approaching Reagan International Airport is told to maintain a holding pattern before being given clearance to land. The formula

$$d(t) = 80\sin(0.75t) + 200$$

can be used to determine the distance of the plane from the airport at time t in minutes. How many minutes does it take for the plane to be 280 miles away from the airport?

93. Biorhythm The function

$$P(t) = 50\sin\dfrac{2\pi}{23}t + 50$$

is used in biorhythm theory to predict an individual's physical potential, as a percentage of the maximum, on a particular day, with $t = 0$ corresponding to birth. Find the day of greatest potential after the person's 21st birthday (day 7,670).

94. Leisure The formula

$$h(t) = 125\sin\left(2\pi t - \dfrac{\pi}{2}\right) + 125$$

represents the height above the ground at time t, in minutes, of a person who is riding the Ferris wheel. During the first turn, how much time does a passenger spend at or above a height of 200 feet?

95. Navigation According to a recent report by the Army Corps of Engineers, on a particular day, the following function

$$d(t) = 4.5 \sin\left(\frac{\pi}{6}t\right) + 7$$

where t is in hours, $t = 0$ corresponds to 2:00 a.m., and $d(t)$ is in feet, may be used to predict the height of the Cape Fear river at one point near its mouth. If your boat needs at least a river height of 5 feet, find the first time interval in which it is not safe for you to navigate that part of the river.

96. Geometry A parallelogram has one side of length 10 inches and a perimeter of 32 inches.

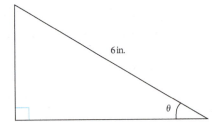

a. Find the lengths of the other three sides of the parallelogram.

b. Write an expression for the area of the parallelogram in terms of the angle θ shown in the figure that has a measure of at most 90°.

c. Show that for $\theta = 90°$, the expression you found in part (b) for the area of the parallelogram, is the formula for the area of a rectangle.

d. For what angle θ does the parallelogram have an area of 48 square inches?

97. Geometry Consider a right triangle with a hypotenuse of 6 inches.

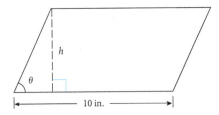

a. Show that the area of the right triangle is given by $A(\theta) = 18 \sin \theta \cos \theta$.

b. For what value of θ is $A(\theta) = 8$ square inches.

c. Without using a graphing utility, find the angle θ for which $A(\theta)$ is maximized. If your calculator is in radian mode, convert your final answer to degrees.

98. Daylight Hours The number of hours of daylight in Anchorage, Alaska, can be approximated by the function

$$d(x) = -7.1 \cos(0.0172x) + 12.27$$

where x is the number of days since January 1.   (*Source*: www.mathdemos.org)

a. For what value(s) of x will Anchorage have 18 hours of daylight?

b. For what value(s) of x will the number of hours of daylight in Anchorage reach a maximum?

99. Daylight Hours The number of hours of daylight in Miami, Florida, can be approximated by the function

$$d(x) = -2.2\cos(0.0175x) + 12.27$$

where x is the number of days since January 1.   (*Source*: www.mathdemos.org)

  a. For what value(s) of x will Miami have 13 hours of daylight?
  b. For what value(s) of x will the number of hours of daylight in Miami reach a maximum?

100. Mechanics The amount of work done on an object by a force F when it causes the object to move in a straight line through a distance d is $W = Fd\cos\theta$, where θ is the angle between the direction of the force and the direction of motion of the object. If F is given in pounds and d is given in feet, then W is in foot pounds.

  a. If $F = 10$ pounds and $d = 5$ feet, for what acute angle θ will the amount of work be 40 foot pounds?
  b. If $d > 0$ and $F > 0$, for what value(s) of θ in $[0°, 360°)$ is $|W|$ maximized?
  c. If $d > 0$ and $F > 0$, for what value(s) of θ in $[0°, 360°)$ is $|W|$ minimized?

Concepts

101. Can you find the solution of $\sin x \cos x = 1$ by setting $\sin x = 1$ and $\cos x = 1$? Explain.

102. By sketching a graph, decide if the equation $\sin x = x$ has an infinite number of solutions.

103. Use a graphing utility to find all the solutions of the equation $\cos x = x$.

104. What is the smallest nonnegative angle x (in radians) such that $\cos\left(\frac{x}{6} + \frac{\pi}{2}\right) = -1$?

105. Find all the values of x (in radians) that satisfy *both* of the equations $\cos^2 x - \sin^2 x = -1$ and $\tan\left(\frac{x}{2}\right) = -1$?

106. Does the equation $\tan^4 x - \sec^4 x = 0$ have any solutions? Why or why not?

Chapter 6 Summary

Section 6.1 Verifying Identities

Basic Identities [Review Exercises 1−4]

Identities as ratios

$$\tan x = \frac{\sin x}{\cos x} \, ; \, \cot x = \frac{\cos x}{\sin x}$$

Reciprocal identities

$$\sec x = \frac{1}{\cos x} \, ; \, \csc x = \frac{1}{\sin x} \, ; \, \cot x = \frac{1}{\tan x}$$

Pythagorean identities

$$\cos^2 x + \sin^2 x = 1 \, ; \, 1 + \tan^2 x = \sec^2 x \, ; \, 1 + \cot^2 x = \csc^2 x$$

Negative-angle identities

$$\cos(-x) = \cos x \, ; \, \sin(-x) = -\sin x \, ; \, \tan(-x) = -\tan x$$

Summary of strategies for verifying identities [Review Exercises 5–14]

1. Write in terms of sine and cosine.
2. Factor expressions when possible.
3. Multiply the numerator and denominator of a fraction to obtain a difference of squares.
4. Working both sides of an equation separately until they equal the same quantity.

Section 6.2 Sum and Difference Identities

Sum and difference identities [Review Exercises 15–30]

The following sum and difference identities are very important and should be memorized.

$$\sin(a + b) = \sin a \cos b + \cos a \sin b$$

$$\sin(a - b) = \sin a \cos b - \cos a \sin b$$

$$\cos(a + b) = \cos a \cos b - \sin a \sin b$$

$$\cos(a - b) = \cos a \cos b + \sin a \sin b$$

$$\tan(a + b) = \frac{\tan a + \tan b}{1 - \tan a \tan b}$$

$$\tan(a - b) = \frac{\tan a - \tan b}{1 + \tan a \tan b}$$

Cofunction identities [Review Exercises 15–30]

The following cofunction identities are quickly derived from the sum and difference identities:

$$\sin\left(\frac{\pi}{2} - a\right) = \cos a \qquad \cos\left(\frac{\pi}{2} - a\right) = \sin a$$

$$\tan\left(\frac{\pi}{2} - a\right) = \cot a \qquad \cot\left(\frac{\pi}{2} - a\right) = \tan a$$

$$\sec\left(\frac{\pi}{2} - a\right) = \csc a \qquad \csc\left(\frac{\pi}{2} - a\right) = \sec a$$

Section 6.3 Multiple-Angle Identities; Sum and Product Identities

Double-angle identities   [Review Exercises 31–34]

These can be quickly derived from the respective sum identities.

$$\sin 2x = 2 \sin x \cos x \qquad \cos 2x = \cos^2 x - \sin^2 x \qquad \tan 2x = \frac{2 \tan x}{1 - \tan^2 x}$$

$$= 1 - 2 \sin^2 x$$

$$= 2 \cos^2 x - 1$$

Power-reducing identities   [Review Exercises 35–36]

These identities are extremely useful in calculus and should be memorized.

$$\sin^2 x = \frac{1 - \cos 2x}{2}$$

$$\cos^2 x = \frac{1 + \cos 2x}{2}$$

$$\tan^2 x = \frac{1 - \cos 2x}{1 + \cos 2x}$$

Half-angle identities   [Review Exercises 37–40]

$$\sin \frac{x}{2} = \pm \sqrt{\frac{1 - \cos x}{2}}$$

$$\cos \frac{x}{2} = \pm \sqrt{\frac{1 + \cos x}{2}}$$

$$\tan \frac{x}{2} = \frac{\sin x}{1 + \cos x} = \frac{1 - \cos x}{\sin x}$$

Product-to-sum identities   [Review Exercises 41–42]

$$\sin a \cos b = \frac{1}{2} \left(\sin(a + b) + \sin(a - b) \right)$$

$$\cos a \sin b = \frac{1}{2} \left(\sin(a + b) - \sin(a - b) \right)$$

$$\cos a \cos b = \frac{1}{2} \left(\cos(a + b) + \cos(a - b) \right)$$

$$\sin a \sin b = \frac{1}{2} \left(\cos(a - b) - \cos(a + b) \right)$$

Sum-to-product identities   [Review Exercises 43–44]

$$\sin a + \sin b = 2 \sin\left(\frac{a+b}{2}\right) \cos\left(\frac{a-b}{2}\right)$$

$$\cos a + \cos b = 2 \cos\left(\frac{a+b}{2}\right) \cos\left(\frac{a-b}{2}\right)$$

$$\sin a - \sin b = 2 \cos\left(\frac{a+b}{2}\right) \sin\left(\frac{a-b}{2}\right)$$

$$\cos a - \cos b = -2 \sin\left(\frac{a+b}{2}\right) \sin\left(\frac{a-b}{2}\right)$$

Section 6.4 Trigonometric Equations

Techniques for solving trigonometric equations   [Review Exercises 45–63]

1. Solve basic trigonometric equations exactly.
2. Find approximate solutions for trigonometric equations with a scientific calculator.
3. Use identities to solve trigonometric equations.
4. Use substitution to solve trigonometric equations.
5. Use a graphing utility to solve trigonometric equations.

Chapter 6 Review Exercises

Section 6.1

In Exercises 1–4, write the given expression in terms of sin x and/or cos x only.

1. $\sec^2 x \cos x$ **2.** $\tan x \cos x$ **3.** $\dfrac{1}{\csc^2 x}$ **4.** $\cot x \sec x$

In Exercises 5–14, verify the given identities.

5. $1 + \cos x = \dfrac{\sin^2 x}{1 - \cos x}$

6. $\sec^4 x - \tan^4 x = \sec^2 x + \tan^2 x$

7. $\sin^4 x = 1 - 2\cos^2 x + \cos^4 x$

8. $\sin^2 x \cos^2 x = \sin^2 x - \sin^4 x$

9. $\cos^4 x - \sin^4 x = 1 - 2\sin^2 x$

10. $\dfrac{1}{1 + \cos x} + \dfrac{1}{1 - \cos x} = 2\csc^2 x$

11. $\dfrac{1 + 2\cos x + \cos^2 x}{1 - \cos^2 x} = \dfrac{1 + \cos x}{1 - \cos x}$

12. $\csc x + \cot x = \dfrac{1 + \cos x}{\sin x}$

13. $\dfrac{\tan^2 x}{\sec x + 1} = \sec x - 1$

14. $\dfrac{\sec x + \tan x}{\cos x - \tan x - \sec x} = -\csc x$

Section 6.2

In Exercises 15–18, write each expression as the sine or cosine of a single angle.

15. $\sin 45° \cos 30° + \cos 45° \sin 30°$ **16.** $\cos 150° \cos 45° - \sin 150° \sin 45°$

17. $\cos \dfrac{\pi}{3} \cos \dfrac{\pi}{4} + \sin \dfrac{\pi}{3} \sin \dfrac{\pi}{4}$ **18.** $\sin \dfrac{\pi}{6} \cos \dfrac{\pi}{4} - \cos \dfrac{\pi}{6} \sin \dfrac{\pi}{4}$

In Exercises 19–22, find the exact values of the sine, cosine, and tangent of the given angles.

19. $-105°$ **20.** $195°$ **21.** $-\dfrac{\pi}{12}$ **22.** $\dfrac{13\pi}{12}$

In Exercises 23–28, given a, b in Quadrant III with $\cos a = -\frac{3}{5}$ and $\sin b = -\frac{5}{13}$, find the exact value of the given functions.

23. $\cos (a + b)$ **24.** $\sin (a + b)$ **25.** $\tan (a + b)$ **26.** $\cot (a + b)$

27. $\sin (a - b)$ **28.** $\cos (b - a)$

29. Mechanics The displacement of a mass suspended on a spring is given by $g(x) = \dfrac{\sqrt{3}}{2} \sin 2x - \dfrac{1}{2} \cos 2x$. Find c in the interval $[0, 2\pi)$ such that $g(x)$ can be written in the form $f(x) = \sin(2x + c)$.

30. Electricity A wave form for an electromagnetic wave is of the form $f(x) = \sin\left(1000\pi x - \dfrac{\pi}{6}\right)$. Write $f(x)$ as a sum of a sine and a cosine function, each of which has no phase shift.

Section 6.3

In Exercises 31–34, let $0 < x < \frac{\pi}{2}$ and let $\cos x = \frac{3}{5}$. Find the exact value of each function.

31. $\sin x$ **32.** $\tan x$ **33.** $\sin 2x$ **34.** $\cos 2x$

In Exercises 35 and 36, write each expression in terms of constants and the first powers of the sine and/or cosine functions.

35. $\sin^2 x \cos^2 x$ **36.** $\sin^3 x \cos^2 x$

In Exercises 37–40, use the half-angle identities to find the exact value of the given function.

37. $\cos \dfrac{3\pi}{12}$ **38.** $\cos\left(-\dfrac{\pi}{8}\right)$ **39.** $\cos 15°$ **40.** $\sin 75°$

In Exercises 41 and 42, write each expression as a sum of two trigonometric functions.

41. $\sin 4x \cos 2x$ **42.** $\sin x \sin 2x$

In Exercises 43 and 44, write each expression as a product of two trigonometric functions.

43. $\cos 2x + \cos 3x$ **44.** $\sin 5x + \sin 2x$

Section 6.4

In Exercises 45–48, find the exact solutions of the given equations, in radians.

45. $2 \cos x - 1 = 0$ **46.** $\tan^2 x - 1 = 0$ **47.** $\sec x = -2$
48. $\sin x + \sin^2 x + \cos^2 x = 1$

In Exercises 49–52, find the approximate solutions of the given equations, in radians, using a calculator. Round your answers to four decimal places.

49. $3 \cos x = -2$ **50.** $5 \sin x = 4$ **51.** $2\sec x = \sec x + 2.5$ **52.** $\tan x = \dfrac{3}{2}$

In Exercises 53–56, find the exact solutions of the given equations, in radians, in the interval $[0, 2\pi)$.

53. $2 \sin^2 x = 1$ **54.** $\sin^3 x + 2 \sin x = 0$ **55.** $\cos^3 x - 2 \cos x = 0$
56. $\csc^2 x + 3 \csc x + 2 = 0$

In Exercises 57 and 58, use a calculator to find the solutions of each equations that lie in the interval $[0, 2\pi)$. Round your answers to four decimal places.

57. $\sin^2 x = \dfrac{3}{5}$ **58.** $\tan^2 x - \tan x = 6$

In Exercises 59 and 60, solve each equation using a graphing utility. Round your answers to four decimal places.

59. $2x + \cos x = 0$ **60.** $x^2 - 2x = \sin x$

61. **Physics** The horizontal range of a projectile fired with an initial velocity of 60 meters per second at an angle θ is given by

$$R = \frac{60^2 \sin \theta \cos \theta}{4.9}$$

where R is in meters. At what acute angle must the projectile be fired so that the range is 280 meters?

62. **Physics** The horizontal range of a projectile fired with an initial velocity of 75 meters per second at an angle θ is given by

$$R = \frac{75^2 \sin \theta \cos \theta}{4.9}$$

where R is in meters. At what acute angle must the projectile be fired so that the horizontal range is maximized?

63. **Geometry** A rectangle of horizontal dimension 8 and vertical dimension 6 is inscribed in a circle of radius r and center at point O, as shown in the figure.

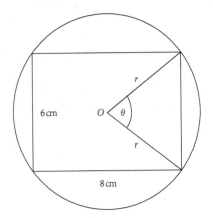

a. What is the value of r?
b. What is the value of $\cos \left(\frac{\theta}{2} \right)$, where θ is the angle shown in the figure?
c. What is the value of θ in degrees?

Chapter 6 Test

In Exercises 1–4, verify each identity.

1. $\dfrac{\cot^2 x}{\csc x + 1} = \csc x - 1$

2. $\dfrac{\cos x}{1 - \sin x} = \sec x + \tan x$

3. $\dfrac{1}{\sec x + 1} + \dfrac{1}{\sec x - 1} = 2 \cot x \csc x$

4. $\sin\left(x + \dfrac{\pi}{2}\right) = \cos x$

5. Write as the sine or cosine of a single angle: $\sin \dfrac{\pi}{4} \cos \dfrac{\pi}{6} + \cos \dfrac{\pi}{6} \sin \dfrac{\pi}{4}$.

In Exercises 6–8, find the exact value of each expression.

6. $\cos\left(-\dfrac{11\pi}{12}\right)$

7. $\cos 165°$

8. $\tan 255°$

In Exercises 9–11, given a, b in Quadrant II with $\cos a = -\dfrac{4}{5}$ and $\sin b = \dfrac{12}{13}$, find the exact value of each expression.

9. $\sin (a + b)$

10. $\tan (a - b)$

11. $\cos (a + b)$

In Exercises 12–14, let $\dfrac{\pi}{2} < x < \pi$ and let $\cos x = -\dfrac{4}{5}$. Find the exact value of each expression.

12. $\sin 2x$

13. $\cos \dfrac{x}{2}$

14. $\tan 2x$

15. Express in terms of constants and the first powers of $\sin 2x$ and/or $\cos 2x$: $\sin^3 x \cos x$.

16. Write as a sum of two trigonometric functions: $\sin 3x \cos 2x$.

17. Write as a product of two trigonometric functions: $\cos 3x + \cos 4x$.

18. Find exact values of all solutions of the equation: $2 \sin x - 1 = 0$.

19. Find exact values of all solutions of the equation: $\tan 2x + \sqrt{3} = 0$.

20. Find exact values of the solutions, in $[0, 2\pi)$, of the equation: $4 \cos^2 x - 1 = 0$.

21. Find exact values of the solutions, in $[0, 2\pi)$, of the equation $\sec^2 x - 3 \sec x + 2 = 0$

22. Use a calculator to find solutions, in $[0, 2\pi)$, of the equation $3 \cos x + 1 = 0$. Round your answer to three decimal places.

23. Use a calculator to find solutions, in $[0, 2\pi)$, of the equation $\tan \left(\dfrac{x}{2}\right) = 2$. Round your answer to three decimal places.

 24. Use a graphing calculator to find solutions of the equation $- 3x = 2 \cos x$. Round your answer to three decimal places.

25. The form of a signal wave is given by the function

$$f(x) = 2 \cos \left(3x - \dfrac{\pi}{2}\right) + 3$$

Find the first positive value of x when the signal wave reaches its minimum value of 1.

26. The function

$$P(t) = 50 \sin \dfrac{2\pi}{33} t + 50$$

is used in biorhythm theory to predict an individual's intellectual potential, as a percentage of the maximum, on a particular day, with $t = 0$ corresponding to birth. Find the first day of 100% potential after the person's 21st birthday (day 7,670).

Chapter 7

Additional Topics in Trigonometry

High precision instruments are used in sports to make accurate measurements in competitions. For instance, a small computer in the measuring device uses trigonometric calculations to compute the distance an athlete throws a discus. See Exercise 37 in Section 7.2.

In this chapter, we will explore additional methods for solving triangles, investigate vectors, and look at alternate ways to represent complex numbers using a trigonometric form and to represent points in the plane using polar form.

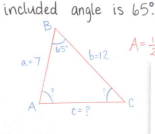

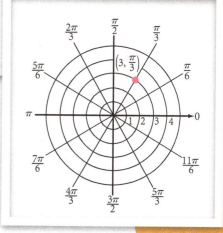

Outline

The Law of Sines

7.1

Objectives

- Solve an oblique triangle using the Law of Sines.

- Solve applied problems using the Law of Sines.

In Chapter 5 you learned how to use right triangles to solve a variety of problems. Many interesting problems involve **oblique triangles**, which are triangles not containing a right angle. In this section and the next, you will learn techniques for solving problems using oblique triangles. Throughout these two sections, we will use the standard notation for triangles. The angles are labeled A, B, and C, and the sides opposite those angles are labeled a, b, and c, respectively, as shown in Figure 1.

The Law of Sines relates the sines of the angles to the sides as follows.

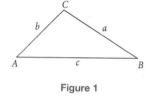

Figure 1

Law of Sines

Let ABC be a triangle with sides a, b, c. (See Figure 1.) Then the following ratios hold:

$$\frac{a}{\sin A} = \frac{b}{\sin B} = \frac{c}{\sin C}$$

The ratios can also be written as

$$\frac{\sin A}{a} = \frac{\sin B}{b} = \frac{\sin C}{c}$$

Note You can use the ratios in either form when solving a triangle, as long as you are consistent.

Proof of Law of Sines   *Case 1*: A is acute. Referring to Figure 2, we have

$$\sin A = \frac{h}{b} \Rightarrow h = b \sin A$$

$$\sin B = \frac{h}{a} \Rightarrow h = a \sin B$$

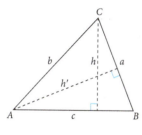

Figure 2

Equating these two expressions for h, we find that

$$b \sin A = a \sin B$$

Dividing by $(\sin A)(\sin B)$, we have

$$\frac{b}{\sin B} = \frac{a}{\sin A} \qquad (1)$$

Note that $\sin A$ and $\sin B$ are both always nonzero, because the angles of a triangle are greater than $0°$ and less than $180°$.

Next, referring to Figure 2, we have

$$\sin B = \frac{h'}{c}; \quad \sin C = \frac{h'}{b}$$

Solving each of these equations for h' and then equating the two expressions for h' gives

$$c \sin B = b \sin C \Rightarrow \frac{c}{\sin C} = \frac{b}{\sin B}$$

Combining this result with Equation (1) gives

$$\frac{a}{\sin A} = \frac{b}{\sin B} = \frac{c}{\sin C}$$

Case 2: A is obtuse. Referring to Figure 3, we have

$$\sin(180° - A) = \frac{h}{b}$$

$$\sin B = \frac{h}{a}$$

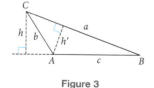

Figure 3

Using the difference identity for sine, $\sin(180° - A) = \sin A$. Thus,

$$\sin A = \frac{h}{b}; \sin B = \frac{h}{a}$$

Following the same procedure as in Case 1 gives:

$$\frac{b}{\sin B} = \frac{a}{\sin A}$$

Continue just as in Case 1 after Equation (1) to obtain

$$\frac{a}{\sin A} = \frac{b}{\sin B} = \frac{c}{\sin C}$$

You can use the Law of Sines to solve an oblique triangle when you are given the following information about the angles and sides.

- **ASA:** A side is common to the two angles (Figure 4).

- **AAS:** A side is opposite to one of the angles (Figure 5).

- **SSA:** Two sides and the angle opposite one of them (Figure 6).

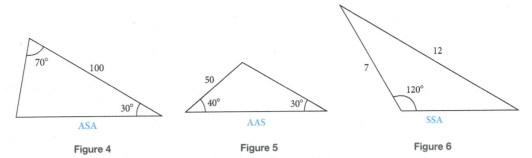

ASA

AAS

SSA

Figure 4   **Figure 5**   **Figure 6**

Oblique triangles can be solved using the Law of Cosines (Section 7.2) if the following is known.

- **SAS:** Two sides and the included angle—the angle formed from the two known sides (Figure 7)

- **SSS:** All three sides (Figure 8)

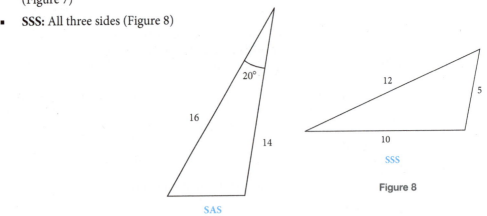

SAS

SSS

Figure 8

Figure 7

Throughout this section, degree measure for angles will be used for solving triangles. We now give examples of the different cases that arise in applying the Law of Sines.

VIDEO EXAMPLES

SECTION 7.1

Example 1 Solving a Triangle: ASA

Solve the triangle in Figure 9.

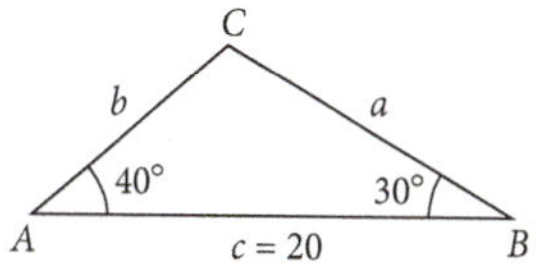

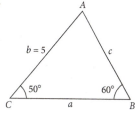

Figure 9

Solution We are given two angles, A and B, and the side common to them, c. To apply the Law of Sines, we must first find C. Because $A + B + C = 180°$, we have

$$C = 180° - (40° + 30°)$$
$$= 110°$$

Because c is known, use it to find a:

$$\frac{a}{\sin A} = \frac{c}{\sin C} \qquad \text{Law of Sines}$$

$$\frac{a}{\sin \mathbf{40°}} = \frac{\mathbf{20}}{\sin \mathbf{110°}} \qquad \text{Substitute } A = 40°, C = 110°, \text{ and } c = 20$$

$$a = \left(\frac{20}{\sin 110°}\right)(\sin 40°) \approx 13.68$$

Similarly, use c to find b:

$$\frac{b}{\sin B} = \frac{c}{\sin C} \qquad \text{Law of Sines}$$

$$\frac{b}{\sin \mathbf{30°}} = \frac{\mathbf{20}}{\sin \mathbf{110°}} \qquad \text{Substitute } B = 30°, C = 110°, \text{ and } c = 20$$

$$b = \left(\frac{20}{\sin 110°}\right)(\sin 30°) \approx 10.64$$

All the angles and sides are known, so the triangle has been solved.

Check It Out 1 Solve the triangle ABC with $A = 20°$, $B = 40°$, and $c = 10$.

Example 2 Solving a Triangle: AAS

Solve the triangle in Figure 10.

Figure 10

Solution We are given two angles, B and C, and the side opposite one of them, b. We will find A, and then apply the Law of Sines to find a.

$$A = 180° - (60° + 50°) = 70°$$

Because b is known, use it to find a.

$$\frac{a}{\sin A} = \frac{b}{\sin B}$$ Law of Sines

$$\frac{a}{\sin 70°} = \frac{5}{\sin 60°}$$ Substitute $A = 70°$, $B = 60°$, and $b = 5$

$$a = \left(\frac{5}{\sin 60°}\right) \sin 70° \approx 5.425$$

Similarly, use b to find c:

$$\frac{c}{\sin C} = \frac{b}{\sin B}$$ Law of Sines

$$\frac{c}{\sin 50°} = \frac{5}{\sin 60°}$$ Substitute $C = 50°$, $B = 60°$, and $b = 5$

$$c = \left(\frac{5}{\sin 60°}\right) \sin 50° \approx 4.423$$

All the angles and sides are known, so the triangle has been solved.

Check It Out 2   Solve the triangle ABC with $C = 40°$, $B = 70°$, and $b = 8$.

Thus far, the ASA and AAS cases have each produced a unique solution. In the SSA case, however, there can be one solution, two solutions, or no solution, depending on the given data. See Figure 11. The case of SSA is also known as the *ambiguous case*.

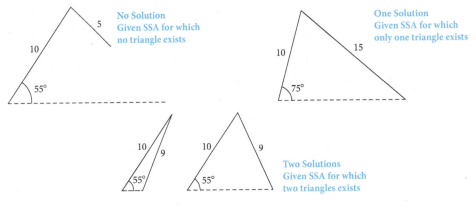

Figure 11

The following three examples explore the ambiguous case, SSA. When solving triangles with SSA, we will use the following identity to determine the number of solutions:

$$\sin(180° - \theta) = \sin \theta$$

Example 3   **One-Solution Case of SSA**

Solve the triangle ABC with $A = 60°$, $a = 6$, and $b = 3$.

Solution We are given two sides and the angle opposite one of them (SSA). Tentatively sketch the triangle as in Figure 12. Our calculations will reveal the number of solutions, if any, for this problem. First, use a and b to find B:

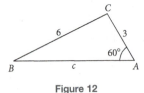

Figure 12

$$\frac{\sin B}{b} = \frac{\sin A}{a}$$ Law of Sines

$$\frac{\sin B}{3} = \frac{\sin 60°}{6}$$ Substitute $b = 3$, $a = 6$, and $A = 60°$

$$\sin B = \left(\frac{\sin 60°}{6}\right)(3)$$

$$\sin B \approx 0.4330$$ Approximate $\sin B$

$$B \approx 25.66°$$ Use $\boxed{\sin^{-1}}$ key on calculator

Thus we have at least one solution. The other possibility for B is the obtuse angle whose sine is also 0.4330, given by $B = 180° - 25.66° = 154.34°$. However,

$$A + B = 60° + 154.34° = 214.34° > 180°.$$

This is not possible because the sum of all three angles of the triangle must equal 180. Thus, the only solution for B is $B \approx 25.66°$. Using this, we determine C:

$$C \approx 180° - (60° + 25.66°) = 94.34°$$

We need only find c to finish solving the triangle. Using a and c, we obtain

$$\frac{c}{\sin C} = \frac{a}{\sin A} \qquad \text{Law of Sines}$$

$$\frac{c}{\sin \mathbf{94.34°}} = \frac{\mathbf{6}}{\sin \mathbf{60°}} \qquad \text{Substitute } C = 94.34° \ a = 6, \text{ and } A = 60°$$

$$c = \left(\frac{6}{\sin 60°}\right)(\sin 94.34°) \approx 6.908$$

 Check It Out 3 Solve the triangle ABC with $A = 70°$, $a = 8$, and $b = 3$.

Example 4   **Two-Solution Case of SSA**

Find angle C in the triangle ABC with $A = 40°$, $c = 10$, and $a = 8$.

Solution We can sketch two possible triangles, as shown in Figure 13. Next, we use the Law of Sines to determine which—if any—of the sketches are valid. Use a and c to find C:

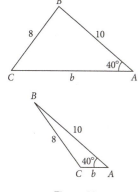

Figure 13

$$\frac{\sin C}{c} = \frac{\sin A}{a} \qquad \text{Law of Sines}$$

$$\frac{\sin C}{\mathbf{10}} = \frac{\sin \mathbf{40°}}{\mathbf{8}} \qquad \text{Substitute } c = 10, a = 8, \text{ and } A = 40°$$

$$\sin C = \left(\frac{\sin 40°}{8}\right)(10)$$

$$\sin C \approx 0.8035 \qquad \text{Approximate } \sin C$$

$$C \approx 53.47° \qquad \text{Use } \boxed{\sin^{-1}} \text{ key on calculator}$$

Thus, one solution for C is the acute angle $C \approx 53.47°$. The other possibility for C is the obtuse angle whose sine is 0.8035, which is $180° - 53.47° = 126.53°$. Because

$$A + C \approx 40° + 126.53° = 166.53° < 180°,$$

this solution works as well. Thus, there are two values of C. Either $C \approx 53.47°$ or $C \approx 126.53°$. Hence, there are two different triangles satisfying the given information.

 Check It Out 4 Solve the triangle ABC with $A = 50°$, $a = 9$ and $c = 11$.

Example 5   **No-Solution Case of SSA**

Show that there is no triangle ABC with $a = 10$, $b = 15$, and $A = 75°$.

Solution If you use the given information to try to sketch a triangle, you will find that it is impossible. To see why, use a and b to find B.

$$\frac{\sin B}{b} = \frac{\sin A}{a} \qquad \text{Law of Sines}$$

$$\frac{\sin B}{15} = \frac{\sin 75}{10} \qquad \text{Substitute } b = 15, A = 75°, \text{ and } a = 10$$

$$\sin B = 15\left(\frac{\sin 75}{10}\right) \approx 1.449$$

Because the sine of an angle cannot be greater than 1, there is no triangle with $a = 10$, $b = 15$, and $A = 75°$.

 Check It Out 5   Show that there is no triangle ABC with $a = 9$, $b = 15$, and $A = 80°$.

Applications of the Law of Sines

The Law of Sines is useful in many problems that involve surveying and measurement, as seen in the next example.

Example 6   **Finding the Height of a Tree**

The angles of elevation from points A and B to the top of a palm tree are 80° and 60° respectively. See Figure 14. Points A and B are 15 feet apart, and the base of the tree lies on the line between them. Find the height of the tree.

Solution From Figure 14, we see that

$$\frac{h}{b} = \sin A = \sin 80°$$

Neither a nor b is known, so we first have to determine C:

$$C = 180° - (80° + 60°) = 40°$$

Then we use c to find b:

$$\frac{b}{\sin B} = \frac{c}{\sin C} \qquad \text{Law of Sines}$$

$$\frac{b}{\sin 60°} = \frac{15}{\sin 40°} \qquad \text{Substitute } B = 60°, c = 15, \text{ and } C = 40°$$

$$b = \left(\frac{15}{\sin 40°}\right)(\sin 60°)$$

$$b \approx 20.21$$

Substituting this result into our original equation, we have

$$\sin 80° = \frac{h}{20.21} \Rightarrow h = 20.21(\sin 80°) \approx 19.90$$

Thus, the height of the tree is approximately 19.90 feet.

 Check It Out 6   Rework Example 6 for the case where $A = 75°$.

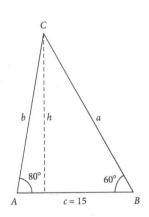

Figure 14

In navigation, the direction traveled by a ship or boat is given in degrees relative to the northerly or southerly direction, followed by east or west. For example, a ship traveling S40°W (which is read as "south 40° west") relative to a location is illustrated in Figure 15. A ship traveling N20°E relative to a location is given in Figure 16. The following problem illustrates an application of the Law of Sines in navigation.

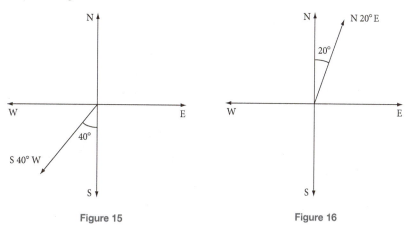

Figure 15   **Figure 16**

Example 7 **Navigation**

A tour boat travels 10 miles due north from its starting point and then travels in the direction S40°W on its second leg of the tour. If the boat is currently 15 miles from its starting point, how far has it traveled on the second leg? Round your final answer to two decimal places.

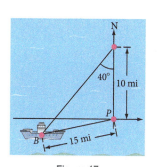

Figure 17

Solution Begin the problem by sketching a figure. Denote the starting point as P and draw a line pointing north. Then draw the second leg of the tour in the direction S40°W. Write in all the given information. See Figure 17.

Use the Law of Sines to first find the measure of angle B.

$$\frac{\sin B}{10} = \frac{\sin 40}{15} \qquad \text{Law of Sines}$$

$$\sin B = \left(\frac{\sin 40}{15}\right) 10 \qquad \text{Multiply both sides of equation by 10}$$

$$\sin B \approx 0.428525 \qquad \text{Evaluate right-hand side}$$

$$B \approx \sin^{-1} 0.428525 \approx 25.3740° \qquad \text{Approximate } B$$

The only other possible solution for B is $180° - 25.3740° = 154.6260°$. The sum of the angles in the triangle then exceeds 180°, which cannot happen. Now, the angle with vertex P has measure $180° - 40° - 25.3740° = 114.6260°$. Applying the Law of Sines again, we have

$$\frac{d}{\sin 114.626°} = \frac{15}{\sin 40°} \qquad \text{Law of Sines}$$

$$d \approx (\sin 114.626°)\left(\frac{15}{\sin 40°}\right)$$

$$d \approx 21.21 \qquad \text{Simplify}$$

Thus, the boat traveled approximately 21.21 miles on the second leg of its tour.

 Check It Out 7 Rework Example 7 if the boat travels in the direction S50°E on its second leg of the tour. Assume all other information stays the same.

Just in Time Exercises

1. True or False: sin 50° = sin 130°

2. True or False: sin 40° = sin 140°

In Exercises 3–6, find two angles θ, 0 < θ < 180° satisfying the given condition.

3. $\sin \theta = \dfrac{1}{2}$ **4.** $\sin \theta = \dfrac{\sqrt{3}}{2}$ **5.** $\sin \theta = 0.8$ (Use calculator) **6.** $\sin \theta = 0.4$ (Use calculator)

Skills

In Exercises 7–10, solve the given triangles. Round all answers to four decimal places.

7.

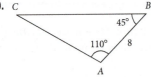

8.

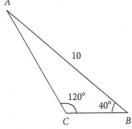

9.

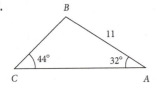

10.

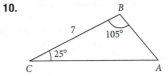

In Exercises 11–22, solve the given triangles. Standard notation is used for labeling triangles. Round all answers to four decimal places.

11. $A = 42°, B = 64°, b = 6$

12. $B = 65°, C = 37°, a = 10$

13. $A = 110°, B = 20°, c = 15$

14. $B = 120°, C = 35°, a = 12$

15. $A = 80°, B = 60°, a = 13$

16. $B = 75°, C = 50°, b = 25$

17. $C = 40°, A = 80°, c = 35$

18. $C = 120°, A = 25°, c = 14$

19. $A = 130.5°, C = 20°, a = 20$

20. $B = 63.7°, C = 48°, b = 33$

21. $C = 52.1°, A = 73°, a = 15$

22. $A = 87.4°, B = 61°, b = 19$

In Exercises 23–28, solve the given triangle. If there is no solution, so state. If there are two solutions, give both of them. Standard notation is used for labeling triangles.

23. $A = 35°, a = 7, b = 5$

24. $A = 25°, a = 7, b = 9$

25. $A = 40°, a = 6, b = 5$

26. $A = 50°, a = 10, b = 8$

27. $B = 70°, b = 10, c = 25$

28. $C = 48°, c = 7, b = 12$

In Exercises 29–32, find the missing length h in each of the given triangles. Round your final answer to the nearest tenth.

29.

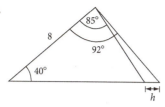

30.

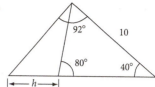

31.

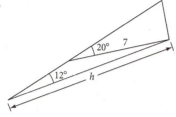

32.

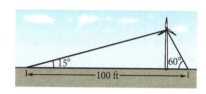

Applications

In Exercises 33–53, round answers to four decimal places unless otherwise stated.

33. **Engineering** The angles of elevation to the top of a tower from points *A* and *B* are 15° and 60°, respectively. Points *A* and *B* are 100 feet apart and at equal elevations, and the base of the tower lies between them. Find the height of the tower.

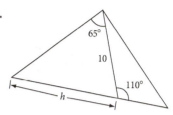

34. **Distance** The locations of point *C* on a small island off the mainland and points *A* and *B* on the shoreline of the mainland are shown below. If *A* = 63° and *B* = 76°, and *A* and *B* are 5 miles apart, find the shortest straight-line distance from point *C* to the mainland shoreline.

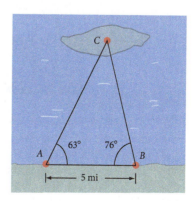

35. **Gardening** Three small circular flower beds, *A*, *B*, and *C*, are arranged so that the angle made by the line segment connecting the center of bed *A* with the center of bed *B* and the line segment connecting the center of bed *A* with bed *C* is 45 degrees. (See the figure.) If the radii of the beds are 2 feet, 3 feet, and 5 feet, respectively, find the distance between the centers of beds *A* and *C*.

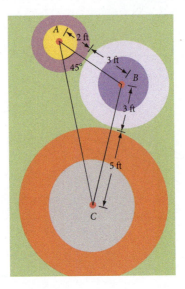

36. **Construction** A homeowner constructs a wooden border for a triangular flower bed. One edge of the bed lies along a north–south line and is 12 feet long. Another edge begins at the southern tip of the first edge and is oriented 65° west of south. If the length of the third edge is 18 feet, what is the perimeter of the flower bed?

37. **Hiking** A group of people go hiking. On the first leg, they hike 2.5 miles due north. The direction of the second and final leg is N36°E. If they end up at a place that is 5.8 miles from their starting point, what is the total distance that they hiked? Sketch a figure first.

38. **Distance** A squirrel runs along an above-ground power line to get a closer look at a peanut on the ground. The angles of depression to the peanut from the squirrel's initial and final locations are 39° and 54°, respectively. If those locations are 9.1 feet apart and the peanut is directly below the power line, how far above the ground is the squirrel?

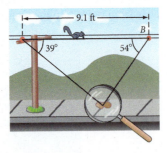

39. **Archery** An archer shoots two arrows at a target. The angle formed from the lines that connect the center of the target to the points at which the arrows hit the target is 109°. If those points are 15 inches apart, and one of them is 7.3 inches from the center, how far from the center is the other point?

40. **Construction** The roof truss represented in the figure is a type of truss that can be used for a cathedral ceiling in a house. Find the missing lengths of sides *b* and *c*. Round your final answer to the nearest tenth of a foot.

41. **Geometry** Marisa has a triangular sign made with her last name on it. She has the sign attached to her lamppost so that visitors can easily identify her house. The lengths of two edges of the sign are 10 inches and 7 inches, and the angle opposite the 10-inch edge is 75°. What is the length of the third edge?

42. **Construction** A flagpole is mounted on a wall in such a way that it makes an angle of 40° with the wall. One end of a brace to support the pole is mounted on the same wall, 2 feet below the point where the pole is mounted, and the other end of the brace is attached to the top of the pole. If the brace makes an angle of 20° with the wall, find the length of the brace.

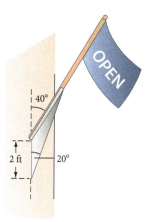

43. **Navigation** A ferry shuttles passengers across a river that runs east. The ferry travels in a straight-line path on each crossing. For the south-to-north crossing, which is 250 feet long and in the direction N25°E, passengers get on at point A and get off at point B. For the north-to-south crossing, which is 230 feet long and oriented in a certain direction west of south, they get on at point B and get off at point C. What is the width w of the river, and what is the distance d between points A and C? Round your answer to the nearest foot.

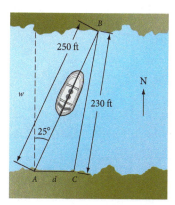

44. **Leisure** Tanya, Chris, and Melody engage in a two-dimensional tug-of-war. After knotting the ends of a rope together, they position themselves at different points of the rope; then they each grip the rope and pull on it. They continually adjust their positions to take up the slack in the rope. At a certain moment, the points at which Tanya and Chris are gripping the rope are 7 feet apart, and the measures of the angles between the corresponding 7-foot segment of the rope and the other two segments are 56° and 78°. What is the length of the rope?

45. **Leisure** Malik, Keisha, and Brian get together for a game of pitch and catch. At a certain moment, Brian is 11 feet away from Malik and 9 feet away from Keisha, and the lines from Keisha to Malik and from Keisha to Brian form an angle of 62°. How far apart are Malik and Keisha?

46. Games A billiard ball traverses a distance of 26 inches on a straight-line path and then collides with another ball, changes direction, and traverses a distance of 18 inches on a different straight-line path before coming to a stop. If an angle of 37° is formed from the lines that connect the initial location of the ball to the final location of the ball and the point of the collision, what are the two possible values of the distance d between the initial and final locations of the ball? Sketch a figure first.

47. Tourist Sites The leaning tower of Pisa was originally 184.5 feet tall when it was perpendicular to the ground. It now leans at an angle of θ from the perpendicular. When the top of the tower is viewed from a distance of 125 feet from the base of the tower, the angle of elevation is 59.7 degrees. Approximate θ to the nearest tenth of a degree.

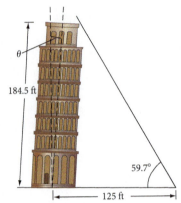

48. History While making a survey of the Maryland–Pennsylvania border, Charles Mason and Jeremiah Dixon indicated in their journal that they measured the distance across the Susquehanna River as follows: They first measured the line BC on the west side of the river, and then measured the angles made by BC with the lines from points B and C to point A on the east side of the river, as shown in the figure below. Find the distance between points A and B.

(*Source:* Surveyors Historical Society at www.surveyhistory.org)

49. Construction A triangular portion of a wooden truss is shown. One edge of the truss is 4 feet long. Find the lengths of the pieces of wood used for the other two edges.

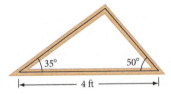

50. Geometry A piece of metal is to be cut in the shape of a parallelogram. The lengths of side AD and diagonal AC are 6.4 cm and 8.2 cm, respectively, and the angle between side AB and diagonal AC is 48°. If the angle between side AB and BC is obtuse, find the area of the parallelogram.

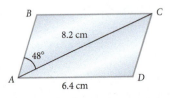

51. **Surveying** A surveyor has determined the dimensions shown in the illustration for a parcel of land. Find the lengths of the missing sides *AD* and *BC*. Round your final answer to the nearest tenth of a yard.

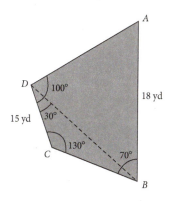

52. **Sewing** Pieces of material for the border of a bedspread are cut out in the shape of a triangle. The pieces all have the same size and shape. Each piece has two equal angles, of measure 51° each, and the edge opposite one of the equal angles is 4 inches long. What are the lengths of the other two edges of each piece?

53. **Geometry** The quadrilateral illustrated has been divided into two triangles with angles as indicated. The length of one side of the quadrilateral is 10 inches. Find the lengths of the other three sides.

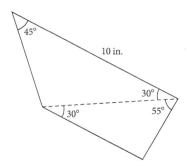

Concepts

54. Verify the Law of Sines for right triangles.

55. Explain how you can use the Law of Sines to solve a right triangle. Is this the best way to solve a right triangle? Explain.

56. Determine the set of positive values of *a* for which there is exactly one triangle *ABC* with *A* = 60° and *b* = 10, where *a* and *b* are the sides opposite angles *A* and *B*, respectively. Then find the set of positive values of *a* for which exactly two such triangles *ABC* exist and the set of positive values of *a* for which no such triangle exists.

57. Can you use the Law of Sines to solve an oblique triangle if you are given only two of the sides and the included angle (SAS) and the two given sides are not of equal length? Explain.

58. Explain why you cannot use the Law of Sines to solve an oblique triangle if you are given only the three sides of the triangle (SSS) and no two of them are of equal length.

The Law of Cosines

Objectives

- Solve an oblique triangle using the Law of Cosines.
- Solve applied problems using the Law of Cosines.
- Know and use alternative formulas for the area of a triangle.

In this section we continue our discussion of the solution of oblique triangles. As in Section 7.1, the sides and angles of a triangle are labeled as shown in Figure 1. In cases where there is not enough information to apply the Law of Sines, the **Law of Cosines** may be used to solve a triangle.

Law of Cosines

Let ABC be a triangle with sides a, b, c. (See Figure 1.) Then,

$$a^2 = b^2 + c^2 - 2bc \cos A$$

$$b^2 = a^2 + c^2 - 2ac \cos B$$

$$c^2 = a^2 + b^2 - 2ab \cos C$$

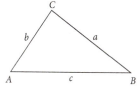

Figure 1

Proof of Law of Cosines To prove the first equation, position triangle ABC on the xy-coordinate plane as shown in Figure 2. We assume that all three angles of the triangle are acute. The case where one of the angles is obtuse is left as an exercise. The coordinates of vertices A, B, and C are $(0, 0)$, $(c, 0)$, and (x, y), respectively.

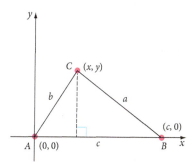

Figure 2

$a = $ Distance from vertex B to vertex C

$a = \sqrt{(x - c)^2 + (y - 0)^2}$ Distance formula

$a = \sqrt{(b \cos A - c)^2 + (b \sin A - 0)^2}$ Substitute $x = b \cos A$ and $y = b \sin A$

$a^2 = (b \cos A - c)^2 + (b \sin A)^2$ Square both sides

$a^2 = b^2 \cos^2 A - 2bc \cos A + c^2 + b^2 \sin^2 A$ Expand

$a^2 = b^2(\cos^2 A + \sin^2 A) + c^2 - 2bc \cos A$ Rearrange terms and factor out b^2

$a^2 = b^2 + c^2 - 2bc \cos A$ Pythagorean identity $\cos^2 A + \sin^2 A = 1$

Relabeling the vertices and repeating the proof gives the other two equations.

The Law of Cosines is used to solve a triangle when either of the following set of information is given:

- **SAS:** Two sides and the included angle
- **SSS:** All three sides

We illustrate the use of the Law of Cosines in the next two examples.

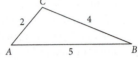 **Example 1**   **Solving a Triangle: SSS**

The sides of triangle ABC in Figure 3 are $a = 4$, $b = 2$, and $c = 5$. Find the angles of the triangle.

Figure 3

Solution We first use the Law of Cosines to find A.

$$a^2 = b^2 + c^2 - 2bc \cos A$$

$$4^2 = 2^2 + 5^2 - 2(2)(5) \cos A \qquad \text{Substitute } a = 4, b = 2, \text{ and } c = 5$$

$$16 = 4 + 25 - 20 \cos A \qquad \text{Simplify}$$

$$-13 = -20 \cos A \qquad \text{Isolate cos } A \text{ term}$$

$$\frac{13}{20} = \cos A \qquad \text{Solve for cos } A$$

$$A = \cos^{-1}\left(\frac{13}{20}\right) \approx 49.46° \qquad \text{Find } A$$

We next use the Law of Cosines to find B:

$$b^2 = a^2 + c^2 - 2ac \cos B$$
$$4 = 16 + 25 - 2(4)(5) \cos B$$
$$-37 = -40 \cos B$$

Thus,

$$\cos B = \frac{37}{40} \quad \Rightarrow \quad B \approx 22.33°$$

To find C, subtract the sum of A and B from $180°$:

$$C \approx 180° - (49.46° + 22.33°)$$

$$C \approx 108.21°$$

Check It Out 1 The sides of triangle ABC are $a = 7$, $b = 4$, and $c = 6$. Find the angles of the triangle.

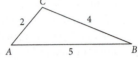 **Example 2**   **Solving a Triangle: SAS**

Find the length of the unknown side of the triangle in Figure 4. Also, find the measures of all the angles.

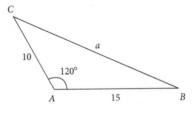

Figure 4

Solution First, use the Law of Cosines to find a:

$$a^2 = b^2 + c^2 - 2bc \cos A$$

$$a^2 = (10)^2 + (15)^2 - 2(10)(15) \cos 120° \qquad \text{Substitute } b = 10, c = 15, \text{ and } A = 120°$$

$$a^2 = 100 + 225 - 300\left(-\frac{1}{2}\right) \qquad \text{Use } \cos 120° = -\frac{1}{2}$$

$$a^2 = 325 + 150 = 475$$

Thus, $a = \sqrt{475} \approx 21.79$.

To find the measure of angle B, we can apply the Law of Sines, because we were already given the measure of one of the angles and we have the lengths of all the sides.

$$\frac{\sin B}{b} = \frac{\sin A}{a} \qquad \text{Law of Sines}$$

$$\frac{\sin B}{10} = \frac{\sin 120°}{\sqrt{475}} \qquad \text{Substitute } b = 10, A = 120°, a = \sqrt{475}$$

$$\sin B = \left(\frac{\sin 120°}{\sqrt{475}}\right) 10 \approx 0.39736 \qquad \text{Solve for } \sin B$$

$$B \approx 23.41° \quad \text{Calculate } \sin^{-1} 0.39736$$

To find C, we can subtract the sum of the two known angles from $180°$.

$$C = 180° - 120° - 23.41° = 36.59°$$

✓ *Check It Out 2*   Find the length of the unknown side of triangle ABC if $b = 9$, $c = 12$, and $A = 105°$.

The Law of Cosines has applications in surveying and navigation, as well as in many problems in engineering. The next example shows how to calculate the distance between a pair of moving boats. The notation for indicating direction, such as N30°W, was discussed in Section 7.1.

Example 3   **Application to Navigation**

Two motorboats start from the same point. One of them travels in the direction N30°W at 45 miles per hour, and the other heads due east at 35 miles per hour. How far apart are the boats after one hour?

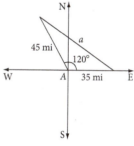

Figure 5

Solution   First, draw a diagram of the situation, as shown in Figure 5.

In one hour, the first motorboat has traveled 45 miles and the second boat 35 miles. Because we want to determine a, use the Law of Cosines to get

$$a^2 = b^2 + c^2 - 2bc \cos A$$

$$a^2 = (45)^2 + (35)^2 - 2(45)(35) \cos 120° \qquad \text{Substitute } b = 45, c = 35, \text{ and } A = 120°$$

$$a^2 = 2025 + 1225 - 3150(-0.5) \qquad \text{Use } \cos 120° = -0.5$$

$$a^2 = 3250 + 1575 = 4825$$

$$a \approx 69.46$$

Thus, the two motorboats are approximately 69.46 miles apart after one hour.

✓ *Check It Out 3*   Rework Example 3 for the case where one boat travels in the direction N45°W at 30 miles per hour, and the other heads due east at 40 miles per hour.

Area of a Triangle: SAS

As you know from geometry, the formula for the area of a triangle is

$$\text{Area} = \frac{1}{2}bh$$

where b is the base of the triangle and h is the height. We can use facts from trigonometry to derive a formula for the area when the height is not known. We only need to know two sides of the triangle and the included angle (SAS).

From Figure 6, we see that

$$\sin C = \frac{h}{a}$$

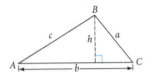

Figure 6

from which it follows that $h = a \sin C$. Substituting this expression into the formula given earlier for the area of a triangle, we obtain

$$\text{Area}(ABC) = \frac{1}{2}bh = \frac{1}{2}b\,(a \sin C) = \frac{1}{2}ab \sin C$$

The same formula holds if C is obtuse, because

$$\sin C = \sin(180° - C) = \frac{h}{a}$$

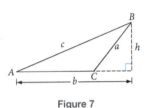

Figure 7

from which it again follows that $h = a \sin C$. (See Figure 7.) The proof that the formula also holds for the area of a right triangle is left as an exercise. By relabeling the vertices and repeating the proof, this formula can be applied whenever any two sides and their included angle are known. Thus,

$$\text{Area}(ABC) = \frac{1}{2}ab \sin C = \frac{1}{2}bc \sin A = \frac{1}{2}ac \sin B$$

Area the of Triangle: SAS

Let ABC be a triangle with sides of lengths a, b, and c. If two of the sides and the included angle are known, then

$$\text{Area}(ABC) = \frac{1}{2}ab \sin C = \frac{1}{2}bc \sin A = \frac{1}{2}ac \sin B$$

Example 4   **Finding Area of Triangle: SAS**

Find the area of a triangle if the lengths of two of its sides are 12 and 7 and their included angle is 65°.

Solution Letting $a = 12$, $b = 7$, and $C = 65°$ in the formula just established, we obtain

$$\text{Area} = \frac{1}{2}ab \sin C = \frac{1}{2}(\mathbf{12})(\mathbf{7}) \sin (\mathbf{65°}) = 42 \sin 65° \approx 38.06$$

✔ *Check It Out 4*   Find the area of a triangle if the lengths of two of its sides are 15 and 22 and their included angle is 108°.

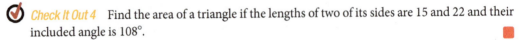

Area of a Triangle: SSS (Heron's Formula)

A third formula for the area of a triangle, known as *Heron's formula*, can be applied whenever the lengths of all three sides of the triangle are known (SSS).

Area of Triangle Using Heron's Formula

Let ABC be a triangle with sides of lengths a, b, and c. Then

$$\text{Area } (ABC) = \sqrt{s(s - a)(s - b)(s - c)}$$

where

$$s = \frac{1}{2}(a + b + c)$$

The proof of this formula is somewhat technical and has been omitted.

Example 5   **Finding Area of Triangle: SSS**

Find the area of a triangle that has sides of length 8, 5, and 9.

Solution   Apply Heron's formula. Letting $a = 8$, $b = 5$, and $c = 9$, compute the value of s.

$$s = \frac{1}{2}(a + b + c) = \frac{1}{2}(8 + 5 + 9) = \frac{1}{2}(22) = 11$$

Now substitute the values of a, b, c, and s into Heron's formula to find the area of the triangle:

$$\text{Area} = \sqrt{s(s - a)(s - b)(s - c)} = \sqrt{11(11 - 8)(11 - 5)(11 - 9)}$$
$$= \sqrt{(11)(3)(6)(2)} = \sqrt{396} \approx 19.90$$

Check It Out 5   Find the area of a triangle that has sides of length 11, 6, and 7.

Skills

In Exercises 1–6, use the Law of Cosines to find the unknown side of each triangle. The triangles are labeled using standard notation. Round your answer to four decimal places.

1. $a = 7, b = 10, C = 80°$ **2.** $b = 10, c = 4, A = 60°$ **3.** $a = 12, c = 8, B = 56°$

4. $a = 10, b = 5, C = 102°$ **5.** $b = 8, c = 4, A = 75°$ **6.** $a = 12, c = 8, B = 115°$

In Exercises 7–22, solve each triangle. The triangles are labeled using standard notation. Round your answer to four decimal places.

7. $b = 10, c = 7, A = 55°$ **8.** $a = 20, c = 12, B = 108°$ **9.** $a = 15, b = 18, C = 37.5°$

10. $b = 14, c = 20, A = 78.4°$ **11.** $a = 5, b = 7, c = 10$ **12.** $a = 4, b = 9, c = 12$

13. $a = 13, b = 17, c = 29$ **14.** $a = 19, b = 15, c = 7$ **15.** $a = 4.7, b = 8.4, c = 5.6$

16. $a = 10.6, b = 8.5, c = 11.3$ **17.** $a = 15, c = 21, B = 100°$ **18.** $a = 30, b = 20, C = 87°$

19. $b = 32, c = 25, A = 98°$ **20.** $c = 27, a = 18, B = 64°$ **21.** $c = 15.2, b = 15.2, A = 169°$

22. $a = 6, c = 16, B = 111°$

In Exercises 23–32, find the area of each triangle. Round your answer to four decimal places.

23.

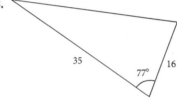

24.

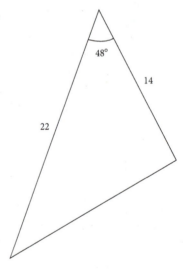

25.

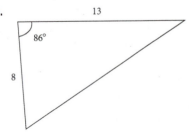

26.

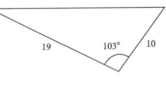

27.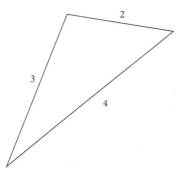

28.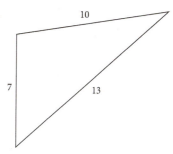

29.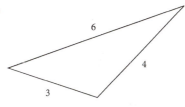

30.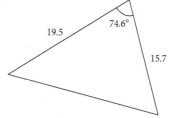

31.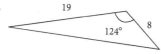

32.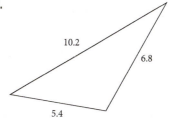

Applications

In Exercises 33–53, round your answer to four decimal places unless otherwise stated.

33. Engineering A triangular truss is to be constructed as shown. Find the length of the wood needed for the third side to the nearest hundredth of a foot.

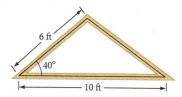

34. Landscaping A landscaper wants to fence in the triangular region as shown. What length of fencing material is needed to completely enclose the region, assuming that none of the fencing goes to waste?

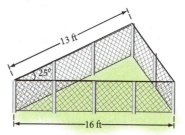

35. **Design** A kite is pictured below. If $C = 110°$, find c, the horizontal dimension of the kite at its widest point.

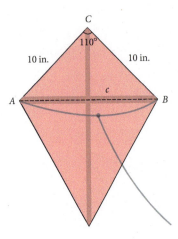

36. **Geometry** A store has designed a triangular sign made with its name on it. The edges of the sign are 11 inches, 14 inches, and 8 inches in length. Find the measure of the angle opposite each edge.

37. **Sports** Modern optical technology can be used to make very accurate measurements in sports competitions. For example, a discus is thrown from Point M to Point Z. A measuring station is set up at Point A. The equipment measures the distances AM and AZ and the angle A, as shown in the figure, and then calculates the distance ZM. (*Source:* Leica Instruments)

    a. How would the equipment's built-in computer calculate the distance ZM?

    b. Find the distance, ZM, to the nearest hundredth of a meter.

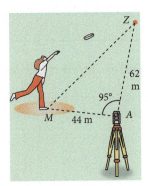

38. **Flight Path** A helicopter has two aid shipments to drop at points A and C, as illustrated. It starts at its base, B, and first flies to A. If the second drop point is at C, find the distance from A to C, to the nearest mile.

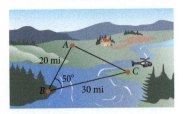

39. **Gardening** A walkway around a flower bed in a park is made up of three straight sections that form the sides of a triangle. If the lengths of the sides are 26 feet, 24 feet, and 21 feet, what is the angle opposite the longest side?

40. Design The lengths of the two sections of a hospital bed are 3 feet and 4 feet. What is the angle between the two sections of the bed when one section is raised up so that the tip of the head of the bed is 6 feet from the tip of the foot?

41. Geometry The lengths of two edges of a triangular bandage are 8 inches and 5 inches, and the angle formed by those two edges is 85°. How long is the third edge of the bandage, and what is the area of the bandage?

42. Shopping A gift shop sells figurines of famous people. Each figurine is mounted on a triangular base. The lengths of the edges of the base are 4, 5, and 6.5 inches. Find the area of the base.

43. Ship Wreckage The crews of two salvage boats have located the wreckage of an old pirate ship. The distances between the boats and the wreckage are indicated. Determine the depth of the wreckage below the surface of the water.

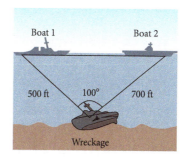

44. Leisure A billiard ball traverses a distance of 15 inches on a straight-line path, and then it collides with another ball, changes direction, and traverses a distance of 8 inches on a different straight-line path before coming to a stop. If the distance between the initial and final locations of the ball is 9 inches, find the measure of the angle formed from the lines that connect the initial location of the ball to the final location of the ball and the point of the collision.

45. Computer Graphics Many computer-graphics packages allow the user to input precise measurements for a drawing. Suppose a graphic artist wants to finish the following drawing of a triangle. In the dialog box shown, what should she input for the length of the third side of the triangle (shown as a dashed line)? What should she input for the angle *A*?

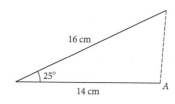

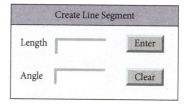

46. Measurement Use the measurements given in the illustration to find the length of the pond.

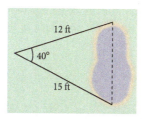

47. **Utilities** The distance from the top of a utility pole to a certain point P on the surrounding level ground is 20 feet, and the angle of elevation from point P to the top of the pole is 38°. What is the distance from the top of the pole to a point on the ground which is 10 feet further away from the base of the pole than P?

48. **Hiking** Sarah and Joycelynn go for a hike. On the first leg, they walk 3.2 miles in the direction E13°S. On the second and final leg, they walk 2.7 miles in the direction E56°S. At the end of the hike, how far are they from their starting point?

49. **Geometry** A square is inscribed in a circle of radius of 15 inches. Find the area of the square.

50. **Area of Polygon** The area of a polygonal region is often found by subdividing the region into triangles. Find the area of the lot shown in the figure to the nearest square foot.

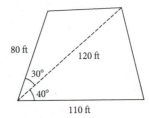

51. **Quilting** A quilt square is to be pieced together of four triangles as illustrated. Determine the lengths AE, BE, and DE to the nearest hundredth of an inch.

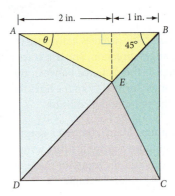

52. **Engineering** The surface of a ramp in a department store warehouse is initially oriented at an angle of 15° with the horizontal. After a load of merchandise is placed on it, the ramp is rotated clockwise about an axis through the lower edge of the ramp until the surface makes an angle of 5° with the horizontal. If the final position of the upper edge of the ramp is at a distance of 3 feet from its initial position as a result of the rotation, what is the length l of the ramp to the nearest tenth of a foot?

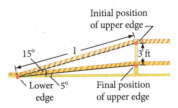

53. **Archery** An archer shoots two arrows at a target. The second arrow lands twice as far from the center of the target as the first arrow. The points at which the arrows hit the target are 6.5 inches apart, and an angle of 76° is formed from the line segments that connect the center of the target to those two points. How far from the center of the target does each of the arrows land?

Concepts

54. Prove that the Law of Cosines holds for a triangle that has an obtuse angle.

55. Show that the Law of Cosines applied to a right triangle yields the Pythagorean Theorem.

56. Explain why you cannot use the Law of Cosines to solve an oblique triangle if you are given only the measures of two angles and one side of the triangle (either AAS or ASA), and no two of the angles of the triangle are of equal measure.

57. Can you use the Law of Cosines to solve an oblique triangle if you are given only two of the sides and the angle opposite one of them (SSA) and the two given sides are not of equal length? Explain.

58. Is it possible for a triangle to have sides $a = 3$, $b = 2$, and $c = 5$? (*Hint:* what happens if you apply the Law of Cosines to this triangle?)

59. If you are given all three sides of a triangle (SSS), how can you tell whether it has a right angle?

60. If you are given all three sides of a triangle (SSS), how can you tell whether it has an obtuse angle?

61. If you are given two sides of a triangle and the included angle (SAS), how can you tell whether the triangle has a right angle if the included angle is acute?

62. If you are given two sides of a triangle and the included angle (SAS), how can you tell whether the triangle has an obtuse angle if the included angle is acute?

63. Show that the formula Area $(ABC) = \dfrac{1}{2} ab \sin C$ holds if ABC is a right triangle.

Polar Coordinates

Objectives

- Graph points in the polar coordinate system in two dimensions.

- Find several pairs of polar coordinates for a given point in two dimensions.

- Convert from polar to rectangular coordinates, and vice versa.

- Convert equations from rectangular form to polar form, and vice versa.

Consider the motion of a carousel horse revolving about a fixed center point. Its motion is circular, and at any given instant, its location can be defined as coordinates (x, y) in the xy-plane such that $x^2 + y^2 = r^2$, where r is the constant distance from the center to the horse. This description uses the rectangular coordinate system. However, circular motion occurs in many applications, and it is simpler to describe with a different coordinate system, known as the **polar coordinate system**.

In two dimensions, the polar coordinate system consists of a point called the **pole** and a ray with its endpoint at the pole, known as the **polar axis**. The pole coincides with the origin of the rectangular coordinate system, and the direction of the polar axis coincides with the direction of the positive x-axis, as shown in Figure 1.

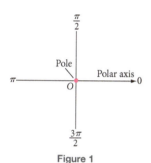

Figure 1

Note Polar grids, with concentric circles representing the radius r, and rays emanating from the origin representing θ are useful for plotting points. Polar grid graph paper can be downloaded from the Internet.

Polar Coordinates of a Point

A point in two dimensions is represented by polar coordinates as (r, θ), where r is a real number and θ is an angle in standard position. The angle θ is usually expressed in radians. Let O denote the pole.

- If $r > 0$, the polar coordinates (r, θ) represent the point that is at distance $|r|$ from O and lies on the ray along the terminal side of θ. See Figure 2.

- If $r < 0$, the polar coordinates (r, θ) represent the point that is at distance $|r|$ from O and lies on the ray *opposite* the terminal side of θ. See Figure 3.

- If θ is any angle, the polar coordinates $(0, \theta)$ represent the pole.

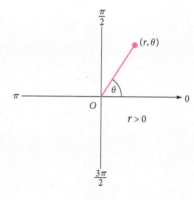

Figure 2

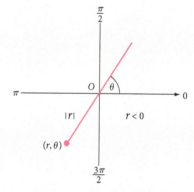

Figure 3

Just in Time
Review angles in
standard position
in Section 5.1.

Example 1   **Plotting Points on a Polar Grid**

Plot the following points on a polar grid.

a. $\left(3, \dfrac{\pi}{3}\right)$ **b.** $\left(-4, \dfrac{5\pi}{6}\right)$ **c.** $\left(\dfrac{3}{2}, -\dfrac{\pi}{4}\right)$

Solution

a. To plot $\left(3, \dfrac{\pi}{3}\right)$, first find the circle of radius 3 on the polar grid. Then, starting at the polar axis, move along this circle in the counterclockwise direction until you intersect the terminal side of the angle $\dfrac{\pi}{3}$. See Figure 4.

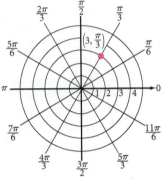

Figure 4

b. To plot $\left(-4, \dfrac{5\pi}{6}\right)$, first locate the terminal side of the angle $\dfrac{5\pi}{6}$. Because $r < 0$, locate the point that is 4 units from the pole and lies on the ray in the *opposite* direction of the terminal side of the angle $\dfrac{5\pi}{6}$. See Figure 5.

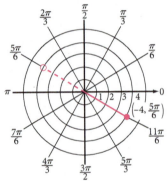

Figure 5

c. To plot $\left(\dfrac{3}{2}, -\dfrac{\pi}{4}\right)$, first find the circle of radius $\dfrac{3}{2}$ on the polar grid. Then, starting at the polar axis, move along this circle in the clockwise direction until you intersect the terminal side of the angle $-\dfrac{\pi}{4}$. See Figure 6.

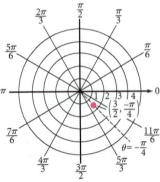

Figure 6

✓ *Check It Out 1*   Plot $\left(-1, \dfrac{\pi}{4}\right)$ on a polar grid.

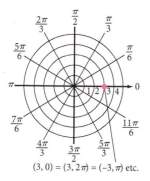

Figure 7

Multiple Representations of Polar Coordinates of a Point

In the rectangular coordinate system, every point has just one pair of coordinates. This is not the case in the polar coordinate system. For instance, $(r, \theta) = (3, 0)$ and $(r, \theta) = (3, 2\pi)$ represent the same point, because 0 and 2π are coterminal angles. The point $(3, 0)$ can also be represented by $(-3, \pi)$ or $(-3, 3\pi)$. See Figure 7. This observation leads to the following general statement about representations of a point in polar coordinates.

Multiple Representations of Polar Coordinates of a Point

The point (r, θ) can also be represented as

$$(r, \theta + 2n\pi) \text{ or } (-r, \theta + (2n + 1)\pi)$$

where n is an integer and $r > 0$.

Example 2   **Finding Multiple Representations of Polar Coordinates of a Point**

For the point with polar coordinates $\left(5, -\frac{3\pi}{4}\right)$, find four additional pairs of polar coordinates (r, θ), two with $r > 0$ and two with $r < 0$.

Solution In order to represent the point $\left(5, -\frac{3\pi}{4}\right)$, with $r > 0$, the value of r must be 5, and $\theta = -\frac{3\pi}{4} + 2n\pi$, n is an integer.

If $n = 1$,

$$(r, \theta) = \left(5, -\frac{3\pi}{4} + 2\pi\right) = \left(5, -\frac{3\pi}{4} + \frac{8\pi}{4}\right) = \left(5, \frac{5\pi}{4}\right)$$

If $n = -1$,

$$(r, \theta) = \left(5, -\frac{3\pi}{4} - 2\pi\right) = \left(5, -\frac{11\pi}{4}\right)$$

In order to represent the point $\left(5, -\frac{3\pi}{4}\right)$, with $r < 0$, the value of r must be -5, and $\theta = -\frac{3\pi}{4} + (2n + 1)\pi$, n is an integer.

If $n = 0$,

$$(r, \theta) = \left(-5, -\frac{3\pi}{4} + \pi\right) = \left(-5, \frac{\pi}{4}\right)$$

If $n = -1$,

$$(r, \theta) = \left(-5, \frac{3\pi}{4} - \pi\right) = \left(-5, -\frac{\pi}{4}\right)$$

See Figure 8.

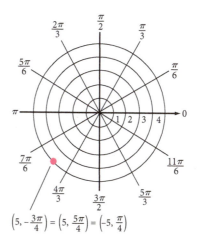

$$\left(5, -\frac{3\pi}{4}\right) = \left(5, \frac{5\pi}{4}\right) = \left(-5, \frac{\pi}{4}\right)$$

Figure 8

✅ *Check It Out 2*   Rework Example 2 for the point with polar coordinates $\left(-4, \frac{6\pi}{5}\right)$. ■

Conversion from Polar Coordinates to Rectangular Coordinates

We next show how to convert the polar coordinates of any point to its rectangular coordinates.

Converting from Polar to Rectangular Coordinates

Given the polar coordinates (r, θ) of a point, use the following equations to convert to its rectangular coordinates (x, y):

$$x = r \cos \theta \qquad y = r \sin \theta$$

Example 3   **Converting from Polar to Rectangular Coordinates**

Find the rectangular coordinates of the point with polar coordinates

a. $\left(3, \dfrac{2\pi}{3}\right)$

b. $\left(-2, \dfrac{\pi}{4}\right)$

Solution

a. Here, $r = 3$ and $\theta = \dfrac{2\pi}{3}$.

$$x = r \cos \theta = 3 \cos \frac{2\pi}{3} = 3\left(-\frac{1}{2}\right) = -\frac{3}{2}$$

$$y = r \sin \theta = 3 \sin \frac{2\pi}{3} = 3\left(\frac{\sqrt{3}}{2}\right) = \frac{3\sqrt{3}}{2}$$

Thus, the rectangular coordinates of the point with polar coordinates $\left(3, \frac{2\pi}{3}\right)$ are $\left(-\frac{3}{2}, \frac{3\sqrt{3}}{2}\right)$. See Figure 9.

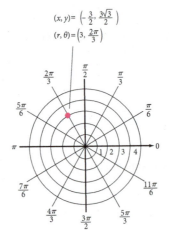

Figure 9

b. Here, $r = -2$ and $\theta = \dfrac{\pi}{4}$.

$$x = r \cos \theta = -2 \cos \frac{\pi}{4} = -2\left(\frac{\sqrt{2}}{2}\right) = -\sqrt{2}$$

$$y = r \sin \theta = -2 \sin \frac{\pi}{4} = -2\left(\frac{\sqrt{2}}{2}\right) = -\sqrt{2}$$

Thus, the rectangular coordinates of the point with polar coordinates $\left(-2, \frac{\pi}{4}\right)$ are $\left(-\sqrt{2}, -\sqrt{2}\right)$. See Figure 10.

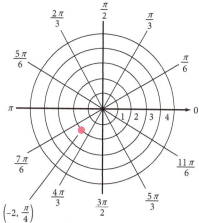

Figure 10

 Check It Out 3 Find the rectangular coordinates of the point with polar coordinates $\left(2, -\frac{5\pi}{6}\right)$.

Converting Rectangular Coordinates to Polar Coordinates

When converting from rectangular to polar coordinates, a point may have multiple representations in polar coordinates. To obtain a unique pair of polar coordinates, (r, θ), we restrict r to be positive and $0 \leq \theta < 2\pi$.

Converting from Rectangular to Polar Coordinates

Given the rectangular coordinates (x, y) of a point other than the origin, its unique pair of polar coordinates (r, θ) with $r > 0$ and $0 \leq \theta < 2\pi$ can be obtained as follows.

- Define r by

$$r = \sqrt{x^2 + y^2}$$

- If $x \neq 0$ and $y \neq 0$, then θ is the unique angle in the interval $[0, 2\pi)$ such that θ satisfies the equation

$$\tan \theta = \frac{y}{x}$$

and its terminal side lies in the same quadrant as the point.

- If $y = 0$, then $r = |x|$, $\theta = 0$ if $x > 0$, and $\theta = \pi$ if $x < 0$. See Figure 11.
- If $x = 0$, then $r = |y|$, $\theta = \frac{\pi}{2}$ if $y > 0$, and $\theta = \frac{3\pi}{2}$ if $y < 0$. See Figure 12.

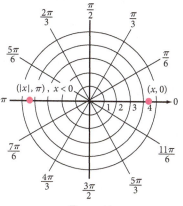

Figure 11

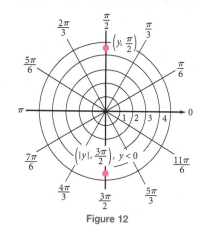

Figure 12

Example 4   **Converting from Rectangular to Polar Coordinates**

Find a pair of polar coordinates (r, θ) of the point with rectangular coordinates $(3, -\sqrt{3})$. Choose r and θ so that $r > 0$ and $0 \le \theta < 2\pi$.

Solution   First, use $r = \sqrt{x^2 + y^2}$ to find r:

$$r = \sqrt{(3)^2 + (-\sqrt{3})^2} = \sqrt{9 + 3} = \sqrt{12} = 2\sqrt{3}$$

Next, use $\tan \theta = \dfrac{y}{x}$ to find θ:

$$\tan \theta = -\frac{\sqrt{3}}{3}$$

Both $\theta = \dfrac{5\pi}{6}$ and $\theta = \dfrac{11\pi}{6}$ satisfy $\tan \theta = -\dfrac{\sqrt{3}}{3}$, and are in $[0, 2\pi)$. Choose $\theta = \dfrac{11\pi}{6}$ because its terminal side lies in the fourth quadrant, the same as the given point $(3, -\sqrt{3})$. Thus the polar coordinates for the point with rectangular coordinates $(3, -\sqrt{3})$ are $\left(2\sqrt{3}, \dfrac{11\pi}{6}\right)$. See Figure 13.

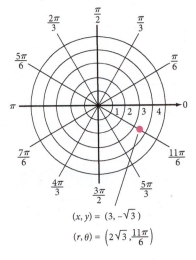

$(x, y) = (3, -\sqrt{3})$

$(r, \theta) = \left(2\sqrt{3}, \dfrac{11\pi}{6}\right)$

Figure 13

Check It Out 4   Rework Example 4 for the point with rectangular coordinates $(4, -4)$.

An equation in **rectangular form** contains the variables x and y that correspond to the rectangular coordinates of points. An equation in **polar form** contains the variables r, θ that correspond to the polar coordinates of points. The following equations can be used to convert an equation in rectangular form to an equation in polar form.

$$x = r \cos \theta \text{ and } y = r \sin \theta$$

Example 5   **Writing an Equation in Polar Form**

Write a polar form of the equation of the circle $x^2 + y^2 - 2y = 8$.

Solution

$x^2 + y^2 - 2y = 8$	Given equation
$(r \cos \theta)^2 + (r \sin \theta)^2 - 2(r \sin \theta) = 8$	Substitute $x = r \cos \theta$ and $y = r \sin \theta$
$r^2 \cos^2 \theta + r^2 \sin^2 \theta - 2r \sin \theta = 8$	Eliminate parentheses
$r^2(\cos^2 \theta + \sin^2 \theta) - 2r \sin \theta = 8$	Factor out r^2
$r^2 - 2r \sin \theta = 8$	Substitute $\cos^2 \theta + \sin^2 \theta = 1$

Thus, a polar form for $x^2 + y^2 - 2y = 8$ is $r^2 - 2r \sin \theta = 8$.

✅ *Check It Out 5*   Write a polar form of the equation of the circle $x^2 + 2x + y^2 = 3$.

To convert from the polar to the rectangular form of an equation, use the following substitutions.

$$r = \sqrt{x^2 + y^2} \qquad \cos\theta = \frac{x}{\sqrt{x^2 + y^2}} \qquad \sin\theta = \frac{y}{\sqrt{x^2 + y^2}} \qquad \tan\theta = \frac{x}{y}$$

Example 6   Converting an Equation to Rectangular Form

Convert the following equations from polar to rectangular form. Identify the equation.

a. $\theta = -\dfrac{\pi}{4}$

b. $r = 6\sin\theta$

Solution

a. Use the substitution $\frac{y}{x} = \tan\theta$ to convert to rectangular form by first taking the tangent of both sides of the equation.

$$\theta = -\frac{\pi}{4} \qquad\qquad \text{Original equation}$$

$$\tan\theta = \tan\left(-\frac{\pi}{4}\right) \qquad \text{Take tangent of both sides}$$

$$\frac{y}{x} = -1 \qquad\qquad \tan\theta = \frac{y}{x}$$

$$y = -x \qquad\qquad \text{Solve for } y$$

Thus the equation $\theta = -\frac{\pi}{4}$ is the same as the equation of the line $y = -x$.

b. Use the substitution $r = \sqrt{x^2 + y^2}$ and $\sin\theta = \frac{y}{r} = \frac{y}{\sqrt{x^2 + y^2}}$.

$$r = 6\sin\theta \qquad\qquad \text{Original equation}$$

$$\sqrt{x^2 + y^2} = 6\frac{y}{\sqrt{x^2 + y^2}} \qquad \text{Substitute } r = \sqrt{x^2 + y^2} \text{ and } \sin\theta = \frac{y}{\sqrt{x^2 + y^2}}$$

$$x^2 + y^2 = 6y \qquad\qquad \text{Multiply both sides by } \sqrt{x^2 + y^2}$$

$$x^2 + y^2 - 6y = 0$$

$$x^2 + y^2 - 6y + 9 = 9 \qquad \text{Complete the square}$$

$$x^2 + (y - 3)^2 = 9 \qquad \text{Write as a perfect trinomial square}$$

Thus the equation $r = 6\sin\theta$ represents a **circle** of radius 3 with center $(0, 3)$.

✅ *Check It Out 6*   Convert $r = 2\sin\theta$ to rectangular form and identify the equation.

Exercises 7.3

 Just in Time Exercises

In Exercises 1–4, determine the quadrant where the terminal side of each angle lies.

1. $\theta = -\frac{5\pi}{4}$ **2.** $\theta = \frac{11\pi}{6}$ **3.** $\theta = \frac{10\pi}{3}$ **4.** $\theta = -\frac{11\pi}{6}$

Skills

In Exercises 5–10, match each of the given polar coordinates with one of the points A–F on the graph.

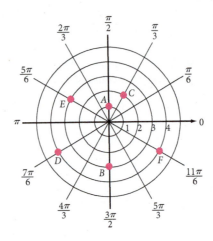

5. $\left(3, \frac{5\pi}{6}\right)$ **6.** $\left(-4, \frac{\pi}{6}\right)$ **7.** $\left(4, -\frac{\pi}{6}\right)$ **8.** $\left(1, -\frac{3\pi}{2}\right)$

9. $\left(-3, \frac{\pi}{2}\right)$ **10.** $\left(-2, \frac{4\pi}{3}\right)$

Paying Attention to Instructions: *Exercises 11 and 12 are intended to give you practice reading and paying attention to the instructions that accompany the problems you are working.*

11. a. Plot using a rectangular grid: $(2, \pi)$. **12. a.** Plot using a rectangular grid: $(-3, \pi)$.

    b. Plot using a polar grid: $(2, \pi)$.   **b.** Plot using a polar grid: $(-3, \pi)$.

In Exercises 13–20, plot the points, given in polar coordinates, on a polar grid.

13. $\left(1, \frac{\pi}{2}\right)$ **14.** $(1, \pi)$ **15.** $\left(-3, \frac{\pi}{3}\right)$ **16.** $\left(-2, \frac{\pi}{6}\right)$

17. $\left(\frac{1}{2}, -\pi\right)$ **18.** $\left(\frac{3}{2}, \frac{-\pi}{2}\right)$ **19.** $\left(0, \frac{3\pi}{2}\right)$ **20.** $(2, 0)$

In Exercises 21–28, convert each of the given pairs of polar coordinates to rectangular coordinates.

21. $\left(3, \frac{\pi}{4}\right)$ **22.** $\left(-2, \frac{\pi}{3}\right)$ **23.** $\left(-1, \frac{\pi}{6}\right)$ **24.** $\left(2, \frac{2\pi}{3}\right)$

25. $\left(-4, \frac{7\pi}{6}\right)$ **26.** $\left(\frac{1}{2}, \frac{3\pi}{2}\right)$ **27.** $\left(\frac{3}{2}, \frac{\pi}{2}\right)$ **28.** $\left(-3, \frac{5\pi}{6}\right)$

In Exercises 29–34, convert each of the given pairs of rectangular coordinates to a pair of polar coordinates (r, θ) with $r > 0$ and $0 \le \theta < 2\pi$.

29. $\left(1, \sqrt{3}\right)$ **30.** $(-4, -4)$ **31.** $\left(-\sqrt{3}, -1\right)$

32. $\left(-2, 2\sqrt{3}\right)$ **33.** $(0, 1)$ **34.** $(-1, 0)$

In Exercises 35–44, for each of the points given in polar coordinates, find two additional pairs of polar coordinates (r, θ), *one with* $r > 0$ *and one with* $r < 0$.

35. $(2, \pi)$ **36.** $\left(-3, \frac{\pi}{2}\right)$ **37.** $\left(4, \frac{3\pi}{2}\right)$ **38.** $(5, 0)$ **39.** $\left(\frac{3}{4}, \frac{\pi}{6}\right)$

40. $\left(-\frac{1}{2}, \frac{4\pi}{5}\right)$ **41.** $\left(-\frac{11}{7}, -5\pi\right)$ **42.** $\left(\frac{7}{2}, 3\pi\right)$ **43.** $\left(1.3, \frac{3\pi}{4}\right)$ **44.** $\left(-2.7, \frac{5\pi}{4}\right)$

In Exercises 45–56, convert each of the given rectangular equations to polar form.

45. $x = 2$ **46.** $y = 3$ **47.** $x + 2y = 4$

48. $3x + y = 1$ **49.** $x^2 + y^2 = 25$ **50.** $x^2 + y^2 = 4$

51. $(x + 1)^2 + y^2 = 1$ **52.** $x^2 + (y + 3)^2 = 9$ **53.** $y = x^2$

54. $y^2 = 3x$ **55.** $y = 2x^2 + x$ **56.** $y = x^2 + 4x$

In Exercises 57–68, convert each of the given polar equations to rectangular form.

57. $r = 3$ **58.** $r = 4$ **59.** $\theta = \frac{\pi}{4}$

60. $\theta = \pi$ **61.** $r \cos \theta = 4$ **62.** $r \sin \theta = 3$

63. $2r \cos \theta + r \sin \theta = 4$ **64.** $r \cos \theta - 3r \sin \theta = 5$ **65.** $r = 2 \cos \theta$

66. $r = 4 \sin \theta$ **67.** $r^2 \cos 2\theta = 4$ **68.** $r^2 \cos 2\theta = 1$

Applications

69. Signal Device A patented device converts a radar signal given in polar coordinates to a format in rectangular coordinates, so that it is better suited to display in a television-type display device. If a radar signal is at the point $\left(3, -\frac{2\pi}{3}\right)$, find the exact values of the corresponding rectangular coordinates in the television display. *(Source:* www.freepatents.com)

70. Navigation A boat departs its starting point and travels 4 miles north and 3 miles west. Determine its current location in polar coordinates, using the starting point as the origin. Use a calculator to approximate θ, in radians, to three decimal places.

Concepts

71. Explain why (r, θ) and $(r, \theta + 2\pi)$ represent the same point in the polar coordinate system.

72. Explain why (r, θ) and $(-r, \theta + \pi)$ represent the same point in the polar coordinate system.

73. List at least two features of the polar coordinate system that are different from those of the rectangular coordinate system.

SPOTLIGHT ON SUCCESS   *Student Instructor Stephanie*

For success, attitude is equally as important as ability.
—Harry F. Banks

Math has always fascinated me. From addition to calculus, I've taken great interest in the material and great pride in my work. Whenever I struggled with concepts, I asked questions and worked problems over and over until they became second nature. I used to assume this was how everyone dealt with concepts they didn't understand. However, in high school, I noticed how easily students got discouraged with mathematics. In my senior year calculus and statistics classes, I was surrounded by bright students who simply gave up on trying to fully understand the material because it seemed confusing or difficult. Even if we shared a similar level of academic ability, the difference between these students' grades and my own reflected a difference in attitude. I noticed many students giving up without really trying to understand the concepts because they lacked confidence and didn't feel they were capable. They began coming to me for help. Though I was glad to help them with the math, I had a greater goal to help them believe they could succeed on their own. Soon the students I tutored gained more understanding and achieved success by simply paying more attention in class and working extra problems outside of class. It was amazing how much improvement I saw in both their confidence levels and their grades. It goes to show that a little extra effort and a positive attitude can truly make a difference.

Objectives

- Graph polar equations.

- Analyze the symmetries of a polar graph.

You are familiar with graphing in the rectangular coordinate system by plotting points, testing for symmetry, and understanding the nature of a specific function. In this section, we will introduce similar techniques for graphing equations in the polar coordinate system.

Lines and Circles

The simplest polar equations to graph are those of the form $r = k$ or $\theta = k$, where k is a constant. The following two examples discuss these special cases.

> **Example 1**   **Graphing the Equation $r = 4$**

Sketch the graph of $r = 4$.

Solution Find and plot a set of points (r, θ) in the polar coordinates system such that $r = 4$. Because θ does not appear in the equation, it can assume any value. The following lists some selected points (r, θ) that satisfy $r = 4$.

$$\left(4, \frac{\pi}{4}\right), \left(4, \frac{2\pi}{3}\right), \left(4, -\frac{\pi}{3}\right), (4, \pi)$$

When you plot these points on a polar grid, you will see that all the points lie on the circle of radius 4, centered at the pole. See Figure 1.

Observe that the equation for a circle centered at the origin with radius 4 is much more compact in polar form than in rectangular form, which would be $x^2 + y^2 = 16$.

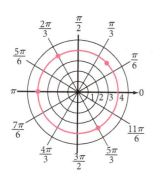

Figure 1

VIDEO EXAMPLES

SECTION 7.4

✓ *Check It Out 1*   Sketch the graph of $r = 3$.

> **Example 2**   **Graphing Lines**

Graph the following equations.

a. $r = 2 \csc \theta$

b. $\theta = \dfrac{\pi}{4}$

Solution

a. Use the equation $r = 2 \csc \theta$ to generate a set of points as shown in Table 1. Plot the labeled points on a polar grid. The graph is a horizontal line, as shown in Figure 2.

θ	$-\dfrac{\pi}{2}$	$\dfrac{\pi}{6}$	$\dfrac{\pi}{3}$	$\dfrac{\pi}{2}$
r	-2	4	≈ 2.31	2
Label	A	B	C	D

Table 1

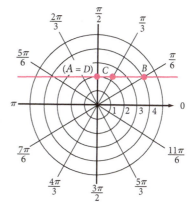

Figure 2

Note that the equation can be rewritten as $r \sin \theta = 2$, because $\csc \theta = \dfrac{1}{\sin \theta}$. Substituting $y = r \sin \theta$, we see that the polar equation $r = 2 \csc \theta$ is equivalent to $y = 2$. This is the rectangular form of the equation of a horizontal line.

b. We first find and plot a set of points (r, θ) in the polar coordinate system such that $\theta = \frac{\pi}{4}$. Because r does not appear in the equation, r can assume any value. The following lists some points (r, θ) satisfying $\theta = \frac{\pi}{4}$.

$$A: \left(2, \frac{\pi}{4}\right), \ B: \left(0, \frac{\pi}{4}\right), \ C: \left(-1, \frac{\pi}{4}\right), \ D: \left(\frac{7}{2}, \frac{\pi}{4}\right)$$

Plotting these points on a polar grid, we see that they lie on the ray $\theta = \frac{\pi}{4}$ or $\theta = \frac{5\pi}{4}$. The points on the ray $\theta = \frac{5\pi}{4}$ are generated by the *negative* values of r. Thus the graph of the equation $\theta = \frac{\pi}{4}$ is a line passing through the pole, making an angle of $\frac{\pi}{4}$ with the polar axis. See Figure 3.

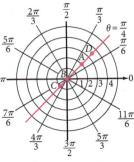

Figure 3

Check It Out 2 Sketch the graph of the equation $\theta = \frac{2\pi}{3}$.

In general, we have the following.

Lines and Circles in Polar Coordinates

Graph of $r = a$, $a > 0$
The graph of the polar equation $r = a$ is a circle of radius a centered at the pole.

Graph of $\theta = \alpha$
The graph of the polar equation $\theta = \alpha$ is a line through the pole, making an angle of α with the polar axis.

Graphs of $r = c \sec \theta$ and $r = d \csc \theta$
The graph of $r = c \sec \theta$ is a vertical line. If $c > 0$, it lies c units to the right of the pole. If $c < 0$, it lies $|c|$ units to the left of the pole.

The graph of $r = d \csc \theta$ is a horizontal line. If $d > 0$, it lies d units above the pole. If $d < 0$, it lies $|d|$ units below the pole.

Symmetries in Polar Coordinates

We can use the following guidelines to test for any symmetries in a polar equation

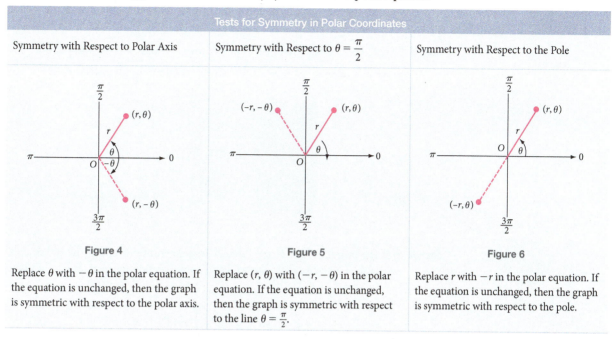

Tests for Symmetry in Polar Coordinates		
Symmetry with Respect to Polar Axis	Symmetry with Respect to $\theta = \dfrac{\pi}{2}$	Symmetry with Respect to the Pole
Figure 4	**Figure 5**	**Figure 6**
Replace θ with $-\theta$ in the polar equation. If the equation is unchanged, then the graph is symmetric with respect to the polar axis.	Replace (r, θ) with $(-r, -\theta)$ in the polar equation. If the equation is unchanged, then the graph is symmetric with respect to the line $\theta = \dfrac{\pi}{2}$.	Replace r with $-r$ in the polar equation. If the equation is unchanged, then the graph is symmetric with respect to the pole.

> **Note** If a polar equation passes a symmetry test, the graph of the equation definitely exhibits that type of symmetry. However, the graph of a polar equation can exhibit a certain type of symmetry even if the equation fails the corresponding symmetry test.

> **Example 3** **Graphing a Circle**

Sketch the graph of the equation $r = 3 \cos \theta$.

Solution

Step 1 Test for symmetry, if any.

- **Symmetry with respect to the polar axis.** Replace θ by $-\theta$ in the equation.

$$r = 3 \cos(-\boldsymbol{\theta}) \qquad \text{Replace } \theta \text{ with } -\theta$$
$$r = 3 \cos \theta \qquad \text{Negative-angle identity for cosine}$$

because we get the original equation ($r = 3 \cos \theta$), the graph is symmetric with respect to the polar axis.

- **Symmetry with respect to the line** $\theta = \dfrac{\pi}{2}$. Replace (r, θ) with $(-r, -\theta)$ in the equation.

$$r = 3 \cos \theta \qquad \text{Original equation}$$
$$-r = 3 \cos(-\boldsymbol{\theta}) \qquad \text{Replace } (r, \theta) \text{ with } (-r, -\theta)$$
$$-r = 3 \cos \theta \qquad \text{Substitute } \cos \theta = \cos(-\theta)$$
$$r = -3 \cos \theta \qquad \text{Simplify}$$

The equation $r = -3 \cos \theta$ is not equivalent to the original equation. The graph may or may not be symmetric with respect to the line $\theta = \dfrac{\pi}{2}$.

- **Symmetry with respect to the pole.** Replace r with $-r$ in the equation.

$$r = 3 \cos \theta \qquad \text{Original equation}$$
$$-r = 3 \cos \theta \qquad \text{Replace } r \text{ with } -r$$
$$r = -3 \cos \theta \qquad \text{Simplify}$$

The equation $r = -3 \cos \theta$ is not equivalent to the original equation. Thus the graph may or may not be symmetric with respect to the pole.

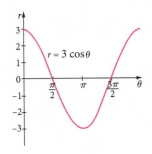

Figure 7

Step 2 Analyze the equation. Because $\cos\theta$ is periodic with period 2π, we consider values of θ in $[0, 2\pi)$. Determine the maximum and minimum values for r and where $r = \theta$ by sketching a graph of $r = 3 \cos\theta$ in *rectangular* coordinates (Figure 7). The maximum value of r is 3 at $\theta = 0$. The minimum value of r is -3, at $\theta = \pi$. For $\theta = \frac{\pi}{2}$ and $\theta = \frac{3\pi}{2}$, $r = 0$.

Step 3 From Step 2, we see that r decreases from $r = 3$ to $r = 0$ for θ in $\left[0, \frac{\pi}{2}\right]$, so we first tabulate values of r for selected angles θ in the interval $\left[0, \frac{\pi}{2}\right]$. See Table 2.

θ	0	$\frac{\pi}{6}$	$\frac{\pi}{4}$	$\frac{\pi}{3}$	$\frac{\pi}{2}$
$r = 3\cos\theta$	3	$3 \cdot \dfrac{\sqrt{3}}{2}$ ≈ 2.60	$3 \cdot \dfrac{\sqrt{2}}{2}$ ≈ 2.12	$\dfrac{3}{2}$	0
Label	A	B	C	D	E

Table 2

Step 4 Plot the points $A - E$ from Table 2. Using symmetry about the polar axis, reflect the points $A - E$ across the polar axis and complete the figure as shown in Figure 8.

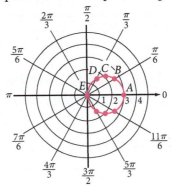

Figure 8

✓ *Check It Out 3*   Rework Example 3 for the equation $r = 3 \sin\theta$.

Roses and Lemniscates

Not all equations of curves in polar form are easy to represent in rectangular form. The following examples will examine the graphs of equations that are relatively simple in polar form, but are more difficult to describe in rectangular form.

Example 4   **Graphing a 4-Petal Rose**

Sketch the graph of the equation $r = 4 \cos(2\theta)$, and determine the type(s) of symmetry exhibited by the graph.

Solution

Step 1   Test for symmetry.

Symmetry with Respect to the Polar Axis	Symmetry with Respect to the Line $\theta = \dfrac{\pi}{2}$	Symmetry with Respect to the Pole
Replacing θ with $-\theta$, $$r = 4 \cos(2(-\theta))$$ $$r = 4 \cos 2\theta$$	Replacing (r, θ) with $(-r, -\theta)$, $$-r = 4 \cos(2(-\theta))$$ $$-r = 4 \cos(2\theta)$$ $$r = -4 \cos(2\theta)$$	Replacing r with $-r$, $$-r = 4 \cos(2\theta)$$ $$r = -4 \cos(2\theta)$$
Note that $\cos(-2\theta) = \cos(2\theta)$, by the negative-angle identity. Because the equation is unchanged, the graph is symmetric with respect to the polar axis.	The result is not equivalent to the original equation, so the graph may or may not be symmetric with respect to the line $\theta = \frac{\pi}{2}$.	The result is not equivalent to the original equation, so the graph may or may not be symmetric with respect to the pole.

Step 2   Analyze the equation. Because $-1 \leq \cos 2\theta \leq 1$, the values of $r = 4 \cos 2\theta$ are in the interval $[-4, 4]$.

Step 3   Sketch the graph piece by piece. Start by tabulating values of r for selected values of θ in $\left[0, \frac{\pi}{2} \right]$. Include a value of θ at which $r = 0$ and for which r is a maximum or minimum. See Table 3.

θ	0	$\dfrac{\pi}{6}$	$\dfrac{\pi}{4}$	$\dfrac{\pi}{3}$	$\dfrac{\pi}{2}$
$r = 4 \cos 2\theta$	4	2	0	-2	-4
Label	A	B	C	D	E

Table 3

Plot the corresponding points (r, θ) and use them to sketch the graph of the equation over the interval $\left[0, \frac{\pi}{2} \right]$. See Figure 9.

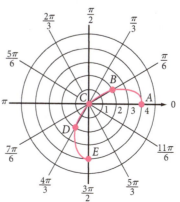

Figure 9

Using Technology

Polar equations can be graphed with a graphing utility using the polar mode. To graph $r = 4\cos(2\theta)$, we used $\theta\min = 0$, $\theta\max = 2\pi$ and $\theta\text{step} = \frac{\pi}{24}$ in the WINDOW menu. A suitable viewing rectangle was obtained with the help of the SQUARE setting. See Figure 11.

Keystroke Appendix:
Section 15

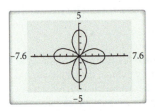

Figure 11

Continue to tabulate and plot another set of points on the interval $\left[\frac{\pi}{2}, \pi\right]$. See Table 4 and Figure 10.

θ	$\frac{2\pi}{3}$	$\frac{3\pi}{4}$	$\frac{5\pi}{6}$	π
$r = 4\cos 2\theta$	-2	0	2	4
Label	F	G	H	I

Table 4

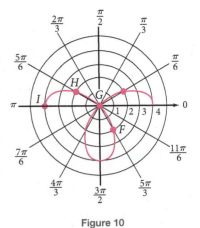

Figure 10

To complete the graph, use symmetry of the graph about the polar axis. See Figure 12.

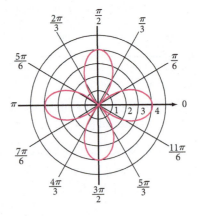

Figure 12

 Check It Out 4   Rework Example 4 for the equation $r = 3\sin(2\theta)$.

A curve with an equation of the form $r = a\cos n\theta$ or $r = a\sin n\theta$, where $a \neq 0$ and n is a nonzero integer, is known as a **rose**. It consists of several closed curves, called **petals**. There are n petals if n is odd, and $2n$ petals if n is even. In either case, the size of the petals is determined by the value of a.

A curve with an equation of the form $r^2 = a\cos 2\theta$ or $r^2 = a\sin 2\theta$, where $a > 0$, is known as a **lemniscate**, which is graphed in the next example.

> **Example 5** **Graphing a Lemniscate**

Sketch the graph of the equation $r^2 = 9 \sin(2\theta)$, and determine the type(s) of symmetry exhibited by the graph.

Solution

Step 1. Test for symmetry.

Symmetry with Respect to the Polar Axis	Symmetry with Respect to $\theta = \dfrac{\pi}{2}$	Symmetry with Respect to the Pole
Replacing θ with $-\theta$, and using the negative angle identity for sine, $$r^2 = 9\sin(2(-\boldsymbol{\theta}))$$ $$r^2 = 9(-\sin 2\theta)$$ $$r^2 = -9\sin 2\theta$$	Replacing (r, θ) with $(-r, -\theta)$, $$(-r)^2 = 9\sin(2(-\boldsymbol{\theta}))$$ $$r^2 = 9(-\sin 2\theta)$$ $$r^2 = -9\sin 2\theta$$	Replacing r with $-r$, $$(-r)^2 = 9\sin 2\theta$$ $$r^2 = 9\sin 2\theta$$
The result is not equivalent to the original equation, so the graph may or may not be symmetric with respect to the polar axis.	The result is not equivalent to the original equation, so the graph may or may not be symmetric with respect to the line $\theta = \frac{\pi}{2}$.	The original equation is unchanged, so the graph is symmetric with respect to the pole.

Step 2 Analyze the equation. For values of θ in the interval $\left[0, \frac{\pi}{2}\right]$, $r^2 = \sin 2\theta \geq 0$. If $\sin 2\theta < 0$, then this equation has no solution. The value of $|r|$ is at a maximum when $\theta = \frac{\pi}{4}$. See Table 5 for selected values of (r, θ).

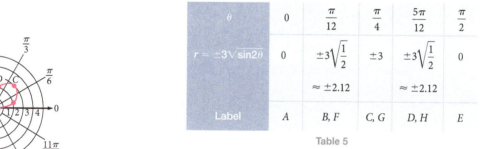

θ	0	$\dfrac{\pi}{12}$	$\dfrac{\pi}{4}$	$\dfrac{5\pi}{12}$	$\dfrac{\pi}{2}$
$r = \pm 3\sqrt{\sin 2\theta}$	0	$\pm 3\sqrt{\dfrac{1}{2}}$	± 3	$\pm 3\sqrt{\dfrac{1}{2}}$	0
		$\approx \pm 2.12$		$\approx \pm 2.12$	
Label	A	B, F	C, G	D, H	E

Table 5

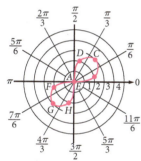

Figure 13

Step 3 Plot the corresponding points (r, θ), and use them to sketch the graph. The negative values of r produce the portion of the graph in the third quadrant (points F, G, and H). The graph is symmetric with respect to the pole, but not with respect to the polar axis or the line $\theta = \frac{\pi}{2}$. See Figure 13.

✅ *Check It Out 5* Rework Example 5 for the equation $r^2 = 9\cos(2\theta)$.

Cardioids and Limaçons

A curve with an equation of the form $r = a + b \cos \theta$ or $r = a + b \sin \theta$, where a and b are nonzero and $|a| = |b|$, is known as a **cardioid.**

Example 6 Graphing a Cardioid

Sketch the graph of the equation $r = 2(1 - \cos \theta)$, and determine the type(s) of symmetry exhibited by the graph.

Solution

Step 1 Test for symmetry

Symmetry with Respect to the Polar Axis	Symmetry with Respect to $\theta = \dfrac{\pi}{2}$	Symmetry with respect to the pole
Replacing θ with $-\theta$, $$r = 2(1 - \cos(-\boldsymbol{\theta}))$$ $$r = 2(1 - \cos\theta)$$	Replacing (r, θ) with $(-r, -\theta)$, $$-\boldsymbol{r} = 2(1 - \cos(-\boldsymbol{\theta}))$$ $$-r = 2(1 - \cos\theta)$$ $$r = -2(1 - \cos\theta)$$	Replacing r with $-r$, $$-r = 2(1 - \cos\theta)$$ $$r = -2(1 - \cos\theta)$$
Note that $\cos(-\theta) = \cos(\theta)$, by the negative-angle identity. Because the equation is unchanged, the graph is symmetric with respect to the polar axis.	The result is not equivalent to the original equation, so the graph may or may not be symmetric with respect to the line $\theta = \frac{\pi}{2}$.	The result is not equivalent to the original equation, so the graph may or may not be symmetric with respect to the pole.

Step 2 Analyze the equation. Because $\cos \theta$ is periodic with period 2π, sketch the graph over any interval of length 2π. The value of r increases from 0 to 2 on the interval $[0, \pi]$ and decreases from 2 to 0 on the interval $[\pi, 2\pi]$

Step 3 Tabulate values of r for selected angles θ in the interval $[0, \pi]$. See Table 6 and Table 7.

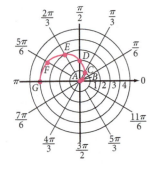

Figure 14

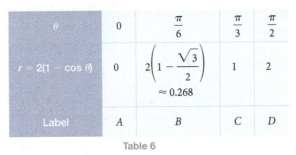

θ	0	$\dfrac{\pi}{6}$	$\dfrac{\pi}{3}$	$\dfrac{\pi}{2}$
$r = 2(1 - \cos \theta)$	0	$2\left(1 - \dfrac{\sqrt{3}}{2}\right)$ ≈ 0.268	1	2
Label	A	B	C	D

Table 6

θ	$\dfrac{2\pi}{3}$	$\dfrac{5\pi}{6}$	π
$r = 2(1 - \cos \theta)$	3	$2\left(1 + \dfrac{\sqrt{3}}{2}\right)$ ≈ 3.732	4
Label	E	F	G

Table 7

Plot the corresponding points, and use them to sketch the graph over the interval $[0, \pi]$. See Figure 14. Then sketch the mirror image of the existing portion of the graph with respect to the polar axis. See Figure 15.

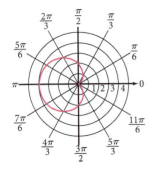

Figure 15

 Check It Out 6 Rework Example 6 for the equation $r = 2(1 - \sin \theta)$.

A curve with an equation of the form $r = a \pm b \cos \theta$ or $r = a \pm b \sin \theta$, where $a > 0$ and $b > 0$, is known as a **limaçon**. If $0 < a < b$, the limaçon has an inner loop. A limaçon with $a = b$ is a cardioid. It has a cusp—rather than a loop—at the pole. A limaçon with $0 < b < a$ does not pass through the pole at all and does not have an inner loop. Some limaçons of this type have a depression.

Example 7   **Graphing a Limaçon**

Use a graphing utility to graph the following equations and determine the type(s) of symmetry of the graphs.

a. $r = 1 - 4 \sin \theta$ **b.** $r = 5 + 3 \cos \theta$

Solution

a. To graph this function, set the calculator to POLAR mode, with θ ranging from 0 to 2π. Use the table of values from your calculator for a range of values of r to determine a window size. Use a window size of $[-6, 6] \times [-6, 2]$ for the graph in Figure 16.

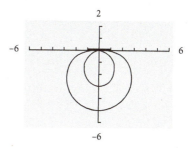

Figure 16

The graph is symmetric with respect to the line $\theta = \frac{\pi}{2}$, but it is not symmetric with respect to the polar axis, nor is it symmetric with respect to the pole. This curve is an example of a **limaçon with an inner loop.**

b. To graph this function, set the calculator to POLAR mode, and use a window size of $[-9, 15] \times [-8, 8]$. See Figure 17. The graph is symmetric with respect to the polar axis, but it is not symmetric with respect to the line $\theta = \frac{\pi}{2}$, nor is it symmetric with respect to the pole. This curve is an example of a **limaçon with a depression.**

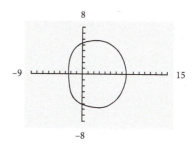

Figure 17

Check It Out 7   Rework Example 7 for the equations $r = 1 - 4 \cos \theta$ and $r = 5 + 3 \sin \theta$.

Using Technology

For more details on window settings and changing modes, consult the keystroke appendix.

Keystroke Appendix:
Section 7 and 15

Summary of Polar Graphs

We now summarize the various types of polar equations and graphs discussed in this section.

Lines in Polar Form		

Line passing through the pole $\theta = \alpha$	Vertical line $r \cos \theta = c$, or $r = c \sec \theta$	Horizontal line $r \sin \theta = d$, or $r = d \csc \theta$
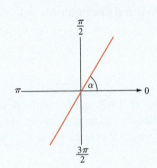	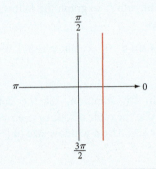	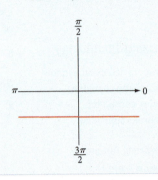

Circle in Polar Form		

Circle centered at the pole (origin) with radius a	Circle with radius $\frac{a}{2}$ and center on the line $\theta = 0$	Circle with radius $\frac{a}{2}$ and center on the line $\theta = \frac{\pi}{2}$
$r = a, a > 0$	$r = \pm a \cos \theta, a > 0$	$r = \pm a \sin \theta, a > 0$
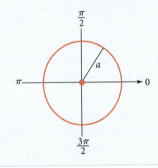	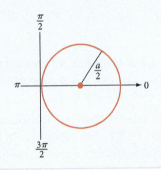	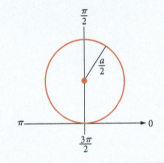

Roses and a Lemniscate

Three-petaled rose

$$r = a \cos(3\theta), a > 0$$
$$r = a \sin(3\theta), a > 0$$

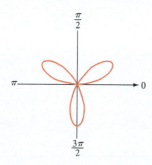

Graph shown is that of $r = \sin(3\theta)$.

Four-petaled rose

$$r = a \cos(2\theta), a > 0$$
$$r = a \sin(2\theta), a > 0$$

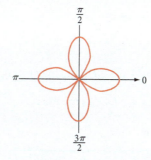

Graph shown is that of $r = \cos(2\theta)$.

Lemniscate

$$r^2 = a^2 \cos(2\theta), a > 0$$
$$r^2 = a^2 \sin(2\theta), a > 0$$

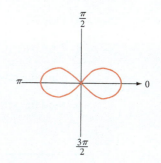

Graph shown is that of $r^2 = \cos(2\theta)$.

Cardiod and Limaçons

Cardiod

$$r = a \pm b \sin(\theta), 0 < a = b$$
$$r = a \pm b \cos(\theta), 0 < a = b$$

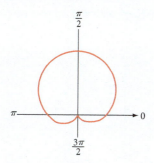

Graph shown is that of $r = 1 + \sin\theta$.

Limaçon with an inner loop

$$r = a \pm b \sin(\theta), 0 < a < b$$
$$r = a \pm b \cos(\theta), 0 < a < b$$

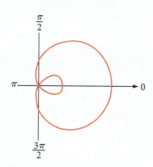

Graph shown is that of $r = 1 + 2\cos\theta$.

Limaçon with no inner loop

$$r = a \pm b \sin(\theta), 0 < b < a$$
$$r = a \pm b \cos(\theta), 0 < b < a$$

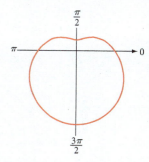

Graph shown is that of $r = 3 - 2\sin\theta$.

Exercises 7.4

1. For what value(s) of θ in $[0, 2\pi]$ does $\sin\theta$ reach a maximum value?

2. For what value(s) of θ in $[0, 2\pi]$ does $\cos\theta$ reach a minimum value?

3. Find θ in $[0, \pi]$ such that $\cos 2\theta = -1$.

4. Find θ in $[0, \pi]$ such that $\sin 2\theta = 1$.

5. Find the zero(s) of $f(\theta) = \cos 2\theta$ in the interval $[0, \pi]$.

6. Find the zero(s) of $f(\theta) = \sin 2\theta$ in the interval $[0, \pi]$.

 Skills

In Exercises 7–22, sketch the graphs of the polar equations.

7. $r = 4$ 8. $r = 2$ 9. $\theta = \frac{\pi}{4}$ 10. $\theta = \frac{2\pi}{3}$ 11. $r = 2\sec\theta$ 12. $r = \csc\theta$

13. $r = -4\csc\theta$ 14. $r = -3\sec\theta$ 15. $r = 2\cos\theta$ 16. $r = -4\sin\theta$ 17. $r = -4\cos\theta$

18. $r = 2\sin\theta$ 19. $r = -\sin\theta$ 20. $r = -2\cos\theta$ 21. $r = \frac{3}{2}\cos\theta$ 22. $r = \sqrt{2}\sin\theta$

In Exercises 23–26, each of the graphs corresponds to an equation of the form $r = a\cos\theta$ or $r = a\sin\theta$, where a is a real number. Determine the equation, including the value of a.

23.
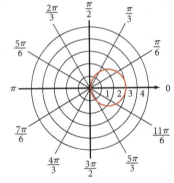

24.
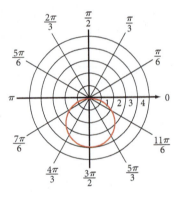

25.
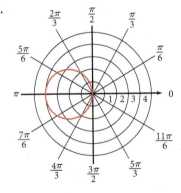

26.
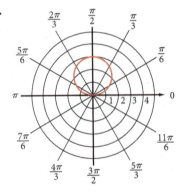

602

In Exercises 27–30, each of the graphs corresponds to an equation of the form r = a sec θ or r = a csc θ, where a is a real number. Determine the equation, including the value of a.

27.

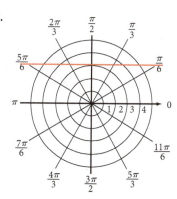

28.

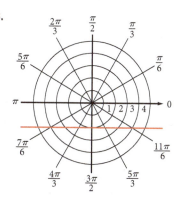

29.

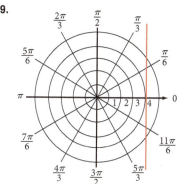

30.

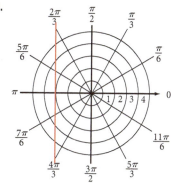

In Exercises 31–46, sketch the graphs of the polar equations.

31. $r = 2 \cos 2\theta$ **32.** $r = 5 \sin 2\theta$ **33.** $r = 1 + \cos \theta$ **34.** $r = 1 + \sin \theta$

35. $r = 2 - 2 \sin \theta$ **36.** $r = 2 + 2 \cos \theta$ **37.** $r = 3 \sin (3\theta)$ **38.** $r = 4 \cos (3\theta)$

39. $r^2 = 4 \cos (2\theta)$ **40.** $r^2 = 9 \sin(2\theta)$ **41.** $r = 4 - 3 \cos \theta$ **42.** $r = 5 + 4 \sin \theta$

43. $r = 2 + \sin \theta$ **44.** $r = 2 - \cos \theta$ **45.** $r = 2 + 3 \cos \theta$ **46.** $r = 2 - 3 \sin \theta$

In Exercises 47–54, use a graphing utility to graph each polar equation.

47. $r = 5 - 4 \sin \theta$ **48.** $r = 3 + 4 \cos \theta$ **49.** $r = \cos\left(\theta - \dfrac{\pi}{4}\right)$ **50.** $r = \sin\left(\theta + \dfrac{\pi}{4}\right)$

51. $r = 3 \sin 4\theta$ **52.** $r = 2 \cos 3\theta$ **53.** $r = 2\theta$ $0 \le \theta \le 4\pi$ **54.** $r = \dfrac{1}{\theta}$ $0 < \theta \le 4\pi$

In Exercises 55–58, use a graphing utility starting to find the smallest value of θ max, with θ min = 0, such that the entire curve is graphed exactly once without retracing.

55. $r = -2 \cos \theta$ **56.** $r = 1 - 2 \cos \theta$ **57.** $r = 3 \cos(2\theta)$

58. $r^2 = \cos(2\theta)$ (*Hint:* Graph as two functions, r_1 and r_2.)

Concepts

59. Write the polar equation $r = 2 - 2\cos\left(\theta + \frac{\pi}{2}\right)$ in terms of just the sine function.

60. Write the polar equation $r = 4\sin\left(\theta - \frac{3\pi}{2}\right)$ in terms of just the cosine function.

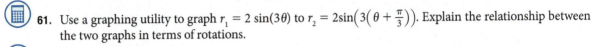

61. Use a graphing utility to graph $r_1 = 2\sin(3\theta)$ to $r_2 = 2\sin\left(3\left(\theta + \frac{\pi}{3}\right)\right)$. Explain the relationship between the two graphs in terms of rotations.

62. Use a graphing utility to graph $r_1 = 1 + \cos\theta$ to $r_2 = 1 + \cos\left(\theta - \frac{\pi}{2}\right)$. Explain the relationship between the two graphs in terms of rotations.

Vectors

7.5

Objectives

- Know vector notation in component form.
- Add and subtract vectors in component form.
- Add and subtract vectors graphically.
- Use vectors to solve applied problems.

When an airplane flies, its speed and direction characterize the *velocity* of the airplane and cannot be represented by a single real number. In physics and engineering, quantities such as velocity and force are represented by a line segment that has a specified length and points in a specified direction. These directed line segments are called **vectors**. This section will give a brief introduction to vectors in two dimensions and their properties.

A vector is depicted graphically by an arrow, with an **initial point**, the "tail," and a **terminal point**, the "head," as shown in Figure 1. The length of the arrow represents the **magnitude** of the vector. Two vectors can have the same magnitude and the same direction, as shown in Figure 2.

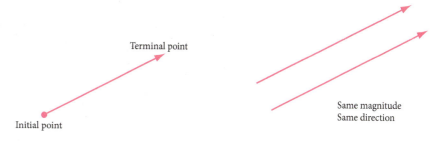

Figure 1

Figure 2

To give a unique representation to a vector, its initial position can be taken to be the origin. This is called the **standard position** of a vector. A vector in standard position is completely and uniquely specified by the coordinates of its terminal point.

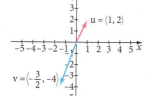

Figure 3

Component Form of a Vector

A vector **v** in two dimensions and in standard position is represented as $\langle v_x, v_y \rangle$, where v_x and v_y are the coordinates of the terminal point. They are referred to as the **x- and y-components**, respectively, of **v**. The **zero vector** is given by $\mathbf{0} = \langle 0, 0 \rangle$.

Pictured in Figure 3 are two vectors, **u** and **v**, in component form.

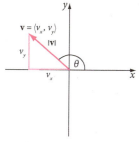

Figure 4

Note Observe that the magnitude of a vector is represented by a single number.

Magnitude and Direction of a Vector

The **magnitude** of a vector $\mathbf{v} = \langle v_x, v_y \rangle$ in standard position is given by

$$\|\mathbf{v}\| = \sqrt{v_x^2 + v_y^2}$$

For **v** nonzero, the **direction angle** of **v** is the angle θ that **v** makes with the positive x-axis. It is given by

$$\cos \theta = \frac{v_x}{\|\mathbf{v}\|} \quad \text{and} \quad \sin \theta = \frac{v_y}{\|\mathbf{v}\|}, \text{ or} \quad \tan \theta = \frac{v_y}{v_x}$$

with θ in the interval $[0°, 360°)$. See Figure 4. The direction of the zero vector is undefined.

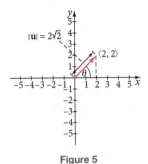

$\|\mathbf{u}\| = 2\sqrt{2}$

Figure 5

VIDEO EXAMPLES

SECTION 7.5

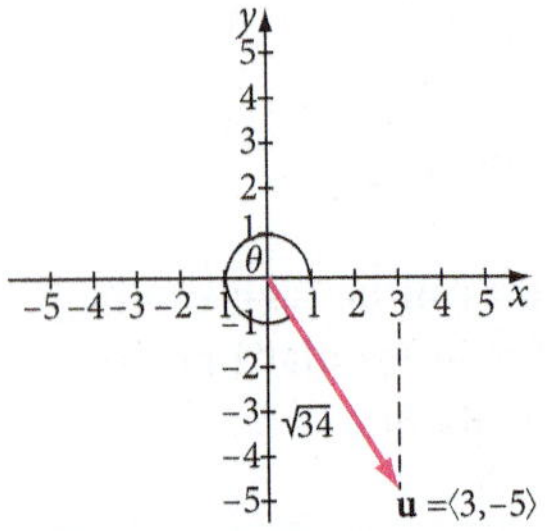

Figure 6

| Example 1 | **Finding Magnitude and Direction of a Vector** |

Find the magnitude and direction of each of the following vectors.

a. $\mathbf{v} = \langle 2, 2 \rangle$ **b.** $\mathbf{u} = \langle 3, -5 \rangle$

Solution

a. Graph the vector as in Figure 5. By the definition of magnitude, we have

$$\|\mathbf{v}\| = \sqrt{(2)^2 + (2)^2} = \sqrt{4 + 4} = \sqrt{8} = 2\sqrt{2}$$

The direction of $\mathbf{v}$ is defined by the angle θ in standard position such that

$$\tan \theta = \frac{v_y}{v_x} = \frac{2}{2} = 1$$

From our knowledge of trigonometric functions of special angles, the direction angle is $\theta = 45°$, because the vector lies in the first quadrant.

b. Graph the vector as in Figure 6. Using the definition of magnitude, we get

$$\|\mathbf{u}\| = \sqrt{3^2 + (-5)^2} = \sqrt{34}$$

The direction of $\mathbf{u}$ is defined by the angle θ such that

$$\tan \theta = \frac{u_y}{u_x} = \frac{-5}{3}$$

Using the $\tan^{-1}$ key on your calculator, we have

$$\tan^{-1}\left(\frac{-5}{3}\right) \approx -59.04°$$

Because the terminal side of θ is in the fourth quadrant, and θ must be in the interval $[0°, 360°)$.

$$\theta = 360° - 59.04° = 300.96°$$

Check It Out 1   Find the magnitude and direction of $\langle -4, 1 \rangle$.

Sometimes, we are given the magnitude of a vector and its direction, and we wish to find its components. By definition of the direction θ of a vector $\mathbf{u} = \langle u_x, u_y \rangle$ with nonzero magnitude,

$$\cos \theta = \frac{u_x}{\|\mathbf{u}\|} \Rightarrow u_x = \|\mathbf{u}\| \cos \theta$$

$$\sin \theta = \frac{u_y}{\|\mathbf{u}\|} \Rightarrow u_y = \|\mathbf{u}\| \sin \theta$$

Components of a Vector

Let $\mathbf{u} = \langle u_x, u_y \rangle$ be a vector with nonzero magnitude. Then $u_x = \|\mathbf{u}\| \cos \theta$ and $u_y = \|\mathbf{u}\| \sin \theta$.

Example 2   Finding Components of a Vector

Find the components of the vector **u** with magnitude 6 and direction angle 117°.

Solution Using the direction angle θ and the magnitude of **u**, we have

$$u_x = \|\mathbf{u}\| \cos \theta = 6 \cos 117° \approx -2.72$$

$$u_y = \|\mathbf{u}\| \sin \theta = 6 \sin 117° \approx 5.35$$

 Check It Out 2   Find the components of the vector with magnitude 22 and direction 37° south of west.

Vector Operations

We now discuss addition and subtraction of vectors and scalar multiplication of vectors.

Addition and Subtraction of Vectors

To add or subtract vectors, add or subtract their components.

$$\mathbf{u} + \mathbf{v} = \langle u_x, u_y \rangle + \langle v_x, v_y \rangle = \langle u_x + v_x, u_y + v_y \rangle$$

$$\mathbf{u} - \mathbf{v} = \langle u_x, u_y \rangle - \langle v_x, v_y \rangle = \langle u_x - v_x, u_y - v_y \rangle$$

Example 3   Adding and Subtracting Vectors

Let $\mathbf{u} = \langle -1, 3 \rangle$ and $\mathbf{v} = \langle 4, -1 \rangle$. Find the following and illustrate the result graphically.

a. u + v   **b. v − u**

Solution

a. Adding the components of **v** to the respective components of **u** gives

$$\mathbf{u} + \mathbf{v} = \langle -1, 3 \rangle + \langle 4, -1 \rangle = \langle -1 + 4, 3 + (-1) \rangle = \langle 3, 2 \rangle$$

See Figure 7.

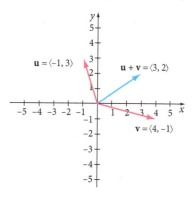

Figure 7

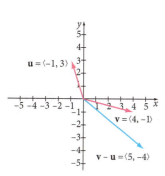

Figure 8

b. Subtracting the components of **u** from the respective components of **v** gives

$$\mathbf{v} - \mathbf{u} = \langle 4, -1 \rangle - \langle -1, 3 \rangle = \langle 4 - (-1), -1 - 3 \rangle = \langle 5, -4 \rangle$$

See Figure 8.

 Check It Out 3   Find $\langle 5, -4 \rangle - \langle -3, 1 \rangle$.

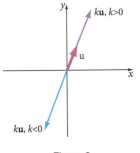

Figure 9

Scalar multiplication

For a real number k and a vector $\mathbf{u}$, the **scalar multiple** $k\mathbf{u}$ is the vector given by

$$k\mathbf{u} = k\langle u_x, u_y \rangle = \langle ku_x, ku_y \rangle$$

Observations

- If $k > 0$, then the vector $k\mathbf{u}$ is in the same direction as $\mathbf{u}$ and has magnitude $k\,\|\mathbf{u}\|$.

- If $k < 0$, then the vector $k\mathbf{u}$ is in the opposite direction as $\mathbf{u}$ and has magnitude $|k|\,\|\mathbf{u}\|$. See Figure 9.

Example 4   **Scalar Multiplication**

Let $\mathbf{u} = \langle 5, -3 \rangle$ and let $\mathbf{v} = \langle -2, 1 \rangle$. Find the following.

a. $3\mathbf{u}$ **b.** $-2\mathbf{v} + \mathbf{u}$

Solution

a. Using the definition of scalar multiplication, we have $3\mathbf{u} = 3\langle 5, -3 \rangle = \langle 15, -9 \rangle$

b. Using the definition of scalar multiplication to get $-2\mathbf{v}$, and then adding $\mathbf{u}$ to the result, gives

$$-2\mathbf{v} + \mathbf{u} = -2\langle -2, 1 \rangle + \langle 5, -3 \rangle = \langle 4, -2 \rangle + \langle 5, -3 \rangle = \langle 4 + 5, -2 - 3 \rangle = \langle 9, -5 \rangle$$

 **Check It Out 4**   Find $4\langle 1, 5 \rangle - \langle 3, -2 \rangle$. ◼

In the special case $k = -1$, we get the vector $-\mathbf{u}$. Note that $\mathbf{u} + (-\mathbf{u}) = \langle 0, 0 \rangle$. The **zero vector**, $\langle 0, 0 \rangle$, has magnitude zero and is denoted by $\mathbf{0}$. Thus $-\mathbf{u} = \langle -u_x, -u_y \rangle$, where u_x and u_y are the components of $\mathbf{u}$.

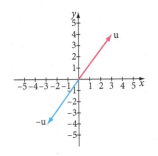

Figure 10

Example 5   **Finding $-\mathbf{u}$**

Let $\mathbf{u} = \langle 3, 4 \rangle$.

a. Find $-\mathbf{u}$ in component form.   **b.** Graph $\mathbf{u}$ and $-\mathbf{u}$.

Solution

a. By the definition of $-\mathbf{u}$,

$$-\mathbf{u} = -\langle 3, 4 \rangle = \langle -3, -4 \rangle$$

b. Graph the vectors $\mathbf{u}$ and $-\mathbf{u}$, as shown in Figure 10. Observe that $-\mathbf{u}$ is directed 180° away from $\mathbf{u}$.

 **Check It Out 5**   Rework Example 5 for the case where $\mathbf{u} = \langle -5, 2 \rangle$. ◼

Vector operations share many of the same properties as operations on real numbers.

Properties of Vector Addition and Scalar Multiplication

Let $\mathbf{u}$, $\mathbf{v}$, and $\mathbf{w}$ be vectors and let k and m be scalars. Then the following hold:

1. $\mathbf{u} + \mathbf{v} = \mathbf{v} + \mathbf{u}$
2. $\mathbf{u} + (\mathbf{v} + \mathbf{w}) = (\mathbf{u} + \mathbf{v}) + \mathbf{w}$

3. $\mathbf{u} + \mathbf{0} = \mathbf{0} + \mathbf{u}$

4. $\mathbf{u} + (-\mathbf{u}) = \mathbf{0}$

5. $k(\mathbf{u} + \mathbf{v}) = k\mathbf{u} + k\mathbf{v}$

6. $(k + m)\mathbf{u} = k\mathbf{u} + m\mathbf{v}$

7. $k(m\mathbf{v}) = (km)(\mathbf{v})$

Unit Vectors

A **unit vector** is a vector of magnitude 1. For instance, $\left\langle \frac{3}{5}, -\frac{4}{5} \right\rangle$ is a unit vector because

$$\left\| \left\langle \frac{3}{5}, -\frac{4}{5} \right\rangle \right\| = \sqrt{\left(\frac{3}{5}\right)^2 + \left(-\frac{4}{5}\right)^2} = \sqrt{\frac{25}{25}} = 1$$

In many applications of vectors, we are interested in a unit vector in the direction of a nonzero vector **v**.

Unit Vector in the Direction of a Given Vector

Let $\mathbf{v} = \langle v_x, v_y \rangle$ be a nonzero vector. Then the unit vector in the direction of **v** is given by

$$\frac{\mathbf{v}}{\|\mathbf{v}\|}, \text{ where } \|\mathbf{v}\| = \sqrt{v_x^2 + v_y^2}$$

Example 6   **Calculating a Unit Vector**

Find the unit vector **u** in the same direction as $\mathbf{v} = \langle -5, 12 \rangle$. Verify that **u** has magnitude 1.

Solution First calculate $\|\mathbf{v}\|$:

$$\|\mathbf{v}\| = \|\langle -5, 12 \rangle\| = \sqrt{(-5)^2 + 12^2} = \sqrt{169} = 13$$

Then the unit vector **u** in the direction of **v** is given by:

$$\mathbf{u} = \frac{\mathbf{v}}{\|\mathbf{v}\|} = \frac{\langle -5, 12 \rangle}{13} = \left\langle -\frac{5}{13}, \frac{12}{13} \right\rangle$$

We now show that the magnitude of **u** is 1 .

$$\|\mathbf{u}\| = \sqrt{\left(-\frac{5}{13}\right)^2 + \left(\frac{12}{13}\right)^2} = \sqrt{\frac{169}{169}} = 1$$

 Check It Out 6   Find the unit vector **u** in the same direction as $\mathbf{v} = \langle 6, 8 \rangle$.

Other Representations of a Vector

A vector can be represented as the sum of vectors that lie along the axes. By the rules of vector addition,

$$\mathbf{u} = \langle u_x, u_y \rangle = \langle u_x, 0 \rangle + \langle 0, u_y \rangle$$

Observe that the vectors $\langle u_x, 0 \rangle$ and $\langle 0, u_y \rangle$ lie along the x- and y-axes, respectively.

A vector can also be represented in terms of unit vectors that lie along the axes. The **standard unit vectors** $\langle 1, 0 \rangle$ and $\langle 0, 1 \rangle$ lie along the positive x-axis and positive y-axis, respectively. By the rules of vector addition and scalar multiplication,

$$\mathbf{u} = \langle u_x, u_y \rangle = \langle u_x, 0 \rangle + \langle 0, u_y \rangle = u_x \langle 1, 0 \rangle + u_y \langle 0, 1 \rangle$$

The unit vectors $\langle 1, 0 \rangle$ and $\langle 0, 1 \rangle$ are also denoted by **i** and **j**, respectively. Thus, **u** can also be written as

$$\mathbf{u} = u_x \mathbf{i} + u_y \mathbf{j}$$

The unit-vector representation of a vector with a negative x-component and a positive y-component is illustrated in Figure 11.

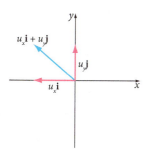

Figure 11

Example 7   **Representing a Vector in Terms of i and j**

Write $\mathbf{u} = \langle -3, 2 \rangle$ in terms of the unit vectors $\mathbf{i}$ and $\mathbf{j}$.

Solution   Because $\mathbf{u} = \langle -3, 2 \rangle$, we have $u_x = -3$ and $u_y = 2$. Thus

$$\mathbf{u} = \langle -3, 2 \rangle = \langle -3, 0 \rangle + \langle 0, 2 \rangle = -3\langle 1, 0 \rangle + 2\langle 0, 1 \rangle = -3\mathbf{i} + 2\mathbf{j}$$

 Check It Out 7   Rework Example 7 for the case where $\mathbf{u} = \langle 6, -5 \rangle$.

Applications of Vectors

Vectors are useful in a number of areas because they can be used for any quantity with direction and magnitude. In navigation, the **velocity** of an airplane or boat gives both speed and direction. Thus, velocity can be represented as a vector, with the speed as its magnitude.

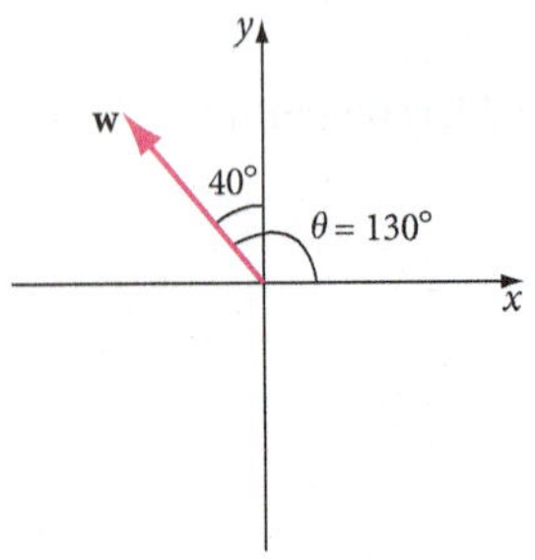

Example 8   **Wind Velocity**

Using a specially designed golf flag, a golfer finds that the wind is blowing through the course at a speed of 12 miles per hour in the direction N40°W. Express the wind velocity in vector form, rounded to two decimal places.   (*Source:* directhitgolfflags.com)

Solution   Let $\mathbf{w}$ denote the velocity of the wind. Then $\|\mathbf{w}\| = 12$ is the speed of the wind. The direction N40°W, 40° west of north, corresponds to the angle $\theta = 90° + 40° = 130°$. (This notation was introduced in Section 7.1.) See Figure 12.

Compute the x and y velocity components of $\mathbf{w}$ as follows.

$$w_x = \|\mathbf{w}\| \cos \theta = 12 \cos 130° \approx -7.71$$

$$w_y = \|\mathbf{w}\| \sin \theta = 12 \sin 130° \approx 9.19$$

Thus, $\mathbf{w} = \langle -7.71, 9.19 \rangle$.

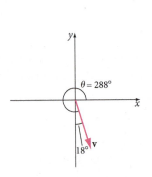

Figure 12

 Check It Out 8   If wind is blowing at a speed of 6 miles per hour in the direction S40°W, express the wind velocity in vector form, rounded to two decimal places.

Example 9   **Combining Boat and Wind Velocities**

A boat travels in the direction S18°E at a speed of 15 miles per hour. It encounters a wind blowing north to south at a speed of 10 miles per hour. Find the resulting speed and direction of the boat.

Solution   The resulting velocity of the boat is obtained by adding the components of the boat velocity to the wind velocity. The magnitude of the resulting velocity vector is the resulting speed.

Step 1 Find the components of boat velocity.   Let $\mathbf{v}$ denote the velocity of the boat. Thus, $\|\mathbf{v}\| = 15$, the speed of the boat. The direction S18°E, 18° east of south, corresponds to the angle $\theta = 270° + 18° = 288°$. See Figure 13. We then have

$$v_x = \|\mathbf{v}\| \cos \theta = 15 \cos 288° \approx 4.64$$

$$v_y = \|\mathbf{v}\| \sin \theta = 15 \sin 288° \approx -14.27$$

So $\mathbf{v} = \langle 4.64, -14.27 \rangle$.

Figure 13

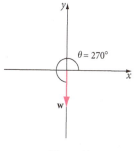

Figure 14

Step 2 **Find the components of wind velocity.** Let **w** denote the velocity of the wind. Because the wind is blowing north to south at 10 miles per hour, $\|\mathbf{w}\| = 10$, and the direction is $\theta = 270°$. See Figure 14. We then have

$$w_x = \|\mathbf{w}\| \cos \theta = 10 \cos 270° = 0$$

$$w_y = \|\mathbf{w}\| \sin \theta = 10 \sin 270° = -10$$

So $\mathbf{w} = \langle 0, -10 \rangle$.

Step 3 **Find the resulting velocity.** To find the resulting velocity, add **v** and **w**.

$$\mathbf{u} = \mathbf{v} + \mathbf{w} = \langle 4.64, -14.27 \rangle = \langle 0, -10 \rangle = \langle 4.64, -24.27 \rangle$$

Step 4 **Calculate the resulting speed and direction.**

$$\text{Speed: } \|\mathbf{u}\| = \sqrt{(4.64)^2 + (-24.27)^2} \approx 24.71$$

The direction θ is defined by

$$\tan \theta = \frac{u_y}{u_x} = \frac{-24.27}{4.64} \approx -5.23$$

Now, $\tan^{-1}(-5.23) = -79.18°$. Because the terminal side of θ is in the fourth quadrant, and θ must be in the interval $[0°, 360°)$, $\theta = 360° - 79.18° = 280.82°$. So the boat travels at a speed of 24.71 miles per hour with $\theta = 280.82°$, or S10.82°E.

✅ *Check It Out 9* Rework Example 9 if the wind is blowing west to east at 10 miles per hour. All other information stays the same.

Exercises 7.5

 Skills

In Exercises 1–6, graph each of the given vectors in the standard position.

1. $\langle 1, 0 \rangle$ **2.** $\langle 4, -1 \rangle$ **3.** $\langle -5, -3 \rangle$ **4.** $\left\langle 0, \dfrac{1}{2} \right\rangle$ **5.** $\left\langle \dfrac{4}{3}, -6 \right\rangle$ **6.** $\langle -2, -5.5 \rangle$

In Exercises 7–12, write each of the given vectors in terms of the unit vectors **i** and **j**

7. $\mathbf{u} = \langle -4, 6 \rangle$ **8.** $\mathbf{v} = \langle 5, -3 \rangle$ **9.** $\mathbf{w} = \langle -2, -1.5 \rangle$ **10.** $\mathbf{v} = \langle -3.5, 4 \rangle$

11. $\mathbf{u} = \left\langle \dfrac{1}{3}, \dfrac{3}{4} \right\rangle$ **12.** $\mathbf{w} = \left\langle -\dfrac{2}{5}, \dfrac{1}{6} \right\rangle$

In Exercises 13–24, for each of the following, find $\mathbf{u} - \mathbf{v}$, $\mathbf{u} + 2\mathbf{v}$, and $-3\mathbf{u} + \mathbf{v}$.

13. $\mathbf{u} = \langle 3, 0 \rangle, \mathbf{v} = \langle 5, 1 \rangle$ **14.** $\mathbf{u} = \langle 6, -2 \rangle, \mathbf{v} = \langle 3, -1 \rangle$

15. $\mathbf{u} = \langle -4, 5 \rangle, \mathbf{v} = \langle 3, -7 \rangle$ **16.** $\mathbf{u} = \langle -2, 6 \rangle, \mathbf{v} = \langle 7, -3 \rangle$

17. $\mathbf{u} = \langle 1.5, 2.5 \rangle, \mathbf{v} = \langle 0, 1 \rangle$ **18.** $\mathbf{u} = \langle 4, 0 \rangle, \mathbf{v} = \langle -1.5, 2.5 \rangle$

19. $\mathbf{u} = \left\langle \dfrac{1}{3}, \dfrac{2}{5} \right\rangle, \mathbf{v} = \langle 1, 2 \rangle$ **20.** $\mathbf{u} = \left\langle \dfrac{1}{4}, \dfrac{1}{2} \right\rangle, \mathbf{v} = \left\langle -\dfrac{1}{2}, \dfrac{3}{4} \right\rangle$

21. $\mathbf{u} = -2\mathbf{i} + 3\mathbf{j}, \mathbf{v} = 4\mathbf{i} - \mathbf{j}$ **22.** $\mathbf{u} = 6\mathbf{i} - 2\mathbf{j}, \mathbf{v} = -5\mathbf{i} + 3\mathbf{j}$

23. $\mathbf{u} = -1.1\mathbf{i} + 4\mathbf{j}, \mathbf{v} = 4\mathbf{i} + 2.4\mathbf{j}$ **24.** $\mathbf{u} = 2.6\mathbf{i} + 5\mathbf{j}, \mathbf{v} = -2\mathbf{i} + 3.7\mathbf{j}$

In Exercises 25–34, find the magnitude and the direction angle θ of each of the given vectors, with $\theta \in [0°, 360°)$.

25. $\mathbf{u} = \langle -1, 2 \rangle$ **26.** $\mathbf{w} = \langle 3, 5 \rangle$ **27.** $\mathbf{v} = \langle 1, 1.5 \rangle$ **28.** $\mathbf{u} = \langle -1.5, 3 \rangle$

29. $\mathbf{v} = \left\langle \dfrac{4}{3}, \dfrac{2}{5} \right\rangle$ **30.** $\mathbf{w} = \left\langle -\dfrac{1}{2}, -\dfrac{1}{4} \right\rangle$ **31.** $\mathbf{v} = 4\mathbf{i} - 2\mathbf{j}$ **32.** $\mathbf{v} = -3\mathbf{i} + 4\mathbf{j}$

33. $\mathbf{v} = 1\mathbf{i} + 2.5\mathbf{j}$ **34.** $\mathbf{v} = -3.2\mathbf{i} + 2\mathbf{j}$

In Exercises 35–40, find a unit vector in the same direction as the given vector.

35. $\mathbf{u} = \langle 3, 4 \rangle$ **36.** $\mathbf{v} = \langle -12, 5 \rangle$ **37.** $\mathbf{w} = \langle 1, 1 \rangle$ **38.** $\mathbf{u} = \langle 3, 2 \rangle$

39. $\mathbf{v} = -2\mathbf{i} + 1\mathbf{j}$ **40.** $\mathbf{u} = 4\mathbf{i} - 3\mathbf{j}$

In Exercises 41–48, express each vector in component form. Round your answers to four decimal places.

41. Magnitude 19; direction $34°$ **42.** Length 7; direction $276°$

43. Magnitude 10; direction $190°$ **44.** Magnitude 8; direction $145°$

45. Magnitude 4.6; points due west   **46.** Length 3.1; direction $16°$ south of east

47. Length 22; points northwest   **48.** Magnitude 59; direction $108°$

Applications

In Exercises 49–57, round your answers to two decimal places.

49. Weather The world's largest weathervane is located in Montague, Michigan. On a July day in 2007, it showed that the wind had a speed of 15 miles per hour in the direction S30°E. Express the wind velocity in component form. *(Source: www.wunderground.com)*

50. Golf A golf ball is hit from a tee with a launch angle of 13.2° and a speed of 140 miles per hour. Express the velocity of the ball in component form. *(Source: www.golf.com)*

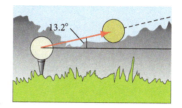

51. Beaufort Scale The Beaufort scale was developed in 1805 by Sir Francis Beaufort of England. It gives a measure for wind intensity based on observed sea and land conditions. For example, a wind speed of 20 knots is classified as a "fresh breeze," and smaller trees sway at this speed. Note that wind speed can also be measured in *knots*, where 1 knot equals 1.15 miles per hour. *(Source: www.noaa.com)*

   a. If the fresh breeze is in the direction S60°W, express the velocity of the breeze in component form. Use knots for the unit of speed.

   b. Express the velocity of the fresh breeze in component form using miles per hour as the unit for speed.

52. Biking Carl rides his bike due east for half an hour at a speed of 12 miles per hour (to be consistent). Then he rides due north for 45 minutes at a speed of 10 mph.

   a. At the end of the trip, how far is Carl from his starting point?

   b. Suppose Carl rides in a single straight-line path, and that his starting point and ending point are the same as before. In what direction does he ride his bike?

53. Distance Wanda goes for a hike. She first walks 2.4 miles in the direction S17°E, and then goes another 1.8 miles in the direction S38°E.

   a. By what east-west distance did Wanda's position change between the time she began the hike and the time she completed it?

   b. By what north-south distance did Wanda's position change?

   c. At the end of the hike, how far is Wanda from her starting point?

   d. Suppose Wanda walks in a single straight-line path and that her starting point and ending point are the same as before. In what direction does she walk?

54. Physics A ball is thrown upward with a velocity of 20 meters per second at an angle of 42° with respect to the horizontal.

a. At the time the ball is thrown, how fast is it moving in the horizontal direction?

b. At the time the ball is thrown, how fast is it moving in the vertical direction?

55. Sailing A sailboat travels on White Lake, Michigan, at a speed of 5 miles per hour in the direction N45°E. It encounters a moderate breeze blowing from south to north at a speed of 12 miles per hour. Find the resulting speed and direction of the sailboat.

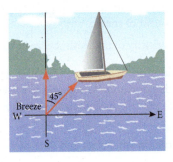

56. Navigation The net velocity of a ship is the vector sum of the velocity originating from the ship's engine and the velocity of the wind. The engine propels the ship at a velocity of 20 miles per hour in the direction S35°E.

a. What are the components of the ship's velocity originating from the engine?

b. If the wind is blowing from north to south at 12 miles per hour, find the magnitude and direction of the net velocity of the ship.

c. Rework part (b) for the case where the wind is blowing from north to south at 15 miles per hour.

Concepts

57. Show that if $\|\mathbf{v}\| = 0$, then $\mathbf{v} = \langle 0, 0 \rangle$.

58. Show that if $\mathbf{u}$ is a nonzero vector, then the vector $\frac{\mathbf{u}}{\|\mathbf{u}\|}$ has magnitude 1.

59. If $\mathbf{u}$ is a nonzero vector, for what values of k does the equation $\|k\mathbf{u}\| = k\|\mathbf{u}\|$ hold? Explain.

Dot Product of Vectors

7.6

Objectives

- Find the dot product of two vectors.
- Determine the angle between two vectors.
- Determine whether two vectors are orthogonal.
- Calculate the projection of a vector onto another vector.
- Calculate the parallel and perpendicular components of a vector.
- Solve applied problems using dot products.

We have already defined multiplication of a vector by a scalar, but we have not yet defined multiplication of two vectors. One type of product of two vectors that is quite useful is called the **dot product.**

Dot Product

The **dot product** of vectors $\mathbf{u} = \langle u_x, u_y \rangle$ and $\mathbf{v} = \langle v_x, v_y \rangle$ is defined as

$$\mathbf{u} \cdot \mathbf{v} = u_x v_x + u_y v_y$$

The dot product of two vectors is a scalar, and is sometimes referred to as the *scalar product* or the *inner product.*

Example 1 **Calculating the Dot Product**

Calculate $\langle -4, 3 \rangle \cdot \langle 2, -5 \rangle$.

Solution Using the definition of the dot product, we have

$$\langle -4, 3 \rangle \cdot \langle 2, -5 \rangle = (-4)(2) + (3)(-5) = -8 - 15 = -23$$

 Check It Out 1 Calculate $\langle 6, -2 \rangle \cdot \langle 3, 4 \rangle$.

The following are properties of the dot product.

Properties of the Dot Product

Let $\mathbf{u}$, $\mathbf{v}$, and $\mathbf{w}$ be vectors, and let k be a real number. Then the following hold:

1. $\mathbf{u} \cdot \mathbf{v} = \mathbf{v} \cdot \mathbf{u}$
2. $\mathbf{u} \cdot (\mathbf{v} + \mathbf{w}) = \mathbf{u} \cdot \mathbf{v} + \mathbf{u} \cdot \mathbf{w}$
3. $\mathbf{0} \cdot \mathbf{u} = 0$
4. $\mathbf{v} \cdot \mathbf{v} = \|\mathbf{v}\|^2$
5. $(k\mathbf{u}) \cdot (\mathbf{v}) = k(\mathbf{u} \cdot \mathbf{v}) = \mathbf{u} \cdot (k\mathbf{v})$

We will prove the first property. The rest are left as exercises at the end of this section. Let $\mathbf{u} = \langle u_x, u_y \rangle$ and $\mathbf{v} = \langle v_x, v_y \rangle$. Then

$$\mathbf{u} \cdot \mathbf{v} = u_x v_x + u_y v_y$$

Now

$$\mathbf{v} \cdot \mathbf{u} = v_x u_x + v_y u_y = u_x v_x + u_y v_y$$

because multiplication is commutative. Thus

$$\mathbf{u} \cdot \mathbf{v} = \mathbf{v} \cdot \mathbf{u}$$

Angle Between Vectors

Another formula for the dot product of nonzero vectors is useful in determining the angle between two vectors and in solving applied problems.

Alternate Formula for Dot Products

$$\mathbf{u} \cdot \mathbf{v} = (\|\mathbf{u}\|)(\|\mathbf{v}\|)\cos\theta,$$

where θ is the smallest nonnegative angle between $\mathbf{u}$ and $\mathbf{v}$, as shown in Figure 1.

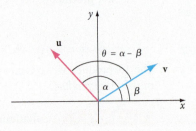

Figure 1

This formula can be derived from the original formula for the dot product:

$$\mathbf{u} \cdot \mathbf{v} = u_x v_x + u_y v_y$$

Let α and β be the directions of $\mathbf{u}$ and $\mathbf{v}$, respectively. Expressing the components of $\mathbf{u}$ and $\mathbf{v}$ in terms of their magnitudes and directions, we have

$$u_x = \|\mathbf{u}\| \cos \alpha \qquad v_x = \|\mathbf{v}\| \cos \beta$$

$$u_y = \|\mathbf{u}\| \sin \alpha \qquad v_y = \|\mathbf{v}\| \sin \beta$$

Substituting these expressions into the formula for the dot product, we obtain

$$\mathbf{u} \cdot \mathbf{v} = (\|\mathbf{u}\| \cos \alpha)(\|\mathbf{v}\| \cos \beta) + (\|\mathbf{u}\| \sin \alpha)(\|\mathbf{v}\| \sin \beta)$$

$$= (\|\mathbf{u}\|)(\|\mathbf{v}\|)(\cos \alpha \cos \beta + \sin \alpha \sin \beta) \qquad \text{Factor out } (\|\mathbf{u}\|)(\|\mathbf{v}\|)$$

$$= (\|\mathbf{u}\|)(\|\mathbf{v}\|)\cos(\alpha - \beta) \qquad \text{Apply difference identity for cosine}$$

$$= (\|\mathbf{u}\|)(\|\mathbf{v}\|) \cos \theta. \qquad \text{Substitute } \theta = \alpha - \beta$$

Example 2 **Finding the Angle Between Two Vectors**

Find the smallest nonnegative angle, in degrees, between the vectors $\langle 3, -7 \rangle$ and $\langle 2, 6 \rangle$. See Figure 2.

Solution First, use the original formula to compute the dot product.

$$\langle 3, -7 \rangle \cdot \langle 2, 6 \rangle = (3)(2) + (-7)(6) = 6 - 42 = -36$$

Next, use the alternative formula for the dot product.

$$\langle 3, -7 \rangle \cdot \langle 2, 6 \rangle = (\|\langle 3, -7 \rangle\|)(\|\langle 2, 6 \rangle\|) \cos\theta$$

where θ is the angle between the vectors. Now, compute the magnitudes of the vectors.

$$\|\langle 3, -7 \rangle\| = \sqrt{3^2 + (-7)^2} = \sqrt{9 + 49} = \sqrt{58}$$

$$\|\langle 2, 6 \rangle\| = \sqrt{2^2 + 6^2} = \sqrt{4 + 36} = \sqrt{40} = \sqrt{4(10)} = 2\sqrt{10}$$

Substituting the magnitudes, we obtain

$$\langle 3, -7 \rangle \cdot \langle 2, 6 \rangle = (\sqrt{58})(2\sqrt{10}) \cos \theta$$

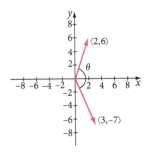

Figure 2

Equating the two expressions for the dot product, we have

$$-36 = \left(\sqrt{58}\right)\left(2\sqrt{10}\right)\cos\theta$$

$$\cos\theta = -\frac{36}{\left(\sqrt{58}\right)\left(2\sqrt{10}\right)} \qquad \text{Isolate } \cos\theta$$

$$\cos\theta = -\frac{18}{\sqrt{580}} \approx -0.7474 \qquad \text{Simplify}$$

$$\theta \approx 138.37° \qquad \text{Use } \boxed{\cos^{-1}} \text{ key on calculator}$$

 Check It Out 2　Find the angle between the vectors $\langle -8, 13 \rangle$ and $\langle 5, -4 \rangle$.

Parallel and Perpendicular Vectors

Two vectors **v** and **w** are said to be **parallel** if the angle θ between the vectors is 0° or 180°. If $\theta = 0°$, the vectors lie in the same direction. If $\theta = 180°$, the vectors lie in opposite directions.

If $\theta = 90°$, then the vectors are **perpendicular**. The term **orthogonal** is also used to mean perpendicular, especially in reference to vectors. When $\theta = 90°$, note that

$$\mathbf{v} \cdot \mathbf{w} = \|\mathbf{v}\| \, \|\mathbf{w}\| \cos\theta = \|\mathbf{v}\| \, \|\mathbf{w}\| \cos 90° = 0$$

We have the following facts about orthogonal vectors.

> ### The Dot Product and Orthogonal Vectors
>
> Let **v** and **w** be two nonzero vectors.
>
> - If **v** and **w** are perpendicular, or orthogonal, then $\mathbf{v} \cdot \mathbf{w} = 0$.
> - If $\mathbf{v} \cdot \mathbf{w} = 0$, then **v** and **w** are perpendicular, or orthogonal.

Example 3　**Determining Orthogonal Vectors**

Determine whether the vectors $\mathbf{v} = \langle -4, 6 \rangle$ and $\mathbf{w} = \langle 3, 2 \rangle$ are orthogonal.

Solution　To determine whether the vectors are perpendicular, check whether their dot product is zero.

$$\mathbf{v} \cdot \mathbf{w} = \langle -4, 6 \rangle \cdot \langle 3, 2 \rangle = (-4)(3) + (6)(2) = -12 + 12 = 0$$

Because the dot product is 0, the vectors are orthogonal, or perpendicular.

 Check It Out 3　Determine whether the vectors $\mathbf{v} = \langle 1, -3 \rangle$ and $\mathbf{w} = \langle -6, -2 \rangle$ are orthogonal.

Vector Projection

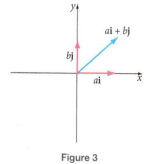

Figure 3

We saw in Section 7.5 that when a vector is written as $\mathbf{v} = a\mathbf{i} + b\mathbf{j}$, the component $a\mathbf{i}$ lies along the horizontal axis, and the component $b\mathbf{j}$ lies along the vertical axis. See Figure 3. Now, we discuss how to find components of a vector **v** along *any* nonzero vector **w** and perpendicular to **w**. We will refer to these components as $\mathbf{v}_1$ and $\mathbf{v}_2$, respectively. See Figure 4.

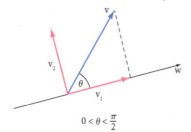

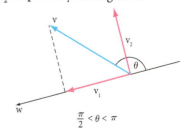

Figure 4

We first calculate $||\mathbf{v}_1||$ using trigonometry:

$$||\mathbf{v}_1|| = ||\mathbf{v}|| \cos \theta$$

Now

$$\mathbf{v} \cdot \mathbf{w} = ||\mathbf{w}|| \, ||\mathbf{v}|| \cos \theta \Rightarrow \cos \theta = \frac{\mathbf{v} \cdot \mathbf{w}}{||\mathbf{w}|| \, ||\mathbf{v}||}$$

Substituting the expression for $\cos \theta$ in the expression for $\mathbf{v}_1$, we have

$$||\mathbf{v}_1|| = ||\mathbf{v}|| \cos \theta = ||\mathbf{v}|| \frac{\mathbf{v} \cdot \mathbf{w}}{||\mathbf{w}|| \, ||\mathbf{v}||} = \frac{\mathbf{v} \cdot \mathbf{w}}{||\mathbf{w}||}$$

Because $\mathbf{v}_1$ is in the direction of $\mathbf{w}$, $\mathbf{v}_1$ can be obtained by scalar multiplication of $||\mathbf{v}_1||$ by a unit vector in the direction of $\mathbf{w}$. Thus

$$\mathbf{v}_1 = ||\mathbf{v}_1|| \left(\frac{\mathbf{w}}{||\mathbf{w}||} \right) \qquad \text{\textcolor{teal}{$\mathbf{v}_1$ in the same direction as $\mathbf{w}$}}$$

$$= \left(\frac{\mathbf{v} \cdot \mathbf{w}}{||\mathbf{w}||} \right) \left(\frac{\mathbf{w}}{||\mathbf{w}||} \right) \qquad \text{\textcolor{teal}{Substitute $||\mathbf{v}_1|| = \dfrac{\mathbf{v} \cdot \mathbf{w}}{||\mathbf{w}||}$}}$$

$$= \frac{\mathbf{v} \cdot \mathbf{w}}{||\mathbf{w}||^2} \mathbf{w}$$

The component $\mathbf{v}_1$ we found above is referred to as the **vector projection** of $\mathbf{v}$ onto $\mathbf{w}$, denoted by $\text{proj}_w \mathbf{v}$.

Vector Projection of v on w

Let $\mathbf{v}$ and $\mathbf{w}$ be nonzero vectors. The vector projection of $\mathbf{v}$ on the vector $\mathbf{w}$ is given by

$$\text{proj}_w \mathbf{v} = \left(\frac{\mathbf{v} \cdot \mathbf{w}}{||\mathbf{w}||^2} \right) \mathbf{w}$$

Example 4   **Finding the Vector Projection of v onto w**

If $\mathbf{v} = \langle 4, 3 \rangle$ and $\mathbf{w} = \langle 3, 1 \rangle$, calculate $\text{proj}_w \mathbf{v}$. See Figure 5.

Solution   First compute

$$\mathbf{v} \cdot \mathbf{w} = \langle 4, 3 \rangle \cdot \langle 3, 1 \rangle = 12 + 3 = 15$$

and

$$||\mathbf{w}||^2 = ((3)^2 + 1^2) = 10$$

Then

$$\text{proj}_w \mathbf{v} = \left(\frac{\mathbf{v} \cdot \mathbf{w}}{||\mathbf{w}||^2} \right) \mathbf{w} = \frac{15}{10} \langle 3, 1 \rangle = \left\langle \frac{9}{2}, \frac{3}{2} \right\rangle$$

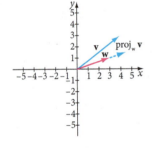

Figure 5

Check It Out 4   If $\mathbf{v} = \langle 4, -1 \rangle$ and $\mathbf{w} = \langle 2, 5 \rangle$, calculate $\text{proj}_w \mathbf{v}$.

Using the vector projection, we can express **v** as the sum of the two orthogonal vector components.

Orthogonal Decomposition of a Vector

Let **v** and **w** be nonzero vectors. Then **v** can be written as

$$\mathbf{v} = \mathbf{v}_1 + \mathbf{v}_2$$

where $\mathbf{v}_1 = \text{proj}_w \, \mathbf{v}$ and $\mathbf{v}_2 = \mathbf{v} - \mathbf{v}_1$. The vectors $\mathbf{v}_1$ and $\mathbf{v}_2$ are orthogonal.

Example 5   **Decomposing a Vector into Components**

Let $\mathbf{v} = \langle 4, 3 \rangle$ and $\mathbf{w} = \langle 3, 1 \rangle$. Decompose **v** into two vectors $\mathbf{v}_1$ and $\mathbf{v}_2$, where $\mathbf{v}_1$ is parallel to **w** and $\mathbf{v}_2$ is orthogonal to **w**. See Figure 6.

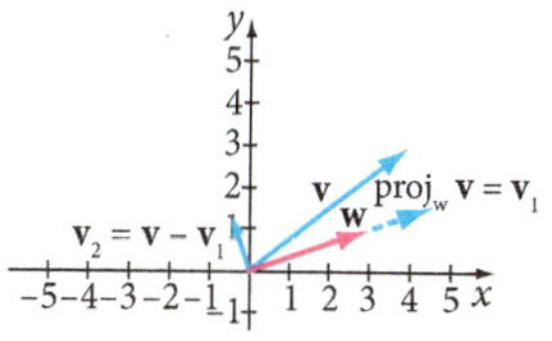

Solution From Example 4,

$$\mathbf{v}_1 = \text{proj}_w \, \mathbf{v} = \left\langle \frac{9}{2}, \frac{3}{2} \right\rangle$$

Now compute $\mathbf{v}_2$.

$$\mathbf{v}_2 = \mathbf{v} - \mathbf{v}_1 = \langle 4, 3 \rangle - \left\langle \frac{9}{2}, \frac{3}{2} \right\rangle \qquad \text{Definition of } v_2$$

$$= \left\langle \frac{8}{2}, \frac{6}{2} \right\rangle - \left\langle \frac{9}{2}, \frac{3}{2} \right\rangle \qquad \text{Rewrite with common denominator}$$

$$= \left\langle -\frac{1}{2}, \frac{3}{2} \right\rangle$$

Thus **v** can be decomposed as

$$\mathbf{v} = \mathbf{v}_1 + \mathbf{v}_2 = \left\langle \frac{9}{2}, \frac{3}{2} \right\rangle + \left\langle -\frac{1}{2}, \frac{3}{2} \right\rangle$$

You can quickly check that $\mathbf{v}_1 + \mathbf{v}_2 = \langle 4, 3 \rangle = \mathbf{v}$.

Check It Out 5 Let $\mathbf{v} = \langle 4, -1 \rangle$ and $\mathbf{w} = \langle 2, 5 \rangle$. Decompose v into two vectors $\mathbf{v}_1$ and $\mathbf{v}_2$, where $\mathbf{v}_1$ is parallel to **w** and $\mathbf{v}_2$ is orthogonal to **w**.

Figure 6

Applications of Dot Products

Dot products have long been used in engineering and physics. Recent applications of them include computer simulations and computer game design.

In physics, the **work** done by a force in moving an object is given by the product of the distance moved with the magnitude of the force component in the direction of motion. See Figure 7. In terms of vectors and dot products, we have the following definition of work.

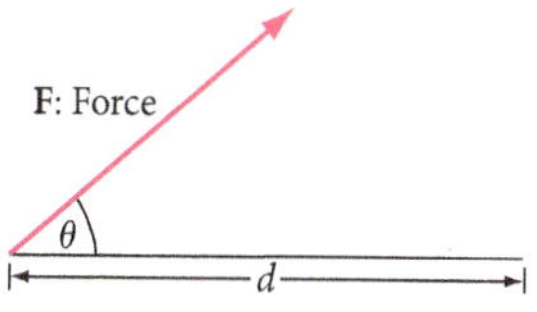

Figure 7

Work Done by a Force

The work W done by a constant force **F** as it moves along the vector **v** is given by

$$W = \mathbf{F} \cdot \mathbf{v} = \|\mathbf{F}\| \, d \cos \theta$$

where $d = \|\mathbf{v}\|$.

Note In most problems involving work, it is usually easier to apply the definition of work using the cosine.

The quantity $\|\mathbf{F}\| \cos \theta$ is the component of the force vector in the direction of motion. In particular, if the force vector is perpendicular to the direction of motion, no work is done because $\theta = 90°$ and so $W = (\|\mathbf{F}\| \cos \theta) \|\mathbf{v}\| = 0$. The units of work are *foot-pounds* in the English system and *Newton-meters* in the metric system.

| **Example 6** | **Work Done by a Force** |

A man is pushing on a lawn mower handle with a force of 35 pounds. If the angle the handle makes with the horizontal is 40°, how much work is done in moving the lawn mower a distance of 150 feet on level ground?

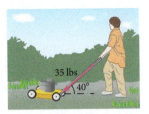

Figure 8

Solution Note that the force is exerted along the handle of the lawn mower, but the direction of motion is horizontal. Thus, only the horizontal component of the force contributes to the work done in moving the lawn mower. See Figure 8. Using the formula for work,

$$W = \|\mathbf{F}\| d \cos \theta$$

$$= (35)(150) \cos(40°) \qquad \|\mathbf{F}\| = 35, d = 150, \theta = 40°$$

$$= 5250(0.7660) \qquad\qquad \cos 40° \approx 0.7660$$

$$= 4021.50$$

Thus, the work done in moving the lawn mower is 4021.50 foot-pounds.

Check It Out 6   Rework Example 6 if the angle made with the horizontal is 35°. Assume all other information stays the same. Round your answer to two decimal places.

Dot products are used extensively in computer graphics and in designing computer video games, because they give algebraic formulations of geometric concepts. Thus, dot products can be programmed more easily than geometric constructions. An elementary example from game design is given next.

Example 7	Video Game Design

A portion of a computer video game consists of a ball colliding with a wall. The origin is taken to be the left bottom-most corner of the computer screen. The ball's location is given by the vector $\mathbf{v} = \langle 6, 10 \rangle$, and the wall makes an angle of 45° with the horizontal. See Figure 9. What is the perpendicular distance from the ball to the wall?

Figure 9

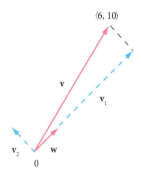

Figure 10

Solution We decompose $\mathbf{v} = \mathbf{v}_1 + \mathbf{v}_2$, so that $\mathbf{v}_1$ lies along the direction of the wall. Then, the magnitude of the perpendicular vector, $\|\mathbf{v}_2\|$, will give the required distance.

First, compute the projection $\mathbf{v}_1$, the component of $\mathbf{v}$ that lies along the wall. To do so, we need a vector $\mathbf{w}$ that lies along the direction of the wall. Because the wall makes an angle of 45° with the horizontal, the vector $\mathbf{w} = \langle \cos 45°, \sin 45° \rangle = \left\langle \frac{\sqrt{2}}{2}, \frac{\sqrt{2}}{2} \right\rangle$ is a unit vector in the same direction as the wall. See Figure 10. Thus

$$\mathbf{v}_1 = \text{proj}_w\,\mathbf{v} = \left(\frac{\mathbf{v} \cdot \mathbf{w}}{\|\mathbf{w}\|^2} \right) \mathbf{w} \qquad \text{Definition of projection}$$

$$= 8\sqrt{2} \left\langle \frac{\sqrt{2}}{2}, \frac{\sqrt{2}}{2} \right\rangle \qquad \mathbf{v} \cdot \mathbf{w} = 8\sqrt{2};\ \|\mathbf{w}\| = 1$$

$$= \langle 8, 8 \rangle \qquad \text{Simplify}$$

Now compute $\mathbf{v}_2$.

$$\mathbf{v}_2 = \mathbf{v} - \mathbf{v}_1 = \langle 6, 10 \rangle - \langle 8, 8 \rangle = \langle -2, 2 \rangle$$

The perpendicular distance is therefore

$$\|\mathbf{v}_2\| = \sqrt{8} = 2\sqrt{2}$$

Check It Out 7 Rework Example 7 if the wall makes an angle of 30° with the horizontal. Assume all other information stays the same.

Exercises 7.6

Skills

In Exercises 1–8, find $\mathbf{v} \cdot \mathbf{w}$.

1. $\mathbf{v} = \langle -3, 4 \rangle$, $\mathbf{w} = \langle 2, -3 \rangle$
2. $\mathbf{v} = \langle 6, -1 \rangle$, $\mathbf{w} = \langle 4, 3 \rangle$
3. $\mathbf{v} = \langle 5, -8 \rangle$, $\mathbf{w} = \langle -2, \frac{1}{2} \rangle$
4. $\mathbf{v} = \langle \frac{3}{2}, -1 \rangle$, $\mathbf{w} = \langle 4, 0 \rangle$
5. $\mathbf{v} = \langle 3, -1 \rangle$, $\mathbf{w} = \langle 1, 3 \rangle$
6. $\mathbf{v} = \langle -2, 0 \rangle$, $\mathbf{w} = \langle 0, 4 \rangle$
7. $\mathbf{v} = \langle -\frac{5}{3}, \frac{4}{5} \rangle$, $\mathbf{w} = \langle \frac{2}{5}, \frac{1}{3} \rangle$
8. $\mathbf{v} = \langle \frac{6}{5}, \frac{1}{2} \rangle$, $\mathbf{w} = \langle -10, 2 \rangle$

In Exercises 9–16, find the smallest nonnegative angle between the vectors $\mathbf{v}$ *and* $\mathbf{w}$. *Round your answer to the nearest tenth of a degree.*

9. $\mathbf{v} = \langle 1, 3 \rangle$, $\mathbf{w} = \langle -2, 0 \rangle$
10. $\mathbf{v} = \langle 0, -1 \rangle$, $\mathbf{w} = \langle 2, 3 \rangle$
11. $\mathbf{v} = \langle -2, 0 \rangle$, $\mathbf{w} = \langle 0, 3 \rangle$
12. $\mathbf{v} = \langle 2, -4 \rangle$, $\mathbf{w} = \langle 6, 3 \rangle$
13. $\mathbf{v} = \langle 4, 3 \rangle$, $\mathbf{w} = \langle 2, -1 \rangle$
14. $\mathbf{v} = \langle 2, 4 \rangle$, $\mathbf{w} = \langle -3, 2 \rangle$
15. $\mathbf{v} = \langle \frac{1}{3}, 1 \rangle$, $\mathbf{w} = \langle 6, -1 \rangle$
16. $\mathbf{v} = \langle -2, \frac{3}{2} \rangle$, $\mathbf{w} = \langle 1, 2 \rangle$

In Exercises 17–24, calculate $\text{proj}_w \mathbf{v}$. *Then, decompose* $\mathbf{v}$ *into* $\mathbf{v}_1$ *and* $\mathbf{v}_2$, *where* $\mathbf{v}_1$ *is parallel to* $\mathbf{w}$ *and* $\mathbf{v}_2$ *is orthogonal to* $\mathbf{w}$.

17. $\mathbf{v} = \langle 2, -4 \rangle$, $\mathbf{w} = \langle 2, 6 \rangle$
18. $\mathbf{v} = \langle 5, -3 \rangle$, $\mathbf{w} = \langle 1, 1 \rangle$
19. $\mathbf{v} = \langle 10, 5 \rangle$, $\mathbf{w} = \langle 2, -1 \rangle$
20. $\mathbf{v} = \langle 1, 2 \rangle$, $\mathbf{w} = \langle -3, 3 \rangle$
21. $\mathbf{v} = \langle 6, 12 \rangle$, $\mathbf{w} = \langle 3, 1 \rangle$
22. $\mathbf{v} = \langle -4, 3 \rangle$, $\mathbf{w} = \langle 1, -3 \rangle$
23. $\mathbf{v} = \langle 4, 5 \rangle$, $\mathbf{w} = \langle -3, 4 \rangle$
24. $\mathbf{v} = \langle 6, -3 \rangle$, $\mathbf{w} = \langle 4, 2 \rangle$

In Exercises 25–32, determine whether the given pairs of vectors are orthogonal.

25. $\mathbf{v} = \langle 1, 2 \rangle$, $\mathbf{w} = \langle -4, 1 \rangle$
26. $\mathbf{v} = \langle 3, 1 \rangle$, $\mathbf{w} = \langle 0, 1 \rangle$
27. $\mathbf{v} = \langle 2, -3 \rangle$, $\mathbf{w} = \langle -6, 4 \rangle$
28. $\mathbf{v} = \langle -5, 2 \rangle$, $\mathbf{w} = \langle 4, -10 \rangle$
29. $\mathbf{v} = \langle \frac{1}{3}, 2 \rangle$, $\mathbf{w} = \langle 6, \frac{5}{2} \rangle$
30. $\mathbf{v} = \langle 3, 5 \rangle$, $\mathbf{w} = \langle \frac{5}{6}, \frac{1}{2} \rangle$
31. $\mathbf{v} = \langle 1, 0 \rangle$, $\mathbf{w} = \langle 0, 3 \rangle$
32. $\mathbf{v} = \langle 2, 0 \rangle$, $\mathbf{w} = \langle 0, 4 \rangle$

In Exercises 33–40, perform each operation, given $\mathbf{u}$, $\mathbf{v}$, *and* $\mathbf{w}$.

$$\mathbf{u} = \langle 3, 2 \rangle, \mathbf{v} = \langle -1, 4 \rangle, \mathbf{w} = \langle -2, -1 \rangle$$

33. $2\mathbf{u} + 3\mathbf{w}$
34. $-4\mathbf{u} + \mathbf{v}$
35. $-\mathbf{u} \cdot (\mathbf{v} + \mathbf{w})$
36. $-2\mathbf{w} \cdot (\mathbf{v} + \mathbf{w})$
37. $3\mathbf{u} + \mathbf{v} - 2\mathbf{w}$
38. $-\mathbf{u} - 2\mathbf{v} + \mathbf{w}$
39. $\text{proj}_w(\mathbf{u} - \mathbf{v})$
40. $\text{proj}_u(\mathbf{v} + \mathbf{w})$

Applications

41. **Work** A parent pulling a wagon in which her child is riding along level ground exerts a force of 20 pounds on the handle. The handle make an angle of 45° with the horizontal. How much work is done in pulling the wagon 100 feet? Round to the nearest foot-pound.

42. **Work** A child pulls a wagon along level ground. He exerts a force of 20 pounds on the handle which makes a 30° angle with the horizontal. Find the work done in pulling the wagon 100 feet. Round to the nearest foot-pound.

43. **Game Design** In a new video game, Mario and Luigi are at positions defined by the vectors $\langle 10, 3 \rangle$ and $\langle x, 15 \rangle$. What must be the value of x so that their position vectors are orthogonal?

44. **Design** The position vectors of a tower and a small garden from the center of a fountain are given by $\langle 50, 60 \rangle$ and $\langle 40, y \rangle$. Find y so that the two position vectors are orthogonal.

45. Distance Two tourboats, *Swan* and *Dolphin*, depart from a lighthouse, with one directly behind another. However, after some time, *Swan* strays off course and ends up at position $\langle 1, 2 \rangle$, relative to the lighthouse. The other boat, *Dolphin*, stays on course with a position vector $\langle 5, 3 \rangle$, relative to the lighthouse. What is the shortest distance that *Swan* must travel to get back on the same course as *Dolphin*, assuming that *Dolphin* keeps traveling in its current direction?

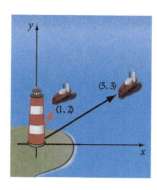

46. Computer Animation An animated figure's location is given by $\langle 5, 2 \rangle$. By what angle must the figure be rotated so that its new location is in the direction of $\langle 4, 3 \rangle$? Round your answer to the nearest tenth of a degree.

47. Work on Incline A wagon is pulled by a child with a force of 20 pounds, at an angle of 30° with the horizontal. Find the work done in pulling the wagon 100 feet up an incline that makes an angle of 10° with the horizontal. Round your answer to the nearest foot-pound.

48. Work on Incline A box weighing 100 pounds is pushed up a hill. The hill makes an angle of 30° with the horizontal. Find the work done against gravity in pushing the box a distance of 60 feet.

For Exercises 49 and 50, use the following information. The horsepower of an engine pulling a cart is determined by the formula

$$P = \frac{1}{550}(\mathbf{F} \cdot \mathbf{v})$$

where **F** is the force, in pounds, exerted on the cart and **v** is the velocity, in feet per second, of the cart as it is moved by the engine.

49. Power Find the horsepower of an engine that is exerting a force of 1500 pounds at an angle of 30° and is moving the cart horizontally at a speed of 10 feet per sec. Round to the nearest tenth of a horsepower.

50. Power Find the horsepower of an engine that is exerting a force of 2000 pounds at an angle of 30° and is moving the cart horizontally at a speed of 15 feet per sec. Round to the nearest tenth of a horsepower.

Concepts

51. Find a so that $\langle 4, a \rangle$ and $\langle -3, 2 \rangle$ are orthogonal.

52. Show that if **v** and **w** are nonzero orthogonal vectors, then $\text{proj}_w \mathbf{v} = 0$.

53. Show that for vectors **u**, **v** and **w**, $\mathbf{u} \cdot (\mathbf{v} + \mathbf{w}) = \mathbf{u} \cdot \mathbf{v} + \mathbf{u} \cdot \mathbf{w}$.

54. Show that for any vector **u**, $\mathbf{0} \cdot \mathbf{u} = 0$.

55. Show that for any vector **v**, $\mathbf{v} \cdot \mathbf{v} = ||\mathbf{v}||^2$. (In advanced mathematics, this relationship is very useful.)

56. Show that for any vector **u** and **v**, and any real number k, $(k\mathbf{u}) \cdot (\mathbf{v}) = k(\mathbf{u} \cdot \mathbf{v}) = \mathbf{u} \cdot (k\mathbf{v})$.

Trigonometric Form of a Complex Number

Objectives

- Write a complex number in trigonometric form.

- Multiply and divide complex numbers in trigonometric form.

- Find powers of complex numbers.

- Find roots of complex numbers.

Just in Time
Review complex numbers in Section 3.5.

Certain arithmetic or algebraic operations on complex numbers can be performed more efficiently if we first express complex numbers in **trigonometric form**, also referred to as **polar form**. To show how we may write complex numbers in trigonometric form, we will first look at how complex numbers appear when graphed in the complex plane.

The complex plane is a two-dimensional coordinate system. The horizontal axis and the vertical axis are known as the **real axis** and the **imaginary axis**, respectively. Figure 1 shows how a complex number of the form $z = a + bi$ is graphed in the complex plane with coordinates (a, b), where a is the real part of z and b is the imaginary part of z.

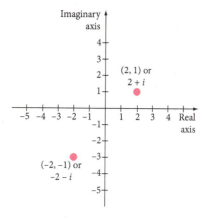

Figure 1

In order to represent the complex coordinates (a, b) in trigonometric form, first refer to Figure 2. The distance of (a, b) from the origin is $r = \sqrt{a^2 + b^2}$ and θ is an angle in standard position such that the point (a, b) lies on its terminal side as shown in Figure 2 and $(a, b) \neq (0, 0)$.

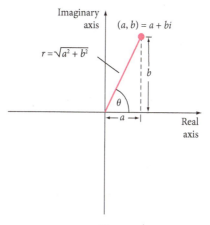

Figure 2

Using the definitions of $\cos \theta$ and $\sin \theta$ from Section 5.4, we obtain

$$a = r \cos \theta \qquad b = r \sin \theta$$

so

$$a + bi = (r \cos \theta) + (r \sin \theta)i$$

These equations hold for every complex number, including $(a, b) = (0, 0)$.

Trigonometric Form of a Complex Number

The complex number $z = a + bi$ can be represented in **trigonometric form**, or **polar form**, as

$$z = r(\cos \theta + i \sin \theta)$$

where $r = \sqrt{a^2 + b^2}$ and θ is the angle in $[0, 2\pi)$ defined as follows:

$$\sin \theta = \frac{b}{r} \text{ and } \cos \theta = \frac{a}{r}$$

The quantity $r = \sqrt{a^2 + b^2}$ is called the **modulus**, or **absolute value** of z.

It is customary to choose θ in the interval $[0, 2\pi)$, as defined above, though any angle θ of the form $\theta + 2n\pi$, n an integer, satisfies the equation $z = r(\cos \theta + i \sin \theta)$.

VIDEO EXAMPLES

SECTION 7.7

Example 1   Complex Numbers in Trigonometric Form

Write the following complex numbers in trigonometric form.

a. 3   **b.** $-i$ **c.** $-1 + i\sqrt{3}$

Using Technology

In the $a + bi$ mode, a graphing utility can compute the absolute value, r, and *theta* of a complex number. Here, θ is in radians. See Figure 4.

Keystroke Appendix:
Section 15

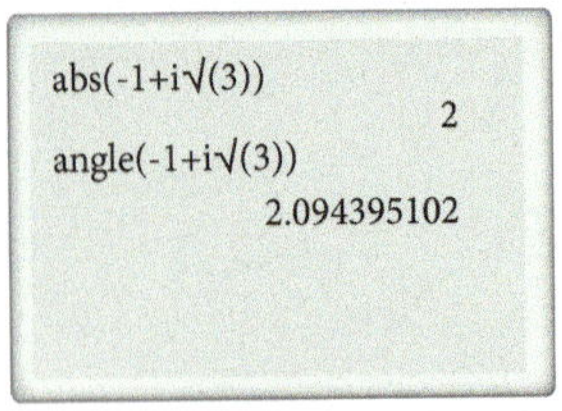

Solution

a. Because $3 = 3 + 0i$, $r = \sqrt{3^2 + (0)^2} = \sqrt{9} = 3$. The number 3 corresponds to the point $(3, 0)$ in the complex plane, which lies on the positive real axis. Thus $\theta = 0$ and we have

$$3 = 3(\cos 0 + i \sin 0)$$

See Figure 3.

```
abs(-1+i√(3))
                    2
angle(-1+i√(3))
          2.094395102
```

Figure 4

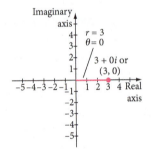

Figure 3

b. Because $-i = 0 - 1i$, $r = \sqrt{0^2 + (-1)^2} = \sqrt{1} = 1$. The number $-i$ corresponds to the point $(0, -1)$ in the complex plane, which lies on the negative imaginary axis, and so $\theta = \frac{3\pi}{2}$. Thus

$$-i = 1\left(\cos \frac{3\pi}{2} + i \sin \frac{3\pi}{2}\right)$$

See Figure 5.

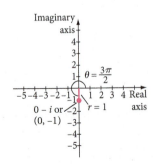

Figure 5

c. For the number $-1 + i\sqrt{3}$, $r = \sqrt{(-1)^2 + (\sqrt{3})^2} = \sqrt{4} = 2$. The number $-1 + i\sqrt{3}$ corresponds to the point $(-1, \sqrt{3})$, which is in the second quadrant. So we need to find the angle θ in the interval $\left(\frac{\pi}{2}, \pi\right)$ that satisfies the equations

$$\sin\theta = \frac{\sqrt{3}}{2} \quad \text{and} \quad \cos\theta = -\frac{1}{2}$$

From our knowledge of special angles, $\theta = \frac{2\pi}{3}$. Thus

$$-1 + i\sqrt{3} = 2\left(\cos\frac{2\pi}{3} + i\sin\frac{2\pi}{3}\right)$$

See Figure 6.

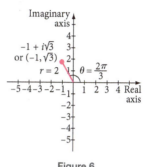

Figure 6

 Check It Out 1   Write the number $2 + 2i$ in trigonometric form.

Multiplication and Division of Complex Numbers

When complex numbers are written in trigonometric form, their products and quotients have convenient representations. The formulas given below can be proved by use of the sum and difference identities for sine and cosine.

> *Note* When finding the trigonometric form of a nonzero complex number, $r = \sqrt{a^2 + b^2}$ is *always* positive.

Multiplication and Division of Complex Numbers in Trigonometric Form

Let $z_1 = r_1(\cos\theta_1 + i\sin\theta_1)$ and $z_2 = r_2(\cos\theta_2 + i\sin\theta_2)$. Then

$$z_1 z_2 = r_1 r_2(\cos(\theta_1 + \theta_2) + i\sin(\theta_1 + \theta_2))$$

$$\frac{z_1}{z_2} = \frac{r_1}{r_2}(\cos(\theta_1 - \theta_2) + i\sin(\theta_1 - \theta_2))$$

These formulas are illustrated in the following example.

Example 2   **Multiplication and Division of Complex Numbers**

Let $w = 3\left(\cos\frac{\pi}{3} + i\sin\frac{\pi}{3}\right)$ and $z = 4\left(\cos\frac{\pi}{6} + i\sin\frac{\pi}{6}\right)$. Leave your answer in trigonometric form. Find

a. wz **b.** $\dfrac{w}{z}$

Solution

a. Using the multiplication formula, we get

$$wz = (3)(4)\left[\cos\left(\frac{\pi}{3} + \frac{\pi}{6}\right) + i\sin\left(\frac{\pi}{3} + \frac{\pi}{6}\right)\right]$$

$$= 12\left[\cos\left(\frac{\pi}{2}\right) + i\sin\left(\frac{\pi}{2}\right)\right]$$

This can also be written as $12[0 + i(1)] = 12i$, in the $a + bi$ form.

b. Using the division formula, we get

$$\frac{w}{z} = \frac{3}{4}\left[\cos\left(\frac{\pi}{3} - \frac{\pi}{6}\right) + i\sin\left(\frac{\pi}{3} - \frac{\pi}{6}\right)\right]$$

$$= \frac{3}{4}\left[\cos\frac{\pi}{6} + i\sin\frac{\pi}{6}\right]$$

$$= \frac{3}{4}\left[\frac{\sqrt{3}}{2} + i\left(\frac{1}{2}\right)\right]$$

This can be written as $\frac{3\sqrt{3}}{8} + \frac{3}{8}i$, in the $a + bi$ form.

 Check It Out 2   With w and z as in Example 2, find $\frac{z}{w}$. ■

Powers of Complex Numbers

When complex numbers are written in trigonometric form, finding their positive integer powers is simplified due to a result known as De Moivre's Theorem, which we will state without proof.

> ### De Moivre's Theorem
>
> Let $r(\cos\theta + i\sin\theta)$ be any complex number and let n be a positive integer. Then
>
> $$[r(\cos\theta + i\sin\theta)]^n = r^n(\cos n\theta + i\sin n\theta)$$

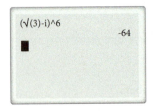

Using Technology

In the $a + bi$ mode, a graphing utility can compute powers of a complex number. See Figure 7.

Keystroke Appendix:
Section 15

Figure 7

| **Example 3**   **Raising a Complex Number to a Positive-Integer Power** |

Use De Moivre's Theorem to find $\left(\left(\sqrt{3} - i\right)\right)^6$.

Solution First write $\sqrt{3} - i$ in trigonometric form. The point in the complex plane that corresponds to $\sqrt{3} - i$ is $\left(\sqrt{3}, -1\right)$.

Thus, $r = \sqrt{\left(\sqrt{3^2}\right) + (-1)^2} = 2$. Find $\theta \in [0, 2\pi)$ using the fact that

$$\cos\theta = \frac{\sqrt{3}}{2} \text{ and } \sin\theta = -\frac{1}{2}$$

From our knowledge of special angles, $\theta = \frac{11\pi}{6}$. Thus

$$\sqrt{3} - i = r(\cos\theta + i\sin\theta) = 2\left(\cos\frac{11\pi}{6} + i\sin\frac{11\pi}{6}\right)$$

Using De Moiver's Theorem,

$$(\sqrt{3} - i)^6 = \left[2\left(\cos\frac{11\pi}{6} + i\sin\frac{11\pi}{6}\right)\right]^6$$

$$= 2^6\left[\cos\left(6 \cdot \frac{11\pi}{6}\right) + i\sin\left(6 \cdot \frac{11\pi}{6}\right)\right]$$

$$= 2^6(\cos 11\pi + i\sin 11\pi)$$

$$= 2^6[-1 + i(0)]$$

$$= 2^6(-1) = 64(-1) = -64$$

 *Check It Out 3*   Find $(-1 - i)^4$. ■

Roots of Complex Numbers

Let z be a nonzero complex number. Then, according to the Fundamental Theorem of Algebra, the equation $u^n = z$, for a positive integer $n \geq 2$, is guaranteed to have n complex solutions. Each solution of this equation is of the form $z^{1/n}$, an nth root of z. The nth roots of a complex number can be found using the following formula, which we state without proof.

Roots of Complex Numbers

For a positive integer $n \geq 2$, the nth roots of a nonzero complex number $z = r(\cos \theta + i \sin \theta)$ are given by

$$\sqrt[n]{r} \left[\cos \left(\frac{\theta + 2k\pi}{n} \right) + i \sin \left(\frac{\theta + 2k\pi}{n} \right) \right], k = 0, 1, 2, \ldots, n - 1,$$

where $\sqrt[n]{r}$ denotes the positive real nth root of r.

We illustrate the application of this formula in the next two examples.

Example 4 **Finding the Square Roots of a Complex Number**

Find the two square roots of $1 - i$.

Solution First, write $1 - i$ in trigonometric form. The point in the complex plane that corresponds to $1 - i$ is $(1, -1)$. Thus, $r = \sqrt{1^2 + (-1)^2} = \sqrt{2}$ and $\theta = \frac{7\pi}{4}$, so

$$1 - i = \sqrt{2} \left(\cos \frac{7\pi}{4} + i \sin \frac{7\pi}{4} \right)$$

Setting n to 2 in the formula for the nth roots of a complex number, we find that the square roots of $1 - i$ have the following form:

$$\sqrt[2]{\sqrt{2}} \left(\cos \frac{\left(\frac{7\pi}{4} \right) + 2k\pi}{2} + i \sin \frac{\left(\frac{7\pi}{4} \right) + 2k\pi}{2} \right), k = 0, 1$$

Note that $= \sqrt[2]{\sqrt{2}} = \sqrt[4]{2}$. For each value of k, a root is generated as follows:

First root $(k = 0)$: $\sqrt[4]{2} \left(\cos \left(\frac{7\pi}{8} \right) + i \sin \left(\frac{7\pi}{8} \right) \right) \approx -1.099 + 0.455i$

Second root $(k = 1)$: $\sqrt[4]{2} \left(\cos \left(\frac{15\pi}{8} \right) + i \sin \left(\frac{15\pi}{8} \right) \right) \approx 1.099 - 0.455i$

The two roots are equally spaced about the circle of radius $r = \sqrt[4]{2} \approx 1.189$. See Figure 8.

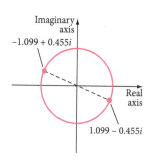

Figure 8

✓ *Check It Out 4*   Find the two square roots of $-i$.

| Example 5 | **Finding the Fifth Roots of a Complex Number** |

Find all the fifth roots of 1.

Solution To determine all the fifth roots of 1, first write 1 in trigonometric form. The point in the complex plane that corresponds to the number 1 is $(1, 0)$. Thus $r = \sqrt{1^2 + 0^2} = 1$ and $\theta = 0$, so

$$1 = 1(\cos 0 + i \sin 0)$$

Setting n to 5 in the formula for the nth roots of a complex number, we find that the roots have the following form:

$$\sqrt[5]{1}\left(\cos \frac{0 + 2k\pi}{5} + i \sin \frac{0 + 2k\pi}{5}\right), \qquad k = 0, 1, 2, 3, 4$$

Note that $\sqrt[5]{1} = 1$. For each value of k, a root is generated as follows

First root $(k = 0)$: $\cos 0 + i \sin 0 = 1 + i(0) = 1$

Second root $(k = 1)$: $\cos \frac{2\pi}{5} + i \sin \frac{2\pi}{5} \approx 0.309 + 0.951i$

Third root $(k = 2)$: $\cos \frac{4\pi}{5} + i \sin \frac{4\pi}{5} \approx -0.809 + 0.588i$

Fourth root $(k = 3)$: $\cos \frac{6\pi}{5} + i \sin \frac{6\pi}{5} \approx -0.809 - 0.588i$

Fifth root $(k = 4)$: $\cos \frac{8\pi}{5} + i \sin \frac{8\pi}{5} \approx 0.309 - 0.951i$

The five roots are equally spaced about the circle of radius $r = 1$. See Figure 9.

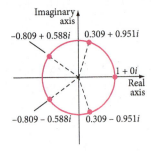

Figure 9

✓ *Check It Out 5* Find the fourth roots of i.

Exercises 7.7

Just in Time Exercises

In Exercises 1–6, evaluate the given expressions.

1. i^3 **2.** $(-2i)^2$ **3.** $-i^4$ **4.** i^5 **5.** $(3+2i)-(4+i)$ **6.** $-1-2i+(5+i)$

Skills

In Exercises 7–12, find r for the given complex numbers.

7. $2-i$ **8.** $3i$ **9.** $1-2i$ **10.** $\sqrt{3}+i$ **11.** $\frac{1}{2}+\frac{3}{4}i$ **12.** $\frac{1}{4}-\frac{2}{5}i$

In Exercises 13–22, express each complex number in trigonometric form.

13. $2i$ **14.** 8 **15.** -4 **16.** $-5i$ **17.** $1-i\sqrt{3}$ **18.** $-2-2i\sqrt{3}$

19. $4-4i$ **20.** $-5+5i$ **21.** $2\sqrt{3}-2i$ **22.** $-3\sqrt{3}+3i$

In Exercises 23–28, multiply or divide as indicated- and leave the answer in trigonometric form.

23. $2\left(\cos\dfrac{\pi}{4}+i\sin\dfrac{\pi}{4}\right)\cdot 4\left(\cos\dfrac{\pi}{3}+i\sin\dfrac{\pi}{3}\right)$ **24.** $3\left(\cos\dfrac{\pi}{6}+i\sin\dfrac{\pi}{6}\right)\cdot 5\left(\cos\dfrac{2\pi}{3}+i\sin\dfrac{2\pi}{3}\right)$

25. $\dfrac{1}{2}\left(\cos\dfrac{5\pi}{4}+i\sin\dfrac{5\pi}{4}\right)\cdot 3\left(\cos\dfrac{4\pi}{3}+i\sin\dfrac{4\pi}{3}\right)$ **26.** $\dfrac{4}{3}\left(\cos\dfrac{\pi}{3}+i\sin\dfrac{\pi}{3}\right)\cdot 2\left(\cos\dfrac{\pi}{4}+i\sin\dfrac{\pi}{4}\right)$

27. $\dfrac{5\left(\cos\dfrac{\pi}{4}+i\sin\dfrac{\pi}{4}\right)}{2\left(\cos\dfrac{\pi}{6}+i\sin\dfrac{\pi}{6}\right)}$ **28.** $\dfrac{6\left(\cos\dfrac{\pi}{12}+i\sin\dfrac{\pi}{12}\right)}{3\left(\cos\dfrac{\pi}{4}+i\sin\dfrac{\pi}{4}\right)}$

In Exercises 29–34, use De Moivre's Theorem to evaluate each expression.

29. $(1+i)^4$ **30.** $(2-2i)^4$ **31.** $\left(\sqrt{3}+i\right)^3$ **32.** $\left(1-\sqrt{3}i\right)^6$ **33.** $\left(-3-3\sqrt{3}i\right)^3$ **34.** $(1-i)^8$

In Exercises 35–38, find the square roots of each complex number. Round all numbers to three decimal places.

35. i **36.** $-2i$ **37.** $1+i\sqrt{3}$ **38.** $-2-2i$

39. Find the fifth roots of -1.   **40.** Find the fifth roots of i.   **41.** Find the fourth roots of -16.

42. Find the cube roots of $8i$.   **43.** Find the fourth roots of $-8i$.   **44.** Find the sixth roots of 1.

In Exercises 45–48, find all the complex solutions of each equation.

45. $z^3+1=0$ **46.** $z^2-i=0$ **47.** $iz^3=1$ **48.** $z^3+iz=0$

Concepts

Let $z = r(\cos \theta + i \sin \theta)$ be a nonzero complex number, and let n be a positive integer greater than 1.

49. Verify that each of the following n numbers is a solution of the equation $u^n = z$.

$$\sqrt[n]{r}\left[\cos\left(\frac{\theta + 2k\pi}{n}\right) + i \sin\left(\frac{\theta + 2k\pi}{n}\right)\right], \qquad k = 0, 1, 2, \ldots, n - 1$$

where $\sqrt[n]{r}$ denotes the positive real number that, when raised to the nth power, gives r. (*Hint*: Use De Moivre's Theorem.)

50. Can two or more of the n solutions of the equation $u^n = z$ be equal?

51. How many solutions of the equation $u^n = z$ are real numbers if n is odd and z is real?

52. How many solutions of the equation $u^n = z$ are real numbers if n is even and z is real?

Chapter 7 Summary

Section 7.1 Law of Sines

Law of Sines   [Review Exercises 1–12]

Let ABC be a triangle with sides a, b, c. See Figure 1. Then the following ratios hold

$$\frac{a}{\sin A} = \frac{b}{\sin B} = \frac{c}{\sin C}$$

The ratios can also be written as

$$\frac{\sin A}{a} = \frac{\sin B}{b} = \frac{\sin C}{c}$$

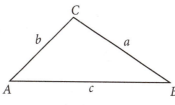

Figure 1

When given the following information, we can solve the triangle using the Law of Sines.

- Two angles and any side (abbreviated AAS or ASA)—always has a solution.
- Two sides and an angle opposite one of the sides (SSA)—can have one, two, or no solutions.

Section 7.2 Law of Cosines

Law of Cosines   [Review Exercises 13–20]

Let ABC be a triangle with sides a, b, c. See Figure 1 above. Then,

$$a^2 = b^2 + c^2 - 2bc \cos A$$

$$b^2 = a^2 + c^2 - 2ac \cos B$$

$$c^2 = a^2 + b^2 - 2ab \cos C$$

The following two situations require the Law of Cosines:

- Two sides and an angle formed by the two given sides (abbreviated SAS)
- Three sides (SSS)

Area of a triangle   [Review Exercises 21, 22]

Let ABC be a triangle with sides a, b and c.

SAS: If two of the sides and the included angle are known, then

$$\text{Area}(ABC) = \frac{1}{2} ab \sin C = \frac{1}{2} bc \sin A = \frac{1}{2} ac \sin B$$

SSS: If all three sides are known, then use Heron's formula:

$$\text{Area}(ABC) = \sqrt{s(s-a)(s-b)(s-c)}$$

where $s = \frac{1}{2}(a + b + c)$.

Section 7.3 **Polar Coordinates**

Polar coordinates of a point   [Review Exercises 23, 24]

In polar coordinates, a point is represented by (r, θ), where r is a real number and θ is an angle in standard position.

Polar to rectangular coordinates   [Review Exercises 25–28]

Given the polar coordinates (r, θ) of a point, use the following equations to convert to its rectangular coordinates (x, y):

$$x = r \cos \theta \qquad y = r \sin \theta$$

Rectangular to polar coordinates   [Review Exercises 29–32]

The point (x, y) can be represented by (r, θ), where

$$r = \sqrt{x^2 + y^2}$$

If $x \neq 0$ and $y \neq 0$, then θ is the unique angle in the interval $[0, 2\pi)$ such that

$$\tan \theta = \frac{y}{x}$$

and its terminal side lies in the same quadrant as the point.
If $y = 0$, then $\theta = 0$ if $x > 0$, and $\theta = \pi$ if $x < 0$. If $x = 0$, then $\theta = \frac{\pi}{2}$ if $y > 0$, and $\theta = \frac{3\pi}{2}$ if $y < 0$.

Converting equation between polar and rectangular forms   [Review Exercises 33–34]

To convert a rectangular form of an equation to polar form, use

$$x = r \cos \theta \quad \text{and} \quad y = r \sin \theta$$

To convert a polar equation to rectangular form, use

$$r = \sqrt{x^2 + y^2}$$

and $\tan \theta = \dfrac{y}{x}$.

Section 7.4 **Graphs of Polar Equations**

Circles and lines   [Review Exercises 35,36]

- The graph of $r = k$ is a circle of radius k centered at the pole.
- The graph of the polar equation $\theta = \alpha$ is a line through the pole, making an angle of α with the polar axis.

Tests for symmetry   [Review Exercises 37–40]

- The graph of a polar equation is **symmetric with respect to the polar axis** if substituting $(r, -\theta)$ for (r, θ) in the equation leaves it unchanged.
- The graph of a polar equation is **symmetric with respect to the line** $\theta = \frac{\pi}{2}$ if substituting $(-r, -\theta)$ for (r, θ) in the equation leaves it unchanged.
- The graph of a polar equation is **symmetric with respect to the pole** if substituting $(-r, \theta)$ for (r, θ) in the equation leaves it unchanged.

Section 7.5 **Vectors**

Magnitude of a vector [Review Exercises 41–44]

The **magnitude** of a vector $\mathbf{v} = \langle v_x, v_y \rangle$ is given by

$$\|\mathbf{v}\| = \sqrt{v_x^2 + v_y^2}$$

Direction angle of a vector [Review Exercises 41–44]

The **direction angle** of $\mathbf{v}$ is the angle θ that $\mathbf{v}$ makes with the positive x-axis such that

$$\cos \theta = \frac{v_x}{\|\mathbf{v}\|} \quad \text{and} \quad \sin \theta = \frac{v_y}{\|\mathbf{v}\|}$$

$$\text{or} \quad \tan \theta = \frac{v_y}{v_x}$$

where θ is in the interval $[0°, 360°)$.

The component form of a vector [Review Exercises 45–48]

A vector $\mathbf{v}$ in two-dimensions and in standard position is represented as $\langle v_x, v_y \rangle$. The quantities v_x and v_y are referred to as components.

To add or subtract vectors, we add or subtract their respective components. The **zero vector** is represented by $\langle 0, 0 \rangle$.

Scalar multiplication of vectors [Review Exercises 45–48]

Let k be a real number. Then the **scalar multiple** of k times a vector $\mathbf{u}$ is given by

$$k\mathbf{u} = k\langle u_x, u_y \rangle = \langle ku_x, ku_y \rangle$$

Unit vector [Review Exercises 49–52]

A unit vector is a vector of length 1. For any nonzero vector $\mathbf{v}$, the vector $\frac{\mathbf{v}}{\|\mathbf{v}\|}$ has length 1.

Using the unit vectors $\mathbf{i} = \langle 1, 0 \rangle$ and $\mathbf{j} = \langle 0, 1 \rangle$, we can write a vector as $v_x \mathbf{i} + v_y \mathbf{j}$.

Section 7.6 **Dot Product of Vectors**

Dot Product [Review Exercises 53–56]

The **dot product** of two vectors $\mathbf{u} = \langle u_x, u_y \rangle$ and $\mathbf{v} = \langle v_x, v_y \rangle$ is defined as

$$\mathbf{u} \cdot \mathbf{v} = u_x v_x + u_y v_y.$$

Angle between vectors [Review Exercises 53–56]

The smallest nonnegative angle between $\mathbf{u}$ and $\mathbf{v}$ is given by

$$\mathbf{u} \cdot \mathbf{v} = (\|\mathbf{u}\|)(\|\mathbf{v}\|) \cos \theta$$

Vector projection [Review Exercises 57–60]

The vector projection of $\mathbf{v}$ on the nonzero vector $\mathbf{w}$ is given by

$$\text{proj}_w \mathbf{v} = \left(\frac{\mathbf{v} \cdot \mathbf{w}}{\|\mathbf{w}\|^2} \right) \mathbf{w}$$

Orthogonal decomposition of a vector [Review Exercises 57–60]

Let $\mathbf{v}$ and $\mathbf{w}$ be nonzero vectors. Then $\mathbf{v}$ can be written as

$$\mathbf{v} = \mathbf{v}_1 + \mathbf{v}_2$$

where $\mathbf{v}_1 = \text{proj}_w \mathbf{v}$ and $\mathbf{v}_2 = \mathbf{v} - \mathbf{v}_1$. The vectors $\mathbf{v}_1$ and $\mathbf{v}_2$ are orthogonal.

Work done by a force

The **work** W done by a constant force $\mathbf{F}$ as it moves along the vector $\mathbf{v}$ is given by

$$W = \mathbf{F} \cdot \mathbf{v} = \|\mathbf{F}\| \, d \cos \theta$$

where $d = \|\mathbf{v}\|$.

Section 7.7 Trigonometric Form of a Complex Number

Trigonometric form of a complex number [Review Exercises 61–66]

The complex number $z = a + bi$ can be represented in the **trigonometric form** as

$$z = r(\cos \theta + i \sin \theta)$$

where $r = \sqrt{a^2 + b^2}$ and $\theta \in [0, 2\pi)$ such that

$$\sin \theta = \frac{b}{r} \quad \text{and} \quad \cos \theta = \frac{a}{r}$$

Multiplication and division of complex numbers in trigonometric form [Review Exercises 67–70]

Let $z_1 = r_1(\cos \theta_1 + i \sin \theta_1)$ and $z_2 = r_2(\cos \theta_2 + i \sin \theta_2)$. Then

$$z_1 z_2 = r_1 r_2(\cos(\theta_1 + \theta_2) + i \sin(\theta_1 + \theta_2))$$

$$\frac{z_1}{z_2} = \frac{r_1}{r_2}(\cos(\theta_1 - \theta_2) + i \sin(\theta_1 - \theta_2))$$

De Moivre's Theorem [Review Exercises 71–74]

Let $r(\cos \theta + i \sin \theta)$ be any complex number and let n be a positive integer. We then have

$$[r(\cos \theta + i \sin \theta)]^n = r^n(\cos n\theta + i \sin n\theta)$$

Roots of complex numbers [Review Exercises 75–78]

The nth roots of a complex number $z = r(\cos \theta + i \sin \theta)$ are given by

$$\sqrt[n]{r}\left[\cos\left(\frac{\theta + 2k\pi}{n}\right) + i \sin\left(\frac{\theta + 2k\pi}{n}\right)\right], \quad k = 0, 1, 2, \ldots, n - 1$$

Chapter 7 Review Exercises

Section 7.1

In Exercises 1–8, solve the given triangles which exist using the Law of Sines. If there is no solution, state so. The standard notation for labeling triangles is used. Round answers to four decimal places.

1. $A = 68°$, $C = 40°$, $a = 25$
2. $B = 55.7°$, $C = 67°$, $b = 43$
3. $C = 52.1°$, $A = 73°$, $a = 15$
4. $A = 78.4°$, $B = 61°$, $b = 19$
5. $A = 65°$, $a = 10$, $b = 8$
6. $A = 125°$, $a = 15$, $b = 12$
7. $A = 35°$, $a = 13$, $b = 15$
8. $A = 56°$, $a = 12$, $b = 7$

9. **Surveying** The angle of elevation to the top of a building from point P on the ground is 32°. The angle of elevation from point Q on the ground to the top of the building is 40°. If points P and Q are 300 feet apart and the building lies between the two points, find the height of the building to four decimal places.

10. **Construction** In the picture of a roof truss shown below, find the lengths a, b, and c, to four decimal places.

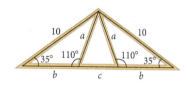

11. **Sports** In 1974, Evel Knievel, Jr., attempted to jump the Snake River Canyon in Idaho using only a ramp and a jet powered sled called a sky cycle. See the diagram. An observer measured angles from points C and again from point A, 309.5 feet away. If he took off from point A and landed at point B, find the distance from A to B across the canyon, to the nearest foot.

(Source: ESPN.com)

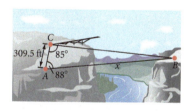

12. **Surveying** Two people, Abe and Carl, are 50 feet apart on the same bank at the narrowest point of the Mississippi river. They each measure the angles to a tree on the opposite bank, as shown in the diagram. Find the distance from Carl to the tree, to the nearest foot.

(Source: www.nps.gov)

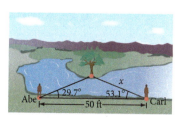

Section 7.2

In Exercises 13–20, solve the given triangles using the Law of Cosines. Round answers to four decimal places.

13. $b = 9, c = 5, A = 35°$ 14. $a = 13, c = 16, B = 68°$

15. $a = 30, b = 20, C = 107°$ 16. $b = 18, c = 14, A = 63.4°$

17. $a = 35, b = 25, C = 95°$ 18. $b = 10, c = 16, A = 120°$

19. $a = 4, b = 6, c = 8$ 20. $a = 7, b = 9, c = 11$

21. Find the area of the triangle in Problem 17. Round your answer to two decimal places.

22. Find the area of the triangle in Problem 18. Round your answer to two decimal places.

Section 7.3

In Exercises 23 and 24, plot the points, given in polar coordinates, on a polar grid.

23. $\left(4, \dfrac{\pi}{6}\right)$ 24. $\left(-2, \dfrac{\pi}{4}\right)$

In Exercises 25–28, convert the points to rectangular coordinates.

25. $\left(4, \dfrac{11\pi}{6}\right)$ 26. $\left(\dfrac{1}{3}, \dfrac{3\pi}{4}\right)$ 27. $\left(-2, \dfrac{3\pi}{2}\right)$ 28. $\left(3, \dfrac{5\pi}{3}\right)$

In Exercises 29–32, convert the points to polar coordinates.

29. $(-1, \sqrt{3})$ 30. $(-5, 5)$ 31. $(0, 3)$ 32. $(-2\sqrt{3}, 2)$

33. Convert the equation $(x - 1)^2 + y^2 = \dfrac{9}{4}$ into polar form.

34. Convert the equation $r \cos \theta = \dfrac{5}{2}$ into rectangular form.

Section 7.4

In Exercises 35–40, graph the polar equations.

35. $r = \dfrac{3}{2}$ 36. $\theta = -\dfrac{\pi}{6}$ 37. $r = 2 \cos \theta$ 38. $r = 4 \sin \theta$

39. $r = 3(1 - \cos \theta)$ 40. $r = 2(1 + \sin \theta)$

Section 7.5

In Exercises 41–44, find the magnitude and direction of each of the given vectors.

41. $\langle 3, 6 \rangle$ 42. $\langle 7, -2 \rangle$ 43. $\langle -5, 7 \rangle$ 44. $\left\langle \dfrac{1}{3}, -\dfrac{4}{3} \right\rangle$

In Exercises 45–48, find $\mathbf{u} - \mathbf{v}$, $\mathbf{u} + 2\mathbf{v}$, $-3\mathbf{u} + \mathbf{v}$.

45. $\mathbf{u} = \langle 0, 4 \rangle, \mathbf{v} = \langle -2, 4 \rangle$ 46. $\mathbf{u} = \langle 7, -3 \rangle, \mathbf{v} = \langle 2, 9 \rangle$

47. $\mathbf{u} = \langle -2.5, 5.5 \rangle, \mathbf{v} = \langle 3, 0 \rangle$ 48. $\mathbf{u} = \langle 5.2, 6.3 \rangle, \mathbf{v} = \langle 2, -1 \rangle$

In Exercises 49 and 50, find a unit vector in the same direction as of each of the given vectors.

49. $\langle -12, 5 \rangle$ **50.** $\langle 3, -2 \rangle$

In Exercises 51 and 52, express the given vectors in the form $a\mathbf{i} + b\mathbf{j}$.

51. $\langle 6, -3 \rangle$ **52.** $\langle -9, -5 \rangle$

Section 7.6

In Exercises 53–56, find $\mathbf{u} \cdot \mathbf{v}$ and the smallest positive angle between them.

53. $\mathbf{u} = \langle 4, -2 \rangle$, $\mathbf{v} = \langle 3, 7 \rangle$ **54.** $\mathbf{u} = \langle -3, 2 \rangle$, $\mathbf{v} = \langle 0, -1 \rangle$

55. $\mathbf{u} = \langle -4, 2 \rangle$, $\mathbf{v} = \left\langle 3, \dfrac{2}{3} \right\rangle$ **56.** $\mathbf{u} = \left\langle -\dfrac{1}{2}, -1 \right\rangle$, $\mathbf{v} = \langle -3, -2 \rangle$

In Exercises 57–60, calculate $\text{proj}_w \mathbf{v}$. Then, decompose $\mathbf{v}$ into $\mathbf{v}_1$ and $\mathbf{v}_2$, where $\mathbf{v}_1$ is parallel to $\mathbf{w}$ and $\mathbf{v}_2$ is orthogonal to $\mathbf{w}$.

57. $\mathbf{v} = \langle 2, -2 \rangle$, $\mathbf{w} = \langle 2, 4 \rangle$ **58.** $\mathbf{v} = \langle -4, 2 \rangle$, $\mathbf{w} = \langle 3, 3 \rangle$

59. $\mathbf{v} = \langle -1, 5 \rangle$, $\mathbf{w} = \langle 5, 1 \rangle$ **60.** $\mathbf{v} = \langle 1, 2 \rangle$, $\mathbf{w} = \langle -3, 3 \rangle$

Section 7.7

In Exercises 61–66, express each complex number in trigonometric form.

61. $-6i$ **62.** 10 **63.** $-1 + i\sqrt{3}$ **64.** $-3 - 3i\sqrt{3}$

65. $-2\sqrt{3} + 2i$ **66.** $-3\sqrt{3} - 3i$

In Exercises 67–70, find $z_1 z_2$ and $\dfrac{z_1}{z_2}$, in trigonometric form, for each pair of complex numbers.

67. $z_1 = 1 + i,\ z_2 = 2 - 2i$ **68.** $z_1 = \sqrt{3} + i,\ z_2 = \sqrt{3} - i$

69. $z_1 = -1 + i\sqrt{3},\ z_2 = 1 - i$ **70.** $z_1 = 1 - i\sqrt{3},\ z_2 = 1 + i\sqrt{3}$

In Exercises 71–74, use De Moivre's Theorem to find the value of each.

71. $(1 - i)^4$ **72.** $(-2 + 2i)^4$ **73.** $\left(\sqrt{3} - i\right)^3$ **74.** $\left(-1 - i\sqrt{3}\right)^3$

In Exercises 75–78, find the square roots of each complex number.

75. $-i$ **76.** $3i$ **77.** $2 - 2i\sqrt{3}$ **78.** $-1 - i$

Chapter 7 Test

In Exercises 1–4, solve the triangle ABC using Law of Sines or Law of Cosines. If no solution exists, state so. Round answers to two decimal places. Standard labeling is used for the triangles.

1. $A = 65°, B = 30°, a = 20$

2. $A = 60°, b = 5, c = 12$

3. $a = 10, b = 8, c = 15$

4. $C = 45°, b = 7, c = 12$

5. Find the area of $\triangle ABC$ if $A = 30°$, $b = 10$ inches, and $c = 20$ inches.

6. Find the polar coordinates corresponding to $(-4\sqrt{3}, -4)$. Express exact value of $\theta \in [0, 2\pi)$.

7. Write the exact values of the rectangular coordinates corresponding to the polar coordinates $\left(5, -\frac{3\pi}{4}\right)$.

8. Write the polar form of the equation $x^2 + (y - 1)^2 = 1$.

9. Write the rectangular form of the following equations.

   a. $r = 2.5$

   b. $\theta = -\dfrac{2\pi}{3}$

10. Graph $r = \dfrac{9}{2}$.

11. Graph $r = 4(1 - \sin\theta)$.

12. Find the magnitude and direction angle θ for the vector $\langle 4, -3 \rangle$. Round θ to the nearest tenth of a degree.

13. Find the components of a vector with magnitude 8 and direction angle 140°. Round your answer to two decimal places.

In Exercises 14–17, let $\mathbf{u} = \langle -3, 1 \rangle$ and $\mathbf{v} = \langle 2, 4 \rangle$. Calculate the given quantities.

14. $\mathbf{u} \cdot \mathbf{v}$

15. $3\mathbf{u} - \mathbf{v}$

16. $\|\mathbf{u}\|$ and $\|\mathbf{v}\|$

17. $\text{proj}_v \mathbf{u}$

18. Multiply and leave your answer in trigonometric form:

$$3\left(\cos\frac{\pi}{6} + i\sin\frac{\pi}{6}\right) \cdot 2\left(\cos\frac{\pi}{4} + i\sin\frac{\pi}{4}\right)$$

19. Divide and leave your answer in trigonometric form: $\dfrac{6\left(\cos\frac{\pi}{4} + i\sin\frac{\pi}{4}\right)}{3\left(\cos\frac{\pi}{6} + i\sin\frac{\pi}{6}\right)}$.

20. Find the perimeter of a triangular flower bed with the following dimensions: two sides are 3 feet and 5 feet, and the angle between the two sides is 50°. Round your answer to two decimal places.

21. A ship travels in the direction N40°W for 10 miles. Find its position in vector notation, rounded to two decimal places.

22. A wagon is pulled on level ground with a force of 20 pounds, with the handle of the wagon making an angle of 25° with the horizontal. Find the work done in pulling the wagon 50 feet. Round your answer to two decimal places.

Systems of Equations and Inequalities

Maintaining good health and nutrition translates into targeting values for calories, cholesterol, and fat, and can be modeled by a system of linear equations. Exercise 44 in Section 8.2 uses a system of linear equations to determine the nutritional value of slices of different types of pizza. In this chapter you will study how systems of equations and inequalities arise, how to solve them, and how they are used in various applications.

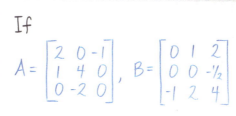

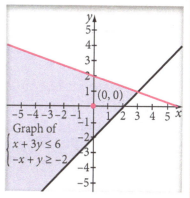

Graph of
$$\begin{cases} x + 3y \le 6 \\ -x + y \ge -2 \end{cases}$$

Outline

Objectives

- Solve systems of linear equations in two variables by elimination.

- Understand what is meant by an inconsistent system of linear equations.

- Understand what is meant by a dependent system of linear equations.

- Relate the nature of the solution(s) of a system of linear equations to their respective graphs.

- Graph and solve systems of linear inequalities in two variables.

- Use systems of equations and systems of inequalities to solve applied problems.

VIDEO EXAMPLES

SECTION 8.1

The general form of a linear equation in the variables x, y is $Ax + By = C$, where A, B, and C are constants such that A and B are not both zero. A **system of linear equations** in x and y variables is a collection of equations written in this form, as shown below.

$$\begin{cases} Ax + By = E \\ Cx + Dy = F \end{cases}$$

where A, B, C, D are not all zero. A solution of a system of linear equations is an ordered pair of numbers, (x, y), satisfying all the equations in the system. Graphically, it is the point of intersection of two lines, if the system has only one solution.

We next introduce a method for solving a system of two linear equations in two variables which can be extended to linear systems with more equations and variables.

Solving Systems of Linear Equations by Elimination

In the **elimination method** for solving linear systems in two variables, we eliminate one of the variables by combining the equations in a suitable manner. Example 1 illustrates this procedure.

Example 1   **Elimination Method to Solve a System of Equations**

Use the method of elimination to solve the following system of equations.

$$\begin{cases} x + y = 30 \\ 3x + 8y = 180 \end{cases}$$

Solution

Step	Example
1. Write the system of equations in the form $\begin{cases} Ax + By = E \\ Cx + Dy = F \end{cases}$	$\begin{cases} x + y = 30 \\ 3x + 8y = 180 \end{cases}$
2. Eliminate x from the second equation. To do this, multiply the first equation by -3, so that the coefficients of x are negatives of one another. Then add the two equations.	$-3(x + y = 30)$ $3x + 8y = 180$ Adding the two equations gives $-3x - 3y = -90$ $+ (3x + 8y = 180)$ $5y = 90$
3. Replace the second equation in the original system with the result of Step 2, and retain the *original* first equation.	$\begin{cases} x + y = 30 \\ 5y = 90 \end{cases}$
4. The second equation has only one variable, y. Solve this equation for y.	$5y = 90$ $y = 18$
5. Substitute this value of y into the first equation and solve for x.	$x + y = 30$ $x + (18) = 30$ $x = 12$
6. Write out the solution.	We have $x = 12$ and $y = 18$. You can check the solution by substituting the values in the original equations.

Check It Out 1   Solve the system of equations.

$$\begin{cases} x + y = 30 \\ 2x + 4y = 100 \end{cases}$$

Nature of Solutions of Systems of Linear Equations in Two Variables

Systems of linear equations can have different types of solutions, summarized as follows.

Nature of solutions of systems of linear equations in two variables

Let a system of linear equations in two variables be given as follows:

$$\begin{cases} Ax + By = E \\ Cx + Dy = F \end{cases}$$

where A, B, C, D are not all zero. The solution of this system will have one of the following properties.

- Exactly one solution (independent system)

- No solution (inconsistent system)

- Infinitely many solutions (dependent system)

Figure 1 summarizes the ways in which two lines can intersect and indicates the nature of the solutions of the corresponding linear systems.

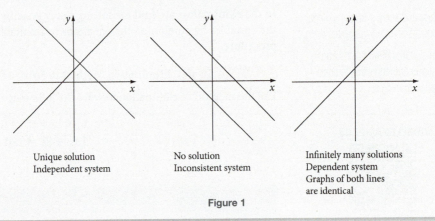

Unique solution
Independent system

No solution
Inconsistent system

Infinitely many solutions
Dependent system
Graphs of both lines
are identical

Figure 1

Example 2   **Solving Systems of Linear Equations**

Use the method of elimination to solve the following systems of linear equations, if a solution exists.

a. $\begin{cases} -3x + 2y = -5 \\ 5x + 4y = 12 \end{cases}$ **b.** $\begin{cases} -2x + y = 4 \\ 4x - 2y = -8 \end{cases}$ **c.** $\begin{cases} -2x + y = 4 \\ -6x + 3y = 10 \end{cases}$

Solution

a. To solve the system

$$\begin{cases} -3x + 2y = -5 \\ 5x + 4y = 12 \end{cases}$$

we can eliminate the x variable by finding a common multiple of the x coefficients, -3 and 5, which is 15. Multiply the first equation by 5, and the second equation by 3, and add them.

$$\begin{array}{rl} -15x + 10y = -25 & \text{5 times the first equation} \\ \underline{15x + 12y = 36} & \text{3 times the second equation} \\ 22y = 11 & \text{5 times the first equation} \\ & \text{plus 3 times the second equation.} \end{array}$$

Solve the equation $22y = 11$ to get $y = \frac{1}{2}$, and retain the original first equation to get

$$\begin{cases} -3x + 2y = -5 \\ y = \dfrac{1}{2} \end{cases}$$

Substituting $\frac{1}{2}$ for y in the first equation and solving for x gives:

$$-3x + 2\left(\frac{1}{2}\right) = -5 \implies -3x + 1 = -5 \implies -3x = -6 \implies x = 2$$

Thus, the solution of this system of equations is $x = 2$, $y = \frac{1}{2}$. You should check that this is indeed the solution.

b. To solve the system

$$\begin{cases} -2x + y = 4 \\ 4x - 2y = -8 \end{cases}$$

we can eliminate the variable x by multiplying the first equation by 2 and then adding the result to the second equation.

$-4x + 2y = 8$	2 times the first equation
$\underline{4x - 2y = -8}$	The second equation
$0 = 0$	2 times the first equation plus the second equation

Replacing the second equation with $0 = 0$ and retaining the original first equation, we can write

$$\begin{cases} -2x + y = 4 \\ 0 = 0 \end{cases}$$

The second equation ($0 = 0$) is true but gives us no useful information. Thus we have a dependent system with infinitely many solutions. Solving the first equation for y in terms of x, we get

$$-2x + y = 4 \implies y = 2x + 4$$

Thus the solution set is the set of all ordered pairs (x, y) such that $y = 2x + 4$, where x is any real number. That is, the solution set consists of all points (x, y) on the line $y = 2x + 4$.

The solution can also be expressed as

$$x = a, y = 2a + 4$$

where a is any real number.

c. To solve the system

$$\begin{cases} -2x + y = 4 \\ -6x + 3y = 10 \end{cases}$$

we can eliminate the variable x by multiplying the first equation by -3 and then adding the result to the second equation.

$6x - 3y = -12$	-3 times the first equation
$\underline{-6x + 3y = 10}$	The second equation
$0 = -2$	-3 times the first equation
	plus the second equation

Replacing the second equation with $0 = -2$ and retaining the original first equation, we can write

$$\begin{cases} -2x + y = 4 \\ 0 = -2 \end{cases}$$

The second equation, $0 = -2$, is false. Thus this system of equations is inconsistent and has no solution.

 **Check It Out 2**   Solve by elimination.

$$\begin{cases} -2x + y = 6 \\ 5x - 2y = 10 \end{cases} \quad (22, 50)$$

Solving a Linear Inequality in Two Variables

An example of a linear inequality in two variables is $2x + y > 0$. A solution of this inequality is any point (x, y) that satisfies the inequality. For example, $(1, 10)$ and $(-1, 4)$ are points that satisfy the inequality. However, $(0, -1)$ does not satisfy it.

> **Definition of a Linear Inequality**
>
> An inequality in the variables x and y is a **linear inequality** if it can be written in the form $Ax + By \leq C$ or $Ax + By < C$ (or $Ax + By \geq C$, $Ax + By > C$), where A, B, and C are constants such that A and B are not both zero. The **solution set** of such a linear inequality is the set of all points (x, y) that satisfy the inequality.

To find the solution set of a linear inequality, we first graph the corresponding equality. This graph divides the xy-plane into two regions. One of these regions will satisfy the inequality. Example 3 illustrates how an inequality in two variables is solved.

Example 3   **Solving a Linear Inequality**

Graph the solution set of each of the following linear inequalities.

a. $x \geq 3$ **b.** $y < 2x$

Solution

a. First graph the line $x = 3$ as a solid line, because it is included in the inequality. This line separates the xy-plane into two half-planes. Choose a point *not* on the line and see if it satisfies the inequality $x \geq 3$. We choose $(0, 0)$ because it is easy to check.

$$x \overset{?}{\geq} 3 \quad \Rightarrow \quad 0 \overset{?}{\geq} 3$$

Because the inequality $0 \geq 3$ is false, the half-plane that contains the point $(0, 0)$ does *not* satisfy the inequality. Thus all the points in the *other* half-plane do satisfy the inequality, and so this region is shaded. See Figure 2.

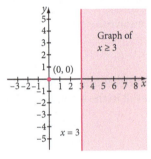

Graph of $x \geq 3$

$(0, 0)$

$x = 3$

Figure 2

b. Graph the line $y = 2x$ as a dashed line, because it is not included in the inequality. Choose a point *not* on the line and see if it satisfies the inequality $y < 2x$. We choose $(2, 0)$ because it is easy to check. Note that we cannot use $(0, 0)$ because it lies on the line $y = 2x$.

$$y \overset{?}{<} 2x \quad \Rightarrow \quad 0 \overset{?}{<} 2(2)$$

Because the inequality $0 < 4$ is true, the half-plane that contains the point $(2, 0)$ is shaded. See Figure 3.

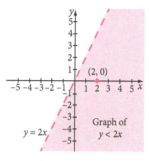

Figure 3

 Check It Out 3   Graph the inequality $y \geq x + 5$.

Solving Systems of Linear Inequalities in Two Variables

The **solution set of a system of linear inequalities** in the variables x and y consists of the set of all points (x, y) in the intersection of the solution sets of the individual inequalities of the system. To find the graph of the solution set of the system, graph the solution set of each inequality and then find the intersection of the shaded regions.

> **Example 4**   **Graphing Systems of Linear Inequalities**

Graph the following system of linear inequalities.

$$\begin{cases} x + 3y \leq 6 \\ -x + y \geq -2 \end{cases}$$

Solution First graph the inequality $x + 3y \leq 6$. Do this by graphing the line $x + 3y = 6$ and then using a test point not on the line to determine which half-plane satisfies the inequality. The test point $(0, 0)$ satisfies the inequality, and so the half-plane containing this point is shaded. All the points on the line $x + 3y = 6$ also satisfy the inequality, so this line is graphed as a solid line. See Figure 4.

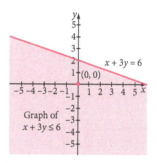

Figure 4

Using Technology

With a graphing utility, you can graph both lines in Example 4 by first solving each equation for y. You then choose the appropriate line marker to shade the portion of the plane above or below the graph of the line. See Figure 6.

Keystroke Appendix:
Section 7

Figure 6

Next graph the inequality $-x + y \geq -2$. Do this by graphing the line $-x + y = -2$ as a solid line and then using a test point not on the line to determine which half-plane satisfies the inequality. The test point $(0, 0)$ satisfies the inequality, and so the half-plane containing this point is shaded. See Figure 5.

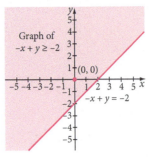

Figure 5

Although we have graphed the two inequalities separately for the sake of clarity, you should graph the second inequality on the same set of coordinate axes as the first inequality, as shown in Figure 7. The region where the two separate shaded regions overlap is the region that satisfies the system of inequalities. Every point in this overlapping region satisfies *both* of the given inequalities.

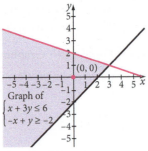

Figure 7

 Check It Out 4　Graph the following system of linear inequalities.

$$\begin{cases} x - 2y \geq 4 \\ -2x + y \leq 3 \end{cases}$$

A Linear Programming Application

In many business and industrial applications, a quantity such as profit or output may need to be optimized subject to certain conditions. These conditions, which are referred to as **constraints,** can often be expressed in the form of linear inequalities. **Linear programming** is a procedure that yields all the solutions of a given system of linear inequalities that correspond to achieving an important goal, such as maximizing profit or minimizing cost. Actual applications, such as airline scheduling, usually involve thousands of variables and numerous constraints, and are solved using sophisticated computer algorithms. In this section, we focus on problems involving only two variables and relatively few constraints.

Example 5 illustrates how a system of linear inequalities can be set up to solve a simple optimization problem.

Example 5　**Maximizing Calories Burned by Exercising**

Cathy can spend at most 30 minutes on the treadmill, in some combination of running and walking. To warm up and cool down, she must spend at least 8 minutes walking. At the walking speed, she burns off 3 calories per minute. At the running speed, she burns off 8 calories per minute. Set up the problem that must be solved to answer the following question: How many minutes should Cathy spend on each activity to maximize the total number of calories burned?

Note   The calorie we reference in Example 5, in the context of nutrition, is a food calorie, which is more accurately a kilocalorie of nutritional energy. The same word, calorie, in the context of chemistry, is 1,000 times smaller than the calorie used in nutrition. It can be confusing. In some contexts, you will see the food calorie written as Calorie (with an uppercase C), to distinguish it from the calorie used in chemistry, but this notation is not universal.

Solution   First define the variables.

x: number of minutes spent walking

y: number of minutes spent running

Using these variables, we formulate inequalities that express the stated constraints. The first constraint is that the total amount of time spent on the treadmill can be at most 30 minutes.

$$x + y \leq 30 \quad \text{First constraint}$$

In addition, we know that Cathy must spend at least 8 minutes walking. This gives us a second constraint.

$$x \geq 8 \quad \text{Second constraint}$$

Because x and y represent time, they must be nonnegative. This gives us two more constraints.

$$x \geq 0, \ y \geq 0 \quad \text{Third and fourth constraints}$$

The system of inequalities that corresponds to this set of constraints is given below. The set of points (x, y) for which *all four constraints* are satisfied is depicted by the shaded region of the graph in Figure 8.

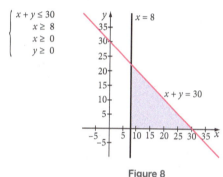

Figure 8

The number of calories burned can be expressed algebraically as

$$C = 3x + 8y$$

Thus the ultimate goal, or *objective,* is to find the point(s) (x, y) within the shaded region at which the value of C attains its maximum value.

 Check It Out 5   Rework Example 5 for the case in which the additional constraint that Cathy can spend at most 15 minutes running is imposed.

To solve optimization problems such as the one that was introduced in Example 5, we need the following important theorem.

Fundamental Theorem of Linear Programming

Given a system of linear inequalities in the variables x and y that express the constraints for an optimization problem, we can define the following.

- The **feasible set** for the optimization problem is the solution set of the system of linear inequalities.

- A **corner point** is a point at which two or more of the boundary lines of the feasible set intersect.

- A **linear objective function** is a function of the form $C = Ax + By$, where A and B are constants that are not both zero.

Then the objective function attains its maximum value (if a maximum value exists) at one or more of the corner points; the minimum value (if it exists) is also attained at one or more of the corner points.

Using this theorem, we can now solve the problem that was set up in Example 6.

Example 6   **Maximizing an Objective Function with Constraints**

Maximize the objective function $C = 3x + 8y$ subject to the following constraints.

$$\begin{cases} x + y \le 30 \\ x \ge 8 \\ x \ge 0 \\ y \ge 0 \end{cases}$$

Solution   The setup for this problem was given in Example 5. By the fundamental theorem of linear programming, we know that the objective function attains its maximum value at one or more of the corner points of the feasible set. Each of the corner points is a point of intersection of the boundary lines, which are illustrated on the graph in Figure 9.

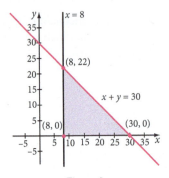

Figure 9

(x, y) (minutes)	$C = 3x + 8y$ (calories)
(8, 0)	24
(8, 22)	200
(30, 0)	90

Table 1

Table 1 lists values of the objective function C at the corner points. From the table, we see that the maximum number of calories that will be burned under the given constraints is 200, and that Cathy will have to spend 8 minutes walking and 22 minutes running to burn this many calories. Note that the minimum amount of time needed to satisfy the walking constraint (8 minutes) turns out to be the value of x in the optimal solution. This makes sense from a physical standpoint, because walking burns off fewer calories per minute than running does.

Check It Out 6   Rework Example 6 for the case in which Cathy is required to spend at least 6 minutes walking.

Exercises 8.1

Just in Time Exercises

In Exercises 1–6, find the intersection of the two lines.

1. $x = -1, y = 2$

2. $x + y = 7, y = 3$

3. $x = 3, y = 2x - 1$

4. $6x + 3y = 9, x - y = 3$

5. $2x - y = 6, x + 2y = 8$

6. $x + y = 4, x - y = 12$

Skills

In Exercises 7–12, verify that each system of equations has the indicated solution.

7. $\begin{cases} x + 5y = -6 \\ -x + 2y = -8 \end{cases}$

Solution: $x = 4, y = -2$

8. $\begin{cases} -x - 2y = 5 \\ -2x + y = -5 \end{cases}$

Solution: $x = 1, y = -3$

9. $\begin{cases} 2x - 3y = -6 \\ -x + 2y = 4 \end{cases}$

Solution: $x = 0, y = 2$

10. $\begin{cases} 2x - y = 2 \\ 6x - 5y = 8 \end{cases}$

Solution: $x = \dfrac{1}{2}, y = -1$

11. $\begin{cases} x - y = 5 \\ -2x + 2y = -10 \end{cases}$

Solution: $x = a, y = a - 5$
(for every real number a)

12. $\begin{cases} 2x - y = 1 \\ 8x - 4y = 4 \end{cases}$

Solution: $x = a, y = -1 + 2a$
(for every real number a)

In Exercises 13–30, use elimination to solve each system of equations. Check your solution.

13. $\begin{cases} x + 2y = 7 \\ -2x + y = 1 \end{cases}$

14. $\begin{cases} -x - y = -7 \\ 3x + 4y = 24 \end{cases}$

15. $\begin{cases} -3x + y = 5 \\ 6x - y = -8 \end{cases}$

16. $\begin{cases} 5x + 3y = -1 \\ -10x + 2y = 26 \end{cases}$

17. $\begin{cases} 2x + y = 5 \\ 4x + 2y = 3 \end{cases}$

18. $\begin{cases} -3x + 4y = 9 \\ 6x - 8y = 3 \end{cases}$

19. $\begin{cases} 3x - y = 9 \\ x + y = -1 \end{cases}$

20. $\begin{cases} 5x - 3y = 23 \\ x + y = -13 \end{cases}$

21. $\begin{cases} x + y = 5 \\ -2x - 2y = -10 \end{cases}$

22. $\begin{cases} -3x + 4y = 9 \\ 9x - 12y = -27 \end{cases}$

23. $\begin{cases} 3x - 6y = 2 \\ y = -3 \end{cases}$

24. $\begin{cases} -2x = -4 \\ -4x + 3y = -3 \end{cases}$

25. $\begin{cases} -2x - 3y = 0 \\ 3x + 5y = -2 \end{cases}$

26. $\begin{cases} 5x - 2y = -3 \\ 3x - y = 1 \end{cases}$

27. $\begin{cases} -3x + 2y = \dfrac{1}{2} \\ 4x + y = 3 \end{cases}$

28. $\begin{cases} 4x + 3y = \dfrac{9}{2} \\ 5x + y = 7 \end{cases}$

29. $\begin{cases} \dfrac{1}{2}x + \dfrac{1}{3}y = -2 \\ \dfrac{1}{4}x + \dfrac{2}{3}y = -\dfrac{5}{2} \end{cases}$

(*Hint*: Clear fractions first to simplify the arithmetic.)

30. $\begin{cases} \dfrac{1}{5}x - \dfrac{3}{2}y = 4 \\ -\dfrac{2}{3}x + \dfrac{1}{2}y = -\dfrac{13}{3} \end{cases}$

(*Hint*: Clear fractions first to simplify the arithmetic.)

In Exercises 31–36, write each system of equations in the form $\begin{cases} Ax + By = E \\ Cx + Dy = F \end{cases}$, and then solve the system.

31. $\begin{cases} 3(x + y) = 1 \\ -2x = -y + 2 \end{cases}$

32. $\begin{cases} 2x = -3y + 4 \\ \dfrac{x + y}{3} = 1 \end{cases}$

33. $\begin{cases} -4y = x + 5 \\ \dfrac{x}{3} + \dfrac{y}{2} = 1 \end{cases}$

34. $\begin{cases} 3y = 7 - x \\ -\dfrac{2x}{3} + \dfrac{3y}{2} = 5 \end{cases}$

35. $\begin{cases} \dfrac{x + 1}{2} + \dfrac{y - 1}{3} = 1 \\ 3x + y = 7 \end{cases}$

36. $\begin{cases} -2x - y = -7 \\ \dfrac{2x}{3} + \dfrac{y + 1}{2} = 1 \end{cases}$

In Exercises 37–40, use a graphing utility to approximate the solution set of each system. If there is no solution, state that the system is inconsistent.

37. $\begin{cases} 0.3x = y - 4 \\ 0.5x + y = 1 \end{cases}$
38. $\begin{cases} 1.9x = y + 2.6 \\ -0.5x - y = 1.7 \end{cases}$
39. $\begin{cases} 1.2x - 0.4y = -2 \\ 0.5x + 1.3y = 3.2 \end{cases}$
40. $\begin{cases} -3.2x + 2.5y = -5.3 \\ 1.6x - 2.8y = 4.7 \end{cases}$

In Exercises 41 and 42, find a system of equations whose solution is indicated graphically.

41.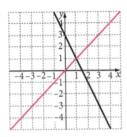

42.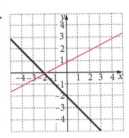

In Exercises 43–52, graph the solution set of each inequality.

43. $x < -2$
44. $y \geq 3$
45. $-2x + y \geq -4$
46. $-3x + y \leq 2$
47. $x > y + 6$

48. $y < 2x + 7$
49. $3x - 4y > 12$
50. $6x - 2y \leq 24$
51. $3x + 5y > 10$
52. $7x \leq 10x + 3$

In Exercises 53–70, graph the solution set of each system of inequalities.

53. $\begin{cases} -2x + y \leq 8 \\ -x + y \geq 2 \end{cases}$
54. $\begin{cases} x + y \leq 10 \\ 3x - y \geq 6 \end{cases}$
55. $\begin{cases} y \geq -2 \\ 5x + 2y \leq 10 \end{cases}$
56. $\begin{cases} x \leq 10 \\ x + y \geq 7 \end{cases}$

57. $\begin{cases} -x + y < 3 \\ x + y > -5 \end{cases}$
58. $\begin{cases} -x + y > 5 \\ x + y < 1 \end{cases}$
59. $\begin{cases} x + y \leq 4 \\ -4x + 2y \geq -1 \end{cases}$
60. $\begin{cases} x \leq 5 \\ x \geq 2 \\ y > 1 \end{cases}$

61. $\begin{cases} y \leq -2 \\ y \geq -5 \\ x \leq -1 \end{cases}$
62. $\begin{cases} -2x + 3y \leq 7 \\ -2x + 9y \geq 1 \\ 2x + 3y \leq 11 \end{cases}$
63. $\begin{cases} x + y \leq 4 \\ -x + y \leq 4 \\ x + 5y \geq 8 \end{cases}$
64. $\begin{cases} 3x - y \geq -1 \\ -2x - y \leq 4 \end{cases}$

65. $\begin{cases} -3x + y \leq -1 \\ 4x + y \leq 6 \end{cases}$
66. $\begin{cases} -x - y \geq 3 \\ 2x - y \leq 1 \end{cases}$
67. $\begin{cases} x \leq 0 \\ -5x + 4y \leq 20 \\ 3x + 4y \geq -12 \end{cases}$
68. $\begin{cases} x \geq 0 \\ 2x + 3y \leq 2 \\ -4x + 3y \geq -4 \end{cases}$

69. $\begin{cases} -\frac{4}{3}x + y \geq -5 \\ -4x + 3y \leq 13 \\ x + y \geq 2 \end{cases}$
70. $\begin{cases} -\frac{3}{2}x + y \geq -3 \\ 2x + y \leq 4 \\ 2x + y \geq -3 \end{cases}$

In Exercises 71–74, solve the optimization problem.

71. Maximize $P = 10x + 8y$ subject to the following constraints.

$$\begin{cases} x \geq 5 \\ y \geq 2 \\ x \leq 9 \\ y \leq 10 \end{cases}$$

72. Maximize $P = 12x + 15y$ subject to the following constraints.

$$\begin{cases} x \geq 3 \\ y \geq 1 \\ x \leq 10 \\ y \leq 14 \end{cases}$$

73. Minimize $P = 16x + 10y$ subject to the following constaints.

$$\begin{cases} y \le 2x \\ x \le 5 \\ x \ge 0 \\ y \ge 0 \end{cases}$$

74. Minimize $P = 20x + 30y$ subject to the following constraints.

$$\begin{cases} 3x + y \le 9 \\ y \ge x \\ y \ge 2 \\ x \ge 0 \end{cases}$$

Applications

In this set of exercises, you will use systems of linear equations and inequalities to study real-world problems.

75. Film Industry The total revenue generated by a film comes from two sources: box-office ticket sales and the sale of merchandise associated with the film. It is estimated that for a very popular film such as *Spiderman* or *Harry Potter,* the revenue from the sale of merchandise is four times the revenue from ticket sales. Assume this is true for the film *Spiderman,* which grossed a total of $3 billion. Find the revenue from ticket sales and the revenue from the sale of merchandise.

(Source: The Economist)

76. Nutrition According to health professionals, the daily intake of fat in a diet that consists of 2000 calories per day should not exceed 50 grams. The total fat content of a meal that consists of a Whopper and a medium order of fries exceeds this limit by 14 grams. Two Whoppers and a medium order of fries have a total fat content of 111 grams. Set up and solve a system of equations to find the fat content of a Whopper and the fat content of a medium order of fries.

(Source: www.burgerking.com)

77. Nutrition The following table lists the caloric content of a typical fast-food meal.

Food (single serving)	calories
Cheeseburger	330
Medium order of fries	450
Medium cola (21 oz)	220

(Source: www.mcdonalds.com)

a. After a lunch that consists of a cheeseburger, a medium order of fries, and a medium cola, you decide to burn off a quarter of the total calories in the meal by some combination of running and walking. You know that running burns 8 calories per minute and walking burns 3 calories per minute. If you exercise for a total of 40 minutes, how many minutes should you spend on each activity?

b. Rework part (a) for the case in which you exercise for a total of only 20 minutes. Do you get a realistic solution? Explain your answer.

78. Criminology In 2004, there were a total of 3.38 million car thefts and burglaries in the United States. The number of burglaries exceeded the number of car thefts by 906,000. Find the number of burglaries and the number of car thefts. *(Source: Federal Bureau of Investigation)*

79. Ticket Pricing An airline charges $380 for a round-trip flight from New York to Los Angeles if the ticket is purchased at least 7 days in advance of travel. Otherwise, the price is $700. If a total of 80 tickets are purchased at a total cost of $39,040, find the number of tickets sold at each price.

80. Utilities In a residential area serviced by a utility company, the percentage of single-family homes with central air conditioning was 4 percentage points higher than 5 times the percentage of homes without central air. What percentage of these homes had central air-conditioning, and what percentage did not?

81. Mixture A chemist wishes to make 10 gallons of a 15% acid solution by mixing a 10% acid solution with a 25% acid solution.

   a. Let x and y denote the total volumes (in gallons) of the 10% and 25% solutions, respectively. Using the variables x and y, write an equation for the total volume of the 15% solution (the mixture).

   b. Using the variables x and y, write an equation for the total volume of acid in the mixture by noting the following fact:

   Volume of acid in 15% solution = volume of acid in 10% solution + volume of acid in 25% solution.

   c. Solve the system of equations from parts (a) and (b), and interpret your solution.

   d. Is it possible to obtain a 5% acid solution by mixing a 10% solution with a 25% solution? Explain without solving any equations.

82. Financial Planning A couple has $20,000 to invest for their child's future college expenses. Their accountant recommends placing at least $12,000 in a high-yield mutual fund and at most $6000 in a low-yield mutual fund.

   a. Use x to denote the amount of money placed into the high-yield fund. Use y to denote the amount of money placed into the low-yield fund. Write a system of linear inequalities that describes the possible amounts in each type of account.

   b. Graph the region that represents all possible amounts the couple could place into each account if they wish to follow the accountant's advice.

83. Financial Planning A couple has $10,000 to invest for their child's wedding. Their accountant recommends placing at least $6000 in a high-yield investment and no more than $4000 in a low-yield investment.

   a. Use x to denote the amount of money placed into the high-yield investment. Use y to denote the amount of money placed into the low-yield investment. Write a system of linear inequalities that describes the possible amounts the couple could invest in each type of venture.

   b. Graph the region that represents all possible amounts the couple could put into each investment if they wish to follow the accountant's advice.

84. Ticket Pricing The gymnasium at a local high school has 1000 seats. The state basketball championship game will be held there next Friday night. Tickets to the game will be priced as follows: $10 if purchased in advance and $15 if purchased at the door. The booster club feels that at least 100 tickets will be sold at the advance ticket price. Total sales of at least $5000 are expected.

   a. Use x to denote the number of tickets sold in advance. Use y to denote the number of tickets sold at the door. Write a system of linear inequalities that describes the possible numbers of tickets sold at each price.

   b. Graph the region that represents all possible combinations of ticket sales.

85. Ticket Pricing The auditorium at the library has 200 seats. Tickets to a lecture by a local author are priced as follows: $6 if purchased in advance and $8 if purchased at the door. According to past records, at least 40 tickets will be sold in advance, and total sales of at least $960 are expected.

   a. Use x to denote the number of tickets sold in advance. Use y to denote the number of tickets sold at the door. Write a system of linear inequalities that describes the possible numbers of tickets sold at each price.

   b. Graph the region that represents all possible combinations of ticket sales.

86. Inventory Control A store sells two brands of mobile phones, Brand A and Brand B. Past sales records indicate that the store must have at least twice as many Brand B phones in stock as Brand A phones. The store must have at least two Brand A phones in stock. The maximum inventory the store can carry is 20 phones.

   a. Use x to denote the number of Brand A phones in stock. Use y to denote the number of Brand B phones in stock. Write a system of linear inequalities that describes the possible numbers of each brand of phone in stock.

   b. Graph the region that represents all possible numbers of Brand A and Brand B phones the store could have in stock.

87. **Inventory Control** A grocery store sells two types of carrots, organic and non-organic. Past sales records indicate that the store must stock at least three times as many pounds of organic carrots as non-organic carrots. The store needs to have at least 5 pounds of non-organic carrots in stock. The maximum inventory of carrots the store can carry is 30 pounds.

    a. Use x to denote the number of pounds of organic carrots in stock. Use y to denote the number of pounds of non-organic carrots in stock. Write a system of linear inequalities that describes the possible numbers of pounds of each type of carrot the store could keep in stock.

    b. Graph the region that represents all possible amounts of organic and non-organic carrots the store could have in stock.

88. **Maximizing Profit** An electronics firm makes a clock radio in two different models: one (model 380) with a battery backup feature and the other (model 360) without. It takes 1 hour and 15 minutes to manufacture each unit of the model 380 radio, and only 1 hour to manufacture each unit of the model 360. At least 500 units of the model 360 radio are to be produced. The manufacturer realizes a profit per radio of $15 for the model 380 and only $10 for the model 360. If at most 2000 hours are to be allocated to the manufacture of the two models combined, how many of each model should be made to maximize the total profit?

89. **Maximizing Profit** A telephone company manufactures two different models of phones: Model 120 is cordless and Model 140 is not cordless. It takes 1 hour to manufacture the cordless model and 1 hour and 30 minutes to manufacture the traditional phone. At least 300 of the cordless models are to be produced. The manufacturer realizes a profit per phone of $12 for Model 120 and $10 for Model 140. If at most 1000 hours are to be allocated to the manufacture of the two models combined, how many of each model should be produced to maximize the total profit?

90. **Minimizing Commuting Time** Bill can't afford to spend more than $90 per month on transportation to and from work. The bus fare is only $1.50 one way, but it takes Bill 1 hour and 15 minutes to get to work by bus. If he drives the 20-mile round trip, his one-way commuting time is reduced to 1 hour, but it costs him $.45 per mile. If he works at least 20 days per month, how often does he need to drive in order to minimize his commuting time and keep within his monthly budget?

91. **Minimizing Commuting Time** Sarah can't afford to spend more than $90 per month on transportation to and from work. The bus fare is only $1.50 one way, but it takes Sarah 1 hour and 15 minutes to get to work by bus. If she drives the 15-mile round trip, her one-way commuting time is reduced to 40 minutes, but it costs her $.40 per mile. If she works at least 20 days a month, how often does she have to drive in order to minimize her commuting time and keep within her monthly budget?

92. **Maximizing Profit** A cosmetics company makes a profit of 15 cents on a tube of lipstick and a profit of 8 cents on a tube of lip gloss. To meet dealer demand, the company needs to produce between 300 and 800 tubes of lipstick and between 100 and 300 tubes of lip gloss. The maximum number of tubes of lipstick and lip gloss the company can produce per day is 800. How many of each type of beauty product should be produced to maximize profit?

93. **Maximizing Profit** A golf club manufacturer makes a profit of $3 on a driver and a profit of $2 on a putter. To meet dealer demand, the company needs to produce between 20 and 50 drivers and between 30 and 50 putters each day. The maximum number of clubs produced each day by the company is 80. How many of each type of club should be produced to maximize profit?

Concepts

94. The sum of money invested in two savings accounts is $1000. If both accounts pay 4% interest compounded annually, is it possible to earn a total of $50 in interest in the first year?

   a. Explain your answer in words.

   b. Explain your answer using a system of equations.

95. The following is a system of three equations in only two variables.

$$\begin{cases} x - y = 1 \\ x + y = 1 \\ 2x - y = 1 \end{cases}$$

   a. Graph the solution of each of these equations.

   b. Is there a single point at which *all three lines* intersect?

   c. Is there one ordered pair (x, y) that satisfies *all three equations*? Why or why not?

96. State the conditions that a system of three equations in two variables must satisfy in order to have just one solution. Give an example of such a system, and illustrate your example graphically.

97. Adult and children's tickets for a certain show sell for $8 each. A total of 1000 tickets are sold, with total sales of $8000. Is it possible to figure out exactly how many of each type of ticket were sold? Why or why not?

Systems of Linear Equations in Three Variables

8.2

Objectives

- Solve systems of linear equations in three variables using Gaussian elimination.

- Understand what is meant by an inconsistent system of linear equations in three variables.

- Understand what is meant by a dependent system of linear equations in three variables.

- Use systems of linear equations in three variables to solve applied problems.

VIDEO EXAMPLES

SECTION 8.2

In this section, we present an elimination method that can be used to solve systems of linear equations in three variables.

Solving Systems of Linear Equations in Three Variables

The general form of a linear equation in the variables x, y, and z is $Ax + By + Cz = D$, where A, B, C, and D are constants such that A, B and C are not all zero. A **system of linear equations** in these variables is a collection of equations that can be put into this form. A solution of such a system of linear equations is an **ordered triple** of numbers (x, y, z) that satisfies all the equations in the system.

We now introduce a process known as **Gaussian elimination,** which can be used to find the solution(s) of a system of linear equations in three variables. Gaussian elimination, named after the German mathematician Karl Friedrich Gauss (1777−1855), is an extension of the method of elimination for solving systems of two linear equations in two variables.

We will first apply this method to a simple system of equations whose solution is easy to find.

Example 1 **Solving a System of Linear Equations in Three Variables**

Solve the following system of equations.

$$\begin{cases} x + y - z = 10 \\ \phantom{x + {}} y + z = 0 \\ \phantom{x + y + {}} z = 2 \end{cases}$$

Solution From the last equation, we immediately have

$$z = 2$$

Note that the middle equation contains only the variables y and z. Since we know that $z = 2$, we can substitute this value of z into the second equation to find y.

$$\begin{array}{ll} y + z = 0 & \text{The middle equation} \\ y + 2 = 0 & \text{Substitute } z = 2 \\ y = -2 & \text{Solve for } y \end{array}$$

Now that we have found both z and y, we can find x by using the first equation. Substituting $y = -2$ and $z = 2$ into the first equation, we have

$$x + (-2) - 2 = 10 \Rightarrow x - 2 - 2 = 10 \Rightarrow x - 4 = 10 \Rightarrow x = 14$$

Thus our solution is $x = 14$, $y = -2$, $z = 2$. You may use substitution to verify that these values of x, y, and z do indeed satisfy all three equations.

✓ *Check It Out 1* Solve the following system of equations.

$$\begin{cases} x - y + z = 10 \\ \phantom{x - {}} y - z = 0 \\ \phantom{x - y + {}} z = -1 \end{cases}$$

The procedure illustrated in Example 1 is known as **back-substitution**. That is, we start with a known value of one variable and successively work backward to find the values of the other variables.

To solve most systems of equations, we must first put the system in a form similar to the one in Example 1 by using algebraic manipulations.

The following operations will be used in solving systems of linear equations by Gaussian elimination.

Operations Used in Manipulating a System of Linear Equations

1. Interchange two equations in the system.
2. Multiply one equation in the system by a nonzero constant.
3. Multiply one equation in the system by a nonzero constant and add the result to another equation in the system.

Whenever these operations are performed on a given system of equations, the resulting system of equations is equivalent to the original system. That is, the two systems will have the same solution(s).

Example 2   **Solving a System of Linear Equations Using Gaussian Elimination**

Solve the system of equations.

$$\begin{cases} -2x + 2y + z = -6 \\ \quad x - 2y + 2z = -1 \\ \quad 3x + 2y - z = \quad 3 \end{cases}$$

Solution Begin by labeling the equations.

$$\begin{cases} -2x + 2y + z = -6 & (1) \\ \quad x - 2y + 2z = -1 & (2) \\ \quad 3x + 2y - z = \quad 3 & (3) \end{cases}$$

Our first goal is to eliminate the variable x from the first two equations. Check to see if the coefficient of x in either of the last two equations is 1. If so, we will interchange that equation with Equation (1) to make the elimination easier.

Step 1 Interchange Equations (1) and (2), since the coefficient of x in Equation (2) is 1.

$$\begin{cases} \quad x - 2y + 2z = -1 & (4) = (2) \\ -2x + 2y + z = -6 & (5) = (1) \\ \quad 3x + 2y - z = \quad 3 & (3) \end{cases}$$

We have relabeled the first two equations to make the subsequent discussion less confusing.

Step 2 Eliminate x from Equation (5).

Multiply Equation (4) by 2 and add the result to Equation (5):

$$\begin{array}{ll} 2x - 4y + 4z = -2 & 2 \cdot (4) \\ -2x + 2y + \ z = -6 & (5) \\ \hline \quad\quad -2y + 5z = -8 & 2 \cdot (4) + (5) \end{array}$$

Replace Equation (5) with the result and relabel it as Equation (6), and retain Equation (4):

$$\begin{cases} x - 2y + 2z = -1 & (4) \\ \quad\ -2y + 5z = -8 & (6) = 2 \cdot (4) + (5) \\ 3x + 2y - z = \quad 3 & (3) \end{cases}$$

Step 3 Eliminate x from Equation (3).

Multiply Equation (4) by -3 and add the result to Equation (3):

$$\begin{array}{ll} -3x + 6y - 6z = 3 & -3 \cdot (4) \\ \quad 3x + 2y - \ z = 3 & (3) \\ \hline \quad\quad\ 8y - 7z = 6 & -3 \cdot (4) + (3) \end{array}$$

Replace Equation (3) with the result and relabel it as Equation (7), and retain Equation (4):

$$\begin{cases} x - 2y + 2z = -1 & (4) \\ \quad -2y + 5z = -8 & (6) \\ \quad 8y - 7z = 6 & (7) = -3 \cdot (4) + (3) \end{cases}$$

Our next goal is to eliminate the variable y from Equation (7). In doing so, we will make use of Equation (6). Note that both Equation (6) and (7) contain only the two variables y and z.

Step 4 Eliminate y from Equation (7).
Multiply Equation (6) by 4 and add the result to Equation (7):

$$\begin{array}{rl} -8y + 20z = -32 & 4 \cdot (6) \\ \underline{8y - 7z = 6} & (7) \\ 13z = -26 & 4 \cdot (6) + (7) \end{array}$$

Replace Equation (7) with the result and relabel it as Equation (8), and retain Equation (6):

$$\begin{cases} x - 2y + 2z = -1 & (4) \\ \quad -2y + 5z = -8 & (6) \\ \quad\quad\quad 13z = -26 & (8) = 4 \cdot (6) + (7) \end{cases}$$

We now have a system of equations that is very similar in structure to the one given in Example 1: Equation (8) contains only the variable z; Equation (6) contains only the variables y and z; and Equation (4) contains all three variables, x, y, z. Use back-substitution to find the value of each variable.

Step 5 Use Equation (8) to solve for z:

$$13z = -26 \Rightarrow z = -2$$

Proceeding as in Example 1, substitute this value into Equation (6) to solve for y:

$$-2y + 5(-2) = -8 \Rightarrow -2y - 10 = -8 \Rightarrow -2y = 2 \Rightarrow y = -1$$

Substitute $z = -2$ and $y = -1$ into Equation (4) to solve for x:

$$x - 2(-1) + 2(-2) = -1 \Rightarrow x + 2 - 4 = -1 \Rightarrow x - 2 = -1 \Rightarrow x = 1$$

Thus the solution is $x = 1$, $y = -1$, $z = -2$. You should verify that these values of x, y, and z satisfy all three equations.

✓ Check It Out 2   Solve the system of equations.

$$\begin{cases} 2x + 2y + z = 1 \\ -3x - 2y + 2z = -5 \\ x + 2y - z = 2 \end{cases}$$

Every linear equation in three variables is represented by a plane in three-dimensional space. Therefore, the solution set of a system of linear equations in three variables can be thought of geometrically as the set of points at which all the planes associated with the equations in the system intersect. Figure 1 summarizes the ways in which three planes can intersect and indicates the nature of the solution(s) of the corresponding system of three linear equations in three variables.

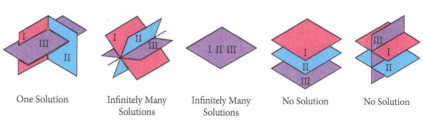

| One Solution | Infinitely Many Solutions | Infinitely Many Solutions | No Solution | No Solution |

Figure 1

The Gaussian elimination method presented in this section can be generalized to find the solutions of systems of linear equations in more than three variables. However, it is difficult to visualize the geometric representation of a linear equation in more than three variables.

Systems of Linear Equations in Three Variables with Either Infinitely Many Solutions or No Solution

When manipulating a system of linear equations, it is possible to end up with an equation of the form $0 = 0$. Such an equation is of no value and you must work with the remaining two equations, as illustrated in the following example.

Example 3   **A System with Infinitely Many Solutions**

Use Gaussian elimination to solve the following system of linear equations.

$$\begin{cases} -3x + y - 2z = 1 \\ 5x - y + 4z = -1 \end{cases}$$

Solution   First, label the equations.

$$\begin{cases} -3x + y - 2z = 1 & (1) \\ 5x - y + 4z = -1 & (2) \end{cases}$$

The coefficients of x in this system of equations are of opposite sign, but neither of them is a multiple of the other. The least common multiple of -3 and 5 is 15, so we will multiply Equation (2) by 3 and then replace Equation (2) with the result.

$$\begin{cases} -3x + y - 2z = 1 & (1) \\ 15x - 3y + 12z = -3 & (3) = 3 \cdot (2) \end{cases}$$

The next step is to eliminate x from Equation (3): Multiply Equation (1) by 5 and add the result to Equation (3).

$$\begin{cases} -3x + y - 2z = 1 & (1) \\ 2y + 2z = 2 & (4) = 5 \cdot (1) + (3) \end{cases}$$

Because there is no third equation, we cannot eliminate y. Instead, we use the second equation to solve for y in terms of z:

$$2y + 2z = 2 \Rightarrow 2y = -2z + 2 \Rightarrow y = -z + 1$$

Now we can substitute $-z + 1$ for y in Equation (1), which will yield an equation in the variables x and z. After simplifying, we will solve for x in terms of z.

$$-3x + (-z + 1) - 2z = 1$$

$$-3x - 3z + 1 = 1$$

$$-3x = 3z \Rightarrow x = -z$$

The solution set is thus the set of all x, y, z such that $x = -z$ and $y = -z + 1$, where z is any real number. Since the expression for at least one of the variables x or y depends explicitly on z, this system of equations is **dependent and has infinitely many solutions.**

Individual solutions can be found by choosing specific values for z. For example, if $z = 0$, then $y = 1$ and $x = 0$. If $z = 2$, then $y = -1$ and $x = -2$.

Check It Out 3   Use Gaussian elimination to solve the following system of linear equations.

$$\begin{cases} -x + y - z = 3 \\ 4x - 5y + z = -2 \end{cases}$$

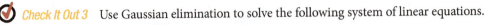

Example 4 **A System of Equations with No Solution**

Use Gaussian elimination to solve the following system of linear equations.

$$\begin{cases} x & + z = 3 \\ 3x + y + 2z = 0 \\ 4x + y + 3z = 5 \end{cases}$$

Solution First, label the equations.

$$\begin{cases} x & + z = 3 & (1) \\ 3x + y + 2z = 0 & (2) \\ 4x + y + 3z = 5 & (3) \end{cases}$$

Next, eliminate x from Equations (2) and (3).

$$\begin{cases} x & + z = & 3 & (1) \\ & y - z = -9 & (4) = -3 \cdot (1) + (2) \\ & y - z = -7 & (5) = -4 \cdot (1) + (3) \end{cases}$$

Finally, eliminate y from Equation (5).

$$\begin{cases} x & + z = & 3 & (1) \\ & y - z = -9 & (4) \\ & 0 = & 2 & (6) = -1 \cdot (4) + (5) \end{cases}$$

The last equation $(0 = 2)$ is false. Thus this system of equations is **inconsistent and has no solution.**

 Check It Out 4 Use Gaussian elimination to solve the following system of linear equations.

$$\begin{cases} x + y - z = -4 \\ 3x & + 2z = 1 \\ -2x + y - 3z = 6 \end{cases}$$

Application of Systems of Linear Equations

Example 5 gives an application of linear system to personal finance.

Example 5 **An Investment Problem**

An investment counselor would like to advise her client about three types of investments: a stock based mutual fund, a corporate bond, and a savings bond. The counselor wants to distribute the total investment amount according to the client's risk tolerance. Risk factors for each investment types range from 1 to 5, with 1 being the most risky, and are summarized in Table 1. The client can tolerate an overall risk level of 3.5, and wants the amount invested in the corporate bond to equal the amount invested in the savings bond. Set up and solve a system of equations to determine the percentage of the total investment that should be allocated to each investment type.

Investment instrument	Risk factor
Stock based mutual fund	2
Corporate bond	4
Savings bond	5

Table 1

Solution First, define the variables:

x: percentage (in decimal form) invested in mutual fund

y: percentage (in decimal form) invested in corporate bond

z: percentage (in decimal form) invested in savings bond

Formulating the first equation

Because the percentage is assumed to be in decimal form, the total percentage should add up to 1:

$$x + y + z = 1$$

Formulating the second equation

To find an expression for the overall risk, mutiply the percentage allocated to each investment by its corresponding risk factor and set it equal to the risk tolerance of 3.5.

$$\text{Overall risk} = 2x + 4y + 5z = 3.5$$

Formulating the third equation

Because the percentage invested in the corporate bond must equal the percentage invested in the savings bond, the third equation is

$$y = z$$

We have the following system of equations.

$$\begin{cases} x + y + z = 1 \\ 2x + 4y + 5z = 3.5 \\ \qquad\quad y = z \end{cases}$$

We now solve the system of equations. Rewrite the system with the variables on the left and the constants on the right, and label the equations.

$$\begin{cases} x + y + z = 1 & (1) \\ 2x + 4y + 5z = 3.5 & (2) \\ \quad\; y - z = 0 & (3) \end{cases}$$

Equation (2) contains the variable x, but Equation (3) does not, so eliminate x from Equation (2) only: Multiply Equation (1) by -2, and add the result to Equation (2).

$$\begin{cases} x + y + z = 1 & (1) \\ \quad\; 2y + 3z = 1.5 & (4) = -2 \cdot (1) + (2) \\ \quad\; y - z = 0 & (3) \end{cases}$$

Since y has a coefficient of 1 in Equation (3), interchange Equations (4) and (3) to make the arithmetic easier.

$$\begin{cases} x + y + z = 1 & (1) \\ \quad\; y - z = 0 & (5) = (3) \\ \quad\; 2y + 3z = 1.5 & (6) = (4) \end{cases}$$

Eliminate y from Equation (6): Multiply Equation (5) by -2, and add the result to Equation (6).

$$\begin{cases} x + y + z = 1 & (1) \\ \quad\; y - z = 0 & (5) \\ \quad\quad\; 5z = 1.5 & (7) = -2 \cdot (5) + (6) \end{cases}$$

Solving the last equation, we get $z = 0.3$. Substituting $z = 0.3$ into the second equation yields $y = 0.3$. Since $x + y + z = 1$, substitution of $y = 0.3$ and $z = 0.3$ gives $x = 0.4$. So 40% of the total investment is allocated to the mutual fund (x), and the corporate bond (y) and the savings bond (z) are each allocated 30% of the total amount.

 Check It Out 5 Solve the following system of equations.

$$\begin{cases} x + y + z = 1 \\ 2x + 3y + 5z = 2 \\ \qquad\qquad z = x \end{cases}$$

Exercises 8.2

In Exercises 1–6, verify that each system of equations has the indicated solution.

1. $\begin{cases} 3x \quad\;\; - z = 3 \\ 2x + y - 2z = 0 \\ 3x - 2y + z = 7 \end{cases}$

Solution: $x = 1, y = -2, z = 0$

2. $\begin{cases} x - 2y - 4z = \;\; 4 \\ -2x \quad\quad - 3z = -7 \\ 5x + y - 2z = \;\; 5 \end{cases}$

Solution: $x = 2, y = -3, z = 1$

3. $\begin{cases} 5x - y + 3z = -13 \\ x - y + 2z = \;\; -9 \\ 4x - y + z = \;\; -5 \end{cases}$

Solution: $x = 0, y = 1, z = -4$

4. $\begin{cases} -3x + 2y - z = -1 \\ 2x \quad\quad + 5z = \;\; 18 \\ -4x + 2y + z = \;\; 8 \end{cases}$

Solution: $x = -1, y = 0, z = 4$

5. $\begin{cases} x + 2y - 3z = \;\; 1 \\ 2x + 3y + z = -3 \end{cases}$

Solution: $x = -9 - 11z, y = 5 + 7z$
(for every real number z)

6. $\begin{cases} x - 3y + z = \;\; 4 \\ -x + 2y - 2z = -7 \end{cases}$

Solution: $x = 13 - 4z, y = 3 - z$
(for every real number z)

In Exercises 7–10, use back-substitution to solve the system of linear equations.

7. $\begin{cases} x - y + z = \;\; 1 \\ -2y - z = \;\; 0 \\ z = -2 \end{cases}$

8. $\begin{cases} x + y - z = -1 \\ -y - 3z = -2 \\ 2z = \;\; 4 \end{cases}$

9. $\begin{cases} -2u + v + 3w = -1 \\ v - w = \;\; 1 \\ 3w = \;\; 9 \end{cases}$

10. $\begin{cases} 2u - 3v + w = -1 \\ v - w = \;\; 1 \\ -2w = \;\; 8 \end{cases}$

In Exercises 11–38, use Gaussian elimination to solve the system of linear equations. If there is no solution, state that the system is inconsistent.

11. $\begin{cases} x + 2y \quad\quad = \;\; 2 \\ -x + y + 3z = \;\; 4 \\ 3y - 3z = -6 \end{cases}$

12. $\begin{cases} x - 2y + z = \;\; 2 \\ -x + 4y + 3z = -8 \\ -6y + 2z = \;\; 4 \end{cases}$

13. $\begin{cases} x + 2y \quad\quad = \;\; 0 \\ -2x + 4y + 8z = \;\; 8 \\ 3x \quad\quad - 3z = -9 \end{cases}$

14. $\begin{cases} x + 3y - z = \;\; 4 \\ -x - 2y \quad = -8 \\ 2x + 4y - z = \;\; 10 \end{cases}$

15. $\begin{cases} 5x + y \quad\quad = \;\; 0 \\ x \quad - z = \;\; 2 \\ 4y + z = -2 \end{cases}$

16. $\begin{cases} 4x - y \quad = 2 \\ y - 2z = 1 \\ 6x \quad - z = 3 \end{cases}$

17. $\begin{cases} x - 4y - 3z = -3 \\ y + z = \;\; 2 \\ x + 3y + 3z = \;\; 0 \end{cases}$

18. $\begin{cases} 4x + y - 2z = \;\; 6 \\ -x - y + z = -2 \\ 3x \quad - z = \;\; 5 \end{cases}$

19. $\begin{cases} 3x + 2y + 3z = \;\; 1 \\ x - y - z = \;\; 1 \\ x + 4y + 5z = -1 \end{cases}$

20. $\begin{cases} -2x \quad\quad + 4z = \;\; 4 \\ x - 2y \quad = \;\; 2 \\ -3x - 2y + 4z = -2 \end{cases}$

21. $\begin{cases} x \quad\quad + 2z = \;\; 0 \\ y + z = -1 \\ x + 8y + 4z = \;\; 1 \end{cases}$

22. $\begin{cases} x + y + 4z = -1 \\ 2x + y + 2z = \;\; 3 \\ 3x \quad - 6z = \;\; 12 \end{cases}$

23. $\begin{cases} 5x \quad\quad + 3z = \;\; 2 \\ x - 4y + z = \;\; 6 \\ x + 8y + z = -10 \end{cases}$

24. $\begin{cases} 4x \quad\quad + z = \;\; 0 \\ 3x - 5y + z = -1 \\ x + 5y \quad = -1 \end{cases}$

25. $\begin{cases} 5x + 6y - 2z = 2 \\ 2x - y + z = 2 \\ x + 4y - 2z = 0 \end{cases}$

26. $\begin{cases} -3u - 2v + w = 1 \\ 2u - v + 2w = 1 \\ u + v - w = 0 \end{cases}$

27. $\begin{cases} 3u - 2v + 2w = -2 \\ -u + 4v + w = -1 \\ 5u + 3v + 5w = \;\; 1 \end{cases}$

28. $\begin{cases} 3r + s + 2t = \;\; 5 \\ -2r - s + t = -1 \\ 4r \quad + 2t = \;\; 6 \end{cases}$

29. $\begin{cases} 4r + s - 2t = 7 \\ 3r - s + t = 6 \\ -6r + 3s - 2t = -1 \end{cases}$

30. $\begin{cases} x + y + 2z = 4 \\ -3x + 2y - z = 3 \end{cases}$

31. $\begin{cases} x - 2y - z = 7 \\ 2x - 3y + z = 10 \end{cases}$

32. $\begin{cases} r - s - t = -3 \\ 3r + s + t = -5 \end{cases}$

33. $\begin{cases} r + 2s = 1 \\ 3r + 5s + 4t = 7 \end{cases}$

34. $\begin{cases} -2x + y - z = 2 \\ 5x - 2y + z = 3 \end{cases}$

35. $\begin{cases} -3x - 2y - z = 7 \\ 3x + 3y + z = -3 \end{cases}$

36. $\begin{cases} x - 2y - 3z = 2 \\ 5x - 3y - z = 3 \\ 6x - 5y - 4z = 5 \end{cases}$

37. $\begin{cases} x - 2y - 5z = -3 \\ 3x - 6y - 7z = 1 \\ -2x + 4y + 12z = -4 \end{cases}$

38. $\begin{cases} x - 4y + 2z = -2 \\ y - z = 2 \\ 3x - 6y + 2z = 3 \end{cases}$

Applications

39. Investments An investor would like to build a portfolio from three specific investment types: a stock-based mutual fund, a high-yield bond, and a certificate of deposit (CD). Risk factors for individual investments can be quantified on a scale of 1 to 5, with 1 being the most risky. The risk factors associated with the particular investment types chosen by this investor are summarized in the following table.

Type of Investment	Risk Factor
Stock-based mutual fund	2
High-yield bond	1
CD	5

The investor can tolerate an overall risk level of 2.7. In addition, the amount of money invested in the mutual fund must equal the sum of the amounts invested in the high-yield bond and the CD. Determine the percentage of the total investment that should be allocated to each investment type.

40. Clothing Princess Clothing, Inc., has the following yardage requirements for making a single blouse, dress, or skirt.

	Fabric (yd)	Lining (yd)	Trim (yd)
Blouse	2	1	1
Dress	4	2	1
Skirt	3	1	0

There are 52 yards of fabric, 24 yards of lining, and 10 yards of trim available. How many blouses, dresses, and skirts can be made, assuming that all the fabric, lining, and trim is used up?

41. Electrical Wiring In an electrical circuit in which two resistors are connected in series, the formula for the total resistance R is $R = R_1 + R_2$, where R_1 and R_2 are the resistances of the individual resistors. Consider three resistors A, B, and C. The total resistance when A and B are connected in series is 55 ohms. The total resistance when B and C arc connected in series is 80 ohms. The sum of the resistances of B and C is four times the resistance of A. Find the resistances of A, B, and C.

42. **Computers** The Hi-Tech Computer Company builds three types of computers: basic, upgrade, and high-power. Component requirements for each model are given below.

	Peripheral Cards	Processors	Disk Drives
Basic	3	1	1
Upgrade	4	1	2
High-power	5	2	2

A recent shipment of parts contained 125 peripheral cards, 40 processors, and 50 disk drives. How many of each computer model can be built if all the parts are to be used up?

43. **Car Rentals** A car rental company structures its rates according to the specific day(s) of the week for which a car is rented. The rate structure is as follows.

Daily rental fee, Level A   Monday through Thursday
Daily rental fee, Level B   Friday
Daily rental fee, Level C   Saturday and Sunday

The total rental fee for each of three different situations is given in the following table.

Days Rented	Total Rental Fee
Thurs., Fri., Sat., Sun.	$140
Wed.,Thurs., Fri.	$125
Fri., Sat., Sun.	$95

What is the daily rental fee at each of the three levels (A, B, and C)?

44. **Nutrition** Joanna and her friends visited a popular pizza place to inquire about the cholesterol content of some of their favorite kinds of pizza. They were told that two slices of pepperoni pizza and one slice of Veggie Delight contain a total of 65 milligrams of cholesterol. Also, the amount of cholesterol in one slice of Meaty Delight exceeds that in a slice of Veggie Delight by 20 milligrams. Finally, the total amount of cholesterol in three slices of Meaty Delight and one slice of Veggie Delight is 120 milligrams. How many milligrams of cholesterol are there in each slice of pepperoni, Veggie Delight, and Meaty Delight?

(*Source*: www.pizzahut.com)

45. **Tourism** In 2003, the top three U.S. states visited by foreign tourists were Florida, New York, and California. The total tourism market share of these three states was 68.7%. Florida's share was equal to that of New York. California's share was lower than Florida's by 1.2 percentage points. What was the market share of each of these three states? (*Source*: U.S. Department of Commerce)

Concepts

46. Perform Gaussian elimination on the following system of linear equations, where a is some unspecified constant.

$$\begin{cases} x - y + z = 1 \\ -x + 2y - 2z = 3 \\ y - z = a \end{cases}$$

a. For what value(s) of a does this system have infinitely many solutions?

b. For what value(s) of a does this system have no solution?

47. The graph of the function $f(x) = ax^2 + bx + c$ is a parabola that passes through the points $(-1, -3)$, $(1, 1)$, and $(-2, -8)$, where a, b, and c are constants to be determined.

a. Because $f(-1) = a(-1)^2 + b(-1) + c = a - b + c$ and the value of $f(-1)$ is given to be -3, one linear equation satisfied by a, b, and c is

$$a - b + c = -3$$

Give two more linear equations satisfied by a, b, and c.

b. Solve the system of three linear equations satisfied by a, b, and c (the equation you were given together with the two equations that you found).

c. Substitute your values of a, b, and c into the expression for $f(x)$ and check that the graph of f passes through the given points.

48. Use the steps outlined in Exercise 47 to find the equation of the parabola that passes through the points $(0, 1)$, $(2, -3)$, and $(-3, -8)$.

49. Use the steps outlined in Exercise 47 to find the equation of the parabola that passes through the point $(2, 6)$ and has $(1, 1)$ as its vertex.

Solving Systems of Equations Using Matrices

8.3

Objectives

- Perform Gaussian elimination using matrices.
- Perform Gauss-Jordan elimination to solve a system of linear equations.
- Solve systems of equations arising from applications.

In the previous section, the operations used for solving systems of linear equations depended only on the coefficients of the variables. We can refine the elimination process so that we keep track of only the coefficients and the constant terms. In this section, we introduce *matrices* precisely for this purpose.

A **matrix** is just a table that consists of rows and columns of numbers, called *entries*. The following is an example of a matrix with three rows and four columns:

$$\begin{bmatrix} -3 & 6 & 2 & 6 \\ 2 & 0 & -1 & 4 \\ 0 & 1 & 8 & \frac{2}{3} \end{bmatrix}$$

A matrix is always enclosed within square brackets, to distinguish it from other types of tables.

Any linear equation can be represented as a row of numbers, consisting of the coefficients and the constant on the right hand side. Consider the following system of equations.

$$\begin{cases} -x + y - 2z = -7 \\ 3x \quad\;\; + z = \;\;5 \\ \quad\;\; 3y + 2z = -2 \end{cases}$$

We can represent this system as a matrix with three rows and four columns. In the first equation, the coefficients of x, y, and z are -1, 1, and -2, respectively, and the constant term is -7. Thus the first row of the matrix will be

$$-1 \quad 1 \quad -2 \quad -7$$

Similarly, we will have the following for the second and third rows, respectively.

$$\begin{matrix} 3 & 0 & 1 & 5 \\ 0 & 3 & 2 & -2 \end{matrix}$$

Since the second equation does not contain the variable y, the coefficient of y in that equation is 0, corresponding to $0y$. Similarly, the coefficient of x in the third equation is 0.

Thus the matrix for this system of equations is

$$\left[\begin{array}{ccc|c} -1 & 1 & -2 & -7 \\ 3 & 0 & 1 & 5 \\ 0 & 3 & 2 & -2 \end{array}\right]$$

The vertical bar inside the matrix separates the coefficients from the constant terms, and the matrix is known as the **augmented matrix** for the corresponding system of equations.

Gaussian Elimination Using Matrices

We now introduce operations on matrices that will help solve the corresponding systems of equations. The following operations are called **elementary row operations;** these operations correspond to the operations you performed on systems of linear equations in Section 8.2.

> ### Elementary row operations on matrices
>
> 1. Interchange two rows of the matrix.
> 2. Multiply one row of the matrix by a nonzero constant.
> 3. Multiply one row of the matrix by a nonzero constant and add the result to another row of the matrix.

Example 1 **Performing Elementary Row Operations**

Perform the indicated row operations (independently of one another, not in succession) on the following augmented matrix.

$$\left[\begin{array}{ccc|c} 0 & 3 & 1 & -1 \\ 1 & -2 & -1 & 2 \\ 0 & 6 & -4 & 5 \end{array}\right]$$

a. Interchange rows 1 and 2.

b. Multiple the first row by -2.

c. Multiply the first row by -2 and add the result to the third row. Retain the original first row.

Solution

Before performing the indicated row operations, label the rows of the matrix.

$$\left[\begin{array}{ccc|c} 0 & 3 & 1 & -1 \\ 1 & -2 & -1 & 2 \\ 0 & 6 & -4 & 5 \end{array}\right] \begin{array}{l} (1) \\ (2) \\ (3) \end{array}$$

a. In the original matrix, interchanging rows (1) and (2) gives

$$\left[\begin{array}{ccc|c} 1 & -2 & -1 & 2 \\ 0 & 3 & 1 & -1 \\ 0 & 6 & -4 & 5 \end{array}\right]$$

b. Simply multiply each entry in row (1) of the original matrix by -2.

$$\left[\begin{array}{ccc|c} \mathbf{0} & \mathbf{-6} & \mathbf{-2} & \mathbf{2} \\ 1 & -2 & -1 & 2 \\ 0 & 6 & -4 & 5 \end{array}\right] \qquad \text{New row} = -2 \cdot \text{row (1)}$$

c. For this computation, there are two steps.

Step 1 Multiply each entry in row (1) by -2, and then add the entries in the resulting row to the corresponding entries in row (3).

$$\begin{array}{rrrr} 0 & -6 & -2 & 2 \\ +0 & 6 & -4 & 5 \\ \hline 0 & 0 & -6 & 7 \end{array}$$

Step 2 Replace row (3) with the result of Step 1 (and relabel the third row), and retain row (1).

$$\left[\begin{array}{ccc|c} 0 & 3 & 1 & -1 \\ 1 & -2 & -1 & 2 \\ \mathbf{0} & \mathbf{0} & \mathbf{-6} & \mathbf{7} \end{array}\right] \qquad \text{New row} = -2 \cdot \text{row (1)} + \text{row (3)}$$

✓ *Check It Out 1*   Rework Example 1, part (c), for the following augmented matrix.

$$\left[\begin{array}{ccc|c} 0 & 0 & -2 & -1 \\ -2 & 4 & 1 & 0 \\ 3 & 4 & -2 & 3 \end{array}\right]$$

By performing elementary row operations, we can convert the augmented matrix for any system of linear equations to a matrix that is in a special form known as **row-echelon form**.

Row-Echelon Form of an Augmented Matrix

1. In each nonzero row (i.e., in each row that has at least one nonzero entry to the left of the vertical bar), the first nonzero entry is 1. This 1 is called the **leading 1**.
2. For any two successive nonzero rows, the leading 1 in the higher of the two rows must be farther to the left than the leading 1 in the lower row.
3. Each zero row (i.e., each row in which every entry to the left of the vertical bar is 0) lies below all the nonzero rows.

Note   We will not necessarily use all three types of elementary row operations in any particular example. You should study all the examples given here to see when a certain type of operation would arise.

The following matrices are in row-echelon form. The leading 1's are given in red, and the stars represent real numbers.

$$
\begin{bmatrix} 1 & * & * & | & * \\ 0 & 1 & * & | & * \\ 0 & 0 & 1 & | & * \end{bmatrix}
\qquad
\begin{bmatrix} 1 & * & * & | & * \\ 0 & 1 & * & | & * \\ 0 & 0 & 0 & | & 0 \end{bmatrix}
\qquad
\begin{bmatrix} 1 & * & * & | & * \\ 0 & 0 & 0 & | & 1 \\ 0 & 0 & 0 & | & 0 \end{bmatrix}
$$

Once the augmented matrix for a system of linear equations is in row-echelon form, back-substitution is used to solve the system.

Example 2   Gaussian Elimination Using Matrices

Apply Gaussian elimination to a matrix to solve the following system of equations.

$$
\begin{cases}
x + y + z = 1 \\
2x + 4y + 5z = 3.5 \\
 y = z
\end{cases}
$$

This is the system of equations we formulated and solved in Example 5 of Section 8.2.

Solution

Step 1   Write the system of equations in the general form, with the variables on the left and the constants on the right. For each variable, make sure that all the terms containing that variable are aligned vertically with one another.

$$
\begin{cases}
x + y + z = 1 \\
2x + 4y + 5z = 3.5 \\
 y - z = 0
\end{cases}
$$

In each of the remaining steps, the corresponding equations are displayed to the right of the matrix.

Step 2   Construct the augmented matrix for this system of equations, and label the rows of the matrix.

$$
\begin{bmatrix} 1 & 1 & 1 & | & 1 \\ 2 & 4 & 5 & | & 3.5 \\ 0 & 1 & -1 & | & 0 \end{bmatrix}
\begin{matrix} (1) \\ (2) \\ (3) \end{matrix}
\qquad
\begin{aligned}
x + y + z &= 1 \\
2x + 4y + 5z &= 3.5 \\
y - z &= 0
\end{aligned}
$$

Using Technology

Steps 3 and 4 of Example 2 are illustrated in Figure 1 using a graphing utility. Note that after each row operation is performed on matrix *A*, the resulting matrix is also called *A*.

Keystroke Appendix: Section 13

Step 3   Eliminate the 2 in the second row, first column: Multiply row (1) by -2 and add the result to row (2).

$$
\begin{bmatrix} 1 & 1 & 1 & | & 1 \\ 0 & 2 & 3 & | & 1.5 \\ 0 & 1 & -1 & | & 0 \end{bmatrix}
\begin{matrix} (1) \\ (2') = -2 \cdot (1) + (2) \\ (3) \end{matrix}
\qquad
\begin{aligned}
x + y + z &= 1 \\
2y + 3z &= 1.5 \\
y - z &= 0
\end{aligned}
$$

Step 4   To obtain a 1 in the second row, second column, swap rows (2′) and (3).

$$
\begin{bmatrix} 1 & 1 & 1 & | & 1 \\ 0 & 1 & -1 & | & 0 \\ 0 & 2 & 3 & | & 1.5 \end{bmatrix}
\begin{matrix} (1) \\ (2'') = (3) \\ (3') = (2') \end{matrix}
\qquad
\begin{aligned}
x + y + z &= 1 \\
y - z &= 0 \\
2y + 3z &= 1.5
\end{aligned}
$$

Step 5   Eliminate the 2 in the third row, second column: Multiply row (2″) by -2 and add the result to row (3′).

$$
\begin{bmatrix} 1 & 1 & 1 & | & 1 \\ 0 & 1 & -1 & | & 0 \\ 0 & 0 & 5 & | & 1.5 \end{bmatrix}
\begin{matrix} (1) \\ (2'') \\ (3'') = -2 \cdot (2'') + (3') \end{matrix}
\qquad
\begin{aligned}
x + y + z &= 1 \\
y - z &= 0 \\
5z &= 1.5
\end{aligned}
$$

Step 6   To get a 1 in the third row, third column, multiply row (3″) by $\frac{1}{5}$.

$$
\begin{bmatrix} 1 & 1 & 1 & | & 1 \\ 0 & 1 & -1 & | & 0 \\ 0 & 0 & 1 & | & 0.3 \end{bmatrix}
\begin{matrix} (1) \\ (2'') \\ (3''') = \frac{1}{5} \cdot (3'') \end{matrix}
\qquad
\begin{aligned}
x + y + z &= 1 \\
y - z &= 0 \\
z &= 0.3
\end{aligned}
$$

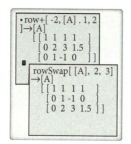

Figure 1

Step 7 Now perform the back-substitution. From the third row, $z = 0.3$. From the second row, $y - z = 0$. Substitute $z = 0.3$ into this equation to solve for y:

$$y - \mathbf{0.3} = 0 \Rightarrow y = 0.3$$

From the first row,

$$x + y + z = 1$$

Substitute $z = 0.3$ and $y = 0.3$ into this equation to solve for x:

$$x + \mathbf{0.3} + \mathbf{0.3} = 1 \Rightarrow x + 0.6 = 1 \Rightarrow x = 0.4$$

The solution to the system of equations is $\boldsymbol{x = 0.4,\ y = 0.3,\ z = 0.3}$, which is the same as the solution found in Example 5 of Section 8.2.

✔ *Check It Out 2* Apply Gaussian elimination to a matrix to solve the following system of equations.

$$\begin{cases} 2x + 2y + z = 1 \\ -3x - 2y + 2z = -5 \\ x + 2y - z = 2 \end{cases}$$

■

Figure 2 gives a schematic illustration of the steps in the Gaussian elimination process. Additional operations may be needed to get the leading 1 in one or more of the nonzero rows of the matrix.

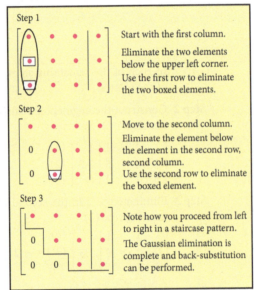

Figure 2 Gaussian Elimination

Example 3   **A System of Equations with Infinitely Many Solutions**

Apply Gaussian elimination to a matrix to solve the following system of equations.

$$\begin{cases} 2x - 4y \quad = -8 \\ 3x \quad - 6z = 4 \end{cases}$$

Solution

Step 1 Construct the augmented matrix for the given system of equations, and label the rows of the matrix.

$$\left[\begin{array}{ccc|c} 2 & -4 & 0 & -8 \\ 3 & 0 & -6 & 4 \end{array} \right] \quad \begin{array}{l} (1) \\ (2) \end{array}$$

Step 2 To get a 1 in the first row, first column, multiply row (1) by $\frac{1}{2}$.

$$\left[\begin{array}{ccc|c} 1 & -2 & 0 & -4 \\ 3 & 0 & -6 & 4 \end{array} \right] \quad \begin{array}{l} (1') = \frac{1}{2} \cdot (1) \\ (2) \end{array}$$

Step 3 Eliminate the 3 in the second row, first column: multiply row (1′) by −3 and add the result to row (2).

$$\begin{bmatrix} 1 & -2 & 0 & | & -4 \\ 0 & 6 & -6 & | & 16 \end{bmatrix} \quad \begin{array}{l} (1') \\ (2') = -3 \cdot (1') + (2) \end{array}$$

Step 4 To get a 1 in the second row, second column, multiply row (2′) by $\frac{1}{6}$.

$$\begin{bmatrix} 1 & -2 & 0 & | & -4 \\ 0 & 1 & -1 & | & \frac{8}{3} \end{bmatrix} \quad \begin{array}{l} (1') \\ (2'') = \frac{1}{6} \cdot (2') \end{array}$$

Because there are only two rows to work with, stop here and perform back-substitution. From the second row,

$$y - z = \frac{8}{3}$$

Solve this equation for y in terms of z.

$$y - z = \frac{8}{3} \Rightarrow y = z + \frac{8}{3}$$

From the first row,

$$x - 2y + 0z = -4$$

Substitute $z + \frac{8}{3}$ for y in this equation and simplify. Then solve for x in terms of z.

$$x - 2\left(z + \frac{8}{3}\right) = -4$$

$$x - 2z - \frac{16}{3} = -4$$

$$x - 2z = \frac{16}{3} - 4$$

$$x - 2z = \frac{16}{3} - \frac{12}{3}$$

$$x - 2z = \frac{4}{3}$$

$$x = 2z + \frac{4}{3}$$

The solution set is $x = 2z + \frac{4}{3}$ and $y = z + \frac{8}{3}$, where z is any real number. Since the expression for at least one of the variables x or y depends explicitly on z, **this system of equations is dependent and has infinitely many solutions.**

 Check It Out 3 Give two specific solutions of the system of equations given in Example 3.

Example 4   **A System of Equations with No Solution**

Apply Gaussian elimination to a matrix to solve the following system of equations.

$$\begin{cases} x + y - z = 1 \\ -x \quad\;\; + z = 2 \\ x + 2y - z = 0 \end{cases}$$

Solution

Step 1 Construct the augmented matrix for the given system of equations, and label the equations.

$$\begin{bmatrix} 1 & 1 & -1 & | & 1 \\ -1 & 0 & 1 & | & 2 \\ 1 & 2 & -1 & | & 0 \end{bmatrix} \quad \begin{array}{l} (1) \\ (2) \\ (3) \end{array}$$

Using Technology

The augmented matrix in Example 3, shown in row-echelon form in Figure 3 differs from the augmented matrix we obtained in Example 3 because the row-echelon form of an augmented matrix is not unique. Fractions are used to display the result so that it is easier to read and easier to check against the result obtained by hand.

Keystroke Appendix:
Section 13

Figure 3

Step 2 Eliminate the -1 and 1 in the second and third rows of the first column. Multiply the first row by 1 and add it to the second row. Multiply the first row by -1 and add it to the third row.

$$\begin{bmatrix} 1 & 1 & -1 & | & 1 \\ 0 & 1 & 0 & | & 3 \\ 0 & 1 & 0 & | & -1 \end{bmatrix} \quad \begin{matrix} (1) \\ (2') = (1) + (2) \\ (3') = -(1) + (3) \end{matrix}$$

Step 3 Eliminate the 1 in the third row, second column: Multiply the second row by -1 and add it to the third row.

$$\begin{bmatrix} 1 & 1 & -1 & | & 1 \\ 0 & 1 & 0 & | & 3 \\ 0 & 0 & 0 & | & -4 \end{bmatrix} \quad \begin{matrix} (1) \\ (2') \\ (3'') = (2') + (3') \end{matrix}$$

Step 4 We stop here because the last row is equivalent to the equation

$$0x + 0y + 0z = -4 \Rightarrow 0 = -4$$

The equation $0 = -4$ is false. Thus the systems of equations has **no solution.**

✓ *Check It Out 4*   Use a matrix to solve the following system of equations.

$$\begin{cases} x & - z = 1 \\ x - y & = 2 \\ 2x - y - z = 4 \end{cases}$$

Gauss-Jordan Elimination

An augmented matrix can be converted to a matrix that is in a special type of row-echelon form known as **reduced row-echelon form.** The distinctive feature of a matrix in reduced row-echelon form is that each leading 1 is the only nonzero entry in its column. The following augmented matrices are in reduced row-echelon form. The leading 1's are given in red.

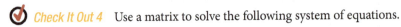

$$\begin{bmatrix} 1 & 0 & 0 & | & -3 \\ 0 & 1 & 0 & | & \frac{1}{4} \\ 0 & 0 & 1 & | & \frac{13}{3} \end{bmatrix} \quad \begin{bmatrix} 1 & 0 & 0 & 0 & | & 2 \\ 0 & 1 & 0 & 0 & | & \frac{1}{2} \\ 0 & 0 & 1 & 0 & | & 0 \end{bmatrix} \quad \begin{bmatrix} 1 & 0 & 4 & | & -3 \\ 0 & 1 & -2 & | & \frac{1}{4} \\ 0 & 0 & 0 & | & 0 \end{bmatrix}$$

When the augmented matrix for a system of linear equations is in reduced row-echelon form, you can read off the solution directly.

The process used to reduce an augmented matrix to reduced row-echelon form is known as **Gauss-Jordan elimination,** illustrated by the following example

Example 5   **Solving a System of Equations Using Gauss-Jordan Elimination**

Use Gauss-Jordan elimination to solve the following system of linear equations.

$$\begin{cases} 3x + y - 10z = -8 \\ x + y - 2z = -4 \\ -2x + 9z = 5 \end{cases}$$

Solution Construct the augmented matrix for this system of equations, and label the rows of the matrix.

$$\begin{bmatrix} 3 & 1 & -10 & | & -8 \\ 1 & 1 & -2 & | & -4 \\ -2 & 0 & 9 & | & 5 \end{bmatrix} \quad \begin{matrix} (1) \\ (2) \\ (3) \end{matrix}$$

First we will reduce the augmented matrix to row-echelon form via ordinary Gaussian elimination. Then we will complete the process of reducing the matrix to reduced row-echelon form by using additional elementary row operations.

Step 1 To get a 1 in the first row, first column, swap rows (1) and (2).

$$\begin{bmatrix} 1 & 1 & -2 & | & -4 \\ 3 & 1 & -10 & | & -8 \\ -2 & 0 & 9 & | & 5 \end{bmatrix} \quad \begin{matrix} (1') = (2) \\ (2') = (1) \\ (3) \end{matrix}$$

Step 2   Eliminate the 3 in the second row, first column: Multiply row (1′) by -3 and add the result to row (2′). Then eliminate the -2 in the third row, first column: multiply row (1′) by 2 and add the result to row (3).

$$\left[\begin{array}{ccc|c} 1 & 1 & -2 & -4 \\ 0 & -2 & -4 & 4 \\ 0 & 2 & 5 & -3 \end{array}\right] \quad \begin{array}{l} (1') \\ (2'') = -3 \cdot (1') + (2') \\ (3') = 2 \cdot (1') + (3) \end{array}$$

Step 3   To get a 1 in the second row, second column, multiply row (2″) by $-\frac{1}{2}$.

$$\left[\begin{array}{ccc|c} 1 & 1 & -2 & -4 \\ 0 & 1 & 2 & -2 \\ 0 & 2 & 5 & -3 \end{array}\right] \quad \begin{array}{l} (1') \\ (2''') = -\dfrac{1}{2} \cdot (2'') \\ (3') \end{array}$$

Step 4   Eliminate the 2 in the third row, second column: Multiply row (2‴) by -2 and add the result to row (3′).

$$\left[\begin{array}{ccc|c} 1 & 1 & -2 & -4 \\ 0 & 1 & 2 & -2 \\ 0 & 0 & 1 & 1 \end{array}\right] \quad \begin{array}{l} (1') \\ (2''') \\ (3'') = -2 \cdot (2''') + (3') \end{array}$$

As far as Gaussian elimination is concerned, we are done and can find the solution by back-substitution. For Gauss-Jordan elimination, we need to continue with the row operations. What remains is elimination of the nonzero entries at the locations indicated in red in the following matrix.

$$\left[\begin{array}{ccc|c} 1 & 1 & -2 & -4 \\ 0 & 1 & 2 & -2 \\ 0 & 0 & 1 & 1 \end{array}\right] \quad \begin{array}{l} (1') \\ (2''') \\ (3'') \end{array}$$

Step 5   Eliminate the 2 in the second row, third column: Multiply row (3″) by -2 and add the result to row (2‴). Then eliminate the -2 in the first row, third column: Multiply row (3″) by 2 and add the result to row (1′).

$$\left[\begin{array}{ccc|c} 1 & 1 & 0 & -2 \\ 0 & 1 & 0 & -4 \\ 0 & 0 & 1 & 1 \end{array}\right] \quad \begin{array}{l} (1'') = 2 \cdot (3'') + (1') \\ (2'''') = -2 \cdot (3'') + (2''') \\ (3'') \end{array}$$

Using Technology

In Figure 4, a graphing utility gives the reduced row-echelon form for the augmented matrix in Example 5.

Keystroke Appendix:
Section 13

Step 6   Eliminate the 1 in the first row, second column: Multiply row (2⁗) by -1 and add the result to row (1″).

$$\left[\begin{array}{ccc|c} 1 & 0 & 0 & 2 \\ 0 & 1 & 0 & -4 \\ 0 & 0 & 1 & 1 \end{array}\right] \quad \begin{array}{l} (1''') = -1 \cdot (2'''') + (1'') \\ (2'''') \\ (3'') \end{array}$$

The matrix is now in reduced row-echelon form and corresponds to the following system of linear equations.

$$\begin{cases} x & & = 2 \\ & y & = -4 \\ & & z = 1 \end{cases}$$

Thus the solution is

$$x = 2, y = -4, z = 1$$

Instead of writing down the system of equations, we could have read the solution from the matrix directly.

```
rref [ [A] ]
   [[ 1  0  0   2]
    [ 0  1  0  -4]
    [ 0  0  1   1]]
```

Figure 4

Check It Out 5   Use Gauss-Jordan elimination to solve the following system of equations.

$$\begin{cases} x - y - 2z = 0 \\ 2x + y + z = 3 \\ 3y + 2z = 3 \end{cases}$$

Example 6 **Finding a Solution from a Reduced Matrix**

Find the solution set of the system of linear equations in the variables x, y, z, and u (in that order) that has the following augmented matrix.

$$\begin{array}{cccc} x & y & z & u \\ \left[\begin{array}{cccc|c} 1 & 0 & 0 & 3 & 0 \\ 0 & 1 & 0 & 5 & 1 \\ 0 & 0 & 1 & -6 & 4 \end{array}\right] \end{array}$$

Solution The given matrix is in reduced row-echelon form. The corresponding system of linear equations consists of *three* equations in *four* variables.

$$\begin{cases} x & + 3u = 0 \qquad \text{From first row} \\ \quad y & + 5u = 1 \qquad \text{From second row} \\ \qquad z - 6u = 4 \qquad \text{From third row} \end{cases}$$

Solving for x, y, and z in terms of u, we have

$$x = -3u, y = -5u + 1$$

and

$$z = 6u + 4$$

The solution set is $x = -3u$, $y = -5u + 1$, and $z = 6u + 4$, where u is any real number. Since the expression for at least one of the variables x, y, or z depends explicitly on u, **this system of equations is dependent and has infinitely many solutions.**

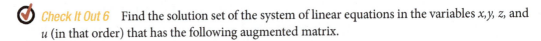 *Check It Out 6* Find the solution set of the system of linear equations in the variables x, y, z, and u (in that order) that has the following augmented matrix.

$$\begin{array}{cccc} x & y & z & u \\ \left[\begin{array}{cccc|c} 1 & 0 & 0 & -2 & 3 \\ 0 & 1 & 0 & 1 & -4 \\ 0 & 0 & 1 & -2 & 0 \end{array}\right] \end{array}$$

Applications

Systems of linear equations have a number of applications in a wide variety of fields. We now solve an applied problem in manufacturing.

Example 7 **Systems of Equations in Manufacturing**

The Quilter's Corner is a company that makes quilted wall hangings, pillows, and bedspreads. The process of quilting involves cutting, sewing, and finishing. For each of the three products, the numbers of hours spent on each task are given in Table 1.

	Cutting (hr)	Sewing (hr)	Finishing (hr)
Wall hanging	1.5	1.5	1
Pillow	1	1	0.5
Bedspread	4	3	2

Table 1

Every week, there are 17 hours available for cutting, 15 hours available for sewing, and 9 hours available for finishing. How many of each item can be made per week if all the available time is to be used?

Solution First, define the variables.

w: number of wall hangings

p: number of pillows

b: number of bedspreads

The total number of hours spent on cutting is given by

Total cutting hours = cutting for wall hanging + cutting for pillow + cutting for bedspread

$$= 1.5w + p + 4b$$

The total amount of time available for cutting is 17 hours, so our first equation is

$$1.5w + p + 4b = 17$$

Since there are 15 hours available for sewing and 9 hours available for finishing, our other two equations are

$$1.5w + p + 3b = 15$$

$$w + 0.5p + 2b = 9$$

Thus the system of equations to be solved is

$$\begin{cases} 1.5w + p + 4b = 17 \\ 1.5w + p + 3b = 15 \\ w + 0.5p + 2b = 9 \end{cases}$$

We will solve this system by applying Gaussian elimination to the corresponding augmented matrix.

$$\begin{bmatrix} 1.5 & 1 & 4 & | & 17 \\ 1.5 & 1 & 3 & | & 15 \\ 1 & 0.5 & 2 & | & 9 \end{bmatrix} \quad \begin{matrix} (1) \\ (2) \\ (3) \end{matrix}$$

Step 1 To get a 1 in the first row, first column, swap rows (1) and (3).

$$\begin{bmatrix} 1 & 0.5 & 2 & | & 9 \\ 1.5 & 1 & 3 & | & 15 \\ 1.5 & 1 & 4 & | & 17 \end{bmatrix} \quad \begin{matrix} (1') = (3) \\ (2) \\ (3') = (1) \end{matrix}$$

Step 2 Eliminate the 1.5 in the second row, first column: Multiply row (1') by -1.5 and add the result to row (2). Then eliminate the 1.5 in the third row, first column: Multiply row (1') by -1.5 and add the result to row (3').

$$\begin{bmatrix} 1 & 0.5 & 2 & | & 9 \\ 0 & 0.25 & 0 & | & 1.5 \\ 0 & 0.25 & 1 & | & 3.5 \end{bmatrix} \quad \begin{matrix} (1') \\ (2') = -1.5 \cdot (1') + (2) \\ (3'') = -1.5 \cdot (1') + (3') \end{matrix}$$

Step 3 To get a 1 in the second row, second column, multiply row (2') by $\frac{1}{0.25}$.

$$\begin{bmatrix} 1 & 0.5 & 2 & | & 9 \\ 0 & 1 & 0 & | & 6 \\ 0 & 0.25 & 1 & | & 3.5 \end{bmatrix} \quad \begin{matrix} (1') \\ (2'') = \frac{1}{0.25} \cdot (2') \\ (3'') \end{matrix}$$

Step 4 Eliminate the 0.25 in the third row, second column: Multiply row (2'') by -0.25 and add the result to row (3'').

$$\begin{bmatrix} 1 & 0.5 & 2 & | & 9 \\ 0 & 1 & 0 & | & 6 \\ 0 & 0 & 1 & | & 2 \end{bmatrix} \quad \begin{matrix} (1') \\ (2'') \\ (3''') = -0.25 \cdot (2'') + (3'') \end{matrix}$$

Step 5   Because the matrix is in row echelon form, perform back-substitution. From the third row, $b = 2$; from the second row, $p = 6$. From the first row,

$$w + 0.5p + 2b = 9$$

Substituting $b = 2$ and $p = 6$ into this equation yields

$$w + 0.5(6) + 2(2) = 9$$

Simplifying and then solving for w, we get

$$w + 0.5(6) + 2(2) = 9 \Rightarrow w + 3 + 4 = 9 \Rightarrow w + 7 = 9 \Rightarrow w = 2$$

Thus **2 wall hangings, 6 pillows, and 2 bedspreads** can be made per week.

 Check It Out 7   Rework Example 7 for the case in which 36 hours are available for cutting, 30 hours are available for sewing, and 19 hours are available for finishing.

Observations

We recommend using Gaussian elimination with back-substitution when working by hand because this method is likely to result in fewer arithmetic errors than Gauss-Jordan elimination. If technology is used, however, Gauss-Jordan elimination, using the **rref** feature of a graphing utility, is much quicker.

Exercises 8.3

 Skills

In Exercises 1–8, construct the augmented matrix for each system of equations. Do not solve the system.

1. $\begin{cases} 4x + y - 2z = 6 \\ -x - y + z = -2 \\ 3x - z = 4 \end{cases}$ 2. $\begin{cases} 3r + s + 2t = -1 \\ -2r - s + t = 3 \\ 4r + 2t = -2 \end{cases}$ 3. $\begin{cases} 3x - 2y + z = -1 \\ x + y - 4z = 3 \\ -2x - y + 3z = 0 \end{cases}$

4. $\begin{cases} -x + 5y - z = 6 \\ x - 4y + 2z = 3 \\ 3x - y + 5z = -1 \end{cases}$ 5. $\begin{cases} 6x - 2y + z = 0 \\ -5x + y - 3z = -2 \\ 2x - 3y + 5z = 7 \end{cases}$ 6. $\begin{cases} -2x - y - z = 5 \\ x + y + z = 0 \\ 3x + 2y + 7z = -8 \end{cases}$

7. $\begin{cases} x + y + 2z = -3 \\ -3x + 2y + z = 1 \end{cases}$ 8. $\begin{cases} -2x + 6z = -1 \\ -3x + 2y + z = 0 \end{cases}$

In Exercises 9–14, perform the indicated row operations (*independently of one another, not in succession*) on the following augmented matrix.

$$\begin{bmatrix} 1 & -2 & 0 & | & -1 \\ 2 & -8 & -2 & | & 1 \\ 3 & 5 & 1 & | & 2 \end{bmatrix}$$

9. Multiply the second row by $\frac{1}{2}$.

10. Multiply the second row by $-\frac{1}{2}$.

11. Switch rows 1 and 2.

12. Switch rows 1 and 3.

13. Multiply the first row by 2 and add the result to the second row.

14. Multiply the first row by -3 and add the result to the third row.

In Exercises 15–22, for each matrix, construct the corresponding system of linear equations. Use the variables listed above the matrix, in the given order. Determine whether the system is consistent or inconsistent. If it is consistent, give the solution(s).

15. $\begin{matrix} x & y \\ \begin{bmatrix} 1 & 0 & | & -7 \\ 0 & 1 & | & 3 \end{bmatrix} \end{matrix}$ 16. $\begin{matrix} x & y \\ \begin{bmatrix} 1 & 0 & | & 3 \\ 0 & 0 & | & 1 \end{bmatrix} \end{matrix}$ 17. $\begin{matrix} x & y \\ \begin{bmatrix} 2 & 0 & | & 6 \\ 0 & 1 & | & 5 \end{bmatrix} \end{matrix}$

18. $\begin{matrix} x & y & z \\ \begin{bmatrix} 3 & -2 & 5 & | & 2 \\ 1 & 0 & -1 & | & -3 \\ 0 & 0 & 0 & | & 8 \end{bmatrix} \end{matrix}$ 19. $\begin{matrix} x & y & z \\ \begin{bmatrix} 1 & 0 & 3 & | & 5 \\ 0 & 1 & -2 & | & -2 \\ 0 & 0 & 0 & | & 0 \end{bmatrix} \end{matrix}$ 20. $\begin{matrix} x & y & z \\ \begin{bmatrix} 1 & 0 & -2 & | & 7 \\ 0 & 1 & 4 & | & 3 \\ 0 & 0 & 0 & | & 0 \end{bmatrix} \end{matrix}$

21. $\begin{matrix} x & y & z & u \\ \begin{bmatrix} 1 & 0 & 0 & -5 & | & 2 \\ 0 & 1 & 0 & -2 & | & -3 \\ 0 & 0 & 1 & 3 & | & 5 \end{bmatrix} \end{matrix}$ 22. $\begin{matrix} x & y & z & u \\ \begin{bmatrix} 1 & 0 & 0 & -4 & | & -3 \\ 0 & 1 & 0 & -2 & | & 1 \\ 0 & 0 & 1 & 3 & | & -10 \end{bmatrix} \end{matrix}$

In Exerciscs 23–52, apply elementary row operations to a matrix to solve the system of equations. If there is no solution, state that the system is inconsistent.

23. $\begin{cases} x + 3y = 10 \\ -2x - 5y = -12 \end{cases}$ 24. $\begin{cases} -x - y = -10 \\ 3x + 4y = 24 \end{cases}$ 25. $\begin{cases} -2x + y = -5 \\ 4x - y = 6 \end{cases}$ 26. $\begin{cases} x + 2y = 5 \\ 2x - 3y = 3 \end{cases}$

27. $\begin{cases} -x + y = 2 \\ 7x - 4y = -2 \end{cases}$ 28. $\begin{cases} 5x + 3y = 7 \\ 2x + y = 2 \end{cases}$ 29. $\begin{cases} x + 2y = 1 \\ x + 5y = -2 \end{cases}$ 30. $\begin{cases} 2x - 3y = -1 \\ -3x + 2y = 6 \end{cases}$

31. $\begin{cases} 5x + 3y = -1 \\ -10x - 6y = 2 \end{cases}$ 32. $\begin{cases} 2x + 4y = 5 \\ -4x - 8y = -10 \end{cases}$ 33. $\begin{cases} 2x - 3y = 7 \\ 4x - 6y = 9 \end{cases}$ 34. $\begin{cases} -x + y = 5 \\ 4x - 4y = 10 \end{cases}$

35. $\begin{cases} x + y + z = 3 \\ x + 2y + z = 5 \\ -2x + y - z = 4 \end{cases}$

36. $\begin{cases} x + y - z = 0 \\ 3x + 2y - z = -1 \\ -2x + y - 2z = -1 \end{cases}$

37. $\begin{cases} x + 3y - 2z = 4 \\ -5x - 3y - 2z = -8 \\ x - y - z = 0 \end{cases}$

38. $\begin{cases} x + 2y - 2z = -7 \\ 2x + 5y - 2z = -10 \\ x - 2y - 3z = -9 \end{cases}$

39. $\begin{cases} 3x - 4y = 14 \\ x - y + 2z = 14 \\ -x + 4z = 18 \end{cases}$

40. $\begin{cases} 2x - 5z = -15 \\ -x - 2y + z = 5 \\ x + y = -1 \end{cases}$

41. $\begin{cases} x + 2y - z = 2 \\ -2x + y - 3z = 6 \\ -x + 3y - 4z = 8 \end{cases}$

42. $\begin{cases} x + 3y = 2 \\ 5x + 12y + 3z = 1 \\ -4x - 9y - 3z = 1 \end{cases}$

43. $\begin{cases} -x + 2y - 3z = 2 \\ 2x + 3y + 2z = 1 \\ 3x + y + 5z = 1 \end{cases}$

44. $\begin{cases} x + 2y + z = -3 \\ 3x + y - 2z = 2 \\ 4x + 3y - z = 0 \end{cases}$

45. $\begin{cases} -x + 4y + 3z = 6 \\ 2x - 8y - 4z = 8 \end{cases}$

46. $\begin{cases} 3u + 5v - 3w = 1 \\ u + 2v - w = -2 \end{cases}$

47. $\begin{cases} 3r + 4s - 8t = 14 \\ 2r - 2s + 4t = 28 \end{cases}$

48. $\begin{cases} -3x + y + 24z = -9 \\ 2x + 2y - 8z = 6 \end{cases}$

49. $\begin{cases} 3x + 4y - 8z = 10 \\ -6x - 8y + 16z = 20 \end{cases}$

50. $\begin{cases} 2u - 3v - 2w = 4 \\ u + 2v + w = -3 \end{cases}$

51. $\begin{cases} z + 2y = 0 \\ z - 5x = -1 \\ 3x + 2y = 3 \end{cases}$

(*Hint*: Be careful with the order of the variables.)

52. $\begin{cases} x + 4z = -3 \\ x - 5y = 0 \\ z + 4y = 2 \end{cases}$

(*Hint*: Be careful with the order of the variables.)

Applications

53. **Mixture** JoAnn's Coffee wants to sell a new blend of three types of coffee: Colombian, Java, and Kona. The company plans to market this coffee in 10-pound bags and has established the following specifications for the product:

    - The amount of Colombian coffee in the blend is to be twice the amount of Java and Kona combined.
    - The amount of Kona coffee in the blend is to be half the amount of Java.

    How many pounds of each type of coffee will go into each bag of the blend?

54. **Sports Equipment** The athletic director of a local high school is ordering equipment for spring sports. He needs to order twice as many baseballs as softballs. The total number of balls he must order is 300. How many of each type should he order?

55. **Take-Out Orders** A boy scout troop orders eight pizzas. Cheese pizzas cost $5 each and pepperoni pizzas cost $6 each. The scout leader paid a total of $43 for the pizzas. How many of each type did he order?

56. **Merchandise Sales** An electronics store carries two brands of video cameras. For a certain week, the number of Brand A video cameras sold was 10 less than twice the number of Brand B cameras sold. Brand A cameras cost $200 and Brand B cameras cost $350. If the total revenue generated that week from the sale of both types of cameras was $16,750, how many of each type were sold?

57. **Merchandise Sales** A grocery store carries two brands of diapers. For a certain week, the number of boxes of Brand A diapers sold was 4 more than the number of boxes of Brand B diapers sold. Brand A diapers cost $10 per box and Brand B diapers cost $12 per box. If the total revenue generated that week from the sale of diapers was $172, how many of each brand did the store sell?

58. Design Sarita is designing a mobile in which three objects will be suspended from a lightweight rod, as illustrated below.

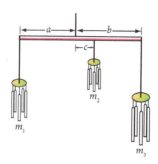

The weights of the objects are given as $m_1 = 2$ ounce, $m_2 = 1$ ounces, and $m_3 = 2.5$ ounces. Use a system of linear equations to determine the values of a, b, and c (in inches) such that the mobile will meet the following three requirements:

- The distance between object 1 and object 3 must be 20 inches.
- To balance the mobile, Sarita must position the objects in such a way that

$$m_1a = m_2b + m_3c$$

- Object 1 must hang three times further from the support than object 2.

59. Investments A financial advisor offers three specific investment types: a stock-based mutual fund, a high-yield bond, and a certificate of deposit (CD). Risk factors for individual instruments can be quantified on a scale of 1 to 5, with 1 being the most risky. The risk factors associated with these particular investment types are summarized in the following table.

Type of Investment	Risk Factor
Stock-based mutual fund	3
High-yield bond	1
CD	5

One of the advisor's clients can tolerate an overall risk level of 3.5. In addition, the client stipulates that the amount of money invested in the mutual fund must equal the sum of the amounts invested in the high-yield bond and the CD. To satisfy the client's requirements, what percentage of the total investment should be allocated to each investment type?

60. Utilities Privately owned, single-family homes in a small town were heated with gas, electricity, or oil. The percentage of homes heated with electricity was 9 times the percentage heated with oil. The percentage of homes heated with gas was 40 percentage points higher than the percentage heated with oil and the percentage heated with electricity combined. Find the percentage of homes heated with each type of fuel.

61. Electrical Engineering An electrical circuit consists of two resistors connected in series. The formula for the total resistance R is given by $R = R_1 + R_2$, where R_1 and R_2 are the resistances of the individual resistors. In a circuit with two resistors A and B connected in series, the total resistance is 60 ohms. The total resistance when B and C are connected in series is 100 ohms. The sum of the resistance of B and C is 2.5 times the resistance of A. Find the resistances of A, B, and C.

Concepts

62. Consider the following system of equations.

$$\begin{cases} x + y = 3 \\ -x + y = 1 \\ 2x + y = 4 \end{cases}$$

Use Gauss-Jordan elimination to find the solution, if it exists. Interpret your answer in terms of the graphs of the given equations.

63. Consider the following system of equations.

$$\begin{cases} x + y = 3 \\ -x + y = 1 \\ 2x + y = 6 \end{cases}$$

Use Gauss-Jordan elimination to show that this system has no solution. Interpret your answer in terms of the graphs of the given equations.

64. Consider the following system of equations.

$$\begin{cases} x + y = 3 \\ -2x - 2y = -6 \\ -x - y = -3 \end{cases}$$

Use Gauss-Jordan elimination to show that this system has infinitely many solutions. Interpret your answer in terms of the graphs of the given equations.

65. Consider the following system of equations.

$$\begin{cases} 6u + 6v - 3w = -3 \\ 2u + 2v - w = -1 \end{cases}$$

a. Show that each of the equations in this system is a multiple of the other equation.

b. Explain why this system of equations has infinitely many solutions.

c. Express w as an equation in u and v.

d. Give two solutions of this system of equations.

66. Consider the following system of equations.

$$\begin{cases} x + y = 1 & (1) \\ x - z = 0 & (2) \\ y + z = 1 & (3) \end{cases}$$

a. Find constants b and c such that Equation (1) can be expressed as

$$\text{Equation (1)} = b\,[\text{Equation (2)}] + c\,[\text{Equation (3)}]$$

b. Solve the system.

c. Give three individual solutions such that $0 < z < 1$.

Operations with Matrices

8.4

Objectives

- Define a matrix.
- Perform matrix addition and scalar multiplication.
- Find the product of two matrices.
- Use matrices and matrix operations in applications.

In Section 8.3, we represented systems of equations by matrices in order to streamline the bookkeeping involved in solving them. In this section, we will study the properties of matrices in more detail.

A **matrix** is a rectangular array of numbers. A matrix that has m rows and n columns has **dimensions** $m \times n$ (pronounced "m by n"); such a matrix is referred to as an $m \times n$ **matrix**.

An uppercase letter of the alphabet, such as A, is typically used to name a matrix. Each **element** of matrix A is referred to as a_{ij}, since its position is in the ith row and jth column. For A, an $m \times n$ matrix, i ranges from 1 through m, and j from 1 through n.

$$A = \begin{bmatrix} a_{11} & a_{12} & \cdots & a_{1n} \\ a_{21} & a_{22} & \cdots & a_{2n} \\ \cdots & \cdots & \cdots & \cdots \\ a_{m1} & a_{m2} & \cdots & a_{mn} \end{bmatrix}$$

The next example will familiarize you with this basic terminology and notation.

VIDEO EXAMPLES

SECTION 8.4

Example 1   Understanding Matrix Notation

Let

$$B = \begin{bmatrix} -2 & 4 & 7 & 0 \\ 0 & 5 & 12 & 6 \\ -8 & -7 & 0 & 1 \end{bmatrix}$$

Find the following.

a. The dimensions of B **b.** The value of b_{31} **c.** The value of b_{13}

Solution

a. Because matrix B has three rows and four coulmns, it has dimensions 3×4.

b. The value of b_{31} (the entry in the third row, first column) is -8.

c. The value of b_{13} (the entry in the first row, third column) is 7.

 Check It Out 1   Let

$$A = \begin{bmatrix} -2 & 4 & -5 \\ -9 & -2 & 1 \\ 12 & 3 & 0 \end{bmatrix}$$

Find the dimensions of A and the values of a_{12} and a_{33}.

Matrix Addition and Scalar Multiplication

Just as with numbers, we can define equality, addition, and subtraction for matrices—but only for matrices that have the *same dimensions.*

Equality of Matrices

Two $m \times n$ matrices A and B are said to be equal, which is denoted by $A = B$, if corresponding entries are equal. That is, for all i, j, $a_{ij} = b_{ij}$.

Addition and Subtraction of Matrices

Let A and B be $m \times n$ matrices. Then the sum of A and B, which is denoted by $A + B$, is the matrix which is formed by simply adding corresponding entries. Note that $A + B$ is also an $m \times n$ matrix. The difference of A and B, which is denoted by $A - B$, is defined analogously.

Another easily defined operation on matrices is the multiplication of an entire matrix by a single number, known as a *scalar*. This operation is known as **scalar multiplication.**

Scalar Multiplication

The product of an $m \times n$ matrix A and a scalar c, which is denoted by cA, is found by multiplying all the entries of A by c. The product is also an $m \times n$ matrix.

Example 2   **Addition, Subtraction and Scalar Multiplication of Matrices**

Let

$$A = \begin{bmatrix} -1 & 2 & 0 \\ 0 & -3.5 & 1 \end{bmatrix} \quad B = \begin{bmatrix} 3 & -2 \\ -1 & 4.2 \\ 2 & -6 \end{bmatrix} \quad C = \begin{bmatrix} 1 & 4 & -5 \\ -2.5 & 0 & 2.3 \end{bmatrix}$$

Perform the following operations, if defined.

a. $A + B$ **b.** $C - A$ **c.** $3C$

Solution

a. The sum $A + B$ is not defined, since A has dimensions 2×3 and B has dimensions 3×2.

b. The difference $C - A$ is defined, since A and C have the same dimensions. To find $C - A$, we subtract the entries of A from the corresponding entries of C:

$$C - A = \begin{bmatrix} 1 - (-1) & 4 - 2 & -5 - 0 \\ -2.5 - 0 & 0 - (-3.5) & 2.3 - 1 \end{bmatrix} = \begin{bmatrix} 2 & 2 & -5 \\ -2.5 & 3.5 & 1.3 \end{bmatrix}$$

c. To compute $3C$, multiply each entry in C by 3.

$$3C = \begin{bmatrix} 3(1) & 3(4) & 3(-5) \\ 3(-2.5) & 3(0) & 3(2.3) \end{bmatrix} = \begin{bmatrix} 3 & 12 & -15 \\ -7.5 & 0 & 6.9 \end{bmatrix}$$

✅ *Check It Out 2*   Let

$$A = \begin{bmatrix} 3 & -2 \\ 1 & 2.5 \\ -0.5 & -3 \end{bmatrix} \quad B = \begin{bmatrix} -3 & 4 \\ 2.1 & 2 \\ 0 & 4 \end{bmatrix}$$

Perform the following operations:

a. $A + B$ **b.** $A - B$ **c.** $4B$

Using Technology

Matrices can be added with a graphing utility. Attempting to add matrices that do not have the same dimensions will yield an error message. See Figure 1 and Figure 2 .

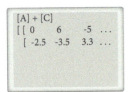

[A] + [C]
[[0   6   -5 ...
 [-2.5 -3.5   3.3 ...

Figure 1

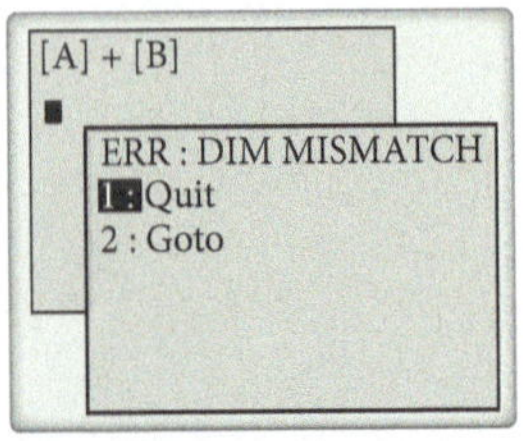

[A] + [B]
■
ERR : DIM MISMATCH
1: Quit
2 : Goto

Figure 2

Matrix Multiplication

We begin our discussion of matrix multiplication with the simplest types of matrices: a row matrix and a column matrix. Multiplication of a row matrix by a column matrix will form the basis for matrix multiplication.

A **row matrix** has just one row and any number of columns. The following is an example of a row matrix with 4 columns.

$$R = \begin{bmatrix} 3 & 2.5 & \dfrac{1}{3} & -2 \end{bmatrix}$$

Similarly, a **column matrix** has just one column and any number of rows. The following is an example of a column matrix with 4 rows.

$$C = \begin{bmatrix} 2 \\ -2 \\ 3 \\ 1.7 \end{bmatrix}$$

Multiplication of a row matrix by a column matrix

Let R be a row matrix with n columns:

$$R = [r_{11} \ r_{12} \ \cdots \ r_{1n}]$$

and let C be a column matrix with n rows:

$$\begin{bmatrix} c_{11} \\ c_{21} \\ \cdots \\ c_{n1} \end{bmatrix}$$

Then the product p is defined as

$$p = r_{11}c_{11} + r_{12}c_{21} + \ldots + r_{1n}c_{n1}$$

The number of columns in the row matrix R *must be the same* as the number of rows in the column matrix C for the product p to be defined. The product p is just a number.

Example 3   **Finding the Product of a Row and Column Matrix**

Find the product p if

$$R = [3 \ 2 \ 3 \ -2] \text{ and } C = \begin{bmatrix} 2 \\ -2 \\ 3 \\ 1 \end{bmatrix}$$

The product p is calculated as follows:

$$p = [3 \ 2 \ 3 \ -2] \begin{bmatrix} 2 \\ -2 \\ 3 \\ 1 \end{bmatrix}$$

$$= [r_{11} \ r_{12} \ r_{13} \ r_{14}] \begin{bmatrix} c_{11} \\ c_{21} \\ c_{31} \\ c_{41} \end{bmatrix}$$

$$= r_{11}c_{11} + r_{12}c_{21} + r_{13}c_{31} + r_{14}c_{41}$$

$$= 3(2) + 2(-2) + (3)(3) + (-2)(1)$$

$$= 6 - 4 + 9 - 2 = 9$$

✓ *Check It Out 3* Find the product p if

$$R = [4 \ -1 \ 2 \ 2] \text{ and } C = \begin{bmatrix} 0 \\ -3 \\ 4 \\ 1 \end{bmatrix}$$

We now define matrix multiplication in general. The idea is to break up the multiplication of a pair of matrices into steps, where each step consists of multiplying some row of the first matrix by some column of the second matrix, using the rule given earlier.

Multiplication of Matrices

Multiplication of an $m \times p$ matrix A by a $p \times n$ matrix B results in a product matrix AB with m rows and n columns.

The entry in the ith row, jth column of the product matrix AB is given by the product of the ith row of A and the jth column of B:

$$(ab)_{ij} = a_{i1}b_{1j} + a_{i2}b_{2j} + \ldots + a_{ik}b_{kj}$$

Example 4 The Product of Two Matrices

Find the product AB for the following matrices.

$$A = \begin{bmatrix} 3 & 4 & 0 & -2 \\ -1 & 0 & -3 & 2 \end{bmatrix}, \qquad B = \begin{bmatrix} 5 & 0 \\ -4 & 1 \\ 2 & 6 \\ 0 & -2 \end{bmatrix}$$

Solution

Step 1 Multiply the *first row of A* by the *first column of B*, using the rule for multiplying a row matrix by a column matrix. The result, which is a single number, will be the entry in the *first row, first column* of the *product matrix.*

$$\begin{bmatrix} 3 & 4 & 0 & -2 \\ * & * & * & * \end{bmatrix} \begin{bmatrix} 5 & * \\ -4 & * \\ 2 & * \\ 0 & * \end{bmatrix} = [3(5) + 4(-4) + 0(2) + (-2)(0)] = \begin{bmatrix} -1 & \\ & \end{bmatrix}$$

Step 2 Multiply the *first row of A* by the *second column of B*. The result will be the entry in the *first row, second column* of the product matrix.

$$\begin{bmatrix} 3 & 4 & 0 & -2 \\ * & * & * & * \end{bmatrix} \begin{bmatrix} * & 0 \\ * & 1 \\ * & 6 \\ * & -2 \end{bmatrix} = [-1 \quad 3(0) + 4(1) + 0(6) + (-2)(-2)] = \begin{bmatrix} -1 & 8 \end{bmatrix}$$

Step 3 Since we have exhausted all the columns of B (for multiplication by the first row of A), we next multiply the *second row of A* by the *first column of B*. The result will be the entry in the *second row, first column* of the product matrix.

$$\begin{bmatrix} * & * & * & * \\ -1 & 0 & -3 & 2 \end{bmatrix} \begin{bmatrix} 5 & * \\ -4 & * \\ 2 & * \\ 0 & * \end{bmatrix} = \begin{bmatrix} -1 & 8 \\ (-1)(5) + 0(-4) + (-3)(2) + 2(0) & \end{bmatrix}$$

$$= \begin{bmatrix} -1 & 8 \\ -11 & \end{bmatrix}$$

Step 4 Multiply the *second row of A* by the *second column of B*. The result will be the entry in the *second row, second column* of the product matrix.

$$\begin{bmatrix} * & * & * & * \\ -1 & 0 & -3 & 2 \end{bmatrix} \begin{bmatrix} * & 0 \\ * & 1 \\ * & 6 \\ * & -2 \end{bmatrix} = \begin{bmatrix} -1 & 8 \\ -11 & (-1)(0) + 0(1) + (-3)(6) + 2(-2) \end{bmatrix}$$

$$= \begin{bmatrix} -1 & 8 \\ -11 & -22 \end{bmatrix}$$

Thus the product matrix is

$$AB = \begin{bmatrix} -1 & 8 \\ -11 & -22 \end{bmatrix}$$

Recall that A has dimensions 2×4 and B has dimensions 4×2. The product AB has dimensions 2×2, where the first 2 arises from the number of rows in A and the second 2 arises from the number of columns in B.

✓ *Check It Out 4* Find the product AB for the following matrices.

$$A = \begin{bmatrix} -2 & 0 & 3 \\ 0 & 4 & -1 \end{bmatrix}, \quad B = \begin{bmatrix} -3 & 2 \\ 1 & 0 \\ -2 & 5 \end{bmatrix}$$

The method that was used to multiply the two matrices in Example 7 can be extended to find the product of any two matrices, provided the number of columns in the first matrix equals the number of rows in the second matrix.

In matrix multiplication, the order of the matrices matters. In general, the product AB *does not equal* the product BA. Furthermore, it is possible that AB is defined but BA is not. For instance, if A is a 2×3 matrix and B is a 3×4 matrix, then AB is defined (since A has three columns and B has three rows) but BA is undefined (since B has four columns and A has only two rows).

Using Technology

Matrices can be multiplied using a graphing utility. Attempting to multiply matrices whose dimensions are incompatible will yield an error message. See Figure 3.

Keystroke Appendix:

Section 13

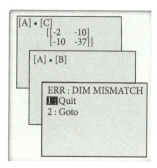

Figure 3

Example 5   Matrix Products

Let

$$A = \begin{bmatrix} 1 & -1 & 0 \\ 2 & -5 & 1 \end{bmatrix}, \quad B = \begin{bmatrix} 6 & -2 \\ 2 & -5 \\ -3 & -1 \\ 1 & -4 \end{bmatrix}, \quad C = \begin{bmatrix} 0 & -4 \\ 2 & 6 \\ 0 & 1 \end{bmatrix}$$

Calculate the following, if defined.

a. AC　　　**b.** BA　　　**c.** AB

Solution

a. For the product AC, we have a 2×3 matrix (A) multiplied by a 3×2 matrix (C). This multiplication is defined because the number of columns of A is the same as the number of rows of C. The result is a 2×2 matrix.

$$AC = \begin{bmatrix} 1 & -1 & 0 \\ 2 & -5 & 1 \end{bmatrix} \begin{bmatrix} 0 & -4 \\ 2 & 6 \\ 0 & 1 \end{bmatrix} = \begin{bmatrix} -2 & -10 \\ -10 & -37 \end{bmatrix}$$

b. For the product BA, we have a 4×2 matrix multiplied by a 2×3 matrix. This multiplication is defined because the number of columns of B is the same as the number of rows of A The result is a 4×3 matrix.

$$BA = \begin{bmatrix} 6 & -2 \\ 2 & -5 \\ -3 & -1 \\ 1 & -4 \end{bmatrix} \begin{bmatrix} 1 & -1 & 0 \\ 2 & -5 & 1 \end{bmatrix} = \begin{bmatrix} 2 & 4 & -2 \\ -8 & 23 & -5 \\ -5 & 8 & -1 \\ -7 & 19 & -4 \end{bmatrix}$$

c. The product AB is not defined, since A has three columns and B has four rows.

Check It Out 5   In Example 8, find the product *CA*, if it is defined.

Example 6 illustrates an application of matrix multiplication to manufacturing.

Example 6   **Application of Matrix Multiplication to Manufacturing.**

A manufacturer of women's clothing makes four different outfits, each of which utilizes some combination of fabrics A, B, and C. The yardage of each fabric required for each outfit is given in matrix *F*.

$$
\begin{array}{c}
\\
\\
\text{Outfit 1}\\
\text{Outfit 2}\\
\text{Outfit 3}\\
\text{Outfit 4}
\end{array}
\begin{array}{ccc}
\text{Fabric A} & \text{Fabric B} & \text{Fabric C}\\
\text{(yd)} & \text{(yd)} & \text{(yd)}
\end{array}
\begin{bmatrix}
1.2 & 0.4 & 1.5\\
0.8 & 0.6 & 2.1\\
1.5 & 0.3 & 1.8\\
2.2 & 0.7 & 1.5
\end{bmatrix} = F
$$

The cost of each fabric (in dollars per yard) is given in matrix *C*.

$$
\begin{array}{c}
\text{Fabric A}\\
\text{Fabric B}\\
\text{Fabric C}
\end{array}
\begin{bmatrix}
8\\
4\\
10
\end{bmatrix} = C
$$

Find the total cost of fabric for each outfit.

Solution   Each row of matrix *F* gives the fabric requirements for a certain outfit. To determine the total cost of fabric for each outfit, we multiply matrix *F* by matrix *C*.

$$
FC = \begin{bmatrix}
1.2 & 0.4 & 1.5\\
0.8 & 0.6 & 2.1\\
1.5 & 0.3 & 1.8\\
2.2 & 0.7 & 1.5
\end{bmatrix}
\begin{bmatrix}
8\\
4\\
10
\end{bmatrix} =
\begin{bmatrix}
26.2\\
29.8\\
31.2\\
35.4
\end{bmatrix}
$$

The total cost of fabric for each outfit is given in Table 1.

Outfit	Total Cost of Fabric
1	$26.20
2	$29.80
3	$31.20
4	$35.40

Table 1

 Check It Out 6   Rework Example 9 for the case in which the cost of each fabric (in dollars per yard) is given by the following matrix.

$$
\begin{array}{c}
\text{Fabric A}\\
\text{Fabric B}\\
\text{Fabric C}
\end{array}
\begin{bmatrix}
6\\
5\\
8
\end{bmatrix}
$$

Exercises 8.4

In Exercises 1–6, use the following matrix.

$$A = \begin{bmatrix} -1 & 2 & 0 & 4 \\ 2.1 & -7 & 9 & 0 \\ 1 & 0 & -\frac{2}{3} & \pi \end{bmatrix}$$

1. Find a_{11}.　　**2.** Find a_{22}.　　**3.** Determine the dimensions of A.

4. Find a_{31}.　　**5.** Find a_{34}.　　**6.** Why is a_{43} not defined?

In Exercises 7–10, indicate whether each statement is True or False. Explain your answers.

7. *Some* matrices that do not have the same dimensions can be added.

8. *Some* matrices that do not have the same dimensions can be multiplied.

9. If A is a 2×4 matrix and B is a 4×3 matrix, then the product AB is a 2×3 matrix.

10. If A is a 2×4 matrix and B is a 4×3 matrix, then the product BA is a 4×4 matrix.

In Exercises 11–28, perform the given operations (if defined) on the matrices.

$$A = \begin{bmatrix} 1 & -3 & \frac{1}{3} \\ 5 & 0 & -2 \end{bmatrix}, \quad B = \begin{bmatrix} 8 & 0 \\ 3 & -2 \\ 2 & -6 \end{bmatrix}, \quad C = \begin{bmatrix} -4 & 5 \\ 0 & 1 \\ -2 & 7 \end{bmatrix}$$

If an operation is not defined, state the reason.

11. $B + C$　**12.** $C - B$　**13.** $2B + C$　**14.** $B + 2C$　**15.** $-3C + B$　**16.** $C - 2B$

17. $A + 2B$　**18.** $3B - C$　**19.** AB　**20.** AC　**21.** BC　**22.** CA

23. $\frac{1}{2}A$　**24.** $\frac{2}{3}C$　**25.** $A(B + C)$　**26.** $(B + C)A$　**27.** $C(AB)$　**28.** $B(AC)$

In Exercises 29–34, for the given matrices A and B, evaluate (if defined) the expressions (a) AB, (b) 3B − 2A, and (c) BA. For any expression that is not defined, state the reason.

29. $A = \begin{bmatrix} -4 & 2 \\ -1 & 0 \end{bmatrix}$; $B = \begin{bmatrix} 1 \\ -3 \end{bmatrix}$

30. $A = \begin{bmatrix} -5 \\ 4 \end{bmatrix}$; $B = \begin{bmatrix} -1 & \frac{1}{2} \\ 0 & -6 \end{bmatrix}$

31. $A = \begin{bmatrix} 0 & 4 \\ -6 & 7 \end{bmatrix}$; $B = \begin{bmatrix} 3 & -7 \\ 2 & -1 \end{bmatrix}$

32. $A = \begin{bmatrix} 9 & -4 \\ 7 & -3 \end{bmatrix}$; $B = \begin{bmatrix} 4 & 0 \\ -7 & 5 \end{bmatrix}$

33. $A = \begin{bmatrix} 3 & 0 & -2 \\ 7 & -6 & -1 \\ 5 & 2 & -1 \end{bmatrix}$; $B = \begin{bmatrix} 4 & -2 \\ 1 & 0 \\ 9 & 3 \end{bmatrix}$

34. $A = \begin{bmatrix} 6 & -7 & 0 \\ 2 & 3 & -4 \\ 1 & 0 & -2 \end{bmatrix}$; $B = \begin{bmatrix} 7 & -2 \\ 3 & 0 \\ 2 & 1 \end{bmatrix}$

In Exercises 35–42, for the given matrices A, B, and C, evaluate the indicated expression.

35. $A = \begin{bmatrix} 6 & -1 \\ 5 & 1 \end{bmatrix}$; $B = \begin{bmatrix} 2 \\ 4 \end{bmatrix}$; $C = \begin{bmatrix} 3 \\ -2 \end{bmatrix}$; $AB + AC$

36. $A = \begin{bmatrix} 8 & 3 \\ -4 & 7 \end{bmatrix}$; $B = \begin{bmatrix} -1 \\ 1 \end{bmatrix}$; $C = \begin{bmatrix} -4 \\ 3 \end{bmatrix}$; $A(C - B)$

37. $A = \begin{bmatrix} 1 & 2 \\ 0 & -5 \end{bmatrix}$; $B = \begin{bmatrix} 7 & -4 \\ -4 & -7 \end{bmatrix}$; $C = \begin{bmatrix} 2 & 1 \\ 0 & -9 \end{bmatrix}$; $CA - B$

38. $A = \begin{bmatrix} 3 & -8 \\ 2 & 4 \end{bmatrix}$; $B = \begin{bmatrix} -6 & 0 \\ 0 & -6 \end{bmatrix}$; $C = \begin{bmatrix} 3 & 5 \\ -2 & 6 \end{bmatrix}$; $(A + 2B)C$

39. $A = \begin{bmatrix} 1 & 3 \\ -4 & 2 \\ 6 & 0 \end{bmatrix}$; $B = \begin{bmatrix} -2 & 1 & 4 \\ 3 & -1 & 1 \end{bmatrix}$; $C = \begin{bmatrix} -4 & 1 \\ 5 & -7 \end{bmatrix}$; $2C + BA$

40. $A = \begin{bmatrix} 5 & -3 \\ -1 & 4 \\ 6 & 0 \end{bmatrix}$; $B = \begin{bmatrix} 0 & 5 & 8 \\ 1 & 0 & -4 \end{bmatrix}$; $C = \begin{bmatrix} 6 & 8 \\ -5 & 3 \end{bmatrix}$; $BA - CC$

41. $A = \begin{bmatrix} 4 & 1 \\ 0 & 2 \\ 5 & 1 \end{bmatrix}$; $B = \begin{bmatrix} 4 & 3 \\ -6 & 2 \\ 3 & -1 \end{bmatrix}$; $C = \begin{bmatrix} 1 & 2 & 3 \\ -2 & -3 & -1 \\ 3 & 1 & 2 \end{bmatrix}$; $C(B - A)$

42. $A = \begin{bmatrix} 3 & 1 \\ 2 & 5 \\ -2 & 1 \end{bmatrix}$; $B = \begin{bmatrix} -5 & -3 \\ 1 & 6 \\ 8 & 3 \end{bmatrix}$; $C = \begin{bmatrix} 2 & 1 & 1 \\ 0 & -1 & 7 \\ 3 & 0 & -3 \end{bmatrix}$; $CB + 2A$

In Exercises 43–46, answer the question pertaining to the matrices

$$A = \begin{bmatrix} a & b \\ c & d \\ e & f \end{bmatrix} \quad and \quad B = \begin{bmatrix} g & h & i \\ j & k & l \end{bmatrix}$$

43. Let $P = AB$, and find p_{11} and p_{33} without performing the entire multiplication of matrix A by matrix B.

44. Let $Q = BA$, and find q_{11} and q_{22} without performing the entire multiplication of matrix B by matrix A.

45. Let $P = AB$, and find p_{32} and p_{23} without performing the entire multiplication of matrix A by matrix B.

46. Let $Q = BA$, and find q_{12} and q_{21} without performing the entire multiplication of matrix B by matrix A.

In Exercises 47–52, find A² (the product AA) and A³ (the product (A²)A).

47. $A = \begin{bmatrix} 2 & -1 \\ 1 & 0 \end{bmatrix}$

48. $A = \begin{bmatrix} 3 & 1 \\ 0 & -1 \end{bmatrix}$

49. $A = \begin{bmatrix} -4 & 0 \\ 0 & 3 \end{bmatrix}$

50. $A = \begin{bmatrix} 1 & 1 \\ -1 & 2 \end{bmatrix}$

51. $A = \begin{bmatrix} 3 & 0 & 0 \\ 0 & 1 & 1 \\ -4 & 1 & 0 \end{bmatrix}$

52. $A = \begin{bmatrix} 2 & -1 & 1 \\ 0 & 1 & 0 \\ 0 & 3 & 2 \end{bmatrix}$

53. If $A = \begin{bmatrix} 0 & 1 \\ a & 0 \end{bmatrix}$ and $B = \begin{bmatrix} 0 & a \\ 1 & 0 \end{bmatrix}$, for what value(s) of a does $AB = \begin{bmatrix} 1 & 0 \\ 0 & 1 \end{bmatrix}$?

54. If $A = \begin{bmatrix} 4a + 5 & -1 \\ -4 & -7 \end{bmatrix}$ and $B = \begin{bmatrix} 7 & 0 \\ -4 & -8 \end{bmatrix}$, for what value(s) of a does $2B - 3A = \begin{bmatrix} 2 & 3 \\ 4 & 5 \end{bmatrix}$?

55. If $A = \begin{bmatrix} 2 & 1 \\ 1 & 3 \end{bmatrix}$ and $B = \begin{bmatrix} 2 & 2a + b \\ b - a & 6 \end{bmatrix}$, for what value(s) of a and b $AB = BA$?

56. If $A = \begin{bmatrix} 1 & 0 & 1 \\ 0 & 0 & 1 \\ 2 & -1 & 0 \end{bmatrix}$ and $B = \begin{bmatrix} 0 & 3 & -1 \\ -1 & 2 & 0 \\ 0 & 0 & 1 \end{bmatrix}$, for what value(s) of a and b does

$$AB = \begin{bmatrix} 0 & 2a + 2b + 1 & 0 \\ 3a + 4b & 0 & 1 \\ 1 & 4 & -2 \end{bmatrix}?$$

57. If $A = \begin{bmatrix} a^2 - 3a + 3 & 1 \\ 0 & 2b + 5 \end{bmatrix}$ and $B = \begin{bmatrix} 0 & 1 \\ 1 & 0 \end{bmatrix}$, for what value(s) of a and b does $AB = \begin{bmatrix} 1 & 1 \\ 1 & 0 \end{bmatrix}$?

58. If $A = \begin{bmatrix} 3 & 16 & 5 \\ 4 & 3 & 6 \end{bmatrix}$ and $B = \begin{bmatrix} 1 & a^2 - 2a - 7 & 2 \\ b^2 - 5b - 4 & 1 & 3 \end{bmatrix}$, for what value(s) of a and b does

$A - 2B = \begin{bmatrix} 1 & 0 & 1 \\ 0 & 1 & 0 \end{bmatrix}$?

Applications

59. Gas Prices At a certain gas station, the prices of regular and high-octane gasoline are $2.40 per gallon and $2.65 per gallon, respectively. Use matrix scalar multiplication to compute the cost of 12 gallons of each type of fuel.

60. Taxi Fares A cab company charges $4.50 for the first mile of a passenger's fare and $1.50 for every mile thereafter. If it is snowing, the fare is increased to $5.50 for the first mile and $1.75 for every mile thereafter. All distances are rounded up to the nearest full mile. Use matrix addition and scalar multiplication to compute the fare for a 6.8-mile trip on both a fair-weather day and a day on which it is snowing.

61. Economics Matrix G gives the U.S. gross domestic product for the years 1999–2001.

$$
\begin{array}{c}
\text{GDP} \\
\text{(billions of \$)}
\end{array}
$$

$$
\begin{array}{c}
1999 \\
2000 \\
2001
\end{array}
\begin{bmatrix}
9274.3 \\
9824.6 \\
10{,}082.2
\end{bmatrix} = G
$$

The finance, retail, and agricultural sectors contributed 20%, 9%, and 1.4%, respectively, to the gross domestic product in those years. These percentages have been converted to decimals and are given in matrix P.

$$
\begin{array}{ccc}
\text{Finance} & \text{Retail} & \text{Agriculture}
\end{array}
$$

$$
\begin{bmatrix} 0.2 & 0.09 & 0.014 \end{bmatrix} = P
$$

a. Compute the product GP.

b. What does GP represent?

c. Is the product PG defined? If so, does it represent anything meaningful? Explain.

62. Tuition Three students take courses at two different colleges, Woosamotta University (WU) and Frostbite Falls Community College (FFCC). WU charges $200 per credit hour and FFCC charges $120 per credit hour. The number of credits taken by each student at each college is given in the following table.

	Credits	
Student	WU	FFCC
1	12	6
2	3	9
3	8	8

Use matrix multiplication to find the total tuition paid by each student.

63. **Furniture** A furniture manufacturer makes three different pieces of furniture, each of which utilizes some combination of fabrics A, B, and C. The yardage of each fabric required for each piece of furniture is given in matrix *F*.

$$
\begin{array}{c}
\\
\\
\text{Sofa} \\
\text{Loveseat} \\
\text{Chair}
\end{array}
\begin{array}{ccc}
\text{Fabric A} & \text{Fabric B} & \text{Fabric C} \\
\text{(yd)} & \text{(yd)} & \text{(yd)}
\end{array}
\left[
\begin{array}{ccc}
10.5 & 2 & 1 \\
8 & 1.5 & 1 \\
4 & 1 & 0.5
\end{array}
\right] = F
$$

The cost of each fabric (in dollars per yard) is given in matrix *C*.

$$
\begin{array}{c}
\text{Fabric A} \\
\text{Fabric B} \\
\text{Fabric C}
\end{array}
\left[
\begin{array}{c}
10 \\
6 \\
5
\end{array}
\right] = C
$$

Find the total cost of fabric for each piece of furniture.

64. **Shopping** Keith and two of his friends, Sam and Cody, take advantage of a sidewalk sale at a shopping mall. Their purchases are summarized in the following table.

Name	Shirt	Sweater	Jacket
Keith	3	2	1
Sam	1	2	2
Cody	2	1	2

The sale prices are $14.95 per shirt, $18.95 per sweater, and $24.95 per jacket. In their state, there is no sales tax on purchases of clothing. Use matrix multiplication to determine the total expenditure of each of the three shoppers.

Concepts

65. Let $A = \begin{bmatrix} 1 & -1 \\ 1 & -1 \end{bmatrix}$ and $B = \begin{bmatrix} 1 & 1 \\ 1 & 1 \end{bmatrix}$. What is the product AB? Is is true that if A and B are matrices such that AB is defined and all the entries of AB are zero, then either all the entries of A must be zero or all the entries of B must be zero? Explain.

66. Show that $A + B = B + A$ for any two matrices A and B for which addition is defined.

67. Let $I = \begin{bmatrix} 1 & 0 \\ 0 & 1 \end{bmatrix}$ and $A = \begin{bmatrix} 2 & -1 \\ 1 & 0 \end{bmatrix}$. Calculate AI and IA. What do you observe?

68. Let $I = \begin{bmatrix} 1 & 0 \\ 0 & 1 \end{bmatrix}$. Show that $IA = AI$, where A is any 2×2 matrix.

69. Let $A = \begin{bmatrix} 1 & 2 \\ 3 & 4 \end{bmatrix}$ and $B = \begin{bmatrix} 0 & 1 \\ 1 & 1 \end{bmatrix}$. Find the products AB and BA. Is it true that if A and B are 2×2 matrices, then $AB = BA$? Explain.

Matrices and Inverses

8.5

Objectives

- Define a square matrix.
- Define an identity matrix.
- Define the inverse of a square matrix.
- Find the inverse of a 2 × 2 matrix.
- Find the inverse of a 3 × 3 matrix.
- Use inverses to solve systems of linear equations.
- Use inverses in applications.

During World War II, Sir Alan Turing, a now-famous mathematician, was instrumental in breaking the code used by the Germans to communicate military secrets. Such a decoding process entails *working backward* from the cryptic information in order to reveal the text of the original message. In Example 6, we will see how an encoded message can be decoded by use of an operation known as *matrix inversion,* which we will study in detail in this section. Before introducing the topic of matrix inversion, we will address two important concepts related to it: **square matrices** and **identity matrices**.

Identity Matrices

The number 1 is called a *multiplicative identity* because $a \cdot 1 = 1 \cdot a = a$ for any non-zero number a. We can extend this idea to **square matrices**—matrices that have the same number of rows and columns. As you might guess, an **identity matrix** will contain some 1's. In fact, the identity matrix for 2 × 2 matrices is

$$I = \begin{bmatrix} 1 & 0 \\ 0 & 1 \end{bmatrix}$$

For any 2 × 2 matrix A, it can be shown that $AI = IA = A$. In Example 1, the property that $IA = A$ is demonstrated for a particular 2 × 2 matrix A.

VIDEO EXAMPLES

SECTION 8.5

> **Example 1**   **Checking a Matrix Identity**
>
> Let
> $$A = \begin{bmatrix} -2 & 3 \\ 4 & 7 \end{bmatrix}$$
>
> Show that $IA = A$.
>
> **Solution** We perform the multiplication of matrix I by matrix A.
> $$IA = \begin{bmatrix} 1 & 0 \\ 0 & 1 \end{bmatrix} \begin{bmatrix} -2 & 3 \\ 4 & 7 \end{bmatrix} = \begin{bmatrix} -2 & 3 \\ 4 & 7 \end{bmatrix}$$
>
> Clearly, the product matrix IA is equal to matrix A.

✓ *Check It Out 1*   For matrix A from Example 1, show that $AI = A$.

For every positive integer $n \geq 2$, the $n \times n$ identity matrix is

$$I = \overbrace{\begin{bmatrix} 1 & 0 & \cdots & 0 \\ 0 & 1 & \cdots & 0 \\ \vdots & \vdots & \ddots & \vdots \\ 0 & 0 & \cdots & 1 \end{bmatrix}}^{n \text{ columns}} \left. \vphantom{\begin{bmatrix} 1 \\ 0 \\ \vdots \\ 0 \end{bmatrix}} \right\} n \text{ rows}$$

Every diagonal entry of the $n \times n$ identity matrix is 1, and every off-diagonal entry is 0.

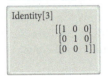
Inverse of a Matrix

For every nonzero number a, the reciprocal of a is called its multiplicative inverse since

$$a \times \frac{1}{a} = 1 = \frac{1}{a} \times a$$

We consider the problem of finding the inverse of a 2 × 2 matrix $A = \begin{bmatrix} a & b \\ c & d \end{bmatrix}$. The goal is to find a 2 × 2 matrix B that satisfies the matrix equation $AB = I$, where I is the 2 × 2 identity matrix. If $B = \begin{bmatrix} x & u \\ y & v \end{bmatrix}$, we want to solve the matrix equation

$$\begin{bmatrix} a & b \\ c & d \end{bmatrix} \begin{bmatrix} x & u \\ y & v \end{bmatrix} = \begin{bmatrix} 1 & 0 \\ 0 & 1 \end{bmatrix}$$

Using the rules of matrix multiplication, we can break up this matrix equation into two separate matrix equations:

$$\begin{bmatrix} a & b \\ c & d \end{bmatrix} \begin{bmatrix} x \\ y \end{bmatrix} = \begin{bmatrix} 1 \\ 0 \end{bmatrix}$$

$$\begin{bmatrix} a & b \\ c & d \end{bmatrix} \begin{bmatrix} u \\ v \end{bmatrix} = \begin{bmatrix} 0 \\ 1 \end{bmatrix}$$

Thus x and y can be found by solving the first system, and u and v can be found by solving the second system. The two systems can be solved simultaneously, by applying Gauss-Jordan elimination to the following augmented matrix:

$$\left[\begin{array}{cc|cc} a & b & 1 & 0 \\ c & d & 0 & 1 \end{array}\right]$$

Example 2 shows how to calculate the inverse of a 2 × 2 matrix.

Example 2   Finding the Inverse of a 2 × 2 Matrix

Find the inverse of $A = \begin{bmatrix} 1 & -3 \\ -2 & 5 \end{bmatrix}$.

Solution We need to find a matrix $B = \begin{bmatrix} x & u \\ y & v \end{bmatrix}$ such that $AB = I$, where I is the 2 × 2 identity matrix. As discussed earlier, we find B by applying Gauss-Jordan elimination to the following augmented matrix:

$$\left[\begin{array}{cc|cc} 1 & -3 & 1 & 0 \\ -2 & 5 & 0 & 1 \end{array}\right] \begin{array}{c} (1) \\ (2) \end{array}$$

The goal is to reduce the part of the augmented matrix which lies to the left of the vertical bar to the 2 × 2 identity matrix. Then, the matrix on the right side of the vertical bar will be the inverse of A.

Step 1 There is already a 1 in the first row, first column, so the first step consists of eliminating the -2 in the second row, first column: Multiply row (1) by 2, and add the result to row (2).

$$\left[\begin{array}{cc|cc} 1 & -3 & 1 & 0 \\ 0 & -1 & 2 & 1 \end{array}\right] \begin{array}{l} (1) \\ (2') = 2 \cdot (1) + (2) \end{array}$$

Step 2 The next step consists of getting a 1 in the second row, second column: Multiply row (2') by -1.

$$\left[\begin{array}{cc|cc} 1 & -3 & 1 & 0 \\ 0 & 1 & -2 & -1 \end{array}\right] \begin{array}{l} (1) \\ (2'') = -1 \cdot (2') \end{array}$$

Step 3 The final step is to eliminate the -3 in the first row, second column: Multiply row $(2'')$ by 3, and add the result to row (1).

$$\left[\begin{array}{cc|cc} 1 & 0 & -5 & -3 \\ 0 & 1 & -2 & -1 \end{array}\right] \quad \begin{array}{l} (1') = 3 \cdot (2'') + (1) \\ (2'') \end{array}$$

The part of the final augmented matrix which lies to the right of the vertical bar is the inverse of matrix A.

$$B = \begin{bmatrix} -5 & -3 \\ -2 & -1 \end{bmatrix}$$

You should check that it is indeed the inverse. That is, $AB = I = BA$, where I is the 2×2 identity matrix.

 Check It Out 2   Use Gauss-Jordan elimination to find the inverse of $A = \begin{bmatrix} -3 & 5 \\ -1 & 2 \end{bmatrix}$.

We now define the inverse of a square matrix in general.

Inverse of an $n \times n$ matrix

For $n \geq 2$, the inverse of an $n \times n$ matrix A is an $n \times n$ matrix B such that

$$AB = I = BA$$

where I is the $n \times n$ identity matrix. If such an inverse exists, it is denoted by A^{-1}.

Example 3   **Checking if Two Matrices Are Inverses**

If

$$A = \begin{bmatrix} 2 & 0 & -1 \\ 1 & 4 & 0 \\ 0 & -2 & 0 \end{bmatrix}, \qquad B = \begin{bmatrix} 0 & 1 & 2 \\ 0 & 0 & -\frac{1}{2} \\ -1 & 2 & 4 \end{bmatrix}$$

show that $AB = I$, where I is the 3×3 identity matrix.

Solution   What we have to show is that

$$AB = \begin{bmatrix} 1 & 0 & 0 \\ 0 & 1 & 0 \\ 0 & 0 & 1 \end{bmatrix}$$

Multiplying A by B, we get

$$AB = \begin{bmatrix} 2 & 0 & -1 \\ 1 & 4 & 0 \\ 0 & -2 & 0 \end{bmatrix} \begin{bmatrix} 0 & 1 & 2 \\ 0 & 0 & -\frac{1}{2} \\ -1 & 2 & 4 \end{bmatrix} = \begin{bmatrix} 0+0+1 & 2+0-2 & 4+0-4 \\ 0+0+0 & 1+0+0 & 2-2+0 \\ 0+0+0 & 0+0+0 & 0+1+0 \end{bmatrix}$$

$$= \begin{bmatrix} 1 & 0 & 0 \\ 0 & 1 & 0 \\ 0 & 0 & 1 \end{bmatrix}$$

 Check It Out 3   Rework Example 3 for the case where

$$A = \begin{bmatrix} -1 & 0 & -2 \\ 1 & 1 & 1 \\ 1 & 1 & 0 \end{bmatrix}, \quad B = \begin{bmatrix} -1 & -2 & 2 \\ 1 & 2 & -1 \\ 0 & 1 & -1 \end{bmatrix}$$

To determine whether a given $n \times n$ matrix A has an inverse, we construct the modified augmented matrix analogous to the 2×2 case. Then perform Gauss-Jordan elimination to reduce the augmented matrix to reduced row-echelon form. Matrix A has an inverse if and only if the final augmented matrix has no row with all zeros to the left of the vertical bar.

The following example illustrates the use of Gauss-Jordan elimination to find the inverse of a 3×3 matrix.

Example 4 **Finding the Inverse of a 3 × 3 Matrix**

Find the inverse of

$$A = \begin{bmatrix} 3 & 3 & 9 \\ 1 & 0 & 2 \\ -2 & 3 & 0 \end{bmatrix}$$

Solution First, construct the augmented matrix with A to the left of the vertical bar and the 3×3 identity matrix to the right of the bar. Label the rows.

$$\begin{bmatrix} 3 & 3 & 9 & | & 1 & 0 & 0 \\ 1 & 0 & 2 & | & 0 & 1 & 0 \\ -2 & 3 & 0 & | & 0 & 0 & 1 \end{bmatrix} \begin{matrix} (1) \\ (2) \\ (3) \end{matrix}$$

Step 1 To get a 1 in the first row, first column, swap rows (1) and (2).

$$\begin{bmatrix} 1 & 0 & 2 & | & 0 & 1 & 0 \\ 3 & 3 & 9 & | & 1 & 0 & 0 \\ -2 & 3 & 0 & | & 0 & 0 & 1 \end{bmatrix} \begin{matrix} (1') = (2) \\ (2') = (1) \\ (3) \end{matrix}$$

Step 2 Eliminate the 3 in the second row, first column: Multiply row (1') by -3 and add the result to row (2'). Then eliminate the -2 in the third row, first column: Multiply row (1') by 2 and add the result to row (3).

$$\begin{bmatrix} 1 & 0 & 2 & | & 0 & 1 & 0 \\ 0 & 3 & 3 & | & 1 & -3 & 0 \\ 0 & 3 & 4 & | & 0 & 2 & 1 \end{bmatrix} \begin{matrix} (1') \\ (2'') = -3 \cdot (1') + (2') \\ (3') = 2 \cdot (1') + (3) \end{matrix}$$

Step 3 To get a 1 in the second row, second column, multiply row (2'') by $\frac{1}{3}$.

$$\begin{bmatrix} 1 & 0 & 2 & | & 0 & 1 & 0 \\ 0 & 1 & 1 & | & \frac{1}{3} & -1 & 0 \\ 0 & 3 & 4 & | & 0 & 2 & 1 \end{bmatrix} \begin{matrix} (1') \\ (2''') = \frac{1}{3} \cdot (2'') \\ (3') \end{matrix}$$

Step 4 Eliminate the 3 in the third row, second column: Multiply row (2''') by -3 and add the result to row (3').

$$\begin{bmatrix} 1 & 0 & 2 & | & 0 & 1 & 0 \\ 0 & 1 & 1 & | & \frac{1}{3} & -1 & 0 \\ 0 & 0 & 1 & | & -1 & 5 & 1 \end{bmatrix} \begin{matrix} (1') \\ (2''') \\ (3'') = -3 \cdot (2''') + (3') \end{matrix}$$

Step 5 Eliminate the 2 in the first row, third column: Multiply row (3'') by -2 and add the result to row (1'). Then eliminate the 1 in the second row, third column: Multiply row (3'') by -1 and add the result to row (2''').

$$\begin{bmatrix} 1 & 0 & 0 & | & 2 & -9 & -2 \\ 0 & 1 & 0 & | & \frac{4}{3} & -6 & -1 \\ 0 & 0 & 1 & | & -1 & 5 & 1 \end{bmatrix} \begin{matrix} (1'') = -2 \cdot (3'') + (1') \\ (2'''') = -1 \cdot (3'') + (2''') \\ (3'') \end{matrix}$$

Thus the inverse of A is

$$A^{-1} = \begin{bmatrix} 2 & -9 & -2 \\ \frac{4}{3} & -6 & -1 \\ -1 & 5 & 1 \end{bmatrix}$$

Check It Out 4 Use Gauss-Jordan elimination to find the inverse of

$$A = \begin{bmatrix} -1 & 0 & -2 \\ 3 & 1 & 5 \\ 1 & 1 & 0 \end{bmatrix}$$

As you can see, finding inverses by hand can get quite tedious. For larger matrices, it is better to use technology. We will illustrate this in Example 6 at the end of this section.

Using Technology

The inverse of a matrix can be found easily using a graphing utility. Converting the entries of the inverse matrix to fractional form makes the output easier to read. See Figure 2.

Keystroke Appendix:
Section 13

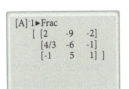

Figure 2

Using Inverses to Solve Systems of Equations

Recall that if a is any nonzero number, we solve an algebraic equation of the form $ax = b$ for x by multiplying both sides of the equation by the reciprocal of a.

$$ax = b \Rightarrow \left(\frac{1}{a}\right)ax = \left(\frac{1}{a}\right)b \Rightarrow x = \left(\frac{1}{a}\right)b$$

If a matrix A has an inverse, we can extend this idea to solve a matrix equation $AX = B$.

$$AX = B \Rightarrow A^{-1}AX = A^{-1}B \Rightarrow X = A^{-1}B$$

Solution of a System of Linear Equations

If a system of linear equations is written in the form $AX = B$, with A an $n \times n$ matrix and A^{-1} its inverse, then

$$X = A^{-1}B$$

The following example shows how to use the inverse of a matrix to solve a system of linear equations.

Example 5   **Solving a System of Equations Using Inverses**

Use the inverse of a matrix to solve the following system of equations.

$$\begin{cases} 3x + 3y + 9z = 6 \\ x + 2z = 0 \\ -2x + 3y = 1 \end{cases}$$

Solution   We first write the system of equations in the form of a matrix equation $AX = B$.

$$\begin{bmatrix} 3 & 3 & 9 \\ 1 & 0 & 2 \\ -2 & 3 & 0 \end{bmatrix} \begin{bmatrix} x \\ y \\ z \end{bmatrix} = \begin{bmatrix} 6 \\ 0 \\ 1 \end{bmatrix}$$

Here,

$$A = \begin{bmatrix} 3 & 3 & 9 \\ 1 & 0 & 2 \\ -2 & 3 & 0 \end{bmatrix}, \quad X = \begin{bmatrix} x \\ y \\ z \end{bmatrix}, \quad B = \begin{bmatrix} 6 \\ 0 \\ 1 \end{bmatrix}$$

Recall that we found the inverse of A in Example 4. Thus, the solution of the matrix equation $AX = B$ is

$$X = A^{-1}B = \begin{bmatrix} 2 & -9 & -2 \\ \frac{4}{3} & -6 & -1 \\ -1 & 5 & 1 \end{bmatrix} \begin{bmatrix} 6 \\ 0 \\ 1 \end{bmatrix} = \begin{bmatrix} 10 \\ 7 \\ -5 \end{bmatrix}$$

Hence the solution of the given system of equations is $x = \mathbf{10}, y = \mathbf{7}, z = \mathbf{-5}$.

 Check It Out 5   Use the inverse of a matrix to solve the following system of equations.

$$\begin{cases} -3x - 2z = -2 \\ x + y + 5z = 3 \\ x + y = -1 \end{cases}$$

Applications

Matrix inverses have a variety of applications. In Example 6 we use the inverse of a matrix to decode a message.

Example 6   **Decoding a Message Using a Matrix Inverse**

Suppose you receive an encoded message in the form of two 4×1 matrices C and D.

$$C = \begin{bmatrix} -30 \\ 109 \\ 120 \\ 14 \end{bmatrix}, \quad D = \begin{bmatrix} 26 \\ -81 \\ -47 \\ -26 \end{bmatrix}$$

Find the original message if the encoding was done using the matrix

$$A = \begin{bmatrix} -1 & 2 & 0 & -2 \\ 3 & -7 & 2 & 6 \\ 2 & -4 & 1 & 7 \\ 1 & -2 & 0 & 1 \end{bmatrix}$$

Solution   To decode the message, we need to find two 4×1 matrices X and Y such that

$$C = AX \quad \text{and} \quad D = AY$$

To do this, we can use the inverse of A:

$$X = A^{-1}C \quad \text{and} \quad Y = A^{-1}D$$

Using a graphing utility, we obtain

$$A^{-1} = \begin{bmatrix} 15 & -2 & 4 & 14 \\ 7 & -1 & 2 & 6 \\ 5 & 0 & 1 & 3 \\ -1 & 0 & 0 & -1 \end{bmatrix}$$

Thus we have the following.

$$X = A^{-1}C = \begin{bmatrix} 15 & -2 & 4 & 14 \\ 7 & -1 & 2 & 6 \\ 5 & 0 & 1 & 3 \\ -1 & 0 & 0 & -1 \end{bmatrix}\begin{bmatrix} -30 \\ 109 \\ 120 \\ 14 \end{bmatrix} = \begin{bmatrix} 8 \\ 5 \\ 12 \\ 16 \end{bmatrix}$$

$$Y = A^{-1}D = \begin{bmatrix} 15 & -2 & 4 & 14 \\ 7 & -1 & 2 & 6 \\ 5 & 0 & 1 & 3 \\ -1 & 0 & 0 & -1 \end{bmatrix}\begin{bmatrix} 26 \\ -81 \\ -47 \\ -26 \end{bmatrix} = \begin{bmatrix} 0 \\ 13 \\ 5 \\ 0 \end{bmatrix}$$

Each positive integer from 1 through 26 corresponds to the letter of a alphabet ($A \leftrightarrow 1$, $B \leftrightarrow 2$, and so on), and 0 corresponds to a space. Applying this correspondence to the matrices X and Y, we find that the decoded message is **HELP ME**.

 **Check It Out 6**   Decode the following message, which was encoded using matrix A from Example 6.

$$\begin{bmatrix} 2 \\ 21 \\ 29 \\ -7 \end{bmatrix}$$

Exercises 8.5

In Exercises 1–6, verify that the matrices are inverses of each other.

1. $\begin{bmatrix} 5 & 2 \\ -3 & -1 \end{bmatrix}, \begin{bmatrix} -1 & -2 \\ 3 & 5 \end{bmatrix}$

2. $\begin{bmatrix} 3 & 4 \\ 5 & 7 \end{bmatrix}, \begin{bmatrix} 7 & -4 \\ -5 & 3 \end{bmatrix}$

3. $\begin{bmatrix} -6 & 5 \\ 4 & -3 \end{bmatrix}, \begin{bmatrix} \frac{3}{2} & \frac{5}{2} \\ 2 & 3 \end{bmatrix}$

4. $\begin{bmatrix} -3 & 2 \\ -4 & 2 \end{bmatrix}, \begin{bmatrix} 1 & -1 \\ 2 & -\frac{3}{2} \end{bmatrix}$

5. $\begin{bmatrix} -1 & 3 & -1 \\ 0 & -5 & 2 \\ 1 & 0 & 0 \end{bmatrix}, \begin{bmatrix} 0 & 0 & 1 \\ 2 & 1 & 2 \\ 5 & 3 & 5 \end{bmatrix}$

6. $\begin{bmatrix} 1 & 1 & 0 \\ -1 & 1 & 0 \\ 1 & 0 & 1 \end{bmatrix}, \begin{bmatrix} \frac{1}{2} & -\frac{1}{2} & 0 \\ \frac{1}{2} & \frac{1}{2} & 0 \\ -\frac{1}{2} & \frac{1}{2} & 1 \end{bmatrix}$

In Exercises 7–22, find the inverse of each matrix.

7. $\begin{bmatrix} 2 & 3 \\ 1 & 1 \end{bmatrix}$

8. $\begin{bmatrix} 4 & 5 \\ 1 & 1 \end{bmatrix}$

9. $\begin{bmatrix} -1 & 3 \\ -1 & 4 \end{bmatrix}$

10. $\begin{bmatrix} 3 & 4 \\ 1 & 2 \end{bmatrix}$

11. $\begin{bmatrix} 5 & 3 \\ 3 & 2 \end{bmatrix}$

12. $\begin{bmatrix} 5 & 3 \\ 4 & 2 \end{bmatrix}$

13. $\begin{bmatrix} 4 & 0 & 5 \\ 0 & 1 & -6 \\ 3 & 0 & 4 \end{bmatrix}$

14. $\begin{bmatrix} 0 & 1 & 0 \\ 3 & 5 & 2 \\ 1 & 2 & 1 \end{bmatrix}$

15. $\begin{bmatrix} -1 & -1 & -1 \\ 3 & 3 & 4 \\ 0 & 1 & 0 \end{bmatrix}$

16. $\begin{bmatrix} 1 & -1 & 0 \\ -2 & 0 & 1 \\ -2 & 5 & -1 \end{bmatrix}$

17. $\begin{bmatrix} 4 & -2 & 1 \\ -2 & 1 & 2 \\ 1 & 2 & 4 \end{bmatrix}$

18. $\begin{bmatrix} 1 & 0 & 0 \\ 0 & 1 & 2 \\ 1 & 0 & 1 \end{bmatrix}$

19. $\begin{bmatrix} 0 & \frac{1}{2} & \frac{1}{2} \\ \frac{1}{2} & 0 & \frac{1}{2} \\ \frac{1}{2} & \frac{1}{2} & 0 \end{bmatrix}$

20. $\begin{bmatrix} 1 & 3 & 0 \\ -1 & -1 & -1 \\ 0 & 3 & -2 \end{bmatrix}$

21. $\begin{bmatrix} 1 & -1 & 0 & 3 \\ 0 & 1 & -2 & 0 \\ -3 & 3 & 1 & -10 \\ 0 & -1 & 2 & 1 \end{bmatrix}$

22. $\begin{bmatrix} 1 & -1 & -2 & 0 \\ 0 & 1 & 0 & 3 \\ -3 & 3 & 7 & -1 \\ 0 & -1 & 0 & -2 \end{bmatrix}$

In Exercises 23 and 24, use one of the matrices given in Exercise 5 to solve the system of equations.

23. $\begin{cases} -x + 3y - z = 6 \\ -5y + 2z = -2 \\ x = 4 \end{cases}$

24. $\begin{cases} -x + 3y - z = 0 \\ -5y + 2z = -3 \\ x = 5 \end{cases}$

In Exercises 25 and 26, use one of the matrices given in Exercise 6 to solve the system of equations.

25. $\begin{cases} x + y = -2 \\ -x + y = 1 \\ x + z = -1 \end{cases}$

26. $\begin{cases} x + y = -4 \\ -x + y = 2 \\ x + z = 6 \end{cases}$

In Exercises 27–46, use matrix inversion to solve the system of equations.

27. $\begin{cases} x - y = -2 \\ -3x + 4y = 5 \end{cases}$

28. $\begin{cases} -x - y = -2 \\ 7x + 6y = 1 \end{cases}$

29. $\begin{cases} 2x + 4y = 1 \\ x + y = -2 \end{cases}$

30. $\begin{cases} x + 2y = -4 \\ -x - y = 5 \end{cases}$

31. $\begin{cases} 3x + 7y = -11 \\ x + 2y = -3 \end{cases}$

32. $\begin{cases} 4x - 3y = 1 \\ 2x - y = -1 \end{cases}$

33. $\begin{cases} 2x - 5y = -7 \\ -3x + 2y = -6 \end{cases}$

34. $\begin{cases} 3x + 2y = -4 \\ 4x + y = 3 \end{cases}$

35. $\begin{cases} x + 2y = 3 \\ 3x + 4y = 3 \end{cases}$

36. $\begin{cases} 7x + 5y = 9 \\ -2x + 3y = -7 \end{cases}$

37. $\begin{cases} x - 3y + 2z = -1 \\ y + z = 4 \\ 2x - 6y + 3z = 3 \end{cases}$

38. $\begin{cases} x - 4y + z = 7 \\ 2x + 9y = -1 \\ y - z = 0 \end{cases}$

39. $\begin{cases} x - y + z = 5 \\ y + 2z = -1 \\ -2x + 3y + z = 6 \end{cases}$

40. $\begin{cases} x + 2z = -3 \\ -2x + y - 7z = 2 \\ x + 3z = 4 \end{cases}$

41. $\begin{cases} 3x - 6y + 2z = -6 \\ x + 2y + 3z = -1 \\ y - z = 5 \end{cases}$

42. $\begin{cases} 2x - y - z = 1 \\ 4x - y + z = -5 \\ x - 3y - 4z = 2 \end{cases}$

43. $\begin{cases} x - 2y - z = \frac{3}{2} \\ 2x - 3y + 2z = -3 \\ -3x + 6y + 4z = 1 \end{cases}$

44. $\begin{cases} x + 5y - 3z = -2 \\ -3x - 16y + 7z = -\frac{1}{2} \\ -x - 5y + 4z = 0 \end{cases}$

45. $\begin{cases} x - y + w = -3 \\ y - 2w = 0 \\ -2x + 2y + z - 3w = 1 \\ -y + 3w = 0 \end{cases}$

46. $\begin{cases} x - y + z + 2w = -3 \\ y - 2z = 0 \\ -2x + 2y - z - w = 1 \\ y - 2z + w = 0 \end{cases}$

In Exercises 47–52, find the inverse of A^2 and the inverse of A^3 (where A^2 is the product AA and A^3 is the product $(A^2)A$).

47. $A = \begin{bmatrix} 1 & 1 \\ 0 & 1 \end{bmatrix}$

48. $A = \begin{bmatrix} 1 & 0 \\ 2 & 1 \end{bmatrix}$

49. $A = \begin{bmatrix} 2 & 1 \\ 0 & -1 \end{bmatrix}$

50. $A = \begin{bmatrix} 1 & 1 \\ 2 & -1 \end{bmatrix}$

51. $A = \begin{bmatrix} 2 & 0 & 0 \\ 0 & 1 & 2 \\ 0 & 0 & 1 \end{bmatrix}$

52. $A = \begin{bmatrix} 1 & 1 & 0 \\ -1 & 1 & 0 \\ 0 & 0 & 1 \end{bmatrix}$

Applications

53. Theater There is a two-tier pricing system for tickets to a certain play: one price for adults, and another for children. One customer purchases 12 tickets for adults and 6 for children, for a total of $174. Another customer purchases 8 tickets for adults and 3 for children, for a total of $111. Use the inverse of an appropriate matrix to compute the price of each type of ticket.

54. Hourly Wage A firm manufactures metal boxes for electrical outlets. The hourly wage for cutting the metal for the boxes is different from the wage for forming the boxes from the cut metal. In a recent week, one worker spent 16 hours cutting metal and 24 hours forming it, and another worker spent 20 hours on each task. The first worker's gross pay for that week was $784, and the second worker grossed $770. Use the inverse of an appropriate matrix to determine the hourly wage for each task.

55. Nutrition Liza, Megan, and Blanca went to a popular pizza place. Liza ate two slices of cheese pizza and one slice of Veggie Delite, for a total of 550 calories. Megan ate one slice each of cheese pizza, Meaty Delite, and Veggie Delite, for a total of 620 calories. Blanca ate one slice of Meaty Delite and two slices of Veggie Delite, for a total of 570 calories. (*Source:* www.pizzahut.com) Use the inverse of an appropriate matrix to determine the number of calories in each slice of cheese pizza, Meaty Delite, and Veggie Delite.

56. Nutrition For a diet of 2000 calories per day, the total fat content should not exceed 60 grams per day. An order of two beef burrito supremes and one plate of nachos supreme exceeds this limit by 2 grams. An order of one beef burrito supreme and one bean tostada contains a total of 28 grams of fat. Also, an order of one beef burrito supreme, one plate of nachos supreme, and one bean tostada contains a total of 54 grams of fat. (*Source:* www.tacobell.com) Use the inverse of an appropriate matrix to determine the fat content (in grams) of each beef burrito supreme, each bean tostada, and each plate of nachos supreme.

57. Quilting A firm manufactures "patriotic" patchwork quilts in three different patterns. The patches are all of the same dimensions and come in three different solid colors: red, white, and blue. For the top layer of each quilt, the firm uses 9 yards of material. For each pattern, the fractions of the total number of squares used for the three colors are given in the table, along with the total cost of the fabric for the top layer of the quilt.

	Fraction of Squares			Total Cost ($)
Pattern	Red	White	Blue	
1	$\frac{1}{4}$	$\frac{5}{12}$	$\frac{1}{3}$	67.50
2	$\frac{1}{3}$	$\frac{1}{3}$	$\frac{1}{3}$	69.00
3	$\frac{1}{4}$	$\frac{1}{2}$	$\frac{1}{4}$	65.25

Use the inverse of an appropriate matrix to determine the cost of the fabric (per yard) for each color.

 Cryptography *In Exercises 58–61, find the decoding matrix for each encoding matrix.*

58. $\begin{bmatrix} 1 & -3 \\ 1 & -2 \end{bmatrix}$ **59.** $\begin{bmatrix} 5 & 7 \\ 2 & 3 \end{bmatrix}$ **60.** $\begin{bmatrix} 1 & 1 & 4 & 1 \\ 2 & -3 & 4 & 1 \\ 3 & -4 & 6 & 2 \\ -1 & 0 & -2 & -1 \end{bmatrix}$ **61.** $\begin{bmatrix} 1 & 0 & -1 & 1 \\ -2 & 3 & 3 & 7 \\ 2 & 0 & -6 & 0 \\ -1 & 1 & 2 & 2 \end{bmatrix}$

 Cryptography *In Exercises 62–65, decode the message, which was encoded using the matrix*

$$\begin{bmatrix} 1 & -2 & 3 \\ -2 & 3 & -4 \\ 2 & -4 & 5 \end{bmatrix}$$

62. $\begin{bmatrix} 29 \\ -47 \\ 45 \end{bmatrix}, \begin{bmatrix} 62 \\ -90 \\ 99 \end{bmatrix}$ **63.** $\begin{bmatrix} 52 \\ -77 \\ 86 \end{bmatrix}, \begin{bmatrix} -24 \\ 38 \\ -53 \end{bmatrix}, \begin{bmatrix} 19 \\ -38 \\ 38 \end{bmatrix}$ **64.** $\begin{bmatrix} -5 \\ 0 \\ -11 \end{bmatrix}, \begin{bmatrix} 20 \\ -36 \\ 38 \end{bmatrix}$ **65.** $\begin{bmatrix} 6 \\ -16 \\ 7 \end{bmatrix}, \begin{bmatrix} 28 \\ -32 \\ 31 \end{bmatrix}$

Concepts

66. Find the inverse of

$$\begin{bmatrix} a & 0 & 0 \\ 0 & b & 0 \\ 0 & 0 & c \end{bmatrix}$$

where a, b, and c are *all* nonzero. Would this matrix have an inverse if $a = 0$? Explain.

67. Find the inverse of

$$\begin{bmatrix} a & a & a \\ 0 & 1 & 0 \\ 0 & 0 & 1 \end{bmatrix}$$

where a is nonzero. Evaluate this inverse for the case in which $a = 1$.

68. Compute $A(BC)$ and $(AB)C$, where

$$A = \begin{bmatrix} 3 & -1 \\ 0 & 2 \end{bmatrix}, B = \begin{bmatrix} 1 & 4 \\ 0 & 1 \end{bmatrix}, \text{ and } C = \begin{bmatrix} -1 & 0 \\ 3 & 1 \end{bmatrix}. \text{ What do you observe?}$$

Exercises 69–74 involve positive-integer powers of a square matrix A. A^2 is defined as the product AA; for $n \geq 3$, A^n is defined as the product $(A^{n-1})A$.

69. Find $(A^2)^{-1}$ and $(A^{-1})^2$, where $A = \begin{bmatrix} 1 & -2 \\ -1 & 3 \end{bmatrix}$. What do you observe?

70. Use the definition of the inverse of a matrix, together with the fact that $(AB)^{-1} = B^{-1}A^{-1}$, to show that $(A^2)^{-1} = (A^{-1})^2$ for every square matrix A.

71. Find $(A^3)^{-1}$ and $(A^{-1})^3$, where $A = \begin{bmatrix} -5 & -1 \\ 4 & 1 \end{bmatrix}$. What do you observe?

72. For $n \geq 3$ and a square matrix A, express the inverse of A^n in terms of A^{-1}. (*Hint:* See Exercise 70.)

73. Let $A = \begin{bmatrix} 4 & 1 \\ 3 & 1 \end{bmatrix}$. Find the inverses of A^2 and A^3 without computing the matrices A^2 and A^3.
(*Hint:* See Exercises 70 and 72.)

74. Let $A = \begin{bmatrix} 0 & 1 \\ 1 & 0 \end{bmatrix}$.

   a. Find A^2, A^3, and A^4.
   b. Find the inverse of A without applying Gauss-Jordan elimination. (*Hint:* Use the answer to part (a).)
   c. For this particular matrix A, what do you observe about A^n for $n = 3, 5, 7, \ldots$?
   d. For this particular matrix A, what do you observe about A^n for $n = 2, 4, 6, \ldots$?

Exercises 75 and 76 involve the use of matrix multiplication to transform one or more points. This technique, which can be applied to any set of points, is used extensively in computer graphics.

75. Let $A = \begin{bmatrix} 0 & 1 \\ 1 & 0 \end{bmatrix}$ and $B = \begin{bmatrix} 2 \\ -1 \end{bmatrix}$.

   a. Calculate the product matrix AB.
   b. On a single coordinate system, plot the point $(2, -1)$ and the point whose coordinates (x, y) are the entries of the product matrix found in part (a). Explain geometrically what the matrix multiplication did to the point $(2, -1)$.
   c. How would you undo the multiplication in part (a)?

76. Consider a series of points (x_0, y_0), (x_1, y_1), (x_2, y_2), … such that, for every nonnegative integer i, the point (x_{i+1}, y_{i+1}) is found by applying the matrix $\begin{bmatrix} 1 & -2 \\ 1 & -3 \end{bmatrix}$ to the point (x_i, y_i).

$$\begin{bmatrix} x_{i+1} \\ y_{i+1} \end{bmatrix} = \begin{bmatrix} 1 & -2 \\ 1 & -3 \end{bmatrix} \begin{bmatrix} x_i \\ y_i \end{bmatrix}$$

   a. Find (x_1, y_1) if $(x_0, y_0) = (2, -1)$.
   b. Find (x_2, y_2) if $(x_0, y_0) = (4, 6)$. (*Hint:* Find (x_1, y_1) first.)
   c. Use the inverse of an appropriate matrix to find (x_0, y_0) if $(x_3, y_3) = (2, 3)$.

Determinants and Cramer's Rule

Objectives

- Calculate the determinant of a matrix.

- Use Cramer's Rule to solve systems of linear equations in two and three variables.

In this section, we introduce **Cramer's Rule,** which gives a formula for the solution of a system of linear equations. We will also discuss the limitations of Cramer's Rule and show that, from a practical standpoint, it is suitable only for solving systems of equations in a few variables. Before presenting Cramer's Rule, we will address an important concept related to it: the determinant of a square matrix.

Determinant of a Square Matrix

We first define the **determinant** of a 2×2 matrix.

Determinant of a 2×2 Matrix

Let A be a 2×2 matrix:

$$A = \begin{bmatrix} a & b \\ c & d \end{bmatrix}$$

where a, b, c, and d represent numbers. Then the **determinant** of A, which is denoted by $|A|$, is defined as

$$|A| = \begin{vmatrix} a & b \\ c & d \end{vmatrix} = ad - bc$$

Note that $|A|$ is just a number, whereas A is a matrix.

VIDEO EXAMPLES

SECTION 8.6

Example 1   **Evaluating a Determinant**

Evaluate $\begin{vmatrix} -3 & -2 \\ 4 & 6 \end{vmatrix}$.

Solution   We see that $a = -3$, $b = -2$, $c = 4$, and $d = 6$. Thus

$$\begin{vmatrix} -3 & -2 \\ 4 & 6 \end{vmatrix} = (-3)(6) - (-2)(4) = -18 - (-8) = -18 + 8 = -10$$

Check It Out 1   Evaluate $\begin{vmatrix} 2 & -3 \\ 5 & -7 \end{vmatrix}$.

For $n > 2$, the determinant of an $n \times n$ matrix is obtained by successively breaking down the computation to those of determinants of smaller matrices. Before explaining how to do this, we define two new terms: **minor** and **cofactor.**

Definition of Minor and Cofactor

Let A be an $n \times n$ matrix, and let a_{ij} be the entry in the ith row, jth column of A.

The **minor** of a_{ij}, denoted by M_{ij}, is the determinant of the matrix obtained by deleting the ith row and the jth column of A.

The **cofactor** of a_{ij}, denoted by C_{ij}, is defined as

$$C_{ij} = (-1)^{i+j} M_{ij}$$

Note that M_{ij} and C_{ij} are a pair of numbers associated with a particular entry of matrix A. Example 2 illustrates how to find minors and cofactors.

Example 2 **Calculating Minors and Cofactors**

Let

$$A = \begin{bmatrix} 0 & 1 & 3 \\ -2 & 5 & 7 \\ 4 & 0 & -1 \end{bmatrix}$$

a. Find M_{12} and C_{12}. **b.** Find M_{13} and C_{13}.

Solution

a. To find M_{12}, let $i = 1$ and $j = 2$.

Steps	Example
1. Since $i = 1$ and $j = 2$, delete the first row and the second column of A. The remaining entries of A are given in red.	$A = \begin{bmatrix} 0 & 1 & 3 \\ -2 & 5 & 7 \\ 4 & 0 & -1 \end{bmatrix}$
2. Find the determinant of the 2 × 2 matrix that is formed by the remaining entries of A.	$M_{12} = \begin{vmatrix} -2 & 7 \\ 4 & -1 \end{vmatrix}$ $= (-2)(-1) - (7)(4)$ $= 2 - 28 = -26$

Because $C_{ij} = (-1)^{i+j}M_{ij}$, the cofactor C_{12} is

$$C_{12} = (-1)^{1+2}M_{12} = (-1)^3(-26) = (-1)(-26) = 26.$$

b. To find M_{13}, let $i = 1$ and $j = 3$. Then delete the first row and the third column of A. The remaining entries of A are shown in boldface.

$$A = \begin{bmatrix} 0 & 1 & 3 \\ \mathbf{-2} & \mathbf{5} & 7 \\ \mathbf{4} & \mathbf{0} & -1 \end{bmatrix}$$

The determinant of the 2 × 2 matrix that is formed by the remaining entries of A is

$$M_{13} = \begin{vmatrix} -2 & 5 \\ 4 & 0 \end{vmatrix} = (-2)(0) - (5)(4) = 0 - 20 = -20$$

The cofactor C_{13} is

$$C_{13} = (-1)^{1+3}M_{13} = (-1)^4(-20) = 1(-20) = -20$$

 Check It Out 2 For matrix A from Example 2, find M_{23} and C_{23}. ■

Now that we have defined minors and cofactors, we can find the determinant of an $n \times n$ matrix.

Finding the Determinant of an $n \times n$ Matrix

Step 1 Choose any row or column of the matrix, preferably a row or column in which at least one entry is zero. This is the row or column by which the determinant will be *expanded*.

Step 2 Multiply each entry in the chosen row or column by its cofactor.

Step 3 The determinant is the *sum* of all the products found in Step 2. It can be shown that the determinant is independent of the choice of row or column made in Step 1.

The determinant of an $n \times n$ matrix A is denoted by $|A|$. Example 3 illustrates the procedure for evaluating the determinant of a 3 × 3 matrix.

Using Technology

Access the MATRIX menu of your graphing utility to compute the determinant of a matrix. Figure 1 shows the determinant of matrix A from Example 3.

Keystroke Appendix:
Section 13

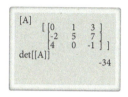

Figure 1

Example 3   Determinant of a 3 × 3 Matrix

Evaluate the determinant of A.

$$A = \begin{bmatrix} 0 & 1 & 3 \\ -2 & 5 & 7 \\ 4 & 0 & -1 \end{bmatrix}$$

Solution We will expand the determinant of A by the first row of A.

$$|A| = \begin{vmatrix} \mathbf{0} & \mathbf{1} & \mathbf{3} \\ -2 & 5 & 7 \\ 4 & 0 & -1 \end{vmatrix} = a_{11}C_{11} + a_{12}C_{12} + a_{13}C_{13}$$ Multiply each element in the first row by its cofactor

$$= (\mathbf{0})C_{11} + \mathbf{1}(26) + 3(-20)$$ $C_{12} = 26$ and $C_{13} = -20$, from Example 2

$$= 0 + 26 - 60 = -34$$

Because $a_{11} = 0$, the product $a_{11}C_{11}$ is zero regardless of the value of C_{11}. So we need not compute C_{11}.

✓ *Check It Out 3*   Evaluate the determinant of A.

$$A = \begin{vmatrix} -1 & 0 & 3 \\ 0 & 2 & 5 \\ -2 & 1 & 0 \end{vmatrix}$$

Cramer's Rule for Systems of Two Linear Equations in Two Variables

We will now see how determinants arise in the process of solving the following system of equations.

$$\begin{cases} ax + by = e \\ cx + dy = f \end{cases}$$

We first solve for the variable x by eliminating the variable y. We assume that a, b, c and d are all nonzero. We multiply the first equation by d and the second equation by $-b$ and then add the resulting equations.

$$\begin{array}{rl} adx + bdy = & ed \\ -bcx - bdy = & -bf \\ \hline adx - bcx \quad = & ed - bf \end{array}$$

Solving for x in the last equation,

$$adx - bcx = ed - bf \Rightarrow (ad - bc)x = ed - bf \Rightarrow x = \frac{ed - bf}{ad - bc}$$

provided $ad - bc \neq 0$. In terms of determinants, we have

$$x = \frac{ed - bf}{ad - bc} = \frac{\begin{vmatrix} e & b \\ f & d \end{vmatrix}}{\begin{vmatrix} a & b \\ c & d \end{vmatrix}}$$

We can solve for y in a similar manner. We now state Cramer's Rule for a 2 × 2 system of equations.

Cramer's Rule for a 2 × 2 System of Linear Equations

The solution of the system of equations

$$\begin{cases} ax + by = e \\ cx + dy = f \end{cases}$$

is

$$x = \frac{\begin{vmatrix} e & b \\ f & d \end{vmatrix}}{\begin{vmatrix} a & b \\ c & d \end{vmatrix}}, \qquad y = \frac{\begin{vmatrix} a & e \\ c & f \end{vmatrix}}{\begin{vmatrix} a & b \\ c & d \end{vmatrix}}$$

provided $\begin{vmatrix} a & b \\ c & d \end{vmatrix} = ad - bc \neq 0.$

The solution can be written more compactly as follows.

$$x = \frac{D_x}{D}, \qquad y = \frac{D_y}{D}$$

where $D_x = \begin{vmatrix} e & b \\ f & d \end{vmatrix},\quad D_y = \begin{vmatrix} a & e \\ c & f \end{vmatrix},\quad D = \begin{vmatrix} a & b \\ c & d \end{vmatrix}$, and $D \neq 0.$

Example 4   **Solving a System of Two Linear Equations Using Cramer's Rule**

Use Cramer's Rule, if applicable, to solve the following system of equations.

$$\begin{cases} 2x - 3y = 6 \\ 5x + y = 7 \end{cases}$$

> **Note** If the quantity $ad - bc$ is nonzero, the system of equations has the unique solution given by Cramer's Rule. If $ad - bc = 0$, Cramer's Rule cannot be applied. If such a system can be solved, it is dependent and has infinitely many solutions.

Solution We first calculate D to determine whether it is nonzero.

$$D = \begin{vmatrix} 2 & -3 \\ 5 & 1 \end{vmatrix} = (2)(1) - (-3)(5) = 2 - (-15) = 2 + 15 = 17$$

Since $D = 17 \neq 0$, we can apply Cramer's Rule to solve this system of equations. Calculating the other determinants, D_x and D_y, we obtain

$$D_x = \begin{vmatrix} 6 & -3 \\ 7 & 1 \end{vmatrix} = (6)(1) - (-3)(7) = 6 - (-21) = 6 + 21 = 27$$

$$D_y = \begin{vmatrix} 2 & 6 \\ 5 & 7 \end{vmatrix} = (2)(7) - (6)(5) = 14 - 30 = -16$$

Thus

$$x = \frac{D_x}{D} = \frac{27}{17}, \qquad y = \frac{D_y}{D} = \frac{-16}{17}$$

✓ **Check It Out 4**   Use Cramer's Rule, if applicable, to solve the following system of equations.

$$\begin{cases} 3x + y = -3 \\ -4x - 2y = 5 \end{cases}$$

Cramer's Rule for Systems of Three Linear Equations in Three Variables

Cramer's Rule for a system of three linear equations in three variables, called a 3×3 system, is similar in form to Cramer's Rule for a 2×2 system. The main difference is the size of the square matrices whose determinants have to be evaluated.

Cramer's Rule for a 3×3 System of Linear Equations

The solution of the system of equations

$$\begin{cases} a_{11}x + a_{12}y + a_{13}z = b_1 \\ a_{21}x + a_{22}y + a_{23}z = b_2 \\ a_{31}x + a_{32}y + a_{33}z = b_3 \end{cases}$$

is

$$x = \frac{D_x}{D}, \quad y = \frac{D_y}{D}, \quad z = \frac{D_z}{D}$$

where

$$D_x = \begin{vmatrix} b_1 & a_{12} & a_{13} \\ b_2 & a_{22} & a_{23} \\ b_3 & a_{32} & a_{33} \end{vmatrix}, \quad D_y = \begin{vmatrix} a_{11} & b_1 & a_{13} \\ a_{21} & b_2 & a_{23} \\ a_{31} & b_3 & a_{33} \end{vmatrix}, \quad D_z = \begin{vmatrix} a_{11} & a_{12} & b_1 \\ a_{21} & a_{22} & b_2 \\ a_{31} & a_{32} & b_3 \end{vmatrix}, \quad \text{and}$$

$$D = \begin{vmatrix} a_{11} & a_{12} & a_{13} \\ a_{21} & a_{22} & a_{23} \\ a_{31} & a_{32} & a_{33} \end{vmatrix}$$

provided $D \neq 0$.

Example 5   **Cramer's Rule for a System of Three Linear Equations**

Use Cramer's Rule to solve the following system of equations.

$$\begin{cases} 3x + y - z = -1 \\ x - y + 2z = 7 \\ -2x + y + z = -2 \end{cases}$$

Solution   We first compute D to determine whether it is nonzero.

$$D = \begin{vmatrix} 3 & 1 & -1 \\ 1 & -1 & 2 \\ -2 & 1 & 1 \end{vmatrix}$$

Expanding the determinant by the first row of the associated matrix, we find that

$$D = \begin{vmatrix} 3 & 1 & -1 \\ 1 & -1 & 2 \\ -2 & 1 & 1 \end{vmatrix} = (-1)^{1+1}(3)\begin{vmatrix} -1 & 2 \\ 1 & 1 \end{vmatrix} + (-1)^{1+2}(1)\begin{vmatrix} 1 & 2 \\ -2 & 1 \end{vmatrix} + (-1)^{1+3}(-1)\begin{vmatrix} 1 & -1 \\ -2 & 1 \end{vmatrix}$$

$$= (1)(3)(-1 - 2) + (-1)(1)[1 - (-4)] + (1)(-1)(1 - 2)$$

$$= 3(-3) - (1)(5) + (-1)(-1) = -9 - 5 + 1 = -13$$

Because $D \neq 0$, we can proceed to compute the other determinants and then use them to compute the solution. We have computed all the determinants by expanding them along the first row of the pertinent matrix.

Using Technology

To check the answer to Example 5, take the matrices whose determinants are D, D_x, D_y, and D_z and enter them into your calculator as matrices A, B, C, and D, respectively. Then compute the quotients of their determinants. See Figure 13.

Keystroke Appendix:

Section 2

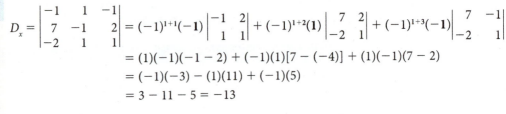

$$D_x = \begin{vmatrix} -1 & 1 & -1 \\ 7 & -1 & 2 \\ -2 & 1 & 1 \end{vmatrix} = (-1)^{1+1}(-1)\begin{vmatrix} -1 & 2 \\ 1 & 1 \end{vmatrix} + (-1)^{1+2}(1)\begin{vmatrix} 7 & 2 \\ -2 & 1 \end{vmatrix} + (-1)^{1+3}(-1)\begin{vmatrix} 7 & -1 \\ -2 & 1 \end{vmatrix}$$

$$= (1)(-1)(-1-2) + (-1)(1)[7-(-4)] + (1)(-1)(7-2)$$

$$= (-1)(-3) - (1)(11) + (-1)(5)$$

$$= 3 - 11 - 5 = -13$$

$$D_y = \begin{vmatrix} 3 & -1 & -1 \\ 1 & 7 & 2 \\ -2 & -2 & 1 \end{vmatrix} = (-1)^{1+1}(3)\begin{vmatrix} 7 & 2 \\ -2 & 1 \end{vmatrix} + (-1)^{1+2}(-1)\begin{vmatrix} 1 & 2 \\ -2 & 1 \end{vmatrix} + (-1)^{1+3}(-1)\begin{vmatrix} 1 & 7 \\ -2 & -2 \end{vmatrix}$$

$$= (1)(3)[7-(-4)] + (-1)(-1)[1-(-4)]$$
$$+ (1)(-1)[-2-(-14)]$$

$$= (3)(11) - (-1)(5) + (-1)(12)$$

$$= 33 + 5 - 12 = 26$$

Figure 2

$$D_z = \begin{vmatrix} 3 & 1 & -1 \\ 1 & -1 & 7 \\ -2 & 1 & -2 \end{vmatrix} = (-1)^{1+1}(3)\begin{vmatrix} -1 & 7 \\ 1 & -2 \end{vmatrix} + (-1)^{1+2}(1)\begin{vmatrix} 1 & 7 \\ -2 & -2 \end{vmatrix} + (-1)^{1+3}(-1)\begin{vmatrix} 1 & -1 \\ -2 & 1 \end{vmatrix}$$

$$= (1)(3)(2-7) + (-1)(1)[-2-(-14)] + (1)(-1)(1-2)$$

$$= (3)(-5) - (1)(12) + (-1)(-1)$$

$$= -15 - 12 + 1 = -26$$

We then have the following result.

$$x = \frac{D_x}{D} = \frac{-13}{-13} = 1$$

$$y = \frac{D_y}{D} = \frac{26}{-13} = -2$$

$$z = \frac{D_z}{D} = \frac{-26}{-13} = 2$$

You should check that these values of x, y, and z satisfy the given system of equations.

 Check It Out 5   Use Cramer's Rule to solve the following system of equations.

$$\begin{cases} -2x + y & = 0 \\ x - y + 2z = -2 \\ y + z = 1 \end{cases}$$

Limitations of Cramer's Rule

Cramer's Rule can be generalized to systems of n linear equations in n variables for all $n \geq 2$ but it uses many more arithmetic operations than does Gaussian elimination. Thus Cramer's Rule is used only for systems with two, three, or four variables, and only in a very limited set of applications, such as computer graphics. For larger systems of equations, Cramer's Rule quickly becomes inefficient, even with extremely fast and powerful supercomputers.

Exercises 8.6

 Skills

In Exercises 1–8, evaluate the determinant of A.

1. $A = \begin{bmatrix} -3 & 1 \\ 2 & 4 \end{bmatrix}$
2. $A = \begin{bmatrix} 5 & 2 \\ -2 & 4 \end{bmatrix}$
3. $A = \begin{bmatrix} \frac{1}{2} & 3 \\ 2 & -6 \end{bmatrix}$
4. $A = \begin{bmatrix} -3 & -\frac{1}{4} \\ 8 & 2 \end{bmatrix}$

5. $A = \begin{bmatrix} 4 & 1 \\ -3 & 8 \end{bmatrix}$
6. $A = \begin{bmatrix} 6 & -2 \\ 5 & 3 \end{bmatrix}$
7. $A = \begin{bmatrix} \frac{1}{3} & -2 \\ 4 & 9 \end{bmatrix}$
8. $A = \begin{bmatrix} 5 & -\frac{2}{5} \\ 10 & 2 \end{bmatrix}$

In Exercises 9–12, find the given minor and cofactor pertaining to the matrix

$$\begin{bmatrix} -3 & 0 & 2 \\ 1 & 5 & -4 \\ 0 & 6 & 5 \end{bmatrix}$$

9. M_{11} and C_{11}
10. M_{23} and C_{23}
11. M_{32} and C_{32}
12. M_{21} and C_{21}

In Exercises 13–22, evaluate the determinant of the matrix.

13. $\begin{bmatrix} 0 & 1 & -2 \\ 5 & -2 & 3 \\ 0 & 6 & 5 \end{bmatrix}$
14. $\begin{bmatrix} -7 & 5 & 0 \\ 0 & 3 & 0 \\ -3 & -2 & 2 \end{bmatrix}$
15. $\begin{bmatrix} -2 & 3 & 5 \\ 6 & -1 & 0 \\ 0 & 1 & -2 \end{bmatrix}$
16. $\begin{bmatrix} -5 & 4 & 9 \\ 1 & 0 & -2 \\ 0 & 7 & 3 \end{bmatrix}$

17. $\begin{bmatrix} 0 & 0 & 0 \\ -7 & 3 & 4 \\ 6 & 3 & 4 \end{bmatrix}$
18. $\begin{bmatrix} -5 & 4 & 9 \\ 1 & 0 & -2 \\ 0 & -5 & 7 \end{bmatrix}$
19. $\begin{bmatrix} 1 & 1 & 1 \\ 2 & 2 & 2 \\ 3 & 3 & 3 \end{bmatrix}$
20. $\begin{bmatrix} 1 & 1 & 1 \\ 1 & 2 & 4 \\ 1 & 3 & 9 \end{bmatrix}$

21. $\begin{bmatrix} -2 & 2 & 0 \\ 0 & -1 & 1 \\ -4 & 5 & 2 \end{bmatrix}$
22. $\begin{bmatrix} 0 & -1 & -3 \\ -2 & 0 & 4 \\ -1 & 0 & 5 \end{bmatrix}$

In Exercises 23–28, solve for x.

23. $\begin{vmatrix} -1 & x \\ 3 & -4 \end{vmatrix} = -2$
24. $\begin{vmatrix} 5 & -1 \\ x & 2 \end{vmatrix} = 13$
25. $\begin{vmatrix} -1 & 0 & 2 \\ 0 & 5 & 3 \\ 0 & x & -2 \end{vmatrix} = -2$
26. $\begin{vmatrix} 5 & 0 & 0 \\ -3 & x & 1 \\ 2 & 8 & -3 \end{vmatrix} = -70$

27. $\begin{vmatrix} 2 & -3 & 5 \\ x & 0 & -4 \\ 3 & 2 & 1 \end{vmatrix} = 39$
28. $\begin{vmatrix} 4 & 3 & x \\ -2 & 8 & 1 \\ 5 & 2 & 1 \end{vmatrix} = 353$

In Exercises 29–46, use Cramer's Rule to solve the system of equations.

29. $\begin{cases} -3x - y = 5 \\ 4x + y = 2 \end{cases}$
30. $\begin{cases} -x + y = -3 \\ -2x + y = -2 \end{cases}$
31. $\begin{cases} 4x - 2y = 7 \\ 3x - y = 1 \end{cases}$
32. $\begin{cases} x - y = -3 \\ 4x + y = 0 \end{cases}$

33. $\begin{cases} x - 2y = 4 \\ -3x + 4y = -8 \end{cases}$
34. $\begin{cases} 7x - y = -8 \\ -x + 3y = 4 \end{cases}$
35. $\begin{cases} 4x + y = -7 \\ 5x + 4y = -6 \end{cases}$
36. $\begin{cases} x + 2y = -3 \\ -x - y = -2 \end{cases}$

37. $\begin{cases} 1.4x + 2y = 0 \\ 3.5x + 3y = -9.7 \end{cases}$
38. $\begin{cases} 2.5x - 0.5y = -7.2 \\ -x + y = 5.6 \end{cases}$
39. $\begin{cases} x + y = 1 \\ x - z = 0 \\ -y + z = 0 \end{cases}$

40. $\begin{cases} x \quad\;\; - z = 0 \\ -y - z = -1 \\ x + y + z = 0 \end{cases}$

41. $\begin{cases} 5x \qquad\; + 3z = \quad 3 \\ -2x + y + z = -1 \\ \quad\; -3y + z = \quad 7 \end{cases}$

42. $\begin{cases} -2x + 3y - z = \quad 7 \\ x \qquad\quad - 2z = -7 \\ \quad\; - 3y + z = -1 \end{cases}$

43. $\begin{cases} 3x - 5y + z = -14 \\ -3x + 7y - 4z = \quad 9 \\ 2x \qquad + z = \quad 6 \end{cases}$

44. $\begin{cases} x \qquad\; + 4z = -3 \\ -2x + y \qquad = -10 \\ x - 2y - z = \quad 7 \end{cases}$

45. $\begin{cases} 3x + y + z = \quad 1 \\ 2x + y - z = -\frac{3}{2} \\ x + 3y - z = -5 \end{cases}$

46. $\begin{cases} x + 2y + z = \quad 0 \\ -x + y + 3z = \quad \frac{5}{2} \\ 4x + y - z = -\frac{3}{2} \end{cases}$

Concepts

47. Without computing the following determinant, explain why its value must be zero.

$$\begin{vmatrix} 1 & 2 & -1 \\ 0 & 0 & 0 \\ 3 & -2 & 1 \end{vmatrix}$$

48. For what value(s) of k can Cramer's Rule be used to solve the following system of equations?

$$\begin{cases} 2x - y = 2 \\ kx + 3y = 4 \end{cases}$$

49. Verify that $x = 1, y = 2, z = 0$ is a solution of the following system of equations.

$$\begin{cases} x + 2y \qquad = \quad 5 \\ 4x + y - z = \quad 6 \\ -2x - 4y \qquad = -10 \end{cases}$$

Even though there is a solution, explain why Cramer's Rule cannot be used to solve this system.

Partial Fractions 8.7

Objective

- Compute the partial fraction decomposition of a rational expression.

You already know how to find the sum and difference of rational expressions. In some situations, however, we need to reverse this procedure. For example, $\frac{5x+1}{x^2-1}$ can be decomposed as $\frac{2}{x+1} + \frac{3}{x-1}$. The technique used to decompose a rational expression into simpler expressions is known as **partial fraction decomposition**. The decomposition procedure depends on the form of the denominator. In this section, we cover four different cases; in each case, it is assumed that the degree of the numerator is less than the degree of the denominator.

> **Case 1: Linear Factors of the Denominator Are All Distinct**
>
> Let $r(x) = \frac{P(x)}{Q(x)}$, where $P(x)$ and $Q(x)$ are polynomials and the degree of $P(x)$ is less than the degree of $Q(x)$, and let $a_1, ..., a_n$ and $b_1, ..., b_n$ be real numbers such that
> $$Q(x) = (a_1 x + b_1)(a_2 x + b_2) \ldots (a_n x + b_n)$$
> Then there exist real numbers $A_1, ..., A_n$ such that the partial fraction decomposition of $r(x)$ is
> $$\frac{A_1}{a_1 x + b_1} + \frac{A_2}{a_2 x + b_2} + \ldots + \frac{A_n}{a_n x + b_n}$$

The constants $A_1, A_2, ..., A_n$ are determined by solving a system of linear equations, as illustrated in the following example.

Example 1 Nonrepeated Linear Factors

Compute the partial fraction decomposition of $\frac{12}{x^2-4}$.

Solution Factoring the denominator, we get $x^2 - 4 = (x+2)(x-2)$. Since both factors are linear and neither of them is a repeated factor,

$$\frac{12}{(x+2)(x-2)} = \frac{A}{x+2} + \frac{B}{x-2}$$

Multiplying both sides of this equation by $(x+2)(x-2)$, we obtain the following.

$$12 = A(x-2) + B(x+2)$$
$$12 = Ax - 2A + Bx + 2B \qquad \text{Expand}$$
$$12 = (A+B)x + (-2A+2B) \qquad \text{Combine like terms}$$

The expression $(A+B)x + (-2A+2B)$ is a polynomial that consists of two terms: a term in x and a constant term. The coefficient of x is $A + B$, and the constant term is $-2A + 2B$. Since these two polynomials are equal, the coefficients of like powers must be equal. Equating coefficients of like powers of x, we obtain

$$0 = A + B \qquad \text{Equate coefficients of } x$$
$$12 = -2A + 2B \qquad \text{Equate constant terms}$$

From the first equation, $A = -B$. Substituting $-B$ for A in the second equation, we have

$$12 = -2(-B) + 2B$$
$$12 = 2B + 2B$$
$$12 = 4B$$
$$3 = B$$

Because $B = 3$ and $A = -B$, we see that $A = -3$. Thus the partial fraction decomposition of $\frac{12}{x^2-4}$ is

$$\frac{-3}{x+2} + \frac{3}{x-2}$$

 Check It Out 1 Compute the partial fraction decomposition of $\frac{-x-1}{x^2-x}$.

Case 2: Denominator Has At Least One Repeated Linear Factor

Let $r(x) = \frac{P(x)}{Q(x)}$, where $P(x)$ and $Q(x)$ are polynomials and the degree of $P(x)$ is less than the degree of $Q(x)$. For every repeated linear factor of $Q(x)$, of the form $(ax+b)^m$, $m \geq 2$, the portion of the partial fraction decomposition of $r(x)$ that pertains to $(ax + b)^m$ is a sum of m fractions of the following form:

$$\frac{A_1}{ax + b} + \frac{A_2}{(ax + b)^2} + \cdots + \frac{A_m}{(ax + b)^m}$$

where $A_1, \ldots, A_m$ are real numbers. The nonrepeated linear factors of $Q(x)$ are treated as in Case 1.

Example 2   **Repeated Linear Factors**

Write the partial fraction decomposition of $\frac{4x^2 - 7x + 1}{x^3 - 2x^2 + x}$.

Solution   Factoring the denominator, we have

$$x^3 - 2x^2 + x = x(x^2 - 2x + 1) = x(x - 1)^2$$

Because every factor is linear and $(x - 1)^2$ is the only repeated linear factor,

$$\frac{4x^2 - 7x + 1}{x(x - 1)^2} = \frac{A}{x} + \frac{B}{x - 1} + \frac{C}{(x - 1)^2}.$$

Multiplying both sides of this equation by the LCD, $x(x - 1)^2$, we obtain the following.

$$\begin{aligned}
4x^2 - 7x + 1 &= A(x - 1)^2 + Bx(x - 1) + Cx \\
&= A(x^2 - 2x + 1) + Bx^2 - Bx + Cx \\
&= Ax^2 - 2Ax + A + Bx^2 - Bx + Cx \\
&= (A + B)x^2 + (-2A - B + C)x + A
\end{aligned}$$

Equating coefficients, we have

$$\begin{aligned}
4 &= A + B & &\text{Equate coefficients of } x^2 \\
-7 &= -2A - B + C & &\text{Equate coefficients of } x \\
1 &= A & &\text{Equate constant terms}
\end{aligned}$$

Because $A = 1$ and $A + B = 4$, we have $B = 3$. Substituting the values of A and B into the second equation leads to the following result.

$$\begin{aligned}
-7 &= -2(1) - 3 + C \\
-7 &= -2 - 3 + C \\
-7 &= -5 + C \\
-2 &= C
\end{aligned}$$

Thus the partial fraction decomposition of $\frac{4x^2 - 7x + 1}{x^3 - 2x^2 + x}$ is

$$\frac{1}{x} + \frac{3}{x - 1} - \frac{2}{(x - 1)^2}$$

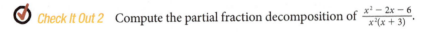 **Check It Out 2**   Compute the partial fraction decomposition of $\frac{x^2 - 2x - 6}{x^2(x + 3)}$.

> **Case 3: Denominator Has At Least One Non-Repeated Irreducible Quadratic Factor**
>
> Let $r(x) = \frac{P(x)}{Q(x)}$, where $P(x)$ and $Q(x)$ are polynomials and the degree of $P(x)$ is less than the degree of $Q(x)$. For every quadratic factor of $Q(x)$, there exist real numbers a, b, and c such that
>
> - $ax^2 + bx + c$ is a factor of $Q(x)$, and
>
> - $ax^2 + bx + c$ cannot be factored over the real numbers, and is called an **irreducible quadratic factor.**
>
> The portion of the partial fraction decomposition of $r(x)$ that pertains to $ax^2 + bx + c$ is a fraction of the form
>
> $$\frac{Ax + B}{ax^2 + bx + c}$$
>
> where A and B are real numbers. Any linear factor of $Q(x)$ is treated as in Case 1 or Case 2.

Example 3   **Checking for Irreducible Quadratic Factors**

Check whether the quadratic polynomial $2x^2 + x + 3$ is irreducible.

Solution The quadratic polynomial $2x^2 + x + 3$ can be factored over the real numbers if and only if the quadratic equation $2x^2 + x + 3 = 0$ has real solutions. To determine whether a quadratic polynomial has real solutions, we evaluate the discriminant, $b^2 - 4ac$, where a and b are the coefficients of the x^2 and x terms, respectively, and c is the constant term. For the polynomial $2x^2 + x + 3$,

$$a = 2, b = 1, c = 3$$

The discriminant is

$$b^2 - 4ac = (\mathbf{1})^2 - 4(\mathbf{2})(\mathbf{3})$$
$$= 1 - 24$$
$$= -23$$

Because $b^2 - 4ac < 0$, the equation

$$2x^2 + x + 3 = 0$$

has no real solutions. Thus $2x^2 + x + 3$ cannot be factored over the real numbers and is therefore irreducible.

 Check It Out 3   Check whether the quadratic polynomial $2x^2 + x - 3$ is irreducible. ■

Example 4   **An Irreducible Quadratic Factor**

Write the partial fraction decomposition of $\frac{3x^2 - 2x + 13}{(x + 1)(x^2 + 5)}$.

Solution Because $x + 1$ is a linear factor and $x^2 + 5$ is an irreducible quadratic factor, we have

$$\frac{3x^2 - 2x + 13}{(x + 1)(x^2 + 5)} = \frac{A}{x + 1} + \frac{Bx + C}{x^2 + 5}$$

Multiplying both sides of this equation by the LCD, $(x + 1)(x^2 + 5)$, we obtain

$$3x^2 - 2x + 13 = A(x^2 + 5) + (Bx + C)(x + 1) \qquad \text{Equation (1)}$$

This equation holds true for any value of x. Because the term $(Bx + C)(x + 1)$ has a linear factor $(x + 1)$, we can easily solve for at least one of the coefficients by substituting the zero of the linear factor for x. Letting $x = -1$ (because -1 is the zero of $x + 1$), we have

$$3(-1)^2 - 2(-1) + 13 = A((-1)^2 + 5) + (B(-1) + C)(-1 + 1)$$
$$3 + 2 + 13 = A(1 + 5) + (-B + C)(0)$$
$$18 = 6A + 0$$
$$\mathbf{3 = A}$$

Substituting $A = 3$ into Equation (1), we get

$$3x^2 - 2x + 13 = 3(x^2 + 5) + (Bx + C)(x + 1)$$
$$3x^2 - 2x + 13 = 3x^2 + 15 + Bx^2 + Cx + Bx + C$$
$$-2x + 13 = Bx^2 + (B + C)x + 15 + C$$

Equating coefficients,

$$\begin{array}{ll} 0 = B & \text{Equate coefficients of } x^2 \\ -2 = B + C & \text{Equate coefficients of } x \\ 13 = 15 + C & \text{Equate constant terms} \end{array}$$

Because $B = 0$ and $B + C = -2$, we see that $C = -2$. Thus the partial fraction decomposition of $\frac{3x^2 - 2x + 13}{(x + 1)(x^2 + 5)}$ is

$$\frac{3}{x + 1} - \frac{2}{x^2 + 5}$$

 Check It Out 4   Compute the partial fraction decomposition of $\frac{x^2 + 3x - 8}{(x^2 + 1)(x - 3)}$.

Case 4: Denominator Has At Least One Irreducible, Repeated Quadratic Factor

Let $r(x) = \frac{P(x)}{Q(x)}$, where $P(x)$ and $Q(x)$ are polynomials and the degree of $P(x)$ is less than the degree of $Q(x)$. For every repeated irreducible quadratic factor of $Q(x)$, of the form $(ax^2 + bx + c)^m$, $m \geq 2$, the portion of the partial fraction decomposition of $r(x)$ that pertains to $(ax^2 + bx + c)^m$ is a sum of m fractions of the following form:

$$\frac{A_1 x + B_1}{ax^2 + bx + c} + \frac{A_2 x + B_2}{(ax^2 + bx + c)^2} + \ldots + \frac{A_m x + B_m}{(ax^2 + bx + c)^m}$$

where $A_1, \ldots, A_m$ and $B_1, \ldots, B_m$ are real numbers. Each of the linear factors of $Q(x)$ is treated as specified in Case 1 or Case 2, as appropriate. The nonrepeated irreducible quadratic factors of $Q(x)$ are treated as in Case 3.

| **Example 5** | **Repeated Irreducible Quadratic Factors** |

Find the partial fraction decomposition of $\dfrac{-2x^3 + x^2 - 6x + 7}{(x^2 + 3)^2}$.

Solution Because $x^2 + 3$ is a repeated irreducible quadratic factor, we have

$$\frac{-2x^3 + x^2 - 6x + 7}{(x^2 + 3)^2} = \frac{Ax+B}{x^2 + 3} + \frac{Cx +D}{(x^2 + 3)^2}.$$

Multiplying both sides of this equation by the LCD, $(x^2 + 3)^2$, we obtain the following.

$$-2x^3 + x^2 - 6x + 7 = (Ax + B)(x^2 + 3) + Cx + D$$
$$-2x^3 + x^2 - 6x + 7 = Ax^3 + Bx^2 + 3Ax + 3B + Cx + D$$
$$-2x^3 + x^2 - 6x + 7 = Ax^3 + Bx^2 + (3A + C)x + 3B + D$$

Equating coefficients,

$-2 = A$	Equate coefficients of x^3
$1 = B$	Equate coefficients of x^2
$-6 = 3A + C$	Equate coefficients of x
$7 = 3B + D$	Equate constant terms

Substituting $A = -2$ into the third equation, we get

$$-6 = 3(-2) + C \Rightarrow -6 = -6 + C \Rightarrow 0 = C$$

Substituting $B = 1$ into the fourth equation, we get

$$7 = 3(1) + D \Rightarrow 7 = 3 + D \Rightarrow 4 = D$$

Thus the partial fraction decomposition is

$$\frac{-2x^3 + x^2 - 6x + 7}{(x^2 + 3)^2} = \frac{-2x + 1}{x^2 + 3} + \frac{4}{(x^2 + 3)^2}$$

Check It Out 5 Compute the partial fraction decomposition of $\dfrac{-2x^2 + 1}{(x^2 + 1)^2}$.

Exercises 8.7

Skills

In Exercises 1–10, write just the form of the partial fraction decomposition. Do not solve for the constants.

1. $\dfrac{3}{x^2 - 2x - 3}$

2. $\dfrac{-1}{x^2 - 3x}$

3. $\dfrac{4x}{(x + 5)^2}$

4. $\dfrac{-2}{x^2(x - 1)}$

5. $\dfrac{x + 3}{(x^2 + 2)(2x + 1)}$

6. $\dfrac{-6x + 7}{(x^2 + x + 1)(x + 5)}$

7. $\dfrac{3x - 1}{x^4 - 16}$

8. $\dfrac{2x^2 - 5}{x^4 - 1}$

9. $\dfrac{x + 6}{3x^3 + 6x^2 + 3x}$

10. $\dfrac{3x - 2}{2x^3 + 4x^2 + 2x}$

In Exercises 11–16, determine whether the quadratic expression is reducible.

11. $x^2 + 5$

12. $x^2 - 9$

13. $x^2 + x + 1$

14. $2x^2 + 2x + 3$

15. $x^2 + 4x + 4$

16. $x^2 + 6x + 9$

In Exercises 17–40, write the partial fraction decomposition of each rational expression.

17. $\dfrac{8}{x^2 - 16}$

18. $\dfrac{-4}{x^2 - 4}$

19. $\dfrac{2}{2x^2 - x}$

20. $\dfrac{3}{x^2 + 3x + 2}$

21. $\dfrac{x}{x^2 + 5x + 6}$

22. $\dfrac{5x - 7}{x^2 - 4x - 5}$

23. $\dfrac{-3x^2 - 3x + 2}{x^3 - x}$

24. $\dfrac{x^2 + 4}{x^3 - 4x}$

25. $\dfrac{-2x + 6}{x^2 - 2x + 1}$

26. $\dfrac{x - 1}{x^2 + 4x + 4}$

27. $\dfrac{-x^2 + 2x + 4}{x^3 + 2x^2}$

28. $\dfrac{-x^2 + 3x - 9}{x^3 - 3x^2}$

29. $\dfrac{-2x^2 - 3x - 4}{(x - 1)(x + 2)^2}$

30. $\dfrac{4x^2 - 5x - 5}{(x - 3)(x + 1)^2}$

31. $\dfrac{4x + 1}{(x + 2)(x^2 + 3)}$

32. $\dfrac{x^2 + 3}{(x - 1)(x^2 + 1)}$

33. $\dfrac{-3x + 3}{(x + 2)(x^2 + x + 1)}$

34. $\dfrac{4x + 4}{(x - 1)(x^2 + x + 1)}$

35. $\dfrac{x^2 + 3}{x^4 - 1}$

36. $\dfrac{6x^2 - 8x + 24}{x^4 - 16}$

37. $\dfrac{-x^2 - 2x - 3}{(x^2 + 2)^2}$

38. $\dfrac{2x^2 - x + 2}{(x^2 + 1)^2}$

39. $\dfrac{x^3 - 3x^2 - x - 3}{x^4 - 1}$

40. $\dfrac{-2x^3 + x^2 + 8x + 4}{x^4 - 16}$

Applications

41. Environment The concentration of a pollutant in a lake t hours after it has been dumped there is given by

$$C(t) = \frac{t^2}{t^3 + 125}, t \geq 0$$

Chemists originally constructed the formula for $C(t)$ by modeling the concentration of the pollutant as a sum of at least two rational functions, where each term in the sum represents a different chemical process. Determine the individual terms in that sum.

Engineering In engineering applications, partial fraction decomposition is used to compute the Laplace transform. The independent variable is usually s.

In Exercises 42 and 43, compute the partial fraction decomposition.

42. $\dfrac{1}{s(s + 1)}$

43. $\dfrac{2}{s(s^2 + 1)}$

Concepts

44. What is wrong with the following decomposition?

$$\frac{x}{x^2(x - 1)^2} = \frac{A}{x^2} + \frac{B}{(x - 1)^2}$$

45. Explain why the following decomposition is incorrect.

$$\frac{1}{x(x^2 + 2x - 3)} = \frac{A}{x} + \frac{Bx + C}{x^2 + 2x - 3}$$

46. Find the partial fraction decomposition of $\dfrac{1}{x(x^2 + a^2)}$.

47. Find the partial fraction decomposition of $\dfrac{1}{(x - c)^2}$.

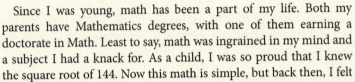

SPOTLIGHT ON SUCCESS   *Student Instructor Penelope*

Never give up on something that you can't go a day without thinking about.
— Sir Winston Churchill

Since I was young, math has been a part of my life. Both my parents have Mathematics degrees, with one of them earning a doctorate in Math. Least to say, math was ingrained in my mind and a subject I had a knack for. As a child, I was so proud that I knew the square root of 144. Now this math is simple, but back then, I felt smart for knowing the answer. That excitement stayed alive through elementary school. Honestly, there were times when I became frustrated, such as when learning division and factoring for the first time, but I was still enthusiastic.

In middle school and high school, I lost my ability to enjoy math. With classes getting progressively more difficult and having to worry about the future, my focus veered away from wanting to learn more to forcing myself to master the material for the grades. The weight of having parents with math backgrounds became quite heavy. This created high standards and expectations from friends and family that turned into expectations that I placed on myself. I felt like I had to do well, to be more than proficient in math so that I would not let them down. It was not until I went to university that I found my motivation and inspiration again.

While working to attain a degree in Business Administration at Cal Poly, I decided to take the Calculus series. In a college setting, math was even more difficult to understand but I did not let that deter me. I persevered and found myself in the Proofs in Mathematics course, where I learned to truly appreciate math again. The class utilized a different way of thinking than I was used to. It was a challenging class; I struggled throughout most of it, but I was eager to learn the new material despite my grades not being what I desired. It was at that moment that I did not care how my grades ended up. Enjoying and understanding what I was learning was my main priority.

Objectives

- Solve systems of nonlinear equations by substitution.
- Solve systems of nonlinear equations by elimination.
- Solve systems of nonlinear equations using technology.
- Using systems of nonlinear equations to solve applied problems.

A system of equations in which one or more of the equations is not linear is called a **system of nonlinear equations.** Unlike systems of linear equations, there are no systematic procedures for solving *all* nonlinear systems. In this section, we give strategies for solving some simple nonlinear systems in just two variables. However, many nonlinear systems cannot be solved algebraically and must be solved using technology, as shown in Example 3.

We first discuss the substitution method, which works well when one of the equations in the system is quadratic and the other equation is linear.

Solving Nonlinear Systems by Substitution

We now outline how to solve a system of nonlinear equations using substitution.

> **Substitution Strategy for Solving a Nonlinear System**
>
> **Step 1** In one of the equations, solve for one variable in terms of the other.
> **Step 2** Substitute the resulting expression into the other equation to get one equation in one variable.
> **Step 3** Solve the resulting equation for that variable.
> **Step 4** Substitute each solution into either of the original equations and solve for the other variable.
> **Step 5** Check your solution.

VIDEO EXAMPLES

SECTION 8.8

Example 1 $\quad$ **Solving a Nonlinear System Using Substitution**

Solve the following system.

$$\begin{cases} x^2 + 4y^2 = 25 \\ x - 2y + 1 = 0 \end{cases}$$

Solution $\quad$ Take the simpler second equation and solve it for x in terms of y. Then substitute the resulting expression for x into the first equation.

$x = 2y - 1$	Solve second equation for x in terms of y
$(2y - 1)^2 + 4y^2 = 25$	Substitute for x in first equation
$4y^2 - 4y + 1 + 4y^2 = 25$	Expand left-hand side
$8y^2 - 4y - 24 = 0$	Collect like terms
$4(2y^2 - y - 6) = 0$	Factor out 4
$2y^2 - y - 6 = 0$	Divide by 4
$(2y + 3)(y - 2) = 0$	Factor

Setting each factor equal to 0 and solving, we have

$$2y + 3 = 0 \Rightarrow y = -\frac{3}{2}; \quad y - 2 = 0 \Rightarrow y = 2$$

Next, substitute each of these values of y into the second equation in the original system. You will obtain a value of x for each value of y.

$x - 2\left(-\dfrac{3}{2}\right) + 1 = 0$	Substitute $y = -\dfrac{3}{2}$
$x = -4$	Solve for x

Thus the first solution is $(x, y) = \left(-4, -\frac{3}{2}\right)$. Next, substitute $y = 2$ into the second equation; this gives $x = 2(2) - 1 = 3$. So the second solution is $(3, 2)$. The two solutions are $(3, 2)$ and $\left(-4, -\frac{3}{2}\right)$. You should check that both solutions satisfy the system of equations.

Check It Out 1   Solve the following system.

$$\begin{cases} x^2 + y^2 = 2 \\ \qquad\; y = -x \end{cases}$$

We next introduce an elimination method that is useful when both of the equations in a system are quadratic.

Solving Nonlinear Systems by Elimination

Next we outline how to solve a system of nonlinear equations using elimination.

Elimination Strategy for Solving a Nonlinear System

Step 1 Eliminate one variable from both equations by suitably combining the equations.

Step 2 Solve one of the resulting equations for the variable that was retained.

Step 3 Substitute each solution into either of the original equations and solve for the variable that was eliminated.

Step 4 Check your solution.

Example 2   **Solving a Nonlinear System by Elimination**

Solve the following system of equations.

$$\begin{cases} \quad x^2 + y^2 = 4 & (1) \\ \qquad\quad y = -2x^2 + 2 & (2) \end{cases}$$

Solution First note that substituting the expression for y into the first equation will give a polynomial equation that has an x^4 term, which is difficult to solve. Instead, we use the elimination technique.

Rewrite the system with the variables on one side and the constants on the other.

$$\begin{cases} x^2 + y^2 = 4 & (1) \\ 2x^2 + y = 2 & (2) \end{cases}$$

Eliminate the x^2 term from Equation (2) by multiplying Equation (1) by -2 and adding the result to Equation (2). This gives the following.

$$
\begin{array}{ll}
-2x^2 - 2y^2 = -8 & \quad -2 \cdot (1) \\
\underline{\quad 2x^2 + y = \quad 2} & \quad (2) \\
-2y^2 + y = -6 &
\end{array}
$$

The last equation is a quadratic equation in y, which can be solved as follows.

$$
\begin{array}{ll}
-2y^2 + y = -6 & \\
-2y^2 + y + 6 = 0 & \quad \text{Add 6 to both sides} \\
2y^2 - y - 6 = 0 & \quad \text{Multiply by } -1 \text{ to write in standard form} \\
(2y + 3)(y - 2) = 0 & \quad \text{Factor}
\end{array}
$$

Setting each factor equal to 0 and solving, we have

$$2y + 3 = 0 \Rightarrow y = -\frac{3}{2}; \quad y - 2 = 0 \Rightarrow y = 2$$

Substitute $y = -\frac{3}{2}$ into Equation (2) and solve for x.

$$y = -2x^2 + 2$$

$$-\frac{3}{2} = -2x^2 + 2 \qquad \text{Substitute } y = -\frac{3}{2}$$

$$-\frac{7}{2} = -2x^2 \qquad \text{Subtract 2 from each side}$$

$$\frac{7}{4} = x^2 \qquad \text{Divide by } -2$$

$$\pm\frac{\sqrt{7}}{2} = x \qquad \text{Take square root}$$

Thus we have two solutions corresponding to $y = -\frac{3}{2}$: $\left(\frac{\sqrt{7}}{2}, -\frac{3}{2}\right)$ and $\left(-\frac{\sqrt{7}}{2}, -\frac{3}{2}\right)$.

Substituting $y = 2$ into Equation (2) and solving for x gives

$$y = -2x^2 + 2 \Rightarrow 2 = -2x^2 + 2 \Rightarrow x = 0$$

Thus we have $(0, 2)$ as another solution. You should check that these solutions satisfy the original system of equations.

The solutions are $\left(\frac{\sqrt{7}}{2}, -\frac{3}{2}\right)$, $\left(-\frac{\sqrt{7}}{2}, -\frac{3}{2}\right)$, and $(0, 2)$.

 Check It Out 2   Solve the following system.

$$\begin{cases} x^2 + y^2 = 4 & \text{(1)} \\ \quad\quad y = x^2 - 1 & \text{(2)} \end{cases}$$

Solving Nonlinear Systems Using Technology

The next example shows how graphing technology can be used to find the solution of a system of equations that is not easy to solve algebraically.

 Example 3   **Using a Graphing Utility to Solve a Nonlinear System of Equations**

Solve the following system of equations.

$$\begin{cases} xy = 1 & \text{(1)} \\ \quad y = -x^2 + 3 & \text{(2)} \end{cases}$$

Solution In Equation (1), solve for y in terms of x, which gives $y = \frac{1}{x}$. This is a rational function with a vertical asymptote at $x = 0$. The graph of the function from Equation (2) is a parabola that opens downward. Enter these functions as Y_1 and Y_2 in a graphing utility, and use a decimal window to display the graphs. The graphs are shown in Figure 1.

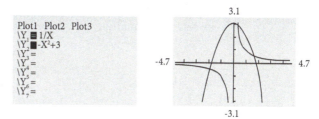

Figure 1

The solution of the system of equations is given by the intersection of the two graphs. From Figure 1, there are three points of intersection. By zooming out, we can make sure that there are no others. Using the INTERSECT feature, we find the intersection points, one by one. The results are shown in Figure 2. The approximate solutions are $(0.3473, 2.879)$, $(1.532, 0.6527)$, and $(-1.879, -0.5321)$.

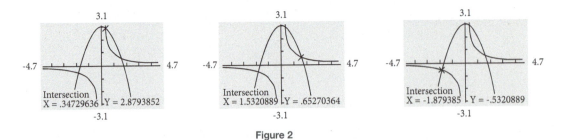

Figure 2

Note that if we had tried to solve this simple system by substitution, we would have obtained

$$\frac{1}{x} = -x^2 + 3$$

Multiplying by x on both sides and rearranging terms, we obtain the cubic equation $x^3 + 3x - 1 = 0$. This equation cannot be solved by any of the algebraic techniques discussed in this text. Hence, even some simple-looking nonlinear systems of equations can be difficult or even impossible to solve algebraically.

 Check It Out 3   Solve the following system of equations.

$$\begin{cases} xy = (1) \\ y = x^2 - 5 \end{cases}$$

Application of Nonlinear Systems

We next examine a geometry application that involves a nonlinear system of equations.

Example 4   **An Application of Geometry**

A right triangle with a hypotenuse of $2\sqrt{15}$ inches has an area of 15 square inches. Find the lengths of the other two sides of the triangle. See Figure 3.

Solution Referring to Figure 3 and using the given information, we can write the following system of equations.

$$\begin{cases} \dfrac{1}{2}xy = 15 & \text{Area of triangle} & (1) \\ x^2 + y^2 = \left(2\sqrt{15}\right)^2 = 60 & \text{Pythagorean Theorem} & (2) \end{cases}$$

We can solve this nonlinear system of equations by substitution. In Equation (1), solve for y.

$$\frac{1}{2}xy = 15 \Rightarrow xy = 30 \Rightarrow y = \frac{30}{x}$$

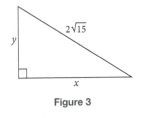

Figure 3

Substituting $y = \frac{30}{x}$ into Equation (2), we have

$$x^2 + y^2 = 60 \qquad \text{Equation (2)}$$

$$x^2 + \left(\frac{30}{x}\right)^2 = 60 \qquad \text{Substitute } y = \frac{30}{x}$$

$$x^2 + \frac{900}{x^2} = 60 \qquad \text{Simplify}$$

$$x^4 + 900 = 60x^2 \qquad \text{Multiply each side by } x^2$$

$$x^4 - 60x^2 + 900 = 0 \qquad \text{Move } x^2 \text{ term to left-hand side}$$

$$(x^2 - 30)(x^2 - 30) = 0 \qquad \text{Factor}$$

Setting the factor $x^2 - 30$ equal to 0 and solving for x gives

$$x^2 - 30 = 0 \Rightarrow x = \pm\sqrt{30}$$

We ignore the negative solution for x, since lengths must be positive. Thus $x = \sqrt{30}$. Using $x^2 + y^2 = 60$, we find that

$$y^2 = 60 - x^2 = 60 - 30 = 30.$$

Therefore, $y = \sqrt{30}$, and so the other two sides of the triangle are both of length $\sqrt{30}$ inches.

Check It Out 4   A right triangle with a hypotenuse of $2\sqrt{13}$ inches has an area of 12 square inches. Find the lengths of the other two sides of the triangle.

Exercises 8.8

 Skills

In Exercises 1–34, find all real solutions of the system of equations. If no real solution exists, so state.

1. $\begin{cases} x^2 + y^2 = 13 \\ \quad\ y = x + 1 \end{cases}$
2. $\begin{cases} x^2 + y^2 = 10 \\ \quad\ y = x + 2 \end{cases}$
3. $\begin{cases} 5x^2 + y^2 = 9 \\ \quad\ y = 2x \end{cases}$
4. $\begin{cases} x^2 + 4y^2 = 16 \\ \quad\ y = \frac{1}{2}x \end{cases}$

5. $\begin{cases} 2x^2 - y = 1 \\ \quad\ y = 5x + 2 \end{cases}$
6. $\begin{cases} x^2 - y = 3 \\ \quad\ y = 3x + 7 \end{cases}$
7. $\begin{cases} 9x^2 + 4y = 4 \\ 3x + 4y = -2 \end{cases}$
8. $\begin{cases} x^2 + y = 4 \\ 2x + y = 1 \end{cases}$

9. $\begin{cases} 3x^2 - 10y = 5 \\ \quad x - \quad y = -2 \end{cases}$
10. $\begin{cases} 2x^2 - 3y = 2 \\ \quad x - 2y = -2 \end{cases}$
11. $\begin{cases} \quad x^2 + 2y = -2 \\ -2x + \quad y = 1 \end{cases}$
12. $\begin{cases} 4x^2 + y = 2 \\ -4x + y = 3 \end{cases}$

13. $\begin{cases} x^2 + y^2 = 8 \\ \quad xy = -4 \end{cases}$
14. $\begin{cases} x^2 + y^2 = 61 \\ \quad xy = 30 \end{cases}$
15. $\begin{cases} x^2 + y^2 = 9 \\ 2x^2 + y = 15 \end{cases}$
16. $\begin{cases} x^2 + y^2 = 16 \\ -3x + y^2 = 6 \end{cases}$

17. $\begin{cases} 5x^2 - 2y^2 = 10 \\ 3x^2 + 4y^2 = 6 \end{cases}$
18. $\begin{cases} 2x^2 + 3y^2 = 4 \\ 6x^2 + 5y^2 = -8 \end{cases}$
19. $\begin{cases} x^2 - y^2 + 2y = 1 \\ 5x^2 - 3y^2 = 17 \end{cases}$

20. $\begin{cases} x^2 - y^2 + 2x = 4 \\ 7x^2 - 5y^2 = 8 \end{cases}$
21. $\begin{cases} x^2 + y^2 = 4 \\ x^2 + 4y^2 = 1 \end{cases}$
22. $\begin{cases} x^2 + y^2 = 6 \\ \quad y = 4 \end{cases}$

23. $\begin{cases} 2x^2 - 5x - 4y^2 = -4 \\ \quad x^2 - 3y^2 = 4 \end{cases}$
24. $\begin{cases} x^2 + y^2 + 4y = 1 \\ x^2 - y^2 = 3 \end{cases}$
25. $\begin{cases} x^2 + (y - 1)^2 = 9 \\ x^2 + \quad y^2 = 4 \end{cases}$

26. $\begin{cases} (x + 2)^2 + y^2 = 6 \\ \quad x^2 + y^2 = -2 \end{cases}$
27. $\begin{cases} x^2 - (y - 3)^2 = 7 \\ x^2 + \quad y^2 = 16 \end{cases}$
28. $\begin{cases} (x+1)^2 - y^2 = -9 \\ \quad x^2 - y^2 = -16 \end{cases}$

29. $\begin{cases} x^2 + 3xy - 2x = -10 \\ \quad 2xy + x = -14 \end{cases}$
30. $\begin{cases} 3x^2 + 2xy + x = 8 \\ \quad xy + x = 3 \end{cases}$
31. $\begin{cases} 2^{3x^2 - y^2} = 4 \\ x^2 + y^2 = 14 \end{cases}$

(*Hint*: Eliminate the *xy* term.)   (*Hint*: Eliminate the *xy* term.)

32. $\begin{cases} x + y^2 = 25 \\ e^{3xy} = 1 \end{cases}$
33. $\begin{cases} \log_{10}(2y^2) + \log_{10}(x^3) = 3 \\ \quad xy = 5 \end{cases}$
34. $\begin{cases} \log_3(y^2) - \log_3 x = 1 \\ 3x - 2y = 3 \end{cases}$

In Exercises 35–40, graph both equations by hand and find their point(s) of intersection, if any.

35. $\begin{cases} x^2 - 9y = -18 \\ -2x + 3y = 3 \end{cases}$
36. $\begin{cases} (x + 2)^2 + 4y = 17 \\ 3x + y = 1 \end{cases}$
37. $\begin{cases} (x - 5)^2 + (y + 3)^2 = 8 \\ x + y = 2 \end{cases}$

38. $\begin{cases} x^2 + y^2 - 8y = -15 \\ 2x - y = -6 \end{cases}$
39. $\begin{cases} (x - 3)^2 + y^2 + 2y = 7 \\ x^2 - 6x + y^2 + 10y = -26 \end{cases}$

40. $\begin{cases} x^2 + 2x + \quad y^2 + 12y = -33 \\ x^2 + 8x + (y + 6)^2 = 9 \end{cases}$

In Exercises 41–48, solve the system using a graphing utility. Round all values to three decimal places.

41. $\begin{cases} y = x^3 - 2x - 1 \\ y = 3x^2 - 2 \end{cases}$

42. $\begin{cases} y = x^3 - 3x + 2 \\ y = 4x^2 - 1 \end{cases}$

43. $\begin{cases} y = -x^2 + 3 \\ y = 3^x \end{cases}$

44. $\begin{cases} y = -x^2 + 2 \\ y = 2^x \end{cases}$

45. $\begin{cases} 2xy = 8 \\ x^2 + y^2 = 3 \end{cases}$

46. $\begin{cases} xy = 1 \\ y = x^2 - 7 \end{cases}$

47. $\begin{cases} 5x^2 - y = 10 \\ 9x^2 + y^2 = 25 \end{cases}$

48. $\begin{cases} 2x^2 - y = 2 \\ 4x^2 + y^2 = 16 \end{cases}$

Applications

49. Design The perimeter of a rectangular garden is 80 feet and the area it encloses is 336 square feet. Find the length and width of the garden.

50. Design The perimeter of a rectangular garden is 54 feet, and its area is 180 square feet. Find the length and width of the garden.

51. Geometry A right triangle with a hypotenuse of $\sqrt{89}$ has an area of 20 square inches. Find the lengths of the other two sides of the triangle.

52. Geometry A right triangle with a hypotenuse of $2\sqrt{10}$ inches has an area of 6 square inches. Find the lengths of the other two sides of the triangle.

53. Manufacturing A manufacturer wants to make a can in the shape of a right circular cylinder with a volume of 6.75π cubic inches and a lateral surface area of 9π square inches. The lateral surface area includes only the area of the curved surface of the can, not the areas of the flat (top and bottom) surfaces. Find the radius and height of the can.

54. Manufacturing A manufacturer wants to make a can in the shape of a right circular cylinder with a volume of 45π cubic inches and a lateral surface area of 30π square inches. The lateral surface area includes only the area of the curved surface of the can, not the area of the flat (top and bottom) surfaces. Find the radius and height of the can.

55. Geometry The volume of a paper party hat, shaped in the form of a right circular cone, is 36π cubic inches. If the radius of the cone is one-fourth the height of the cone, find the radius and the height.

56. Geometry The volume of a super-size ice cream cone, shaped in the form of a right circular cone, is 8π cubic inches. If the radius of the cone is one-third the height of the cone, find the radius and the height of the cone.

57. Number Theory The sum of the squares of two positive integers is 85. If the squares of the integers differ by 13, find the integers.

58. Number Theory The sum of the squares of two positive integers is 74. If the squares of the integers differ by 24, find the integers.

59. Landscaping The area of a rectangular property is 300 square feet. Its length is three times its width. There is a rectangular swimming pool centered within the property. The dimensions of the property are twice the corresponding dimensions of the pool. The portion of the property that lies outside the pool is paved with concrete. What are the dimensions of the property and of the pool? What is the area of the paved portion?

60. Landscaping The area of a rectangular property is 1800 square feet; its length is twice its width. There is a rectangular swimming pool centered within the property. The dimensions of the property are one and one-third times the corresponding dimensions of the pool. The portion of the property that lies outside the pool is paved with concrete. What are the dimensions of the property and of the pool? What is the area of the paved portion?

Concepts

61. For what value(s) of b does the following system of equations have two distinct, real solutions?

$$\begin{cases} y = -x^2 + 2 \\ y = x + b \end{cases}$$

62. Consider the following system of equations.

$$\begin{cases} y = x^2 + 6x - 4 \\ y = b \end{cases}$$

For what value(s) of b do the graphs of the equations in this system have

a. exactly one point of intersection?

b. exactly two points of intersection?

c. no point of intersection?

63. Explain why the following system of equations has no solution.

$$\begin{cases} (x + y)^2 = 36 \\ xy = 18 \end{cases}$$

(*Hint:* Expand the expression $(x + y)^2$.)

64. Give a graphical explanation of why the following system of equations has no real solutions.

$$\begin{cases} x^2 + y^2 = 1 \\ y = x^2 + 3 \end{cases}$$

65. Consider the following system of equations.

$$\begin{cases} x^2 + y^2 = r^2 \\ (x - h)^2 + y^2 = r^2 \end{cases}$$

Let r be a fixed, positive number. For what value(s) of h does this system have

a. exactly one real solution?

b. exactly two real solutions?

c. infinitely many real solutions?

d. no real solution?

(*Hint:* Visualize the graph of the two equations.)

Chapter 8 Summary

Section 8.1 Systems of Linear Equations and Inequalities in Two Variables

Solving a system by elimination [Review Exercises 1–4]

Eliminate one of the variables from the two equations to get one equation in one variable. Solve this equation and substitute the result into one of the original equations. The second variable is solved for by substitution.

Solutions of a linear inequality [Review Exercises 5–8]

An inequality is a **linear inequality** if it can be written in the form $Ax + By \leq C$ or $Ax + By < C$, where A, B, and C are real numbers, not all zero. (The symbols $\leq$ and $<$ may be replaced with $\geq$ or $>$.) The **solution set** of a linear inequality is the set of points (x, y) that satisfy the inequality.

Solutions of a system of linear inequalities [Review Exercises 9,10]

The **solution set of a system of linear inequalities** is the set of points (x, y) consisting of the intersection of the solution sets of the individual inequalities. To find the solution set of the system, graph the solution set of each inequality and find the intersection of the shaded regions.

Fundamental theorem of linear programming [Review Exercises 11–14]

The **fundamental theorem of linear programming** states that the maximum or minimum of a linear objective function occurs at a corner point of the boundary of the solution set of a system of linear inequalities (the feasible set).

Section 8.2 Systems of Linear Equations in Three Variables

Gaussian elimination [Review Exercises 15–19]

Gaussian elimination is used to solve a system of linear equations in three or more variables. The following operations are used in the elimination process for solving systems of equations.

Operations for manipulating systems of equations

- Interchange two equations in the system.
- Multiply one equation in the system by a nonzero constant.
- Multiply one equation in the system by a nonzero constant and add the result to another equation in the system.

Section 8.3 Solving Systems of Equations Using Matrices

Augmented matrix for a linear system [Review Exercises 20, 21]

In an **augmented matrix** corresponding to a system of equations, the rows of the left-hand side are formed by taking the coefficients of the variables in each equation, and the constants appear on the right-hand side.

Gaussian elimination using matrices [Review Exercises 22–25]

The following operations are used when performing Gaussian elimination on matrices.

Elementary row operations on matrices

- Interchange two rows of the matrix.
- Multiply a row of the matrix by a nonzero constant.
- Multiply a row of the matrix by a nonzero constant and add the result to another row.

These operations are used to solve a system of linear equations in matrix form.

We can continue the elementary row operations in the previous example so that the first 1 in each row has a 0 above and below it.

Section 8.4 Operations on Matrices

A **matrix** is a rectangular array of real numbers with m rows and n columns. The **dimension** of such a matrix is $m \times n$. A matrix is usually referred to by a capital letter.

Two matrices can be added or subtracted only if their dimensions are the same. The resulting matrix, which will be of the same size, is obtained by adding (subtracting) the corresponding elements of the two matrices.

The product of an $m \times n$ matrix A and a scalar c is found by multiplying all the elements of A by c.

The product of an $m \times p$ matrix A and a $p \times n$ matrix B is n product matrix AB with m rows and n columns.

Furthermore, the element in the ith row, jth column of the product matrix AB is given by the product of the ith row of A and the jth column of B.

Section 8.5 Matrices and Inverses

The inverse of an $n \times n$ matrix A, denoted by A^{-1}, is another $n \times n$ matrix such that

$$AA^{-1} = A^{-1}A = I$$

where I is the $n \times n$ identity matrix.

If a system of equations can be expressed in the form $AX = B$, with A an $n \times n$ matrix and A^{-1} its inverse, then we have the following:

$$AX = B \Rightarrow A^{-1}AX = A^{-1}B \Rightarrow X = A^{-1}B$$

Section 8.6 Determinants and Cramer's Rule

The determinant of a 2×2 matrix $A = \begin{bmatrix} a & b \\ c & d \end{bmatrix}$ is defined as

$$|A| = \begin{vmatrix} a & b \\ c & d \end{vmatrix} = ad - bc$$

The determinants of larger matrices can be found by cofactor expansion.

Cramer's Rule is a set of formulas for solving systems of linear equations with exactly one solution.

Minors and cofactors   [Review Exercises 49, 50, 53, 54]

Let A be an $n \times n$ matrix, and let a_{ij} be the entry in the ith row, jth column of A.

The **minor** of a_{ij} (denoted by M_{ij}) is the determinant of the matrix that is obtained by deleting the ith row and jth column of A.

The **cofactor** of a_{ij} (denoted by C_{ij}) is defined as $C_{ij} = (-1)^{i+j} M_{ij}$.

Section 8.7 Partial Fractions

Partial fraction decomposition   [Review Exercises 55–58]

Partial fraction decomposition is a technique used to decompose a rational expression into simpler expressions. There are four cases:

1. The denominator can be factored over the real numbers as a product of distinct linear factors.

2. The denominator can be factored over the real numbers as a product of linear factors, at least one of which is a repeated linear factor.

3. The denominator can be factored over the real numbers as a product of distinct irreducible quadratic factors and possibly one or more linear factors.

4. The denominator can be factored over the real numbers as a product of irreducible quadratic factors, at least one of which is a repeated quadratic factor, and possibly one or more linear factors.

Section 8.8 Systems of Nonlinear Equations

Systems of nonlinear equations   [Review Exercises 59–63]

A system of equations in which one or more of the equations is not linear is called a **system of nonlinear equations**. Systems of nonlinear equations can be solved using **substitution** or **elimination**. Systems of nonlinear equations can also be solved using graphing utilities.

Chapter 8 Review Exercises

Section 8.1

In Exercises 1–4, solve each system of equations by elimination and check your solution.

1. $\begin{cases} -x - y = -7 \\ 3x + 4y = 24 \end{cases}$ **2.** $\begin{cases} -3x + 4y = 9 \\ 6x - 8y = 3 \end{cases}$ **3.** $\begin{cases} x + y = 5 \\ -2x - 2y = -10 \end{cases}$

4. $\begin{cases} 3x - 6y = 2 \\ y = -3 \end{cases}$

In Exercises 5–8, graph the solution set of each inequality.

5. $y > 3x - 7$ **6.** $x < -y + 4$ **7.** $2x - y \geq 1$ **8.** $x - 3y \leq 6$

In Exercises 9 and 10, graph the solution set of each system of inequalities.

9. $\begin{cases} x \leq 8 \\ 2x - y \geq 6 \end{cases}$ **10.** $\begin{cases} x \leq 5 \\ x \geq 2 \\ y > 1 \end{cases}$

In Exercises 11 and 12, solve the optimization problem.

11. Maximize $P = 5x + 7y$ subject to the following constraints.

$$\begin{cases} x \geq 1 \\ y \geq 2 \\ x \leq 7 \\ y \leq 10 \end{cases}$$

12. Maximize $P = 10x + 20y$ subject to the following constraints.

$$\begin{cases} x + 3y \leq 12 \\ y \leq x \\ y \geq 1 \\ x \geq 0 \end{cases}$$

13. Networks Telephone and computer networks transmit messages from one relay point to another. Suppose 450 messages are relayed to a single point on a network, and at that point the messages are distributed between two separate lines. The capacity of one line is 3.5 times the capacity of the other. Find the number of messages carried by each line.

14. Maximizing Revenue The cows on a dairy farm yield a total of 2400 gallons of milk per week. The farmer receives \$1 per gallon for the portion that is sold as milk. The rest is used in the production of cheese, which the farmer sells for \$5 per pound. It takes 1 gallon of milk to make a pound of cheese. If at least 25% of the milk is to be sold as milk, how many gallons of milk should go into the production of cheese in order to maximize the farmer's total revenue?

Section 8.2

In Exercises 15–18, solve the system of linear equations using Gaussian elimination.

15. $\begin{cases} x - 2y + z = 2 \\ -x + 4y + 3z = -8 \\ -6y + 2z = 4 \end{cases}$ **16.** $\begin{cases} 4x - y = 2 \\ y - 2z = 1 \\ 6x - z = 3 \end{cases}$ **17.** $\begin{cases} x + y + 2z = 4 \\ -3x + 2y - z = 3 \end{cases}$

18. $\begin{cases} 2x + y - z = 2 \\ 5x - 2y + z = 3 \end{cases}$

19. Nutrition The members of the Nutrition Club at Grand State University researched the lunch menu at their college cafeteria. They found that a meal consisting of a soft taco, a tostada, and a side dish of rice totaled 600 calories. Also, a meal consisting of two soft tacos and a tostada totaled 580 calories. Finally, a meal consisting of just a soft taco and rice totaled 400 calories. (*Source:* www.tacobell.com) Determine the numbers of calories in a soft taco, a tostada, and a side dish of rice.

Section 8.3

In Exercises 20 and 21, represent the system of equations in augmented matrix form. Do not solve the system.

20. $\begin{cases} 4x + y + z = 0 \\ -y + 2z = -1 \\ x + z = 3 \end{cases}$ **21.** $\begin{cases} x + 2y - 5z = 3 \\ 3x + z = -1 \end{cases}$

In Exercises 22–25, solve the system of equations using matrices and row operations. If there is no solution, so state.

22. $\begin{cases} -x - y = -10 \\ 3x + 4y = 24 \end{cases}$ **23.** $\begin{cases} x + 2y + z = -3 \\ 3x + y - 2z = 2 \\ 4x + 3y - z = 0 \end{cases}$ **24.** $\begin{cases} x + y - z = 0 \\ 3x + 2y - z = -1 \\ -2x + y - 2z = -1 \end{cases}$

25. $\begin{cases} -x + 4y + 3z = 8 \\ 2x - 8y - 4z = 3 \end{cases}$

In Exercises 26 and 27, for each augmented matrix, construct the corresponding system of linear equations. Use the variables listed above the matrix, in the given order. Determine whether the system is consistent or inconsistent. If it is consistent, give the solution(s).

26. $\begin{array}{ccc} x & y & z \\ \end{array}$
$\left[\begin{array}{ccc|c} 1 & 0 & -2 & 3 \\ 0 & 1 & 1 & 5 \\ 0 & 0 & 0 & 0 \end{array}\right]$

27. $\begin{array}{ccc} x & y & z \\ \end{array}$
$\left[\begin{array}{ccc|c} 1 & 0 & 0 & 7 \\ 0 & 1 & 0 & 3 \\ 0 & 0 & 1 & 2 \end{array}\right]$

28. Manufacturing Princess Clothing, Inc., has the following yardage requirements for making a single blazer, pair of trousers, or skirt.

	Fabric (yd)	Lining (yd)	Trim (yd)
Blazer	2.5	2	0.5
Trousers	3	2	1
Skirt	3	1	0

There are 42 yards of fabric, 25 yards of lining, and 7 yards of trim available. How many blazers, pairs of trousers, and skirts can be made, assuming that all the fabric, lining, and trim is used up?

Section 8.4

In Exercises 29–32, use the following matrix A.

$$A = \begin{bmatrix} 0 & -3 & 1 & 8 \\ -4.2 & -5 & 8 & 4 \\ 5 & 1.9 & \frac{4}{5} & \sqrt{2} \end{bmatrix}$$

29. Find a_{11}.   **30.** Find a_{22}.   **31.** Determine the dimensions of A.   **32.** Find a_{31}.

In Exercises 33–38, let A, B, and C be as given. Perform like indicated operation(s), if defined. If not defined, state the reason.

$$A = \begin{bmatrix} 4 & -1 & \frac{1}{2} \\ 0 & 3 & -1 \end{bmatrix}, \qquad B = \begin{bmatrix} -4 & 1 \\ 2 & -3 \\ 2 & -6 \end{bmatrix}, \qquad C = \begin{bmatrix} 3 & 5 \\ 1 & -1 \\ 2 & -3 \end{bmatrix}$$

33. $B + C$ **34.** $C - B$ **35.** $2B + C$ **36.** $B - 3C$ **37.** AB **38.** AC

Section 8.5

In Exercises 39–44, find the inverse of each matrix.

39. $\begin{bmatrix} 4 & 5 \\ 1 & 1 \end{bmatrix}$ **40.** $\begin{bmatrix} 5 & 3 \\ 3 & 2 \end{bmatrix}$ **41.** $\begin{bmatrix} -4 & 1 \\ -3 & 1 \end{bmatrix}$ **42.** $\begin{bmatrix} 4 & 3 \\ 5 & 4 \end{bmatrix}$

43. $\begin{bmatrix} 1 & -1 & 0 \\ -2 & 0 & 1 \\ -2 & 5 & -1 \end{bmatrix}$ **44.** $\begin{bmatrix} 4 & -2 & 1 \\ -2 & 1 & 2 \\ 1 & 2 & 4 \end{bmatrix}$

In Exercises 45 and 46, use inverses to solve the system of equations.

45. $\begin{cases} x + 2y = -4 \\ -x - y = 5 \end{cases}$ **46.** $\begin{cases} x - 3y + 2z = -1 \\ y + z = 4 \\ 2x - 6y + 3z = 3 \end{cases}$

Section 8.6

In Exercises 47–50, find the determinant of the matrix.

47. $A = \begin{bmatrix} 4 & 2 \\ -3 & 1 \end{bmatrix}$

48. $A = \begin{bmatrix} -2 & -3 \\ 1 & 5 \end{bmatrix}$

49. $A = \begin{bmatrix} -7 & 5 & 0 \\ 0 & 3 & 0 \\ -3 & -2 & 2 \end{bmatrix}$

50. $A = \begin{bmatrix} -2 & 3 & 5 \\ 6 & -1 & 0 \\ 0 & 1 & -2 \end{bmatrix}$

In Exercises 51–54, use Cramer's Rule to solve the system of equations.

51. $\begin{cases} -x - y = -2 \\ 2x + y = 0 \end{cases}$

52. $\begin{cases} -3x + 2y = 1 \\ -2x - y = 2 \end{cases}$

53. $\begin{cases} -2x + y + z = 0 \\ y + z = 4 \\ -3y + z = 1 \end{cases}$

54. $\begin{cases} x + 3y - z = 3 \\ x - z = 0 \\ x - y = 2 \end{cases}$

Section 8.7

In Exercises 55–58, write the partial fraction decomposition of each rational expression.

55. $\dfrac{-3x + 8}{x^2 + 5x + 6}$

56. $\dfrac{-2x^2 - x + 1}{x^3 - x^2}$

57. $\dfrac{9}{(x + 2)(x^2 + 5)}$

58. $\dfrac{-x^3 - x + x^2 + 2}{x^4 + 4x^2 + 4}$

Section 8.8

In Exercises 59–62, find all real solutions of the system of equations. If no real solution exists, so state.

59. $\begin{cases} x^2 + y^2 = 9 \\ y = x + 1 \end{cases}$

60. $\begin{cases} x^2 - 2y^2 = 4 \\ 2x^2 + 5y^2 = 12 \end{cases}$

61. $\begin{cases} (x + 3)^2 + (y - 4)^2 = 25 \\ x^2 - 8x + (y + 3)^2 = 9 \end{cases}$

62. $\begin{cases} (x - 1)^2 + (y + 1)^2 = 0 \\ x^2 + y^2 + 2y = 0 \end{cases}$

63. Geometry The sum of the areas of two squares is 549 square inches. The length of a side of the larger square is 3 inches more than the length of a side of the smaller square. Find the length of a side of each square.

Chapter 8 Test

In Exercises 1 and 2, solve the system of equations by elimination and check your solution.

1. $\begin{cases} x + 2y = 4 \\ 4x - y = -11 \end{cases}$

2. $\begin{cases} -2x + 3y = -8 \\ 5x - 2y = 9 \end{cases}$

3. Graph the solution set of the following system of inequalities.

$$\begin{cases} x \geq 1 \\ y \leq 4 \\ 3x + 2y \geq 6 \end{cases}$$

4. Maximize $P = 15x + 10y$ subject to the following constraints.

$$2x + y \leq 8$$
$$y \leq 2x$$
$$y \leq 1$$
$$y \geq 0$$
$$x \geq 0$$

In Exercises 5 and 6, solve the system of linear equations by using Gaussian elimination.

5. $\begin{cases} 2x - y + z = -3 \\ -3x + z = 11 \\ x + 3y = -6 \end{cases}$

6. $\begin{cases} -x + 2y - z = 0 \\ 2x - y = 2 \end{cases}$

In Exercises 7 and 8, use matrices A, B, and C given below. Perform the indicated operations, if defined. If an operation is not defined, state the reason.

$$A = \begin{bmatrix} 5 & 0 & 3 \\ 2 & -4 & 3 \end{bmatrix}, \quad B = \begin{bmatrix} -3 & 4 \\ -1 & 2 \\ 7 & -3 \end{bmatrix}, \quad C = \begin{bmatrix} 8 & 3 \\ -6 & 5 \\ 5 & -2 \end{bmatrix}$$

7. $B - 2C$

8. CA

In Exercises 9 and 10, find the inverse of the matrix.

9. $\begin{bmatrix} 3 & 5 \\ 1 & 2 \end{bmatrix}$

10. $\begin{bmatrix} 1 & 4 & 2 \\ -2 & 1 & 0 \\ 0 & 2 & 1 \end{bmatrix}$

11. Solve the following system of equations using an inverse.

$$\begin{cases} x + 3y - z = 0 \\ x + 4y + z = -2 \\ 2x + 6y - z = 1 \end{cases}$$

12. Find the determinant of $A = \begin{bmatrix} 2 & -3 & 1 \\ 0 & 5 & 2 \\ -4 & 2 & 0 \end{bmatrix}$.

In Exercises 13 and 14, use Cramer's Rule to solve the system of equations.

13. $\begin{cases} x + 3y = 6 \\ x + y = 2 \end{cases}$

14. $\begin{cases} x - 2y - z = 0 \\ -x - y = 3 \\ x + z = -1 \end{cases}$

15. Write the partial fraction decomposition of

$$\frac{-x^2 + 2x + 2}{x^3 + 2x^2 + x}$$

16. Write the partial fraction decomposition of

$$\frac{5x^2 - 7x + 6}{(x^2 + 1)(x - 3)}$$

In Exercises 17 and 18, find all real solutions of the system of equations. If no real solution exists, so state.

17. $\begin{cases} x^2 + y^2 = 4 \\ y - 3x = 0 \end{cases}$

18. $\begin{cases} (x + 1)^2 + (y - 1)^2 = 9 \\ x^2 + 2x + y^2 = 9 \end{cases}$

19. A computer manufacturer produces two types of laptops, the CX100 and the FX100. It takes 2 hours to manufacture each unit of the FX100 and only 1 hour to manufacture each unit of the CX100. At least 100 of the FX100 models are to be produced. At most 800 hours are to be allocated to the manufacture of the two models combined. If the company makes a profit of $100 on each FX100 model and a profit of $150 on each CX100 model, how many of each model should the company produce to maximize its profit?

20. Ten airline tickets were purchased for a total of $14,200. Each first-class ticket cost $2500. Business class and coach tickets cost $1500 and $300, respectively. Twice as many coach tickets were purchased as business class tickets. How many of each type of ticket were purchased?

21. A cell phone company charges $0.10 per minute during prime time and $0.05 per minute for non-prime time. The numbers of minutes used by three different customers are as follows.

	Prime-Time Minutes	Non-Prime-Time Minutes
Customer 1	200	400
Customer 2	300	200
Customer 3	150	300

Perform an appropriate matrix multiplication to find the total amount paid by each customer.

Chapter 9

Conic Sections

Astronomical objects emit radiation, which is not detectable by optical telescopes. These objects are usually studied using a radio telescope, which consists of a dish with a parabolic cross-section. Because a parabola exhibits a reflective property, the incoming radio waves are reflected off the surface of the dish and focused onto a single point. Exercise 68 in Section 9.1 asks you to find the equation of a parabola for the cross-section of a reflecting telescope. This chapter will examine a class of equations known as the conic sections, which are widely used in engineering, astronomy, physics, architecture, and navigation because they possess reflective and other unique properties.

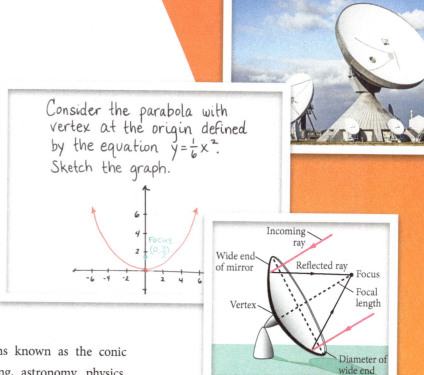

Outline

Objectives

- Define a parabola.
- Find the focus, directrix, and axis of symmetry of a parabola.
- Determine the equation of a parabola and write it in standard form.
- Translate a parabola in the *xy*-plane.
- Sketch a parabola.
- Understand the reflective property of a parabola.

This chapter covers a group of curves known as the *conic sections*, often just known as the conics. These curves are formed by intersecting a right circular cone with a plane. The circular cone consists of two halves, called **nappes**, which intersect in only one point, called the **vertex.**

Figure 1 shows how the conic sections are formed.

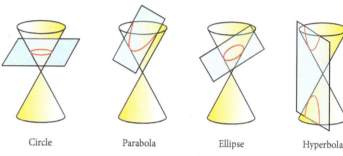

| Circle | Parabola | Ellipse | Hyperbola |

Figure 1

Observations

- If the plane is perpendicular to the axis of the cone but does not pass through the vertex, the conic is a **circle.**
- If the plane is tilted and intersects only one nappe of the cone, the conic is a single curve. If the curve is closed, the conic is an **ellipse;** otherwise, it is a **parabola.**
- If the plane is parallel to the axis of the cone and does not pass through the vertex, it intersects the cone in a pair of non-intersecting open curves, called a **hyperbola.** Each curve is called a **branch** of the hyperbola.

In this section, we will define parabolas differently than we did in Chapter 2. Later on, we will make connections between the two definitions.

Definition of a Parabola

A **parabola** is the set of all points (x, y) in a plane such that the distance of (x, y) from a fixed line is equal to the distance of (x, y) from a fixed point that is not on the fixed line. The fixed line and the fixed point, which lie in the plane of the parabola, are called the **directrix** and the **focus**, respectively.

The **axis of symmetry** of a parabola is the line that is perpendicular to the directrix and passes through the focus of the parabola. The **vertex** of a parabola is on the axis of symmetry, midway between the focus and the directrix. See Figure 2.

The only parabolas discussed in this section are those whose axis of symmetry is either vertical or horizontal.

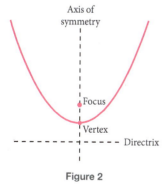

Figure 2

Parabolas with Vertex at the Origin

Suppose a parabola has its vertex at the origin and the y-axis as its axis of symmetry. Suppose also that the equation of its directrix is $y = -p$, for some $p > 0$. The coordinates of the focus are $(0, p)$, since the vertex is midway between the focus and directrix.

Let P be any point on the parabola with coordinates (x, y), and let dist(PF) denote the distance from P to the focus F. To find the distance from P to the directrix, we draw a perpendicular line segment from P to the line $y = -p$ (the directrix). Let M be the point of intersection of that line segment and the directrix, and let dist(PM) denote the distance from P to M. Then the coordinates of M are $(x, -p)$, as shown in Figure 3.

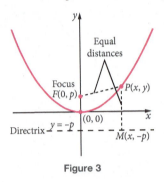

Figure 3

Just in Time
Review the distance formula in Section 2.2.

To derive the equation of the parabola, we apply the definition of a parabola along with the distance formula. From the definition of a parabola, we have

$$\text{Distance from } P \text{ to focus } F = \text{distance from } P \text{ to directrix}$$

That is,

$$\text{dist}(PF) = \text{dist}(PM)$$

Applying the distance formula,

$$\text{dist}(PF) = \sqrt{(x - 0)^2 + (y - p)^2} = \sqrt{x^2 + (y - p)^2} \qquad \text{Distance from } P \text{ to focus } F$$

$$\text{dist}(PM) = \sqrt{(x - x)^2 + (y - (-p))^2} = \sqrt{0^2 + (y + p)^2} \qquad \text{Distance from } P \text{ to directrix}$$

$$\sqrt{x^2 + (y - p)^2} = \sqrt{(y + p)^2} \qquad \text{Equate dist}(PF) \text{ and dist}(PM)$$

$$x^2 + y^2 - 2py + p^2 = y^2 + 2py + p^2 \qquad \text{Square both sides and expand}$$

$$x^2 = 4py \qquad \text{Combine like terms}$$

Thus the standard form of the equation of an upward-opening parabola with vertex at the origin is $x^2 = 4py$, where $p > 0$. The derivations of the standard forms of the equations of parabolas with vertex at the origin, but open downward, to the right, or to the left, are similar.

Their main features are summarized as follows.

Parabola with Vertex at the Origin

	Vertical Axis of Symmetry	Horizontal Axis of Symmetry
Graph		
	Figure 4	Figure 5
Equation	$x^2 = 4py$	$y^2 = 4px$
Direction of opening	Upward if $p > 0$, downward if $p < 0$	To the right if $p > 0$, to the left if $p < 0$
Vertex	$(0, 0)$	$(0, 0)$
Focus	$(0, p)$	$(p, 0)$
Directrix	The line $y = -p$	The line $x = -p$
Axis of symmetry	y-axis	x-axis

We will use the following observations when solving problems involving parabolas.

Observations

- A parabola opens toward the focus and away from the directrix.
- The axis of symmetry of a parabola is perpendicular to the directrix.
- The axis of symmetry intersects the parabola in only one point, namely, the vertex.
- The directrix does *not* intersect the parabola, and it is *not* part of the parabola.
- The focus of a parabola lies on the axis of symmetry, but it is *not* part of the parabola.
- If a parabola opens to the right or to the left, the variable y in the equation of the parabola is not a function of x, since the graph of such an equation does not pass the vertical line test.

Example 1   **Parabola with Vertex at the Origin**

Consider the parabola with vertex at the origin defined by the equation $y = \frac{1}{6}x^2$.

a. Find the coordinates of the focus.

b. Find the equations of the directrix and the axis of symmetry.

c. Find the value(s) of a for which the point $(a, 4)$ is on the parabola.

d. Sketch the parabola, and indicate the focus and the directrix.

Solution

a. First write the equation in standard form by multiplying both sides by 6:

$$y = \frac{1}{6}x^2 \Rightarrow x^2 = 6y$$

This equation is in the form $x^2 = 4py$. To find the focus, we must first calculate p.

$$x^2 = 6y = 4py \Rightarrow 6 = 4p \Rightarrow \boldsymbol{p = \frac{3}{2}}$$

Because the parabola is of the form $x^2 = 4py$ and $p > 0$, the parabola opens upward. The focus is $(0, p) = \left(0, \frac{3}{2}\right)$.

b. The equation of the directrix is $y = -p$; substituting the value of p, we obtain $y = -\frac{3}{2}$. The axis of symmetry of this parabola is the y-axis, which is given by the equation $x = 0$.

c. Since the point $(a, 4)$ is on the parabola, we will substitute its coordinates into the equation $y = \frac{1}{6}x^2$.

$$\boldsymbol{4 = \frac{1}{6}a^2} \qquad \text{Point } (a, 4) \text{ satisfies equation of parabola}$$

$$24 = a^2 \qquad \text{Multiply both sides by 6}$$

$$a = \pm\sqrt{24} = \pm 2\sqrt{6} \quad \text{Take square root of both sides}$$

Thus the points $\left(2\sqrt{6}, 4\right)$ and $\left(-2\sqrt{6}, 4\right)$ lie on the parabola.

d. The vertex of the parabola is at $(0, 0)$. Plot the vertex, along with the points $\left(2\sqrt{6}, 4\right)$ and $\left(-2\sqrt{6}, 4\right)$ from part (c). Draw a parabola through these three points. To find more points, construct a table (see Table 1). Then indicate the focus and the directrix, as shown in Figure 6.

x	y
-6	6
-2	$\frac{2}{3}$
0	0
2	$\frac{2}{3}$
6	6

Table 1

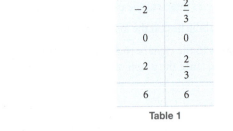

Figure 6

✓ *Check It Out 1*   Find the coordinates of the focus of the parabola $y = -\frac{1}{8}x^2$, as well as the equations of the directrix and the axis of symmetry. Sketch the parabola.   ◼

Example 2 illustrates how to obtain the equation of a parabola given a description of its features.

> ### Example 2   Finding the Equation of a Parabola
>
> Determine the equation in standard form of the parabola with vertex at the origin and directrix $x = 3$. Sketch the parabola, and indicate the focus and the directrix.
>
> **Solution** The directrix, $x = 3$, is a vertical line. The vertex of this parabola is at the origin, which is to the left of the directrix. Thus the parabola must open to the left, since a parabola always opens *away* from the directrix. The focus is at $(-3, 0)$, since the vertex is midway between the focus and the directrix. With this information, we can make a preliminary sketch, as shown in Figure 7. Because the focus is at $(-3, 0) = (p, 0)$, we have $p = -3$. As a check, we know that p must be negative if the parabola opens to the left. Next we find the equation of the parabola.

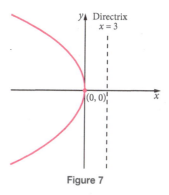

Figure 7

Using Technology

The APPS button in a graphing calculator will bring up a CONICS menu. Choose the option for Parabola, and enter the equation to get a graph of the parabola. See Figure 9.

Keystroke Appendix:

Section 17

$$y^2 = 4px \qquad \text{Standard form of equation of parabola opening to the left}$$

$$y^2 = 4(-3)x = -12x \qquad \text{Substitute } p = -3$$

The equation of the parabola is $y^2 = -12x$. To get a more accurate sketch, find points (x, y) that are on the parabola.

$$x = -3 \Rightarrow y^2 = -12(-3) = 36 \Rightarrow y = \pm 6$$

$$x = -2 \Rightarrow y^2 = -12(-2) = 24 \Rightarrow y = \pm\sqrt{24} \approx \pm 4.899$$

The graph is shown in Figure 8,

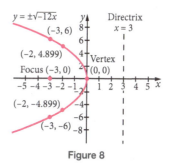

Figure 8

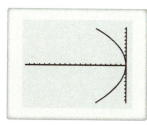

Figure 9

 Check It Out 2 Determine the equation in standard form of the parabola with vertex at the origin and directrix $x = -2$.

Parabolas with Vertex at the Point (h, k)

Just in Time
Review translations of graphs in Section 2.7

Recall the following facts about translations, or shifts, of graphs.

- If the variable x in an equation of the form $x^2 = 4py$ or $y^2 = 4px$ is replaced by $(x - h)$, the graph of the equation is translated horizontally by $|h|$ units. The translation is to the right if $h > 0$, and to the left if $h < 0$.

- If the variable y in an equation of the form $x^2 = 4py$ or $y^2 = 4px$ is replaced by $(y - k)$, the graph of the equation is translated vertically by $|k|$ units. The translation is upward if $k > 0$, and downward if $k < 0$.

The standard form of the equation of a parabola with vertex at the origin and a vertical axis of symmetry is $x^2 = 4py$, where $p \neq 0$. Using the given facts about translations, the equation of a parabola with vertex at the point (h, k) and a vertical axis of symmetry is $(x - h)^2 = 4p(y - k)$, where $p \neq 0$. A similar equation can be written for a parabola with vertex at the point (h, k) and a horizontal axis of symmetry.

Their main features are summarized as follows.

Parabola with Vertex at (h, k)

	Vertical Axis of Symmetry	Horizontal Axis of Symmetry
Graph	$(x - h)^2 = 4p(y - k), p > 0$ (above) $(x - h)^2 = 4p(y - k), p < 0$ (below) **Figure 10**	$(y - k)^2 = 4p(x - h), p > 0$ (above) $(y - k)^2 = 4p(x - h), p < 0$ (below) **Figure 11**
Equation	$(x - h)^2 = 4p(y - k)$	$(y - k)^2 = 4p(x - h)$
Direction of opening	Upward if $p > 0$, downward if $p < 0$	To the right if $p > 0$, to the left if $p < 0$
Vertex	(h, k)	(h, k)
Focus	$(h, k + p)$	$(h + p, k)$
Directrix	The line $y = k - p$	The line $x = h - p$
Axis of symmetry	The line $x = h$	The line $y = k$

Example 3   **Finding the Equation of a Parabola with Vertex at (h, k)**

Determine the equation in standard form of the parabola with directrix $y = 7$ and focus at $(-3, 3)$. Sketch the parabola.

Solution Use the given information to determine the orientation of the parabola and the location of the vertex. Plotting the information as you proceed can be very helpful.

Step 1 First find the axis of symmetry of the parabola, which is perpendicular to its directrix. Here, the directrix is the horizontal line $y = 7$, so the axis of symmetry is vertical. Because the focus lies on the axis of symmetry, the value of h must be equal to the x-coordinate of the focus, which is -3. Thus the equation of the axis of symmetry is $x = -3$.

Step 2 Next, find the vertex. The vertex (h, k) is midway between the focus and the directrix and lies on the axis of symmetry. Because $(-3, 7)$ is the point on the directrix that lies on the axis of the symmetry of the parabola, we obtain the coordinates of the vertex by finding the midpoint of the segment that joins the focus to the point $(-3, 7)$.

$$\text{Vertex: } (h, k) = \left(\frac{-3 + (-3)}{2}, \frac{3 + 7}{2} \right) = (-3, 5)$$

Just in Time
Review equations of horizontal and vertical lines in Section 2.1.

Step 3 Find the orientation of the parabola to find the correct form of its equation. Since the vertex lies below the directrix, the parabola opens downward, away from the directrix.

A preliminary sketch of the parabola is given in Figure 12. The standard form of the equation of the directrix of a parabola with a vertical axis of symmetry and vertex (h, k) is $y = k - p$. We can use this equation to find the value of p.

$y = k - p$	General equation of directrix in terms of k and p
$y = 7$	Given equation of directrix
$k - p = 7$	Equate expressions for y
$5 - p = 7$	Substitute $k = 5$
$p = -2$	Solve for p

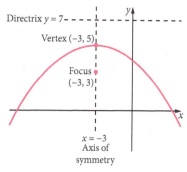

Figure 12 Preliminary sketch

Substituting $(h, k) = (-3, 5)$ and $p = -2$ into the equation $(x - h)^2 = 4p(y - k)$ gives us the equation of the parabola in standard form.

$$(x + 3)^2 = -8(y - 5)$$

To generate other points on the parabola, we solve for y to obtain

$$y = \frac{(x + 3)^2 - 40}{-8}$$

$$= -\frac{1}{8}(x + 3)^2 + 5$$

Substituting $x = -7$ yields $y = 3$. Because the parabola is symmetric with respect to the line $x = -3$, $(1, 3)$ is also a point on the parabola. Table 2 shows the coordinates of additional points on the parabola. The graph of the parabola is shown in Figure 13.

x	y
-7	3
-5	4.5
-3	5
-1	4.5
1	3

Table 2

Directrix $y = 7$

Vertex $(-3, 5)$

$(-7, 3)$ Focus $(-3, 3)$ $(1, 3)$

$y = -\frac{1}{8}(x + 3)^2 + 5$

$x = -3$
Axis of symmetry

Figure 13

✓ *Check It Out 3* Determine the equation in standard form of the parabola with directrix $y = 9$ and focus at $(-2, 5)$. Sketch the parabola.

Just in Time
Review
completing
the square in
Section 1.5.

Writing the Equation of a Parabola in Standard Form

Sometimes, the equation of a parabola is not given in standard form. In such a case, we complete the square on the given equation to put it in standard form. The procedure is outlined below.

Transforming the Equation of a Parabola into Standard Form

Step 1 Gather the y terms on one side of the equation and the x terms on the other side.

Step 2 Put the constant(s) on the same side as the variable that is raised to only the first power.

Step 3 Complete the square on the variable that is raised to the second power.

The above procedure is illustrated in Example 4.

Example 4   **Completing the Square to Write the Equation of a Parabola**

Find the vertex and focus of the parabola defined by the equation $-4x + 3y^2 + 12y - 8 = 0$. Determine the equation of the directrix and sketch the parabola.

Solution The equation of the parabola first must be rewritten in standard form. We can proceed as follows.

$$-4x + 3y^2 + 12y - 8 = 0 \qquad \text{Given equation}$$
$$3y^2 + 12y = 4x + 8 \qquad \text{Keep } y \text{ terms on left side}$$
$$3(y^2 + 4y) = 4x + 8 \qquad \text{Factor out 3 on left side}$$

Find the number a that will make $y^2 + 4y + a$ a perfect square.

$$3(y^2 + 4y + \mathbf{4}) = 4x + 8 + \mathbf{12} \qquad \text{Complete the square}$$

The square was completed on the expression *inside* the parentheses by using $\left(\frac{4}{2}\right)^2 = 4$. In order to keep the equation balanced, $12 = 3 \cdot 4$ was added to the *right* side as well.

$$3(y + 2)^2 = 4x + 20 = 4(x + 5) \qquad \text{Simplify and factor}$$
$$(y + 2)^2 = \frac{4}{3}(x + 5) \qquad \text{Write equation in standard form}$$

By comparing the standard form of the equation of this parabola with the general form $(y - k)^2 = 4p(x - h)$, we can see that $(h, k) = (-5, -2)$ and $4p = \frac{4}{3} \Rightarrow p = \frac{1}{3}$. Because $p > 0$, the parabola opens to the right. The graph is shown in Figure 14.

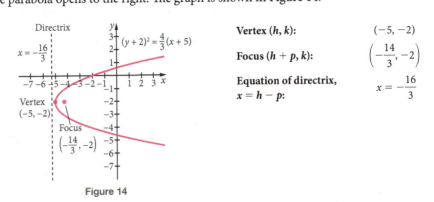

Vertex (h, k): $(-5, -2)$

Focus $(h + p, k)$: $\left(-\dfrac{14}{3}, -2\right)$

Equation of directrix,
$x = h - p$: $x = -\dfrac{16}{3}$

Figure 14

✓ **Check It Out 4** Find the vertex and focus of the parabola defined by the equation $x^2 - 4x - 4y = 0$. Also give the equation of the directrix. ■

Applications

Parabolas are found in many applications, such as the design of flashlights, fluorescent lamps, suspension bridge cables, and the study of the paths of projectiles. Also, parabolas exhibit a remarkable reflective property. Rays of light that emanate from the focus of a mirror with a parabolic cross-section will bounce off the mirror and travel parallel to its axis of symmetry.

Example 5 examines the design of a headlight reflector.

Example 5   **Headlight Design**

The cross-section of a headlight reflector is in the shape of a parabola. The reflector is 6 inches in diameter and 5 inches deep, as illustrated in Figure 15.

a. Find an equation of the parabola, using the position of the vertex of the parabola as the origin of your coordinate system.

b. The bulb for the headlight is positioned at the focus. Find the position of the bulb.

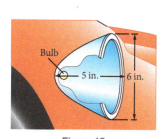

Figure 15

Solution

a. Setting up the problem with respect to a suitable coordinate system is the first task. Using the vertex of the parabola as the origin, and orienting the coordinate axes in such a way that the focus of the parabola lies on the positive x-axis, we obtain the coordinate system shown in Figure 16.

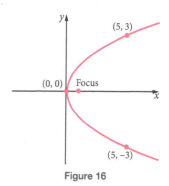

Figure 16

The points $(5, 3)$ and $(5, -3)$ are on the parabola, since the reflector is 5 inches deep and the cross-sectional diameter is 6 inches. Because the axis of symmetry is horizontal, the standard form of the equation of the parabola is $y^2 = 4px$. Because the point $(5, 3)$ lies on the parabola, we can solve for p by substituting $(x, y) = (5, 3)$ into this equation.

$$4(p)(5) = (3)^2 \Rightarrow 20p = 9 \Rightarrow p = \frac{9}{20}$$

Thus the equation of the parabola is $y^2 = \frac{9}{5}x$.

b. The coordinates of the focus are $(p, 0) = \left(\frac{9}{20}, 0\right)$, so the bulb should be placed $\frac{9}{20}$ of an inch away from the vertex.

Check It Out 5   Rework Example 5 for the case in which the reflector is 8 inches in diameter and 6 inches deep.

Exercises 9.1

1. True or False: The distance between two points (a, b) and (c, d) is given by the formula
$$d = \sqrt{(a - c)^2 + (b - d)^2}$$

2. Find the distance between the points $(1, 3)$ and $(-5, 3)$.

3. Find the distance between the points $(2, 6)$ and $(-8, 6)$.

4. Write the equation of the function $f(x)$ that is obtained by shifting the graph of $g(x) = x^2$ to the left 3 units.

5. Write the equation of the function $f(x)$ that is obtained by shifting the graph of $g(x) = x^2$ to the right 1 unit.

6. True or False: The graph of the equation $x = 3$ is a horizontal line.

7. True or False: The graph of the equation $y = -2$ is a horizontal line.

8. Write the equation of the vertical line that passes through the point $(2, -5)$.

In Exercises 9 and 10, complete the square.

9. $x^2 + 6x$

10. $x^2 - 10x$

Skills

In Exercises 11–14, match each graph (labeled a, b, c, and d) with the appropriate equation.

a.
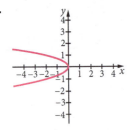

b.
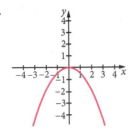

c.
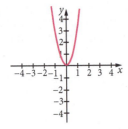

d.
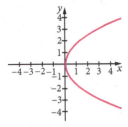

11. $y = 3x^2$

12. $y^2 = 3x$

13. $x^2 = -2y$

14. $y^2 = -\dfrac{x}{2}$

In Exercises 15–38, find the vertex and focus of the parabola that satisfies the given equation. Write the equation of the directrix, and sketch the parabola.

15. $y^2 = 12x$　　　　**16.** $x^2 = -16y$　　　　**17.** $y^2 = 8x$　　　　**18.** $x^2 = 8y$

19. $y = -\dfrac{1}{4}x^2$　　　**20.** $y^2 = -4x$　　　　**21.** $8x = 3y^2$　　　　**22.** $-4y = 3x^2$

23. $5y^2 = 4x$　　　　**24.** $-3x^2 = 16y$　　　　**25.** $y^2 = -12(x - 2)$　　**26.** $x^2 = 8(y + 3)$

27. $(x - 5)^2 = -4(y + 1)$　　**28.** $(y + 3)^2 = 8(x + 2)$　　**29.** $(y - 4)^2 = -(x - 1)$

30. $(x + 7)^2 = -(y + 2)$　　**31.** $y^2 - 4y + 4x = 0$　　**32.** $x^2 + 2x - 4y = 0$

33. $x^2 - 2x = -y$　　　**34.** $y^2 + 6y = x$　　　**35.** $x^2 + 6x + 4y + 25 = 0$

36. $y^2 - 4y + 2x - 6 = 0$　　**37.** $x^2 - 5y + 1 = 8x$　　**38.** $y^2 + 4x + 22 = -10y$

In Exercises 39–62, determine the equation in standard form of the parabola that satisfies the given conditions.

39. Directrix $x = -1$; vertex at $(0, 0)$　　　　**40.** Directrix $x = -3$; vertex at $(0, 0)$

41. Directrix $y = -3$; vertex at $(0, 0)$　　　　**42.** Directrix $y = 2$; vertex at $(0, 0)$

43. Focus at $(0, 2)$; vertex at $(0, 0)$　　　　**44.** Focus at $(0, -3)$; vertex at $(0, 0)$

45. Focus at $(-2, 0)$; vertex at $(0, 0)$　　　　**46.** Focus at $(4, 0)$; vertex at $(0, 0)$

47. Opens upward; distance of focus from x-axis is 4; vertex at $(0, 0)$

48. Opens downward; distance of directrix from x-axis is 6; vertex at $(0, 0)$

49. Opens to the left; distance of directrix from y-axis is 5; vertex at $(0, 0)$

50. Opens to the right; distance of focus from y-axis is 3; vertex at $(0, 0)$

51. Vertex at $(-2, 1)$; directrix $y = -2$　　　**52.** Vertex at $(4, -1)$; directrix $x = 6$

53. Focus at $(0, 4)$; directrix $y = -4$　　　**54.** Focus at $(0, -5)$; directrix $y = 5$

55. Horizontal axis of symmetry; vertex at $(0, 0)$; passes through the point $(1, 3)$

56. Vertical axis of symmetry; vertex at $(0, 0)$; passes through the point $(-2, 6)$

57. Vertical axis of symmetry; vertex at $(4, 3)$; passes through the point $(5, 2)$

58. Horizontal axis of symmetry; vertex at $(-7, -5)$; passes through the point $(2, -1)$

59. Opens upward; vertex at $(4, 1)$; passes through the point $(8, 3)$

60. Opens to the right; vertex at $(-3, 2)$; passes through the point $(5, 10)$

61. Opens downward; vertex at $(5, 3)$; passes through the point $(2, 0)$

62. Opens to the left; vertex at $(2, -4)$; passes through the point $(-1, 3)$

Applications

63. **Satellite Technology** Suppose that a satellite receiver with a parabolic cross-section is 36 inches across and 16 inches deep. How far from the vertex must the receptor unit be located to ensure that it is at the focus of the parabola?

64. **Landscaping** A sprinkler set at ground level shoots water upward in a parabolic arc. If the highest point of the arc is 9 feet above the ground, and the water hits the ground at a maximum of 6 feet from the sprinkler, what is the equation of the parabola in standard form?

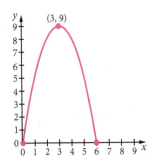

65. **Design** A monument to a famous orator is in the shape of a parabola. The base of the monument is 96 inches wide. The vertex of the parabola is at the top of the monument, which is 144 inches above the base.
    a. How high is the parabola above the locations on the base of the monument that are 32 inches from the ends?
    b. What locations on the base of the monument are 63 inches below the parabola?

66. **Audio Technology** A microphone with a parabolic cross-section is formed by revolving the portion of the parabola $10y = x^2$ between the lines $x = 7$ and $x = -7$ about its axis of symmetry. The sound receiver should be placed at the focus for best reception. Find the location of the sound receiver.

67. **Road Paving** The surface of a footpath over a hill is in the shape of a parabola. The straight-line distance between the ends of the path, which are at the base of the hill, is 60 feet. The path is 5 feet higher at the middle than at the ends. How high above the base of the hill is a point on the path that is located at a horizontal distance of 12 feet from an end of the path?

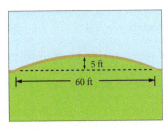

68. **Parabolic Mirror** A reflecting telescope has a mirror with a parabolic cross-section (see the accompanying schematic, which is not drawn to scale). The border of the wide end of the mirror is a circle of radius 50 inches, and the *focal length* of the mirror, which is defined as the distance of the focus from the vertex, is 300 inches.

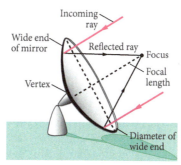

Schematic of Parabolic Mirror

a. How far is the vertex from the center of the wide end of the mirror?

b. Astronomers characterize reflecting telescopes by their *focal ratio*, which is the ratio of the focal length to the aperture (the diameter of the wide end of the mirror). What is the focal ratio of this telescope?

Concepts

69. A parabola has directrix $y = -2$ and passes through the point $(4, -7)$. Find the distance of the point $(4, -7)$ from the focus of the parabola.

70. Find the distance of the point $(1, -4)$ from the axis of symmetry of the parabola with vertex $(3, 2)$, if the axis of symmetry is horizontal.

71. Find the distance of the point $(0, 5)$ from the axis of symmetry of the parabola with focus $(-2, 1)$ and directrix $x = 4$.

72. Find the distance of the point $(11, 7)$ from the focus of the parabola with vertex $(5, 3)$ and directrix $x = 2$.

73. A parabola has axis of symmetry parallel to the y-axis and passes through the points $(-7, 4)$, $(-5, 5)$, and $(3, 29)$. Determine the equation of this parabola. (*Hint:* Use the general equation of a parabola from Chapter 3 together with the information you learned in Chapter 6 about solving a system of equations.)

74. What point in the xy-plane is the mirror image of the point $(2, 8)$ with respect to the axis of symmetry of a parabola that has vertex $(-5, 3)$ and opens to the right? (*Hint:* Use what you learned in Chapter 3 about the mirror image of a point with respect to a line.)

75. What point in the xy-plane is the mirror image of the point $(4, -6)$ with respect to the axis of symmetry of a parabola that has focus $(-1, 1)$ and directrix $x = -5$? (*Hint:* Use what you learned in Chapter 3 about the mirror image of a point with respect to a line.)

SPOTLIGHT ON SUCCESS   *Student Instructor Nathan*

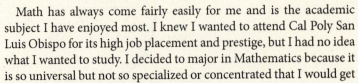

Keep steadily before you the fact that all true success depends at last upon yourself.

—*Theodore T. Hunger*

Math has always come fairly easily for me and is the academic subject I have enjoyed most. I knew I wanted to attend Cal Poly San Luis Obispo for its high job placement and prestige, but I had no idea what I wanted to study. I decided to major in Mathematics because it is so universal but not so specialized or concentrated that I would get stuck in a field that I did not enjoy. I felt that if I kept studying math and its related fields, I would set myself up to be successful later in life, as math is the foundation for engineering, physics, and other science related fields. I have not looked back on my decision. I know it will be a degree that I am proud to have achieved.

I appreciate the consistency that math offers in its problems and in its solutions. I like that math can be simplified into smaller easier-to-understand parts, and its answers are almost always definite. It provides challenges that I enjoy solving, like completing a puzzle piece by piece. In the end, I am able to enjoy the success I have put together for myself.

The Ellipse

Objectives

- Define an ellipse.
- Find the foci, vertices, and major and minor axes of an ellipse.
- Determine the equation of an ellipse and write it in standard form.
- Translate an ellipse in the *xy*-plane.
- Sketch an ellipse.

When a planet revolves around the sun, it traces out an oval-shaped orbit known as an *ellipse*. Once we derive the basic equation of an ellipse, we can use it to study a variety of applications, from planetary motion to elliptical domes.

Definition of an Ellipse

An **ellipse** is the set of all points (x, y) in a plane such that the sum of the distances of (x, y) from two fixed points in the plane is equal to a fixed positive number d, where d is greater than the distance between the fixed points. Each of the fixed points is called a **focus** of the ellipse. Together, they are called the **foci**.

The **major axis** of an ellipse is the line segment that passes through the two foci and has both of its endpoints on the ellipse. The midpoint and the endpoints of the major axis are called the **center** and the **vertices**, respectively, of the ellipse. The **minor axis** of an ellipse is the line segment that is perpendicular to the major axis, passes through the center of the ellipse, and has both of its endpoints on the ellipse. See Figure 1.

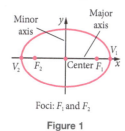

Foci: F_1 and F_2

Figure 1

The only ellipses discussed in this section are those whose major axes are either horizontal or vertical.

Equation of an Ellipse Centered at the Origin

Suppose an ellipse is centered at the origin, with its foci F_1 and F_2 on the *x*-axis. See Figure 2. If c is the distance from the center to either focus, then F_1 and F_2 are located at $(c, 0)$ and $(-c, 0)$, respectively. Now let (x, y) be any point on the ellipse, and let d_1 and d_2 be the distances from (x, y) to F_1 and F_2, respectively. See Figure 2.

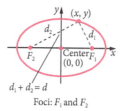

Foci: F_1 and F_2

Figure 2

By the definition of an ellipse, there is some fixed positive number d such that $d_1 + d_2 = d$. The individual distances d_1 and d_2 will change from one point on the ellipse to another, but their sum d will not.

By symmetry, the distance from the center of the ellipse to either vertex is $\frac{d}{2}$. Thus the coordinates of the vertices are $(a, 0)$ and $(-a, 0)$, where $a = \frac{d}{2}$.

Just in Time
Review the
distance formula
in Section 2.2.

Writing $d = 2a$, the equation $d_1 + d_2 = 2a$ is satisfied by all the points on the ellipse. Applying the distance formula, to this equation, we obtain

$$d_1 + d_2 = 2a$$

$$\sqrt{(x - c)^2 + (y - 0)^2} + \sqrt{(x + c)^2 + (y - 0)^2} = 2a$$

Isolating the second radical and squaring both sides gives

$$\sqrt{(x + c)^2 + y^2} = 2a - \sqrt{(x - c)^2 + y^2}$$

$$(x + c)^2 + y^2 = 4a^2 - 4a\sqrt{(x - c)^2 + y^2} + (x - c)^2 + y^2$$

Isolating the radical, we get

$$4a\sqrt{(x - c)^2 + y^2} = (x - c)^2 - (x + c)^2 + 4a^2 = 4a^2 - 4cx$$

Dividing by 4 and then squaring both sides gives

$$a^2[(x - c)^2 + y^2] = a^4 - 2a^2cx + c^2x^2$$

Next, expand the left-hand side and combine like terms.

$$a^2[(x^2 - 2cx + c^2) + y^2] = a^4 - 2a^2cx + c^2x^2$$

$$a^2x^2 - 2a^2cx + a^2c^2 + a^2y^2 = a^4 - 2a^2cx + c^2x^2$$

$$a^2x^2 - c^2x^2 + a^2y^2 = a^4 - a^2c^2$$

Factoring both sides, we get

$$(a^2 - c^2)x^2 + a^2y^2 = a^2(a^2 - c^2)$$

Because $a > c > 0$, $a^2 > c^2$. Let $b^2 = a^2 - c^2$. Then

$$b^2x^2 + a^2y^2 = a^2b^2$$

Dividing by a^2b^2, we have

$$\frac{x^2}{a^2} + \frac{y^2}{b^2} = 1$$

This is the standard form of the equation of an ellipse with center at $(0, 0)$ and foci on the x-axis at $(c, 0)$ and $(-c, 0)$, where c is related to a and b via the equation $b^2 = a^2 - c^2$. Thus $c^2 = a^2 - b^2$. Also, $a^2 - b^2 > 0$ and so $a^2 > b^2$. Since $a, b > 0$, $a > b$.

The minor axis of the ellipse is a segment of the y-axis, and so the x-coordinate of every point on the minor axis is zero. The endpoints of the minor axis lie on the ellipse, so their y-coordinates can be found by substituting $x = 0$ into the equation of the ellipse.

$$\frac{0^2}{a^2} + \frac{y^2}{b^2} = 1 \Rightarrow \frac{y^2}{b^2} = 1 \Rightarrow y^2 = b^2 \Rightarrow y = \pm b$$

Thus the endpoints of the minor axis are $(0, b)$ and $(0, -b)$.

The derivation of the standard form of the equation of an ellipse with center at $(0, 0)$ and foci on the y-axis is similar. Their main features are summarized next.

Ellipses with Center at the Origin

	Horizontal Major Axis	Vertical Major Axis
Graph	 **Figure 3**	 **Figure 4**
Equation	$\dfrac{x^2}{a^2} + \dfrac{y^2}{b^2} = 1, \, a > b > 0$	$\dfrac{x^2}{b^2} + \dfrac{y^2}{a^2} = 1, \, a > b > 0$
Center	$(0, 0)$	$(0, 0)$
Foci	$(-c, 0), (c, 0), \, c = \sqrt{a^2 - b^2}$	$(0, -c), (0, c), \, c = \sqrt{a^2 - b^2}$
Vertices	$(-a, 0), (a, 0)$	$(0, -a), (0, a)$
Major axis	Segment of x-axis from $(-a, 0)$ to $(a, 0)$	Segment of y-axis from $(0, -a)$ to $(0, a)$
Minor axis	Segment of y-axis from $(0, -b)$ to $(0, b)$	Segment of x-axis from $(-b, 0)$ to $(b, 0)$

We will use the following observations to solve problems involving ellipses centered at the origin.

Observations

Assume that $a > b > 0$, where $\frac{1}{a^2}$ and $\frac{1}{b^2}$ are the coefficients of the quadratic terms in the standard form of the equation of an ellipse. Then the following statements hold.

- The major axis of the ellipse is horizontal if the x^2 term in the equation is the term with a coefficient equal to $\frac{1}{a^2}$.

- The major axis of the ellipse is vertical if the y^2 term in the equation is the term with a coefficient equal to $\frac{1}{a^2}$.

- The vertices of the ellipse are the endpoints of the major axis, and are a distance of a units from the center. The length of the major axis is $2a$.

- The distance of either endpoint of the minor axis from the center of the ellipse is b, and the length of the minor axis is $2b$.

- The foci of the ellipse lie on the major axis, but they are *not* part of the ellipse. The distance of either focus from the center is $c = \sqrt{a^2 - b^2}$.

- The center of the ellipse lies on the major axis, midway between the foci, but the center is *not* part of the ellipse.

- In the equation of an ellipse, the variable y is not a function of x, because the graph of such an equation does not pass the vertical line test.

Example 1   **Ellipse Centered at the Origin**

Consider the ellipse that is centered at the origin and defined by the equation

$$16x^2 + 25y^2 = 400$$

a. Write the equation of the ellipse in standard form, and determine the orientation of the major axis.

b. Find the coordinates of the vertices and the foci.

c. Sketch the ellipse, and indicate the vertices and foci.

Solution

a. Divide the given equation by 400 to get a 1 on one side.

$$\frac{x^2}{25} + \frac{y^2}{16} = 1$$

Since $25 > 16$ and $a^2 > b^2$, we have

$$a^2 = 25, \quad b^2 = 16, \quad c^2 = a^2 - b^2 = 25 - 16 = 9$$

Using the fact that a, b, and c are all positive, we obtain

$$a = 5, b = 4, c = 3$$

The x^2 term is the term with a coefficient equal to $\frac{1}{a^2}$, so the major axis is horizontal.

b. Because the major axis is horizontal and the ellipse is centered at the origin, the vertices are

$$(a, 0) = (5, 0) \quad \text{and} \quad (-a, 0) = (-5, 0)$$

The foci also lie on the major axis, so they are

$$(c, 0) = (3, 0) \quad \text{and} \quad (-c, 0) = (-3, 0)$$

c. First plot the endpoints of the major and minor axes and then sketch an ellipse through these points. The endpoints of the major axis are the vertices, $(\pm 5, 0)$, and the endpoints of the minor axis are $(0, \pm b) = (0, \pm 4)$. See Figure 5. To obtain additional points on the ellipse, solve for y in the original equation to get
$y = \frac{\pm\sqrt{400 - 16x^2}}{5}$. Then make a table of values. See Table 1.

x	y
−5	0
−3	±3.2
0	±4
3	±3.2
5	0

Table 1

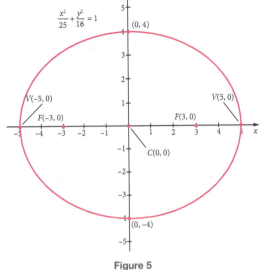

Figure 5

✓ *Check It Out 1*   Determine the orientation of the major axis of the ellipse given by the equation $\frac{x^2}{4} + \frac{y^2}{9} = 1$. Find the vertices and foci and sketch the ellipse.

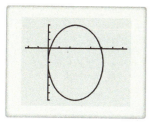

Example 2   Finding the Equation of an Ellipse

Determine the equation in standard form of the ellipse with center at the origin, one focus at $\left(0, -\frac{3}{2}\right)$, and one vertex at $(0, 2)$. Sketch the ellipse.

Solution

Step 1 Determine the orientation of the ellipse. Because the focus and the vertex lie on the y-axis, the major axis is a segment of the y-axis, and so the ellipse is oriented vertically.

Step 2 Next find the values of a and b. Since one of the vertices is at $(0, 2)$, the distance from the center to either vertex is 2. Thus **$a = 2$.**

Recall that a and b are related by the equation $b^2 = a^2 - c^2$, where c is the distance of either focus from the center of the ellipse. One of the foci is at $\left(0, -\frac{3}{2}\right)$, so $c = \frac{3}{2}$. Thus

$$b^2 = a^2 - c^2 = (2)^2 - \left(\frac{3}{2}\right)^2 = 4 - \frac{9}{4} = \frac{7}{4} \Rightarrow b = \frac{\sqrt{7}}{2}$$

We take b to be the positive square root of $\frac{7}{4}$ because b was defined to be positive.

Step 3 Find the equation. Because the major axis is vertical, the equation of the ellipse is $\frac{x^2}{b^2} + \frac{y^2}{a^2} = 1$, where $a^2 = 4$ and $b^2 = \frac{7}{4}$. We thus obtain

$$\frac{x^2}{\frac{7}{4}} + \frac{y^2}{4} = 1$$

Step 4 Sketch the ellipse by plotting the vertices and the endpoints of the minor axis, as shown in Figure 6.

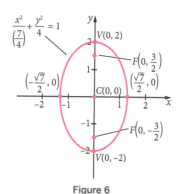

Figure 6

✓ *Check It Out 2*   Determine the equation in standard form of the ellipse with center at the origin, one focus at $(4, 0)$, and one vertex at $(5, 0)$.

Equation of an Ellipse with Center at the Point (h, k)

Recall the following facts about translations.

- If the variable x in an equation of the form $\frac{x^2}{a^2} + \frac{y^2}{b^2} = 1$ or $\frac{x^2}{b^2} + \frac{y^2}{a^2} = 1$ is replaced by $(x - h)$, the graph of the equation is translated horizontally by $|h|$ units. The translation is to the right if $h > 0$, and to the left if $h < 0$.

- If the variable y in an equation of the form $\frac{x^2}{a^2} + \frac{y^2}{b^2} = 1$ or $\frac{x^2}{b^2} + \frac{y^2}{a^2} = 1$ is replaced by $(y - k)$, the graph of the equation is translated vertically by $|k|$ units. The translation is upward if $k > 0$, and downward if $k < 0$.

The standard form of the equation of an ellipse with center at the origin and a horizontal major axis is $\frac{x^2}{a^2} + \frac{y^2}{b^2} = 1$, where $a > b > 0$. Using the given facts about translations, the standard form of the equation of an ellipse with center at the point (h, k) and a horizontal major axis is $\frac{(x - h)^2}{a^2} + \frac{(y - k)^2}{b^2} = 1$, where $a > b > 0$. A similar equation can be written for an ellipse with center at the point (h, k) and a vertical major axis. Their main features are summarized as follows.

Ellipses with Center at (h, k)

	Horizontal Major Axis	Vertical Major Axis
Graph	Figure 8	Figure 9
Equation	$\dfrac{(x-h)^2}{a^2} + \dfrac{(y-k)^2}{b^2} = 1, a > b > 0$	$\dfrac{(x-h)^2}{b^2} + \dfrac{(y-k)^2}{a^2} = 1, a > b > 0$
Center	(h, k)	(h, k)
Foci	$(h-c, k), (h+c, k), c = \sqrt{a^2 - b^2}$	$(h, k-c), (h, k+c), c = \sqrt{a^2 - b^2}$
Vertices	$(h-a, k), (h+a, k)$	$(h, k-a), (h, k+a)$
Major axis	Segment of the line $y = k$ from $(h-a, k)$ to $(h+a, k)$	Segment of the line $x = h$ from $(h, k-a)$ to $(h, k+a)$
Minor axis	Segment of the line $x = h$ from $(h, k-b)$ to $(h, k+b)$	Segment of the line $y = k$ from $(h-b, k)$ to $(h+b, k)$

Example 3   **Finding the Equation of an Ellipse with Vertex at (h, k)**

Determine the equation in standard form of the ellipse with foci at $(4, 1)$ and $(4, -5)$ and one vertex at $(4, 3)$. Sketch the ellipse.

Solution

Step 1 Determine the orientation. The x-coordinate of each focus is 4, so the foci lie on the vertical line $x = 4$. Thus the major axis of the ellipse is vertical, and the standard form of its equation is $\dfrac{(x-h)^2}{b^2} + \dfrac{(y-k)^2}{a^2} = 1$.

Step 2 Next determine the center (h, k) and the values of a and b. Because the center is the midpoint of the line segment that has its endpoints at the foci, we obtain

$$\text{Center} = (h, k) = \left(\frac{4+4}{2}, \frac{1+(-5)}{2} \right) = (4, -2)$$

The quantity a is the distance from either vertex to the center. Using the vertex at $(4, 3)$ and the distance formula, we find that the distance from $(4, 3)$ to $(4, -2)$ is

$$a = \sqrt{(4-4)^2 + (3-(-2))^2} = 5$$

Recall that a and b are related by the equation $b^2 = a^2 - c^2$, where c is the distance from the center to either focus. Using the distance formula to compute the distance from the center to the focus at $(4, 1)$ gives

$$c = \sqrt{(4-4)^2 + (-2-1)^2} = 3$$

Thus

$$b^2 = a^2 - c^2 = (5)^2 - (3)^2 = 25 - 9 = 16 \Rightarrow b = 4$$

We take b to be the positive square root of 16 because b was defined to be positive.

Step 3 To find the equation, substitute the values of h, k, a, and b into the standard form of the equation.

$$\frac{(x - h)^2}{b^2} + \frac{(y - k)^2}{a^2} = 1 \qquad \text{Standard form of equation}$$

$$\frac{(x - 4)^2}{16} + \frac{(y + 2)^2}{25} = 1 \qquad \text{Substitute } h = 4, k = -2, a = 5, b = 4$$

The vertices are

$$(h, k + a) = (4, -2 + 5) = \mathbf{(4, 3)} \text{ and } (h, k - a) = (4, -2 - 5) = \mathbf{(4, -7)}$$

The first vertex checks with the given information.

The endpoints of the minor axis are

$$(h + b, k) = (4 + 4, -2) = \mathbf{(8, -2)} \text{ and } (h - b, k) = (4 - 4, -2) = \mathbf{(0, -2)}$$

Step 4 To sketch the ellipse, we plot the vertices and the endpoints of the minor axis. These four points can be used to sketch the ellipse, as shown in Figure 10.

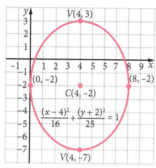

Figure 10

Check It Out 3 Determine the equation in standard form of the ellipse with foci at (3, 2) and (3, −2) and one vertex at (3, 3).

Example 4 **Completing the Square to Write the Equation of an Ellipse**

Write the equation in standard form of the ellipse defined by the equation $5x^2 - 30x + 25y^2 + 50y + 20 = 0$. Sketch the ellipse.

Solution To obtain the standard form of the equation, first complete the squares on both x and y. As the first step in doing this, move the constant to the right side of the equation.

$$5x^2 - 30x + 25y^2 + 50y = -20$$

Next factor out the common factor in the x terms, and then do the same for the y terms. Then complete the square on each expression in parentheses.

$$5(x^2 - 6x + 9) + 25(y^2 + 2y + 1) = -20 + \mathbf{45} + \mathbf{25}$$

The squares were completed using $\left(\frac{-6}{2}\right)^2 = 9$ and $\left(\frac{2}{2}\right)^2 = 1$, respectively. On the right side, we added $45 = (5 \cdot 9)$ and $25 = (25 \cdot 1)$. Factoring the left side of the equation and simplifying the right side gives

$$5(x - 3)^2 + 25(y + 1)^2 = 50$$

Dividing by 50 on both sides, we obtain

$$\frac{(x - 3)^2}{10} + \frac{(y + 1)^2}{2} = 1$$

This is the standard form of the equation of the ellipse, which implies that the center of the ellipse is (3, −1). Because $10 > 2$ and $a^2 > b^2$, we see that $a^2 = 10$ and $b^2 = 2$. Thus $a = \sqrt{10}$ and $b = \sqrt{2}$. The major axis is horizontal, since the $(x - 3)^2$ term is the term with a coefficient equal to $\frac{1}{a^2}$.

Just in Time
Review completing the square in Section 1.5.

The vertices are

$$(h + a, k) = (3 + \sqrt{10}, -1) \approx (6.162, -1) \text{ and}$$
$$(h - a, k) = (3 - \sqrt{10}, -1) \approx (-0.162, -1)$$

The endpoints of the minor axis are

$$(h, k + b) = (3, -1 + \sqrt{2}) \approx (3, 0.414) \text{ and}$$

$$(h, k - b) = (3, -1 - \sqrt{2}) \approx (3, -2.414)$$

Finally, we plot the vertices and the endpoints of the minor axis and use these four points to sketch the ellipse. See Figure 11.

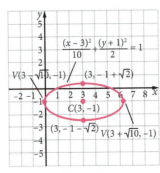

Figure 11

Check It Out 4 Write the equation in standard form of the ellipse defined by the equation $4x^2 - 8x + y^2 = 8$.

Applications

Ellipses occur in many science and engineering applications. The orbit of each planet in our solar system is an ellipse with the sun at one of the foci. The orbit of our moon is an ellipse with Earth at one focus, as are the orbits of some comets.

| **Example 5** | **Elliptical Orbit of Halley's Comet** |

The orbit of Halley's comet is elliptical, with the sun at one of the foci. The length of the major axis of the orbit is approximately 36 astronomical units (AU), and the length of the minor axis is approximately 9 AU (1 AU $\approx$ 92,600,000 miles). Find the equation in standard form of the path of Halley's comet, using the origin as the center of the ellipse and a segment of the x-axis as the major axis.

Solution The major axis is a segment of the x-axis and its length is 36. This length can be expressed as $2a$, where a is the distance from either vertex of the ellipse to the center. Equating $2a$ and 36, we find that $a = 18$. Thus $a^2 = 324$.

The minor axis is a segment of the y-axis and its length is 9. This length can be expressed as $2b$, where b is the distance from either endpoint of the minor axis to the center of the ellipse. Equating $2b$ and 9, we obtain $b = 4.5$. Thus $b^2 = 20.25$.

Because the ellipse is centered at the origin, the equation in standard form of the orbit is

$$\frac{x^2}{324} + \frac{y^2}{20.25} = 1$$

Check It Out 5 In Example 6, find the distance (in AU) from the sun to the center of the ellipse. (*Hint:* The sun is one of the foci of the elliptical orbit of Halley's comet.)

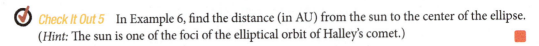

Exercises 9.2

1. True or False: The distance between the points (a, b) and $(0, 4)$ is given by $d = \sqrt{a^2 + (b - 4)^2}$.

2. True or False: The distance between the points (a, b) and $(2, -7)$ is given by $d = \sqrt{(a + 7)^2 + (b - 2)^2}$.

3. Find the distance between the points $(3, 4)$ and $(1, -2)$.

4. Find the distance between the points $(5, -1)$ and $(5, -8)$.

In Exercises 5 and 6, what number must be added to complete the square?

5. $x^2 - 12x$

6. $y^2 + 24y$

Skills

In Exercises 7–10, match each graph (labeled a, b, c, and d) with the appropriate equation.

a.
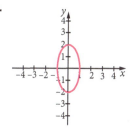

b.
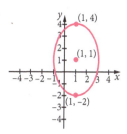

c.
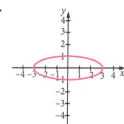

d.
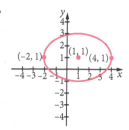

7. $\dfrac{x^2}{9} + y^2 = 1$

8. $x^2 + \dfrac{y^2}{4} = 1$

9. $\dfrac{(x - 1)^2}{9} + \dfrac{(y - 1)^2}{4} = 1$

10. $\dfrac{(x - 1)^2}{4} + \dfrac{(y - 1)^2}{9} = 1$

In Exercises 11–34, find the center, vertices, and foci of the ellipse that satisfies the given equation, and sketch the ellipse.

11. $\dfrac{x^2}{25} + \dfrac{y^2}{16} = 1$

12. $\dfrac{x^2}{9} + \dfrac{y^2}{25} = 1$

13. $\dfrac{x^2}{16} + \dfrac{y^2}{25} = 1$

14. $\dfrac{x^2}{25} + \dfrac{y^2}{9} = 1$

15. $\dfrac{x^2}{9} + \dfrac{y^2}{4} = 1$

16. $\dfrac{x^2}{9} + \dfrac{y^2}{16} = 1$

17. $\dfrac{4x^2}{9} + \dfrac{9y^2}{16} = 1$

18. $\dfrac{16x^2}{49} + \dfrac{25y^2}{81} = 1$

19. $\dfrac{(x + 3)^2}{9} + \dfrac{(y + 1)^2}{16} = 1$

20. $\dfrac{(x + 5)^2}{25} + \dfrac{(y - 3)^2}{16} = 1$

21. $\dfrac{(x - 1)^2}{100} + \dfrac{(y + 1)^2}{36} = 1$

22. $\dfrac{(x + 6)^2}{64} + \dfrac{y^2}{100} = 1$

23. $\dfrac{(x-1)^2}{16} + \dfrac{(y+2)^2}{9} = 1$ **24.** $\dfrac{(x-6)^2}{36} + \dfrac{(y-3)^2}{25} = 1$ **25.** $3x^2 + 4y^2 = 12$

26. $25x^2 + 12y^2 = 300$ **27.** $5x^2 + 2y^2 = 10$ **28.** $4x^2 + 7y^2 = 28$

29. $4x^2 + y^2 - 24x - 8y + 48 = 0$ **30.** $x^2 + 9y^2 + 6x - 36y + 36 = 0$

31. $5x^2 + 9y^2 - 20x + 54y + 56 = 0$ **32.** $9x^2 + 16y^2 + 36x - 16y - 104 = 0$

33. $25x^2 + 16y^2 - 200x + 96y + 495 = 0$ **34.** $9x^2 + 4y^2 + 90x - 16y + 216 = 0$

 In Exercises 35–40, use a graphing utility to graph the given equation.

35. $3x^2 + 7y^2 = 20$ **36.** $8x^2 + 3y^2 = 15$

37. $\dfrac{x^2}{6} + \dfrac{y^2}{11} = 1$ **38.** $\dfrac{x^2}{4} + \dfrac{y^2}{13} = 1$

39. $\dfrac{(x-1)^2}{5} + \dfrac{(y+3)^2}{6} = 1$ **40.** $\dfrac{(x+4)^2}{7} + \dfrac{(y-1)^2}{3} = 1$

In Exercises 41–48, determine the equation in standard form of the ellipse centered at the origin that satisfies the given conditions.

41. One vertex at $(6, 0)$; one focus at $(3, 0)$

42. One vertex at $(7, 0)$; one focus at $(2, 0)$

43. Minor axis of length $2\sqrt{39}$; foci at $(-5, 0)$, $(5, 0)$

44. Minor axis of length 8; foci at $(0, -5)$, $(0, 5)$

45. Minor axis of length 7; major axis of length 9; major axis vertical

46. Minor axis of length 6; major axis of length 14; major axis horizontal

47. One endpoint of minor axis at $(2, 0)$; major axis of length 18

48. One endpoint of minor axis at $(0, -4)$: major axis of length 12

In Exercises 49–56, determine the equation in standard form of the ellipse that satisfies the given conditions.

49. Center at $(-2, 4)$; one vertex at $(-6, 4)$; one focus at $(1, 4)$

50. Center at $(2, 1)$; one vertex at $(7, 1)$; one focus at $(-2, 1)$

51. Vertices at $(-3, 4)$, $(-3, -2)$; foci at $(-3, 3)$, $(-3, -1)$

52. Vertices at $(5, 6)$, $(5, -4)$; foci at $(5, 4)$, $(5, -2)$

53. Major axis of length 8; foci at $(4, 1)$, $(4, -3)$

54. Center at $(-9, 3)$; one focus at $(-5, 3)$; one vertex at $(-3, 3)$

55. One endpoint of minor axis at $(7, -4)$; center at $(7, -8)$; major axis of length 12

56. One endpoint of minor axis at $(-6, 1)$; one vertex at $(-3, -4)$; major axis of length 10

In Exercises 57–60, determine the equations in standard form of two different ellipses that satisfy the given conditions.

57. Center at $(0, 0)$; major axis of length 9; minor axis of length 5

58. Center at $(3, 2)$; major axis of length 11; minor axis of length 6

59. One endpoint of minor axis at $(5, 0)$; minor axis of length 8; major axis of length 12; major axis vertical

60. Major axis of length 10; minor axis of length 4; one vertex at $(-5, -3)$; major axis horizontal

Applications

61. Medical Technology A lithotripter is a device that breaks up kidney stones by propagating shock waves through water in a chamber that has an elliptical cross-section. High-frequency shock waves are produced at one focus, and the patient is positioned in such a way that the kidney stones are at the other focus. On striking a point on the boundary of the chamber, the shock waves are reflected to the other focus and break up the kidney stones. Find the coordinates of the foci if the center of the ellipse is at the origin, one vertex is at $(6, 0)$, and one endpoint of the minor axis is at $(0, -2.5)$.

62. Design An arched bridge over a 20-foot stream is in the shape of the top half of an ellipse. The highest point of the bridge is 5 feet above the base. How high is a point on the bridge that is 5 feet (horizontally) from one end of the base of the bridge?

63. Astronomy The orbit of the moon around Earth is an ellipse, with Earth at one focus. If the major axis of the orbit is 477,736 miles and the minor axis is 477,078 miles, find the maximum and minimum distances from Earth to the moon.

64. Reflective Property An ellipse has a reflective property. A sound wave that passes through one focus of an ellipse and strikes some point on the ellipse will be reflected through the other focus. This reflective property has been used in architecture to design whispering galleries, such as the one in the Capitol building in Washington, D.C. A weak whisper at one focus can be heard clearly at the other focus, but nowhere else in the room. The dome of the Capitol has an elliptical cross-section. If the length of the major axis is 400 feet and the highest point of the dome is 60 feet above the floor, where should two senators stand to hear each other whisper? Assume that the ellipse is centered at the origin.

65. Physics A laser is located at one focus of an ellipse. A sheet of metal, which is only a fraction of an inch wide and serves as a reflecting surface, lines the entire ellipse and is located at the same height above the ground as the laser. A very narrow beam of light is emitted by the laser. When the beam strikes the metal, it is reflected toward the other focus of the ellipse. If the foci are 20 feet apart and the shorter dimension of the ellipse is 12 feet, how great a distance is traversed by the beam of light from the time it is emitted by the laser to the time it reaches the other focus?

66. Graphic Design A graphic artist draws a schematic of an elliptically-shaped logo for an IT firm. The shorter dimension of the logo is 10 inches. If the foci are 8 inches apart, what is the longer dimension of the logo?

67. **Leisure** One of the holes at a miniature golf course is in the shape of an ellipse. The teeing-off point and the cup are located at the foci, which are 12 feet apart. The minor axis of the ellipse is 8 feet long. If on the first shot a player hits the ball off to either side at ground level, and the ball rolls until it reaches some point on the (elliptical) boundary, the ball will bounce off and proceed toward the cup. Under such circumstances, what is the total distance traversed by the ball if it just barely reaches the cup?

68. **Construction** A cylindrically-shaped vent pipe with a diameter of 4 inches is to be attached to the roof of a house. The pipe first has to be cut at an angle that matches the slope of the roof, which is $\frac{3}{4}$. Once that has been done, the lower edge of the pipe will be elliptical. If the shorter dimension of the lower edge of the pipe is to be parallel to the base of the roof, what is the longer dimension?

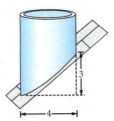

Concepts

This set of exercises will draw on the ideas presented in this section and your general math background.

69. For the ellipse with equation $\frac{x^2}{16} + \frac{y^2}{9} = 1$, find the distance from either endpoint of the major axis to either endpoint of the minor axis.

70. For the ellipse with equation $\frac{(x+2)^2}{25} + \frac{(y-3)^2}{144} = 1$, find the distance from either endpoint of the major axis to either endpoint of the minor axis.

71. Write the equation that is satisfied by the set of points whose distances from the points $(3, 0)$ and $(-3, 0)$ add up to 8.

72. Write the equation that is satisfied by the set of points whose distances from the points $(0, 5)$ and $(0, -5)$ average to 6.

73. What is the set of all points that satisfy an equation of the form $\frac{x^2}{a^2} + \frac{y^2}{b^2} = 1$ if $a = b > 0$?

74. The *eccentricity* of an ellipse is defined as $e = \frac{c}{a} \left(= \frac{\sqrt{a^2 - b^2}}{a} \right)$, where a, b, and c are as defined in this section. Since $0 < c < a$, the value of e lies between 0 and 1. In ellipses that are long and thin, b is small compared to a, so the eccentricity is close to 1. In ellipses that are nearly circular, b is almost as large as a, so the eccentricity is close to 0. What is the eccentricity of the ellipse with equation $\frac{x^2}{9} + \frac{y^2}{25} = 1$? Does this ellipse have a greater or lesser eccentricity than the ellipse with equation $\frac{x^2}{16} + \frac{y^2}{25} = 1$?

The Hyperbola

Objectives

- Define a hyperbola.

- Find the foci, vertices, transverse axis, and asymptotes of a hyperbola.

- Determine the equation of a hyperbola and write it in standard form.

- Translate a hyperbola in the *xy*-plane.

- Sketch a hyperbola.

The last of the conic sections that we will discuss is the hyperbola. We first give its definition and derive its equation. The rest of this section will analyze various equations of hyperbolas, and conclude with an application to sculpture.

Definition of a Hyperbola

A **hyperbola** is the set of all points (x, y) in a plane such that the absolute value of the difference of the distances of (x, y) from two fixed points in the plane is equal to a fixed positive number d, where d is less than the distance between the fixed points. Each of the fixed points is called a **focus** of the hyperbola. Together, they are called the **foci**.

A hyperbola consists of a pair of non-intersecting, open curves, each of which is called a **branch** of the hyperbola. The two points on the hyperbola that lie on the line passing through the foci are known as the **vertices** of the hyperbola. The **transverse axis** of a hyperbola is the line segment that has the vertices as its endpoints. See Figure 1.

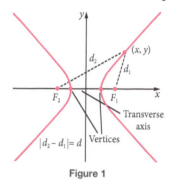

Figure 1

The only hyperbolas discussed in this section are those whose transverse axes are either horizontal or vertical.

Equation of a Hyperbola Centered at the Origin

Suppose a hyperbola is centered at the origin, with its foci F_1 and F_2 on the *x*-axis. If c is the distance from the center to either focus, then F_1 and F_2 are located at $(c, 0)$ and $(-c, 0)$, respectively. Let (x, y) be any point on the hyperbola, and let d_1 and d_2 be the distances from (x, y) to F_1 and F_2, respectively. See Figure 2.

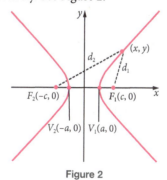

Figure 2

Now d is the distance between the vertices, and the vertices are symmetric with respect to the center of the hyperbola, so the distance from the center of the hyperbola to either vertex is $\frac{d}{2}$. Thus the coordinates of the vertices are $(a, 0)$ and $(-a, 0)$, where $a = \frac{d}{2}$.

Just in Time
Review the
distance formula
in Section 2.2.

Because $a = \frac{d}{2}$, we can write $d = 2a$. Hence the equation $|d_2 - d_1| = 2a$ is satisfied by all points on the hyperbola.

$$|d_2 - d_1| = 2a \quad \Rightarrow \quad d_2 - d_1 = \pm 2a$$

$$\sqrt{(x + c)^2 + y^2} - \sqrt{(x - c)^2 + y^2} = \pm 2a$$

Isolating the first radical and squaring both sides gives

$$\sqrt{(x + c)^2 + y^2} = \pm 2a + \sqrt{(x - c)^2 + y^2}$$

$$(x + c)^2 + y^2 = 4a^2 \pm 4a\sqrt{(x - c)^2 + y^2} + (x - c)^2 + y^2$$

We then switch sides and isolate the radical.

$$\pm 4a\sqrt{(x - c)^2 + y^2} = (x + c)^2 - (x - c)^2 - 4a^2$$

$$= 4cx - 4a^2$$

Next, we divide by 4 and square both sides.

$$a^2[(x - c)^2 + y^2] = c^2x^2 - 2a^2cx + a^4$$

Expanding the left-hand side and combining like terms, we get

$$a^2x^2 - 2a^2cx + a^2c^2 + a^2y^2 = c^2x^2 - 2a^2cx + a^4$$

$$a^2x^2 + a^2c^2 + a^2y^2 = c^2x^2 + a^4$$

Switching sides and isolating the constant terms then gives

$$c^2x^2 + a^4 = a^2x^2 + a^2c^2 + a^2y^2$$

$$c^2x^2 - a^2x^2 - a^2y^2 = a^2c^2 - a^4$$

Factoring both sides leads to the equation

$$(c^2 - a^2)x^2 - a^2y^2 = a^2(c^2 - a^2)$$

Because $c > a$ and a and c are both positive, we can write $c^2 > a^2$. Letting $b^2 = c^2 - a^2$ in the last equation and then dividing by a^2b^2, we obtain

$$b^2x^2 - a^2y^2 = a^2b^2$$

$$\frac{x^2}{a^2} - \frac{y^2}{b^2} = 1$$

This is the standard form of the equation of a hyperbola with center at $(0, 0)$ and foci on the x-axis at $(c, 0)$ and $(-c, 0)$, where c is related to a and b via the equation $b^2 = c^2 - a^2$. Thus $c^2 = a^2 + b^2$.

Solving the equation $\frac{x^2}{a^2} - \frac{y^2}{b^2} = 1$ for y, we obtain

$$y = \pm \frac{b}{a}\sqrt{x^2 - a^2}$$

As x becomes very large, the quantity $x^2 - a^2$ approaches x^2. Therefore, the graph of the equation $y = \frac{b}{a}\sqrt{x^2 - a^2}$ approaches the line $y = \frac{b}{a}x$, and the graph of the equation $y = -\frac{b}{a}\sqrt{x^2 - a^2}$ approaches the line $y = -\frac{b}{a}x$. These lines, $y = \frac{b}{a}x$ and $y = -\frac{b}{a}x$, are called the **asymptotes** of the hyperbola, and pass through the center of the hyperbola.

The standard equation of a hyperbola with center at $(0, 0)$ and foci on the y-axis at $(0, c)$ and $(0, -c)$ is $\frac{y^2}{a^2} - \frac{x^2}{b^2} = 1$, where $a^2 + b^2 = c^2$, using a similar derivation.

Hyperbolas with Center at the Origin

	Horizontal Transverse Axis	Vertical Transverse Axis
Graph	Figure 3	Figure 4
Equation	$\dfrac{x^2}{a^2} - \dfrac{y^2}{b^2} = 1, a, b > 0$	$\dfrac{y^2}{a^2} - \dfrac{x^2}{b^2} = 1, a, b > 0$
Center	$(0, 0)$	$(0, 0)$
Foci	$(-c, 0), (c, 0), c = \sqrt{a^2 + b^2}$	$(0, -c), (0, c), c = \sqrt{a^2 + b^2}$
Vertices	$(-a, 0), (a, 0)$	$(0, -a), (0, a)$
Transverse axis	Segment of x-axis from $(-a, 0)$ to $(a, 0)$	Segment of y-axis from $(0, -a)$ to $(0, a)$
Asymptotes	The lines $y = \dfrac{b}{a}x$ and $y = -\dfrac{b}{a}x$	The lines $y = \dfrac{a}{b}x$ and $y = -\dfrac{a}{b}x$

Observations

- The transverse axis of a hyperbola is horizontal if the coefficient of the x^2 term is positive in the standard form of the equation.

- The transverse axis of a hyperbola is vertical if the coefficient of the y^2 term is positive in the standard form of the equation.

- The vertices of a hyperbola are a distance of a units from the center.

- Each branch of a hyperbola opens *away* from the center and *toward* one of the foci.

- The center of a hyperbola lies on the transverse axis, midway between the vertices and midway between the foci, but the center is *not* part of the hyperbola.

- In the equation of a hyperbola, the variable y is not a function of x, since the graph of such an equation does not pass the vertical line test.

VIDEO EXAMPLES

SECTION 9.3

Example 1   **Hyperbola Centered at the Origin**

Consider the hyperbola centered at the origin and defined by the equation

$$16x^2 - 9y^2 = 144$$

a. Write the equation of the hyperbola in standard form, and determine the orientation of the transverse axis.

b. Find the coordinates of the vertices and foci.

c. Find the equations of the asymptotes.

d. Sketch the hyperbola, and indicate the vertices, foci, and asymptotes.

Solution

a. Divide the equation by 144 to get a 1 on one side, and then simplify.

$$\frac{16x^2}{144} - \frac{9y^2}{144} = 1 \quad \Rightarrow \quad \frac{x^2}{9} - \frac{y^2}{16} = 1$$

Because the y^2 term is subtracted from the x^2 term, the values of a^2 and b^2 are 9 and 16, respectively, and the transverse axis is a segment of the x-axis. Thus the transverse axis is horizontal, so the hyperbola opens to the side.

b. The vertices are the endpoints of the transverse axis, so the vertices and foci lie on the x-axis. Using $a^2 = 9$ and $b^2 = 16$, we find that

$$c^2 = a^2 + b^2 = 16 + 9 = 25$$

Using the fact that a, b, and c are all positive, we have

$$a = 3, \quad b = 4, \quad \text{and} \quad c = 5$$

Vertices: $(\pm a, 0) \Rightarrow (\mathbf{3}, 0)$ and $(-\mathbf{3}, 0)$
Foci: $(\pm c, 0) \Rightarrow (\mathbf{5}, 0)$ and $(-\mathbf{5}, 0)$

c. The equations of the asymptotes are given by

$$y = \frac{b}{a}x = \frac{4}{3}x \quad \text{and} \quad y = -\frac{b}{a}x = -\frac{4}{3}x$$

d. To sketch the hyperbola, proceed as follows.

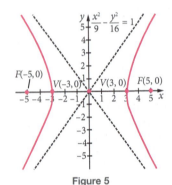

Figure 5

- Plot and label the vertices and the foci.

- Draw a rectangular box that is centered at $(0, 0)$ and has horizontal and vertical dimensions of $2a = 6$ and $2b = 8$, respectively.

- Graph the asymptotes, each of which is a line that passes through the center of the hyperbola *and* through one pair of opposite corners of the box.

- Sketch the hyperbola so that it passes through the vertices and approaches the asymptotes. See Figure 5.

Check It Out 1   For the hyperbola given by the equation $\frac{x^2}{4} - \frac{y^2}{4} = 1$, find the coordinates of the vertices and foci and the equations of the asymptotes. Sketch the hyperbola.

In the next example, we derive the equation of a hyperbola given one of its foci and one of its vertices.

Example 2   **Finding the Equation of a Hyperbola**

Determine the equation in standard form of the hyperbola with center at $(0, 0)$, one focus at $(0, 4)$, and one vertex at $(0, -1)$. Find the other focus and the other vertex. Sketch a graph of the hyperbola by finding and plotting some additional points that lie on the hyperbola.

Solution   The given focus and vertex are on the y-axis. Because the transverse axis is a segment of the y-axis, the equation of the hyperbola is of the form $\frac{y^2}{a^2} - \frac{x^2}{b^2} = 1$. Because one of the foci is at $(0, 4)$, $\mathbf{c = 4}$. Now, a is the distance from $(0, 0)$ to the vertex. Since there is a vertex at $(\mathbf{0, -1})$, $\mathbf{a = 1}$. Substituting the values of a and c, we find that

$$b^2 = c^2 - a^2$$
$$= (4)^2 - (1)^2$$
$$= 16 - 1 = 15$$

Thus the equation of the hyperbola in standard form is

$$\frac{y^2}{a^2} - \frac{x^2}{b^2} = \frac{y^2}{1} - \frac{x^2}{15} = 1$$

Because the center is at $(0, 0)$, the other focus is at $(0, -c) = (0, -4)$ and the other vertex is at $(0, a) = (0, 1)$.

To find some additional points on the hyperbola, we solve the equation of the hyperbola for y to obtain

$$y = \pm\sqrt{1 + \frac{x^2}{15}}$$

The expression for y is defined for all real numbers x. Substituting $x = 5$ gives

$$y = \pm\sqrt{1 + \frac{(5)^2}{15}} = \pm\sqrt{1 + \frac{5}{3}} = \pm\sqrt{\frac{8}{3}} \Rightarrow y \approx \pm 1.633$$

Similarly, substituting $x = -5$ gives $y \approx \pm 1.633$. We can sketch the hyperbola by plotting the two vertices, $(0, \pm 1)$, along with the four points $(5, \pm 1.633)$ and $(-5, \pm 1.633)$. See Figure 6.

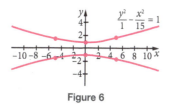

Figure 6

Check It Out 2   Determine the equation in standard form of the hyperbola with center at $(0, 0)$, one focus at $(0, 4)$, and one vertex at $(0, 3)$.

Equation of Hyperbola with Center at the point (h, k)

The main features and properties of hyperbolas with center at the point (h, k) and a horizontal or vertical transverse axis are summarized next. Their graphs are shown in Figures 7 and 8.

Hyperbolas with Center at (h, k)

	Horizontal Transverse Axis	Vertical Transverse Axis
Graph	Figure 7	Figure 8
Equation	$\dfrac{(x-h)^2}{a^2} - \dfrac{(y-k)^2}{b^2} = 1$, $a, b > 0$	$\dfrac{(y-k)^2}{a^2} - \dfrac{(x-h)^2}{b^2} = 1$, $a, b > 0$
Center	(h, k)	(h, k)
Foci	$(h - c, k), (h + c, k), c = \sqrt{a^2 + b^2}$	$(h, k - c), (h, k + c), c = \sqrt{a^2 + b^2}$
Vertices	$(h - a, k), (h + a, k)$	$(h, k - a), (h, k + a)$
Transverse axis	Segment of the line $y = k$ from $(h - a, k)$ to $(h + a, k)$	Segment of the line $x = h$ from $(h, k - a)$ to $(h, k + a)$
Asymptotes	The lines $y = \dfrac{b}{a}(x - h) + k$ and $y = -\dfrac{b}{a}(x - h) + k$	The lines $y = \dfrac{a}{b}(x - h) + k$ and $y = -\dfrac{a}{b}(x - h) + k$

Example 3   **Finding the Equation of a Hyperbola with Vertex at** (h, k)

Determine the equation in standard form of the hyperbola with foci at $(4, 3)$ and $(4, -7)$ and a transverse axis of length 6. Find the asymptotes and sketch the hyperbola.

Solution

Step 1 Determine the orientation of the hyperbola. Because the foci are located on the line $x = 4$, the transverse axis is vertical. You can see this by plotting the foci. Thus the standard form of the equation of the hyperbola is

$$\frac{(y - k)^2}{a^2} - \frac{(x - h)^2}{b^2} = 1$$

Step 2 Next we must determine the center (h, k) and the values of a and b. Because the center is the midpoint of the line segment that has its endpoints at the foci, the coordinates of the center are

$$(h, k) = \left(\frac{4 + 4}{2}, \frac{3 + (-7)}{2}\right) = (4, -2)$$

To find a, we note that the length of the transverse axis is given as 6. The distance from either vertex to the center is half the length of the transverse axis, so $a = 3$.

Because c is the distance from either focus to the center, we obtain

$$c = \text{distance from } (4, 3) \text{ to } (4, -2) = 3 - (-2) = \mathbf{5}$$

To find b, we substitute the values of a and c.

$$b^2 = c^2 - a^2 = (\mathbf{5})^2 - (\mathbf{3})^2 = 25 - 9 = 16 \Rightarrow b = 4$$

Substituting the values of h, k, a^2, and b^2, we find that the equation of the hyperbola is

$$\frac{(y - k)^2}{a^2} - \frac{(x - h^2)}{b^2} = \frac{(y + 2)^2}{9} - \frac{(x - 4)^2}{16} = 1$$

Step 3 Now we can find the asymptotes and vertices. The asymptotes are

$$y = \frac{a}{b}(x - h) + k = \frac{3}{4}(x - 4) - 2 = \frac{3}{4}x - 5 \text{ and}$$

$$y = -\frac{a}{b}(x - h) + k = -\frac{3}{4}(x - 4) - 2 = -\frac{3}{4}x + 1$$

The vertices are located at $(4, -2 + 3) = (\mathbf{4, 1})$ and $(4, -2 - 3) = (\mathbf{4, -5})$.

Step 4 Sketch the hyperbola as follows.

- Plot and label the vertices.

- Draw a rectangular box that is centered at $(h, k) = (4, -2)$ and has horizontal and vertical dimensions of $2b = 8$ and $2a = 6$, respectively.

- Graph the asymptotes, each of which is a line that passes through the center of the hyperbola *and* through one pair of opposite corners of the box.

- Sketch the hyperbola so that it passes through the vertices and approaches the asymptotes. See Figure 9.

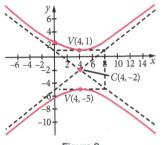

Figure 9

Using Technology

The APPS button in a graphing calculator will bring up a CONICS menu. Choose the option for Hyperbola, and enter the equation to get a graph of the hyperbola. See Figure 10.

Keystroke Appendix:

Section 17.

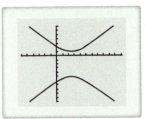

Figure 10

Check It Out 3   Determine the equation in standard form of the hyperbola with foci at $(4, 3)$ and $(-4, 3)$ and a transverse axis of length 4.

Example 4   **Completing the Square to Write the Equation of a Hyperbola**

Write the equation in standard form of the hyperbola defined by the equation $x^2 - 4y^2 + 2x - 24y = 39$. Find the vertices and the foci, and sketch the hyperbola.

Solution   First, rewrite the given equation by grouping the x terms and the y terms separately and then factoring -4 out of the y terms.

$$(x^2 + 2x) - 4(y^2 + 6y) = 39$$ Group x and y terms and factor -4 out of y terms

$$(x^2 + 2x + \mathbf{1}) - 4(y^2 + 6y + \mathbf{9}) = 39 + \mathbf{1} - \mathbf{36}$$ Complete the square on both x and y

$$(x + 1)^2 - 4(y + 3)^2 = 4$$ Factor

$$\frac{(x + 1)^2}{4} - \frac{(y + 3)^2}{1} = 1$$ Write in standard form

From the standard form of the equation, we see that the center of the hyperbola is at $(h, k) = (-1, -3)$ and that $a^2 = 4$ and $b^2 = 1$. Thus $a = 2$ and $b = 1$.

Since the $(y - k)^2$ term is subtracted from the $(x - h)^2$ term, the transverse axis is horizontal. Thus the vertices lie on the line $y = -3$.

Vertices: $(h + \mathbf{a}, k) = (-1 + \mathbf{2}, -3) = (1, -3)$ and $(h - \mathbf{a}, k) = (-1 - \mathbf{2}, -3) = (-3, -3)$

To find the foci, we determine c, the distance from the center to either focus.

$$c^2 = a^2 + b^2 = 4 + 1 = 5 \Rightarrow c = \sqrt{5}$$

Foci: $(h + \mathbf{c}, k) = \left(-1 + \sqrt{\mathbf{5}}, -3\right)$ and $(h - \mathbf{c}, k) = \left(-1 - \sqrt{\mathbf{5}}, -3\right)$

Just in Time
Review completing the square in Section 1.5.

The asymptotes are

$$y = \frac{b}{a}(x - h) + k = \frac{1}{2}(x + 1) - 3 \quad \text{and} \quad y = -\frac{b}{a}(x - h) + k = -\frac{1}{2}(x + 1) - 3$$

To sketch the hyperbola, first plot and label the center of the hyperbola. Then draw a rectangular box that is centered at $(h, k) = (-1, -3)$ and has horizontal and vertical dimensions of $2a = 4$ and $2b = 2$, respectively. Graph the asymptotes, each of which is a line that passes through the center of the hyperbola *and* through one pair of opposite corners of the box. Sketch the hyperbola so that it passes through the vertices and approaches the asymptotes. See Figure 11.

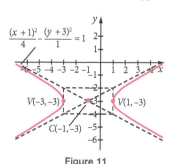

Figure 11

 Check It Out 4   Write the equation in standard form of the hyperbola defined by the equation $x^2 - y^2 + 2x - 4y = 12$.

An Application of a Hyperbola

In Example 5, we examine a sculpture that is designed in the shape of a hyperbola.

Example 5   **Application to Sculpture**

The front face of a wire frame sculpture is in the shape of the branches of a hyperbola that opens to the side. The transverse axis of the hyperbola is 40 inches long. If the base of the sculpture is 60 inches below the transverse axis and one of the asymptotes has a slope of $\frac{3}{2}$, how wide is the sculpture at the base?

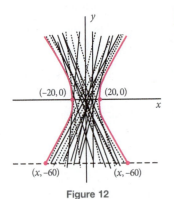

Figure 12

Solution First sketch a hyperbola with transverse axis on the x-axis, centered at (0, 0). See Figure 12. We want to find the x-coordinates of two points on the hyperbola such that the y-coordinate of those two points is -60. The distance between the two points will give the width of the sculpture at the base. Because the length of the transverse axis is **40** inches, the vertices are **(20, 0)** and **(−20, 0)**.

The next step is to find the equation of the hyperbola. Because the transverse axis is horizontal, we have the equation

$$\frac{x^2}{a^2} - \frac{y^2}{b^2} = 1$$

Now we must find a and b. The point (20, 0) is a vertex, so $a = 20$. Using the fact that the slope of an asymptote of the hyperbola is $\frac{3}{2}$, we obtain

$$\frac{b}{a} = \frac{b}{20} = \frac{3}{2} \quad \Rightarrow \quad b = 30$$

Thus the equation of the hyperbola is

$$\frac{x^2}{20^2} - \frac{y^2}{30^2} = 1$$

Now, if the base of the sculpture is 60 inches below the transverse axis, we need to find x such that $(x, -60)$ is on the hyperbola.

$$\frac{x^2}{20^2} - \frac{y^2}{30^2} = 1 \qquad \text{Equation of hyperbola}$$

$$\frac{x^2}{20^2} - \frac{(-60)^2}{30^2} = 1 \qquad \text{Substitute } y = -60$$

$$\frac{x^2}{400} - \frac{3600}{900} = 1 \qquad \text{Simplify}$$

$$\frac{x^2}{400} - 4 = 1$$

$$x^2 - 1600 = 400 \qquad \text{Multiply both sides of equation by 400}$$

$$x^2 = 2000 \quad \Rightarrow \quad x = \pm 20\sqrt{5}$$

The distance between $\left(-20\sqrt{5}, -60\right)$ and $\left(20\sqrt{5}, -60\right)$ is $40\sqrt{5}$. Thus the sculpture is $40\sqrt{5} \approx 89.44$ inches wide at the base.

 Check It Out 5   In Example 5, if the top of the sculpture is 30 inches above the transverse axis, how wide is the sculpture at the top?

Exercises 9.3

 Just in Time Exercises

In Exercises 1–3, find the distance between the given points.

1. (c, d) and (w, v) **2.** $(2, 1)$ and $(-5, -1)$ **3.** $(-3, 5)$ and $(0, 5)$

In Exercises 4–6, complete the square.

4. $y^2 + 8y$ **5.** $x^2 + 22x$ **6.** $x^2 - 2x$

 Skills

This set of exercises will reinforce the skills illustrated in this section.

In Exercises 7–10, match each graph (labeled a, b, c, and d) with the appropriate equation.

a.
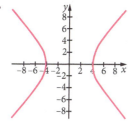

b.
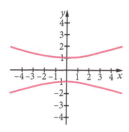

c.
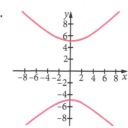

d.
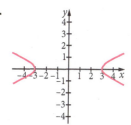

7. $\dfrac{x^2}{9} - y^2 = 1$ **8.** $y^2 - \dfrac{x^2}{9} = 1$ **9.** $y^2 - x^2 = 25$ **10.** $x^2 - y^2 = 16$

In Exercises 11–34, find the center, vertices, foci, and asymptotes of the hyperbola that satisfies the given equation, and sketch the hyperbola.

11. $\dfrac{x^2}{16} - \dfrac{y^2}{9} = 1$ **12.** $\dfrac{y^2}{16} - \dfrac{x^2}{9} = 1$ **13.** $\dfrac{y^2}{9} - \dfrac{x^2}{16} = 1$

14. $\dfrac{4x^2}{9} - \dfrac{9y^2}{16} = 1$ **15.** $\dfrac{y^2}{64} - \dfrac{x^2}{36} = 1$ **16.** $\dfrac{y^2}{4} - \dfrac{x^2}{4} = 1$

17. $x^2 - y^2 = 9$ **18.** $4x^2 - 9y^2 = 36$ **19.** $\dfrac{(x + 3)^2}{16} - \dfrac{(y + 1)^2}{9} = 1$

20. $\dfrac{(x - 6)^2}{9} - \dfrac{(y + 4)^2}{16} = 1$ **21.** $\dfrac{(x + 5)^2}{25} - \dfrac{(y + 1)^2}{16} = 1$ **22.** $\dfrac{(y - 4)^2}{25} - \dfrac{(x - 2)^2}{9} = 1$

23. $y^2 - (x + 4)^2 = 1$ **24.** $\dfrac{x^2}{16} - \dfrac{(y + 5)^2}{4} = 1$ **25.** $\dfrac{(y - 3)^2}{9} - \dfrac{(x + 1)^2}{25} = 1$

26. $\dfrac{(y - 3)^2}{25} - \dfrac{(x + 1)^2}{9} = 1$ **27.** $\dfrac{(x + 2)^2}{144} - \dfrac{(y - 3)^2}{25} = 1$ **28.** $\dfrac{(y - 3)^2}{25} - \dfrac{(x + 2)^2}{144} = 1$

29. $36x^2 - 16y^2 = 225$ **30.** $9y^2 - 16x^2 = 100$ **31.** $9x^2 + 54x - y^2 = 0$

32. $y^2 - 9x^2 - 18x = 18$ **33.** $8x^2 - 32x - y^2 - 6y = 41$ **34.** $4y^2 - 16y - x^2 + 12x = 29$

In Exercises 35–48, determine the equation in standard form of the hyperbola that satisfies the given conditions.

35. Vertices at $(3, 0)$, $(-3, 0)$; foci at $(4, 0)$, $(-4, 0)$

36. Vertices at $(0, 2)$, $(0, -2)$; foci at $(0, 3)$, $(0, -3)$

37. Foci at $(4, 0)$, $(-4, 0)$; asymptotes $y = \pm 2x$

38. Foci at $(0, 5)$, $(0, -5)$; asymptotes $y = \pm x$

39. Foci at $(2, 0)$, $(-2, 0)$; passes through the point $(2, 3)$

40. Foci at $(5, 0)$, $(-5, 0)$; passes through the point $(3, 0)$

41. Vertices at $(-4, 0)$, $(4, 0)$; passes through the point $(8, 2)$

42. Vertices at $(0, 5)$, $(0, -5)$; passes through the point $\left(12, 5\sqrt{2}\right)$

43. Foci at $(-3, -6)$, $(-3, -2)$; slope of one asymptote is 1

44. Foci at $(4, -2)$, $(-2, -2)$; slope of one asymptote is $\dfrac{\sqrt{5}}{2}$

45. Vertices at $(5, -2)$, $(1, -2)$; slope of one asymptote is $\dfrac{5}{2}$

46. Vertices at $(4, 6)$, $(-4, 6)$; slope of one asymptote is -2

47. Transverse axis of length 10; center at $(1, -4)$; one focus at $(9, -4)$

48. Transverse axis of length 6; one vertex at $(7, 8)$; one focus at $(7, 5)$

 In Exercises 49–54, use a graphing utility to graph the given equation.

49. $x^2 - 5y^2 = 10$ **50.** $3y^2 - 2x^2 = 15$ **51.** $\dfrac{x^2}{5} - \dfrac{y^2}{7} = 1$

52. $\dfrac{y^2}{10} - \dfrac{x^2}{12} = 1$ **53.** $\dfrac{(x + 1)^2}{15} - \dfrac{(y - 3)^2}{3} = 1$ **54.** $\dfrac{(y - 4)^2}{8} - \dfrac{x^2}{13} = 1$

In Exercises 55–58, determine the equations in standard form of two different hyperbolas that satisfy the given conditions.

55. Center at $(0, 0)$; transverse axis of length 12; slope of one asymptote is 4.

56. Center at $(-3, -6)$; distance of one vertex from center is 5; distance of one focus from center is 7.

57. Transverse axis of length 6; transverse axis vertical; one vertex at $(-1, 1)$; distance of one focus from nearest vertex is 4.

58. Transverse axis of length 12; transverse axis horizontal; one vertex at $(6, 5)$; slope of one asymptote is -5.

Applications

In this set of exercises, you will use hyperbolas to study real-world problems.

59. Art The front face of a small paperweight sold by a modern art store is in the shape of a hyperbola, and includes both of its branches. The transverse axis of the paperweight has a length of 4 inches and is oriented horizontally. If the base of the paperweight is 2 inches below the transverse axis and one of the asymptotes has a slope of 1, how wide is the paperweight at the base?

60. Physics Because positively-charged particles repel each other, there is a limit on how close a small, positively-charged particle can get to the nucleus of a heavy atom. (A nucleus is positively charged.) As a result, the smaller particle follows a hyperbolic path in the neighborhood of the nucleus. If the asymptotes of the hyperbola have slopes of ± 1, what is the overall change in the direction of the path of the smaller particle as it first approaches the nucleus of the heavy atom and ultimately recedes from it?

61. Astronomy The path of a certain comet is known to be hyperbolic, with the sun at one focus. Assume that a space station is located 13 million miles from the sun and at the center of the hyperbola, and that the comet is 5 million miles from the space station at its point of closest approach. Find the equation of the hyperbola if the coordinate system is set up so that the sun lies on the x-axis and the origin coincides with the center of the hyperbola.

Concepts

This set of exercises will draw on the ideas presented in this section and your general math background.

62. For the hyperbola defined by the equation $25y^2 - 16x^2 = 400$, solve for x in terms of y and then use your expression for x to determine the equations of the asymptotes.

63. Find the domain of the function $f(x) = \sqrt{\dfrac{16x^2 - 144}{9}}$. This is the top half of the hyperbola discussed in Example 1.

64. The hyperbolas $\dfrac{x^2}{a^2} - \dfrac{y^2}{b^2} = 1$ and $\dfrac{y^2}{a^2} - \dfrac{x^2}{b^2} = 1$ are called conjugate hyperbolas. Show that one hyperbola is the reflection of the other in the line $y = x$.

65. Let a, b, and c be as defined in this section. For a rectangle with a length of $2a$ and a width of $2b$, express the length of a diagonal of the rectangle in terms of c alone.

66. What are the slopes of the asymptotes of a hyperbola that satisfies an equation of the form $\dfrac{x^2}{a^2} - \dfrac{y^2}{b^2} = 1$ if $a = b > 0$? At what angle do the asymptotes intersect?

67. Write the equation for the set of points whose distances from the points $(3, 2)$ and $(3, -8)$ differ by 6.

68. The *eccentricity* of a hyperbola $e = \dfrac{c}{a} = \left(\dfrac{\sqrt{a^2 + b^2}}{a} \right)$, where a, b, and c are as defined in this section. Because $0 < a < c$, the value of e is greater than 1. In hyperbolas that are highly curved at the vertices, b is small compared to a, so the eccentricity is close to 1. In hyperbolas that are nearly flat at the vertices, b is much larger than a, so the eccentricity is large. What is the eccentricity of the hyperbola with equation $\dfrac{x^2}{9} - \dfrac{y^2}{16} = 1$? Does this hyperbola have a greater or lesser eccentricity than the hyperbola with equation $\dfrac{x^2}{9} - \dfrac{y^2}{81} = 1$?

69. There are hyperbolas other than the types studied in this section. For example, some hyperbolas satisfy an equation of the form $xy = c$, where c is a nonzero constant. In which quadrant(s) of the coordinate plane does the hyperbola with equation $xy = 10$ lie? The hyperbola with equation $xy = -10$?

Rotation of Axes; General Form of Conic Sections

Objectives

- Use the change-of-coordinates formula.

- Identify conics from the general equation.

- Write and graph a conic equation in standard form using rotation of axes.

In Sections 9.1–9.3, we studied conic sections whose axes were parallel to the x- or y-axis. We continue our study of conics with ones whose axes are rotated. For instance, Figure 1 shows the graph of an ellipse whose major and minor axes are rotated by an angle of 30° counterclockwise. To study these types of rotated conics, we first introduce the idea of rotation of a coordinate system.

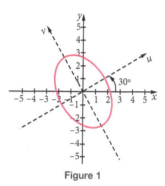

Figure 1

Change of Coordinates by Rotating Axes

We discuss how points are represented in a coordinate system that is a rotation of the x- and y-axes about the origin. Consider a point P with coordinates (x, y). We want to write (x, y) in terms of its coordinates (u, v) in the uv-coordinate system, obtained by rotating the x- and y-axes about the origin by a fixed angle θ. From Figure 2, we have

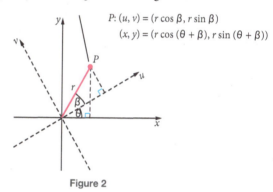

$$P: (u, v) = (r \cos \beta, r \sin \beta)$$
$$(x, y) = (r \cos (\theta + \beta), r \sin (\theta + \beta))$$

Figure 2

$$x = r \cos(\beta + \theta) \qquad y = r \sin(\beta + \theta)$$

where r is the distance of P from the origin. Using the sum identity for cosine yields

$$x = r \cos \beta \cos \theta - r \sin \beta \sin \theta$$
$$= u \cos \theta - v \sin \theta \qquad \qquad u = r \cos \beta, v = r \sin \beta$$

Similarly,

$$y = r \sin \beta \cos \theta + r \sin \theta \cos \beta \qquad \text{Use sum identity for sine}$$
$$= v \cos \theta + u \sin \theta \qquad \qquad u = r \cos \beta, v = r \sin \beta$$
$$= u \sin \theta + v \cos \theta$$

We can derive similar formulas for u and v in terms of x and y. Thus we have the following change-of-coordinates formula.

Change of Coordinates

Suppose that the xy-coordinate system is rotated by a positive, acute angle θ about the origin to produce a new uv-coordinate system. See Figure 3.

- Given the point (u, v), it can be written in the xy-coordinate system as follows:

$$x = u \cos \theta - v \sin \theta \qquad (1)$$
$$y = u \sin \theta + v \cos \theta$$

- Given the point (x, y), it can be written in the uv-coordinate system as follows:

$$u = x \cos \theta + y \sin \theta \qquad (2)$$
$$v = -x \sin \theta + y \cos \theta$$

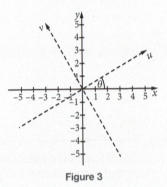

Figure 3

| Example 1 | **Change of Coordinates** |

Let a uv-coordinate system be obtained by rotating the x- and y-axes by an angle of $\theta = \frac{\pi}{3}$.

a. Express the point $(u, v) = (1, 0)$ in the xy-coordinate system.

b. Express the point $(x, y) = (1, 2)$ in the uv-coordinate system.

Solution

a. Using Formula (1), with $u = 1$, $v = 0$, and $\theta = \frac{\pi}{3}$, we get

$$x = u \cos \theta - v \sin \theta \qquad\qquad y = u \sin \theta + v \cos \theta$$

$$= (1) \cos \frac{\pi}{3} - (0) \sin \frac{\pi}{3} \qquad\qquad = (1) \sin \frac{\pi}{3} + (0) \cos \frac{\pi}{3}$$

$$= \frac{1}{2} \qquad\qquad\qquad\qquad\qquad = \frac{\sqrt{3}}{2}$$

Thus $(1, 0)$ in uv coordinates corresponds to $\left(\frac{1}{2}, \frac{\sqrt{3}}{2}\right)$ in xy coordinates.

b. Using Formula (2), with $x = 1$, $y = 2$, and $\theta = \frac{\pi}{3}$, we get

$$u = x \cos \theta + y \sin \theta \qquad\qquad v = -x \sin \theta + y \cos \theta$$

$$= (1) \cos \frac{\pi}{3} + (2) \sin \frac{\pi}{3} \qquad\qquad = -(1) \sin \frac{\pi}{3} + (2) \left(\cos \frac{\pi}{3}\right)$$

$$= \frac{1}{2} + 2\left(\frac{\sqrt{3}}{2}\right) \qquad\qquad\qquad = -\frac{\sqrt{3}}{2} + 2\left(\frac{1}{2}\right)$$

$$= \frac{1 + 2\sqrt{3}}{2} \qquad\qquad\qquad\qquad = \frac{2 - \sqrt{3}}{2}$$

Thus $(1, 2)$ in xy coordinates corresponds to $\left(\frac{1 + 2\sqrt{3}}{2}, \frac{2 - \sqrt{3}}{2}\right)$ in uv coordinates.

Check It Out 1   Write the xy coordinates corresponding to $(0, 1)$ in the uv coordinate system in Example 1.

General Equation of Conic Sections

In the previous sections of this chapter, we studied conic sections with equations of the form

$$Ax^2 + Cy^2 + Dx + Ey + F = 0$$

As we saw there, the graph of this equation was always an ellipse, a parabola, or a hyperbola, with vertical or horizontal axes. If we introduce an xy term, we get the most general form of this equation:

$$Ax^2 + Bxy + Cy^2 + Dx + Ey + F = 0$$

We now give conditions on the equation that determine the type of conic section.

> ### Identifying Conics from the General Equation
>
> The general equation of a conic section is
>
> $$Ax^2 + Bxy + Cy^2 + Dx + Ey + F = 0$$
>
> with A, B not both equal to zero. The graph of this equation is
> - a parabola if $B^2 - 4AC = 0$.
> - a hyperbola if $B^2 - 4AC > 0$.
> - an ellipse if $B^2 - 4AC < 0$.

Example 2   **Classifying a Conic Section**

Identify the conic section given by the equation $x^2 + 3xy - y^2 + 7x - 4 = 0$.

Solution   Here $A = 1$, $B = 3$, $C = -1$, so

$$B^2 - 4AC = (3)^2 - 4(1)(-1) = 13 > 0$$

Thus the equation is that of a hyperbola.

 Check It Out 2   Identify the conic section given by the equation $2x^2 + 4xy + 2y^2 + 7x - 4 = 0$.

Rotated Conic Sections

We next discuss how to sketch the graph when the equation has an xy term present. It is possible to change the coordinates to remove the xy term, so that they take the form of the equations seen in Sections 9.1–9.3. This change of coordinates can be accomplished by choosing an appropriate angle θ of rotation.

> ### Rotation of Axes to Eliminate an xy Term
>
> The general equation for a conic section,
>
> $$Ax^2 + Bxy + Cy^2 + Dx + Ey + F = 0$$
>
> can be rewritten as
>
> $$A'u^2 + C'v^2 + D'u + E'v + F' = 0$$
>
> by rotating the x- and y-axes through an angle θ, where
>
> $$\cot 2\theta = \frac{A - C}{B}$$
>
> The equation is written in the new variables u and v by making the following substitutions in the original equation.
>
> $$x = u \cos \theta - v \sin \theta \qquad y = u \sin \theta + v \cos \theta$$

Example 3   **Graphing a Rotated Conic**

Identify the conic section given by the equation $xy - 2 = 0$. Write the equation in the uv-coordinate system so that there is no uv term in the equation, and sketch its graph.

Solution Because $B^2 - 4AC = 1^2 - 4(0)(0) = 1$, the conic section is a hyperbola. It has an xy term, so it is a rotated conic. In order to express the equation in a form similar to $\frac{x^2}{a^2} - \frac{y^2}{b^2} = 1$, we first find the angle of rotation of the axes:

$$\cot 2\theta = \frac{A - C}{B} = 0 \Rightarrow \cot 2\theta = 0$$

Thus

$$2\theta = \frac{\pi}{2} \Rightarrow \theta = \frac{\pi}{4}$$

We then have

$$x = u \cos \frac{\pi}{4} - v \sin \frac{\pi}{4} \qquad\qquad y = u \sin \frac{\pi}{4} + v \cos \frac{\pi}{4}$$

$$= u\left(\frac{\sqrt{2}}{2}\right) - v\left(\frac{\sqrt{2}}{2}\right) \qquad\qquad = u\left(\frac{\sqrt{2}}{2}\right) + v\left(\frac{\sqrt{2}}{2}\right)$$

$$= \frac{\sqrt{2}}{2}(u - v) \qquad\qquad\qquad = \frac{\sqrt{2}}{2}(u + v)$$

Substituting into the equation $xy - 2 = 0$ yields

$$\left(\frac{\sqrt{2}}{2}(u - v)\right)\left(\frac{\sqrt{2}}{2}(u + v)\right) - 2 = 0$$

$$\frac{1}{2}(u^2 - v^2) - 2 = 0$$

$$\frac{1}{2}u^2 - \frac{1}{2}v^2 = 2 \qquad\text{Move constant to right side}$$

$$\frac{u^2}{4} - \frac{v^2}{4} = 1 \qquad\text{Divide by 2 on both sides}$$

This is a hyperbola in the uv plane with asymptotes $v = u$ and $v = -u$ and with vertices $(u_1, v_1) = (2, 0)$ and $(u_2, v_2) = (-2, 0)$. To obtain the vertices in terms of x and y, use the substitution

$$x = \frac{\sqrt{2}}{2}(u - v) \text{ and } y = \frac{\sqrt{2}}{2}(u + v)$$

to get

Vertex 1: $x = \dfrac{\sqrt{2}}{2}(2 - 0) = \sqrt{2}, \qquad y = \dfrac{\sqrt{2}}{2}(2 + 0) = \sqrt{2}$

Vertex 2: $x = \dfrac{\sqrt{2}}{2}(-2 - 0) = -\sqrt{2}, \qquad y = \dfrac{\sqrt{2}}{2}(-2 + 0) = -\sqrt{2}$

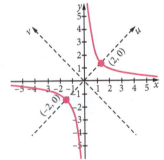

Figure 4

The vertices are $\left(\sqrt{2}, \sqrt{2}\right)$ and $\left(-\sqrt{2}, -\sqrt{2}\right)$ in the xy-coordinate system. See Figure 4.

Check It Out 3   Rework Example 3 for the equation $xy = 1$.

Example 4   **Graphing a Rotated Conic**

Identify the conic section given by the equation $13x^2 - 6\sqrt{3}xy + 7y^2 - 16 = 0$. Write the equation in the uv-coordinate system so that there is no uv term in the equation, and sketch its graph.

Solution Because $B^2 - 4AC = 108 - (4)(13)(7) < 0$, the conic is an ellipse. It has an xy term, so by rotation of the axes,

$$\cot 2\theta = \frac{A - C}{B} = \frac{13 - 7}{-6\sqrt{3}} = \frac{1}{-\sqrt{3}} = \frac{-\sqrt{3}}{3}$$

Thus $2\theta = \frac{2\pi}{3} \Rightarrow \theta = \frac{\pi}{3}$. We then have

$$x = u\cos\frac{\pi}{3} - v\sin\frac{\pi}{3} \qquad\qquad y = u\sin\frac{\pi}{3} + v\cos\frac{\pi}{3}$$

$$= \frac{1}{2}u - \frac{\sqrt{3}}{2}v \qquad\qquad\qquad = \frac{\sqrt{3}}{2}u + \frac{1}{2}v$$

Substituting into the original equation, $13x^2 - 6\sqrt{3}xy + 7y^2 - 16 = 0$, we get

$$13x^2 = \frac{13}{4}u^2 - \frac{13}{2}uv\sqrt{3} + \frac{39}{4}v^2$$

$$-6\sqrt{3}xy = \frac{-9}{2}u^2 + 3uv\sqrt{3} + \frac{9}{2}v^2$$

$$7y^2 = \frac{21}{4}u^2 + \frac{7}{2}uv\sqrt{3} + \frac{7}{4}v^2$$

Thus the original equation becomes

$$4u^2 + 16v^2 - 16 = 0$$

In standard form, it is written as

$$\frac{u^2}{4} + \frac{v^2}{1} = 1$$

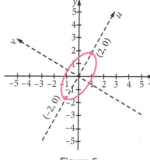

Figure 5

In the uv-coordinate system, the endpoints of the minor axis are $(u, v) = (0, -1)$ and $(u, v) = (0, 1)$, and the vertices are $(u, v) = (-2, 0)$ and $(u, v) = (2, 0)$. Substituting in the expression for x and y, we have the following points in the xy-coordinate system:

$$(x, y) = \left(\frac{\sqrt{3}}{2}, \frac{-1}{2}\right); (x, y) = \left(\frac{-\sqrt{3}}{2}, \frac{1}{2}\right); (x, y) = (-1, -\sqrt{3}); (x, y) = (1, \sqrt{3})$$

The conic section is shown in Figure 5.

Check It Out 4 Rework Example 4 for the equation $7x^2 + 6\sqrt{3}xy + 13y^2 - 16 = 0$.

Example 5   **Graphing a Rotated Conic**

Identify the conic section given by the equation

$$2x^2 + 4xy + 2y^2 + \frac{\sqrt{2}}{2}x - \frac{\sqrt{2}}{2}y = 0$$

Write the equation in the uv-coordinate system so that there is no uv term in the equation, and sketch its graph.

Solution Because $B^2 - 4AC = 16 - 4(2)(2) = 0$, the conic section is a parabola. It has an xy term, so it is a rotated conic. To find the angle of rotation of the axes, we use

$$\cot 2\theta = \frac{A - C}{B} = 0 \quad\Rightarrow\quad \cot 2\theta = 0$$

Thus $2\theta = \frac{\pi}{2} \Rightarrow \theta = \frac{\pi}{4}$. We then have

$$x = u\cos\frac{\pi}{4} - v\sin\frac{\pi}{4} \qquad\qquad y = u\sin\frac{\pi}{4} + v\cos\frac{\pi}{4}$$

$$= \frac{\sqrt{2}}{2}(u - v) \qquad\qquad\qquad = \frac{\sqrt{2}}{2}(u + v)$$

Substituting into the original equation,

$$2x^2 + 4xy + 2y^2 + \frac{\sqrt{2}}{2}x - \frac{\sqrt{2}}{2}y = 0$$

we get

$$2x^2 = u^2 - 2uv + v^2$$

$$4xy = 2u^2 - 2v^2$$

$$2y^2 = u^2 + 2uv + v^2$$

$$\frac{\sqrt{2}}{2}x = \frac{1}{2}(u - v)$$

and

$$\frac{-\sqrt{2}}{2}y = -\frac{1}{2}(u + v)$$

Thus, the original equation becomes $4u^2 - v = 0$. In the uv-coordinate system, the vertex of the parabola is $(0, 0)$, which has the v-axis as the axis of symmetry. See Figure 6.

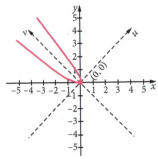

Figure 6

✓ *Check It Out 5*   Rework Example 5 for the equation $x^2 + 2x\sqrt{3}y + 3y^2 + 8\sqrt{3}x - 8y = 0$. ■

Exercises 9.4

 Skills

In Exercises 1–8, the coordinates of a point in the uv-coordinate system are given. Find the coordinates of the point in the xy-coordinate system, which has been rotated θ radians from the xy-coordinate system.

1. $(1, 0), \theta = \dfrac{\pi}{4}$ **2.** $(0, -1), \theta = \dfrac{\pi}{4}$ **3.** $(1, 2), \theta = \dfrac{\pi}{3}$ **4.** $(-3, 1), \theta = \dfrac{\pi}{3}$

5. $(-2, 4), \theta = \dfrac{\pi}{6}$ **6.** $(-1, -2), \theta = \dfrac{\pi}{6}$ **7.** $(-1, -1), \theta = \dfrac{\pi}{4}$ **8.** $(2, 1), \theta = \dfrac{\pi}{4}$

In Exercises 9–16, identify the conic section given by each of the equations by using the general form of the conic equations.

9. $3x^2 + 2x - 4y = 0$ **10.** $-y^2 + 2y + x = 0$ **11.** $x^2 - 2x + y^2 - 1 = 0$

12. $3x^2 - 2y^2 - 2 = 0$ **13.** $2x^2 - y^2 + 2xy + 3 = 0$ **14.** $x^2 + xy + 2y^2 + 4y = 0$

15. $x^2 - 2xy + y^2 + 3y = 1$ **16.** $3x^2 + 6xy - y^2 + 6x + 4 = 0$

In Exercises 17–24, find the vertex, or vertices, of the given conic sections in the uv-coordinate system, obtained by rotating the x- and y-axes by $\theta = \dfrac{\pi}{6}$.

17. $v^2 = 16u$ **18.** $v^2 = -8u$ **19.** $\dfrac{u^2}{1} + \dfrac{v^2}{4} = 1$ **20.** $\dfrac{u^2}{16} + \dfrac{v^2}{9} = 1$

21. $\dfrac{u^2}{9} - \dfrac{v^2}{4} = 1$ **22.** $\dfrac{u^2}{1} - \dfrac{v^2}{16} = 1$ **23.** $u^2 - 9v^2 = 36$ **24.** $4u^2 + v^2 = 36$

For each of the equations in Exercises 25–34:

  a. _Identify the conic section._
  b. _Write the equation in the uv-coordinate system so that there is no uv term in the equation._
  c. _Find the vertex, or vertices, in the uv-coordinate system, and graph._

25. $xy = 4$ **26.** $xy = 3$

27. $3x^2 + 10xy + 3y^2 - 8 = 0$ **28.** $-4x^2 + 10xy - 4y^2 - 9 = 0$

29. $5x^2 - 8xy + 5y^2 - 9 = 0$ **30.** $5x^2 + 6xy + 5y^2 - 8 = 0$

31. $x^2 - 10\sqrt{3}xy + 11y^2 - 16 = 0$ **32.** $3x^2 - 4\sqrt{3}xy + 7y^2 - 9 = 0$

33. $x^2 - 2\sqrt{3}xy + 3y^2 + \dfrac{\sqrt{3}}{2}x + \dfrac{1}{2}y = 0$ **34.** $\dfrac{3}{4}x^2 + \dfrac{\sqrt{3}}{2}xy + \dfrac{1}{4}y^2 + 4x - 4\sqrt{3}y = 0$

Concepts

35. Derive Equation (2) of the change-of-coordinates formula.

36. Suppose you are given the equation $Ax^2 + y^2 - 3 = 0$. What values of A will give the equation of a hyperbola? What values of A will give the equation of an ellipse?

37. Suppose you are given the equation $x^2 + xy + Cy^2 + x - 3 = 0$. What value of C will give the equation of a parabola?

38. Without trying to graph, show that there are _no_ real number values x and y such that $x^2 + 2y^2 + 1 = 0$.

Polar Equations of Conic Sections

Objectives

- Identify a conic from its equation.
- Graph a conic given in polar form.

Note The ratio e here denotes eccentricity and should not be confused with the number $e = 2.718 \ldots$.

Recall that the definition of a parabola was given in terms of a line (the **directrix**) and a point not on the line (a **focus**). It is also possible to give a single definition of all three conic sections in terms of a directrix and a focus. We will use this definition to find the equations of conic sections in polar coordinates. The advantage of this approach is that it enables us to unify the derivation of the equations of all three conic sections.

Focus-Directrix Definition of a Conic Section

Let L be a fixed line, and let F denote a fixed point not on the line L. Let e be a given positive real number. The set of all points P such that

$$\frac{\text{dist}(P, F)}{\text{dist}(P, L)} = e$$

where $\text{dist}(P, F)$ is the distance from P to F and $\text{dist}(P, L)$ is the perpendicular distance from P to the line L, is a conic with a focus at F.

- If $0 < e < 1$, the conic is an **ellipse**.
- If $e = 1$, the conic is a **parabola**.
- If $e > 1$, the conic is a **hyperbola**.

The ratio e is known as the **eccentricity** of the conic.

Just in Time
Review polar equations and coordinates in Sections 7.3 and 7.4.

Polar Equation of a Conic Using the focus-directrix definition, we now derive the polar equation of a conic whose directrix is the vertical line lying to the left of the pole and whose focus is at the pole. Refer to Figure 1. From the definition of eccentricity,

$$\frac{\text{dist}(P, F)}{\text{dist}(P, L)} = e$$

$$\frac{r}{\text{dist}(P, L)} = e$$

$$\frac{r}{d + \text{dist}(F, T)} = e \qquad \text{dist}(P, L) = d + \text{dist}(F, T)$$

$$\frac{r}{d + r \cos \theta} = e \qquad \text{dist}(F, T) = r \cos \theta$$

$$r = e(d + r \cos \theta)$$

$$r = ed + er \cos \theta$$

$$r - er \cos \theta = ed \qquad \text{Combine like terms}$$

$$r(1 - e \cos \theta) = ed \qquad \text{Factor out } r$$

$$r = \frac{ed}{1 - e \cos \theta} \qquad \text{Divide by } 1 - e \cos \theta$$

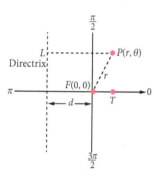

Figure 1

Other cases can be derived similarly. The polar equations of conic sections are given in Table 1.

Equation	$0 < e < 1$	$e = 1$	$e > 1$
$r = \dfrac{ed}{1 - e\cos\theta}$	Ellipse with vertices at $\theta = 0$ and $\theta = \pi$	Parabola opening to the right	Hyperbola with horizontal transverse axis
$r = \dfrac{ed}{1 + e\cos\theta}$	Ellipse with vertices at $\theta = 0$ and $\theta = \pi$	Parabola opening to the left	Hyperbola with horizontal transverse axis
$r = \dfrac{ed}{1 - e\sin\theta}$	Ellipse with vertices at $\theta = \frac{\pi}{2}$ and $\theta = \frac{3\pi}{2}$	Parabola that opens upward	Hyperbola with vertical transverse axis
$r = \dfrac{ed}{1 + e\sin\theta}$	Ellipse with vertices at $\theta = \frac{\pi}{2}$ and $\theta = \frac{3\pi}{2}$	Parabola that opens downward	Hyperbola with vertical transverse axis

Table 1 Polar Equations of Conics

Figures 2, 3, and 4 show some different types of conics in polar coordinates.

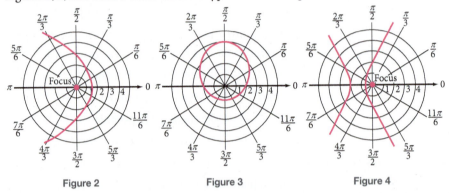

Figure 2   Figure 3   Figure 4

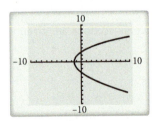
Using Technology

When graphing polar equations of conics, the calculator should be in POLAR and RADIAN modes. See Figure 5.

Keystroke Appendix:
Section 15

Example 1   **Identifying and Graphing the Polar Equation of a Conic**

Identify and graph the conic section given by the equation $r = \dfrac{3}{1 - \cos\theta}$.

Solution

Step 1 First, determine the eccentricity to identify the conic. Because the equation is already in one of the forms given in Table 1, the coefficient of $\cos\theta$ is $e = 1$. Thus the conic is a parabola opening to the right.

Step 2 Now determine the coordinates of the vertex. Because the vertex of the parabola lies on the ray $\theta = \pi$, set $\theta = \pi$ to find r. Thus

$$r = \frac{3}{1 - \cos\pi} = \frac{3}{1 - (-1)} = \frac{3}{2}$$

and $\left(\frac{3}{2}, \pi\right)$ is the vertex.

Step 3 Find other points on the parabola. Setting $\theta = \frac{\pi}{2}$ gives

$$r = \frac{3}{1 - \cos\left(\frac{\pi}{2}\right)} = \frac{3}{1 - 0} = 3$$

Thus $\left(3, \frac{\pi}{2}\right)$ is one point. Similarly, setting $\theta = \frac{3\pi}{2}$ gives $\left(3, \frac{3\pi}{2}\right)$ as the third point.

Step 4 Graph the conic as shown in Figure 6.

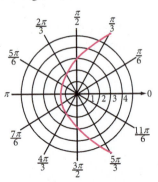

Figure 6

Figure 5

✓ *Check It Out 1*   Identify and graph the conic section given by the equation $r = \dfrac{2}{1 + \cos\theta}$. ■

Example 2   **Identifying and Graphing the Polar Equation of a Conic**

Identify and graph the conic section given by the equation $r = \dfrac{4}{2 - \sin\theta}$.

Solution

Step 1 First write the equation in one of the forms in Table 1. Comparing with those equations, we see that the denominator must be put in the form $1 - e\sin\theta$. Thus, dividing the numerator and denominator by 2 gives

$$r = \frac{2}{1 - \dfrac{1}{2}\sin\theta}$$

Step 2 Determine the eccentricity to identify the conic. Because $e = \dfrac{1}{2} < 1$, the equation is that of an ellipse. Also, from Table 1, its major axis is along $\theta = \dfrac{\pi}{2}$.

Step 3 Find the vertices. The ellipse has its vertices at $\theta = \dfrac{\pi}{2}$ and $\theta = \dfrac{3\pi}{2}$. Substituting into the equation gives

$$r = \frac{4}{2 - \sin\left(\dfrac{\pi}{2}\right)} = \frac{4}{2 - 1} = 4$$

Thus one vertex is $\left(4, \dfrac{\pi}{2}\right)$. We used the original equation only to avoid arithmetic with fractions. Similarly, the second vertex is given by $\left(\dfrac{4}{3}, \dfrac{3\pi}{2}\right)$.

Step 4 We will need additional points to graph the ellipse. The easiest points to find are those where the ellipse intersects the horizontal axis. This occurs at $\theta = 0$ and $\theta = \pi$. Setting $\theta = 0$ gives

$$r = \frac{4}{2 - \sin 0} = 2$$

Setting $\theta = \pi$ gives

$$r = \frac{4}{2 - \sin\pi} = 2$$

Thus two additional points are $(2, 0)$ and $(2, \pi)$.

Step 5 Use the information above to sketch the ellipse as in Figure 7.

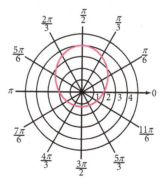

Figure 7

✓ *Check It Out 2*   Identify and graph the conic section given by the equation $r = \dfrac{2}{2 + \cos\theta}$. ■

> **Example 3**   **Identifying and Graphing the Polar Equation of a Conic**

Identify and graph the conic section given by the equation $r = \frac{9}{3 + 6 \sin \theta}$.

Solution

Step 1 First write the equation in one of the forms in Table 1. Comparing with those equations, we see that the denominator must be put in the form $1 + e \sin \theta$. Thus, dividing the numerator and denominator by 3 gives

$$r = \frac{3}{1 + 2 \sin \theta}$$

Step 2 Determine the eccentricity to identify the conic. Because $e = 2 > 1$, the equation is that of a hyperbola. Also, from Table 1, its transverse axis is vertical.

Step 3 Find the vertices. The hyperbola has its vertices at $\theta = \frac{\pi}{2}$ and $\theta = \frac{3\pi}{2}$. Substituting into the equation gives

$$r = \frac{3}{1 + 2 \sin \left(\dfrac{\pi}{2} \right)} = \frac{3}{1 + 2} = 1$$

Thus one vertex is $\left(1, \frac{\pi}{2} \right)$. Similarly, the second vertex is given by $\left(-3, \frac{3\pi}{2} \right)$.

Step 4 We will need additional points to graph the hyperbola. The easiest points to find are those where the hyperbola intersects the horizontal axis. This occurs at $\theta = 0$ and $\theta = \pi$. Setting $\theta = 0$ gives

$$r = \frac{3}{1 + 2 \sin 0} = 3$$

Setting $\theta = \pi$ gives

$$r = \frac{3}{1 + 2 \sin \pi} = 3$$

Thus two additional points are $(3, 0)$ and $(3, \pi)$. These points are not sufficient to graph the upper branch of the hyperbola. Plot additional points by choosing $\theta = \frac{4\pi}{3}$ and $\theta = \frac{5\pi}{3}$.

θ	$\dfrac{4\pi}{3}$	$\dfrac{5\pi}{3}$
r	-4.098	-4.098

Step 5 Use the information to sketch the hyperbola as in Figure 8.

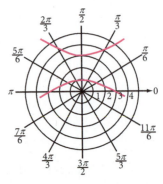

Figure 8

✓ *Check It Out 3* Identify and graph the conic section given by the equation $r = \frac{4}{2 - 4 \sin \theta}$. ■

Exercises 9.5

Just in Time Exercises

1. True or False: $(2, \pi)$ and $(-2, 0)$ represent the same point.

2. True or False: $\left(3, \frac{\pi}{2}\right)$ and $\left(3, \frac{3\pi}{2}\right)$ represent the same point.

3. The graph of $r = 5$ is a _____.   4. The graph of $\theta = \pi$ is a _____.

Skills

In Exercises 5–12, identify the conic section given by each of the equations.

5. $r = \dfrac{3}{1 + 6 \sin \theta}$ 6. $r = \dfrac{1}{1 + 0.75 \cos \theta}$ 7. $r = \dfrac{2}{1 - \sin \theta}$ 8. $r = \dfrac{1}{1 + 2 \sin \theta}$

9. $r = \dfrac{6}{3 - 3 \sin \theta}$ 10. $r = \dfrac{8}{8 - 4 \sin \theta}$ 11. $r = \dfrac{3}{1 - 0.6 \sin \theta}$ 12. $r = \dfrac{12}{4 - 8 \cos \theta}$

In Exercises 13–26, identify and graph the conic section given by each of the equations.

13. $r = \dfrac{1}{1 - \sin \theta}$ 14. $r = \dfrac{1}{1 + \sin \theta}$ 15. $r = \dfrac{4}{1 + 2 \cos \theta}$ 16. $r = \dfrac{6}{1 - 2 \sin \theta}$

17. $r = \dfrac{1}{1 - 0.5 \cos \theta}$ 18. $r = \dfrac{2}{1 + 0.5 \sin \theta}$ 19. $r = \dfrac{6}{6 + 3 \sin \theta}$ 20. $r = \dfrac{16}{8 - 4 \cos \theta}$

21. $r = \dfrac{18}{6 + 12 \cos \theta}$ 22. $r = \dfrac{20}{5 + 10 \sin \theta}$ 23. $r = \dfrac{6}{6 - 8 \sin \theta}$ 24. $r = \dfrac{4}{2 + 5 \cos \theta}$

25. $r = \dfrac{4}{3 - 3 \cos \theta}$ 26. $r = \dfrac{15}{5 + 2 \sin \theta}$

In Exercises 27–32, use an appropriate identity to transform each of the equations to one of the forms shown in Table 1. Then identify the conic section.

27. $r = \dfrac{1}{1 - \sin\left(\theta - \dfrac{\pi}{2}\right)}$ 28. $r = \dfrac{1}{2 + \cos\left(\theta + \dfrac{\pi}{2}\right)}$ 29. $r = \dfrac{1}{3 + \sin(\theta + \pi)}$

30. $r = \dfrac{3}{1 + 3 \cos\left(\theta - \dfrac{\pi}{2}\right)}$ 31. $r = \dfrac{4}{1 + 4 \cos(\theta + \pi)}$ 32. $r = \dfrac{1}{1 - \sin\left(\theta + \dfrac{\pi}{2}\right)}$

In Exercises 33–36, graph each equation using a graphing utility.

33. $r = \dfrac{\sqrt{2}}{1 + \sin \theta}$ 34. $r = \dfrac{5.6}{1 + 0.7 \sin \theta}$

35. $r = \dfrac{5}{1 - 3 \cos\left(\theta + \dfrac{\pi}{6}\right)}$

How is this graph different from the ones studied in this section?

36. $r = \dfrac{1}{1 + 0.7 \sin\left(\theta + \dfrac{\pi}{4}\right)}$

How is this graph different from the ones studied in this section?

Concepts

37. Find the equation of a parabola in polar form whose focus is at the pole and whose vertex is $(1, \pi)$.

38. Find the equation of an ellipse in polar form that has eccentricity 0.5 and a directrix with equation $x = 2$.

39. Find the equation of an ellipse in polar form that has eccentricity 0.3 and a directrix with equation $y = -3$.

 40. Graph the following equations on the same screen. What do you observe as e gets close to 0?

a. $r = \dfrac{1}{1 + 0.4 \sin \theta}$ **b.** $r = \dfrac{1}{1 + 0.2 \sin \theta}$ **c.** $r = \dfrac{1}{1 + 0.1 \sin \theta}$ **d.** $r = \dfrac{1}{1 + 0.01 \sin \theta}$

9.6 Parametric Equations

Objectives

- Graph parametric equations.
- Convert parametric equations to rectangular form.
- Solve applied problems using parametric equations.

t	0	1	2	3
x	0	2	4	6
y	0	3	6	9

Table 1

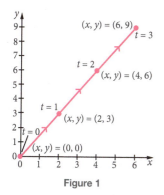

Figure 1

In this section, we examine equations where the x and y coordinates of a set of points are described in terms of another variable, called a *parameter*. That is, the x- and y-coordinates are given by **parametric equations.** This representation is especially useful when representing moving objects, such as satellites, aircraft carriers, or even subatomic particles. The reason is that parametric equations take into account not only an object's position but also the direction of its motion. To see an application of parametric equations, we can examine the motion of a remotely controlled vehicle, as in Example 1.

Example 1   **Motion of a Vehicle**

An engineer tracking the motion of a remotely controlled vehicle creates a data table as follows. For each value of time t, she records the x- and y-coordinates of the vehicle relative to a fixed coordinate system. Her data is given in Table 1. Plot just the x- and y- values as coordinate points in the xy-plane. Indicate the corresponding value of t. As t increases, show the direction in which the vehicle is moving.

Solution   The points are plotted in Figure 1. From the graph, the vehicle appears to be moving in a straight line. Note that in addition to the xy-coordinates, there is also information in the graph about the *direction* of the vehicle. This extra information, coming from the variable t, is not available when we graph a function of the form $y = f(x)$.

If it is known that x and y are linear functions of t, the x-coordinate can be written as a function of t—in this case, as $x = 2t$. Similarly, the y-coordinate can be written as a function of t as $y = 3t$.

 Check It Out 1   If the vehicle continues in the same direction, what will its xy-coordinates be at $t = 4$?

In Example 1, we saw that the x and y coordinates describing the motion of a vehicle could be separately described using a variable t. Expressing x and y in terms of t is referred to as a **parametric representation** of x and y. The variable t is called a **parameter.** We now give the formal definition of parametric equations.

Parametric Equations

Let f and g be continuous functions of t on an interval $[t_1, t_2]$. If

$$x = f(t) \quad \text{and} \quad y = g(t)$$

then the set of all such points (x, y) is called a **plane curve.** The set of equations defining x and y are called **parametric equations.** The variable t is called a **parameter.**

To graph parametric equations, you make a table of values of t, x, and y. Then plot the xy-coordinate pairs, and indicate the direction of increasing t by arrows. We next give examples of parametric equations and their graphs.

Example 2   **Graph of a Set of Parametric Equations**

Let $x = t - 1$ and $y = t^2$ for $-2 \le t \le 3$.

a. Make a table of values for x and y with $t = -2, -1, 0, 1, 2, 3$.

b. Plot the (x, y) points in the table, and sketch a smooth curve through the points. Use arrows to indicate the direction of increasing t.

Solution

a. Table 2 is filled in by substituting the given values of t, one by one, for the expressions for x and y.

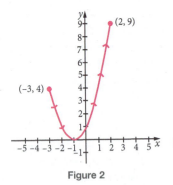

Figure 2

t	-2	-1	0	1	2	3
$x = t - 1$	-3	-2	-1	0	1	2
$y = t^2$	4	1	0	1	4	9

Table 2

b. Now plot each coordinate pair (x, y) in the table, and connect the points with a smooth curve. As t increases, the curve is traced from left to right, as shown in Figure 2. This is indicated by the direction of the arrows.

Note that the curve begins at $(-3, 4)$ and ends at $(2, 9)$ because t is restricted to a specific interval.

 Check It Out 2   Sketch a graph of the curve given by the equations $x = -t$ and $y = t^2$ for $0 \leq t \leq 3$.

Using Technology

Parametric equations can be graphed by setting the calculator to PAR mode. To graph the equations in Example 3, enter the equations for $X_{1T} = T - 1$ and $Y_{1T} = T^2$ in the $Y =$ editor and then set Tmin $= -2$ and Tmax $= 3$ in the graphing window. The x and y dimensions of the window are $[-10, 10] \times [-10, 10]$. The graph shows the parabola with x restricted to the interval $[-3, 2]$. See Figure 3.

Keystroke Appendix:
Section 16

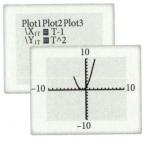

Figure 3

Converting a Parametric Equation to Rectangular Form

To get an idea of the shape of the graph of a set of parametric equations, it is useful to convert them to rectangular form; that is, eliminate the t parameter to get an equation in x and y only. Keep in mind that when we eliminate the parameter t, the information regarding direction and motion is completely lost. Also, if the values of t are constrained, so will the values of x and y be restricted in the resulting equation. The following two examples illustrate these ideas.

Example 3   **Eliminating t in Parametric Equations**

Write the set of parametric equations $x = t - 1$ and $y = t^2$, where $-2 \leq t \leq 3$, in an equivalent rectangular form.

Solution   Solve for t in the equation for x to get $t = x + 1$. Substitute in the equation for y to obtain

$$y = t^2 = (x + 1)^2$$

Because $x = t - 1$ and $-2 \leq t \leq 3$, x is restricted to the interval $[-3, 2]$. Thus $y = (x + 1)^2$, $-3 \leq x \leq 2$, is an equivalent rectangular form. The graph of the set of parametric equations is the portion of the parabola $y = (x + 1)^2$ with x restricted to the interval $[-3, 2]$. See Figure 4.

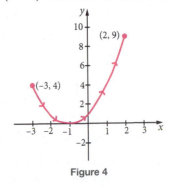

Figure 4

 Check It Out 3   Write the set of parametric equations $x = t - 2$ and $y = t^2$, where $-1 \leq t \leq 2$, in an equivalent rectangular form.

| Example 4 | **Writing Parametric Equations in Rectangular Form** |

Let $x = -t$, $y = \sqrt{4 - t^2}$; $-2 \leq t \leq 2$.

a. Sketch the curve given by this set of parametric equations.

b. Find an equivalent rectangular equation for these parametric equations, and note any restrictions on the variables.

Solution

a. We start by making a table of values of t, x, and y. See Table 3.

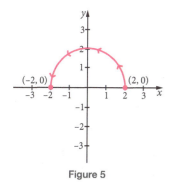

(−2, 0)   (2, 0)

Figure 5

t	−2	−1	0	1	2
$x = -t$	2	1	0	−1	−2
$y = \sqrt{4 - t^2}$	0	$\sqrt{3}$	2	$\sqrt{3}$	0

Table 3

Note A rectangular equation is useful for sketching the graph of a set of parametric equations. However, it may not always be possible to eliminate the parameter t to produce a rectangular equation. Also, the rectangular equation does not give additional information about motion and direction that t provides.

Note that as t increases, a semicircle is traced out counterclockwise, as shown in Figure 5.

b. Because $x = -t$, substitute in the expression for $y = \sqrt{4 - t^2}$ to obtain

$$y = \sqrt{4 - x^2} \qquad \text{Substitute } t = -x$$
$$y^2 = 4 - x^2 \qquad \text{Square both sides}$$
$$x^2 + y^2 = 4 \qquad \text{Add } x^2 \text{ to each side}$$

This is the equation of a circle of radius 2 with center $(0, 0)$.

The rectangular equation does *not* tell you that $y \geq 0$. This information is obtained only from the parametric representation—that is, from $y = \sqrt{4 - t^2} \geq 0$. Thus the complete representation of the parametric equations in rectangular form is

$$x^2 + y^2 = 4, \quad y \geq 0$$

 Check It Out 4 Sketch the curve given by the equations $x = t^2$, $y = \sqrt{2 - t^2}$; $0 \leq t \leq 1$ and find an equivalent rectangular equation.

The functions describing x and y in a set of parametric equations are often trigonometric functions, as illustrated in the next example.

| Example 5 | **Parametric Representation of an Ellipse** |

Sketch the curve given by the parametric equations $x = 2\cos t$; $y = \sin t$; $0 \leq t \leq \pi$. Convert to rectangular coordinates to verify that it is the top half of an ellipse.

Solution Make a table of values for t, x, and y. See Table 4. Plotting these values, we see that the top half of an ellipse is obtained, as shown in Figure 6. The direction of increasing t is counterclockwise.

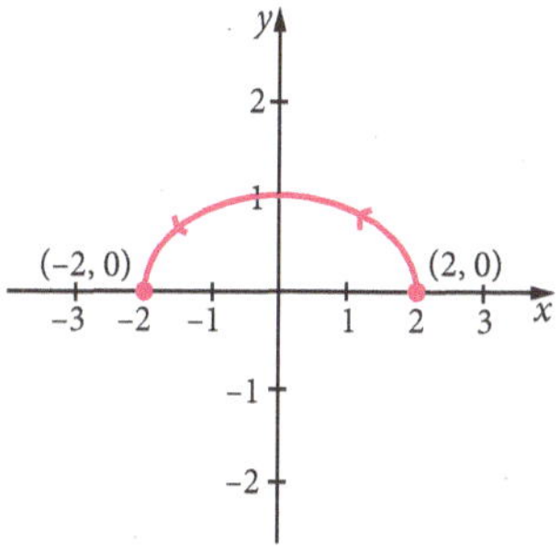

(−2, 0)   (2, 0)

Figure 6

t	0	$\dfrac{\pi}{4}$	$\dfrac{\pi}{2}$	$\dfrac{3\pi}{4}$	π
x	2	$\sqrt{2}$	0	$-\sqrt{2}$	−2
y	0	$\dfrac{\sqrt{2}}{2}$	1	$\dfrac{\sqrt{2}}{2}$	0

Table 4

To verify that this is an ellipse, note that

$$\frac{x}{2} = \cos t \text{ and } y = \sin t$$

Thus

$$\left(\frac{x}{2}\right)^2 + y^2 = (\cos t)^2 + (\sin t)^2 = 1$$

by the Pythagorean identity. The equation of the ellipse in standard rectangular form is

$$\frac{x^2}{4} + \frac{y^2}{1} = 1$$

with $0 \le y \le 1$.

 Check It Out 5 Sketch the curve given by the parametric equations $x = 2 \cos t$; $y = 2 \sin t$; $0 \le t \le \pi$. ■

Applications of Parametric Equations

At the beginning of this section, we discussed the position (x, y) of an automated vehicle at time t by gathering data. Another type of motion that can be modeled by parametric equations is that of an object propelled upward at an angle θ with an initial velocity v_0. This type of motion is called **projectile motion**.

Equations for Projectile Motion

The parametric equations for the path of a projectile are

$$x = v_0(\cos \theta)t \quad \text{and} \quad y = -16t^2 + v_0(\sin \theta)t + h$$

where v_0 is the initial speed in feet per second, θ is the angle of inclination, and h is the initial height of the projectile.

These equations neglect the effect of air resistance and are usually derived in a first year physics course.

Example 6 **Projectile Motion**

A professional tennis player hits a tennis ball with an initial speed of 140 feet per second at a height of 2 feet and an angle of 10° with respect to the horizontal. See Figure 7.

(*Source:* http://www.aroundhawaii.com/speed_training.html)

a. Find the parametric equations that give the position of the ball as a function of time.

b. When is the ball at its maximum height? What is its maximum height?

Figure 7

c. What is the horizontal distance traveled?

Solution

a. From the equations of projectile motion, we have

$$x = v_0 (\cos \theta)t$$
$$= (140)(\cos 10°)t \qquad \text{Use } v_0 = 140 \text{ and } \theta = 10°$$
$$= 137.87t \qquad \text{Answer rounded to two decimal places}$$

$$y = -16t^2 + v_0(\sin \theta)t + h$$
$$= -16t^2 + (140)(\sin 10°)t + 2 \qquad \text{Use } v_0 = 140, \theta = 10° \text{ and } h = 2$$
$$= -16t^2 + 24.31t + 2 \qquad \text{Answer rounded to two decimal places}$$

b. To find the maximum height, first note that the vertical position, given by y, is a quadratic function in t. Thus the ball is at its maximum height at the vertex:

$$t = \frac{-b}{2a} = \frac{-24.31}{2(-16)} \approx 0.76 \text{ second}$$

Using the expression for y, we find that the maximum height is

$$y = -16(0.76)^2 + 24.31(0.76) + 2 = 11.23 \text{ feet}$$

c. First solve the equation $y = 0$ to find out how long the ball is in the air.

$$-16t^2 + 24.31t + 2 = 0 \qquad\qquad \text{Set } y = 0$$

$$t = \frac{-24.31 \pm \sqrt{(24.31)^2 - 4(-16)(2)}}{2(-16)} \qquad \text{Use quadratic formula}$$

$$= \frac{-24.31 \pm \sqrt{24.31^2 + 128}}{-32}$$

$$= 1.60 \text{ seconds}$$

We disregard the negative value of t because it is not realistic. The horizontal distance traveled is

$$x(t) = 137.87(1.60) = 220.592 \text{ feet}$$

Check It Out 6   A golf ball is hit from the ground with an initial speed of 150 feet per second at an angle of 30° with respect to the horizontal. Find the parametric equations that give the position of the ball as a function of time.

Skills

In Exercises 1–10, sketch the graph of the parametric equations. Indicate the direction of increasing t.

1. $x = t, y = t - 2, 0 \leq t \leq 3$

2. $x = 2t, y = -4t, -2 \leq t \leq 2$

3. $x = -t, y = 2t + 3, -1 \leq t \leq 2$

4. $x = t, y = -t + 5, -3 \leq t \leq 3$

5. $x = -t, y = t^2 + 1, -1 \leq t \leq 2$

6. $x = t, y = 3t^2 - 1, -3 \leq t \leq 3$

7. $x = t + 1, y = -2t^2 - 1, -3 \leq t \leq 3$

8. $x = t - 1, y = t^2 + 2, 0 \leq t \leq 4$

9. $x = t^2, y = t, -2 \leq t \leq 2$

10. $x = t^2, y = 2t, 0 \leq t \leq 4$

In Exercises 11–20, eliminate the parameter t to find an equivalent equation with y in terms of x. Give any restrictions on x. Sketch the corresponding graph, indicating the direction of increasing t.

11. $x = t - 2, y = -t^2, -3 \leq t \leq 2$

12. $x = t + 1, y = -2t^2, -3 \leq t \leq 2$

13. $x = e^{-t}, y = e^t, t \geq 0$

14. $x = e^t, y = e^{2t}, t \geq 0$

15. $x = 3 \cos t, y = 3 \sin t, 0 \leq t \leq 2\pi$

16. $x = 4 \cos t, y = 4 \sin t, 0 \leq t \leq 2\pi$

17. $x = 3 \cos t, y = 4 \sin t, 0 \leq t \leq 2\pi$

18. $x = 2 \cos t, y = \sin t, 0 \leq t \leq 2\pi$

19. $x = \cos t, y = \sec t, 0 \leq t < \dfrac{\pi}{2}$

20. $x = \sin t, y = \csc t, 0 < t \leq \dfrac{\pi}{2}$

 In Exercises 21–24, use a graphing utility to graph the parametric equations and answer the given questions.

21. $x = \cos t, y = (\cos t)^2, 0 \leq t \leq \pi$. Graph again for $0 \leq t \leq 2\pi$. What do you observe?

22. $x = t \cos t, y = t \sin t, 0 \leq t \leq 4\pi$. What type of shape is produced?

23. $x = 2(t - \sin t), y = 2(1 - \cos t), 0 \leq t \leq 2\pi$. Will y ever be negative? Explain.

24. $x = 2 \sin t, y = 4 \cos t, 0 \leq t \leq 2\pi$. Is the direction of increasing t clockwise or counterclockwise?

In Exercises 25–28, show that each of the pairs of parametric equations gives the same rectangular representation but different graphs and restrictions on x and/or y.

25. **a.** $x = t, y = t, 0 \leq t \leq 2$

b. $x = t^2, y = t^2, 0 \leq t \leq 2$

26. **a.** $x = \dfrac{1}{t}, y = \dfrac{1}{t^2}, 1 \leq t \leq 2$

b. $x = t - 1, y = (t - 1)^2, 1 \leq t \leq 2$

27. **a.** $x = t + 1, y = t + 2, -2 \leq t \leq 1$

b. $x = t^2, y = t^2 + 1, -2 \leq t \leq 1$

28. **a.** $x = -t, y = -t - 2, -3 \leq t \leq 2$

b. $x = t^2, y = t^2 - 2, -3 \leq t \leq 2$

Applications

29. **Projectile Motion** Sara kicks a soccer ball from the ground with an initial velocity of 120 feet per second at an angle of 30° to the horizontal.

    a. Find the parametric equations that give the position of the ball as a function of time.

    b. When is the ball at its maximum height, to the nearest hundredth of a second? What is its maximum height, to the nearest tenth of a foot?

    c. How far did the ball travel? Round your answer to the nearest foot.

30. **Projectile Motion** Jim kicks a football from the ground with an initial velocity of 140 feet per second at an angle of 45° to the horizontal.

    a. Find the parametric equations that give the position of the ball as a function of time.

    b. When is the ball at its maximum height, to the nearest hundredth of a second? What is its maximum height, to the nearest tenth of a foot?

    c. How far did the ball travel? Round your answer to the nearest foot.

31. **Projectile Motion** Suppose Jacob hits a baseball from a tee that is 3 feet tall. He hits the ball with an initial velocity of 100 feet per second at an angle of 60° to the horizontal. What is the maximum distance the ball will travel? Round your answer to the nearest foot.

32. **Projectile Motion** Suppose Jacob hits a baseball from a tee that is 3 feet tall. He hits the ball with an initial velocity of 120 feet per second at an angle of 50° to the horizontal. What is the maximum distance the ball will travel? Round your answer to the nearest foot.

33. **Leisure** Sara's favorite ride at the fair is the carousel. Her favorite horse to ride makes a path that is a circle of radius 10 feet as the carousel spins around. She gets on the horse when it is at point (10, 0), and the carousel turns in a counterclockwise direction. Find parametric equations of the form $x = a\cos(bt)$ and $y = a\sin(bt)$, with a and b to be determined, that describe her motion if she requires 15 seconds for one complete revolution.

34. **Geometry** In an amusement park, Jason rides a go-cart on an elliptical track. The equation

$$x^2 + \frac{y^2}{16} = 1$$

    may be used to describe the shape of the track.

    a. Find parametric equations of the form $x = a\cos(bt)$ and $y = c\sin(bt)$, with a, b, and c to be determined, if he starts at the point (1, 0), travels in a counterclockwise direction, and requires 4 minutes to make one complete loop.

    b. What are Jason's coordinates at $t = 1$ minute?

35. **Sports** A ball is kicked from the ground at an angle of 30° to the horizontal and lands 350 feet away 4 seconds later. Find the initial velocity of the ball to the nearest whole number.

36. **Sports** A golf ball is hit from the ground at an angle θ with an initial speed of 114 feet per second and lands 300 feet away after 3 seconds. Find the angle θ with the horizontal when the ball was hit. Round to the nearest degree.

37. Video Game Design A video game developer gives a parametric representation of the motion of one of the game's characters, at time t, as

$$x = f(t) \quad \text{and} \quad y = g(t)$$

where the table of values for f and g are as given.

t	0	2	4	6	8
$x = f(t)$	0	2	2	0	0

t	0	2	4	6	8
$y = g(t)$	0	0	2	2	0

Sketch the motion of the game character in the xy-plane, indicating the direction of increasing t. Assume that the path between successive points is a straight line.

38. Robotics A robot has x- and y-coordinates at time t given by the parametric equations

$$x = f(t) \quad \text{and} \quad y = g(t)$$

where the table of values for f and g are as given.

t	0	1	2	3
$x = f(t)$	0	2	1	0

t	0	1	2	3
$y = g(t)$	0	0	2	0

Sketch the motion of the robot in the xy-plane, indicating the direction of increasing t. Assume that the path between successive points is a straight line.

Concepts

39. Graph the parametric equations $x = \frac{1}{t}, y = \frac{1}{t+1}$ for $0 < t \le 10$, indicating direction of increasing t. Will $(0, 0)$ ever be reached? Explain.

40. Explain why the following statement is wrong. "The graph of $x = \frac{1}{t}, y = \frac{2}{t}, t \ne 0$, is the same as that of $y = 2x$ because $y = 2\left(\frac{1}{t}\right) = 2x$."

41. Explain why there is a difference between the graphs of the following sets (a) and (b).

a. $x = t, y = t + 1, 0 \le t \le 1$

b. $x = -t, y = -t + 1, 0 \le t \le 1$

Chapter 9 Summary

Section 9.1 The Parabola

Parabola with vertical axis of symmetry

[Review Exercises 1–18]

Equation	$(x - h)^2 = 4p(y - k)$
Opening	Upward if $p > 0$, downward if $p < 0$
Vertex	(h, k)
Focus	$(h, k + p)$
Directrix	$y = k - p$
Axis of symmetry	$x = h$

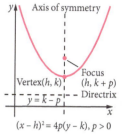

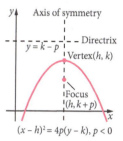

Parabola with horizontal axis of symmetry

[Review Exercises 1–18]

Equation	$(y - k)^2 = 4p(x - h)$
Opening	To the right if $p > 0$, to the left if $p < 0$
Vertex	(h, k)
Focus	$(h + p, k)$
Directrix	$x = h - p$
Axis of symmetry	$y = k$

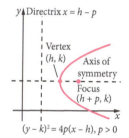

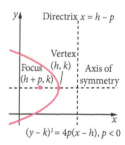

Section 9.2 **The Ellipse**

Ellipse with horizontal major axis   [Review Exercises 19–38]

Equation	$\dfrac{(x-h)^2}{a^2} + \dfrac{(y-k)^2}{b^2} = 1, a > b > 0$
Center	(h, k)
Foci	$(h-c, k), (h+c, k), c = \sqrt{a^2 - b^2}$
Vertices	$(h-a, k)$ and $(h+a, k)$
Major axis	Parallel to x-axis between $(h-a, k)$ and $(h+a, k)$
Minor axis	Parallel to y-axis between $(h, k-b)$ and $(h, k+b)$

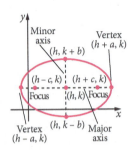

Ellipse with vertical major axis   [Review Exercises 19–38]

Equation	$\dfrac{(x-h)^2}{b^2} + \dfrac{(y-k)^2}{a^2} = 1, a > b > 0$
Center	(h, k)
Foci	$(h, k-c), (h, k+c), c = \sqrt{a^2 - b^2}$
Vertices	$(h, k-a)$ and $(h, k+a)$
Major axis	Parallel to y-axis between $(h, k-a)$ and $(h, k+a)$
Minor axis	Parallel to x-axis between $(h-b, k)$ and $(h+b, k)$

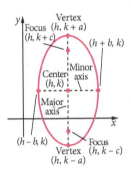

Section 9.3 **The Hyperbola**

Hyperbola with horizontal transverse axis   [Review Exercises 39–54]

Equation	$\dfrac{(x-h)^2}{a^2} - \dfrac{(y-k)^2}{b^2} = 1,$ $a, b > 0$
Center	(h, k)
Foci	$(h-c, k), (h+c, k),$ $c = \sqrt{a^2 + b^2}$
Vertices	$(h-a, k), (h+a, k)$
Transverse axis	Parallel to x-axis between $(h-a, k)$ and $(h+a, k)$
Asymptotes	$y = \dfrac{b}{a}(x-h) + k$ and $y = -\dfrac{b}{a}(x-h) + k$

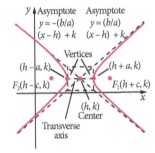

Hyperbola with vertical transverse axis [Review Exercises 39–54]

Equation $$\frac{(y-k)^2}{a^2} - \frac{(x-h)^2}{b^2} = 1,$$

$a, b > 0$

Center (h, k)

Foci $(h, k - c), (h, k + c),$

$c = \sqrt{a^2 + b^2}$

Vertices $(h, k - a), (h, k + a)$

Transverse axis Parallel to y-axis between

$(h, k - a)$ and $(h, k + a)$

Asymptotes $y = \dfrac{a}{b}(x - h) + k$ and

$y = -\dfrac{a}{b}(x - h) + k$

Section 9.4 Rotation of Axes; General Form of Conic Sections

General equation of a conic section [Review Exercises 59–62]

Equation:

$$Ax^2 + Bxy + Cy^2 + Dx + Ey + F = 0$$

The graph of this equation is
- a parabola if $B^2 - 4AC = 0$.
- a hyperbola if $B^2 - 4AC > 0$.
- an ellipse if $B^2 - 4AC < 0$.

Rotation of axes [Review Exercises 55–62]

The general equation for a conic section can be rewritten without a uv term as

$$A'u^2 + C'v^2 + D'u + E'v + F' = 0$$

by rotating the x- and y-axes through an angle θ, where

$$\cot 2\theta = \frac{A - C}{B}$$

Section 9.5 Polar Equations of Conic Sections

Polar equation of a conic section [Review Exercises 63–66]

The polar equation of a conic section with a focus at the pole is given by $r = \dfrac{ed}{1 \pm e \cos \theta}$
or $r = \dfrac{ed}{1 \pm e \sin \theta}$.

- If $0 < e < 1$, the conic is an **ellipse**.
- If $e = 1$, the conic is a **parabola**.
- If $e > 1$, the conic is a **hyperbola**.

The ratio e is known as the **eccentricity** of the conic.

Section 9.6 **Parametric Equations**

Parametric equations   [Review Exercises 67–73]

Let f and g be continuous functions of t on an interval $[t_1, t_2]$. Then

$$x = f(t) \quad \text{and} \quad y = g(t)$$

are **parametric equations** for x and y. The variable t is called a **parameter** and can be eliminated to get an equation in x and y.

Projectile motion   [Review Exercises 67–73]

The parametric equations for **projectile motion** are

$$x = v_0(\cos \theta)t$$

$$y = -16t^2 + v_0(\sin \theta)t + h$$

where v_0 is the initial speed in feet per second, θ is the angle of inclination, and h is the initial height of the projectile.

Chapter 9 Review Exercises

Section 9.1

In Exercises 1–10, find the vertex, focus, and directrix of the parabola. Sketch the parabola.

1. $x^2 = 12y$

2. $(x - 7)^2 = y - 1$

3. $y^2 = 4x$

4. $(y - 4)^2 = -(x + 2)$

5. $x^2 - 6x = y$

6. $x^2 + 2x + 3y = 5$

7. $y^2 + 10y + 5x = 0$

8. $y^2 - 8y + 4 = 4x$ **9.** $7y^2 = x$

 10. $x^2 + 3x - y = 0$

In Exercises 11–16, find the equation in standard form of the parabola satisfying the given conditions. Sketch the parabola.

11. Vertical axis of symmetry; vertex at $(0, 0)$; passes through the point $(-2, -6)$

12. Focus at $(0, 2)$; directrix $y = -2$ **13.** Focus at $(0, -5)$; directrix $y = 5$

14. Horizontal axis of symmetry; vertex at $(0, 0)$; passes through the point $(1, 6)$

15. Vertex at $(2, 0)$; directrix $x = -2$ **16.** Vertex at $(1, -4)$; directrix $x = 6$

17. Gardening The foliage in a planter takes the form of a parabola with a base that is 10 feet across. If the plant at the center is 6 feet tall and the heights of the plants taper off to zero toward each end of the base, how far from the center are the plants that are 4 feet tall?

18. Water Flow As a stream of water goes over the edge of a cliff, it forms a waterfall that takes the shape of one-half of a parabola. The equation of the parabola is $y = -64x^2$, where x and y are measured with respect to the edge of the cliff. How high is the cliff if the water hits the ground below at a horizontal distance of 3 feet from the edge of the cliff?

Section 9.2

In Exercises 19–30, find the vertices and foci of the ellipse given by the equation. Sketch the ellipse.

19. $\dfrac{x^2}{4} + \dfrac{y^2}{1} = 1$

20. $\dfrac{(x - 1)^2}{64} + \dfrac{y^2}{25} = 1$

21. $\dfrac{(x + 2)^2}{25} + \dfrac{(y - 1)^2}{16} = 1$

22. $\dfrac{x^2}{4} + \dfrac{y^2}{9} = 1$

23. $\dfrac{(x + 2)^2}{9} + \dfrac{y^2}{16} = 1$

24. $\dfrac{(x - 1)^2}{4} + \dfrac{(y - 2)^2}{9} = 1$

 25. $3x^2 + y^2 = 7$

 26. $2x^2 + 3y^2 = 10$

27. $x^2 - 4x + 6y^2 - 20 = 0$

28. $4y^2 - 32y + 3x^2 + 12x + 40 = 0$

29. $4y^2 - 24y + 7x^2 + 42x + 71 = 0$

30. $3x^2 + 18x + 2y^2 + 8y + 17 = 0$

In Exercises 31–36, find the equation in standard form of the ellipse centered at (0, 0).

31. One vertex at $(0, 5)$; one focus at $(0, 3)$

32. Major axis of length 8; foci at $(0, -2), (0, 2)$

33. One vertex at $(0, 6)$; one focus at $(0, 4)$

34. One vertex at $(5, 0)$; one focus at $(3, 0)$

35. Major axis of length 10; foci at $(-3, 0), (3, 0)$

36. One vertex at $(-6, 0)$; one focus at $(1, 0)$

37. **Astronomy** The orbit of Pluto around the sun is an ellipse with a major axis of about 80 astronomical units (AU), where 1 AU is approximately 92,600,000 miles. If the distance between the foci is one-fourth the length of the major axis, how long (in AU) is the minor axis?

38. **Sporting Equipment** The head of a tennis racket is in the shape of an ellipse. The ratio of the length of its major axis to the length of its minor axis is $\frac{17}{13}$. If the longer dimension of the head of the racket is 15 inches, how far is each focus from the center?

Section 9.3

In Exercises 39–48, find the center, vertices, foci, and asymptotes for each hyperbola. Sketch the hyperbola.

39. $\dfrac{y^2}{36} - \dfrac{x^2}{25} = 1$ 40. $\dfrac{(y-1)^2}{25} - \dfrac{x^2}{36} = 1$ 41. $\dfrac{(y+2)^2}{4} - \dfrac{(x-1)^2}{4} = 1$

42. $\dfrac{x^2}{9} - \dfrac{y^2}{9} = 1$ 43. $\dfrac{x^2}{4} - \dfrac{(y+2)^2}{9} = 1$ 44. $\dfrac{(x-1)^2}{4} - \dfrac{(y-3)^2}{1} = 1$

45. $16y^2 - x^2 - 4x = 20$ 46. $y^2 + 10y - 4x^2 - 16x + 5 = 0$

 47. $8x^2 - y^2 = 64$ 48. $7x^2 + 42x - 5y^2 - 50y = 307$

In Exercises 49–53, find the standard form of the equation of each hyperbola.

49. Foci at $(0, 6), (0, -6)$; transverse axis of length 4

50. Vertices at $(-2, -6), (-2, 2)$; one focus at $(-2, -9)$

51. Transverse axis of length 6; center at $(3, -4)$; one focus at $(3, 1)$

52. Vertices at $(3, 0), (-3, 0)$; slope of one asymptote is -4

53. Transverse axis of length 8; center at $(-4, 2)$; one focus at $(2, 2)$

54. **Landscaping** The borders of the plantings in the two sections of a mathematician's garden form the branches of a hyperbola whose vertices are 16 feet apart. There are only two rosebushes in the garden, one at each focus of the hyperbola. If the transverse axis is horizontal and the slope of one of the asymptotes is $\frac{7}{8}$, how far apart are the rosebushes?

Section 9.4

55. If $(x, y) = (2, 0)$, find the coordinates in the uv-system, obtained by rotating x- and y-axes by $\frac{\pi}{3}$.

56. If $(x, y) = (0, -3)$, find the coordinates in the uv-system, obtained by rotating the x- and y-axes by $\frac{\pi}{3}$.

57. If the uv-system is obtained by rotating the x- and y-axes by $\frac{\pi}{6}$, and $(u, v) = (-4, 0)$, find the coordinates in the xy-system.

58. If the uv-system is obtained by rotating the x- and y-axes by $\frac{\pi}{6}$, and $(u, v) = (0, -2)$, find the coordinates in the xy-system.

For each of the equations in Exercises 59–62:

(a) Identify the conic section.
(b) Write the equation in the uv-coordinate system so that there is no uv term in the equation.
(c) Find the vertex or vertices, and graph.

59. $xy - 6 = 0$

60. $-2x^2 + 2\sqrt{3}\,xy - 6 = 0$

61. $\frac{3}{2}x^2 - xy + \frac{3}{2}y^2 - 4 = 0$

62. $\frac{1}{2}x^2 + xy + \frac{1}{2}y^2 - 2\sqrt{2}\,x + 2\sqrt{2}\,x = 0$

Section 9.5

In Exercises 63–66, identify and graph the conic section given by each of the equations.

63. $r = \dfrac{4}{1 - \sin\theta}$ **64.** $r = \dfrac{8}{1 + 3\sin\theta}$ **65.** $r = \dfrac{6}{2 - 2\cos\theta}$ **66.** $r = \dfrac{4}{1 - 2\cos\theta}$

Section 9.6

In Exercises 67–73, eliminate the parameter t to find an equivalent equation with y in terms of x. Give any restrictions on x. Sketch the corresponding graph, indicating the direction of increasing t.

67. $x = -t, y = 2t + 1, t \geq 0$

68. $x = t - 1, y = -t, -3 \leq t \leq 3$

69. $x = -t + 2, y = t^2, -2 \leq t \leq 2$

70. $x = t^2, y = -t, -2 \leq t \leq 2$

71. $x = 2\sin t, y = 2\cos t, 0 \leq t \leq 2\pi$ **72.** $x = \sin t, y = 2\cos t, 0 \leq t \leq \pi$

73. Projectile Motion A tennis ball is hit from a height of 4 feet above the ground with an initial velocity of 120 feet per second at an angle of 30° to the horizontal.

  a. Find parametric equations that give the position of the ball as a function of time.

  b. When is the ball at its maximum height, to the nearest thousandth of a second? What is its maximum height, to the nearest hundredth of a foot?

Chapter 9 Test

In Exercises 1 and 2, find the vertex, focus, and directrix of the parabola. Sketch the parabola.

1. $(y + 3)^2 = -12(x + 1)$

2. $y^2 - 4y + 4x = 0$

In Exercises 3 and 4, find the equation, in standard form, of the parabola satisfying the given conditions.

3. Vertical axis of symmetry; vertex at $(0, -2)$; passes through the point $(4, 0)$.

4. Focus at $(3, 0)$; directrix $x = -3$

In Exercises 5 and 6, find the vertices and foci of each ellipse. Sketch the ellipse.

5. $\dfrac{(x + 2)^2}{16} + \dfrac{(y - 3)^2}{25} = 1$

6. $\dfrac{x^2}{9} + \dfrac{(y + 1)^2}{4} = 1$

In Exercises 7 and 8, find the equation of the ellipse in standard form.

7. $4x^2 + 8x + y^2 = 0$

8. Foci at $(0, 3)$, $(0, -3)$; one vertex at $(0, -5)$

In Exercises 9 and 10, find the center, vertices, foci, and asymptotes for each hyperbola. Sketch the hyperbola.

9. $\dfrac{y^2}{4} - x^2 = 1$

10. $\dfrac{(x + 2)^2}{9} - \dfrac{y^2}{16} = 1$

In Exercises 11 and 12, find the standard form of the equation of the hyperbola.

11. Foci at $(3, 0)$, $(-3, 0)$; transverse axis of length 2

12. Vertices at $(0, 1)$, $(0, -1)$; slope of one asymptote is -3

In Exercises 13 and 14:

(a) Identify the conic section.
(b) Write the equation in the uv-coordinate system so that there is no uv term in the equation.
(c) Find the vertex or vertices in uv coordinates, and graph.

13. $\dfrac{1}{2}x^2 - xy + \dfrac{1}{2}y^2 - \sqrt{2}\,x - \sqrt{2}\,y = 0$

14. $3x^2 + 4\sqrt{3}\,xy + 7y^2 - 9 = 0$

15. Identify and graph the equation $r = \dfrac{6}{3 - 6\sin\theta}$.

16. Identify and graph the equation $r = \dfrac{4}{4 + 2\cos\theta}$.

In Exercises 17 and 18, eliminate the parameter t to find an equivalent equation with y in terms of x. Give any restrictions on x. Sketch the corresponding graph, indicating the direction of increasing t.

17. $x = -t + 2$, $y = -t^2 + 1$, $-2 \le t \le 2$

18. $x = -3\cos t$, $y = \sin t$, $0 \le t \le \pi$

19. A museum plans to design a whispering gallery whose cross-section is the top half of an ellipse. Because of the reflective properties of ellipses, a whisper from someone standing at one focus can be heard by a person standing at the other focus. If the dimensions of the gallery are as shown, how far from the gallery's center should two people stand so that they can hear each other's whispers?

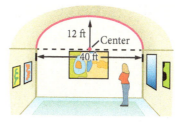

20. A small microphone with a parabolic cross-section measures 2 inches across and 4 inches deep. The sound receiver should be placed at the focus for best reception. How far from the vertex of the parabola should the sound receiver be placed?

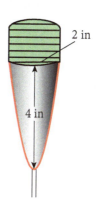

21. **Projectile Motion** A ball is thrown upward from a height of 3 feet above the ground with an initial velocity of 100 feet per second at an angle of 30° to the horizontal.
   a. Find parametric equations that give the position of the ball as a function of time.
   b. Find the horizontal distance traveled by the ball, to the nearest foot.

More Topics in Algebra

In a bicycle race with many participants, you might ask how many different possibilities exist for the first-place and second-place finishes. Determining the possibilities involves a counting technique known as the multiplication principle. See Example 1 in Section 10.4. In this chapter you will study various topics in algebra, including sequences, counting methods, probability, and mathematical induction. These concepts are used in a variety of applications, such as finance, sports, pharmacology, and biology, as well as in advanced courses in mathematics.

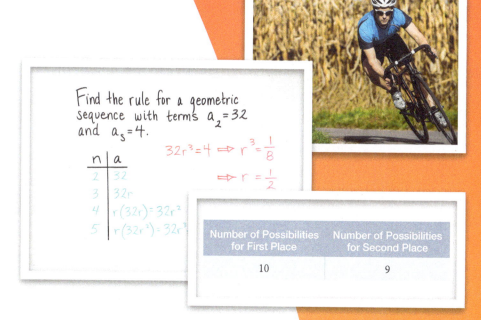

Find the rule for a geometric sequence with terms $a_2 = 32$ and $a_5 = 4$.

n	a
2	32
3	$32r$
4	$r(32r) = 32r^2$
5	$r(32r^2) = 32r^3$

$32r^3 = 4 \Rightarrow r^3 = \dfrac{1}{8}$

$\Rightarrow r = \dfrac{1}{2}$

Number of Possibilities for First Place	Number of Possibilities for Second Place
10	9

Outline

Sequences

Objectives

- Define and identify an arithmetic sequence.
- Define and identify a geometric sequence.
- Compute the terms of an arithmetic or a geometric sequence.
- Apply arithmetic and geometric sequences to word problems.

In this section, we examine sequences of numbers. They occur in a wide variety of contexts, such as music and biology. In Examples 5 and 6, we investigate some specific applications to music and biology. But first, we introduce notation and definitions for our work with sequences.

Definition of a Sequence

A **sequence** is a function $f(n)$ whose domain is the set of all nonnegative integers (i.e., $n = 0, 1, 2, 3, \ldots$) and whose range is a subset of the set of all real numbers. The numbers $f(0), f(1), f(2), \ldots$ are called the **terms** of the sequence.

For a nonnegative number n, it is conventional to denote the term that corresponds to n by a_n rather than by $f(n)$. We shall use both notations in our discussion.

Next we will study some special types of sequences for which a general rule exists for the function f.

Arithmetic Sequences

A sequence in which we set the starting value a_0 to a certain number and *add* a fixed number d to any term of the sequence to get the next term is known as an **arithmetic sequence**.

Definition of an Arithmetic Sequence

Each term of an **arithmetic sequence** is given by the rule

$$a_n = a_0 + nd, n = 0, 1, 2, 3, \ldots$$

where a_0 is the starting value of the sequence and d is the common difference between successive terms. That is,

$$d = a_1 - a_0 = a_2 - a_1 = a_3 - a_2 = \ldots$$

VIDEO EXAMPLES

SECTION 10.1

> **Example 1** **Using the First Few Terms to Find a Rule**
>
> Table 1 gives the first four terms of an arithmetic sequence.
>
> | n | 0 | 1 | 2 | 3 |
> |-----|-----|-----|-----|-----|
> | a_n | -1 | -3 | -5 | -7 |
>
> **Table 1**
>
> **a.** Find the rule corresponding to the given sequence.
>
> **b.** Use your rule to find a_{10}.
>
> **Solution**
>
> **a.** We first find the common difference d between successive terms of the sequence. Note that
>
> $$a_1 - a_0 = a_2 - a_1 = a_3 - a_2 = -2$$
>
> Thus, $d = -2$. Since $a_0 = -1$, the general rule is then given by
>
> $$a_n = a_0 + dn = -1 + (-2)n \Rightarrow a_n = -1 - 2n$$
>
> **b.** To find a_{10}, simply substitute $n = 10$ into the rule, which gives
>
> $$a_{10} = -1 - 2(10) = -21$$

Check It Out 1   Check that the rule generated in part (a) of Example 1 gives the correct values for the four terms listed in Table 1. ■

Example 2   Using Any Two Terms to Find a Rule

Suppose two terms of an arithmetic sequence are $a_8 = 10$ and $a_{12} = 26$. What is the rule for this sequence?

Solution Because this is an arithmetic sequence, we know that the difference between any two successive terms is a constant, d. We can use the given terms to construct Table 2.

Thus $10 + 4d = 26$. To find d, we solve this equation.

$$10 + 4d = 26 \quad \Rightarrow \quad d = 4$$

The rule for the sequence is then $a_n = a_0 + 4n$. We still need to find a_0, the starting value. We could construct another table and work our way backward, but that would be too tedious. Instead, we use the information available to us to find an equation in which a_0 is the only unknown.

$a_n = a_0 + 4n$	Known rule so far
$\mathbf{10} = a_0 + 4(\mathbf{8})$	Substitute $n = 8$ and $a_8 = 10$
$10 = a_0 + 32$	Simplify
$-22 = a_0$	Solve for a_0

Thus the rule for this sequence is given by

$$a_n = -22 + 4n$$

 Check It Out 2

a. Check that the rule in Example 2 is correct.

b. Find the rule for the arithmetic sequence with terms $a_9 = 15$ and $a_{12} = 6$. ■

Geometric Sequences

We saw that arithmetic sequences are formed by setting the starting value a_0 equal to a certain number and then *adding* a fixed number d to any term of the sequence to get the next term. Another way to generate a sequence is to set the starting value a_0 equal to a certain number and then *multiply* any term of the sequence by a fixed number r to get the next term. A simple example is when we start with the number 2 and keep multiplying by 2, which gives the sequence 2, 4, 8, 16, …. A sequence generated in this manner is called a **geometric sequence**.

Definition of a Geometric Sequence

A **geometric sequence** is defined by the rule

$$a_n = a_0 r^n, \; n = 0, 1, 2, 3, \ldots$$

where a_0 is the starting value of the sequence and r is the *fixed number* by which any term of the sequence is multiplied to obtain the next term. If $r \neq 0$, this is equivalent to saying that r is the *fixed ratio* of successive terms:

$$r = \frac{a_1}{a_0} = \frac{a_2}{a_1} = \cdots = \frac{a_i}{a_{i-1}} = \cdots$$

Table 2

n	a_n
8	10
9	$10 + d$
10	$10 + 2d$
11	$10 + 3d$
12	$10 + 4d = 26$

Table 2

 Using Technology

A graphing utility in SEQUENCE mode can be used to graph the terms of a sequence. Figure 1 shows the graph of the sequence $a_n = -1 - 2n$, $n = 0, 1, 2, 3, \ldots$.

Keystroke Appendix: Section 14

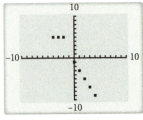

Figure 1

Using Technology

To see how quickly a geometric sequence can grow, you can use a graphing utility to plot the terms $a_n = 3(2)^n$, $n = 0, 1, 2, 3, \ldots$. See Figure 2.

Keystroke Appendix
Section 14

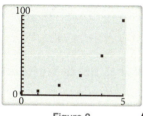

Figure 2

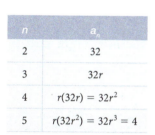

n	a_n
2	32
3	$32r$
4	$r(32r) = 32r^2$
5	$r(32r^2) = 32r^3 = 4$

Table 3

Example 3 **Using Two Consecutive Terms to Find a Rule**

Find the rule for the geometric sequence with terms $a_2 = 12$ and $a_3 = 24$.

Solution Because a_2 and a_3 are successive terms of the sequence, $a_3 = ra_2$. To find r, we solve this equation.

$$r = \frac{a_3}{a_2} = \frac{24}{12} = 2$$

The rule for the sequence is then $a_n = a_0(2)^n$.

Next we find a_0, the starting value. We use the information available to write an equation in which a_0 is the only unknown.

$a_n = a_0(2)^n$ Known rule so far

$12 = a_0(2)^2$ Substitute $n = 2$ and $a_2 = 12$

$12 = 4a_0$ Simplify

$3 = a_0$ Solve for a_0

Hence, the rule for this sequence is

$$a_n = 3(2)^n$$

 Check It Out 3

a. Verify that the rule in Example 3 is correct.

b. Find the rule for the geometric sequence with terms $a_3 = 54$ and $a_4 = 162$.

Example 4 **Using Two Nonconsecutive Terms to Find a Rule**

Find the rule for the geometric sequence with terms $a_2 = 32$ and $a_5 = 4$.

Solution Unlike in Example 3, we do not have successive terms here. To find the ratio r between successive terms, we can set up a table to summarize the information we are given. See Table 3.

From the last row, we have

$$32r^3 = 4 \Rightarrow r^3 = \frac{1}{8} \Rightarrow r = \frac{1}{2}$$

Thus the rule for this sequence is $a_n = a_0\left(\frac{1}{2}\right)^n$. We still need to find a_0, the starting value. Use the information available to set up an equation in which a_0 is the only unknown:

$a_n = a_0\left(\frac{1}{2}\right)^n$ Known rule so far

$32 = a_0\left(\frac{1}{2}\right)^2$ Substitute $n = 2$ and $a_2 = 32$

$32 = a_0\left(\frac{1}{4}\right)$ Simplify

$a_0 = 128$ Solve for a_0

Hence, the rule for this geometric sequence is

$$a_n = 128\left(\frac{1}{2}\right)^n$$

Note that the ratio $r = \frac{1}{2} < 1$. The terms of this sequence get smaller as n gets larger.

 Check It Out 4

a. Verify that the rule in Example 4 is correct.

b. Find the rule for the geometric sequence with terms $a_3 = 3$ and $a_6 = 81$.

Applications of Arithmetic and Geometric Sequences

Sequences are used in a wide variety of situations. In particular, they are used to model processes in which there are discrete jumps from one stage to the next. The following examples will examine applications from music and biology.

Example 5   A Vibrating String

A string of a musical stringed instrument vibrates in many modes, called *harmonics*. Associated with each harmonic is its *frequency*, which is expressed in units of cycles per second, or Hertz (Hz). The harmonic with the lowest frequency is known as the fundamental mode; all the other harmonics are known as overtones. Table 4 lists the frequencies corresponding to the first four harmonics of a string of a certain instrument.

Harmonic	Frequency (Hz)
Fundamental	55
Second	110
Third	165
Fourth	220

Table 4

a. What type of sequence do the frequencies form: arithmetic, geometric, or neither?

b. Find the frequency corresponding to the fifth harmonic.

c. Find the frequency corresponding to the nth harmonic.

Solution

a. Because the *difference* between successive harmonics is constant at 55 Hz, the sequence is arithmetic.

b. From Table 4, the fourth harmonic is 220 Hz. Thus the fifth harmonic is

$$220 + d = 220 + 55 = 275 \text{ Hz}$$

c. The common difference is $d = 55$. Denote the frequency of the nth harmonic by $f_n, n = 1, 2, 3, \ldots$. From Table 4, $f_1 = 55, f_2 = 110 = 55(2), f_3 = 165 = 55(3)$, and so on. Thus we have

$$f_n = 55n, \quad n = 1, 2, 3, \ldots$$

Harmonic	Frequency (Hz)
Fundamental	35
Second	70
Third	140
Fourth	280

Table 5

Note that this sequence begins at $n = 1$.

 Check It Out 5   Rework Example 5 for the case in which the frequencies of the first four harmonics are as shown in Table 5. ■

Example 6   DNA Fragments

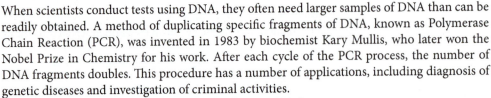

When scientists conduct tests using DNA, they often need larger samples of DNA than can be readily obtained. A method of duplicating specific fragments of DNA, known as Polymerase Chain Reaction (PCR), was invented in 1983 by biochemist Kary Mullis, who later won the Nobel Prize in Chemistry for his work. After each cycle of the PCR process, the number of DNA fragments doubles. This procedure has a number of applications, including diagnosis of genetic diseases and investigation of criminal activities.

Suppose a biologist begins with a sample of 1000 DNA fragments.

a. What type of sequence is generated by the repeated application of the PCR process?

b. How many DNA fragments will there be after five cycles of the PCR process?

c. How many DNA fragments will there be after n cycles of the PCR process?

d. Laboratories typically require millions of DNA fragments to conduct proper tests. How many cycles of the PCR process are needed to produce one million fragments?

e. Each cycle of the PCR process takes approximately 30 minutes. How long will it take to generate one million DNA fragments?

Solution

a. Because the number of DNA fragments doubles after each cycle of the PCR process, the resulting sequence of the number of DNA fragments is geometric.

b. The doubling of DNA fragments after each cycle is summarized in Table 6.

Number of Cycles	Number of Fragments
0	1000
1	$2(1000) = 2000$
2	$2(2000) = 2^2(1000) = 4000$
3	$2^3(1000) = 8000$
4	$2^4(1000) = 16,000$
5	$2^5(1000) = 32,000$

Table 6

From the table, the number of DNA fragments after five cycles is

$$1000(2)^5 = 32,000$$

c. Let $F(n)$ denote the number of DNA fragments after n cycles. Because the fixed, or common ratio is 2 and the initial number of fragments is 1000, we can write

$$F(n) = 1000(2)^n$$

d. To compute the number of cycles needed to produce one million DNA fragments, solve the following equation for n.

$$1000(2)^n = 1,000,000 \quad \text{Set } F(n) \text{ equal to 1,000,000}$$

$$2^n = 1000 \quad \begin{array}{l}\text{Isolate exponential term: divide}\\\text{both sides of equation by 1000}\end{array}$$

Because n appears as an exponent, use logarithms to solve for n.

$$\log 2^n = \log 1000 \qquad \text{Take common log of both sides}$$

$$n \log 2 = \log 1000$$

$$n = \frac{\log 1000}{\log 2} = \frac{3}{\log 2} \qquad \text{Solve for } n$$

$$n \approx 9.966$$

The number of cycles must be a whole number. Thus **10 cycles** of the PCR process are needed to produce at least one million DNA fragments.

e. Each cycle lasts 30 minutes, so it will take **10(30) = 300 minutes, or 5 hours,** to produce one million DNA fragments.

✓ *Check It Out 6* How long will it take to produce one million fragments of DNA if the biologist in Example 6 starts with 2000 fragments of DNA?

Exercises 10.1

 ## Just in Time Exercises

1. True or False: If an amount P is invested in an account that pays interest at rate r and the interest is compounded n times per year, then after t years the amount in the account will be

$$A(t) = P\left(1 + \frac{r}{n}\right)^{nt}$$

2. If interest on an account is compounded quarterly, what is the value of n in the compound interest formula?

3. A bank is advertising an account with an interest rate of 4%, compounded semiannually. A customer opens an account with an initial deposit of $100 and makes no more deposits or withdrawals. What will the account balance be at the end of 2 years?

4. A bank is advertising an account with an interest rate of 6%, compounded yearly. A customer opens an account with an initial deposit of $100 and makes no more deposits or withdrawals. What will the account balance be at the end of 2 years?

 ## Skills

In Exercises 5–20, find the terms a_0, a_1, and a_2 for each sequence.

5. $a_n = 4 + 6n$
6. $a_n = 3 + 5n$
7. $a_n = -5 + 3n$
8. $a_n = -3 + 2n$

9. $a_n = -4 - 4n$
10. $a_n = -3 - 3n$
11. $a_n = 8 - 2n$
12. $a_n = 5 - 3n$

13. $a_n = 7(4^n)$
14. $a_n = 3(2^n)$
15. $a_n = 5(3^n)$
16. $a_n = 6(5^n)$

17. $a_n = -2(3^n)$
18. $a_n = -4(6^n)$
19. $a_n = -3(5^n)$
20. $a_n = -5(2^n)$

In Exercises 21–32, find the rule for the arithmetic sequence having the given terms.

21.

n	0	1
a_n	-3	3

22.

n	0	1
$f(n)$	-1	4

23.

n	6	7
$g(n)$	8	12

24.

n	3	4
b_n	-8	-9

25.

n	10	12
b_n	10	16

26.

n	8	11
h_n	9	3

27. The common difference d is 5 and $a_9 = 55$.   28. The common difference d is 2 and $a_5 = 24$.

29. The common difference d is -2 and $a_8 = 5$.   30. The common difference d is -3 and $a_{10} = 7$.

31. The common difference d is $\frac{1}{2}$ and $b_6 = 13$.   32. The common difference d is $\frac{1}{4}$ and $c_8 = 7$.

In Exercises 33–44, find the rule for the geometric sequence having the given terms.

33.

n	0	1
a_n	3	6

34.

n	0	1
a_n	4	8

35.

n	2	3
$h(n)$	32	128

36.

n	3	4
$g(n)$	-8	-16

37.

n	4	7
b_n	$\frac{3}{16}$	$\frac{3}{128}$

38.

n	3	6
h_n	$\frac{2}{27}$	$\frac{2}{729}$

39. The common ratio r is 2 and $a_5 = 128$.   40. The common ratio r is 3 and $a_4 = -162$.

41. The common ratio r is $\frac{3}{2}$ and $b_4 = \frac{81}{4}$.   42. The common ratio r is $\frac{4}{3}$ and $c_3 = -\frac{64}{27}$.

43. The common ratio r is 5 and $a_4 = 2500$.   44. The common ratio r is 4 and $a_6 = 12{,}288$.

In Exercises 45–50, fill in the missing terms of each arithmetic *sequence.*

45.

n	0	1	2	3
a_n		5	8	

46.

n	0	1	2	3
a_n		5	6	

47.

n	0	1	2	3
a_n		2	6	

48.

n	0	1	2	3
a_n		−2	1	

49.

n	0	1	2	3
a_n		3	1	

50.

n	0	1	2	3
a_n		−9	−15	

In Exercises 51–56, fill in the missing terms of each geometric *sequence.*

51.

n	0	1	2	3
a_n		3	9	

52.

n	0	1	2	3
a_n		5	25	

53.

n	0	1	2	3
a_n		10	50	

54.

n	0	1	2	3
a_n		12	36	

55.

n	0	1	2	3
a_n		$\frac{1}{4}$	$\frac{1}{8}$	

56.

n	0	1	2	3
a_n		$\frac{1}{9}$	$\frac{1}{27}$	

In Exercises 57–68, state whether the sequence is arithmetic or geometric.

57. $1, 3, 5, 7, \ldots$

58. $4, 10, 16, 22, \ldots$

59. $2, 6, 18, 54, \ldots$

60. $8, 5, 2, -1, \ldots$

61. $-7, -11, -15, -19, \ldots$

62. $3, 15, 75, 375, \ldots$

63. $\frac{1}{2}, \frac{1}{6}, \frac{1}{18}, \ldots$

64. $0.929, 0.939, 0.949, \ldots$

65. $\frac{111}{1000}, \frac{115}{1000}, \frac{119}{1000}, \ldots$

66. $0.9, 0.81, 0.729, \ldots$

67. $0.4, 0.8, 1.6, 3.2, \ldots$

68. $0.4, 0.9, 1.4, 1.9, \ldots$

Applications

69. Investment An income-producing investment valued at $2000 pays interest at an annual rate of 6%. Assume that the interest is taken out as income and therefore is not compounded.

   a. Make a table in which you list the initial investment along with the total value of the investment-related assets (initial investment plus total interest earned) at the end of each of the first 4 years.

   b. What is the total value of the investment-related assets after n years?

70. Investment An income-producing investment valued at $3000 pays interest at an annual rate of 4.5%. Assume that the interest is taken out as income and therefore is not compounded.

   a. Make a table in which you list the initial investment along with the total value of the investment-related assets (initial investment plus total interest earned) at the end of each of the first 4 years.

   b. What is the total value of the investment-related assets after n years?

71. Knitting Knitting, whether by hand or by machine, uses a sequence of stitches and proceeds row by row. Suppose you knit 100 stitches for the bottommost row and increase the number of stitches in each row thereafter by 4. This is a standard way to make the sleeve portion of a sweater.

   a. What type of sequence does the number of stitches in each row produce: arithmetic, geometric, or neither?

   b. Find a rule that gives the number of stitches in the nth row. Note that this sequence starts at $n = 1$.

   c. How many rows must be knitted to end with a row of 168 stitches?

72. Knitting New trends in knitting involve creating vibrant patterns with geometric shapes. Suppose you want to knit a large right triangle. You start with 85 stitches and decrease each row thereafter by 2 stitches.

   a. What type of sequence does the number of stitches in each row produce: arithmetic, geometric, or neither?

   b. Find a rule that gives the number of stitches for the nth row. Note that this sequence starts at $n = 1$.

   c. How many rows must be knitted to end with a row of just one stitch?

73. **Music** In music, the frequencies of a certain sequence of tones that are an octave apart are

    55 Hz, 110 Hz, 220 Hz, ...

    where Hz (Hertz) is a unit of frequency (1 Hz = 1 cycle per second).
    a. Is this an arithmetic or a geometric sequence? Explain.
    b. Compute the next two terms of the sequence.
    c. Find a rule for the frequency of the nth tone.

74. **Sports** The men's and women's U.S. Open tennis tournaments are elimination tournaments. Each tournament starts with 128 players in 64 separate matches. After the first round of competition, 64 players are left. The process continues until the final championship match has been played.
    a. What type of sequence gives the number of players left after each round?
    b. How many rounds of competition are there in each tournament?

75. **Salary** An employee starting with an annual salary of $40,000 will receive a salary increase of 4% at the end of each year. What type of sequence would you use to find her salary after 6 years on the job? What is her salary after 6 years?

76. **Salary** An employee starting with an annual salary of $40,000 will receive a salary increase of $2000 at the end of each year. What type of sequence would you use to find his salary after 5 years on the job? What is his salary after 5 years?

77. **Biology** A cell divides into two cells every hour.
    a. How many cells will there be after 4 hours if we start with 10,000 cells?
    b. Is this a geometric sequence or an arithmetic sequence?
    c. How long will it take for the number of cells to equal 1,280,000?

 78. **Salary** Joan is offered two jobs with differing salary structures. Job A has a starting salary of $30,000 with an increase of 4% per year. Job B has a starting salary of $35,000 with an increase of $500 per year. During what years will Job A pay more? During what years will Job B pay more?

79. **Geometry** A sequence of square boards is made as follows. The first board has dimensions 1 inch by 1 inch, the second has dimensions 2 inches by 2 inches, the third has dimensions 3 inches by 3 inches, and so on.
    a. What type of sequence is formed by the *perimeters* of the boards? Explain.
    b. Write a rule for the sequence formed by the *areas* of the boards. Is the sequence arithmetic, geometric, or neither? Explain your answer.

 80. **Social Security** The following table gives the average monthly Social Security payment, in dollars, for retired workers for the years 2000 to 2003.   (*Source*: Social Security Administration)

Year	2000	2001	2002	2003
Amount	843	881	917	963

    a. Is this sequence better approximated by an arithmetic sequence or a geometric sequence? Explain.
    b. Use the regression capabilities of your graphing calculator to find a suitable function that models this data. Make sure that n represents the number of years after 2000.

 81. **Recreation** The following table gives the amount of money, in billions of dollars, spent on recreation in the United States from 1999 to 2002.   (*Source*: Bureau of Economic Analysis)

Year	1999	2000	2001	2002
Amount ($ billions)	546.1	585.7	603.4	633.9

    Assume that this sequence of expenditures approximates an arithmetic sequence.
    a. If n represents the number of years since 1999, use the linear regression capabilities of your graphing calculator to find a function of the form $f(n) = a_0 + nd$, $n = 0, 1, 2, 3,...$, that models these expenditures.
    b. Use your model to project the amount spent on recreation in 2007.

Concepts

82. Is 4, 4, 4,… an arithmetic sequence, a geometric sequence, or both? Explain.

83. What are the terms of the sequence generated by the expression $a_n = a_0 + nd$, $d = 0$?

84. What are the terms of the sequence generated by the expression $a_n = a_0 r^n$, $r = 1$?

85. You are given two terms of a sequence: $a_1 = 1$ and $a_3 = 9$.
   a. Find the rule for this sequence, assuming it is arithmetic.
   b. Find the rule for this sequence, assuming it is geometric.
   c. Find $a_4, …, a_8$ for the sequence in part (a).
   d. Find $a_4, …, a_8$ for the sequence in part (b).
   e. Which sequence grows faster, the one in part (a) or the one in part (b)? Explain your answer.

86. Consider the sequence

$$1, 10, 100, 1000, 10{,}000, ….$$

Is this an arithmetic sequence or a geometric sequence? Explain. Now take the common logarithm of each term in this sequence. Is the new sequence arithmetic or geometric? Explain.

87. Find the next two terms in the geometric sequence whose first three terms are $(1 + x)$, $(1 + x)^2$, and $(1 + x)^3$. What is the common ratio r in this case?

88. Suppose a, b, and c are three consecutive terms in an arithmetic sequence. Show that $b = \frac{a + c}{2}$.

89. If $a_0, a_1, a_2,…$ is a geometric sequence, what kind of sequence is $a_0^3, a_1^3, a_2^3,…$? Explain your reasoning.

Sums of Terms of Sequences

- Find the sum of terms of an arithmetic sequence.

- Understand and work with summation notation.

- Find the sum of terms of a geometric sequence.

- Find the sum of an infinite geometric series.

- Apply sums of arithmetic and geometric sequences to word problems.

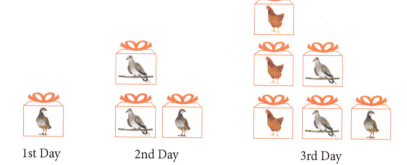

1st Day   2nd Day   3rd Day

On the first day of Christmas, my true love sent to me
a partridge in a pear tree.
On the second day of Christmas, my true love sent to me
two turtle doves and a partridge in a pear tree.
On the third day of Christmas, my true love sent to me
three French hens, two turtle doves, and a partridge in a pear tree.
...

(*Source*: William Henry Husk, Songs of the Nativity, 1868.)

In the holiday song *Twelve Days of Christmas*, what is the total number of gifts the lucky person will receive on the twelfth day? This is an example of finding the *sum* of numbers in a sequence. This section will examine sums of numbers generated by special types of sequences.

Sum of Terms of an Arithmetic Sequence

From Section 10.1, the sequence 1, 2,..., 12, is an arithmetic sequence since each term differs by 1. In this section, we derive a general formula for such a sequence. First, we show how to obtain a formula for the sum of the first 12 numbers in the sequence 1, 2, 3,..., 12. We then generalize that technique to find the sum of the first n positive integers. This formula is useful in finding the sum of terms in any arithmetic sequence.

From Figure 1, we see that the first and last (twelfth) numbers in the sequence add to 13. The second and eleventh numbers also add to 13.

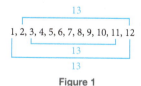

Figure 1

The same is true of four more sums: the sum of the third and tenth numbers, the sum of the fourth and ninth numbers, the sum of the fifth and eighth numbers, and the sum of the sixth and seventh numbers each add to 13. We thus have six pairs of numbers, each of which adds to 13. The sum is

$$1 + 2 + 3 + \cdots + 12 = (1 + 12) + (2 + 11) + (3 + 10) + \cdots + (6 + 7)$$
$$= (6)(13) = 78$$

Thus, the fortunate person receiving the gifts will receive a total of 78 gifts on the twelfth day. To find the total number of gifts received on the nth day, where n is even, we find the sum $1 + 2 + 3 + \cdots + n$. See Figure 2.

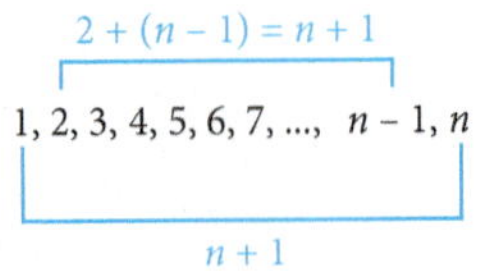

Figure 2

Applying the technique from the previous discussion gives

$$1 + 2 + 3 + \cdots + (n - 1) + n$$

$$= (1 + n) + (2 + (n - 1)) + \cdots + \left(\frac{n}{2} + \left(\frac{n}{2} + 1\right)\right) \qquad \text{See Figure 2}$$

$$= \underbrace{(n + 1) + (n + 1) + \cdots + (n + 1)}_{\frac{n}{2} \text{ times}} \qquad \text{Evaluate within parentheses}$$

$$= \frac{n}{2}(n + 1) = \frac{n(n + 1)}{2} \qquad \text{There are } \frac{n}{2} \text{ pairs of numbers}$$

The number of gifts given on the nth day is $\dfrac{n(n + 1)}{2}$.

Because we are given that n is even, we have $\frac{n}{2}$ pairs of numbers to sum in this calculation.

If n is odd, you can use this technique to find the sum of the first n positive integers, but then you will have a leftover term to deal with. However, it can be shown that the same formula holds for the sum regardless of whether n is even or odd. This leads to the following result.

> **Sum of Numbers from 1 to n**
>
> Let n be a positive integer. Then
>
> $$1 + 2 + 3 + \cdots + n = \frac{n(n + 1)}{2}$$

We can use this formula to find the general formula for the sum of the first n terms of *any* arithmetic sequence. Let $a_j = a_0 + jd$, $j = 0, 1, 2, \ldots$, be an arithmetic sequence, and let n be a positive integer. To find the sum of the terms a_0 through a_{n-1}, proceed as follows.

We wish to find

$$S_n = a_0 + a_1 + a_2 + \cdots + a_{n-1}$$

The quantities that we know are a_0, d, and n. If we can obtain a formula that contains just these quantities, we can compute the sum of any number of terms very easily. From the general form for a_j, we have

$$S_n = a_0 + a_1 + a_2 + \cdots + a_{n-1}$$

$$= a_0 + (a_0 + d) + (a_0 + 2d) + \cdots + (a_0 + (n - 1)d)$$

$$= \underbrace{a_0 + a_0 + \cdots + a_0}_{n \text{ times}} + d + 2d + \cdots + (n - 1)d \qquad \text{Collect like terms}$$

$$= (n)a_0 + d(1 + 2 + \cdots + (n - 1)) \qquad \text{Factor out } d$$

$$= (n)a_0 + d\left(\frac{(n - 1)(n)}{2}\right) \qquad \begin{array}{l}\text{Use formula for} \\ 1 + 2 + \cdots + (n - 1)\end{array}$$

$$= \frac{n}{2}(2a_0 + d(n - 1)) \qquad \text{Factor out } \frac{n}{2}$$

We can stop at this point, since we have found a formula for the sum that contains only the known quantities, a_0, d, and n. We can also write another formula for S_n by using the fact that

$$a_{n-1} = a_0 + d(n-1).$$ This gives

$$
\begin{aligned}
S_n &= a_0 + a_1 + a_2 + \cdots + a_{n-1} \\
&= \frac{n}{2}(2a_0 + d(n-1)) && \text{Use formula for sum} \\
&= \frac{n}{2}(a_0 + (a_0 + d(n-1))) \\
&= \frac{n}{2}(a_0 + a_{n-1}) && \text{Substitute } a_{n-1} = a_0 + d(n-1)
\end{aligned}
$$

We now have the following formula for the sum of the first n terms of an arithmetic sequence.

Sum of the First n Terms of an Arithmetic Sequence

Let $a_j = a_0 + jd$, $j = 0, 1, 2,\ldots$, be an arithmetic sequence, and let n be a positive integer. The sum of the first n terms of the sequence, a_0 through a_{n-1}, is given by

$$S_n = a_0 + a_1 + \cdots + a_{n-1} = \frac{n}{2}(2a_0 + d(n-1)) = \frac{n}{2}(a_0 + a_{n-1})$$

You can use either of these forms, depending on the information you are given.

VIDEO EXAMPLES

SECTION 10.2

 **Using Technology**

Using the SUM and SEQUENCE features of your graphing utility, you can calculate the sum in Example 1. See Figure 3.

Keystroke Appendix:
Section 14

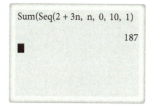

Sum(Seq(2 + 3n, n, 0, 10, 1)

           187

Figure 3

Example 1   **Calculating the Sum from the First Few Terms**

Find the sum of the first 11 terms of the arithmetic sequence

$$2, 5, 8, 11, 14, \ldots$$

Solution For the given sequence, the first term, a_0, equals 2 and the common difference, d, equals 3. The sequence can be written as

$$a_j = 2 + 3j,\ j = 0, 1, 2, 3, \ldots$$

The sum of the first 11 terms is then found by using the sum formula.

$$
\begin{aligned}
S_n &= \frac{n}{2}(2a_0 + d(n-1)) && \text{We know } a_0,\ d,\ \text{and } n \\
S_{11} &= \frac{11}{2}(2(2) + 3(11-1)) && \text{Substitute } a_0 = 2,\ n = 11,\ \text{and } d = 3 \\
&= \frac{11}{2}(34) = 187 && \text{Simplify}
\end{aligned}
$$

✓ **Check It Out 1**   Find the sum of the first 20 terms of the arithmetic sequence

$$2, 4, 6, 8, 10, \ldots$$

The next example uses an alternate formula for finding the sum of an arithmetic sequence.

Example 2   **Calculating the Sum from the Initial and Final Terms**

Find the sum of all even numbers between 2 and 100, inclusive.

Solution Here, we are given the initial term and the final term of the sequence. Thus, the formula

$$S_n = \frac{n}{2}(a_0 + a_{n-1})$$

can be used. There are 50 even numbers between 2 and 100, inclusive. Substituting $a_0 = 2$, $a_{49} = 100$, and $n = 50$, we have

$$S_{50} = \frac{50}{2}(a_0 + a_{49}) = \frac{50}{2}(2 + 100) = 2550$$

✓ **Check It Out 2**   Find the sum of all odd numbers between 3 and 51, inclusive.

Summation Notation

It can be very cumbersome to write out the sum of the terms of a sequence (such as the sum $1 + 2 + \cdots + n$) in the form of a string of numbers separated by plus signs. Fortunately, there is a shorthand notation, known as **summation notation**, that can be used to indicate a sum. The capital Greek letter sigma, Σ, is used to denote a sum. In addition, a variable known as an **index**, ranging over some set of consecutive integers, keeps track of the terms in the sum.

Consider the following example of summation notation.

$$\text{The sum } 1 + 2 + \cdots + n \text{ would be written as } \sum_{i=1}^{n} i.$$

The details of this notation are illustrated below.

$$\text{Index } i \text{ ends at } n$$
$$\downarrow$$
$$\text{Summation symbol} \rightarrow \sum_{i=1}^{n} i^2 \leftarrow \text{Expression to be evaluated and summed}$$
$$\uparrow$$
$$\text{Index } i \text{ starts at } 1$$

The best way to understand summation notation is to work with it. The next two examples illustrate the use of summation notation.

Example 3   Expanding a Sum Given in Summation Notation

Expand the following sums given in summation notation. Do not evaluate.

a. $\displaystyle\sum_{i=2}^{5} 3i$ **b.** $\displaystyle\sum_{i=3}^{6} 7$

Solution

a. We see that i goes from 2 to 5. Make a table of values corresponding to each term in the sum. From Table 1,

$$\sum_{i=2}^{5} 3i = 6 + 9 + 12 + 15$$

i	$3i$
2	6
3	9
4	12
5	15

Table 1

b. To expand $\displaystyle\sum_{i=3}^{6} 7$, we first observe that i goes from 3 to 6. Every term in this sum is 7. The terms do not depend on i, the index, which simply "counts" the terms. Therefore,

$$\sum_{i=3}^{6} 7 = \underset{i=3}{7} + \underset{i=4}{7} + \underset{i=5}{7} + \underset{i=6}{7}$$

 Check It Out 3   Expand the sum given by $\displaystyle\sum_{i=3}^{7} 4i$. Do not evaluate.

Example 4   Writing an Expanded Sum in Summation Notation

Write the sum $-2 -4 -6 -8 - \cdots -20$ using summation notation and evaluate.

Solution   Note that each term in the sum can be written as $-2i$, and that i ranges from 1 to 10. Therefore, this sum can be written as

$$\sum_{i=1}^{10} -2i$$

When using summation notation, the index i is always incremented in steps of 1 as it goes from its starting value to its ending value.

Using the formula for the sum of terms in an arithmetic sequence,

$$S_{10} = \frac{10}{2}[-2 + -20] = -110$$

 Check It Out 4   Write the sum $4 + 8 + 12 + \cdots + 36$ using summation notation, and evaluate.

Sum of Terms of a Geometric Sequence

Note We have used the letter j here for the index. The actual letter used for the index does not matter, as long as we are consistent— the variable used for the index must be the same as the variable used for the expression that is to be evaluated and summed.

Just as with arithmetic sequences, it is possible to find the sum of a finite number of terms of a geometric sequence.

Recall that a geometric sequence is given by the rule $a_j = a_0 r^j$, $j = 0, 1, 2, \ldots$. We want to examine the sum of the first n terms. That is,

$$\sum_{j=0}^{n-1} a_0 r^j = a_0 + a_0 r + a_0 r^2 + \cdots + a_0 r^{n-1}$$

To find a formula for this sum, we first write the sum as follows.

$$S_n = \sum_{j=0}^{n-1} a_0 r^j = a_0(1 + r + r^2 + r^3 + \cdots + r^{n-1})$$

It turns out that

$$1 - r^n = (1 - r)(1 + r + r^2 + r^3 + \cdots + r^{n-1})$$

You can easily check this equation for $n = 1, 2, 3$. It holds true for all other positive integers n as well. Therefore, if $r \neq 1$, we can write

$$\frac{1 - r^n}{1 - r} = 1 + r + r^2 + r^3 + \cdots + r^{n-1}$$

Substituting this expression into the expression for S_n, we then have the formula for the sum of terms of a geometric sequence.

The Sum of Terms of a Geometric Sequence

Let n be a positive integer, and a_0 and r be real numbers, $r \neq 1$.

$$S_n = \sum_{j=0}^{n-1} a_0 r^j = a_0(1 + r + r^2 + r^3 + \cdots + r^{n-1})$$

$$= a_0\left(\frac{1 - r^n}{1 - r}\right)$$

Example 5   **Direct Application of the Summation Formula**

Find the sum of the first five terms of the sequence $a_j = \dfrac{1}{2}(3)^j, j = 0, 1, 2, \ldots$.

Solution Here, $a_0 = \dfrac{1}{2}$, $r = 3$, and $n = 5$. Using the formula for the sum of terms of a geometric sequence, we have

$$S_5 = \frac{1}{2}(1 + 3 + 3^2 + 3^3 + 3^4)$$

$$= \frac{1}{2}\left(\frac{1 - 3^5}{1 - 3}\right)$$

$$= \frac{1}{2}\left(\frac{1 - 243}{-2}\right) = \frac{121}{2}$$

Check It Out 5   Find the sum of the first six terms of the sequence $a_j = 1024\left(\dfrac{1}{4}\right)^j, j = 0, 1, 2, \ldots$.

| **Example 6** | **Finding the Sum if the Formula Cannot Be Directly Applied** |

Find the sum: $\displaystyle\sum_{j=1}^{5} 3(2)^{j-1}$.

Solution Note that this sum is not quite in the form to which our formula for the sum of terms of a geometric sequence can be applied, because
- the value of the index begins with 1 instead of 0; and
- the exponent is $j - 1$ rather than j.

To see how we can approach this problem, first write out the terms in the sum.

$$\sum_{j=1}^{5} 3(2)^{j-1} = 3(2)^0 + 3(2)^1 + 3(2)^2 + 3(2)^3 + 3(2)^4$$
$$= 3 + 3(2) + 3(2)^2 + 3(2)^3 + 3(2)^4$$
$$= 3\left(\frac{1 - 2^5}{1 - 2}\right) \qquad\qquad a_0 = 3, n = 5, r = 2$$
$$= 93$$

 Check It Out 6 Find the sum: $\displaystyle\sum_{i=2}^{5} 5(3)^{i-2}$. ■

When computing a sum of terms of a sequence, it often helps to write out the first few terms and the last few terms. Then it should be clear what should be substituted into the formula for the sum. This is particularly true for applications.

Infinite Geometric Series

The sum of the terms of an infinite geometric sequence is called an **infinite geometric series.** If the terms of the geometric sequence are increasing—for instance, if $r > 1$—then the sum will increase to infinity. However, it can be shown that if $|r| < 1$, then the infinite geometric series has a finite sum.

The Sum of an Infinite Geometric Series

If $|r| < 1$, then the infinite geometric series

$$a_0 + a_0 r + a_0 r^2 + a_0 r^3 + \cdots + a_0 r^{n-1} + \cdots$$

has the sum

$$S = \sum_{i=0}^{\infty} a_0 r^i = \frac{a_0}{1 - r}$$

| **Example 7** | **Sum of an Infinite Geometric Series** |

Determine whether each of the following infinite geometric series has a sum. If so, find the sum.

a. $2 + 4 + 8 + 16 + \cdots$

b. $1 + \dfrac{1}{2} + \dfrac{1}{4} + \dfrac{1}{8} + \dfrac{1}{16} + \cdots$

Solution

a. For this geometric series, $r = 2$ because each term is twice the previous term. Thus $|r| = 2$. The series does not have a sum because $|r| > 1$.

b. For this geometric series, each term is one-half the previous term. Thus $r = \dfrac{1}{2}$.

Because $|r| = \dfrac{1}{2}$ is less than 1, the series has a sum. We use the formula for the sum of an infinite geometric series to find the sum.

$$S = \frac{a_0}{1 - r} = \frac{1}{1 - \left(\dfrac{1}{2}\right)} = \frac{1}{\dfrac{1}{2}} = 2$$

 Check It Out 7 Find the following sum: $2 + \dfrac{1}{2} + \dfrac{1}{8} + \dfrac{1}{32} + \cdots$ ■

Applications

The notion of summing terms of a particular sequence occurs in a variety of applications. Two such applications are discussed in the next two examples.

| Example 8 | Seating Capacity

An auditorium has 30 seats in the front row. Each subsequent row has two seats more than the row directly in front of it. If there are 12 rows in the auditorium, how many seats are there altogether?

Solution From the statement of the problem, we see that the second row must have 32 seats, the third row must have 34 seats, and so on. Therefore, this is an arithmetic sequence with $a_0 = 30$ and common difference $d = 2$. Also, $n = 12$ because there are 12 rows of seats. We then have

$$S_n = \frac{n}{2}(2a_0 + d(n - 1)) \qquad \text{We know } a_0, d, \text{ and } n$$

$$S_{12} = \frac{12}{2}(2(\mathbf{30}) + \mathbf{2}(\mathbf{12} - 1)) \qquad \text{Substitute } a_0 = 30, n = 12, \text{ and } d = 2$$

$$= \frac{12}{2}(60 + 2(11)) = 492 \qquad \text{Simplify}$$

Thus there are 492 seats in the auditorium.

 Check It Out 8 Rework Example 8 for the case in which there are 25 seats in the front row, each subsequent row has two seats more than the row directly in front of it, and there are 10 rows.

Annuities are investments in which a fixed amount of money is invested each year. The interest earned on the investment is compounded annually. Example 9 discusses a specific case of an annuity.

| Example 9 | Retirement Funds

Suppose $2000 is deposited initially (and at the *end* of each year) into a retirement account that pays 5% interest compounded annually. What is the total amount in the account at the end of 10 years?

Solution The interest earned on the $2000 that is deposited in any given year will not begin to be paid until a year later. Thus, by the end of the tenth year, only 9 years' worth of interest will have been paid on the amount deposited at the end of the first year, and only 1 year's worth of interest will have been paid on the amount deposited at the end of ninth year. No interest will have been paid on the amount deposited at the end of the tenth year, since that deposit will just have been made.

This information is summarized in Table 2. The compound interest formula has been used to calculate the total value of each $2000 deposit for 10 years after the account was opened.

Years After Opening Account	0	1	$\cdots$	9	10
Deposit ($)	2000	2000	$\cdots$	2000	2000
Value at End of Tenth Year ($)	$2000(1.05)^{10}$	$2000(1.05)^9$	$\cdots$	$2000(1.05)$	2000

Table 2

To find the total amount, we must add all these amounts:

Amount at end of 10 years $= 2000 + 2000(1.05) + \cdots + 2000(1.05)^9 + 2000(1.05)^{10}$

Using $a_0 = 2000$, $n = 11$, and $r = 1.05$ in the geometric sum formula, we find that this sum is

$$S_{11} = 2000\left(\frac{1 - (1.05)^{11}}{1 - 1.05}\right)$$

$$= 28{,}413.57$$

Thus there will be **$28,413.57** in the account at the end of 10 years.

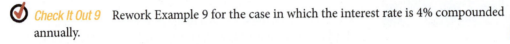 **Check It Out 9**   Rework Example 9 for the case in which the interest rate is 4% compounded annually.

Exercises 10.2

 Just in Time Exercises

1. True or False: If a bank is advertising an account that pays 7% interest compounded quarterly, interest will be computed 6 times per year.

2. True or False: A bank is advertising an account that pays 6% interest, compounded monthly. A customer opens an account of this type and leaves money in the account for 5 years. Over the 5 years, interest will be deposited into the account 60 times.

3. A bank is advertising an account with an interest rate of 2%, compounded semiannually. A customer opens an account with an initial deposit of $1000 and makes no more deposits or withdrawals. What will the account balance be at the end of 2 years?

4. A bank is advertising an account with an interest rate of 5%, compounded quarterly. A customer opens an account with an initial deposit of $1000 and makes no more deposits or withdrawals. What will the account balance be at the end of 2 years?

 Skills

In Exercises 5–10, find the sum of the first 14 terms of each arithmetic sequence.

5. $3, 6, 9, 12, 15, \ldots$

6. $4, 8, 12, 16, 20, \ldots$

7. $-6, -1, 4, 9, \ldots$

8. $-8, -5, -2, 1, 4, \ldots$

9. $2, 7, 12, 17, 22, \ldots$

10. $6, 13, 20, 27, \ldots$

In Exercises 11–16, find the sum of the first eight terms of each geometric sequence.

11. $3, 6, 12, 24, \ldots$

12. $4, 8, 16, 32, \ldots$

13. $6, 3, \dfrac{3}{2}, \dfrac{3}{4}, \ldots$

14. $12, 4, \dfrac{4}{3}, \dfrac{4}{9}, \ldots$

15. $2, 3, \dfrac{9}{2}, \dfrac{27}{4}, \ldots$

16. $12, 16, \dfrac{64}{3}, \dfrac{256}{9}, \ldots$

In Exercises 17–42, find the sum.

17. $3 + 6 + 9 + \cdots + 90$

18. $1 + 5 + 9 + \cdots + 53$

19. $10 + 13 + 16 + \cdots + 55$

20. $8 + 12 + 16 + \cdots + 72$

21. $4 + 9 + 14 + \cdots + (5n + 4)$

22. $1 + 3 + 5 + \cdots + (2n + 1)$

23. $1 + 5 + 25 + \cdots + 78{,}125$

24. $1 + 4 + 16 + \cdots + 1024$

25. $2 + 6 + 18 + \cdots + 1458$

26. $7 + 14 + 28 + \cdots + 896$

27. Sum of the odd integers from 5 to 125, inclusive

28. Sum of the odd integers from 35 to 105, inclusive

29. Sum of the odd integers from 27 to 115, inclusive

30. Sum of the odd integers from 51 to 205, inclusive

31. Sum of the even integers from 4 to 130, inclusive

32. Sum of the even integers from 8 to 160, inclusive

33. Sum of the even integers from 10 to 102, inclusive

34. Sum of the even integers from 20 to 200, inclusive

35. $\displaystyle\sum_{i=0}^{6} (2i)$

36. $\displaystyle\sum_{i=0}^{5} (-3i)$

37. $\displaystyle\sum_{i=0}^{10} (5 + 2i)$

38. $\displaystyle\sum_{i=0}^{7} (2 + 4i)$

39. $\displaystyle\sum_{i=0}^{4} \left(\dfrac{1}{2}\right)^{i}$

40. $\displaystyle\sum_{k=0}^{5} \left(\dfrac{2}{3}\right)^{k}$

41. $\displaystyle\sum_{i=0}^{4} 8(3^{i})$

42. $\displaystyle\sum_{i=0}^{5} 5(2^{i})$

In Exercises 43–54, (a) Write using summation notation, and (b) find the sum.

43. $2 + 4 + 6 + \cdots + 40$ **44.** $1 + 4 + 7 + \cdots + 58$ **45.** $a + 2a + 3a + \cdots + 60a$

46. $2z + 4z + 6z + \cdots + 20z$ **47.** $2 + 4 + 8 + \cdots + 1024$ **48.** $3 + 9 + 27 + \cdots + 59{,}049$

49. $a + a^2 + a^3 + \cdots + a^{40}$ **50.** $2z + 6z^3 + 18z^5 + \cdots + 486z^{11}$

51. The sum of the first 25 terms of the sequence defined by $a_n = 2.5n$, $n = 0, 1, 2, \ldots$

52. The sum of the first 50 terms of the sequence defined by $a_n = 6.5n$, $n = 0, 1, 2, \ldots$

53. The sum of the first 45 terms of the sequence defined by $a_n = (0.5)^n$, $n = 0, 1, 2, \ldots$

54. The sum of the first 60 terms of the sequence defined by $a_n = (0.4)^n$, $n = 0, 1, 2, \ldots$

In Exercises 55–66, evaluate the sum. For each sum, state whether it is arithmetic or geometric. Depending on your answers, state the value of d or r.

55. $\displaystyle\sum_{k=0}^{6} (2k + 1)$ **56.** $\displaystyle\sum_{k=0}^{8} (3k - 1)$ **57.** $\displaystyle\sum_{k=0}^{20} (0.5k)$

58. $\displaystyle\sum_{k=0}^{18} (0.25k)$ **59.** $\displaystyle\sum_{k=0}^{6} 2^{k-1}$ **60.** $\displaystyle\sum_{k=0}^{7} \left(\frac{3}{4}\right)^{k+1}$

61. $\displaystyle\sum_{k=5}^{10} (0.5)^{k-4}$ **62.** $\displaystyle\sum_{k=4}^{9} (0.25)^{k-3}$ **63.** $\displaystyle\sum_{k=1}^{5} 2(0.5)^k$

64. $\displaystyle\sum_{k=1}^{9} 3(0.25)^k$ **65.** $\displaystyle\sum_{k=0}^{5} (3(k+2) + 3(k-1))$ **66.** $\displaystyle\sum_{k=0}^{6} (2(2k+4) - 2(k+1))$

In Exercises 67–76, determine whether the infinite geometric series has a sum. If so, find the sum.

67. $6 + 3 + \dfrac{3}{2} + \dfrac{3}{4} + \cdots$ **68.** $8 + 4 + 2 + 1 + \cdots$ **69.** $9 + 3 + 1 + \dfrac{1}{3} + \dfrac{1}{9} + \cdots$

70. $12 + 3 + \dfrac{3}{4} + \dfrac{3}{12} + \cdots$ **71.** $4 + 8 + 16 + 32 + \cdots$ **72.** $\dfrac{1}{2} + \dfrac{3}{2} + \dfrac{9}{2} + \dfrac{27}{2} + \cdots$

73. $\displaystyle\sum_{k=0}^{\infty} 2(0.5)^k$ **74.** $\displaystyle\sum_{k=0}^{\infty} 3(0.25)^k$ **75.** $\displaystyle\sum_{k=0}^{\infty} 3\left(\frac{1}{5}\right)^k$ **76.** $\displaystyle\sum_{k=0}^{\infty} 5\left(\frac{1}{8}\right)^k$

Applications

77. Stacking Displays A store clerk is told to stack cookie boxes in a pyramid pattern for a store display, as pictured.

Etc.

a. If the clerk has 55 boxes, how many boxes must be placed in the bottom row if all the boxes are to be displayed at one time?

b. The store manager gives the clerk 15 more boxes and tells her to use all 70 boxes to build a display in the same pyramid pattern. The clerk replies that would be impossible. Explain why she is correct.

c. The clerk offers to start with 70 boxes and make a display in the same pyramid pattern—and to do it in such a way that as few boxes as possible will be left over. How many boxes will be in the bottom row? How many boxes will be left over?

78. **Communication** Many large corporations have in place an emergency telephone chain in which one employee in each division is designated to be the first called in the case of an emergency. That employee then calls three employees within the division, each of whom in turn calls three employees, and so on. The chain stops once all the employees have been notified of the emergency.

   a. Write the first five terms of the sequence that represents the number of people called at each step of the chain. (The first step consists of just the "designated" employee being called.) Is this an arithmetic sequence or a geometric sequence?

   b. Use an appropriate formula to answer the following question: How many steps of the chain are needed to notify all the employees of a corporate division with 600 employees?

   c. Explain why this method of notification is very efficient.

79. **Education Savings** The parents of a newborn child decide to start saving for the child's college education. At the end of each calendar year, they put $1500 into an Educational Savings Account (ESA) that pays 6% interest compounded annually. What will be the total amount in the account 18 years after they make their initial deposit?

80. **Retirement Savings** Maria is a recent college graduate who wants to take advantage of an individual retirement account known as a Roth IRA. In order to build savings for her retirement, she wants to put $2500 at the end of each calendar year into an IRA that pays 5.5% interest compounded annually. If she stays with this plan, what will be the total amount in the account 40 years after she makes her initial deposit?

81. **Physics** A ball dropped to the floor from a height of 10 feet bounces back up to a point that is three-fourths as high. If the ball continues to bounce up and down, and if after each bounce it reaches a point that is three-fourths as high as the point reached on the previous bounce, calculate the total distance the ball travels from the time it is dropped to the time it hits the floor for the third time.

82. **Television Piracy** The loss of revenue to an industry due to piracy can be staggering. For example, a newspaper article reported that the pay television industry in Asia lost nearly $1.3 billion in potential revenue in 2003 because of the use of stolen television signals. The loss was projected to grow at a rate of 10% per year. *(Source: The Financial Times)*

   a. Assuming the projection was accurate, how much did the pay television industry in Asia lose in the years 2004, 2005, and 2006?

   b. Assuming the projected trend has prevailed to the present time and will continue into the future, what is the projected loss in revenue for the year that is *n* years after 2003?

   c. Find the total loss of revenue for the years 2003 to 2012, inclusive.

83. **Literature** The following poem (*As I Was Going to St. Ives*, circa 1730) refers to the name of a quaint old village in Cornwall, England. *(Source: www.rhymes.org.uk)*

   > As I was going to St. Ives
   > I met a man with seven wives.
   > Every wife had seven sacks,
   > Every sack had seven cats,
   > Every cat had seven kits.
   > Kits, cats, sacks, and wives,
   > How many were going to St. Ives?

   a. Use the sum of a *sequence* of numbers to express the number of people and objects (combined) that the author of this poem encountered while going to St. Ives. Do not evaluate the sum. Is this the sum of terms of an arithmetic sequence or a geometric sequence? Explain.

   b. Use an appropriate formula to find the sum from part (a).

84. **Dimensions** A carpet warehouse needs to calculate the diameter of a rolled carpet given its length, width, and thickness. If the diameter of the carpet roll can be predicted ahead of time, the warehouse will know how much to order so as not to exceed warehouse capacity. Assume that the carpet is rolled lengthwise. The cross-section of the carpet roll is then a spiral. To simplify the problem, approximate the spiral cross-section by a set of *n* concentric circles whose radii differ by the thickness *t*. Calculate the number of circles *n* using the fact that the sum of the circumferences of the *n* circles must equal the given length. How can you find the diameter once you know *n*?

Concepts

85. The first term of an arithmetic sequence is 4. The sum of the first three terms of the sequence is 24. Use summation notation to express the sum of the first eight terms of this sequence, and use an appropriate formula to find the sum.

86. Find the following sum:

$$1 + 2\left(1 + \frac{1}{2}\right) + 3\left(1 + \frac{1}{3}\right) + \cdots + 50\left(1 + \frac{1}{50}\right)$$

(*Hint:* Expand first.)

87. Given two terms of an arithmetic sequence, $a_2 = 14$ and $a_6 = 2$, find $\sum_{k=1}^{8} a_k$. (*Hint:* First find d and a_0.)

88. Given two terms of a geometric sequence, $a_0 = -1$ and $a_3 = 27$, find $\sum_{j=0}^{5} a_j$.

89. For a geometric sequence, find a_0 if $\sum_{j=0}^{4} a_j = 3$ and $r = \frac{1}{2}$.

Objectives

- Generate terms of a general sequence.

- Find a rule for a sequence given a few terms.

- Generate terms of a recursively defined sequence.

- Find a rule for a recursively defined sequence.

- Calculate partial sums of terms of a sequence.

- Apply general sequences to word problems.

VIDEO EXAMPLES

SECTION 10.3

The number of spirals in the head of a sunflower forms a sequence that does not fit the pattern of an arithmetic or a geometric sequence. This is only one example of the many different types of sequences that can be studied. In this section, we will discuss sequences in a more general setting. The Fibonacci sequence illustrated by the sunflower pattern will be discussed in Example 6.

Sequences Given by a Rule

Recall that the terms of arithmetic and geometric sequences are generated by specific types of rules. We can generate other types of sequences simply by using other kinds of rules. The following examples illustrate some of the types of sequences that can be generated in this way.

Example 1   **Generating a Sequence from a Rule**

Find the first four terms of each of the following sequences.

a. $a_n = n^2$, $n = 0, 1, 2, 3, \ldots$

b. $f(n) = \dfrac{1}{n + 1}$, $n = 0, 1, 2, 3, \ldots$

Solution

a. To find the first four terms, successively substitute $n = 0, 1, 2, 3$ into the formula $a_n = n^2$, which gives

$$a_0 = (0)^2 = 0, \ a_1 = (1)^2 = 1, \ a_2 = (2)^2 = 4, \ a_3 = (3)^2 = 9$$

b. Substitute $n = 0, 1, 2, 3$ into $f(n) = \dfrac{1}{n + 1}$, which gives

$$a_0 = \frac{1}{0 + 1} = 1, \ a_1 = \frac{1}{1 + 1} = \frac{1}{2}, \ a_2 = \frac{1}{2 + 1} = \frac{1}{3}, \ a_3 = \frac{1}{3 + 1} = \frac{1}{4}$$

✅ *Check It Out 1*   Find the first four terms of the sequence defined by $a_n = 1 + 2n^2$, $n = 0, 1, 2, 3, \ldots$

Example 2   **Using the First Few Terms to Find the Rule**

Assuming that the pattern continues, find a rule for the sequence whose first four terms are as given.

a. $1, 8, 27, 64$

b. $1, \dfrac{1}{4}, \dfrac{1}{9}, \dfrac{1}{16}$

Solution

a. The given terms are all perfect cubes. Thus the rule for this sequence is

$$a_n = n^3, \quad n = 1, 2, 3, \ldots$$

Note that this sequence starts with $n = 1$.

b. Examining the terms, we see that each term is the reciprocal of a perfect square. Thus the rule is

$$a_n = \frac{1}{n^2}, \quad n = 1, 2, 3, \ldots$$

Check It Out 2   Find a rule for the sequence whose first four terms are

$$1, \frac{1}{8}, \frac{1}{27}, \frac{1}{64}$$

Alternating Sequences

In an **alternating sequence,** the terms *alternate* between positive and negative numbers. The next example involves finding a rule for a simple sequence of this type.

Example 3   **Finding the Rule for an Alternating Sequence**

Assuming that the given pattern continues, find a rule for the sequence whose terms are given by

$$1, -1, 1, -1, 1, -1, \ldots$$

Solution The terms of this sequence consist only of 1 and -1, with the two numbers alternating. One way to write a rule for this sequence is

$$a_n = (-1)^n, \quad n = 0, 1, 2, 3, \ldots$$

This rule works because -1 raised to an even power will equal 1, while -1 raised to an odd power will equal -1. Since the value of n alternates between even and odd numbers, the rule $a_n = (-1)^n$ produces the sequence

$$1, -1, 1, -1, 1, -1, \ldots$$

Check It Out 3   Find a rule for the sequence whose terms are given by

$$-1, 1, -1, 1, -1, 1, \ldots$$

Example 4   **Generating and Graphing Terms of a Sequence**

Let $f(n) = (-1)^n(n^2 + 1)$. Fill in Table 1 and plot the first five terms of the sequence.

n	0	1	2	3	4
$f(n)$					

Table 1

Solution To fill in the table, substitute $n = 0, 1, 2, 3, 4$ (in succession) into the expression for $f(n)$.

$$f(0) = (-1)^0 \; ((0)^2 + 1) = \quad (1)(1) = \quad 1$$
$$f(1) = (-1)^1 \; ((1)^2 + 1) = \quad (-1)(2) = \quad -2$$
$$f(2) = (-1)^2 \; ((2)^2 + 1) = \quad (1)(5) = \quad 5$$
$$f(3) = (-1)^3 \; ((3)^2 + 1) = (-1)(10) = -10$$
$$f(4) = (-1)^4 \; ((4)^2 + 1) = \quad (1)(17) = \quad 17$$

See Table 2.

n	0	1	2	3	4
$f(n)$	1	−2	5	−10	17

Table 2

The first five terms of the sequence are plotted in Figure 1. Note that the dots are *not* connected. Because the function *f* is a sequence, its domain consists of the set of all nonnegative integers. Thus, *f* is not defined for any number that lies *between* two consecutive nonnegative integers.

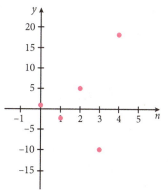

Figure 1

Check It Out 4 Let $f(n) = (-1)^n \left(\frac{1}{n}\right)$, $n = 1, 2, 3, \ldots$. Fill in Table 3 and plot the first five terms of the sequence.

n	1	2	3	4	5
f(n)					

Table 3

Recursively Defined Sequences

Some sequences are defined by expressing the *n*th term a_n as a function of one or more of the preceding terms rather than as a function of *n* alone, and specifying the value of the first term of the sequence, or the values of the first several terms. A sequence generated in this manner is called a **recursively defined sequence**. Example 5 illustrates such a sequence.

Example 5 **Generating Terms of a Recursively Defined Sequence**

Find the first four terms of the sequence defined recursively by

$$a_0 = 1, \quad a_n = 3 + a_{n-1}, \quad n = 1, 2, 3, \ldots$$

What type of sequence is generated? Find a rule for the *n*th term of the sequence that depends only on *n*.

Solution The general rule for a_n is defined using a_{n-1}, the term just preceding a_n. The first four terms are

$$a_0 = 1 \qquad\qquad\qquad \text{Given}$$
$$a_1 = 3 + a_0 = 3 + 1 = 4 \qquad \text{Rule for } a_n \text{ with } n = 1$$
$$a_2 = 3 + a_1 = 3 + 4 = 7 \qquad \text{Rule for } a_n \text{ with } n = 2$$
$$a_3 = 3 + a_2 = 3 + 7 = 10 \qquad \text{Rule for } a_n \text{ with } n = 3$$

The terms of this sequence are thus

$$1, 4, 7, 10, \ldots$$

This is an arithmetic sequence with a common difference of 3. Using the formula for the *n*th term of an arithmetic sequence, we can write

$$a_n = a_0 + nd = 1 + 3n$$

Check It Out 5   Consider the recursively defined sequence

$$b_0 = 1, \quad b_n = 1 - b_{n-1}, \quad n = 1, 2, 3, \ldots$$

a. Find the first five terms of the sequence.

b. What is the range of the function f that corresponds to this sequence (i.e., the function $f(n) = b_n, n = 1, 2, 3, \ldots)$?

The next example deals with the Fibonacci sequence, which was discussed briefly at the beginning of this section.

| **Example 6** | **Generating Terms of the Fibonacci Sequence** |

Find the first five terms of the Fibonacci sequence, defined recursively as follows.

$$f_0 = 1, \quad f_1 = 1, \quad f_n = f_{n-1} + f_{n-2}, \quad n = 2, 3, \ldots$$

Solution   The recursive definition of the Fibonacci sequence states that the term f_n (for $n \geq 2$) is the sum of the *two preceding terms*. We can compute the first five terms as follows.

$$f_0 = 1; \quad f_1 = 1 \qquad \text{Given}$$
$$f_2 = f_1 + f_0 = 1 + 1 = 2 \qquad \text{Substitute } n = 2$$
$$f_3 = f_2 + f_1 = 2 + 1 = 3 \qquad \text{Substitute } n = 3$$
$$f_4 = f_3 + f_2 = 3 + 2 = 5 \qquad \text{Substitute } n = 4$$

Hence, the first five terms of the Fibonacci sequence are

$$1, 1, 2, 3, 5$$

 Check It Out 6   Generate the terms $f_5, f_6, f_7,$ and f_8 of the Fibonacci sequence.

Partial Sums

It is sometimes necessary to find the sum of two or more consecutive terms of a given sequence of numbers. In Section 10.2, we found sums of terms of arithmetic and geometric sequences. In this section, we extend the discussion to include sums of terms of other types of sequences.

Suppose we want to denote the sum of the first n terms of the sequence $a_0, a_1, a_2, \ldots$. Using the summation notation from Section 10.2, we have the following.

> ### nth Partial Sum of a Sequence
>
> Let n be a positive integer. The sum of the first n terms of the sequence $a_0, a_1, a_2, \ldots$ (the terms a_0 through a_{n-1}) is denoted by $\displaystyle\sum_{i=0}^{n-1} a_i$. In other words,
>
> $$\sum_{i=0}^{n-1} a_i \text{ means } a_0 + a_1 + \cdots + a_{n-1}$$
>
> This sum is referred to as the **nth partial sum** of the sequence.

Some partial sums will be computed in the following examples.

| **Example 7** | **Direct Calculation of a Partial Sum of a Sequence** |

Compute each sum.

a. $\displaystyle\sum_{i=0}^{4} i^2$

b. $\displaystyle\sum_{i=0}^{3} (-1)^i (2i + 1)$

Solution

a. We consecutively substitute 0, 1, 2, 3, and 4 for i in the expression i^2 and then add the results.

$$\sum_{i=0}^{4} i^2 = 0^2 + 1^2 + 2^2 + 3^2 + 4^2 = 0 + 1 + 4 + 9 + 16 = 30$$

b. By consecutively substituting 0, 1, 2, and 3 for i in the expression $(-1)^i(2i + 1)$ and then adding the results, we obtain

$$\sum_{i=0}^{3} (-1)^i(2i + 1) = (-1)^0(2(0) + 1) + (-1)^1(2(1) + 1) + (-1)^2(2(2) + 1) + (-1)^3(2(3) + 1)$$

$$= 1(1) + (-1)(3) + (1)(5) + (-1)(7) = 1 - 3 + 5 - 7 = -4$$

 Check It Out 7 Compute the following sums.

a. $\displaystyle\sum_{j=0}^{5} (j + 1)$

b. $\displaystyle\sum_{k=0}^{4} (6k - 5)$

Applications

Sequences are extremely useful in studying events that happen at regular intervals. This is usually done with a recursive sequence. We illustrate this idea in the next example.

Example 8   **Drug Dose**

An initial dose of 40 milligrams of the pain reliever acetaminophen is given to a patient. Subsequent doses of 20 milligrams each are administered every 5 hours. Just before each 20-milligram dose is given, the amount of acetaminophen in the patient's bloodstream is 25% of the total amount in the bloodstream just after the previous dose was administered.

a. Let a_0 represent the initial amount of the drug in the bloodstream and, for $n \geq 1$, let a_n represent the amount in the bloodstream immediately after the nth 20-milligram dose is given. Make a table of values for a_0 through a_6.

b. Plot the values you tabulated in part (a). What do you observe?

c. With the aid of the values you tabulated in part (a), find a recursive definition of a_n.

Solution

a. We construct Table 4 as follows.

n	a_n, Amount in Bloodstream (mg)	Comments
0	40	Initial amount
1	$(0.25)(40) + 20 = 30$	Amount remaining from prior dose + new dose
2	$(0.25)(30) + 20 = 27.5$	Amount remaining from prior dose + new dose
3	$(0.25)(27.5) + 20 = 26.875$	Amount remaining from prior dose + new dose
4	$(0.25)(26.875) + 20 \approx 26.72$	Amount remaining from prior dose + new dose
5	$(0.25)(26.72) + 20 \approx 26.68$	Amount remaining from prior dose + new dose
6	$(0.25)(26.68) + 20 \approx 26.67$	Amount remaining from prior dose + new dose

Table 4

b. The values of a_n, the amount of acetaminophen in the bloodstream after the nth 20-milligram dose, are plotted in Figure 2.

Note that as n increases, the amount of the drug in the bloodstream seems to approach a constant amount of approximately 26.67 milligrams. This is known as reaching **steady state**.

c. Examining the table of values closely, we see that

$$a_0 = 40 \text{ and } a_n = (0.25)(\text{previous amount}) + 20 = 0.25a_{n-1} + 20$$

Figure 2

 Check It Out 8 Rework Example 8 for the case in which the initial dose of acetaminophen is 40 milligrams, subsequent doses of 20 milligrams are given every 5 hours thereafter, and, just before each 20-milligram dose is administered, the amount of the drug in the patient's bloodstream is only 20% of the total amount in the bloodstream just before the previous dose was given.

Exercises 10.3

Skills

In Exercises 1–18, find the first five terms of the sequence.

1. $a_n = -4n + 6$, $n = 0, 1, 2, 3, \ldots$

2. $a_n = 6n + 2$, $n = 0, 1, 2, 3, \ldots$

3. $a_n = -4\left(\dfrac{1}{3}\right)^n$, $n = 0, 1, 2, 3, \ldots$

4. $a_n = -\left(\dfrac{5}{2}\right)^n$, $n = 0, 1, 2, 3, \ldots$

5. $b_n = n^2 + 4$, $n = 0, 1, 2, 3, \ldots$

6. $b_n = -6n^3 + 1$, $n = 0, 1, 2, 3, \ldots$

7. $f(n) = \dfrac{n}{2n^2 + 1}$, $n = 1, 2, 3, \ldots$

8. $f(n) = \dfrac{n}{-3n^2 - 1}$, $n = 1, 2, 3, \ldots$

9. $a_n = (-1)^n e^{n+1}$, $n = 0, 1, 2, 3, \ldots$

10. $a_n = (-1)^n 2^{n+1}$, $n = 0, 1, 2, 3, \ldots$

11. $g(n) = \dfrac{2^n + 1}{n^2 + 1}$, $n = 1, 2, 3, \ldots$

12. $g(n) = \dfrac{3^n - 1}{n^2}$, $n = 1, 2, 3, \ldots$

13. $a_n = \sqrt{2n + 4}$, $n = 0, 1, 2, 3, \ldots$

14. $a_n = \dfrac{25}{n}$, $n = 1, 2, 3, \ldots$

15. $a_n = n^2 + 2$, $n = 0, 1, 2, 3, \ldots$

16. $a_n = 2 + n$, $n = 0, 1, 2, 3, \ldots$

17. $a_n = 2n^3$, $n = 0, 1, 2, 3, \ldots$

18. $a_n = (2n^3)^{1/3}$, $n = 0, 1, 2, 3, \ldots$

In Exercises 19–28, find a rule for each sequence whose first four terms are given. Assume that the given pattern will continue.

19. $-2, -6, -10, -14, \ldots$

20. $-3, 2, 7, 12, \ldots$

21. $1, \dfrac{1}{2}, \dfrac{1}{4}, \dfrac{1}{8}, \ldots$

22. $1, \dfrac{1}{3}, \dfrac{1}{9}, \dfrac{1}{27}, \ldots$

23. $1, \sqrt{2}, \sqrt{3}, 2, \ldots$

24. $\sqrt{2}, 2, \sqrt{6}, \sqrt{8}, \ldots$

25. $1, 0.4, 0.16, 0.064, \ldots$

26. $1, \dfrac{1}{4}, \dfrac{1}{16}, \dfrac{1}{64}, \ldots$

27. $\sqrt{3}, \sqrt{6}, 3, \sqrt{12}, \ldots$

28. $2, 2\sqrt{2}, 2\sqrt{3}, 4$

In Exercises 29–36, find the first four terms of the recursively defined sequence.

29. $a_0 = 6$; $a_n = a_{n-1} - 2$, $n = 1, 2, 3, \ldots$

30. $a_0 = -3$; $a_n = a_{n-1} + 2.5$, $n = 1, 2, 3, \ldots$

31. $b_1 = 4$; $b_n = \dfrac{1}{4}b_{n-1}$, $n = 2, 3, 4, \ldots$

32. $b_1 = 2$; $b_n = \dfrac{3}{2}b_{n-1}$, $n = 2, 3, 4, \ldots$

33. $a_0 = -1$; $a_n = a_{n-1} + n$, $n = 1, 2, 3, \ldots$

34. $a_0 = 4$; $a_n = a_{n-1} - n$, $n = 1, 2, 3, \ldots$

35. $b_1 = \sqrt{3}$; $b_n = \sqrt{b_{n-1} + 3}$, $n = 2, 3, 4, \ldots$

36. $b_1 = \sqrt{3}$; $b_n = \sqrt{\dfrac{b_{n-1}}{3}}$, $n = 2, 3, 4, \ldots$

In Exercises 37–40, find the first four terms of the recursively defined sequence. Find the rule for a_n in terms of just n.

37. $a_0 = -4$; $a_n = a_{n-1} + 2$, $n = 1, 2, 3, \ldots$

38. $a_0 = 3$; $a_n = a_{n-1} + 1.5$, $n = 1, 2, 3, \ldots$

39. $b_1 = 6$; $b_n = \dfrac{1}{2}b_{n-1}$, $n = 2, 3, 4, \ldots$

40. $b_1 = 7$; $b_n = \dfrac{3}{4}b_{n-1}$, $n = 2, 3, 4, \ldots$

Applications

41. Games A popular electronic game originally sold for $200. The price of the game was adjusted annually by just enough to keep up with inflation. Assume that the rate of inflation was 4% per year.

   a. For $n \geq 0$, let p_n be the price of the game n years after it was put on the market. Define p_n recursively.

   b. Find an expression for p_n in terms of just n.

   c. What was the price of the game 3 years after it was put on the market?

   d. How many years after the game was put on the market did the price first exceed $250?

   e. How many years after the game was put on the market did the price first exceed $300?

42. Compensation A certain company rewards its employees with annual bonuses that grow with the number of years the employee remains with the company. At the end of the first full year of employment, an employee's bonus is $1000. At the end of each full year beyond the first, the employee receives $1000 plus 50% of the previous year's bonus.

   a. What bonuses does an employee receive after each of the first four full years of employment?

   b. For $n \geq 0$, let b_n be the bonus received after n full years of employment. Define b_n recursively.

43. Manufacturing One product line offered by a window manufacturer consists of rectangular windows that are 36 inches in height. The widths of the windows range from 18 inches to 54 inches, in increments of half an inch.

   a. Give an expression for the perimeter P_n of the nth window in this product line, where P_0 is the perimeter of the smallest window, P_1 is the perimeter of the second-smallest window, and so on.

   b. Give an expression for the area A_n of the nth window in this product line, where A_0 is the area of the smallest window, A_1 is the area of the second-smallest window, and so on.

   c. What are the dimensions, perimeter, and area of the sixth-smallest window in this product line?

   d. If the smallest window sells for $250 and the prices of the windows are graduated at the rate of $10 for every half-inch of additional width, what is the total cost (ignoring sales tax) of one 19.5-inch-wide window and five 36-inch-wide windows?

44. Horticulture The Morales family bought a Christmas tree. As soon as they got the tree home and set it up, they put 3 quarts of water into the tree holder. Every day thereafter, they awoke to find that half of the water from the previous day was gone, so they added a quart of water.

   a. For $n \geq 0$, let w_n be the volume of water in the tree holder (just after water was added) n days after the tree was set up in the home of the Morales family. Define w_n recursively.

   b. How many days after the tree was initially set up did the family awake to find that the water level had dipped below the 2.1-quart mark for the first time?

   c. When, if ever, did the family awake to find that the water level had dipped below the 1-quart mark for the first time? Explain.

45. Fundraising During a recent month, students contributed money at school for the benefit of flood victims in another part of the country. One enterprising student, Matt, asked his aunt to donate money on his behalf. She agreed that on each day that Matt contributed, she would match his donation plus donate 10 cents more. There were 21 school days during the month in question. From the second school day on, Matt donated 3 cents more than he gave on the previous school day. In total, Matt and his aunt contributed $17.22.

   a. How much money did Matt contribute on the first school day of the month in question?

   b. What was Matt's total contribution for that month?

   c. How much did Matt's aunt donate on his behalf?

46. **Distribution** Kara gives a pencil to every child who comes to her house trick-or-treating on Halloween. The first year she did this, she bought 120 pencils, which turned out to be one-third more pencils than she needed. Kara kept the extras to hand out the next year. The second year, she bought x new pencils (to add to the supply she had left over from the first year). One-fourth of all the pencils she had to give to trick-or-treaters the second year (the new pencils plus the extras from the first year) were left over. The third year, she again bought x new pencils, and one-fifth of the total number available for handout that year were left over.

   a. How many trick-or-treaters went to Kara's house the first year, and how many pencils were left over that year?

   b. Give expressions (in terms of x) for the total number of pencils available for handout the second year and the number of children who came to Kara's house trick-or-treating that year.

   c. Give expressions (in terms of x) for the total number of pencils available for handout the third year and the number of trick-or-treaters who went to Kara's house that year.

   d. If Kara had 14 pencils left over the third year, what is the value of x?

   e. Use the value of x that you found in part (d) to determine the number of children who came to Kara's house trick-or-treating the second year and the number who came the third year.

Concepts

47. Find the smallest positive integer n such that $\left| \sum_{i=0}^{n-1} b_i \right| \geq 81$ if $b_n = 3n^2 - 2$, $n = 0, 1, 2, \dots$

48. For $n \geq 1$, let V_n be the volume of a cube that is n units on a side. Using summation notation, give an expression for the sum of V_n over the first six positive integers n, and find the sum.

49. A sequence $a_0, a_1, a_2, \dots$ has the property that $a_n = 3a_{n-1} + 2$ for $n = 1, 2, 3, \dots$. If $a_3 = 134$, what is the value of a_0?

50. A sequence $b_0, b_1, b_2, \dots$ has the property that $b_n = c\left(\frac{n+3}{n+2}\right)b_{n-1}$ for $n = 1, 2, 3, \dots$, where c is a positive constant to be determined. Find c if $b_2 = 25$ and $b_4 = 315$.

51. If $a_n = 1 - (a_{n-1})^3$ for $n = 1, 2, 3, \dots$, for what value(s) of a_0 is the sequence $a_0, a_1, a_2, \dots$ an alternating sequence?

52. If $a_n = \sqrt{a_{n-1}} + \frac{1}{1000}$ for $n = 1, 2, 3, \dots$, for what value(s) of a_0 are all the terms of the sequence $a_0, a_1, a_2, \dots$ defined?

Counting Methods

10.4

Objectives

- Use the multiplication principle for counting.

- Use permutations in a counting problem.

- Use combinations in a counting problem.

- Distinguish between permutations and combinations.

VIDEO EXAMPLES

SECTION 10.4

Many applications involve counting the number of ways certain events can occur. For example, consider a bicycle race with many participants. One could ask how many possibilities exist for the first-place and second-place finishes. Answering this question involves the use of *counting strategies*, which we discuss in this section.

Multiplication Principle

Basic to almost all counting problems is a simple rule that relies on multiplication. We can think of this rule in terms of setting up slots and filling each slot with one of the factors in the multiplication. In some cases, different slots can contain the same number; in other cases, each slot must contain a unique number. We illustrate this concept in the next two examples.

Example 1 **Filling Slots with Numbers: No Repetition**

Suppose 10 members of a cycling club are practicing for a race. From among these members, how many possibilities are there for the first-place and second-place finishes?

Solution Make two slots, one for first place and one for second place, and decide how many possibilities there are for each slot. From the given information, there are 10 possibilities for the first-place finish, since any of the 10 participants could win the race. For *each* possible first-place winner, there are only 9 possibilities for the second-place finish, since no one can finish in both first place *and* second place. See Table 1.

Number of Possibilities for First Place	Number of Possibilities for Second Place
10	9

Table 1

To get the total number of possibilities for the first-place and second-place finishes, we must *multiply* the number of possibilities for the first slot by the number of possibilities for the second slot. Because there are 10 of the former and 9 of the latter, we write

$$\text{Total} = \text{First-place possibilities} \times \text{Second-place possibilities}$$
$$= (10)(9) = 90$$

Thus there are 90 different ways of filling the first- and second-place positions.

Check It Out 1 Find the total number of possibilities for the first-place, second-place, and third-place finishes in a horse race in which 12 horses compete.

Example 2 **Filling Slots with Numbers That Can Be Repeated**

You wish to form a three-digit number, with each digit ranging from 1 to 9. You may use the same digit more than once. How many three-digit numbers can you make?

Solution We use the same strategy as in Example 1, in the sense of filling slots. Now, however, we have *three* slots to fill, one for each of the three digits in the number we are forming. There are nine possibilities for each slot, since

- each digit is to be from the set {1, 2, …, 9}; and
- any number in that set may fill any slot.

See Table 2.

Possibilities for First Digit	Possibilities for Second Digit	Possibilities for Third Digit
9	9	9

Table 2

By the multiplication technique used in Example 1, there are 81 possibilities for just the first two digits. For each of these 81 two-digit numbers, there are nine possibilities for the third digit.

Total = Possibilities for first × Possibilities for second × Possibilities for third
= 9 × 9 × 9 = **729**

Thus there are 729 different three-digit numbers that can be formed from the digits 1 through 9.

Notice how this example differs from Example 1. Here, we can use the same digit more than once, and so we have nine possibilities for each slot. In the first example, we had to reduce the number of possibilities for the second slot by 1.

 Check It Out 2   How many three-digit numbers can be formed if the first digit can be anything but zero and there is no restriction on the second and third digits? ■

The preceding two examples illustrate an important rule known as the **multiplication principle**.

Multiplication Principle

Suppose there are n slots to fill. Let a_1 be the number of possibilities for the first slot, a_2 the number of possibilities for the second slot, and so on. Then the total number of ways in which the n slots can be filled is

$$a_1 \times a_2 \times a_3 \times \cdots \times a_n$$

Permutations

Certain types of counting situations happen so frequently that they have a specific name attached to them. For example, one may be interested in knowing the number of ways in which a particular set of objects can be ordered or arranged. Each way of ordering or arranging a certain set of objects is called a **permutation** of those objects.

Example 3   **Filling Just As Many Slots As There Are Objects**

How many different four-block towers can be built with a given set of four different colored blocks?

Solution Think of the problem as that of arranging four blocks in four slots, from bottom to top.

←— 1 possibility
←— 2 possibilities
←— 3 possibilities
←— 4 possibilities

Using the multiplication principle and the fact that any block can be used only once, we see that there are $(4)(3)(2)(1) = \textbf{24}$ different towers that can be built.

 Check It Out 3   In how many ways can the letters of the word TRAIN be arranged? ■

The product of the first n natural numbers occurs so frequently that it has a special name, **n factorial**, which is written as $n!$.

Definition of n factorial ($n!$)

Let n be a positive integer. Then

$$n! = n(n-1)(n-2)\ldots(3)(2)(1)$$

The quantity $0!$ is defined to be 1.

Using this definition, we can say that there are $4!$ ways to build the four-block tower in Example 3.

| Example 4 | **Using Factorial Notation**

Use factorial notation to calculate the number of ways in which seven people can be arranged in a row for a photograph.

Solution There are seven slots available for the seven people. The first slot has seven possibilities, the second slot has only six possibilities (no one can be in both the first position *and* the second position for the photograph), and so on. Using the multiplication principle, we see that

$$\text{Number of arrangements} = (7)(6)(5) \cdots (1) = 7! = 5040$$

Check It Out 4 Use factorial notation to give the number of ways in which 16 people can form a line while waiting to get on a ride at an amusement park. ■

Arrangements also exist in which the number of slots is less than the number of available objects. How do we use factorials to count the arrangements in this case? The next example explores such a situation.

| Example 5 | **Filling Fewer Slots Than There Are Objects**

A store manager has six different candy boxes, but spaces for only four of them on the shelf. In how many ways can the boxes be arranged horizontally on the shelf?

Solution Since there are four "slots" on the shelf, we can create the following diagram:

| one of 6 boxes | one of 5 boxes | one of 4 boxes | one of 3 boxes |

Therefore, the number of ways in which the boxes can be arranged is

$$(6)(5)(4)(3) = 360$$

Check It Out 5 A total of seven people want to hone their archery skills, and there are only three targets available for them to practice on. In how many ways can the targets be assigned if each target is to be used by only one archer? ■

Recall that a permutation is an arrangement of objects. Motivated by Example 5, we now give the formal definition of a permutation.

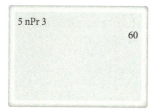
Definition of Permutation

Given *n distinct* objects and *r* slots to fill, the number of **r-permutations of n objects** (i.e., the number of permutations of *r* objects from the collection of *n* objects) is denoted by $P(n, r)$.

$$P(n, r) = (n)(n - 1) \cdots (n - r + 1)$$

Using the definition of factorial, the expression for $P(n, r)$ can also be written as

$$P(n, r) = \frac{n!}{(n - r)!}$$

Note that the definition of $P(n, r)$ is really just a new name for an application of the multiplication principle. Other common notations for $P(n, r)$ include $P_{n,r}$ and $_nP_r$.

The following example illustrates the use of permutation notation.

| Example 6 | **Using Permutation Notation**

A fast food restaurant holds a promotion in which each customer scratches off three boxes on a ticket. Each box contains a picture of one of five items: a burger, a bag of fries, a shake, a pie, or a salad. The item in the first box is free. The item in the second box can be purchased at a 50% discount, and the item in the third box can be bought at 25% off. Using permutation notation, determine the number of different tickets that are possible if each picture can be used only once per ticket.

Solution On any given ticket, there are pictures of three of the five items, in some order or other. Thus each ticket represents a permutation of three objects from a collection of five objects. Therefore, the number of different tickets is

$$P(5, 3) = (5)(4)(3) = 60$$

Note that the second formula for $P(5, 3)$ gives the same answer:

$$P(5, 3) = \frac{5!}{(5 - 3)!} = \frac{5!}{2!} = \frac{120}{2} = 60$$

Use whichever formula you find to be more helpful.

Check It Out 6 Use permutation notation to find the number of ways in which eight teachers' aides at an elementary school can be assigned to cafeteria duty during a 5-day week. Only one aide has cafeteria duty each day, and no one takes cafeteria duty more than once during the week.

Certain problems can be solved only by combining different counting strategies. Thus it is important to read problems carefully and think them through, and not just memorize formulas. The next example illustrates this point.

Example 7   **Filling Two Sets of Slots That Are Independent of Each Other**

A group photograph is taken with four children in the front row and five adults in the back row. How many different photographs are possible?

Solution Break up the problem into two parts.

1. There are 4! ways to arrange the children in the front row.
2. There are 5! ways to arrange the adults in the back row.

How can we use these two pieces of information?

Note that for each arrangement of the children in the front, there are 5! ways to arrange the adults in the back. Since there are 4! ways to arrange the children in the front, we have the following.

Total number of photographs = (ways to arrange kids)(ways to arrange adults)

$$= 4! \, 5! = (24)(120) = 2880$$

Check It Out 7 There are two special parking areas in a small company: one with three spaces for SUVs and one with five spaces for cars. There are three employees who drive their SUVs to work and five who drive their cars. How many different arrangements of employee vehicles in the special lots are possible?

Combinations

In a permutation, we are counting the number of ways in which a certain set of objects can be *ordered*. However, the way in which objects are ordered is not always relevant. The next example illustrates this concept.

Example 8   **Selecting Objects When Order Is Irrelevant**

You have three textbooks on your desk: history (H), English (E), and mathematics (M). You choose two of them to put in your book bag. In how many ways can you do this?

Solution Let us first list all the different ways in which the books can be chosen. See Table 3.

Ways to Arrange History, English	Ways to Arrange History, Math	Ways to Arrange English, Math
HE	HM	EM
EH	MH	ME

Table 3

In the book bag, the order doesn't matter. The only thing that counts is which two books are in the bag. Therefore, HE and EH in Table 3 count as only one possibility, and similarly for HM and MH and for EM and ME. Therefore,

$$\text{Number of ways} = \frac{\text{Ways to arrange any two of three books}}{\text{Ways to arrange any set of two books}}$$

$$= \frac{(3)(2)}{2!} = \frac{6}{2} = 3$$

Thus there are **3** ways to choose two books to go into the book bag.

✓ *Check It Out 8*   Rework Example 8 for the case in which you have four textbooks (history, English, mathematics, and biology) and you put two of them in your book bag. ■

Definition of Combination

Note Other common notations for $C(n, r)$ include $C_{n,r}$ and $_nC_r$. Note that $\binom{n}{r}$ is not the fraction $\left(\frac{n}{r}\right)$.

Let r and n be positive integers with $r \le n$. A selection of r objects from a set of n objects, without regard to the order of the selected objects, is called a **combination.** A combination is denoted by

$$C(n, r) \quad \text{or} \quad \binom{n}{r}$$

To compute $C(n, r)$ or $\binom{n}{r}$, we use the following strategy.

$$\binom{n}{r} = \text{Ways to choose } r \text{ objects from } n \text{ objects}$$

$$= \frac{\text{Ways to fill } r \text{ slots}}{\text{Ways to arrange any set of } r \text{ elements}}$$

$$= \frac{(n)(n - 1) \cdots (n - r + 1)}{r!}$$

Alternatively, $C(n, r) = \dfrac{P(n, r)}{r!} = \dfrac{n!}{(n - r)! \, r!}$

We use the idea of combinations in the next two examples.

Example 9   **Using Combination Notation**

Suppose a cycling club has 10 members. From among the members of the club, how many ways are there to select a two-person committee?

Solution   Within the committee, the order of the members chosen does not matter. All that matters is that there are two committee members. Thus, the number of ways to choose the committee members is

$$C(10, 2) = \frac{(10)(9)}{2!} = 45$$

We can also use the formula

$$C(10, 2) = \frac{10!}{8! \, 2!} = 45$$

In Example 1, we computed the number of possibilities for first and second place in a race. Those are two *distinct* positions. There, order mattered; in this example it does not.

✓ *Check It Out 9*   A physics lab is equipped with 12 workstations. There are 15 students in one of the physics classes. Assume that all the workstations are in use during the lab period, and that each workstation is being used by a student from the class. If only one student is stationed at each workstation, how many possibilities are there for the particular group of students using the workstations? ■

Using Technology

Using the Probability menu, you can calculate $C(10, 2)$ using your graphing utility. See Figure 2.

Keystroke Appendix:
Section 4

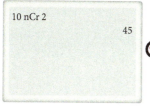

10 nCr 2
                    45

Figure 2

Example 10   **Selecting Objects from Two Different Sets**

A six-member student board is formed with three male students and three female students. If there are five male candidates and six female candidates, how many different student boards are possible?

Solution We need to break this problem into separate parts.

In how many ways can three male board members be chosen from the five male candidates?
Because order does not matter, the number of ways this can be done is

$$\binom{5}{3} = \frac{(5)(4)(3)}{3!} = 10$$

In how many ways can three female board members be chosen from the six female candidates?
Because order does not matter, the number of ways this can be done is

$$\binom{6}{3} = \frac{6!}{(6-3)!\,3!} = \frac{(6)(5)(4)}{3!} = 20$$

Here we used the second formula to calculate the number of combinations.

For *each* group of three male board members, there are 20 possible groups of three females. Using the multiplication principle, the total number of possible student boards is calculated as follows.

Total number of student boards = (Number of three-male groups) × (Number of three-female groups)

$$= \binom{5}{3}\binom{6}{3} = (10)(20) = 200$$

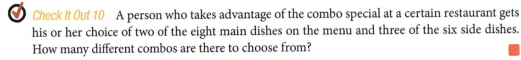*Check It Out 10*   A person who takes advantage of the combo special at a certain restaurant gets his or her choice of two of the eight main dishes on the menu and three of the six side dishes. How many different combos are there to choose from?

In exercises 1–22, evaluate.

1. 4!　　　**2.** 6!　　　**3.** $\dfrac{5!}{2!}$　　**4.** $\dfrac{5!}{3!}$　　**5.** $\dfrac{6!}{4!}$　　**6.** $\dfrac{7!}{3!}$

7. $P(4, 3)$　　**8.** $P(5, 3)$　　**9.** $P(7, 5)$　　**10.** $P(8, 4)$　　**11.** $P(8, 5)$　　**12.** $P(9, 4)$

13. $C(4, 3)$　　**14.** $C(5, 3)$　　**15.** $C(8, 5)$　　**16.** $C(9, 4)$　　**17.** $\dbinom{8}{6}$　　**18.** $\dbinom{9}{7}$

19. $\dbinom{8}{8}$　　**20.** $\dbinom{8}{0}$　　**21.** $\dbinom{100}{99}$　　**22.** $\dbinom{100}{1}$

23. Write out all possible three-letter arrangements of the letters B, C, Z.

24. Write out all possible two-letter arrangements of letters selected from B, C, Z.

25. John, Maria, Susan, and Angelo want to form a subcommittee consisting of only three of them. List all the different subcommittees possible.

26. List all the possible ways in which two marbles from a set of three marbles labeled 1, 2, and 3 can be chosen.

27. How many different photographs are possible if four children line up in a row?

28. How many different photographs are possible if six college students line up in a row?

29. How many different photographs are possible if two children sit in the front row and three adults sit in the back row?

30. How many different photographs are possible if four children sit in the front row and two adults sit in the back row?

31. How many different three-person committees can be formed in a club with 12 members?

32. How many different four-person committees can be formed in a club with 12 members?

33. In how many different ways can Sara give her friend Brittany two pieces of candy from a bag containing 10 different pieces of candy?

34. In how many different ways can Jason give his friend Dylan three pieces of candy from a bag containing eight different pieces of candy?

35. In how many different ways can four people be chosen to receive a prize package from a group of 20 people at the grand opening of a local supermarket?

36. In how many different ways can five people be chosen to receive a prize package from a group of 50 people at the grand opening of a local supermarket?

37. There are eight books in a box. There is space on a bookshelf for only three books. How many different three-book arrangements are possible?

Applications

38. **Investments** A stock analyst plans to include in her portfolio stocks from four of the 10 top-performing companies featured in a finance journal. In how many ways can she do this?

39. **Book List** An editor received a short list of 20 books that his company is considering for publication. If he can only choose six of these books to be published this year, in how many different ways can he choose?

40. **Collectibles** A doll collector has a collection of 22 different dolls. She wants to display four of them on her living room shelf. In how many different ways can she display the dolls?

41. **Collectibles** If the collector in Exercise 40 decides to give one of her dolls to each of her four nieces, in how many different ways can she give the dolls to her nieces?

42. **Sports** The manager of a baseball team of 12 players wants to assign infield positions (first base, second base, third base, catcher, pitcher, and shortstop). In how many different ways can the manager make the assignments if each of the players can play any infield position?

43. **Crafts** Mary is making a gift basket for her friend Kate's birthday. She is planning to include two different eye shadow packs, two different lipsticks, and one blush. The store in which she is shopping has 10 different eye shadow packs, five different lipsticks, and three different blushes available for purchase. In how many different ways can Mary make up her gift basket?

44. **Sports** Adnan is purchasing supplies for a weekend fishing trip. He needs to buy three different lures, one spool of line, and one rod-and-reel. The store in which he is shopping has 25 different lures, three different spools of line, and seven different rod-and-reels. In how many different ways can Adnan purchase what he needs for his trip?

45. **Horse Racing** In how many different orders can 15 horses in the Kentucky Derby finish in the top three spots if there are no ties?

46. **Modern Dance** The choreographer Twyla Tharp has 11 male and 11 female dancers in her dance company. Suppose she wants to arrange a dance consisting of a lead pair of a male and a female dancer. In how many ways can she do this, assuming all dancers are qualified for the lead? (*Source*: www. twylatliarp.org)

47. **Sports** The Lake Wobegon Little League has to win four of their seven games to earn an "above average" certificate of distinction. In how many ways can this be done?

48. **Hardware** A combination lock can be opened by turning the dial of the lock to three predetermined numbers ranging from 0 to 35. A number can be used more than once. How many different three-number arrangements are possible? Why is a combination lock *not* a good name for this type of lock?

49. **License Plates** How many different license plates can be made by using two letters, followed by three digits, followed by one letter?

50. **Airline Routes** There are five different airline routes from New York to Minneapolis and seven different airline routes from Minneapolis to Los Angeles. How many different trips are possible from New York to Los Angeles with a connection in Minneapolis?

51. **Lottery** The lottery game Powerball is played by choosing six different numbers from 1 through 53, and an extra number from 1 through 44 for the "Powerball." How many different combinations are possible?

(*Source*: Iowa State Lottery)

52. **Card Game** A standard card deck has 52 cards. How many five-card hands are possible from a standard deck?

53. **Card Game** A standard card deck has 52 cards. A bridge hand has 13 cards. How many bridge hands are possible from a standard deck?

54. **Card Game** How many five-card hands consisting of all red cards are possible from a standard deck of 52 cards?

55. **Photography** A wedding photographer lines up four people plus the bride and groom for a photograph. If the bride and groom stand side-by-side, how many different photographs are possible?

56. **Computer Security** A password for a computer system consists of six characters. Each character must be a digit or a letter of the alphabet. Assume that passwords are *not* case-sensitive. How many passwords are possible? How many passwords are possible if a password must contain at least one digit? (*Hint for second part*: How many passwords are there containing just letters?)

57. **Computer Security** Rework Exercise 56 for the case in which the passwords *are* case-sensitive.

58. **Retail Store** The Woosamotta University bookstore sells "W. U." T-shirts in four sizes: S, M, L, and XL. Both the blue and yellow shirts are available in all four sizes, but the red shirts come in only small and medium. What is the minimum number of W. U. T-shirts the bookstore should stock if it wishes to have available at least one of each size and color?

59. **Car Options** You are shopping for a new car and have narrowed your search to three models, two colors, and four optional features. These are detailed below.

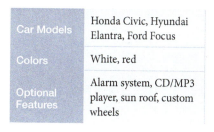

Car Models	Honda Civic, Hyundai Elantra, Ford Focus
Colors	White, red
Optional Features	Alarm system, CD/MP3 player, sun roof, custom wheels

How many different cars are there to choose from

a. if you want *no* optional features?

b. if you want *all four* optional features?

c. if you want *any two* of the optional features?

d. if you want *at most two* of the optional features?

60. **Board Game** In the board game Mastermind, one of two players chooses at most four pegs to place in a row of four slots, and then hides the colors and positions of the pegs from his opponent. Each peg comes in one of six colors, and the player can use a color more than once. Also, one or more of the slots can be left unfilled.

a. How many different ways are there to arrange the pegs in the four-slot row? In this game, the order in which the pegs are arranged matters.

b. The Mastermind website states: "With 2401 combinations possible, it's a mind-bending challenge every time!" Is *combination* the appropriate mathematical term to use here? Explain. This is an instance of how everyday language and mathematical language can be contradictory.   (*Source*: www.pressman.com)

Concepts

61. Given n points ($n \geq 3$) such that no three of them lie on the same line, how many different line segments can be drawn connecting exactly two of the n points?

62. Write out all the different four-digit numbers possible using the numbers 1, 1, 2, 3. Why is your number of possibilities *not* equal to 4!?

63. A diagonal of a polygon is defined as a line segment with endpoints at a pair of nonadjacent vertices of the polygon. How many diagonals does a pentagon have? an octagon? an n-gon (that is, a polygon with n sides)?

64. How many different six-letter arrangements are there of the letters in the word PIPPIN? This exercise involves a slightly different strategy than the strategies discussed in the Examples.
- First draw six slots for six letters. In how many ways can you put the three P's in the slots?
- You have three slots left over. In how many ways can you place the two I's?
- The last slot, by default, will contain the N.
- Put together the information outlined above to come up with a solution.

65. Use the strategy outlined in Exercise 64 to find the number of different 11-letter arrangements of the letters in the word MISSISSIPPI.

Objectives

- Define and identify outcomes and events.
- Calculate probabilities of equally likely events.
- Define and identify mutually exclusive events.
- Calculate probabilities of mutually exclusive events.
- Calculate probabilities of complements of events.
- Apply probabilities to problems involving collected data.

The notion of chance is something we deal with every day. Weather forecasts, Super Bowl predictions, and auto insurance rates all incorporate elements of chance. The mathematical study of chance behavior is called **probability**. This section will cover some basic ideas in the study of probability.

In order to study these notions further, we need to introduce two new terms, **outcome** and **event**.

Definition of Outcome and Event

- An **outcome** is a possible result of an experiment. Here, an experiment simply denotes an activity that yields random results.

- An **event** is a collection of outcomes.

Example 1   **Rolling a Six-Sided Die**

What are the possible outcomes when a six-sided die is rolled and the number on the top face is recorded? What is the event that the number on the top face is even?

Solution   Each face of a die is marked with a unique number from 1 to 6, so the possible outcomes are

$$\{1, 2, 3, 4, 5, 6\}$$

The event that the number on the top face is even is

$$\{2, 4, 6\}$$

Check It Out 1   What is the event that a six-sided die is rolled and the number on the top face is at least 2 but no greater than 5?

An event is generally given in set notation. The set consisting of all possible outcomes for a certain situation under consideration is called the **sample space**, which is often denoted by the letter S. For example, the sample space in Example 1 is $S = \{1, 2, 3, 4, 5, 6\}$.

Example 2   **Tossing a Coin Three Times in Succession**

A coin is tossed three times, and the sequence of heads and tails that occurs is recorded.

a. What is the sample space for this experiment?

b. What is the event that *at least* two heads occur?

c. What is the event that *exactly* two heads occur?

Solution

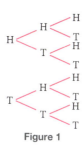

Figure 1

a. Using the multiplication principle, we know that there are a total of eight possible outcomes. Next we list the possible outcomes. A tree diagram is helpful for doing this. It lists the possibilities for the first toss and then branches out to list the possibilities for the second and third tosses, respectively. See Figure 1.

From the tree diagram, we see that the sample space is

$$\{HHH, HHT, HTH, HTT, THH, THT, TTH, TTT\}$$

b. The event that *at least* two heads occur is

$$\{HHH, HHT, HTH, THH\}$$

c. The event that *exactly* two heads occur is

$$\{HHT, HTH, THH\}$$

Note that the event in part (c) is different from the event in part (b). This illustrates the importance of paying careful attention to the *wording* of probability problems.

 Check It Out 2   A coin is tossed four times and the sequence of heads and tails that occurs is recorded.

a. What is the sample space for this experiment?

b. What is the event that *exactly three* tails occur?

c. What is the event that *at most two* heads occur?

We are now in a good position to give a formal definition of probability.

> *Note*   Remember these important points:
>
> If $P(E) = 1$, then the event E is certain to happen.
> If $P(E) = 0$, then the event E will *not* happen.

Definition of Probability

Assume that a sample space S consists of equally likely outcomes. Then the probability of an event E, denoted by $P(E)$, is defined as

$$P(E) = \frac{\text{Number of outcomes in } E}{\text{Number of outcomes in } S}$$

Note that $P(E)$ must satisfy

$$0 \leq P(E) \leq 1$$

Throughout this textbook, we have been discussing functions. The probability of an event can also be thought of as a function. Because an event is described as a set, the probability function has a particular set as its domain and the numbers in the interval [0, 1] as its range.

We now calculate the probabilities of some of the events described in the previous examples.

Example 3   **Probability of Winning a Specific Prize**

During a certain episode, the television game show *Wheel of Fortune* had a wheel with 24 sectors. One sector was marked "Trip to Hawaii" and two of the other sectors were marked "Bonus." What is the probability of winning a trip to Hawaii, assuming the wheel is equally likely to stop at any one of the 24 sectors?

Solution   We know that the total number of outcomes is 24, since there are 24 different sectors in which the wheel can stop. Let E be the event that the wheel stops in the "trip to Hawaii" sector. Calculate $P(E)$ as follows.

$$P(E) = \frac{\text{Number of outcomes in } E}{\text{Number of outcomes in } S} = \frac{1}{24}$$

Using mathematical terminology helps us to generalize our ideas and make them precise.

 Check It Out 3   Find the probability of winning a "Bonus" in the game described in Example 3.

Example 4   **Probability of a Specific Event in a Coin-Tossing Experiment**

Suppose a coin is tossed three times. What is the probability of obtaining at least two heads?

Solution   Here, E is the event of obtaining at least two heads. To find the probability of E, we use the answers to parts (a) and (b) of Example 2.

$$P(E) = \frac{\text{Number of ways to get at least two heads}}{\text{Number of outcomes in } S} = \frac{4}{8} = \frac{1}{2}$$

Check It Out 4   What is the probability of obtaining at most three heads when a coin is tossed four times?

Mutually Exclusive Events

Many applications entail finding a probability that involves two events. If the events have no overlap, they are said to be **mutually exclusive.** Next we examine some pairs of events that may or may not be mutually exclusive.

> **Example 5**　**Deciding Whether Events Are Mutually Exclusive**

Decide which of the following pairs of events are mutually exclusive.

a. "Drawing a queen" and "drawing a king" from a deck of 52 cards.

b. "Drawing a queen" and "drawing a spade" from a deck of 52 cards.

Solution

a. A card cannot be both a king and a queen. Therefore, the event of drawing a king and the event of drawing a queen have no overlap and are mutually exclusive.

b. The event of drawing a queen overlaps with the event of drawing a spade because drawing the queen of spades is an element of both events. Therefore, these two events are *not* mutually exclusive.

 Check It Out 5　Decide which of the following pairs of events are mutually exclusive.

a. When rolling a die, "rolling a 2" and "rolling the smallest possible even number".

b. "Drawing a club" and "drawing a red card" from a deck of 52 cards.

When two events are mutually exclusive, the probability of one or the other occurring is easy to compute—you simply add up the two respective probabilities.

Computing the Probability of Mutually Exclusive Events
Let F and G be two *mutually exclusive events*. Then $$P(F \text{ or } G) = P(F) + P(G)$$

> **Example 6**　**Combining the Probabilities of Mutually Exclusive Events**

Find the probability of drawing a queen or a king from a deck of 52 cards.

Solution　Drawing a king and drawing a queen are mutually exclusive events, since they cannot both happen on one draw. See Example 6 for details. Therefore,

$$P(\text{queen or king}) = P(\text{queen}) + P(\text{king})$$
$$= \frac{4}{52} + \frac{4}{52}$$
$$= \frac{8}{52} = \frac{2}{13}$$

The probability of drawing a queen or a king from a deck of 52 cards is $\frac{2}{13}$.

 Check It Out 6　Find the probability of tossing three heads or three tails in three tosses of a coin.

Example 7   **Rolling a Pair of Six-Sided Dice**

Figure 2 shows all the possibilities for the numbers on the top faces when rolling a pair of dice.

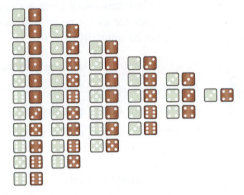

Figure 2

Find the following.

a. The probability of rolling a sum of 5

b. The probability of rolling a sum of 6 or 7

c. The probability of rolling a sum of 13

Solution We examine Figure 2 to help us answer the questions. Note that there are 36 different possible outcomes when rolling the two dice. All 36 outcomes are equally likely.

a. Probability of rolling a sum of 5: All the possibilities resulting in a sum of 5 are listed in the fourth row of the figure. Written in set notation, the event of rolling a sum of 5 is

$$\{(4, 1), (1, 4), (3, 2), (2, 3)\}$$

Because all of the possibilities are equally likely, the probability of this event is

$$P(\text{rolling a sum of 5}) = \frac{4}{36} = \frac{1}{9}$$

b. Probability of rolling a sum of 6 or 7: These two events are mutually exclusive, since they cannot happen at the same time. From the figure, there are five ways to roll a sum of 6 and six ways to roll a sum of 7. Therefore,

$$P(\text{rolling a sum of 6 or 7}) = P(\text{rolling a sum of 6}) + P(\text{rolling a sum of 7})$$

$$= \frac{5}{36} + \frac{6}{36} = \frac{11}{36}$$

c. Probability of rolling a sum of 13: It is not possible to roll a sum of 13 with two six-sided dice whose faces are uniquely numbered from 1 to 6. Therefore,

$$P(\text{rolling a sum of 13}) = \frac{0}{36} = 0$$

Check It Out 7 When rolling two dice, find the probability of rolling a sum that is an even number less than or equal to 6.

Complement of an Event

This section will deal with the probability of an event *not* happening, which is known as the **complement** of the event. Complements of events occur all the time in daily life. Example 8 explores one such everyday application.

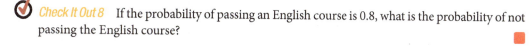

Definition of the Complement of an Event

The set of all outcomes in a sample space that do *not* belong to event E is called the **complement** of E and is denoted by E'.

By definition, E and E' are mutually exclusive. Thus, because either E or E' is certain to happen, we can write

$$P(E \text{ or } E') = P(E) + P(E') = 1$$

This gives us a way to calculate $P(E')$:

$$P(E') = 1 - P(E)$$

Complements are very helpful in determining certain types of probabilities.

Example 8   **Calculating the Probability of the Complement of an Event**

Suppose you are told that the probability of rain today is 0.6. What is the probability that it will *not* rain?

Solution   Because the total probability must be 1, the probability of *not* raining is

$$P(\text{no rain}) = 1 - P(\text{rain}) = 1 - \mathbf{0.6} = 0.4$$

✓ *Check It Out 8*   If the probability of passing an English course is 0.8, what is the probability of not passing the English course?

Example 9   **Using the Complement of an Event to Find a Probability**

Refer to Figure 2 in Example 7, which lists all the possible outcomes of rolling two dice. What is the probability of rolling a sum of *at least* 4?

Solution   It is much easier to figure out the number of ways in which the sum is *not* at least 4. The sum is *not* at least 4 when the sum is equal to 2 or 3. There are only three ways in which this can happen. Thus,

$$P(not \text{ rolling a sum of at least 4}) = \frac{3}{36}$$

Using complements,

$$P(\text{rolling a sum of at least 4}) = 1 - P(not \text{ rolling a sum of at least 4})$$

$$= \frac{33}{36} = \frac{11}{12}$$

✓ *Check It Out 9*   In tossing a coin four times, what is the probability of getting at least two tails?

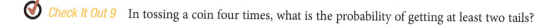

Calculating Probabilities from Percentages

In real life, probabilities are often calculated from data that is expressed in terms of percentages. In fact, probabilities are often quoted in terms of percentages. To be consistent with the mathematical definition of probability, we will convert all percentages to their equivalent values between 0 and 1.

Example 10   **Converting Percentages to Probabilities**

Every spring, the National Basketball Association holds a lottery to determine which team will get first pick of its number 1 draft choice from a pool of college players. The teams with poorer records have a higher chance of winning the lottery than those with better records. Table 1 lists the percentage chance that each team had of getting first pick of its number 1 draft choice for the year 2002.   (*Source*: National Basketball Association)

Team	Chance of Winning First Pick (%)
Golden State Warriors	22.5
Chicago Bulls	22.5
Memphis Grizzlies	15.7
Denver Nuggets	12.0
Houston Rockets	8.9
Cleveland Cavaliers	6.4
New York Knicks	4.4
Atlanta Hawks	2.9
Phoenix Suns	1.5
Miami Heat	1.4
Washington Wizards	0.7
L.A. Clippers	0.6
Milwaukee Bucks	0.5

Table 1

Find the probability of

a. the Bulls or the Warriors getting first draft pick.

b. the Clippers *not* getting first draft pick.

Solution   First note that these probabilities are not equal—the teams with the poorer records have a higher chance of getting first pick. Also, convert the percentages into their respective decimal equivalents.

a. The probability of the Bulls or the Warriors getting first draft pick is

$$P(\text{Bulls or Warriors}) = P(\text{Bulls}) + P(\text{Warriors}) = 0.225 + 0.225 = 0.45$$

We have used the fact that the two events are mutually exclusive, since the two teams cannot both get first draft pick.

b. Using the formula for computing the complement of an event, the probability of the Clippers *not* getting first draft pick is

$$P(not\ \text{Clippers}) = 1 - P(\text{Clippers}) = 1 - 0.006 = 0.994$$

Thus the probability is very high that the Clippers will *not* get first draft pick.

 Check It Out 10   Use Table 1 from Example 10 find the probability of

a. the Atlanta Hawks or the L.A. Clippers getting first draft pick.

b. the New York Knicks *not* getting first draft pick.

Exercises 10.5

In Exercises 1–4, consider the following experiment: toss a coin twice and record the sequence of heads and tails.

1. What is the sample space (for tossing a coin twice)?

2. What is the event that you get at least one head?

3. What is the complement of the event that you get at least one head?

4. Calculate the probability of the event in Exercise 3.

In Exercises 5–8, consider the following experiment: roll a die and record the number on the top face.

5. What is the event that the number on the top face is odd?

6. What is the complement of the event that the number on the top face is odd?

7. What is the probability that the number on the top face is greater than or equal to 5?

8. What is the probability that the number on the top face is less than 1?

In Exercises 9–12, consider the following experiment: draw a single card from a standard deck of 52 cards.

9. What is the event that the card is a spade?

10. What is the complement of the event that the card is a spade? Describe in words only.

11. What is the probability that the card drawn is the ace of spades?

12. What is the probability that the card drawn is the 2 of clubs?

In Exercises 13–16, consider the following experiment: pick one coin out of a bag that contains one quarter, one dime, one nickel, and one penny.

13. Give the sample space (for picking one coin out of the bag).

14. What is the complement of the event that the coin you pick has a value of 10 cents?

15. What is the probability of picking a nickel?

16. What is the probability of picking a quarter or a penny?

In Exercises 17–22, answer True or False.

17. When rolling a die, "rolling a 2" and "rolling an even number" are mutually exclusive events.

18. When randomly picking a card from a standard deck of 52 cards, "picking a queen" and "picking a jack" are mutually exclusive events.

19. Consider the roll of a die. The complement of the event "rolling an even number" is "rolling a 1, a 3, or 5".

20. Consider randomly picking a card from a standard deck of 52 cards. The complement of the event "picking a black card" is "picking a heart".

21. When picking one coin at random from a bag that contains one quarter, one dime, one nickel, and one penny, "picking a coin with a value of more than one cent" and "picking a penny" are mutually exclusive events.

22. Consider picking one coin from a bag that contains one quarter, one dime, one nickel, and one penny. The complement of the event "picking a quarter or a nickel" is "picking a dime or a nickel".

Applications

23. **Coin Toss** A coin is tossed four times and the number of heads that appear is counted. Fill in the following table listing the probabilities of obtaining various numbers of heads. What do you observe? Are all of these outcomes equally likely?

Number of Heads	Probability
0	
1	
2	
3	
4	

24. **Cards** If a card is drawn from a standard deck of 52 cards, what is the probability that it is a heart?

25. **Cards** If a card is drawn from a standard deck of 52 cards, what is the probability that it is an ace?

26. **Candy Colors** Students in a college math class counted 29 packages (1.5 ounces each) of plain M&M'S and recorded the following color distribution.

Color	Red	Blue	Green	Yellow	Brown	Orange	Total
Number of M&M'S	278	157	261	265	549	139	1649

If one M&M is drawn at random from the total, find the following probabilities.
   a. The probability of getting a red candy
   b. The probability of getting a red or a green candy
   c. The probability of *not* getting a blue candy

27. **Cards** What is the probability of drawing the 4 of clubs from a standard deck of 52 cards?

28. **Card Game** During the play of a card game, you have seen 20 of the 52 cards in the deck and none of them is the 4 of clubs. You need the 4 of clubs to win the game. What is the probability that you will win the game on the next card drawn?

29. **Cards** What is the probability of drawing a face card (a face card is a jack, queen, or king) from a standard deck of 52 cards?

30. **Cards** What is the probability of drawing a red face card (a face card is a jack, queen, or king) from a standard deck of 52 cards?

In Exercises 31–34, use counting principles from Section 10.4 to calculate the number of outcomes.

31. **Dice Games** A pair of dice, one blue and one green, are rolled and the number showing on the top of each die is recorded. What is the probability that the sum of the numbers on the two dice is 7?

32. **Dice Games** Refer to Exercise 31. What is the probability that the sum is 10?

33. **Movie Theater Seating** A group of friends, five girls and five boys, wants to go to the movies on Friday night. The friends select, at random, two of their group to go to the ticket office to purchase the tickets. What is the probability that the two selected are both boys?

34. **Movie Theater Seating** Refer to Exercise 33. What is the probability that the two selected are a boy and a girl?

Phone Numbers *Exercises 35–38 involve dialing the last four digits of a phone number that has an area code of 907 and an exchange of 316. The exchange consists of the first three digits of the seven-digit phone number.*

35. How many outcomes are there for dialing the last four digits of a phone number?

36. How many possible outcomes are in the event that the first three (of the last four) digits you dial are 726, in that order?

37. What is the probability that the (last four) digits you dial are different from one another?

38. What is the probability that all of the (last four) digits you dial are different from all the digits of the area code *and* different from all the digits of the exchange? Assume each digit can be repeated.

39. Refer to Exercise 33. What is the probability that Ann (who is one of the five girls) is selected?

40. The 10 friends in Exercise 33 all have different last names. The seats they purchased for the movie are numbered 1 through 10. If the tickets are distributed among the friends at random, what is the probability the friends will be seated in alphabetical order from seat 1 to seat 10?

41. **Card Game** During the play of a card game, you see 20 of 52 cards in the deck drawn and discarded and none of them is a black 4. You need a black 4 to win the game. What is the probability that you will win the game on the next card drawn?

42. **Roulette** A roulette wheel has 38 sectors. Two of the sectors are green and are numbered 0 and 00, respectively, and the other 36 sectors are equally divided between red and black. The wheel is spun and a ball lands in one of the 38 sectors.

    a. What is the probability of the ball landing in a red sector?

    b. What is the probability of the ball landing in a green sector?

    c. If you bet $1 on a red sector and the ball lands in a red sector, you will win another $1. Otherwise, you will lose the dollar that you bet. Do you think this is a fair game? That is, do you have the same chance of wining as you do of losing? Why or why not?

43. **Dart Game** Many probabilities are computed by using ratios of areas. This exercise illustrates such a scenario. What is the probability of hitting the shaded inner region of the dart board in the figure if all of the points within the larger circle are equally likely to be hit by the dart? You may assume that the dart will never land anywhere outside the larger circle.

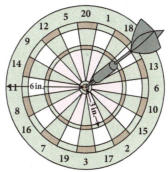

44. **Card Probabilities** Suppose five cards are drawn from a standard deck of 52 cards. Find the probability that all five cards are black. (*Hint*: Use the counting principles from Section 10.4.)

45. **Card Probabilities** If a card is drawn from a standard deck of 52 cards, the probability of drawing a king or a spade is *not* $\frac{17}{52}$. Explain. What is the correct answer?

Slot Machines *A slot machine has four reels, with 10 symbols on each reel. Assume that there is exactly one cherry symbol on each reel. Use this information and the counting principles from Section 10.4 when solving Exercises 46–48.*

46. What is the probability of getting four cherries?

47. What is the probability of getting exactly three cherries?

48. What is the probability of getting at least one cherry?

Concepts

49. Consider rolling a pair of dice. Which, if either, of the following events has a higher probability: "rolling a sum that is odd" or "rolling a sum that is even"?

50. Each card in a standard deck of 52 cards belongs to one of four different suits: hearts, diamonds, spades, or clubs. There are 13 cards in each suit. Consider a scenario in which you draw five cards from the deck, one at a time, and record only the suit to which each card drawn belongs.
    a. Describe the sample space.
    b. What is the probability that the set of five cards you draw consists of two spades, one heart, one diamond, and one club (drawn in any order)?
    c. What is the probability that exactly two of the five cards you draw are from the same suit?

51. In a telephone survey, people are asked whether they have seen each of four different films. Their answers for each film (yes or no) are recorded.
    a. What is the sample space?
    b. What is the probability that a respondent has seen exactly two of the four films?
    c. Assuming that all outcomes are equally likely, what is the probability that a respondent has seen all four films?

52. Assume that the probability of winning $5 in the lottery (on one lottery ticket) for any given week is $\frac{1}{50}$, and consider the following argument. "Henry buys a lottery ticket every week, but he hasn't won $5 in any of the previous 49 weeks, so he is assured of winning $5 this week." Is this a valid argument? Explain.

In Exercises 53 and 54, consider a bag that contains eight coins: three quarters, two dimes, one nickel, and two pennies.

53. Assume that two coins are chosen from the bag.
    a. How many ways are there to choose two coins from the bag?
    b. What is the probability of choosing two coins of equal value?

54. Assume that two coins are picked out of the bag, one at a time, and the first coin is put back into the bag before the second coin is chosen.
    a. How many outcomes are there? (*Hint*: Count the possibilities for the first coin and the possibilities for the second coin.)
    b. What is the probability of picking two coins of equal value?

The Binomial Theorem

Objectives

- Calculate the variable parts of terms in a binomial expansion.
- Calculate binomial coefficients.
- Expand a binomial using the Binomial Theorem.
- Find the *i*th term of a binomial expansion.
- Relate combinations to the Binomial Theorem.

When applying algebraic techniques in order to solve a problem, it is sometimes necessary to write a quantity of the form $(a + b)^n$ as the sum of its terms. Because $a + b$ is a binomial, this process is called a **binomial expansion**. You already know the following:

$$(a + b)^1 = a + b$$
$$(a + b)^2 = a^2 + 2ab + b^2$$

Building on these expansions, we can write $(a + b)^3$ as

$$(a + b)^3 = (a + b)(a + b)^2$$

We then can expand $(a + b)^2$, multiply $(a + b)$ by the result, and simplify.

$$(a + b)^3 = (a + b)(a + b)^2$$
$$= (a + b)(a^2 + 2ab + b^2)$$
$$= a^3 + 3a^2b + 3ab^2 + b^3$$

As the exponent on the binomial increases to numbers larger than 3, multiplying out the entire expression to find all the terms becomes more tedious. Fortunately, in this section we discuss a way to find all the terms of the expansion without having to multiply repeatedly.

Observations:

Note the following about the expansion of $(a + b)^n$, where n is a positive integer.
- The first term of the expansion is a^n and the last term is b^n.
- For each successive term after a^n, the exponent on b increases by 1 while the exponent on a decreases by 1.
- For any term in the expansion, the sum of the exponents on a and b is n.

VIDEO EXAMPLES

SECTION 10.6

Example 1   **Finding the Variable Parts of the Terms in a Binomial Expansion**

Consider the expansion of $(a + b)^4$.

a. Write down the variable parts of all the terms that occur in the expansion.

b. What is the sum of the exponents on a and b for each term of the expansion?

Solution

a. First write

$$(a + b)^4 = (a + b)(a + b)(a + b)(a + b)$$

From the second bulleted item in the preceding list of observations, the variable parts of all the terms that occur in the expansion are

$$a^4, a^3b, a^2b^2, ab^3, b^4$$

b. The sum of the exponents on a and b for each term of the expansion is 4.

 Check It Out 1   Consider the expansion of $(x - y)^5$.

a. Write down the variable parts of all the terms that occur in the expansion.

b. What is the sum of the exponents on x and y for each term of the expansion?

The Binomial Theorem

Recall the definition of n factorial from Section 10.4.

Definition of n Factorial

Let n be a positive integer. Then

$$n! = n(n-1)(n-2)\cdots(3)(2)(1)$$

The quantity $0!$ is defined to be 1.

Definition of Binomial Coefficient

Let n and r be nonnegative integers with $r \le n$. In the expansion of $(a+b)^n$, the coefficient of the term $a^{n-r}b^r$ is

$$\binom{n}{r} = \frac{n!}{(n-r)!r!}$$

Example 2   **Calculating Factorials and Binomial Coefficients**

Evaluate the following expressions.

a. $5!$ **b.** $\binom{6}{3}$ **c.** $\binom{5}{2}$

Solution

a. $5! = 5 \cdot 4 \cdot 3 \cdot 2 \cdot 1 = 120$

b. Applying the formula for binomial coefficients,

$$\binom{6}{3} = \frac{6!}{(6-3)!3!} = \frac{6!}{3!3!}$$

$$= \frac{6 \cdot 5 \cdot 4 \cdot 3 \cdot 2 \cdot 1}{(3 \cdot 2 \cdot 1)(3 \cdot 2 \cdot 1)} = 20$$

c. $\binom{5}{2} = \frac{5!}{(5-2)!2!} = \frac{5!}{3!2!}$

$$= \frac{5 \cdot 4 \cdot 3 \cdot 2 \cdot 1}{(3 \cdot 2 \cdot 1)(2 \cdot 1)} = 10$$

 Check It Out 2   Evaluate $\binom{6}{4}$.

Now that we have a method for finding each of the coefficients in the expansion of $(a+b)^n$, we present the Binomial Theorem.

Binomial Theorem

Let n be a positive integer. Then

$$(a+b)^n = \sum_{i=0}^{n}\binom{n}{i}a^{n-i}b^i$$

$$= \binom{n}{0}a^n + \binom{n}{1}a^{n-1}b + \binom{n}{2}a^{n-2}b^2 + \cdots + \binom{n}{n}b^n$$

Virtually any variable or constant can take the place of a and b in the Binomial Theorem. The next two examples illustrate the use of this theorem.

Example 3   **Expanding a Binomial Raised to the Fourth Power**

Expand: $(3 + 2y)^4$.

Solution We apply the Binomial Theorem with $a = 3$, $b = 2y$, and $n = 4$.

$$(3 + 2y)^4 = \binom{4}{0}(3)^4 + \binom{4}{1}(3)^3(2y) + \binom{4}{2}(3)^2(2y)^2$$

$$+ \binom{4}{3}(3)(2y)^3 + \binom{4}{4}(2y)^4 \qquad \text{Substitute } a = 3, b = 2y, \text{ and } n = 4$$

$$= (1)(81) + (4)(27)(2y) + (6)(9)(4y^2)$$

$$+ (4)(3)(8y^3) + (1)(16y^4) \qquad \binom{4}{0} = \binom{4}{4} = 1, \binom{4}{1} = \binom{4}{3} = 4, \text{ and } \binom{4}{2} = 6$$

$$= 81 + 216y + 216y^2 + 96y^3 + 16y^4 \qquad \text{Simplify}$$

$$= 16y^4 + 96y^3 + 216y^2 + 216y + 81$$

✓ *Check It Out 3* Expand: $(2y - x)^4$.

Example 4   **Expanding a Binomial Raised to the Fifth Power**

Expand: $(2z - y)^5$.

Solution First write $2z - y$ as $2z + (-y)$. We apply the Binomial Theorem with $a = 2z$, $b = -y$, and $n = 5$.

$$(2z - y)^5 = (2z + (-y))^5 = \binom{5}{0}(2z)^5 + \binom{5}{1}(2z)^4(-y)^1 + \binom{5}{2}(2z)^3(-y)^2$$

$$+ \binom{5}{3}(2z)^2(-y)^3 + \binom{5}{4}(2z)^1(-y)^4 + \binom{5}{5}(-y)^5$$

$$= (1)(32z^5) + (5)(16z^4)(-y) + (10)(8z^3)(y^2) + (10)(4z^2)(-y^3)$$

$$+ (5)(2z)(y^4) + (1)(-y^5)$$

$$= 32z^5 - 80z^4y + 80z^3y^2 - 40z^2y^3 + 10zy^4 - y^5 \qquad \text{Simplify}$$

✓ *Check It Out 4* Expand: $(3u + 2v)^5$.

The *i*th Term of a Binomial Expansion

In many instances, we may be interested only in a particular term or terms of a binomial expansion. In such cases, we can use the following formula, which is a direct result of examining the individual terms of the Binomial Theorem.

> ### The *i*th Term of a Binomial Expansion
>
> Let n and i be positive integers such that $1 \le i \le n + 1$. Then the ith term of $(a + b)^n$ is given by
>
> $$\binom{n}{i - 1}a^{n - i + 1}b^{i - 1}$$
>
> Note that the exponent on b is one less than the number of the term.

Example 5   **Finding a Specific Term of a Binomial Expansion**

Find the fourth term in the expansion of $(3x + 5)^6$.

Solution Using the formula for the ith term of a binomial expansion with $a = 3x, b = 5, n = 6$, and $i = 4$, we obtain

$$\binom{6}{4 - 1}(3x)^{6 - 4 + 1}(5)^{4 - 1} = \binom{6}{3}(3x)^3(5)^3$$

$$= (20)(27)x^3(125) = 67{,}500x^3$$

Check It Out 5   Find the second term in the expansion of $(3x + 5)^6$.

Exercises 10.6

Skills

In Exercises 1–4, write down the variable parts of the terms in the expansion of the binomial.

1. $(a + b)^5$ **2.** $(a + b)^6$ **3.** $(x + y)^7$ **4.** $(x + y)^8$

In Exercises 5–16, evaluate each expression.

5. $4!$ **6.** $6!$ **7.** $\dfrac{3!}{2!}$ **8.** $\dfrac{4!}{3!}$

9. $\dbinom{6}{2}$ **10.** $\dbinom{5}{3}$ **11.** $\dbinom{7}{5}$ **12.** $\dbinom{7}{4}$

13. $\dbinom{10}{10}$ **14.** $\dbinom{10}{0}$ **15.** $\dbinom{100}{0}$ **16.** $\dbinom{100}{100}$

In Exercises 17–28, use the binomial theorem to expand the expression.

17. $(x + 2)^4$ **18.** $(x - 3)^3$ **19.** $(2x - 1)^3$ **20.** $(2x + 3)^4$

21. $(3 + y)^5$ **22.** $(4 - z)^4$ **23.** $(x - 3z)^4$ **24.** $(2z + y)^3$

25. $(x^2 + 1)^3$ **26.** $(x^2 - 2)^3$ **27.** $(y - 2x)^4$ **28.** $(z + 4x)^5$

In Exercises 29–42, use the Binomial Theorem to find the indicated term or coefficient.

29. The coefficient of x^3 when expanding $(x + 4)^5$

30. The coefficient of y^2 when expanding $(y - 3)^5$

31. The coefficient of x^5 when expanding $(3x + 2)^6$

32. The coefficient of y^4 when expanding $(2y + 1)^7$

33. The coefficient of x^6 when expanding $(x + 1)^8$

34. The coefficient of y^7 when expanding $(y - 3)^{10}$

35. The third term in the expansion of $(x - 4)^6$

36. The fourth term in the expansion of $(x + 3)^6$

37. The sixth term in the expansion of $(x + 4y)^5$

38. The seventh term in the expansion of $(a + 2b)^6$

39. The fifth term in the expansion of $(3x - 2)^6$

40. The fifth term in the expansion of $(3x + 1)^8$

41. The fourth term in the expansion of $(4x - 2)^6$

42. The fourth term in the expansion of $(3x - 1)^8$

Concepts

43. Show that $\dbinom{n}{r} = \dbinom{n}{n - r}$, where $0 \le r \le n$, with n and r integers.

44. Show that $\dbinom{n}{0} = 1$.

45. Evaluate the following.

$$\dbinom{4}{0}\left(\dfrac{1}{3}\right)^4 + \dbinom{4}{1}\left(\dfrac{1}{3}\right)^3\left(\dfrac{2}{3}\right) + \dbinom{4}{2}\left(\dfrac{1}{3}\right)^2\left(\dfrac{2}{3}\right)^2 + \dbinom{4}{3}\left(\dfrac{1}{3}\right)\left(\dfrac{2}{3}\right)^3 + \dbinom{4}{4}\left(\dfrac{2}{3}\right)^4$$

Objective

- Prove a statement by mathematical induction.

Many mathematical facts are established by first observing a pattern, then making a conjecture about the general nature of the pattern, and finally *proving* the conjecture. In order to *prove* a conjecture, we use existing facts, combine them in such a way that they are relevant to the conjecture, and proceed in a logical manner until the truth of the conjecture is established.

For example, let us make a conjecture regarding the sum of the first n even integers. First, we look for a pattern:

$$2 = 2$$
$$2 + 4 = 6$$
$$2 + 4 + 6 = 12$$
$$2 + 4 + 6 + 8 = 20$$
$$2 + 4 + 6 + 8 + 10 = 30$$

n	Sum of First n Even Integers
1	2
2	6
3	12
4	20
5	30

Table 1

From the equations above, we can build Table 1. The numbers in the "sum" column in the table can be factored as follows: $2 = 1 \cdot 2$, $6 = 2 \cdot 3$, $12 = 3 \cdot 4$, $20 = 4 \cdot 5$, and $30 = 5 \cdot 6$. Noting the values of n to which the factorizations correspond, we make our conjecture:

The sum of the first n even integers is $n(n + 1)$.

According to our calculations, this conjecture holds true for n up to and including 5. But does it hold true for all n? To establish the pattern for all values of n, we must *prove* the conjecture. Simply substituting various values of n is not feasible because we would have to verify the statement for infinitely many n. A more practical proof technique is needed. We next introduce a proof method called **mathematical induction**, which is typically used to prove statements such as this.

Mathematical Induction

Before giving a formal definition of mathematical induction, we take our discussion of the sum of the first n even integers and introduce some new notation that we will need in order to work with this type of proof.

First, the conjecture is given a name: P_n. The subscript n means that the conjecture depends on n. Stating our conjecture, we write

P_n: The sum of the first n even integers is $n(n + 1)$

For some specific values of n, the conjecture reads as follows:

P_8: The sum of the first 8 even integers is $8 \cdot 9 = 72$.

P_{12}: The sum of the first 12 even integers is $12 \cdot 13 = 156$.

P_k: The sum of the first k even integers is $k(k + 1)$.

$P_{k + 1}$: The sum of the first $k + 1$ even integers is $(k + 1)(k + 2)$.

We now state the principle of mathematical induction, which we will need to complete the proof of our conjecture.

The Principle of Mathematical Induction

Let n be a natural number and let P_n be a statement that depends on n. If
1. P_1 is true, and
2. for all positive integers k, $P_{k + 1}$ can be shown to be true if P_k is assumed to be true,
   then P_n is true for all natural numbers n.

The underlying scheme behind proof by induction consists of two key pieces:

1. Proof of the base case: proving that P_1 is true
2. Use the assumption that P_k is true for a general value of k to show that P_{k+1} is true.

Taken together, these two pieces prove that P_n holds true for every natural number n. The assumption that P_k is true is known as the **induction hypothesis.**

In proving statements by induction, we often have to take an expression containing the variable k and replace k with $k + 1$. Example 1 illustrates this process.

VIDEO EXAMPLES

SECTION 10.7

| **Example 1** | **Replacing k with $k + 1$ in an Algebraic Expression** |

Replace k with $k + 1$ in the following.

a. $3^k - 1$

b. $\dfrac{k(k + 1)(2k + 1)}{6}$

Solution

a. Replacing k by $k + 1$, we obtain

$$3^{k+1} - 1$$

b. Replacing k by $k + 1$ and simplifying, we obtain

$$\frac{(k + 1)((k + 1) + 1)(2(k + 1) + 1)}{6} = \frac{(k + 1)(k + 2)(2k + 3)}{6}$$

 Check It Out 1   Replace k by $k + 1$ in $2k(k + 2)$.

We now return to the conjecture we made at the beginning of this section, and prove it by induction.

| **Example 2** | **Proving a Formula by Induction** |

Prove the following formula by induction:

$$2 + 4 + \cdots + 2n = n(n + 1)$$

Solution This is just the statement that we conjectured earlier, but in the form of an equation. Recall that we denoted this statement by P_n, so we denote the proposed equation by P_n as well.

First we must prove that P_n is true for $n = 1$. We do this by replacing every n in P_n with a 1, and then demonstrating that the result is true.

$$P_1: 2(1) = 1(1 + 1)$$

Since $2(1) = 1(1 + 1)$, we see that P_1 is true.

Next we state P_k and assume that P_k is true.

$$P_k: 2 + 4 + \cdots + 2k = k(k + 1)$$

Finally, we state P_{k+1} and use the assumption that P_k is true to prove that P_{k+1} holds true as well.

$$P_{k+1}: 2 + 4 + \cdots + 2k + 2(k + 1) = (k + 1)(k + 2)$$

To prove P_{k+1}, we start with the expression on the left side of P_{k+1} and show that it is equal to the expression on the right side.

$$2 + 4 + \cdots + 2k + 2(k + 1) \qquad \text{Left-hand side of } P_{k+1}$$

$$= k(k + 1) + 2(k + 1) \qquad \text{Induction hypothesis: } P_k \text{ is true}$$

$$= k^2 + k + 2k + 2 \qquad \text{Expand}$$

$$= k^2 + 3k + 2 \qquad \text{Combine like terms}$$

$$= (k + 1)(k + 2) \qquad \text{Factor}$$

We see that the result, $(k + 1)(k + 2)$, is the expression on the right side of P_{k+1}. Thus, by mathematical induction, P_n is true for all natural numbers n.

 *Check It Out 2*   Prove the following formula by mathematical induction:

$$1 + 3 + 5 + \cdots + (2n - 1) = n^2$$

Example 3   **Proving a Summation Formula by Induction**

Prove the following formula by induction:

$$1 + 2 + 3 + \cdots + n = \frac{n(n + 1)}{2}$$

Solution   First denote the proposed equation by P_n and prove that it holds true for $n = 1$. Replacing every n with a 1, we get

$$P_1: 1 = \frac{1(1 + 1)}{2}$$

Clearly this is true, so P_1 holds.

Next state P_k and assume that P_k is true.

$$P_k: 1 + 2 + 3 + \cdots + k = \frac{k(k + 1)}{2}$$

Finally, state P_{k+1} and use the assumption that P_k is true to prove that P_{k+1} holds true as well.

$$P_{k+1}: 1 + 2 + 3 + \cdots + k + (k + 1) = \frac{(k + 1)(k + 2)}{2}$$

Show that the expression on the left side of P_{k+1} is equal to the expression on the right-hand side.

$$\mathbf{1 + 2 + 3 + \cdots + k + k + 1} \qquad \text{Left side of } P_{k+1}$$

$$= \frac{k(k + 1)}{2} + k + 1 \qquad \text{Induction hypothesis: } P_k \text{ is true}$$

$$= \frac{k(k + 1) + 2(k + 1)}{2} \qquad \text{Use common denominator}$$

$$= \frac{k^2 + k + 2k + 2}{2} \qquad \text{Expand}$$

$$= \frac{k^2 + 3k + 2}{2} \qquad \text{Combine like terms}$$

$$= \frac{(k + 1)(k + 2)}{2} \qquad \text{Factor}$$

We see that the result, $\dfrac{(k + 1)(k + 2)}{2}$, is the expression on the right side of P_{k+1}. Thus, by mathematical induction, P_n is true for all natural numbers n.

Check It Out 3   Prove by induction:

$$2 + 5 + 8 + \cdots + (3n - 1) = \frac{1}{2}n(3n + 1)$$

Example 4   **Proving a Formula for Partial Sums by Induction**

Prove by induction:

$$1 + 2 + 2^2 + 2^3 + \cdots + 2^{n-1} = 2^n - 1$$

Solution   First denote the proposed equation by P_n and prove that it holds true for $n = 1$ by replacing every n with a 1.

$$P_1: 1 = 2^1 - 1$$

It is easy to see that P_1 is true.

Next state P_k and assume that P_k is true.

$$P_k: 1 + 2 + 2^2 + 2^3 + \cdots + 2^{k-1} = 2^k - 1$$

Finally, state P_{k+1} and use the induction hypothesis (the assumption that P_k is true) to show that P_{k+1} holds true as well.

$$P_{k+1}: 1 + 2 + 2^2 + 2^3 + \cdots + 2^{k-1} + 2^k = 2^{k+1} - 1$$

$1 + 2 + 2^2 + 2^3 + \cdots + 2^{k-1} + 2^k$	Left side of P_{k+1}
$= 2^k - 1 + 2^k$	Induction hypothesis: P_k is true
$= 2(2^k) - 1$	Combine like terms
$= 2^{k+1} - 1$	Simplify

We see that the result, $2^{k+1} - 1$, is the expression on the right side of P_{k+1}. Thus, by mathematical induction, P_n is true for all natural numbers n.

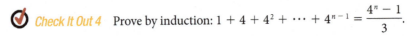 **Check It Out 4**   Prove by induction: $1 + 4 + 4^2 + \cdots + 4^{n-1} = \dfrac{4^n - 1}{3}$.   ■

You may wonder how we get the formulas to prove by induction in the first place. Many of these formulas are arrived at by first examining patterns and then coming up with a general formula using various mathematical facts. A complete discussion of how to obtain these formulas is beyond the scope of this book.

Exercises 10.7

 Skills

In Exercises 1–4, replace k by k + 1 in each expression.

1. $k(k + 1)(k + 2)$ **2.** $3^k - 1$ **3.** $\dfrac{k}{k + 1}$ **4.** $\dfrac{3}{1 + k^2}$

In Exercises 5–25, prove the statement by induction.

5. $3 + 5 + \cdots + (2n + 1) = n(n + 2)$

6. $2 + 6 + 10 + \cdots + (4n - 2) = 2n^2$

7. $1 + 4 + 7 + \cdots + (3n - 2) = \dfrac{n(3n - 1)}{2}$

8. $5 + 4 + 3 + \cdots + (6 - n) = \dfrac{1}{2}n(11 - n)$

9. $7 + 5 + 3 + \cdots + (9 - 2n) = -n^2 + 8n$

10. $3 + 9 + 15 + \cdots + (6n - 3) = 3n^2$

11. $2 + 5 + 8 + \cdots + (3n - 1) = \dfrac{1}{2}n(3n + 1)$

12. $1^2 + 2^2 + 3^2 + \cdots + n^2 = \dfrac{n(n + 1)(2n + 1)}{6}$

13. $1^3 + 2^3 + \cdots + n^3 = \dfrac{n^2(n + 1)^2}{4}$

14. $1 + 2 + 2^2 + \cdots + 2^{n-1} = 2^n - 1$

15. $1^2 + 3^2 + \cdots + (2n - 1)^2 = \dfrac{n(2n - 1)(2n + 1)}{3}$

16. $\dfrac{1}{1 \cdot 2} + \dfrac{1}{2 \cdot 3} + \dfrac{1}{3 \cdot 4} + \cdots + \dfrac{1}{n(n + 1)} = \dfrac{n}{n + 1}$

17. $1 \cdot 2 + 2 \cdot 3 + 3 \cdot 4 + \cdots + n(n + 1) = \dfrac{n(n + 1)(n + 2)}{3}$

18. $1 + 3 + 3^2 + \cdots + 3^{n-1} = \dfrac{3^n - 1}{2}$

19. $1 + 5 + 5^2 + \cdots + 5^{n-1} = \dfrac{5^n - 1}{4}$

20. $1 + r + r^2 + \cdots + r^{n-1} = \dfrac{r^n - 1}{r - 1}$, r a positive integer, $r \neq 1$

21. $3^n - 1$ is divisible by 2.

22. $n^3 - n + 3$ is divisible by 3.

23. $n^2 + 3n$ is divisible by 2.

24. $n^2 + n$ is even.

25. $2^n > n$

Concepts

Induction is not the only method of proving that a statement is true. Exercises 26–29 suggest alternate methods for proving statements.

26. By factoring $n^2 + n$, n a natural number, show that $n^2 + n$ is divisible by 2.

27. By factoring $a^3 - b^3$, a and b positive integers, show that $a^3 - b^3$ is divisible by $a - b$.

28. Prove that $1 + 4 + 7 + \cdots + (3n - 2) = \dfrac{n(3n - 1)}{2}$ by using the formula for the sum of terms of an arithmetic sequence.

29. Prove that $1 + 4 + 4^2 + \cdots + 4^{n-1} = \dfrac{4^n - 1}{3}$ by using the formula for the sum of terms of a geometric sequence.

Chapter 10 Summary

Section 10.1 Sequences

Sequence

A **sequence** is a function $f(n)$ whose domain is the set of all nonnegative integers and whose range is a subset of the set of all real numbers. The numbers $f(0), f(1), f(2), \ldots$ are called the **terms** of the sequence.

Definition of an arithmetic sequence [Review Exercises 1–4, 9]

Each term of an **arithmetic sequence** is given by the rule

$$a_n = a_0 + nd, \quad n = 0, 1, 2, 3, \ldots$$

where a_0 is the starting value of the sequence and d is the *common difference* between successive terms.

Definition of a geometric sequence [Review Exercises 5–8]

A **geometric sequence** is defined by the rule

$$a_n = a_0 r^n, \quad n = 0, 1, 2, 3, \ldots$$

where a_0 is the initial value of the sequence and $r \neq 0$ is the *fixed ratio* between successive terms.

Section 10.2 Sums of Terms of Sequences

Sum of the first *n* terms of an arithmetic sequence [Review Exercises 10–13]

Let $a_j = a_0 + jd, j = 0, 1, 2, \ldots$, be an arithmetic sequence, and let n be a positive integer. The sum of the first n terms from a_0 to a_{n-1} is given by

$$S_n = a_0 + a_1 + \cdots + a_{n-1}$$

$$= \frac{n}{2}(2a_0 + d(n-1))$$

$$= \frac{n}{2}(a_0 + a_{n-1})$$

Summation notation [Review Exercises 14, 15]

The **summation symbol** is indicated by the Greek letter Σ (sigma).

$$\text{Summation symbol} \rightarrow \sum_{i=1}^{n} i \leftarrow \text{Expression to be evaluated and summed}$$

Index i ends at n

Index i starts at 1

Sum of the first *n* terms of a geometric sequence   [Review Exercises 14–21, 26]

Let $a_j = a_0 r^j$, $j = 0, 1, 2, \ldots$, be a geometric sequence. The sum of the first n terms from a_0 to a_{n-1} is given by

$$S_n = \sum_{j=0}^{n-1} a_0 r^j = a_0 \left(\frac{1 - r^n}{1 - r} \right)$$

Sum of an infinite geometric series   [Review Exercises 22–25]

If $|r| < 1$, then the infinite geometric series $a_0 + a_0 r + a_0 r^2 + a_0 r^3 + \cdots + a_0 r^{n-1} + \cdots$ has the sum

$$S = \sum_{i=0}^{\infty} a_0 r^i = \frac{a_0}{1 - r}$$

Section 10.3 General Sequences and Series

Sequences given by a rule   [Review Exercises 27–36]

A **general rule for the terms** a_n of a sequence can be used to generate terms of the sequence.

Alternating sequences   [Review Exercises 35, 36]

In an **alternating sequence**, the terms alternate between positive and negative values.

Recursively defined sequences   [Review Exercises 37–40]

Recursively defined sequences define the nth term by using the preceding terms, along with the first term or first several terms.

Partial sums   [Review Exercises 41, 42]

The **nth partial sum** of a sequence $\{a_n\}$ is given by

$$\sum_{i=0}^{n-1} a_i$$

Section 10.4 Counting Methods

Multiplication principle   [Review Exercises 45–47]

Suppose there are n slots to fill. Let a_1 be the number of possibilities for the first slot, a_2 the number of possibilities for the second slot, and so on, with a_k representing the number of possibilities for the kth slot. Then the total number of ways in which the n slots can be filled is

$$\text{Total possibilities} = a_1 \times a_2 \times a_3 \times \cdots \times a_n$$

Definition of *n*! (*n* factorial)   [Review Exercises 43, 44]

Let n be a positive integer. Then

$$n! = n(n - 1)(n - 2) \cdots (3)(2)(1)$$

The quantity $0! = 1$.

Definition of permutation [Review Exercises 43, 45–47, 50]

Given n *distinct* objects and r slots to fill, the number of **r-permutations of n objects** is given by $P(n, r) = \frac{n!}{(n-r)!}$. In a permutation, order matters.

Definition of combination [Review Exercises 48, 49, 51, 52]

Let r and n be positive integers with $r \leq n$. When the order of the objects does not matter, the selection of r objects at a time from a set of n objects is called a **combination**:

$$C(n, r) = \frac{n!}{(n-r)!\, r!}$$

Section 10.5 Probability

Basic terminology [Review Exercises 53–58]

- An **outcome** is any possibility resulting from an experiment. Here, an experiment simply denotes an activity yielding random results.

- An **event** is a collection (set) of outcomes.

- A **sample space** is the set consisting of all possible outcomes for a certain situation under consideration.

Definition of probability [Review Exercises 59–62]

Assume that a sample space S consists of equally likely outcomes. Then the **probability** of an event E, denoted by $P(E)$, is defined as

$$P(E) = \frac{\text{Number of outcomes in } E}{\text{Number of outcomes in } S}$$

Note that $0 \leq P(E) \leq 1$.

Mutually exclusive events [Review Exercises 56, 60]

Two events F and G are **mutually exclusive** if they have no overlap. In this case,

$$P(F \text{ or } G) = P(F) + P(G)$$

Complement of an event [Review Exercise 57]

The set of all outcomes in a sample space that do *not* belong to event E is called the **complement** of E and is denoted by E'.

The probability of E' is $P(E') = 1 - P(E)$.

Section 10.6 The Binomial Theorem

Binomial coefficients [Review Exercises 63–70]

Let n and r be nonnegative integers with $r \leq n$. In the expansion of $(a + b)^n$, the **coefficient** of the term $a^{n-r} b^r$ is

$$\binom{n}{r} = \frac{n!}{(n-r)!\, r!}$$

The Binomial Theorem   [Review Exercises 63–70]

Let n be a positive integer. Then

$$(a + b)^n = \sum_{i=0}^{n} \binom{n}{i} a^{n-i} b^i$$

$$= \binom{n}{0} a^n + \binom{n}{1} a^{n-1} b + \binom{n}{2} a^{n-2} b^2 + \cdots + \binom{n}{n} b^n$$

The ith term of $(a + b)^n$ is given by $\binom{n}{i-1} a^{n-i+1} b^{i-1}$, where n and i are positive integers, with $1 \le i \le n + 1$.

Section 10.7 Mathematical Induction

Principle of mathematical induction   [Review Exercises 71–76]

Let n be a natural number and let P_n be a statement that depends on n. If

1. P_1 is true, and

2. for all positive integers k, P_{k+1} can be shown to be true if P_k is assumed to be true
   then P_n is true for all natural numbers n.

Chapter 10 Review Exercises

Section 10.1

In Exercises 1–4, find a rule for an arithmetic sequence that fits the given information.

1.

n	0	1
a_n	7	9

2.

n	0	1
$f(n)$	2	−1

3. The common difference d is 4 and $a_2 = 9$.

4. The common difference d is −2 and $a_6 = 10$.

In Exercises 5–8, find a rule for a geometric sequence that fits the given information.

5.

n	0	1
a_n	5	10

6.

n	0	1
a_n	8	4

7. The common ratio r is 3 and $a_3 = 54$.

8. The common ratio r is $\frac{1}{3}$ and $b_2 = 9$.

9. Compensation Carolyn has worked as an accountant for the same firm for the past 8 years. Her annual starting salary with her current employer was $36,000. Each year, on the anniversary of her first day on the job with this firm, she has been given a raise of $1500. What was her annual salary just after her fifth anniversary with the company?

Section 10.2

In Exercises 10–13, find the sum of the terms of the arithmetic sequence using the summation formula.

10. $1 + 5 + 9 + \cdots + 61$

11. $2 + 7 + 12 + \cdots + 102$

12. The sum of the first 20 terms of the sequence defined by

$$a_n = 4 + 5n, n = 0, 1, 2, \ldots$$

13. The sum of the first 15 terms of the sequence defined by

$$a_n = 2 - 3n, n = 0, 1, 2, \ldots$$

In Exercises 14–17, find the sum of the terms of the geometric sequence using the summation formula.

14. $4 + 8 + 16 + 32 + 64 + 128 + 256$

15. $0.6 + 1.8 + 5.4 + 16.2$

16. $\displaystyle\sum_{i=0}^{5} \left(\frac{1}{2}\right)^n$

17. $\displaystyle\sum_{i=0}^{7} (3)^n$

In Exercises 18–21, (a) write using summation notation and (b) find the sum using the summation formula.

18. The first 25 terms of the series defined by $a_n = 3n, n = 0, 1, 2, 3, \ldots$

19. The first 10 terms of the series defined by $a_n = 2.5n, n = 0, 1, 2, 3, \ldots$

20. The first eight terms of the series defined by $a_n = 2^n, n = 0, 1, 2, 3, \ldots$

21. $a + a^2 + a^3 + \cdots + a^{16}, a \neq 0, a \neq 1$

In Exercises 22–25, determine whether the infinite geometric series has a sum. If so, find the sum.

22. $1 + \dfrac{1}{3} + \dfrac{1}{9} + \cdots$

23. $27 + 9 + 3 + 1 + \dfrac{1}{3} + \cdots$

24. $1 + 1.1 + 1.21 + 1.331 + \cdots$

25. $3 + 2 + \dfrac{4}{3} + \dfrac{8}{9} + \cdots$

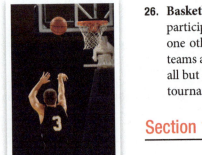

26. Basketball Playoffs An elimination basketball tournament is held, with 32 teams participating. All the teams play in the first round, with each team playing against just one other team. The losing teams in the first round are eliminated, and the winning teams advance to the second round. This process continues for additional rounds, until all but one team have been eliminated. How many rounds of games were played in the tournament?

Section 10.3

In Exercises 27–36, find the first five terms of the sequence.

27. $a_n = -2n + 5, n = 0, 1, 2, 3, \ldots$

28. $a_n = 4n + 1, n = 0, 1, 2, 3, \ldots$

29. $a_n = -3\left(\dfrac{1}{2}\right)^n, n = 0, 1, 2, 3, \ldots$

30. $a_n = -\left(\dfrac{2}{3}\right)^n, n = 0, 1, 2, 3, \ldots$

31. $b_n = -n^2 + 1, n = 0, 1, 2, 3, \ldots$

32. $b_n = 3n^3 - 1, n = 0, 1, 2, 3, \ldots$

33. $f(n) = \dfrac{n}{n^2 + 1}, n = 1, 2, 3, \ldots$

34. $f(n) = \dfrac{1}{2n^2 - 1}, n = 1, 2, 3, \ldots$

35. $a_n = (-1)^n \left(\dfrac{1}{3}\right)^n, n = 1, 2, 3, \ldots$

36. $a_n = (-1)^n (2n + 3), n = 1, 2, 3, \ldots$

In Exercises 37–40, find the first four terms of the recursively defined sequence.

37. $a_0 = 4; a_n = a_{n-1} + 3, n = 1, 2, 3, \ldots$

38. $a_0 = -2; a_n = a_{n-1} + 0.5, n = 1, 2, 3, \ldots$

39. $a_0 = -1; a_n = a_{n-1} + 2n, n = 1, 2, 3, \ldots$

40. $b_1 = \sqrt{2}; b_n = \sqrt{b_{n-1} + 2}, n = 2, 3, 4, \ldots$

In Exercises 41 and 42, find the partial sum.

41. $\displaystyle\sum_{i=0}^{3} 2i^2$

42. $\displaystyle\sum_{i=0}^{3} (-1)^i (2i + 1)$

Section 10.4

In Exercises 43 and 44, evaluate.

43. a. $\dfrac{4!}{2!}$ 　　 **b.** $P(5, 4)$ 　　 **c.** $P(4, 4)$ 　　 **d.** $P(6, 3)$

44. a. $\dfrac{5!}{3!}$ 　　 **b.** $C(5, 4)$ 　　 **c.** $\dbinom{4}{4}$ 　　 **d.** $C(6, 3)$

45. How many different photographs are possible if five adults line up in a row?

46. How many different photographs with exactly four college students are possible if there are six college students to select from? Here, order matters.

47. How many different photographs are possible if four children stand in the front row and three adults stand in the back row?

48. How many three-person committees can be formed in a club with 10 members?

49. How many five-letter words, including nonsense words, can be made from the letters in the word TABLE?

50. You have three English books, two history books, and three physics books.
    a. How many ways are there to arrange all these books on a shelf?
    b. How many ways are there to arrange the books if they must be grouped together by subject?

51. **Graduation** All 69 graduates of a middle school attended their commencement ceremony, and they all posed for the class photograph. They formed six rows for the photo, with one row in back of another. The number of graduates in each row beyond the first (front) row increased by 1. How many graduates were in the back row of the photo?

52. **Sports** There are 11 swimmers on a swim team. How many different ways could a swim team be formed from a pool of 18 swimmers, assuming that every swimmer is qualified to be on the team?

Section 10.5

In Exercises 53–58, use this scenario: A coin is tossed three times and the sequence of heads and tails is recorded.

53. List all the possible outcomes for this problem.

54. What is the event that you get at least one head?

55. What is the sample space for this problem?

56. Are the events of getting all tails and getting all heads mutually exclusive?

57. What is the complement of the event that you get at least one head?

58. Calculate the probability of the event that you get at least one head.

59. A bag consists of three red marbles, two white marbles, and four green marbles. If one marble is chosen randomly, what is the probability that the marble is white?

60. You are dealt one card from an ordinary deck of 52 cards. What is the probability that it is an ace of hearts or an ace of spades?

61. **Card Game** Kim takes part in a card game in which every player is dealt a hand of five cards from a standard 52-card deck. What is the probability that the hand dealt to her consists of five red cards, if she is dealt the first five cards?

62. **Cryptography** A secret code is made up of a sequence of four letters of the alphabet. What is the probability that all four letters of the code are identical?

Section 10.6

In Exercises 63–66, expand each expression using the Binomial Theorem.

63. $(x + 3)^4$ **64.** $(2x + 1)^3$ **65.** $(3x + y)^3$ **66.** $(3z - 2w)^4$

In Exercises 67–70, use the Binomial Theorem to find the indicated term or coefficient.

67. The coefficient of x^4 when expanding $(x + 3)^5$

68. The coefficient of y^3 when expanding $(2y + 1)^5$

69. The second term in the expansion of $(x + 2y)^3$

70. The third term in the expansion of $(y - z)^4$

Section 10.7

In Exercises 71–76, use mathematical induction to prove the statement.

71. $1 + \dfrac{1}{2} + \dfrac{1}{4} + \cdots + \dfrac{1}{2^n} = 2 - \dfrac{1}{2^n}$

72. $1 + 3 + 6 + \cdots + \dfrac{n(n + 1)}{2} = \dfrac{n(n + 1)(n + 2)}{6}$

73. $(n + 1)^2 + n$ is odd.

74. $n^3 + 2n$ is divisible by 3.

75. $3 + 6 + 9 + \cdots + 3n = \dfrac{3}{2}n(n + 1)$

76. $1 + 5 + 9 + \cdots + (4n - 3) = n(2n - 1)$

1. Find a rule for an arithmetic sequence with $a_0 = 8$ and $a_1 = 11$.

2. Find a rule for an arithmetic sequence with common difference $d = 5$ and $a_4 = 27$.

3. Find a rule for a geometric sequence with $a_0 = 15$ and $a_1 = 5$.

4. Find a rule for a geometric sequence with common ratio $r = \frac{2}{3}$ and $a_2 = 2$.

5. Find the following sum using a formula for the sum of terms: $5 + 8 + 11 + \cdots + 104$.

6. Find the sum of the first 30 terms of the sequence defined by $a_n = -2 + 4n$, $n = 0, 1, 2, \ldots$.

7. Write the following sequence using summation notation, and find its sum: $a_n = 5n$, $n = 0, 1, 2, 3, \ldots, 9$.

8. Find the first four terms of the sequence defined by $a_n = -2n^3 + 2$, $n = 0, 1, 2, 3, \ldots$.

9. Find the first five terms of the sequence defined by $a_n = (-1)^n \left(\frac{1}{4}\right)^n$, $n = 0, 1, 2, 3, \ldots$.

10. Write the first five terms of the following recursively defined sequence: $a_0 = 4$; $a_n = a_{n-1} - n$, $n = 1, 2, 3, \ldots$.

11. Evaluate the following partial sum: $\sum_{i=0}^{3} (3i + 2)$

12. Evaluate the following.

    a. $\dfrac{5!}{2!}$ b. $C(6, 4)$ c. $P(7, 1)$ d. $P(6, 4)$

13. How many committees of three people are possible in a club with 12 members?

14. How many six-letter words, including nonsense words, can be made with the letters in the word SAMPLE?

15. If you have four mathematics books, two biology books, and three chemistry books, how many ways are there to arrange the books on one row of a bookshelf if they must be grouped by subject?

In Exercises 16–18, consider a bag containing four red marbles, five blue marbles, and two white marbles. One marble is drawn randomly from the bag and its color is recorded.

16. List all the possible outcomes for this problem.

17. What is the probability that a white marble is drawn?

18. What is the probability that a blue marble is *not* drawn?

19. You are dealt one card from an ordinary deck of 52 cards. What is the probability that it is the queen of hearts or the jack of spades?

20. Expand using the Binomial Theorem: $(3x + 2)^4$.

21. Prove by induction: $2 + 6 + 10 + \cdots + (4n - 2) = 2n^2$.

22. A clothing store sells a "fashion kit" consisting of three pairs of pants, four shirts, and two jackets. How many different outfits, each consisting of a pair of pants, a shirt, and a jacket, can be put together using this kit?

23. A license plate number consists of two letters followed by three nonzero digits. How many different license plates are possible, assuming no letter or digit can be used more than once?

Answers

Chapter 1

1.1 Exercises

1. $-1, 0, 10, 40$ **3.** $-1.67, -1, 0, 0.5, \frac{4}{5}, 10, 40$

5. $0, 10, 40$ **7.** $-1, 0$

9. Associative property of multiplication
11. Distributive property of multiplication over addition
13. Commutative property of multiplication

15.

17.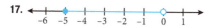

19.

21.

23.

25. $[-3, 4]$ **27.** $[0, \infty)$ **29.** $(-2, 4)$ **31.** $(1, \infty)$

	Interval	
	Inequality	Notation	Graph
33.	$4 \le x \le 10$	$[4, 10]$	
35.	$-3 \le x < 0$	$[-3, 0)$	
37.	$2 \le x \le 12$	$[2, 12]$	

39. 3.2 **41.** 253 **43.** $\frac{5}{4}$ **45.** $\frac{4}{3}$ **47.** 6 **49.** -4.5

51. 7 **53.** -3 **55.** 6 **57.** 9 **59.** 4.5 **61.** 12.2 **63.** 3

65. $\frac{7}{15}$ **67.** -5 **69.** 4 **71.** 14 **73.** -240 **75.** $-\frac{6}{11}$

77. -9 **79.** $-\frac{43}{3}$ **81.** $\frac{1}{4}$ **83.** $\frac{19}{5}$ **85.** $5, 6, 7, 8, 9$

87. $[25, 36]$ **89.** $20{,}602$ ft   **91.** No; it can be zero.
93. -7 and 1.

1.2 Exercises

1. -9 **3.** $\frac{1}{64}$ **5.** -1 **7.** $\frac{16}{9}$ **9.** $-16x^4y^8$ **11.** $\frac{1}{4x^{10}}$

13. $-2a^4b^{10}$ **15.** $\frac{1}{16x^2y^4}$ **17.** $\frac{x^3y^2}{3}$ **19.** $\frac{4}{y}$ **21.** $\frac{4}{x^6y^2}$

23. $\frac{9t^6}{s^4}$ **25.** 5.1×10^{-3} **27.** 5.6×10^3

29. 5.67×10^{-5} **31.** 1.76×10^6 **33.** 371 **35.** 0.028
37. $596{,}000$ **39.** $43{,}670{,}000$ **41.** 9.03×10^7
43. 2×10^{-1} **45.** 7 **47.** $\frac{1}{2}$ **49.** 343 **51.** $\frac{2}{5}$
53. $4\sqrt{2}$ **55.** $5\sqrt[3]{2}$ **57.** 16 **59.** $\frac{-2\sqrt{2}}{5}$ **61.** $\frac{\sqrt{15}}{5}$

63. $\frac{\sqrt[3]{21}}{3}$ **65.** $\frac{5\sqrt{6}}{21}$ **67.** $\frac{-2\sqrt[3]{2}}{3}$ **69.** $xy^2\sqrt{x}$

71. $x^2y\sqrt[3]{x^2y}$ **73.** $3xy^2\sqrt{5x}$ **75.** $2xy\sqrt[3]{3x^2}$ **77.** $19\sqrt{2}$

79. $-2\sqrt{6} + \sqrt{3}$ **81.** -4 **83.** $-1 - \sqrt{5}$

85. $\sqrt{3} + \sqrt{2}$ **87.** $\frac{1}{3^{2/3}}$ **89.** 5 **91.** $\frac{1}{7^{3/4}}$ **93.** $\frac{1}{4^{7/12}}$

95. $\frac{1}{xy^{3/2}}$ **97.** $\frac{1}{x^{7/6}}$ **99.** 3.68×10^4 **101.** 3.54×10^6

103. $10\sqrt{2}$ in.   **105.** 816 species
107. Defined for $-\infty < x < \infty$ and $y \ne 0$.
109. if $x = 1$ and $y = 2$, then $\sqrt{5} \ne 3$.

1.3 Exercises

1. $5y + 26; 1$ **3.** $3t^2 - 2t + 5; 2$ **5.** $-4s^2 - 6s + 7; 2$
7. $6z + 14$ **9.** $-11x^2 - 17x + 8$
11. $z^5 - 4z^4 + z^3 - 15z^2 + 8z + 7$
13. $25v^5 + 3v^4 - 16v^3 + 3v^2 - 6v + 5$
15. $9t^3 - 21t^2 + 21t + 23$ **17.** $2s^2 + s$
19. $-18z^3 - 15z$ **21.** $7t^3 - 3t^2 - 9t$
23. $7z^4 + 63z^3 - 56z^2$ **25.** $y^2 + 11y + 30$
27. $-v^2 - 9v + 36$ **29.** $-28v^2 - 59v - 30$
31. $u^3 + 3u^2 - 9u - 27$
33. $x^2 + 8x + 16$ **35.** $s^2 + 12s + 36$
37. $25t^2 + 40t + 16$ **39.** $v^2 - 81$ **41.** $-81s^2 + 49$
43. $v^4 - 9$ **45.** $25y^4 - 16$ **47.** $16z^4 - 25$ **49.** $x^2 - 4z^2$
51. $-t^3 - 11t^2 - 29t + 6$ **53.** $x^3 + x^2 - 13x + 14$
55. $-20u^4 + 59u^3 - 14u^2 - 85u + 63$
57. $2x(x^2 + 3x - 4)$ **59.** $-2t^2(t^4 + 2t^3 - 5)$
61. $-5x^3(x^4 - 2x^2 + 3)$ **63.** $(x + 1)(3 + x)$
65. $(s - 5)(s + 3)(s - 3)$ **67.** $(3u + 1)(2u + 1)(2u - 1)$
69. $(x + 1)(x + 3)$ **71.** $(x + 2)(x - 8)$ **73.** $3(s + 4)(s + 1)$
75. $-6(t + 2)(t - 6)$ **77.** $9(u - 2)(u - 1)$
79. $(x + 4)(x - 4)$ **81.** $(2x + 1)^2$
83. $(y + 4)(y^2 - 4y + 16)$ **85.** $(2y + 1)(4y^2 - 2y + 1)$
87. $(z + 7)(z + 6)$ **89.** $(x + 6)^2$ **91.** $-(y - 2)^2$
93. $(z - 8)^2$ **95.** $-(2y - 1)(y - 3)$ **97.** $(3y + 2)^2$
99. $3(2z + 3)(z - 2)$ **101.** $5(3t + 1)(t - 5)$
103. $-5(2u + 1)(u + 4)$ **105.** $(v + 2)(v - 2)$
107. $(7 + s)(7 - s)$ **109.** $(2 + 5t)(2 - 5t)$ **111.** $t^2(t - 16)$
113. $4u(3u - 5)(u + 2)$ **115.** $-5t(2t - 3)(t + 1)$
117. $-5z(3z + 4)(z - 1)$ **119.** $(2y + 3)(y + 2)(y - 2)$
121. $(2z - 3)(4 + 3z)(4 - 3z)$ **123.** $3y^2(y + 4)(y + 2)$
125. $7x^3(x + 3)(x - 3)$ **127.** $-6s^3(s^2 + 5)$
129. $8(x + 2)(x^2 - 2x + 4)$ **131.** $(1 - 2y)(1 + 2y + 4y^2)$
133. $24x$ **135. a.** $4s$ **b.** $8s$
137. a. $1000r^2 + 2000r + 1000$ **b.** $\$1102.50$
139. No. If the coefficients are additive inverses, then the third degree term would disappear and the resulting polynomial would not be of degree 3.
141. 0 **143.** $a = 0$
145. $x^2 + 6x + 9 = (x + 3)^2$. Answers may vary.

1.4 Exercises

1. $\frac{19}{8}$ **3.** $\frac{x - 2}{6}, x \ne -2$ **5.** $\frac{x + 1}{x + 3}, x \ne -3, 3$

7. $x^2(x - 1), x \ne -1$ **9.** $\frac{x^2 + x + 1}{x + 1}, x \ne -1, 1$ **11.** $\frac{1}{xy}$

13. $\frac{1}{(x - 3)(x + 2)}$ **15.** $\frac{6}{x + 6}$ **17.** $\frac{2(x - 2)}{x(x + 2)(x + 5)}$

19. $\frac{(x^2 - x + 1)(2x + 1)}{x + 2}$ **21.** $\frac{5}{(x - 4)(x + 1)}$

23. $\frac{(x - 2)(x - 1)}{x + 1}$ **25.** $\frac{x^2 + 2x + 4}{(x + 2)(x - 2)}$ **27.** $\frac{2x + 3}{x^2}$

29. $\frac{5x + 3}{(x + 1)(x - 1)}$ **31.** $\frac{2x(x^2 - 3x + 6)}{3(x + 3)(x - 3)}$

33. $\frac{2(x + 2)}{(x + 1)(x - 1)}$ **35.** $-\frac{x(x - 3)(x - 1)}{(x + 4)(x - 4)}$ **37.** $\frac{1}{1 - x}$

39. $\dfrac{2x - 11}{(x + 2)(x - 2)}$ **41.** $\dfrac{-8x - 7}{(x - 3)(x + 2)}$ **43.** $\dfrac{x}{x - 1}$

45. $\dfrac{y(y + x)}{x - 2y^2}$ **47.** $\dfrac{rst}{st + rt + rs}$ **49.** $\dfrac{x}{1 - x}$

51. $\dfrac{-2(x + 1)}{(x - 3)(5x - 1)}$ **53.** $\dfrac{-1}{x(x + h)}$ **55.** $\dfrac{x + 4}{4(x - 2)}$

57. $\dfrac{a + ba - b^2}{a + b}$

59. \$ 3.50; average cost for 100 booklets

61. $t = \dfrac{12}{7}$hr   **63.** $x = 1, y = 1,$ for example

65. Not true for $x = 0$

1.5 Exercises

1. $x = 1$ **3.** $x = -3$ **5.** $x = -3$ **7.** $x = -9$ **9.** $x = \dfrac{19}{2}$

11. $x = -\dfrac{18}{13}$ **13.** $x = 2$ **15.** $x = 9$ **17.** $x = 2$ **19.** $x = \dfrac{4}{5}$

21. $x = 6$ **23.** $x = \dfrac{60}{19}$ **25.** $x = 10$ **27.** $x = \dfrac{40}{49}$

29. $x = \dfrac{1}{\pi}$ **31.** $y = 5 - x$ **33.** $y = 3 + 2x$

35. $y = \dfrac{5}{2} - \dfrac{5}{4}x$ **37.** $y = 5 - 4x$ **39.** $x = 5, x = -5$

41. $x = 4, x = 3$ **43.** $x = 2, x = -2$ **45.** $x = \dfrac{2}{3}, x = -\dfrac{1}{2}$

47. $x = \dfrac{1}{2}$ **49.** $t = -1, t = \dfrac{3}{2}$ **51.** $x = -1, x = -3$

53. $x = 1 + \sqrt{5}, x = 1 - \sqrt{5}$ **55.** $x = 1, x = -2$

57. $x = -2 + \dfrac{3\sqrt{2}}{2}, x = -2 - \dfrac{3\sqrt{2}}{2}$

59. $x = -1 + \sqrt{2}, x = -1 - \sqrt{2}$

61. $x = \dfrac{1}{2} + \dfrac{\sqrt{3}}{2}, x = \dfrac{1}{2} - \dfrac{\sqrt{3}}{2}$

63. $x = -\dfrac{1}{2} + \dfrac{\sqrt{13}}{2}, x = -\dfrac{1}{2} - \dfrac{\sqrt{13}}{2}$

65. No real solution   **67.** $l = 20 + 10\sqrt{3}, l = 20 - 10\sqrt{3}$

69. $t = 4 + \sqrt{22}, t = 4 - \sqrt{22}$

71. $x = -\dfrac{4}{3} + \dfrac{2\sqrt{10}}{3}, x = -\dfrac{4}{3} - \dfrac{2\sqrt{10}}{3}$

73. $x = 2, x = -2$ **75.** $x = 1$ **77.** $x = -1, x = -\dfrac{1}{2}$

79. $x = 1 + \sqrt{10}, x = 1 - \sqrt{10}$

81. $x = -\dfrac{1}{2} + \dfrac{\sqrt{13}}{2}, x = -\dfrac{1}{2} - \dfrac{\sqrt{13}}{2}$

83. 8; 2 real solutions   **85.** -8; no real solutions
87. $40x - 200 = 800$; 25 Blu-ray players
89. 18 in.   **91.** Near end of 1998
93. $x = 5 + \sqrt{10}$ in. or $x = 5 - \sqrt{10}$ in.
95. 5 ft. No, you get a negative width, which is unrealistic.
97. $a \neq -1$ **99.** $k > 0$

1.6 Exercises

1. No   **3.** Yes   **5.** $[-1, \infty)$ **7.** $[4, \infty)$ **9.** $(10, \infty)$

11. $\left[-\dfrac{3}{2}, \infty\right)$ **13.** $(2, \infty)$ **15.** $(-\infty, 1]$ **17.** $\left(-\infty, -\dfrac{13}{5}\right]$

19. $[3, \infty)$ **21.** $(-4, \infty)$ **23.** $\left[\dfrac{18}{7}, \infty\right)$ **25.** $\left[-\dfrac{3}{2}, 1\right]$

27. $(1, 5)$ **29.** $(-3, 3)$ **31.** $(5, 20); q > 5$

33. $(40, 600); q > 40$
35. a. $C = 2q + 750$ **b.** $R = 4q$ **c.** $q > 375$
37. At least 62.5 mins   **39.** At least $[89, 100]$
41. a. $A = 4.95 + 0.07x$;   **b.** $B = 0.10x$;
   c. 165 mins; \$ 16.50
43. $70 \leq x \leq 89.1$ **45.** $4 \geq 0$ is always true, regardless of x.

1.7 Exercises
1. $\{-10, 2\}$ **3.** $\{6, -2\}$ **5.** $\left\{-\dfrac{17}{4}, \dfrac{23}{4}\right\}$ **7.** $\{-9, -1\}$

9. $\{-3, 2\}$ **11.** $\{-3, 5\}$ **13.** $\left\{\dfrac{3}{2}, \dfrac{7}{2}\right\}$

15. $\{-3, -\sqrt{7}, \sqrt{7}, 3\}$ **17.** No   **19.** Yes
21. $x > -3$

23. $-1 \leq s \leq 2$

25. $-\dfrac{4}{3} \leq x \leq \dfrac{4}{3}$

27. $x < -7$ or $x > 7$

29. $(-\infty, -4) \cup (4, \infty)$

31. $[-7, 1]$

33. $(-\infty, 4) \cup (16, \infty)$

35. $(-\infty, 2) \cup (5, \infty)$

37. $\left[-\dfrac{8}{3}, 4\right]$

39. $[-22, -2]$

41. $(-37, 23)$

43. $(-\infty, -4] \cup [12, \infty)$

45. $\left(-\dfrac{17}{3}, 1\right)$

47. $(-\infty, \infty)$

49 $(3.999, 4.001)$

51. $|x + 7| = 3$　**53.** $|x - 8| < 5$　**55.** $|x + 6.5| > 8$
57. $-30 < T < 10$
　　This means the temperature is always less than $10°$F but greater than $-30°$F.
59. $|x| < 30$, the origin is Omaha.
61. Let $T =$ the temperature. Then $|T - 68| \leq 1.5$.
63. $|-3(x + 2)|$ will produce strictly nonnegative values for all real x, whereas $-3|x + 2|$ will produce strictly nonpositive values for all real x.
65. By definition, $|x - k| = x - k$ or $k - x$. Similarly, $|k - x| = k - x$ or $x - k$
67. The lowest value of $|x|$ is 0 when $x = 0$, so there is no value of x for which $|x| < 0$.

1.8 Exercises

1. 3　**3.** $x^2 + 2x + 4$　**5.** $x = \sqrt{7}, x = -\sqrt{7}$
7. $x = \pm\sqrt{3}, x = \pm\sqrt{7}$　**9.** $s = \pm\dfrac{\sqrt{6}}{3}$
11. $x = \pm\sqrt{2}$　**13.** $x = -1, x = \sqrt[3]{5}$
15. $t = -\dfrac{\sqrt[3]{18}}{3}, t = -\sqrt[3]{4}$　**17.** $x = \dfrac{30}{19}$
19. $x = \dfrac{4}{3}$　**21.** $x = -\dfrac{1}{2}, x = \dfrac{1}{5}$　**23.** $x = 3$
25. $x = \dfrac{13 \pm \sqrt{249}}{8}$　**27.** $x = 0$　**29.** No solution
31. $x = 0, x = 3$　**33.** $x = 22$　**35.** $x = \pm 4$
37. $x = -8, x = 2$　**39.** $x = \dfrac{5 + \sqrt{13}}{2}$
41. $x = 4$　**43.** $x = 3, x = 11$　**45.** $x = 122$
47. $x = 1, x = 9$　**49.** $x = -1, x = \dfrac{1}{27}$
51. $\$120$　**53.** 12 hr　**55.** 4.681 km

57. $(\sqrt{x + 1} - 2)^2 = (x + 1) - 4\sqrt{x + 1} + 4$

59. Numerator is never zero

Chapter 1 Review Exercises

1. $-5, 3, 8$　**3.** $3, 8$　**5.** Associative property of addition
7.

9.

11. 10　**13.** -3.7　**15.** -5　**17.** $\dfrac{6y^4}{x^2}$　**19.** $\dfrac{4x^4}{y^3}$　**21.** $\dfrac{16x^{12}}{y^6}$

23. -4.67×10^6　**25.** $30,010$　**27.** 6.4×10^2

29. 1.6×10^{-2}g　**31.** $3x\sqrt{5x}$　**33.** $\dfrac{-2\sqrt[3]{12}}{5}$　**35.** $-\sqrt[3]{3} - 4$
37. $1 - \sqrt{3}$　**39.** -5　**41.** 9　**43.** $20x^{7/6}y^{3/2}$　**45.** $\dfrac{2}{x^{1/3}y}$
47. $6y^3 + 13y^2 + 24y - 12$　**49.** $9t^5 + 3t^4 - 10$
51. $12u^2 - 37u - 10$　**53.** $-24z^2 + 91z - 72$
55. $6z^3 + 7z^2 + 19z + 40$　**57.** $9x^2 - 4$　**59.** $x^2 - 10x + 25$
61. $4z^2(2z + 1)$　**63.** $(y + 4)(y + 7)$　**65.** $(3x + 5)(x - 4)$
67. $(5x + 2)(x - 2)$　**69.** $(3u + 7)(3u - 7)$　**71.** $z(z^2 + 8)$
73. $2(x + 1)^2$　**75.** $4(x + 2)(x^2 - 2x + 4)$　**77.** $x + 3, x \neq 3$
79. $\dfrac{(x - 2)(x + 1)}{x - 1}$　**81.** $\dfrac{(x + 2)^2}{x(x - 2)}$　**83.** $\dfrac{5x + 1}{(x + 1)(x - 3)}$
85. $\dfrac{-2x^4 + 11x^2 + 7x + 78}{(x + 3)(x - 3)}$　**87.** $\dfrac{a + b}{a}$　**89.** $x = -3$
91. $x = 1$　**93.** $y = 2x + 5$　**95.** $x = 2, x = -2$

97. $x = 3, x = -4$　**99.** $x = \dfrac{1}{2}, x = -\dfrac{4}{3}$
101. $x = -1 + \sqrt{6}, x = -1 - \sqrt{6}$　**103.** $x = 2 \pm \dfrac{\sqrt{14}}{2}$
105. $x = 1 \pm \sqrt{3}$　**107.** $x = \dfrac{-3 \pm \sqrt{105}}{8}$
109. $x = 5, x = -2$　**111. a.** 2007　**b.** No　**113.** $\left(-\infty, \dfrac{9}{2}\right]$
115. $\left(-\infty, \dfrac{20}{3}\right]$　**117.** $[1, 16]$　**119.** $\{-13, 1\}$
121. $\left\{-\dfrac{23}{8}, \dfrac{17}{8}\right\}$　**123.** $\{-1, 9\}$
125. $(-13, 1)$

127. $[-3, 2]$

129. $\left[-\dfrac{11}{2}, -\dfrac{3}{2}\right]$

131. $x = \sqrt[3]{7}, x = -\sqrt[3]{3}, x = -\sqrt[3]{3}$　**133.** $x = \dfrac{3}{2}, x = -\dfrac{1}{3}$
135. $x = -3, x = -7$　**137.** $x = 0$　**139.** $x = 7$　**141.** $x = 3$

Chapter 1 Test

1. $-1, 1.55, 41$　**2.** Distributive Property
3.

4. 10.3　**5.** $\dfrac{9}{7}$　**6.** -25　**7.** 8.903×10^6　**8.** $-36x^4y^{10}$
9. $\dfrac{64x^{27}}{y^{12}}$　**10.** $4x\sqrt{3x}$　**11.** $-30x^{8/15}$　**12.** $\dfrac{9x^{1/3}}{y^3}$
13. $\sqrt[3]{2}$　**14.** 25　**15.** $\dfrac{-7 + 7\sqrt{5}}{4}$　**16.** $(5 + 7y)(5 - 7y)$
17. $(2x + 5)^2$　**18.** $(3x - 5)(2x + 1)$
19. $x(2x + 3)(2x - 3)$　**20.** $(3x - 7)(x + 5)$
21. $2(x + 2)(x^2 - 2x + 4)$　**22.** $\dfrac{2(2x + 1)}{(x + 3)(x - 2)}$
23. $\dfrac{x - 1}{(x + 2)(x - 2)}$　**24.** $\dfrac{19 - 7x}{(x + 2)(x - 2)}$
25. $\dfrac{3x^2 + 6x - 4}{(2x + 1)(x - 1)(x + 3)}$　**26.** $\dfrac{6 - 8x}{3x + 2}$　**27.** $x = \dfrac{11}{4}$
28. $\left[-\dfrac{3}{5}, \dfrac{9}{5}\right)$　**29.** $\left\{-\dfrac{19}{18}, \dfrac{11}{18}\right\}$　**30.** $\left\{-\dfrac{1}{2}, 4\right\}$
31. $x = 6, x = -1$　**32.** $x = 1 \pm \dfrac{\sqrt{10}}{2}$
33. $x = -\dfrac{1}{6} \pm \dfrac{\sqrt{13}}{6}$　**34.** $x = \dfrac{1}{2} \pm \dfrac{\sqrt{7}}{2}$
35. $x = \dfrac{4}{3}; x = -1$　**36.** $x = -\dfrac{1}{2} \pm \dfrac{1}{2}\sqrt{11}$
37. $(-\infty, 1] \cup [4, \infty)$　**38.** $\left(-\dfrac{9}{5}, 1\right)$
39. $x = \pm\dfrac{2\sqrt{3}}{3}$　**40.** $x = \dfrac{4}{7}$　**41.** $x = 5$　**42.** 2.85×10^{-5}g
43. 100 min　**44.** $t = 4$ sec

Chapter 2

2.1 Exercises

1. $y = -6$ **3.** $y = 2$ **5.** $y = -\dfrac{1}{2}x + 2$

7. a. Yes, of the form $y = mx + b$, $m = 3$ and $b = 1$.
b. Does not represent a line.
c. Yes, of the form $y = mx + b$ where $m = -5$ and $b = 0$.
d. Does not represent a line.

9. -7 **11.** 0 **13.** $\dfrac{1}{2}$ **15.** undefined **17.** undefined

19. 1 **21.** $\dfrac{2}{9}$ **23.** 1

25.

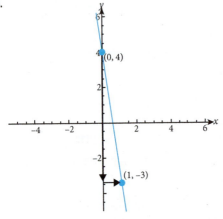

27. On line **29.** Not on line **31.** $y + 1 = -5(x - 2)$

33. $y + 4 = \dfrac{1}{3}(x + 1)$ **35.** $y - 5 = 0(x - 4)$

37. $y - 3.6 = 0.9(x - 1.5)$ **39.** $y + 1 = \dfrac{3}{5}\left(x - \dfrac{1}{2}\right)$

41. $y = \dfrac{1}{2}x + \dfrac{1}{3}$, $m = \dfrac{1}{2}$, y-int: $\left(0, \dfrac{1}{3}\right)$

43. $y = -\dfrac{3}{4}x + \dfrac{13}{4}$, $m = -\dfrac{3}{4}$, y-int: $\left(0, \dfrac{13}{4}\right)$

45. $x = 3$

47. $y = -2x + 6$
$m = -2$
x-intercept: $(3, 0)$
y-intercept: $(0, 6)$

49. $y = \dfrac{4}{3}x + \dfrac{2}{3}$
$m = \dfrac{4}{3}$
x-intercept: $\left(-\dfrac{1}{2}, 0\right)$
y-intercept: $\left(0, \dfrac{2}{3}\right)$

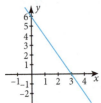

51. $y = -2x + 15$
$m = -2$
x-intercept: $\left(\dfrac{15}{2}, 0\right)$
y-intercept: $(0, 15)$

53. $y = \dfrac{5}{3}x + 3$
$m = \dfrac{5}{3}$
x-intercept: $\left(-\dfrac{9}{5}, 0\right)$
y-intercept: $(0, 3)$

55. $y = 2$
$m = 0$
y-intercept: $(0, 2)$

57. $y = x + 1$
59. $y = \dfrac{2}{5}x - \dfrac{6}{5}$
61. $x = 2$

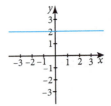

63.

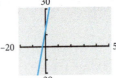

65.

67.

69. $y = -x + 2$ **71.** $y = \dfrac{1}{3}x + 1$ **73.** $y = -\dfrac{1}{2}x + \dfrac{5}{2}$

75. $y = -3x + 8$ **77.** $y = \dfrac{3}{4}x - \dfrac{3}{2}$ **79.** $y = -6x + 3$

81. $x = 4$ **83.** $y = -1$ **85.** $y = 0.5$ **87.** $y = -3x - 1$

89. $y = -\dfrac{1}{2}x + 1$ **91.** $y = -\dfrac{2}{3}x - 1$ **93.** $y = -\dfrac{3}{2}x - 1$

95. $y = -\dfrac{1}{2}x$ **97.** $y = 2x - 3$ **99.** $x = 0$ **101.** $x = -2$

103. a. $y = 50x + 650$
b. $m = 50$. An additional \$50 is earned for each additional computer sold.
$b = 650$. Amount earned if no computers are sold is \$650.

105. a. 40.2 billion US dollars
b. R-intercept $(0, 27.46)$ means in 2008 the global box office revenue was 27.46 billion US dollars.

107. a. 980 million handbags
     b. (0, 500), 500 million handbags were sold in 2009.
     c. 2013
109. $P = 1.50t + 25.50$ **111.** $F = \dfrac{9}{5}C + 32$
113. The graph does not appear in the viewing window.

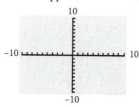

The maximum and minimum values of x and y shown in the viewing window must be manually changed to accommodate the values of this equation.

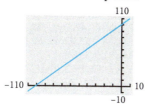

115. $y = -\dfrac{4}{5}x - 1$ **117.** $y = \dfrac{5}{3}x + \dfrac{28}{3}$

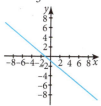 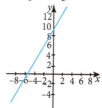

2.2 Exercises

1. False   **3.** $(x + 4)^2$ **5.** $d = 7\sqrt{5}$; $\left(-1, \dfrac{15}{2}\right)$

7. $d = 6\sqrt{2}$; $(-7, 17)$ **9.** $d = 2\sqrt{13}$; $(3, 2)$

11. $d = 14$; $(6, 4)$ **13.** $d = \dfrac{\sqrt{17}}{4}$; $\left(-\dfrac{3}{8}, \dfrac{1}{2}\right)$

15. $d = \sqrt{(b_1 - a_1)^2 + (b_2 - a_2)^2}$; $\left(\dfrac{a_1 + b_1}{2}, \dfrac{a_2 + b_2}{2}\right)$

17. $x^2 + y^2 = 25$ **19.** $(x + 1)^2 + y^2 = 9$

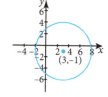 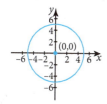

21. $(x - 3)^2 + (y + 1)^2 = 25$ **23.** $(x - 1)^2 + y^2 = \dfrac{9}{4}$

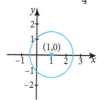 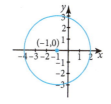

25. $(x - 1)^2 + (y - 1)^2 = 3$ **27.** $x^2 + y^2 = 10$

29. $(x - 2)^2 + y^2 = 25$
31. $(x - 1)^2 + (y + 2)^2 = 25$
33. $\left(x - \dfrac{1}{2}\right)^2 + y^2 = \dfrac{37}{4}$
35. 36
37. $\dfrac{25}{4}$

39. $\dfrac{9}{4}$ **41.** Center: $(0, 0)$; radius: 6
43. Center: $(1, -2)$; radius: 6   **45.** Center: $(8, 0)$; radius: $\dfrac{1}{2}$
47. Center: $(3, -2)$; radius: 4   **49.** Center: $(1, -1)$; radius: 3
51. Center: $(3, 2)$; radius: $3\sqrt{2}$
53. Center: $\left(\dfrac{1}{2}, 0\right)$; radius: $\dfrac{3}{2}$ **55.** $x^2 + y^2 = 4$
57. $(x + 1)^2 + (y + 1)^2 = 4$
59. Center: $(0, 0)$; radius: 2.5

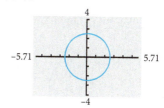

61. Center: $(3.5, 0)$; radius: $\sqrt{10}$

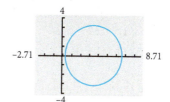

63.   **65.**

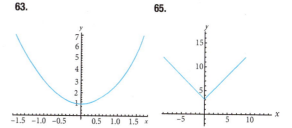

67.

69.

71. a. $x^2 + y^2 = 900$ **b.** $x^2 + y^2 = 1369$
73. a. $A: (0, 0)$ $B: (5, 0)$ $C: (10, 0)$ $D: (15, 0)$ $E: (20, 0)$
   $F: (15, 12)$ $G: (10, 12)$ $H: (5, 12)$
   b. 118 ft.

75. $(x + 1)^2 + (y - 3)^2 = 5$

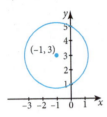

77. $(x + 5)^2 + (y + 1)^2 = 64$
79. $(0, -2 - 3\sqrt{7}), (0, -2 + 3\sqrt{7})$

2.3 Exercises
1. False   **3.** $x > -\dfrac{1}{2}$

5. $f(3) = 18$ **7.** $f(3) = -\dfrac{17}{2}$ **9.** $f(3) = 11$
   $f(-1) = -2$ $f(-1) = 3$
   $f(0) = 3$ $f(-1) = \dfrac{11}{2}$ $f(0) = 2$
            $f(0) = 2$

11. $f(3) = -36$ **13.** $f(3) = \sqrt{13}$ **15.** $f(3) = \dfrac{4}{3}$
   $f(-1) = -4$ $f(-1) = 1$
   $f(0) = -6$ $f(0) = 2$ $f(-1) = 0$
               $f(0) = -\dfrac{1}{3}$

17. $f(a) = 4a + 3$ **19.** $f(a) = -a^2 + 4$
   $f(a + 1) = 4a + 7$ $f(a + 1) = -a^2 - 2a + 3$
   $f\left(\dfrac{1}{2}\right) = 5$ $f\left(\dfrac{1}{2}\right) = \dfrac{15}{4}$

21. $f(a) = \sqrt{3a - 1}$ **23.** $f(a) = \dfrac{1}{a + 1}$
   $f(a + 1) = \sqrt{3a + 2}$ $f(a + 1) = \dfrac{1}{a + 2}$
   $f\left(\dfrac{1}{2}\right) = \dfrac{\sqrt{2}}{2}$ $f\left(\dfrac{1}{2}\right) = \dfrac{2}{3}$

25. $g(-x) = \sqrt{6}$ **27.** $g(-x) = -2x - 3$
   $g(2x) = \sqrt{6}$ $g(2x) = 4x - 3$
   $g(a + h) = \sqrt{6}$ $g(a + h) = 2a + 2h - 3$

29. $g(-x) = 3x^2$ **31.** $g(-x) = -\dfrac{1}{x}$
   $g(2x) = 12x^2$
   $g(a + h) = 3a^2 + 6ah + 3h^2$ $g(2x) = \dfrac{1}{2x}$
                  $g(a + h) = \dfrac{1}{a + h}$

33. $g(-x) = -x^2 + 3x + 5$
   $g(2x) = -4x^2 - 6x + 5$
   $g(a + h) = -a^2 - 2ah - h^2 - 3a - 3h + 5$

35. $f(-2)$: undefined; $f(0) = 1/2; f(1) = 1/2$
37. $f(-2) = -2; f(0) = 1; f(1) = 1$
39. $f(-2) = -1; f(0) = 2; f(1) = 4$
41. $f(-2)$: undefined; $f(0) = 0; f(1) = 3$
43. a. $g(5) = -1$ **b.** $g(0) = 2$
   c. No. The table does not include an output value $g(t)$
      for the input $t = 3$.
45. a. 3   **b.** 21   **c.** 4   **47.** Yes   **49.** No   **51.** Yes   **53.** Yes
55. Domain: $(-\infty, \infty)$ **57.** Domain: $(-\infty, -1)\cup(-1, \infty)$
59. Domain: $(-\infty, 3)\cup(3, \infty)$
61. Domain: $(-\infty, -2)\cup(-2, 2)\cup(2, \infty)$
63. Domain: $(-\infty, 2]$ **65.** Domain: $(-\infty, \infty)$
67. Domain: $(-7, \infty)$ **69.** $36\pi \approx 113.097$ in³.
71. Earnings is $1600 when 30 items are sold.
73. a. Let $t =$ time (in hours)
      distance = rate × time
         $d(t) = 45t$
   b. $d(2) = 45(2) = 90$
      The car travels 90 miles in two hours.
   c. Domain: $[0, \infty)$
75. a. D(The Hunger Games) ≈ 408; $408 (million)
      grossed by The Hunger Games.
   b. Domain: {The Hobbit: An Unexpected Journey,
      The Dark Knight Rises, Marvel's The Avengers, The
      Amazing Spider-Man, The Twilight Saga: Breaking
      Dawn Part 2, The Hunger Games}
77. $A(\omega) = 3\omega^2$
79. a. $h(0) = 100$
      The height of the ball 0 seconds after it is dropped is
      100 feet.
   b. $h(2) = 36$ The height of the ball 2 seconds after it is
      dropped is 36 feet.
81. a. This table represents a function because for each
      year (input), there is only one value for per capita
      consumption (output).
   b. $S(2005) = 59.2$
      In 2005, the per capita consumption of high-fructose
      corn syrup was 59.2 pounds.
   c. Steady increase in its use from 1970-2000
   d. Use started to decrease.
83. a. It is a flat rate for the first 20 minutes
   b. $C(t) = \begin{cases} 1, & 0 < t \le 20 \\ 1 + 0.07(t - 20), & t > 20 \end{cases}$
   c. $1, $1, $1.70
85. $a = -3$ **87.** Domain: $(-\infty, \infty)$

2.4 Exercises
1. −2   **3.** 2   **5.** 3
7. No. For the value $x = 1$ there are two values of y.
9. No. For the value $x = 0$ there are two values of y.

11. No. For the value $x = -5$ there are two values of y.

13.

x	-4	-2	0	2	4
$f(x)$	-2	-3	-4	-5	-6

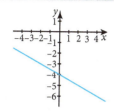

15.

x	0	2	$9/2$	8	18
$f(x)$	0	2	3	4	6

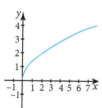

17. Domain: $(-\infty, \infty)$, range: $(-\infty, \infty)$

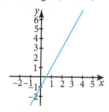

19. Domain: $(-\infty, \infty)$, range: $(-\infty, \infty)$

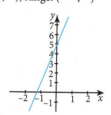

21. Domain: $(-\infty, \infty)$, range: $(-\infty, \infty)$

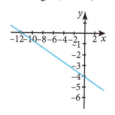

23. Domain: $(-\infty, \infty)$, range: $(-\infty, \infty)$

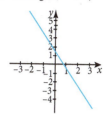

25. Domain: $(-\infty, \infty)$, range: $\{4\}$

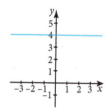

27. Domain: $(-\infty, \infty)$, range: $[0, \infty)$

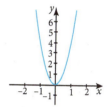

29. Domain: $(-\infty, \infty)$, range: $(-\infty, 4]$

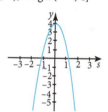

31. Domain: $[-4, \infty)$, range: $[0, \infty)$

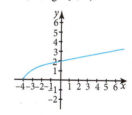

33. Domain: $[0, \infty)$, range: $[0, \infty)$

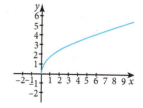

35. Domain: $[0, \infty)$, range: $[-2, \infty)$

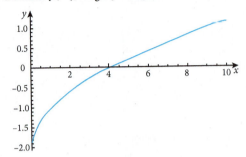

37. Domain: $(-\infty, \infty)$, range: $[0, \infty)$

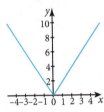

39. Domain: $(-\infty, \infty)$, range: $(-\infty, 0]$

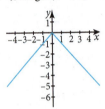

41. Domain: $(-\infty, \infty)$, range: $[0, \infty)$

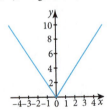

43. Domain: $(-\infty, 0]$, range: $[0, \infty)$

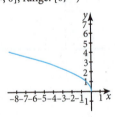

45. Domain: $(-\infty, \infty)$, range: $(-\infty, \infty)$

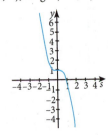

47.

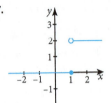

49.

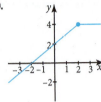

51.

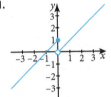

53.

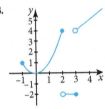

55. Not a function; fails vertical line test
57. Function; passes vertical line test
59. Domain: $(-\infty, \infty)$, range: $(-\infty, 1]$
61. Domain: $(-\infty, \infty)$, range: $[-1, 2]$
63. Domain: $(-\infty, \infty)$; range: $\{2\} \cup [3, \infty)$

65.

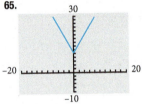

67.

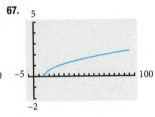

69.

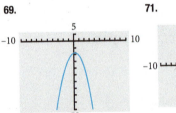

71.

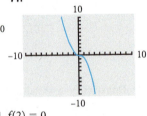

73. a. $f(-1) = 1.5, f(0) = 1, f(2) = 0$
 b. Domain: $(-\infty, \infty)$
 c. x-intercept: $(2, 0)$, y-intercept: $(0, 1)$
75. a. $f(-1) = 1, f(0) = 2, f(2) = 0$
 b. Domain: $(-\infty, \infty)$
 c. x-intercept: $(-2, 0)$, $(2, 0)$ y-intercept: $(0, 2)$
77. a. $f(-1) = -1.7, f(0) = -2, f(2) = -2.4$
 b. Domain: $[-4, \infty)$
 c. x-intercept: $(-4, 0)$, y-intercept: $(0, -2)$
79. Yes **81.** No **83.** Yes
85. a. \$18.7 billion **b.** Approx. \$1 billion
87. Radius measures distance and is a positive number.

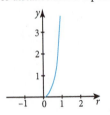

89. $d(t) = 55t$, t = time traveled. The values of t must be greater than or equal to zero.

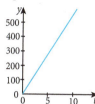

91. a.

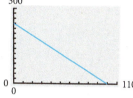

b. The t-intercept represents the time the when the coastal region has completely eroded.

93. $f(x)$ is independent of x, whereas $g(x)$ increases as x increases.

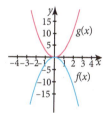

95. $f(x)$ opens downward, whereas $g(x)$ opens upward.

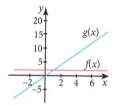

97. This is not a function because each element of the domain greater than zero has two distinct values in the range.

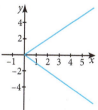

99. Range of $f(x)$: $[0, \infty)$, range of $g(x)$: $[-3, \infty)$

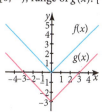

101. Range of $f(x)$: $[0, \infty)$, range of $g(x)$: $[0, \infty)$

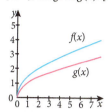

2.5 Exercises

1. Even **3.** Neither **5.** Odd **7.** $[-2, 2]$ **9.** $(0, 4)$

11. $(0, 1)$ **13.** Even **15.** $(-3, -2), (3, 4)$ **17.** $(0, 1)$ **19.** $\dfrac{1}{2}$

21. 3 **23.** $(0, 1)$ **25.** -4 **27.** Neither **29.** Even **31.** Even

33. Even **35.** Neither **37.** Neither **39.** 6 **41.** 4 **43.** -3

45. 3 **47** 2 **49.** $\sqrt{3} - 2$ **51.** 2 **53.** $2a + h$

55. Neither **57.** Odd **59.** Even

61. This is a decreasing function : if $x_1 < x_2$, then $d(x_1) > d(x_2)$.

63. -300 dollars per year; slope is constant, so average rate will be the same

65. 10,000; 3,000; is not the same

67. The average rate of change appears to get closer and closer to a single value, -4.

Interval	Average Rate of Change
$[1, 2]$	-3
$[1.9, 2]$	-3.9
$[1.95, 2]$	-3.95
$[1.99, 2]$	-3.99
$[1.999, 2]$	-3.999

69. No

2.6 Exercises

1. Rational expression; zero

3. $(-\infty, \infty)$

5. $(-\infty, -3] \cup [3, \infty)$

7. a. $(f + g)(x) = 2x - 2$; $(-\infty, \infty)$

   b. $(f - g)(x) = 4x - 8$; $(-\infty, \infty)$

   c. $(fg)(x) = -3x^2 + 14x - 15$; $(-\infty, \infty)$

   d. $\left(\dfrac{f}{g}\right)(x) = \dfrac{3x - 5}{-x + 3}$; $(-\infty, 3) \cup (3, \infty)$

9. a. $(f + g)(x) = x^2 + x - 2$; $(-\infty, \infty)$

   b. $(f - g)(x) = -x^2 + x - 4$; $(-\infty, \infty)$

   c. $(fg)(x) = x^3 - 3x^2 + x - 3$; $(-\infty, \infty)$

   d. $\left(\dfrac{f}{g}\right)(x) = \dfrac{x - 3}{x^2 + 1}$; $(-\infty, \infty)$

11. a. $(f + g)(x) = \dfrac{3x - 1}{x(2x - 1)}$; $(-\infty, 0) \cup \left(0, \dfrac{1}{2}\right) \cup \left(\dfrac{1}{2}, \infty\right)$

   b. $(f - g)(x) = \dfrac{x - 1}{x(2x - 1)}$; $(-\infty, 0) \cup \left(0, \dfrac{1}{2}\right) \cup \left(\dfrac{1}{2}, \infty\right)$

   c. $(fg)(x) = \dfrac{1}{x(2x - 1)}$; $(-\infty, 0) \cup \left(0, \dfrac{1}{2}\right) \cup \left(\dfrac{1}{2}, \infty\right)$

   d. $\left(\dfrac{f}{g}\right)(x) = \dfrac{2x - 1}{x}$; $(-\infty, 0) \cup \left(0, \dfrac{1}{2}\right) \cup \left(\dfrac{1}{2}, \infty\right)$

13. a. $(f + g)(x) = \sqrt{x} - x + 1; [0, \infty)$

b. $(f - g)(x) = \sqrt{x} + x - 1; [0, \infty)$

c. $(fg)(x) = -x\sqrt{x} + \sqrt{x}; [0, \infty)$

d. $\left(\dfrac{f}{g}\right)(x) = \dfrac{\sqrt{x}}{-x + 1}; [0, 1) \cup (1, \infty)$

15. a. $(f + g)(x) = \dfrac{2x|x| + 5|x| + 1}{2x + 5}; \left(-\infty, -\dfrac{5}{2}\right) \cup \left(-\dfrac{5}{2}, \infty\right)$

b. $(f - g)(x) = \dfrac{2x|x| + 5|x| - 1}{2x + 5}; \left(-\infty, -\dfrac{5}{2}\right) \cup \left(-\dfrac{5}{2}, \infty\right)$

c. $(fg)(x) = \dfrac{|x|}{2x + 5}; \left(-\infty, -\dfrac{5}{2}\right) \cup \left(-\dfrac{5}{2}, \infty\right)$

d. $\left(\dfrac{f}{g}\right)(x) = 2x|x| + 5|x|; \left(-\infty, -\dfrac{5}{2}\right) \cup \left(-\dfrac{5}{2}, \infty\right)$

17. 1   **19.** 3   **21.** $-\dfrac{8}{3}$ **23.** -7 **25.** -3 **27.** -7 **29.** 3

31. 0   **33.** -2 **35.** 3   **37.** 0   **39.** No. No value for $g(3)$

41. 210   **43.** -6 **45.** 6   **47.** $\sqrt{6}$ **49.** -18 **51.** $-\dfrac{9}{4}$

53. $(f \circ g)(x) = -x^2 - 2x$
domain: $(-\infty, \infty)$
$(g \circ f)(x) = -x^2 + 2$
domain: $(-\infty, \infty)$

55. $(f \circ g)(x) = x$
domain: $(-\infty, \infty)$
$(g \circ f)(x) = x$
domain: $(-\infty, \infty)$

57. $(f \circ g)$
$(x) = 3x^2 + 16x + 20$
domain: $(-\infty, \infty)$
$(g \circ f)(x) = 3x^2 + 4x + 2$
domain: $(-\infty, \infty)$

59. $(f \circ g)(x) = \dfrac{1}{2x + 5}$
domain: $\left(-\infty, -\dfrac{5}{2}\right) \cup \left(-\dfrac{5}{2}, \infty\right)$
$(g \circ f)(x) = \dfrac{5x + 2}{x}$
domain: $(-\infty, 0) \cup (0, \infty)$

61. $(f \circ g)(x) = \dfrac{3}{4x^2 + 1}$
domain: $(-\infty, \infty)$
$(g \circ f)(x) = \dfrac{18}{(2x + 1)^2}$
domain: $\left(-\infty, -\dfrac{1}{2}\right) \cup \left(-\dfrac{1}{2}, \infty\right)$

63. $(f \circ g)(x) = \sqrt{-3x - 3}$
domain: $(-\infty, -1]$
$(g \circ f)(x) = -3\sqrt{x + 1} - 4$
domain: $[-1, \infty)$

65. $(f \circ g)(x) = \left|\dfrac{2x}{x - 1}\right|$
domain: $(-\infty, 1) \cup (1, \infty)$
$(g \circ f)(x) = \dfrac{2|x|}{|x| - 1}$
domain: $(-\infty, -1) \cup (-1, 1) \cup (1, \infty)$

67. $(f \circ g)(x) = x^2$
domain: $(-\infty, \infty)$
$(g \circ f)(x) = x^2 - 2x + 2$
domain: $(-\infty, \infty)$

69. $(f \circ g)(x) = \dfrac{x^2 + 1}{x^2 - 1}$
domain: $(-\infty, -1) \cup (-1, 1) \cup (1, \infty)$
$(g \circ f)(x) = \left|\dfrac{x^2 + 1}{x^2 - 1}\right|$
domain: $(-\infty, -1) \cup (-1, 1) \cup (1, \infty)$

71. $(f \circ g)(x) = \dfrac{9x^2 - 6x + 1}{13x^2 - 2x + 2}$
domain: $\left(-\infty, \dfrac{1}{3}\right) \cup \left(\dfrac{1}{3}, \infty\right)$
$(g \circ f)(x) = \dfrac{x^2 + 3}{2 - x^2}$
domain: $(-\infty, -\sqrt{2}) \cup (-\sqrt{2}, \sqrt{2}) \cup (\sqrt{2}, \infty)$

73. $g(x) = 3x - 1, f(x) = x^2$

75. $g(x) = 4x^2 - 1, f(x) = \sqrt[3]{x}$

77. $g(x) = 2x + 5, f(x) = \dfrac{1}{x}$

79. $g(x) = x^2 + 1, f(x) = \sqrt{x} + 5$

81. $g(x) = 5x + 7, f(x) = \sqrt[3]{x} - 2$

83. $(f \circ f)(-1) = -1$

85. $(f \circ f)(t) = -t^4$
domain: $(-\infty, \infty)$

87. $(f \circ f)(2) = 22$

89. $(f \circ f)(t) = 9t + 4$
domain: $(-\infty, \infty)$

91. 3   **93.** $-2x - h + 1$ **95.** $\dfrac{-1}{(x + h - 3)(x - 3)}$

97. $P(t) = 81 - 10t$

99. $(np)(3) = \$2167.50$
This is the revenue generated by the sale of the book during the third month after its release.

101. $(f \circ R)(t) = 30 + 1.5t$
This function represents the GlobalEx revenue in euros.

103. $(C \circ A)(r) = 25.8064\pi r^2$
This function represents the surface area of a sphere in square centimeters based on its radius in inches.

105. No, not in general. Let $f(x) = x + 5$ and $g(x) = 2x$;
$(fg)(x) \ne (f \circ g)(x)$.

107. $(f + g)(x) = (a + c)(x) + (b + d)$; linear function
$(f - g)(x) = (a - c)(x) + (b - d)$; linear function

2.7 Exercises

1. The graph of $g(t)$ is the graph of $f(t) = t^2$ shifted up by 1 unit.

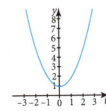

3. The graph of $f(x)$ is the graph of $g(x) = \sqrt{x}$ shifted down by 2 units.

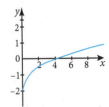

5. The graph of $h(x)$ is the graph of $f(x) = |x|$ shifted right by 2 units.

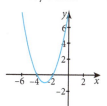

7. The graph of $F(s)$ is the graph of $f(s) = s^2$ shifted left by 5 units.

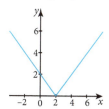

9. The graph of $f(x)$ is the graph of $g(x) = \sqrt{x}$ shifted right by 4 units.

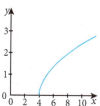

11. The graph of $H(x)$ is the graph of $g(x) = |x|$ shifted right by 2 units and up by 1 unit.

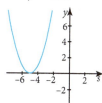

13. The graph of $S(x)$ is the graph of $g(x) = x^2$ shifted left by 3 units and down by 1 unit.

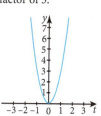

15. The graph of $H(t)$ is the graph of $g(t) = t^2$ stretched vertically away from the x-axis by a factor of 3.

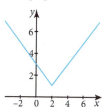

17. The graph of $S(x)$ is the graph of $f(x) = |x|$ reflected across the x-axis and vertically stretched away from the x-axis by a factor of 4.

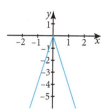

19. The graph of $H(s)$ is the graph of $f(s) = |s|$ reflected across the x-axis and shifted down by 3 units.

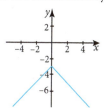

21. The graph of $h(x)$ is the graph of $f(x) = |x|$ reflected across the x-axis, compressed vertically toward the x-axis by a factor of $\frac{1}{2}$, shifted left by 1 unit, and shifted down by 3 units.

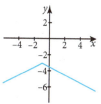

23. The graph of $g(x)$ is the graph of $f(x) = x^2$ reflected across the x-axis, vertically stretched away from the x-axis by a factor of 3, shifted left by 2 units, and shifted down by 4 units.

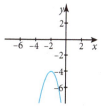

25. The graph of $f(x)$ is the graph of $g(x) = |x|$ compressed horizontally toward the y-axis, scaled by a factor of $\frac{1}{2}$.

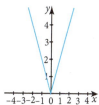

27. The graph of $f(x)$ is the graph of $g(x) = x^2$ compressed horizontally toward the y-axis, scaled by a factor of $\frac{1}{2}$.

29. The graph of $g(x)$ is the graph of $f(x) = \sqrt{x}$ compressed horizontally towards the y-axis and scaled by a factor of $\frac{1}{3}$.

31. The graph is shifted left by 1 unit; $g(x) = |x + 1|$.
33. The graph is reflected across the x-axis and shifted up by 1 unit; $g(x) = -|x| + 1$.
35. The graph is shifted right by 2 units; $g(x) = (x - 2)^2$.
37. The graph is shifted right by 1 unit and down by 2 units; $g(x) = (x - 1)^2 - 2$.
39. $g(t) = |t + 4| - 3$ **41.** $g(t) = -3(t - 1)^2$
43. $h(x) = (2x)^2 - 4$

45.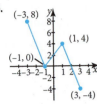
(−3, 8)
(1, 4)
(−1, 0)
(3, −4)

47.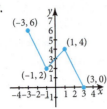
(−3, 6)
(1, 4)
(−1, 2)
(3, 0)

49.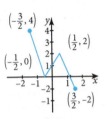
$\left(-\frac{3}{2}, 4\right)$
$\left(\frac{1}{2}, 2\right)$
$\left(-\frac{1}{2}, 0\right)$
$\left(\frac{3}{2}, -2\right)$

51.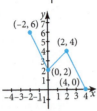
(−2, 6)
(2, 4)
(0, 2)
(4, 0)

53.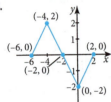
(−4, 2)
(−6, 0)
(2, 0)
(−2, 0)
(0, −2)

55.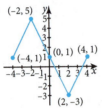
(−2, 5)
(0, 1) (4, 1)
(−4, 1)
(2, −3)

57.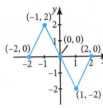
(−1, 2)
(0, 0)
(−2, 0) (2, 0)
(1, −2)

59.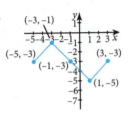
(−3, −1)
(−5, −3) (3, −3)
(−1, −3)
(1, −5)

61. $k = -2$ **63.** $k = 1$

65.

x	$f(x)$	$g(x) = f(x) - 3$
−2	36	33
−1	25	22
0	16	13
1	9	6
2	4	1

67. The graph of $f(x)$ is a translation of $h(x)$ to the left by 3.5 units, whereas $g(x)$ is a translation of $h(x)$ up by 3.5 units.

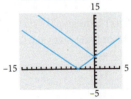

69. The graph of $f(x - 4.5)$ is a translation of $f(x)$ to the right by 4.5 units.

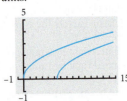

71. The graphs are different. The graph of $-2f(x)$ is a vertical scaling of $f(x)$ by a factor of 2 away from the x-axis, and then reflected across the x-axis. The graph of $f(-2x)$ is a horizontal scaling of $f(x)$ by a factor of $\frac{1}{2}$ toward the y-axis, and then reflected across the y-axis.

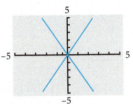

73. The graph of g is the graph of f shifted to the right by 7 units.

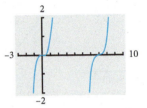

75. Let T be the sales tax.
$T(x) = 6\%$ of $P(x)$
$= 0.06\,P(x)$

77. a. $C(x) = 450 + 3x$
  b. The graph of the decreased cost function has a lower y-intercept but the same slope as the graph of the original cost function.

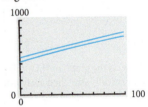

79. a. $h = 0 \Rightarrow s(t) = -16t^2 + 30t$
    $s(1) = -16(1)^2 + 30(1) = 14$
    The ball is 14 ft high after 1 sec.
  b. $h = 20 \Rightarrow s(t) = -16t^2 + 30t + 20$ and
    $s(1) = -16(1)^2 + 30(1) + 20 = 34$
    The ball is 34 ft high after 1 sec.
  c. The value of $s(t)$ shifted up by h feet when $h > 0$ and shifted down by $|h|$ feet when $h < 0$.

81. From the x-coordinates, the shift is 2−(−3) = 5 left by 5 units. Thus, the function g is $g(x) = (x + 5)^2$.

83. a. $g(x) = 2|x| + 3$
  b. $g(x) = 2(|x| + 3) = 2|x| + 6$
  c. In part (b), the y-intercept is 3 units higher than in part (a) because scaling by 2 after adding 3 units effectively scales the original equation by 2 and then moves it up 6 units. In part (a), the function is moved up by 3 units after it is scaled by 2.

2.8 Exercises

1. $f(x) = 4x - 7$ **3.** $f(x) = 2.4x - 9.94$

5. $f(x) = 15x + 150$ **7.** $k = 4, y = 4x$

9. $k = 10, y = \dfrac{10}{x}$ **11.** $k = 3.5, y = 3.5x$

13. $k = 15, y = \dfrac{15}{x}$ **15.** $k = \dfrac{1}{6}, y = \dfrac{1}{6}x$

17. $k = 6, y = \dfrac{6}{x}$ **19.** $k = 5, y = 5x$ **21.** $k = 98, y = \dfrac{98}{x}$

23. $C(n) = 4.50 + 0.07n$ **25.** $C = \dfrac{5}{9}(F - 32)$ or $C = \dfrac{5}{9}F - \dfrac{160}{9}$

27. $k = \dfrac{5}{4}$; volume = 75 cc.   **29.** 12.5 ft.   **31.** $3\dfrac{1}{3}$ days

33. a. $C(m) = 19.95 + 0.99m$
 b. $m = 0.99$. This represents the increase in the total cost for each additional mile driven. y-intercept: $(0, 19.95)$. This represents the cost for renting the truck for one day when 0 miles are driven.
 c. \$75.39

35. a.

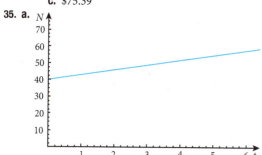

 b. $m = \dfrac{18}{6} = 3$
 c. The slope represents the number of additional visitors each month, in millions.
 d. $v(t) = 3t + 40$
 e. 70 million visitors
37. a. $g(t) = 1.1t + 18.2$
 b. $m = 1.1$; consumption increases by 1.1 gals. per year
 c. $(0, 18.2)$; gallons consumed in 2001
39. a. $P(t) = 3t + 20$
 b. 38%.
 c. 2018.
 d. No, because the model predicts a percentage of consumers over 100% in 2038.
 e. This model cannot accurately predict the buying habits over a long period of time.
41. a. 6250 vehicles per year.
  Mathematically, this is the slope of a linear model of the traffic volume.
 b. $T(t) = 6250t + 175,000$
  212,500 vehicles.
43. a. The percentage is decreasing over time.
 b. $R(t) = -0.114t + 20.34$

45. No, because direct variation refers to an equation of the form $y = kx$, not $y = kx + b, b \neq 0$.

Chapter 2 Review Exercises

1. $\dfrac{5}{2}$ **3.** $-\dfrac{12}{7}$

5. $y = -2x + 7$

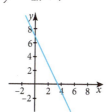

7. $y = \dfrac{3}{2}x + 3$

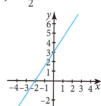

9. $y = -\dfrac{1}{5}x - \dfrac{23}{5}$

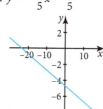

11. $y = -x + 1$

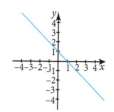

13. $y = -x + 2$

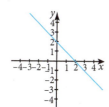

15. $x = -2$

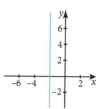

17. \$7000

19. $d = \sqrt{34}$
 midpoint: $\left(-\dfrac{3}{2}, \dfrac{1}{2}\right)$

21. $d = \sqrt{41}$
 midpoint: $\left(\dfrac{17}{2}, 11\right)$

23. $(x + 1)^2 + (y - 2)^2 = 36$

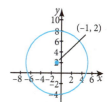

25. $x^2 + (y + 1)^2 = \dfrac{1}{4}$

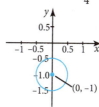

27. Center: $(3, 2)$;
 radius: $\sqrt{\dfrac{1}{4}} = \dfrac{1}{2}$

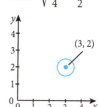

29. Center: $(1, -1)$;
radius: $\sqrt{9} = 3$

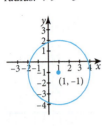

31.

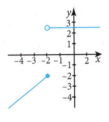

33. $x^2 + y^2 = 484$

35. a. 31 **b.** 7 **c.** $2a^2 - 1$ **d.** $2a^2 + 4a + 1$

37. a. $2\sqrt{3}$ **b.** 0 **c.** $\sqrt{a^2 - 4}$ **d.** $\sqrt{a^2 + 2a - 3}$

39. $f(0) = 0$ **41.** $f(3) = 3$ **43.** Domain: $(-\infty, \infty)$

45. Domain: $(-\infty, 2)\cup(2, \infty)$ **47.** Domain: $(-\infty, \infty)$

49. Domain: $(-\infty, \infty)$

51. $s(3) = 113.097$ units^2. $r > 0$, since the radius of a sphere must have a positive value.

53. Domain: $(-\infty, \infty)$;   **55.** Domain: $(-\infty, \infty)$;
Range: $(-\infty, \infty)$ Range: $(-\infty, 3]$

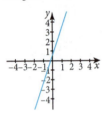

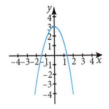

57. Domain: $(-\infty, \infty)$;   **59.**
Range: $(-\infty, 0]$

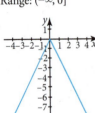

61. a. No; does not pass the vertical line test.
b. Yes; passes the vertical line test.
c. No; does not pass the vertical line test.
d. Yes; passes the vertical line test.

63. Yes **65.** Odd **67.** Even **69.** Odd **71.** Neither

73. Increasing on $(-2, 2)$ and $(3, 4)$

75. Constant on $(4, 5)$ **77.** 1 **79.** -2 **81.** 2

83. a. $(f + g)(x) = 4x^2 + x + 2$ **b.** $(f - g)(x) = 4x^2 - x$
Domain: $(-\infty, \infty)$ Domain: $(-\infty, \infty)$

c. $(fg)(x) = 4x^3 + 4x^2 + x + 1$
Domain: $(-\infty, \infty)$

d. $\left(\dfrac{f}{g}\right)(x) = \dfrac{4x^2 + 1}{x + 1}$
Domain: $(-\infty, -1)\cup(-1, \infty)$

85. a. $(f + g)(x) = \dfrac{x^2 + 2x + 1}{2x(x^2 + 1)}$
Domain: $(-\infty, 0)\cup(0, \infty)$

b. $(f - g)(x) = \dfrac{x^2 - 2x + 1}{2x(x^2 + 1)}$
Domain: $(-\infty, 0)\cup(0, \infty)$

c. $(fg)(x) = \dfrac{1}{2x(x^2 + 1)}$
Domain: $(-\infty, 0)\cup(0, \infty)$

d. $\left(\dfrac{f}{g}\right)(x) = \dfrac{x^2 + 1}{2x}$
Domain: $(-\infty, 0)\cup(0, \infty)$

87. a. $(f + g)(x) = \dfrac{3x^3 - 12x^2 + 2}{x - 4}$
Domain: $(-\infty, 4)\cup(4, \infty)$

b. $(f - g)(x) = \dfrac{-3x^3 + 12x^2 + 2}{x - 4}$
Domain: $(-\infty, 4)\cup(4, \infty)$

c. $(fg)(x) = \dfrac{6x^2}{x - 4}$
Domain: $(-\infty, 4)\cup(4, \infty)$

d. $\left(\dfrac{f}{g}\right)(x) = \dfrac{2}{3x^2(x - 4)}$
Domain: $(-\infty, 0)\cup(0, 4)\cup(4, \infty)$

89. 0 **91.** 5 **93.** 7 **95.** $\dfrac{5}{8}$ **97.** -13 **99.** -19

101. $(f \circ g)(x) = -x^2 + 4x$
domain: $(-\infty, \infty)$
$(g \circ f)(x) = -x^2 + 2$
domain: $(-\infty, \infty)$

103. $(f \circ g)(x) = -x^2 + 9x - 18$
domain: $(-\infty, \infty)$
$(g \circ f)(x) = -x^2 + 3x - 3$
domain: $(-\infty, \infty)$

105. $(f \circ g)(x) = \dfrac{1}{(x + 2)(x - 1)}$
domain: $(-\infty, -2)\cup(-2, 1)\cup(1, \infty)$
$(g \circ f)(x) = \dfrac{x - 1}{(x - 2)^2}$
domain: $(-\infty, 2)\cup(2, \infty)$

107. $(f \circ g)(x) = \dfrac{|x|}{|x| + 3}$
domain: $(-\infty, \infty)$
$(g \circ f)(x) = \left|\dfrac{x}{x + 3}\right|$
domain: $(-\infty, -3)\cup(-3, \infty)$

109. 4

111.

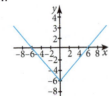

113.

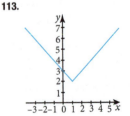

115.

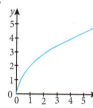

117.

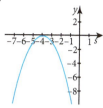

119.

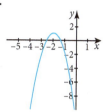

121.

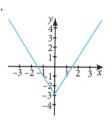

123. $g(x) = |x - 3| + 1$ **125.** $g(x) = 2(x + 1)^2$

127. $2f(x) - 3$ **129.** $f(2x)$

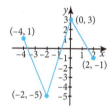

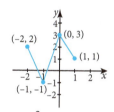

131. $y_1(x)$ is $f(x)$ shifted right 1.5 units. y_2 is $f(x)$ shifted down 1.5 units.

133. $f(x) = \dfrac{2}{3}x + 4$

135. $y = 2.5x$

137. $y = \dfrac{54}{x}$

139. $R(x) = 6x$; $R(800) = \$4,800$

141. a. $v(t) = -50t + 400$
  b. $\$400$
  c. In 8 years, 2016

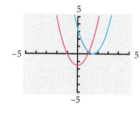

Chapter 2 Test

1. a. $\dfrac{7}{2}$ **b.** Undefined

2. $y = -4x - 1$ **3.** $x + y - 3 = 0$ **4.** $y = -2x + 6$

5. $y = -4x - 12$ **6.** $y = -5$ **7.** $x = 7$

8. $\sqrt{26}, \left(-\dfrac{3}{2}, \dfrac{5}{2}\right)$ **9.** $(x - 2)^2 + (y - 5)^2 = 36$

10. Center: $(-1, 2)$;
    radius: $\sqrt{9} = 3$

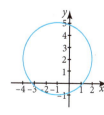

11. No, y is not a function of x, fails the vertical line test.

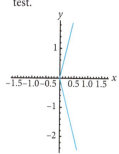

12. a. -8 **b.** $a^2 + 4a - 3$ **c.** 3 **d.** 4

13. Domain: $(-\infty, \infty)$ **14.** Domain: $(-\infty, 5) \cup (5, \infty)$

15. Domain: $(-\infty, \infty)$ **16.** Domain: $[-3, \infty)$

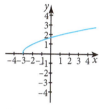

17. Domain: $(-\infty, \infty)$; range: $(-\infty, 4]$

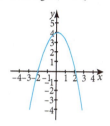

18. No   **19.** Yes

20. a. $f(x) = \begin{cases} -x + 1, \text{ if } x \le 1 \\ x - 1, \text{ if } x > 1 \end{cases}$

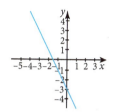

  b. $f(x) = |x - 1|$

21. Neither   **22.** Even   **23.** Odd

24. a. $(0, 2)$ **b.** $(-5, 0)$ **c.** $(2, 5)$

25. $-\dfrac{1}{6}$ **26.** $8x + 4h$ **27.** $f(x) = x^2 + 4x - 1$; $(-\infty, \infty)$

28. $f(x) = -x^2 - 1$; $(-\infty, \infty)$ **29.** 3 **30.** 0 **31.** -1 **32.** -3

33. $(fg)(x) = 2x^3 + 3x^2 - 2x$; $(-\infty, \infty)$

34. $\left(\dfrac{f}{g}\right)(x) = \dfrac{x^2 + 2x}{2x - 1}$;

    Domain: $\left(-\infty, \dfrac{1}{2}\right) \cup \left(\dfrac{1}{2}, \infty\right)$

35. $(f \circ g)(x) = \dfrac{1}{2x^2 - 2} = \dfrac{1}{2(x - 1)(x + 1)}$

    Domain: $(-\infty, -1) \cup (-1, 1) \cup (1, \infty)$

36.

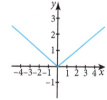

37.

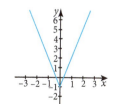

38.

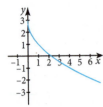

39.

40. $g(x) = |x + 2| + 1$ **41.** $g(x) = (2x)^2 - 1$

42. $k = \dfrac{9}{2}, y = \dfrac{9}{2}x$ **43.** $k = 70, y = \dfrac{70}{x}$

44. a. $v(t) = 300{,}000 + 15{,}000t$ **b.** In 8 years, 2014

Chapter 3

3.1 Exercises

1. up   **3.** $(x - 8)(x - 5)$ **5.** $\dfrac{9}{4}$

7. Domain for both: $(-\infty, \infty)$
range for both: $(-\infty, 0]$

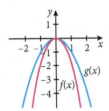

9. Domain for both: $(-\infty, \infty)$
range for $f(x)$: $[1, \infty)$; range for $g(x)$: $[-1, \infty)$

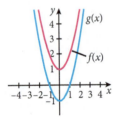

11. a. Domain $(-\infty, \infty)$; Range $(-\infty, 1]$
b. -2 **c.** 1   **d.** $-3, -1$
13. b   **15.** a   **17.** e   **19.** f

21. Vertex: $(-2, -1)$

23. Vertex: $(-1, -1)$

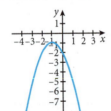

25. Vertex: $(-4, -2)$

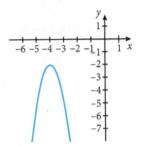

27. $g(x) = (x + 1)^2 + 4$
vertex: $(-1, 4)$;

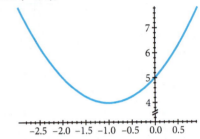

29. $w(x) = -(x - 3)^2 + 13$
vertex: $(3, 13)$;

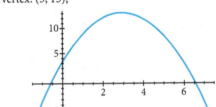

31. $h(x) = \left(x + \dfrac{1}{2}\right)^2 - \dfrac{13}{4}$
vertex: $\left(-\dfrac{1}{2}, -\dfrac{13}{4}\right)$;

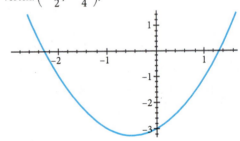

33. $f(x) = 3(x + 1)^2 - 7$
vertex: $(-1, -7)$;

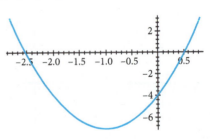

35. 8; The graph of f has two x-intercepts, and the equation $f(x) = 0$ has two real solutions.

37. -7; The graph of f has no x-intercepts, and the equation $f(x) = 0$ has no real solutions.

39. 0; The graph of f has one x-intercept, and the equation $f(x) = 0$ has one real solution.

41. vertex: $(1, 1)$; axis of symmetry: $x = 1$; y-intercept: $(0, -1)$; another point: $(2, -1)$; x-intercepts: $\left(1 - \frac{\sqrt{2}}{2}, 0\right)$, $\left(1 + \frac{\sqrt{2}}{2}, 0\right)$

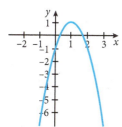

43. vertex: $(2, 1)$; axis of symmetry: $x = 2$; y-intercept: $(0, -3)$; another point: $(4, -3)$; x-intercepts: $(1, 0)$, $(3, 0)$

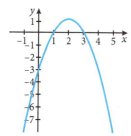

45. vertex: $\left(\frac{3}{2}, \frac{11}{4}\right)$; axis of symmetry: $x = \frac{3}{2}$; y-intercept: $(0, 5)$; another point: $(3, 5)$; No x-intercepts

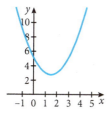

47. vertex: $(0, 100)$; axis of symmetry: $t = 0$; y-intercept: $(0, 100)$ is the same as the vertex so choose two other points: $(2, 36)$, $(-2, 36)$. t-intercepts: $\left(-\frac{5}{2}, 0\right)$, $\left(\frac{5}{2}, 0\right)$

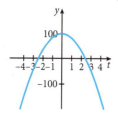

49. vertex: $(3, -4)$; axis of symmetry: $x = 3$; y-intercept: $(0, 5)$; another point: $(2, -3)$; x-intercepts: $(1, 0)$, $(5, 0)$

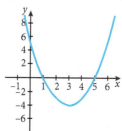

51. vertex: $\left(\frac{3}{4}, \frac{17}{8}\right)$; axis of symmetry: $s = \frac{3}{4}$; y-intercept: $(0, 1)$; another point: $(1, -2)$; s-intercepts: $\left(\frac{1}{4}(3 + \sqrt{17}), 0\right)$, $\left(\frac{1}{4}(3 - \sqrt{17}), 0\right)$

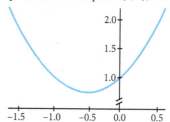

53. vertex: $\left(-\frac{1}{2}, \frac{3}{4}\right)$; axis of symmetry: $t = -\frac{1}{2}$; y-intercept: $(0, 1)$; another point: $(1, 3)$; no t-intercepts

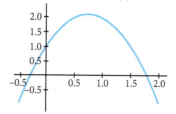

55. vertex: $\left(\frac{3}{2}, -\frac{23}{12}\right)$; axis of symmetry: $t = \frac{3}{2}$; y-intercept: $\left(0, \frac{1}{3}\right)$; another point: $\left(3, \frac{1}{3}\right)$; t-intercepts: $\left(\frac{1}{6}(9 + \sqrt{69}), 0\right)$, $\left(\frac{1}{6}(9 - \sqrt{69}), 0\right)$

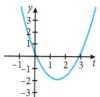

57. vertex: (5, 17); axis
of symmetry: $x = 5$;
increasing on: $(-\infty, 5)$;
decreasing on: $(5, \infty)$;
range: $(-\infty, 17]$

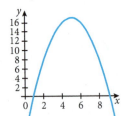

59. vertex: $(-1, -5)$; axis
of symmetry: $x = -1$;
increasing on: $(-1, \infty)$;
decreasing on: $(-\infty, -1)$;
range: $[-5, \infty)$

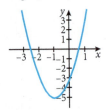

61. vertex: $(1, -2)$; axis
of symmetry: $x = 1$;
increasing on: $(-\infty, 1)$;
decreasing on: $(1, \infty)$;
range: $(-\infty, -2]$

63. vertex: $\left(-1, -\frac{9}{4}\right)$; axis
of symmetry: $x = -1$;
increasing on: $(-1, \infty)$;
decreasing on: $(-\infty, -1)$;
range: $\left[-\frac{9}{4}, \infty\right)$

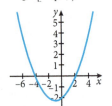

65. $f(x) = -(x - 3)(x + 1)$ **67.** $f(x) = (x + 1)(x - 4)$
69. Vertex: (0, 20)

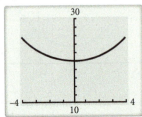

71. Vertex: $(-0.3536, 0.8232)$

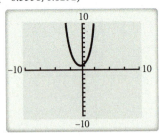

73. Vertex: (1.25, 145), or $\left(\frac{5}{4}, 145\right)$

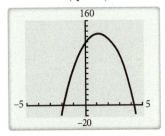

75. a. Axis of symmetry: $x = 2$
b. Another point: $(1, -1)$
c.

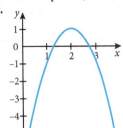

77. $b = 4$
79. If $k < -\frac{1}{4}$, the function has no real zeros (and the graph
has no x-intercepts). If $k = -\frac{1}{4}$, the function has one
real zero (and the graph has one x-intercept). If $k > -\frac{1}{4}$,
the function has two real zeros.

81. a.

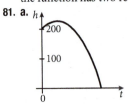

b. maximum height of 225 ft at 1.25 seconds.
83. At 49°F, 83.8%, and at 47°F, 2.2%.
85. a. 8.9534 million   **b.** 21.335 million
c. vertex: (7.99, 7.1876). About 8 years after 1981
(1989), the attendance at Broadway shows was at a
minimum of 7.18
d. No, because 2025 is too many years past the data
used to create the model.
87. $\frac{200}{3}$ ft × 50 ft.

89. a. There is a constant change of 0.125 in. diameter but
there is not a corresponding change in load.
b. 9,620 lb

91. a. $d(t) = a(t - 3)^2 + 144$
b. $a = -16$
c. $d(t) = -16(t - 3)^2 + 144$
d. Yes, (3, 144) is the vertex, $d(0) = 0$

93. a.

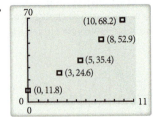

b. $f(x) = 0.1954x^2 + 3.6454x + 11.8844$
c.

Year	Actual	Predicted
1995	11.8	11.8
1998	24.6	24.58
2000	35.4	35.0
2003	52.9	53.56
2005	68.2	67.88

d. About 60.52 million tax returns

e. The number compares quite well, being an error of about 1% in prediction.

95. A linear function has a constant rate of change (slope) but a quadratic function does not. A linear function increases (or decreases) over its entire domain, but a quadratic function decreases and then increases (or increases and then decreases) over its domain.

97. a. $(-6, 0)$.   **b.** $f(x) = -\dfrac{2}{25}(x+1)^2 + 2$

c.

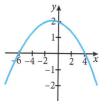

99. a. Vertex $(2, 8) \Rightarrow h = 2, k = 8$

b. $f(x) = a(x-2)^2 + 8$ **c.** $a = -2$

d. $f(x) = -2(x-2)^2 + 8$

e.

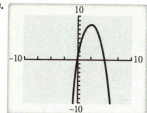

3.2 Exercises

1. left, two   **3.** $(2, 0), (0, 4)$

5. Symmetry with respect to y-axis

7. a. $-2x, x+3, x-3$ **b.** $0, -3, 3$

 c. $(0, 0), (-3, 0), (3, 0)$

9. Yes; no breaks, no corners, end behavior like a polynomial.

11. No; has a corner.   **13.** Yes; 3

15. No. This function cannot be written as a sum of terms in which t is raised to a nonnegative-integer power.

17. Yes; 0   **19.** Yes; 3

21. $f(t) \to -\infty$ as $t \to -\infty$ and $f(t) \to \infty$ as $t \to \infty$.

23. $f(x) \to \infty$ as $x \to -\infty$ and $f(x) \to -\infty$ as $x \to \infty$.

25. $H(x) \to -\infty$ as $x \to -\infty$ and $H(x) \to -\infty$ as $x \to \infty$.

27. $g(x) \to \infty$ as $x \to -\infty$ and $g(x) \to -\infty$ as $x \to \infty$.

29. $f(s) \to -\infty$ as $s \to -\infty$ and $f(s) \to \infty$ as $s \to \infty$.

31.

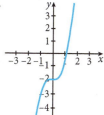

33.

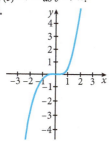

35.

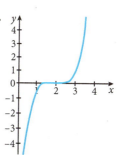

37.

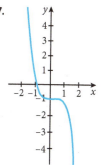

39.

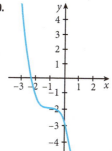

41.

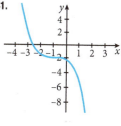

43. $y = -5x^3$

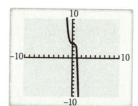

45. $y = 1.5x^5$

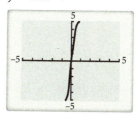

47. a. $y = -2x^3$

b. y-intercept: $(0, 0)$; x-intercepts: $(0, 0), (-2, 0), (2, 0)$

c. $(-\infty, -2) \cup (0, 2)$.   **d.** $(-2, 0) \cup (2, \infty)$.

e.

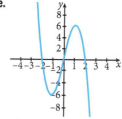

49. a. $y = x^3$
  b. y-intercept: $(0, 12)$; x-intercepts: $(3, 0), (-4, 0), (1, 0)$
  c. $(-4, 1) \cup (3, \infty)$ **d.** $(-\infty, -4) \cup (1, 3)$
  e.

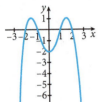

51. a. $y = -\dfrac{1}{2}x^4$
  b. y-intercept: $(0, -2)$; x-intercepts: $(-2, 0), (2, 0)$,
    $(-1, 0), (1, 0)$ **c.** $(-2, -1) \cup (1, 2)$
  d. $(-\infty, -2) \cup (-1, 1) \cup (2, \infty)$
  e.

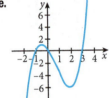

53. a. $y = x^3$
  b. y-intercept: $(0, 0)$; x-intercepts: $(0, 0), (-1, 0), (3, 0)$
  c. $(-1, 0) \cup (3, \infty)$.
  d. $(-\infty, -1) \cup (0, 3)$
  e.

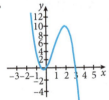

55. a. $y = -2x^3$
  b. y-intercept: $(0, 0)$; x-intercepts: $(0, 0), \left(-\dfrac{1}{2}, 0\right), (3, 0)$
  c. $\left(-\infty, -\dfrac{1}{2}\right) \cup (0, 3)$
  d. $\left(-\dfrac{1}{2}, 0\right) \cup (3, \infty)$
  e.

57. a. $y = -x^4$
  b. y-intercept: $(0, -6)$; x-intercepts: $(-1, 0), (1, 0)$,
    $(2, 0), (-3, 0)$
  c. $(-3, -1) \cup (1, 2)$
  d. $(-\infty, -3) \cup (-1, 1) \cup (2, \infty)$
  e.

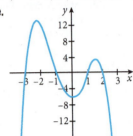

59. a. $y = 2x^3$
  b. y-intercept: $(0, 0)$; x-intercepts: $(0, 0), (-3, 0)$
  c. $(-3, 0) \cup (0, \infty)$
  d. $(-\infty, -3)$
  e.

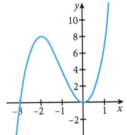

61. a. $y = 2x^4$
  b. y-intercept: $(0, -3)$; x-intercepts: $\left(-\dfrac{1}{2}, 0\right), (3, 0)$
  c. $\left(-\infty, -\dfrac{1}{2}\right) \cup (3, \infty)$ **d.** $\left(-\dfrac{1}{2}, 3\right)$
  e.

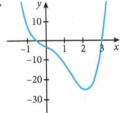

63. a. $(-1, 0)$ **b.** $(0, 3)$ **c.** Odd   **d.** Positive
65. a. $(-1, 0), (0, 0), (1, 0)$ **b.** $(0, 0)$ **c.** Odd   **d.** Negative
67. $x = 2$ has multiplicity 2, graph touches the x-axis;
    $x = -5$ has multiplicity 5, graph passes through the x-axis
69. $t = 0$ has multiplicity 2, graph touches the t-axis;
    $t = 1$ has multiplicity 1, graph passes through the t-axis;
    $t = -2$ has multiplicity 1, graph passes through the
    t-axis
71. $x = -1$ has multiplicity 2, graph touches the x-axis
73. $s = 0$ has multiplicity 1, graph passes through the s-axis;
    $s = -1$ has multiplicity 2, graph touches the s-axis
75. y-axis symmetry; even   **77.** No symmetry; neither
79. Origin symmetry; odd   **81.** y-axis symmetry; even
83. $x = -1$, odd multiplicity; $x = \dfrac{3}{2}$, odd multiplicity;
    $x = 0$, odd multiplicity
85. $x = 0$, even multiplicity; $x = 4$, even multiplicity

87. a. $f(x) \to \infty$ $x \to \infty$; $f(x) \to -\infty$ as $x \to -\infty$ **b.** $(0,0)$
  c. $x = 0$, multiplicity 2; $x = 1$, multiplicity 1; $(0,0)$, $(1,0)$
  d. No symmetry
  e. Positive: $(1, \infty)$; negative: $(-\infty, 0) \cup (0, 1)$

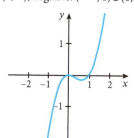

89. a. $f(x) \to \infty$ as $x \to \infty$; $f(x) \to -\infty$ as $x \to -\infty$ **b.** $(0, 8)$
  c. $x = 2$, multiplicity 2; $x = -2$, multiplicity 1; $(2, 0)$, $(-2, 0)$
  d. No symmetry
  e. Positive: $(-2, 2) \cup (2, \infty)$; negative: $(-\infty, 2)$

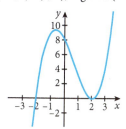

91. a. $g(x) \to \infty$ as $x \to \infty$; $g(x) \to \infty$ as $x \to -\infty$
    b. $(0, -6)$
  c. $x = -1$, multiplicity 2; $x = 2$, multiplicity 1;
    $x = -3$, multiplicity 1; $(-1, 0)$, $(2, 0)$, $(-3, 0)$
    d. No symmetry
    e. Positive: $(-\infty, -3) \cup (2, \infty)$;
      negative: $(-3, -1) \cup (-1, 2)$

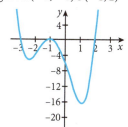

93. a. $g(x) \to -\infty$ as $x \to \infty$; $g(x) \to -\infty$ as $x \to -\infty$
    b. $(0, -18)$ **c.** $x = -1$, multiplicity 2; $x = 3$, multiplicity 2;
    $(-1, 0)$, $(3, 0)$
  d. No symmetry
  e. Negative: $(-\infty, -1) \cup (-1, 3) \cup (3, \infty)$

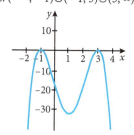

95. a. $f(x) \to \infty$ as $x \to \infty$; $f(x) \to -\infty$ as $x \to -\infty$ **b.** $(0,0)$
  c. $x = 0$, multiplicity 1; $x = -2$, multiplicity 2; $(0, 0)$,
  $(-2, 0)$
  d. No symmetry
  e. Positive: $(0, \infty)$; negative: $(-\infty, -2) \cup (-2, 0)$

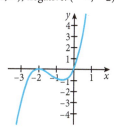

97. a. $h(x) \to -\infty$ as $x \to -\infty$; $h(x) \to -\infty$ as $x \to \infty$
  b. $(0,0)$
  c. $x = 0$, multiplicity 2; $x = 1 + \sqrt{2}$, multiplicity 1;
    $x = 1 - \sqrt{2}$, multiplicity 1; $(0, 0)$, $(1 + \sqrt{2}, 0)$,
    $(1 - \sqrt{2}, 0)$
  d. No symmetry
  e. Positive: $(1 - \sqrt{2}, 0) \cup (0, 1 + \sqrt{2})$;
    negative: $(-\infty, 1 - \sqrt{2}) \cup (1 + \sqrt{2}, \infty)$

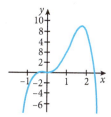

99. $f(x) = (x + 2)(x - 5)(x - 6)$
101. $f(x) = (x - 2)^2(x - 4)^2$ **103.** $f(x) = (x - 2)(x + 3)^2$
105. $f(x) = (x + 2)(x + 1)(x - 5)^3$

107. a. $(-1.5321, 0)$, $(-0.3473, 0)$, $(1.8794, 0)$
  b. Positive: $(-\infty, -1.5321) \cup (-0.3473, 1.8794)$;
    negative: $(-1.5321, -0.3473) \cup (1.8794, \infty)$
  c. Local maximum: $(1, 3)$; local minimum: $(-1, -1)$
  d. No symmetry

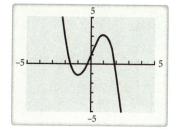

109. a. $(0.7167, 0), (-2.1069, 0)$
  b. Positive: $(-\infty, -2.1069) \cup (0.7167, \infty)$;
  negative: $(-2.1069, 0.7167)$
  c. Local minimum: $(-1.5, -2.6875)$
  d. No Symmetry

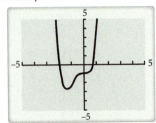

111. a. The graph appears parabolic in the standard window.

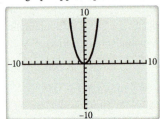

  b.

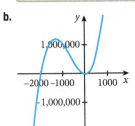

  c.

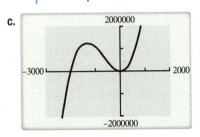

113. a. $h = 10 - 2r$
  b. $V = \pi r^2(10 - 2r)$
  c. $0 < r < 5$, so that the height is a positive value
115. a. $V = h(h + 3)^2$
  b.

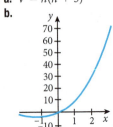

  c. $h > 0$

117. a. $5778, \$9378, \$4402$ **b.** 2002 is too far outside the range of the data used to model the equation.   **c.** 1943

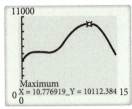

119. The graph below represents the function
  $f(x) = (x + 2)(x - 1)^2$

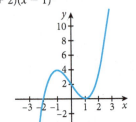

121. $f(x) = (x - 2)(x - 1)$; No, because the graph crosses the x-axis at $(2, 0)$ and $(1, 0)$ each factor could have any odd positive integer power.

123. a.

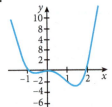

  b. Local minimum: $(-0.693, -0.397)$ and $(1.443, -2.833)$
  c. $f(s) = s^2(s + 1)(s - 2)$
  d. Graphing calculator

125. a.

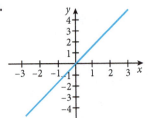

  b. No local extrema
  c. $q(x) = x$
  d. Graphing calculator

3.3 Exercises
  1. $2x + 3$ **3.** $x^2 - 4$ **5.** $x^2 - 5x + 12 - \dfrac{28}{x + 2}$
  7. $-x^3 - \dfrac{1}{3}x^2 + \dfrac{2}{9}x + \dfrac{2}{27} - \dfrac{52}{27(3x - 1)}$
  9. $x^5 - x^4 + x^3 - x^2 + x - 1 + \dfrac{2}{x + 1}$
  11. $x + 2 + \dfrac{2x - 1}{x^2 - 2}$ **13.** $x^2 - x + 1 + \dfrac{3x^2 - 3x + 2}{x^3 + x - 1}$
  15. $x^2 + x + 1 = (x + 1)\,x + 1$
  17. $3x^3 + 2x - 8 = (x - 4)(3x^2 + 12x + 50) + 192$
  19. $x^6 - 3x^5 + x^4 + 0x^3 - 2x^2 - 5x + 6$

$= (x^2 + 2)(x^4 - 3x^3 - x^2 + 6x) - 17x + 6$

21. $f(3) = 11; f(5) = 95$ **23.** $f(-1) = 21; f(2) = -108$

25. $f(3) = 201; f(-2) = -4$ **27.** $f\left(\dfrac{1}{2}\right) = \dfrac{9}{16}$

29. No. $\dfrac{p(x)}{q(x)}$ has a nonzero remainder.

31. Yes. $\dfrac{p(x)}{q(x)}$ has no remainder.

33. No. $\dfrac{p(x)}{q(x)}$ has a nonzero remainder.

35. No. $\dfrac{p(x)}{q(x)}$ has a nonzero remainder.

37. Yes. $\dfrac{p(x)}{q(x)}$ has no remainder.

39. $x - 1$ **41.** 8 **43.** 0 **45.** 3

3.4 Exercises

1. integers **3.** False **5.** $x = 10$
7. Neither **9.** $x = \sqrt{3}$
11. $p(2) = (2)^3 - 5(2)^2 + 8(2) - 4 = 0$;
  $p(x) = (x - 2)^2(x - 1)$
13. $p(1) = -(1)^4 - (1)^3 + 18(1)^2 + 16(1) - 32 = 0$;
  $p(x) = -(x - 1)(x + 2)(x + 4)(x - 4)$
15. $p\left(\dfrac{2}{3}\right) = 3\left(\dfrac{2}{3}\right)^3 - 2\left(\dfrac{2}{3}\right)^2 + 3\left(\dfrac{2}{3}\right) - 2 = 0$;
  $p(x) = (3x - 2)(x^2 + 1)$
17. $p\left(-\dfrac{1}{3}\right) = 3\left(-\dfrac{1}{3}\right)^3 + \left(-\dfrac{1}{3}\right)^2 + 24\left(-\dfrac{1}{3}\right) + 8 = 0$;
  $p(x) = (3x + 1)(x^2 + 8)$

	Function	Zero	x-intercept	Factor
19.	$f(x)$	-2	$(-2, 0)$	$x + 2$
21.	$h(x)$	-4	$(-4, 0)$	$x + 4$

23. Real zeros: $x = -1, -3, 2$

25. Real zeros: $x = 0, -1, -3, 4$

27. Real zeros: $s = 2, -2, -\dfrac{3}{2}, \dfrac{3}{2}$

29. Real zeros: $x = -3, \dfrac{1}{4}$

31. Real zeros: $x = -3, 1, \pm\sqrt{2}$

33. Real zeros: $x = 3, -4$

35. $x = -1$ **37.** $x = -1, 3, 4$ **39.** $x = -2, \dfrac{1}{2}, 3$

41. $x = -1, 1$

43. Positive zeros: 3 or 1; negative zeros: 1
45. Positive zeros: 3 or 1; negative zeros: 0
47. Positive zeros: 2 or 0; negative zeros: 2 or 0
49. Positive zeros: 2 or 0; negative zeros: 3 or 1
51. Positive zeros: 2 or 0; negative zeros: 2 or 0

53. $x = 4$ **55.** $x = -\dfrac{1}{2}, \dfrac{3}{2}, 2$

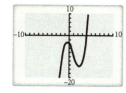 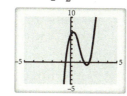

57. $x = -\dfrac{1}{2}, 1, 3$

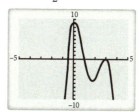

59. $x \approx -3.8265$

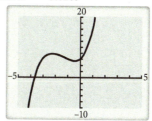

61. 12 in. $\times$ 7 in. $\times$ 2 in.
63. Radius $\approx$ 5.5047 in.; height $\approx$ 10.5047 in.

65. Answers may vary. Example:
  $y = (x + 1)^3; y = -(x + 1)(x^2 + 1)$

3.5 Exercises

1. $3x^2 - 7x - 6$ **3.** $x^2 + (\sqrt{2} - \sqrt{3})x - \sqrt{6}$
5. $x = 0, x = 3$ **7.** $x = -\dfrac{1}{2}, x = 1$
9. $4i$ **11.** $2i\sqrt{3}$ **13.** $\dfrac{2}{5}i$
15. Real part: 2; imaginary part: 0
17. Real part: 0; imaginary part: $-\pi$
19. Real part: $1 + \sqrt{5}$; imaginary part: 0
21. Real part: 1; imaginary part: $\sqrt{5}$
23. -2 **25.** $-1 - i$ **27.** $3 + \sqrt{2}$ **29.** -1
31. $x + y = 2 + 2i$ $x - y = -2 + 4i$
  $xy = 3 + 6i$ $\dfrac{x}{y} = -\dfrac{3}{5} + \dfrac{6}{5}i$
33. $x + y = -1 + 2i$ $x - y = -5 + 8i$
  $xy = 9 + 19i$ $\dfrac{x}{y} = -\dfrac{21}{13} + \dfrac{1}{13}i$
35. $x + y = 7 - 3i$ $x - y = 1 - 7i$
  $xy = 22 - 7i$ $\dfrac{x}{y} = \dfrac{2}{13} - \dfrac{23}{13}i$
37. $x + y = \dfrac{7}{10} - \dfrac{5}{3}i$ $x - y = \dfrac{3}{10} - \dfrac{13}{3}i$
  $xy = \dfrac{41}{10} + \dfrac{1}{15}i$ $\dfrac{x}{y} = -\dfrac{1755}{818} - \dfrac{285}{409}i$
39. $x + y = -\dfrac{5}{6} - i\sqrt{5}$ $x - y = \dfrac{1}{6} + 3i\sqrt{5}$
  $xy = \dfrac{61}{6} + \dfrac{\sqrt{5}}{6}i$ $\dfrac{x}{y} = -\dfrac{118}{243} - \dfrac{14\sqrt{5}}{243}i$
41. $x + y = -\dfrac{5}{2} + 2i$ $x - y = -\dfrac{7}{2}$
  $xy = -\dfrac{5}{2} - \dfrac{5}{2}i$ $\dfrac{x}{y} = -\dfrac{2}{5} + \dfrac{14}{5}i$
43. $\pm 4i$ **45.** $\pm 2i\sqrt{2}$ **47.** $\pm i\sqrt{10}$
49. a. no real zeros **b.** $x = \pm i\sqrt{2}$

51. a. no real solutions

b. $x = -\dfrac{1}{2} - i\dfrac{\sqrt{3}}{2}, x = -\dfrac{1}{2} + i\dfrac{\sqrt{3}}{2}$

53. $x = -\dfrac{3\sqrt{2}}{2}i, \dfrac{3\sqrt{2}}{2}i$

55. $x = -\dfrac{1}{2} + \dfrac{\sqrt{3}}{2}i, -\dfrac{1}{2} - \dfrac{\sqrt{3}}{2}i$

57. $t = \dfrac{1 + 2\sqrt{7}}{3}, \dfrac{1 - 2\sqrt{7}}{3}$

59. $x = -4, \dfrac{4}{3}$

61. $x = -1 + i\sqrt{2}$ and $x = -1 - i\sqrt{2}$

63. $x = \dfrac{1}{3} + \dfrac{\sqrt{11}}{3}i$ and $x = \dfrac{1}{3} - \dfrac{\sqrt{11}}{3}i$

65. $x = \dfrac{1}{5} + \dfrac{\sqrt{14}}{5}i$ and $x = \dfrac{1}{5} - \dfrac{\sqrt{14}}{5}i$

67. $x = -\dfrac{1}{5} + \dfrac{\sqrt{14}}{5}i$ and $x = -\dfrac{1}{5} - \dfrac{\sqrt{14}}{5}i$

69. $x = \dfrac{4}{3} + \dfrac{4\sqrt{2}}{3}i$ and $x = \dfrac{4}{3} - \dfrac{4\sqrt{2}}{3}i$

71. $t = \dfrac{1}{8} + \dfrac{\sqrt{7}}{8}i$ and $t = \dfrac{1}{8} - \dfrac{\sqrt{7}}{8}i$

73. $x = -\dfrac{3}{4} + \dfrac{\sqrt{15}}{4}i$ and $x = -\dfrac{3}{4} - \dfrac{\sqrt{15}}{4}i$

75. $x = -1 + 5i$ and $x = -1 - 5i$.

77. $z + \bar{z} = (a + bi) + (a - bi) = (a + a) + (b - b)i = 2a$
and $z - \bar{z} = (a + bi) - (a - bi)$
$= (a - a) + (b + b)i = 2bi$.

79. $z + \bar{z} = 2a$ so $\dfrac{z + \bar{z}}{2} = \dfrac{2a}{2} = a$, which is the real part of z.

81. $b^2 - 4ac = 4 - 4a$ **83.** $a = 1$ **85.** $x \approx 2.28 \pm 2.191i$

87. $t \approx 0.333 \pm 1.158i$

89. a. Because the graph does not intersect the x-axis, there are no real zeros of the function.

b. $f(x) = a(x - h)^2 + k \Rightarrow f(x) = ax^2 + 4$ where $a > 0$ so $f(x) = 2x^2 + 4$ is a possible function. Using any $a > 0$ gives a possible function.

c. zeros: $x = -i\sqrt{2}, i\sqrt{2}$

91. a. $(x + i)(x - i) = x^2 - xi + xi - i^2 = x^2 + 1$

b. zeros: $x = -i, i$

c. $(x + i)(x - i)$ represent the factorization of $x^2 + 1$, and the solutions of $(x + i)(x - i) = 0$ provide the zeros of $f(x) = x^2 + 1$.

d. $(x + 3i)(x - 3i)$ **e.** $(x + ci)(x - ci)$

93. a. line of symmetry is $x = 0$ because $f(x) = f(-x)$ and so $(x, f(x))$ and $(-x, f(-x))$ are the same distance from $(0, f(0))$ for all x.

b. Minimum: 1, which occurs at $x = 0$

c.

d. The function has two nonreal zeros, since its graph doesn't intersect the x-axis.

3.6 Exercises

1. $x = 1$, multiplicity 3; $x = 4$, multiplicity 5

3. $s = \pi$, multiplicity 10; $s = -\pi$, multiplicity 3

5. $x = \dfrac{3}{2}, 1; p(x) = (2x - 3)(x - 1)$

7. $x = -\pi, \pi; p(x) = (x + \pi)(x - \pi)$

9. $x = \pm 3i; p(x) = (x + 3i)(x - 3i)$

11. $x = \pm\sqrt{3}; p(x) = (x + \sqrt{3})(x - \sqrt{3})$

13. $x = 0, \pm i\sqrt{3}; p(x) = x(x - i\sqrt{3})(x + i\sqrt{3})$

15. $x = \pm\sqrt{3}, \pm i\sqrt{3};$
$p(x) = (x + \sqrt{3})(x - \sqrt{3})(x + i\sqrt{3})(x - i\sqrt{3})$

17. $(x - 2)(x^2 + 1)$ **19.** $(x - 5)(2x - 3)(x + 2)$

21. $(x - 3)(x - 2)(x^2 + 1)$ **23.** $(x - 2)(x + i)(x - i)$

25. $(x + 5)(x - 1)(x + 2i)(x - 2i)$

27. $(x - 3)(x - 2)(x + i)(x - i)$.

29. $(x - 2)(2x - 1)(x + i\sqrt{3})(x - i\sqrt{3})$

31. $p(x) = x^2 - x - 2$ **33.** $p(x) = x^3 - 2x^2 + x$

35. $p(x) = 9x^4 - 24x^3 + 22x^2 - 8x + 1$ **37.** $1 - i$

39. Each of these zeros would have to be of even multiplicity, and so the function would have to be at least a quartic function.

41. 3; $x = -1$ is a zero of odd multiplicity (at least 1), $x = 1$ is a zero of even multiplicity (at least 2); the least degree is $1 + 2 = 3$.
$p(x) = (x + 1)(x - 1)^2 = x^3 - x^2 - x + 1$

3.7 Exercises

1. False   **3.** False

5. $(-\infty, -6)\cup(-6, \infty)$; vertical asymptote: $x = -6$; horizontal asymptote: $y = 0$

7. $(-\infty, -2)\cup(-2, 2)\cup(2, \infty)$; vertical asymptotes: $x = \pm 2$; horizontal asymptote: $y = 0$

9. $(-\infty, -2)\cup(-2, 2)\cup(2, \infty)$; vertical asymptotes: $x = \pm 2$; horizontal asymptote: $y = \dfrac{-1}{-2} = \dfrac{1}{2}$

11. $(-\infty, 2)\cup(2, \infty)$; vertical asymptote: $x = 2$; horizontal asymptote: $y = 0$

13. $(-\infty, -1)\cup(-1, \infty)$; vertical asymptote: $x = -1$; horizontal asymptote: none

15. $(-\infty, -3)\cup\left(-3, \dfrac{1}{2}\right)\cup\left(\dfrac{1}{2}, \infty\right)$; vertical asymptote: $x = -3, \dfrac{1}{2}$; horizontal asymptote: $y = 0$

17. $(-\infty, \infty)$; vertical asymptote: none; horizontal asymptote: $y = 0$

19. $(-\infty, 2)\cup(2, \infty)$; vertical asymptote: $x = 2$; horizontal asymptote: $y = 3$; x-intercept: $(0, 0)$ y-intercept: $(0, 0)$

21. $(-\infty, -2)\cup(-2, 1)\cup(1, \infty)$; vertical asymptote: $x = -2$, 1; horizontal asymptote: $y = 0$; x-intercept: $(-3, 0)$; y-intercept: $\left(0, -\dfrac{3}{2}\right)$

23. a. $f(x) \to -\infty$ as $x \to -1$ from the left
$f(x) \to \infty$ as $x \to -1$ from the right.

x	-1.5	-1.1	-1.01	-0.99	-0.9	-0.5
$f(x)$	-4	-20	-200	200	20	4

b. $f(x)$ gets close to zero.

x	10	50	100	1000
$f(x)$	$\dfrac{2}{11}$	$\dfrac{2}{51}$	$\dfrac{2}{101}$	$\dfrac{2}{1001}$

c. $f(x)$ gets close to zero.

x	−1000	−100	−50	−10
f(x)	$-\dfrac{2}{999}$	$-\dfrac{2}{99}$	$-\dfrac{2}{49}$	$-\dfrac{2}{9}$

25. a. $f(x) \to -\infty$ as $x \to 0$ from the left; $f(x) \to -\infty$ as $x \to 0$ from the right.

x	−0.5	−0.1	−0.01	0.01	0.1	0.5
f(x)	−2	−98	−9998	−9998	−98	−2

b. $f(x) \to 2$ as $x \to \infty$.

x	10	50	100	1000
f(x)	1.99	1.9996	1.9999	1.999999

c. $f(x) \to 2$ as $x \to -\infty$.

x	−1000	−100	−50	−10
f(x)	1.999999	1.9999	1.9996	1.99

27. Vertical asymptote: $x = 2$; horizontal asymptote: $y = 0$; y-intercept: $\left(0, -\dfrac{1}{2}\right)$.

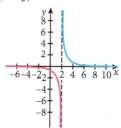

29. Vertical asymptote: $x = -6$; horizontal asymptote: $y = 0$; y-intercept: $(0, -2)$

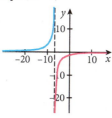

31. Vertical asymptote: $x = 4$; horizontal asymptote: $y = 0$; y-intercept: $(0, 2)$

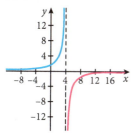

32. Vertical asymptote: $x = 3$; horizontal asymptote: $y = 0$; y-intercept: $(0, 4)$

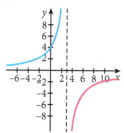

33. Vertical asymptote: $x = -1$; horizontal asymptote: $y = 0$; y-intercept: $(0, 3)$

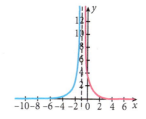

35. Vertical asymptote: $x = -4$; horizontal asymptote: $y = -1$; x-intercept: $(3, 0)$; y-intercept: $\left(0, \dfrac{3}{4}\right)$

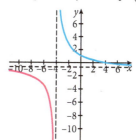

37. Vertical asymptotes: $x = 1, -4$; horizontal asymptote: $y = 0$; x & y-intercept: $(0, 0)$

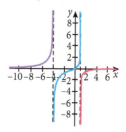

39. Vertical asymptotes: $x = -1, 2$; horizontal asymptote: $y = 3$; x & y-intercept: $(0, 0)$

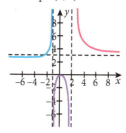

41. Vertical asymptotes: $x = -\frac{1}{2}$, 3; horizontal asymptote: $y = 0$; x-intercept: $(1, 0)$; y-intercept: $\left(0, \frac{1}{3}\right)$

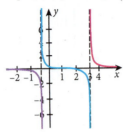

43. Vertical asymptotes: $x = \pm 1$; horizontal asymptote: $y = 1$; x-intercepts: $(-3, 0)$, $(2, 0)$; y-intercept: $(0, 6)$

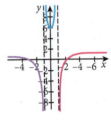

45. Vertical asymptote: none; horizontal asymptote: $y = 0$; x-intercept: none; y-intercept: $(0, 1)$

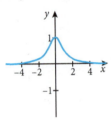

47. Vertical asymptote: $x = -4$; horizontal asymptote: none; slant asymptote: $y = x - 4$; x-intercept: $(0, 0)$; y-intercept: $(0, 0)$

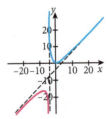

49. Vertical asymptote: $x = 3$; horizontal asymptote: none; slant asymptote: $y = -x - 3$; x-intercept: $(0, 0)$; y-intercept: $(0, 0)$

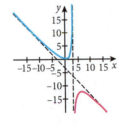

51. Vertical asymptote: $x = 0$; horizontal asymptote: none; slant asymptote: $y = -x$; x-intercepts: $(-2, 0)$, $(2, 0)$; $x \neq 0$ so there is no y-intercept

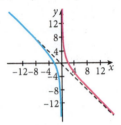

53. Vertical asymptote: $x = 1$; horizontal asymptote: none; slant asymptote: $y = x + 2$; x-intercepts: none; y-intercept: $(0, -1)$

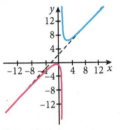

55. Vertical asymptote: $x = -1$; horizontal asymptote: none; slant asymptote: $y = 3x + 2$; x-intercepts: $\left(\frac{1}{3}, 0\right)$, $(-2, 0)$; y-intercept: $(0, -2)$

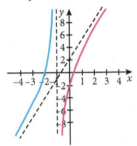

57. Vertical asymptotes: $x = 0$, -3; horizontal asymptote: none; slant asymptote: $y = x - 3$; x-intercepts: $(-1, 0)$; $x \neq 0$ so there is no y-intercept

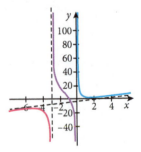

59. Vertical asymptote: $x = 3$; horizontal asymptote: $y = 0$;
x-intercept: none; y-intercept: $(0, -1)$

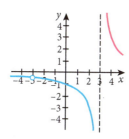

61. Vertical asymptote: $x = -3$; horizontal asymptote:
$y = 1$; x-intercept: $(-2, 0)$; y-intercept: $\left(0, \frac{2}{3}\right)$

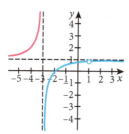

63. Vertical asymptote: none; horizontal asymptote: none
x-intercept: $(-5, 0)$; y-intercept: $(5, 0)$

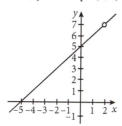

65. a. ≈ 1.2308 milligrams per liter
b. $C(t) = 0$. The concentration will near zero as time
increases.
67. $C(x) = \dfrac{15}{x} + 0.25; C(50) = \dfrac{15}{50} + 0.25 = \$0.55$
69. a. $C(x) = \dfrac{30}{x}, 0 < x < 250$
b. $C(x) = \dfrac{30}{x} + \dfrac{0.60(x - 250)}{x}, x > 250.$
c. \$0.40
71. 1.75 oz
73. a. $V(x) = x(5 - 2x)(3 - 2x); 0 < x < \dfrac{3}{2}.$
b. $S(x) = (5 - 2x)(3 - 2x) + 2x(5 - 2x) + 2x(3 - 2x)$
c. $r(x) = \dfrac{x(5 - 2x)(3 - 2x)}{(5 - 2x)(3 - 2x) + 2x(5 - 2x) + 2x(3 - 2x)}$
d.

x	0.2	0.4	0.6	0.8	1.0
r(x)	0.1612	0.2574	0.3027	0.3061	0.2727

e. It increases as x increases, then decreases as x
increases; 0.7

f. $x \approx 0.7170$

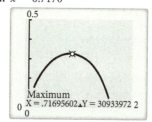

75. $r(x) = \dfrac{-2x(x - 2)}{(x - 1)^2}$

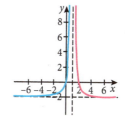

77. $r(x) = \dfrac{1}{x}$
79. It doesn't show the behavior of the graph near the
vertical asymptote $x = 10$.

3.8 Exercises

1. $[-1, 1]$ **3.** $\left(-\infty, -\dfrac{5}{2}\right] \cup [1, \infty)$ **5.** $\left(-\dfrac{1}{2}, 1\right)$

7. $\left(-\infty, -\dfrac{2}{5}\right] \cup [2, \infty)$ **9.** $[-5, 0] \cup [3, \infty)$

11. $(-\infty, -4) \cup (0, 4)$ **13.** $(-2, 0) \cup (2, \infty)$ **15.** $(-\infty, -2)$

17. $[-2, 1] \cup [4, \infty)$ **19.** $(-\infty, -1.5175) \cup (1.5175, \infty)$
21. $(-\infty, -2] \cup [-1, 2]$ **23.** $(-\infty, -2] \cup [0, 2]$

25. $(-\infty, -1) \cup (0, 3)$ **27.** $[-2, 1)$ **29.** $(-\infty, -2] \cup [2, 3)$

31. $(-\infty, -1] \cup [0, \infty)$ **33.** $(-\infty, -2) \cup (1, 2)$

35. $(-\infty, 0) \cup \left(\dfrac{1}{2}, 1\right]$ **37.** $(-\infty, 1) \cup \left[\dfrac{5}{2}, \infty\right)$

39. $(-\infty, -2) \cup [1, \infty)$ **41.** $\left(-\infty, -\dfrac{1}{2}\right)$

43. $(-\infty, -3) \cup (-1, 3)$ **45.** $(-\infty, -4) \cup \left[\dfrac{1}{5}, 3\right)$

47. $x \geq 4$ **48.** $x \geq 2.2939$ **49.** $(1, 3)$
51. Each term doesn't necessarily have to be less than 2 for
the product of the terms to be less than 2.
53. Example: $p(x) = x(x - 1)(x - 3)$

Chapter 3 Review Exercises

1. Vertex: $(-3, -1)$

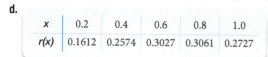

3. Vertex: $(0, -1)$

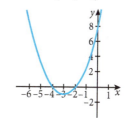

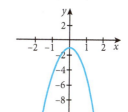

5. $f(x) = (x - 2)^2 - 1$

vertex: $(2, -1)$; minimum point

7. $f(x) = 4(x + 1)^2 - 5$

vertex: $(-1, -5)$; minimum point

9. $h = \frac{-b}{2a} = \frac{2}{2(1)} = 1$ $k = f(1) = (1)^2 - 2(1) + 1 = 0$

x-intercept: $(1, 0)$; y-intercept: $(0, 1)$

vertex: $(1, 0)$; axis of symmetry: $x = 1$

increasing on: $(1, \infty)$; decreasing on: $(-\infty, 1)$

range: $[0, \infty)$

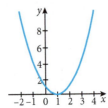

11. $h = \frac{-b}{2a} = \frac{3}{2(-1)} = -\frac{3}{2}$

$k = f\left(-\frac{3}{2}\right) = -\left(-\frac{3}{2}\right)^2 - 3\left(-\frac{3}{2}\right) + 1 = \frac{13}{4}$

vertex: $\left(-\frac{3}{2}, \frac{13}{4}\right)$; axis of symmetry: $x = -\frac{3}{2}$

x-intercepts: $\left(\frac{-3 - \sqrt{13}}{2}, 0\right), \left(\frac{-3 + \sqrt{13}}{2}, 0\right)$;

y-intercept: $(0, 1)$; range: $\left(-\infty, \frac{13}{4}\right]$

increasing on: $\left(-\infty, -\frac{3}{2}\right)$; decreasing on: $\left(-\frac{3}{2}, \infty\right)$

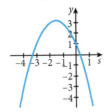

13. $h = \frac{-b}{2a} = \frac{2}{2\left(\frac{1}{2}\right)} = 2$

$k = g(2) = \frac{1}{2}(2)^2 - 2(2) + 5 = 3$

vertex: $(2, 3)$; axis of symmetry: $x = 2$ increasing on:

$(2, \infty)$; decreasing on: $(-\infty, 2)$

x-intercept: none; y-intercept: $(0, 5)$; range: $[3, \infty)$

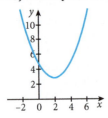

15. $b^2 - 4ac = 20 > 0$

The graph of f has two x-intercepts, and the equation $f(x) = 0$ has two real solutions.

17. $b^2 - 4ac = 0$

The graph of f has one x-intercept, and the equation $f(x) = 0$ has one real solution.

19. 60 ft $\times$ 30 ft   **21.** Yes; degree: 3; $-1, -6, 5; -1$ **23.** No

25. $f(x) \to \infty$ as $x \to -\infty$; $f(x) \to -\infty$ as $x \to \infty$.

27. $H(s) \to -\infty$ as $s \to -\infty$; $H(s) \to -\infty$ as $s \to \infty$.

29. $h(s) \to -\infty$ as $s \to -\infty$; $h(s) \to \infty$ as $s \to \infty$.

31. a. $f(x) \to \infty$ as $x \to -\infty$; $f(x) \to -\infty$ as $x \to \infty$

b. y-intercept: $(0, 8)$; x-intercepts: $(1, 0), (-2, 0), (-4, 0)$

c. $(-\infty, -4) \cup (-2, 1)$ **d.** $(-4, -2) \cup (1, \infty)$

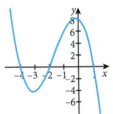

33. a. $f(t) \to -\infty$ as $t \to -\infty$; $f(t) \to \infty$ as $t \to \infty$

b. y-intercept: $(0, 0)$; t-intercepts: $(0, 0), \left(\frac{1}{3}, 0\right), (-4, 0)$

c. $(-4, 0) \cup \left(\frac{1}{3}, \infty\right)$ **d.** $(-\infty, -4) \cup \left(0, \frac{1}{3}\right)$

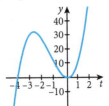

35. a. $f(x) \to -\infty$ as $x \to -\infty$; $f(x) \to \infty$ as $x \to \infty$

b. y-intercept: $(0, 0)$; x-intercepts: $(0, 0), \left(\frac{1}{2}, 0\right), (-1, 0)$

c. $(-1, 0) \cup \left(\frac{1}{2}, \infty\right)$.   **d.** $(-\infty, -1) \cup \left(0, \frac{1}{2}\right)$

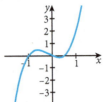

37. $x = -2$, multiplicity 3, crosses the x-axis; $x = -7$, multiplicity 2, touches the x-axis

39. $t = 0$, multiplicity 2, touches the t-axis; $t = -1$, multiplicity 1, crosses the t-axis; $t = 2$, multiplicity 1, crosses the t-axis

41. $x = 0$, multiplicity 1, crosses the x-axis; $x = -1$, multiplicity 2, touches the x-axis

43. a. x-intercepts: $(0, 0), \left(-\frac{1}{2}, 0\right)$; y-intercept: $(0, 0)$

b. $x = 0$, multiplicity 2; $x = -\frac{1}{2}$, multiplicity 1

c. $f(x) \to -\infty$ as $x \to -\infty$; $f(x) \to \infty$ as $x \to \infty$

d. Positive: $\left(-\frac{1}{2}, 0\right) \cup (0, \infty)$; negative: $\left(\infty, -\frac{1}{2}\right)$

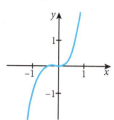

45. a. x-intercepts: $\left(\frac{1}{2}, 0\right)$, $(4, 0)$; y-intercept: $(0, -1)$

b. $x = \frac{1}{2}$, multiplicity 2; $x = 4$, multiplicity 1

c. $f(x) \to -\infty$ as $x \to -\infty$; $f(x) \to \infty$ as $x \to \infty$

d. Positive: $(4, \infty)$; negative: $\left(\infty, \frac{1}{2}\right) \cup \left(\frac{1}{2}, 4\right)$

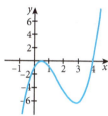

47. a. t-intercepts: $(-2, 0)$, $(1, 0)$; y-intercept: $(0, -2)$

b. $t = -2$, multiplicity 1; $t = 1$, multiplicity 1

c. $f(t) \to \infty$ as $t \to -\infty$; $f(t) \to \infty$ as $t \to \infty$

d. Positive: $(-\infty, -2) \cup (1, \infty)$; negative: $(-2, 1)$

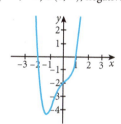

49. a. x-intercepts: $(0, 0)$, $(-3, 0)$, $(6, 0)$; y-intercept: $(0, 0)$

b. $x = 0$, multiplicity 2; $x = -3$, multiplicity 1; $x = 6$, multiplicity 1

c. $g(x) \to \infty$ as $x \to -\infty$; $g(x) \to \infty$ as $x \to \infty$

d. Positive: $(-\infty, -3) \cup (6, \infty)$; negative: $(-3, 0) \cup (0, 6)$

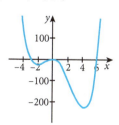

51. a. $V(x) = x(11 - 2x)(8 - 2x)$

b. $0 < x < 4$, since you must cut more than 0 inch and less than half of the shortest side of the box.

c. About 1.5252 inches

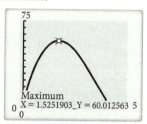

53. $-4x^2 + x - 7 = (x - 4)(-4x - 15) - 67$

55. $x^5 - x^4 + x^2 - 3x + 1 = (x^2 + 3)$
$(x^3 - x^2 - 3x + 4) + 6x - 11$

57. $r = 0$; yes, no remainder   **59.** $r = 0$; yes, no remainder

61. $p(x) = (x - 2)(x + 1)(x - 5)$

63. $p(3) = -(3)^4 + (3)^3 + 4(3)^2 + 5(3) + 3 = 0$
$p(x) = (x - 3)(-x^3 - 2x^2 - 2x - 1)$
$= -(x - 3)(x^3 + 2x^2 + 2x + 1)$
Possible rational zeros of $q(x) = x^3 + 2x^2 + 2x + 1$ are
± 1; $p(x) = -(x - 3)(x^3 + 2x^2 + 2x + 1)$
$= -(x - 3)(x + 1)(x^2 + x + 1)$

65. Real zeros: $x = 1, \frac{3}{2}, -1$ **67.** Real zero: $x = 3$

69. $x = 1, -\frac{1}{2}, -5$ **71.** $x = 1, 2, 4$

73. Positive zeros: 2 or 0; negative zeros: 2 or 0

75. $h(h + 3)(h + 4) = 60$; $h = 2$

77. Real part: 0; imaginary part: $-\frac{3}{2}$

79. $-1 - \sqrt{-5} = -1 - i\sqrt{5}$
Real part: -1; imaginary part: $-\sqrt{5}$

81. $3 + i$ **83.** $\left(1 - \sqrt{2}\right) - 3i$

85. $x + y = -1 + 5i$
$x - y = 7 - i$
$xy = -18 + i$ $\dfrac{x}{y} = -\dfrac{6}{25} - \dfrac{17}{25}i$

87. $x + y = \left(-3 - \sqrt{2}\right) + i$
$x - y = \left(3 - \sqrt{2}\right) + i$
$xy = 3\sqrt{2} - 3i$ $\dfrac{x}{y} = \dfrac{\sqrt{2}}{3} - \dfrac{1}{3}i$

89. $x + y = 1 - \left(1 + \sqrt{3}\right)i$
$x - y = -1 + \left(\sqrt{3} - 1\right)i$
$xy = -\sqrt{3} - i$ $\dfrac{x}{y} = \dfrac{\sqrt{3}}{4} - \dfrac{1}{4}i$

91. $x = \dfrac{1}{4} \pm \dfrac{\sqrt{7}}{4}i$. These are also the zeros of
$f(x) = -2x^2 + x - 1$.

93. $x = \dfrac{1}{4} \pm \dfrac{\sqrt{8\sqrt{13} - 1}}{4}$. These are also the zeros of
$f(t) = 2t^2 - t - \sqrt{13}$.

95. $x = i, -i, 4$; $p(x) = (x - 4)(x - i)(x + i)$

97. $x = \pm 3, \pm i$; $p(x) = (x + 3)(x - 3)(x + i)(x - i)$

99. a. $f(x) \to \infty$ as $x \to 3$ from the left and right.

x	2.5	2.9	2.99	3.01	3.1	3.5
$f(x)$	4	100	10,000	10,000	100	4

b. $f(x) \to 0$ as $x \to \infty$.

x	10	50	100	1000
f(x)	0.0204	0.0004527	0.0001063	0.000001006

c. $f(x) \to 0$ as $x \to \infty$.

x	−1000	−100	−50	−10
f(x)	0.0000009940	0.00009426	0.0003560	0.005917

101. Vertical asymptote: $x = -5$; horizontal asymptote: $y = \dfrac{3}{1} = 3$; x-intercept: $(0, 0)$; y-intercept: $(0, 0)$

103. Vertical asymptote: $x = -1, 1$; horizontal asymptote: $y = \dfrac{2}{1} = 2$; x-intercept: $(0, 0)$; y-intercept: $(0, 0)$

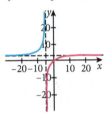
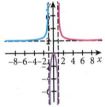

105. Vertical asymptote: $x = -2, 2$; horizontal asymptote: $y = \dfrac{1}{1} = 1$; x-intercepts: $(-\sqrt{2}, 0)$, $(\sqrt{2}, 0)$; y-intercept: $\left(0, \dfrac{1}{2}\right)$

107. Vertical asymptote: $x = 1$; horizontal asymptote: none; slant asymptote: $y = x + 1$; x-intercept: $(-2, 0)$, $(2, 0)$; y-intercept: $(0, 4)$

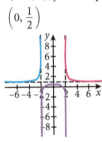
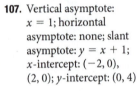

109. $\left[-5, -\dfrac{1}{4}\right]$ **111.** $\left(-\dfrac{4}{3}, \dfrac{1}{2}\right)$ **113.** $(-3, -1) \cup (0, 3)$

115. $(-\infty, -3] \cup [-2, 1]$ **117.** $(-\infty, -1) \cup (-1, 1]$

119. $\left(0, \dfrac{1}{3}\right)$ **121.** $s^2(s - 1) \geq 48$; $s \geq 4$

Chapter 3 Test

1. $f(x) = 2(x - 1)^2 - 1$
Vertex: $(1, -1)$; minimum point

2. Vertex: $(1, 2)$; axis of symmetry: $x = 1$ range: $(-\infty, 2]$

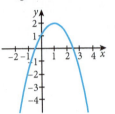

3. Vertex: $(-2, -2)$; axis of symmetry: $x = -2$ range: $[-2, \infty)$

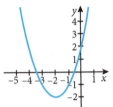

4. Vertex: $(2, 4)$; axis of symmetry: $x = 2$; range: $(-\infty, 4]$

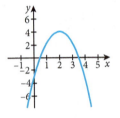

5. Vertex: $(-1, -3)$; axis of symmetry: $x = -1$; x-intercepts: $(0, 0)$, $(-2, 0)$ y-intercept: $(0, 0)$ increasing on $(-1, \infty)$; decreasing on $(-\infty, -1)$

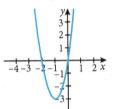

6. Degree: 5; coefficients: 3, 4, −1, 7; leading: 3
7. $p(x) \to -\infty$ as $x \to -\infty$;
$p(x) \to -\infty$ as $x \to \infty$.
8. $p(x) = -2x^2(x^2 - 9)$
$= -2x^2(x + 3)(x - 3)$
Zeros: $x = 0$, multiplicity 2; $x = -3$, multiplicity 1; $x = 3$, multiplicity 1; x-intercepts: $(0, 0)$, touches the x-axis at this point; $(-3, 0)$ and $(3, 0)$, crosses the x-axis at these points
9. a. x-intercepts: $(0, 0)$, $(2, 0)$, $(-1, 0)$; y-intercept: $(0, 0)$
b. $x = 0$, multiplicity 1; $x = 2$, multiplicity 1; $x = -1$, multiplicity 1
c. $f(x) \to \infty$ as $x \to -\infty$; $f(x) \to -\infty$ as $x \to \infty$
d. Positive: $(-\infty, -1) \cup (0, 2)$; negative: $(-1, 0) \cup (2, \infty)$

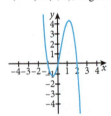

10. a. x-intercepts: $(-1, 0)$, $(2, 0)$; y-intercept: $(0, 4)$
b. $x = -1$, multiplicity 1; $x = 2$, multiplicity 2
c. $f(x)$ behaves like $g(x) = x^3$ so $f(x) \to -\infty$ as $x \to -\infty$ and $f(x) \to \infty$ as $x \to \infty$.
d. Positive: $(-1, 2) \cup (2, \infty)$; negative: $(-\infty, -1)$

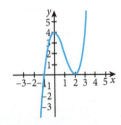

11. a. x-intercepts: $(0, 0)$, $(-1, 0)$; y-intercept: $(0, 0)$
  b. $x = 0$, multiplicity 1; $x = -1$, multiplicity 2
  c. $f(x) \to \infty$ as $x \to -\infty$; $f(x) \to -\infty$ as $x \to \infty$
  d. Positive: $(-\infty, -1) \cup (-1, 0)$; negative: $(0, \infty)$

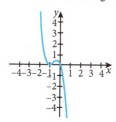

12. a. x-intercepts: $(0, 0)$, $\left(-\frac{1}{2}, 0\right)$, $(-2, 0)$; y-intercept: $(0, 0)$
  b. $x = 0$, multiplicity 2; $x = -\frac{1}{2}$, multiplicity 1;
    $x = -2$, multiplicity 1
  c. $f(x) \to \infty$ as $x \to -\infty$; $f(x) \to \infty$ as $x \to \infty$.
  d. Positive: $(-\infty, -2) \cup \left(-\frac{1}{2}, 0\right) \cup (0, \infty)$
    negative: $\left(-2, -\frac{1}{2}\right)$

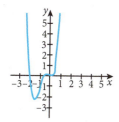

13. $3x^4 - 6x^2 + x - 1 = (x^2 + 1)(3x^2 - 9) + x + 8$
14. $-2x^5 + x^4 - 4x^2 + 3$
  $= (x - 1)(-2x^4 - x^3 - x^2 - 5x - 5) - 2$
15. The remainder is 0. Yes, $x - 2$ is a factor of $p(x)$ since $p(2) = 0$ (by the Factor Theorem).
16. $p(x) = x^4 - 3x^3 - x + 3$
  $= (x - 3)(x^3 - 1)$
  $= (x - 3)(x - 1)(x^2 + x + 1)$
17. Real zeros: $x = 2, \pm\sqrt{3}$
18. Real zeros: $x = -2, -\frac{1}{2}, -1$
19. $x = -1, \frac{1}{2}, -2$ **20.** $x = 3, -1, 1$
21. $p(x) = -x^5 + 4x^4 - 3x^2 + x + 8$; 2 sign variations
  $p(-x) = -(-x)^5 + 4(-x)^4 - 3(-x)^2 + (-x) + 8$
  $= x^5 + 4x^4 - 3x^2 - x + 8$; 2 sign variations
  Positive zeros: 2 or 0; negative zeros: 2 or 0
22. Real part: 4; imaginary part: $-\sqrt{2}$
23. $-1 + 2i$ **24.** $-2 + 11i$ **25.** $\frac{4}{13} + \frac{7}{13}i$
26. $x = -1 + i\sqrt{2}, x = -1 - i\sqrt{2}$
27. $x = \frac{1}{4} + \frac{\sqrt{7}}{4}i, x = \frac{1}{4} - \frac{\sqrt{7}}{4}i$
28. Zeros: $x = 0, \pm 2i, \pm 2$
  $p(x) = x(x + 2)(x - 2)(x + 2i)(x - 2i)$
29. Zeros: $x = -2, 3, \pm 2i$
  $p(x) = (x + 2)(x - 3)(x + 2i)(x - 2i)$

30. Vertical asymptote: $x = -3$; horizontal asymptote: $y = 0$; x-intercept: none; y-intercept: $(0, -1)$

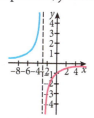

31. Vertical asymptote: $x = 2$; horizontal asymptote: $y = \frac{-2}{1} = -2$; x-intercept: $(0, 0)$; y-intercept: $(0, 0)$

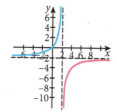

32. Vertical asymptote: $x = -\frac{1}{2}, 2$; horizontal asymptote: $y = 0$; x-intercept: none; y-intercept: $(0, -1)$

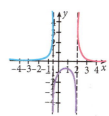

33. $\left(-\frac{5}{3}, 3\right)$ **34.** $(-\infty, -3] \cup [-2, 2]$
35. $(-2, -1) \cup (5, \infty)$ **36.** $\left(-\frac{1}{3}, \frac{1}{3}\right]$
37. Height: $h = 4$ inches.
38. a. $C(x) = \frac{50 + 0.25x}{x} = \frac{50}{x} + 0.25$
  b. $C(250) = \frac{50}{250} + 0.25 = \$0.45$

Chapter 4

4.1 Exercises
1. True **3.** $(f \circ g)(x) = (x - 1)^2$ **5.** x
7. a. 27 **b.** $\sqrt[3]{3}$ **c.** $27\sqrt[3]{3}$ **d.** 3
9. $f(g(x)) = (x + 2) - 2 = x$
  $g(f(x)) = (x - 2) + 2 = x$
11. $f(g(x)) = 6\left(\frac{1}{6}x\right) = x$
  $g(f(x)) = \frac{1}{6}(6x) = x$
13. $f(g(x)) = -3\left(-\frac{1}{3}x + \frac{8}{3}\right) + 8 = x - 8 + 8 = x$
  $g(f(x)) = -\frac{1}{3}(-3x + 8) + \frac{8}{3} = x - \frac{8}{3} + \frac{8}{3} = x$
15. $f(g(x)) = (\sqrt[3]{x + 2})^3 - 2 = x + 2 - 2 = x$
  $g(f(x)) = \sqrt[3]{(x^3 - 2) + 2} = \sqrt[3]{x^3} = x$
17. $f(g(x)) = \frac{1}{\left(\frac{1}{x}\right)} = x; g(f(x)) = \frac{1}{\left(\frac{1}{x}\right)} = x$

19. The function is one-to-one because there are no values of a and b in the chart such that $f(a) = f(b)$, where $a \neq b$.

21. The function is not one-to-one because there are values of a and b in the chart such that $f(a) = f(b)$, where $a \neq b$. Here, $f(0) = f(2) = 9$.

23. Not one-to-one   **25.** One-to-one   **27.** Not one-to-one

29. One-to-one   **31.** Not one-to-one,   **33.** One-to-one

35. $f^{-1}(x) = -\dfrac{3}{2}x$ **37.** $f^{-1}(x) = -\dfrac{1}{4}x + \dfrac{1}{20}$

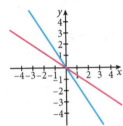 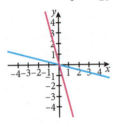

39. $f^{-1}(x) = \sqrt[3]{x+6}$ **41.** $g^{-1}(x) = \sqrt{8-x}$

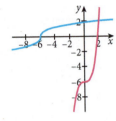 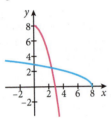

43. $f^{-1}(x) = \sqrt[3]{\dfrac{7-x}{2}}$ **45.** $f^{-1}(x) = \sqrt[5]{\dfrac{9-x}{4}}$

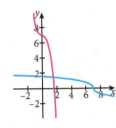 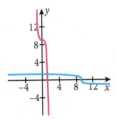

47. $f^{-1}(x) = 3x$ **49.** $g^{-1}(x) = 1 + \sqrt{x}$

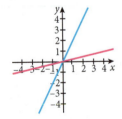 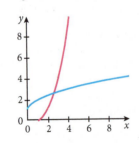

51. $f^{-1}(x) = x^2 - 3, x \geq 0$ **53.** $f^{-1}(x) = \dfrac{x}{x-2}$

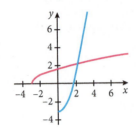 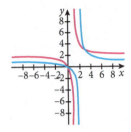

55. $f^{-1}(x) = \dfrac{1}{2}x$

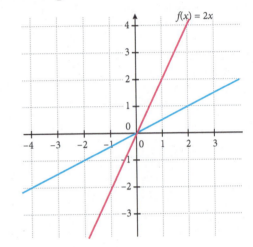

57. $f^{-1}(x) = -\dfrac{1}{3}x + 1$

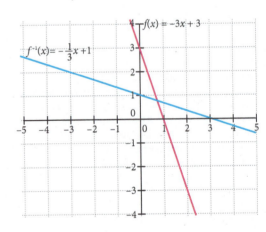

59. $f^{-1}(x) = \dfrac{\sqrt[3]{x}}{2}$

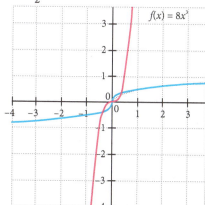

$f(x) = 8x^3$

61. $f^{-1}(x) = \sqrt{x-2}$

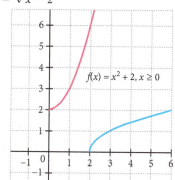

$f(x) = x^2 + 2,\ x \geq 0$

63. Domain of f: $[-3, 3]$
Range of f: $[0, 4]$
Domain of f^{-1}: $[0, 4]$
Range of f^{-1}: $[-3, 3]$

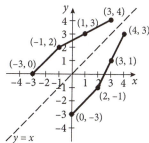

65. Domain of f: $[-3, 2]$
Range of f: $[-5, 5]$
Domain of f^{-1}: $[-5, 5]$
Range of f^{-1}: $[-3, 2]$

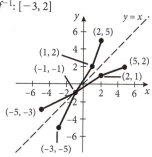

67. $f^{-1}(1) = -2$
69. $f^{-1}(-2) = 2$ so $f^{-1}(f^{-1}(-2)) = f^{-1}(2) = -1$
71. $f^{-1}(1) = 0$
73. $f^{-1}(-3) = -2$ so $f^{-1}(f^{-1}(-3)) = f^{-1}(-2) = -1.5$
75. $f(x) = 4x$, where x is the number of gallons and $f(x)$ is the number of quarts in x gallons. The inverse, $f^{-1}(x) = \frac{1}{4}x$, gives the number of gallons, where x is the number of quarts.
77. Solve for q: $q = 1000 - 10p$
79. a. Range of f: even integers in the interval $[32, 54]$.
   b. $f^{-1}(s) = s - 30$ This function converts a woman's dress size in France to American size.
81. a. $x = -2$ **b.** $g(2) = 1$
   c.

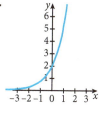

83. No. If f is symmetric with respect to the y-axis, then $f(a) = f(-a)$. This statement does not meet the criteria for a one-to-one function; in this case, if $f(a) = f(b)$, then $a \neq b$.
85. The simplest function that is its own inverse is $y = x$.

4.2 Exercises

1. 64   **3.** $\dfrac{1}{9}$ **5.** 32   **7.** 1.2806   **9.** $4^{1.6} = 9.1896$
11. $3^{\sqrt{2}} = 4.7288$ **13.** $e^3 = 20.0855$ **15.** $e^{-2.5} = 0.0821$
17. a. $f(3) = 8; g(3) = 9$ **b.** $f(-2) = \dfrac{1}{4}; g(-2) = 4$
   c. f **d.** g

19.

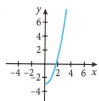

21.

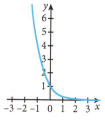

23.

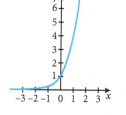

25.

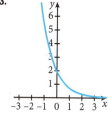

27.

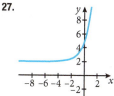

29.

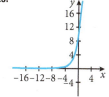

31.

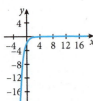

33.

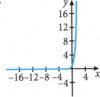

35.

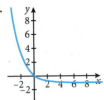

37. a. y-intercept: $(0, -1)$
b. Domain: $(-\infty, \infty)$; Range: $(-\infty, 0)$
c. Horizontal Asymptote: $y = 0$
d. $f(x) \to 0$ as $x \to -\infty$; $f(x) \to -\infty$ as $x \to \infty$

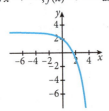

39. a. y-intercept: $(0, 2)$
b. Domain: $(-\infty, \infty)$; Range: $(-\infty, 3)$
c. Horizontal Asymptote: $y = 3$
d. $f(x) \to 3$ as $x \to -\infty$; $f(x) \to -\infty$ as $x \to \infty$

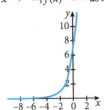

41. a. y-intercept: $(0, 7)$
b. Domain: $(-\infty, \infty)$; Range: $(0, \infty)$
c. Horizontal Asymptote: $y = 0$
d. $f(x) \to 0$ as $x \to -\infty$; $f(x) \to \infty$ as $x \to \infty$

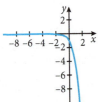

43. a. y-intercept: $(0, -1)$
b. Domain: $(-\infty, \infty)$; Range: $(-4, \infty)$
c. Horizontal Asymptote: $y = -4$
d. $f(x) \to \infty$ as $x \to -\infty$; $f(x) \to -4$ as $x \to \infty$

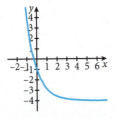

45. a. y-intercept: $(0, -3)$
b. Domain: $(-\infty, \infty)$; Range: $(-\infty, 1)$
c. Horizontal Asymptote: $y = 1$
d. $f(x) \to 1$ as $x \to -\infty$; $f(x) \to -\infty$ as $x \to \infty$

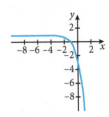

47. This graph does not represent an exponential function since it does not include a horizontal asymptote.
49. This graph does not represent an exponential function since it includes a vertical asymptote.
51. $x \approx 1.4650$

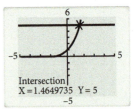

53. $x \approx -3.3219$

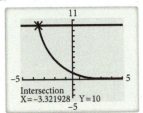

55. $x \approx 11.5525$

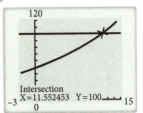

57. a.

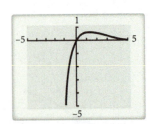

b. Domain: $(-\infty, \infty)$; range: $(-\infty, 0.3679]$
c. x-intercept: $(0, 0)$; y-intercept: $(0, 0)$
d. $f(x) \to -\infty$ as $x \to -\infty$; $f(x) \to 0$ as $x \to \infty$

59. $A = \$2007.34$ **61.** $A = \$2023.28$ **63.** $A = \$1795.83$
65. $A = \$1793.58$

67. $S(t) = 10,000(1.05)^t$

Years at Work	Annual Salary
0	$10,000.00
1	$10,500.00
2	$11,025.00
3	$11,576.25
4	$12,155.06

69. $V(t) = 20,000(0.9)^t$

Years Since Purchase	Value
0	$20,000
1	$18,000
2	$16,200
3	$14,580
4	$13,122

71. $V(t) = 18,000(0.7)^t$

Years Since Purchase	Value
1	$12,600.00
2	$8820.00
3	$6174.00
4	$4321.80
5	$3025.26

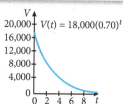

73. $A = 1348.35$. The bond would be worth $1348.35. If the bonds continued paying interest, their value would have no upper limit. For instance, after 80 years, a bond purchased for $1000 would be worth $10,924.90, nearly 11 times its purchase price.

75. a. $C(0) = 0$. This answer makes sense because the drug has not yet been ingested.
  b. $C(1) = 4.5(1)e^{-1.275(1)} \approx 3.4181$ mg/L
  c.

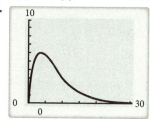

  d. $C(t) \to 0$ as $t \to \infty$. This makes sense because the concentration of a drug decreases over time after reaching a maximum value.

e. $C(t)$ reaches its maximum 3.6 hours after the drug is administered.

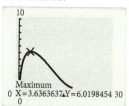

f. $C(t)$ reaches 3 mg/L for the second time about 9.8 hours after the drug is administered.

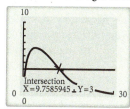

77. Because when $a = 1$ for $f(x) = Ca^x$, the function is a horizontal line.

79. a. $C = 12$
  b. Based on the given points, $f(x)$ is decreasing because as x increases, $f(x)$ decreases
  c. $a = \dfrac{1}{2}$

81. $f(x) = 2^x$, has no vertical asymptotes because the function is defined for every value in its domain, which has no restrictions – that is, $x \in (-\infty, \infty)$

4.3 Exercises

1. $13^{1/2}$ **3.** $e^{1/3}$ **5.** True   **7.** False   **9.** 3.6×10^{-2}

11.

Logarithmic Statement	Exponential Statement
$\log_3 1 = 0$	$3^0 = 1$
$\log 10 = 1$	$10^1 = 10$
$\log_5 \dfrac{1}{5} = -1$	$5^{-1} = \dfrac{1}{5}$
$\log_a x = b, a > 0$	$a^b = x, a > 0$

13.

Exponential Statement	Logarithmic Statement
$3^4 = 81$	$\log_3 81 = 4$
$5^{1/3} = \sqrt[3]{5}$	$\log_5 \sqrt[3]{5} = \dfrac{1}{3}$
$6^{-1} = \dfrac{1}{6}$	$\log_6 \dfrac{1}{6} = -1$
$a^v = u, a > 0$	$\log_a u = v, a > 0$

15. 4   **17.** $\dfrac{1}{3}$ **19.** 2   **21.** $\dfrac{1}{3}$ **23.** $x + y$ **25.** k **27.** $\dfrac{1}{2}$

29. -4 **31.** -2 **33.** $x^2 + 1$ **35.** 1.2041   **37.** 0.3466

39. 3.1461   **41.** -3.2189 **43.** 0.2031   **45.** -0.4307

47. 3.5850   **49.** 2.5750   **51.** 8   **53.** $\sqrt[3]{3}$

55. $x = \sqrt[3]{216} = 6$

57. $x > 0 \Rightarrow$ Domain: $(0, \infty)$; vertical asymptote: $x = 0$; x-intercept: $(1, 0)$; y-intercept: none

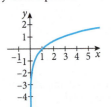

59. $x > 0 \Rightarrow$ Domain: $(0, \infty)$; vertical asymptote: $x = 0$; x-intercept: $(1, 0)$; y-intercept: none

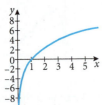

71. $|x| > 0$ for all real x except $x \neq 0 \Rightarrow$ Domain: $(-\infty, 0) \cup (0, \infty)$; vertical asymptote: $x = 0$; x-intercepts: $(-1, 0)$, $(1, 0)$; y-intercept: none

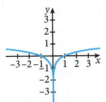

61. $x > 0 \Rightarrow$ Domain: $(0, \infty)$; vertical asymptote: $x = 0$; x-intercept: $(1000, 0)$; y-intercept: none

63. $x + 1 > 0 \Rightarrow x > -1$; Domain: $(-1, \infty)$; vertical asymptote: $x = -1$; x-intercept: $(0, 0)$; y-intercept: $(0, 0)$

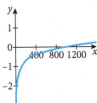

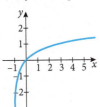

73. $\log 7 \approx 0.85$. The graph corresponds to $f(x) = 10^x$, so $\log f(x) = \log 10^x = x$. In this case, when $f(x) = 7$, $x \approx 0.85$

75. c   **77.** b

79. $\log 7 = t$; $t \approx 0.8451$

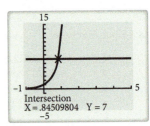

65. $x + 4 > 0 \Rightarrow x > -4$; Domain: $(-4, \infty)$; vertical asymptote: $x = -4$; x-intercept: $(-3, 0)$; y-intercept: $(0, \ln 4)$

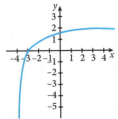

81. $\log 5 = x$; $x \approx 0.6990$

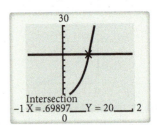

83. 4   **85.** $63{,}095{,}734.45 I_0$ **87.** $794.33{:}1$.   **89.** 4

67. $x - 1 > 0 \Rightarrow x > 1$; Domain: $(1, \infty)$; vertical asymptote: $x = 1$; x-intercept: $(2, 0)$; y-intercept: none

69. $t > 0$; Domain: $(0, \infty)$; vertical asymptote: $t = 0$; t-intercept: $(1, 0)$; y-intercept: none

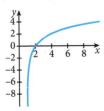

91. a. Bubble sort: $100^2 = 10{,}000$ operations heap sort: $100 \log 100 = 100 \cdot 2 = 200$ operations

b.

n	Operations, n^2
5	25
10	100
15	225
20	400

The corresponding increase would be 300 operations

c.

n	Operations, $n\log n$
5	3.49
10	10
15	17.64
20	26.02

The corresponding increase would be 16 operations

d. The heap sort is more efficient because n^2 grows faster than $n\log n$.

e. n^2 grows faster than $n\log n$

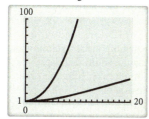

93. Because $10^2 = 100$ and $10^3 = 1000$, the value of x for which $10^x = 400$ is between 2 and 3, as is the value of $\log 400$.

95. $\log 1000 = x; f(x) = 10^x; f(x) = 1000 \therefore x = 3$

97. $\log 0.5 = x; f(x) = 10^x; f(x) = 0.5 \therefore x = -0.3010$

99. $f(x) = 3\log x$ **101.** The graphs of the two functions are identical.

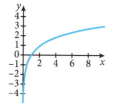
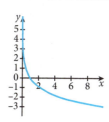

4.4 Exercises

1. $x^{1/3}$ **3.** $x^{3/4}$ **5.** 1.5441   **7.** -0.3980 **9.** 0.1505

11. 2.097   **13.** $\log x + 3\log y$ **15.** $\frac{1}{3}\log x + \frac{1}{4}\log y$

17. $\log y + \frac{1}{4}\log x$ **19.** $2\log x + 5\log y - 1$

21. $\frac{2}{3}\ln x - 2$ **23.** $\frac{1}{2}\log_a (x^2 + y) - 3$

25. $3\log_a x - \frac{3}{2}\log_a y - \frac{5}{2}\log_a z$

27. $\frac{1}{3}\log x + \log y - \frac{5}{3}\log z$

29. $\log 2.1$ **31.** $\log 3x\sqrt{y}$

33. $\ln \frac{4}{e}$ **35.** $\log (100x^3)$ **37.** $\log_4 (2x)$

39. $\ln (x - 3)$ **41.** $\log (x\sqrt{x - 1})$ **43.** $\log \frac{64x^6}{y^4}$

45. $1 + b$ **47.** $3b$ **49.** $-b$ **51.** $\sqrt{2}$ **53.** $x + 2$

55. $3x + 1$ **57.** $x^2 + 1$ **59.** $\frac{2}{5}$

61. $f(x) = 1 + g(x)$

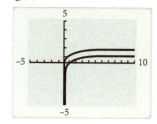

63. $f(x) = 2 + g(x)$

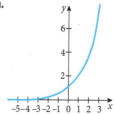

65. 10,000:1   **67.** 7.2   **69.** $10^{-3.4}$ **71.** 13.01 dB

73. a.

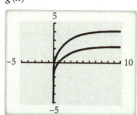

b. Domain of f: $(-\infty, \infty)$; range of f: $(0, \infty)$

c.

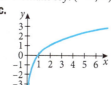

d. Domain of f^{-1}: $(0, \infty)$; range of f^{-1}: $(-\infty, \infty)$

75. a. Domain of f: $(0, \infty)$; domain of g: $(-\infty, \infty)$

b. The functions agree for all real values of $x > 0$.

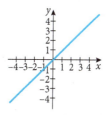

77. No, because there is no real value of y for which a^y is equal to a negative number, in this case $x = -3$.

4.5 Exercises

1. True   **3.** True   **5.** $x = 3$ **7.** $x = 4$ **9.** $x = -2$

11. $x = \ln 9 \approx 2.197$ **13.** $x = \dfrac{\ln 5 - \ln 3}{\ln 1.3} \approx 1.947$

15. $x = \dfrac{\log 16}{1 + \log 2} \approx 0.926$

17. $x = \dfrac{-\ln 3}{2\ln 3 + \ln 2} \approx -0.380$ **19.** $x = \dfrac{\ln 2}{0.04} \approx 17.329$

21. $x = \ln 5 \approx 1.609$ **23.** $x = \dfrac{\ln 11 - \ln 2}{\ln 0.8} \approx -7.640$

25. $x = \pm\sqrt{\ln 5 - 1} \approx \pm 0.781$

27. $x \approx 1.302$

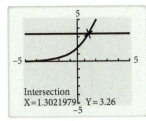

29. $x \approx 0.524$

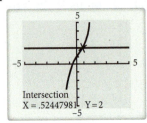

31. $x = 1$ **33.** $x = e^2 + 1 \approx 8.389$ **35.** $x = 7$
37. $x = 5$ **39.** $x = \sqrt{2}, -\sqrt{2}$ **41.** $x = -5, x = 2$
43. $x = -1, x = 4$ **45.** No solution   **47.** $x = 0, x = 3$
49. $x = \dfrac{-7 + \sqrt{89}}{4}$ **51.** $x = -1, x = 5$
53. $x \approx 3.186$

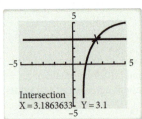

55. $x \approx 5.105, x \approx -3.105$

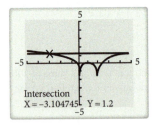

57. $t = \dfrac{\ln 2}{0.06} \approx 11.55$ years   **59.** $t = \dfrac{\ln 2}{0.0575} \approx 12.05$ years

61. $t = \dfrac{\ln 2}{0.075} \approx 9.24$ years

63. $r = \dfrac{\ln \frac{4}{3}}{5} = 0.05754; r = 5.75\%$

65. $r = \dfrac{\ln 1.5}{8} = 0.05068; r = 5.07\%$

67. $r = \dfrac{\ln \frac{20}{17}}{5} = 0.03250; r = 3.25\%$

69. $k = \dfrac{\ln 2}{12} \approx 0.0578; t = \dfrac{\ln 4}{0.0578} \approx 24$ hr;

$t = \dfrac{\ln 8}{0.0578} \approx 36$ hr

71. a. 2297.1   **b.** $t = \dfrac{\ln 2}{0.3316} \approx 2.09$ yr

73. $x = e^{\frac{49.202}{26.203}} \approx 6.54$ weeks

75. $[\text{H}^+] = 10^{-6.2}$ moles per liter

77. $P_1 = 10^{0.07}(75\text{W}) \approx 88\text{W}$ **79.** \$5000

81. A little under 12 years

83. The two equations share a solution, but the first equation has an additional negative solution because the logarithm is taken after x is squared. This allows for a nonpositive solution.

85. The step incorrectly uses the properties of addition of logarithms of the same base. Instead, the step should read $\log x + \log(x + 1) = 0 \Rightarrow \log(x(x + 1)) = 0 \Rightarrow x(x + 1) = 1$.

87. $x = 10^e$ **89.** $x = -23, x = 27$

4.6 Exercises

1. 10   **3.** $t \approx 0.6931$ **5.** $4e \approx 10.8731$ **7.** $t \approx 0.6391$

9. $\dfrac{3}{110}$ **11.** 0.3000   **13.** -1 **15.** $x = e^2 \approx 7.3891$

17. c   **19.** a

21. a. $A(t) = A_0 e^{-0.0001216t}$ **b.** $t \approx 9035$ years

23. a. $P(t) = 282e^{0.008611t}$ **b.** About 2025

25. a. $P(t) = 175{,}000e^{0.06714t}$
b. The model assumes exponential growth and does not take into account other factors such as economic conditions.

27. a. 20 deer   **b.** 32 deer

29. a. $A(t) = 5e^{-0.087t}$ **b.** $B(t) = 5e^{-0.231t}$ **c.** 5.87; 2.21
d. The solution produced with water that has a pH of 6.0 should be used because more malathion will remain after several days than will remain in the solution produced with water that has a pH of 7.0.
e.

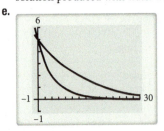

The graphs show that one function decreases more quickly than the other.

31. a.

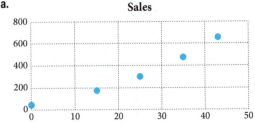

$f(x) = 53.366(1.064)^x$
b. It is a growth model.
c. \$870.238 billion
d. No. Eventually, the revenue growth will slow down, or may even decline.

33. a. A logistic function suits this data well because the number of cases is confined to the upper limit of the population.

b. $f(x) = \dfrac{264.8933}{1 + 6.0071e^{-0.6329x}}$

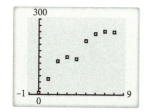

c. c signifies the greatest value that the model can produce.

d. According to this model, the greatest number of SARS cases is about 265 cases. By July, that number will have been effectively achieved.

35. a. As time increases, the percentage of women who smoke during pregnancy decreases.

b. $p(t) = -1.3705\ln t + 14.823$; a must be negative so that the value of $a\ln t + b$ decreases as t increases.

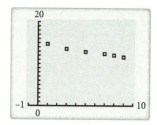

c. $\approx 11.2\%$

37. The term refers to the upper bound of the function as x increases, which is c. The function can be used to model the maximum number of individuals that a given environment can support, which is referred to as that environment's "carrying capacity."

39. If a and c are positive, the numerator is clearly positive, and the denominator is the sum of a positive number and the product of two positive numbers (since e^{-bx} is positive for any value of b). Thus, $\frac{c}{1 + ae^{-bx}}$ is the quotient of two positive numbers, so $f(x) > 0$.

Chapter 4 Review Exercises

1. $f(g(x)) = f\left(\dfrac{x-7}{2}\right) = 2\left(\dfrac{x-7}{2}\right) + 7 = x - 7 + 7 = x$

$g(f(x)) = g(2x+7) = \dfrac{(2x+7)-7}{2} = \dfrac{2x}{2} = x$

3. $f(g(x)) = f\left(\dfrac{\sqrt[3]{x}}{2}\right) = 8\left(\dfrac{\sqrt[3]{x}}{2}\right)^3 = 8\left(\dfrac{x}{8}\right) = x$

$g(f(x)) = g(8x^3) = \dfrac{\sqrt[3]{8x^3}}{2} = \dfrac{2x}{2} = x$

5. $f^{-1}(x) = -\dfrac{5}{4}x$ **7.** $f^{-1}(x) = \dfrac{6-x}{3}$

9. $f^{-1}(x) = \sqrt[3]{x - 8}$ **11.** $g^{-1}(x) = \sqrt{8 - x}$

13. $f^{-1}(x) = -x - 7$ **15.** $f^{-1}(x) = \sqrt[3]{1 - x}$

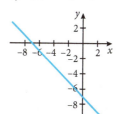   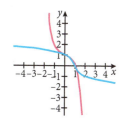

17. y-intercept: $(0, -1)$; other points: $(1, -4)$, $(2, -16)$ domain: $(-\infty, \infty)$; range: $(-\infty, 0)$; $f(x) \to 0$ as $x \to -\infty$ and $f(x) \to -\infty$ as $x \to \infty$

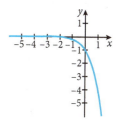

19. y-intercept: $(0, 1)$; other points: $\left(1, \dfrac{2}{3}\right), \left(-1, \dfrac{3}{2}\right)$; domain: $(-\infty, \infty)$; range: $(0, \infty)$; $g(x) \to \infty$ as $x \to -\infty$ and $g(x) \to 0$ as $x \to \infty$

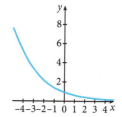

21. y-intercept: $(0, 4)$; other points: $(1, 4e), \left(-1, \dfrac{4}{e}\right)$ domain: $(-\infty, \infty)$; range: $(0, \infty)$; $f(x) \to 0$ as $x \to -\infty$ and $f(x) \to \infty$ as $x \to \infty$

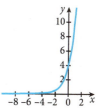

23. y-intercept: $(0, 3)$; other points: $\left(1, \dfrac{2}{e} + 1\right), (-1, 2e + 1)$ domain: $(-\infty, \infty)$; range: $(1, \infty)$; $g(x) \to \infty$ as $x \to -\infty$ and $g(x) \to 1$ as $x \to \infty$

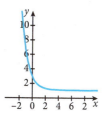

25. $2021.03   **27.** $2065.69

29.

Number of Years After Purchase	Value
1	$12,750.00
2	$9562.50
3	$7171.88
4	$5378.91
5	$4034.18

$v(t) = 17,000(0.75)^t$

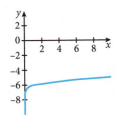

31.

Logarithmic Statement	Exponential Statement
$\log_3 9 = 2$	$3^2 = 9$
$\log 0.1 = -1$	$10^{-1} = 0.1$
$\log_5 \dfrac{1}{25} = -2$	$5^{-2} = \dfrac{1}{25}$

33. 4   **35.** 2   **37.** $\dfrac{1}{2}$ **39.** $\dfrac{1}{3}$ **41.** $x + 2$ **43.** 1.2041

45. 1.0397   **47.** 1.3277   **49.** -0.1606

51. Domain: $(0, \infty)$;
y-intercept: none;
x-intercept: $(10^6, 0)$;
vertical asymptote: $x = 0$

53. $x > 0 \Rightarrow$ Domain:
$(0, \infty)$; y-intercept:
none; x-intercept: $(1, 0)$;
vertical asymptote: $x = 0$

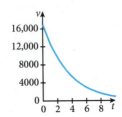

55. 1   **57.** 1.3222   **59.** 0.2219   **61.** $\dfrac{1}{4}\log x + \dfrac{1}{3}\log y$

63. $3\log_a x - \dfrac{3}{2}\log_a y - \dfrac{5}{2}\log_a z$

65. $\dfrac{1}{3}\ln x + \ln y - \dfrac{5}{3}\ln z$

67. $\ln x$ **69.** $\log\left(x^3\sqrt[4]{x - 1}\right)$ **71.** $\log_3 x^{23/6}$

73. $x = 4$ **75.** $x = -2$ **77.** $x = \dfrac{\ln 4}{0.04}$ **79.** $x = -\dfrac{1}{2}$

81. $x = \dfrac{\ln 4 - 1}{2}$ **83.** $x = 1$ **85.** $x = e^4$

87. $x = \dfrac{5}{2}, x = -2$ **89.** $x = 0, x = 3$ **91.** $x = -4, x = 1$

93. 4; 0.0022   **95.** 14.8629; 33.1888   **97.** 0.0020; 0.1999

99. $P(t) = 2.5(1.07)^t$, where $P(t)$ is the purchasing power in trillions of dollars.

101. a. 45 trout
  b. 409 trout
  c. As t increases, the number of trout in the pond increases, but at an increasingly slower rate. The graph is asymptotic to $N(t) = 450$.

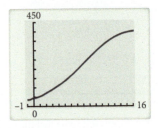

  d. About 14.3 months

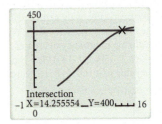

Chapter 4 Test

1. Since $f(g(x)) = x$ and $g(f(x)) = x$, they are inverses of each other.

2. $f^{-1}(x) = \sqrt[3]{\dfrac{x + 1}{4}}$

3. $f^{-1}(x) = \sqrt{x + 2}$

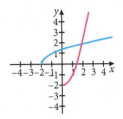

4. $f(x) \to 1$ as $x \to -\infty$ and $f(x) \to -\infty$ as $x \to \infty$

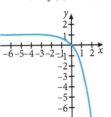

5. $f(x) \to \infty$ as $x \to -\infty$ and $f(x) \to -3$ as $x \to \infty$

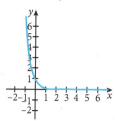

6. $f(x) \to \infty$ as $x \to -\infty$ and $f(x) \to 0$ as $x \to \infty$

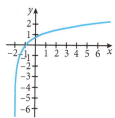

7. $\log_6 \dfrac{1}{216} = -3 \Leftrightarrow 6^{-3} = \dfrac{1}{216}$

8. $2^5 = 32 \Leftrightarrow \log_2 32 = 5$

9. $\log_8 \dfrac{1}{64} = \log_8 \dfrac{1}{8^2} = \log_8 8^{-2} = -2$ **10.** $\ln e^{3.2} = 3.2$

11. $\log_7 4.91 = \dfrac{\ln 4.91}{\ln 7} \approx 0.8178$

12. Domain: $(-2, \infty)$; y-intercept: $(0, \ln 2)$; x-intercept: $(-1, 0)$; vertical asymptote: $x = -2$

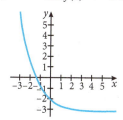

13. $\dfrac{2}{3}\log x + \dfrac{4}{3}\log y$ **14.** $2 + 2\ln x + \ln y$

15. $\ln [x(x + 2)]$ **16.** $\log_2 x^2$ **17.** $x = \dfrac{1}{2}$

18. $x \approx 1.4139$ **19.** $x \approx -0.6137$ **20.** $t \approx 6.9315$

21. $x = 0$ **22.** $x = -5, x = 2$ **23.** \$4042.05

24. \$4934.71 **25.** $f(t) = 900(0.6)^t$

26. $I = 10^{6.2} I_0$ or about $(1.585 \times 10^6)I_0$

27. a. 30 students
b. 114 students

28. $P(t) = 28{,}000e^{0.06677t}$

Chapter 5

5.1 Exercises

1. III **3.** IV **5.** F **7.** II

9.

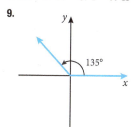

11.

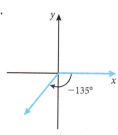

13.

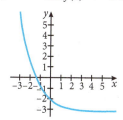

15.

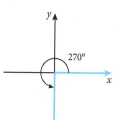

17.

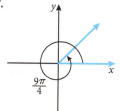

19.

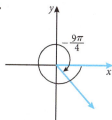

21. $500°, 860°$ **23.** $305°, 665°$ **25.** $420°, 780°$

27. $570°, 930°$ **29.** $\dfrac{7\pi}{3}, \dfrac{13\pi}{3}$ **31.** $-\dfrac{13\pi}{6}, \dfrac{11\pi}{6}$

33. $500°, -220°$ **35.** $295°, -425°$ **37.** $\dfrac{13\pi}{6}, -\dfrac{11\pi}{6}$

39. $33°$ **41.** $42°$ **43.** $75°$ **45.** $75°$ **47.** $91°$ **49.** $50°$

51. $\dfrac{4\pi}{3}$ **53.** $-\dfrac{5\pi}{6}$ **55.** $\dfrac{3\pi}{2}$ **57.** $-\dfrac{3\pi}{2}$ **59.** 4π **61.** $\dfrac{13\pi}{6}$

63. $540°$ **65.** $1°$ **67.** $-72°$ **69.** $60°, \dfrac{\pi}{3}$ **71.** $270°, \dfrac{3\pi}{2}$

73. $15°$ **75.** 9.8 inches **77.** 433π miles

79. $\dfrac{2\pi}{9}$ feet per second

81. 1120.45 degrees per second; 19.56 radians per second

83. 12.6 feet per minute **85.** $34.38°$ **87.** $\dfrac{15}{2\pi}$ feet

89. 12.6 feet per second

91. a. 15 degrees per hour; $\dfrac{\pi}{12}$ radians per hour
b. 2600π miles **c.** 325π miles per hour

93. 1 radian $= 57.2958°$ **94.** $\dfrac{\pi}{12}$ units^2 or square units

95. $-\dfrac{\pi}{2}$ **97.** $\dfrac{2\pi}{3}$ **99.** -4π

5.2 Exercises

1. Yes **3.** No **5.** III **7.** I **9. a.** $\dfrac{\pi}{3}$ **b.** $\dfrac{11\pi}{3}$ **11.** $\dfrac{\pi}{6}$

13. $\dfrac{\pi}{6}$ **15.** $\dfrac{\pi}{4}$ **17.** $\dfrac{\pi}{6}$ **19.** $\dfrac{\pi}{8}$ **21.** $\dfrac{\pi}{5}$

23. $\sin 5\pi = 0$; $\cos 5\pi = -1$; $\tan 5\pi = 0$;
$\cot 5\pi =$ undefined; $\csc 5\pi =$ undefined;
$\sec 5\pi = -1$

25. $\sin (-3\pi) = \sin (-\pi) = -\sin \pi = 0$;
$\cos (-3\pi) = \cos (-\pi) = \cos \pi = -1$;
$\tan (-3\pi) = \dfrac{0}{-1} = 0$; $\cot (-3\pi)$ is undefined;
$\csc (-3\pi)$ is undefined; $\sec (-3\pi) = -1$

27. $\sin \left(-\dfrac{\pi}{3}\right) = -\dfrac{\sqrt{3}}{2}$; $\cos \left(-\dfrac{\pi}{3}\right) = \dfrac{1}{2}$;
$\tan \left(-\dfrac{\pi}{3}\right) = -\sqrt{3}$; $\cot \left(-\dfrac{\pi}{3}\right) = -\dfrac{\sqrt{3}}{3}$;
$\csc \left(-\dfrac{\pi}{3}\right) = -\dfrac{2\sqrt{3}}{3}$; $\sec \left(-\dfrac{\pi}{3}\right) = 2$

29. $\sin\left(-\dfrac{\pi}{4}\right) = -\dfrac{\sqrt{2}}{2}; \cos\left(-\dfrac{\pi}{4}\right) = \dfrac{\sqrt{2}}{2};$
$\tan\left(-\dfrac{\pi}{4}\right) = -1; \cot\left(-\dfrac{\pi}{4}\right) = -1;$
$\csc\left(-\dfrac{\pi}{4}\right) = -\sqrt{2}; \sec\left(-\dfrac{\pi}{4}\right) = \sqrt{2}$

31. $\sin\dfrac{11\pi}{4} = \dfrac{\sqrt{2}}{2}; \cos\dfrac{11\pi}{4} = -\dfrac{\sqrt{2}}{2}; \tan\dfrac{11\pi}{4} = -1;$
$\cot\dfrac{11\pi}{4} = -1; \csc\dfrac{11\pi}{4} = \sqrt{2}; \sec\dfrac{11\pi}{4} = -\sqrt{2}$

33. $\sin\dfrac{13\pi}{6} = \dfrac{1}{2}; \cos\dfrac{13\pi}{6} = \dfrac{\sqrt{3}}{2}; \tan\dfrac{13\pi}{6} = \dfrac{\sqrt{3}}{3};$
$\cot\dfrac{13\pi}{6} = \sqrt{3}; \csc\dfrac{13\pi}{6} = 2; \sec\dfrac{13\pi}{6} = \dfrac{2\sqrt{3}}{3}$

35. $\sin\dfrac{13\pi}{3} = \dfrac{\sqrt{3}}{2}; \cos\dfrac{13\pi}{3} = \dfrac{1}{2}; \tan\dfrac{13\pi}{3} = \sqrt{3};$
$\cot\dfrac{13\pi}{3} = \dfrac{\sqrt{3}}{3}; \csc\dfrac{13\pi}{3} = \dfrac{2\sqrt{3}}{3}; \sec\dfrac{13\pi}{3} = 2$

37. $\sin\left(-\dfrac{10\pi}{3}\right) = \dfrac{\sqrt{3}}{2}; \cos\left(-\dfrac{10\pi}{3}\right) = -\dfrac{1}{2};$
$\tan\left(-\dfrac{10\pi}{3}\right) = -\sqrt{3}; \cot\left(-\dfrac{10\pi}{3}\right) = -\dfrac{\sqrt{3}}{3};$
$\csc\left(-\dfrac{10\pi}{3}\right) = \dfrac{2\sqrt{3}}{3}; \sec\left(-\dfrac{10\pi}{3}\right) = -2$

39. -0.9900 **41.** -0.1411 **43.** 0.2588 **45.** -2.1850
47. -3.0777 **49.** -0.5463 **51.** -3.2361 **53.** -1.0017
55. -1.6709 **57.** 14.1368 **59.** -1.0555

61. $\cos t = \dfrac{\sqrt{3}}{2}; \tan t = \dfrac{\sqrt{3}}{3}$

63. $\sin t = -\dfrac{\sqrt{2}}{2}; \cos t = -\dfrac{\sqrt{2}}{2}$

65. $\sin t = \dfrac{\sqrt{3}}{2}; \tan t = -\sqrt{3}$

67. $\cos t = 0.8; \tan t = -0.75$

69. $\sin t = \dfrac{12}{13}; \tan t = -\dfrac{12}{5}$

71. $\sin t = -\dfrac{2}{\sqrt{5}} = -\dfrac{2\sqrt{5}}{5}; \cos t = \dfrac{1}{\sqrt{5}} = \dfrac{\sqrt{5}}{5}$

73. $\sin\left(-\dfrac{2\pi}{3}\right) = -\dfrac{\sqrt{3}}{2}$ **75.** $\sec\left(-\dfrac{4\pi}{3}\right) = -2$

77. $\tan\left(-\dfrac{7\pi}{3}\right) = -\sqrt{3}$ **79.** $\sin\dfrac{\pi}{2} + \cos\pi = 0$

81. $3\sin\dfrac{\pi}{4} + 2\cos\dfrac{3\pi}{4} = \dfrac{\sqrt{2}}{2}$ **83.** $\sin\dfrac{\pi}{4}\cos\dfrac{\pi}{4} = \dfrac{1}{2}$

85. $\tan\dfrac{\pi}{4}\sec\dfrac{\pi}{4} = \sqrt{2}$ **87.** $\csc\dfrac{\pi}{2} - 4\cot\dfrac{\pi}{2} = 1$

89. $\tan\dfrac{\pi}{3} - \cos\dfrac{\pi}{6} = \dfrac{\sqrt{3}}{2}$ **91.** $\sin t = 0; \cos t = -1$

93. $\sin t = \dfrac{3}{5}; \cos t = -\dfrac{4}{5}$ **95.** $\sin t = -\dfrac{5}{13}; \cos t = -\dfrac{12}{13}$

97. $s = \dfrac{7\pi}{4}$ **99.** $s = \dfrac{2\pi}{3}$ **101.** $s = 0$ **103.** $s = \dfrac{\pi}{3}$

105. $1; 1$ **107.** $0; \dfrac{\sqrt{3}}{2}$

109. No; for example, when $t = \dfrac{\pi}{2}$,
$\sin\left(\dfrac{\pi}{2} + \pi\right) = \sin\dfrac{3\pi}{2} = -1$ and $\sin\dfrac{\pi}{2} + \sin\pi = 1$

111. $t = \dfrac{\pi}{4}, \dfrac{5\pi}{4}$ **113.** $\cos(t + \pi) = -\cos t$

115. $\sin^2 t + \cos^2 t = 1$
$\dfrac{\sin^2 t}{\sin^2 t} + \dfrac{\cos^2 t}{\sin^2 t} = \dfrac{1}{\sin^2 t}$
$1 + \cot^2 t = \csc^2 t$

5.3 Exercises
1. 5 **3.** $\sqrt{13}$ **5.** $3\sqrt{3}$

7. $\sin\theta = \dfrac{4}{5}; \cos\theta = \dfrac{3}{5}; \tan\theta = \dfrac{4}{3}; \csc\theta = \dfrac{5}{4}; \sec\theta = \dfrac{5}{3};$
$\cot\theta = \dfrac{3}{4}$

9. $\sin\theta = \dfrac{12}{37}; \cos\theta = \dfrac{35}{37}; \tan\theta = \dfrac{12}{35}; \csc\theta = \dfrac{37}{12};$
$\sec\theta = \dfrac{37}{35}; \cot\theta = \dfrac{35}{12}$

11. $\sin\theta = \dfrac{5\sqrt{41}}{41}; \cos\theta = \dfrac{4\sqrt{41}}{41}; \tan\theta = \dfrac{5}{4};$
$\sec\theta = \dfrac{\sqrt{41}}{4}; \csc\theta = \dfrac{\sqrt{41}}{5}; \cot\theta = \dfrac{4}{5}$

13. $\cos\theta = \dfrac{4\sqrt{17}}{17}; \sin\theta = \dfrac{\sqrt{17}}{17}; \tan\theta = \dfrac{1}{4};$
$\sec\theta = \dfrac{\sqrt{17}}{4}; \csc\theta = \sqrt{17}; \cot\theta = 4$

15. $\sin\theta = \dfrac{4}{5}; \tan\theta = \dfrac{4}{3}; \csc\theta = \dfrac{5}{4}; \sec\theta = \dfrac{5}{3}; \cot\theta = \dfrac{3}{4}$

17. $\sin\theta = \dfrac{3\sqrt{13}}{13}; \cos\theta = \dfrac{2\sqrt{13}}{13}; \csc\theta = \dfrac{\sqrt{13}}{3};$
$\sec\theta = \dfrac{\sqrt{13}}{2}; \cot\theta = \dfrac{2}{3}$

19. $\cos\theta = \dfrac{4}{5}; \tan\theta = \dfrac{3}{4}; \csc\theta = \dfrac{5}{3}; \sec\theta = \dfrac{5}{4};$
$\cot\theta = \dfrac{4}{3}$

21. $\cos\theta = \dfrac{3\sqrt{13}}{13}; \sin\theta = \dfrac{2\sqrt{13}}{13}; \tan\theta = \dfrac{2}{3};$
$\sec\theta = \dfrac{\sqrt{13}}{3}; \csc\theta = \dfrac{\sqrt{13}}{2}$

23. $\sin\theta = \dfrac{1}{2}; \cos\theta = \dfrac{\sqrt{3}}{2}; \tan\theta = \dfrac{\sqrt{3}}{3}; \cot\theta = \sqrt{3};$
$\sec\theta = \dfrac{2\sqrt{3}}{3}$

25. $\cos\theta = \dfrac{4\sqrt{17}}{17}; \sin\theta = \dfrac{\sqrt{17}}{17}; \tan\theta = \dfrac{1}{4};$
$\sec\theta = \dfrac{\sqrt{17}}{4}; \csc\theta = \sqrt{17}$

27. $\dfrac{\sqrt{3}}{2}$ **29.** $\dfrac{2\sqrt{3}}{3}$ **31.** $\sqrt{3}$ **33.** $\dfrac{2\sqrt{3}}{3}$ **35.** 0.6157

37. 2.1445 **39.** 0.1219 **41.** 1.0353 **43.** 0.4245
45. $\cos 55°$ **47.** $\cot 50°$ **49.** $\csc 43°$ **51.** $\tan 23°$
53. $2; 2\sqrt{2}; 45°$ **55.** $3.6397; 10.6418; 70°$
57. $3.1058; 11.5911; 75°$ **59.** $30°$ **61.** $45°$ **63.** $45°$
65. $\csc\alpha = \dfrac{1}{a}; \cos(90° - \alpha) = a$ **67.** 10 feet

69. 548.7404 feet   **71.** 88.5849 feet   **73.** 14.8349 feet

75. 51.0415 feet   **77.** $7.1250°$ **79.** 271.8683 m   **81.** $11.3099°$
83. 84 in   **85.** 0.7265 **87.** 277.1757 feet
89. 320.4294 feet   **91.** 55.0672 feet
93. $L \approx 109.8697$ feet; $H \approx 143.0187$ feet
95. a. 7.19 in, 10.27 in, 12.54 in   **b.** 36.93 in^2

97. $\csc(90° - \theta) = \dfrac{1}{\sin(90° - \theta)} = \dfrac{1}{\cos\theta} = \sec\theta$

99. Since sine and cosine are cofunctions, the values of them will switch for $[45°, 90°]$.

5.4 Exercises

1. 410°; −310°   **3.** 610°; −110°   **5.** $\sqrt{17}$

7. $\sin\theta = \dfrac{\sqrt{2}}{2}$; $\cos\theta = \dfrac{\sqrt{2}}{2}$; $\tan\theta = 1$; $\csc\theta = \sqrt{2}$; $\sec\theta = \sqrt{2}$; $\cot\theta = 1$

9. $\sin\theta = -\dfrac{\sqrt{3}}{2}$; $\cos\theta = -\dfrac{1}{2}$; $\tan\theta = \sqrt{3}$; $\csc\theta = -\dfrac{2\sqrt{3}}{3}$; $\sec\theta = -2$; $\cot\theta = \dfrac{\sqrt{3}}{3}$

11. $\sin\theta = \dfrac{2\sqrt{5}}{5}$; $\cos\theta = -\dfrac{\sqrt{5}}{5}$; $\tan\theta = -2$; $\csc\theta = \dfrac{\sqrt{5}}{2}$; $\sec\theta = -\sqrt{5}$; $\cot\theta = -\dfrac{1}{2}$

13. $\sin\theta = \dfrac{4\sqrt{17}}{17}$; $\cos\theta = \dfrac{\sqrt{17}}{17}$; $\tan\theta = 4$; $\csc\theta = \dfrac{\sqrt{17}}{4}$; $\sec\theta = \sqrt{17}$; $\cot\theta = \dfrac{1}{4}$

15. Quadrant IV   **17.** Quadrant III   **19.** Quadrant I

21. 60°   **23.** 60°   **25.** 75°   **27.** 40°   **29.** $\dfrac{\pi}{4}$ **31.** $\dfrac{\pi}{3}$

33. $\dfrac{1}{2}$ **35.** $-\dfrac{\sqrt{2}}{2}$ **37.** $-\dfrac{1}{2}$ **39.** $\dfrac{\sqrt{3}}{2}$ **41.** 1

43. $-\dfrac{\sqrt{3}}{3}$ **45.** $-\dfrac{2\sqrt{3}}{3}$ **47.** 0   **49.** $\sqrt{3}$

51. $-\dfrac{\sqrt{3}}{2}$ **53.** −1   **55.** $-\sqrt{2}$ **57.** Undefined

59. 1   **61.** $\dfrac{\sqrt{2}+1}{2}$ **63.** 0   **65.** $-\sqrt{3}-1$ **67.** $\dfrac{4\sqrt{3}}{3}$

69. −2   **71.** $\dfrac{\sqrt{3}-1}{2}$ **73.** 0   **75.** $\dfrac{\sqrt{6}+\sqrt{2}}{4}$ **77.** 0

79. $\sin\theta = -\dfrac{4}{5}$ **81.** $\sec\theta = -\dfrac{13}{12}$ **83.** $\tan\theta = 2\sqrt{2}$

85. $\cos\theta = \dfrac{2\sqrt{6}}{5}$ **87.** $\sec\theta = \dfrac{5}{3}$ **89.** 0.3907

91. 0.2521   **93.** −1.4281   **95.** 1.4826   **97.** −1.4242

99. $\dfrac{20\sqrt{3}}{3}$ feet   **101.** 50 miles

103. $\sec 270° = \dfrac{1}{\cos 270°} = \dfrac{1}{0}$, which is undefined.

105. Using (x, y), $\cos\theta = \dfrac{x}{\sqrt{x^2+y^2}}$. Using (kx, ky), $k > 0$,

$\cos\theta = \dfrac{kx}{\sqrt{(kx)^2+(ky)^2}} = \dfrac{kx}{\sqrt{k^2x^2+k^2y^2}}$

$= \dfrac{kx}{\sqrt{k^2}\cdot\sqrt{x^2+y^2}} = \dfrac{kx}{k\sqrt{x^2+y^2}}$, since $k > 0$.

So $\cos\theta = \dfrac{x}{\sqrt{x^2y^2}}$, as in previous answer.

5.5 Exercises

1. Upward   **3.** To the left   **5.** Vertical   **7.** Horizontal

9.

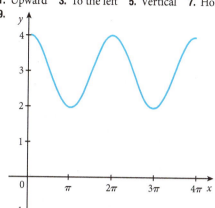

11.

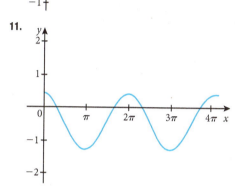

13.

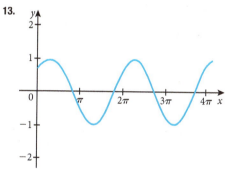

15.

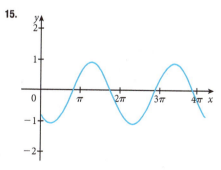

17.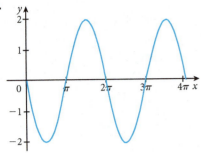

19.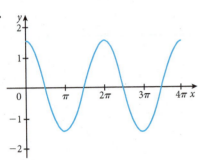

21.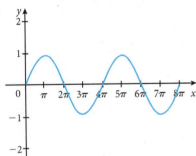

23.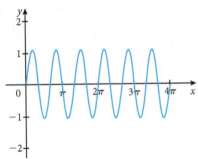

25.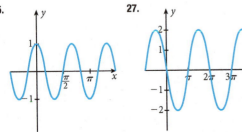

27.

29.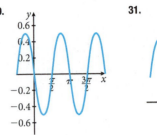

31.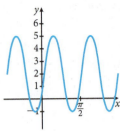

33.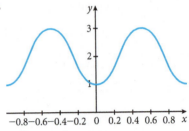

35.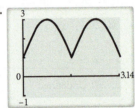

37. No.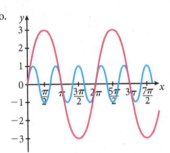

39. No.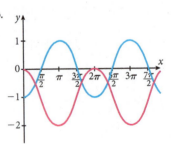

41. No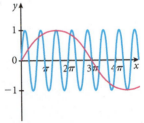

43. $\sin\left(x - \dfrac{\pi}{2}\right) + 1$ **45.** $\sin(x + \pi) - 1$ **47.** $3 \sin 2x$

49. $-2 \sin 2x$ **51.** $-\cos \dfrac{x}{2}$ **53.** $3 \sin \dfrac{x}{2}$

55. c   **57.** a   **59.** f   **61.** $\dfrac{7\sqrt{2}}{2}$ units; 7 units

63. \$135,000; toy sales in April; 12 months   **65.** $\dfrac{\pi}{6}$

67. 25; $\dfrac{\pi}{2}$; $\dfrac{2}{\pi}$ **69.** 280 miles   **71. a.** 23 days   **b.** 18.4%

73.

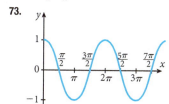

5.6 Exercises

1.

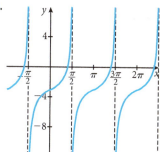

3.   **5.**

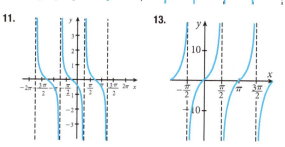

7.   **9.**

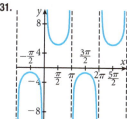

11.   **13.**

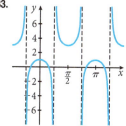

15.   **17.**

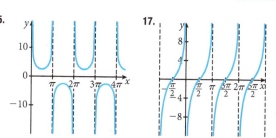

19.   **21.**

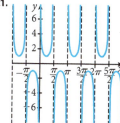

23.

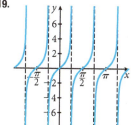

25.   **27.**

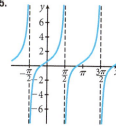

29.

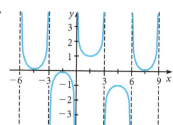

31.   **33.**

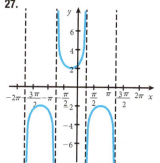

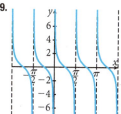

35. $\cot(2x)$　**37.** $\tan\left(\dfrac{x}{3}\right)$

39. $f(x) = \tan(x)$ and $f(x) = -\tan(x)$ have the same period and asymptotes but are reflections of each other across the x-axis.

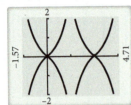

41. $f(x) = 2\sec(x)$ is a horizontal and vertical stretch of $f(x) = \sec(2x)$.

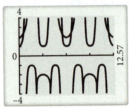

43. $f(x) = \cot(3\pi x)$ is a horizontal compression of $f(x) = \cot\left(\dfrac{\pi}{3}x\right)$

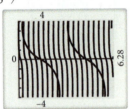

45. $d(x) = 5\cot(x)$; 7.1407 feet

47. $h(x) = 8\tan\left(x + \dfrac{\pi}{4}\right)$; 21.9798 feet

49. D: $\mathbb{R}$ except for $\dfrac{n\pi}{2}$, n is an integer; R: $(-\infty, 2]\cup[4, \infty)$

51. $x = \dfrac{n\pi}{2}$, n is an integer

5.7 Exercises

1. 1　**3.** 3　**5.** $\dfrac{\pi}{2}$　**7.** π　**9.** 0　**11.** $\dfrac{\pi}{3}$　**13.** $\dfrac{5\pi}{6}$　**15.** $-\dfrac{\pi}{3}$

17. -0.2527　**19.** 0.7227　**21.** 1.3734　**23.** 1.4455

25. 2.8240　**27.** -1.3734　**29.** $\dfrac{\pi}{3}$　**31.** 0.3　**33.** 4

35. $\dfrac{5}{13}$　**37.** $\dfrac{5}{13}$

39.

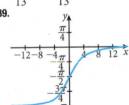

41.

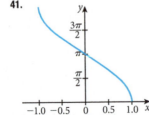

43.

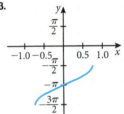

45. The domain of arccos x is $[-1, 1]$.

47. The domain of arcsin x is $[-1, 1]$.

49. $\dfrac{\pi}{4}$　**51.** $\dfrac{5\pi}{6}$　**53.** $-\dfrac{\pi}{6}$　**55.** $\dfrac{\pi}{6}$　**57.** 1.1593

59. -0.2815　**61.** 0.3948　**63.** -0.3029　**65.** 75.2564°

67. 48.5904°; minimum to stay within the necessary ratio

69. 38.6598°　**71.** $t = \dfrac{1}{2}\sin^{-1}\left(\dfrac{d}{5}\right)$; $[-5, 5]$

73. 30.2940°　**75.** -0.20　**77.** -0.93　**79.** 1.3　**81.** 1.4

Chapter 5 Review Exercises

1.

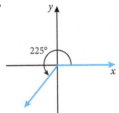

3.

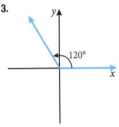

5.

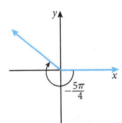

7. 55°; 145°　**9.** $\dfrac{4\pi}{3}$　**11.** $-\dfrac{5\pi}{6}$　**13.** 150°　**15.** 2°

17. $\dfrac{4\pi}{3}$; 240°　**19.** $\dfrac{\pi}{3}$　**21.** $\dfrac{\pi}{3}$

23. $\sin\left(-\dfrac{5\pi}{2}\right) = -1$; $\cos\left(-\dfrac{5\pi}{2}\right) = 0$; $\tan\left(-\dfrac{5\pi}{2}\right) =$ undefined

25. $\sin\dfrac{7\pi}{3} = \dfrac{\sqrt{3}}{2}$; $\cos\dfrac{7\pi}{3} = \dfrac{1}{2}$; $\tan\dfrac{7\pi}{3} = \sqrt{3}$

27. $\sin t = \dfrac{1}{2}$; $\tan t = -\dfrac{\sqrt{3}}{3}$

29. $\sin t = -\dfrac{\sqrt{2}}{2}$; $\cos t = \dfrac{\sqrt{2}}{2}$　**31.** -1

33. 10 units; 6 seconds

35. $\cos\theta = \dfrac{\sqrt{91}}{10}$; $\tan\theta = \dfrac{3\sqrt{91}}{91}$; $\csc\theta = \dfrac{10}{3}$; $\sec\theta = \dfrac{10\sqrt{91}}{91}$; $\cot\theta = \dfrac{\sqrt{91}}{3}$

37. $\sin\theta = \dfrac{10\sqrt{149}}{149}$; $\cos\theta = \dfrac{7\sqrt{149}}{149}$; $\tan\theta = \dfrac{10}{7}$;

$\csc\theta = \dfrac{\sqrt{149}}{10}$; $\sec\theta = \dfrac{\sqrt{149}}{7}$

39. 18.8073; 21.3005   **41.** About 81 ft.   **43.** About 11 ft.

45. $-\dfrac{12}{13}$ **47.** -1 **49.** $45°$; $\dfrac{\sqrt{2}}{2}$ **51.** $60°$; -2

53. 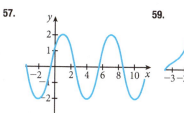   **55.**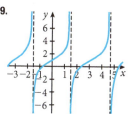

57. 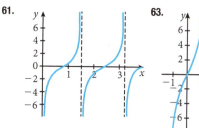   **59.**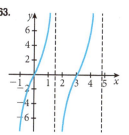

61. 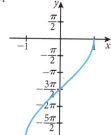   **63.**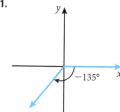

65. $-\dfrac{\pi}{2}$ **67.** $\dfrac{\pi}{2}$ **69.** $\dfrac{\pi}{4}$ **71.** 7   **73.** 0.4   **75.** $\dfrac{5}{13}$

77. 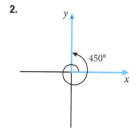   **79.** 9.4623°

Chapter 5 Test

1. 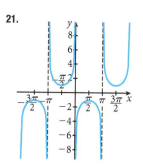   **2.** (image)

3. 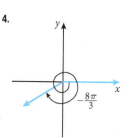   **4.** (image)

5. $\dfrac{5\pi}{12}$ **6.** $-\dfrac{7\pi}{4}$ **7.** $10°$ **8.** $-6°$

9. $\dfrac{\pi}{6}$; $\sin\left(-\dfrac{5\pi}{6}\right) = -\dfrac{1}{2}$; $\cos\left(-\dfrac{5\pi}{6}\right) = -\dfrac{\sqrt{3}}{2}$

10. $\dfrac{\pi}{3}$; $\sin\left(\dfrac{8\pi}{3}\right) = \dfrac{\sqrt{3}}{2}$; $\cos\left(\dfrac{8\pi}{3}\right) = -\dfrac{1}{2}$

11. $\sin t = -\dfrac{\sqrt{2}}{2}$; $\tan t = 1$

12. $\cos\theta = \dfrac{\sqrt{21}}{5}$, $\tan\theta = \dfrac{2\sqrt{21}}{21}$, $\cot\theta = \dfrac{\sqrt{21}}{2}$,

$\sec\theta = \dfrac{5\sqrt{21}}{21}$, $\csc\theta = \dfrac{5}{2}$

13. $\sin\theta = \dfrac{3\sqrt{10}}{10}$, $\cos\theta = \dfrac{\sqrt{10}}{10}$, $\cot\theta = \dfrac{1}{3}$, $\sec\theta = \sqrt{10}$,

$\csc\theta = \dfrac{\sqrt{10}}{3}$

14. $a = \dfrac{3\sqrt{3}}{2}$; $b = \dfrac{3}{2}$

15. $\cos\theta = -\dfrac{\sqrt{10}}{10}$; $\csc\theta = \dfrac{\sqrt{10}}{3}$

16. $60°$; $\sin 240° = -\dfrac{\sqrt{3}}{2}$; $\sec 240° = -2$

17. (image)   **18.** (image)

19. (image)   **20.**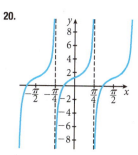

21. (image)   **22.**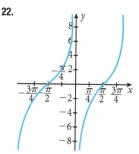

23. $\dfrac{\pi}{2}$ **24.** 8   **25.** $\dfrac{3}{5}$

26.

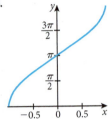

27. 180 degrees per minute; $100\pi \approx 314$ ft/min

28. 19.5°   **29.** 14.1 in. × 14.1 in.   **30. a.** 33 days   **b.** 95.5%

Chapter 6

6.1 Exercises

1. $(2x + 3)(2x - 3)$ **3.** $9(x + 3)(x - 3)$ **5.** $\dfrac{1 - 3x}{1 - x^2}$

7. $\dfrac{2}{x - 1}$ **9.** $\cos \pi = -1 \neq 0.5$

11. $\sin\left(\dfrac{\pi}{2} + \pi\right) = -1 \neq 1 = \sin\left(\dfrac{\pi}{2}\right) + \sin(\pi)$

13. $\sin\left(2\left(\dfrac{\pi}{2}\right)\right) = 0 \neq 2 = 2\sin\left(\dfrac{\pi}{2}\right)$

15. $\dfrac{\cos x}{\sin^2 x}$ **17.** $\dfrac{\sin x}{\cos^2 x}$ **19.** $\dfrac{1 - \cos^2 x}{\cos^2 x}$ or $\dfrac{\sin^2 x}{\cos^2 x}$

21. $\sin x\,(1 + \cos x)$ **23.** $(1 + \sin x)(1 - \sin x)$

25. $(\sin x + \cos x)(\sin x - \cos x)$

27. $\tan x \csc x = \dfrac{\sin x}{\cos x}\dfrac{1}{\sin x} = \dfrac{1}{\cos x} = \sec x$

29. $\sin x + \cos x \cot x = \sin x + \cos x \dfrac{\cos x}{\sin x}$
$= \dfrac{\sin^2 x + \cos^2 x}{\sin x} = \dfrac{1}{\sin x} = \csc x$

31. $\sec^2 x(1 - \sin^2 x) = \dfrac{1}{\cos^2 x}(\cos^2 x) = 1$

33. $\sin^2(-x) + \cos^2(-x) = (-\sin x)^2 + \cos^2 x$
$= \sin^2 x + \cos^2 x = 1$

35. $(\sec^2 x - 1)\cot^2 x = (\tan^2 x)\cot^2 x$
$= \dfrac{\sin^2 x}{\cos^2 x}\dfrac{\cos^2 x}{\sin^2 x} = 1$

37. $\dfrac{\cot x}{\csc x} = \dfrac{\dfrac{\cos x}{\sin x}}{\dfrac{1}{\sin x}} = \cos x$

39. $\sin^2 x \sec x = (1 - \cos^2 x)\sec x = \sec x - \cos x$

41. $(\cos x + \sin x)^2 - 2\sin x \cos x$
$= \cos^2 x + 2\sin x \cos x + \sin^2 x - 2\sin x \cos x$
$= \cos^2 x + \sin^2 x = 1$

43. $\sec x + \tan x = \dfrac{1}{\cos x} + \dfrac{\sin x}{\cos x} = \dfrac{1 + \sin x}{\cos x}$

45. $\cos^3 x = \cos x\,(1 - \sin^2 x) = \cos x - \cos x \sin^2 x$

47. $\sec x \cos^3 x = \dfrac{1}{\cos x}\cos^3 x = \cos^2 x = 1 - \sin^2 x$

49. $\dfrac{1}{1 + \cos x} + \dfrac{1}{1 - \cos x} = \dfrac{1 - \cos x + 1 + \cos x}{1 - \cos^2 x}$
$= \dfrac{2}{\sin^2 x} = 2\csc^2 x$

51. $\dfrac{\sin x - \sin(-x)}{1 - \cos^2 x} = \dfrac{2\sin x}{\sin^2 x} = \dfrac{2}{\sin x} = 2\csc x$

53. $\dfrac{\sec^2 x}{1 + \sin x} = \dfrac{\sec^2 x}{1 + \sin x}\dfrac{1 - \sin x}{1 - \sin x}$
$= \dfrac{\sec^2 x - \sec x\dfrac{\sin x}{\cos x}}{1 - \sin^2 x}$
$= \dfrac{\sec^2 x - \sec x \tan x}{\cos^2 x}$

55. $\cos^2 x - \sin^2 x = 1 - \sin^2 x - \sin^2 x = 1 - 2\sin^2 x$

57. $\tan x + \cot x = \dfrac{\sin x}{\cos x} + \dfrac{\cos x}{\sin x}$
$= \dfrac{\sin^2 x}{\cos x \sin x} + \dfrac{\cos^2 x}{\cos x \sin x}$
$= \dfrac{1}{\cos x \sin x} = \sec x \csc x$

59. $\cos^4 x - \sin^4 x = (\cos^2 x + \sin^2 x)(\cos^2 x - \sin^2 x)$
$= \cos^2 x - \sin^2 x$

61. $\dfrac{\tan^2 x - 1}{1 + \tan^2 x} = \dfrac{\tan^2 x - 1}{\sec^2 x} = \cos^2 x\left(\dfrac{\sin^2 x}{\cos^2 x} - 1\right)$
$= \sin^2 x - \cos^2 x = \sin^2 x - (1 - \sin^2 x)$
$= 2\sin^2 x - 1$

63. $\csc^2 x + \sec^2 x = \dfrac{1}{\sin^2 x} + \dfrac{1}{\cos^2 x} = \dfrac{\cos^2 x + \sin^2 x}{\sin^2 x \cos^2 x}$
$= \dfrac{1}{\sin^2 x \cos^2 x} = \csc^2 x \sec^2 x$

65. $\sec^4 x - \tan^4 x = (\sec^2 x - \tan^2 x)(\sec^2 x + \tan^2 x)$
$= \sec^2 x + \tan^2 x$

67. $\dfrac{1}{1 + \cos x} = \dfrac{1}{1 + \cos x}\dfrac{1 - \cos x}{1 - \cos x} = \dfrac{1 - \cos x}{1 - \cos^2 x}$
$= \dfrac{1 - \cos x}{\sin^2 x} = \dfrac{1}{\sin^2 x} - \dfrac{\cos x}{\sin^2 x}$
$= \csc^2 x - \cot x \csc x$

69. $(\csc x - \cot x)^2 = \left(\dfrac{1}{\sin x} - \dfrac{\cos x}{\sin x}\right)^2 = \dfrac{(1 - \cos x)^2}{1 - \cos^2 x}$
$= \dfrac{(1 - \cos x)(1 - \cos x)}{(1 + \cos x)(1 - \cos x)} = \dfrac{1 - \cos x}{1 + \cos x}$

71. $\cot x + \tan x = \dfrac{\cos x}{\sin x} + \dfrac{\sin x}{\cos x}$
$= \dfrac{\cos^2 x}{\sin x \cos x} + \dfrac{\sin^2 x}{\sin x \cos x}$
$= \dfrac{\cos^2 x + \sin^2 x}{\sin x \cos x} = \dfrac{1}{\sin x \cos x}$
$= \dfrac{\cos x}{\sin x \cos^2 x} = \sec^2 x \cot x$

73. $\dfrac{\sec^2 x - 1}{\tan x} = \dfrac{\tan^2 x}{\tan x} = \dfrac{\sin x}{\cos x} = \dfrac{\sec x}{\csc x}$

75. $\dfrac{\sin x}{\csc x - \cot x} = \dfrac{\sin x}{\dfrac{1}{\sin x} - \dfrac{\cos x}{\sin x}} = \dfrac{\sin x}{\dfrac{1 - \cos x}{\sin x}}$
$= \dfrac{\sin^2 x}{1 - \cos x} = \dfrac{\sin^2 x\,(1 + \cos x)}{(1 - \cos x)(1 + \cos x)}$
$= \dfrac{(1 - \cos^2 x)(1 + \cos x)}{1 - \cos^2 x} = 1 + \cos x$

77. $a \csc^2 x\,(1 + \cos x)(1 - \cos x) = a \csc^2 x\,(1 - \cos^2 x)$
$= a \csc^2 x \sin^2 x = a$

79. $\ln|\tan x| = \ln\left|\dfrac{1}{\cot x}\right| = \ln 1 - \ln|\cot x|$
$= -\ln|\cot x|$

81. No

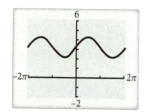 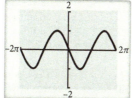

$y = \sin x + \pi$ $y = \sin(x + \pi)$

83. Yes

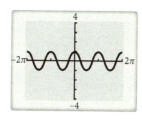

85. No

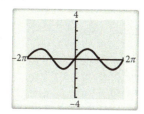 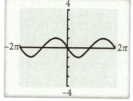

$y = \sin x$ $y = \sin(x - \pi)$

87. $\sin x$ can take on negative values, whereas the right-hand expression cannot.

89. No, $\cos\frac{\pi}{2} = 0$, so $\tan\frac{\pi}{2}$ is undefined.

6.2 Exercises

1. $\frac{1}{2}$ **3.** $-\frac{\sqrt{2}}{2}$ **5.** $\frac{\sqrt{3}}{2}$ **7. a.** $\frac{1+\sqrt{3}}{2}$ **b.** 0

9. 0 **11.** $-\frac{3}{4}$ **13.** $\frac{\sqrt{2}}{4}$ **15.** $-\frac{\sqrt{6}}{4}$ **17.** $-\frac{1}{2}$

19. $\frac{\sqrt{2}-\sqrt{6}}{4}$

21. $\sin\frac{11\pi}{12} = \frac{\sqrt{6}-\sqrt{2}}{4}$; $\cos\frac{11\pi}{12} = \frac{-\sqrt{2}-\sqrt{6}}{4}$;

$\tan\frac{11\pi}{12} = \frac{-\sqrt{3}+1}{1+\sqrt{3}}$, or $\sqrt{3}-2$

23. $\sin\left(-\frac{\pi}{12}\right) = \frac{\sqrt{2}-\sqrt{6}}{4}$; $\cos\left(-\frac{\pi}{12}\right) = \frac{\sqrt{2}+\sqrt{6}}{4}$;

$\tan\left(-\frac{\pi}{12}\right) = \frac{1-\sqrt{3}}{1+\sqrt{3}}$, or $\sqrt{3}-2$

25. $\sin\frac{3\pi}{4} = \frac{\sqrt{2}}{2}$; $\cos\frac{3\pi}{4} = -\frac{\sqrt{2}}{2}$; $\tan\frac{3\pi}{4} = -1$

27. $\sin\left(-\frac{13\pi}{12}\right) = \frac{-\sqrt{2}+\sqrt{6}}{4}$;

$\cos\left(-\frac{13\pi}{12}\right) = \frac{-\sqrt{2}-\sqrt{6}}{4}$;

$\tan\left(-\frac{13\pi}{12}\right) = \frac{1-\sqrt{3}}{1+\sqrt{3}}$, or $\sqrt{3}-2$

29. $\sin\left(-\frac{7\pi}{12}\right) = \frac{-\sqrt{2}-\sqrt{6}}{4}$;

$\cos\left(-\frac{7\pi}{12}\right) = \frac{\sqrt{2}-\sqrt{6}}{4}$;

$\tan\left(-\frac{7\pi}{12}\right) = \frac{-1-\sqrt{3}}{1-\sqrt{3}}$, or $\sqrt{3}+2$

31. $\sin 240° = \frac{-\sqrt{3}}{2}$; $\cos 240° = -\frac{1}{2}$; $\tan 240° = \sqrt{3}$

33. $\sin 75° = \frac{\sqrt{2}+\sqrt{6}}{4}$; $\cos 75° = \frac{\sqrt{6}-\sqrt{2}}{4}$;

$\tan 75° = \frac{\sqrt{3}+3}{3-\sqrt{3}}$, or $\sqrt{3}+2$

35. $\sin 105° = \frac{\sqrt{2}+\sqrt{6}}{4}$; $\cos 105° = \frac{\sqrt{2}-\sqrt{6}}{4}$;

$\tan 105° = \frac{1+\sqrt{3}}{1-\sqrt{3}}$, or $-2-\sqrt{3}$

37. $\sin(-195°) = \frac{\sqrt{6}-\sqrt{2}}{4}$;

$\cos(-195°) = \frac{-\sqrt{2}-\sqrt{6}}{4}$;

$\tan(-195°) = \frac{1-\sqrt{3}}{1+\sqrt{3}}$, or $\sqrt{3}-2$

39. $\cos\left(\frac{\pi}{2}-a\right) = \cos\frac{\pi}{2}\cos a + \sin\frac{\pi}{2}\sin a$

$= 0 + \sin a = \sin a$

41. $\sec\left(\frac{\pi}{2}-a\right) = \dfrac{1}{\cos\left(\frac{\pi}{2}-a\right)}$

$= \dfrac{1}{\cos\frac{\pi}{2}\cos a + \sin\frac{\pi}{2}\sin a}$

$= \dfrac{1}{\sin a} = \csc a$

43. $-\frac{16}{35}$ **45.** $\frac{63}{16}$ **47.** $\frac{3}{5}$ **49.** $\frac{2\sqrt{2}+\sqrt{15}}{12}$

51. $\frac{32\sqrt{2}-9\sqrt{15}}{7}$ **53.** $\frac{-2\sqrt{2}}{3}$

55. $f(x) = \sin\left(x+\frac{\pi}{4}\right)$ **57.** $f(x) = \sqrt{2}\sin\left(x+\frac{3\pi}{4}\right)$

59. $f(x) = 2\sin\left(x+\frac{5\pi}{3}\right)$ **61.** $-\sin x$

63. $-\sin x$ **65.** $-\cot x$

67. $\cos(x+x) = \cos x\cos x - \sin x\sin x = \cos^2 x - \sin^2 x$

69. $\tan(\pi-x) = \dfrac{\tan\pi - \tan x}{1+\tan\pi\tan x} = -\tan x$

71. $\tan\left(x+\frac{\pi}{4}\right) = \dfrac{\tan x + \tan\frac{\pi}{4}}{1-\tan x\tan\frac{\pi}{4}} = \dfrac{\tan x + 1}{1-\tan x}$

73. $\cos\left(x-\frac{\pi}{2}\right) = \cos x\cos\frac{\pi}{2} + \sin x\sin\frac{\pi}{2}$

$= \cos\frac{\pi}{2}\cos x + \sin\frac{\pi}{2}\sin x$

$= \cos\left(\frac{\pi}{2}-x\right)$

75. $\sin\left(\frac{\pi}{3}-x\right) = \sin\frac{\pi}{3}\cos x - \cos\frac{\pi}{3}\sin x$

$= \cos x\sin\frac{\pi}{3} - \sin x\cos\frac{\pi}{3}$

$= -\sin x\cos\frac{\pi}{3} + \cos x\sin\frac{\pi}{3}$

$= -\sin\left(x-\frac{\pi}{3}\right)$

77. $\cos(x+y) + \cos(x-y)$
$= \cos x \cos y - \sin x \sin y + \cos x \cos y + \sin x \sin y$
$= 2 \cos x \cos y$

79. $\sin(x+y) \sin(x-y)$
$= (\sin x \cos y + \cos x \sin y) \cdot (\sin x \cos y - \cos x \sin y)$
$= \sin^2 x \cos^2 y - \cos^2 x \sin^2 y$

81. $\dfrac{\sqrt{3}}{2}$ **83.** $-\dfrac{3}{5}$ **85.** 7 **87.** $c = \dfrac{5\pi}{6}$

89. $A = \dfrac{\sqrt{2}}{2}, B = 300\pi, C = \dfrac{\sqrt{2}}{2}, D = 300\pi$

91. $\dfrac{\cos 2x(\cos 2h - 1) - \sin 2x \sin 2h}{h}$

93. $A = 10\sqrt{2}; c = \dfrac{\pi}{4}$

95. $\sin(2x) = \sin(x+x) = \sin x \cos x + \sin x \cos x$
$= 2 \sin x \cos x$

97. $2 \cos a \sin b$

99. $\sin(a + (-b)) = \sin a \cos(-b) + \cos a \sin(-b)$
$= \sin a \cos b - \cos a \sin b$

6.3 Exercises

1. $\dfrac{2\sqrt{2}}{3}$ **3.** $\dfrac{4\sqrt{2}}{9}$

5. $\dfrac{4\sqrt{2}}{7}$ **7.** $\dfrac{9}{7}$

9. $\sin 2x = \dfrac{24}{25}; \cos 2x = -\dfrac{7}{25}; \tan 2x = -\dfrac{24}{7}$

11. $\sin 2x = \dfrac{\sqrt{3}}{2}; \cos 2x = \dfrac{1}{2}; \tan 2x = \sqrt{3}$

13. $\sin 2x = \dfrac{4\sqrt{2}}{9}; \cos 2x = -\dfrac{7}{9}; \tan 2x = -\dfrac{4\sqrt{2}}{7}$

15. $\sin 2x = \dfrac{5\sqrt{11}}{18}; \cos 2x = \dfrac{7}{18}; \tan 2x = \dfrac{5\sqrt{11}}{7}$

17. $\dfrac{\cos x + \cos x \cos 2x}{2}$ **19.** $\dfrac{\sin x + \sin x \cos 2x}{2}$

21. $\dfrac{\cos x - \cos x \cos 4x}{8}$ **23.** $\dfrac{1 - \cos 2x}{2}$

25. $\dfrac{\sqrt{2 + \sqrt{3}}}{2}$ **27.** $-\dfrac{\sqrt{2 + \sqrt{3}}}{2}$ **29.** $-\dfrac{\sqrt{2 + \sqrt{2}}}{2}$

31. 1 **33.** $\dfrac{\sqrt{10}}{10}$ **35.** $\dfrac{1}{3}$ **37.** 3 **39.** $\dfrac{2\sqrt{5}}{5}$

41. -2 **43.** $-\dfrac{1}{2}$ **45.** $\dfrac{1}{2}(\sin 7x + \sin x)$

47. $\dfrac{1}{2}(\cos 4x - \cos 8x)$ **49.** $\dfrac{1}{2}(\cos 2x - \cos 4x)$

51. $-2 \cos \dfrac{5x}{2} \sin \dfrac{x}{2}$ **53.** $2 \sin 4x \sin x$

55. $-2 \sin \dfrac{7x}{2} \sin \dfrac{3x}{2}$

57. $\sec^2 x = \dfrac{1}{\cos^2 x} = \dfrac{1}{\dfrac{1 + \cos 2x}{2}} = \dfrac{2}{1 + \cos 2x}$

59. $2 \cos^2 2x = 2\left(\dfrac{1 + \cos 2(2x)}{2}\right) = 1 + \cos 4x$

61. $\sec^2 2x = \dfrac{1}{\cos^2 2x} = \dfrac{1}{\dfrac{(1 + \cos 4x)}{2}} = \dfrac{2}{1 + \cos 4x}$

63. $\cos 6x = \cos 2(3x) = 1 - 2 \sin^2(3x) = 1 - 2 \sin^2 3x$
65. $\sin 6x = \sin 2(3x) = 2 \sin 3x \cos 3x$
67. $\sin(2x + \pi) = \sin 2x \cos \pi + \cos 2x \sin \pi$
$= -\sin 2x = -2 \sin x \cos x$

69. $\cos 4x = 1 - 2 \sin^2 2x = 1 - 2(\sin 2x)^2$
$= 1 - 2(2 \sin x \cos x)^2$
$= 1 - 8 \sin^2 x \cos^2 x = 1 - 8 \sin^2 x (1 - \sin^2 x)$
$= 1 - 8 \sin^2 x + 8 \sin^4 x$

71. $\sin 3x = \sin(2x + x) = \sin 2x \cos x + \cos 2x \sin x$
$= 2 \sin x \cos^2 x + (2 \cos^2 x - 1) \sin x$
$= \sin x (4 \cos^2 x - 1)$

73. $\cos 3x + \cos x = \cos(2x + x) + \cos x$
$= \cos 2x \cos x - \sin 2x \sin x + \cos x$
$= \cos 2x \cos x - 2 \sin^2 x \cos x + \cos x$
$= \cos 2x \cos x + \cos x (1 - 2 \sin^2 x)$
$= 2 \cos x \cos 2x = 2 \cos x (2 \cos^2 x - 1)$
$= 4 \cos^3 x - 2 \cos x$

75. $\tan\left(-\dfrac{x}{2}\right) = -\tan \dfrac{x}{2} = -\left(\dfrac{1 - \cos x}{\sin x}\right)$
$= \cot x - \csc x$

77. $-\dfrac{7}{25}$ **79.** $\dfrac{\sqrt{3}}{2}$ **81.** $\dfrac{1}{10}$

83. $\dfrac{1}{3}$ **85.** 93.4471 meters

87. a. $b = 2s \sin\left(\dfrac{\alpha}{2}\right)$ **b.** $h = s \cos\left(\dfrac{\alpha}{2}\right)$

c. $A = s^2 \sin\left(\dfrac{\alpha}{2}\right) \cos\left(\dfrac{\alpha}{2}\right)$

d. $A = \dfrac{1}{2} s^2 \sin(\alpha)$ **e.** $\dfrac{72}{5}$

89. $A = -1$
91. $\cos(a+b) = \cos a \cos b - \sin a \sin b$
$\cos(a-b) = \cos a \cos b + \sin a \sin b$
$\cos(a+b) + \cos(a-b) = 2 \cos a \cos b$
$\dfrac{\cos(a+b) + \cos(a-b)}{2} = \cos a \cos b$

6.4 Exercises

1. $\sin^2 x$ **3.** $\tan^2 x$ **5.** $x = 4, -1$

7. a. $\sin x \cdot \dfrac{\cos x}{\sin x} = \cos x$ **b.** 0

9. $x = \dfrac{7\pi}{6} + 2n\pi, \dfrac{11\pi}{6} + 2n\pi$, where n is an integer

11. $x = \dfrac{\pi}{4} + n\pi$, where n is an integer

13. $x = \dfrac{\pi}{6} + 2n\pi, \dfrac{5\pi}{6} + 2n\pi$, where n is an integer

15. $x = \dfrac{\pi}{2} + 2n\pi, \dfrac{3\pi}{2} + 2n\pi$, where n is an integer
or $x = \dfrac{\pi}{2} + n\pi$

17. $x = \dfrac{\pi}{12} + n\pi, \dfrac{5\pi}{12} + n\pi$, where n is an integer

19. $x = \dfrac{\pi}{4} + \dfrac{n\pi}{2}$, where n is an integer

21. $x = \dfrac{3\pi}{8} + \dfrac{n\pi}{2}$, where n is an integer

23. $x \approx 2.4189 + 2n\pi, 3.8643 + 2n\pi$, where n is an integer
25. $x \approx 1.2490 + n\pi$, where n is an integer
27. $x \approx 3.7851 + 2n\pi, 5.6397 + 2n\pi$, where n is an integer
29. $x \approx 0.7297 + 2n\pi, 2.4119 + 2n\pi$, where n is an integer
31. $x = 0, \dfrac{\pi}{2}, \pi$ **33.** $x = 0, \dfrac{\pi}{2}, \pi, \dfrac{3\pi}{2}$

35. $x = \dfrac{\pi}{2}, \pi, \dfrac{3\pi}{2}$ **37.** $x = 0, \pi$ **39.** $x = \dfrac{\pi}{4}, \dfrac{5\pi}{4}$

41. $x = \dfrac{\pi}{2}, \dfrac{3\pi}{2}$ **43.** $x = \dfrac{\pi}{2}, \dfrac{3\pi}{2}$ **45.** $x = 0, \pi$

47. $x = \dfrac{\pi}{2}, \dfrac{3\pi}{2}$ **49.** $x = \dfrac{\pi}{3}, \dfrac{5\pi}{3}$

51. $x = \dfrac{\pi}{4}, \dfrac{3\pi}{4}, \dfrac{5\pi}{4}, \dfrac{7\pi}{4}$ **53.** $x = \pi$ **55.** $x = \dfrac{\pi}{2}, \dfrac{3\pi}{2}$

57. $x = \dfrac{\pi}{2}, \dfrac{3\pi}{2}$ **59.** $x = 0, \dfrac{\pi}{3}, \dfrac{5\pi}{3}$ **61.** $x = \dfrac{\pi}{4}, \dfrac{5\pi}{4}$

63. $x = \dfrac{\pi}{3}, \pi, \dfrac{5\pi}{3}$ **65.** $x = \dfrac{3\pi}{4}, \dfrac{7\pi}{4}, \dfrac{\pi}{6}, \dfrac{7\pi}{6}$

67. $x \approx 0.4636, 2.6779, 3.6052, 5.8195$

69. $x \approx 3.4814, 5.9433$ **71.** $x \approx 0.8411, \pi, 5.4421$

73. $x \approx \dfrac{\pi}{2}, 3.3943, \dfrac{3\pi}{2}, 6.0305$

75. $x \approx 0, 1.1071, \pi, 4.2487$

77. $x \approx 0.2014, 2.9402, 3.3943, 6.0305$

79. $x \approx 0.9828, 2.6779, 4.1244, 5.8195$

81. $x \approx 0.6155, 2.5261, 3.7571, 5.6677$

83. No solution   **85.** $x \approx 1.2744, 4.5611$

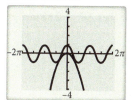 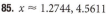

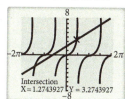

87. $x \approx 0.3927, 1.9635, 3.5343, 5.1051$

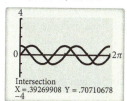

89. $x \approx 0.6662, 2.4754$

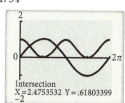

91. $\theta \approx 0.3218$ or about $18.43°$ **93.** Day 7676

95. Approximately 9 A.M. to 1 P.M.

97. a. $\sin\theta = \dfrac{h}{6}, \cos\theta = \dfrac{b}{6}, A = \dfrac{1}{2}bh$

$\qquad = \dfrac{1}{2}\,36\sin\theta\cos\theta = 18\sin\theta\cos\theta$

b. $\theta = 31.3670°$ **c.** $\theta = 45°$

99. a. $x \approx 109, 250$ **b.** 180 days

101. No, there is no value of x for which $\sin x$ and $\cos x$ are both equal to 1. This is a misstatement of the rule that if $ab = 0$, then $a = 0$, $b = 0$, or $a = b = 0$.

103. $x = 0.7391$

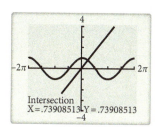

105. $\dfrac{3\pi}{2} + 2n\pi$, n an integer

Chapter 6 Review Exercises

1. $\dfrac{1}{\cos x}$ **3.** $\sin^2 x$

5. $1 + \cos x = (1 + \cos x)\dfrac{1 - \cos x}{1 - \cos x}$

$\qquad = \dfrac{1 - \cos^2 x}{1 - \cos x} = \dfrac{\sin^2 x}{1 - \cos x}$

7. $\sin^4 x = (\sin^2 x)^2 = (1 - \cos^2 x)^2 = 1 - 2\cos^2 x + \cos^4 x$

9. $\cos^4 x - \sin^4 x = (\cos^2 x - \sin^2 x)(\cos^2 x + \sin^2 x)$

$\qquad = (\cos^2 x - \sin^2 x) = 1 - \sin^2 x - \sin^2 x = 1 - 2\sin^2 x$

11. $\dfrac{1 + 2\cos x + \cos^2 x}{1 - \cos^2 x} = \dfrac{(1 + \cos x)^2}{(1 - \cos x)(1 + \cos x)}$

$\qquad = \dfrac{1 + \cos x}{1 - \cos x}$

13. $\dfrac{\tan^2 x}{\sec x + 1} = \dfrac{\sec^2 x - 1}{\sec x + 1} = \dfrac{(\sec x - 1)(\sec x + 1)}{\sec x + 1}$

$\qquad = \sec x - 1$

15. $\sin 75°$ **17.** $\cos\dfrac{\pi}{12}$

19. $\sin(-105°) = \dfrac{-\sqrt{2} - \sqrt{6}}{4};$

$\qquad \cos(-105°) = \dfrac{\sqrt{2} - \sqrt{6}}{4};$

$\qquad \tan(-105°) = \dfrac{-1 - \sqrt{3}}{1 - \sqrt{3}}$, or $\sqrt{3} + 2$

21. $\sin\left(-\dfrac{\pi}{12}\right) = \dfrac{\sqrt{2} - \sqrt{6}}{4}$; $\cos\left(-\dfrac{\pi}{12}\right) = \dfrac{\sqrt{2} + \sqrt{6}}{4}$;

$\qquad \tan\left(-\dfrac{\pi}{12}\right) = \dfrac{1 - \sqrt{3}}{1 + \sqrt{3}}$, or $\sqrt{3} - 2$

23. $\dfrac{16}{65}$ **25.** $\dfrac{63}{16}$ **27.** $\dfrac{33}{65}$ **29.** $f(x) = \sin\left(2x + \dfrac{11\pi}{6}\right)$

31. $\dfrac{4}{5}$ **33.** $\dfrac{24}{25}$ **35.** $\dfrac{1 - \cos 4x}{8}$ **37.** $\dfrac{\sqrt{2}}{2}$

39. $\dfrac{\sqrt{2 + \sqrt{3}}}{2}$ **41.** $\dfrac{1}{2}(\sin 6x + \sin 2x)$

43. $2\cos\left(\dfrac{5x}{2}\right)\cos\left(\dfrac{x}{2}\right)$

45. $x = \dfrac{\pi}{3} + 2n\pi$, $x = \dfrac{5\pi}{3} + 2n\pi$, where n is an integer

47. $x = \dfrac{2\pi}{3} + 2n\pi$, $x = \dfrac{4\pi}{3} + 2n\pi$, where n is an integer

49. $x \approx 2.3005 + 2n\pi$, $x \approx 3.9827 + 2n\pi$, where n is an integer

51. $x \approx 1.1593 + 2n\pi$, $x \approx 5.1239 + 2n\pi$, where n is an integer

53. $x = \dfrac{\pi}{4}, \dfrac{3\pi}{4}, \dfrac{5\pi}{4}, \dfrac{7\pi}{4}$ **55.** $x = \dfrac{\pi}{2}, \dfrac{3\pi}{2}$

57. $x \approx 0.8861, 2.2556, 4.0277, 5.3971$

59. $x \approx -0.4502$

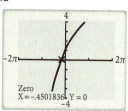

61. $\theta \approx 0.4334$ or about $24.83°$

63. a. 5 cm   **b.** $\dfrac{4}{5}$ **c.** $73.7398°$

Chapter 6 Test

1. $\dfrac{\cot^2 x}{\csc x + 1} = \dfrac{\csc^2 x - 1}{\csc x + 1} = \dfrac{(\csc x - 1)(\csc x + 1)}{\csc x + 1}$
$= \csc x - 1$

2. $\dfrac{\cos x}{1 - \sin x} \cdot \dfrac{1 + \sin x}{1 + \sin x} = \dfrac{\cos x(1 + \sin x)}{1 - \sin^2 x}$
$= \dfrac{\cos x(1 + \sin x)}{\cos^2 x} = \dfrac{1 + \sin x}{\cos x} = \sec x + \tan x$

3. $\dfrac{1}{\sec x + 1} + \dfrac{1}{\sec x - 1} = \dfrac{\sec x - 1 + \sec x + 1}{\sec^2 x - 1}$
$= \dfrac{2 \sec x}{\tan^2 x} = 2 \cot^2 x \sec x = 2 \cot x \csc x$

4. $\sin\left(x + \dfrac{\pi}{2}\right) = \sin x \cos \dfrac{\pi}{2} + \cos x \sin \dfrac{\pi}{2} = \cos x$

5. $\sin \dfrac{5\pi}{12}$ **6.** $\dfrac{-\sqrt{2} - \sqrt{6}}{4}$ **7.** $\dfrac{-\sqrt{6} - \sqrt{2}}{4}$

8. $\sqrt{3} + 2$ **9.** $-\dfrac{63}{65}$ **10.** $\dfrac{33}{56}$ **11.** $-\dfrac{16}{65}$ **12.** $-\dfrac{24}{25}$

13. $\dfrac{\sqrt{10}}{10}$ **14.** $-\dfrac{24}{7}$ **15.** $\dfrac{1}{4}(1 - \cos 2x)\sin 2x$

16. $\dfrac{1}{2}(\sin 5x + \sin x)$ **17.** $2 \cos \dfrac{7x}{2} \cos \dfrac{x}{2}$

18. $\dfrac{\pi}{6} + 2n\pi, \dfrac{5\pi}{6} + 2n\pi$, where n is an integer

19. $\dfrac{\pi}{3} + \dfrac{n\pi}{2}$, where n is an integer

20. $\dfrac{\pi}{3}, \dfrac{2\pi}{3}, \dfrac{4\pi}{3}, \dfrac{5\pi}{3}$ **21.** $0, \dfrac{\pi}{3}, \dfrac{5\pi}{3}$ **22.** 1.911, 4.373

23. 2.214   **24.** -0.564 **25.** $\dfrac{\pi}{2}$ **26.** Day 9

Chapter 7

7.1 Exercises

1. True   **3.** 30°; 150°   **5.** 53.13°; 126.87°
7. $B = 104°; a \approx 8.3913; b \approx 15.3648$
9. $C = 25°; a \approx 17.7880; b \approx 13.3853$
11. $C = 74°; a \approx 4.4669; c \approx 6.4170$
13. $C = 50°; a \approx 18.4002; b \approx 6.6971$
15. $C = 40°; b \approx 11.4320; c \approx 8.4851$
17. $B = 60°; a \approx 53.6231; b \approx 47.1554$
19. $B = 29.5°; b \approx 12.9516; c \approx 8.9957$
21. $B = 54.9°; b \approx 12.8330; c \approx 12.3771$
23. One solution: $B = 24.1858°; C = 120.8142°; c \approx 10.4813$
25. One solution: $B = 32.3884°; C = 107.6116°; c \approx 8.8968$
27. No solution   **29.** 1.0   **31.** 11.5   **33.** 23.2051 feet
35. 10.7119 feet   **37.** 6.0882 miles   **39.** 10.9410 inches
41. 9.1794 feet   **43.** $w = 227$ feet; $d = 145$ feet
45. 11.8313 feet   **47.** 5.5°   **49.** 2.3031 feet and 3.0759 feet
51. $AD \approx 14.7$ yards; $BC \approx 21.9$ yards
53. Counterclockwise from the left: approximately 5.1764

inches, 6.0195 inches, and 3.6742 inches.
55. The Law of Sines works for a right triangle as well as for any other, and it is easy to use because $\sin 90° = 1$ The Pythagorean Theorem and the trig ratios are more efficient for right triangles.
57. No. There is an infinite number of three angle sets that can solve this equation.

7.2 Exercises

1. $c \approx 11.1664$ **3.** $b \approx 10.0317$ **5.** $a \approx 7.9646$
7. $a \approx 8.2885; B \approx 81.2265°; C \approx 43.7735°$
9. $c \approx 10.9813; A \approx 56.2575°; B \approx 86.2425°$
11. $A \approx 27.6604°; B \approx 40.5358°; C \approx 111.8037°$
13. $A \approx 12.9298°; B \approx 17.0140°; C \approx 150.0562°$
15. $A \approx 31.9475°; B \approx 108.9679°; C \approx 39.0846°$
17. $b \approx 27.8460; A \approx 32.0388°; C \approx 47.9612°$
19. $a = 43.2629; B \approx 47.0935°; C \approx 34.9065°$
21. $B \approx 5.5°; C \approx 5.5°; a \approx 30.2600$
23. 272.8236   **25.** 51.8733   **27.** 2.9047   **29.** 5.3327
31. 63.0069   **33.** 6.64 feet   **35.** $c \approx 16.3830$ inches
37. a. Use the Law of Cosines   **b.** 79.09 meters
39. 70.2271°   **41.** ≈ 9.0569 in, ≈ 19.9239 square inches
43. 371.3456 feet   **45.** 6.7804 centimeters; 94.2368°
47. 28.5518 feet   **49.** 450 square inches
51. $AE = 2.24$ inches; $BE = 1.41$ inches; $DE = 2.83$ inches
53. 3.2370 inches; 6.4739 inches
55. $c^2 = b^2 + a^2 - 2ab \cos C$
    $C = 90°, \cos C = 0$
    $c^2 = b^2 + a^2 - 0$
    $a^2 + b^2 = c^2$
57. No, a direct application of the Law of Cosines results in a single equation with an unknown side and an unknown angle, and cannot be solved uniquely.
59. If the sum of the squares of the two smaller sides is equal to the square of the largest side.
61. If the quotient of the side lengths is equal to the cosine of the angle.
63. If C is a right angle, then the area formula is $\dfrac{1}{2} ab \sin 90° = \dfrac{1}{2} ab.$

7.3 Exercises

1. Quadrant II   **3.** Quadrant III   **5.** E **7.** F **9.** B
11. a.

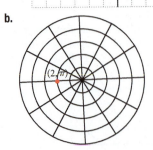

b.

13. $\left(1, \dfrac{\pi}{2}\right)$

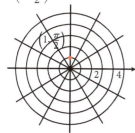

15. $\left(-3, \dfrac{\pi}{3}\right)$

17. $\left(\dfrac{1}{2}, -\pi\right)$

19. $\left(0, \dfrac{3\pi}{2}\right)$

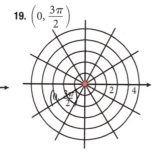

21. $\left(\dfrac{3\sqrt{2}}{2}, \dfrac{3\sqrt{2}}{2}\right)$ **23.** $\left(-\dfrac{\sqrt{3}}{2}, -\dfrac{1}{2}\right)$ **25.** $(2\sqrt{3}, 2)$

27. $\left(0, \dfrac{3}{2}\right)$ **29.** $\left(2, \dfrac{\pi}{3}\right)$ **31.** $\left(2, \dfrac{7\pi}{6}\right)$ **33.** $\left(1, \dfrac{\pi}{2}\right)$

35. $(2, -\pi); (-2, 0)$ **37.** $\left(4, \dfrac{7\pi}{2}\right); \left(-4, \dfrac{\pi}{2}\right)$

39. $\left(\dfrac{3}{4}, \dfrac{13\pi}{6}\right); \left(-\dfrac{3}{4}, \dfrac{7\pi}{6}\right)$ **41.** $\left(\dfrac{11}{7}, 0\right); \left(-\dfrac{11}{7}, \pi\right)$

43. $\left(1.3, \dfrac{11\pi}{4}\right); \left(-1.3, \dfrac{7\pi}{4}\right)$ **45.** $r\cos\theta = 2$

47. $r\cos\theta + 2r\sin\theta = 4$ **49.** $r = 5$ **51.** $r = -2\cos\theta$
53. $\sin\theta = r\cos^2\theta$ **55.** $\sin\theta = 2r\cos^2\theta + \cos\theta$
57. $x^2 + y^2 = 9$ **59.** $y = x$ **61.** $x = 4$ **63.** $2x + y = 4$
65. $(x-1)^2 + y^2 = 1$ **67.** $x^2 - y^2 = 4$

69. $\left(-\dfrac{3}{2}, -\dfrac{3\sqrt{3}}{2}\right)$

71. The coordinates (r, θ) and $(r, \theta + 2\pi)$ are equivalent in the polar coordinate system, since the terminal side of $\theta + 2\pi$ is the same as the terminal side of θ.

73. Two features of the polar coordinate system that are different from those of the rectangular coordinate system: (1) The coordinates of a point in the polar coordinate system are expressed in terms of real numbers and measures of angles, whereas in the rectangular coordinate system, the coordinates are expressed in terms of real numbers only. (2) The rectangular coordinates of a point are unique, whereas the polar coordinates of a point are not.

7.4 Exercises

1. $\theta = \dfrac{\pi}{2}$ **3.** $\theta = \dfrac{\pi}{2}$ **5.** $\theta = \dfrac{\pi}{4}, \dfrac{3\pi}{4}$

7.

9.

11.

13.

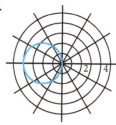

15.

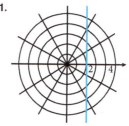

17.

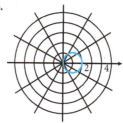

19.

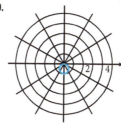

21.

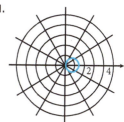

23. $r = 3\cos\theta$ **25.** $r = -4\cos\theta$
27. $r = \dfrac{3}{\sin\theta}$ **29.** $r = \dfrac{4}{\cos\theta}$

31.

33.

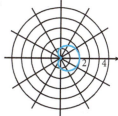

35.

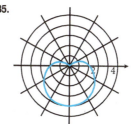

37.

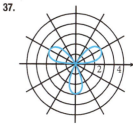

39. 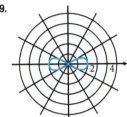 **41.**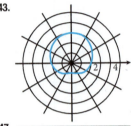

43. 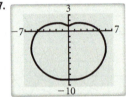 **45.**

47. 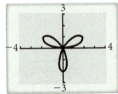 **49.**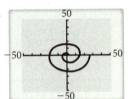

51. 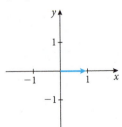 **53.**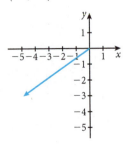

55. π **57.** 2π **59.** $r = 2 + 2\sin\theta$

61.

The graph of r_2 is obtained by rotating the graph of r_1 by $\frac{\pi}{3}$ units clockwise.

7.5 Exercises

1. $\langle 1, 0\rangle$

3. $\langle -5, -3\rangle$

5. $\left\langle \frac{4}{3}, -6\right\rangle$

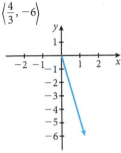

7. $-4\mathbf{i} + 6\mathbf{j}$ **9.** $-2\mathbf{i} - 1.5\mathbf{j}$ **11.** $\frac{1}{3}\mathbf{i} + \frac{3}{4}\mathbf{j}$

13. $\langle -2, -1\rangle, \langle 13, 2\rangle, \langle -4, 1\rangle$

15. $\langle -7, 12\rangle, \langle 2, -9\rangle, \langle 15, -22\rangle$

17. $\langle 1.5, 1.5\rangle, \langle 1.5, 4.5\rangle, \langle -4.5, -6.5\rangle$

19. $\left\langle -\frac{2}{3}, -\frac{8}{5}\right\rangle, \left\langle \frac{7}{3}, \frac{22}{5}\right\rangle, \left\langle 0, \frac{4}{5}\right\rangle$

21. $-6\mathbf{i} + 4\mathbf{j}, 6\mathbf{i} + \mathbf{j}, 10\mathbf{i} - 10\mathbf{j}$

23. $-5.1\mathbf{i} + 1.6\mathbf{j}, 6.9\mathbf{i} + 8.8\mathbf{j}, 7.3\mathbf{i} - 9.6\mathbf{j}$

25. $\sqrt{5}, 116.57°$ **27.** $\sqrt{3.25}, 56.31°$ **29.** $\frac{2\sqrt{109}}{15}, 16.70°$

31. $2\sqrt{5}, 333.43°$ **33.** $\sqrt{7.25}, 68.20°$ **35.** $\left\langle \frac{3}{5}, \frac{4}{5}\right\rangle$

37. $\left\langle \frac{\sqrt{2}}{2}, \frac{\sqrt{2}}{2}\right\rangle$ **39.** $-\frac{2\sqrt{5}}{5}\mathbf{i} + \frac{\sqrt{5}}{5}\mathbf{j}$

41. $\langle 15.7517, 10.6247\rangle$ **43.** $\langle -9.8481, -1.7365\rangle$

45. $\langle -4.6, 0\rangle$ **47.** $\langle -15.5563, 15.5563\rangle$

49. $\langle 7.50, -12.99\rangle$

51. a. $\langle -17.32, -10\rangle$ **b.** $\langle -19.92, -11.50\rangle$

53. a. 1.81 miles **b.** 3.71 miles **c.** 4.13 miles **d.** S26°E

55. 15.93 mph; $\theta = 77.18°$ or N12.82°E

57. $||\mathbf{v}|| = 0 \rightarrow \sqrt{v_x^2 + v_y^2} = 0$. Thus $v_x^2 + v_y^2 = 0$.
If either $v_x, v_y,$ or both are nonzero, $v_x^2 + v_y^2 > 0$.
Thus $v_x = v_y = 0$.

59. Let $\mathbf{u} = \langle a, b\rangle$
Since $||k\mathbf{u}|| = ||\langle ka, kb\rangle||$
$= \sqrt{(ka)^2 + (kb)^2} = \sqrt{k^2a^2 + k^2b^2}$
$= \sqrt{k^2(a^2 + b^2)} = \sqrt{k^2} \cdot \sqrt{a^2b^2}$
$= |k| \cdot ||\mathbf{u}||$
Therefore, $||k\mathbf{u}|| = |k| ||\mathbf{u}|| = k||\mathbf{u}||$ iff $|k| = k$ and this occurs for $k \geq 0$ (or when k is non-negative).

7.6 Exercises

1. -18 **3.** -14 **5.** 0 **7.** $-\frac{2}{5}$ **9.** $108.4°$ **11.** $90°$

13. $63.4°$ **15.** $81.0°$ **17.** $\mathbf{v}_1 = \langle -1, -3\rangle; \mathbf{v}_2 = \langle 3, -1\rangle$

19. $\mathbf{v}_1 = \langle 6, -3\rangle; \mathbf{v}_2 = \langle 4, 8\rangle$

21. $\mathbf{v}_1 = \langle 9, 3\rangle; \mathbf{v}_2 = \langle -3, 9\rangle$

23. $\mathbf{v}_1 = \left\langle -\frac{24}{25}, \frac{32}{25}\right\rangle; \mathbf{v}_2 = \left\langle \frac{124}{25}, \frac{93}{25}\right\rangle$

25. No **27.** No **29.** No **31.** Yes **33.** $\langle 0, 1\rangle$ **35.** 3

37. $\langle 12, 12\rangle$ **39.** $\left\langle \frac{12}{5}, \frac{6}{5}\right\rangle$ **41.** 1414 foot-pounds

43. -4.5 **45.** 1.20 miles **47.** 1879 foot-pounds

49. 23.6 horsepower **51.** $a = 6$

7.5 Exercises

1. $\langle 1, 0\rangle$

3. $\langle -5, -3\rangle$

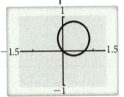

53.
$$\mathbf{v} + \mathbf{w} = \langle v_x + w_x, v_y + w_y \rangle$$
$$\mathbf{u} \cdot (\mathbf{v} + \mathbf{w}) = u_x(v_x + w_x) + u_y(v_y + w_y)$$
$$= (u_x v_x + u_x w_x) + (u_y v_y + u_y w_y)$$
$$\mathbf{u} \cdot \mathbf{v} + \mathbf{u} \cdot \mathbf{w}$$
$$= u_x v_x + u_y v_y + u_x w_x + u_y w_y$$
Thus $\mathbf{u} \cdot (\mathbf{v} + \mathbf{w}) = \mathbf{u} \cdot \mathbf{v} + \mathbf{u} \cdot \mathbf{w}$.

55. $\mathbf{v} \cdot \mathbf{v} = v_x \cdot v_x + v_y \cdot v_y$
$$= v_x^2 + v_y^2 = \|\mathbf{v}\|^2$$

7.7 Exercises

1. $-i$ **3.** -1 **5.** $-1 + i$ **7.** $\sqrt{5}$ **9.** $\sqrt{5}$ **11.** $\dfrac{\sqrt{13}}{4}$

13. $2\left(\cos\dfrac{\pi}{2} + i\sin\dfrac{\pi}{2}\right)$ **15.** $4(\cos\pi + i\sin\pi)$

17. $2\left(\cos\dfrac{5\pi}{3} + i\sin\dfrac{5\pi}{3}\right)$

19. $4\sqrt{2}\left(\cos\dfrac{7\pi}{4} + i\sin\dfrac{7\pi}{4}\right)$

21. $4\left(\cos\dfrac{11\pi}{6} + i\sin\dfrac{11\pi}{6}\right)$ **23.** $8\left(\cos\dfrac{7\pi}{12} + i\sin\dfrac{7\pi}{12}\right)$

25. $\dfrac{3}{2}\left(\cos\dfrac{31\pi}{12} + i\sin\dfrac{31\pi}{12}\right)$ **27.** $\dfrac{5}{2}\left(\cos\dfrac{\pi}{12} + i\sin\dfrac{\pi}{12}\right)$

29. -4 **31.** $8i$ **33.** 216

35. $0.707 + 0.707i; -0.707 - 0.707i$

37. $1.225 + 0.707i; -1.225 - 0.707i$

39. $-1; -0.309 - 0.951i; 0.809 - 0.588i;$
$0.809 + 0.588i; -0.309 + 0.951i$

41. $1.414 + 1.414i; -1.414 + 1.414i; -1.414 - 1.414i;$
$1.414 - 1.414i$

43. $1.554 - 0.644i; -1.554 + 0.644i; 0.644 + 1.554i;$
$-0.644 - 1.554i$

45. $-1; 0.500 + 0.866i; 0.500 - 0.866i$

47. $-0.866 - 0.500i; 0.866 - 0.500i; i$

49. Let $z = r(\cos\theta + i\sin\theta)$. By DeMoivre's Theorem,
$$\left(\sqrt[n]{r}\left[\cos\dfrac{\theta + 2k\pi}{n} + i\sin\dfrac{\theta + 2k\pi}{n}\right]\right)^n$$
$$= r(\cos(\theta + 2k\pi) + i\sin(\theta + 2k\pi))$$
$$= r(\cos\theta + i\sin\theta) = z \text{ because cosine and sine are}$$
periodic with period $2k\pi$, k an integer.

51. One

Chapter 7 Review Exercises

1. $B = 72°$, $b \approx 25.6437$, $c \approx 17.3317$

3. $B = 54.9°$, $b \approx 12.8330$, $c \approx 12.3771$

5. $B \approx 46.4727°$, $C \approx 68.5273°$, $c \approx 10.2679$

7. $B \approx 41.4387°$, $C \approx 103.5613°$, $c \approx 22.0329$
or $B \approx 138.5613°$, $C \approx 6.4387°$, $c \approx 2.5416$

9. 107.4465 feet

11. 2530 feet

13. $a \approx 5.6812$, $B \approx 114.6819°$, $C \approx 30.3181°$

15. $c \approx 40.6306$, $A \approx 44.9182°$, $B \approx 28.0818°$

17. $c \approx 44.7496$, $A \approx 51.1832°$, $B \approx 33.8168°$

19. $A \approx 28.9550°$, $B \approx 46.5675°$, $C \approx 104.4775°$

21. 435.84

23. $\left(4, \dfrac{\pi}{6}\right)$

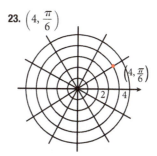

25. $(2\sqrt{3}, -2)$ **27.** $(0, 2)$ **29.** $\left(2, \dfrac{2\pi}{3}\right)$ **31.** $\left(3, \dfrac{\pi}{2}\right)$

33. $r^2 - 2r\cos\theta = \dfrac{5}{4}$

35. 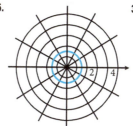 **37.**

39.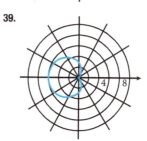

41. $3\sqrt{5} \approx 6.71; 63.43°$ **43.** $\sqrt{74} \approx 8.60; 125.54°$

45. $\langle 2, 0 \rangle; \langle -4, 12 \rangle; \langle -2, -8 \rangle$

47. $\langle -5.5, 5.5 \rangle; \langle 3.5, 5.5 \rangle; \langle 10.5, -16.5 \rangle$ **49.** $\left\langle -\dfrac{12}{13}, \dfrac{5}{13}\right\rangle$

51. $6i - 3j$ **53.** $-2; 93.4°$ **55.** $-\dfrac{32}{3}; 140.9°$

57. $\mathbf{v_1} = \left\langle -\dfrac{2}{5}, -\dfrac{4}{5}\right\rangle; \mathbf{v_2} = \left\langle \dfrac{12}{5}, -\dfrac{6}{5}\right\rangle$

59. $\mathbf{v_1} = \langle 0, 0 \rangle; \mathbf{v_2} = \langle -1, 5 \rangle$ **61.** $6\left(\cos\dfrac{3\pi}{2} + i\sin\dfrac{3\pi}{2}\right)$

63. $2\left(\cos\dfrac{2\pi}{3} + i\sin\dfrac{2\pi}{3}\right)$ **65.** $4\left(\cos\dfrac{5\pi}{6} + i\sin\dfrac{5\pi}{6}\right)$

67. $4(\cos 2\pi + i\sin 2\pi); \dfrac{1}{2}\left(\cos\dfrac{\pi}{2} + i\sin\dfrac{\pi}{2}\right)$

69. $2\sqrt{2}\left(\cos\dfrac{5\pi}{12} + i\sin\dfrac{5\pi}{12}\right);$
$\sqrt{2}\left(\cos\dfrac{11\pi}{12} + i\sin\dfrac{11\pi}{12}\right)$

71. -4 **73.** $-8i$ **75.** $0.707 - 0.707i; -0.707 + 0.707i$

77. $1.732 - i; -1.732 + i$

Chapter 7 Test

1. $C = 85°$, $b \approx 11.03$, $c \approx 21.98$

2. $a \approx 10.44$, $B \approx 24.50°$, $C \approx 95.50°$

3. $A \approx 38.05°$, $B \approx 29.54°$, $C \approx 112.41°$

4. $a \approx 15.88$, $A \approx 110.64°$, $B \approx 24.36°$

5. 50 square inches

6. $\left(8, \dfrac{7\pi}{6}\right)$ **7.** $\left(-\dfrac{5\sqrt{2}}{2}, -\dfrac{5\sqrt{2}}{2}\right)$

8. $r = 2\sin\theta$

9. a. $x^2 + y^2 = 6.25$ **b.** $y = \sqrt{3}x$

10.

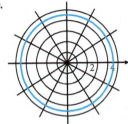

11.

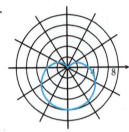

12. Magnitude: 5; $\theta \approx 323.1°$ **13.** $\langle -6.13, 5.14\rangle$

14. -2 **15.** $\langle -11, -1\rangle$ **16.** $\|\mathbf{u}\| = \sqrt{10}; \|\mathbf{v}\| = 2\sqrt{5}$

17. $\left\langle -\dfrac{1}{5}, -\dfrac{2}{5}\right\rangle$ **18.** $6\left(\cos\dfrac{5\pi}{12} + i\sin\dfrac{5\pi}{12}\right)$

19. $2\left(\cos\dfrac{\pi}{12} + i\sin\dfrac{\pi}{12}\right)$ **20.** 11.84 feet

21. $\langle -6.43, 7.66\rangle$ **22.** 906.31 foot-pounds

Chapter 8

8.1 Exercises

1. $(-1, 2)$ **3.** $(3, 5)$ **5.** $(4, 2)$ **7.** $\begin{cases} 4 + 5(-2) = -6 \\ -4 + 2(-2) = -8 \end{cases}$

9. $\begin{cases} 2(0) - 3(2) = -6 \\ -0 + 2(2) = 4 \end{cases}$ **11.** $\begin{cases} a - (a - 5) = 5 \\ -2a + 2(a - 5) = -10 \end{cases}$

13. $x = 1, y = 3$ **15.** $x = -1, y = 2$ **17.** No solution

19. $x = 2, y = -3$ **21.** Dependent system

23. $x = -\dfrac{16}{3}, y = -3$ **25.** $x = 6, y = -4$

27. $x = \dfrac{1}{2}, y = 1$ **29.** $x = -2, y = -3$

31. $\begin{cases} 3x + 3y = 1 \\ -2x + y = 2 \end{cases}$; $x = -\dfrac{5}{9}, y = \dfrac{8}{9}$

33. $\begin{cases} x + 4y = -5 \\ 2x + 3y = 6 \end{cases}$; $x = \dfrac{39}{5}, y = \dfrac{-16}{5}$

35. $\begin{cases} 3x + 2y = 5 \\ 3x + y = 7 \end{cases}$; $x = 3, y = -2$

37. $x = -3.75, y = 2.875$

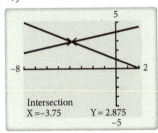

Intersection
X = -3.75   Y = 2.875

39. $x = -0.75, y = 2.75$

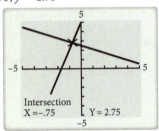

Intersection
X = -.75   Y = 2.75

41. $\begin{cases} x - y = 0 \\ 2x + y = 3 \end{cases}$

43. $x < -2$

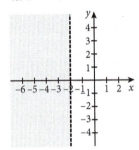

45. $-2x + y \geq -4$

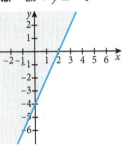

47. $x > y + 6$

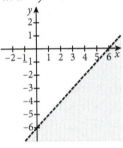

49. $3x - 4y > 12$

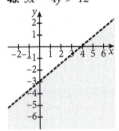

51. $3x + 5y > 10$

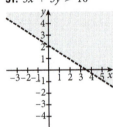

53. $\begin{cases} -2x + y \leq 8 \\ -x + y \geq 2 \end{cases}$

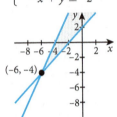

$(-6, -4)$

55. $\begin{cases} y \geq 2 \\ 5x + 2y \leq 10 \end{cases}$

$\left(\dfrac{14}{5}, -2\right)$

57. $\begin{cases} -x + y < 3 \\ x + y > -5 \end{cases}$

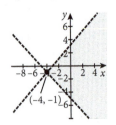

$(-4, -1)$

59. $\begin{cases} x + y \le 4 \\ -4x + 2y \ge -1 \end{cases}$

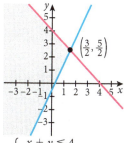

$\left(\frac{3}{2}, \frac{5}{2}\right)$

61. $\begin{cases} y \le -2 \\ y \ge -5 \\ x \le -1 \end{cases}$

$(-1, -2)$

$(-1, -5)$

63. $\begin{cases} x + y \le 4 \\ -x + y \le 4 \\ x + 5y \ge 8 \end{cases}$

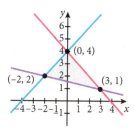

$(0, 4)$

$(-2, 2)$ $(3, 1)$

65. $\begin{cases} -3x + y \le -1 \\ 4x + y \le 6 \end{cases}$

$(1, 2)$

67. $\begin{cases} x \le 0 \\ -5x + 4y \le 20 \\ 3x + 4y \ge -12 \end{cases}$

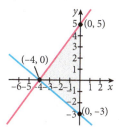

$(0, 5)$

$(-4, 0)$

$(0, -3)$

69. $\begin{cases} -\frac{4}{3}x + y \ge -5 \\ -4x + 3y \le 13 \\ x + y \ge 2 \end{cases}$

$(-1, 3)$

$(3, -1)$

71. $P = 170$ **73.** $P = 0$

75. Tickets: \$600 million; merchandise: \$2.4 billion

77. a. Running: 26 minutes; walking: 14 minutes
   b. Running: 38 minutes; walking: -18 minutes. No.
   Time must be nonnegative.

79. \$380: 53 tickets; \$700: 27 tickets

81. a. $x + y = 10$ **b.** $0.10x + 0.25y = 1.5$
   c. $x = \frac{20}{3}$ gallons, $y = \frac{10}{3}$ gallons; $\frac{20}{3}$ gallons of 10%
   solution and $\frac{10}{3}$ gallons of 25% solution are needed
   for a 15% solution.
   d. No. The acidity of the less acidic of the two solutions used
   to make the mixture is greater than the acidity desired.

83. a. $\begin{cases} x + y \le 10{,}000 \\ x \ge 6000 \\ y \le 4000 \\ y \ge 0 \end{cases}$ **b.**

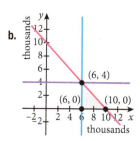

$(6, 4)$

$(6, 0)$ $(10, 0)$

thousands

85. a. $\begin{cases} x + y \le 200 \\ x \ge 40 \\ 6x + 8y \ge 960 \\ y \ge 0 \end{cases}$
   b.

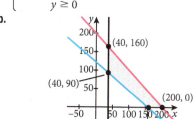

$(40, 160)$

$(40, 90)$

$(200, 0)$

$(160, 0)$

87. a. $\begin{cases} x + y \le 30 \\ x \ge 3y \\ y \ge 5 \\ x \ge 0 \end{cases}$ **b.**

$(0, 30)$

$\left(22\frac{1}{2}, 7\frac{1}{2}\right)$

$(0, 5)$ $(15, 5)$

89. 1000 of Model 120; 0 of Model 140   **91.** 10 days
93. 50 drivers, 30 putters
95. a.

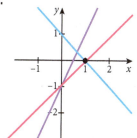

   b. No   **c.** No. There is no point of intersection.
97. No. The resulting equations form a dependent system.

8.2 Exercises

1. $\begin{cases} 3(1) - (0) = 3 \\ 2(1) + (-2) - 2(0) = 0 \\ 3(1) - 2(-2) + (0) = 7 \end{cases}$

3. $\begin{cases} 5(0) - (1) + 3(-4) = -13 \\ (0) - (1) + 2(-4) = -9 \\ 4(0) - (1) + (-4) = -5 \end{cases}$

5. $\begin{cases} (-9 - 11z) + 2(5 + 7z) - 3z = 1 \\ 2(-9 - 11z) + 3(5 + 7z) + z = -3 \end{cases}$

7. $x = 4, y = 1, z = -2$ **9.** $u = 7, v = 4, w = 3$
11. $x = 2, y = 0, z = 2$ **13.** $x = -4, y = 2, z = -1$
15. $x = 0, y = 0, z = -2$ **17.** $x = -6, y = -9, z = 11$
19. $x = \frac{2}{3} + \frac{1}{6}y, z = -\frac{1}{3} - \frac{5}{6}y$
21. $x = 3, y = \frac{1}{2}, z = -\frac{3}{2}$ **23.** $x = 0, y = -\frac{4}{3}, z = \frac{2}{3}$
25. $x = 2, y = -3, z = -5$ **27.** $u = 4, v = 2, w = -5$
29. $r = 3, s = 11, t = 8$ **31.** $x = -1 - 5z, y = -4 - 3z$
33. $r = 9 - 8t, s = -4 + 4t$ **35.** $y = 4, z = -15 - 3x$

37. Inconsistent

39. Mutual fund: 50%, high-yield bond: 20%, CD: 30%
41. A: 20 ohms, B: 35 ohms, C: 45 ohms
43. Level A: \$45, Level B: \$35, Level C: \$30

45. Florida: 23.3%, New York: 23.3%, California: 22.1%

47. a. $a + b + c = 1; 4a - 2b + c = -8$

   b. $a = -1, b = 2, c = 0$ **c.** $f(x) = -x^2 + 2x$

49. $f(x) = 5x^2 - 10x + 6$

8.3 Exercises

1. $\begin{bmatrix} 4 & 1 & -2 & | & 6 \\ -1 & -1 & 1 & | & -2 \\ 3 & 0 & -1 & | & 4 \end{bmatrix}$ **3.** $\begin{bmatrix} 3 & -2 & 1 & | & -1 \\ 1 & 1 & -4 & | & 3 \\ -2 & -1 & 3 & | & 0 \end{bmatrix}$

5. $\begin{bmatrix} 6 & -2 & 1 & | & 0 \\ -5 & 1 & -3 & | & -2 \\ 2 & -3 & 5 & | & 7 \end{bmatrix}$ **7.** $\begin{bmatrix} 1 & 1 & 2 & | & -3 \\ -3 & 2 & 1 & | & 1 \end{bmatrix}$

9. $\begin{bmatrix} 1 & -2 & 0 & | & -1 \\ 1 & -4 & -1 & | & \frac{1}{2} \\ 3 & 5 & 1 & | & 2 \end{bmatrix}$ **11.** $\begin{bmatrix} 2 & -8 & -2 & | & 1 \\ 1 & -2 & 0 & | & -1 \\ 3 & 5 & 1 & | & 2 \end{bmatrix}$

13. $\begin{bmatrix} 1 & -2 & 0 & | & -1 \\ 4 & -12 & -2 & | & -1 \\ 3 & 5 & 1 & | & 2 \end{bmatrix}$

15. $\begin{cases} 1x + 0y = -7 \\ 0x + 1y = 3 \end{cases}$; Consistent

   $x = -7, y = 3$

17. $\begin{cases} 2x + 0y = 6 \\ 0x + 1y = 5 \end{cases}$; Consistent

   $x = 3, y = 5$

19. $\begin{cases} x + 3z = 5 \\ y - 2z = -2 \end{cases}$; Consistent

   $x = 5 - 3z, y = -2 + 2z$

21. $\begin{cases} x - 5u = 2 \\ y - 2u = -3 \\ z + 3u = 5 \end{cases}$; Consistent

   $x = 2 + 5u, y = -3 + 2u, z = 5 - 3u$

23. $x = -14, y = 8$ **25.** $x = \frac{1}{2}, y = -4$

27. $x = 2, y = 4$ **29.** $x = 3, y = -1$

31. $y = -\frac{1}{3} - \frac{5}{3}x$

33. Inconsistent system   **35.** $x = -3, y = 2, z = 4$

37. $x = 1, y = 1, z = 0$ **39.** $x = 2, y = -2, z = 5$

41. $x = -y, z = y - 2, y$ is a real number

43. Inconsistent system   **45.** $x = 4y + 24, z = 10$

47. $r = 10, s = 2t - 4$ **49.** Inconsistent system

51. $x = -1, y = 3, z = -6$

53. $6\frac{2}{3}$ pounds of Colombian, $2\frac{2}{9}$ pounds of Java, $1\frac{1}{9}$ pounds of Kona

55. 5 cheese pizzas, 3 pepperoni pizzas

57. 10 boxes of Brand A, 6 boxes of Brand B

59. Mutual fund: 50%, bond: 12.5%, CD: 37.5%

61. $A = 40$ ohms, $B = 20$ ohms, $C = 80$ ohms

63.

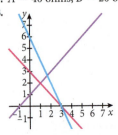

65. a. The first equation is the second equation multiplied by 3. The second equation is the first multiplied by $\frac{1}{3}$.

   b. Because the equations are multiples of each other, the values that satisfy one will satisfy the other.

   c. $w = 2u + 2v + 1$

   d. $(1, 1, 5), (-1, 1, 1)$

8.4 Exercises

1. -1 **3.** 3×4 **5.** π

7. False. For addition, matrices must have the same dimensions so that corresponding entries can be added.

9. True. The product of an $m \times n$ matrix and an $n \times p$ matrix is an $m \times p$ matrix.

11. $\begin{bmatrix} 4 & 5 \\ 3 & -1 \\ 0 & 1 \end{bmatrix}$ **13.** $\begin{bmatrix} 12 & 5 \\ 6 & -3 \\ 2 & -5 \end{bmatrix}$ **15.** $\begin{bmatrix} 20 & -15 \\ 3 & -5 \\ 8 & -27 \end{bmatrix}$

17. Not defined. The dimensions do not allow for addition.

19. $\begin{bmatrix} -\frac{1}{3} & 4 \\ 36 & 12 \end{bmatrix}$

21. Not defined. The number of columns of the first matrix does not match the number of rows of the second.

23. $\begin{bmatrix} \frac{1}{2} & -\frac{3}{2} & \frac{1}{6} \\ \frac{5}{2} & 0 & -1 \end{bmatrix}$ **25.** $\begin{bmatrix} -5 & \frac{25}{3} \\ 20 & 23 \end{bmatrix}$ **27.** $\begin{bmatrix} \frac{544}{3} & 44 \\ 36 & 12 \\ \frac{758}{3} & 76 \end{bmatrix}$

29. a. $\begin{bmatrix} -10 \\ -1 \end{bmatrix}$

   b. The dimensions do not allow for subtraction.

   c. The dimensions do not allow for multiplication in the given order.

31. a. $\begin{bmatrix} 8 & -4 \\ -4 & 35 \end{bmatrix}$ **b.** $\begin{bmatrix} 9 & -29 \\ 18 & -17 \end{bmatrix}$ **c.** $\begin{bmatrix} 42 & -37 \\ 6 & 1 \end{bmatrix}$

33. a. $\begin{bmatrix} -6 & -12 \\ 13 & -17 \\ 13 & -13 \end{bmatrix}$

   b. The dimensions do not allow for subtraction.

   c. The dimensions do not allow for multiplication in the given order.

35. $\begin{bmatrix} 28 \\ 27 \end{bmatrix}$ **37.** $\begin{bmatrix} -5 & 3 \\ 4 & 52 \end{bmatrix}$ **39.** $\begin{bmatrix} 10 & -2 \\ 23 & -7 \end{bmatrix}$

41. $\begin{bmatrix} -18 & -4 \\ 20 & -2 \\ -10 & 2 \end{bmatrix}$ **43.** $ag + bj; ei + fl$

45. $eh + fk; ci + dl$

47. $\begin{bmatrix} 3 & -2 \\ 2 & -1 \end{bmatrix}, \begin{bmatrix} 4 & -3 \\ 3 & -2 \end{bmatrix}$ **49.** $\begin{bmatrix} 16 & 0 \\ 0 & 9 \end{bmatrix}, \begin{bmatrix} -64 & 0 \\ 0 & 27 \end{bmatrix}$

51. $\begin{bmatrix} 9 & 0 & 0 \\ -4 & 2 & 1 \\ -12 & 1 & 1 \end{bmatrix}, \begin{bmatrix} 27 & 0 & 0 \\ -16 & 3 & 2 \\ -40 & 2 & 1 \end{bmatrix}$

53. $a = -1, 1$ **55.** $a = 0; b = 4$ **57.** $a = 2, 1; b = -2$

59. Regular: $28.80; high-octane: $31.80

61. a. $\begin{bmatrix} 1854.86 & 834.687 & 129.8402 \\ 1964.92 & 884.214 & 137.5444 \\ 2016.44 & 907.398 & 141.1508 \end{bmatrix}$

b. It represents the breakdown of the contribution from each of these three sectors per year.

c. Yes. It represents the total contribution of these three sectors over the given years.

63. Sofa: $122; loveseat: $94; chair: $48.50

65. $AB = \begin{bmatrix} 0 & 0 \\ 0 & 0 \end{bmatrix}$. No, as demonstrated by the given matrices.

67. $AI = \begin{bmatrix} 2 & -1 \\ 1 & 0 \end{bmatrix}\begin{bmatrix} 1 & 0 \\ 0 & 1 \end{bmatrix} = \begin{bmatrix} 2 & -1 \\ 1 & 0 \end{bmatrix}$

$IA = \begin{bmatrix} 1 & 0 \\ 0 & 1 \end{bmatrix}\begin{bmatrix} 2 & -1 \\ 1 & 0 \end{bmatrix} = \begin{bmatrix} 2 & -1 \\ 1 & 0 \end{bmatrix}$

$AI = IA$

69. $AB = \begin{bmatrix} 2 & 3 \\ 4 & 7 \end{bmatrix}$; $BA = \begin{bmatrix} 3 & 4 \\ 4 & 6 \end{bmatrix}$. No. The corresponding entries of the resulting matrices are computed using different rows and columns.

8.5 Exercises

1. $\begin{bmatrix} 5 & 2 \\ -3 & -1 \end{bmatrix}\begin{bmatrix} -1 & -2 \\ 3 & 5 \end{bmatrix} = \begin{bmatrix} 1 & 0 \\ 0 & 1 \end{bmatrix}$; Yes

3. $\begin{bmatrix} -6 & 5 \\ 4 & -3 \end{bmatrix}\begin{bmatrix} \frac{3}{2} & \frac{5}{2} \\ 2 & 3 \end{bmatrix} = \begin{bmatrix} 1 & 0 \\ 0 & 1 \end{bmatrix}$; Yes

5. $\begin{bmatrix} -1 & 3 & 1 \\ 0 & 5 & 2 \\ 1 & 0 & 0 \end{bmatrix}\begin{bmatrix} 0 & 0 & 1 \\ 2 & 1 & 2 \\ 5 & 3 & 5 \end{bmatrix} = \begin{bmatrix} 1 & 0 & 0 \\ 0 & 1 & 0 \\ 0 & 0 & 1 \end{bmatrix}$; Yes

7. $\begin{bmatrix} -1 & 3 \\ 1 & -2 \end{bmatrix}$ **9.** $\begin{bmatrix} -4 & 3 \\ -1 & 1 \end{bmatrix}$ **11.** $\begin{bmatrix} 2 & -3 \\ -3 & 5 \end{bmatrix}$

13. $\begin{bmatrix} 4 & 0 & -5 \\ -18 & 1 & 24 \\ -3 & 0 & 4 \end{bmatrix}$ **15.** $\begin{bmatrix} -4 & -1 & -1 \\ 0 & 0 & 1 \\ 3 & 1 & 0 \end{bmatrix}$ **17.** $\begin{bmatrix} 0 & -\frac{2}{5} & \frac{1}{5} \\ -\frac{2}{5} & -\frac{3}{5} & \frac{2}{5} \\ \frac{1}{5} & \frac{2}{5} & 0 \end{bmatrix}$

19. $\begin{bmatrix} -1 & 1 & 1 \\ 1 & -1 & 1 \\ 1 & 1 & -1 \end{bmatrix}$ **21.** $\begin{bmatrix} 7 & 0 & 2 & -1 \\ 6 & 3 & 2 & 2 \\ 3 & 1 & 1 & 1 \\ 0 & 1 & 0 & 1 \end{bmatrix}$

23. $x = 4, y = 18, z = 44$

25. $x = -\frac{3}{2}, y = -\frac{1}{2}, z = \frac{1}{2}$ **27.** $x = -3, y = -1$

29. $x = -\frac{9}{2}, y = \frac{5}{2}$ **31.** $x = 1, y = -2$

33. $x = 4, y = 3$ **35.** $x = -3, y = 3$

37. $x = 36, y = 9, z = -5$ **39.** $x = -47, y = -35, z = 17$

41. $x = 4, y = 2, z = -3$ **43.** $x = -49, y = -28, z = 5.5$

45. $x = -3, y = 0, z = -5, w = 0$

47. $\begin{bmatrix} 1 & -2 \\ 0 & 1 \end{bmatrix}, \begin{bmatrix} 1 & -3 \\ 0 & 1 \end{bmatrix}$ **49.** $\begin{bmatrix} \frac{1}{4} & -\frac{1}{4} \\ 0 & 1 \end{bmatrix}, \begin{bmatrix} \frac{1}{8} & \frac{3}{8} \\ 0 & -1 \end{bmatrix}$

51. $\begin{bmatrix} \frac{1}{4} & 0 & 0 \\ 0 & 1 & -4 \\ 0 & 0 & 1 \end{bmatrix}, \begin{bmatrix} \frac{1}{8} & 0 & 0 \\ 0 & 1 & -6 \\ 0 & 0 & 1 \end{bmatrix}$ **53.** Adult: $12; child: $5

55. Cheese: 200 calories; Meaty Delite: 270 Calories; Veggie Delite: 150 calories

57. Red: $72; white: $54; blue: $81

59. $\begin{bmatrix} 3 & -7 \\ -2 & 5 \end{bmatrix}$ **61.** $\begin{bmatrix} 1.5 & -1.5 & 0.5 & 4.5 \\ 0.5 & -2.5 & 1.5 & 8.5 \\ 0.5 & -0.5 & 0 & 1.5 \\ 0 & 1 & -0.5 & -3 \end{bmatrix}$

63. $\begin{bmatrix} 16 \\ 9 \\ 18 \end{bmatrix}, \begin{bmatrix} 1 \\ 20 \\ 5 \end{bmatrix}, \begin{bmatrix} 19 \\ 0 \\ 0 \end{bmatrix}$; PIRATES **65.** $\begin{bmatrix} 19 \\ 14 \\ 5 \end{bmatrix}, \begin{bmatrix} 5 \\ 26 \\ 25 \end{bmatrix}$; SNEEZY

67. $\begin{bmatrix} \frac{1}{a} & -1 & -1 \\ 0 & 1 & 0 \\ 0 & 0 & 1 \end{bmatrix}$. If $a = 1$, $\begin{bmatrix} 1 & -1 & -1 \\ 0 & 1 & 0 \\ 0 & 0 & 1 \end{bmatrix}$

69. $(A^2)^{-1} = (A^{-1})^2 = \begin{bmatrix} 11 & 8 \\ 4 & 3 \end{bmatrix}$

71. $(A^3)^{-1} = (A^{-1})^3 = \begin{bmatrix} -13 & -17 \\ 68 & 89 \end{bmatrix}$

73. $(A^2)^{-1} = \begin{bmatrix} 4 & -5 \\ -15 & 19 \end{bmatrix}$; $(A^3)^{-1} = \begin{bmatrix} 19 & -24 \\ -72 & 91 \end{bmatrix}$

75. a. $\begin{bmatrix} -1 \\ 2 \end{bmatrix}$

b. The coordinates were transposed from (a, b) to (b, a).

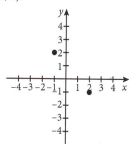

c. Multiply the product by A.

8.6 Exercises

1. -14 **3.** -9 **5.** 35 **7.** 11 **9.** $49; 49$ **11.** $10; -10$

13. -85 **15.** 62 **17.** 0 **19.** 0 **21.** 6 **23.** $x = 2$

25. $x = -4$ **27.** $x = -1$ **29.** $x = 7, y = -26$

31. $x = -\frac{5}{2}, y = -\frac{17}{2}$ **33.** $x = 0; y = -2$

35. $x = -2; y = 1$ **37.** $x \approx -6.9286, y = 4.85$

39. $x = \frac{1}{2}, y = \frac{1}{2}, z = \frac{1}{2}$ **41.** $x = 0, y = -2, z = 1$

43. $x = \frac{25}{32}, y = \frac{133}{32}, z = \frac{71}{16}$ **45.** $x = \frac{1}{2}, y = -\frac{3}{2}, z = 1$

47. The entries of the second row are all zero, so the sum of the products of these entries and their cofactors will be zero.

49. $\begin{cases} (1) + 2(2) = 5 \\ 4(1) + (2) - (0) = 6 \\ -2(1) - 4(2) = -10 \end{cases}$

$D = \begin{bmatrix} 1 & 2 & 0 \\ 4 & 1 & -1 \\ -2 & -4 & 0 \end{bmatrix} = 0.$

Cramer's Rule applies only for $D \ne 0$.

8.7 Exercises

1. $\dfrac{Ax + B}{x^2 - x - 3}$ **3.** $\dfrac{A}{x + 5} + \dfrac{B}{(x + 5)^2}$

5. $\dfrac{Ax + B}{x^2 + 2} + \dfrac{C}{2x + 1}$

7. $\dfrac{A}{x - 2} + \dfrac{B}{x + 2} + \dfrac{Cx + D}{x^2 + 4}$

9. $\dfrac{A}{3x} + \dfrac{B}{x+1} + \dfrac{C}{(x+1)^2}$

11. Irreducible   **13.** Irreducible   **15.** Reducible

17. $\dfrac{-1}{x+4} + \dfrac{1}{x-4}$ **19.** $\dfrac{-2}{x} + \dfrac{4}{2x-1}$

21. $\dfrac{-2}{x+2} + \dfrac{3}{x+3}$ **23.** $\dfrac{-2}{x} + \dfrac{1}{x+1} + \dfrac{-2}{x-1}$

25. $\dfrac{-2}{x-1} + \dfrac{4}{(x-1)^2}$ **27.** $\dfrac{2}{x^2} + \dfrac{-1}{x+2}$

29. $\dfrac{-1}{x-1} + \dfrac{-1}{x+2} + \dfrac{2}{(x+2)^2}$ **31.** $\dfrac{-1}{x+2} + \dfrac{x+2}{x^2+3}$

33. $\dfrac{3}{x+2} + \dfrac{-3x}{x^2+x+1}$ **35.** $\dfrac{-1}{x+1} + \dfrac{1}{x-1} + \dfrac{-1}{x^2+1}$

37. $\dfrac{-1}{x^2+2} + \dfrac{-2x-1}{(x^2+2)^2}$

39. $\dfrac{3}{2(x+1)} + \dfrac{-3}{2(x-1)} + \dfrac{x}{x^2+1}$

41. $\dfrac{1}{3(t+5)} + \dfrac{2t-5}{3(t^2-5t+25)}$ **43.** $\dfrac{2}{s} + \dfrac{-2s}{s^2+1}$

45. $\dfrac{1}{x(x^2+2x-3)} = \dfrac{A}{x} + \dfrac{B}{x+3} + \dfrac{C}{x-1}$ **47.** $\dfrac{1}{(x-c)^2}$

8.8 Exercises

1. $(2,3), (-3,-2)$ **3.** $(1,2), (-1,-2)$

5. $\left(-\dfrac{1}{2}, -\dfrac{1}{2}\right), (3,17)$ **7.** $\left(1, -\dfrac{5}{4}\right), \left(-\dfrac{2}{3}, 0\right)$

9. $(5,7), \left(-\dfrac{5}{3}, \dfrac{1}{3}\right)$ **11.** $(-2,-3)$ **13.** $(2,-2), (-2,2)$

15. $\left(2\sqrt{2}, -1\right), \left(-2\sqrt{2}, -1\right), \left(-\dfrac{3\sqrt{3}}{2}, \dfrac{3}{2}\right), \left(\dfrac{3\sqrt{3}}{2}, \dfrac{3}{2}\right)$

17. $(-\sqrt{2}, 0), (\sqrt{2}, 0)$

19. $(-2,-1), (2,-1), (5,6), (-5,6)$ **21.** No solution

23. $(4,2), (4,-2), \left(\dfrac{7}{2}, -\dfrac{\sqrt{11}}{2}\right), \left(\dfrac{7}{2}, \dfrac{\sqrt{11}}{2}\right)$

25. $(0,-2)$ **27.** $(4,0), (-4,0), \left(\sqrt{7}, 3\right), \left(-\sqrt{7}, 3\right)$

29. $(-2,3), \left(\dfrac{11}{2}, -\dfrac{39}{22}\right)$

31. $\left(2, \sqrt{10}\right), \left(-2, \sqrt{10}\right), \left(2, -\sqrt{10}\right), \left(-2, -\sqrt{10}\right)$

33. $\left(20, \dfrac{1}{4}\right)$

35. $(3,3)$ **37** $(3,-1), (7,-5)$

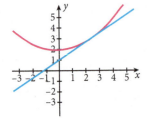   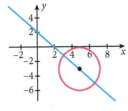

39. $(1,-3), (5,-3)$

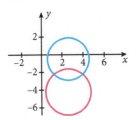

41. $(0.343, -1.646), (-0.834, 0.088)$

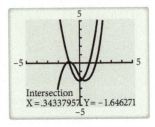

43. $(0.789, 2.378), (-1.686, 0.157)$

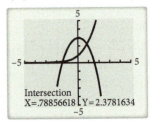

45. No solution

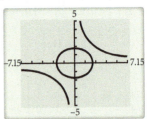

47. $(-1.123, -3.695), (-1.542, 1.895), (1.123, -3.695),$
$(1.542, 1.895)$

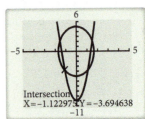

49. 28 feet $\times$ 12 feet   **51.** 8 inches, 5 inches

53. $r = \dfrac{3}{2}$ inches, $h = 3$ inches

55. Radius $= 3$ inches, height $= 12$ inches

57. 6 and 7

59. Property: 30 feet $\times$ 10 feet; pool: 15 feet $\times$ 5 feet; area of paved portion: 225 square feet

61. $b < \dfrac{9}{4}$

63. If $x, y \neq 0$, then $x^2, y^2 \neq 0$. If $xy = 18$, then $2xy = 36$. Expanding the first equation, we get $x^2 + 2xy + y^2 = 36$. Because $2xy = 36$, we can subtract these terms from both sides of the equation. Then $x^2 + y^2 = 0$, which contradicts the assumption that $x, y \neq 0$.

65. a. $h = \pm 2r$ **b.** $0 < h < 2r$ or $-2r < h < 0$ **c.** $h = 0$
d. $h > 2r$ or $h < -2r$

Chapter 8 Review Exercises

1. $x = 4, y = 3$ **3.** $y = 5 - x$, x a real number
5.

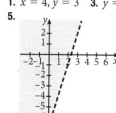

7.

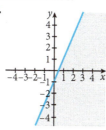

9.

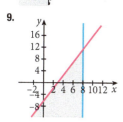

11. $P = 105$
13. $100, 350$ **15.** $x = 1, y = -1, z = -1$
17. $x = 1 - z, y = 3 - z$
19. Soft taco: 190; tostada: 200; rice: 210

21. $\begin{bmatrix} 1 & 2 & -5 & | & 3 \\ 3 & 0 & 1 & | & -1 \end{bmatrix}$ **23.** No solution

25. $y = -\dfrac{41}{8} + \dfrac{1}{4}x$

$z = \dfrac{19}{2}$

27. Consistent; $x = 7, y = 3, z = 2$ **29.** 0 **31.** 3×4

33. $\begin{bmatrix} -1 & 6 \\ 3 & -4 \\ 4 & -9 \end{bmatrix}$ **35.** $\begin{bmatrix} -5 & 7 \\ 5 & -7 \\ 6 & -15 \end{bmatrix}$ **37.** $\begin{bmatrix} -17 & 4 \\ 4 & -3 \end{bmatrix}$

39. $\begin{bmatrix} -1 & 5 \\ 1 & -4 \end{bmatrix}$ **41.** $\begin{bmatrix} -1 & 1 \\ -3 & 4 \end{bmatrix}$ **43.** $\begin{bmatrix} 5 & 1 & 1 \\ 4 & 1 & 1 \\ 10 & 3 & 2 \end{bmatrix}$

45. $x = -6, y = 1$ **47.** 10 **49.** -42 **51.** $x = -2, y = 4$
53. $x = 2, y = \dfrac{3}{4}, z = \dfrac{13}{4}$ **55.** $\dfrac{14}{x + 2} - \dfrac{17}{x + 3}$

57. $\dfrac{1}{x + 2} - \dfrac{x - 2}{x^2 + 5}$ **59.** $(1.56, 2.56), (-2.56, -1.56)$

61. $(0, 0), (1, 1)$ **63.** 15 inches, 18 inches

Chapter 8 Test

1. $x = -2, y = 3$ **2.** $x = 1, y = -2$
3.

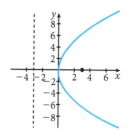

4. $P = 62.5$ at $(3.5, 1)$ **5.** $x = -3, y = -1, z = 2$
6. $x = \dfrac{4}{3} + \dfrac{1}{3}z, y = \dfrac{2}{3} + \dfrac{2}{3}z$, z any real number

7. $\begin{bmatrix} -19 & -2 \\ 11 & -8 \\ -3 & 1 \end{bmatrix}$ **8.** $\begin{bmatrix} 46 & -12 & 33 \\ -20 & -20 & -3 \\ 21 & 8 & 9 \end{bmatrix}$

9. $\begin{bmatrix} 2 & -5 \\ -1 & 3 \end{bmatrix}$ **10.** $\begin{bmatrix} 1 & 0 & -2 \\ 2 & 1 & -4 \\ -4 & -2 & 9 \end{bmatrix}$

11. $x = 13, y = -4, z = 1$ **12.** 36 **13.** $x = 0, y = 2$
14. $x = -\dfrac{7}{4}, y = -\dfrac{5}{4}, z = \dfrac{3}{4}$

15. $-\dfrac{3}{x + 1} + \dfrac{1}{(x + 1)^2} + \dfrac{2}{x}$ **16.** $\dfrac{2x - 1}{x^2 + 1} + \dfrac{3}{x - 3}$

17. $\left(\dfrac{\sqrt{10}}{5}, \dfrac{3\sqrt{10}}{5} \right), \left(-\dfrac{\sqrt{10}}{5}, -\dfrac{3\sqrt{10}}{5} \right)$

18. $(2, 1), (-4, 1)$
19. 600 CX100 models; 100 FX100 models; maximum profit: $100,000
20. 4 first class; 2 business class; 4 coach
21.

	Total
Customer 1	$40
Customer 2	$40
Customer 3	$30

Chapter 9

9.1 Exercises

1. True **3.** 10 **5.** $f(x) = (x - 1)^2$ **7.** True
9. $x^2 + 6x + 9$ **11.** c **13.** b
15. Vertex: $(0, 0)$; focus: $(3, 0)$; directrix: $x = -3$

17. Vertex: $(0, 0)$; focus: $(2, 0)$; directrix: $x = -2$

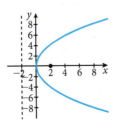

19. Vertex: $(0, 0)$; focus: $(0, -1)$; directrix: $y = 1$

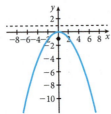

21. Vertex: $(0, 0)$; focus: $\left(\frac{2}{3}, 0\right)$; directrix: $x = -\frac{2}{3}$

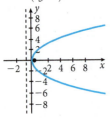

23. Vertex: $(0, 0)$; focus: $\left(\frac{1}{5}, 0\right)$; directrix: $x = -\frac{1}{5}$

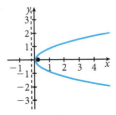

25. Vertex: $(2, 0)$; focus: $(-1, 0)$; directrix: $x = 5$

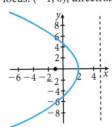

27. Vertex: $(5, -1)$; focus: $(5, -2)$; directrix: $y = 0$

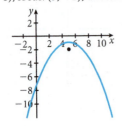

29. Vertex: $(1, 4)$; focus: $\left(\frac{3}{4}, 4\right)$; directrix: $x = \frac{5}{4}$

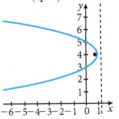

31. Vertex: $(1, 2)$; focus: $(0, 2)$; directrix: $x = 2$

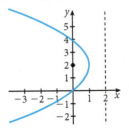

33. Vertex: $(1, 1)$; focus: $\left(1, \frac{3}{4}\right)$; directrix: $y = \frac{5}{4}$

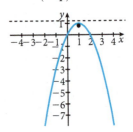

35. Vertex: $(-3, -4)$; focus: $(-3, -5)$; directrix: $y = -3$

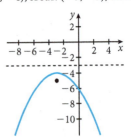

37. Vertex: $(4, -3)$; focus: $\left(4, -\frac{7}{4}\right)$; directrix: $y = -\frac{17}{4}$

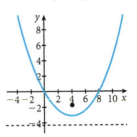

39. $y^2 = 4x$ **41.** $x^2 = 12y$ **43.** $x^2 = 8y$ **45.** $y^2 = -8x$

47. $x^2 = 16y$ **49.** $y^2 = -20x$ **51.** $(x + 2)^2 = 12(y - 1)$

53. $x^2 = 16y$ **55.** $y^2 = 9x$ **57.** $(x - 4)^2 = -(y - 3)$

59. $(x - 4)^2 = 8(y - 1)$ **61.** $(x - 5)^2 = -3(y - 3)$

63. 5.0625 inches from the vertex

65. a. 128 inches   **b.** 12 inches from the ends   **67.** 3.2 feet

69. 5   **71.** 4   **73.** $(x + 7)^2 = 4(y - 4)$ **75.** $(4, 8)$

9.2 Exercises

1. True **3.** $2\sqrt{10}$ **5.** 36 **7.** c **9.** d

11. Center: $(0, 0)$; vertices: $(5, 0)$ and $(-5, 0)$; foci: $(3, 0)$ and $(-3, 0)$

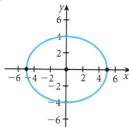

13. Center: $(0, 0)$; vertices: $(0, 5)$ and $(0, -5)$; foci: $(0, 3)$ and $(0, -3)$

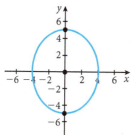

15. Center: $(0, 0)$; vertices: $(3, 0)$ and $(-3, 0)$; foci: $(\sqrt{5}, 0)$ and $(-\sqrt{5}, 0)$

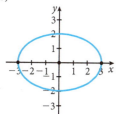

17. Center: $(0, 0)$; vertices: $\left(\frac{3}{2}, 0\right)$ and $\left(-\frac{3}{2}, 0\right)$; foci: $\left(\frac{\sqrt{17}}{6}, 0\right)$ and $\left(-\frac{\sqrt{17}}{6}, 0\right)$

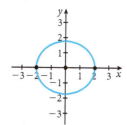

19. Center: $(-3, -1)$; vertices: $(-3, -5)$ and $(-3, 3)$; foci: $(-3, -1 + \sqrt{7})$ and $(-3, -1 - \sqrt{7})$

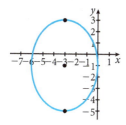

21. Center: $(1, -1)$; vertices: $(11, -1)$ and $(-9, -1)$; foci: $(-7, -1)$ and $(9, -1)$

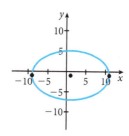

23. Center: $(1, -2)$; vertices: $(-3, -2)$ and $(5, -2)$; foci: $(1 - \sqrt{7}, -2)$ and $(1 + \sqrt{7}, -2)$

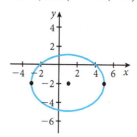

25. Center: $(0, 0)$; vertices: $(-2, 0)$ and $(2, 0)$; foci: $(-1, 0)$ and $(1, 0)$

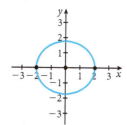

27. Center: $(0, 0)$; vertices: $(0, \sqrt{5})$ and $(0, -\sqrt{5})$; foci: $(0, \sqrt{3})$ and $(0, -\sqrt{3})$

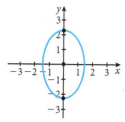

29. Center: $(3, 4)$; vertices: $(3, 6)$ and $(3, 2)$; foci: $(3, 4 - \sqrt{3})$ and $(3, 4 + \sqrt{3})$

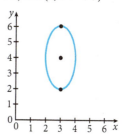

31. Center: $(2, -3)$; vertices: $(5, -3)$ and $(-1, -3)$; foci: $(4, -3)$ and $(0, -3)$

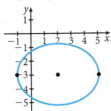

33. Center: $(4, -3)$; vertices: $(4, -4.75)$ and $(4, -1.25)$; foci: $(4, -4.05)$ and $(4, -1.95)$

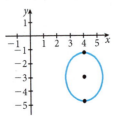

35.

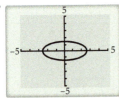

37.

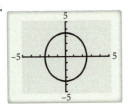

39.

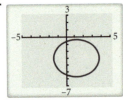

41. $\dfrac{x^2}{36} + \dfrac{y^2}{27} = 1$ **43.** $\dfrac{x^2}{64} + \dfrac{y^2}{39} = 1$ **45.** $\dfrac{4x^2}{49} + \dfrac{4y^2}{81} = 1$

47. $\dfrac{x^2}{4} + \dfrac{y^2}{81} = 1$ **49.** $\dfrac{(x + 2)^2}{16} + \dfrac{(y - 4)^2}{7} = 1$

51. $\dfrac{(x + 3)^2}{5} + \dfrac{(y - 1)^2}{9} = 1$

53. $\dfrac{(x - 4)^2}{12} + \dfrac{(y + 1)^2}{16} = 1$

55. $\dfrac{(x - 7)^2}{36} + \dfrac{(y + 8)^2}{16} = 1$

57. $\dfrac{4x^2}{81} + \dfrac{4y^2}{25} = 1; \dfrac{4x^2}{25} + \dfrac{4y^2}{81} = 1$

59. $\dfrac{(x - 9)^2}{16} + \dfrac{y^2}{36} = 1; \dfrac{(x - 1)^2}{16} + \dfrac{y^2}{36} = 1$

61. $(\sqrt{29.75}, 0), (-\sqrt{29.75}, 0)$

63. Minimum distance: 226,335 miles; maximum distance: 251,401 miles

65. $4\sqrt{34}$ feet **67.** $4\sqrt{13}$ feet **69.** 5 **71.** $\dfrac{x^2}{16} + \dfrac{y^2}{7} = 1$

73. $y = \pm\sqrt{a^2 - x^2}$, a circle

9.3 Exercises

1. $d = \sqrt{(c - w)^2 + (d - v)^2}$ **3.** 3

5. $x^2 + 22x + 121$

7. d **9.** c

11. Center: $(0, 0)$; vertices: $(4, 0)$ and $(-4, 0)$; foci: $(-5, 0)$ and $(5, 0)$; asymptotes: $y = \dfrac{3}{4}x$ and $y = -\dfrac{3}{4}x$

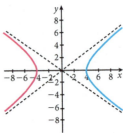

13. Center: $(0, 0)$; vertices: $(0, 3)$ and $(0, -3)$; foci: $(0, -5)$ and $(0, 5)$; asymptotes: $y = \dfrac{3}{4}x$ and $y = -\dfrac{3}{4}x$

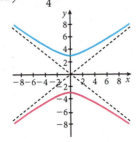

15. Center: $(0, 0)$; vertices: $(0, 8)$ and $(0, -8)$; foci: $(0, -10)$ and $(0, 10)$; asymptotes: $y = \dfrac{4}{3}x$ and $y = -\dfrac{4}{3}x$

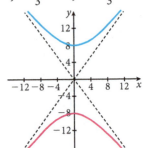

17. Center: $(0, 0)$; vertices: $(3, 0)$ and $(-3, 0)$; foci: $\left(-3\sqrt{2}, 0\right)$ and $\left(3\sqrt{2}, 0\right)$; asymptotes: $y = x$ and $y = -x$

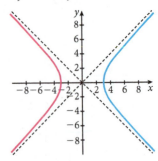

19. Center: $(-3, -1)$; vertices: $(-7, -1)$ and $(1, -1)$;
   foci: $(2, -1)$ and $(-8, -1)$;
   asymptotes: $y = \dfrac{3}{4}x + \dfrac{5}{4}$ and $y = -\dfrac{3}{4}x - \dfrac{13}{4}$

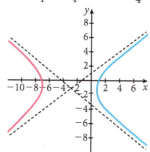

21. Center: $(-5, -1)$; vertices: $(-10, -1)$ and $(0, -1)$;
   foci: $\left(-5 + \sqrt{41}, -1\right)$ and $\left(-5 - \sqrt{41}, -1\right)$;
   asymptotes: $y = \dfrac{4}{5}x + 3$ and $y = -\dfrac{4}{5}x - 5$

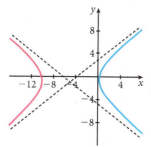

23. Center: $(-4, 0)$; vertices: $(-4, 1)$ and $(-4, -1)$;
   foci: $\left(-4, \sqrt{2}\right)$ and $\left(-4, -\sqrt{2}\right)$;
   asymptotes: $y = x + 4$ and $y = -x - 4$

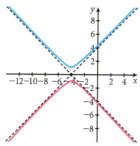

25. Center: $(-1, 3)$; vertices: $(-1, 0)$ and $(-1, 6)$;
   foci: $\left(-1, 3 + \sqrt{34}\right)$ and $\left(-1, 3 - \sqrt{34}\right)$;
   asymptotes: $y = \dfrac{3}{5}x + \dfrac{18}{5}$ and $y = -\dfrac{3}{5}x + \dfrac{12}{5}$

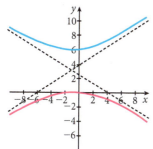

27. Center: $(-2, 3)$; vertices: $(10, 3)$ and $(-14, 3)$;
   foci: $(-15, 3)$ and $(11, 3)$;
   asymptotes: $y = \dfrac{5}{12}x + \dfrac{23}{6}$ and $y = -\dfrac{5}{12}x + \dfrac{13}{6}$

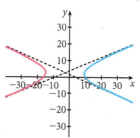

29. Center: $(0, 0)$; vertices: $\left(\dfrac{5}{2}, 0\right)$ and $\left(-\dfrac{5}{2}, 0\right)$;
   foci: $\left(-\dfrac{5\sqrt{13}}{4}, 0\right)$ and $\left(\dfrac{5\sqrt{13}}{4}, 0\right)$;
   asymptotes: $y = \dfrac{3}{2}x$ and $y = -\dfrac{3}{2}x$

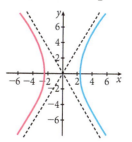

31. Center: $(-3, 0)$; vertices: $(-6, 0)$ and $(0, 0)$;
   foci: $\left(-3 - 3\sqrt{10}, 0\right)$ and $\left(-3 + 3\sqrt{10}, 0\right)$;
   asymptotes: $y = 3x + 9$ and $y = -3x - 9$

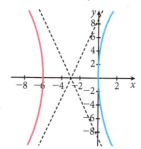

33. Center: $(2, -3)$; vertices: $\left(2 + 2\sqrt{2}, -3\right)$ and
   $\left(2 - 2\sqrt{2}, -3\right)$; foci: $\left(2 - 6\sqrt{2}, -3\right)$ and
   $\left(2 + 6\sqrt{2}, -3\right)$; asymptotes: $y = 2\sqrt{2}x - 4\sqrt{2} - 3$
   and $y = -2\sqrt{2}x + 4\sqrt{2} - 3$

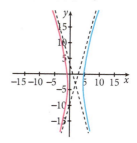

35. $\dfrac{x^2}{9} - \dfrac{y^2}{7} = 1$ **37.** $\dfrac{5x^2}{16} - \dfrac{5y^2}{64} = 1$

39. $\dfrac{x^2}{1} - \dfrac{y^2}{3} = 1$ **41.** $\dfrac{x^2}{16} - \dfrac{3y^2}{4} = 1$

43. $\dfrac{(y+4)^2}{2} - \dfrac{(x+3)^2}{2} = 1$ **45.** $\dfrac{(x-3)^2}{4} - \dfrac{(y+2)^2}{25} = 1$

47. $\dfrac{(x-1)^2}{25} - \dfrac{(y+4)^2}{39} = 1$

49. 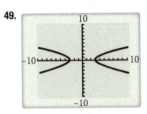   **51.**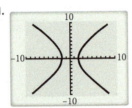

53.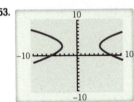

55. $\dfrac{x^2}{36} - \dfrac{y^2}{576} = 1; \dfrac{y^2}{36} - \dfrac{4x^2}{9} = 1$

57. $\dfrac{(y-4)^2}{9} - \dfrac{(x+1)^2}{40} = 1; \dfrac{(y+2)^2}{9} - \dfrac{(x+1)^2}{40} = 1$

59. $4\sqrt{2}$ inches

61. $\dfrac{x^2}{25{,}000{,}000{,}000{,}000} - \dfrac{y^2}{144{,}000{,}000{,}000{,}000} = 1$

63. $x \geq 3$ or $x \leq -3$

65. $2c$ **67.** $\dfrac{(y+3)^2}{9} - \dfrac{(x-3)^2}{16} = 1$

69. The hyperbola $xy = 10$ lies in the first and third quadrants. The hyperbola $xy = -10$ lies in the second and fourth quadrants.

9.4 Exercises

1. $\left(\dfrac{\sqrt{2}}{2}, \dfrac{\sqrt{2}}{2}\right)$ **3.** $\left(\dfrac{1 - 2\sqrt{3}}{2}, \dfrac{\sqrt{3} + 2}{2}\right)$

5. $\left(-\sqrt{3} - 2, -1 + 2\sqrt{3}\right)$ **7.** $(0, -\sqrt{2})$ **9.** Parabola

11. Ellipse   **13.** Hyperbola   **15.** Parabola   **17.** $(0, 0)$
19. $(0, -2), (0, 2)$ **21.** $(-3, 0), (3, 0)$ **23.** $(-6, 0), (6, 0)$

25. a. Hyperbola   **b.** $\dfrac{u^2}{8} - \dfrac{v^2}{8} = 1$

  c. Vertices: $(-2\sqrt{2}, 0), (2\sqrt{2}, 0)$

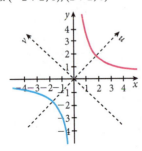

27. a. Hyperbola   **b.** $\dfrac{u^2}{1} - \dfrac{v^2}{4} = 1$

  c. Vertices: $(-1, 0), (1, 0)$

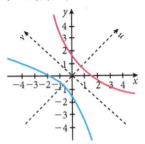

29. a. Ellipse   **b.** $\dfrac{u^2}{9} + \dfrac{v^2}{1} = 1$

  c. Vertices: $(-3, 0), (3, 0)$

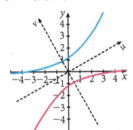

31. a. Hyperbola   **b.** $\dfrac{v^2}{1} - \dfrac{u^2}{4} = 1$

  c. Vertices: $(0, -1), (0, 1)$

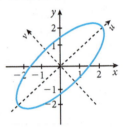

33. a. Parabola   **b.** $v^2 = \dfrac{-1}{4}u$ **c.** Vertex: $(0, 0)$

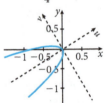

35. $(x = u \cos\theta - v \sin\theta)(-\sin\theta)$
  $(y = u \sin\theta + v \cos\theta)(\cos\theta)$
  $\overline{-x \sin\theta + y \cos\theta = v(\sin^2\theta + \cos^2\theta)}$
  $-x \sin\theta + y \cos\theta = v$
  Similarly, eliminate v to get $u = x \cos\theta + y \sin\theta$.

37. $C = \dfrac{1}{4}$

9.5 Exercises

1. True   **3.** Circle   **5.** Hyperbola   **7.** Parabola
9. Parabola   **11.** Ellipse

13. Parabola
$$r = \frac{1}{1 - \sin \theta}$$

15. Hyperbola
$$r = \frac{4}{1 + 2 \cos \theta}$$

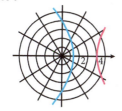

17. Ellipse
$$r = \frac{1}{1 - 0.5 \cos \theta}$$

19. Ellipse
$$r = \frac{6}{6 + 3 \sin \theta}$$

21. Hyperbola
$$r = \frac{18}{6 + 12 \cos \theta}$$

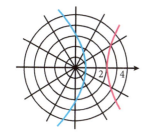

23. Hyperbola
$$r = \frac{6}{6 - 8 \sin \theta}$$

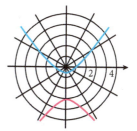

25. Parabola
$$r = \frac{4}{3 - 3 \cos \theta}$$

27. $r = \dfrac{1}{1 + \cos \theta}$; parabola

29. $r = \dfrac{\frac{1}{3}}{1 - \frac{1}{3} \sin \theta}$; ellipse

31. $r = \dfrac{4}{1 - 4 \cos \theta}$; hyperbola

33.

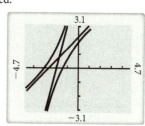

35. It is rotated.

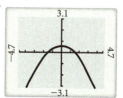

36. It is rotated.

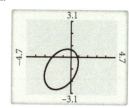

37. $r = \dfrac{2}{1 - \cos \theta}$ **39.** $r = \dfrac{9}{10 - 3 \sin \theta}$

9.6 Exercises

1.

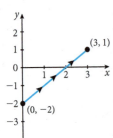

3.

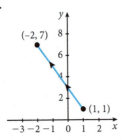

5.

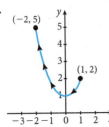

7.

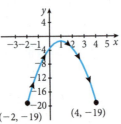

9.

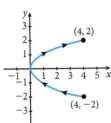

11. $y = -(x + 2)^2, -5 \leq x \leq 0$

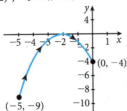

13. $y = \dfrac{1}{x}, 0 < x \leq 1$

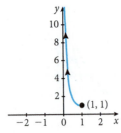

15. $x^2 + y^2 = 9, -3 \leq x \leq 3$

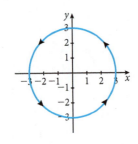

17. $\dfrac{x^2}{9} + \dfrac{y^2}{16} = 1, -3 \leq x \leq 3$

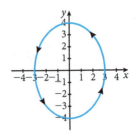

19. $y = \dfrac{1}{x}, 0 < x \leq 1$

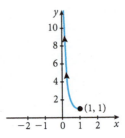

21. Same graph for both intervals. Point moves from $(1, 1)$ to $(-1, 1)$ and back again.

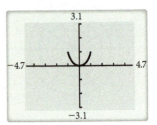

23. No; $-1 \leq \cos t \leq 1$, so $0 \leq y \leq 4$.

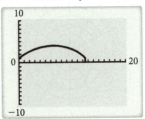

25. Common rectangular equation: $y = x$
a. Graph and restrictions:

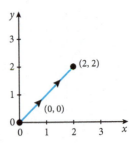

$0 \leq x \leq 2, 0 \leq y \leq 2$

b. Graph and restrictions:

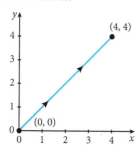

$0 \le x \le 4, 0 \le y \le 4$

27. Common rectangular equation: $y = x + 1$
   a. Graph and restrictions:

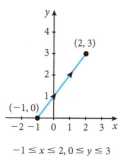

$-1 \le x \le 2, 0 \le y \le 3$

   b. Graph and restrictions:

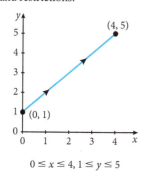

$0 \le x \le 4, 1 \le y \le 5$

29. a. $x = (60\sqrt{3})t$ and $y = -16t^2 + 60t$
   b. Time ≈ 1.88 seconds; maximum height ≈ 56.3 feet
   c. 390 feet
31. The maximum distance the ball will travel is approximately 275 feet.
33. $x = 10 \cos\left(\dfrac{2\pi}{15}t\right)$ and $y = 10 \sin\left(\dfrac{2\pi}{15}t\right), 0 \le t \le 15$

35. 101 feet per second
37.

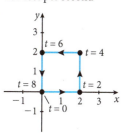

39. No. There is no value of t for which $x(t) = y(t) = 0$.

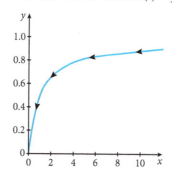

41. Despite the two having the common rectangular equation $y = x + 1$, the restrictions on x and y are different. For part (a), $0 \le x \le 1$, $1 \le y \le 2$; for part (b), $-1 \le x \le 0$, $0 \le y \le 1$.

Chapter 9 Review Exercises

1. Vertex: $(0, 0)$; focus: $(0, 3)$; directrix: $y = -3$

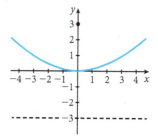

3. Vertex: $(0, 0)$; focus: $(1, 0)$; directrix: $x = -1$

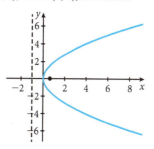

5. Vertex: $(3, -9)$; focus: $(3, -8.75)$; directrix: $y = -9.25$

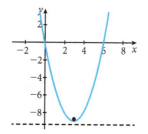

7. Vertex: $(5, -5)$; focus: $(3.75, -5)$; directrix: $x = 6.25$

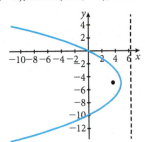

9. Vertex: $(0, 0)$; focus: $\left(\dfrac{1}{28}, 0\right)$; directrix: $x = -\dfrac{1}{28}$

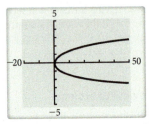

11. $x^2 = -\dfrac{2}{3}y$

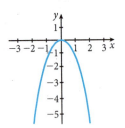

13. $x^2 = -20y$

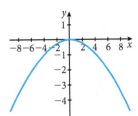

15. $y^2 = 16(x - 2)$

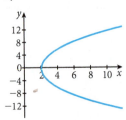

17. $\dfrac{5\sqrt{3}}{3}$ feet

19. Vertices: $(2, 0)$ and $(-2, 0)$;
foci: $\left(-\sqrt{3}, 0\right)$ and $\left(\sqrt{3}, 0\right)$

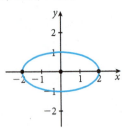

21. Vertices: $(3, 1)$ and $(-7, 1)$; foci: $(1, 1)$ and $(-5, 1)$

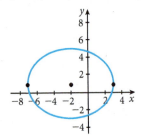

23. Vertices: $(-2, 4)$ and $(-2, -4)$;
foci: $\left(-2, -\sqrt{7}\right)$ and $\left(-2, \sqrt{7}\right)$

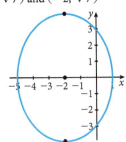

25. Vertices: $\left(0, \sqrt{7}\right)$ and $\left(0, -\sqrt{7}\right)$;
foci: $\left(0, \dfrac{\sqrt{42}}{3}\right)$ and $\left(0, -\dfrac{\sqrt{42}}{3}\right)$

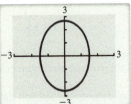

27. Vertices: $\left(2 + 2\sqrt{6}, 0\right)$ and $\left(2 - 2\sqrt{6}, 0\right)$;
foci: $\left(2 + 2\sqrt{5}, 0\right)$ and $\left(2 - 2\sqrt{5}, 0\right)$

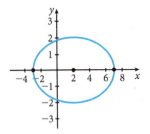

29. Vertices: $\left(-3, 3 - \sqrt{7}\right)$ and $\left(-3, 3 + \sqrt{7}\right)$;
foci: $\left(-3, 3 - \sqrt{3}\right)$ and $\left(-3, 3 + \sqrt{3}\right)$

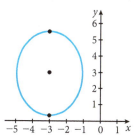

31. $\dfrac{x^2}{16} + \dfrac{y^2}{25} = 1$ **33.** $\dfrac{x^2}{20} + \dfrac{y^2}{36} = 1$ **35.** $\dfrac{x^2}{25} + \dfrac{y^2}{16} = 1$

37. $20\sqrt{15}$ Au

39. Center: $(0, 0)$; vertices: $(0, -6)$ and $(0, 6)$;
foci $\left(0, -\sqrt{61}\right)$ and $\left(0, \sqrt{61}\right)$; asymptotes: $y = \pm\dfrac{6}{5}x$

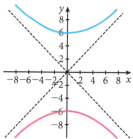

41. Center: $(1, -2)$; vertices: $(1, -4)$ and $(1, 0)$;
foci: $\left(1, -2 - 2\sqrt{2}\right)$ and $\left(1, -2 + 2\sqrt{2}\right)$;
asymptotes: $y = x - 3$ and $y = -x - 1$

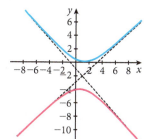

43. Center: $(0, -2)$; vertices: $(-2, -2)$ and $(2, -2)$;
foci: $\left(-\sqrt{13}, -2\right)$ and $\left(\sqrt{13}, -2\right)$;
asymptotes: $y = \dfrac{3}{2}x - 2$ and $y = -\dfrac{3}{2}x - 2$

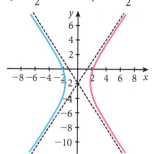

45. Center: $(-2, 0)$; vertices: $(-2, -1)$ and $(-2, 1)$;
foci: $\left(-2, -\sqrt{17}\right)$ and $\left(-2, \sqrt{17}\right)$;
asymptotes: $y = \dfrac{1}{4}x + \dfrac{1}{2}$ and $y = -\dfrac{1}{4}x - \dfrac{1}{2}$

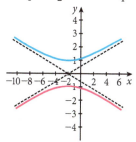

47. Center: $(0, 0)$; vertices: $\left(-2\sqrt{2}, 0\right)$ and $\left(2\sqrt{2}, 0\right)$;
foci: $\left(-6\sqrt{2}, 0\right)$ and $\left(6\sqrt{2}, 0\right)$;
asymptotes: $y = 2\sqrt{2}x$ and $y = -2\sqrt{2}x$

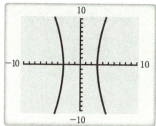

49. $\dfrac{y^2}{4} - \dfrac{x^2}{32} = 1$ **51.** $\dfrac{(y + 4)^2}{9} - \dfrac{(x - 3)^2}{16} = 1$

53. $\dfrac{(x + 4)^2}{16} - \dfrac{(y - 2)^2}{20} = 1$ **55.** $\left(1, -\sqrt{3}\right)$

57. $\left(-2\sqrt{3}, -2\right)$

59. a. Hyperbola **b.** $\dfrac{u^2}{12} - \dfrac{v^2}{12} = 1$
  c. Vertices: $\left(-2\sqrt{3}, 0\right)$, $\left(2\sqrt{3}, 0\right)$

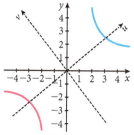

61. a. Ellipse **b.** $\dfrac{u^2}{4} + \dfrac{v^2}{2} = 1$ **c.** Vertices: $(-2, 0)$, $(2, 0)$

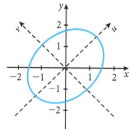

63. Parabola

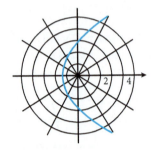

65. Parabola

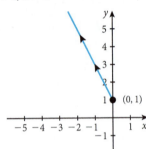

67. $y = -2x + 1, x \leq 0$

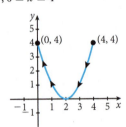

69. $y = (2 - x)^2, 0 \leq x \leq 4$

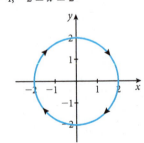

71. $x^2 + y^2 = 4, -2 \leq x \leq 2$

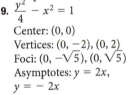

73. a. $x = 60\sqrt{3}\, t$ and $y = -16t^2 + 60t + 4$
   b. Time $= 1.875$ Sec.; Maximum height $= 60.25$ ft.

Chapter 9 Test

1. $(y + 3)^2 = -12(x + 1)$
   Vertex: $(-1, -3)$
   Focus: $(-4, -3)$
   Directrix: $x = 2$

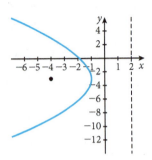

2. $y^2 - 4y + 4x = 0$
   Vertex: $(1, 2)$
   Focus: $(0, 2)$
   Directrix: $x = 2$

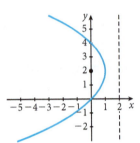

3. $x^2 = 8(y + 2)$ **4.** $y^2 = 12x$

5. $\dfrac{(x + 2)^2}{16} + \dfrac{(y - 3)^2}{25} = 1$
   Vertices: $(-2, -2), (-2, 8)$
   Foci: $(-2, 0), (-2, 6)$

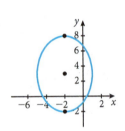

6. $\dfrac{x^2}{9} + \dfrac{(y + 1)^2}{4} = 1$
   Vertices: $(-3, -1), (3, -1)$
   Foci: $(-\sqrt{5}, -1), (\sqrt{5}, -1)$

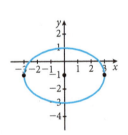

7. $\dfrac{(x + 1)^2}{1} + \dfrac{y^2}{4} = 1$ **8.** $\dfrac{x^2}{16} + \dfrac{y^2}{25} = 1$

9. $\dfrac{y^2}{4} - x^2 = 1$
   Center: $(0, 0)$
   Vertices: $(0, -2), (0, 2)$
   Foci: $(0, -\sqrt{5}), (0, \sqrt{5})$
   Asymptotes: $y = 2x$,
   $y = -2x$

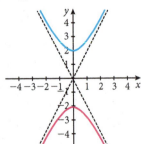

10. $\dfrac{(x+2)^2}{9} - \dfrac{y^2}{16} = 1$

Center: $(-2, 0)$

Vertices: $(-5, 0), (1, 0)$

Foci: $(-7, 0), (3, 0)$

Asymptotes: $y = \dfrac{4}{3}(x+2)$,

$y = -\dfrac{4}{3}(x+2)$

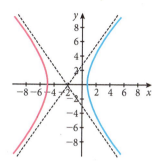

11. $\dfrac{x^2}{1} - \dfrac{y^2}{8} = 1$ **12.** $\dfrac{y^2}{1} - \dfrac{9x^2}{1} = 1$

13. a. Parabola   **b.** $v^2 = 2u$ **c.** Vertex: $(0, 0)$

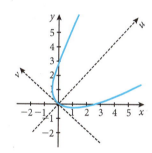

14. a. Ellipse   **b.** $\dfrac{u^2}{1} + \dfrac{v^2}{9} = 1$ **c.** Vertices: $(0, -3), (0, 3)$

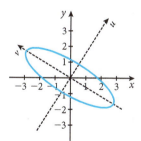

15. Hyperbola

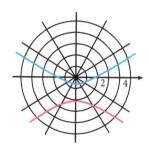

16. Ellipse

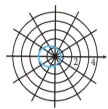

17. $y = -(2-x)^2 + 1, 0 \le x \le 4$

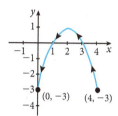

18. $\dfrac{x^2}{9} + \dfrac{y^2}{1} = 1, -3 \le x \le 3$

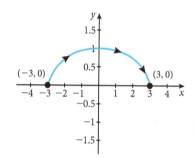

19. 16 feet from the center

20. $\dfrac{1}{16}$ inch from the vertex

21. a. $x = 50\sqrt{3}\,t$ and $y = -16t^2 + 50t + 3$
    b. 276 feet

Chapter 10

10.1 Exercises

1. True   **3.** \$108.24   **5.** 4, 10, 16   **7.** $-5, -2, 1$

9. $-4, -8, -12$ **11.** 8, 6, 4   **13.** 7, 28, 112   **15.** 5, 15, 45

17. $-2, -6, -18$ **19.** $-3, -15, -75$ **21.** $a_n = -3 + 6n$

23. $g(n) = -16 + 4n$ **25.** $b_n = -20 + 3n$

27. $a_n = 10 + 5n$ **29.** $a_n = 21 - 2n$ **31.** $b_n = 10 + \dfrac{1}{2}n$

33. $a_n = 3(2)^n$ **35.** $h(n) = 2(4)^n$ **37.** $b_n = 3\left(\dfrac{1}{2}\right)^n$

39. $a_n = 4(2)^n$ **41.** $b_n = 4\left(\dfrac{3}{2}\right)^n$ **43.** $a_n = 4(5)^n$

45.

n	0	1	2	3
a_n	2	5	8	11

47.

n	0	1	2	3
a_n	-2	2	6	10

49.

n	0	1	2	3
a_n	5	3	1	-1

51.

n	0	1	2	3
a_n	1	3	9	27

53.

n	0	1	2	3
a_n	2	10	50	250

55.

n	0	1	2	3
a_n	$\frac{1}{2}$	$\frac{1}{4}$	$\frac{1}{8}$	$\frac{1}{16}$

57. Arithmetic   **59.** Geometric   **61.** Arithmetic
63. Geometric   **65.** Arithmetic   **67.** Geometric
69. a.

Year	Total Value ($)
1	2120
2	2240
3	2360
4	2480

   b. $V = 2000 + 120n$
71. a. Arithmetic   **b.** $a_n = 100 + 4(n-1)$ **c.** 18 rows
73. a. Geometric   **b.** 440 Hz, 880 Hz   **c.** $a_n = 55(2)^{n-1}$
75. Geometric; $50,613
77. a. 160,000   **b.** Geometric   **c.** 7 hours
79. a. Arithmetic; each term is increased by 4.
   b. Neither; it is a sequence of squared terms.
81. a. $f(n) = 550.1 + 28.11n$ **b.** $775 billion
83. $a_0, a_0, a_0, \ldots$
85. a. $a_n = -3 + 4n$ **b.** $a_n = \frac{1}{3}(3)^n$
   c. $a_4 = 13, a_5 = 17, a_6 = 21, a_7 = 25, a_8 = 29$
   d. $a_4 = 27, a_5 = 81, a_6 = 243, a_7 = 729,$
      $a_8 = 2187$
   e. The sequence in part (b) because it is geometric
87. $(1 + x)^4, (1 + x)^5; r = (1 + x)$
89. A geometric sequence; each term can be rewrittten in
   the form $a_n^3 = a_0^3(r^3)^n$.

10.2 Exercises

1. False   **3.** $1040.60   **5.** 315   **7.** 371   **9.** 483
11. 765   **13.** $11\frac{61}{64}$ **15.** $98\frac{33}{64}$ **17.** 1395   **19.** 520
21. $\frac{n}{2}(5n + 8)$ **23.** 97,656   **25.** 2186   **27.** 3965
29. 3195   **31.** 4288   **33.** 2632   **35.** 42   **37.** 165   **39.** $1\frac{15}{16}$
41. 968   **43. a.** $\sum_{i=0}^{19} 2 + 2i$ **b.** 420   **45. a.** $\sum_{i=1}^{60} ia$ **b.** 1830a
47. a. $\sum_{i=1}^{10} 2^i$ **b.** 2046   **49. a.** $\sum_{i=1}^{40} a^i$ **b.** $a\left(\frac{1-a^{40}}{1-a}\right)$
51. a. $\sum_{i=0}^{24} 2.5i$ **b.** 750   **53. a.** $\sum_{i=0}^{44} (0.5)^i$ **b.** 2
55. 49, arithmetic, $d = 2$ **57.** 105, arithmetic, $d = 0.5$

59. $\frac{127}{2}$, geometric, $r = 2$ **61.** $\frac{63}{64}$, geometric, $r = 0.5$
63. 1.9375, geometric, $r = 0.5$ **65.** 108, arithmetic, $d = 6$
67. 12   **69.** $\frac{27}{2}$ **71.** No   **73.** 4   **75.** $\frac{15}{4}$
77. a. 10
   b. The next two possible arrangements can be made
      with 66 and 78 boxes.
   c. 11; 4
79. $50,639.99   **81.** 36.25 feet
83. a. $\sum_{i=0}^{4} 7^i$; geometric   **b.** 2801
85. $\sum_{i=0}^{7} 4 + 4i$; 144   **87.** 52   **89.** $\frac{48}{31}$

10.3 Exercises

1. $6, 2, -2, -6, -10$
3. $-4, -\frac{4}{3}, -\frac{4}{9}, -\frac{4}{27}, -\frac{4}{81}$
5. $4, 5, 8, 13, 20$
7. $\frac{1}{3}, \frac{2}{9}, \frac{3}{19}, \frac{4}{33}, \frac{5}{51}$
9. $e, -e^2, e^3, -e^4, e^5$
11. $\frac{3}{2}, 1, \frac{9}{10}, 1, \frac{33}{26}$
13. $2, \sqrt{6}, \sqrt{8}, \sqrt{10}, \sqrt{12}$
15. $2, 3, 6, 11, 18$ **17.** $0, 2, 16, 54, 128$
19. $a_n = -2 - 4n, n = 0, 1, 2, 3, \ldots$
21. $a_n = \frac{1}{2^n}, n = 0, 1, 2, 3, \ldots$
23. $a_n = \sqrt{n}, n = 1, 2, 3, \ldots$
25. $a_n = (0.4)^n, n = 0, 1, 2, 3, \ldots$
27. $a_n = \sqrt{3n}, n = 1, 2, 3, \ldots$
29. $6, 4, 2, 0$
31. $4, 1, \frac{1}{4}, \frac{1}{16}$
33. $-1, 0, 2, 5$
35. $\sqrt{3}, \sqrt{\sqrt{3} + 3}, \sqrt{\sqrt{\sqrt{3} + 3} + 3},$
   $\sqrt{\sqrt{\sqrt{\sqrt{3} + 3} + 3} + 3}$
37. $-4, -2, 0, 2; a_n = -4 + 2n, n = 0, 1, 2, 3, \ldots$
39. $6, 3, \frac{3}{2}, \frac{3}{4}; b_n = 6\left(\frac{1}{2}\right)^{n-1}; n = 1, 2, 3, \ldots$
41. a. $p_1 = 200, p_n = 1.04(p_{n-1})$ **b.** $p_n = 200(1.04)^n$
   c. $224.97   **d.** 6 years   **e.** 11 years
43. a. $P_n = 2(36) + 2(18 + 0.5n) = 108 + n$
   b. $A_n = (36)(18 + 0.5n) = 648 + 18n$
   c. 36 inches $\times$ 20.5 inches, 113 inches, 738 square
      inches
   d. $3330
45. a. $0.06   **b.** $7.56   **c.** $9.66
47. $n = 5$ **49.** 4
51. $a_0 < 0$

10.4 Exercises

1. 24   **3.** 60   **5.** 30   **7.** 24   **9.** 2520   **11.** 6720   **13.** 4
15. 56   **17.** 28   **19.** 1   **21.** 100
23. BCZ, BZC, CZB, CBZ, ZBC, ZCB
25. John, Maria, Susan; Maria, Susan, Angelo; John, Susan,
   Angelo; John, Maria, Angelo
27. 24   **29.** 12   **31.** 220   **33.** 45   **35.** 4845   **37.** 336

39. 38,760 **41.** 175,560 **43.** 1350 **45.** 2730 **47.** 35
49. 17,576,000 **51.** 1,010,129,120 **53.** 635,013,559,600
55. 240 (if the bride and groom can switch positions)
57. 56,800,235,584; 37,029,625,920
59. a. 6 **b.** 6 **c.** 36 **d.** 66
61. $(n-1) + (n-2) + \cdots + 1$
63. 5; 20; $2(n-3) + (n-4) + (n-5) + \cdots + 1$ **65.** 34,650

10.5 Exercises

1. {HH, HT, TH, TT} **3.** {TT} **5.** {1, 3, 5} **7.** $\dfrac{1}{3}$
9. {2 of spades, 3 of spades, 4 of spades, 5 of spades, 6 of spades, 7 of spades, 8 of spades, 9 of spades, 10 of spades, jack of spades, queen of spades, king of spades, ace of spades}
11. $\dfrac{1}{52}$ **13.** {quarter, dime, nickel, penny} **15.** $\dfrac{1}{4}$
17. False **19.** True **21.** True
23.

Number of Heads	Probability
0	$\dfrac{1}{16}$
1	$\dfrac{1}{4}$
2	$\dfrac{3}{8}$
3	$\dfrac{1}{4}$
4	$\dfrac{1}{16}$

If n = number of heads, $P(n) = P(4-n)$. No.

25. $\dfrac{1}{13}$ **27.** $\dfrac{1}{52}$ **29.** $\dfrac{3}{13}$ **31.** $\dfrac{1}{6}$ **33.** $\dfrac{2}{9}$ **35.** 10,000
37. 0.504 **39.** $\dfrac{1}{9}$ **41.** $\dfrac{1}{16}$ **43.** $\dfrac{1}{4}$
45. This answer overcounts a card, since there is one spade that is a king. $\dfrac{4}{13}$
47. $\dfrac{9}{2,500}$ **49.** Neither
51. a. {YYYY, YYYN, YYNY, YNYY, NYYY, NNYY, YYNN, NYNY, YNYN, NYYN, YNNY, YNNN, NYNN, NNYN, NNNY, NNNN}
   b. $\dfrac{3}{8}$ **c.** $\dfrac{1}{16}$
53. a. 28 **b.** $\dfrac{5}{28}$

10.6 Exercises

1. $a^5, a^4b, a^3b^2, a^2b^3, ab^4, b^5$
3. $x^7, x^6y, x^5y^2, x^4y^3, x^3y^4, x^2y^5, xy^6, y^7$
5. 24 **7.** 3 **9.** 15 **11.** 21 **13.** 1 **15.** 1
17. $x^4 + 8x^3 + 24x^2 + 32x + 16$
19. $8x^3 - 12x^2 + 6x - 1$
21. $243 + 405y + 270y^2 + 90y^3 + 15y^4 + y^5$
23. $x^4 - 12x^3z + 54x^2z^2 - 108xz^3 + 81z^4$
25. $x^6 + 3x^4 + 3x^2 + 1$
27. $y^4 - 8y^3x + 24y^2x^2 - 32yx^3 + 16x^4$
29. 160 **31.** 2916 **33.** 28 **35.** $240x^4$ **37.** $1024y^5$
39. $2160x^2$ **41.** $-10,240x^3$
43. $\dbinom{n}{r} = \dfrac{n!}{(n-r)!r!} = \dfrac{n!}{(n-r)!(n-(n-r))!} = \dbinom{n}{n-r}$
45. 1

10.7 Exercises

1. $(k+1)(k+2)(k+3)$ **3.** $\dfrac{k+1}{k+2}$
5. $P_1: 2(1) + 1 = 1(1+2)$
Assume $P_k: 3 + 5 + \cdots + (2k+1) = k(k+2)$.
Prove $P_{k+1}: 3 + 5 + \cdots + (2k+1)$
$\qquad\qquad + (2(k+1) + 1) = (k+1)(k+3)$.
$3 + 5 + \cdots + (2k+1) + (2(k+1) + 1)$
$\qquad = k(k+2) + (2(k+1) + 1)$
$\qquad = k^2 + 2k + 2k + 2 + 1$
$\qquad = k^2 + 4k + 3$
$\qquad = (k+1)(k+3)$

7. $P_1: 3(1) - 2 = \dfrac{1(3(1)-1)}{2}$
Assume $P_k: 1 + 4 + 7 + \cdots + (3k-2) = \dfrac{k(3k-1)}{2}$.
Prove $P_{k+1}: 1 + 4 + 7 + \cdots + (3k-2) + (3(k+1)-2)$
$\qquad\qquad = \dfrac{(k+1)(3(k+1)-1)}{2}$.
$1 + 4 + 7 + \cdots + (3k-2) + (3(k+1)-2)$
$\qquad = \dfrac{k(3k-1)}{2} + (3(k+1)-2)$
$\qquad = \dfrac{k(3k-1)}{2} + 3k + 1$
$\qquad = \dfrac{3k^2 + 5k + 2}{2}$
$\qquad = \dfrac{(k+1)(3k+2)}{2}$
$\qquad = \dfrac{(k+1)(3(k+1)-1)}{2}$

9. $P_1: 9 - 2(1) = -(1)^2 + 8(1)$
Assume $P_k: 7 + 5 + 3 + \cdots + (9-2k) = -k^2 + 8k$.
Prove $P_{k+1}: 7 + 5 + 3 + \cdots + (9-2k) + (9-2(k+1))$
$\qquad\qquad = -(k+1)^2 + 8(k+1)$.
$7 + 5 + 3 + \cdots + (9-2k) + (9-2(k+1))$
$\qquad = -k^2 + 8k + (9-2(k+1))$
$\qquad = -k^2 + 6k + 7$
$\qquad = -(k^2 + 2k + 1) + 8k + 8$
$\qquad = -(k+1)^2 + 8(k+1)$

11. $P_1: 3(1) - 1 = \dfrac{1}{2}(1)(3(1)+1)$
Assume $P_k: 2 + 5 + 8 + \cdots + (3k-1) = \dfrac{1}{2}k(3k+1)$.
Prove $P_{k+1}: 2 + 5 + 8 + \cdots + (3k-1) + (3(k+1)-1)$
$\qquad\qquad = \dfrac{1}{2}(k+1)(3(k+1)+1)$.
$2 + 5 + 8 + \cdots + (3k-1) + (3(k+1)-1)$
$\qquad = \dfrac{1}{2}k(3k+1) + (3(k+1)-1)$
$\qquad = \dfrac{1}{2}(3k^2 + k) + (3k+2)$
$\qquad = \dfrac{1}{2}(3k^2 + k) + \dfrac{1}{2}(6k+4)$
$\qquad = \dfrac{1}{2}(3k^2 + k + 6k + 4)$
$\qquad = \dfrac{1}{2}(3k^2 + 7k + 4)$
$\qquad = \dfrac{1}{2}(k+1)(3k+4)$
$\qquad = \dfrac{1}{2}(k+1)(3(k+1)+1)$

13. $P_1: (1)^3 = \dfrac{1^2((1)+1)^2}{4}$

   Assume $P_k: 1^3 + 2^3 + \cdots + k^3 = \dfrac{k^2(k+1)^2}{4}$.

   Prove $P_{k+1}: 1^3 + 2^3 + \cdots + k^3 + (k+1)^3$
   $$= \dfrac{(k+1)^2(k+2)^2}{4}.$$

   $1^3 + 2^3 + \cdots + k^3 + (k+1)^3 = \dfrac{k^2(k+1)^2}{4} + (k+1)^3$

   $= \dfrac{k^2(k+1)^2 + 4(k+1)^3}{4}$

   $= \dfrac{(k+1)^2(k^2 + 4(k+1))}{4}$

   $= \dfrac{(k+1)^2(k^2 + 4k + 4)}{4}$

   $= \dfrac{(k+1)^2(k+2)^2}{4}$

15. $P_1: (2(1)-1)^2 = \dfrac{1(2(1)-1)(2(1)+1)}{3}$

   Assume $P_k: 1^2 + 3^2 + \cdots + (2k-1)^2$
   $$= \dfrac{k(2k-1)(2k+1)}{3}.$$

   Prove $P_{k+1}: 1^2 + 3^2 + \cdots + (2k-1)^2 + (2(k+1)-1)^2$
   $= \dfrac{(k+1)(2(k+1)-1)(2(k+1)+1)}{3}$.

   $1^2 + 3^2 + \cdots + (2k-1)^2 + (2(k+1)-1)^2$

   $= \dfrac{k(2k-1)(2k+1)}{3} + (2(k+1)-1)^2$

   $= \dfrac{k(4k^2-1)}{3} + \dfrac{3(2k+1)^2}{3}$

   $= \dfrac{4k^3 - k + 3(4k^2 + 4k + 1)}{3}$

   $= \dfrac{4k^3 + 12k^2 + 11k + 3}{3}$

   $= \dfrac{(k+1)(2k+1)(2k+3)}{3}$

   $= \dfrac{(k+1)(2(k+1)-1)(2(k+1)+1)}{3}$

17. $P_1: 1(1+1) = \dfrac{1(1+1)(1+2)}{3}$

   Assume $P_k: 1\cdot 2 + 2\cdot 3 + 3\cdot 4 + \cdots + k(k+1)$
   $$= \dfrac{k(k+1)(k+2)}{3}.$$

   Prove $P_{k+1}:$

   $1\cdot 2 + 2\cdot 3 + 3\cdot 4 + \cdots + k(k+1) + (k+1)(k+2)$
   $= \dfrac{(k+1)(k+2)(k+3)}{3}.$

   $1\cdot 2 + 2\cdot 3 + 3\cdot 4 + \cdots + k(k+1) + (k+1)(k+2)$

   $= \dfrac{k(k+1)(k+2)}{3} + (k+1)(k+2)$

   $= \dfrac{k(k+1)(k+2) + 3(k+1)(k+2)}{3}$

   $= \dfrac{(k+1)(k+2)(k+3)}{3}$

19. $P_1: 5^{1-1} = \dfrac{5^1 - 1}{4}$

   Assume $P_k: 1 + 5 + 5^2 + \cdots + 5^{k-1} = \dfrac{5^k - 1}{4}$.

   Prove $P_{k+1}: 1 + 5 + 5^2 + \cdots + 5^{k-1} + 5^k = \dfrac{5^{k+1} - 1}{4}$.

   $1 + 5 + 5^2 + \cdots + 5^{k-1} + 5^k = \dfrac{5^k - 1}{4} + 5^k$

   $= \dfrac{5^k - 1 + 4(5^k)}{4}$

   $= \dfrac{5^k(1+4) - 1}{4}$

   $= \dfrac{5^{k+1} - 1}{4}$

21. $P_1: 3^1 - 1$ is divisible by 2.

   Assume $P_k: 3^k - 1$ is divisible by 2.
   Prove $P_{k+1}: 3^{k+1} - 1$ is divisible by 2.
   $3^{k+1} - 1 = 3(3^k) - 1$
   $\qquad\qquad = 3(3^k - 1) - 1 + 3$
   $\qquad\qquad = 3(3^k - 1) + 2$
   Because $3^k - 1$ is divisible by 2, $3(3^k - 1)$ is divisible
   by 2, and so $3(3^k - 1) + 2$ is divisible by 2.

23. $P_1: 1^2 + 3(1)$ is divisible by 2.
   Assume $P_k: k^2 + 3k$ is divisible by 2.
   Prove $P_{k+1}: (k+1)^2 + 3(k+1)$ is divisible by 2.
   $(k+1)^2 + 3(k+1) = k^2 + 2k + 1 + 3k + 3$
   $\quad = (k^2 + 3k) + (2k + 1 + 3) = (k^2 + 3k) + 2(k+2)$
   Both $(k^2 + 3k)$ and $2(k+2)$ are divisible by 2,
   so $(k^2 + 3k) + 2(k+2)$ is divisible by 2.

25. $P_1: 2^1 > 1$
   Assume $P_k: 2^k > k$.
   Prove $P_{k+1}: 2^{k+1} > k + 1$.
   $2^{k+1} = 2(2^k) > 2(k) = k + k \ge k + 1$

27. $a^3 - b^3 = (a-b)(a^2 + ab + b^2)$. This product is
   divisible by $a - b$.

29. For $1 + 4 + 4^2 + \cdots + 4^{n-1}, a_0 = 1, r = 4,$ and
   $a_0\left(\dfrac{1 - r^n}{1 - r}\right) = \left(\dfrac{1 - 4^n}{1 - 4}\right) = \dfrac{4^n - 1}{3}.$

Chapter 10 Review Exercises

1. $a_n = 7 + 2n$ **3.** $a_n = 1 + 4n$ **5.** $a_n = 5(2)^n$

7. $a_n = 2(3)^n$ **9.** \$43,500   **11.** 1092   **13.** -285 **15.** 24

17. 3280   **19. a.** $\displaystyle\sum_{i=0}^{9} 2.5i$ **b.** 112.5

21. a. $\displaystyle\sum_{i=1}^{16} a^i$ **b.** $\dfrac{a(1 - a^{16})}{1 - a}$ **23.** 40.5   **25.** 9

27. $5, 3, 1, -1, -3$ **29.** $-3, -\dfrac{3}{2}, -\dfrac{3}{4}, -\dfrac{3}{8}, -\dfrac{3}{16}$

31. $1, 0, -3, -8, -15$ **33.** $\dfrac{1}{2}, \dfrac{2}{5}, \dfrac{3}{10}, \dfrac{4}{17}, \dfrac{5}{26}$

35. $-\dfrac{1}{3}, \dfrac{1}{9}, -\dfrac{1}{27}, \dfrac{1}{81}, -\dfrac{1}{243}$ **37.** $4, 7, 10, 13$

39. $-1, 1, 5, 11$ **41.** 28   **43. a.** 12   **b.** 120   **c.** 24   **d.** 120

45. 120   **47.** 144   **49.** 120   **51.** 14

53. {HHH, HHT, HTH, THH, TTH, THT, HTT, TTT}

55. {HHH, HHT, HTH, THH, TTH, THT, HTT, TTT}

57. {TTT}   **59.** $\dfrac{2}{9}$ **61.** $\dfrac{253}{9996}$

63. $x^4 + 12x^3 + 54x^2 + 108x + 81$

65. $27x^3 + 27x^2y + 9xy^2 + y^3$ **67.** 15 **69.** $6x^2y$

71. $P_1: 1 + \dfrac{1}{2^1} = 2 - \dfrac{1}{2^1}$

Assume $P_k: 1 + \dfrac{1}{2} + \dfrac{1}{4} + \cdots + \dfrac{1}{2^k} = 2 - \dfrac{1}{2^k}$.

Prove $P_{k+1}: 1 + \dfrac{1}{2} + \dfrac{1}{4} + \cdots + \dfrac{1}{2^k} + \dfrac{1}{2^{k+1}}$
$$= 2 - \dfrac{1}{2^{k+1}}.$$

$$1 + \dfrac{1}{2} + \dfrac{1}{4} + \cdots + \dfrac{1}{2^k} + \dfrac{1}{2^{k+1}} = 2 - \dfrac{1}{2^k} + \dfrac{1}{2^{k+1}}$$
$$= 2 - \dfrac{2}{2^{k+1}} + \dfrac{1}{2^{k+1}}$$
$$= 2 - \dfrac{1}{2^{k+1}}$$

73. $P_1: (1+1)^2 + 1$ is odd.
Assume $P_k: (k+1)^2 + k$ is odd.
Prove $P_{k+1}: (k+2)^2 + (k+1)$ is odd.
$(k+2)^2 + (k+1) = k^2 + 4k + 4 + (k+1)$
$$= k^2 + 5k + 5$$
$$= k^2 + 2k + 1 + k + 2k + 4$$
$$= [(k+1)^2 + k] + 2(k+2)$$
The sum of an odd number and an even number is odd.

75. $P_1: 3(1) = \dfrac{3}{2}(1)(1+1)$
Assume $P_k: 3 + 6 + 9 + \cdots + 3k = \dfrac{3}{2}k(k+1)$.
Prove $P_{k+1}: 3 + 6 + 9 + \cdots + 3k + 3(k+1)$
$$= \dfrac{3}{2}(k+1)(k+2).$$
$$3 + 6 + 9 + \cdots + 3k + 3(k+1) = \dfrac{3}{2}k(k+1) + 3(k+1)$$
$$= (k+1)\left(\dfrac{3}{2}k + 3\right)$$
$$= \dfrac{3}{2}(k+1)(k+2)$$

Chapter 10 Test

1. $a_n = 8 + 3n$ **2.** $a_n = 7 + 5n$

3. $a_n = 15\left(\dfrac{1}{3}\right)^n$ **4.** $a_n = \dfrac{9}{2}\left(\dfrac{2}{3}\right)^n$ **5.** 1853 **6.** 1680

7. $\displaystyle\sum_{i=0}^{9} 5i = 225$ **8.** $a_0 = 2, a_1 = 0, a_2 = -14, a_3 = -52$

9. $a_0 = 1, a_1 = -\dfrac{1}{4}, a_2 = \dfrac{1}{16}, a_3 = -\dfrac{1}{64}, a_4 = \dfrac{1}{256}$

10. $a_0 = 4, a_1 = 3, a_2 = 1, a_3 = -2, a_4 = -6$

11. 26 **12. a.** 60 **b.** 15 **c.** 7 **d.** 360

13. 220 **14.** 720 **15.** 1728 **16.** {red, blue, white}

17. $\dfrac{2}{11}$ **18.** $\dfrac{6}{11}$ **19.** $\dfrac{1}{26}$

20. $81x^4 + 216x^3 + 216x^2 + 96x + 16$

21. P_1 is true: $2 = 2(1)^2$. Assume P_k is true. Then
$P_{k+1}: 2 + 6 + 10 + \cdots + 4k - 2 + 4(k+1) - 2$
$= 2k^2 + 4(k+1) - 2 = 2k^2 + 4k + 2 = 2(k+1)^2$.
Since P_{k+1} is of the form $2(k+1)^2$, the proposition holds true for all natural numbers n.

22. 24 **23.** 327,600

Index

Photo Credit